Sturkie's Avian Physiology
FIFTH EDITION

Sturkie's Avian Physiology

FIFTH EDITION

Edited by

G. Causey Whittow

Department of Physiology
John A. Burns School of Medicine
University of Hawaii at Manoa
Honolulu, Hawaii

Academic Press

San Diego   London   Boston   New York   Sydney   Tokyo   Toronto

Cover photograph: Adult wing of *Garrulus glandarius* after complete postbrachial moult in autumn. Photo by Raffael Winkler. From *Moult and Ageing in European Passerines* by Lukas Jenni and Raffael Winkler (Fig. 470, p. 157, Academic Press, 1994).

This book is printed on acid-free paper.

Copyright © 2000, 1986, 1976 by ACADEMIC PRESS

All Rights Reserved.
No part of this publication may be reproduced or transmitted in any form or by any means, electronic or mechanical, including photocopy, recording, or any information storage and retrieval system, without permission in writing from the publisher.

Requests for permission to make copies of any part of the work should be mailed to: Permissions Department, Harcourt Inc., 6277 Sea Harbor Drive, Orlando, Florida, 32887-6777

Academic Press
A Harcourt Science and Technology Company
525 B Street, Suite 1900, San Diego, California 92101-4495, USA
http://www.apnet.com

Academic Press
24-28 Oval Road, London NW1 7DX, UK
http://www.hbuk.co.uk/ap/

Library of Congress Catalog Card Number: 99-60592

International Standard Book Number: 0-12-747605-9

PRINTED IN THE UNITED STATES OF AMERICA
03 04 EB 9 8 7 6 5 4 3 2

To Paul D. Sturkie,
Emeritus Professor of Physiology,
Rutgers University.
Author of the first edition
of Avian Physiology, editor
of the second, third, and fourth editions.

Contents

Contributors xi
Preface to the Fifth Edition xiii

1. Sensory Physiology: Vision
ONUR GÜNTÜRKÜN

I. Introduction 1
II. Structure and Functions of the Eye 1
III. Central Processing—Anatomy and Function 7
References 14

2. The Avian Ear and Hearing
REINHOLD NECKER

I. Introduction 21
II. Structure of the Ear 21
III. Function of the Ear 24
IV. Auditory Nerve Fibers 27
V. Central Pathways 28
VI. Behavioral Aspects 33
VII. Summary and Conclusions 35
References 35

3. The Chemical Senses in Birds
J. RUSSELL MASON AND LARRY CLARK

I. Chemosensory Systems 39
II. Chemesthesis 40
III. Gustation 43
IV. Olfaction 46
References 51

4. The Somatosensory System
REINHOLD NECKER

I. Introduction 57
II. Types of Receptors and Afferent Fibers 57
III. Central Processing 62
IV. Behavioral Aspects 65
V. Summary and Conclusions 66
References 67

5. Functional Organization of the Spinal Cord
REINHOLD NECKER

I. Introduction 71
II. Gross Anatomy 71
III. Cytoarchitectonic Organization of the Spinal Gray 72
IV. Peripheral Input to the Spinal Cord 74
V. Functional Aspects of Laminae and Nuclear Groups 75
VI. Ascending Pathways 78
VII. Descending Pathways 79
VIII. Summary and Conclusions 79
References 80

6. Motor Control System
JACOB L. DUBBELDAM

I. Introductory Remarks 83
II. The Control of Eye and Head Movements: The Oculomotor System 84
III. The Motor Control of Jaw and Tongue Movements 87
IV. Other Premotor and Motor Systems in the Brainstem 91
V. Structure and Function of the Cerebellum 92
VI. Telencephalic Centers for Motor Control 94
VII. Concluding Remarks 97
References 98

7. The Autonomic Nervous System of Avian Species
WAYNE J. KUENZEL

I. Introduction 101
II. Components 102
III. Peripheral Motor Components of the Autonomic Nervous System 102
IV. Central Components of the Autonomic Nervous System 105
V. Functional Neural Pathways of the Autonomic Nervous System 112
VI. Shifts in Homeostasis: An Autonomic Hypothesis Explaining the Regulation of Annual Cycles of Birds 117
VII. Consequences of Genetic Selection for Rapid Growth in Broilers and Turkeys: An Imbalanced Autonomic Nervous System 118
References 118

8. Skeletal Muscle
A. L. HARVEY AND I. G. MARSHALL

I. Introduction 123
II. Development of Avian Muscle 123
III. Muscle Fiber Types 126
IV. Innervation 128
V. Electrical Properties of Muscle Fibers 129
VI. Contractile Properties 129
VII. Neuromuscular Transmission 130
VIII. Uses of Avian Muscle in Neuromuscular Pharmacology 133
References 135

9. The Cardiovascular System
FRANK M. SMITH, NIGEL H. WEST, AND DAVID R. JONES

I. Introduction 141
II. Heart 142
III. General Circulatory Hemodynamics 154
IV. The Vascular Tree 157
V. Blood 176
VI. Control of the Cardiovascular System 181
References 223

10. Respiration
F. L. POWELL

I. Overview 233
II. Anatomy of the Avian Respiratory System 234
III. Ventilation and Respiratory Mechanics 238
IV. Pulmonary Circulation 243
V. Gas Transport by Blood 244
VI. Pulmonary Gas Exchange 249
VII. Tissue Gas Exchange 254
VIII. Control of Breathing 256
References 259

11. Renal and Extrarenal Regulation of Body Fluid Composition
DAVID L. GOLDSTEIN AND ERIK SKADHAUGE

I. A Review of Reviews 265
II. Introduction 265
III. Intake of Water and Solutes 266
IV. The Kidneys 267
V. Extrarenal Organs of Osmoregulation: Introduction 282
VI. The Lower Intestine 283
VII. Salt Glands 288
VIII. Evaporative Water Loss 291
References 291

12. Gastrointestinal Anatomy and Physiology
D. MICHAEL DENBOW

I. Anatomy of the Digestive Tract 299
II. Anatomy of the Accessory Organs 305
III. Motility 305
IV. Neural and Hormonal Control of Motility 310
V. Secretions and Digestion 313
VI. Absorption 317
VII. Age-Related Effects on Gastrointestinal Function 320
VIII. Food Intake Regulation 320
References 321

13. Energy Balance
CHARLES R. BLEM

I. Introduction 327
II. The Measurement of Energy Exchange 327
III. Energy Costs of Activity 331
IV. Energy Storage and Production 334
V. Daily Energy Budgets and Energetics of Free-Living Birds 337
VI. Geographic Variation in Energy Balance 338
VII. Energy Requirements of Populations and Communities 338
VIII. Summary 338
References 339

14. Regulation of Body Temperature
W. R. DAWSON AND G. C. WHITTOW

I. Introduction 344
II. Body Temperature 344
III. Heat Balance 348
IV. Changes in Bodily Heat Content (S) 348
V. Heat Production (H) 349
VI. Heat Transfer within the Body 355
VII. Heat Loss 356
VIII. Heat Exchange under Natural Conditions 364
IX. Behavioral Thermoregulation 364
X. Control Mechanisms 367
XI. Thermoregulation at Reduced Body Temperatures 372
XII. Development of Thermoregulation 375
XIII. Summary 379
References 379

15. Flight
P. J. BUTLER AND C. M. BISHOP

I. Introduction 391
II. Scaling 392
III. Energetics of Bird Flight 393
IV. The Flight Muscles of Birds 406
V. Development of Locomotor Muscles and Preparation for Flight 413
VI. Metabolic Substrates and Fuel Deposits 415
VII. The Cardiovascular System 416
VIII. The Respiratory System 418
IX. Migration and Long-Distance Flight Performance 425
X. Flight at High Altitude 428
References 430

16. Introduction to Endocrinology: Pituitary Gland
COLIN G. SCANES

I. Introduction 437
II. Anatomy of the Hypothalamic–Hypophyseal Complex 438
III. Gonadotropins 438
IV. Thyrotropin 443
V. Growth Hormone 444
VI. Prolactin 447
VII. Adrenocorticotropic Hormone 449
VIII. Other Adenohypophyseal Peptides 450
IX. Neurohypophysis 451
X. Arginine Vasotocin and Mesotocin 451
References 452

17. Thyroids
F. M. ANNE MCNABB

I. Anatomy, Embryology, and Histology of Thyroid Glands 461
II. Thyroid Hormones 462
III. Hypothalamic–Pituitary–Thyroid Axis 466
IV. Mechanism of Action of Thyroid Hormones 466
V. Effects of Thyroid Hormones 467
VI. Thyroid Interactions with Other Hormones 468
VII. Environmental Influences on Thyroid Function 469
References 469

18. The Parathyroids, Calcitonin, and Vitamin D
CHRISTOPHER G. DACKE

I. Introduction 473
II. Parathyroid Hormone and Related Peptides 474
III. Calcitonin and the Ultimobranchial Glands 479
IV. The Vitamin D System 481
V. Prostaglandins and Other Factors 483
VI. Conclusions 484
References 485

19. Adrenals
R. V. CARSIA AND S. HARVEY

I. Anatomy 489
II. Adrenocortical Hormones 492
III. Physiology of Adrenocortical Hormones 514
IV. Adrenal Chromaffin Hormones 519
References 522

20. Pancreas
ROBERT L. HAZELWOOD

I. Glucoregulation: Why Glucoregulate? 539
II. Organs of Importance in Glucoregulation 540
III. Avian Carbohydrate Metabolism: Different from Mammals? 541
IV. Central Role of the Pancreatic Organ 542
V. Mechanisms of Pancreatic Hormone Action: Molecular Events 544
VI. Pancreatic–Enteric Regulation of Carbohydrate Metabolism 548

VII. Carbohydrate Metabolism in Other Tissues 551
VIII. Altered Avian Carbohydrate Metabolism 552
References 554

21. The Pineal Gland, Circadian Rhythms, and Photoperiodism

EBERHARD GWINNER AND MICHAELA HAU

I. Anatomy of the Pineal Gland 557
II. Pineal Hormones 558
III. Physiological Effects of the Pineal Gland 560
References 565

22. Reproduction in the Female

A. L. JOHNSON

I. Anatomy of the Female Reproductive System 569
II. Breeding and Ovulation–Oviposition Cycles 575
III. Ovarian Hormones 577
IV. Hormonal and Physiologic Factors Affecting Ovulation 580
V. Oviposition 584
VI. Composition and Formation of Yolk, Albumen, Organic Matrix, and Shell 586
References 591

23. Reproduction in Male Birds

JOHN D. KIRBY AND DAVID P. FROMAN

I. Introduction 597
II. Reproductive Tract Anatomy 597
III. Ontogeny of the Reproductive Tract 600
IV. Development and Growth of the Testis 603
V. Hormonal Control of Testicular Function 605
VI. Spermatogenesis and Extragonadal Sperm Maturation 607
References 612

24. Incubation Physiology

HIROSHI TAZAWA AND G. CAUSEY WHITTOW

I. Introduction 617
II. Composition of the Freshly Laid Egg 617
III. Changes in the Composition of the Egg during Incubation 618
IV. Heat Transfer to the Egg 620
V. Development of Physiological Functions 621
VI. Requirements and Procedures for Incubation 629
References 632

25. Physiology of Growth and Development

LARRY A. COGBURN, JOAN BURNSIDE, AND COLIN G. SCANES

I. Introduction 635
II. Somatotropic Axis 636
III. Thyrotropic Axis 641
IV. Gonadotropic Axis 642
V. Lactotropic Axis 642
VI. Adrenocorticotropic Axis 644
VII. Pancreatic Hormones 644
VIII. Growth Factors 644
IX. Models in Avian Growth 648
References 649

26. Immunophysiology

B. GLICK

I. Introduction 657
II. Cytoarchitecture and Development of the Immune System 657
III. Regulation of Immune Response 662
References 667

Index 671

Contributors

Numbers in parentheses indicate the pages on which the authors' contributions begin.

C. M. Bishop (391) School of Biological Sciences, University of Wales–Bangor, Gwynedd LL57 2UW United Kingdom

Charles R. Blem (327) Department of Biology, Virginia Commonwealth University, Richmond, Virginia 23284

Joan Burnside (635) Department of Animal and Food Sciences, College of Agricultural Sciences, Delaware Agricultural Experiment Station, University of Delaware, Newark, Delaware 19717

P. J. Butler (391) School of Biological Sciences, The University of Birmingham, Birmingham B15 2TT, United Kingdom

R. V. Carsia (489) Department of Cell Biology, University of Medicine and Dentistry of New Jersey, School of Osteopathic Medicine, Stratford, New Jersey 08084

Larry Clark (39) United States Department of Agriculture, Animal and Plant Health Inspection Service, Wildlife Services, National Wildlife Research Center and Monell Chemical Senses Center, Philadelphia, Pennsylvania 19104

Larry A. Cogburn (635) Department of Animal and Food Sciences, College of Agricultural Sciences, Delaware Agricultural Experiment Station, University of Delaware, Newark, Delaware 19717

Christopher G. Dacke (473) Pharmacology Division, School of Pharmacy and Biomedical Science, University of Portsmouth, Portsmouth PO1 2DT, United Kingdom

W. R. Dawson (343) Department of Biology, The University of Michigan at Ann Arbor, Ann Arbor, Michigan 48109

D. Michael Denbow (299) Department of Animal and Poultry Sciences, Virginia Tech, Blacksburg, Virginia 24061

Jacob L. Dubbeldam (83) Institute of Evolutionary and Ecological Sciences, Section of Dynamic Morphology, Van der Klaauw Laboratorium, Leiden University, 2300 RA Leiden, The Netherlands

David P. Froman (597) Department of Animal Sciences, Oregon State University, Corvallis, Oregon 97331

B. Glick[1] (657) Department of Poultry Science, College of Agricultural Sciences, Clemson University, Clemson, South Carolina 29634

David L. Goldstein (265) Department of Biological Sciences, Wright State University, Dayton, Ohio 45435

Onur Güntürkün (1) AE Biopsychologie, Fakultät für Psychologie, Ruhr-Universität Bochum, D-44780 Bochum, Germany

Eberhard Gwinner (557) Max-Planck-Institut für Verhaltensphysiologie, D-82346 Andechs, Germany

A. L. Harvey (123) Strathclyde Institute for Drug Research, University of Strathclyde, Glasgow G1 1XW, United Kingdom

S. Harvey (489) Department of Physiology, University of Alberta, Edmonton, Alberta, Canada T69 2H7

Michaela Hau (557) Max-Planck-Institut für Verhaltensphysiologie, D-82346 Andechs, Germany

Robert L. Hazelwood (539) Department of Biology, University of Houston, Houston, Texas 77204

A. L. Johnson (569) Department of Biological Sciences, University of Notre Dame, Notre Dame, Indiana 46556

[1]Retired; see note on p. 657.

David R. Jones (141) Department of Zoology, University of British Columbia, Vancouver, British Columbia, Canada V6T 1Z4

John D. Kirby (597) Department of Poultry Science, University of Arkansas, Fayetteville, Arkansas 72701

Wayne J. Kuenzel (101) Department of Animal and Avian Sciences, University of Maryland, College Park, Maryland 20742

I. G. Marshall (123) Strathclyde Institute for Drug Research, University of Strathclyde, Glasgow G1 1XW, United Kingdom

J. Russell Mason (39) United States Department of Agriculture, Animal and Plant Health Inspection Service, Wildlife Services, National Wildlife Research Center and Monell Chemical Senses Center, Philadelphia, Pennsylvania 19104

F. M. Anne McNabb (461) Department of Biology, Virginia Polytechnic Institute and State University, Blacksburg, Virginia 24061

Reinhold Necker (21, 57, 71) Fakultät für Biologie, Lehrstuhl für Tierphysiologie, Ruhr-Universität Bochum, D-44780 Bochum, Germany

F. L. Powell (233) Division of Physiology, Department of Medicine, School of Medicine, University of California at San Diego, La Jolla, California 92093

Colin G. Scanes (437, 635) Department of Animal Science, Cook College, Rutgers–The State University, New Brunswick, New Jersey 08903, and College of Agriculture, Iowa State University, Ames, Iowa 50011

Erik Skadhauge (265) Department of Anatomy and Physiology, Royal Veterinary and Agricultural University, DK-Frederiksberg C, DK-1870 Denmark

Frank M. Smith (141) Department of Anatomy and Neurobiology, Dalhousie University, Halifax, Nova Scotia, Canada B3H 4H7

Hiroshi Tazawa (617) Department of Electrical and Electronic Engineering, Muroran Institute of Technology, Muroran 050, Japan

Nigel H. West (141) Department of Physiology, University of Saskatchewan, Saskatoon, Saskatchewan, Canada S7N 5E5

G. C. Whittow (343, 617) Department of Physiology, John A. Burns School of Medicine, University of Hawaii at Manoa, Honolulu, Hawaii 96822

Preface to the Fifth Edition

When the first edition of *Avian Physiology* appeared in 1954, it broke new ground—it was the first book on avian physiology ever published. The next three editions sustained the book's distinctiveness; it remained the only comprehensive, single-volume textbook on the physiology of birds. The present edition recognizes the retirement of the original author and editor, Dr. Paul D. Sturkie, by being titled *Sturkie's Avian Physiology*. The volume continues to break new ground. The treatment of the nervous system has been greatly expanded. A new chapter on flight has been added, occupying a central place in the book, as befits the most conspicuous feature of birds. There are new chapters also on incubation and growth and development, reflecting the characteristic development of the avian egg outside the body and the great deal of information now available. Most of the authors of this edition are new. However, the book retains the unique feature of earlier editions; it covers wild and domestic birds within the compass of one volume, making it suitable for use as a text.

G.C.W.

CHAPTER 1

Sensory Physiology: Vision

ONUR GÜNTÜRKÜN
AE Biopsychologie
Fakultät für Psychologie
Ruhr-Universität Bochum
44780 Bochum
Germany

I. Introduction 1
II. Structure and Functions of the Eye 1
   A. Eye Shape, Stereopsis, and Acuity 2
   B. Retina 4
III. Central Processing—Anatomy and Function 7
   A. Centrifugal Pathway 7
   B. Tectofugal Pathway 9
   C. Thalamofugal Pathway 12
   References 14

I. INTRODUCTION

Birds are the most visually dependent class of vertebrates and the phrase of Rochon-Duvigneaud (1943) that a pigeon is nothing else but two eyes with wings is probably valid for most avian species. Man, a highly visual primate, sees the world with the information transmitted by about one million fibers within each of his optic nerves. This is only 40% of the number of retinal axons counted in a single optic nerve of pigeons and chicks (Binggeli and Paule, 1969; Rager and Rager, 1978). The acuity of many birds of prey surpasses that of other living beings (Fox *et al.*, 1976) and even the unspecialized pigeon excels relative to humans in its ability to discriminate luminances (Hodos *et al.*, 1985) and discern subtle color differences (Emmerton and Delius, 1980). Food-storing birds like Clark's nutcracker store 33,000 seeds in about 6,600 caches to survive in winter (Vander Wall and Balda, 1977). Pigeons acquire visual concepts of, for example, "animals" (Roberts and Mazmanian, 1988), "same versus different" (Wright *et al.*, 1988), and even cartoon figures such as "Charlie Brown" (Cerella, 1980). They communicate using visual symbols (Lubinski and MacCorquodale, 1984) and are able to rank optic patterns by using transitive inference logic (von Fersen *et al.*, 1992). If we, on the basis of countless evidence, assume that the visual system of amniotes has evolved only once (Shimizu and Karten, 1993), the avian visual system is a remarkable model to explore its morphology, its modes of operations, and the unanticipated complexity of its function.

II. STRUCTURE AND FUNCTIONS OF THE EYE

Avian eyes take up a considerable volume of the bird's head and are very large in relation to brain size (Figure 1). In general terms, the structure of their eyes is not much different from that of other vertebrates. Incoming light has to pass through four media: the cornea, the anterior chamber, the lens, and the vitrous body, before reaching the retina, where photoreceptors convert light energy into electric impulses by bleaching of visual pigments. All four optic media are remarkably transparent, transmitting wavelengths down to at least 310 nm in the near-ultraviolet range (Emmerton *et al.*, 1980).

The avian retina is completely avascularized to prevent shadows and light scattering. This arrangement is associated with the presence of an unusual nutritional device specific for birds—the pecten. This black pig-

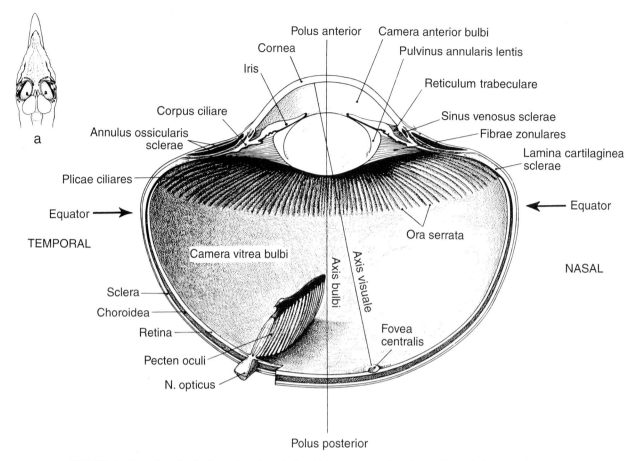

FIGURE 1 Drawing of a horizontal section of the chicken eye showing the position of the eyes within the head. (From H. Evans, 1996.)

mented and manifoldly pleated structure projects from the ventral retina above the exit of the optic nerve toward the lens and is completely made up of blood vessels and extravascular pigmented stromal cells. There is evidence that it also has a nutritive function. This is shown by the presence of an oxygen gradient from the pecten to the retina, the passing of nutrients from the pecten into the vitreous, and the observation that fluorescent markers pass from the pecten into the vitreous (Bellhorn and Bellhorn, 1975). Also, Pettigrew *et al.* (1990) posit that the inertia of the pecten during saccadic eye movements could be used like a shaker to propel oxygen and nutrients within the eye.

A. Eye Shape, Stereopsis, and Acuity

The eyeshapes of birds are a result of ecological requirements (Figure 2). Generally, acuity can be maximized by increasing the anterior focal length of an eye; the optic image is then spread over a larger retinal surface and thus over a larger number of photoreceptors (Martin, 1993). Increasing the number of photorecep-

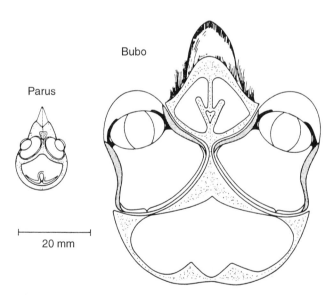

FIGURE 2 Horizontal section through the head of the black-capped chickadee (*Parus atricapillus*) and the great owl (*Bubo virginianus*). (From *Perception and Motor Control in Birds,* Form and function in the optical structure of bird eyes, G. R. Martin, pp. 5–34, Fig. 1.2, 1994, © Springer-Verlag.)

tors also makes it possible to connect several receptors to single bipolar cells and thus to maximize visual detection even under low light conditions. Since an increase in eye size is advantageous, birds, which rely heavily on vision, generally have the largest absolute and relative eyes within the animal kingdom. The eye of the ostrich, for example, has an axial length of 50 mm, the largest of any land vertebrate and twice that of the human eye (Walls, 1942). The tube-shaped eyecups of birds of prey, which create an extremely large image on the retina, represent another extreme version of biological optimization to achieve high acuity. These eyes generally also have a low retinal convergence ratio (receptors per ganglion cell) so that the receptor inputs are not pooled to increase visual resolution (Snyder et al., 1977). However, these optimizations are limited by trade-offs for brightness sensitivity. Retinae in which receptors are not pooled function only optimally at high light intensities and, indeed, resolution of birds of prey deteriorates at dusk (Reymond, 1985).

Visual acuity measurements in pigeons (Columba livia) have shown that the acuity in the frontal field depends on stimulus time (Bloch and Martinoya, 1982), wavelength of light (Hodos and Leibowitz, 1977), luminance (Hodos et al., 1976; Hodos and Leibowitz, 1977), and age of the pigeon (Hodos et al., 1991a). Under favorable conditions 1-year-old pigeons reach a frontal acuity of 12.7 c/deg, increase this value to 16–18 c/deg at 2 years, and decline to 3 c/deg at 17 years (Hodos et al., 1985, 1991b). The frontal binocular visual field of pigeons is represented in the superiotemporal area dorsalis, while the lateral monocular visual field is observed via the area centralis (both lack a true foveal depression). These two retinal regions seem to subserve different visual functions with differing capacities for optic resolutions. Behavioral studies show that many avian species, including pigeons, fixate distant objects preferentially with their lateral and monocular field (pigeon: Blough, 1971; dove: Friedman, 1975; kestrel: Fox et al., 1976; eagle: Reymond, 1985; passerine birds: Bischof, 1988; Kirmse, 1990). This behavior is often pronounced; birds orient themselves sideways in order to achieve a lateral orientation to the inspected object. This behavior, together with the fact that retinal ganglion cell densities reach peak values in the central fovea, suggest that resolution is maximal in the lateral visual field. However, the acuity of young pigeons is 12.6 c/deg in their lateral visual field and thus identical with the values obtained for frontal vision in same aged subjects (Hahmann and Güntürkün, 1993). However, lateral acuity measurements are naturally obtained under monocular conditions, while frontal acuity is generally tested binocularly. In humans, binocular sensitivity can almost double that of one-eyed viewing (Pirenne, 1943). This same effect is known in pigeons and possibly depends on probability summation of the input of both eyes (DiStefano et al., 1987; Kusmic et al., 1991). The power of this mechanism is visible when pigeons are frontally tested under monocular conditions. Their acuity then drops to a mean of 6.5 c/deg and thus to less than half the value obtained under binocular conditions (Güntürkün and Hahmann, 1994). If only monocular data are used to compare frontal and lateral acuity, resolution in the lateral field (12.6 c/deg) is considerably higher than in the frontal field (6.5 c/deg). These psychophysical data are in perfect accord with the observations that many bird species prefer to use their lateral visual field for a detailed inspection of distant objects.

These acuity data are easily surpassed by some birds of prey. The wedge-tailed eagle Aquila audax reaches a maximum acuity of 143 c/deg, more than two times higher than the human optic resolution measured under identical conditions (Reymond, 1985). These values are even surpassed by the American kestrel Falco sparverius. The acuity threshold of this falcon was measured to be 160 c/deg, which would enable this animal to discriminate 2-mm insects from 18-m-high treetops (Fox et al., 1976). In both studies, these birds of prey were reported to be considerably luminance dependent with acuity dropping to 58 c/deg at 2 cd/m^2 in the wedge-tailed eagle (Reymond, 1985). Thus, while visual adaptations allow for high acuity they necessitate a loss of optical sensitivity. Not all birds of prey, however, reach high acuity values. The nocturnal barn owl Tyto alba, which heavily relies on auditory cues to detect prey, reaches an acuity of only 8.4 c/deg as predicted from its retinal ganglion cell density (Wathey and Pettigrew, 1989).

The ability to focus the eye to see objects at various distances sharply is called accomodation; it is achieved by alterations in corneal curvature and by lens deformation and constitutes one of the most important mechanisms of achieving high visual resolution. In addition to these dynamic accommodation mechanisms, some birds possess static mechanisms which keep objects along the ground in focus, irrespective of their distance. This is achieved by asymmetries of the eye such that it is emmetropic in its superior parts but increasingly myopic with decreasing elevation (Fitzke et al., 1985). As a result, objects along the horizon or in the upper visual field are in focus together with objects at various distances on the ground. The degree of this lower-field myopia seems to adjust to the height of the head of the animal so that cranes can also benefit from its effect (Hodos and Erichsen, 1990). The presence of a lower field myopia would not be advantageous for raptors which pursue and capture their mobile prey, the prey often being seen with their lower field of view. Consequently, Murphy *et al.* (1995) demonstrated that raptors lack lower-field myopia.

B. Retina

1. Oil Droplets, Photoreceptors, and Color Vision

Differing from those of placentalia, avian eyes are characterized by the presence of oil droplets within the distal end of the inner segment of their cones. Microspectrophotometric studies show that oil droplets act as cut-off filters and absorb light below their characteristic wavelength of transmission (Emmerton, 1983b). Colored oil droplets thus provide a protective shield against UV light, similar to the yellowish lenses of mammals. Additionally they probably act as lenses which focus light onto the photoreceptor, thus increasing the quantum reception of visual pigments (Young and Martin, 1984). A detailed inspection shows at least five different-colored types of oil droplets depending on the presence, mixture, and concentration of different carotenoids: red, orange, greenish-yellow, pale, and transparent (Varela *et al.*, 1993).

The spectral sensitivity of an avian cone is the result of the relation between the spectral transmittance of the oil droplets and the spectral absorptance of the visual pigments. This condition creates the possibility that birds can increase the number of their chromatic channels by varying the combinations of oil droplets and cone pigments. Indeed, there is evidence that at least some bird species have two absorption maxima operating with one visual pigment which is associated with two different oil droplets (Jane and Bowmaker, 1988). Birds studied up to now have at least three to four cone pigments which, together with their associated oil droplets, create spectral sensitivity maxima reaching from 370 to 580 nm (Chen and Goldsmith, 1986).

Another feature that increases the complexity of color perception in birds is the differential distribution of oil droplets across the retina. This hererogeneous distribution reaches an extreme in pigeons where the dorsotemporal "red field," with large numbers of red and orange droplets, is clearly separated from the remaining "yellow field," which is characterized by a high density of greenish-yellow droplets (Galifret, 1968). Bowmaker (1977) showed that the transmission curves of oil droplets in the red field are shifted 10 nm toward longer wavelengths. These data may indicate differences in color perception between different retinal areas in pigeons and, indeed, behavioral experiments demonstrate that colors backprojected onto two pecking keys are treated differently by pigeons when both are seen with the red field or when one is viewed with the red and the other with the yellow field (Delius *et al.*, 1981).

The authors suggest that their results are due to a subjective discrepancy, as the birds perceived the two keys illuminated with light of identical spectral composition as being of different color when one was seen with the yellow and the other with the red field. However, probably the most important differentiation of color perception between retinal areas is related to UV sensitivity. Remy and Emmerton (1989) showed in a behavioral study with head-fixed pigeons that UV sensitivity is high in the yellow and low in the red field. Emmerton (1983a) additionally demonstrated that pigeons perform excellent pattern discrimination in UV. Thus, pigeons and several other avian species may use their UV sensitivity to view objects such as plumage or fruits reflecting UV light (Burkhardt, 1989).

2. Neuronal Wiring

The basic design of all vertebrate retinae is essentially the same and those of birds are no exception. Light passes through the neural retina and is transduced in the outer segments of photoreceptors to electrical signals which are relayed via bipolar cells to the ganglion cells and thus to the brain. Horizontal intraretinal interactions are provided by horizontal and amacrine cells which in birds are also partly responsible for long intraretinal projections. But imposed on this basic uniformity, there is wide variation in details (Thompson, 1991) (Figure 3).

In the diurnal pigeon, rods and principal members of the double cones terminate in the outer sublayer of the outer plexiform layer (OPL), the straight single cones in the middle sublayer, and the oblique single cones terminate exclusively in the inner sublayer of the OPL (Mariani and Leure-du Pree, 1978; Mariani, 1987; Nalbach *et al.*, 1993). According to morphological criteria, Mariani (1987) distinguished four types of horizontal and eight types of bipolar cells with each bipolar cell showing a distinct type of termination within the five sublayers of the inner plexiform layer (IPL). The diversity of amacrine cells described by Golgi techniques in the early 1980s (Mariani, 1983) turned out to be an extreme oversimplification as shown by immunocytochemical studies within the past decade. These experiments revealed amacrine cells specific for substance P (Ehrlich *et al.*, 1987), tyrosine hydroxylase (Keyser *et al.*, 1990), enkephalin (Britto and Hamassaki-Britto, 1992), glucagon (Keyser *et al.*, 1988), somatostatin Morgan *et al.*, 1983), 5-hydroxytryptamine (Kiyama *et al.*, 1985), avian pancreatic polypeptide (Katayama *et al.*, 1984), choline acetyltransferase (Millar *et al.*, 1987), neuropeptide Y (Verstappen *et al.*, 1986), neurotensin-related hexapeptide LANT-6 (Reiner, 1992), and GABA (Hamassaki-Britto *et al.*, 1991). Some of the substance

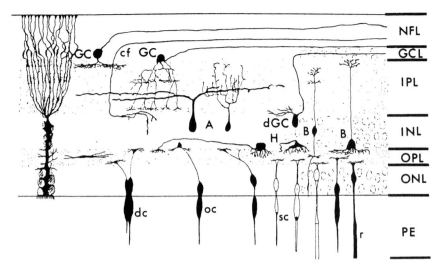

FIGURE 3 Schematic drawing of the avian retina. A, amacrine cell; B, bipolar cell; cf, centrifugal fiber; dc, double cone; dGC, displaced ganglion cell; GC, ganglion cell; GCL, ganglion cell layer; H, horizontal cell; INL, inner nuclear layer; IPL, inner plexiform layer; oc, oblique cone; ONL, outer nuclear layer; OPL, outer plexiform layer; PE, pigment epithelium containing the outer parts of the photoreceptors; r, rod; and sc, single straight cone. (From Nalbach *et al.*, 1993, *Vision, Brain and Behavior*, The MIT Press. © 1993 The MIT Press.)

P and/or glucagon-positive amacrine cells are the "bullwhip" neurons with long thin processes directed toward the posterodorsal pole of the retina (Ehrlich *et al.*, 1987; Keyser *et al.*, 1988). Catsicas *et al.* (1987a) could show that some amacrine cells, of which the bullwhip neurons probably represent a subclass, are localized within the central and ventral retina and project toward the superiodorsal retina. They suggested that the intraretinal connections may be involved in a system for switching attention between the upper and lower halves of the visual field, which could be modulated by centrifugal axons entering the retina from the contralateral tegmentum (Fritzsch *et al.*, 1990). It is interesting to note that these experiments demonstrate a one-way route from central and ventral retinal areas to the red field, but not vice versa. Mallin and Delius (1983) showed that pigeons can transfer information about discriminatory cues from the central retina to the red field, but not from the red field to the area centralis. Behaviorally, this asymmetry makes sense since pigeons spot seeds from a distance (central retina) and approach to peck them after making a final inspection in the binocular field (superiodorsal red field). The reverse behavioral pattern never occurs. There may be a neural basis for this behavioral constraint.

A subpopulation of ganglion cells is located within the inner nuclear layer (INL) and they are thus called "displaced ganglion cells" (DGCs) (Brecha and Karten, 1981). Medium-sized and large DGCs have dendrites which arborize for considerable distances in the outermost lamina of the IPL (Britto *et al.*, 1988), are predominantly distributed in the peripheral retina (Prada *et al.*, 1989; Prada *et al.*, 1992), and project to the avian accessory optic nucleus (Fite *et al.*, 1981; Yang *et al.*, 1989). A part of the DGCs are substance P positive (Britto and Hamassaki-Britto, 1991), while others are cholinergic (Britto *et al.*, 1988). Further aspects of the accessory optic system will be discussed in Chapter 4. Additionally, a population of DGCs appears to exist in the avian retina, which exhibit smaller soma sizes, are located centrally in the retina, and whose central connections are uncertain (Hayes and Holden, 1983).

Cajal (1892) described two main types of ganglion cells in the chicken retina: mono- and polystratified neurons. More modern attempts to classify avian retina ganglion cells into categories similar to that developed by Boycott and Wässle (1974) and Fukuda and Stone (1974) in cats did not lead to unequivocal results (Ikushima *et al.*, 1986). Hayes and Holden (1980) suggested, on the basis of perikaryal morphology and electrophysiological properties (Holden, 1978), that retinal ganglion cells projecting to the optic tectum would be comparable to W-cells. Studies in owls (Bravo and Pettigrew, 1981) and pigeons (Remy and Güntürkün, 1991) demonstrated that indeed the tectum receives its input from a large number of very small and a few very large ganglion cells while the GLd is characterized by its afferents from medium-sized and very large retinal neurons. These conditions suggest similarities to the differential sizes and central projections of cat alpha, beta, and gamma cells (Illing and Wässle, 1981). It should, however, be remarked that these assumptions rest on observations of soma diameters and projections and do not include any data on dendritic morphology and axonal diameters. Additionally, electrophysiological studies demonstrated various ganglion cell properties in avian retinae with important deviations from the usual schema known from mammals (Maturana and Frenk, 1963, Miles, 1972a).

As outlined for amacrine cells, immunocytochemical analyses have demonstrated a very large number of diverse ganglion cells specific for certain transmitters or neuromodulators. Among them are neurons positive for cholecystokinin (Britto and Hamassaki-Britto, 1991), tyrosine hydroxylase (Keyser *et al.*, 1990), substance P (Britto and Hamassaki-Britto, 1991), dopamine (Karten *et al.*, 1990), GABA (Hamassaki-Britto *et al.*, 1991), LANT6 (Reiner, 1992), enkephalin (Britto and Hamassaki-Britto, 1992), and glutamate (Morino *et al.*, 1991). Such diversity in ganglion cell transmitters/modulators implies a far more heterogeneous influence of retinal axons on central targets than previously imagined and may require revision of the broadly held concept that ganglion cell classifications based on frequency coding and dendritic morphology provide sufficient information on the type of central effects of retinal inputs.

The retinae of birds are characterized by a large variation of different regional specializations (Figure 4). In pigeons the density of cells in the outer nuclear layer (ONL) and the inner nuclear layer (INL) as well as the ganglion cell layer (GCL) increase in the area centralis and the dorsotemporal red field, while a streak of slightly increased ganglion cell densities connects these two areas of enhanced vision (Galifret, 1968). This arrangement is typical for granivorous birds (pigeons: Binggeli and Paule, 1969; quail: Budnik *et al.*, 1984; but see chicks: Ehrlich, 1981) which probably have to switch

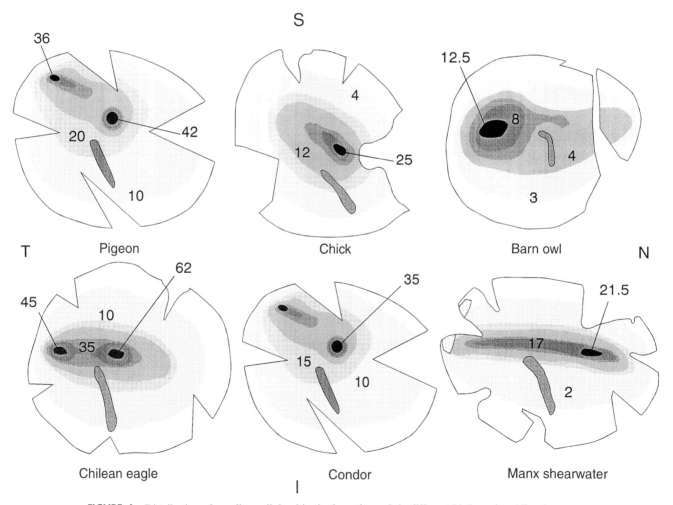

FIGURE 4 Distribution of ganglion cell densities in the retinae of six different bird species. All retinae are drawn to the same size and gray shadings indicate neuronal densities. Drawings according to Binggeli and Paule (1969; pigeon), Ehrlich (1981; chick), Wathey and Pettigrew, 1989; barn owl), Inzunza *et al.* (1991; chilean eagle and condor), and Hayes et al. (1991; Manx shearwater). The dark gray stippled structure is the pecten. The dotted region in the most superiotemporal retina of Manx shearwater is the area giganto cellularis. Numbers indicate 1.000/mm^2. Abbreviations: I, inferior; N, nasal; S, superior; and T, temporal.

between monocular lateral and binocular frontal vision. With regard to synaptic interactions, the dorsotemporal red field seems to be the most complex one, while the area centralis displays a very low synaptic complexity (Yazulla, 1974). Golgi studies make it likely that this is probably due to the area centralis being specialized for precise point-to-point interactions by a midget-like system that does not require extensive horizontal processing by amacrine cells (Lockhart, 1979, Quesada et al., 1988).

A streak with two real foveae can be found in birds of prey like the American kestrel or the Chilean eagle *Buteo fuscenses,* which attack birds or rodents directly from flight and have to combine panoramic sight with excellent stereoscopic vision (Inzunza et al., 1991). The density of ganglion cells reaches up to 65,000 mm^2 in the central fovea of these animals, which surpasses foval values from mammals with high acuity (human: 38,000 mm^2, Curcio and Allen, 1990; macaque: 33,000 mm^2, Perry and Cowey, 1985). Carrion-eating birds like the condor *Vultur gryphus* or the black vulture *Coragyps atratus* pursue their prey from the ground and are not in need of high stereoscopic vision. Consequently they not only have reduced ganglion cell densities within their visual streak but they also have lost their temporal fovea (Inzunza et al., 1991). A single temporal fovea characterizes nocturnal predators like owls (Oehme, 1961, Wathey and Pettigrew, 1989), which have to summate light from both eyes under dim conditions. Diurnal birds living in open country generally have a pronounced streak aligned with the horizon (Duijm, 1985, Kirmse, 1990). According to phylogenetic conditions (Nalbach et al., 1993) or the ecological habitat, specializations within this streak can be found. A prominent example is the area giganto cellularis along the ora serrata of the dorsotemporal retina in procellariiform seabirds (Hayes et al., 1991). These are pelagic seabirds which come ashore only to breed and spend most of their life wandering close to the surface of the oceans, often within the troughs of the waves. According to Hayes et al. (1991), the location of this specialized retinal area and the morphology of its cells suggest a function in the detection of prey due to relative movements within a small binocular field projecting below and around the bill tip (Martin, 1993).

III. CENTRAL PROCESSING—ANATOMY AND FUNCTION

A. Centrifugal Pathway

The centrifugal visual system of birds originates in two different mesencephalic cell groups: the isthmo-optic nucleus (ION), a folded bilaminate structure in the dorsolateral midbrain tegmentum, and the nucleus of the ectopic isthmo-optic neurons (EION), a loosely scattered array of cells with reticular appearance surrounding the ION (Hayes and Webster, 1981, Wolf-Oberhollenzer, 1987). Both structures are part of a closed loop consisting of a projection from the retinal ganglion cells to the contralateral tectum, the efferents of which in turn project both to the ipsilateral ION and EION, whence back-projections lead to the contralateral retina (Clarke, 1992) (Figure 5). All projections within this system seem to be topographically organized (McGill et al., 1966a,b; Catsicas et al., 1987b). Weidner et al. (1987) showed in a comparative study in different bird species important differences between raptors and ground-feeding birds. In seed- or fruit-eating birds, the ION was always large, well differentiated, and laminated. In raptors, the ION was small, poorly differentiated, and reticular in appearance. The authors suggested from their observations that the centrifugal system is probably involved in pecking and visual food selection among static stimuli at a short viewing distance.

The cell bodies of quail tecto-ION neurons are located in layer 9 of the tectum and with their dendrites branching outside the retinorecipient superficial layers (Uchijama and Watanabe, 1985). Thus, tecto-ION neurons have to receive their retinal input via intratectal mechanisms. Uchijama et al. (1987) could demonstrate that electrical stimulations of the Wulst elicited ION neurons, indicating a forebrain influence on activity patterns within this structure. The situation seems to be slightly different in pigeons and chicks, where tecto-ION neurons reach up to layer 2 with their dendrites and could thus pick up direct retinal input (Woodson et al., 1991).

The ION consists of a highly convoluted lamina in which two perikaryal layers are separated by a neuropil in which the dendrites from opposing layers branch toward the middle of the two layers (Güntürkün, 1987) (Figure 6). Afferent axons of presumably tectal origin pass through this dendritic field and synapse topographically on small dendritic appendages and spines providing virtually all excitatory synapses in the ION (Cowan, 1970; Angaut and Repérant, 1978). Additionally, large numbers of inhibitory synapses on ION dendrites are found, which partly originate from a small number of GABAergic neurons within the ION (Miceli et al., 1995). Axons from ION cells emerge at opposing ends of the two laminae and proceed, together with those from the EION, to the contralateral retina. The number of efferent axons within the optic nerve is supposed to be about 12,000 in the pigeon, of which the ION contributes about 10,000 (Cowan, 1970; Weidner et al., 1987; Wolf-Oberhollenzer, 1987). Since the tecto-ION and the tecto-EION pathways also consist of about 12,000 neurons, a 1:1 ratio of tectal and centrifugal

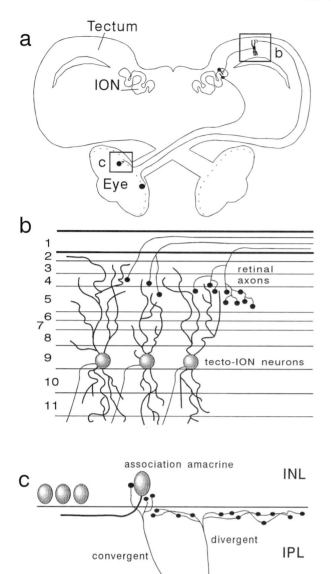

FIGURE 5 Schematic view of different aspects of the avian centrifugal system. (a) Overview of the centrifugal system with retinal ganglion cells projecting to the contralateral tectum, the tectal cells constituting the tecto-ION projection, and the ION neurons, which project back to the contralateral tectum. Components of the schema are not drawn to scale. The EION, which would take a position surrounding the ION, is not depicted. b and c give the position of the drawings with the same letters. (b) Schematic view of the tectum with layers 1 to 11 showing some retinal axons and the position of tecto-ION neurons as described by Woodson *et al.* (1991); (c) schematic view of the retina showing the two types of centrifugal axons. Abbreviations: ION, n. isthmo-opticus; and OFL, optic fiber layer or as in the legend to Figure 3.

neurons is likely (Woodson *et al.*, 1991). The centrifugal axons terminate near the IPL/INL border in the horizontal and ventral retina, barely penetrating the red field (Hayes and Holden, 1983; Catsicas *et al.*, 1987b; Fritzsch *et al.*, 1990). They are composed of two distinct types, with very different degrees of topographic localization. The "convergent" type of axon probably stems from the ION and generally gives rise to a single restricted type of terminal fiber, which forms a dense pericellular nest covering the perikaryon of a single association amacrine cell (Maturana and Frenk, 1965; Dowling and Cowan, 1966; Fritzsch *et al.*, 1990; Uchijama and Ito, 1993; Uchiyama *et al.*, 1995). Association amacrines have long intraretinal axons, are mainly located in the horizontal plus ventral retina, and project dorsally (Catsicas *et al.*, 1987a). In pigeons their projections are directed toward the red field (Ehrlich *et al.*, 1987). The fibers from the ION could thus be involved in a mechanism for switching attention between the upper and the lower field of view (Catsicas *et al.*, 1987a). In contrast, the "divergent" centrifugal axons from the EION give rise to several terminal branches, each constituting an extensive and highly branched arbor of up to 1 mm^2 in the IPL, such that the total termination field of these axons must be several square millimeters (Chmielewski *et al.*, 1990; Fritzsch *et al.*, 1990).

Electrophysiological data are only available for the ION. Miles (1972b) and Holden and Powell (1972) demonstrated that a large number of ION units show a preference for target movements in the anterior visual field and accomodate rapidly to repetitive stimulations, indicating a role in the analysis of transient and dynamic features of the visual environment. Miles (1972c) additionally demonstrated an effect of ION stimulation on the disinhibition of retinal ganglion cell surrounds and activation of ganglion cell centers. This would indicate a role in the modulation of local contrast and luminance sensitivities. Most ION cells have their receptive fields in the inferior anterior visual field and are thus related to the upper posterior parts of the retina, where paradoxically, ION terminals are virtually absent (Hayes and Holden, 1983; Catsicas *et al.*, 1987a).

Several authors tried to establish the functional importance of the ION and EION in behavioral studies. Hodos and Karten (1974), Jarvis (1974), Shortess and Klose (1977), and Knipling (1978) observed only mild or no deficits in visual intensity and pattern discrimination experiments after bilateral centrifugal lesions. However, using a different approach Rogers and Miles (1972) demonstrated profound deficits in the detection of suddenly occuring moving stimuli and the perception of grain on the black squares of a checkerboard. These authors suggested that the centrifugal system may play a role in detecting moving objects and in enhancing contrast under dim light conditions through a mechanism of dynamic adaption at the retinal level. A study of Hahmann and Güntürkün (1992) could not confirm

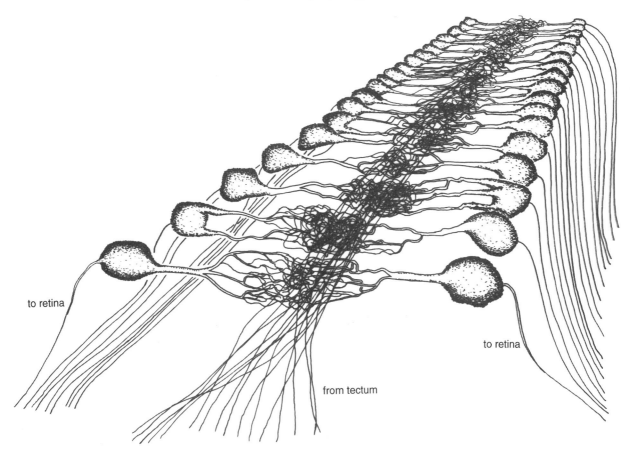

FIGURE 6 Simplified model of the cellular organization of the ION. Only two opposing cell rows separated by dendritic arborizations are shown. Axons entering the system are of presumably tectal origin. (From *Cell Tissue Res.*, A Golgi study of the isthmic nuclei in the pigeon (*Columba livia*), O. Güntürkün, **248,** 439–448, Fig. 5, 1987, © Springer-Verlag.)

the first result but provided additional evidence for the second. Three different behavioral experiments, each testing different aspects of visual analysis, were performed in this study. In the first two experiments, a grain–grit discrimination task and a visual acuity determination, stimuli were presented in the frontal binocular visual field. A third experiment investigated the early detection of moving objects, introduced into the monocular lateral visual field. After bilateral ION and EION lesions a multiple linear regression analysis was employed to correlate the postoperative performance in all three tasks with the amount of structure loss within ION and EION. Deficits in the grain–grit discrimination procedure were a function of the ION lesion extent and did not depend on EION damage. Thus, these two structures could be functionally differentiated. However, neither the ION nor the EION seemd to be involved in visual acuity performance or the early detection of large shadows moving through the visual field. These data support the hypothesis that at least the ION in pigeons is involved through its projections onto the association amacrines in pecking and food selection among static stimuli in the inferior frontal visual field (superiotemporal retina). The electrophysiological study of Miles (1972c) demonstrated an effect of ION stimulation on the disinhibition of retinal ganglion cell receptive field surrounds and the facilitation of the exciatory field centers. The first mechanism would sacrifice contrast sensitivity for responsiveness to a wide range of target forms and would thus confer improved detectability. The second mechanism, on the other hand, would increase sensitivity to small objects without loosing constraints on shape and size, thus facilitating the discriminative capacity of the visual system. Both mechanisms would enable birds to adapt to local optic background variations within the context of feeding or to "highlight" the object of choice as supposed by Uchiyama (1989).

B. Tectofugal Pathway

The tectofugal pathway is composed of optic nerve axons which decussate virtually completely in the chiasma opticum and end in the optic tectum (TO). The tectum projects bilaterally to the thalamic nucleus rotundus (Rt), which itself sends efferent fibers to the ipsilateral ectostriatum (E). Ectostriatal cells project to a surrounding shelf area, the ectostriatal belt (Eb), from where intratelencephalic projections lead to different forebrain structures (Figure 7).

In probably most avian species the majority of retinal ganglion cells project to the optic tectum. The exact proportion is difficult to estimate but according to the data of Bravo and Pettigrew (1981) in the barn owl *Tyto alba* and Remy and Güntürkün (1991) in pigeons, 75–95% of ganglion cells have axons leading to the tectum. With regard to these numbers, the burrowing owl *Speotyto cunicularia* is an exception. This bird is supposed to rely heavily on its thalamofugal pathway and consequently seems to have less than 50% tectally projecting ganglion cells (Bravo and Pettigrew, 1981).

Retinal axons, which constitute the first of the 15 tectal laminae, innervate only superficial layers 2–7 and reach their highest synaptic density in layer 5 (Hayes and Webster, 1985). The retinal projection onto the tectum is strictly topographical in all species studied with the inferior retina projecting to the dorsal tectum while the posterior tectum is reached by the nasal retina (Clarke and Whitteridge, 1976; Frost *et al.*, 1990a; Remy and Güntürkün, 1991). The tectal representation of the foveae or the areas of enhanced vision are considerably expanded (Clarke and Whitteridge, 1976; Frost *et al.*, 1990a). Single-unit recordings in the optic tectum demonstrate that the visual receptive fields of neurons in the superficial layers are small (0.5–4°) but increase to up to 150° in deeper laminae (Jassik-Gerschenfeld *et al.*, 1975; Frost *et al.*, 1981). Despite this modulation in the z-axis, changes in the x- and y-axes also occur: Frost *et al.* (1990a) could demonstrate that receptive field sizes increase from foveal to peripheral representations in the tectum of the American kestrel. Most responses to stationary targets have typical on–off characteristics, and in a large number of cells the activating area is surrounded by an inhibitory region (Hughes and Pearlman, 1974). Using moving bars of monochromatic light, Letelier (1983) could show that 30% of the recorded tectal units had specific wavelength preferences, mostly for short wavelengths. The majority of cells (70%) are movement sensitive with about 30% of them having directional preferences (Jassik-Gerschenfeld and Guichard, 1972). Directionally responsive units are either narrowly tuned or, more commonly, they respond to a wide range of directions, with the majority being inhibited by backward movement (Frost and DiFranco,

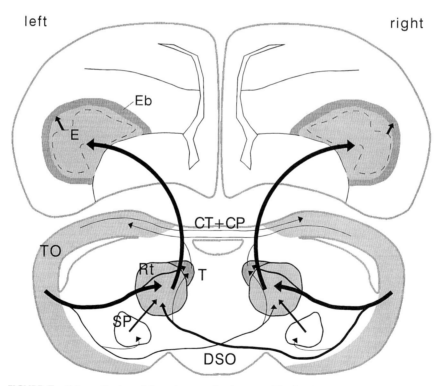

FIGURE 7 Schematic view of the avian tectofugal system. The broken lines within E gives the borderline between the subcomponents as described by Hellmann *et al.* (1995). Abbreviations: CP, commissura posterior; CT, commissura tectalis; DSO, dorsal supraoptic decussation; E, ectostriatum; Eb, ectostriatal belt; Rt, n. rotundus; SP, n. subpretectalis; T, n. triangularis; and TO, tectum opticum.

1976). If the stimulus is positioned on a random-dot background, tectal cells are activated when the background is moved out of phase to the stimulus, while a profound inhibition is produced by an in-phase motion of background and stimulus (Frost and Nakayama, 1983). These data strongly suggest that tectal cells play an important function in figure–ground segregation through discontinuities in velocity (Frost et al., 1990b). If kinematograms (motion equivalents of random dot stereograms) are presented, tectal cells in deep laminae respond only to virtual "objects" shearing above the surface of the background and not to virtual "holes" in the background through which a further texture is visible (Frost et al., 1988). Thus, deep tectal units seem to prefer moving objects rather than movement per se.

Tectal cells also respond selectively to the spatial frequency of drifting sine wave gratings, with most neurons having their optima between 0.45 and 0.6 c/deg (Jassik-Gerschenfeld and Hardy, 1979). Most of these cells are more selective to spatial frequencies than they are to single bar stimuli (Jassik-Gerschenfeld and Hardy, 1980). Birds therefore appear to be able to perform Fourier analysis of patterns in visual space. Recently, Neuenschwander and Varela (1994) could additionally prove the presence of visually triggered gamma oscillations in the pigeon's tectum. This oscillatory activity had characteristics similar to those reported in the mammalian neocortex in the context of synchronization of unit responses as a putative physiological basis of perceptual binding (Singer, 1993).

In pigeons, up to layer 12, most, if not all, cells are purely visually driven, while deeper neurons are often bi- or multimodal, integrating visual, auditory, and somatosensory afferents (Cotter, 1976). This seems to be different in the barn owl, in which the majority of superficial and deep tectal units are bimodal and have their auditory and visual receptive fields in the same space coordinates (Knudsen, 1982).

Visual information is transmitted either directly by axodendritic contacts or with interneurons to the cells of layer 13, which project to the Rt in the thalamus (Hardy et al., 1985). Within Rt they end in synaptic glomerulilike structures constituted by the end-claws of several dendrites and bundles of tectal fibers (Thin et al., 1992; Tömböl et al., 1992). As demonstrated by Bischof and Niemann (1990) and Güntürkün et al. (1993a), the tectal projection is bilaterally organized with the efferents to the contralateral TO crossing through the dorsal supraoptic decussation. Accordingly, Engelage and Bischof (1988) could demonstrate the existence of ipsi- and contralaterally evoked potentials in the Rt of zebra finches. The Rt seems to consist of several distinct subfields as shown by histochemical (Martinez-de-la Torre et al., 1990) and electrophysiological results (Revzin, 1979, Wang et al., 1993). In anterior sections Rt units seem to be specialized successively from dorsal to ventral to color, luminance, and 2D motion, while in posterior sections looming cells can be found in the most dorsal portion of the Rt (Wang et al., 1993). These looming cells seem to signal time to collision with an activity peak about 1 sec before virtual collision with the object (Wang and Frost, 1992).

Rt- and T-cells project ipsilaterally and in a topographic manner onto the ectostriatum (E) in the forebrain, from where projections lead to the surrounding ectostriatal belt (Eb) (Benowitz and Karten, 1976). Due to a Wulst projection onto Eb, this structure seems to be the first forebrain entity in which thalamo- and tectofugal systems interact (Ritchie, 1979). Kimberly et al. (1971) established that E-cells have properties similar to those of Rt, that is, most respond preferentially to moving stimuli with wide receptive fields. Engelage and Bischof (1988) revealed ipsi- and contralaterally evoked potentials in the E of zebra finches and even showed an intraectostriatal differentiation in the current source density profile of ipsi-, contra-, and binocularly evoked potentials (Engelage and Bischof, 1989). Hellmann et al. (1995) demonstrated that the E can be parceled according to the long-term activity pattern of its neurons into at least two components which might reflect ocular dominance areas within this structure.

Lesions of tectum or Rt cause pronounced deficits in pattern (TO: Jarvis, 1974; Hodos and Karten, 1974; Rt: Hodos and Karten, 1966; E: Hodos and Karten, 1970), intensity (TO: Hodos and Karten, 1974; Rt: Hodos and Karten, 1966; E: Hodos and Karten, 1970), or color discrimination (Rt: Hodos, 1969). Psychophysical techniques confirmed the drastic elevation of acuity or intensity thresholds after tectofugal lesions (Rt: Hodos and Bonbright, 1974; Macko and Hodos, 1984; E: Hodos et al., 1984, 1988). The data of Güntürkün and Hahmann (1998) make it likely that the tectofugal system operates according to asymmetric principles. Their unilateral lesions of the Rt revealed that only structure loss within the left Rt correlates significantly with right- or left-sided acuity losses, while right-sided Rt lesions had no impact on monocular acuity. These behavioral data thus confirm the anatomical results of Güntürkün and Melsbach (1992), who demonstrated that left-sided Rt injection of retrograde tracers revealed a twice-as-numerous contingent of contralaterally projecting tectal neurons than after right-sided injections. Since each tectum represents the input from the contralateral eye, asymmetries in the contralateral tectal afferents could create asymmetries in the degree of the visual bilateral integration at the rotundal level. Despite these left–right asymmetries, behavioral studies additionally show a frontal–lateral difference within the tectofugal system.

According to Güntürkün and Hahmann (1998), Rt lesions interfere with acuity in the frontal but not in the lateral visual field. At the same time GLd lesions attenuate lateral but not frontal acuity. Thus, frontal and lateral visual acuity seem to depend on tecto- and thalamofugal mechanisms, respectively.

C. Thalamofugal Pathway

The thalamofugal pathway consists of the retinal projection onto the n. geniculatis lateralis pars dorsalis (GLd), a group of nuclei in the contralateral dorsal thalamus, and the bilateral projection of the GLd onto the Wulst ("bulge") in the anteriodorsal forebrain (Güntürkün et al., 1993b). Most people agree that the avian thalamofugal pathway corresponds due to its anatomical, physiological, and functional properties to the mammalian geniculostriate system (Shimizu and Karten, 1993) (Figure 8).

While the tectofugal pathway receives afferents from the complete extent of the retina, the retinal location of ganglion cells projecting onto the GLd differs in various species. In birds of prey, ganglion cells in the temporal retina subserving frontal vision project primarily onto the GLd (Bravo and Pettigrew, 1981; Bravo and Inzunza, 1983). Consequently, many neurons in the visual Wulst of owls, kestrels, and vultures possess binocular visual fields and detect retinal disparity (Pettigrew, 1979; Porciatti et al., 1990). In pigeons, however, mainly ganglion cells outside the "red field" of superiotemporal retina have efferents to the GLd (Remy and Güntürkün, 1991). The paucity of afferents from the red field should render the thalamofugal pathway of pigeons largely frontally blind, an assumption supported by electrophysiological results (Miceli et al., 1979). Thus, while the GLd of several birds of prey seems to be specialized for the frontal binocular visual field, the GLd of pigeons mainly receives afferents from the lateral monocular field. This functionally important difference is not the result of the laterally placed eyes of pigeons, since the kestrel, a diurnal raptor that has lateral eyes, is also characterized by an overrepresentation of the frontal binocular visual field within its thalamofugal pathway (Pettigrew, 1978). The "frontal blindness" of the pigeon's thalamofugal system is very likely the reason for the virtual absence of behavioral deficits in a variety of discrimination tasks after GLd or Wulst lesions (Güntürkün, 1991). Generally in these experiments the pigeons were required to perform discriminative pecking responses to patterns presented upon response keys. Pigeons pecking a key fixate it with their red field (Goodale, 1983). Since the red field has only limited projections onto the GLd, thalamofugal lesions are likely to produce minimal deficits when tested with this procedure. When using discriminations of laterally presented stimuli, GLd lesions produce severe deficits (Güntürkün and Hahmann, 1998).

The differing ecological demands of seed-eating versus hunting birds are probably the reason for the different thalamofugal specialization to only one visual field. Pigeons and many other seed- or fruit-eating birds fixate novel or complex and distant stimuli laterally and only switch to frontal binocular vision to peck the scrutinized object (Bischof, 1988; Bloch et al., 1988). Thus, in these species visual detection and analysis is mainly performed by those parts of the neural apparatus which represent the lateral visual field, while the frontal binocular area is only involved during the last visually guided sequences before and within pecking bouts. The lateral specialization of the thalamofugal pathway in pigeons could therefore be related to the fact that it is mainly the lateral visual field which requires fine analysis of the visual scenery. The frontal specialization of the thalamofugal system in birds of prey could be related to their more complex feeding habits which require them to specify the distance of objects with great precision through flow-field variables while moving with high speed (Davies and Green, 1990). Although eagles and falcons fixate distant objects mainly laterally (Reymond, 1985, 1987) they switch to frontal vision when approaching prey. The need for complex and fast visual information analysis of moving objects could explain the specialization of the thalamofugal pathway to the frontal visual field in birds of prey.

The GLd is composed of six components, of which only four constitute the core portion since they are retinorecipient and project onto the visual Wulst: n. dorsolateralis anterior thalami, pars lateralis (DLL), n. dorsolateralis anterior thalami, pars magnocellularis (DLAmc), n. lateralis dorsalis nuclei optici principalis thalami (LdOPT), and the n. suprarotundus (SpRt), with DLL and DLAmc being the two largest substructures (Güntürkün and Karten, 1991). Avian GLd neurons are also characterized by relatively small receptive fields (1° in owls, 2°–4° in pigeons, 3° in chicks), by center-surround organization and by a low adaptation to stimulus repetition (Pateromichelakis, 1981; Britten 1987). Aditionally, in pigeons and chicks many directionally selective cells with large receptive fields have been encountered (Wilson, 1980; Britten, 1987).

The GLd-Wulst projection is bilateral and topographically organized in all species studied, but the relative contribution of both sides probably depends on the orientation of this system to the frontal or the lateral field of view (Bagnoli et al., 1990; Miceli et al., 1990; Güntürkün et al., 1993b). In owls with their essentially frontal eyes and the frontal thalamofugal orientation, the ipsi- and contralateral sides contribute an approximately equal number of fibers to the forebrain projec-

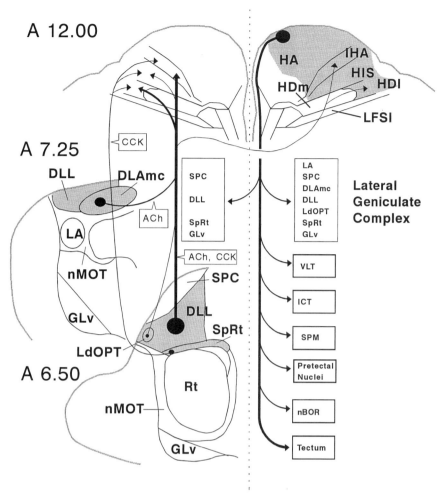

FIGURE 8 Schematic view of the avian thalamofugal pathway. The various core components of the dorsolateral geniculate complex (gray shading) are depicted to the left together with the ascending projections onto the ipsi- and contralateral Wulst and the neurotransmitters/modulators of the relay neurons. Note that the SpRt projects only ipsilaterally. The sizes of the black circles indicate the relative number of neurons contributing to the depicted projections. At the top the subdivisions of the Wulst are given with the nomenclature and the presumed extent of the visual components to the right (gray shading). To the right, the descending Wulst projections via the TSM are depicted. The broken line is the partition between the hemispheres. Each section represents a frontal plane of the atlas of Karten and Hodos (1967). Abbreviations: ACh, acetylcholine; CCK, cholecystokinin; DLAmc, n. dorsolateralis anterior thalami, pars magnocellularis; DLL, n. dorsolateralis anterior thalami, pars lateralis; GLv, n. geniculatus lateralis, pars ventralis; HA, hyperstriatum accessorium; HDl, lateral portion of the hyperstriatum dorsale; HDm, medial portion of the hyperstriatum dorsale; HIS, Hyperstriatum intercalatus superior; ICT, n. intercalatus thalami; IHA, n. intercalatus hyperstriati accessorii; LA, n. lateralis anterior; LdOPT, n. lateralis dorsalis nuclei optici principalis thalami; LFSl, lateral portion of the lamina frontalis superior; nBOR, n. of the basal optic root, pars dorsalis; nMOT, n. marginalis tractus optici; Rt, n. rotundus; SPC, n. superficialis parvocellularis; SPM, n. spiriformis medialis; SpRt, n. suprarotundus; and VLT, n. ventrolateralis thalami. (After Güntürkün et al., 1993b, in *Vision, Brain and Behavior,* The MIT Press. © 1993 The MIT Press.)

tion (Bagnoli et al., 1990). In lateral-eyed birds like pigeons, only a few contralateral projections can be found (Hahmann et al., 1994).

The Wulst can be subdivided into a rostral somatosensory, a medial hippocampal, and a caudal visual division. The visual Wulst is organized from dorsal to ventral in four laminae: hyperstriatum accessorium (HA), intercalated nucleus of the hyperstriatum accessorium (IHA), hyperstriatum intercalatus superior (HIS), and hyperstriatum dorsale (HD). These subdivisions are based on the cytoarchitectonics of the Wulst and do not reflect the full complexity of the structure, since Shimizu and Karten (1990) were able to distinguish at least eight subdivions using immunocytochemical techniques. The granular IHA and probably also lateral HD are the major recipients of the cholinergic and colecystokinergic GLd input (Watanabe et al., 1983; Güntürkün and Karten, 1991). Electrophysiological studies demonstrate similarities between the visual Wulst of birds of prey and the striate cortex of mammals. In the visual Wulst of raptors most neurons are primarily concerned with binocular visual processing, are selectively tuned to stereoscopic depth cues, are sensitive to visual experience during the neonatal period, and have small receptive fields of about 1° (Pettigrew and Konishi, 1976a,b; Pettigrew, 1979). This is not the case for species like pigeons, chickens, and zebra finches in which binocular neurons are rare or in which ipsilaterally evoked visual responses are very weak and irregular (Bredenkötter and Bischof, 1990a,b). Additionally, the receptive fields encountered are considerably larger in nonraptors (pigeons: 2°, Revzin, 1969; chickens: 10°–20°, Wilson, 1980).

Raptors especially need stereoscopic depth cues. However, the distance of a visual target cannot be directly determined from its retinal location but has to be computed by comparing inputs from both eyes. The fundamental problem for a binocular system is to find the correct correspondence; that is, to identify the pair of image segments that belong to the same visual target (Pettigrew, 1993). With repeated stimuli like from a certain spatial frequency the phase ambiguity has thus to be resolved by the nervous system. Wagner and Frost (1993) proposed a solution to this problem by assuming that disparity sensitive neurons in the Wulst of barn owls might be tuned to a characteristic disparity. Indeed, they found that in many disparity-sensitive neurons the reaction peak to visual noise at a certain disparity did not change when using stimuli of different spatial frequencies (Wagner and Frost, 1994). Thus, disparity-sensitive cells in the barn owl's Wulst have a characteristic disparity which could be used to detect the depth plane of a stimulus which exhibits the appropriate combinations of spatial frequency and interocular phase.

The extratelencephalic Wulst efferents project primarily to the GLd, pretectal nuclei, the basal optic root nucleus, and the tectum opticum (Miceli et al., 1987). Within GLd, the terminal fields partially overlap with those areas which both receive direct retinal input and project to the visual Wulst. The Wulst thus modulates its own GLd- input either by direct excitation of relay neurons or by inhibition of GABAergic interneurons (Watanabe, 1987). The Wulst projection onto the tectum is probably of great functional importance. Leresche et al. (1983) could demonstrate that many tectal cells depend for their receptive field properties upon input from an intact Wulst. Cryogenic block of the Wulst caused a reversible response depression of a majority of tectal cells and drastically diminished the directional tuning of half of the directionally selective neurons. Thus, the visual properties of tectal cells are not solely a reflection of the retinal afferents and the intratectal circuitry, but also depend on the thalamofugal input.

References

Angaut, P., and Repérant, J. (1978). A light and electron microscopic study of the nucleus isthmo-optic in the pigeon. *Arch. Anat. Mikr.* **67,** 63–78.

Bagnoli, P., Fontanesi, G., Casini, G., and Porciatti, V. (1990). Binocularity in the little owl, *Athene noctua.* I. Anatomical investigation of the thalamo-Wulst pathway. *Brain Behav. Evol.* **35,** 31–39.

Bellhorn, R. W., and Bellhorn, M. S. (1975). The avian pecten. *Ophtalmic Res.* **7,** 1–7.

Benowitz, L. I., and Karten, H. J. (1976). Organization of tectofugal visual pathway in pigeon: Retrograde transport study. *J. Comp. Neurol.* **167,** 503–520.

Binggeli, R. L., and Paule, W. J. (1969). The pigeon retina: Quantitative aspects of the optic nerve and ganglion cell layer. *J. Comp. Neurol.* **137,** 1-18.

Bischof, H. J. (1988). The visual field and visually guided behavior in the zebra finch (*Taeniopygia guttata*). *J. Comp. Physiol. A* **163,** 329–337.

Bischof H. J., and J. Niemann, (1990). Contralateral projection of the optic tectum in the zebra finch (*Taenopygia guttata castanotis*). *Cell Tissue Res.* **262,** 307–313.

Bloch, S., and Martinoya, C. (1982). Comparing frontal and lateral viewing in the pigeon. I. Tachistoscopic visual acuity as a function of distance. *Behav. Brain Res.* **5,** 231–244.

Bloch, S., Jäger, R., Lemeignant, M., and Martinoya, C. (1988). Correlations between ocular saccades and headmovements in walking pigeons, *J. Physiol.* (*London*) **406,** 173.

Blough, P. M. (1971). The visual acuity of the pigeon for distant targets. *J. Exp. Anal. Behav.* **15,** 57–67.

Bowmaker, J. K. (1977). The visual pigments of oil droplets and spectral sensitivity of the pigeon. *Vision Res.* **17,** 1129–1138.

Boycott, B. B., and Wässle, H. (1974). The morphological types of ganglion cells of the domestic cat's retina. *J. Physiol.* **240,** 397–419.

Bravo, H., and Inzunza, O. (1983). Estudio anatomico en las vias visuales parallelas en falconiformes. *Arch. Biol. Med. Exp.* **16,** 283–289.

Bravo, H., and Pettigrew, J. D. (1981). The distribution of neurons projecting from the retina and visual cortex to the thalamus and tectum opticum of the barn owl, *Tyto alba,* and the burrowing owl, *Speotyto cunicularia. J. Comp. Neurol.* **199,** 419–441.

Brecha, N. C., and Karten, H. J. (1981). Organization of avian accessory optic system. *Ann. N.Y. Acad. Sci.* **374,** 215–229.

Bredenkötter, M., and Bischof, H.-J. (1990a). Differences between ipsilaterally and contralaterally evoked potentials in the visual Wulst of the zebra finch. *Vis. Neurosci.* **5,** 155–163.

Bredenkötter, M., and Bischof, H.-J. (1990b). Ipsilaterally evoked responses of the zebra finch visual Wulst are reduced during ontogeny. *Brain Res.* **515,** 343–346.

Britten, K. H. (1987). Receptive fields of neurons of the principal optic nucleus of the pigeon (*Columba livia*). Ph.D. thesis, SUNY, Stony Brook.

Britto, L. R. G., and Hamassaki-Britto, D. E. (1991). A subpopulation of displaced ganglion cells of the pigeon retina exhibits substance P-like immunoreactivity. *Brain Res.* **546**, 61–68.

Britto, L. R. G., and Hamassaki-Britto, D. E. (1992). Enkephalin-immunoreactive ganglion cells in the pigeon retina. *Vis. Neurosci.* **9**, 389–398.

Britto, L., and Hammassaki, D. E. (1991). A subpopulation of displaced ganglion cells of the pigeon retina exhibits substance P-like immunoreactivity. *Brain Res.* **546**, 61–68.

Britto, L. R. G., Keyser, K. T., Hamassaki, D. E., and Karten, H. J. (1988). Catecholaminergic subpopulation of retinal displaced ganglion cells projects to the accessory optic nucleus in the pigeon (*Columba livia*). *J. Comp. Neurol.* **269**, 109–117.

Britto, L. R. G., Hamassaki, D. G., Keyser, K. T., and Karten, H. J. (1989). Neurotransmitters, receptors, and neuropeptides in the accessory optic system: An immunohistochemical survey in the pigeon (*Columba livia*). *Visual Neurosci.* **3**, 463–475.

Budnik, V., Mpodozis, J., Varela, F. J., and Maturana, H. R. (1984). Regional specialization of the quail retina: ganglion cell density and oil droplet distribution. *Neurosci. Lett.* **51**, 145–150.

Burkhardt, D. (1989). UV vision: A bird's eye view of feathers. *J. Comp. Physiol. A* **164**, 787–796.

Cajal, S. R. (1892). La rétine des vertébrés. *La Cellule* **9**, 119–257.

Catsicas, S., Catsicas, M., and Clarke, P. G. H. (1987a). Long-distance intraretinal connections in birds. *Nature* **326**, 186–187.

Catsicas, S., Thanos, S., and Clarke, P. G. H. (1987b) Major role for neuronal death during brain development: Refinement of topographical connections. *Proc. Natl. Acad. Sci. USA* **84**, 8165–8168.

Cerella, J. (1980). The Pigeon's analysis of pictures. *Pattern Recog.* **12**, 1–6.

Chen, D. M., and Goldsmith, T. H. (1986). Four spectral classes of cone in the retinas of birds. *J. Comp. Physiol. A* **159**, 473–479.

Chmielewski, C. E., Dorado, M. E., Quesada, A., Géniz-Gálvéz, J.-M., and Prada, F. A. (1988). Centrifugal fibers in the chick retina. *Anat. Histol. Embryol.* **17**, 319–327.

Clarke, P. G. H. (1992). Neuron death in the developing avian isthmo-optic nucleus, and its relation to the establishment of functional circuitry. *J. Neurobiol.* **23**, 1140–1158.

Clarke, P. G. H., and Whitteridge D. (1976). Projection of retina, including Red Area onto optic tectum of pigeon. *Quart. J. Exp. Psychol.* **61**, 351–358.

Cotter, J. R. (1976). Visual and nonvisual units recorded from optic tectum of *Gallus domesticus*. *Brain Behav. Evol.* **13**, 1–21.

Cowan, W. M. (1970). Centrifugal fibres to avian retina. *Br. Med. Bull.* **26**, 112–119.

Curcio, C. A., and Allen, K. A. (1990). Topography of ganglion cells in human retina. *J. Comp. Neurol.* **300**, 5–25.

Davies, M. N. O., and Green, P. R. (1990). Optic flow-field variables trigger landing in hawk but not in pigeons. *Naturwissenschaften* **77**, 142–144.

Delius, J. D., Jahnke-Funk, E., and Hawker, A. (1981). Stimulus display geometry and colour discrimination learning by pigeons. *Curr. Psych. Res.* **1**, 203–214.

DiStefano, M., Kusmic, C., and Musumeci, D. (1987). Binocular interactions measured by choice reaction times in pigeons. *Behav. Brain Res.* **25**, 161–165.

Dowling, J. E., and Cowan, W. M. (1966). Electron microscope study of normal and degenerating centrifugal fibre terminals in pigeon retina. *Z. Zellforsch.* **71**, 14–28.

Duijm, M. (1958). On the position of a ribbon like central area in the eyes of some birds. *Arch. Neerl. Zool.* **13**, 128–145.

Ehrlich, D. (1981). Regional specialization of the chick retina as revealed by the size and density of neurons in the ganglion cell layer. *J. Comp. Neurol.* **195**, 643–657.

Ehrlich, D., Keyser, K. T., and Karten, H. J. (1987). Distribution of substance P-like immunoreactive retinal ganglion cells and their pattern of termination in opt tect of chick. *J. Comp. Neurol.* **266**, 220–233.

Emmerton, J. (1983a). Pattern discrimination in the near-ultraviolet by pigeons. *Percept. Psychophys.* **34**, 555–559.

Emmerton, J. (1983b). Vision. *In* "Physiology and Behaviour of the Pigeon" (M. Abs, ed.), pp. 245–266. Academic Press, London.

Emmerton, J., and Delius, J. D. (1980). Wavelength discrimination in the "visible" and ultraviolet spectrum by pigeons. *J. Comp. Physiol.* **141**, 47–52.

Emmerton, J., Schwemer, J., Muth, I., and Schlecht, P. (1980). Spectral transmission of the ocular media of the pigeon (*Columba livia*). *Invest. Ophtalmol. Visual Sci.* **19**, 1382–1387.

Engelage, J., and Bischof, H. J. (1988). Enucleation enhances ipsilateral flash evoked responses in ectostriatum of zebra fish. *Exp. Brain Res.* **70**, 79–89.

Engelage, J., and Bischof, H. J. (1989). Flash evoked potentials in ectostriatum of zebra finch: current source density analysis. *Exp. Br. Res.* **74**, 563–572.

Evans, H. E. (1996). Anatomy of the budgerigar and other birds. *In* "Diseases of Cage and Aviary Birds" (W. J. Rosskopf and R. W. Woerpel, eds.). Williams & Wilkins Co., Baltimore.

Fite, K. V., Brecha, N., Karten, H. J., and Hunt, S. P. (1981). Displaced ganglion cells and the accessory optic system of pigeon. *J. Comp. Neurol.* **195**, 279–288.

Fitzke, F. W., Hayes, B. P., Hodos, W., Holden, A. L., and Low, J. C. (1985). Refractive sectors in the visual field of the pigeon eye. *J. Physiol.* (*London*) **369**, 33–44.

Fox, R., Lehmkuhle, S. W., and Westendorf, D. H. (1976). Falcon visual acuity. *Science* **192**, 263–265.

Friedman, M. B. (1975). How birds use their eyes. *In* "Neural and Endocrine Aspects of Behaviour in Birds" (P. Wright, P. Cryl, and D. M. Volwes, eds.), pp. 182–204. Elsevier, Amsterdam.

Fritzsch, B., Crapon DeCaprona, M.-D., and Clarke, P. G. H. (1990). Development of two morphological types of retinopetal fibers in chick embryos, as shown by the diffusion along axons of a carbocyanine dye in the fixed retina. *J. Comp. Neurol.* **301**, 1-17.

Frost, B. J., and diFranco, D. E. (1976). Motion characteristics of single units in pigeon optic tectum. *Vision Res.* **16**, 1229–1234.

Frost, B. J., and Nakayama, K. (1983). Single visual neurons code opposing motion independent of direction. *Science* **220**, 744–745.

Frost, B. J., Cavanaugh, P., and Morgan, B. (1988). Deep tectal cells in pigeon respond to kinematograms. *J. Comp. Physiol. A* **162**, 639–647.

Frost, B. J, Scilley, P. L., and Wong, S. C. P. (1981). Moving Background patterns reveal double-opponency of directionally specific pegeon tectal neurons. *Exp. Brain Res.* **43**, 173–185.

Frost, B. J., Wise, L. Z., Morgan, B., and Bird, D. (1990a). Retinotopic representation of the bifoveate eye of the kestrel (*Falco sparverius*) on the optic tectum. *Vis. Neurosci.* **5**, 231–239.

Frost, B. J., Wylie, D. R., and Wang, Y.-C. (1990b). The processing of object and self-motion in the tectofugal and accessory optic pathways of birds. *Vision Res.* **30**, 1677–1688.

Fukuda, Y., Stone, J. (1974). Retinal distribution and central projections of Y-, X-, and W-cells of the cat's retina. *J. Neurophysiol.* **37**, 749–772.

Galifret, Y. (1968). Les diverses aires fonctionelles de la retine di pigeon. *Z. Zellforsch. Mikrosk. Anat.* **86**, 535–545.

Goodale, M. A. (1983).Visually guided pecking in the pigeon (*Columba livia*). *Brain Behav. Evol.* **22**, 22–41.

Güntürkün, O. (1987). A Golgi study of the isthmic nuclei in the pigeon (*Columba livia*). *Cell Tissue Res.* **248**, 439–448.

Güntürkün, O. (1991). The functional organization of the avian visual system. In "Neural and Behavioural Plasticity: The Use of the Domestic Chick as a Model" (R. J. Andrew, ed.), pp. 92–105. Oxford Univ. Press, Oxford.

Güntürkün, O., and Hahmann, U. (1994). Cerebral asymmetries and visual acuity in pigeons, *Behav. Brain Res.* **60,** 171–175.

Güntürkün, O., and Hahmann, U. (1998). Functional subdivisions of the ascending visual pathways in the pigeon. *Behav. Brain Res.,* in press.

Güntürkün, O., and Karten, H. J. (1991). An immunocytochemical analysis of the lateral geniculate complex in the pigeon (*Columba livia*). *J. Comp. Neurol.* **314,** 721–749.

Güntürkün, O., and Melsbach, G. (1992). Asymmetric tecto-rotundal projections in pigeons: Cues to the origin of lateralized binocular integration. In "Rhythmogenesis in Neurons and Networks" (N. Elsner and D. W. Richter, eds.), p. 364. Thieme, Stuttgart.

Güntürkün, O., Melsbach, G., Hörster, W., and Daniel S. (1993a). Different sets of afferents are demonstrated by the two fluorescent tracers Fast Blue and Rhodamine. *J. Neurosci. Meth.* **49,** 103–111.

Güntürkün, O., Miceli, D., and Watanabe, M. (1993b). Anatomy of the Avian Thalamofugal Pathway. In "Vision, Brain and Behavior in Birds" (H. P. Zeigler and H.-J. Bischof, eds.), p. 115–135. MIT Press, Cambridge.

Hahmann, U., and Güntürkün, O. (1992). Visual discrimination deficits after lesions of the centrifugal visual system in pigeons. *Visual Neurosci.* **9,** 225–234.

Hahmann, U., and Güntürkün, O. (1993). The visual acuity for the lateral visual field in the pigeon (*Columba livia*). *Vision Res.* **33,** 1659–1664.

Hahmann, U., Güntürkün, O., and Shimizu, T. (1994). Immunohistochemical analysis of the thalamofugal-Wulst projection of the pigeon (*Columba livia*). In "Göttingen Neurobiology Report" (N. Elsner and H. Breer, eds.), p. 488. Thieme, Stuttgart.

Hamassaki-Britto, D. E, Brzozowska-Prechtl, A., and Karten, H. J. (1991). GABA-like immunoreactive cells containing nicotinic acetylcholine receptors in the chick retina. *J. Comp. Neurol.* **313,** 394–408.

Hardy, O., Leresche, N., and Jassik-Gerschenfeld, D. (1985). Morphology and laminar distribution of electrophysiologically identified cells in the pigeon's optic tectum: An intracellular study. *J. Comp. Neurol.* **233,** 390–404.

Hayes, B., Martin, G. R., and Brooke, M. de L. (1991). Novel area serving binocular vision in the retinae of Procellariiform seabirds. *Brain. Behav. Evol.* **37,** 79–84.

Hayes, B. P., and Holden, A. L. (1980). Size classes of ganglion cells in the central yellow field of the pigeon retina. *Exp. Brain Res.* **39,** 269–275.

Hayes, B. P., and Holden, A. L. (1983). The distribution of displaced ganglion cells in the retina of the pigeon. *Exp. Brain Res.* **49,** 181–188.

Hayes, B. P., and Webster, K. E. (1981). Neurons situated outside the isthmo-optic nucleus and projecting to the eye in adult birds. *Neurosci. Lett.* **26,** 107–112.

Hayes, B. P., and Webster, K. E. (1985). Cytoarchitectural fields and retinal termination: an axonal transport study of laminar organization in the avian optic tectum. *Neuroscience* **16,** 641–657.

Hellmann, B., Waldmann, C., and Güntürkün, O. (1995). Cytochrome oxidase activity reveals parcellations of the pigeon's ectostriatum. *Neuroreport.* **6,** 881–885.

Hodos, W. (1969). Color discrimination deficits after lesions of the nucleus rotundus in pigeons. *Brain Behav. Evol.* **2,** 185–200.

Hodos, W., and Bonbright, J. L. (1974). Intensity difference thresholds in pigeons after lesions of the tectofugal and thalamofugal visual pathways. *J. Comp. Physiol. Psychol.* **87,** 1013–1031.

Hodos, W., and Erichsen, J. T. (1990). Lower field myopia in birds: An adaptation that keeps the ground in focus. *Vision Res.* **30,** 653–657.

Hodos, W., and Karten, H. J. (1966). Brightness and pattern discrimination deficits in the pigeon after lesions of nucleus rotundus. *Exp. Brain Res.* **2,** 151–167.

Hodos, W., and Karten, H. J. (1970). Visual intensity and pattern discrimination deficts after lesions of ectostriatum in pigeons. *J. Comp. Neurol.* **140,** 53–68.

Hodos, W., and Karten, H. J. (1974). Visual intensity and pattern discrimination deficits after lesions of the optic lobe in pigeons. *Brain. Behav. Evol.* **9,** 165–194.

Hodos, W., and Leibowitz, R. W. (1977). Near-field visual acuity of pigeons: Effects of scotopic adaption and wavelength. *Vision Res.* **17,** 463–467.

Hodos, W., Bessette, B. B., Macko, K. A., and Weiss, S. R. B. (1985). Normative data for pigeon vision. *Vision Res.* **25,** 1525–1527.

Hodos, W., Leibowitz, R. W., and Bonbright, J. C. Jr. (1976). Near-field visual acuity of pigeons: Effects of head location and stimulus luminance. *J. Exp. Anal. Behav.* **25,** 129–141.

Hodos, W., Macko, K. A., and Bessette, B. B. (1984). Near-field acuity changes after visual system lesions in pigeons. II. Telencephalon. *Behav. Brain Res.* **13,** 15–30.

Hodos, W., Miller, R. F., and Fite, K. V. (1991a). Age-dependent changes in visual acuity and retinal morphology in pigeons. *Vision Res.* **31,** 669–677.

Hodos, W., Miller, R. F., Fite, K. V., Porciatti, V., Holden, A. L., Lee, J.-Y., and Djamgoz, M. B. A. (1991b). Life-span changes in the visual acuity and retina in birds, In "The Changing Visual System" (P. Bagnoli and W. Hodos, eds.), pp. 137–148. Plenum, New York.

Hodos, W., Weiss, S. R. B., and Bessette, B. B. (1988). Intensity difference thresholds after lesions of ectostriatum in pigeons. *Behav. Brain Res.* **30,** 43–53.

Holden, A. L. (1978). Antidromic invasion of ganglion cells in the pigeon retina. *Vision Res.* **18,** 1357–1365.

Holden, A. L., and Powell, T. P. S. (1972). The functional organization of the isthmo-optic nucleus in the pigeon. *J. Physiol.* **233,** 419–447.

Hughes, C. P., and Pearlman, A. L. (1974). Single unit receptive fields and the cellular layers of the pigeon optic tectum. *Brain Res.* **80,** 365–377.

Ikushima, M., Watanabe, M., and Ito, H. (1986). Distribution and morphology of retinal ganglion cells in the Japanese quail. *Brain Res.* **376,** 320–334.

Illing, R. B., and Wässle, H. (1981). The retinal projection to the thalamus in the cat: A quantitative investigation and a comparison with the retinotectal pathway. *J. Comp. Neurol.* **202,** 265–285.

Inzunza, O., Bravo, H., Smith, R. L., and Angel, M. (1991). Topography and morphology of retinal ganglion cells in Falconiforms: A study on predatory and carrion-eating birds. *Anat. Rec.* **229,** 271–277.

Jane, S. D., and Bowmaker, J. K. (1988). Tetrachromic colour vision in the duck (*Anas platyrhynchos L.*): Microspectrophotometry of visual pigments and oil droplets. *J. Comp. Physiol. A* **62,** 225–235.

Jarvis, C. D. (1974). Visual discrimination and spatial localization deficits after lesions of the tectufugal pathway in pigeons. *Brain Behav. Evol.* **9,** 195–228.

Jassik-Gerschenfeld, D., and Guichard, J. (1972). Visual receptive fields of single cells in the pigeon's optic tectum. *Brain Res.* **40,** 303–317.

Jassik-Gerschenfeld, D., and Hardy, O. (1979). Single neuron responses to moving sine-wave gratings in the pigeon optic tectum. *Vision Res.* **19,** 993–999.

Jassik-Gerschenfeld, D., and Hardy, O. (1980). Single-cell responses to bar width and to sine-wave grating frequency in the pigeon. *Vision Res.* **21,** 745–747.

Jassik-Gerschenfeld, D., Guichard, J., and Tessier, Y. (1975). Localization of directionally selective and movement sensitive cells in the optic tectum of the pigeon. *Vision Res,* **15,** 1037–1038.

Karten, H. J., and Hodos, W. (1967). "A Stereotaxic Atlas of the Brain of the Pigeon (*Columba livia*)." The Johns Hopkins Press, Baltimore.

Karten, H. J., Keyser, K. T., and Brecha, N. C. (1990). Biochemical and morphological heterogeneity of retinal ganglion cells. *In* "Vision and the Brain" (B. Cohen and I. Bodis-Wollner, eds.), pp. 19–33. New York, Raven.

Katayama, Y., Kiyama, H., Avai, Y., and Tohyama, M. (1984). Immunoreactive avian pancreatic polypeptide in the chicken retina: Overall distribution. *Brain Res.* **310,** 164–167.

Keyser, K. T., Britto, L. R. G., Woo, J.-I., Park, D. H., and Ioh, T. H. (1990). Presumtive catecholaminergic ganglion cells in the pigeon retina. *Vis. Neurosci.* **4,** 225–235.

Keyser, K. T., Karten, H. J., and Ehrlich, D. (1988). "Bullwhip" association amacrine cells in the pigeon retina: morphology and histochemical heterogeneity. *Invest. Ophtalmol. Vis. Sci. (Suppl.)* **29,** 196.

Kimberly, R. P., Holden, A. L., and Bamborough, P. (1971). Response characteristics of pigeon forebrain cells to visualstimulation. *Vision Res.* **11,** 475–478.

Kirmse, W. (1990). Kritische Übersicht zur selektiven Sensomotorik des Blickens und multifoveales Spähen bei Vögeln. *Zoologisch. Jahrbuch Physiol.* **94,** 217–228.

Kiyama, H., Katayama-Kumoi, Y., Steinbusch, H., Powell, J. F., Smith, A. D., and Tohyama, M. (1985). Three dimensional analysis of retina neuropeptides and amines in the chick. *Brain Res. Bull.* **15,** 155–165.

Knipling, R. R. (1978). No deficit in near-field visual acuity of pigeons after transection of the isthmo-optic tract. *Brain Res.* **22,** 813–816.

Knudsen, E. I. (1982). Auditory and visual maps of space in the optic tectum of the owl. *J. Neurosci.* **2,** 1177–1194.

Kusmic, C., Musumeci, D., and Spinelli, R. (1991). Binocular probability summation in a choice reaction-time task in pigeons. *NeuroReport* **2,** 615–618.

Leresche, N., Hardy, O., and Jassik-Gerschenfeld, D. (1983). Receptive field properties of single cells in the pigeon's optic tectum during cooling of the "visual Wulst." *Brain Res.* **267,** 225–236.

Letelier, J. (1983). Respuestas cromáticas en el tectum óptico de la paloma. Tesis de Licenciatura, Fakultad de Ciencias, Universitad de Chile, Santiago, Chile.

Lockhart, M. (1979). Quantitative morphological investigations of retinal cells in the pigeon: a golgi, light microscopic study. *In* "Neural Mechanisms of Behavior in the Pigeon" (A. M. Granda, and J. H. Maxwell, eds.), pp. 371–394. Plenum, New York.

Lubinski, D., and MacCorquodale, K. (1984). "Symbolic communication" between two pigeons (Columba livia) without unconditioned reinforcement. *J. Comp. Psychol.* **98,** 372–380.

Macko, K. A., and Hodos, W. (1984). Near-field acuity after visual system lesions in pigeons I. Thalamus. *Behav. Brain Res.* **13,** 1–14.

Mallin, H. D., and Delius, J. D. (1983). Inter- and intraocular transfer of colour discriminations with mandibulation as an operant in the head-fixed pigeon. *Behav. Anal. Lett.* **3,** 297–309.

Mariani, A. P. (1983). A morphological basis for verticality detectors in the pigeon retina: asymmetric amacrine cells. *Naturwissenschaften* **70,** 368–369.

Mariani, A. P. (1987). Neuronal and synaptic organization of the outer plexiform layer of the pigeon retina. *Am. J. Anat.* **179,** 25–39.

Mariani, A. P., and Leure-du Pree, A. E. (1978). Photoreceptors and oildroplet colors in the red area of the pigeon retina. *J. Comp. Neurol.* **182,** 821–838.

Martin, G. R. (1993). Producing the Image. *In* "Vision, Brain and Behavior in Birds" (H. P. Zeigler and H.-J. Bischof, eds.), pp. 5–24, MIT Press, Cambridge.

Martin, G. R. (1994). Form and Function in the Optical Structure of Bird Eyes. *In* "Perception and Motor Control in Birds" (M. N. O. Davies and P. R. Green, eds.), pp. 5–34, Springer-Verlag, Berlin.

Martinez-de-la-Torre, M., Martinez, S., and Puelles, L. (1990). Acetylcholinesterase–histochemical differential staining of subdivisions within the nucleus rotundus in the chick. *Anat. Embryol.* **181,** 129–135.

Maturana, H. R., and Frenk, S. (1963). Directional movement and horizontal edge detctors in the pigeon retina. *Science* **142,** 977.

Maturana, H. R., and Frenk, S. (1965). Synaptic connections of the centrifugal fibers in the pigeon retina. *Science* **150,** 359–361.

McGill, J. I., Powell, T. P. S., and Cowan, W. M. (1966a). The retinal representation upon the optic tectum and isthmo-otpic nucleus in the pigeon. *J. Anat.* **100,** 5–34.

McGill, J. I., Powell, T. P. S., and Cowan, W. M. (1966b). The organisation of the projection of the centrifugal fibers to the retina in the pigeon. *J. Anat.* **100,** 35–40.

Miceli, D., Gioanni, H., Repérant, J., and Peyrichoux, J. (1979). The avian visual Wulst. I. An anatomical study of afferent and efferent pathways; II. An electrophysiological study of the functional properties of single neurons. *In* "Neural Mechanisms of Behavior in Birds" (A. M. Granda, and Maxwell, J. H. eds.). Plenum, New York.

Miceli, D., Marchand, L., Repérant, J., and Rio, J.-P. (1990). Projections of the dorsolateral anterior complex and adjacent thalamic nuclei upon the visual Wulst in the pigeon. *Brain Res.* **518,** 317–323.

Miceli, D., Repérant, J., Rio, J.-R. and Medina, M. (1995). GABA immunoreactivity in the nucleus isthmo-opticus of the centrifugal visual system in the pigeon: A light and electron microscopic study. *Vis. Neurosci.* **12,** 425–441.

Miceli, D., Repérant, J., Villalobos, J., and Dionne, L. (1987). Extratelencephalic projections of the avian visual Wulst: A quantitative autoradiographic study in the pigeon Columba livia. *J. Hirnforsch.* **28,** 45–57.

Miles, F. A. (1972a). Centrifugal control of the avian retina. I. Receptive field properties of retinal ganglion cells. *Brain Res.* **48,** 65–92.

Miles, F. A. (1972b). Centrifugal control of the avian retina. II. Receptive field properties of cells in the isthmo-optic nucleus. *Brain Res.* **48,** 93–113.

Miles, F. A. (1972c). Centrifugal control of the avian retina. III. Effects of electrical stimulation of the isthmo-optic tract on the receptive field properties of reinal ganglion cells. *Brain Res.* **48,** 115–129.

Millar, T. J., Ishimoto, I., Boelen, M., Epstein, M. L., Johnson, C. D., and Morgan, I. G. (1987). The toxic effects of ethylcoline mustard aziridinium ion on cholinergic cells in the chicken retina. *J. Neurosci.* **7,** 343–356.

Morgan, I. G., Oliver, J., and Chubb, I. W. (1983). The development of amacrine cells containing somatostatin-like immunoreactivity in chicken retina. *Dev. Brain Res.* **8,** 71–76.

Morino, P., Bahro, M., Cuénod, M., and Streit, P. (1991). Glutamate-like immunoreactivity in the pigeon optic tectum and effects of retinal ablation. *Eur. J. Neurosci.* **3,** 366–378.

Murphy, C. J., Howland, M., and Howland, H. C. (1995). Raptors lack lower field myopia. *Vision Res.* **35,** 1153–1155.

Nalbach, H.-O., Wolf-Oberhollenzer, F., and Remy, M. (1993). Exploring the Image, *In* "Vision, Brain and Behavior in Birds" (H. P. Zeigler and H.-J. Bischof, eds.), pp. 25–46. MIT Press, Cambridge.

Neuenschwander, S., and Varela, F. J. (1993). Visually triggered neuronal oscillations in the pigeon: An autocorrelation of tectal activity. *Eur. J. Neurosci.* **5,** 870–881.

Oehme, H. (1962). Das Auge von Mauersegler, Star und Amsel. *J. Ornithol.* **103**, 187–212.

Pateromichelakis, S. (1981). Response properties of visual units in the anterior dorsolateral thalamus of the chick (*Gallus domesticus*). *Experientia* **37**, 279–280.

Perry, V. H., and Cowey, A. (1985). The ganglion cell and cone distributions in the monkey's retina: implications for central magnification factors. *Vision Res.* **25**, 1795–1810.

Pettigrew, J. D. (1978). Comparison of the retinotopic organization of the visual Wulst in nocturnal and diurnal raptors, with a note on the evolution of frontal vision. *In* "Frontiers of Visual Science" (S. J. Cool. and E. L. Smith, eds.), pp. 328–335. Springer-Verlag, New York.

Pettigrew, J. D. (1979). Binocular visual processing in the owl's telencephalon. *Proc. R. Soc. London* **204**, 435–455.

Pettigrew, J. D. (1993). Two ears and two eyes. *Nature* **364**, 756–757.

Pettigrew, J. D., and Konishi, M. (1976a). Neurons selective for orientation and binocular disparity in the visual Wulst of the barn owl (*Tyto alba*). *Science* **193**, 675–678.

Pettigrew, J. D., and Konishi, M. (1976b). Effects of monocular deprivation on binocular neurons in the owl's visual Wulst. *Nature* **264**, 753–754.

Pettigrew, J. D., Wallman, J., and Wildsoet, C. F. (1990). Saccadic oscillations facilitate ocular perfusion from the avian pecten. *Nature* **343**, 362–363.

Pirenne, M. H. (1943). Binocular and uniocular thresholds in vision. *Nature* **152**, 698–699.

Porciatti, V., Fontanesi, G., Rafaelli, A., and Bagnoli, P. (1990). Binocularity in the little owl, *Athene noctua*. II. Properties of visually evoked potentials from the Wulst in response to monocular and binocular stimulation with sine wave gratings. *Brain Behav. Evol.* **35**, 40–48.

Prada, F. A., Chmielewski, C. E., Dorado, M. E., Prada, C., and Génis-Gálvez, J. M. (1989). Displaced ganglion cells in the chick retina. *Neurosci. Res.* **6**, 329–339.

Prada, C., Medina, J. I., López, R., Génis-Gálvez, J. M., and Prada, F. A. (1992). Development of retinal displaced ganglion cells in the chick: Neurogenesis and morphogenesis. *J. Neurosci.* **12**, 3781–3788.

Quesada, A., Prada, F. A., and Génis-Gálvez, J. M. (1988). Bipolar cells in the chicken retina. *J. Morphol.* **197**, 337–351.

Rager, G., and Rager, U. (1978). Systems-matching by degeneration. I. A quantiavie electron microscopic study of the generation and degeneration of retinal ganglion cells in the chicken. *Exp. Brain Res.* **33**, 65–78.

Reiner, A. (1992). The neurotensin-related hexapeptide LANT6 is found in retinal ganglion cells and in their central projections in pigeons. *Vis. Neurosci.* **9**, 217–223.

Remy, M., and Emmerton, J. (1989). Behavioral spectral sensitivities of different retinal areas in pigeons. *Behav. Neurosci.* **103**, 170–177.

Remy, M., and Güntürkün, O. (1991). Retinal afferents to the tectum opticum and the n. opticus principalis thalami in the pigeon. *J. Comp. Neurol.* **305**, 57–70.

Reymond, L. (1985). Spatial visual acuity of the eagle *Aquila audax*: A behavioral, optical and anatomical investigation. *Vision Res.* **25**, 1477–1491.

Reymond, L. (1987). Spatial visual acuity of the falcon, *Falco berigora*: A behavioural, optical and anatomical investigation. *Vision Res.* **27**, 1859–1874.

Revzin, A. M. (1969). A specific visual projection area in the hyperstriatum of the pigeon. *Brain Res.* **15**, 246–249.

Revzin, A. M. (1979). Functional localization in the nucleus rotundus. *In* "Neural Mechanisms of Behavior in Birds" (A. M. Granda and J. H. Maxwell, eds.), pp. 165–175. Plenum, New York.

Ritchie, T. L. C. (1979). Intratelencephlic visual connections and their relationship to the archistriatum in the pigeon (*Columba livia*). Unpublished Ph.D. Thesis, University of Virginia.

Rochon-Duvigneaud, A. (1943). Les yeux et la vision des vertebrés, Paris: Masson.

Roberts, W. A., and Mazmanian, D. S. (1988). Concept learning at different levels of abstraction by pigeons, monkeys, and people. *J. Exp. Psychol. Anim. Behav. Process* **14**, 247–260.

Rogers, J. J., and Miles, F. A. (1972). Centrifugal control of the avian retina. V. Effects of lesions of the isthmo-optic nucleus on visual behaviour. *Brain Res.* **48**, 147–156.

Shimizu, T., and Karten, H. J. (1990). Immunohistochemical analysis of the visual Wulst of the pigeon (*Columba livia*). *J. Comp. Neurol.* **300**, 346–369.

Shimizu, T., and Karten, H. J. (1993). The avian visual system and the evolution of the neocortex. *In* "Vision, Brain and Behavior in Birds" (H. P. Zeigler and H.-J. Bischof, eds.). MIT Press, Cambridge.

Shortess, G. K., and Klose, E. F. (1977). Effects of lesions involving efferent fibers to the retina in pigeon (*Columba livia*). *Physiol. Behav.* **18**, 409–414.

Singer, W. (1993). Synchronization of cortical activity and its putative role in information processing and learning. *Annu. Rev. Physiol.* **55**, 349–374.

Snyder, A. W., Laughlin, S. B., and Stavenga, D. G. (1977). Information capacity of eyes. *Vision Res.* **17**, 1163–1175.

Thin, N. D., Egedy, G., and Tömböl, T. (1992). Golgi study on neurons and fibers in nucleus rotundus of the thalamus in chicks. *J. Hirnforsch.* **33**, 203–214.

Thompson, I. (1991). Considering the evolution of vertebrate neural retina. *In* "Vision and Visual Dysfunction" (J. R. Cronly-Dillon and R. L. Gregory, eds.), Vol. 2, pp. 136–151. Macmillan, New York.

Tömböl, T., Ngo, T. D., and Egedy, G. (1992). EM and EM golgi study on structure of nucleus rotundus in chicks. *J. Hirnforsch.* **33**, 215–234.

Uchiyama, H. (1989). Centrifugal pathways to the retina: Influence of the optic tectum. *Vis. Neurosci.* **3**, 183–206.

Uchiyama, H., and Ito, H. (1993). Target cells for the isthmo-optic fibers in the retina of the Japanese quail. *Neurosci. Lett.* **154**, 35–38.

Uchiyama, H., Ito, H., and Tauchi, M. (1995). Retinal neurones specific for centrifugal modulation of vision. *NeuroReport* **6**, 889–892.

Uchiyama, H., and Watanabe, M. (1985). Tectal neurons projecting to the isthmo-optic nucleus in the japanese quail. *Neurosci. Lett.* **58**, 381–385.

Uchiyama, H., Matsutani, S., and Watanabe, M. (1987) Activation of the isthmo-optic neurons by the visual Wulst stimulation. *Brain Res.* **406**, 322–325.

Vander Wall, S. B., and Balda, R. P. (1977). Coadaptation of the Clark's nutcracker and the pinyon pine for efficient seed harvest and dispersal. *Ecol. Mon.* **47**, 89–111.

Varela, F. J., Palacios, A. G., and Goldsmith, T. H. (1993). Color Vision of Birds. *In* "Vision, Brain and Behavior in Birds" (H. P. Zeigler and H.-J. Bischof, eds.), pp. 77–98. MIT Press, Cambridge.

Verstappen, A., Van Reeter, O., Vaudry, H., and Pelletier, G. (1986). Demonstration of a neuropeptide Y (NPY)-like immunoreactivity in the pigeon retina. *Neurosci. Lett.* **70**, 193–197.

von Fersen, L., Wynne, C. D. L., Delius, J. D., and Staddon, J. E. R. (1992). Transitive inference formation in pigeons. *J. Exp. Psychol. Anim. Behav. Process* **17**, 334–341.

Wagner, H., and Frost, B. (1993). Disparity-sensitive cells in the owl have a characteristic disparity. *Nature* **364**, 796–798.

Wagner, H., and Frost, B. (1994). Binocular responses of neurons in the barn owl's visual Wulst. *J. Comp. Physiol. A* **174**, 661–670.

Walls, G. L. (1942). "The Vertebrate Eye and Its Adaptive Radiations." Cranbook Institute of Science, Bloomfield Hills, MI.

Wang, Y., and Frost, B. J. (1992). Time to collision is signalled by neurons in the nucleus rotundus of pigeons. *Nature* **356,** 236–238.

Wang, Y., Jiang, S., and Frost, B. J. (1993). Visual processing in pigeon nucleus rotundus: Luminance, color, motion, and looming subdivisions. *Vis. Neurosci.* **10,** 21–30.

Watanabe, M. (1987). Synaptic organization of the nucleus dorsolateralis anterior thalami in the Japanese quail (*Coturnix coturnix japonica*). *Brain Res.* **401,** 279–291.

Watanabe, M., Ito, H., and Masai, H. (1983). Cytoarchitecture and visual receptive neurons in the Wulst of the Japanese quail (*Coturnix coturnix japonica*). *J. Comp. Neurol.* **213,** 188–198.

Wathey, J. C., and Pettigrew, J. D. (1989). Quantitative analysis of the retinal ganglion cell layer and optic nerve of the barn owl, *Tyto alba*. *Brain Behav. Evol.* **33,** 279–292.

Weidner, C., Repérant, J., Desroches, A. M., Miceli, D., and Vesselkin, N. P. (1987). Nuclear origin of the centrifugal visual pathway in birds of prey. *Brain Res.* **436,** 153–160.

Wilson, P. (1980). The organization of the visual hyperstriatum in the domestic chick. II. Receptive field properties of single units. *Brain Res.* **188,** 333–345.

Wolf-Oberhollenzer, F. (1987). A study of the centrifugal projections to the pigeon retina using two fluorescent markers. *Neurosci. Lett.* **73,** 16–20.

Woodson, W., Reiner, A., Anderson, K., and Karten, H. J. (1991). Distribution, laminar location, and morphology of tectal neurons projecting to the isthmo-optic nucleus and the nucleus isthmi, pars parvocellularis in the pigeon (*Columba livia*) and chick (*Gallus domesticus*): A retrograde labelling study. *J. Comp. Neurol.* **305,** 470–488.

Wright, A. A., Cook, R. G., Rivera, J. J., Sands, S. F., and Delius, J. D. (1988). Concept learning by pigeons: Matching-to-sample with trial-unique video picture stimuli. *Anim. Learn. Behav.* **16,** 436–444.

Yang, G., Millar, T. J., and Morgan, I. G. (1989) Co-lamination of cholinergic amacrine cell and displaced ganglion cell dendrites in the chicken retina. *Neurosci. Lett.* **103,** 151–156.

Yazulla, S. (1974). Intraretinal differentiation in the synaptic organization of the inner plexiform layer of the pigeon retina. *J. Comp. Neurol.* **153,** 309–324.

Young, S. R., and Martin, G. R. (1984). Optics of retinal oil droplets: a model of light collection and polarization detction in the avian retina. *Vision Res.* **24,** 129–137.

CHAPTER 2

The Avian Ear and Hearing

REINHOLD NECKER
Ruhr-Universität Bochum
Fakultät für Biologie
Lehrstuhl für Tierphysiologie
D-44780 Bochum
Germany

I. Introduction 21
II. Structure of the Ear 21
   A. External Ear 21
   B. Middle Ear 22
   C. Inner Ear (Cochlea) 22
III. Function of the Ear 24
   A. Transfer Function of the Middle Ear 24
   B. Oscillations of the Basilar Membrane 24
   C. Hair Cell Activation 24
IV. Auditory Nerve Fibers 27
V. Central Pathways 28
   A. Anatomy 28
   B. Electrophysiology 29
VI. Behavioral Aspects 33
   A. Behavioral Audiograms 33
   B. Frequency Discrimination 33
   C. Intensity Discrimination 33
   D. Time Resolution 33
   E. Sound Localization 34
   F. Echolocation 35
VII. Summary and Conclusions 35
   References 35

I. INTRODUCTION

Both sensitive hearing and vocal communication play an important role in a bird's life. Among nonmammalian vertebrates birds possess the most highly evolved auditory system. Although the mammalian ear is more specialized, hearing performances of birds are not necessarily inferior to those of most mammalian species. However, hearing range in birds is restricted to an upper limit of about 20 kHz (Schwartzkopff, 1973) but the normal range does not exceed 10 kHz significantly. As in some mammalian species (e.g., elephants) some birds are able to hear infrasound.

Hearing depends on the structure and function of the sense organ, the ear, and on processing in the central nervous system (CNS). Air-borne sound waves enter the ear by the external acoustic meatus to impinge on the tympanic membrane (eardrum). Movements of the eardrum are transmitted to the inner ear via one ossicle of the middle ear, the avian columella. The inner ear is part of the labyrinth which contains various sensory patches of which the basilar papilla serves auditory functions. Soundwave-driven movements of the basilar papilla result in mechanical stimulation of the sensory receptors located on the papilla, the hair cells. Excitation of hair cells is transmitted to the auditory nerve whose fibers convey the information to the CNS where further processing occurs. This finally results in audition and behavioral responses.

II. STRUCTURE OF THE EAR

A. External Ear

In most birds the external ear is inconspicuous, lacking auricles. The tube-like short external acoustic meatus is normally surrounded by specialized feathers which are adapted to minimize flight turbulences but which do not obstruct sound transmission. There may be ear flaps and specialized feathers which serve to reflect and amplify sound as is true, for example, for the ear and facial ruff in the barn owl (*Tyto alba*). In addition, in some owls their left and right ears may differ in height and direction of their external openings which is the basis for detection of elevation above ground of a sound source (Volman, 1994).

The tympanic membrane at the end of the external acoustic meatus protrudes cone-like into the meatus. This is caused by the processes of the extracolumellar cartilage (Figure 1).

B. Middle Ear

The tympanic cavity of the middle ear contains the auditory ossicle. The single ossicle consists of two segments, a cartilaginous extracolumella and the bony columella proper (Figure 1). The extracolumella inserts on the ear drum whereas the footplate at the proximal end of the columella fits into the oval (vestibular) window of the cochlea. There is a middle ear muscle, M. columellae, which inserts at the extracolumellar end and which upon activation through the facial nerve may increase tension of the columella and the tympanic membrane (Kühne and Lewis, 1985).

The middle ear cavity communicates with air cavities of the surrounding skull bones and the right and left tympanic cavities communicate with each other via interconnecting air sinuses. This may have functional implications for sound localization in some birds (Hill *et al.*, 1980; Coles *et al.*, 1980).

The tympanic cavity contains a sensory patch with hair cells, the paratympanic organ, whose function is so far unknown. There is some evidence that it may detect barometric pressures (von Bartheld, 1994).

C. Inner Ear (Cochlea)

1. Gross Morphology

The inner ear consists of a cochlear organ and a vestibular organ. The bony labyrinth includes a membraneous labyrinth which contains the sensory epithelia. Whereas maculae (m. utriculus, m. sacculus) and cupulae of the semicircular canals are part of the vestibular organ and serve equilibrium, the cochlear organ contains the basilar papilla as a sensory epithelium for hearing and a macula lagena with a still unknown function. The cochlea is not coiled as in mammals but slightly curved approaching the midline at the base of the cranium. The bony cochlea has two openings facing the tympanic cavity, the oval (vestibular) window and the round (cochlear) window

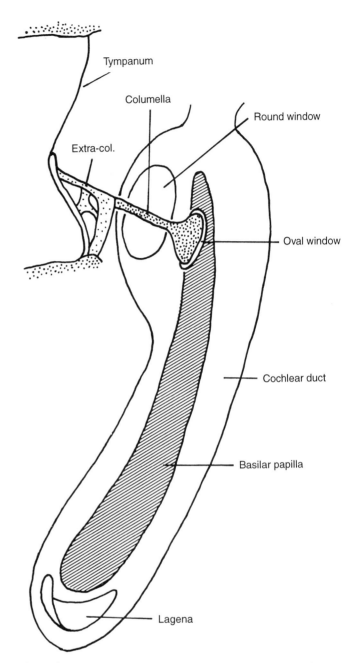

FIGURE 1 Schematic drawing of the middle ear and of the cochlea of birds.

(Figure 1). The length of the cochlea ranges from less than 3 mm in small songbirds to about 12 mm in owls, usually being less than 7 mm even in large birds such as the swan (for comparison, cochlea of mice = 7 mm; Schwartzkopff, 1968).

The cochlea is divided by a cartilagenous frame which spans the basilar membrane (extracellular material) which supports the cochlear duct. This duct is composed of different epithelial specializations and encloses the endolymphatic space of the scala media (Figure 2A). It

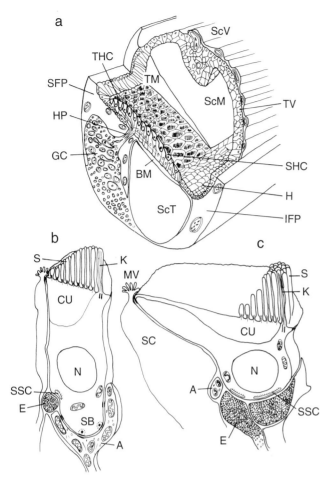

FIGURE 2 Transverse section of the cochlear duct (A) and of tall hair cells (B) and short hair cells (C). A, afferent nerve ending; BM, basilar membrane; CU, cuticular plate; E, efferent nerve ending; GC, cochlear ganglion; H, hyaline cells; HP, habenula perforata; IFP, inferior fibrocartilaginous plate (abneural limbus); K, kinocilium; MV, microvilli; N, nucleus; S, stereocilium (stereovillus); SB, synaptic ball structure; SC, supporting cell; ScM, scala media; ScT, scala tympani; ScV, scala vestibuli; SFP, superior fibrocartilaginous plate (neural limbus); SHC, short hair cell; SSC, subsynaptic cisterna; THC, tall hair cell; TM, tectorial membrane; and TV, tegmentum vasculosum. After Takasaka and Smith (1971) with permission.

cently a separate papilla chaotica near the apical end of the basilar papilla has been described in the pigeon (Schermuly et al., 1991).

The basilar papilla of birds consists of hair cells and of supporting cells. In larger birds it widens from basal to apical (Figure 1). There is no organ of Corti as in mammals with a separation into inner and outer hair cells. Instead up to 50 hair cells make up a transverse row (Figure 2A). Hair cells are characterized by numerous cilia (up to about 100) which protrude from a cuticular plate at the apical side of the cell (Figures 2B and 2C). These stereocilia or stereovilli resemble large microvilli with dense actin filaments. In birds there is usually one kinocilium (Takasaka and Smith, 1971) which may be absent in part of the hair cells in the chicken (Tanaka and Smith, 1978). This kinocilium has the typical microtubuli organization and a basal body found in other cilia also. The arrangement of microtubuli and dynein arms is, however, irregular, which points to an impeded function as a motile organ (Smith, 1985). In a bundle stereocilia increase in heights toward the kinocilium (Figures 2B and 2C). This means a morphological polarization. This orientation may change in dependence on the location on the basilar papilla. There is a change in polarization over the width of the basilar papilla which seems to be similar in different species (Gleich and Manley, 1988; Fischer et al., 1988; Manley et al., 1993; Gleich et al., 1994).

The cochlear ganglion is located at the superior side of the cochlea. This defines a neural (superior) and an abneural (inferior) edge of the basilar papilla. Different types of hair cells can be distinguished depending at least in part on the location on the basilar papilla. Tall hair cells (THC) are elongated and short hair cells (SHC) are larger in width than in height (Figure 2). There is a continuum from tall to short. THCs are located on the cartilagenous neural limbus which does not move during sound stimulation (Fischer et al., 1992). SHCs are located on the free basilar membrane and thus exposed to movements. This is comparable to the location of inner and outer hair cells in the mammalian cochlea.

A thick extracellular tectorial membrane extends from the columnar cells at the superior edge of the cochlear duct and covers the hair cells (Figure 2A). The inferior side of the membrane has a honeycomblike structure into which the longest stereocilia protrude. Furthermore, there are cavities and channels which allow the endolymph to reach the apical surface of the hair cells (Smith, 1985). Because of its unilateral attachment, movements of the basilar membrane result in shearing forces at the tectorial membrane–stereocilia interface.

communicates with the sacculus by the sacculocochlear duct. There are perilymphatic spaces on each side of the cochlear duct, a narrow scala vestibuli on the vestibular window side, and a wider scala tympani on the cochlear window side. Scala vestibuli communicates with scala tympani both at the base and at the apex (helicotrema). There is a widening of the scala tympani near the round window (Recessus scalae tympani).

Main epithelial elements of the cochlear duct are the basilar papilla and the tegmentum vasculosum with inner and outer hyaline cells at the transition zone between both tissues. The macula lagena at the apical end of the cochlear duct (Figure 1) has some similarity to the maculae of the vestibular organ (Smith, 1985). Re-

The richly folded tegmentum vasculosum protrudes into the cochlear duct (Figure 2). It contains dark and light cells. Dark cells contain numerous mitochondria and microvilli at the apical side. The tegmentum vasculosum which is probably homologuous to the mammalian stria vascularis is richly supplied with blood vessels (Schmidt, 1964). Because this epithelial specialization is rich in Na^+-K^+-ATPase (Kuypers and Bonting, 1970), it may be involved in secretory processes and/or in the maintenance of the endocochlear potential (see below).

There is both an afferent and an efferent innervation of hair cells (Figure 2). THCs are contacted by 1 to 4 afferent fibers, which is much less than in mammalian inner hair cells (15–20). Efferent innervation is scarce in THC but abundant in SHC where an afferent innervation may even be lacking (Takasaka and Smith, 1971; Fischer, 1992; Fischer et al., 1992). The efferent innervation originates from neurons in the brain stem which are located near the superior olive, and there is both a crossed and an uncrossed efferent pathway (Whitehead and Morest, 1981; Schwarz et al., 1981).

III. FUNCTION OF THE EAR

A. Transfer Function of the Middle Ear

The middle ear has to transform large amplitudes and velocities of sound waves at the tympanic membrane to small amplitudes but great forces at the oval window. This is achieved by the transmission of the oscillations of the large tympanic membrane to the small oval window. The area of the tympanic membrane is always considerably larger than that of the columellar footplate by ratios which vary between about 15 and 40 in various species of birds (Schwartzkopff, 1968). This means a substantial impedance matching of sound transmission which is about the same as in mammals.

The transfer function of the middle ear has been measured with the Mössbauer technique (Saunders and Johnstone, 1972; Saunders, 1985; Gummer et al., 1989a,b). Above 2 kHz amplitudes of both the tympanic membrane and columellar footplate decrease steeply at more than −10 dB/octave (guinea pig: −1 dB/octave). This limits high frequency hearing to about 10 kHz (see Figure 6). The reason for the decrease in amplitude is an energy loss in the flexing motion of the processes of the cartilaginous extracolumella (Manley and Gleich, 1992).

The activation of the single middle ear muscle results in a reduction of the electrical response of the cochlea (cochlear microphonics). However, this effect depends on frequency, and attenuation mainly occurs at frequencies higher or lower than the most sensitive hearing range (Oeckinghaus and Schwartzkopff, 1983). This may help to select a frequency range; for example, that of vocal communication in songbirds. In owls there is an acoustic reflex of middle ear muscle (activation at high sound intensities) which may serve to keep the sound level in the sensitive range (Kühn and Lewis, 1985).

B. Oscillations of the Basilar Membrane

Movements of the columellar footplate force oscillations of inner ear fluids which finally result in movements of the basilar membrane. Waves travel along the basilar membrane from proximal to distal, getting shorter at the distal end. The location of maximal amplitude of these travelling waves depends on the frequency (Békésy, 1944). Because of the shorter basilar membrane in birds the frequency dependent maxima are in closer proximity than in mammals. The very short basilar membranes of small songbirds probably vibrate as a whole, as do those of larger birds (chicken) below 100 Hz. The existence of traveling waves in the pigeon was confirmed recently with the Mössbauer technique (Gummer et al., 1987). The frequency of maximal response shifted along the basilar membrane with about 0.6 mm/octave (basal part of the basilar membrane only).

C. Hair Cell Activation

Hair cell activation results from shearing forces of the tectorial membrane acting on the stereocilia during oscillations of the basilar membrane. Hair cells are polarized electrically with a negative charge inside. Potential difference across the apical surface of the cells is increased by a positive potential in the endolymphatic space (scala media). This anoxia-sensitive endocochlear potential (EP) is lower in birds (less than +20 mV; Schmidt and Fernandez, 1962; Necker, 1970; Runhaar et al., 1991) than in mammals (+80 mV). There is evidence that EP increases the sensitivity of hair cells (Vossieck et al., 1991). EP is probably due to electrogenic pumps located in the stria vascularis in mammals (Kuijpers and Bonting, 1970) or tegmentum vasculosum in birds. The pumping results in an unusual high concentration of K^+ ions in the extracellular endolymph (about 160 mM/liter in the chicken; Runhaar et al., 1991) which is about the same as the intracellular concentration. The lower value of EP despite a mammalianlike K^+ concentration and an insensitivity to furosemide (Schermuly *et al.*, 1990) point to a modified generation compared to mammals.

The current view of the activation of the hair cells can be summarized as follows (Ashmore, 1991). It has been shown that there are so-called "tip-links" connecting neighboring stereovilli (Figure 3A). These tip links are assumed to open mechanosensitive channels since they are arranged in a way to be stretched during bending the bundle of stereovilli toward the kinocilium (Figure 3B). The opening of the channels probably results

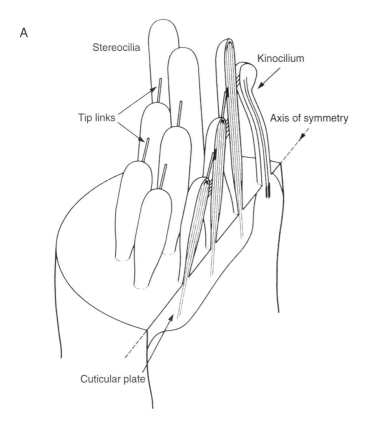

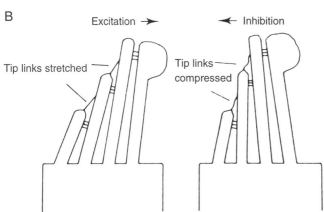

FIGURE 3 Schematic drawings of the organization of the tip links of stereocilia (stereovilli) under rest (A) and when activated (B). After Pickles *et al.* (1990) and Pickles and Corey (1992) with permission.

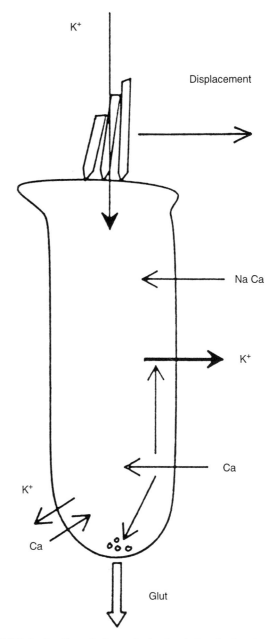

FIGURE 4 Ion flows during excitation in a hair cell. glut, glutamate (transmitter at afferent synapse). (Reproduced, with permission, from the *Annual Review of Physiology*, Volume 53, © 1991, by Annual Reviews Inc.)

mainly in an inward K^+ current and a depolarization of the cell membrane (Figure 4). The role of Ca^{2+} in the tranduction process is not yet clear although it is necessary for excitation. The depolarization primarily causes an activation of voltage dependent K^+ channels at the basolateral membrane of the cell to remove the excess K^+ entered during excitation. Opening of voltage-gated Ca^{2+} channels results in an influx of calcium which triggers the release of transmitter vesicles. The transmitter (probably glutamate) depolarizes the postsynaptic afferent terminals and results in an increase in the firing rate in the afferent fiber.

In nonmammalian vertebrates including birds, Ca^{2+} influx activates Ca^{2+}-sensitive K^+ channels (Figure 4) which results in hyperpolarization of the cell. This is the basis of oscillations of the membrane potential which are thought to underly electrical tuning found in hair cells of lower vertebrates (Ashmore, 1991). In the chicken oscillation frequency depends on the location of the hair cells on the basilar membrane in such a way that higher frequencies are observed in more basal parts (Fuchs et al., 1988).

Mammalian outer hair cells have been shown to change their length with a change in membrane polarization (electromechanical transduction) and it has been postulated that this process contributes to an increase in sensitivity and sharpness of tuning of inner hair cells (Ashmore, 1991, 1994). These events are thought to result in otoacoustic emissions of the ear (sound recorded at the ear after stimulation). Both otoacoustic emissions and electromechanical events have been demonstrated in birds (Manley et al., 1987, Brix and Manley, 1994). It is, however, not clear whether these events have the same origin as in mammalian hair cells.

Extracellular recordings from the fluid spaces surrounding the basilar papilla show that there are two different types of responses related to sound stimulation: (1) cochlear microphonics (CM), which are a mirror image of the stimulus, and (2) a DC component superimposed on the AC events called summating potential (SP) which is mainly directed toward depolarization of the hair cells (Figure 5; Necker, 1970; Pierson and Dallos, 1976). It is unclear whether electrical or mechanical nonlinearities contribute to SP.

Cochlear microphonics have been used to study the sensitivity of the avian ear. The dependence of CM on frequency is best described by isopotential curves—the sound pressure level (SPL) required to produce a constant CM amplitude. As shown in Figure 6 sensitivity decreases rapidly above about 3 kHz as in middle ear transfer function.

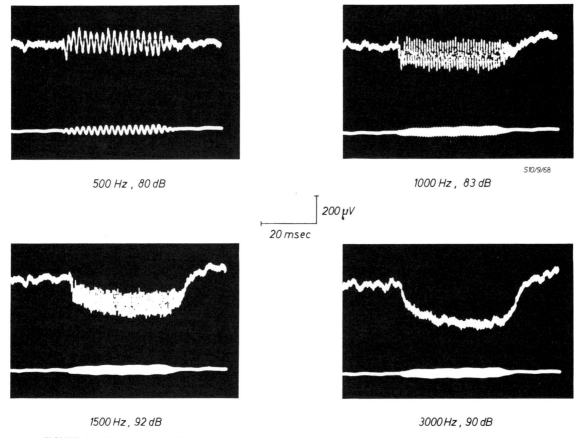

FIGURE 5 Examples of cochlear microphonics and summating potentials in the pigeon inner ear. Sound frequency and intensity as indicated. (After Z. vergl. Physiol., Zur Entstehung der Cochleapotentiale von Vögeln: Verhalten bei O$_2$-Mangel, Cyanidvergiftung und Unterkühlung sowie Beobachtung über die rämliche Verteilung, R. Necker, **69,** 367–425, Fig. 7, 1970 © Springer-Verlag.)

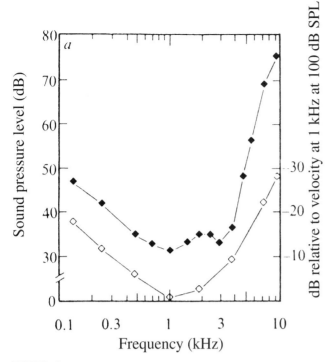

FIGURE 6 Average 1 μm r.m.s. cochlear microphonics isopotential curve (◆, left ordinate) in the frequency range 0.1 to 10 kHz. For comparison (◇, right ordinate) an inverted representation of an average middle ear tranfer function is shown. (Reprinted with permission from *Nature* (**262**, 599–601). Copyright (1976) Macmillian Magazines Limited.)

The efferent innervation of hair cells has been shown to result in a hyperpolarization via cholinergic mechanisms (Ashmore, 1991, Fuchs and Murrow, 1992). This is in accordance with an increase in CM amplitude after activation of the olivocochlear bundle (Desmedt and Delwaide, 1965). However, nerve fiber activity decreases, which points to a desensitization of the system (e.g., inhibition of transmitter release). The function of the efferent innervation is not clear at present (Ashmore, 1991; Kaiser and Manley, 1994).

IV. AUDITORY NERVE FIBERS

There is a tonotopic arrangement in that fibers with low characteristic frequency (CF) innervate apical HCs and fibers with high CF, basal HCs (Gleich, 1989; Manley *et al.*, 1989). Whereas the frequency distribution is below 1 mm/octave in most birds, there is a specialization in owls in that frequencies in the range of 5–10 kHz occupy the basal 6 mm of the 11-mm-long basilar papilla (Köppl *et al.*, 1993). This means 6 mm/octave, which points to an auditory "fovea" as found in echolocating bats. Fibers responding to infrasound in the pigeon contact specialized areas (probably papilla chaotica) at the apical end of the basilar papilla (Schermuly and Klinke, 1990a). Lagenar afferent fibers seem not to be involved in hearing (Manley *et al.*, 1991).

Cochlear afferents have high rates of spontaneous activity (up to 150 Imp./sec) and rates may increase up to 400 Imp./sec during activation (Sachs *et al.*, 1980). Some of the fibers show periodic discharges with preferred intervals which are near but not identical to the characteristic frequency (Temchin, 1988; Hill, 1990; Manley and Gleich, 1992). The basis of spontaneous periodic discharges is thought to be electrical tuning of the hair cells (see above). Since electrical tuning works only at low frequencies and is not found in all hair cells, other mechanisms of tuning must be present.

Tuning curves show a rapid increase of threshold intensities for frequencies above and below CF (Figure 7). CFs range from below 100 Hz up to about 6 kHz in most birds and up to 9 kHz in owls (Manley and Gleich, 1992). Frequencies outside CF may cause inhibitory responses. Such inhibitions (single tone suppression) are found in oscillating fibers only (Temchin, 1988). Two-tone suppression (suppression of the response by a second tone of nearby frequency) is common as in mammals.

Phase-locking (spikes occur at a distinct phase of the stimulus wave) has been observed up to frequencies of 4 kHz in many birds but up to 9 kHz in owls (Sachs *et al.*, 1980; Sullivan and Konishi, 1984; Gleich and Narins, 1988). Phase-locking (periodicity) may be an additional mechanism of frequency discrimination besides the place principle evident from the distribution of CF along the basilar papilla.

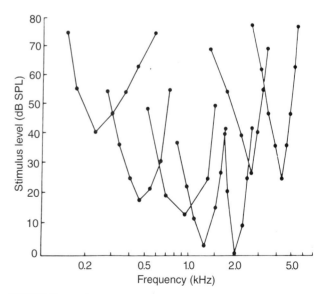

FIGURE 7 Tuning curves of auditory nerve fibers. Each curve is an isorate contour corresponding to a discharge rate 50% above spontaneous activity. (After Sachs *et al.* (1974) *Brain Res.* **70**, 431–447, with permission from Elsevier Science.)

Infrasound-sensitive primary afferents have been described for the pigeon (Schermuly and Klinke, 1990b). Spontaneous activity is similar to normal afferents. There is no change in mean firing rate during tonal stimulation. However, activity is modulated at the stimulus frequency in the range 1–100 Hz (Figure 8).

V. CENTRAL PATHWAYS

A. Anatomy

The central pathways as described by Boord (1969) are shown in Figure 9. More recent details are outlined schematically in Figure 10. The cochlear afferents join the nervus octavus (NVIII) and enter the medulla oblongata where the fibers branch to reach both the nucleus angularis (NA) and the nucleus magnocellularis (NM). The NM projects bilaterally to the nucleus laminaris (NL), which is the first nucleus to receive an input from both ears. The NA projects mainly contralaterally to the midbrain nucleus mesencephalicus lateralis pars dorsalis (MLD) to give off collaterals to the superior olive and to the nuclei of the lateral lemniscus (dorsal, intermediate, and ventral nucleus of the lateral lemniscus: LLD, LLI, LLV; Arends and Zeigler, 1986). The NL has a similar projection as the NA. The superior olive (OS) projects to the MLD. The nuclei of the lateral

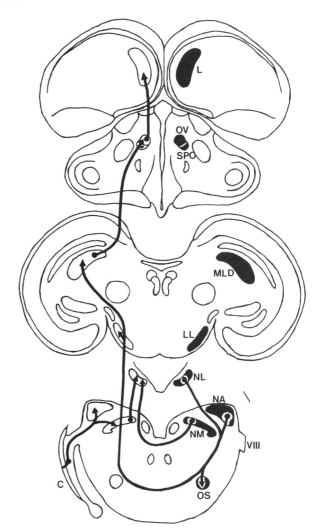

FIGURE 9 Diagram showing the location of auditory nuclei and the main ascending auditory pathway in the pigeon. C, cochlea; L, Field L; LL, nucleus of lateral lemniscus; MLD, nucleus mesencephalicus lateralis pars dorsalis; NA, nucleus angularis; NL, nucleus laminaris; NM, nucleus magnocellularis; OS, oliva superior; Ov, nucleus ovoidalis; SPO, nucleus semilunaris parovoidalis; and VIII, auditory nerve. (Based on Boord (1969), used with permission.)

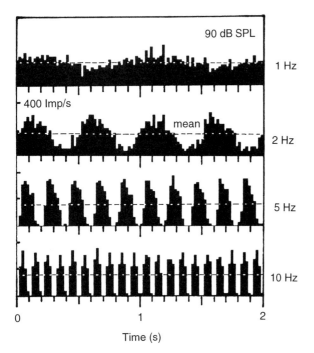

FIGURE 8 Modulation of spontaneous activity in infrasound-sensitive primary afferent fibers by tones of 1 to 10 Hz in the pigeon. (After, *J. Comp. Physiol. A*, Infrasound sensitive neurones in the pigeon cochlear ganglion, L. Schermuly and R. Klinke, **166**, 355–363, Fig. 2, 1990b, © Springer-Verlag.)

lemniscus have differential projections, LLI projecting to the forebrain nucleus basalis area (nucleus of the somatosensory system), and LLD and LLV projecting to the MLD and directly, mainly, to the thalamic nucleus semilunaris parvovoidalis (SPO; Wild, 1987). The midbrain MLD projects ipsilaterally to the thalamic nucleus ovoidalis (Ov) including SPO with collaterals to the contralateral Ov via the dorsal supraoptic decussation (DSOD). The MLD of both sides are interconnected via the intertectal commissure. The Ov projects ipsilaterally to Field L of the caudal neostriatum of the telencephalon. Field L has been divided into three laminae (L1, L2, L3; Bonke *et al.*, 1979) and it is L2 where the fibers from the Ov mainly terminate (Boord, 1969; Conlee and Parks, 1986; Wild, 1987; Carr, 1992).

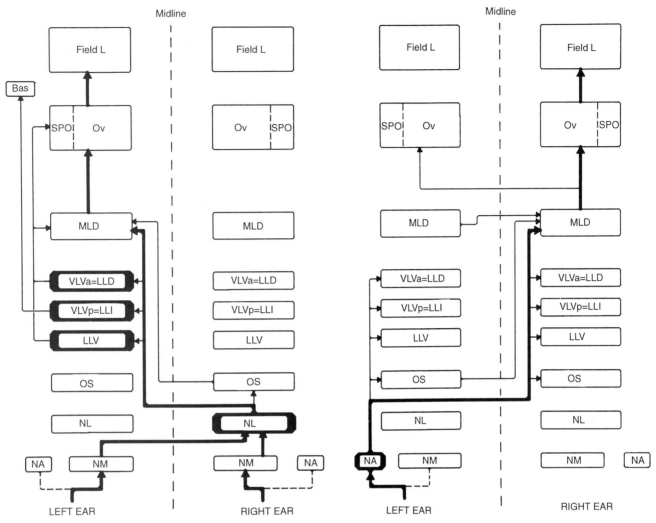

FIGURE 10 Diagram of ascending auditory pathways originating from nucleus laminaris and nuclei of the lateral lemniscus (left side) or nucleus angularis (right side). Thick lines mean main pathways. Bas, nucleus basalis; LLD/LLI/LLV, nucleus lemniscus lateralis pars dorsalis/pars intermedialis/pars ventralis; VLVa/VLVp, nucleus ventralis lemnisci lateralis pars anterior/pars posterior. See the legend to Fig. 9 for further abbreviations.

In the owl there are a few modifications of the general scheme in that the NL and NA reach different subnuclei of the ventral nucleus (VLV) of the lateral lemniscus and of the MLD. In more recent investigations the MLD in the owl is called colliculus inferior (IC) by analogy with the mammalian mesencephalic auditory nucleus (Knudsen, 1983). There is strong evidence that the NA and its projections process intensity cues whereas the NL processes time cues so that a time pathway and an intensity pathway have been postulated (Figure 11; Volman, 1994). Whereas time cues from both ears are processed first in the NL, intensity cues first appear in the posterior part of the VLV (VLVp). The VLVp projects to the shelf region of the central nucleus of the IC (ICc) whereas the NL projects to the ICc. The ICc and its shelf region both project to an external subdivision of the IC (ICx) where the information of both pathways is combined.

B. Electrophysiology

Neurons in the central auditory pathway have been characterized with regard to frequency (tuning curves), intensity, phase, and response to monaural or binaural stimulation. Binaural stimulation may be free field stimulation or dichotic stimulation with independent variations of intensity and phase at both ears. Topographic distribution of CFs within a nucleus indicates tonotopic organization.

The avian cochlear nuclei include the nucleus angularis, nucleus magnocellularis, and nucleus laminaris, which is, however, a third-order nucleus probably ho-

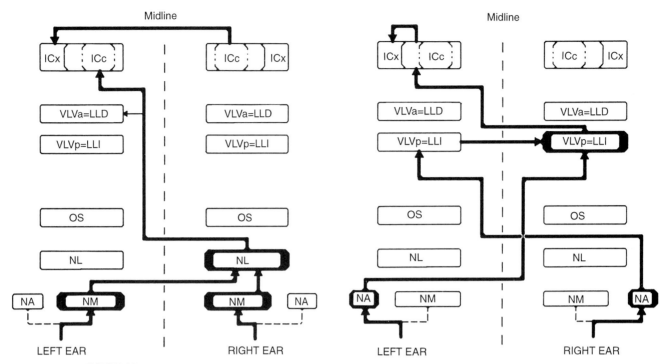

FIGURE 11 Time pathway (left side) and intensity pathway (right side) in the barn owl. ICc: core of inferior colliculus (MLD in the legend to Figure 10); ICx, external nucleus of IC. See the legends to Figures 9 and 10 for further abbreviations.

mologous to the medial superior olive (MSO) of mammals. The NA and NM correspond to the mammalian dorsal and ventral cochlear nuclei.

1. Cochlear Nuclei

Most neurons are spontaneously active both in the NA and the NM, with a higher spontaneous activity in the NM (Sachs and Sinnott, 1978; Warchol and Dallas, 1990). Intensity functions are monotonic in the NM but often reach a maximum in the NA. Tuning curves are more complex in the NA than in the NM. However, the general appearance is a V-shaped tuning curve with inhibitory side bands (Figure 12). In the NA additional inhibition may occur within the excitatory area (Sachs and Sinnott, 1978).

Characteristic frequencies range from 200 to about 5000 Hz in the chicken (Rubel and Parks, 1975, Warchol and Dallos 1990) and from 1000 to 9000 Hz in the owl (Sullivan and Konishi, 1984). In the chicken NA there are neurons which are broadly tuned at low frequencies or which decrease threshold down to 10 Hz (Warchol and Dallos, 1989). These neurons are assumed to process infrasound. Both the NA and the NM show a tonotopic organization which is dorsoventral in the NA and caudorostral in the NM (increasing CF). Phase-locking is better in the NM (time pathway) than in the NA and

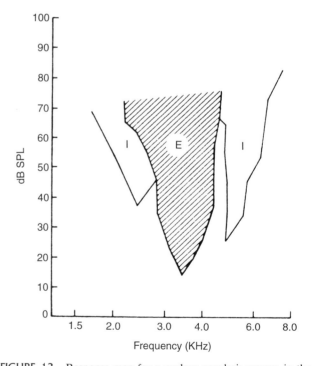

FIGURE 12 Response map for a nucleus angularis neuron in the blackbird. E, excitatory regions; I, inhibitory regions. (After, *J. Comp. Physiol.*, Responses to tones of single cells in nucleus magnocellularis and nucleus angularis of the red-wing blackbird (*Agelaius phoeniceus*), M. B. Sachs and J. M. Sinnott, **126**, 347–361, Fig. 3, 1978, © Springer-Verlag.)

can be observed up to 2000 Hz in the chicken but up to 9000 Hz in the owl.

2. Nucleus Laminaris

NL neurons have a similar tonotopic organization as NM neurons (Rubel and Parks, 1975). The CF of a neuron is the same independent of the ear stimulated. CFs range from 1 to 7 kHz in the owl (Carr and Konishi, 1990). Activity depends strongly on the interaural time delay (ITD) between both ears being maximal at ITDs of 0–300 μsec in the owl (Figure 13). However, at higher frequencies there are multiple maxima, i.e. information is ambiguous (Carr and Konishi, 1990). A prominent feature of NL neurons is phase-locking which increases with binaural stimulation. There is a dorsoventral map of ITDs (Sullivan and Konishi, 1986). It is assumed that delay lines and coincidence detection in NL are the basis for time difference or phase difference detection between both ears (Carr and Konishi, 1990; Overholt et al., 1992).

3. Nuclei of the Lateral Lemniscus

In owls the VLVp of the nuclei of the lateral lemniscus has been studied because it is the first nucleus in the intensity pathway where information from both ears converge (Manley et al., 1988). The neurons in this nucleus are excited (E) by contralateral stimulation and inhibited (I) by ipsilateral stimulation (EI type of response). Changing interaural intensity difference (IID) results in a monotonic change in response in such a way that response increases with reducing ipsilateral intensity. Similar responses are found in the MLD (see Figure 14). Absolute sound pressure level has not much effect, it is mainly the difference to which the cells are sensitive. There is a parallel shift in the response curves of different neurons (see Figure 14). This means that different neurons may detect different interaural intensity differences. There is a topographic organization in that ispsilateral inhibition decreases from dorsal to ventral. There is also a tonotopic organization which is, however, from rostral (low frequencies) to caudal (high frequencies). The CFs range from 2 to 10 kHz, most being 5 to 10 kHz.

4. The Midbrain Auditory Nucleus

The MLD has been investigated in different species. Although it is not clear whether there is a subdivision into an external nucleus (ICx) in other birds than owls, it seems that the main or central nucleus has a similar function in all birds. Neurons are spontaneously active and the most common response is an excitation from contralateral and an inhibition from ipsilateral (EI type, Figure 14). However, excitation to ipsilateral and contralateral stimulation (EE) and no response to ipsilateral stimulation (E0) have been observed in the chicken and in the pigeon also (Coles and Aitkin, 1979; Lewald 1990). EI units and EE units are assumed to detect IIDs and ITDs, respectively. EE units often show phase-locking and have low CFs. In the barn owl ITD-sensitive neurons have similar characteristics as neurons in the NM, including phase ambiguity.

Tuning curves of MLD neurons are V-shaped and some have inhibitory bands of various degree of complexity (Biedermann-Thorson, 1967) which points to a specialization for detecting complex sounds (Scheich et al., 1977). Tonotopy is from dorsal (low CF) to ventral (high CF) in diurnal raptors (Calford et al., 1985). Neurons at the rostrodorsal margin, which is at the low CF side, respond to infrasound in the guinea fowl

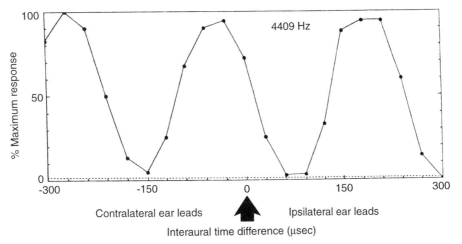

FIGURE 13 Response of a nucleus laminaris neuron to interaural time differences in the barn owl. Note multiple maxima at integer intervals of tone period (226 μs). (After Carr and Konishi (1990), *The Journal of Neuroscience,* used with permission.)

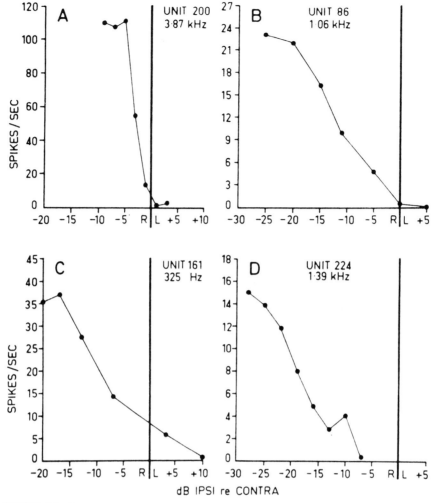

FIGURE 14 Response of four units in the chicken nucleus mesencephalicus lateralis pars dorsalis (MLD) to interaural intensity differences expressed as dB ipsi re contra (decreasing intensity on the ipsilateral side; i.e., dominance of contralateral stimulation increases responses). (After, *J. Comp. Physiol.*, The response properties of auditory neurones in the midbrain of the domestic foul (*Gallus gallus*) to monaural and binaural stimuli, R. B. Coles and L. M. Aitkin, **134**, 241–251, Fig. 6, 1979, © Springer-Verlag.)

(Theurich *et al.*, 1984). In these neurons activity is modulated at the soundwave amplitudes at frequencies as low as 2–10 Hz.

In the barn owl neurons with just one maximal response to ITD and with a peak at a distinct IID appear only in the ICx (Figure 15; Knudsen and Konishi, 1978). Such neurons are space-specific neurons; that is, neurons which under free-field conditions respond only to sound coming from a restricted area of the space. The response of space-specific neurons is independent of frequency and intensity and it has been shown that IID sensitivity depends on NA and ITD sensitivity on NM (Takahashi *et al.*, 1984). There is an orderly representation of azimuth and elevation in the ICx. Space-specific neurons are found in the optic tectum also and the auditory map matches the visual map (Knudsen, 1982, 1984).

5. Thalamic Nucleus Ovoidalis

The Ov has been studied much less than the MLD. There is spontaneous activity and both excitation and inhibitory sidebands are present, sometimes in complex forms (Biedermann-Thorson, 1970; Bigalke-Kunz *et al.*, 1987). There is a clear tonotopic organization from dorsal (high CF) to ventral (low CF; Bigalke-Kunz *et al.*, 1987).

6. Telencephalic Auditory Field L

In Field L activity is highest in layer L2, which is the input area from Ov. Tuning curves are V-shaped with inhibitory bands (Leppelsack, 1974; Biedermann-Thorson, 1970). CFs range from 0.2 to 6 kHz. Neurons

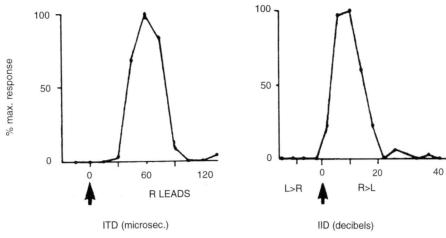

FIGURE 15 Response of neurons in the external nucleus of the inferior colliculus (ICx) to interaural time differences (ITD, on the left) and to interaural intensity differences (IID, on the right). (After Takahashi *et al.* (1984), *The Journal of Neuroscience,* used with permission.)

responding preferentially to species-specific sound, especially in L1 and L3, have been described (Leppelsack and Vogt, 1976; Bonke *et al.,* 1979). However, recently it has been questioned whether such a specialization really exists (Schäfer *et al.,* 1992; Knipschild *et al.,* 1992). There is a distinct tonotopic organization with increasing frequencies from dorsolateral to ventromedial (Rübsamen and Dörrscheidt, 1986; Bonke *et al.,* 1979; Müller and Scheich, 1985; Heil and Scheich, 1991). Phase-locking may occur up to 600 Hz in the starling (Leppelsack, 1974). Binaural interactions have not been studied systematically although most neurons repond more strongly to contralateral stimulation and all to stimulation of both sides (Leppelsack, 1974). In the barn owl space-specific neurons similar to those in ICx have been described (Knudsen and Konishi, 1977).

In the pigeon there is a second auditory area in the telencephalon which is anatomically connected to one of the nuclei of the lateral lemniscus (see Figure 10). Auditory neurons were identified near nucleus basalis which is the telencephalic representation of the beak. It seems that nucleus basalis and its surroundings represent a multisensory area involved in pecking of the pigeon (Delius *et al.,* 1979; Schall and Delius, 1986; Schall *et al.,* 1986).

VI. BEHAVIORAL ASPECTS

A. Behavioral Audiograms

Audiograms (frequency-dependent threshold of hearing) have been assessed in a variety of species (Dooling 1980, 1992). The best hearing is in the range of 1 to 4 kHz and there is a steep increase in the threshold up to 10 kHz, which is the upper limit of hearing in birds. The low-frequency slope is about -10 to -15 dB/octave and the high-frequency slope is about 30 to 50 dB/octave. Intensities at the best frequencies are 5 to 10 dB SPL in most birds (Sound Pressure Level: O dB = 20 μPascal) but about -15 to -20 dB in owls; that is, hearing in owls is at least 20 dB more sensitive. In the pigeon the low-frequency slope is very flat (-8 dB/octave; blackbird: -15 dB/octave; Hienz *et al.,* 1977) and hearing extends into the infrasound range down to 0.1 Hz (Figure 16; Kreithen and Quine, 1979).

B. Frequency Discrimination

Birds are well equipped to discriminate frequency differences and frequency difference limens (frequency difference, Δf, which just can be discriminated) are similar to mammals or humans. Weber fractions ($\Delta f/f$) range from 0.004 to 0.018 in various species (untrained human: 0.004; Sinnott *et al.,* 1980; Kuhn *et al.,* 1980).

C. Intensity Discrimination

Difference limen, to discriminate two sounds of different intensity, is as sensitive as in mammals or humans being in the range of 1–3 dB (Hienz *et al.,* 1980; Klump and Baur, 1990; Lewald, 1987b). Because of the small head of most birds sound attenuation between both ears is small. This is important for sound localization.

D. Time Resolution

Time resolution of avian hearing is again in the range of humans (Dooling, 1980). Two sounds separated by a gap (gap detection) are recognized as separate if the gap exceeds values of 2 to 10 msec (Wilkinson and

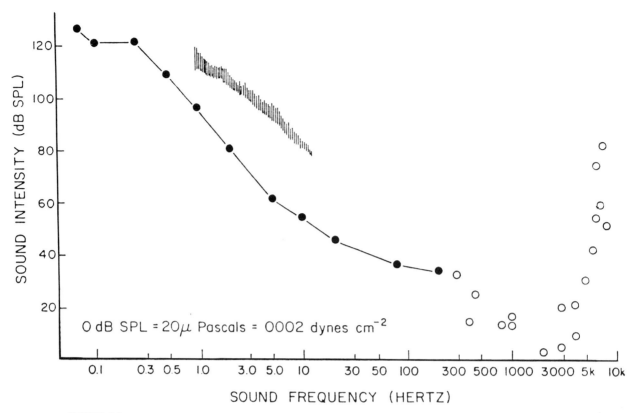

FIGURE 16 Behavioral audiogram of the pigeon in the low-frequency and infrasound range (filled circles). Open circles are pigeon data in the high-frequency range from other investigations. Hatched area indicates natural infrasound intensities. (After, *J. Comp. Physiol.*, Infrasound detection by the homing pigeon: A behavioral audiogram, M. L. Kreithen and D. B. Quine, **129**, 1–4, Fig. 2, 1979, © Springer-Verlag.)

Howe, 1975; Klump and Maier, 1989). This has been confirmed at the neural level by recordings from Field L neurons (Buchfellner *et al.,* 1989). A group of neurons strongly responded to gaps of a few milliseconds. Gap detection should be important for conspecific vocal communication especially in duetting birds.

The discrimination of ongoing interaural time difference (phase difference) between both ears has been tested as yet only in the pigeon in the behavioral context (Lewald, 1987b). At frequencies below 1 kHz, time differences as low as 10 μsec (phase angle difference of 0.9°) can be discriminated but this threshold increases rapidly up to above 100 μsec (phase angle: 144°) with increasing frequency. These results agree with similar data from mammals and humans.

E. Sound Localization

Sound localization depends on the function of two ears and location of a sound source is extracted from binaural disparities. A variety of disparities are used: arrival time (interaural time difference, ITD), phase difference (IPD), intensity difference (IID), spectrum of frequencies (ISD) at both ears, especially in natural sounds. Monaural cues may also be used (Knudsen, 1980). This requires moving the head in the sound field (estimation of elevation of a sound source by humans). The generally small heads of birds impose some limitations. Given a head diameter of 2 cm, the maximum time difference between both ears is about 60 μsec. Phase differences between both ears depend on frequency and at higher frequencies (short wavelength) the same phase may occur at both ears (phase ambiguity). This is, however, not a severe problem for birds because of the limited hearing range: wavelength at 8 kHz is 4.2 cm, about double that of "average" head diameter. It seems that phase-locking in central neurons plays an important role in detection of ongoing time differences. This limits time resolution to frequencies below 2 kHz in normal birds but includes most of the hearing range in the barn owl.

The duplex theory of sound localization assumes that sound localization at low frequencies (no sound attenuation by the head) uses phase differences whereas at high frequencies (phase ambiguity) intensity differences are used (sound shadow of the head increases with frequency). This concept seems to fit for nonspecialized birds (Lewald, 1990) but is probably not valid for owls.

Most birds are able to localize sound in azimuth but not in elevation. Minimum resolvable angle of location of a sound source is about 6° in the pigeon (Lewald, 1987a) but may be much larger in songbirds (e.g., 20° in the great tit; Klump et al., 1986). In the barn owl this angle is as low as 1 to 2° (Knudsen and Konishi, 1979; Knudsen et al., 1979). Owls are able to localize sound both in azimuth and in elevation and minimum localization error is similar in both directions. Detection of azimuth depends on ongoing time differences up to frequencies of 8 kHz and detection of elevation depends on intensity differences which arise from facial ruff and asymmetrical ears. However, detection of elevation is much better with noise than with tones which suggests that the frequency spectrum difference between both ears (ISD) is important also.

F. Echolocation

Echolocation occurs only in two families of birds, the Steatornidae, which are represented by the oilbird (*Steatornis caripensis*), and swiftlets of the family Apodidae. Echolocation is used for orientation in dark caves. Hearing ranges are not unusual with the best frequencies between 1 to 3 kHz in oilbirds (Konishi and Knudsen, 1979) and about 1 to 5 kHz in swiftlets (Coles et al., 1987). The energy of sounds used for echolocation is highest in the range of best hearing.

VII. SUMMARY AND CONCLUSIONS

The auditory system of birds has been studied in some detail at all levels both anatomically and physiologically. Furthermore, hearing performances are well investigated in behavioral experiments. Compared to mammals there are some differences in the structure of the avian ear. Although tuning of auditory nerve fibers is as sharp as in mammals the mechanisms which contribute to this tuning seem to be different but are not yet fully understood. Central pathways and central processing are as elaborate as in mammals. Although there is no neocortex in the avian brain the telencephalic auditory area shows layers which may be compared to the layers typical for the mammalian cortex. Barn owls have a unique sensitivity to hear and localize sound sources (prey). This is due to a number of now well-investigated specializations in the auditory system. The study of this highly evolved system contributed much to the understanding of hearing mechanisms in general.

References

Arends, J. J. A., and Zeigler, H. P. (1986). Anatomical identification of an auditory pathway from a nucleus of the lateral lemniscal system to the frontal telencephalon (nucleus basalis) of the pigeon. *Brain Res.* **398,** 375–381.

Ashmore, J. F. (1991). The electrophysiology of hair cells. *Annu. Rev. Physiol.* **53,** 465–476.

Ashmore, J. F. (1994). The cellular machinery of the cochlea. *Exp. Physiol.* **79,** 113–134.

Biederman-Thorson, M. (1967). Auditory responses of neurones in the lateral mesencephalic nucleus (inferior colliculus) of the barbary dove. *J. Physiol.* **193,** 695–705.

Biederman-Thorson, M. (1970). Auditory responses of the units in the ovoid nucleus and cerebrum (field L) of the ring dove. *Brain Res.* **24,** 247–256.

Bigalke-Kunz, B., Rübsamen, R., and Dörrscheidt, G. J. (1987). Tonotopic organization and functional characterization of the auditory thalamus in a songbird, the European starling. *J. Comp. Physiol. A* **161,** 255–265.

Bonke, B. A., Bonke, D., and Scheich, H. (1979a). Connectivity of the auditory forebrain nuclei in the guinea fowl (*Numida meleagris*). *Cell Tissue Res.* **200,** 101–121.

Bonke, D., Scheich, H., and Langner, G. (1979b). Responsiveness of units in the auditory neostriatum of the guinea fowl (*Numida meleagris*) to species-specific calls and synthetic stimuli. *J. Comp. Physiol.* **132,** 243–255.

Boord, R. L. (1969). The anatomy of the avian auditory system. *Ann. N.Y. Acad. Sci.* **167,** 186–198.

Brix, J., and Manley, G. A. (1994). Mechanical and electromechanical properties of the stereovillar bundles of isolated and cultured hair cells of the chicken. *Hearing Res.* **76,** 147–157.

Buchfellner, E., Leppelsack, H.-J., Klump, G.M., and Häusler, U. (1989). Gap detection in the starling (*Sturnus vulgaris*). II. Coding of gaps by forebrain neurons. *J. Comp. Physiol. A* **164,** 539–549.

Calford, M. B., Wise, L. Z., and Pettigrew, J. D. (1985). Coding of sound location and frequency in the auditory midbrain of diurnal birds of prey, families Accipitridae and Falconidae. *J. Comp. Physiol.* **157,** 149–160.

Carr, C. E. (1992). Evolution of the central auditory system in reptiles and birds. *In* "The evolutionary biology of hearing" (D. B. Webster, R. R. Fay, and A. N. Popper, eds.), pp. 511–543. Springer-Verlag, New York/Heidelberg/Berlin.

Carr, C. E., and Konishi, M. (1990). A circuit for detection of interaural time differences in the brain stem of the barn owl. *J. Neurosci.* **10,** 3227–3246.

Coles, R. B., and Aitkin, L. M. (1979). The response properties of auditory neurones in the midbrain of the domestic fowl (*Gallus gallus*) to monaural and binaural stimuli. *J. Comp. Physiol.* **134,** 241–251.

Coles, R. B., Lewis, D. B., Hill, K. G., Hutchings, M. E., and Gower, D. M. (1980). Directional hearing in the Japanese quail (*Coturnix coturnix japonica*). II. Cochlear physiology. *J. Exp. Biol.* **86,** 153–170.

Coles, R. B., Konishi, M., and Pettigrew, J. D. (1987). Hearing and echolocation in the Australian grey swiftlet, *Collocalia spodiopygia*. *J. Exp. Biol.* **129,** 365–371.

Conlee, J. W., and Parks, T. N. (1986). Origin of ascending auditory projections to the nucleus mesencephalicus lateralis pars dorsalis in the chicken. *Brain Res.* **367,** 96–113.

Dallos, P. (1985). Response characteristics of mammalian cochlear hair cells. *J. Neurosci.* **5,** 1591–1608.

Delius, J. D., Runge, T. E., and Oeckinghaus, H. (1979). Short-latency auditory projection to the frontal telencephalon of the pigeon. *Exp. Neurol.* **63**, 594–609.

Desmedt, J. E., and Delwaide, P. (1963). Activation of the efferent cochlear bundle in the pigeon. *J. Acoust. Soc. Am.* **35**, 809.

Dooling, R. (1980). Behavior and psychophysics of hearing in birds. *In* "Comparative Studies of Hearing in Vertebrates" (A. N. Popper and R. R. Fay, eds.), pp. 261–288. Spinger-Verlag, New York.

Dooling, R. (1992). Hearing in birds. *In* "The Evolutionary Biology of Hearing" (D. B. Webster, R. R. Fay, and A. N. Popper, eds.), pp. 545–559. Springer-Verlag, New York.

Fischer, F. P. (1992). Quantitative analysis of the innervation of the chicken basilar papilla. *Hearing Res.* **61**, 167–178.

Fischer, F. P., Köppl, C., and Manley, G. A. (1988). The basilar papilla of the barn owl *Tyto alba*: A quantitative morphological SEM analysis. *Hearing Res.* **34**, 87–102.

Fischer, F. P., Miltz, C., Singer, I., and Manley, G. A. (1992). Morphological gradients in the starling basilar papilla. *J. Morphol.* **213**, 225–240.

Fuchs, P. A., and Murrow, B. W. (1992). Cholinergic inhibition of short (outer) hair cells of the chick's cochlea. *J. Neurosci.* **12**, 800–809.

Fuchs, P. A., Nagai, T., and Evans, M. G. (1988). Electrical tuning in hair cells isolated from the chick cochlea. *J. Neurosci.* **8**, 2460–2467.

Gleich, O. (1989). Auditory primary afferents in the starling: correlation of function and morphology. *Hearing Res.* **37**, 255–268.

Gleich, O., and Manley, G. A. (1988). Quantitative morphological analysis of the sensory epithelium of the starling and pigeon basilar papilla. *Hearing Res.* **34**, 69–85.

Gleich, O., and Narins, P. M. (1988). The phase response of primary auditory afferents in a songbird (*Sturnus vulgaris* L.). *Hearing Res.* **32**, 81–92.

Gleich, O., Manley, G. A., Mandl, A., and Dooling, R. J. (1994). Basilar papilla of the canary and zebra finch: a quantitative scanning electron microscopical description. *J. Morphol.* **221**, 1–24.

Gummer, A. W., Smolders, J. W. T., and Klinke, R. (1987). Basilar membrane motion in the pigeon measured with the Mössbauer technique. *Hearing Res.* **29**, 63–92.

Gummer, A. W., Smolders, J. W. T., and Klinke, R. (1989a). Mechanics of a single-ossicle ear. I. The extra-stapedius of the pigeon. *Hearing Res.* **39**, 1–14.

Gummer, A. W., Smolders, J. W. T., and Klinke, R. (1989b). Mechanics of a single-ossicle ear. II. The columella footplate of the pigeon. *Hearing Res.* **39**, 15–26.

Heil, P., and Scheich, H. (1991). Functional organization of the avian cortex analogue. I. Topographic representation of isointensity bandwidth. *Brain Res.* **539**, 110–120.

Hienz, R. D., Sinnott, J. M., and Sachs, M. B. (1977). Auditory sensitivity of the redwing blackbird (*Agelaius phoeniceus*) and brown-headed cowbird (*Molothrus ater*). *J. Comp. Physiol. Psychol.* **91**, 1365–1376.

Hienz, R. D., Sinnott, J. M., and Sachs, M. B. (1980). Auditory intensity discrimination in blackbirds and pigeons. *J. Comp. Physiol. Psychol.* **94**, 993–1002.

Hill, K. G. (1989). Comparative aspects of cochlear function: Avian mechanisms. *Neurol. Neurobiol.* **56**, 73–79.

Hill, K. G., Lewis, D. B., Hutchings, M. E., and Coles, R. B. (1980). Directional hearing in the Japanese quail (*Coturnix coturnix japonica*). I. Acoustic properties of the auditory system. *J. Exp. Biol.* **86**, 135–151.

Kaiser, A., and Manley, G. A. (1994). Physiology of single putative cochlear efferents in the chicken. *J. Neurophysiol.* **72**, 2966–2979.

Klump, G.M., and Baur, A. (1990). Intensity discrimination in the European starling (*Sturnus vulgaris*). *Naturwissenschaften* **77**, 545–547.

Klump, G. M., and Maier, E. H. (1989). Gap detection in the starling (*Sturnus vulgaris*). I. Psychophysical thresholds. *J. Comp. Physiol. A* **164**, 531–538.

Klump, G. M., Windt, W., and Curio, E. (1986). The great tit's (*Parus major*) auditory resolution in azimuth. *J. Comp. Physiol. A* **158**, 383–390.

Knipschild, M., Dörrscheidt, G. J., and Rübsamen, R. (1992). Setting complex tasks to single units in the avian auditory forebrain. I. Processing of complex artificial stimuli. *Hearing Res.* **57**, 216–230.

Knudsen, E. I. (1980). Sound localization in birds. *In* "Comparative Studies of Hearing in Vertebrates" (A. N. Popper, and R. R. Fay, eds.), pp. 289–322. Springer-Verlag, New York.

Knudsen, E. I. (1982). Auditory and visual maps of space in the optic tectum of the owl. *J. Neurosci.* **2**, 1177–1194.

Knudsen, E. I. (1983). Subdivisions of the inferior colliculus in the barn owl (*Tyto alba*). *J. Comp. Neurol.* **218**, 174–186.

Knudsen, E. I. (1984). Auditory properties of space-tuned units in owl's optic tectum. *J. Neurophysiol.* **52**, 709–723.

Knudsen, E. I., and Konishi, M. (1977). Receptive fields of auditory neurons in the owl. *Science* **198**, 1278–1280.

Knudsen, E. I., and Konishi, M. (1978). A neural map of auditory space in the owl. *Science* **200**, 795–797.

Knudsen, E. I., and Konishi, M. (1979). Mechanisms of sound localization in the barn owl (*Tyto alba*). *J. Comp. Physiol.* **133**, 13–21.

Knudsen, E. I., Blasdel, G. G., and Konishi, M. (1979). Sound localization by the barn owl (*Tyto alba*) measured with the search coil technique. *J. Comp. Physiol.* **133**, 1–11.

Konishi, M., and Knudsen, E. I. (1979). The oilbird: Hearing and echolocation. *Science* **204**, 425–427.

Köppl, C., Gleich, O., and Manley, G. A. (1993). An auditory fovea in the barn owl cochlea. *J. Comp. Physiol. A* **171**, 695–704.

Kreithen, M. L., and Quine, D. B. (1979). Infrasound detection by the homing pigeon: A behavioral audiogram. *J. Comp. Physiol.* **129**, 1–4.

Kuhn, A., Leppelsack, H.-J., and Schwartzkopff, J. (1980). Measurement of frequency discrimination in the starling (*Sturnus vulgaris*) by conditioning of heart rate. *Naturwissenschaften* **67**, 102.

Kühne, R., and Lewis, B. (1985). External and middle ears. *In* "Form and Function in Birds" (A. S. King and J. McLeland, eds.), Vol. 3, pp. 227–271. Academic Press, London.

Kuijpers, W., and Bonting, S. L. (1970). The cochlear potentials. II. The nature of the cochlear endolymphatic resting potential. *Pflügers Arch.* **320**, 359–372.

Leppelsack, H.-J. (1974). Funktionelle Eigenschaften der Hörbahn im Feld L des Neostriatum caudale des Staren (*Sturnus vulgaris* L., Aves). *J. Comp. Physiol.* **88**, 271–320.

Leppelsack, H.-J., and Vogt, M. (1976). Responses of auditory neurons in the forebrain of a songbird to stimulation with species-specific sounds. *J. Comp. Physiol.* **107**, 263–274.

Lewald, J. (1987a). The acuity of sound localization in the pigeon (*Columba livia*). *Naturwissenschaften* **74**, 296–297.

Lewald, J. (1987b). Interaural time and intensity difference thresholds of the pigeon (*Columba livia*). *Naturwissenschaften* **74**, 449–451.

Lewald, J. (1990). Neural mechanisms of directional hearing in the pigeon. *Exp. Brain Res.* **82**, 423–436.

Manley, G. A., and Gleich, O. (1992). Evolution and specialization of function in the avian auditory periphery. *In* "The Evolutionary Biology of Hearing" (D. B. Webster, R. R. Fay, and A. N. Popper, eds.), pp. 561–580. Springer-Verlag, New York.

Manley, G. A., Gleich, O., Kaiser, A., and Brix, J. (1989). Functional differentiation of sensory cells in the avian auditory periphery. *J. Comp. Physiol. A* **164,** 289–296.

Manley, G. A., Haeseler, C., and Brix, J. (1991). Innervation patterns and spontaneous activity of afferent fibers to the lagenar macula and apical basilar papilla of the chick's cochlea. *Hearing Res.* **56,** 211–226.

Manley, G. A., Köppl, C., and Konishi, M. (1988). A neural map of interaural intensity differences in the brain stem of the barn owl. *J. Neurosci.* **8,** 2665–2676.

Manley, G. A., Schulze, M., and Oeckinghaus, H. (1987). Otoacoustic emissions in a song bird. *Hearing Res.* **26,** 257–266.

Manley, G. A., Schwabedissen, G., and Gleich, O. (1993). Morphology of the basilar papilla of the budgerigar *Melopsittacus undulatus*. *J. Morphol.* **218,** 153–165.

Müller, S. C., and Scheich, H. (1985). Functional organization of the avian auditory field L: A comparative 2DG study. *J. Comp Physiol. A* **156,** 1–12.

Necker, R. (1970). Zur Entstehung der Cochleapotentiale von Vögeln: Verhalten bei O_2-Mangel, Cyanidvergiftung und Unterkühlung sowie Beobachtung über die räumliche Verteilung. *Z. vergl. Physiol.* **69,** 367–425.

Oeckinghaus, H., and Schwartzkopff, J. (1983). Electrical and acoustical activation of the middle ear muscle in a songbird. *J. Comp. Physiol.* **150,** 61–67.

Overholt, E. M., Rubel, E. W., and Hyson, R. L. (1992). A circuit for coding interaural time differences in the chick brainstem. *J. Neurosci.* **12,** 1698–1708.

Pickles, J. O., and Corey, D. P. (1992). Mechanoelectrical transduction by hair cells. *Trends Neurosci.* **15,** 254–259.

Pickles, J. O., Brix, J., and Gleich, O. (1990). The search for the morphological basis of mechano-transduction in cochlear hair cells. *In* "Information Processing in Mammalian Auditory and Tactile Systems," pp. 20–43. A. R. Liss, New York.

Pierson, M., and Dallos, P. (1976). Re-examination of avian cochlear potentials. *Nature* **262,** 599–601.

Rubel, E. W., and Parks, T. N. (1975). Organization and development of brain stem auditory nuclei of the chicken: Tonotopic organization of N. magnocellularis and N. laminaris. *J. Comp. Neurol.* **164,** 411–434.

Rübsamen, R., and Dörrscheidt, G. J. (1986). Tonotopic organization of auditory forebrain in a songbird, the European starling. *J. Comp. Physiol. A* **158,** 639–646.

Runhaar, G., Schedler, J., and Manley, G. A. (1991). The potassium concentration in the cochlear fluids of the embryonic and posthatching chick. *Hearing Res.* **56,** 227–238.

Sachs, M. B., and Sinnott, J. M. (1978). Responses to tones of single cells in nucleus magnocellularis and nucleus angularis of the redwing blackbird (*Agelaius phoeniceus*). *J. Comp. Physiol.* **126,** 347–361.

Sachs, M. B., Young, E. D., and Lewis, R. H. (1974). Discharge pattern of single fibers in the pigeon auditory nerve. *Brain Res.* **70,** 431–447.

Sachs, M. B., Woolf, N. K., and Sinnott J. M. (1980). Response properties of neurons in the avian auditory system: comparisons with mammalian homologues and consideration of the neural encoding of complex stimuli. *In* "Comparative Studies of Hearing in Vertebrates" (A. N. Popper and R. R. Fay, eds.), pp. 323–353. Springer-Verlag, New York.

Saunders, J. C. (1985). Auditory structure and function in the bird middle ear: An evaluation by SEM and capacitive probe. *Hearing Res.* **18,** 253–268.

Saunders, J. C., and Johnstone, B. M. (1972). A comparative analysis of middle-ear function in non-mammalian vertebrates. *Acta Otolaryng.* **73,** 353–361.

Schäfer, M., Rübsamen, R., Dörrscheidt, G. J., and Knipschild, M. (1992). Setting complex tasks to single units in the avian auditory forebrain. II. Do we really need natural stimuli to describe neuronal response characteristics? *Hearing Res.* **57,** 231–244.

Schall, U., and Delius, J. D. (1986). Sensory inputs to the nucleus basalis prosencephali, a feeding-pecking centre in the pigeon. *J. Comp. Physiol. A* **159,** 33–41.

Schall, U., Güntürkün, O., and Delius, J. D. (1986). Sensory projections to the nucleus basalis prosencephali of the pigeon. *Cell Tissue Res.* **245,** 539–546.

Scheich, H., Langner, G., and Koch, R. (1977). Coding of narrow-band and wide-band vocalizations in the auditory midbrain nucleus (MLD) of the guinea fowl (*Numida meleagris*). *J. Comp. Physiol.* **117,** 245–265.

Schermuly, L., and Klinke, R. (1990a). Origin of infrasound sensitive neurones in the papilla basilaris of the pigeon: an HRP study. *Hearing Res.* **48,** 69–78.

Schermuly, L., and Klinke, R. (1990b). Infrasound sensitive neurones in the pigeon cochlear ganglion. *J. Comp. Physiol. A* **166,** 355–363.

Schermuly, L., Vossieck, T., and Klinke, R. (1990). Furosemide has no effect on endocochlear potential and tuning properties of primary afferent fibres in the pigeon inner ear. *Hearing Res.* **50,** 295–298.

Schermuly, L., Topp, G., and Klinke, R. (1991). A previously unknown hair cell epithelium in the pigeon cochlea: the papilla chaotica. *Hearing Res.* **53,** 49–56.

Schmidt, R. S. (1964). Blood supply of pigeon inner ear. *J. Comp. Neurol.* **123,** 187–204.

Schmidt, R. S., and Fernandez, C. (1962). Labyrinthine DC potentials in representative vertebrates. *J. Cell. Comp. Physiol.* **59,** 311–322.

Schwartzkopff, J. (1968). Structure and function of the ear and of the auditory brain areas in birds. Ciba Foundation Symposium on Hearing Mechanisms in Vertebrates, pp. 41–59.

Schwartzkopff, J. (1973). Mechanoreception. *In* "Avian Biology" (D. S. Farner, J. R. King, and K. C. Parkes, eds.), pp. 417–477. Academic Press, New York.

Schwarz, I. E., Schwarz, D. W. F., Fredrickson, J. M., and Landolt, J. P. (1981). Efferent vestibular neurons: A study employing retrograde tracer methods in the pigeon (*Columba livia*). *J. Comp. Neurol.* **196,** 1–12.

Sinnott, J. M., Sachs, M. B., and Hienz, R. D. (1980). Aspects of frequency discrimination in passerine birds and pigeons. *J. Comp. Physiol. Psychol.* **94,** 401–415.

Smith, C. A. (1985). Inner ear. *In* "Form and Function in Birds," Vol. 3, pp. 273–310. Academic Press, London.

Sullivan, W. E., and Konishi, M. (1984). Segregation of stimulus phase and intensity coding in the cochlear nucleus of the barn owl. *J. Neurosci.* **4,** 1787–1799.

Sullivan, W. E., and Konishi, M. (1986). Neural map of interaural phase difference in the owl's brainstem. *Proc. Natl. Acad. Sci. USA* **83,** 8400–8404.

Takahashi, T., Moiseff, A., and Konishi, M. (1984). Time and intensity cues are processed independently in the auditory system of the owl. *J. Neurosci.* **4,** 1781–1786.

Takasaka, T., and Smith, C. A. (1971). The structure and innervation of the pigeon's basilar papilla. *J. Ultrastructure Res.* **35,** 20–65.

Tanaka, K., and Smith, C. A. (1978). Structure of the chicken's inner ear: SEM and TEM study. *Am. J. Anat.* **153,** 251–272.

Temchin, A. N. (1988). Unusual discharge patterns of single fibers in the pigeon's auditory nerve. *J. Comp. Physiol. A* **163,** 99–115.

Theurich, M., Langner, G., and Scheich, H. (1984). Infrasound responses in the midbrain of the guinea fowl. *Neurosci. Lett.* **49,** 81–86.

Volman, S. F. (1994). Directional hearing in owls: Neurobiology, behaviour and evolution. *In* "Perception and Motor Control in Birds" (M. N. O. Davies and P. R. Green, eds.), pp. 292–314. Springer-Verlag, New York.

Von Bartheld, C. S. (1994). Functional morphology of the paratympanic organ in the middle ear of birds. *Brain Behav. Evol.* **44**, 61–73.

Von Bekesy, G. (1944). Über die mechanische Frequenzanalyse in der Schnecke verschiedener Tiere. *Akust. Z.* **9**, 3–11.

Vossieck, T., Schermuly, L., and Klinke, R. (1991). The influence of DC-polarization of the endocochlear potential on single fibre activity in the pigeon cochlear nerve. *Hearing Res.* **56**, 93–100.

Warchol, M. E., and Dallos, P. (1989). Neural response to very low-frequency sound in the avian cochlear nucleus. *J. Comp. Physiol. A* **166**, 83–95.

Warchol, M. E., and Dallos, P. (1990). Neural coding in the chick cochlear nucleus. *J. Comp. Physiol. A* **166**, 721–734.

Whitehead, M. C., and Morest, D. K. (1981). Dual populations of efferent and afferent cochlear axons in the chicken. *Neuroscience* **11**, 2351–2365.

Wild, J. M. (1987). Nuclei of the lateral lemniscus project directly to the thalamic auditory nuclei in the pigeon. *Brain Res.* **408**, 303–307.

Wilkinson, R., and Howse, P. E. (1975). Time resolution of acoustic signals by birds. *Nature* **258**, 320–321.

CHAPTER 3

The Chemical Senses in Birds

J. RUSSELL MASON AND LARRY CLARK
United States Department of Agriculture
Animal and Plant Health Inspection Service Wildlife Services
National Wildlife Research Center
and Monell Chemical Senses Center
Philadelphia, Pennsylvania 19104

I. Chemosensory Systems 39
II. Chemesthesis 40
   A. Trigeminal and Somatosensory Chemoreceptors 40
   B. Innervation of Chemesthetic Receptors 40
   C. Behavioral Responses to Chemical Stimuli 40
   D. Structure-Activity Relationships for Aromatic Stimuli 42
   E. Responses to Respiratory Stimuli 42
   F. Nasal and Respiratory Irritation and Interaction of the Olfactory and Trigeminal Systems 42
   G. Summary 43
III. Gustation 43
   A. Taste Receptors 43
   B. Innervation of Taste Receptors 43
   C. Taste Behavior 43
   D. Response to Sweet 43
   E. Response to Salt 44
   F. Response to Sour 44
   G. Response to Bitter 44
   H. Response to Other Tastes 46
   I. Temperature and Taste 46
   J. Summary 46
IV. Olfaction 46
   A. Morphology of Olfactory System 46
   B. Innervation of Olfactory Receptors 47
   C. Olfactory Neuronal Response 48
   D. Laboratory Detection and Discrimination Capabilities 48
   E. Olfactory Performance in the Field 49
   F. Summary 50
References 51

I. CHEMOSENSORY SYSTEMS

The chemical senses are commonly thought to fall into three classes: (1) olfaction (smell), (2) gustation (taste), and (3) chemesthesis (the common chemical sense). In birds, as in most other vertebrates, olfaction is usually thought to be a telereceptor, capable of receiving airborne chemical stimuli in extreme dilution over relatively great distances. Olfactory receptors are located in the nasal conchae. Gustation, on the other hand, usually requires more intimate contact between the source(s) of chemical stimuli and receptors. Gustatory receptors are located in the taste buds of the oral cavity. Chemesthesis is usually reserved for nonspecific stimuli, which are often irritating or painful. Chemoreceptive fibers are concentrated in exterior mucous membranes, although they occur throughout the animal.

Traditional emphasis in describing responsiveness to chemical stimuli has been placed on taste and smell. This emphasis is misplaced. Trigeminal chemoreception (chemesthesis) also may be involved. The sensory afferents of the trigeminal and olfactory nerves are in close proximity in the nasal cavity, and the trigeminal and gustatory nerves are in close proximity in the oral cavity. Most chemicals can stimulate multiple sensory afferents, although circumstances may favor detection by one sensory system over others. Except in the case of electrophysiological studies in which specific nerve function in response to specific chemical stimulus can be documented, attributing specific sensory mediation of a chemostimulant is not possible.

II. CHEMESTHESIS

Chemesthesis is the perception of chemically induced pain. A major component of the chemesthetic system is the trigeminal nerve (TN). The TN is the principal somatic sensory nerve of the head, and its primary function is the coding of mechanical and thermal stimuli. However, the trigeminal nerve also contains chemoreceptive fibers that mediate the detection of chemical irritants (Silver and Maruniak, 1980). The somatosensory system is the primary somatic sensory system of the rest of the body. Like the TN, the somatosensory system primarily codes for mechanical and thermal stimuli, but it does have sensory afferents that are chemosensitive (Kitchell and Erikson, 1983), though little is known about this system in birds. Sensitivity to chemical irritants is adaptive because animals can avoid noxious stimuli before actual physical damage occurs.

A. Trigeminal Chemoreceptors

Chemosensitive fibers of the avian trigeminal and somatosensory systems are similar to mammalian sensory afferents. Most are unmyelinated C-type polymodal nociceptors with conduction velocities of 0.3–1 m/sec. However, some myelinated A-delta high-threshold mechanoreceptors with conduction velocities of 5–40 m/sec also respond to chemical stimuli. The discharge patterns and conduction velocities for the chicken (*Gallus gallus* var *domesticus*), duck (*Anas platyrhyncos*), and pigeon (*Columba livia*) are similar to those observed in mammals (Gentle, 1989; Necker, 1974).

The underlying physiological and biochemical processes of chemically induced pain appear to be similar for birds and mammals. Endogenous pain-promoting substances such as substance P, 5HT, histamine, bradykinin, and acetylcholine evoke pain-related behaviors in chickens, pigeons, and guinea pigs (Gentle and Hill, 1987; Gentle and Hunter, 1993; Szolcsanyi *et al.*, 1986). Prostaglandins that modulate the pain response in mammals also subserve this function in starlings (*Sturnus vulgaris*), and their effects can be abolished by prostaglandin biosynthase inhibitors, such as aspirinlike analgesics (Clark, 1995a). However, there are profound differences in how birds and mammals respond to exogenous chemical stimuli. In mammals, chemicals such as capsaicin are potent trigeminal irritants. These irritants deplete substance P from afferent terminals and the dorsal root ganglion, producing an initial sensitization followed by a desensitization to further chemical stimulation (Szolcsanyi, 1982). In contrast, birds are insensitive to capsaicin (Mason and Maruniak, 1983; Szolcsanyi *et al.*, 1986). Peripheral presentation of capsaicin to pigeons and chickens does not cause release of substance P in avian sensory afferents (Pierau *et al.*, 1986; Sann *et al.*, 1987; Szolcsanyi *et al.*, 1986).

B. Innervation of Chemesthetic Receptors

The trigeminal nerve is the VIth cranial nerve in birds, arising from the rostrolateral medulla near the caudal surface of the optic lobe (Getty, 1975; Schrader, 1970). The TN travels along with the trochlear nerve (IV), entering a fossa in the floor of the cranial cavity where the trigeminal ganglion (TG) is found. The TG is subdivided into a smaller medial ophthalmic region and a larger lateral maxillomandibular region from which the nerve splits into three branches. In the chicken, the ophthalmic branch of the TN innervates the frontal region, the eyeball, upper eyelid, conjunctiva, glands in the orbit, the rostrodorsal part of the nasal cavity, and the tip of the upper jaw. The ophthalmic branch has a communicating ramus with the trochlear nerve which serves for motor control of the eye region. This aspect can provide for reflexive response to irritating stimuli to the ocular region. The larger medial ramus accompanies the olfactory nerve into the nasal fossa via the medial orbitonasal foramen. The maxillary branch of the TN provides sensory input from the integument of the crown, temporal region, rostral part of the external ear, upper and lower eyelids, the region between the nostrils and eye, conjunctival mucosa, the mucosal part of the palate, and the floor and medial wall of the nasal cavity. The mandibular branch of the TN provides sensory input from the skin and rhamphotheca of the lower jaw, intermandibular skin, wattles, oral mucosa of the rostral floor of the mouth, and the palate near the angle of the mouth.

C. Behavioral Responses to Chemical Stimuli

Although the morphological organization of the peripheral trigeminal system in birds is not very different from that found in mammals (Dubbeldam and Karten, 1978; Dubbeldam and Veenman, 1978), profound functional differences appear to exist (Mason *et al.*, 1989; Norman *et al.*, 1992; Mason and Otis, 1990; Mason *et al.*, 1991a,b). Birds rarely avoid mammalian irritants, even though the avian trigeminal system is responsive to chemical stimuli (Walker *et al.*, 1979; Mason and

Silver, 1983). For example, cedar waxwings (*Bombycilla cedrorum;* Norman *et al.,* 1992) are indifferent to ≥1000 ppm capsaicin, the pungent principle in *Capsicum* peppers, whereas mammals typically avoid much lower concentrations: 100 ppm capsaicin is typically avoided by rodents (Figure 1). Nevertheless, it is interesting to note that birds can be trained to avoid mammalian irritants (Mason and Clark, 1995a) and that some trigeminal input appears to mediate the response (Mason and Clark, 1995b).

Many aromatic structures are aversive to birds (Avery and Decker, 1991; Clark and Shah, 1991a, 1993; Crocker and Perry, 1990; Crocker *et al.,* 1993; Kare, 1961; Mason *et al.,* 1989). Several lines of evidence suggest that a variety of compounds have intrinsic properties that cause them to be aversive on a purely sensory basis. First, the aversive quality is unlearned; that is avoidance occurs upon initial contact (Clark and Shah, 1991b). Second, there is no evidence that consumption is altered by gastrointestinal feedback—intake of fluid treated with these sensory repellents is constant over time (Clark and Mason, 1993). Third, birds seem unable to associate the aversive quality of the stimulus with other chemosensory cues,

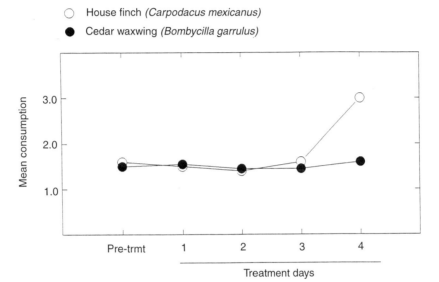

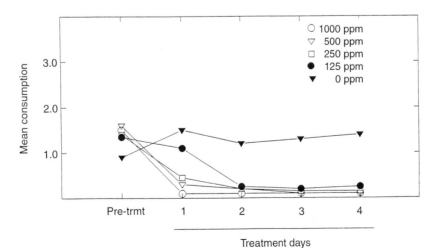

FIGURE 1 Responses of house finches (*Carpodacus mexicanus*), cedar waxwings (Bombycilla cedrorum), and house mice (Mus musculus) to capsaicin adulterated chow. (Modified from Norman *et al.* (1992) with permission.)

suggesting that conditioned flavor avoidance learning does not occur (Clark, 1995b; Mason et al., 1989). Fourth, birds do not habituate to the stimulus—avoidance persists in the absence of reinforcement (Clark and Shah, 1994; Mason et al., 1989).

D. Structure–Activity Relationships for Aromatic Stimuli

The structure–activity relationships of aromatic avian repellents have been elucidated. An aromatic parent structure is critical for repellency. Factors that affect the delocalization of electrons around the aromatic structure contribute to modifying the repellent effect. Thus, acidic substituents to the benzene ring generally detract from repellency, and this is amplified if the acidic function is contained within the electron-withdrawing group. Electron donation to the benzene ring enhances repellency. Heteroatoms that distort the plane of the aromatic structure tend to lessen repellency (Clark and Shah, 1991a, 1994; Clark et al., 1991; Mason et al., 1991a; Shah et al., 1991, 1992) (Figure 2).

E. Responses to Respiratory Stimuli

Changes in carbon dioxide concentration in the nasopharynx region can cause species-specific changes in reflexive breathing in birds (Hiestand and Randall, 1941). However, concentrations of carbon dioxide that are sufficiently high to be irritating to mammals have no effect on blood pressure, heart rate, tidal volume, breathing frequency, upper airway resistance or lower airway resistance in geese (*Anser anser* and *Cygnopsis cygnoid*; Callanan et al., 1974). Similarly, geese respond differently than mammals to exposure to sulfur dioxide, but in a similar manner when exposed to ammonia and phenyl diguanide (Callanan et al., 1974).

F. Nasal and Respiratory Irritation and Interaction of the Olfactory and Trigeminal Systems

The trigeminal nerve is important in the perception of odors (Keverne et al., 1986; Silver and Maruniak, 1980; Tucker, 1971). Electrophysiological evidence shows that the trigeminal nerve is responsive to odors, albeit generally less sensitive than the olfactory nerve (Tucker, 1963). Behavioral assays yield similar results. Pigeons trained to respond to odors fail to respond after olfactory nerve transection. However, odor responding can be reinstated if the odor concentration is increased (Henton, 1969; Henton et al., 1966; Michelsen, 1960). Walker et al. (1979, 1986) found that odor sensitivity of pigeons decreased by 2–4 log units (vapor saturation) after olfactory nerve transection.

Although olfaction can modulate responding to chemical irritants, it is relatively unimportant. Clark

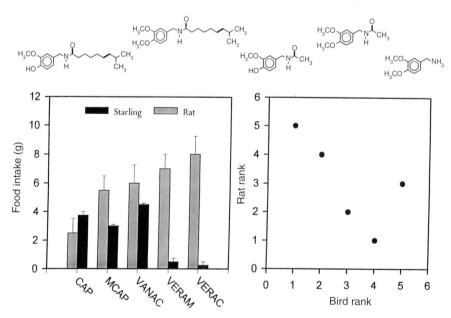

FIGURE 2 (Left) Consumption of food adulterated with capsaicin derivatives for rats and starlings. Codes are CAP, capsaicin; MCAP, methyl capsaicin; VANAC, vanillyl acetamide; VERAM, veratryl amine; VERAC, veratryl acetamide. Structures shown are in order presented in panel codes. (Right) The rank order of food intake for rats and starlings, demonstrating an inverse relationship between palitability.

(1995a) and Mason et al. (1989) showed that avoidance of repellent anthranilates was partially a consequence of olfactory cues. When the olfactory nerves of starlings were transected, avoidance of the anthranilate repellents was mildly suppressed. When the ophthalmic branches of the trigeminal nerve were cut, the starlings became insensitive to the repellent effects of the anthranilates (Mason et al., 1989).

G. Summary

The anatomical configuration and the physiological and biochemical processes of chemosensory afferents of the avian trigeminal and somatosensory systems are similar in birds and mammals. However, there are significant differences in sensitivity to exogenous chemical stimuli between these two taxa. Structure–activity studies suggest that these differences may reflect different receptor mechanisms in peripheral afferents. Confirmation using molecular and pharmacological techniques is needed to clarify this possibility.

III. GUSTATION

A. Taste Receptors

In comparison to other vertebrates, birds have few taste buds (Table 1). They are distributed throughout the oral mucosa, but most often in close association with salivary gland openings (Berkhoudt, 1985). The greatest numbers are on the caudal surface of the tongue and the pharyngeal floor (Kare, 1971; Gentle, 1975; Kare and Rogers, 1976). Ontogenetic changes in taste bud number occur (Duncan, 1960). Adult chickens have twice the number of taste buds of day-old chicks (Lindenmaier and Kare, 1959; Saito, 1966). However, within adults, the number of taste buds declines with age (Botezat, 1910; Duncan, 1960; Lalonde and Eglitis, 1961).

Saliva is critical for the transport of taste stimuli to receptors (Belman and Kare, 1961). This is particularly true for birds, since avian taste buds do not open directly into the oral cavity via taste pores (Berkhoudt, 1985). Although the role of saliva on avian taste responding has not been extensively studied, there is evidence that changes in salivary flow rate affect taste related behaviors. Gentle and Dewar (1981) and Gentle et al. (1981) reported significant declines in taste avoidance by chicks that were vitamin A and zinc deficient. These deficiencies lower salivary flow rate.

B. Innervation of Taste Receptors

The lingual branch of the glossopharyngeal nerve was once considered the only gustatory nerve in birds (Kitchell et al., 1959; Duncan, 1960; Halpern, 1963; Kadono et al., 1966; Landolt, 1970). However, more recent investigations show that the palatine branch of the facial nerve (Krol and Dubbeldam, 1979) and the chorda tympani (Berkhoudt, 1985; Gentle, 1979, 1983) also transmit gustatory information.

C. Taste Behavior

Simple evaluations of ingestion are the most common laboratory method used to measure the sensitivity of birds to taste stimuli, although operant methods have been used (Mariotti and Fiore, 1980). Usually, the test stimuli are presented in aqueous solution, and animals choose between mixtures and distilled water. Chickens show a characteristic response to aversive oral stimulation typified by persistent tongue and beak movements and head-shaking and beak-wiping behaviors (Gentle 1973, 1976, 1978). No characteristic responses to presentations of neutral or appetitive oral stimuli have been observed (Gentle, 1978; Gentle and Harkin, 1979).

D. Response to Sweet

Many species show modest preferences for natural sugars mixed with drinking water (Brindley, 1965; Brindley and Prior, 1968; Duncan, 1960; Engelmann, 1934, 1937, 1960; Gentle 1972, 1975; Gunthur and Wagner, 1971; Harriman and Milner, 1969; Rensch and Neunzig, 1925; Warren and Vince, 1963). Strong preferences are exhibited by parrots, budgerigars, hummingbirds, and other nectar-feeders (Bradley, 1971; Hainesworth and Wolf, 1976; Kare and Rogers, 1976; Stromberg and Johnsen, 1990).

TABLE 1 Absolute Number of Taste Buds in Various Animals[a]

Species	Number	Source
Chicken	24	Lindemaier and Kare (1959)
Bullfinch	46	Duncan (1960)
Starling	200	Bath (1906)
Japanese quail	62	Warner et al. (1967)
Lizard	550	Schwenk (1985)
Kitten	473	Elliot (1937)
Bat	800	Moncrieff (1951)
Human	9,000	Cole (1941)
Pig	15,000	Moncrieff (1951)
Rabbit	17,000	Moncrieff (1951)
Catfish	100,000	Hyman (1942)

[a] Modified from Kare and Mason (1986).

A variety of granivores and some omnivores reject sugars, perhaps for physiological reasons. For example, red-winged blackbirds select pure water over sucrose (Rogers and Maller, 1973; Martinez del Rio et al., 1988). Common grackles (*Quiscula quiscula*), European starlings, cedar waxwings, and robins (*Turdus migratorius*), also reject sucrose, although other sugars (e.g., fructose, glucose) are preferred (Schuler, 1980, 1983). Brugger and Nelms (1991), Brugger (1992), and Brugger et al. (1992) have suggested that rejection occurs because these birds lack the enzyme sucrase. Ingestion of sucrose by sucrase-deficient birds causes sickness, due to malabsorption (Martinez del Rio, 1990; Martinez del Rio et al., 1988; Martinez del Rio and Stevens, 1989; Brugger and Nelms, 1991).

Besides taste, osmotic pressure, viscosity, melting point, nutritive quality, digestibility, and toxicity are all involved in birds' response to tastes. Some have suggested that visual properties and surface texture sometimes take precedence over all other qualities in the birds' selection of food (Engelmann, 1957; Kare and Rogers, 1976; Morris, 1955; Kear, 1960; Mason and Reidinger, 1983a,b). Across species, no physical or chemical characteristic has been shown to reliably predict how a bird on an adequate diet will respond to the taste of a solution (Kare and Medway, 1959).

E. Response to Salt

Sodium chloride rejection thresholds for 58 species ranged from 0.35% in a parrot to 37.5% in the pine siskin (*Carduelis pinus;* Rench and Neunzig, 1925). Salt-eating has been reported for a number of species (Reeks, 1920; Mousley, 1921, 1946; Pierce, 1921; McCabe, 1927; Gorsuch, 1934; Aldrich, 1939; Marshall, 1940; Peterson, 1942; Calhoun, 1945; Packard, 1946; Bleitz, 1958; Duncan, 1964; Cade, 1964; Dawson et al., 1965; Mason and Espaillat, 1990). Numerous finches of the family Carduelidae have notorious appetites for salt. Cross-bills can be caught in traps baited with salt alone (Welty, 1975; Willoughby, 1971). Cade (1964) suggests that finches, which have 0.001–0.03% sodium in their diets (Altmann and Dittmer, 1968), are chronically sodium deficient.

The presence of a nasal salt gland is associated with salt acceptance taste thresholds. Birds without such glands generally refuse concentrations of salt that are hypertonic to their body fluids (Bartholomew and Cade, 1958; Bartholomew and MacMillian, 1960). However, rejection thresholds in no-choice tests do not always predict responding in choice situations. When given a choice, gulls (*Larus spp.*) (with salt glands) select pure water over saline solution (Harriman, 1967; Harriman and Kare, 1966). Similarly, penguins prefer fresh water after having been at sea for extended periods (Warham, 1971). Preference could reflect the toxic effects of chronic exposure to saline or salt waters. Mallards possess salt glands (Shoemaker, 1972), but hatching success and duckling survival is influenced by the salinity of drinking water in the natal marsh (Mitcham and Wobeser, 1988). The order of acceptability of ionic series by birds does not appear to fit into the lyotropic or sensitivity series reported for other animals.

F. Response to Sour

Birds are tolerant of acidic and alkaline solutions (Fuerst and Kare, 1962; Table 2), and some species exhibit preferences for acid over plain tapwater (Brindley and Prior, 1968). Not surprisingly, species differences exist. Rensch and Neunzig (1925) and Engelmann (1934) reported that pigeons were more sensitive than ducks or fowl. Engelmann (1950) also reported that chicks were more sensitive than adults. Berkhoudt (1985) reports that hooded crows (*Corvus corone*) are profoundly sensitive to hydrochloric acid and speculates that this sensitivity might be linked to the assessment of the quality of carrion as potential food. Although the ecological reason(s) for acid tolerance in some avian species remains unclear, one possibility is that it permits the exploitation of certain otherwise unpalatable food resources. For example, even though starlings prefer insect prey to fruit, juvenile starlings are less successful in capturing animal prey than are adults (Stevens, 1985). Accordingly, juveniles eat large amounts of fruit because it is readily available. Much of this fruit is unripe and sour.

G. Response to Bitter

Avian responsiveness to bitter is enigmatic. In some cases, compounds evoke similar responses in mammals and birds (e.g., quinine hydrochloride; Engelmann, 1934; Gentle, 1975). In others, compounds that are extremely bitter to humans (e.g., sucrose octaacetate) are readily accepted by birds (Halpern, 1963; Heinroth, 1938). This acceptance may reflect physiological insensitivity (Kitchell et al., 1959, Landolt, 1970). There is evidence that acceptance may decrease as individuals age (Brindley, 1965; Cane and Vince, 1968).

The bitter phenolic compounds produced by some plants (Robinson, 1983) and utilized by various species of pharmacophagus insects (Nishida and Fukami, 1990) may serve as defenses against birds (e.g., Greig-Smith, 1988; Rodriguez and Levin, 1976). There is abundant evidence that the tannin content of fruits and grain is

TABLE 2 The Influence of pH on Fluid Preferences of the Chick[a]

Substance	pH1[b]	pH2[c]	Versus	Percentage intake[d]
Acetic acid	2.9	3.2	Water	16.1
Acetic acid	4.1	4.5	Water	53.3
Acetic acid	4.9	7.3	Water	50.0
Acetate buffer	4.0	4.1	Acetate buffer, pH 6	47.8
Acetate buffer	4.0	4.1	Acetate buffer, pH 5	38.0
Acetate buffer	4.0	4.1	Water	52.1
Acetate buffer	5.1	5.1	Water	57.6
Acetate buffer	5.1	5.1	Acetate buffer, pH 6	38.0
Acetate buffer	6.0	6.0	Water	54.6
Acetate buffer	6.0	6.1	Water	54.2
Acetate buffer	6.0	6.1	Phosphate buffer, pH 7	53.0
Acetate buffer	6.0	6.1	Phosphate buffer, pH 6	53.9
Acetate buffer	6.0	6.1	Veronal buffer, pH 7	52.4
Glycine buffer	2.3	2.3	Glycine stock	61.2
Glycine buffer	3.0	3.2	Glycine stock	52.8
Glycine stock	5.4	6.6	Water	50.7
Glycine buffer	7.2	7.0	Glycine stock	48.9
Glycine buffer	9.0	7.8	Glycine stock	49.2
Glycine buffer	10.0	8.7	Glycine stock	48.8
Glycine buffer	11.0	9.0	Glycine stock	49.8
Hydrochloric acid	1.1	1.1	Water	4.0
Hydrochloric acid	1.5	1.6	Water	18.6
Hydrochloric acid	2.1	2.1	Water	36.5
Hydrochloric acid	1.6	1.6	Glycine stock	24.7
Hydrochloric acid	1.7	1.7	Glycine stock	16.4
Hydrochloric acid	2.0	2.0	Glycine stock	16.4
Hydrochloric acid	2.1	2.1	Glycine stock	39.8
Hydrochloric acid	1.7	1.7	Water	14.8
Hydrochloric acid	2.0	2.0	Water	50.0
Hydrochloric acid	3.0	3.1	Sulfuric acid, pH 3	49.4
Hydrochloric acid	3.0	3.2	Water	59.1
Hydrochloric acid	3.1	3.1	Sodium hydroxide, pH 10.2	53.3
Hydrochloric acid	4.1	7.4	Water	48.8
Lactic acid	2.3	2.3	Water	14.6
Lactic acid	2.9	3.0	Water	60.6
Lactic acid	4.1	6.7	Water	50.2
Nitric acid	1.1	1.1	Water	8.1
Nitric acid	2.0	2.0	Water	62.0
Nitric acid	3.0	3.2	Water	52.5
Phosphate buffer	6.0	6.0	Water	52.3
Phosphate buffer	6.0	6.0	Phosphate buffer, pH 7	53.6
Phosphate buffer	6.0	6.0	Veronal buffer, pH 7	48.0
Phosphate buffer	7.0	7.2	Water	49.0
Potassium hydroxide	11.1	9.0	Water	48.3
Potassium hydroxide	11.1	10.1	Sodium hydroxide, pH 11	47.9
Potassium hydroxide	12.0	11.2	Water	36.4
Potassium hydroxide	13.0	12.1	Water	2.7
Sodium hydroxide	10.2	9.2	Water	45.0
Sodium hydroxide	11.1	9.5	Water	46.8
Sodium hydroxide	12.2	11.2	Water	33.3
Sodium hydroxide	13.0	12.4	Water	1.8
Sulfuric acid	1.2	1.3	Water	15.2
Sulfuric acid	1.5	1.5	Water	35.4
Sulfuric acid	1.9	1.9	Water	54.2
Sulfuric acid	2.0	2.0	Sulfuric acid, pH 3	45.7
Sulfuric acid	3.1	3.2	Water	55.7
Sulfuric acid	4.1	6.9	Water	51.2
Veronal buffer	7.0	7.0	Water	51.8

[a] From Fuerst and Kare (1962).
[b] pH1 = initial.
[c] pH2 = after 24 hr.
[d] Percentage intake = (volume of test fluid consumed/total consumption) × 100. Each intake percentage is the mean of 18 daily values.

associated with resistance to bird damage (Bullard et al., 1981; Greig-Smith et al., 1983; Mason et al., 1984), and laboratory preference tests show that consumption is negatively correlated with tannin concentration (Mason and Espaillat, 1990). Other phenolic substances (e.g., phenylpropanoids, including coniferyl and cinnamyl derivatives; Crocker and Perry, 1990; Jakubas et al., 1992) produce analogous effects. Jakubas and his colleagues (Jakubas et al., 1992) suggest that it may be possible to genetically engineer crops to produce analogs of coniferyl alcohol as an inherent defense against pests and pathogens. The occurrence of coniferyl alcohol is widespread in higher plants because it is the primary precursor of lignin (Hahlbrock and Scheel, 1989; Lewis and Yamamoto, 1990). It may be possible to localize production of these compounds to specific plant tissues (Collins, 1986; Jakubas et al., 1992; McCallum and Walker, 1990). By localizing the production of repellent phenylpropanoids to specific plant tissues, autotoxic effects could be minimized along with the impact of these compounds on the nutritional value and palatability of the grain.

H. Response to Other Tastes

Apart from responses to simple tastes, reactions to more complex substances and synthetic flavors have been reported (Kare et al., 1957; Romoser et al., 1958; Kare and Medway, 1959; Kare and Pick, 1960; Deyoe et al., 1962). In general, birds are more sensitive to such stimuli in drinking than in feeding tests.

Very few experiments have dealt with natural taste compounds. However, there is evidence that several species of shorebirds can discriminate between clean sand and sand that had contained worms (Gerritsen et al., 1984; van Heezik et al., 1983). Conceivably, these birds were detecting amino acids in mucus secretions of the worms. Espaillat and Mason (1990) reported that both European starlings and red-winged blackbirds detect and show preferences toward diets adulterated with L-alanine (Figure 3). Whether or not L-alanine sensitivity reflects sensitivity to other free amino acids or to protein is unknown. However, L-alanine and similar substances (e.g., L-glutamine) occur as free amino acids in vegetable matter, fruits, and meat (Hac et al., 1949; Maeda et al., 1958; Baker and Baker 1983). These substances could aid in food search and selection. At least for starlings, assimilation efficiency increases as the overall protein content of the diet increases (Twedt, 1984).

There is also some evidence that taste sensitivity may assist in the rejection of potentially dangerous natural substances. Berkhoudt (1985) reports that a great-crested grebe (*Podiceps cristatus*) apparently used taste cues to reject minnows with slime infections of the epidermis.

I. Temperature and Taste

The domestic fowl is acutely sensitive to the temperature of water. Acceptability decreases as the temperature of the water increases above the ambient. Fowl can discriminate a temperature difference of only 5°F, and usually reject the higher temperature. Similar results have been reported for red-winged blackbirds (Mason and Maruniak, 1983). Chickens suffer from acute thirst rather than drink water 10°F above their body temperature. Because the response to temperature may take precedence over all chemical stimulants (Kare and Rogers, 1976), temperature should be eliminated as a variable in taste studies of the fowl. The ecological reason(s) underlying the interaction between taste and temperature remains obscure.

J. Summary

Kare and Beauchamp (1976), in discussing the comparative aspects of the sense of taste in birds and mammals, pointed out that most of the work on the basic mechanism of taste has been conducted with mammals. This mammalian work has suggested that the initial interaction of a taste stimulus and a receptor cell occurs on the microvilli of taste receptor cells. Although stimulus–receptor interactions in avian taste are probably similar to those described for mammals, this has not been demonstrated.

Birds have a sense of taste. However, no pattern, whether chemical, physical, nutritional, or physiologic, can be correlated consistently with the bird's taste behavior. The behavioral, ecologic, and chemical context of a taste stimulant can influence the birds' response. The observed response, particularly to sweet and bitter, indicates that the bird does not share human taste experiences. The supposition that there is a difference in degree between individual birds and an absolute difference between some species appears warranted.

IV. OLFACTION

A. Morphology of Olfactory System

Olfactory receptors are located in the olfactory epithelium in the caudal conchae where each receptor cell is surrounded by a cluster of supporting cells. The receptor nerve dendrite passes through these cells to the lumen, ending in a knob bristling with 6–15 cilia. The length of the cilia vary with species. Black vultures have cilia of 40–50 μm, while that of the domestic fowl is about 7–10 μm (Shibuya and Tucker, 1967). To gain access to receptors, odor molecules must diffuse through a mucous membrane. The cilia of the sensory cells have no transport function. Rather, the secretions covering

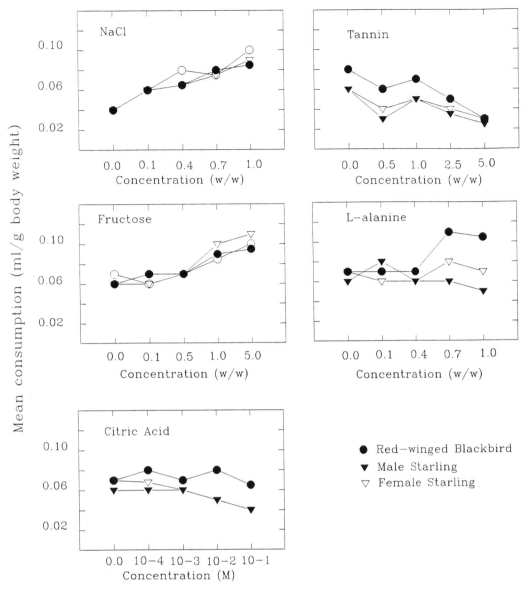

FIGURE 3 Mean consumption of sodium chloride, fructose, citric acid, tannin, and L-alanine per gram of body weight by male red-winged blackbirds (*Agelaius phoeniceus*) and male and female european starlings (*Sturnus vulgaris*). (From Espaillat and Mason (1990) with permission.)

the cilia provide rapid flow for transport of odor molecules and must constantly be replaced to avoid receptor habituation. Olfactory gland secretions are removed by traction of the surrounding respiratory cilia.

The nasal conchae are important structures that influence nasal air flow dynamics and direct odors to the olfactory epithelium (Bang, 1960, 1961, 1963, 1964, 1965, 1966; Bang and Cobb, 1968). The extent of scrolling of the caudal conchae is correlated with the relative size of the olfactory bulb (Bang and Wenzel, 1986). Furthermore, olfactory thresholds and relative size of the olfactory bulb are inversely related at the taxonomic ordinal level; that is, orders with high olfactory thresholds have relatively small olfactory bulbs (Clark *et al.*, 1993; Table 3, Figure 4). These patterns suggest that the elaborated olfactory systems belong to species with demonstrated reliance on odor cues in the field (Stager, 1964; Hutchison and Wenzel, 1980).

B. Innervation of Olfactory Receptors

Birds have a fully developed olfactory bulb, but lack an accessory olfactory system—the vomeronasal organ and accessory olfactory bulb (Rieke and Wenzel, 1975, 1978). However, the latter has been identified in the early embryonic development of some birds (Matthes,

TABLE 3 Summary of Mean Ratios of Ipsilateral Olfactory Bulb Diameter to Cerebral Hemisphere Diameter and Their Standard Errors (SE) for Several Orders of Birds[a]

Order	N	Ratio	SE	Order	N	Ratio	SE
Anseriformes	4	19.4	1.5	Psittaciformes	2	8.0	1.4
Apodiformes	8	12.3	1.9				
Apterygiformes	1	34.0	0.0	Falconiformes	5	17.4	2.6
Caprimulgiformes	3	23.3	0.7	Charadriiformes	9	16.4	0.9
Ciconiiformes	2	20.9	0.6	Galliformes	3	14.2	1.4
Columbiformes	2	20.0	1.4	Piciformes	5	11.4	1.3
Cuculiformes	4	19.5	0.6	Passeriformes	25	13.3	0.7
Gaviiformes	1	20.0	0.0	Pelecaniformes	4	12.1	1.6
Gruiformes	14	22.2	0.9	Coraciiformes	5	14.5	1.6
Podicipediformes	2	24.5	1.8	Sphenisciformes	1	17.0	0.0
Procellariiformes	10	29.1	1.4	Strigiformes	2	18.5	0.4

[a] Data adapted from Bang and Cobb (1968). Sample sizes are in terms of number of species (N).

1934). The olfactory bulb is composed of concentric structures, where the incoming olfactory nerve fibers constitute the outer layer. The branching terminals penetrate to the adjacent, glomerular layer, where they connect with dendrites of mitral and tufted cells in spherical arborizations called glomeruli. The perikarya of these cells are in the deeper mitral cell layer, where their axon leave to project to many areas of the forebrain. There are many interneuronal connections in the layers between the mitral and glomerular regions. There are no direct connections between contralateral bulbs (Rieke and Wenzel, 1978).

C. Olfactory Neuronal Response

Single and multiunit electrophysiological responses to odor stimuli are typically taken as definitive evidence of olfactory capacity. Electrophysiological recordings of units and nerve fibers from mammals, amphibians, reptiles, and birds respond to odor stimuli in a similar fashion, irrespective of the gross anatomical development of the species' olfactory apparatus (Tucker, 1965; Shibuya and Tonosaki, 1972). In black vultures, the electroolfactogram (EOG) appears during inspiration and less so at expiration. The EOG also coincides with peak spike activity (Shibuya and Tucker, 1967). The spike duration of 3–4 msec is similar to that observed for the tortoise and frog (Gesteland et al., 1963; Shibuya and Shibuya, 1963). Because olfactory nerve fibers are unmyelinated, conduction velocities are slow, about 1.5 m/sec (Macadar et al., 1980). As is the case for mammals, continuous presentation of the stimulus to the bird's receptor field will result in physiological adaptation of the nerve units. Recovery can be achieved within a few minutes of rest. In terms of nerve function, species with even the most vestigial olfactory anatomies compare favorably with those with more developed anatomies in terms of olfactory detection thresholds (Tucker, 1965).

Olfactory nerve sections have been used to verify that spontaneous and trained behaviors are based upon odor cues. Transected olfactory nerves grow back within 30 days of transection and recover full physiologic capacity to respond to odor stimuli (Tucker et al., 1974). Healed nerves often were smaller, have neuromas, and are enmeshed in scar tissue. However, electrophysiological recordings and autonomic reflex responses to odorant did not differ between controls and nerves cut 6 months or more before (Tucker, 1971; Tucker et al., 1974).

D. Laboratory Detection and Discrimination Capabilities

Physiological responses (e.g., change in respiration or heart rate) to novel odor stimuli have been observed (Wenzel and Sieck, 1972). However, habitua-

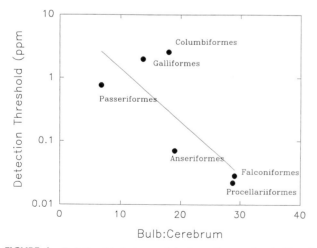

FIGURE 4 Relationship between detection olfactory threshold and relative size of the olfactory bulb for different orders of birds. (From Clark and Shah (1993) with permission.)

tion to the stimulus under this paradigm is always a difficulty.

Various operant and classical conditioning paradigms have also been employed to determine olfactory ability (Michelsen, 1959; Henton et al., 1966; Henton, 1969). Positive or negative reinforcement is used to make the olfactory stimulus a "biologically" relevant cue, irrespective of whether the odor cue is of natural relevance to a species. Overall, classical conditioning techniques have proven to be relatively poor assays for olfactory discrimination in birds (Calvin et al., 1957), but conditioned suppression variants have proven to be quite reliable (Henton et al., 1966; Clark and Mason, 1987). A generally successful assay for determining olfactory detection and discrimination thresholds is cardiac conditioning (Walker et al., 1986; Clark and Mason, 1989; Clark and Smeraski, 1990; Clark, 1991a; Clark et al., 1993). In this procedure, the odor (the conditional stimulus) is paired with an aversive experience; for example, shock (the unconditional stimulus). Heart rate is compared pre- and poststimulus presentation throughout training, when a criterion level of cardiac acceleration is achieved as a result of the stimulus–shock pairing, tests can proceed on detection or discrimination tasks. Most birds that have been tested have shown olfactory capabilities comparable to mammals (Davis, 1973), and even passerines, with the least developed olfactory system, demonstrate behavioral responsiveness to odors (Clark and Mason, 1987; Clark and Smeraski, 1990; Clark, 1991a; Clark et al., 1993) (Table 4).

E. Olfactory Performance in the Field

The use of olfactory cues for locating food has been documented for a number of species. Turkey vultures are attracted to ethyl mercaptan fumes (Stager, 1964, 1967) and can locate decomposed carcasses in the absence of visual cues (Houston, 1987). Procellariiformes can use odor cues as navigational aids in locating food from considerable distances (Table 5). Black-footed albatrosses (*Diomedea nigripes*) are attracted to bacon drippings from distances as great as 20 miles (Miller, 1942). Using cardiac conditioning techniques for estimating odor detection thresholds, field observations, and detailed atmospheric models of odor dispersion, Clark and Shah (1992) estimated that the Leach's storm petrel (*Oceanodroma leucorhoa*) is capable of detecting and homing in on an odor target for distances from 1 to 12 km.

Procellariiformes also appear to rely on olfactory cues to locate their burrows, showing differential return rates to their nest sites as a function of surgical manipulation: control (C), sham surgery (SS), and olfactory nerve section (ONS). For Leach's storm petrel the return rates were C = 91%, SS = 74%, and ONS = 0% (Grubb, 1974). For the wedge-tailed shearwater the return rates were C = 90%, SS = 70%, and ONS = 25% (Shallenberger, 1975).

Pigeons can use odor cues for orientation and navigation (Papi, 1986; Wallraff, 1991; Waldvogel, 1989). However, reliance on odor cues for orientation is dependent upon the atmospheric predictability of the cues experienced during the bird's development and early training experience (Wiltschko et al., 1987). Pigeons can obtain positional information when atmospheric odors are derived from boundary-layer free airspace in an open landscape. However, positional information is obscured when the atmosphere sample is derived from close to ground level (Wallraff et al., 1993). When regional odor maps cannot be relied upon because of atmospheric instability, pigeons use a variety of alternative cues, such as visual, magnetic, and polarized light to orient themselves (Waldvogel, 1987).

A number of species have now been shown to be capable of using olfactory cues to locate food. Ravens (*C. corax;* Harriman and Berger, 1986), magpies (Buitron and Nuechterlein, 1985), jays, crows (Goodwin, 1955), chickadees (*Parus atricapillus;* Jarvi and Wiklund, 1984), hummingbirds (Goldsmith and Goldsmith, 1982; Ioale and Papi, 1989), honey guides (Archer and Glen, 1969), and kiwis (Wenzel, 1968) have all been shown to be capable of using olfactory cues to locate and discriminate between foods.

There are several intriguing studies suggesting that odor recognition is important in the reproductive behavior of birds. Male mallards decreased social displays and sexual behavior toward females when their olfactory nerves were sectioned (Balthazart and Schoffeniels, 1979). When unfamiliar fruit odors were applied to squabs of the ring dove (*Streptopelia risoria*), parents decreased parental care, resulting in higher mortality of scented squabs. Bilateral olfactory nerve cuts eliminated the differential feeding of the scented and control squabs (Cohen, 1981). Olfactory recognition of parents and/or home sites may be advantageous to young as well. Just as in mammals (Corey, 1978), domestic chicks show neophobia to familiar nests treated with novel odors (Jones, 1988) and demonstrate a preference for familiar nest odors (Jones and Faure, 1982; Wurdinger, 1982).

There is also evidence that starlings may use olfaction to select nest material used in the fumigation of ectoparasites and pathogens (Clark and Mason, 1985, 1987, 1988; Clark 1991b) or in the selection of material used in

TABLE 4 Summary of Behavioral Olfactory Threshold Data for Different Species of Birds

Species	Ratio[a]	Stimulus	Threshold (ppm) Min	Max	Source[b]
Rock dove *Columba livia*	18.0	n-Amyl acetate	0.31	29.80	5,6,9,10
		Benzaldehyde	0.47	0.75	10
		Butanethiol	13,820		7
		Butanol	0.17	1.30	10
		n-Butyl acetate	0.11	2.59	5,10
		Butyric acid	2.59		5
		Ethanethiol	10,080		7
		Heptane	0.29	0.38	8
		Hexane	1.53	2.98	8
		Pentane	16.45	20.76	8
Chicken *Gallus gallus*	15.0	Heptane	0.31	0.57	8
		Hexane	0.64	1.00	8
		Pentane	1.58	2.22	8
Northern bobwhite *Colinus virginianus*	—	Heptane	2.14	3.49	8
		Hexane	3.15	4.02	8
		Pentane	7.18	10.92	8
Black-billed magpie *Pica pica*	—	Butanethiol	13,416		7
		Ethanethiol	8,400		7
European starling *Sturnus vulgaris*	9.7	Cyclohexanone	2.50		3
Cedar waxwing *Bombycilla cedrorum*	—	Cyclohexanone	6.80	86.46	1
Tree swallow *Tachycineta bicolor*	15.0	Cyclohexanone	73.42		1
Brown-headed cowbird *Molothrus ater*	7.0	Ethyl butyrate	0.76		2
Catbird *Dumetella carolinensis*	—	Cyclohexanone	35.14		4
Eastern phoebe *Sayornis phoebe*	—	Cyclohexanone	35.61		4
European goldfinch *Carduelis carduelis*	—	Cyclohexanone	13.05		4
Great tit *Parus major*	—	Cyclohexanone	34.10		4
Black-capped chickadee *Parus atricapillus*	3.0	Cyclohexanone	59.95		5

[a] The ratio of the longest axis of the olfactory bulb to that of the ipsilateral cerebral hemisphere.
[b] Sources: (1) Clark (1991a); (2) Clark and Mason (1989); (3) Clark and Smeraski (1990); (4) Clark et al. (1993). Reprinted by permission of the publisher from (Cedar thresholds in passerines, Clark et al.), *Comp. Biochem. Physiol.*, **104A,** 305–312. Copyright 1993 by Elsevier Science Inc.; (5) Henton (1969); (6) Henton et al. (1996); (7) Snyder and Peterson (1979); (8) Stattelman et al. (1975); (9) Walker et al. (1979); (10) Walker et al. (1986).

"anting" behavior, which is postulated to be a grooming response to rid the bird of ectoparasites (Clark et al., 1990). Multiunit recordings from olfactory nerves indicate starlings respond to a number of natural plant odors and are capable of making discriminations between complex sets of odors (Clark and Mason, 1987). However, olfactory discrimination by starlings shows a strong correlation with breeding season (specifically nest-building), suggesting hormonal influence on detection and discrimination ability in this species (Clark and Smeraski, 1990).

F. Summary

All evidence indicates that the extent of olfactory development in birds is on par with that found in mammals. Some species, such as passerines, have relatively poorly developed olfactory capacities, though nonethe-

TABLE 5 Summary of Olfactory Orientation toward a Prey-Odorized Target for Seabirds

Taxa	Percentages[a] Sea water	Cod liver oil	Source
Albatrosses			
Diomedea exulans	12	0	Lequette *et al.* (1989)
Phoebetria palpebrata	0	14	Lequette *et al.* (1989)
Pelicanoididae			
Pelecanoides sp.	0	0	Lequette *et al.* (1989)
Procellariidae			
Pagodroma nivea	—	78	Jouventin and Robin (1984)
Pachyptila spp.	0	0	Lequette *et al.* (1989)
Procellaria aequinoctialis	3	58	Lequette *et al.* (1989)
Macronectes spp.	16	30	Lequette *et al.* (1989)
Daption capense	10	54	Lequette *et al.* (1989)
Daption capense	0	82	Jouventin and Robin (1984)
Puffinus gravis	5	95	Grubb (1972)
Puffinus grisenus	67	33	Grubb (1972)
Oceanitidae			
Oceanodroma leucorhoa	0	100	Grubb (1972)
Oceanites oceanicus	24	76	Grubb (1972)
Oceanites oceanicus	13	77	Jouventin and Robin (1984)
Oceanites oceanicus	0	87	Lequette *et al.* (1989)
Fregetta tropica	0	95	Lequette *et al.* (1989)
Nonprocellariiformes			
Larus dominicanus	11	0	Lequette *et al.* (1989)
Phalacrocorax atricpes	0	0	Lequette *et al.* (1989)
Sterna spp.	9	0	Lequette *et al.* (1989)

[a] Values are the percentage of the birds observed that were attracted to the target (control or cod liver oil-soaked sponge).

less show some degree of olfactory acuity. Other species, such as procellariiformes, have olfactory systems acutely sensitive to odor cues. Relative to mammals, few systematic physiological and behavioral studies are available. This gap in knowledge is unfortunate because there is a well-developed anatomical database on the avian olfactory system.

References

Aldrich, E. C. (1939). Notes on the salt-feeding of the Red Crossbill. *Condor* **41**, 172–173.

Altmann, P. L., and Dittmer, D. S. (1968). "Metabolism." Federation of American Societies for Experimental Biology, Bethesda.

Archer, A. L., and Glen, R. M. (1969). Observations on the behavior of two species of honey-guides *Indicator variegatus* (Lesson) and *Indicator exilis* (Cassin). *Los Ang. Cty. Mus. Contrib. Sci.* **160**, 1–6.

Avery, M. L., and Decker, F. G. (1991). Repellency of fungicidal rice seed treatments to red-winged blackbirds. *J. Wildl. Manage.* **55**, 327–334.

Baker, H. S., and Baker, I. (1983). Floral nectar sugar constituents in relation to pollinator type. *In* "Handbook of Pollination Biology" (C. E. Jones and R. J. Little, eds.), pp. 117–141. Scientific and Academic Editions, New York.

Balthazart, J., and Schoffeniels, E. (1979). Pheromones are involved in the control of sexual behaviour in birds. *Naturwissenschaften* **66**, 55–56.

Bang, B. G. (1960). Anatomical evidence for olfactory function some species of birds. *Nature* **188**, 547–549.

Bang, B. G. (1961). The surface pattern of the nasal mucosa and its relation to mucous flow—A study of chicken and herring gull nasal mucosae. *J. Morphol.* **109**, 57–72.

Bang, B. G. (1963). Comparative studies of the nasal organs of birds: A study of 28 species of birds of West Bengal. *PAVO* **1**, 79–89.

Bang, B. G. (1964). The nasal organs of the Black and Turkey Vultures: A comparative study of the cathartid species *Coragyps atratus atratus* and *Carthartes aura septentrionalis* (with notes on *Cathartes aura falklandica, Pseudogyps bengalensis,* and *Neophron percnopterus*). *J. Morphol.* **115**, 153–184.

Bang, B. G. (1965). Anatomical adaptations for olfaction in the snow petrel. *Nature* **205**, 513–515.

Bang, B. G. (1966). The olfactory apparatus of tube-nosed birds (Procellariiformes). *Acta Anat.* **65**, 391–415.

Bang, B. G., and Cobb, S. (1968). The size of the olfactory bulb in 108 species of birds. *Auk* **85**, 55–61.

Bang, B., and Wenzel, B. M. (1986). Nasal cavity and olfactory system. *In* "Form and Function in Birds III" (A. S. King and J. McLellan, eds.), pp. 195–225. Academic Press, London.

Bartholomew, G. A., and Cade, T. J. (1958). Effects of sodium chloride on the water consumption of house finches. *Physiol. Zool.* **31**, 304–310.

Bartholomew, G. A., and MacMillan, R. E. (1960). The water requirements of mourning doves and their use of sea water and NaCl solutions. *Physiol. Zool.* **33,** 171.

Bath, W. (1906). Die Geschmaksorgane der Vogel und Krokodile. *Arch. Biontol.* **1,** 5–74.

Belman, A. L., and Kare, M. R. (1961). Character of salivary flow in the chicken. *Poult. Sci.* **40,** 1377.

Berkhoudt, H. (1985). Structure and function of avian taste receptors. *In* "Form and Function in Birds III" (A. S. Levy and J. McLelland, eds.), pp. 463–496. Academic Press, New York.

Bleitz, D. (1958). Attraction of birds to salt licks placed for mammals. *Wilson Bull.* **70,** 92.

Botezat, E. (1910). Morphologie, Physiologie und phylogenetische Bedeutung der Geschmacksorgane der Vögel. *Anat. Anz.* **36,** 428–461.

Bradley, R. M. (1971). Tongue topography. *In* "Handbook of Sensory Physiology" (L. M. Beidler, ed.). Springer-Verlag, Berlin.

Brindley, L. D. (1965). Taste discrimination in bobwhite and Japanese quail. *Anim. Behav.* **13,** 507–512.

Brindley, L. D., and Prior, S. (1968). Effects of age on taste discrimination in the bobwhite quail. *Anim. Behav.* **16,** 304–307.

Brugger, K. E. (1992). Repellency of sucrose to American robins (Turdus migratorius). *J. Wildl. Manage.* **56,** 794–799.

Brugger, K. E., and Nelms, C. O. (1991). Sucrose avoidance by American Robins (Turdus migratorius): Implications to control of bird damage in fruit crops. *Crop Protect.* **10,** 455–460.

Brugger, K. E., Nol, P., and Phillips, C. I. (1992). Sucrose repellency to European Starlings: Will high sucrose cultivars deter bird damage to fruit? *Ecol. Appl.* **3,** 256–261.

Buitron, D., and Nuechterlein, G. L. (1985). Experiments on olfactory detection of food caches by Black-billed Magpies. *Condor* **87,** 92–95.

Bullard, R. W., York, J. O., and Kilburn, S. R. (1981). Polyphenolic changes in ripening bird-resistant sorghums. *J. Agric. Food Chem.* **29,** 972–981.

Cade, T. J. (1964). Water and salt balance in granivorous birds. *In* "Thirst" (M. J. Wayner, ed.), pp. 237–256. Pergammon Press, Oxford.

Calhoun, J. B. (1945). English Sparrow eating salt. *Auk* **62,** 455.

Callanan, D., Dixon, M., Widdicombe, J. G., and Wise, J. C. M. (1974). Responses of geese to inhalation of irritant gases and injection of phenyl diguanide. *Resp. Physiol.* **22,** 157–166.

Calvin, A. D., Williams, C. N., and Westmoreland, N. (1957). Olfactory sensitivity in the domestic pigeon. *Am. J. Psychol.* **188,** 100–105.

Cane, V. R., and Vince, M. A. (1968). Age and learning in quail. *Br. J. Psychol.* **59,** 37–46.

Clark, L. (1991a). Odor detection thresholds in Tree Swallows and Cedar Waxwings. *Auk* **108,** 177–180.

Clark, L. (1991b). The nest protection hypothesis: The adaptive use of plant secondary compounds by European starlings. *In* "Bird-Parasite Interactions: Ecology, Evolution, and Behaviour" (J. E. Loye and B. Zuk, eds.), pp. 205–221. Oxford Univ. Press, Oxford.

Clark, L. (1995). Modulation of avian responsiveness to chemical irritants: Effects of prostaglandin E1 and analgesics. *J. Exp. Zool.* **271,** 432–440.

Clark, L. (1996). Trigeminal repellents do not promote conditional odor avoidance in european starlings. *Wilson Bull.* **108,** 36–52.

Clark, L., and Mason, J. R. (1985). Use of nest material as insecticidal and anti-pathogenic agents by the European starling. *Oecologia* **67,** 169–176.

Clark, L., and Mason, J. R. (1987). Olfactory discrimination of plant volatiles by the European Starling. *Anim. Behav.* **35,** 227–235.

Clark, L., and Mason, J. R. (1988). Effect of biologically active plants used as nest material and the derived benefit to starling nestlings. *Oecologia* **77,** 174–180.

Clark, L., and Mason, J. R. (1989). Sensitivity of Brown-headed Cowbirds to volatiles. *Condor* **91,** 922–932.

Clark, L., and Mason, J. R. (1993). Interactions between sensory and postingestional repellents in starlings: Methyl anthranilate and sucrose. *Ecol. Appl.* **3,** 262–270.

Clark, L., and Shah, P. (1991a). Nonlethal bird repellents: In search of a general model relating repellency and chemical structure. *J. Wildl. Manage.* **55,** 538–545.

Clark, L., and Shah, P. (1991b). Chemical bird repellents: Applicability for deterring use of waste water. "Issues and Technology in the Management of Impacted Wildlife" (S. Foster, ed.). Thorne Ecological Institute, Boulder.

Clark, L., and Shah, P. S. (1992). Information content of prey odor plumes: What do foraging Leach's storm petrels know? *In* "Chemical Signals in Vertebrates" (R. L. Doty and D. Muller-Schwarze, eds.), pp. 421–428. Plenum Press, New York.

Clark, L., and Shah, P. S. (1993). Chemical bird repellents: Possible use in cyanide ponds. *J. Wildl. Manage.* **57,** 657–664.

Clark, L., and Shah, P. S. (1994). Tests and refinements of a general structure-activity model for avian repellents. *J. Chem. Ecol.* **20,** 321–339.

Clark, L., and Smeraski, C. A. (1990). Seasonal shifts in odor acuity by starlings. *J. Exp. Zool.* **177,** 673–680.

Clark, C. C., Clark, L., and Clark, L. (1990). "Anting" behavior by common grackles and European starlings. *Wilson Bull.* **102,** 167–169.

Clark, L., Shah, P. S., and Mason, J. R. (1991). Chemical repellency in birds: Relationship between chemical structure and avoidance response. *J. Exp. Zool.* **260,** 310–322.

Clark, L., Avilova, K. V., and Bean, N. J. (1993). Odor thresholds in passerines. *Comp. Biochem. Physiol. A.* **104,** 305–312.

Cohen, J. (1981). Olfaction and parental behavior in Ring Doves. *Biochem. Sys. Ecol.* **9,** 351–354.

Cole, E. C. (1941). "Comparative Histology." Blakiston, Philadelphia.

Collins, F. W. (1986). Oat phenolics: Structure, occurrence, and function. *In* "Oats Chemistry and Technology" (F. H. Webster, ed.), pp. 227–295. American Association of Cereal Chemists, St. Paul, MN.

Corey, D. T. (1978). The determinants of exploration and neophobia. *Neurosci. Biobehav. Rev.* **2,** 235–253.

Crocker, D. R., and Perry, S. M. (1990). Plant chemistry and bird repellents. *Ibis* **132,** 300–308.

Crocker, D. R., Perry, S. M., Wilson, M., Bishop, J. D., and Scanlon, C. D. (1993). Repellency of cinnamic acid derivatives to captive rock doves. *J. Wildl. Manage.* **57,** 113–122.

Davis, R. G. (1973). Olfactory psychophysical parameters in man, rat, dog, and pigeon. *J. Comp. Physiol. Psychol.* **85,** 221–232.

Dawson, W. R., Shoemaker, V. H., Tordoff, H. B., and Borut, A. (1965). Observations on the metabolism of sodium chloride in the Red Crossbill. *Auk* **82,** 606–623.

Deyoe, C. W., Davies, R. E., Krishnan, R., Khaund, R. K., and Couch, J. R. (1962). Studies on the taste preference of the chick. *Poult. Sci.* **41,** 781–784.

Dubbeldam, J. L., and Karten, H. J. (1978). The trigeminal system in the pigeon (Columba livia). I. Projections of the Gasserian ganglion. *J. Comp. Neurol.* **180,** 661.

Dubbledam, J. L., and Veenman, C. L. (1978). Studies on the somatotopy of the trigeminal system in the mallard, *Anas platyrhyncos* L: The Ganglion Trigeminale. *Netherlands J. Zool.* **28,** 150–160.

Duncan, C. J. (1960). The sense of taste in birds. *Ann. Appl. Biol.* **48,** 409–414.

Duncan, C. J. (1964). The sense of taste in the feral pigeon: The response to acids. *Anim. Behav.* **12,** 77–83.

Elliot, R. (1937) Total distribution of taste buds on the tongue of the kitten at birth. *J. Comp. Neurol.* **66**, 361–366.

Engelmann, C. (1934). Versuche über den Geschmackssinn von Taube, Ente und Huhn. *A. vergl. Physiol.* **20**, 626–645.

Engelmann, C. (1937). Vom Geschmackssinn des Huhns. *Forshc. Fortschr.* **13**, 425–426.

Engelmann, C. (1950). Über den Geschmackssinn des Huhns IX. *Z. Tierpsychol.* **7**, 84–111.

Engelmann, C. (1957). "So leben Hühner, Tauben, Gänse." Neumann-Verlag, Radebeul, Germany.

Engelman, C. (1960). Weitere Versuche über die Futterwahl des Wassergeflügels: Über die Schmeckempfindlichkeit der Gänse. *Arch. Geflügelzucht Kleintierk.* **9**, 91–104.

Espaillat, J. E., and Mason, J. R. (1990). Differences in taste preference between Red-winged Blackbirds and European Starlings. *Wilson Bull.* **102**, 292–299.

Fuerst, F. F., and Kare, M. R. (1962). The influence of pH on fluid tolerance and preferences. *Poult. Sci.* **41**, 71–77.

Gentle, M. J. (1972). Taste preferences in the chicken (Gallus domesticus). *Br. Poult. Sci.* **13**, 141–155.

Gentle, M. J. (1973). Diencephalic stimulation and mouth movement in the chicken. *Br. Poult. Sci.* **14**, 167–171.

Gentle, M. J. (1975). Gustatory behavior of the chicken and other birds. *In* "Neural and Endocrine Aspects of Behaviors in Birds" (P. Wright, P. E. Caryl, and D. M. Vowles, eds.). Elsevier Scientific, Amsterdam.

Gentle, M. J. (1976). Quinine hydrochloride acceptability after water deprivation in *Gallus domesticus*. *Chem. Senses Flav.* **2**, 121–128.

Gentle, M. J. (1978). Extra-lingual chemoreceptors in the chicken (*Gallus domesticus*). *Chem. Senses Flav.* **3**, 325–329.

Gentle, M. J. (1979). Single unit responses from the solitary complex following oral stimulation in the chicken. *J. Comp. Physiol.* **130**, 259–264.

Gentle, M. J. (1983). The chorda tympani nerve and taste in the chicken. *Experientia* **39**, 1002–1003.

Gentle, M. J. (1989). Cutaneous sensory afferents recorded from the nervus intramandibularis of *Gallus gallus* var domesticus. *J. Comp. Physiol. A* **164**, 763–774.

Gentle, M. J., and Dewar, W. A. (1981). The effects of vitamin A deficiency on oral gustatory behavior in chicks. *Br. Poult. Sci.* **22**, 275–279.

Gentle, M. J., and Harkin, C. (1979). The effect of sweet stimuli on oral behavior in the chicken. *Chem. Senses Flav.* **4**, 183–190.

Gentle, M. J., and Hill, F. L. (1987). Oral lesions in the chicken: Behavioural responses following nociceptive stimulation. *Physiol. Behav.* **40**, 781–783.

Gentle, M. J., and Hunter, L. N. (1993). Neurogenic inflammation in the chicken (*Gallus gallus* var *domesticus*). *Comp. Biochem. Physiol. C* **105**, 459–462.

Gentle, M. J., Dewar, W. A., and Wight, P. A. L. (1981). The effects of zinc deficiency on oral behavior and taste bud morphology in chicks. *Br. Poult. Sci.* **22**, 265–273.

Gerritsen, A. F. C., van Heezik, Y. M., and Swennen, C. (1984). Chemoreception in two further Calidris species (*C. maritima* and *C. canutus*) with a comparison of the relative importance of chemoreception during foraging in Calidris species. *Netherlands J. Zool.* **33**, 485–496.

Gesteland, R. C., Lettvin, J. Y., Pitts, W. H., and Rojas, A. (1963). Odor specificities of frog's olfactory receptors. *In* "Olfaction and Taste" (Y. Zotterman, ed.), pp. 19–44. Pergammon Press, New York.

Getty R. (1975). "Sisson and Grossman's The Anatomy of the Domestic Animals." W. B. Saunders Company, Philadelphia.

Goldsmith, K. M., and Goldsmith, T. H. (1982). Sense of smell in the black-chinned hummingbird. *Condor* **84**, 237–238.

Goodwin, D. (1955). Jays and carrion crows recovering hidden food. *Br. Birds* **48**, 181–183.

Gorsuch, D. M. (1934). Life history of the Gambel Quail in Arizona. *Univ. Arizona Biol. Sci. Bull.* **2**, 12–17.

Greig-Smith, P. W. (1988). Bullfinches and ash trees: Assessing the role of plant chemicals in controlling damage by herbivores. *J. Chem. Ecol.* **14**, 1889–1903.

Greig-Smith, P. W., Wilson, M. F., Blunden, C. A., and Wilson, G. M. (1983). Bud-eating by bullfinches, Pyrrhula pyrrhula in relation to the chemical constituents of two pear cultivars. *Ann. Appl. Biol.* **103**, 335–343.

Grubb, T. C. (1972). Smell and foraging in shearwaters and petrels. *Nature* **237**, 404–405.

Grubb, T. C. (1974). Olfactory navigation to the nesting burrow in Leach's Petrel (*Oceanodroma leucorrhoa*). *Anim. Behav.* **22**, 192–202.

Gunther, W. C., and Wagner, M. W. (1971). Preferences for natural and artificial sweeteners in heat-stressed chicks of different ages. *Proc. Ind. Acad. Sci.* **81**, 401–409.

Hac, L. R., Long, M. L., and Blish, M. J. (1949). The occurrence of free L-glutamic acid in various foods. *Food Technol.* **3**, 351–354.

Hahlbrock, K., and Scheel, D. (1989). Physiology and molecular biology of phenylpropanoid metabolism. *Annu. Rev. Plant Physiol. Plant Mol. Biol.* **40**, 347–369.

Hainsworth, F. R., and Wolf, L. L. (1976). Nectar characteristics and food selection by hummingbirds. *Oecologia* **25**, 101–113.

Halpern, B. P. (1963). Gustatory nerve responses in the chicken. *Am. J. Physiol.* **203**, 541–544.

Harriman, A. E. (1967). Laughing gull offered saline in preference and survival tests. *Physiol. Zool.* **40**, 273.

Harriman, A. E., and Berger, R. H. (1986). Olfactory acuity in the Common Raven (*Corvus corax*). *Physiol. Behav.* **36**, 257–262.

Harriman, A. E., and Kare, M. R. (1966). Aversion to saline solutions in starlings, purple grackles, and herring gulls. *Physiol. Zool.* **39**, 123–126.

Harriman, A. E., and Milner, J. S. (1969). Preference for sucrose solutions by Japanese quail (*Coturnix coturnix japonica*) in two-bottle drinking test. *Am. Midl. Nat.* **81**, 575–578.

Heinroth, O. (1938). "Aus dem Leben der Vogel." Springer-Verlag, Berlin.

Henton, W. W. (1969). Conditioned suppression to odorous stimuli in pigeons. *J. Exp. Anal. Behav.* **12**, 175–185.

Henton, W. W., Smith, J. C., and Tucker, D. (1966). Odor discrimination in pigeons. *Science* **153**, 1138–1139.

Hiestand, W. A., and Randall, W. C. (1941). Species differentiation in the respiration of birds following carbon dioxide administration and the location of inhibitory receptors in the upper respiratory tract. *J. Cell. Comp. Physiol.* **17**, 333–340.

Houston, D. C. (1987). Scavenging efficiency of turkey vultures in tropical forests. *Condor* **88**, 318–323.

Hutchison, L. V., and Wenzel, B. M. (1980). Olfactory guidance in foraging by procellariiforms. *Condor* **82**, 314–319.

Hyman, L. H. (1942). "Comparative Vertebrate Anatomy." Univ. of Chicago Press, Chicago.

Ioale, P., and Papi, F. (1989). Olfactory bulb size, odor discrimination and magnetic insensitivity in hummingbirds. *Physiol. Behav.* **45**, 995–999.

Jakubas, W. J., Shah, P. S., Mason, J. R., and Norman, D. M. (1992). Avian repellency of coniferyl and cinnamyl derivatives. *Ecol. Appl.* **2**, 147–156.

Jarvi, T., and Wiklund, C. (1984). A note on the use of olfactory cues by the Great Tit *Parus major* in food choice. *Faunanorv. Ser. C. Cinclus* **139**.

Jones, R. B. (1988). Food neophobia and olfaction in domestic chicks. *Bird Behav.* **7**, 78–81.

Jones, R. B., and Faure, J. M. (1982). Domestic chick prefer familiar soiled substrate in an otherwise novel environment. *IRCS Med. Sci.* **13**, 847.

Jouventin, P., and Robin, J. P. (1984). Olfactory experiments on some Antarctic birds. *Emu* **85**, 46–48.

Kadono, H., Okado, T., and Ohno. K., (1966). Neurophysiological studies of the sense of taste in the chicken. *Res. Bull. Fac. Agric. Gifu Univ.* **22**, 149–159.

Kare, M. R. (1961). Comparative aspects of taste. *In* "Physiological and Behavioral Aspects of Taste" (M. R. Kare and B. P. Halpern, eds.), pp. 13–23. University of Chicago Press, Chicago.

Kare, M. R. (1971). Comparative Study of Taste. *In* "Handbook of Sensory Physiology" (L. M. Beidler, ed.). Springer-Verlag, Berlin.

Kare, M. R., and Beauchamp, G. K. (1976). Taste, smell and hearing. *In* "Duke's Physiology of Domestic Animals" (M. J. Swenson, ed.). Comstock, Ithaca.

Kare, M. R., and Medway, W.(1959). Discrimination between carbohydrates by the fowl. *Poult. Sci.* **38**, 1119–1127.

Kare, M. R., and Pick, H. L. (1960). The influence of the sense of taste on feed acceptability. *Poult. Sci.* **39**, 697–705.

Kare, M. R., and Rogers, J. G. (1976). Sense organs: Taste. *In* "Avian Physiology" (P. D. Sturkie, ed.). Springer-Verlag, Berlin.

Kare, M. R., and Scott, M. L. (1962). Nutritional value and feed acceptability. *Poult. Sci.* **44**, 276.

Kare, M. R., Black, R., and Allison, E. G. (1957). The sense of taste in the fowl. *Poult. Sci.* **36**, 129–138.

Kear, J. (1960). Food selection in certain finches with special reference to interspecific differences. Ph.D. Thesis, Cambridge University.

Keverne, E. B., Murphy, C. L., Silver, W. L., Wysocki, C. J., and Meredith, M. (1986). Non-olfactory chemoreceptors of the nose: recent advances in understanding the vomeronasal and trigeminal systems. *Chem. Senses,* **11**, 119–133.

Kitchell, R. L., and Erickson, H. H. (1983). "Animal Pain, Perception and Alleviation." American Physiological Society, Bethesda.

Kitchell, R. L., Strom, L., and Zotterman, Y. (1959). Electrophysiological studies of thermal and taste reception in chickens and pigeons. *Acta Physiol. Scand.* **46**, 133–151.

Krol, C. P. M., and Dubbeldam, J. L. (1979). On the innervation of taste buds by n. facialis in the mallard, *Anas platyrnchos* L. *Netherlands J. Zool.* **29**, 267–274.

Lalonde, E. R., and Eglitis, J. A. (1961). Number and distribution of taste buds on the epiglottis, pharynx, larynx, soft palate and uvula in a human newborn. *Anat. Rec.* **140**, 91–95.

Landolt, J. P. (1970). Neural properties of pigeon lingual chemoreceptors. *Physiol. Behav.* **5**, 1151–1160.

Lequette, B., Verheyden, C., and Jouventin, P. (1989). Olfaction in subantarctic seabirds: Its phylogenetic and ecological significance. *Condor* **91**, 732–735.

Lewis, N. G., and Yamamoto, E. (1990). Lignin: occurrence, biogenesis and biodegradation. *Annu. Rev. Plant Physiol. Plant Mol. Biol.* **41**, 455–496.

Lindenmaier, P., and Kare, M. R. (1959). The taste end organs of the chicken. *Poult. Sci.* **38**, 545–550.

Macadar, A. W., Rausch, L. J., Wenzel, B. M., and Hutchison, L. V. (1980). Electrophysiology of the olfactory pathway in the pigeon. *J. Comp. Physiol.* **137**, 39–46.

Maeda, S., Eguchi, S., and Sasaki, H. (1958). The content of free L-glutamic acid in various foods. *J. Home Econ. Jpn.* **9**, 163–167.

Mariotti, G., and Fiore, L. (1980). Operant conditioning studies of taste discrimination in the pigeon (*Columba livia*). *Physiol. Behav.* **24**, 163–168.

Marshall, W. H. (1940). More notes on salt-feeding of Red Crossbills. *Condor* **42**, 218–219.

Martinez del Rio, C. (1990). Dietary, phylogenetic, and ecological correlates of intestinal sucrase and maltase activity in birds. *Physiol. Zool.* **63**, 987–1011.

Martinez del Rio, C., and Stevens, B. R. (1989). Physiological constraint on feeding behavior: Intestinal membrane disaccharidase of the starling. *Science,* **243**, 794–796.

Martinez del Rio, C., Stevens, B. R., Daneke, D. E., and Andreadis, P. T. (1988). Physiological correlates of preference and aversion for sugars in three species of birds. *Physiol. Zool.* **61**, 222–229.

Mason, J. R., and Clark, L. (1995a). Mammalian irritants as chemical stimuli for birds: The importance of training. *Auk* **112**, 511–515.

Mason, J. R., and Clark, L. (1995b). Capsaicin detection in trained starlings: The importance of olfaction and trigeminal chemoreception. *Wilson Bull.* **107**, 165–169.

Mason, J. R., and Espaillat, J. E. (1990). Differences in taste preference between Red-winged Blackbirds and European Starlings. *Wilson Bull.* **102**, 292–299.

Mason, J. R., and Maruniak, J. A. (1983). Behavioral and physiological effects of capsaicin in red-winged blackbirds. *Pharmacol. Biochem. Behav.* **19**, 857–862.

Mason, J. R., and Otis, D. L. (1990). Aversiveness of six potential irritants on consumption by Red-winged Blackbirds (Agelaius phoeniceus) and European Starlings (Sturnus vulgaris). *In* "Chemical Senses II: Irritation" (B. G. Green, J. R. Mason, and M. R. Kare, eds.), pp. 309–323. Marcel Dekker, New York.

Mason, J. R., and Reidinger, R. F. (1983a). Exploitable characteristics of neophobia and food aversions for improvements in rodent and bird control. *In* "Test Methods for Vertebrate Pest Control and Management Materials" (D. Kaukienen, ed.), pp. 20–42. American Society for Testing and Materials, Philadelphia.

Mason, J. R., and Reidinger, R. F. (1983b). Importance of color for methiocarb-induced taste aversions in red-winged blackbirds. *J. Wildl. Manage.* **47**, 383–393.

Mason, J. R., and Silver, W. L. (1983). Trigeminally mediated odor aversions in starlings. *Brain Res.* **269**, 196–199.

Mason, J. R., Dolbeer, R. A., Arzt, A. H., Reidinger, R. F., and Woronecki, P. P. (1984). Taste preferences of male Red-winged Blackbirds among dried samples of ten corn hybrids. *J. Wildl. Manage.* **48**, 611–616.

Mason, J. R., Adams, M. A. and Clark, L. (1989). Anthranilate repellency to starlings: Chemical correlates and sensory perception. *J. Wildl. Manage.* **53**, 55–64.

Mason, J. R., Bean, N. J., Shah, P. S., and Clark, L. (1991a). Taxon-specific differences in responsiveness to capsaicin and several analogues: Correlates between chemical structure and behavioral aversiveness. *J. Chem. Ecol.* **17**, 2539–2551.

Mason, J. R., Clark, L., and Shah, P. S. (1991b). Taxonomic differences between birds and mammals in their responses to chemical irritants. *In* "Chemical Signals in Vertebrates" (R. Doty and D. Muller-Schwarze, eds.), pp. 291–296. Plenum Press, New York.

Matthes, E. (1934). "Geruchsorgan, Lubosch Handbuch der vergleichenden Anatomie der Wirbeltiere, Groppert, Kallius, Vol 11." Urban und Schwarzbeig, Berlin.

McCabe, T. T. (1927). Bird banding near Barkerville, British Columbia. *Condor* **24**, 206–207.

McCallum, J. A., and Walker, J. R. (1990). Phenolic biosynthesis during grain development in wheat: Changes in phenylalanine ammonia-lyase activity and soluble phenolic content. *J. Cereal Sci.* **11**, 35–49.

Michelsen, W. J. (1959). Procedure for studying olfactory discrimination in pigeons. *Science* **130**, 630–631.

Michelsen, W. J. (1960). Reply to Calvin. *Science* **130**, 632.

Miller, L. (1942). Some tagging experiments with black-footed albatrosses. *Condor* **44**, 3-9.

Mitcham, S. A., and Wobeser, G. (1988). Toxic effects of natural saline waters on mallard ducklings. *J. Wildl. Disease* **24**, 45-50.

Moncreiff, R. W. (1951). "The Chemical Senses." Hill, London.

Morris, D. (1955). The seed preferences of certain finches under controlled conditions. *Avic. Mag.* **61**, 271-287.

Mousley, H. (1921). Goldfinches and Purple Finches wintering at Hatley, Stanstead County, Quebec. *Auk* **38**, 606.

Mousley, H. (1946). English Sparrow eating salt. *Auk* **63**, 89.

Necker, R. (1974). Dependence of mechanoreceptor activity on skin temperature in sauropsid. II. Pigeon and duck. *J. Comp. Physiol.* **92**, 75-83.

Nishida, R., and Fukami, H. (1990). Sequestration of distasteful compounds by some pharmacophagus insects. *J. Chem. Ecol.* **16**, 151-164.

Norman, D. L., Mason, J. R., and Clark, L. (1992). Capsaicin effects on consumption of food by cedar waxwings and house finches. *Wilson Bull.* **104**, 549-551.

Packard, F. M. (1946). Some observations of birds eating salt. *Auk* **63**, 89.

Papi, F. (1986). Pigeon navigation: Solved problems and open questions. *Monit. Zool. Ital.* **20**, 471-517.

Peterson, J. G. (1942). Salt feeding habits of the House Finch. *Condor* **44**, 73.

Pierau, F.-K., Sann, H., and Harti, G. (1986). Resistance of birds to capsaicin and differences in their substance P (SP) system. *Proc. Int. Union Physiol. Sci.* **16**, 207-211.

Pierce, F. J. (1921). Birds and salt. *Bird Lore* **23**, 90-91.

Reeks, E. (1920). House finches eat salt. *Bird Lore* **22**, 286.

Rensch, B., and Neunzig, R. (1925). Experimentalle Untersuchungen uber den Geschmackssinn der Vogel, II. *J. Ornithol.* **73**, 633.

Rieke, G. K., and Wenzel, B. M. (1975). The ipsilateral olfactory projection field in the pigeon. In "Olfaction and Taste" (V. D. Denton and J. P. Coghlan, eds.), pp. 361-368. Academic Press, New York.

Rieke, G. K., and Wenzel, B. M. (1978). Forebrain projections of the pigeon olfactory bulb. *J. Morphol.* **158**, 41-55.

Robinson, T. V. (1983). "The Organic Constituents of Higher Plants." Cordus, North Amherst, MA.

Rodriguez, E., and Levin, D. A. (1976). Biochemical parallels of repellents and attractants in higher plants and arthropods. In "Recent Advances in Phytochemistry: Biochemical Interactions between Plants and Insects" (J. W. Wallace and R. L. Mansell, eds.), pp. 214-270. Plenum Press, New York.

Rogers, J. G., and Maller, O. (1973). Effect of salt on the response of birds to sucrose. *Physiol. Psychol.* **1**, 199.

Romoser, G. L., Bossard, E. H., and Combs, G. F. (1958). Studies on the use of certain flavors in the diet of chick. *Poult. Sci.* **37**, 631-633.

Saito, I. (1966). Comparative anatomical studies of the oral organs of the poultry. V. Structure and distribution of taste buds of the fowl. *Bull. Fac. Agric. Miyazaki Univ.* **13**, 95-102.

Sann, H., Harti, G., Pierau, F-K., and Simon, E. (1987). Effect of capsaicin upon afferent and efferent mechanisms of nociception and temperature regulation in birds. *Can. J. Physiol. Pharmacol.* **65**, 1347-1354.

Schrader, E. (1970). Die Topographie der Kopfnerven vom Huhn. Ph.D. Dissertation. Freien Univ. Berlin.

Schuler, W. (1980). Learned responses to sugars in a songbird: Learning supplements physiological adaptations. *Verh. Dtsch. Zool. Ges.* **366**, 366.

Schuler, W. (1983). Responses to sugars and their behavioural mechanisms in the starling (*Sturnus vulgaris*). *Behav. Ecol. Sociobiol.* **13**, 243-251.

Schwenk, K. (1985). Occurence, distribution, and function significance of taste buds in lizards. *Copeia* **1**, 91-101.

Shah, P. S., Clark, L., and Mason, J. R. (1991). Prediction of avian repellency from chemical structure: The aversiveness of vanillin, vanillyl alcohol and veratryl alcohol. *Pesticide Biochem. Physiol.* **40**, 169-175.

Shah, P. S., Mason, J. R., and Clark, L. (1992). Avian chemical repellency: A structure-activity approach and implications. In "Chemical Signals in Vertebrates" (R. L. Doty and D. Muller-Schwarze, eds.), pp. 291-296. Plenum Press, New York.

Shallenberger, R. J. (1975). Olfactory use in the wedge-tailed shearwater (*Puffinus pacificus*) on Manana Island, Hawaii. In "Olfaction and Taste" (D. A. Denton and J. P. Coghlan, eds.), pp. 355-359. Academic Press, New York.

Shibuya, T., and Shibuya, S. (1963). Olfactory epithelium: Unitary responses in the tortoise. *Science* **140**, 495-496.

Shibuya, T., and Tonosaki, K. (1972). Electrical responses of single olfactory receptor cells in some vertebrates. In "Olfaction and Taste" (D. Schneider, ed.), pp. 102-108. Wissenschaftliche Verlagsgellschaft MBH, Stuttgart.

Shibuya, T., and Tucker, D. (1967). Single unit responses of olfactory receptors in vultures. In "Olfaction and Taste" (T. Hayashi, ed.), pp. 219-220. Pergamon Press, Oxford.

Shoemaker, V. H. (1972). Osmoregulation and excretion in birds. In "Avian Biology" (D. S. Farner and J. R. King, eds.), pp. 527-574. Academic Press, New York.

Sieck, M. H., and Wenzel, B. M. (1969). Electrical activity of the olfactory bulb of the pigeon. *Electoencephalogr. Clin. Neurophysiol.* **26**, 62-69.

Silver, W. L., and Maruniak, J. A. (1980). Trigeminal chemoreception in the nasal and oral cavities. *Chem. Senses* **6**, 295-305.

Snyder, G. K., and Peterson, T. T. (1979). Olfactory sensitivity in the black-billed magpie and in the pigeon. *Comp. Biochem. Physiol. A* **62**, 921-925.

Stager, K. E. (1964). The role of olfaction in food location by the Turkey Vulture (Cathartes aura). *Los Ang. Cty. Mus. Contrib. Sci.* **81**, 1-63.

Stager, K. E. (1967). Avian olfaction. *Am. Zool.* **7**, 415-420.

Stattelman, A. J., Talbot, R. B., and Coulter, D. B. (1975). Olfactory thresholds of pigeons (*Columba livia*), quail (*Colinus virginianus*) and chickens (*Gallus gallus*). *Comp. Biochem. Physiol. A* **50**, 807-809.

Stevens, J. (1985). Foraging success of adult and juvenile Starlings Sturnus vulgaris: A tentative explanation for the preference of juveniles for cherries. *Ibis* **127**, 341-347.

Stromberg, M. R., and Johnsen, P. B. (1990). Hummingbird sweetness preferences: Taste or viscosity. *Condor* **32**, 606-612.

Szolcsanyi, J. (1982). Capsaicin type pungent agents producing pyrexia. In "Handbook of Experimental Pharmacology" (A. S. Milton, ed.), pp. 437-478. Springer-Verlag. Berlin.

Szolcsanyi, J., Sann, H., and Pierau, F.-K. (1986). Nociception is not impaired by capsaicin. *Pain* **27**, 247-260.

Tucker, D. (1963). Olfactory, vomeronasal and trigeminal receptor responses to odorants. In "Olfaction and Taste" (Y. Zotterman, ed.), pp. 45-69. Pergamon, New York.

Tucker, D. (1965). Electrophysiological evidence for olfactory function in birds. *Nature* **207**, 34-36.

Tucker, D. (1971). Nonolfactory responses from the nasal cavity: Jacobson's organ and the trigeminal system. In "Handbook of Sensory Physiology IV: Chemical Senses, Olfaction" (L. M. Beidler, ed.), pp. 151-181. Springer-Verlag, Berlin.

Tucker, D., Graziadei, P. C., and Smith, J. C. (1974). Recovery of olfactory function in pigeons after bilateral transection of the olfactory nerves. *In* "Olfaction and Taste" (D. A. Denton and J. P. Coghlan, eds.), pp. 369–373. Academic Press, New York.

Twedt, D. J. (1984). The effect of dietary protein and feed size on the assimilation efficiency of Starlings. *Denver Wildl. Res. Center Bird Dam. Res. Rep.* **335,** 1–10.

van Heezik, Y. M., Gerritsen, A. F. C., and Swennen, C. (1983). The influence of chemoreception on the foraging behavior of two species of sandpiper, Calidris alba (Pallas) and Calidris alpina (L.). *Netherlands J. Sea Res.* **17,** 47–56.

Waldvogel, J. A. (1987). Olfactory navigation in homing pigeons: Are the current models atmospherically realistic? *Auk* **104,** 369–379.

Waldvogel, J. A. (1989). Olfactory orientation by birds. *In* "Current Ornithology" (D. E. Power, ed.), pp. 369–379. Plenum Press, New York.

Walker, J. C., Tucker, D., and Smith, J. C. (1979). Odor sensitivity mediated by trigeminal nerve in the pigeon. *Chem. Senses Flav.* **4,** 107–116.

Walker, J. C., Walker, D. B., Tambiah, C. R., and Gilmore, K. S. (1986). Olfactory and nonolfactory odor detection in pigeons: Elucidation by a cardiac acceleration paradigm. *Physiol. Behav.* **38,** 575–580.

Wallraff, H. G. (1991). Conceptual approaches to avian navigation systems. *In* "Orientation in Birds" (P. Berthold, ed.), pp. 128–165. Birkhauser-Verlag, Basel.

Wallraff, H. G., Kiepenheuer, J., and Streng, A. (1993). Further experiments on olfactory navigation and non-olfactory pilotage by homing pigeons. *Behav. Ecol. Sociobiol.* **32,** 387–390.

Warham, J. (1971). Aspects of breeding behaviors in the Royal Penguin, Endyptes chrysolophys schlegeli. *Notornis* **18,** 91.

Warner, R.L., McFarland, L. Z., and Wilson, W.O. (1967). Microanatomy of the upper digestive tract of the Japanese quail. *Am. J. Vet. Res.* **28,** 1537–1540.

Warren, R. P., and Vince, M. A. (1963). Taste discrimination in the Great Tit (*Parus major*). *J. Comp. Physiol. Psychol.* **56,** 910–913.

Welty, J. C. (1975). "The Life of Birds." W. B. Saunders, Philadelphia.

Wenzel, B. M. (1968). The olfactory prowess of the Kiwi. *Nature* **220,** 1133–1134.

Wenzel, B. M., and Sieck, M. H. (1972). Olfactory perception and bulbar electrical activity in several avian species. *Physiol. Behav.* **9,** 287–293.

Willoughby, E. J. (1971). Drinking responses of the red crossbill (Loxia curvirostra) to solutions of NaCl, $MgCl_2$, and CaCl2. *Auk* **84,** 828–830.

Wiltschko, W., Wiltschko, R., and Walcott, C. (1987). Pigeon homing: Different effects of olfactory deprivation in different countries. *Behav. Ecol. Sociobiol.* **21,** 333–342.

Wurdinger, I. (1982). Olfaction and home learning in juvenile geese (Anser- and Branta-species). *Biol. Behav.* **5,** 347–351.

CHAPTER 4

The Somatosensory System

REINHOLD NECKER
Ruhr-Universität Bochum
Fakultät für Biologie
Lehrstuhl für Tierphysiologie
D-44780 Bochum
Germany

I. Introduction 57
II. Types of Receptors and Afferent Fibers 57
    A. Mechanoreceptors 58
    B. Thermoreceptors 61
    C. Nociceptors 62
III. Central Processing 62
    A. Somatosensory Pathways 62
    B. Electrophysiological Investigations 64
IV. Behavioral Aspects 65
    A. Mechanoreception 65
    B. Thermoreception 66
    C. Pain 66
V. Summary and Conclusion 66
    References 67

spinal system innervates the body surface and the extremities (wings and legs).

Cutaneous receptors are the peripheral (dendritic) endings of spinal or cranial ganglion cells. These endings are specialized for being excited by mechanical, thermal, or noxious stimuli. Accordingly the skin includes the senses of mechanoreception (touch), thermoreception, and nociception, which serve quite different functions.

The information taken up by the receptors is conveyed up to the telencephalon via relays in the brainstem. Both receptors and central pathways have been studied with anatomical and electrophysiological means. However, our knowledge of the somatosensory system of birds is very limited compared to what is known of the mammalian counterpart.

I. INTRODUCTION

The contact of the body surface with the environment is sensed by a variety of receptors located in the skin. This chapter deals with the exteroreceptive cutaneous sensory system. Deep receptors located in the viscera, muscles, and joints are sometimes also included into the somatosensory system. However, since they serve quite different functions (e.g., gastrointestinal motility, circulation, respiration, and motor control) they will not be included here.

The somatosensory system may be divided into two parts, the trigeminal system and the spinal system. The trigeminal system primarily innervates the beak. The

II. TYPES OF RECEPTORS AND AFFERENT FIBERS

In the avian skin there are both free nerve endings and encapsulated endings (sensory corpuscles; Andres and von Düring, 1973; 1990; Andres, 1974; Gottschaldt, 1985). Whereas free nerve endings are thought to serve mainly thermoreception and nociception, sensory corpuscles are mechanoreceptors.

Free nerve endings are supplied with unmyelinated and thinly myelinated axons (C-fibers and Aδ-fibers or group IV and group III fibers), corpuscular cutaneous mechanoreceptors are supplied with thickly myelinated fibers of the Aβ-type (group II; group I or Aα-fibers supply proprioreceptors). Accordingly, different groups of fiber diameters and conduction velocities have been described (Necker and Meinecke, 1984). C-fibers have fiber diameters of less than 1 μm and conduction velocities (CV) of less than 2 m/sec. Aδ-fibers have mean diameters of about 2 μm and mean CVs of about 5 m/sec. There are two groups of large myelinated Aβ-fibers which have mean diameters of about 4 and 7 μm and mean CVs of about 15 and 35 m/sec, respectively.

A. Mechanoreceptors

Four main types of mechanosensitive sensory corpuscles may be distinguished in birds: Herbst corpuscles, Merkel cell receptors, Grandry corpuscles, and Ruffini endings. Although free nerve endings may function as mechanoreceptors (Iggo and Andres, 1982), electrophysiological evidence is lacking in birds.

1. Morphology and Distribution of Cutaneous Mechanoreceptors

a. Herbst Corpuscles

Herbst corpuscles are lamellated sensory receptors comparable to the Pacinian corpuscles in mammals. The flattened axon ending is enlarged at its tip and is surrounded by a central inner bulb (Figure 1f). The inner bulb cells are of Schwann cell origin and form two opposing rows. The Schwann cell membrane adjacent to the axon forms a complicated network of interdigitated lamellae. Numerous fingerlike processes of the axon protrude into and form contacts with the lamellae. These axon processes are thought to be the sites where the mechanical stimulus is tranduced into excitation of the sensory membrane (Gottschaldt et al., 1982). The inner bulb is surrounded by a capsule space which contains cells of endoneural origin and collagen fibers which form perforated concentric lamellae. The capsular space is enclosed by an outer capsule whose dense lamellae are of perineural origin. The myelinated afferent fiber looses its myelin sheath before entering the inner bulb.

Herbst corpuscles are the most widely distributed receptors in the skin of birds. They are located in the deep dermis and they are found in the beak, in the legs, and in the feathered skin (Gottschaldt, 1985). There is a conspicuous assembly of sometimes more than one hundred Herbst corpuscles on the interosseous membrane of the leg ("Herbstscher Strang" [strand of Herbst corpuscles]; Schildmacher, 1931). In aquatic birds like ducks and geese, in some shorebirds, and in the chicken there are bill tip organs with numerous Herbst corpuscles (Bolze, 1968; Gottschaldt and Lausmann, 1975; Berkhoudt, 1980; Gentle and Breward, 1986). In feathered skin they are associated with the feather follicles and with muscles of the feathers (Stammer, 1961; Winkelmann and Myers, 1961; Ostmann et al., 1963; Andres and von Düring, 1990). There is a decreasing number from head to tail to neck to wing with relatively few corpuscles on the back and fewest on the abdomen, and flying birds are supplied with a larger number than nonflying birds (Stammer, 1961).

b. Merkel Cell Receptors

Merkel cell receptors of the avian skin share some similarity to the intraepidermal Merkel cell receptors of mammals (Andres and von Düring, 1973). They are, however, located in the dermis. The basic morphology of Merkel cell receptors is a Merkel cell and a disc-like axon ending contacting this cell (Merkel cell neurite complex; Figure 1b). Merkel cells are characterized by their clear cytoplasm which typically contains dense-cored granula. Fingerlike processes interdigitate with the surrounding Schwann cells. It is still a matter of speculation whether Merkel cells function as secondary sensory cells like hair cells in the inner ear. There are symmetrical membrane thickenings of the axon membrane and of the apposed Merkel cell membrane which resemble desmosomes rather than synaptic contacts (Toyoshima and Shimamura, 1991).

Merkel cells may occur as single cells as well as in groups and may even be organized as corpuscles with stacked arrangement similar to Grandry corpuscles of aquatic birds. Such corpuscles lack, however, a perineural sheath. Merkel cells have been found predominantly in the beak and tongue of various nonaquatic birds (Botezat, 1906; Saxod, 1978; Gentle and Breward, 1986; Toyoshima and Shimamura, 1991; Halata and Grim, 1993) but they have been described for the toe skin (Ide and Munger, 1978) and also for the feathered skin (Andres and von Düring, 1990).

c. Grandry Corpuscles

Grandry corpuscles occur in aquatic birds (anseriforms) only (Gottschaldt, 1985). As with Merkel cell receptors there is an intimate contact between Grandry cell and nerve ending. Grandry cells are thought to be of neural crest origin and are described as ganglion cell-like. Typically two or more Grandry cells are stacked with discoid axon endings in between the cells (Figure 1a). There is an ongoing debate whether Grandry cells and Merkel cells are two varieties of the same cell (Toyoshima, 1993). Grandry cells in aquatic birds are generally larger than Merkel cells in nonaquatic birds. Except for size, Merkel cells and Grandry cells share most structural specializations. Whereas Merkel cell corpuscles lack a sheath, Grandry corpuscles are always encapsulated by a single-layered capsule of perineural origin.

Grandry cells occur in the dermis of the bill of ducks and geese (Saxod, 1978; Berkhoudt, 1980; Gottschaldt

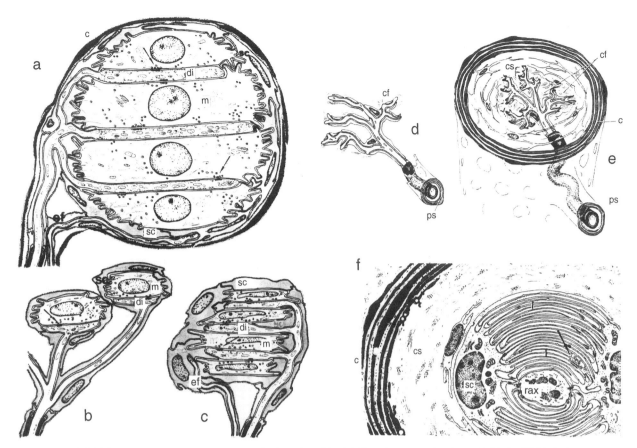

FIGURE 1 Types of mechanoreceptors in the avian skin. (a) Grandry corpuscle of aquatic birds; (b) Merkel cell receptors; (c) Merkel cell corpuscle; (d) free stretch receptor ending; (e) Ruffini corpuscle; (f) Herbst corpuscle. Abbreviations: c, capsule; cs, capsule space; cf, collagen fibers; di, disk-like afferent nerve ending; ef, efferent fiber; m, Merkel cell; ps, perineural sheath; rax, receptor axon; sc, Schwann cell. After Andres (1974) with permission.

and Lausmann, 1974) and they are numerous in bill tip organs, which are accumulations of sensory receptors in connective tissue filled channels of the horny premaxillary plate of the bill.

d. Ruffini Endings

Ruffini endings are well known and well studied in mammals but there are only few reports of this type of mechanoreceptor in the avian skin. Ruffini corpuscles are the encapsulated modification of free stretch receptors (Figures 1d and 1e; Andres and von Düring, 1990). There is an extensive ramification of the axon endings which are in contact with bundles of collagen fibers. The contact zones are probably the tranducer sites. A capsule consisting of layers of perineural cells may lack in avian Ruffini endings (Gottschaldt, 1985).

Ruffini endings have been identified so far only in the bill of geese (Gottschaldt *et al.*, 1982) and in the beak of the Japanese quail (Halata and Grim, 1993). There are, however, numerous Ruffini corpuscles in joint capsules (Halata and Munger, 1980). There is electrophysiolocical (Reinke and Necker, 1992a) but no morphological evidence of Ruffini endings in the feathered skin.

2. Electrophysiology of Mechanoreceptors

Electrophysiologically mechanoreceptors have been characterized by their response to a standard ramp-like stimulus with a plateau (Figure 2). There are rapidly adapting (RA) and slowly adapting (SA) responses which may be further subdivided (Iggo and Gottschaldt, 1974). Type SAI receptors respond both to the ramp and to the plateau with a sustained firing at irregular spike intervals (random or Poisson distribution of intervals). This type of receptor may detect amplitudes of a stimulus (strength of touch or pressure). In mammals the morphological basis of this type of response is the intraepidermal Merkel cell receptor. Type SAII receptors also have a sustained firing rate. However, there is often spontaneous activity and a regular firing (normal or Gaussian distribution of intervals). In mammals the

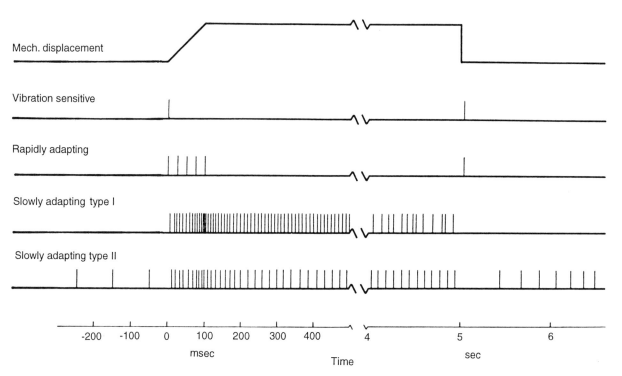

FIGURE 2 Types of responses of mechanoreceptors to a ramp-and-hold stimulus (uppermost trace). After Iggo and Gottschaldt (1974) with permission.

Ruffini corpuscle has been identified as an SAII receptor and the most effective stimulus consists of lateral stretching of the skin. RA receptors respond to a change of stimulus intensity only, for example, during the ramp. Response varies with steepness (velocity of ramp). The morphological basis for these velocity detectors are the Meissner corpuscles in glabrous skin and lanceolate endings in hairy skin of mammals; neither type of receptor exists in birds. A very rapidly adapting type of response shows spikes only at the beginning and/or end of the ramp; that is, during the acceleration phase of the ramp stimulus. This type of receptor responds best to vibration. The morphological basis of the vibration receptor in mammals is a lamellated corpuscle, the Pacinian corpuscle. All types of responses have been observed in birds also (Necker, 1983; Gottschaldt, 1985). However, the correlation of structure and function is less clear in birds than in mammals.

There is no doubt that Herbst corpuscles are vibration-sensitive receptors (Dorward, 1970; Dorward and McIntyre, 1971; Gottschaldt, 1974; Shen and Xu, 1994). Vibration receptors are usually characterized by strong phase coupling; that is, there is one spike per stimulus cycle. In the cycle histogram it can be seen that the spikes of successive cycles fall within a limited phase angle range of the full 360° cycle (Figure 3; Reinke and Necker, 1992b).

Herbst corpuscles are most sensitive to rather high frequencies (Dorward and Mcintyre, 1971; Gregory, 1973; Leitner and Roumy, 1974a; Hörster, 1990; Reinke and Necker, 1992b; Shen and Xu, 1994). Thresholds are rather high below 100 Hz but decrease in the frequency range 300 Hz up to 1000 Hz (Figure 4). In the high-frequency range threshold amplitudes may be less than 0.1 µm which is in the range of human vibration sensitivity.

The morphological basis of velocity-sensitive rapidly adapting responses is less clear although RA responses are very common. In the bill of aquatic birds RA responses are most likely based on Grandry corpuscles (Gottschaldt, 1974). RA responses in the chicken beak have been ascribed to Merkel cell (Grandry) corpuscles (Gentle, 1989). The morphological basis of RA responses in the feathered skin (Dorward, 1970; Necker and Reiner, 1980; Necker 1985c, Reinke and Necker, 1992a) is even less clear. Assuming, however, that avian Merkel cell receptors differ in location and hence in function from the mammalian counterpart (Andres and von Düring, 1990), and considering that there is a continuum of rapidly adapting to slowly adapting responses (Dorward, 1970; Necker, 1985c), one might argue that RA receptors in the feathered skin are due to Merkel cell receptor activation. This means that avian Merkel cell receptors may have both rapidly adapting and slowly adapting characteristics.

Both types of SA responses have been described in birds. SAI responses have most clearly been demonstrated so far only in the feathered skin (Dorward, 1970;

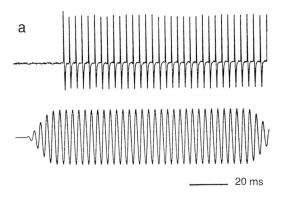

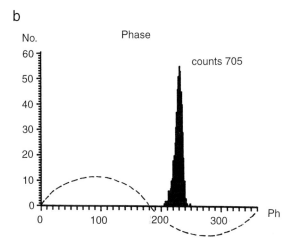

FIGURE 3 Response of a vibration-sensitive afferent fiber. (a) Original traces of action potentials (above) and a 300-Hz vibration stimulus (below). (b) Cycle histogram of the same recording shows the occurence of action potentials at a distinct phase of the stimulus cycle (0° to 360°). (After *J. Comp. Physiol. A,* Spinal dorsal column afferent fiber composition in the pigeon: An electrophysiological investigation, H. Reinke and R. Necker, **171**, 397–403, 1992, © Springer-Verlag.)

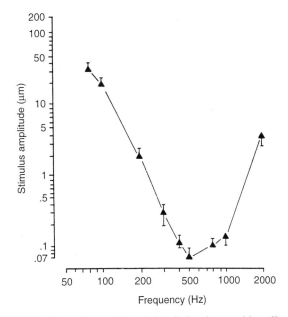

FIGURE 4 Dependence of threshold of vibration-sensitive afferent fibers in the interosseous nerve of the pigeon on vibration frequency. (After *J. Comp. Physiol. A,* Response characteristics of Herbst corpuscles in the interosseous region of the pigeon's hind limb, J. X. Shen and Z. M. Xu, **175**, 667–674, Fig. 7, 1994, © Springer-Verlag.)

Necker 1985c; Brown and Fedde, 1993). Response characteristics are very similar to those of mammalian Merkel cell receptors. This includes a high dynamic sensitivity during mechanical stimulation and cold sensitivity (Necker, 1985c). The location near filoplume follicles (Necker, 1985c) agrees with the anatomical demonstration of groups of Merkel cells in the follicle wall (Andres and von Düring, 1990).

Type SAII responses seem to be common in the beak skin (Necker, 1974a,b; Gottschaldt, 1974; Gottschaldt *et al.,* 1982) and the Ruffini endings are most likely the morphological basis (Gottschaldt *et al.,* 1982). SAII reponses have been observed in the feathered skin of the pigeon also, and the most effective stimulus was lateral stretch of skin, as in mammals (Reinke and Necker, 1992a). SA responses of unclear origin (probably Ruffini endings or free stretch receptors) have been described with wing afferents in the chicken (Brown and Fedde, 1993). These receptors increase activity with increasing elevation of covert feathers.

B. Thermoreceptors

Thermoreceptors are thought to be free nerve endings (Hensel, 1973). This seems to hold for avian thermoreceptors also since the conduction velocity of thermoreceptive afferents has been shown to be in the range of Aδ- and C-fibers (mean: 2 m/sec; Gentle 1989). Thermoreceptors are characterized by spontaneous activity at normal skin temperature which increases during cooling (cold receptors) or during warming (warm receptors). Typically, rapid temperature changes result in an excitatory overshoot. Thermoreceptors in the avian skin (mainly beak and tongue) have been described repeatedly (Kitchell *et al.,* 1959; Leitner and Roumy, 1974b; Gregory, 1973; Necker, 1972, 1973; Gentle, 1987, 1989; Schäfer *et al.,* 1989). Most fibers were cold afferents and there are only few demonstrations of warm receptors (Necker, 1972, 1973; Gentle, 1987, 1989).

Both the dynamic overshoot and the static temperature-dependent activity of cold receptors in the beak and tongue is lower than in mammalian cold receptors (Figures 5 and 6). As in mammals there is a maximum static and dynamic activity at about 25° to 30°C. There is more indirect (Necker, 1977) than direct evidence of cold sensitivity of the feathered skin (Necker and Reiner, 1980; Necker, 1985b). Spontaneous activity of

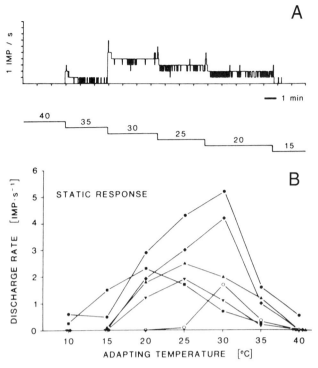

FIGURE 5 Activity of cold receptor neurons in the trigeminal ganglion of the pigeon. (A) Response to cooling steps (temperature as indicated): (B) dependence of static activity of six fibers on adapting temperature. (After Schäfer et al. (1989), *Brain Res.* **501,** 66–72, with permission from Elsevier Science.)

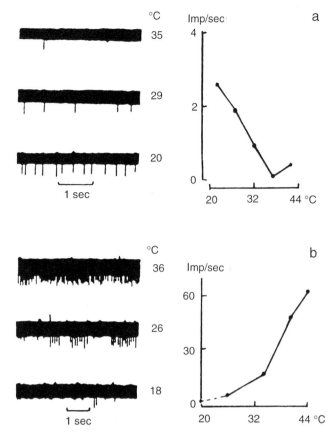

FIGURE 6 Dependence of activity of a cold receptor (a) and of a warm receptor (b) on beak skin temperature in the pigeon. Original spike traces on the left and static activity on the right. (After *J. Comp. Physiol. A,* Response of trigeminal ganglion neurons to thermal stimulation of the beak of pigeons, R. Necker, **78,** 307–314, 1972, © Springer-Verlag.)

warm receptors is high and there is an increase of static activity with increasing temperature in the range of 25° to 45°C (Figure 6).

C. Nociceptors

Nociceptors respond to stimuli which threaten to damage the skin. Both mechanical stimuli (pin prick, squeezing) and thermal stimuli (heat above about 45°C) are effective in exciting nociceptor afferents. Different types of nociceptors have been described: high threshold mechanoreceptors, heat nociceptors, and polymodal nociceptors (activated by heat, mechanical stimuli, and chemical agents like bradykinin; Burgess and Perl, 1973). All of these types seem to occur both in the feathered skin (Necker and Reiner, 1980) and in the beak skin (Gentle, 1989).

Nociceptors have no or only little spontaneous activity. High-threshold (nociceptive) mechanoreceptors increase activity with increasing force of mechanical stimulation (Figure 7a). Heat nociceptors increase activity when skin temperature exceeds about 45°C, and there is an increasing activation up to temperatures above 50°C (Figure 7b). All of these responses show slow adaptation. Nociceptors seem to be quite numerous both in the beak skin and in the feathered skin (Necker and Reiner, 1980; Gentle, 1989).

III. CENTRAL PROCESSING

A. Somatosensory Pathways

1. Trigeminal System

The trigeminal nerve consists of three branches, the ramus ophthalmicus, which innervates the orbita, the nasal area, and rostral part of the upper beak; the ramus maxillaris, which innervates the upper beak; and the ramus mandibularis which innervates the lower beak. The ophthalmic nerve and the maxillary nerve are pure sensory nerves whereas the mandibular nerve is a mixed sensory and motor nerve (Barnikol, 1953). Sensory components of the facial nerve and the glossopharyngeal nerve may join the trigeminal system (Dubbeldam *et al.,* 1979; Dubbeldam, 1984a; Bout and Dubbeldam, 1985).

The somata of the trigeminal nerve are located in the trigeminal ganglion (g. gasseri). The central root enters the brainstem and afferent fibers either ascend in the ascending trigeminal tract (TTA), which ends in the main sensory nucleus of the fifth cranial nerve (PrV,

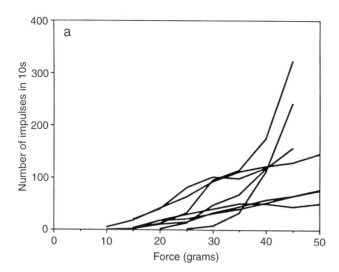

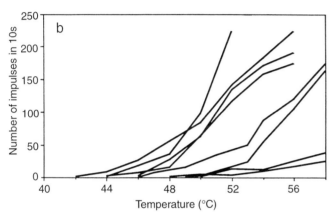

FIGURE 7 Dependence of activity of nociceptors on increasing force (a) and increasing temperature (b). (After *J. Comp. Physiol. A*, Cutaneous sensory afferents recorded from the nervus intramandibularis of *Gallus gallus* var. *domesticus*, M. J. Gentle, **164**, 763–774, 1989, © Springer-Verlag.)

nucleus principalis nervi trigemini), or descend in the descending trigeminal tract (TTD), which extends caudally to the upper spinal cervical segments (Karten and Hodos, 1967; Dubbeldam and Karten, 1978; Dubbeldam, 1980). In the pigeon there is a lateral component of TTD (lTTD) which mainly terminates in the external cuneate nucleus of the medulla. There is a somatotopic projection of the three divisions of the trigeminal nerve to PrV in such a way that the mandibular branch projects dorsally, the maxillary branch intermedially, and the ophthalmic branch ventrally.

The TTD is accompanied by the nucleus of the TTD (nTTD) which can be devided into several subnuclei from rostral to caudal: pars oralis (or) near PrV, pars interpolaris (ip), pars caudalis (cd), and spinal dorsal horn (dh) in the upper cervical spinal cord (Figure 8). In the nTTD fibers of the three branches of the trigeminal nerve terminate in a topographic order in all subnuclei (Dubbeldam and Karten, 1978; Arends and Dubbel-

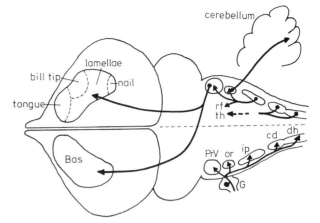

FIGURE 8 Schematic outline of central pathways of the trigeminal system. Afferents to the trigeminal nuclei on the left side of the brain, efferents on the right side. Bas, nucleus basalis; cd, pars caudalis of the descending trigeminal system (TTD); dh, cervical dorsal horn; G, trigeminal ganglion; ip, pars interpolaris of TTD, or, pars oralis of TTD; PrV, nucleus principalis nervi trigemini; rf, projection to the reticular formation; th, projection to the thalamus. (After Dubbeldam (1984b) with permission from S. Karger AG, Basel.)

dam, 1984). Pars oralis is the only trigeminal nucleus to project to the cerebellum (Arends *et al.*, 1984; Arends and Zeigler, 1989). Pars interpolaris has mainly intranuclear connections to oralis and PrV. Pars caudalis and cervical dorsal horn may project up to the thalamus, joining the medial lemniscus (Arends *el al.*, 1984). The more caudal subnuclei have projections to the neighboring reticular formation which may be important for motor control (see Chapter 6).

It has long been known that PrV projects via the quintofrontal tract to the nucleus basalis (Bas) or nucleus prosencephali trigeminalis of the telencephalon, bypassing the thalamus (Cohen and Karten, 1974). There is both an ipsilateral and a contralateral projection and in the mallard a topographic representation of the branches of the trigeminal nerve could be demonstrated (Dubbeldam *et al.*, 1981, Figure 8). Nucleus basalis projects to the nearby frontal neostriatum which seems to be at the origin of a network connecting the beak sensory system to the motor system, a circuit important for feeding (see Chapter 6).

2. Spinal System

Whereas the peripheral branches of spinal ganglion cells innervate receptors in the periphery the central branches enter the spinal cord by the dorsal root. Collaterals either ascend in the dorsal column or terminate in the grey substance (for details of the spinal cord see Chapter 5).

There are several ascending pathways in the spinal cord. The main pathways described for the mammalian spinal cord (dorsal column, spinoolivary, spinocerebellar, spinoreticular, spinomesencephalic, and spinothalamic) are found in the bird also (see Chapter 5). However, there is only sparse direct projection to the thalamus.

The dorsal column pathway (mechanoreception) is outlined in Figure 9. Primary afferent fibers ascend in the dorsal column to terminate in the dorsal column nuclei (nuclei gracilis et cuneatus, GC; nucleus cuneatus externus, CE) of the medulla oblongata (van den Akker, 1970; Wild, 1985; Necker and Schermuly, 1985; Schulte and Necker, 1994). The medially located nucleus gracilis (leg afferents) and the laterally located nucleus cuneatus (wing afferents) can be separated in caudal parts of the medulla only and there is an overlapping projection in rostral GC and in the external cuneate nucleus (van den Akker, 1970; Wild, 1985). The CE does not correspond to the mammalian analog (no relay of forelimb muscle spindle afferents). In addition to the primary afferent fibers there are secondary afferents from laminae IV and V neurons of the dorsal horn (see Chapter 5).

The dorsal column nuclei all project to the thalamus via the medial lemniscus. There is a crossed and a smaller uncrossed pathway and fibers give off collaterals to the inferior olive, to the intercollicular area ventral and medial to the auditory nucleus mesencephali lateralis pars dorsalis (MLD), and to nucleus spiriformes medialis (Wild, 1989). The main somatosensory thalamic nucleus has now been identified as the nucleus dorsalis intermedius ventralis anterior (DIVA) first described in the owl (Karten *et al.,* 1978) and then confirmed in the pigeon (Wild, 1987; Funke, 1989b; Schneider and Necker, 1989). A smaller contingent of dorsal column nuclei afferents reaches the nucleus dorsolateralis posterior (DLP). There is a differential projection of GC and CE whose significance is unknown as yet (Wild, 1989). The same thalamic targets are reached by spinal afferents (Schneider and Necker, 1989).

The thalamic somatosensory nuclei have different ipsilateral projections to the telencephalon. The DLP projects to a somatosensory area in the medial caudal neostriatum (NC) near the auditory field L (see hearing Chapter 2) and to the rostrally adjacent intermediate neostriatum (NI). The NC further projects to the overlying hyperstriatum ventrale (HV; Funke, 1989b). The main thalamic projection is from DIVA to a somatosensory area far rostral in the telencephalon in the Wulst, a rostromedial bulge in the avian telencephalon. The rostral part of the Wulst serves somatosensory representation whereas the caudal part serves vision. This part of the brain belongs to the hyperstriatum accessorium (HA) and it is the intermediate part (IHA) which receives DIVA afferents (Wild, 1987; Funke, 1989b). Both telencephalic areas are interconnected (Figure 9). Both spinal somatosensory representation areas have connections to descending (motor) systems (see Chapter 6).

B. Electrophysiological Investigations

1. Thermoreception and Nociception

Nociception and thermoreception have been studied so far only in the spinal dorsal horn. As in mammals, nociceptive and thermoreceptive neurons are mainly found in lamina I of the spinal dorsal horn both of the cervical and of the lumbar enlargements (Necker, 1985b; Woodbury, 1992). There is no significant direct projection of these neurons to the thalamus (see Chapter 5). However, thermoreceptive and nociceptive information may reach higher levels of the brain via relays in the brainstem reticular formation (Necker, 1989; Günther and Necker, 1995).

2. Mechanoreception

a. Trigeminal System and Beak Representation

In the trigeminal system the beak is represented both in PrV and in nTTD. In both nuclei a dorsoventral somatotopic organization has been confirmed with electrophysiological recordings in the pigeon (Zeigler and Witkovsky, 1968; Silver and Witkovsky, 1973). Units were rapidly adapting or slowly adapting and some responded to opening and/or closing the beak. Tongue stimulation was ineffective which confirms the lack of glossopharyngeal projections to the trigeminal system in the pigeon (Dubbeldam, 1984a).

There is evidence of responses of neurons in the somatosensory thalamus to beak stimulation (Delius and

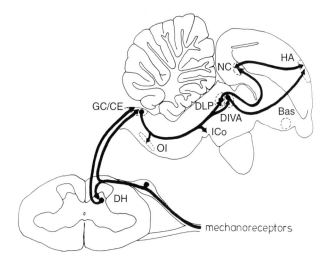

FIGURE 9 Schematic outline of central pathways of the mechanoreceptive spinal system. Bas, nucleus basalis; DH, dorsal horn; DIVA, nucl. dorsalis intermedius ventralis anterior; DLP, nucleus dorsolateralis posterior; GC/CE, nuclei gracilis et cuneatus/nucleus cuneatus externus; HA, hyperstriatum accessorius; ICo, nucleus intercollicularis; NC, neostriatum caudale; OI, oliva inferior.

Bennetto, 1972; Witkovsky *et al.,* 1973) and it seems that DLP is the main site of thalamic beak representation (Korzeniewska, 1987). Because of the lack of collaterals from the quintofrontal tract this information may reach the thalamus via the descending trigeminal system.

Processing in the nucleus basalis has been studied both in the pigeon and in the mallard. In both species receptive fields were often rather small especially at the tip of the beak and there was a somatotopic organization (Figure 8; Witkovsky *et al.,* 1973; Berkhoudt *et al.,* 1981). The tongue was represented in the mallard but not in the pigeon as expected from the anatomical studies. In contrast to PrV all units showed rapid adaptation.

In sandpipers and snipes (family Scolopacidae) an enormous enlargement of nucleus basalis forming a bulge on the basolateral surface of the brain has been observed (Pettigrew and Frost, 1985). Electrophysiological recordings revealed an overrepresentation of the bill tip. By analogy to the fovea in the eye a tactile fovea has been postulated for these birds.

b. Spinal System and Body Representation

The spinal dorsal horn has been studied electrophysiologically in detail in the pigeon cervical enlargement (Necker, 1985a,b, 1990) and in the chicken lumbosacral enlargement (Woodbury, 1992). There is a somatotopic organization similar to that in mammals; for example, distal parts of the extremities are represented medially and proximal parts laterally (see Chapter 5).

Mechanoreceptive neurons are located in lamina IV of the dorsal horn both in the pigeon and in the chicken. Both slowly adapting and rapidly adapting responses have been observed but there is no clear evidence of input from Herbst corpuscles in the cervical cord of the pigeon although it seems to be present in the lumbosacral enlargement of the chicken. In the ascending dorsal column many primary afferent fibers respond to vibratory stimuli (Reinke and Necker, 1992b).

In the dorsal column nuclei there is evidence for a separate representation of leg and wing at least in caudal parts of these nuclei (Necker, 1991). Many neurons in the GC show Herbst corpuscle input and there are also slowly adapting responses (Reinke and Necker, 1996). CE seems to process primarily deep input (joint receptors). However, input from muscle spindles was not found. Together with the finding that the CE does not project to the cerebellum (Wild, 1989) this supports the assumption that the avian CE is not homologous to the mammalian external cuneate nucleus.

Delius and Bennetto (1972) were the first to record somatosensory responses from the avian thalamus in the DLP/DIVA region. Whereas the DLP turned out to be a multimodal nucleus processing both somatosensory and visual and auditory stimuli DIVA seems to be a specific somatosensory relay (Korzeniewska 1987; Korzeniewska and Güntürkün, 1989). A detailed analysis showed that most DIVA neurons respond specifically to body stimulation (Schneider and Necker, 1996). Receptive fields were often large, normally covering the whole extremity and some including both extremities. The smallest receptive fields were located on the toes. A somatotopic organization was largely missing although the area with predominant wing responses could be separated from a more rostral area with predominant leg responses.

The telencephalic areas (NC/NI/HV, IHA) were studied at the single-unit level (Funke, 1989a). Both areas disclosed poor somatotopic organization. Receptive fields were smaller in IHA than in NC. Accordingly, there was a faint somatotopy in the Wulst (HA) area with rostral parts of the body being represented superficially and caudal parts in deeper layers. In the owl a detailed representation of the toes was found in the Wulst area (Karten *et al.,* 1978). There was no somatotopic representation of the body in the NC/NI and adjacent HV and bimodal input (auditory/somatosensory) was common in this caudal area (Funke, 1989a). These differences in the two areas suggest that the HA area may be compared to SI and NC/NI to SII of the mammalian somatosensory cortical representations.

IV. BEHAVIORAL ASPECTS

A. Mechanoreception

Cutaneous mechanoreceptors are involved in a variety of behavioral responses. Most evident is a contribution of beak receptors to feeding (see Chapter 6). In this context it has to be kept in mind that the avian beak serves as a prehensile organ comparable to the human hand. Parrots and birds of prey use both feet and beak for the handling of food items. Interestingly both beak and feet (toes) are the only parts of the avian body which are represented in great detail in the CNS.

In the feet the conspicuous strand of Herbst corpuscles on the interosseous membrane is exquisitely suitable to detect vibrations of the ground (Schwartzkopff, 1949), perhaps even earthquakes (Shen and Xu, 1994).

Mechanoreceptors in the feathered skin can detect disorders of the plumage (Necker, 1985a). This may trigger preening although this behavior does not necessarily depend on sensory input (Delius, 1988). Vibrations of the plumage occur during flight and mechanoreceptors may detect turbulences in the air stream and in this way influence flight control. Air stream evoked stimulation of mechanoreceptors in the feathered skin is important for flight pattern (Gewecke and Woike, 1978) and for flight reflexes (Bilo and Bilo, 1978). It is interesting that information from feather mechanoreceptors is conveyed directly from the spinal dorsal horn to the cerebellum (see Chapter 5).

B. Thermoreception

Thermoreception clearly serves temperature regulation in this homeothermic class of vertebrates (see Chapter 14). This is valid both for autonomic and behavioral thermoregulation. There is, however, not much evidence for behavioral responses apart from thermoregulation. In incubator birds like the mallee fowl (family Megapodiidae) thermoreceptors in the beak seem to be used for controlling the temperature of the incubator mound (Frith, 1959).

C. Pain

The behavior of birds to noxious stimuli has been studied employing pinch, feather plucking, heat, and pain-producing substances (Wooley and Gentle, 1987; Gentle and Hill, 1987; Szolcsanyi et al., 1986; Gentle and Hunter, 1990; Gentle, 1992). Pinching and a hot plate evoke reflex motor behavior (withdrawal reflex). Noxious heat (comb only), however, results in passive immobility which was observed during feather removal also (note that feather pecking is common in commercially reared poultry). This points to two types of responses to noxious stimulation, namely reflex/escape (flight–fight response) and immobility (conservation-withdrawal) which may be mediated by different types of nociceptors (Gentle, 1992). Apart from motor responses, changes in blood pressure and heart rate were observed during noxious stimulation and there was a parallel increase in heart rate and nociceptor activity above about 47°C in pigeons standing on a hot plate (Figure 10; Necker, unpublished data). These responses may well be mediated by the spinal lamina I–brainstem pathway (see above). Reflex responses occur only at very high temperatures (above 60°C) in hot plate experiments (Sufka and Hughes, 1990, own observation).

The time course of pain-related behavior after beak amputation (trimming) studied as the number of pecks in a visual stimulus task includes a pain-free period of about 1 day which is followed by long-lasting guarding behavior (Gentle et al., 1991). This means that birds may experience chronic pain (Gentle, 1992).

V. SUMMARY AND CONCLUSIONS

The avian somatosensory system is largely comparable to the mammalian one although there are much less detailed investigations. This includes the function and morphology of cutaneous receptors, organization of central pathways, and central processing. Altogether, a major difference between the avian and the mammalian somatosensory system is the lack of a detailed somatotopy in the brain. This poor somatotopy may be due

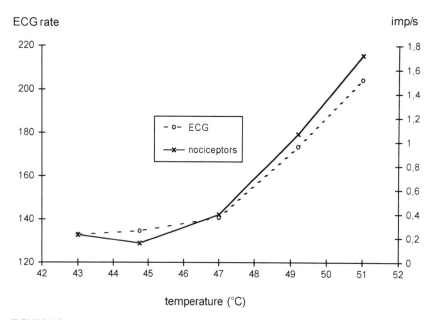

FIGURE 10 Dependence of nociceptor activity and rate of electrocardiogram (ECG, beats per minute) on hot plate temperature. Nociceptor activity after *J. Comp. Physiol.*, Temperature-sensitive mechanoreceptors, thermoreceptors and heat nociceptors in the feathered skin of pigeons, R. Necker and B. Reiner, **135**, 201–207, Fig. 8, 1980, © Springer-Verlag; ECG unpublished own data.

to the feather cover which does not allow punctuate stimulation of skin where the receptors are located. There is, however, a detailed beak and toe representation. This corresponds to a skilful use of these organs in handling of food or other objects. Whereas there is a double representation of the body in the telencephalon as in mammals, the beak (head) is represented only once (Bas), and there is no relay in the thalamus. The meaning of this difference and of the nonadjacent location of all somatosensory areas (see Figure 9) is unclear and awaits further investigations.

References

Andres, K. H. (1974). Morphological criteria for the differentiation of mechanoreceptors in vertebrates. *Abh. Rhein. Westf. Akad. Wiss.* **53,** 135–151.

Andres, K. H., and von Düring, M. (1973). Morphology of cutaneous receptors. *In* "Handbook of Sensory Physiology: Somatosensory System" (A. Iggo, ed.), Vol. II, pp. 3–28. Springer-Verlag, Berlin/Heidelberg/New York.

Andres, K. H., and von Düring, M. (1990). Comparative and functional aspects of the histological organization of cutaneous receptors in vertebrates. *In* "The Primary Afferent Neuron—A Survey of Recent Morpho-Functional Aspects" (W. Zenker and W. Neuhuber, eds.), pp. 1–16. Plenum, New York.

Arends, J. J. A., and Dubbeldam, J. L. (1984). The subnuclei and primary afferents of the descending trigeminal system in the mallard (*Anas platyrhynchos* L.). *Neuroscience* **13,** 781–795.

Arends, J. J. A., and Zeigler H. P. (1989). Cerebellar connections of the trigeminal system in the pigeon (*Columba livia*). *Brain Res.* **487,** 69–78.

Arends, J. J. A., Woelders-Blok, A., and Dubbeldam J. L. (1984). The efferent connections of the nuclei of the descending trigeminal tract in the mallard (*Anas platyrhynchos* L.). *Neuroscience* **13,** 797–817.

Barnikol, A. (1953). Zur Morphologie des Nervus trigeminus der Vögel unter besonderer Berücksichtigung der Accipitres, Cathartidae, Striges und Anseriformes. *Z. wiss. Zool.* **157,** 285–332.

Berkhoudt, H. (1980). The morphology and distribution of cutaneous mechanoreceptors (Herbst and Grandy corpuscles) in bill and tongue of the mallard (*Anas platyrhynchos* L.). *Netherlands J. Zool.* **30,** 1–34.

Berkhoudt, H., Dubbeldam, J. L., and Zeilstra, C. (1981). Studies on the somatotopy of the trigeminal system in the mallard, *Anas platyrhynchos* L. IV. Tactile representation in the nucleus basalis. *J. Comp. Neurol.* **196,** 407–420.

Bilo, D., and Bilo, A. (1978). Wind stimuli control vestibular and optokinetic reflexes in the pigeon. *Naturwissenschaften* **65,** 161–162.

Bolze, G. (1968). Anordnung und Bau der Herbstschen Koerperchen in Limicolenschnaebeln im Zusammenhang mit der Nahrungsfindung. *Zool. Anzeiger* **181,** 313–355.

Botezat, E. (1906). Die Nervenendapparate in den Mundteilen der Vögel und die einheitliche Endigungsweise der peripheren Nerven bei den Wirbeltieren. *Z. wiss. Zool.* **84,** 205–360.

Bout, R. G., and Dubbeldam, J. L. (1985). An HRP study of the central connections of the facial nerve in the mallard (*Anas platyrhynchos* L.). *Acta Morphol. Neerl.-Scand.* **23,** 181–193.

Brown, R. E., and Fedde, M. R. (1993). Airflow sensors in the avian wing. *J. Exp. Biol.* **179,** 13–30.

Burgess, P. R., and Perl, E. R. (1973). Cutaneous mechanoreceptors and nociceptors. *In* "Handbook of Sensory Physiology: Somatosensory system" (A. Iggo, ed.), Vol. II, pp. 29–78. Springer-Verlag/Berlin/Heidelberg/New York.

Cohen, D. H., and Karten, H. J. (1974). The structural organization of avian brain: An overview.. *In* "Birds: Brain and Behavior" (I. J. Goodman and M. Schein, eds.), pp. 29–73. Academic Press, New York.

Delius, J. D. (1988). Preening and associated comfort behaviour in birds. *Ann. N.Y. Acad. Sci.* **525,** 40–55.

Delius, J. D., and Bennetto, K. (1972). Cutaneous sensory projections to the avian forebrain. *Brain Res.* **37,** 205–221.

Dorward, P. K. (1970). Response patterns of cutaneous mechanoreceptors in the domestic duck. *Comp. Biochem. Physiol.* **35,** 720–735.

Dorward, P. K., and McIntyre, A. K. (1971). Responses of vibration-sensitive receptors in the interosseous region of the duck's hind limb. *J. Physiol.* **219,** 77–87.

Dubbeldam, J. L. (1980). Studies on the somatotopy of the trigeminal system in the mallard, *Anas platyrhynchos* L. II. Morphology of the principal sensory nucleus. *J. Comp. Neurol.* **191,** 557–571.

Dubbeldam, J. L. (1984a). Afferent connections of nervus facialis and nervus glossopharyngeus in the pigeon (*Columba livia*) and their role in feeding behaviour. *Brain Behav. Evol.* **24,** 47–57.

Dubbeldam, J. L. (1984b). Brainstem mechanisms for feeding in birds: interaction or plasticity: A functional-anatomical consideration of the pathways. *Brain Behav. Evol.* **25,** 85–98.

Dubbeldam, J. L., and Karten, H.J. (1978). The trigeminal system in the pigeon (*Columba livia*). I. Projections of the Gasserian ganglion. *J. Comp. Neurol.* **180,** 661–678.

Dubbeldam, J. L., Brauch, C. S. M., and Don, A. (1981). Studies on the somatotopy of the trigeminal system in the mallard, *Anas platyrhynchos* L. III. Afferents and organization of the nucleus basalis. *J. Comp. Neurol.* **196,** 391–405.

Dubbeldam, J. L., Brus, E. R., Menken, S. B. J., and Zeilstra, S. (1979). The central projections of the glossopharyngeal and vagus ganglia in the mallard, *Anas platyrhynchos* L. *J. Comp. Neurol.* **183,** 149–168.

Frith, H. J. (1959). Incubator birds. *Sci. Am.* **201,** 52–58.

Funke, K. (1989a). Somatosenory areas in the telencephalon of the pigeon. I. Response characteristics. *Exp. Brain Res.* **76,** 603–619.

Funke, K. (1989b). Somatosensory areas in the telencephalon of the pigeon. II. Spinal pathways and afferent connections. *Exp. Brain Res.* **76,** 620–638.

Gentle, M. (1987). Facial nerve sensory responses recorded from the geniculate ganglion of *Gallus gallus* var. *domesticus*. *J. Comp. Physiol.* A **160,** 683–691.

Gentle, M. J. (1989). Cutaneous sensory afferents recorded from the nervus intramandibularis of *Gallus gallus* var. *domesticus*. *J. Comp. Physiol.* A **164,** 763–774.

Gentle, M. J. (1992). Pain in birds. *Anim. Welfare* **1,** 235–247.

Gentle, M. J., and Breward, J. (1986). The bill tip organ of the chicken (*Gallus gallus* var. *domesticus*). *J. Anat.* **145,** 79–85.

Gentle, M. J., and Hill, F. L. (1987). Oral lesions in the chicken: behavioral responses following nociceptive stimulation. *Physiol. Behav.* **40,** 781–783.

Gentle, M. J., and Hunter, L. N. (1990). Physiological and behavioural responses associated with feather removal in *Gallus gallus* var. *domestica*. *Res. Vet. Sci.* **50,** 95–101.

Gentle, M. J., Hunter, L. N., and Waddington, D. (1991). The onset of pain related behaviours following partial beak amputation in the chicken. *Neurosci. Lett.* **128,** 113–116.

Gewecke, M., and Woike, M. (1978). Breast feathers as an air-current sense organ for the control of flight behaviour in a songbird (*Carduelis spinus*). *Z. Tierpsychol.* **47,** 293–298.

Gottschaldt, K.-M. (1974). The physiological basis of tactile sensibility in the beak of geese. *J. Comp. Physiol.* **95**, 29–47.

Gottschaldt, K.-M. (1985). Structure and function of avian somatosensory receptors. *In* "Form and Function in Birds" (A. S. King and J. McLelland, eds.), Vol. 3, pp. 375–461. Academic Press, London.

Gottschaldt, K.-M., and Lausmann, S. (1974). The peripheral morphological basis of tactile sensibility in the beak of geese. *Cell Tiss. Res.* **153**, 477–476.

Gottschaldt, K.-M., Fruhstorfer, H., Schmidt, W., and Kräft, I. (1982). Thermosensitivity and its possible fine-structural basis in mechanoreceptors in the beak skin of geese. *J. Comp. Neurol.* **205**, 219–245.

Gregory, J. E. (1973). An electrophysiological investigation of the receptor apparatus of the duck's bill. *J. Physiol.* **22**, 151–164.

Günther, S., and Necker, R. (1995). Spinal distribution and brainstem projection of lamina I neurons in the pigeon. *Neurosci. Lett.* **186**, 111–114.

Halata, Z., and Grim, M. (1993). Sensory nerve endings in the break skin of Japanese quail. *Anat. Embryol.* **187**, 131–138.

Halata, Z., and Munger, B. L. (1980). The ultrastructure of Ruffini and Herbst corpuscles in the articular capsule of domestic pigeon. *Anat. Rec.* **198**, 681–692.

Hensel, H. (1973). Cutaneous thermoreceptors. *In* "Handbook of Sensory Physiology: Somatosensory system" (A. Iggo, ed.), Vol. II, pp. 79–110. Springer-Verlag, Berlin/Heidelberg/New York.

Hörster, W. (1990). Histological and electrophysiological investigations on the vibration-sensitive receptors (Herbst corpuscles) in the wing of the pigeon (*Columba livia*). *J. Comp. Physiol. A* **166**, 663–673.

Ide, C., and Munger, B. L. (1978). A cytologic study of Grandy corpuscle development in chicken toe skin. *J. Comp. Neurol.* **179**, 301–324.

Iggo, A., and Andres, K. H. (1982). Morphology of cutaneous receptors. *Ann. Rev. Neurosci.* **5**, 1–31.

Iggo, A., and Gottschaldt, K. M. (1974). Cutaneous mechanoreceptors in simple and in complex sensory structures. *Abh. Rhein. Westf. Akad. Wiss.* **53**, 153–174.

Karten, H. J., and Hodos, W. (1967). "A Stereotaxic Atlas of the Brain of the Pigeon." John Hopkins Press, Baltimore.

Karten, H. J., Konishi, M., and Pettigrew, J. (1978). Somatosensory representation in the anterior wulst of the owl. *Soc. Neurosci. Abstr.* **5**, 554.

Kitchell, R. L., Ström L. and, Zotterman, Y. (1959). Electrophysiological studies of thermal and taste reception in chickens and pigeons. *Acta Physiol. Scand.* **46**, 133–151.

Korzeniewska, E. (1987). Multisensory convergence in the thalamus of the pigeon (*Columba livia*). *Neurosci. Lett.* **80**, 55–60.

Korzeniewska, E., and Güntürkün, O. (1990). Sensory properties and afferents of the N. dorsolateralis posterior thalami of the pigeon. *J. Comp. Neurol.* **292**, 457–479.

Leitner, L. M., and Roumy, M. (1974a). Mechanosensitive units in the upper bill and in the tongue of the domestic duck. *Pflügers Arch.* **346**, 141–150.

Leitner, L. M., and Roumy, M. (1974b). Thermosensitive units in the tongue and in the skin of the duck's bill. *Pflügers Arch.* **346**, 151–156.

Necker, R. (1972). Response of trigeminal ganglion neurons to thermal stimulation of the beak of pigeons. *J. Comp. Physiol.* **78**, 307–314.

Necker, R. (1973). Temperature sensitivity of thermoreceptors and mechanoreceptors on the beak of pigeon. *J. Comp. Physiol.* **87**, 379–391.

Necker, R. (1974). Temperature sensitivity of slowly-adapting mechanoreceptors on the beaks of pigeons. *Abh. Rhein. Westf. Akad. Wiss.* 53, 115–121.

Necker, R. (1977). Thermal sensitvity of different skin areas in pigeons. *J. Comp. Physiol. A* **116**, 239–246.

Necker, R. (1983). Somatosensory system. *In* "Physiology and Behaviour of the Pigeon" (M. Abs, ed.), pp. 169–191. Academic Press, London.

Necker, R. (1985a). Projection of a cutaneous nerve to the spinal cord of the pigeon. I. Evoked field potentials. *Exp. Brain Res.* **59**, 338–343.

Necker, R. (1985b). Projection of a cutaneous nerve to the spinal cord of the pigeon. II. Responses of dorsal horn neurons. *Exp. Brain Res.* **59**, 344–352.

Necker, R. (1985c). Observations on the function of a slowly-adapting mechanoreceptor associated with filoplumes in the feathered skin of pigeons. *J. Comp. Physiol. A* **156**, 391–394.

Necker, R. (1989). Cells of origin of spinothalamic, spinotectal, spinoreticular and spinocerebellar pathways in the pigeon as studied by the retrograde transport of horseradish peroxidase. *J. Hirnforsch.* **30**, 33–43.

Necker, R. (1990). Sensory representation of the wing in the spinal dorsal horn of the pigeon. *Exp. Brain Res.* **81**, 403–412.

Necker, R. (1991). The dorsal column nuclei in the pigeon: Electrophysiological investigations. *In* "Synapse—Transmission—Modulation" (N. Elsner and H. Penzlin, eds.), p. 44. Proceedings of th 19th Göttingen Neurobiology Conferences. Thieme, Stuttgart.

Necker, R., and Meinecke, C. C. (1984). Conduction velocities and fiber diameters in a cutaneous nerve of the pigeon. *J. Comp. Physiol. A* **154**, 817–824.

Necker, R., and Reiner, B. (1980). Temperature-sensitive mechanoreceptors, thermoreceptors and heat nociceptors in the feathered skin of pigeons. *J. Comp. Physiol.* **135**, 201–207.

Necker, R., and Schermuly, C. (1985). Central projections of the radial nerve and of one of its cutaneous branches in the pigeon. *Neurosci. Lett.* **58**, 271–276.

Ostmann, O. W., Ringer, R. K., and Tetzlaff, M. (1963). The anatomy of the feather follicle and its immediate surroundings. *Poult. Sci.* **42**, 958–969.

Pettigrew, J. D., and Frost, B. J. (1985). A tactile fovea in the Scolopacidae? *Brain Behav. Evol.* **26**, 185–195.

Reinke, H., and Necker, R. (1992a). Response properties of cutaneous low-threshold mechanoreceptors in the feathered skin of the pigeon. *In* "Rhythmogenesis in Neurons and Networks" (N. Elsner and D. W. Richter, eds.), p. 122. Proceedings of the 20th Göttingen Neurobiology Conferences. Thieme, Stuttgart.

Reinke, H., and Necker, R. (1992b). Spinal dorsal column afferent fiber composition in the pigeon: An electrophysiological investigation. *J. Comp. Physiol. A* **171**, 397–403.

Reinke, H., and Necker, R. (1996). Coding of vibration by neurones of the dorsal column nuclei in the pigeon. *J. Comp. Physiol. A* **179**, 263–276.

Saxod, R. (1978). Development of cutaneous sensory receptors in birds. *In* "Handbook of Sensory Physiology: Development of Sensory Systems" (M. Jacobson, ed.), Vol. IX, pp. 337–417.

Schäfer, K., Necker, R., and Braun, H. A. (1989). Analysis of avian cold receptor function. *Brain Res.* **501**, 66–72.

Schildmacher, H. (1931). Untersuchungen über die Funktion der Herbstschen Körperchen. *J. Ornithol.* **79**, 374–415.

Schneider, A., and Necker, R. (1996). Electrophysiological investigations of the somatosensory thalamus of the pigeon. *Exp. Brain Res.* **109**, 377–383.

Schneider, A., and Necker, R. (1989). Spinothalamic projections in the pigeon. *Brain Res.* **484**, 139–149.

Schulte, M., and Necker, R. (1994). Projection of wing nerves to spinal cord and brain stem of the pigeon as studied by transganglionic transport of Fast Blue. *J. Brain Res.* **35,** 313–325.

Schwartzkopff, J. (1949). Über Sitz und Leistung von Gehör und Vibrationssinn bei Vögeln. *Z. vergl. Physiol.* **31,** 527–608

Shen, J. X., and Xu, Z. M. (1994). Response characteristics of Herbst corpuscles in the interosseous region of the pigeon's hind limb. *J. Comp. Physiol. A* **175,** 667–674.

Silver, R., and Witkovsky, P. (1973). Functional characteristics of single units in the spinal trigeminal nucleus of the pigeon. *Brain Behav. Evol.* **8,** 287–303.

Sufka, K. J., and Hughes, R. A. (1990). Dose and temporal parameters of morphine-induced hyperalgesia in domestic fowl. *Physiol. Behav.* **47,** 385–387.

Stammer, A. (1961). Die Nervenendorgane der Vogelhaut. *Acta Universit. Szeged. Acta Biol. biol. Sci. natur.* **7,** 115–131.

Szolczanyi, J., Sann, H., and Pierau, F.-K. (1986). Nociception in pigeons is not impaired by capsaicin. *Pain* **27,** 247–260.

Toyoshima, K. (1993). Are Merkel and Grandry cells two varieties of the same cell in birds? *Arch. Histol. Cytol.* **56,** 167–175.

Toyoshima, K., and Shimamura, A. (1991). Ultrastructure of Merkel corpuscles in the tongue of the finch, *Lonchura striata. Cell Tiss. Res.* **264,** 427–436.

Van den Akker, L. M. (1970). "An Anatomical Outline of the Spinal Cord of the Pigeon." Van Gorcum, Assen.

Wild, J. M. (1985). The avian somatosensory system. I. Primary spinal afferent input to the spinal cord and brainstem in the pigeon (*Columba livia*). *J. Comp. Neurol.* **240,** 377–395.

Wild, J. M. (1987). The avian somatosensory system: Connections of body representation in the forebrain of the pigeon. *Brain Res.* **412,** 205–223.

Wild, J. M. (1989). Avian somatosensory system: II. Ascending projections of the dorsal column and external cuneate nuclei in the pigeon. *J. Comp. Neurol.* **287,** 1–18.

Winkelmann, R. K., and Myers, T. T. (1961). The histochemistry and morphology of the cutaneous sensory end-organs of the chicken. *J. Comp. Neurol.* **117,** 27–35.

Witkovsky, P., Zeigler, H. P., and Silver, R. (1973). The nucleus basalis of the pigeon: A single-unit analysis. *J. Comp. Neurol.* **147,** 119–128.

Woodbury, C. J. (1992). Physiological studies of cutaneous inputs to dorsal horn laminae I-IV of adult chickens. *J. Neurophysiol.* **67,** 241–254.

Woolley, S. C., and Gentle, M. J. (1987). Physiological and behavioural responses in the hen (*Gallus domesticus*) to nociceptive stimulation. *Comp. Biochem. Physiol. A* **88,** 27–31.

Zeigler, H. P., and Witkovsky, P. (1968). The main sensory trigeminal nucleus in the pigeon: A single-unit analysis. *J. Comp. Neurol.* **134,** 255–263.

CHAPTER 5

Functional Organization of the Spinal Cord

REINHOLD NECKER
Ruhr-Universität Bochum
Fakultät für Biologie
Lehrstuhl für Tierphysiologie
D-44780 Bochum
Germany

I. Introduction 71
II. Gross Anatomy 71
III. Cytoarchitectonic Organization of the Spinal Gray 72
IV. Peripheral Input to the Spinal Cord 74
V. Functional Aspects of Laminae and Nuclear Groups 75
   A. Lamina I 75
   B. Laminae II and III 75
   C. Lamina IV (Nucleus Proprius) 75
   D. Lamina V 75
   E. Laminae VI, VII, VIII 76
   F. Lamina IX (Motoneurons) 76
   G. Clarke's Column (Magnocellular Column) 76
   H. Column of Terni 77
   I. Paragriseal Cells 77
VI. Ascending Pathways 78
VII. Descending Pathways 79
VIII. Summary and Conclusions 79
   References 80

I. INTRODUCTION

The general organization of the avian spinal cord resembles that of all other vertebrates. There are, however, some specializations which birds largely share with the phylogenetically related class of reptiles. Compared to the mammalian cord the most outstanding deviations are the lack of a cauda equina and a filum terminale and the occurance of a sinus rhomboidalis or lumbosacralis. At the microscopic level there are some significant differences in the organization of cell groups and pathways. Birds differ from most other vertebrates in their kind of locomotion—bipedal walk and flight. It has to be kept in mind that specializations in the spinal cord may result from adaptations to this peculiar kind of locomotion.

This chapter deals with the structure and function of the spinal cord of adult birds. There is a considerable number of anatomical and functional studies of the development of the chicken cord as a model of embryonic development. Since most of these studies were aimed at understanding the development of the spinal cord or central nervous system in general, they are not included here.

II. GROSS ANATOMY

The spinal cord consists of a number of segments which may be grouped into cervical, thoracic, lumbar, sacral, and coccygeal segments according to the distribution of vertebrae along the axis of the vertebral column (Figure 1a). Each segment gives rise to a pair of spinal nerves. The number of segments varies from species to species. The pigeon has 39 segments (14 cervical, 6 thoracic, 4 lumbar, and 15 sacrococcygeal; Huber, 1936) and the ostrich 51 segments (15 cervical, 8 thoracic, 19 lumbosacral, and 9 coccygeal; Streeter, 1904). There are two distinct enlargements, the cervical enlargement (C11 to T1 or segments 11 to 15 in the pigeon), whose

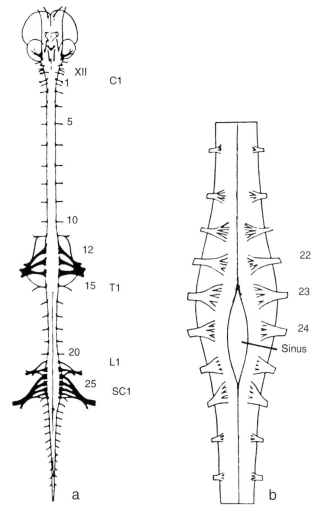

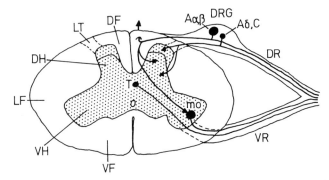

FIGURE 1 (a) Outline of the spinal cord of the pigeon. Cervical segments: 1–14 (C1–C14), thoracic segments: 15–20 (T1–T6), lumbar segments: 21–24 (L1–L4), sacrococcygeal segments: 25–39 (SC1–SC15). (b) Lumbosacral enlargement with sinus. (After Nerve roots and nuclear groups in the spinal cord of the pigeon, J. F. Huber, *J. Comp. Neurol.*, Copyright © 1936 John Wiley & Sons, Inc. Reprinted by permission of John Wiley & Sons, Inc.)

FIGURE 2 Cross-sectional anatomy of the spinal cord including afferent and efferent peripheral connections. Aδ, C, small myelinated and unmyelinated fibers; DF, dorsal funiculus; DH, dorsal horn; DR, dorsal root; DRG, dorsal root ganglion; LF, lateral funiculus; LT, Lissauers tract; mo, motoneurons; T, column of Terni; VF, ventral funiculus; and VR, ventral root.

segments innervate the wing via the brachial nerve plexus, and the lumbosacral enlargement (L1 to SC2 or segments 21 to 26 in the pigeon), whose segments innervate the legs via the lumbosacral plexus. The lumbosacral enlargement contains the very conspicuous rhomboid or lumbosacral sinus with the gelatinous or glycogen body as a peculiarity of all birds (Figure 1b).

In cross sections of the spinal cord a central gray substance with surrounding white matter can be distinguished (Figure 2). The gray substance which contains the cell bodies has been divided into a dorsal horn, a ventral horn, and an intermediate gray in between both horns. The white matter has been divided into a dorsal funiculus or dorsal column, a lateral funiculus, and a ventral funiculus. Peripheral afferent nerve fibers have their somata near the spinal cord in the spinal or dorsal root ganglia. The central processes of the ganglia cells enter the spinal cord in the dorsal roots where they usually bifurcate into an ascending and a descending branch. Sympathetic and parasympathetic preganglionic neurons are located near the central canal. The axons of these neurons and those of the motoneurons in the ventral horn leave the spinal cord in the ventral roots. The ventral roots join the dorsal roots distal to the dorsal root ganglion to give rise to the spinal nerves.

The lumbosacral sinus results from an incomplete or lack of fusion of the dorsal part of the cord (see Figure 4). The sinus is filled by the gelatinous or glycogen body. The glycogen body consists of glial cells with a high content of glycogen (Welsch and Wächtler, 1969). It is supplied with blood vessels and unmyelinated fibers (Paul, 1971). Its function is unknown. A less conspicuous glycogen body surrounding the central canal of cervical segments 14 and 15 has been described in the chicken (Sansone and Lebeda, 1976).

III. CYTOARCHITECTONIC ORGANIZATION OF THE SPINAL GRAY

In the dorsal horn a marginal layer, a substantia gelatinosa, a nucleus proprius, and, in the enlargements, a magnocellular column can easily be discriminated. The intermediate gray shows no specializations, and there is the group of large motoneuron somata in the ventral horn, especially of the segments innervating the limbs. A more detailed description of the spinal gray which follows a scheme of lamination originally described for the cat exists for the spinal cord of the pigeon (Leonard and Cohen, 1975a) and of the chicken (Brinkmann and Martin, 1973; Martin, 1979). According to this scheme

(Figure 3) there are nine laminae labeled I to IX. Cells near the central canal may be grouped into lamina X.

The head of the dorsal horn contains laminae I to IV. Lamina I is characterized by small cells with few large cells (cells of Waldeyer) bordering the gray substance. Lamina II contains many small cells. Because of its gelatinous appearance in histological sections this lamina have been named Substantia gelatinosa. Lamina III has fewer cells than lamina II. There is a difference between the pigeon and chicken in that lamina III is not ventral but medial to lamina II in the chicken (Brinkmann and Martin, 1973). Lamina IV represents the nucleus proprius of the dorsal horn and consists of a distinct group of medium-sized multipolar cells. Lamina V in the neck of the dorsal horn contains cells of various sizes. In the enlargements the very large cells of Clarke's column (ClC) or magnocellular column form a distinct group of neurons within lamina V.

Laminae VI to VIII occupy the intermediate gray and part of the ventral horn. Its delineation is uncertain. There are both small and large cells. The motoneurons in the ventral horn make up lamina IX. This lamina has two subdivisions in the enlargements, a lateral column (motor supply of limb muscles) and a medial group corresponding to lamina IX of all other segments (motor supply of axial muscles; Huber, 1936). In thoracic and sacral segments there is a column of Terni dorsal to the central canal which contains the preganglionic neurons of the autonomic nervous system.

In birds there are several groups of neurons outside the gray substance (paragriseal cells, Figure 4). Lateral to the head of the dorsal horn there are neurons which seem to correspond to the lateral spinal nucleus of mammals (Necker, 1990a). A functional equivalent of the mammalian lateral cervical nucleus seems to be absent in birds (van den Akker, 1970). In thoracic segments neurons are found in Lissauers tract (entrance zone of dorsal root fibers; see Figure 2) outside the gray (van den Akker, 1970; Günther and Necker, 1995). In the lumbosacral cord there are numerous paragriseal cells in the lateral and ventral funiculi (Huber, 1936; Leonard and Cohen, 1975a).

Marginal neurons at the lateral border of the spinal cord and in close proximity to the dentate ligament (specialization of the meninges) are found throughout the length of the spinal cord (Huber, 1936). There are major marginal nuclei of Hofmann-Kölliker or accessory lobes of Lachi in the lumbosacral enlargement and minor marginal nuclei in all other segments (Figure 4; Kölliker, 1902; Huber, 1936). The accessory lobes form distinct protuberances from the ventrolateral funiculi at segments L1 to SC4 in the pigeon. They are supported

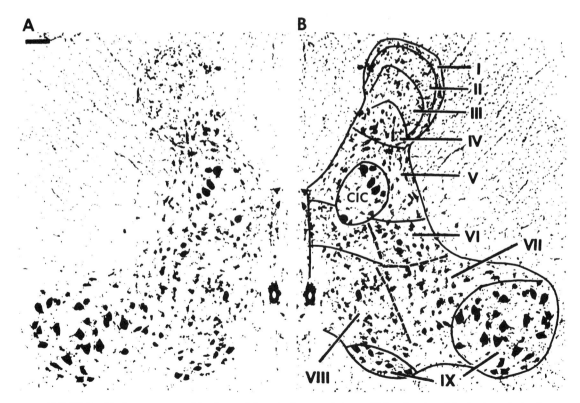

FIGURE 3 Nuclear groups and laminae of the spinal gray substance. [After A cytoarchitectonic analysis of the spinal cord of the pigeon (*Columba livia*), R. B. Leonard and D. H. Cohen, *J. Comp. Neurol.*, Copyright © 1975 John Wiley & Sons, Inc. Reprinted by permission of John Wiley & Sons, Inc.]

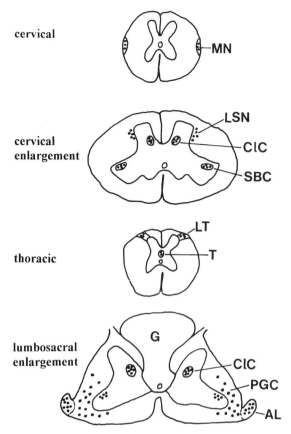

FIGURE 4 Distribution of special cell groups along the spinal cord. AL, accessory lobes; ClC, Clarke's column; G, glycogen body; LSN, lateral spinal nucleus; LT, Lissauers tract neurons; MN, marginal nuclei; PGC, paragriseal cells; SBC, spinal border cells; and T, column of Terni.

by the dentate ligament which has at this level of the spinal cord both a longitudinal and a transverse component (Schroeder and Murray, 1987). The accessory lobes lie between the roots as do the minor nuclei. Aside from nerve cell somata the minor marginal nuclei contain few glial material but the accessory lobes contain similar cells as the glycogen body.

IV. PERIPHERAL INPUT TO THE SPINAL CORD

When entering the spinal cord the dorsal roots separate into a bundle of coarse fibers which runs medially and a lateral bundle of fine fibers which enters through Lissauers tract (Nieuwenhus, 1964). This separation can be shown electrophysiologically (Figure 5): recordings from the medial cord show short-latency responses from fast-conducting large fibers only (N1 in Figure 5), whereas lateral recordings show in addition long-latency responses of small myelinated (N3 in Figure 5) and unmyelinated fibers (Necker, 1985a).

The fibers bifurcate into an ascending branch and a descending branch (van den Akker, 1970; Leonard and

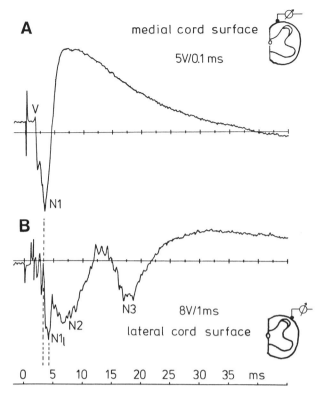

FIGURE 5 Evoked potentials recorded from (A) the medial cord surface or (B) lateral cord surface. Electrical stimulation of a cutaneous nerve. Note long latency peaks in the lower trace. (After *Exp. Brain Res.*, Projection of a cutaneous nerve to a spinal cord of the pigeon. I. Evoked field potentials, R. Necker, **59**, 338–343, Fig. 2, 1995a, © Springer-Verlag.)

Cohen, 1975b). Most medial fibers terminate in the dorsal horn but some reach the ventral horn (see Figure 2). Most fibers terminate in adjacent segments but part of the ascending fibers reach the dorsal column nuclei in the brain stem. Although there is no separation into a fasciculus gracilis (hindlimb afferents) and a fasciculus cuneatus (forelimb afferents), at cervical levels leg afferents always assume a medial position in the dorsal columns (van den Akker, 1970). Small fibers terminate in superficial layers (laminae I and II) of the dorsal horn but there is also a bundle running ventrally lateral to the dorsal horn to reach deep layers of the dorsal horn (Ohmori *et al.*, 1987; Ohmori and Necker, 1995).

The projection of peripheral nerves to the spinal cord has been studied repeatedly (van den Akker, 1970; Leonard and Cohen 1975b; Necker and Schermuly, 1985; Wild, 1985; Woodbury and Scott, 1991; Schulte and Necker, 1994). All these investigations show that there is a somatotopic organization of this projection which is similar to that found in mammals. The extremities are represented in such a way that distal parts (toes) are represented medially and proximal parts (upper arm, thigh) laterally. There is, however, considerable anatomical overlap of projection fields of individual nerves (Schulte and Necker, 1994).

V. FUNCTIONAL ASPECTS OF LAMINAE AND NUCLEAR GROUPS

A. Lamina I

Lamina I receives an input primarily from small fibers which are thought to innervate nociceptors and thermoreceptors. Accordingly, latencies of responses are often long and many lamina I neurons respond specifically to noxious stimulation (nociceptive specific neurons, NS) and a few to thermal stimulation (Necker, 1985b; Woodbury, 1992). There are, however, also neurons which were activated by light mechanical stimulation (low threshold mechanoreceptive neurons, LTM) or by both light and noxious mechanical or noxious heat stimulation (wide dynamic range neurons, WDR); that is, the whole spectrum of neuron types found in the mammalian cord (Brown, 1981).

The large lamina I neurons project bilaterally to the brainstem and seem to terminate predominantly in the lateral reticular formation, in the nucleus tractus solitarius, and in the parabrachial area (Necker, 1989; Günther and Necker, 1995).

B. Laminae II and III

The small cells in laminae II and III do not project beyond a few segments and are thought to belong to the propriospinal system. As in mammals small peripheral fibers terminate in lamina II (Woodbury and Scott, 1990) which thus seems to be involved in nociception. Large cutaneous fibers terminate in lamina III (and IV) where they probably contact dendrites of lamina IV neurons as in the mammalian dorsal horn (Brown, 1981). There are no detailed investigations of these laminae in birds.

C. Lamina IV (Nucleus Proprius)

Lamina IV or nucleus proprius is a well-circumscribed group of medium-sized multipolar neurons which receive an afferent input from large fibers of cutaneous mechanoreceptors. Accordingly, response latencies are short and nearly all neurons respond to light mechanical stimulation of the skin (LTM neurons; Necker, 1985b; Woodbury, 1992). There is a distinct topographic organization in lamina IV with distal parts of the limbs being represented medially and proximal parts laterally (Figure 6; Necker 1990b). As in mammals this physiologically defined somatotopy shows less overlap than the anatomical one (Schulte and Necker, 1994).

The majority of lamina IV neurons projects in the dorsal column or dorsolateral funiculus to the dorsal column nuclei (Funke, 1988; Necker, 1991). There is,

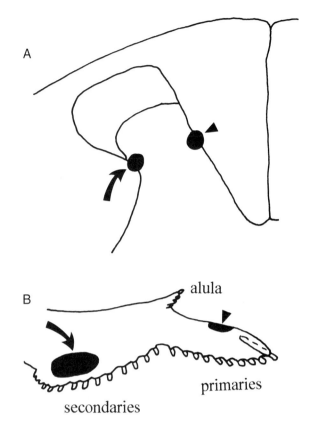

FIGURE 6 Somatotopic representation of the wing in the spinal dorsal horn of the pigeon. (A) Two recording sites (medial and lateral). (B) Receptive fields on the wing corresponding to the recording sites shown in A (same arrows). (After *Exp. Brain Res.*, Sensory representation of the wing in the spinal dorsal horn of the pigeon, R. Necker, **81**, 403–412, Fig. 11, 1990b, © Springer-Verlag.)

however, a difference between cervical and lumbar lamina IV neurons. Whereas cervical neurons project predominantly to the dorsal column nuclei, those of lumbar segments do not project beyond the cervical enlargement (Necker, 1991). A distinct group of cervical lamina IV neurons has descending projections. This means a reciprocal innervation of cervical and lumbosacral enlargements by lamina IV neurons. This scheme of innervation which seems to be peculiar to birds, may be important for coordination of wing and leg movements (Necker, 1990a, 1994).

A further group of lamina IV neurons located in most caudal cervical segments and in thoracic segments projects to the cerebellum (see Figure 9; Necker, 1992). This pathway, which is not present in mammals, may be important for control of flight.

D. Lamina V

Lamina V includes Clarke's column at the enlargements which will be considered separately below. Medial lamina V receives an input from large afferent fibers and medial lamina V neurons, especially those of lumbar segments, project ipsilaterally to the dorsal column nuclei (Necker, 1991). Lateral lamina V neurons receive an input from fine fibers (Ohmori et al., 1987; Ohmori and Necker, 1995) and project to the reticular formation (Necker, 1989). Some lamina V neurons have descending projections. There are no detailed electrophysiological investigations of this lamina. However, it is likely that there is convergence of sensory inputs especially in lateral lamina V as is true in mammals where lamina V neurons often show wide dynamic range responses (WDR) and are thought to be part of the pain system.

E. Laminae VI, VII, and VIII

There is not much known about these laminae of the intermediate gray except that they have both ascending projections to the reticular formation (Necker, 1989) as well as descending projections (Necker, 1990a). Most cells are probably interneurons of the sensorimotor interface and of descending systems (termination of, e.g., the rubrospinal tract, see below)

F. Lamina IX (Motoneurons)

Motoneurons which innervate different muscles are arranged in distinct columns with some overlap which extend for a few segments (Figure 7; Landmesser, 1978; Hollyday, 1980; Ohmori et al., 1982, 1984a; Martin and Hrycyshyn, 1981; Sokoloff et al., 1989).

Recent investigations showed that both the lateral and the medial motor column contain neurons which project to the cerebellum (see Figure 9; Necker, 1989, 1992). Although there is no clear histological distinction, a dorsal group of lateral lamina IX neurons in the enlargements has been identified as cells of origin of a spinocerebellar pathway. These cells may be compared to the spinal border cells (SBC, see Figure 4) of the mammalian cord which are the cells of origin of the ventral spinocerebellar tract (Matsushita et al., 1979). The physiology of this group of cells has not yet been studied in birds.

Lamina IX receives a peripheral input most probably from proprioceptor afferents. This input seems to be distributed both to the motoneurons and to the spinal border cells. There is electrophysiological evidence of monosynaptic connections between muscle afferents and motoneurons (Rabin 1975a). This means that birds have not only polysynaptic reflexes but monosynaptic reflexes like mammals also.

G. Clarke's Column (Magnocellular Column)

Clarke's column receives an input from peripheral large fiber afferents, most probably from proprioceptors as has been shown electrophysiologically for the

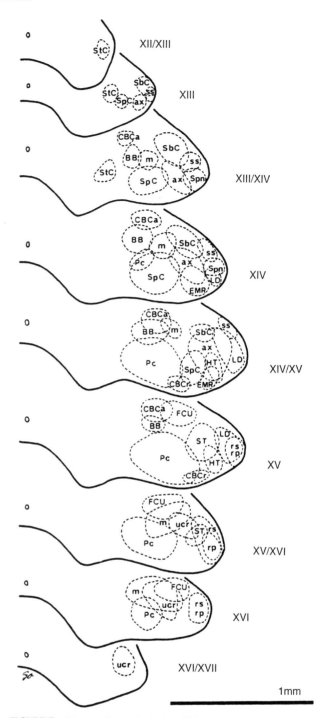

FIGURE 7 Motor columns in the brachial cord of the chicken. Abbreviations of muscles (lowercase abbreviations are nerves): BB, M. biceps brachii; CBCa, CBCp, M. coracobrachialis caudalis, cranialis; EMR, M. extensor metacarpi radialis; FCU, M. flexor carpi ulnaris; HT, M. humerotriceps; LD, M. latissimus dorsi; Pc, M. pectoralis, SbC, M. subcoracoideus; SpC, M. supracoracoideus; Spn, M. supinator; ST, M. scapulotriceps; and StC, M. sternocoracoideus. After Ohmori et al. (1982) with permission.

cervical enlargement (Necker, 1990b). Cervical Clarke's column neurons project ipsilaterally to the cerebellum whereas those of the lumbar cord cross to the contralateral side and recross at the level of the brainstem to terminate in the ipsilateral cerebellum (see Figure 9; Vielvoye, 1977; Necker, 1992). Cervical Clarke's column seems to correspond to the external cuneate nucleus in the brainstem of mammals since both nuclei process proprioreceptor afferents from the forelimb. It is unclear why the location is so different. Clarke's column in the lumbar cord seems to be the equivalent of the mammalian Clarke's column. However, the course of the pathway arising from these cells is uncrossed in mammals. In both classes of vertebrates, however, the projection is to the ispilateral cerebellum.

H. Column of Terni

The cells of this medial column receive an input from visceral afferent fibers (Ohmori et al., 1987) and from the hypothalamus (Cabot et al., 1982). They have been shown to innervate visceral organs (Cabot and Cohen, 1977; Ohmori et al., 1984).

I. Paragriseal Cells (See Figure 4)

1. Lateral Spinal Nucleus (LSN)

Input and function of LSN is unknown. The projection is similar to that of lamina I neurons; that is, to the brainstem reticular formation and adjacent structures (Necker, 1989; Günther and Necker, 1995). There is also a significant descending projection (Necker, 1990a). In mammals LSN has been suggested to be part of the pain system.

2. Lissauer Tract Neurons

These neurons may have migrated from lamina I toward their input (dorsal root entrance zone). It is most likely that the neurons have a peripheral input from visceral or cardiac afferents (Cabot and Cohen, 1977). The projection is to the nucleus tractus solitarius and it seems that rostral parts of the solitarius complex receive spinal input from these neurons only (Günther and Necker, 1995). No electrophysiological investigations have been done but it has been suggested that the thoracic Lissauer tract neurons are involved in visceronociception.

3. Lumbar Paragriseal Cells

There is probably no peripheral input to these numerous paragriseal cells in the lumbosacral enlargement. Recent investigations showed that cells of the accessory lobes have terminals near these paragriseal cells which might be one of their inputs (Necker, 1997). Since they project to the cerebellum (Necker, 1992; see Figure 9) they seem to be involved in sensorimotor control. However, their function has not yet been investigated.

4. Marginal Nuclei

Of all paragriseal cell groups only the minor marginal nuclei have been shown to have an input from peripheral nerves (van den Akker, 1970). This input is probably from axial muscle afferents since radial nerve afferents do not reach the marginal nuclei (Necker and Schulte, 1994). The axons of the marginal cells cross to the contralateral side where they ascend and descend in the ventrolateral cord for a few segments (Necker, 1997). The exact terminations and functional implications are not yet known.

So far no input for the accessory lobes in the lumbosacral cord could be demonstrated. However, the course of the axons of these cells has been shown both for the chicken (Matsushita, 1968; Eide, 1996) and the pigeon (Necker, 1994, 1997). Axons cross to the contralateral side to terminate in the ventromedial gray of rostral and caudal neighboring segments (Figure 8). Some axons or collaterals reach the ventrolateral funiculus where they contact spinocerebellar paragriseal cells (Necker, 1997). It has been suggested that the accessory lobes may function as mechanoreceptors (Schroeder and Murray,

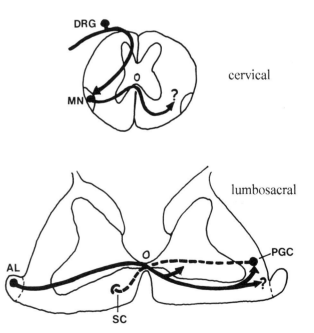

FIGURE 8 Schematic representation of afferents and efferents of the cervical marginal nuclei (top) and of the lumbosacral accessory lobes (bottom). AL, accessory lobes; DRG, dorsal root ganglion cell; MN, marginal nucleus; PGC, paragriseal cell; and SC, spinocerebellar tract.

1987). However, since there have been no electrophysiological investigations the function of these peculiar marginal nuclei remains enigmatic.

VI. ASCENDING PATHWAYS

In the mammalian cord dorsal column, spinocervical, spinothalamic, spinomesencephalic, spinoreticular, spinoolivary, and spinocerebellar pathways can be separated. All these pathways are present in birds also (Necker, 1989, 1991, 1992).

Ascending pathways have been studied anatomically at the level of the spinal cord (Kühn and Trendelenburg, 1911; van den Akker, 1970; Vielvoye. 1977) and there is one electrophysiological investigation also (Oscarsson et al., 1963). However, although different tracts have been identified the investigations were largely unable to show the cells of origin in the spinal cord and the terminations in the brain which define the function of the tracts. The cells of origin have been studied recently by investigations based on neuroanatomical tracer techniques and most projections have been mentioned above. Here, a summary will be given.

The most prominent and distinct ascending pathways are the dorsal columns and the spinocerebellar pathways in the lateral funiculus. Most tracts terminate at the level of the medulla oblongata and there are few projections to the mescencephalic or thalamic level (Necker, 1989).

The dorsal column consists of primary afferent fibers and of axons of secondary afferents originating in cervical lamina IV and in lumbar medial lamina V neurons. Since most primary afferents terminate in the spinal gray, the extent of the dorsal column is reduced significantly beyond the enlargements (Streeter, 1904). Fibers in the dorsal column are afferents of cutaneous mechanoreceptors and of joint receptors (Reinke and Necker, 1992). The termination of this tract is in the dorsal column nuclei in the medulla oblongata (nuclei gracilis et cuneatus and nucleus cuneatus externus).

The spinocerebellar pathways travel in the lateral funiculus. Four different pathways can be distiguished from dorsal to ventral (Figure 9; Necker, 1992). Fibers from cervical lamina IV neurons some of which project to the cerebellum, assume the most dorsal pathway. This is followed ventrally by fibers from cervical Clarke's column. Fibers from lumbar Clarke's column, spinal border cells, and paragriseal spinocerebellar cells all cross to the contralateral side where they first course in the ventral funiculus and then shift dorsally to join the cervical Clarke's column fibers ventrally. The most ventral pathway consists of fibers from cervical spinal border cells whose axons first course medially and then ventrally into the ventral funiculus with a dorsalward

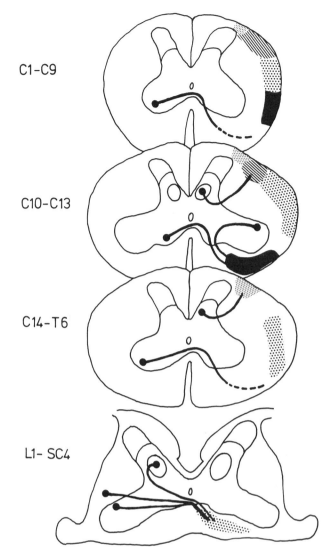

FIGURE 9 Cells of origin and spinal pathways of spinocerebellar tracts of the pigeon. (After *Anat. Embryol.*, Spinal neurons projecting to an anterior or posterior cerebellum in the pigeon, R. Necker, **185**, 325–334, Fig. 9, 1992, © Springer-Verlag.)

shift during their ascent. The course of spinocerebellar cells located in the medial motor column is unclear.

There is a rostral (lobules I to VI) and a caudal somatosensory area in the cerebellum (lobule IX). The spinocerebellar pathways terminate in both areas except for the spinal border cells, which project to anterior cerebellum only. Whereas the main input to the anterior cerebellum seems to be from the proprioreceptive system (Clarke's columns, SBC), that to the posterior cerebellum is dominated by lamina IV neurons; that is, cutaneous mechanoreception (Necker, 1992).

The axons of lamina I neurons predominantly course in the dorsolateral funiculus (Günther and Necker, 1995). Lateral lamina V, intermediate gray, and ventral horn neurons probably project in more ventral parts of the lateral funiculus and lateral parts of the ventral funiculus. All these pathways largely terminate in medullar nuclei (reticular formation, inferior olive, nucleus tractus solitarius, and parabrachial area). This means that a significant spinothalamic pathway is lacking in birds.

VII. DESCENDING PATHWAYS

A variety of brainstem nuclei has been shown to project to the cervical cord in pigeons (Cabot *et al.,* 1982) or lumbar cord in chickens (Webster and Steeves, 1991). These include from rostral to caudal: paraventricular nucleus in the hypothalamus (PVM); interstitial nucleus of Cajal (IS); intercollicular nucleus (ICo) and nucleus ruber (Ru) in the midbrain; nuclei coeruleus (LoC) and subcoeruleus (Scd, Scv); large parts of the mesencephalic, pontine, and medullar reticular formation; caudal raphe nuclei; and vestibular nuclei (mainly lateral vestibular or Deiters nucleus). For more details see Figure 10.

The main projection of the PVM as part of the autonomic nervous system is to the column of Terni in an ipsilateral dorsolateral pathway (Cabot *et al.,* 1982). The rubrospinal tract as a main motor pathway is located contralaterally in the medial dorsolateral funiculus and terminals are found in laminae V to VII; in the intermediate gray (Wild *et al.,* 1979).

There are several raphe–spinal projections of presumably serotonergic neurons in the lateral and ventral funiculus to laminae I and II, laminae V–VII, central areas including column of Terni, and lamina IX (motoneurons). Reticular formation and vestibular projections course in the ventral funiculus and terminals have been demonstrated in the medial ventral gray (Janzik, 1966).

Long descending tracts from the forebrain (septomesencephalic tract, occipitomesencephalic tract) reach first cervical segments only (see Dubbeldam, Chapter 6). This is even true for parrots with their pedal dexterity (Webster *et al.,* 1990).

On the whole, the cells of origin of descending projections are well known but their termination is less well studied and there is not much known about functional aspects of descending input to the spinal gray and to the motoneurons. Lesion studies suggest that the descending pathway from the medullar reticular formation is essential for walking in ducks (Webster and Steeves, 1991). Disynaptic vestibular input to neck motoneurons has been demonstrated electrophysiologically; wing and leg motoneurons were not affected by the same stimuli (Rabin, 1975b). This latter finding could be explained by later findings which showed that vestibular reflexes are greatly fascilitated by activation of cutaneous mechanoreceptors as is the case during flight (Bilo and Bilo, 1978, 1983).

VIII. SUMMARY AND CONCLUSIONS

Although cytoarchitectonics and pathways of the spinal cord of birds have been studied in some detail there is still a limited amount of knowledge of functional aspects.

There seems to be a rather clear separation of nociception (and probably thermoreception), cutaneous mechanoreception, and proprioception. Noxious stimuli are processed in superficial layers of the dorsal horn (laminae I and II). Information from cutaneous mechanoreceptors is processed by lamina IV and medial lamina V neurons. Muscle spindle afferents reach Clarke's column, spinal border cells, and lamina IX. The physiology of the remaining parts of the gray substance has not yet been investigated.

Bipedal locomotion and flight suggest specializations in the avian spinal cord and there are indeed some peculiarities in birds compared to mammals. These can be summarized as follows (see Necker, 1994): Spinocerebellar systems differ from the mammalian scheme in that there are "Clarke's columns" and "spinal border cells" in both enlargements. However, since the projection to the cerebellum is similar, the different locations do not necessarily mean functional differences between the two classes of vertebrates. Electrophysiological investigations should clarify this issue. Some lamina IV neurons (cutaneous mechanoreception) have projections that differ from the mammalian ones. These modifications (reciprocal innervation of the enlargements,

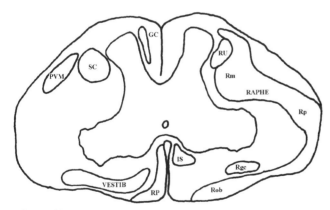

FIGURE 10 Descending pathways in the pigeon cervical enlargement (based on Cabot *et al.,* 1982). GC, dorsal column nuclei; IS, n. interstitialis Cajal; PVM, n. paraventricularis; Rgc, n. reticularis gigantocellularis; Rm, Rp, Rob, nucll. raphe magnus, pallidus, obscurus; RP, n. reticularis pontis; RU, n. ruber; SC, n. subcoeruleus; and VESTIB, nn. vestibularis. Reprinted from *Progress in Brain Research: Descending Pathways to the Spinal Cord,* Vol. 57, Cabot *et al.,* Avian bulbospinal pathways: anterograde and retrograde studies of cells of origin, funicular trajectories and laminar terminations, pp. 79–108, Copyright (1982), with permission from Elsevier Science.

projection to the cerebellum) may well be understood as adaptations to flight.

Because of the location in the lumbosacral cord the numerous paragriseal cells and the accessory lobes seem to play a role in bipedal walking. Functional evidence is, however, lacking. It seems that the marginal nuclei outside the lumbosacral enlargement are involved in proprioception (probably interneurons in sensorimotor propriospinal circuits). However, the meaning of the location of the marginal nuclei far outside the gray substance is completely unknown.

Sensorimotor circuits in the spinal cord have been studied in great detail in the mammal (Jankowska, 1992). Such investigations are nearly completely lacking in birds. There are some reflex studies, however, mainly in embryonic stages (Bekoff, 1992). Despite being involved in bipedal walking and bird flight the contribution of the spinal cord to this and other sensorimotor control systems is far from being elucidated sufficiently.

References

Bekoff, A. (1992). Neuroethological approaches to the study of motor development in chicks: Achievements and challenges. *J. Neurobiol.* **23,** 1486–1505.

Bilo, D., and Bilo, A. (1978). Wind stimuli control vestibular and optokinetic reflexes in the pigeon. *Naturwissenschaften* **65,** 161–162.

Bilo, D., and Bilo, A. (1983). Neck flexion related activity of flight control muscles in the flow-stimulated pigeon. *J. Comp. Physiol.* **153,** 111–122.

Brinkmann, R., and Martin, A. H. (1973). A cytoarchitectonic study of the spinal cord of the domestic fowl *Gallus gallus domesticus.* I. Brachial region. *Brain Res.* **56,** 43–62.

Brown, A. H. (1981). "Organization in the Spinal Cord." Springer-Verlag, Berlin.

Cabot, J. B., and Cohen, D. H. (1977). Anatomical and physiological characterization of avian sympathetic cardiac afferents. *Brain Res.* **131,** 89–101.

Cabot, J. B., Reiner A., and Bogan, N. (1982). Avian bulbospinal pathways: anterograde and retrograde studies of cells of origin, funicular trajectories and laminar terminations. *In* "Progress in Brain Research: Descending Pathways to the Spinal Cord" (H. G. J. M. Kuypers and G. F. Martin, eds.), Vol. 57, pp. 79–108. Elsevier, Amsterdam.

Eide, A. L. (1996). The axonal projections of the Hofmann nuclei in the spinal cord of the late stage chicken embryo. *Anat. Embryol.* **193,** 543–557.

Funke, K. (1988). Spinal projections to the dorsal column nuclei in pigeons. *Neurosci. Lett.* **91,** 295–300.

Günther, S., and Necker, R. (1995). Spinal distribution and brainstem projection of lamina I neurons in the pigeon. *Neurosci. Lett.* **186,** 111–114.

Hollyday, M. (1980). Organization of motor pools in the chick lumbar lateral motor column. *J. Comp. Neurol.* **194,** 143–170.

Hollyday, M., and Jacobson, R.D. (1990). Location of motor pools innervating chick wing. *J. Comp. Neurol.* **302,** 575–588.

Huber, J. F. (1936). Nerve roots and nuclear groups in the spinal cord of the pigeon. *J. Comp. Neurol.* **65,** 43–91.

Jankowska, A. E. (1992). Interneuronal relay in spinal pathways from proprioceptors. *Progr. Neurobiol.* **38,** 335–378.

Janzik, H. H. (1966). Der Vorderstrang der Hühner. *Verh. Anat. Ges.* **61,** 351–355.

Kölliker, A. (1902). Über die oberflächlichen Nervenkerne im Marke der Vögel und Reptilien. *Z. wiss. Zool.* **77,** 125–179.

Kühn, A., and Trendelenburg, W. (1911). Die exogenen und endogenen Bahnen des Rückenmarks der Taube mit der Degenerationsmethode untersucht. *Arch. Anat. Entwicklungsgesch.* 35–48.

Landmesser, L. (1978). The distribution of motoneurones supplying chick hind limb muscles. *J. Physiol.* (*London*) **284,** 371–389.

Leonard, R. B., and Cohen, D. H. (1975a). A cytoarchitectonic analysis of the spinal cord of the pigeon (*Columba livia*). *J. Comp. Neurol.* **163,** 159–180.

Leonard, R. B., and Cohen, D. H. (1975b). Spinal terminal fields of dorsal root fibers in the pigeon (*Columba livia*). *J. Comp. Neurol.* **163,** 181–192.

Martin, A. H. (1979). A cytoarchitectonic scheme for the spinal cord of the domestic fowl, *Gallus gallus domesticus*: Lumbar region. *Acta Morphol. Neerl. Scand.* **17,** 105–117.

Martin, A. H, and Hrycyshyn, A. W. (1981). Neurons for flight, a horseradish peroxidase study. *Exp. Neurol.* **72,** 252–256.

Matsushita, M. (1968). Zur Cytoarchitektonik des Hühnerrückenmarks nach Silberimpregnation. *Acta Anat.* **70,** 238–259.

Matsushita, M., Hosoya, Y., and Ikeda, M. (1979). Anatomical organization of the spinocerebellar system of the cat, as studied by retrograde transport of horseradish peroxidase. *J. Comp. Neurol.* **184,** 81–192.

Necker, R. (1985a). Projection of a cutaneous nerve to the spinal cord of the pigeon. I. Evoked field potentials. *Exp. Brain Res.* **59,** 338–343.

Necker, R. (1985b). Projection of a cutaneous nerve to the spinal cord of the pigeon. II. Responses of dorsal horn neurons. *Exp. Brain Res.* **59,** 344–352.

Necker, R. (1989). Cells of origin of spinothalamic, spinotectal, spinoreticular and spinocerebellar pathways in the pigeon as studied by the retrograde transport of horseradish peroxidase. *J. Hirnforsch.* **30,** 33–43.

Necker, R. (1990a). Cells of origin of ascending and descending as well as branching fibers in the cervical spinal cord of the pigeon. *Neurosci. Lett.* **119,** 1–4.

Necker, R. (1990b). Sensory representation of the wing in the spinal dorsal horn of the pigeon. *Exp. Brain Res.* **81,** 403–412.

Necker, R. (1991). Cells of origin of avian postsynaptic dorsal column pathways. *Neurosci. Lett.* **126,** 91–93.

Necker, R. (1992). Spinal neurons projecting to anterior or posterior cerebellum in the pigeon. *Anatomy and Embryology* **185,** 325–334.

Necker, R. (1994). Sensorimotor aspects of flight control in birds: specializations in the spinal cord. *Eur. J. Morphol.* **32,** 207–211.

Necker, R. (1997). Projections of the marginal nuclei in the spinal cord of the pigeon. *J. Comp. Neurol.* **377,** 95–104.

Necker, R., and Schermuly, C. (1985). Central projections of the radial nerve and of one of its cutaneous branches in the pigeon. *Neurosci. Lett.* **58,** 271–276.

Nieuwenhuys, R. (1964). Comparative anatomy of the spinal cord. *In* "Progres in Brain Research: Organization of the Spinal Cord" (J. C. Eccles and J. P. Schadé, eds.), Vol. 11, pp. 1–57. Elsevier, Amsterdam.

Ohmori, Y., and Necker, R. (1995). Central projections of primary afferents from the interosseous nerve in the pigeon. *Brain Res. Bull.* **38,** 269–274.

Ohmori, Y., Watanabe, T., and Fujioka, T. (1982). Localization of the motoneurons innervating the forelimb muscles in the spinal cord of the domestic fowl. *Zbl. Vet. Med. C. Anat. Histol. Embryol.* **11**, 124–137.

Ohmori, Y., Watanabe, T., and Fujioka, T. (1984a). Localization of motoneurons innervating the hindlimb muscles in the spinal cord of the domesic fowl. *Zbl. Vet. Med. C. Anat. Histol. Embryol.* **13**, 141–155.

Ohmori, Y., Watanabe, T., and Fujioka, T. (1984b). Location of parasympathetic preganglionic neurons in the sacral spinal cord of the domestic fowl. *Jap. J. Zootech. Sci.* **55**, 792–794.

Ohmori, Y., Watanabe, T., and Fujioka, T. (1987). Projections of visceral and somatic primary afferents to the sacral spinal cord of the domestic fowl revealed by transganglionic transport of horseradish peroxidase. *Neurosci. Lett.* **74**, 175–179.

Oscarsson, O., Rosen, I., and Uddenberg, N. (1963). Organization of ascending tracts in the spinal cord of the duck. *Acta Physiol. Scand.* **59**, 143–153.

Paul, E. (1971). Neurohistologische und fluoreszenzmikroskopische Untersuchungen über die Innervation des Glykogenkörpers der Vögel. *Z. Zellforsch.* **112**, 516–525.

Rabin, A. (1975a). Electrophysiology of spinal motoneurons in the pigeon. *Brain Res.* **84**, 351–356.

Rabin, A. (1975b). Labyrintine and vestibulospinal effects on spinal motoneurons in the pigeon. *Exp. Brain Res.* **22**, 431–448.

Reinke, H., and Necker, R. (1992). Spinal dorsal column afferent fiber composition in the pigeon: an electrophysiological investigation. *J. Comp. Physiol. A* **171**, 397–403.

Sansone, F. M., and Lebeda, F. J. (1976). A brachial glycogen body in the spinal cord of the domestic chicken. *J. Morphol.* **148**, 23–32.

Schroeder, D. M., and Murray, R. G. (1987). Specializations within the lumbosacral spinal cord of the pigeon. *J. Morphol.* **194**, 41–53.

Schulte, M., and Necker, R. (1994). Projection of wing nerves to spinal cord and brain stem of the pigeon as studied by transganglionic transport of Fast Blue. *J. Brain Res.* **35**, 313–325.

Sokoloff, A., Deacon, T., and Goslow, G. E., Jr. (1989). Musculotopic innervation of the primary flight muscles, the pectoralis (pars thoracicus) and supracoracoideus, of the pigeon (*Columba livia*): A WGA-HRP study. *Anat. Rec.* **225**, 35–40.

Streeter, G. L. (1904). The structure of the spinal cord of the ostrich. *Amer. J. Anat.* **3**, 1–27.

Van den Akker, L.M. (1970). An Anatomical Outline of the Spinal Cord of the Pigeon. Van Gorcum, *Assen.*

Vielvoye, G. J. (1977). Spinocerebellar tracts in the white leghorn (*Gallus domesticus*). Ph.D. thesis, Leiden

Webster, D. M. S., and Steeves, J. D. (1991). Funicular organization of avian brainstem-spinal projections. *J. Comp. Neurol.* **312**, 467–476.

Webster, D. M. S., Rogers, L. J., Pettigrew, J. D., and Steeves, J. D. (1990). Origins of descending spinal pathways in prehensile birds: do parrots have a homologue to the corticospinal tract of mammals? *Brain Behav. Evol.* **36**, 216–226.

Welsch, U., and Wächtler, K. (1969). Zum Feinbau des Glykogenkörpers im Rückenmark der Taube. *Z. Zellforsch.* **97**, 160–168.

Wild, J. M. (1985). The avian somatosensory system. I. Primary spinal afferent input to the spinal cord and brainstem in the pigeon (*Columba livia*). *J. Comp. Neurol.* **240**, 377–395.

Wild, J. M, Cabot, J. B., Cohen, D. H., and Karten, H. J. (1979). Origin, course and terminations of the rubrospinal tract in the pigeon (*Columba livia*). *J. Comp. Neurol.* **187**, 639–654.

Woodbury, C. J. (1992). Physiological studies of cutaneous inputs to dorsal horn laminae I-IV of adult chickens. *J. Neurophysiol.* **67**, 241–254

Woodbury, C. J., and Scott, S. A. (1991). Somatotopic organization of hindlimb skin sensory inputs to the dorsal horn of hatchling chicks (*Gallus g. domesticus*). *J. Comp. Neurol.* **314**, 237–256.

CHAPTER

6

Motor Control System

JACOB L. DUBBELDAM
Institute of Evolutionary and Ecological Sciences
Section of Dynamic Morphology
Van der Klaauw Laboratorium
Leiden University
2300 RA Leiden
The Netherlands

I. Introductory Remarks 83
II. The Control of Eye and Head Movements: The Oculomotor System 84
   A. Oculomotor Nuclei and Eye Muscles 84
   B. Vestibulocollic and Tectobulbospinal Control of Head and Neck Movements 87
III. The Motor Control of Jaw and Tongue Movements 87
   A. Jaw Opening and Closing Muscles and Their Motor Nuclei 87
   B. Intrinsic and Extrinsic Tongue Muscles 89
   C. Bulbar Premotor Centers and Proprioceptive Control 89
IV. Other Premotor and Motor Systems in the Brainstem 91
   A. Locomotion Centers 91
   B. Centers for Vocalization and Respiration 91
V. Structure and Function of the Cerebellum 92
VI. Telencephalic Centers for Motor Control 94
   A. Eminentia Sagittalis (Wulst) and Septomesencephalic Tract 94
   B. Archistriatum and Occipitomesencephalic Tract 95
   C. Paleostriatal Complex and Ansa Lenticularis 97
VII. Concluding Remarks 97
   References 97

I. INTRODUCTORY REMARKS

All animals are active; each activity depends upon the well coordinated action of groups of muscles. The nervous system is responsible for this coordination. Motor control and the organization of the motor control systems in birds are the subjects of this chapter.

Motor control systems have a hierarchical organization; that is, they comprise centers in "lower" and "higher" parts of the central nervous system. Each level has its specific function: motor centers in the central nervous system activate muscles; higher centers control and coordinate the activity of the respective motor centers. Such higher centers can be premotor regions in the brainstem, but parts of the cerebellum and the telencephalon are also involved. There is also a hierarchy of motor activities: reflexes, fixed or—perhaps better—modal action patterns and complex behavior. More complex activities require more complex control systems.

These few observations already suffice to make clear that there is not just *one* motor control system: activities of different complexity and using varying mechanical systems each require their own motor control system, even though some of these systems may share nervous centers. Further it is important to keep in mind that the muscular systems of the various mechanical systems may have a different embryological origin and thus may be under the control of differently organized neuronal systems. For example, the extrinsic eye muscles have a somatic origin and thus are innervated by somatic motor nerves, whereas the muscles of the jaws and some of the tongue muscles are considered "visceral" and innervated by visceromotor nerves (e.g., Romer and Parssons, 1977). We will begin to consider several motor systems which have their motor control systems in the brainstem; after that, higher hierarchical systems will be discussed. In this terminology, the medulla oblongata, mesencephalon, and diencephalon are part of the brainstem, whereas the cerebellum and telencephalon contain the higher systems.

Sensory information from various sources is indispensable to guide activities. This can be information from exteroceptive sensory systems such as the visual system, tactile (somatosensory) system, and auditory system, but also proprioception; that is, direct registration of the bird's own activity of muscles and of movements of bony elements. Some of the sensory systems are dealt with in Chapters 1–3. Finally, even though quite a lot of work has been done on the organization of the motor and premotor systems in birds, there are still more questions than answers. Some of the current uncertainties will pop up in the following sections.

II. THE CONTROL OF EYE AND HEAD MOVEMENTS: THE OCULOMOTOR SYSTEM

A sitting bird is able to survey a large part of its surroundings using both head and eye movements. Several types of eye movements can be distinguished such as saccades (fast flicks elicited by a sudden stimulus) and smooth pursuit (following a moving target). The combined movements of head and eyes are also of importance when, for example, fixing objects like food particles (e.g., Zeigler et al., 1980; Zweers, 1982). Head movements are effectuated by the neck muscles and are under the control of motoneurons in the cervical cord. Emphasis in this section will be upon eye movements and the neural substrate for their control. However, the two types of movements cannot be regarded entirely independently, as they may share parts of their premotor systems.

A. Oculomotor Nuclei and Eye Muscles

Six muscles are responsible for all movements of the eye (Figure 1). These extrinsic eye muscles are innervated by branches of the oculomotor, trochlear and abducens nerves. Each muscle has its own motor cell group (Table 1). The two muscles that turn the eye to dorsal are innervated by contralateral centers, the other muscles by ipsilateral cell groups. It is interesting to note that apparently the cells innervating the contralateral m. rectus superior migrate during ontogeny from one side of the brain to the other, thus forming an oculomotor decussation (Heaton, 1981). Several regions send

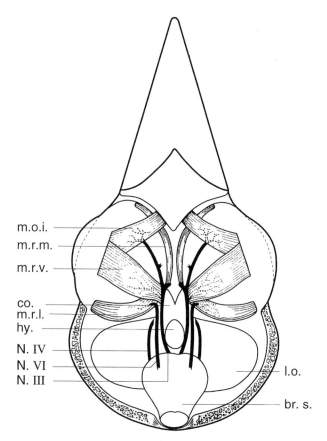

FIGURE 1 Ventral view of a bird's head showing the eyes, extrinsic eye muscles, and nerve branches innervating these muscles. Abbreviations: br.s., brainstem; co., chiasma opticum; hy., hypothalamus; l.o., lobus opticus; m.o.i., musculus obliquus internus; m.r.l., musc.rectus lateralis; m.r.m., musc.rectus medialis; m.r.v., musc.rectus ventralis; N.III, N.IV, N.VI, oculomotor nerve, trochlear nerve, abducens nerve. (Modified after Bubien-Waluszewska (1981) with permission.)

projections to these motor nuclei, such as the vestibular nuclei and several parts of the reticular formation in the mesencephalon and pontine region (Labandeira-

TABLE 1 The Extrinsic Eye Muscles and Their Innervating Motor Nuclei

Muscle	Motor nucleus[a]
Rectus inferior	Ipsilateral Om p. dorsolateralis (Omdl)
Rectus medialis	Ipsilateral Om p. dorsomedials (Omdm)
Rectus superior	Contralateral Om p. ventromedialis (medial part) (Omvm,med)
Rectus lateralis	Ipsilateral m VI (VI)
Obliquus superior	Contralateral m IV (IV)
Obliquus inferior	Ipsilateral Om p. ventromedialis (lateral part) (Omvm,lat)

[a]Om, nucleus oculomotorius; m IV, n. trochlearis; m VI, n. abducens (Heaton and Wayne, 1983; Labandeira-Garcia et al., 1987; Sohal and Holt, 1978).

Garcia *et al.*, 1989). Two systems serve to stabilize the visual field.

1. Optokinetic Nystagmus

The optokinetic nystagmus (OKN) is found in all classes of vertebrates including birds. It is a visuomotor reflex that serves to stabilize the retinal image with respect to movements of the animal or of its environment. Under experimental conditions OKN is a stereotyped pattern of eye motion consisting of a pursuit movement (slow phase) followed by a rapid movement in the opposite direction (saccadelike resetting fast phase) (Fite, 1979). The reflex arch underlying OKN consists of displaced ganglion cells occurring in the peripheral part of the retina (see Chapter 1) and projecting to part of the contralateral accessory optic system, the nucleus of the basal optic root (Figure 2: nBOR; Reiner *et al.*, 1979). The cells of this center project directly to oculomotor neurons: a dorsal part of nBOR is sensitive to dorsal movements in the visual field and projects to the ventromedial part of the ipsilateral oculomotor nucleus (Omvm). The nBOR proper is sensitive to ventral movements and projects to the dorsolateral part of the contralateral Om (Omdl) and trochlear nucleus (Brecha and Karten, 1979; McKenna and Wallman, 1985). This is a disynaptic feedback from retinal ganglion cells to the oculomotor system; that is, a closed-loop system: visual information is used directly to compensate for movements of the visual field. The nBOR cells are particularly sensitive to slow, wholefield visual motion. A small number of units respond to binocular stimulation, some with a preference for motion in the same direction, others for motion in opposite directions in the two eyes, enabling the bird to distinguish translational and rotational movements (Wylie and Frost, 1990).

In addition to the direct retinal input nBOR receives afferents from several visual centers—pretectal nuclei and the ventral lateral geniculate nucleus (see Chapter 1)—from the telencephalon and from several other areas. The variety of input is reflected in a wide variety of transmitters and neuropeptides in this center (Britto *et al.*, 1989). One of the important neurotransmitters is GABA (gamma-aminobutyric acid), which has a role in modulating the directional selectivity of nBOR cells. The source of this GABA-ergic input is uncertain. There is evidence that the visual Wulst (see Chapter 1) influences the directional selectivity of nBOR units (Britto *et al.*, 1990). The nBOR itself has direct projections to folia IXc,IXd, and the paraflocculus of the cerebellum as well as indirect cerebellar projections via the oliva inferior (Brecha *et al.*, 1980). The possible role of this cerebellar input will be discussed in a later section.

Finally, the nucleus interstitialis of Cajal is a relay between nBOR and motoneurons in the cervical cord.

The n. lentiformis mesencephali (LM; Figure 2B) is the second cell group of the accessory optic system. Units in this center are sensitive for rather fast movements (20–60°/sec), predominantly ($\pm 60\%$ of the units) in the horizontal plane (Winterson and Brauth, 1985). This cell group does not project directly to the oculomotor nuclei, but has reciprocal connections with nBOR (Figure 2C). Further, it sends efferents to the cerebellum, partly directly and partly indirectly via the lateral pons (Clarke, 1977). Bodnarenko *et al.* (1988) mapped the metabolic activity in the lentiform nucleus using [^{14}C]-2-deoxy-D-glucose: when part of the eye was occluded, only part of the nucleus was labeled; the lentiform nucleus receives a retinotopically organized input.

> Cells use glucose when active. Deoxy-D-glucose is taken up by the cells but cannot be digested. It remains in the cells and can—if radioactively labeled—be visualized in histological sections using autoradiography. This technique can be used to estimate differences in activity between cell groups.

In case of binocular stimulation in the horizontal plane the OKN appears to be asymmetrical, as the amplitude is larger after stimulation in the temporal–nasal direction than after stimulation in the nasal–temporal direction (Gioanni *et al.*, 1981). Suppression of the GABA activity in the accessory optic nuclei destroys this asymmetry; it has been suggested that these nuclei code the visual signal according to coordinates corresponding to vestibular input (Bonaventure *et al.*, 1992).

2. The VestibuloOculomotor Reflex

When a bird's body orientation is changing with respect to the horizontal position, the animal is perfectly able to keep its head in a stable horizontal position. The vestibular nuclei record all movements of the head. This information is used to correct both head and eye movements. Direct vestibulocollic motor and vestibulo-oculomotor pathways are instrumental in this correction. Vestibular cell groups send fibers either to the neck muscle motor neurons, to the motor nuclei of the eye muscles, or to both (Figure 3). In particular, one vestibular cell group, the n. tangentialis, receives very large afferents from the vestibular sense organ and sends direct projections to the ipsilateral dorsomedial oculomotor nucleus (Omdm) and possibly to the remaining contralateral cell groups, thus forming a fast reflex pathway (Wold, 1978; Cox and Peusner, 1990). However, there is not yet unanimity about the precise pattern of vestibular projections to the oculomotor complex (e.g., Arends *et al.*, 1991). Nevertheless it can be assumed

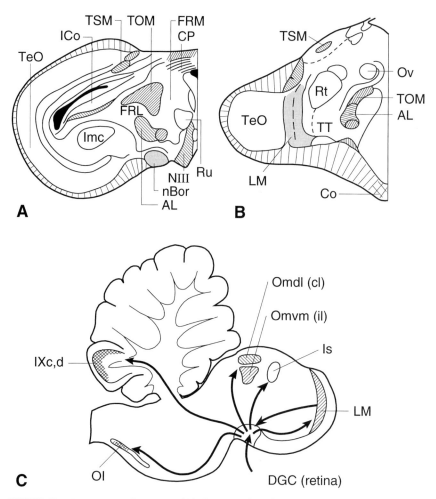

FIGURE 2 Accessory optic system. (A) Cross section through pigeon brain showing the position of the nucleus of the basal optic root (nBor). (B) Cross section showing the position of the nucleus lentiformis (LM). (C) Sagittal view of brainstem plus cerebellum showing the input in nBor from displaced retinal ganglion cells (DGC) and its output to oculomotor centers (Omdl and Omvm), interstitial nucleus of Cajal (Is), to the oliva inferior (OI), and to folia IXc,d of the cerebellum. Note the reciprocal connections between nBor and LM. Further abbreviations: AL, TOM, and TSM, telencephalic fiber systems (compare to Figures 10 and 12); Co, chiasma opticum; CP, commissura posterior; FRL, FRM, lateral/medial reticular formation; ICo, nucleus intercollicularis; Imc, nu. isthmi, pars magnocellularis; NIII, oculomotor nerve; Ov, nu. ovoidalis; Rt, nu. rotundus; Ru, nu. ruber; TeO, tectum opticum. (Modified after, Projections of the nucleus of the basal optic root in the pigeon: An autoradiographic and horseradish peroxidase study, N. Brecha, H. J. Karten, and S. P. Hunt, *J. Comp. Neurol.* Copyright © 1980 John Wiley & Sons, Inc. Reprinted by permission of John Wiley & Sons, Inc.)

that nuclei responsible for horizontal eye movements (Omdm and mVI) receive vestibular input from one side (the ipsilateral or contralateral vestibular nuclei, respectively); those responsible for vertical and oblique movements receive bilateral input (Table 2). The effect is that vestibular input will always cause the two eyes to move in the same direction. For example, stimulation in the right horizontal vestibular ampulla causes a twitch of the two eyes to the left (du Lac and Lisberger, 1992). This is an example of a feedforward system: the vestibular information is used to correct eye movements before visual feedback can reach the oculomotor nuclei. This is due to the fact that the vestibular pathway has less synaptic interruptions from receptor to oculomotor cell than has the visual pathway (see retinal structure, Chapter 1). Moreover, the tangential vestibular nucleus, source of the vestibulooculomotor projection receives input via a system of large (i.e., fast) fibers. This picture may be too simple, as little is known about the kind of neurotransmitters used in the vestibulooculomotor system. Recently, Wentzel *et al.* (1995) showed that in rabbits the ipsilateral projection from the superior vestibular nucleus is predominantly GABA-ergic (i.e., inhibitory), whereas other vestibular nuclei exert an excitatory effect.

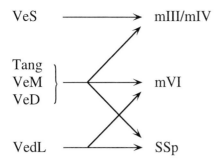

FIGURE 3 Scheme of vestibular projections to oculomotor centers (mIII, mIV, mIV) and the motor neck system (SSp). VeD, VeDL, VeM, VeS, nucleus vestibularis descendens/dorsolateralis/medialis/superior; Tang, nu. (vestibularis) tangentialis.

As in the optokinetic system, a second vestibuloocular control pathway includes part of the cerebellum. Cerebellar projections to the vestibular nuclei arise from several regions including the flocculus. When a single electrical pulse is given simultaneously to the vestibulum and flocculus, the effect of the vestibulum stimulation (see above) is suppressed (du Lac and Lisberger, 1992). This cerebellar involvement may serve to refine the vestibular control, possibly by a learning process. The vestibulooculomotor reflex can undergo rapid, but long-lasting adaptive changes under experimental conditions in both newly hatched and adult animals (Wallman *et al.*, 1982; chicken).

B. Vestibulocollic and Tectobulbospinal Control of Head and Neck Movements

It is clear that visual orientation depends upon eye movements as well as upon head movements. The motor pools innervating the neck muscles, responsible for head movements, are part of the motor system of the cervical (spinal) cord and will not be discussed here. However, the brainstem contains premotor centers of the neck motor system using the same type of information as the oculomotor system. Visual information from the mesencephalic (optic) tectum enters one of these premotor centers (Tellegen and Dubbeldam, 1994) via a crossing of the tectobulbar pathway to the n. gigantocellularis of the reticular formation (Reiner and Karten, 1982). A second visual input originates from the accessory optic system and reaches neck motor neurons via the n. interstitialis of Cajal (see section on Optokinetic Reflex). Vestibular nuclei have direct projections to the cervical motor pools (Arends *et al.*, 1991), as well as indirect via the parvocellular reticular nucleus. In particular, the direct pathway has much in common with the vestibulooculomotor system, both arising from the same group of vestibular nuclei (Figure 3; Arends *et al.*, 1991). We will return to the role of the reticular formation in the next sections.

III. THE MOTOR CONTROL OF JAW AND TONGUE MOVEMENTS

A. Jaw Opening and Closing Muscles and Their Motor Nuclei

The jaw muscles have a visceral origin (are part of the pharyngeal region). Therefore, the nerves innervating these muscles and the corresponding motor nuclei are considered visceromotor elements (Romer and Parssons, 1977). In contrast to the situation for the oculomotor nerves, the two nerves leave the brainstem dorsolaterally and not ventrally. A complication in birds is that they possess a so-called kinetic skull. This means that not only the lower jaw, but also the upper jaw moves to open the beak. Movements of the upper beak depend upon the presence of the quadrate bone; this has articulations with the skull and with the lower jaw. Movements of the beaks are achieved by four groups of muscles, two groups of beak openers and two groups of beak closers (Figure 4A). One opener group consists of the depressor muscles running from the caudal part of the lower jaw to the skull. This muscle complex is innervated by motor neurons of the facial nerve and is responsible for lowering the mandible. The second group of beak opener muscles, the protractors, is attached to the quadrate bone and to the skull; contraction of these muscles causes a rotation of the quadrate to medial and rostral.

TABLE 2 The Oculomotor Nuclei, the Vestibular Input, and the Direction of Eye Movements Caused by Each of the Nuclei

Motor center	Side-innervated muscle	Side origin (tangent.input)[a]	Side origin (vest.input)[b]	Direction (eye motion)[c]
Omdl	Ipsilateral	Contralateral	Contra-ipsilateral	↓
Omdm	Ipsilateral	Ipsilateral	Ipsilateral	→m
Omvm,med	Contralateral	Contralateral	Contra-/ipsilateral	↑
Omvm,lat	Ipsilateral	Contralateral	Contra-/ipsilateral	↙
IV	Contralateral	Contralateral	Contra-/ipsilateral	↘
VI	Ipsilateral	Contralateral	Contra-/ipsilateral	←m

[a]Several authors, see text.
[b]According to Correia *et al.* (1983).
[c]m, medial.

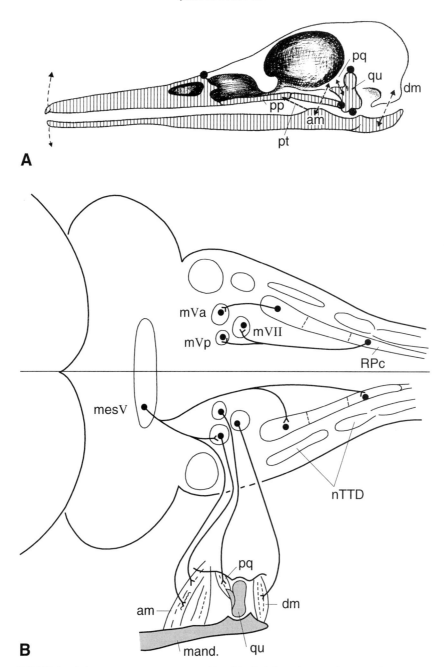

FIGURE 4 (A) Jaw mechanism in a duck (simplified); (B) horizontal view of brainstem with motor nuclei, reticular premotor regions and the proprioceptive mesencephalic trigeminal system. Abbreviations: am, adductor muscle of mandible; dm, depressor muscle of mandible; mand, mandibula; mesV, trigeminal mesencephalic nucleus; mVa, mVp, trigeminal motor nuclei innervating jaw closing (a) and opening (p) muscles; mVII, facial motor nucleus; nTTD, nuclei of the descending trigeminal system; pp, pterygoid muscle; pq, protractor muscle of qu; qu, quadrate bone; RPc, parvocellular reticular formation.

Two bony elements (pterygoid and palatinum) transmit this movement to the upper beak (Figure 4A) pushing its tip upward. Adductor muscles connecting lower jaw and skull cause a closing movement of the lower jaw; the pterygoid muscles at the ventral side of the head are responsible for the backmovement of the quadrate bone and closing movement of the upper beak. The openers of the upper beak and the two groups of beak closing muscles are innervated by motor neurons of the trigeminal nerve (e.g., Wild and Zeigler, 1980).

This description is a gross simplification of a very complex mechanism (with many variations among the avian species), but it is sufficient to illustrate that four motor centers are needed to effectuate the movements of the beak. These motor centers can be identified by tracing methods: tracers are injected in the muscles and transported to the motor centers, where they can be visualized. Electrostimulation in such a motor center results in a short contraction (twitch) of the corresponding muscle. The electrical activity of the muscle can be recorded directly by electromyography.

> Electromyography: Recording of voltage gradients produced by currents in the muscle tissue during activation of muscle fibers by the innervating motor neurons.

Electromyography can be used to record the normal pattern of activity of a group of muscles; for example, during the pecking of a free-moving bird. During normal movements, beak-opening and -closing motor neurons fire alternately, the periods of activity partly overlapping (Figure 5). Myograms show that the activity of facial and trigeminal innervated beak-opening muscles precedes the activity of the two trigeminal innervated beak-closing muscles during fast pecking movements. Jaw-closing muscles may be active during part of the opening movement and determine maximal opening amplitude (Bout and Zeigler, 1994). A superimposed system is required to generate this pattern of alternating activities; the activity of the motor nuclei is under control of a premotor system. Before having a closer look at this presumed premotor system (or systems) we will have a look at another complex kinetic system: that of the tongue.

B. Intrinsic and Extrinsic Tongue Muscles

Movements of the tongue are caused by two groups of muscles. The extrinsic muscles are part of the visceral musculature, comparable to the jaw muscles. These extrinsic muscles are innervated by a trigeminal motor center (intermandibular muscle), a facial center (stylohyoid and serpihyoid muscles: tongue protractors), and a glossopharyngeal center (geniohoid muscle: tongue retractor; Dubbeldam and Bout, 1990). The second group, the intrinsic muscles, have a somatic origin and are innervated by part of the dorsal motor nucleus of the 12th cranial nerve, the hypoglossal nerve. Notwithstanding their different embryological origin, the two groups of muscles behave in a similar way. The positions of the motor nuclei innervating the various muscles are wide apart. As jaws and tongue move in close harmony during feeding, drinking, and comparable actions, a premotor system is needed not only to coordinate the activity the tongue motor centers, but also the activity of these centers and that of the jaw motor centers.

C. Bulbar Premotor Centers and Proprioceptive Control

The motor centers of the jaw muscles receive input from two sources. One source is the mesencephalic trigeminal nucleus; this is a sensory center innervating the muscle spindles in the jaw muscles (Figure 4B). Muscle spindles are receptors sensitive to stretch of muscles and are part of the proprioceptive system (e.g., Manni et al., 1965). A muscle spindle consists of a few small, intrafusal muscle fibers surrounded by a capsule. The intrafusal fibers are innervated by two types of afferent nerve fibers (Ia and II fibers) showing dynamic and static responses, respectively (for details of muscle spindles; see Maier, 1992). Two further categories of motor neurons can be distinguished, α-motor neurons and γ-motor neurons. The latter category innervates the intrafusal muscle fibers and thus can influence the sensitivity of the muscle spindles. The γ-cells are smaller than the α-motor neurons innervating the extrafusal ("normal") muscle fibers. The two categories of neurons mingle in the motor nuclei.

The cells of the mesencephalic trigeminal nucleus project directly to the α-motoneurons of the jaw closers thus forming a monosynaptic reflex (Passatore et al., 1979). The muscle spindles are distributed in such a way that they are in the best position to record changes of length of the muscles (Bout and Dubbeldam, 1991); jaw openers don't have muscle spindles. Stretch of a muscle spindle causes activation of the motoneurons innervating this muscle through the monosynaptic reflex. The contribution of this reflex in different kinds of movements is not clear. It has been suggested that the muscle spindles may have a role in the control of the gape of the beak (Dubbeldam, 1984).

The reticular formation is the most important source of input to the motor nuclei; that is, it can be considered the premotor system of the jaw motor system (Berkhoudt et al., 1982, Arends and Dubbeldam, 1982). The reticular formation is characterized by the presence of many dispersed interconnected interneurons. Roughly, it consists of three longitudinal zones in the brain stem: a dorsolateral and a ventromedial parvocellular zone (both containing small cells) and a ventral gigantocellular zone (with large cells). The two parvocellular zones contain the jaw premotor system. The dorsolateral zone can be subdivided in a rostral, an intermediate, and a caudal compartment (Bout and Dubbeldam, 1994). The rostral compartment sends ipsilateral projections to the jaw-closing motor nuclei, the caudal one to the jaw-opening centers. The two compartments receive input

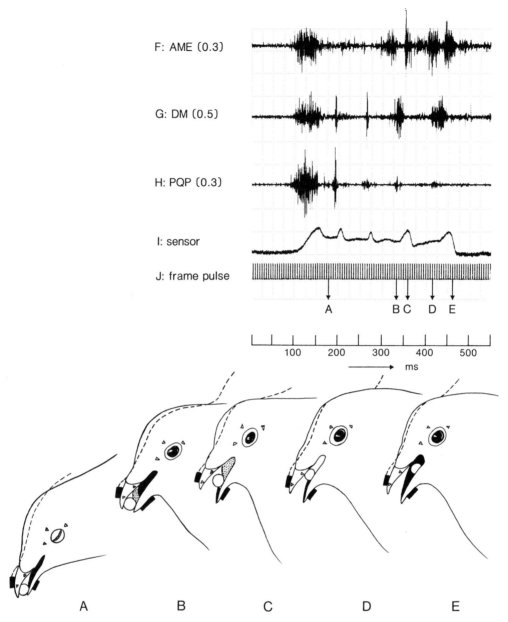

FIGURE 5 Electromyogram of pea-eating pigeon. (A–E), Sequence of beak movements during grasp and transport; (F–H), electromyographic registration of an adductor (AME), a depressor (DM) and a protractor (PQP) muscle. (I) Distance between beak-tips recorded by a sensor (black blocks on upper and lower beak). (J) Frame-pulses to synchronize film pictures (A–E) and electromyogram. (From Ontogeny of feeding in pigeons: A cinefluoroscopic and electromyographic analysis, E. M. S. J. van Gennip and H. Berkhoudt, *J. Exp. Zool.*, Copyright © 1994 John Wiley & Sons, Inc. Reprinted by permission of John Wiley & Sons, Inc.)

from the mesencephalic trigeminal nucleus; proprioceptive information from the muscle spindles can thus be used to control indirectly the motor activity of both the jaw-opener and jaw-closer motoneurons. The ventromedial parvocellular zone is not compartmentalized. Cells over the whole length of this zone project bilaterally to the motor centers of jaw-opening and jaw-closing muscles (Bout and Dubbeldam, 1994).

Often pecking—the basic mode of feeding in birds—is considered a stereotyped pattern of movements; the term fixed action pattern expresses this stereotypy. However, it has been recognized that this pattern is less stereotyped than initially assumed. For example, Zeigler *et al.* (1980) demonstrated that in pigeons the gape of the beak during pecking is tuned to the size of the food particles. Partly, this tuning depends upon visual information, but also tactile sense is important both for this tuning and for monitoring the transport of food through the oropharyngeal cavity (e.g., Zweers, 1982). Tactile (trigeminal) input reaches the dorsolateral premotor system via the nuclei of the descending sensory trigeminal system (Arends *et al.*, 1984; Chapter 4). The adjustment of the gape requires a continuous tuning of the pattern of motor activity: there is a modal—modifiable—rather than a fixed motor pattern. Both the dorsolateral zone and the ventromedial zone receive direct input from a telencephalic sensorimotor region (Dubbeldam and den Boer, 1994); the most lateral part receives a cerebellar input (Arends and Zeigler, 1989, 1991b). This strongly suggests that the premotor regions are under control of higher centers (see below).

Several models of pattern generating neuronal configurations have been developed. So far, however, no generally accepted model exists; moreover, none of these models has been designed specifically for the situation in birds. Although there is no direct evidence in birds, work in mammals suggests that the parvocellular premotor areas do have a crucial role in generating jaw motor patterns. There is further evidence that the ventromedial zone has a comparable function for tongue muscle motor centers and craniocervical muscle motor centers, thus integrating the activity of the various muscle systems (Tellegen and Dubbeldam, 1994). Such a coordination is a prerequisite when pecking, preening, and other behavioral activities are considered to be the effects of integrated motor patterns of jaw, tongue, and neck muscles.

IV. OTHER PREMOTOR AND MOTOR SYSTEMS IN THE BRAINSTEM

A. Locomotion Centers

The reticular formation contains premotor neurons not only of motor systems of the head, but also of motor systems of the spinal cord; among others, those of the locomotor systems (flying and walking). Steeves *et al.* (1987) could demonstrate that electrical stimulation in the gigantocellular formation of the medulla oblongata, as well as in the subtrigeminal region of the caudal medulla, evokes locomotory movements in decerebrate birds (Figure 6; goose and duck). Earlier, we saw that the gigantocellular region contains also premotor neurons of the neck muscle system—a coincidence of premotor elements for locomotion and head movements seems quite obvious, when looking at a walking chicken. However, this is not proof of such a relationship!

B. Centers for Vocalization and Respiration

Birds use their syrinx to produce sounds. The syrinx is an organ at the transition of the trachea and bronchiae. Vocalization depends upon the control of a few small syringeal muscles and the precise regulation of the stream of (expiratory) air passing through the trachea. This requires a close coordination of the motor centers of the syringeal and respiratory muscles. The caudal

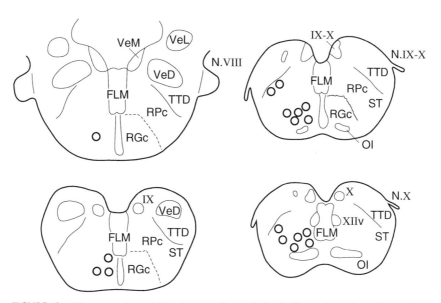

FIGURE 6 Diagrams of caudal brainstem. Open circles indicate sites, where locomotion could be evoked by electrostimulation. Abbreviations: N.VIII, N.IX, N.X: eighth, ninth, and tenth cranial nerves; OI, inferior olive; RGc, RPc, gigantocellular/parvocellular reticular formation; ST, subtrigeminal reticular nucleus; TTD, descending trigeminal system; Ved, VeL, VeM, vestibular nuclei; IX, IX–X, X, motor centers of NIX, NX. (Reprinted from *Brain Res.* **401**, J. D. Steeves, G. N. Sholomenko, and D. M. S. Webster, Stimulation of the pontomedullary reticular formation initiates locomotion in decerebrate birds, pp. 205–212, Copyright (1987), with permission from Elsevier Science.)

part of the dorsal hypoglossal motor nucleus innervates the syringeal muscles; the motor cells of the respiratory muscles are part of the motor system of the spinal cord. However, several cell groups have a role in the control of respiration: a dorsomedial cell group in the mesencephalic nucleus intercollicularis (Figure 12) and the n. retroambiguus in the medulla oblongata. The latter appears to be a premotor center of motor neurons innervating abdominal expiratory muscles (Wild, 1993a). The nucleus ambiguus is a cell group innervating laryngeal muscles and thus may be also of importance for vocalizations. In songbirds these centers are all under direct telencephalic control (Wild, 1993b); we will return to this topic when discussing the telencephalic systems.

V. STRUCTURE AND FUNCTION OF THE CEREBELLUM

Birds possess a well-developed cerebellum, but, as in mammals, its precise role in the control of motor activity is not yet well understood. However, there is little doubt that the ease and precision of motor performance depend upon an intact cerebellum. Before having a closer look at this aspect we will consider the organization of the cerebellum. Two major cell regions can be distinguished, the cerebellar cortex and the central cerebellar nuclei. The cerebellar cortex has essentially the same structure in all vertebrate groups. It is a three-layered structure with an outer molecular layer, a monolayer of Purkyne cells (P-cells) and an inner granular layer. In Nissl-stained sections the molecular layer looks rather "empty," as it contains relatively few cell bodies. The P-cells have large cell bodies and large dendritic trees extending to the outer surface of the molecular layer. Each tree lies in a single plane, the dendrites of all P-cells forming parallel planes (Figure 7). The granular layer consists of densely packed, small cell bodies, the granular cells. The cerebellum has two systems of afferent fibers, the climbing fibers and the mossy fibers (Figure 7). Many sources send mossy fibers to the granular layer, where each fiber makes contact with several granular cells. Each granular cell sends its axon into the molecular layer, where it splits into two branches running parallel to the surface of the cerebellum. All axons of the granular cells or parallel fibers run parallel to each other and perpendicular to the plane of the dendrites of the P-cells (Figure 7). In this way each parallel fiber makes contacts with many P-cells and each P-cell receives input from many parallel fibers.

The oliva inferior—a lamellar cell group in the medulla oblongata—is the source of the second input system, that of the climbing fibers. Each of these fibers "climbs" in the dendritic tree of one P-cell and each P-cell receives one climbing fiber. The inferior olive projects to the cortex in a strictly ordered way; that is, each particular part of the olive has its own projection area in the cortex (Furber, 1983). The effect is that longitudinal zones of olivocerebellar projections, so-called modules, can be recognized (Arends and Voogd,

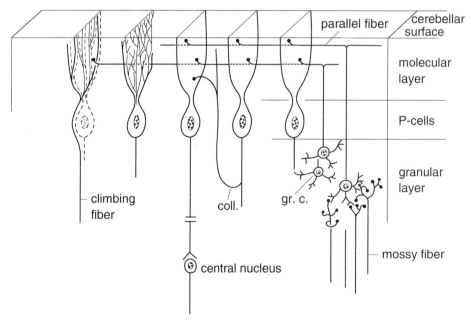

FIGURE 7 Organization of cerebellar cortex. The molecular layer contains the dendrites of the P-cells, the terminals of the climbing fibers, and the parallel fibers; the Purkyne (P-)cells form a monolayer of large perikarya; the granular layer contains the numerous granular cells (gr.c) and terminals of the mossy fibers. The mossy fibers, climbing fibers, and axons of the P-cells form the white core of the cerebellar lobes. Further details in the text.

1989). Previously, three longitudinal zones have been described in an electrostimulation study (Goodman *et al.*, 1964), but later studies revealed a more complex pattern—we will return to this issue.

The axons of the P-cells form the output of the cerebellar cortex; the central cerebellar nuclei and some vestibular nuclei are the targets of these fibers. The input of both the climbing fibers and the mossy fibers plus parallel fibers is excitatory; the P-cells have an inhibitory effect on the activity of the central nuclei. Collaterals of the P-cell axons reenter the molecular layer and have an inhibitory effect on neighboring P-cells. In mammals, three more cell types are found in the outer layers of the cerebellar cortex in addition to the P-cells—stellate, basket, and Golgi cells—which all have inhibitory effects. Probably, such cells also occur in the avian cerebellum, but no details are available in the literature. The study of neurotransmitters may shed some more light. GABA is the inhibitory neurotransmitter of these interneurons in mammals; in the pigeon GABA-A receptors have been found in both the granular and molecular layer, GABA-B receptors only in the latter (Albin *et al.*, 1991). GABA-A is coupled to an inhibitory chloride channel and GABA-B to a G-protein-coupled receptor that modulates potassium and calcium channels. This strongly suggests that interneurons modulate the activity of P-cells in birds also.

Marr (1969) was the first to develop a model ascribing a learning function to the cerebellum. He defined a module as consisting of a number of microzones. Each zone can be considered a three-step computing device with a receptor part consisting of the mossy fibers, an associative part (the granular cells + parallel fibers), and an effector (the P-cells). Such a microzone may control a specific motor function. The role of the climbing fibers would be the refining of the synaptic relationship between granular cells and P-cells resulting in an optimization of the control. In the original theory this role of climbing fibers was restricted to the early development; in this respect the theory did not hold. More recent studies ascribe a specific role to the climbing fibers in the fine tuning of P-cell activity. It has been demonstrated that the neurons of the inferior olive operate as "a distributed system, whose collective activity is rhythmic and temporally related to specific parameters of movement" (Welsh *et al.*, 1995). In the vision of these authors the rhythmic olivary input may cause groups of P-cells to fire synchronously; the axons of these P-cells may converge upon specific motor zones within the central cerebellar nuclei and, through these, control specific motor activities. The cerebellar nuclei do not project directly to motor centers, but their output reaches several vestibular nuclei and many parts of the reticular formation, through two large fiber systems (Figure 8;

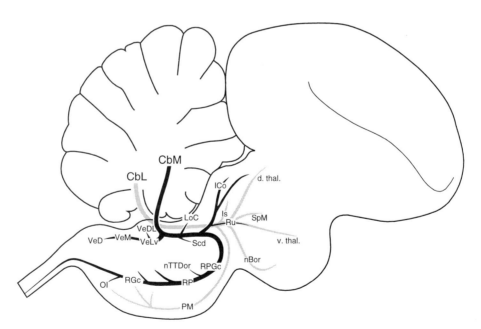

FIGURE 8 Simplified overview of the projections from the lateral and medial central cerebellar nuclei (CbL, CbM). Abbreviations: d.thal, v.thal, dorsal/ventral thalamic nuclei; ICo, nucleus intercollicularis; Is, interstitial nucleus of Cajal; LoC, locus ceruleus; nTTDor. nucleus oralis of descending trigeminal tract; OI, inferior olive; PM, medial pontine nucleus; RGc, RP, RPGC, reticular cell groups; Ru, red nucleus (nu.ruber); Scd, nucleus subceruleus dorsalis; SpM, medial spiriform nucleus; VeD, VeDL, VeM, VeLv, vestibular cell groups. [Modified after Organization of the cerebellum in the pigeon (*Columba livia*): II. Projections of the cerebellar nuclei, J. J. A. Arends and H. P. Zeigler, *J. Comp. Neurol.*, Copyright © 1991 John Wiley & Sons, Inc. Reprinted by permission of John Wiley & Sons, Inc.]

Arends and Zeigler, 1991b). We already saw that parts of this reticular formation serve as premotor systems for various motor activities. Some of the cell groups receiving cerebellar input, such as the red nucleus (n. ruber; Figures 8 and 11), are the source of fiber systems descending into the spinal cord and thus participate in the control of motor activities of the spinal motor centers (see Chapter 6).

Even though there is no definite theory on cerebellar function, it is clear that it does have an important role in the control of learned movements. We already mentioned the convergence of input from the accessory optic and vestibular systems in folia IXc and IXd that may be important to tune eye movements to movements of the head. Another example is the trigeminal system: one of the nuclei of the descending system (n. oralis) sends input to folia VIII and IXa of the cerebellum and a second one (n. interpolaris) to the oliva inferior and from there to the cerebellar cortex (Arends *et al.*, 1984; Arends and Zeigler, 1989). Here, trigeminal information reaches the cerebellar cortex via two pathways and may thus be used in the control of beak movements such as pecking.

VI. TELENCEPHALIC CENTERS FOR MOTOR CONTROL

The telencephalon comprises the highest hierarchical centers for the processing of sensory information and the control of motor activity. A telencephalon consists of an outer pallial and a deeper subpallial region. The cortex of the mammalian forebrain contains neurons of pallial origin, and the basal ganglion or corpus striatum contains the subpallial cell populations. For a long time the avian forebrain was assumed to consist of a hypertrophied striatum and a rudimentary cortex (e.g., Kappers *et al.*, 1963). This belief is reflected in the nomenclature, with many names still ending with -striatum. More recently, the interpretation of the avian forebrain has changed thoroughly. For the following overview it is sufficient to subdivide the telencephalic hemisphere roughly into three regions (Figure 9). One region forms a slight dorsal bulge and is called the eminentia sagittalis or Wulst; it is bordered by a shallow furrow, the vallecula. The second and largest part is often indicated as the dorsal ventricular ridge; in fact, this term reflects the embryological origin of this region. The paleostriatal complex is the deepest part of the telencephalon. The first two regions contain areas that are recipients of ascending sensory systems (Chapter 4); these regions are of pallial origin and may contain the same populations of cells that form the cortex in mammals (Nauta and Karten, 1970). The paleostriatal complex is a sub-

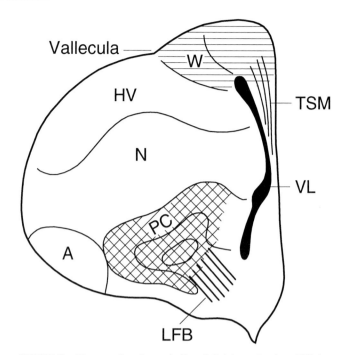

FIGURE 9 Three main telencephalic subdivisions: the dorsal Wulst (W), the dorsal ventricular ridge with neostriatum (N), hyperstriatum ventrale (HV) and archistriatum (A), and the paleostriatal complex (PC). The output system of PC is part of the lateral forebrain bundle (LFB), that of W is the septomesencephalic tract (TSM) descending medial to the lateral ventricle (VL).

pallial structure and corresponds to the mammalian basal ganglion (Karten and Dubbeldam, 1973). It receives input from the overlying telencephalic parts as well as from the mesencephalon, but no ascending sensory input. Each of these regions includes the source of a large descending or extratelencephalic fiber system. We will have a closer look at these systems and see what role each may have in the control of motor activity.

A. Eminentia Sagittalis (Wulst) and Septomesencephalic Tract

The Wulst is one of the few telencephalic areas with a clear cortical appearance. In the pigeon and owl four layers can be distinguished, from dorsal to ventral the hyperstriatum accessorium (HA), the nucleus intercalatus of HA (IHA), the hyperstriatum intercalatum supremum (HIS), and the hyperstriatum dorsale (HD). The four layers are not obvious in all avian species. The IHA is the main recipient of thalamic (i.e., sensory) input, whereas HA is the origin of the septomesencephalic tract. The rostral part of the Wulst receives somatosensory input, the caudal part visual input (Chapter 4). The septomesencephalic fibers arising from the rostral part of HA form the basal branch of this tract, the fibers

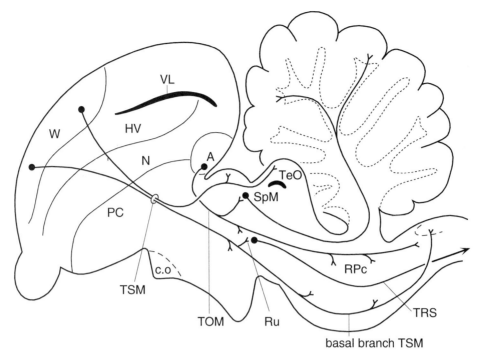

FIGURE 10 Sagittal view of brain with the trajectories of two extratelencephalic fiber systems. The *septomesencephalic tract* (TSM) derives from the Wulst (W); its dorsal branch projects to the dorsal thalamus and optic tectum (TeO), the basal branch has (among others) projections to the nu. ruber (Ru) and medial reticular formation. Ru is source of the rubrospinal tract (TRS) with projections in brainstem and spinal cord (see Chapter 5). The *occipitomesencephalic tract* (TOM) derives from archistriatum (A) and has (among others) projections to the medial spiriform nucleus (SpM, a relay to the cerebellar cortex) and to the parvocellular reticular formation (i.e., the premotor system of various motor systems).

from the caudal part form the dorsal branch (Figure 10; Karten, 1971; Verhaart 1971). The latter branch projects to cell groups in the dorsal thalamus and to the optic tectum (Karten *et al.*, 1973). The Wulst modifies through this pathway the excitability of cells in the optic tectum (Leresche *et al.*, 1983). The basal branch splits off after descending through the septum of the forebrain and runs caudally along the ventral margin of the brainstem. In its caudal course it sends projections to the prerubral area and, in the owl, to the nucleus ruber, to the medial spiriform nucleus, to the medial reticular formation and pontine nucleus and, in the owl and the parrot, also to the gracilis–cuneatus complex (Karten, 1971: owl; Zecha, 1962: parrot). The reticular formation and (pre-)rubral region are premotor centers for spinal functions, whereas the medial spiriform and pontine nuclei are relays for cerebellar input (Figure 10). Taking these connections into account the basal branch appears to convey information from telencephalic centers to motor control centers. In parrots this system is large and can be followed into the cervical cord. It has been speculated that it is comparable to the mammalian pyramidal or corticospinal tract, its importance in parrots having to do with the high prehensile abilities of their feet. However, there is no experimental evidence to support this claim (Webster *et al.*, 1990).

B. Archistriatum and Occipitomesencephalic Tract

The archistriatum is one of the regions derived from the dorsal ventricular ridge, other parts are the neostriatum and hyperstriatum ventrale (Figure 9). The neostriatum embraces telencephalic end stations of ascending sensory systems—the nucleus trigeminalis prosencephali or n. basalis (tactile sense head region), the ectostriatum of the visual system, and field L of the auditory system (see Chapters 1–2). Each of these sensory centers is, via relays in the neostriatum and hyperstriatum ventrale, connected with the archistriatum, a region in the temporal pole of the telencephalon. Such circuits have been described for the visual system (Ritchie, 1979) and the trigeminal system (Figure 11; Wild *et al.*, 1985; Dubbeldam and Visser, 1987) and can be considered to be sensorimotor circuits. Interruption of the visual circuit (e.g., bilateral ablation of the lateral neostriatum)

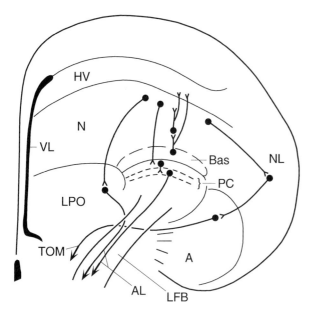

FIGURE 11 Example of intratelencephalic sensorimotor circuits: cells in nucleus basalis (trigeminal system, see Chapter 3) project to neostriatum (N) and hyperstriatum ventrale (HV). N-cells project to the paleostriatal complex (PC) and via a relay in lateral N (NL) to the archistriatum (A). PC and the lateral part of the lobus parolfactorius (LPO) are source of the ansa lenticularis (AL), A of the occipitomesencephalic tract (TOM). Modified after Dubbeldam and den Boer-Visser (1994).

1. Vocalization in Songbirds: A Special Case

Song in oscine passeriformes is a learned motor skill; the young birds have to learn the song from a tutor, generally a parent. In a series of studies Nottebohm and his coworkers clarified the organization of the neuronal substrate of this vocalization system (Figure 12). Two circuits can be distinguished (Nottebohm, 1993); one is essential for the maintenance of normal song (vocal control), the second has special importance for the learning process. The first circuit consists of a neostriatal region, often called the "high vocalization center" (HVC), with a core and a belt area and a nucleus robustus in the archistriatum (RA; Figure 12). HVC receives auditory input from the field L complex (see Chapter 2) and probably multisensory input from the thalamic uvaeform nucleus, partly directly and partly indirectly via an "interface" nucleus (not shown in figure). The second circuit is from HVC to area X in the parolfactory lobe, from here to the medial part of the dorsolateral thalamic nucleus (DLM), then to the lateral magnocellular nucleus of neostriatum (lMAN), and finally from here again to RA. Fibers from RA join the TOM and terminate directly on several centers connected with the respiration system (e.g., nucleus intercollicularis, ICo; Figure 12) and on the motor center of the syrinx (mXII; e.g., Wild, 1993b). Through these connections RA is in the right position to coordinate the activity of both respiratory and vocalization motor centers. Respiration has an important role in producing sounds. Manogue and Paton (1982) demonstrated that

causes a visuomotor deficit during feeding in pigeons: birds can still direct their peck, but have difficulties with grasping and ingesting grains (Jäger, 1990). Comparable effects can be found after disruption of the trigeminoarchistriatal pathway: feeding behavior deficits were found after bilateral damage to the ascending sensory trigeminal system (Zeigler and Karten, 1973). Sometimes this intratelencephalic circuit is called a "feeding circuit" (e.g., Dubbeldam, 1985).

The archistriatum can be subdivided into several parts; one of these is considered a sensorimotor region and is source of one of the large descending systems, the *occipitomesencephalic tract* (Zeier and Karten, 1971). In previous sections it was already mentioned that this fiber system has projections to, among other areas, the medial spiriform nucleus (a relay to the cerebellar cortex) and to the premotor systems of the jaw, tongue, and neck system (Figure 10). Apparently, disruption of the intratelencephalic pathways causes disturbance of motor performance through these projections to the premotor system. To demonstrate the importance of the intratelencephalic pathways we will have a closer look at a specific system, the vocalization system in songbirds.

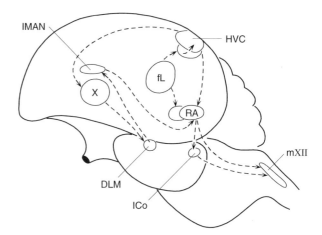

FIGURE 12 Vocalization circuits in the brain of a song bird. Explanation in text. Abbreviations: DLM, medial part of dorsolateral thalamic nucleus; fL, field L (auditory center); HVC, "high-vocalization center"; ICo, intercollicular nucleus; lMAN, lateral magnocellular nucleus of anterior neostriatum; mXII, hypoglossal motor nucleus (innervating syrinx); RA, nucleus robustus of archistriatum; X, area X in parolfactory lobe.

the effect of electrostimulation in HVC or RA on the activity of the syringeal motor nucleus is influenced by the respiration: the effect was much larger during expiration than during inspiration; apparently, respiratory centers in the brainstem have a gating function; that is, they control the activity of the syringeal part of the hypoglossal motor center.

The intratelencephalic vocalization system presents a unique system and many fascinating phenomena have been discovered about the properties of, in particular, the circuit including HVC and RA. A few will be mentioned. A first observation concerns the occurrence of auditory activity in several centers of the vocal control system; an important function may be auditory feed back, at least during learning (e.g., Doupe and Konishi, 1991). Deafening adult zebra finches does not disturb their song. Sutter and Margoliash (1994), studying synchronous activity of HVC units in zebra finches in response to autogenous song, found no clear topographic distribution of units responding to specific syllables. Most units reacted within a narrow range of time. McCashland (1987), recording single- and multiunit activity from the interface nucleus, HVC and RA, found that the units in these centers are sequentially activated before the onset of song. The correspondence of the pattern of activity in these cell areas and the timing of song elements suggests that these nuclei "generate or relay learned temporal cues for song" (McCashland, 1987). Because of their stereotyped songs zebra finches are very suitable for studying the effects of electrophysiological interference during song. In a series of experiments using chronically implanted electrodes into either the HVC, the RA, or the connecting fiber system, Vu *et al.* (1994) demonstrated that only stimulation in HVC affects the temporal pattern of song, whereas stimulation in RA may disrupt a syllable, but not the temporal pattern of song. Their conclusion is that RA is part of the premotor pathway, but that HVC is part of a telencephalic network responsible for producing the central motor program of the song. This network controls the syringeal activity directly via the projection from RA to the motor nucleus innervating the syringeal muscles (Figure 12). However exciting these observations may be, many details about the further organization of this network still have to be filled in.

C. Paleostriatal Complex and Ansa Lenticularis

The deepest ventromedial part of the avian telencephalon is the paleostriatal complex consisting of a paleostriatum augmentatum (PA), paleostriatum primitivum (PP), and a nucleus intrapeduncularis (INP), whereas also the lobus olfactorius (LPO) is considered part of this complex (Figures 9 and 10; Karten and Dubbeldam, 1973). It is a subpallial structure comparable to the basal ganglion of the mammals. Large parts of the Wulst and dorsal ventricular ridge send input to LPO and PA (Veenman *et al.*, 1995). PA projects to PP; the latter structure and LPO are the source of the third large extratelencephalic pathway, the ansa lenticularis. This fiber system has projections to several nuclei, including the lateral spiriform nucleus and the nucleus tegmenti pedunculopontinus (review in Reiner *et al.*, 1984). These nuclei have a substantial input to the optic tectum; probably this system plays an important role in visuomotor control. The n. tegmenti pedunculopontinus projects back to the paleostriatal complex; at least part of this projection is dopaminergic (uses the neurotransmitter dopamine). In this respect this cell group is comparable to the mammalian substantia nigra; in man, disturbance of this dopaminergic projection is the cause of pathological motor disturbances such as Parkinson's disease. A comparable phenomenon has not been described in birds. However, injections with neurotoxic substances (e.g., kainic acid) in the paleostriatal complex of pigeons do cause disturbances such as irregular jerky head movements and abnormal head postures (Rieke, 1980). In particular, in birds with damage to PP and INP, rotating movements toward the side of the damage were observed.

Through the multiple projections the paleostriatal complex receives not only input from the visual system, but also from other sensory systems. The input from the various sources may be used to coordinate head position or, more generally, body position with object location in space (Reiner *et al.*, 1984; Dubbeldam and den Boer-Visser, 1994). In particular, the lateral spiriform nucleus and deep tectal layers may be instrumental in the control of this type of movements.

VII. CONCLUDING REMARKS

It is clear that motor control is not the function of one specific region of the brain; on the contrary, large parts of the brain are involved. Each region appears to have a specific role, but there are still many unanswered questions about the nature of these roles. A further complication is that motor control systems for different activities may share part of their circuitry. This may be due partly to the need to coordinate the activity of certain motor systems. Anyhow, the picture arising from the preceding overview makes clear that the central nervous system is not composed of a number of independent parallel organized systems, but is an integrated system serving many functions.

Acknowledgments

Drs. R. G. Bout and A. J. Tellegen critically read an earlier version of this chapter and assisted in the final preparation of the manuscript. Mr. M. Brittijn prepared the final figures.

References

Albin, R. L., Sakurai, S. Y., Malowiec, R. L., and Gilman, S. (1991). Excitatory and inhibitory amino acid neurotransmitter sites in the cerebellar cortex of the pigeon (*Columba livia*). *J. Chem. Neuroanat.* **4,** 429–437.

Arends, J. J. A., and Dubbeldam, J. L. (1982). Exteroceptive and proprioceptive afferents of the trigeminal and facial motor nuclei in the mallard (*Anas platyrhynchos* L.). *J. Comp. Neurol.* **209,** 313–329.

Arends, J. J. A., and Voogd, J. (1989). Topographical aspects of the olivocerebellar system in the pigeon. *Exp. Brain Res.* **17**(Suppl.), 52–57.

Arends, J. J. A., and Zeigler, H. P. (1989). Cerebellar connections of the trigeminal system in the pigeon (*Columba livia*). *Brain Res.* **487,** 69–78.

Arends, J. J. A., and Zeigler, H. P. (1991). Organization of the cerebellum in the pigeon (*Columba livia*): II. Projections of the cerebellar nuclei. *J. Comp. Neurol.* **306,** 245–272.

Arends, J. J. A., Woelders-Blok, A., and Dubbeldam, J. L. (1984). The efferent connections of the nuclei of the descending trigeminal tract in mallard (*Anas platyrhynchos* L.). *Neuroscience* **13,** 797–817.

Arends, J. J. A., Allen, R. W., and Zeigler, H. P. (1991). Organization of the cerebellum in the pigeon (*Columba livia*): III. Corticovestibular connections with eye and neck premotor areas. *J. Comp. Neurol.* **306,** 273–289.

Berkhoudt, H., Klein, B. G., and Zeigler, H. P. (1982). Afferents to the trigeminal and facial motor nuclei in pigeon (*Columba livia* L.): Central connections of jaw motoneurons. *J. Comp. Neurol.* **209,** 301–312.

Bodnarenko, S. R., Rojas, X., and McKenna, O. C. (1988). Spatial organization of the retinal projection to the avian lentiform nucleus of the mesencephalon. *J. Comp. Neurol.* **269,** 431–447.

Bonaventure, N., Kim, M. S., and Jardon, B. (1992). Effects on the chicken monocular OKN of unilateral microinjections of $GABA_A$ antagonist into the mesencephalic structures responsible for OKN. *Exp. Brain Res.* **90,** 63–71.

Bout, R. G., and Dubbeldam, J. L. (1991). Functional morphological interpretation of the distribution of muscle spindles in the jaw muscles of the mallard (*Anas platyrhynchos*). *J. Morphol.* **210,** 215–226.

Bout, R. G., and Dubbeldam, J. L. (1994). The reticular premotor neurons of the jaw muscle motor nuclei in the mallard (*Anas platyrhynchos* L.). *Eur. J. Morphol.* **32,** 134–137.

Bout, R. G., and Zeigler, H. P. (1994). Jaw muscle (EMG) activity and amplitude scaling of jaw movements during eating in pigeon (*Columba livia*). *J. Comp. Physiol. A.* **174,** 433–442.

Brecha, N., and Karten, H. J. (1979). Accessory projections upon oculomotor nuclei and vestibulocerebellum. *Science* **203,** 913–916.

Brecha, N., Karten, H. J., and Hunt, S. P. (1980). Projections of the nucleus of the basal optic root in the pigeon: An autoradiographic and horseradish peroxidase study. *J. Comp. Neurol.* **189,** 615–670.

Britto, L. R. G., Hamassaki, D. E., Keyser, K. T., and Karten, H. J. (1989). Neurotransmitters, receptors, and neuropeptides in the accessory optic system: an immunohistochemical survey in the pigeon (*Columba livia*). *Vis. Neurosci.* **3,** 463–475.

Britto, L. R. G., Gasparotto, O. C., and Hamassaki, D. E. (1990). Visual telencephalon modulates directional selectivity of accessory optic neurons in pigeons. *Vis. Neurosci.* **4,** 3–10.

Bubień-Waluszewska, A. (1981). The cranial nerves. *In* "Form and function in birds" (A. S. King and J. McLelland, eds.), Vol. 2, pp. 385–438. Academic Press, London/New York.

Clarke, P. G. H. (1977). Some visual and other connections to the cerebellum of the pigeon. *J. Comp. Neurol.* **174,** 535–552.

Correia, M. J., Eden, A. R., Westlund, K. N., and Coulter, J. D. (1983). A study of the ascending and descending vestibular pathways in the pigeon (*Columba livia*) using anterograde transneuronal autoradiography. *Brain Res.* **278,** 53–61.

Cox, R. G., and Peusner, K. D. (1990). Horseradish peroxidase labelling of the efferent and afferent pathways of the avian tangential vestibular nucleus. *J. Comp. Neurol.* **296,** 324–341.

Doupe, A. J., and Konishi, M. (1991). Song-selective auditory circuits in the vocal control system of the zebra finch. *Proc. Natl. Acad. Sci. USA* **88,** 11339–11343.

Dubbeldam, J. L. (1984). Brainstem mechanisms for feeding in birds: Interaction or plasticity. *Brain Behav. Evol.* **25,** 85–98.

Dubbeldam, J. L. (1985). Neuronal circuits of the feeding system in birds. *Fortschr. Zool.* **30,** 273–275.

Dubbeldam, J. L., and den Boer-Visser, A. M. (1994). Organization of "feeding circuits" in birds: pathways for the control of beak and head movements. *Eur. J. Morphol.* **32,** 127–133.

Dubbeldam, J. L., and Bout, R. G. (1990). The identification of the motor nuclei innervating the tongue muscles in the mallard (*Anas platyrhynchos*): An HRP study. *Neurosci. Lett.* **119,** 223–227.

Dubbeldam, J. L., and Visser, A. M. (1987). The organization of the nucleus basalis-neostriatum complex of the mallard (*Anas platyrhynchos* L.) and its connections with the archistriatum and the paleostriatum complex. *Neuroscience* **21,** 487–517.

du Lac, S., and Lisberger, S. G. (1992). Eye movements and brainstem neuronal responses evoked by cerebellar and vestibular stimulation in chicks. *J. Comp. Physiol. A.* **171,** 629–638.

Fite, K. V. (1979). Optokinetic nystagmus and the pigeon visual system. *In* "Neural Mechanisms of Behavior in the Pigeon" (A. M. Granda and J. H. Maxwell, eds.), pp. 395–407. Plenum, New York/London.

Furber, S. E. (1983). The organization of the olivocerebellar projection in the chicken. *Brain Behav. Evol.* **22,** 198–211.

Gioanni, H., Rey, J., Villalobos, J., Bouyer, J. J., and Gioanni, Y. (1981). Optokinetic nystagmus in the pigeon (*Columba livia*). *Exp. Brain Res.* **44,** 362–370.

Goodman, D. C., Horel, J. A., and Freemon, F. R. (1964). Functional localization in the cerebellum of the bird and its bearing on the evolution of cerebellar function. *J. Comp. Neurol.* **123,** 45–54.

Heaton, M. B. (1981). The development of the oculomotor nuclear complex in the Japanese quail embryo. *J. Comp. Neurol.* **198,** 633–648.

Heaton, M. B., and Wayne, D. B. (1983). Patterns of extraocular innervation by the oculomotor complex in the chick. *J. Comp. Neurol.* **216,** 245–252.

Jäger, R. (1990). Visuomotor feeding perturbations after lateral telencephalic lesions in pigeons. *Behav. Brain Res.* **40,** 73–80.

Karten, H. J. (1971). Efferent connections of the Wulst of the owl. *Anat. Rec.* **169,** 353.

Karten, H. J., Hodos, W., Nauta, W. J. H., and Revzin, A. M. (1973). Neural connections of the "visual Wulst" of the avian telencephalon. Experimental studies in the pigeon (*Columba livia*) and owl (*Speotyto cunicularia*). *J. Comp. Neurol.* **150**, 253–278.

Labandeira-Garcia, J. L., Guerra-Seijas, M. J., Segade, L. A. G., and Suarez-Nuñez, J. M. (1987) Identification of abducens motoneurons, accessory abducens motoneurons, and abducens internuclear neurons in the chick by retrograde transport of horseradish peroxidase. *J. Comp. Neurol.* **259**, 140–149.

Labandeira-Garcia, J. L., Guerra-Seijeas, M. J., Labandeira-Garcia, J. A., and Jorge-Barreiro, F. J. (1989). Afferent connections of the oculomotor nucleus in the chick. *J. Comp. Neurol.* **282**, 523–534.

Leresche, N., Hardy, O., and Jassik-Gerschenfeld, D. (1983). Receptive field properties of single cells in the pigeon's optic tectum during cooling of the "visual Wulst." *Brain Res.* **267**, 225–236.

Maier, A. (1992). The avian muscle spindle. *Anat. Embryol.* **186**, 1–26.

Manni, E., Bortolami, R., and Azzena, G. B. (1965). Jaw muscle proprioception and mesencephalic trigeminal cells in birds. *Exp. Neurol.* **12**, 320–328.

Manogue, K. R., and Paton, J. A. (1982). Respiratory gating of activity in the avian vocal control system. *Brain Res.* **247**, 383–387.

Marr, D. (1969). A theory of cerebellar cortex. *J. Physiol.* **202**, 437–470.

McCashland, J. S. (1987). Neural control of bird song production. *J. Neurosci.* **7**, 23–39.

McKenna, O. C., and Wallman, J. (1985). Accessory optic system and pretectum of birds: Comparisons with those of other Vertebrates. *Brain Behav. Evol.* **26**, 91–116.

Nauta, W. J. H., and Karten, H. J. (1970). A general profile of the vertebrate brain, with sidelights on the ancestry of cerebral cortex. *In* "The Neurosciences: The Second Study Program" (F. O. Schmidt, ed.), pp 7–26. Rockefeller Univ. Press, New York.

Nottebohm, F. (1993). The search for neural mechanisms that define the sensitive period for song learning in birds. *Neth. J. Zool.* **43**, 193–234.

Passatore, M., Bortolami, R., and Manni, E. (1979). Somatotopic arrangement of the proprioceptive afferents from the jaw muscles in the mesencephalic trigeminal nucleus of the duck. *Arch. Ital. Biol.* **117**, 123–139.

Reiner, A., Brauth, S. E., and Karten, H. J. (1984). Evolution of the amniote basal ganglia. *Trends Neurosci.* **7**, 320–325.

Reiner, A., Brecha, N., and Karten, H. J. (1979). A specific projection of retinal displaced ganglion cells to the nucleus of the basal optic root in the chicken. *Neuroscience* **4**, 1679–1688.

Reiner, A, and Karten, H. J. (1982). Laminar distribution of the cells of origin of the descending tectofugal pathways in the pigeon (*Columba livia*). *J. Comp. Neurol.* **204**, 165–187.

Rieke, G. K. (1980). Kainic acid lesions of pigeon paleostriatum: A model for study of movement disorders. *Physiol. Behav.* **24**, 683–687.

Romer, A. S., and Parssons, T. S. (1977). "The Vertebrate Body." Saunders, Philadelphia.

Sohal, G. S., and Holt, R. K. (1978). Identification of the trochlear motoneurons by retrograde transport of horseradish peroxidase. *Exp. Neurol.* **59**, 509–514.

Steeves, J. D., Sholomenko, G. N., and Webster, D. M. S. (1987). Stimulation of the pontomedullary reticular formation initiates locomotion in decerebrate birds. *Brain Res.* **401**, 205–212.

Tellegen, A. J., and Dubbeldam, J. L. (1994) Location of premotor neurons of the motor nuclei innervating craniocervical muscles in the mallard (*Anas platyrhynchos* L.). *Eur. J. Morphol.* **32**, 138–141.

van Gennip, E. M. S. J., and Berkhoudt, H. (1994). Ontogeny of feeding in pigeons: A cinefluoroscopic and electromyographic analysis. *J. Exp. Zool.* **269**, 489–506.

Veenman, C. L., Wild, J. M., and Reiner, A. (1995). Organization of the avian "corticostriatal" projection system: A retrograde and anterograde pathway tracing study in pigeons. *J. Comp. Neurol.* **354**, 87–126.

Verhaart, W. J. C. (1971). Forebrain bundles and fibre systems in the avian brainstem. *J. Hirnforsch.* **13**, 39–64.

Vu, E. T., Mazurek, M. E., and Kuo, Y.-C. (1994). Identification of a forebrain motor programming network for the learned song of zebra finches. *J. Neurosci.* **14**, 6924–6934.

Wallman, J., Velez, J., Weinstein, B., and Green, A. E. (1982). Avian vestibuloocular reflex: Adaptive plasticity and developmental changes. *J. Neurophysiol.* **48**, 952–967.

Webster, D. M. S., Rogers, L. J., Pettigrew, J. D., and Steeves, J. D. (1990). Origins of descending spinal pathways in prehensile birds: Do parrots have a homologue to the corticospinal tract of mammals? *Brain Behav. Evol.* **36**, 216–226.

Welsh, J. P., Lang, E. J., Sugihara, I., and Llinás, R. (1995). Dynamic organization of motor control within the olivocerebellar system. *Nature* **374**, 453–457.

Wentzel, P. R., de Zeeuw, C. L., Holstege, J. C., and Gerrits, N. M. (1995). Inhibitory synaptic inputs to the oculomotor nucleus from vestibulo-ocular-reflex-related nuclei in the rabbit. *Neuroscience* **65**, 161–174.

Wild, J. M. (1993a). The avian nucleus retroambigualis: A nucleus for breathing, singing and calling. *Brain Res.* **606**, 119–124.

Wild, J. M. (1993b). Descending projections of the songbird nucleus robustus archistriatalis. *J. Comp. Neurol.* **338**, 225–241.

Wild, J. M., and Zeigler, H. P. (1980). Central representation and somatotopic organization of the jaw muscles within the facial and trigeminal nuclei in the pigeon (*Columba livia*). *J. Comp. Neurol.* **192**, 175–201.

Wild, J. M., Arends, J. J. A., and Zeigler, H. P. (1985). Telencephalic connections of the trigeminal system in the pigeon (*Columba livia*): A trigeminal sensorimotor circuit. *J. Comp. Neurol.* **234**, 441–464.

Winterson, B. J., and Brauth, S. E. (1985). Direction-selective single units in the nucleus lentiformis mesencephali of the pigeon (*Columba livia*). *Exp. Brain Res.* **60**, 215–226.

Wold, J. E. (1978). The vestibular nuclei in the domestic hen (*Gallus domesticus*). III. Ascending projections to the mesencephalic eye motor nuclei. *J. Comp. Neurol.* **179**, 393–406.

Wylie, D. R., and Frost, B. J. (1990). Binocular neurons in the nucleus of the basal optic root (nBOR) of he pigeon are selective for either translational or rotational visual flow. *Vis. Neurosci.* **5**, 489–495.

Zecha, A. (1962) The "pyramidal tract" and other telencephalic efferents in birds. *Acta Morphol. Neerl.-Scand.* **5**, 194–195.

Zeier, H., and Karten, H. J. (1971). The archistriatum of the pigeon: Organization of afferent and efferent connections. *Brain Res.* **31**, 313–326.

Zeigler, H. P., and Karten, H. J. (1973). Brain mechanisms and feeding behavior in the pigeon (*Columba livia*). II. Analysis of feeding behavior deficits following lesions of quintofrontal structures. *J. Comp. Neurol.* **152**, 83–102.

Zeigler, H. P., Levitt, P., and Levine, R. R. (1980) Eating in the pigeon: Response topography, stereotopy and stimulus control. *J. Comp. Physiol. Psychol.* **94**, 783–794.

Zweers, G. A. (1982) Pecking of the pigeon (*Columba livia*). *Behaviour* **81**, 173–230.

CHAPTER 7

The Autonomic Nervous System of Avian Species

WAYNE J. KUENZEL

Department of Animal and Avian Sciences
University of Maryland
College Park, Maryland 20742

I. Introduction 101
II. Components 102
    A. General Description 102
III. Peripheral Motor Components of the Autonomic Nervous System 102
    A. Sympathetic Nervous System 102
    B. Parasypathetic Nervous System 104
IV. Central Components of the Autonomic Nervous System 105
    A. Sympathetic Nervous System 105
    B. Parasympathetic Nervous System 105
    C. Hypothalamopituitary and Hypothalomo-ANS Systems 108
    D. Other Systems Involving Circumventricular Organs (CVOs) 111
V. Functional Neural Pathways of the Autonomic Nervous System 112
    A. An Autonomic Pathway Regulating Food Intake 112
    B. Visceral Forebrain System 114
    C. Limbic System 115
VI. Shifts in Homeostasis: An Autonomic Hypothesis Explaining the Regulation of Annual Cycles of Birds 117
VII. Consequences of Genetic Selection for Rapid Growth in Broilers and Turkeys: An Imbalanced Autonomic Nervous System 118
    References 118

I. INTRODUCTION

The autonomic nervous system (ANS) of birds, as in mammals, has traditionally been described as a network of nerves innervating smooth and cardiac muscle, striated intrinsic muscle of the eyes, and some types of secretory cells (Bennett, 1974). It is usually presented anatomically as a motor or effector system that innervates a large number of body organs known as viscera within the thoraco-abdominal-pelvic body cavity and cranium. Functionally it is an important neural system intimately involved in many basic, physiological functions and behaviors critical to the survival of an organism, and hence, its species. Specific examples of its direct regulation of physiological processes include respiration, cardiovascular function, thermoregulation, osmotic balance, gastrointestinal movements, and circadian rhythms. Examples of the latter are the regulation of feeding, sexual behavior, and diurnal and circannual rhythmic behaviors displayed by many, if not all, avian species. Manifestation of the behaviors just listed require emotions, motivation, and/or cognitive function for appropriate execution. It is clear that the ANS responsible for regulating the previously listed physiological functions and behaviors involves more than an extensive network of ganglia and peripheral nerves outside the confines of the central nervous system. In effect the system regulates processes that involve conscious acts as well as basic ones which function automatically; therefore, it must include all major parts of the nervous system. The purpose of this chapter is to describe both central and peripheral components of the avian ANS and functions controlled by the system, with particular emphasis on the ability of birds to display homeostasis and dynamic shifts in homeostatic regulation throughout their annual cycles to maximize their fitness.

II. COMPONENTS

A. General Description

The principal components of the ANS include the sympathetic (thoracolumbar), parasympathetic (craniosacral), and enteric nervous systems (Langley, 1921). The sympathetic and parasympathetic systems innervate the same end organs but, in general, liberate different neural transmitter substances and have generally opposing actions on visceral functions. It has been suggested that the terms sympathetic and parasympathetic should be discarded with respect to their use in the avian nervous system since deviations from the traditional anatomical separation have been reported (Bennett, 1974). On the other hand, it is the opinion of this author and others (Akester, 1979) that the traditional scheme found in the mammalian literature is a very useful framework from which students can develop an anatomical and functional understanding of the nervous system to justify its retention. Therefore it should be noted that the following is clearly a simplification. The scheme, however, should serve to synthesize past data and, more importantly, suggest hypotheses which can be tested in the future.

The principal components of the ANS reside mostly in the viscera, hence they are regarded as predominantly peripheral and motor. Figure 1 shows a schematic diagram of the principal peripheral or motor components of the ANS. The sympathetic nervous system is depicted on the left while the parasympathetic nervous system is shown on the right. Note that both the sympathetic and parasympathetic nervous systems consist of a number of two-neuron complexes: a preganglionic and postganglionic neuron. The former resides within the central nervous system (brain or spinal cord). Therefore the preganglionic neurons should be regarded as central components of the ANS. A number of ganglia have been identified and examined in birds. The cranial (superior) cervical ganglion (g.) is an important sympathetic g. that contains neurons which project to the pineal as well as to other organs (Figure 1). Major parasympathetic g. include the ciliary, ethmoid, sphenopalatine, geniculate, and chorda tympani (Figure 1). Detailed descriptions of each can be found in Akester (1979) and Bubien-Waluszewska (1981). A second constituent (not shown) is a set of sensory (visceral afferent) fibers from the heart, lungs, gastrointestinal tract, and other viscera that projects to specific subnuclei within the nucleus (n.) tractus solitarius (nTS). The nTS will be discussed in a later section, as it is a major interface between the peripheral and central nervous system regarding the ANS. A third constituent of the ANS is a series of mono- and multisynaptic connections to the preganglionic neurons of the ANS which reside within the brain and comprise functional neural pathways of the ANS. Two prominent systems directly connect to and therefore are constituents of the ANS: the visceral forebrain system and the limbic system. The two will be discussed near the end of the chapter. A final component of the ANS is a series of circumventricular organs (CVOs) that will be referred to as the neuroendocrine/paracrine portion of the ANS. Circumventricular organs contain specialized ependymal cells, neurons, and neuropeptides, the latter of which can either be released into or taken up from the cerebrospinal fluid. In addition to neurons that make point-to-point contact via synapses, CVOs have elements that can release substances into the cerebrospinal fluid or vasculature and affect target structures located at a distance thereby serving more of a paracrine and neuroendocrine role. Two CVOs that have received considerable attention are the median eminence (the interface between the brain and pituitary gland) and the pineal gland. Refer to Chapter 16 on the pituitary gland and Chapter 21 on the pineal for more detail on the neuroendocrinology and function of these important organs. Figure 2 shows the location of CVOs currently identified in avian species.

III. PERIPHERAL MOTOR COMPONENTS OF THE AUTONOMIC NERVOUS SYSTEM

A. Sympathetic Nervous System

The sympathetic nervous system of birds and mammals consists primarily of a chain of ganglia on each side of the spinal cord at the level of the thoracic and lumbar regions (Figure 1). Since the thoracic and lumbar vertebrae are highly fused in birds due to their adaptation for flight (Gill, 1995), individual vertebra are not numbered and therefore ganglia found within the paravertebral chain are not numbered as well. Individual ganglia, however, have been clearly and carefully described by Hsieh (1951) and Akester (1979). A general scheme that occurs in both mammals and birds is that the preganglionic component is a short neuron that originates in the spinal cord and projects to a ganglion. Within the ganglion, a major neurotransmitter released by the terminal region of the neuron is acetylcholine (ACh). Acetylcholine is synthesized from choline and acetyl CoA as shown in Figure 3. Choline is a vitamin and is required in the diet for adequate ACh production and for the prevention of perosis or twisted leg problems in poultry (Scott et al., 1969). The enzyme acetylcholinesterase is available to inactivate ACh such that postganglionic neurons do not continuously respond to the presence of ACh (Kandel et al., 1991). Nicotinic recep-

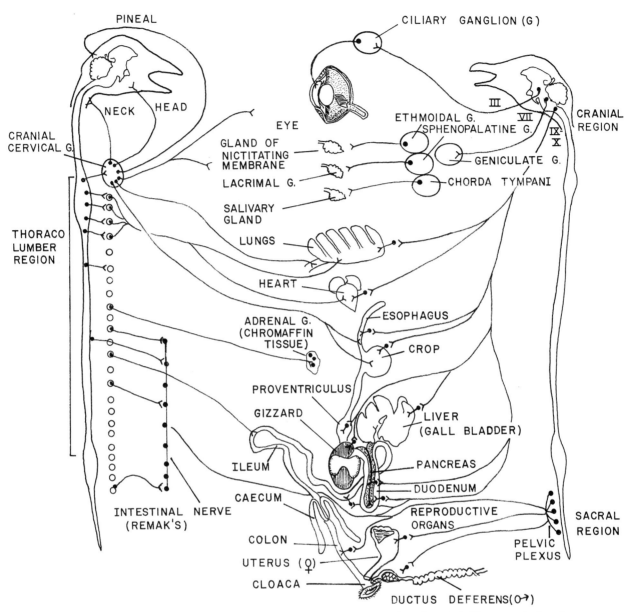

FIGURE 1 The efferent component of the autonomic nervous system (ANS) of birds. The sympathetic and parasympathetic arms of the ANS are shown on the left and right sides, respectively.

tors are present on postganglionic neurons found within each ganglion. Postganglionic neurons usually have longer processes compared to preganglionic ones and project to a target organ within the viscera. A major neurotransmitter released from postganglionic neurons is norepinephrine (NE) and the receptors specific for NE are α- or β-receptors (Cooper *et al.*, 1991; Kandel *et al.*, 1991). The precursor of NE is an essential amino acid, tyrosine, that must be provided in poultry rations (Scott *et al.*, 1969). Tyrosine is then converted to a series of catecholamines beginning with L-dopa and ending with epinephrine. The metabolic pathway for NE synthesis can be found in Figure 4 (modified from Cooper *et al.*, 1991). As depicted in Figure 1, there are exceptions to the generalized scheme of a short preganglionic neuron projecting to a chain of ganglia found on either side of the spinal cord and a long postganglionic neuron. In addition, avian species have specialized ganglia found within the viscera. One unusual and important group of ganglia found in birds is known as the intestinal or

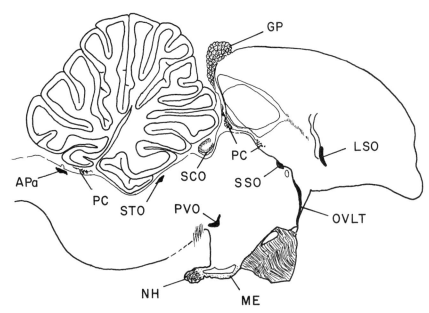

FIGURE 2 Location of circumventricular organs in a sagittal view of the chick brain near midline. APa, area postrema; GP, pineal gland; LSO, lateral septal organ; ME, median eminence; NH, neurohypophysis; OVLT, organum vasculosum of the lamina terminalis; PC, choroid plexus; PVO, paraventricular organ; SCO, subcommissural organ; SSO, subseptal organ; and STO, subtrochlear organ.

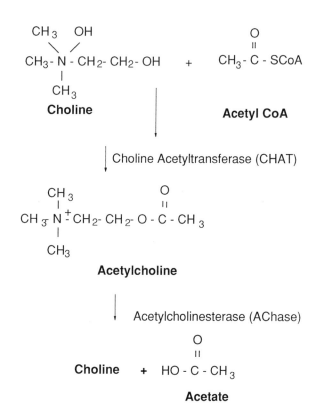

FIGURE 3 The synthesis and degradation of acetylcholine within the autonomic nervous system.

Remak's nerve (Figure 1). The ganglionated, intestinal nerve is found along the entire length of the alimentary canal from the duodenum to cloaca. It is a mixed sympathetic and parasympathetic nerve (Akester, 1979).

B. Parasympathetic Nervous System

The parasympathetic nervous system consists of nerves originating either from the brain or sacral region of the spinal cord (right side of Figure 1) that project to ganglia usually located either near or in end organs. Preganglionic neurons from the brain are found in motor nuclei of the oculomotor (III), facial (VII), glossopharyngeal (IX), and vagus (X) nerves (Akester, 1979; Dubbeldam, 1993). Preganglionic neurons in the sacral region are found in the nucleus pudendus or n. pelvinus. Parasympathetic neurons in the sacral region innervate reproductive organs, colon, and cloaca (Ohmori et al. 1984). Their fibers form part of the pelvic plexus (Figure 1) and project to specific body organs. In contrast to the preganglionic nerves of the sympathetic nervous system, preganglionic neurons of the parasympathetic nervous system have long processes and terminate in a ganglion juxtapositioned to the targeted visceral organ. Similar to the sympathetic nervous system, the preganglionic nerve cell releases ACh. Within each ganglion, postganglionic neurons contain nicotinic receptors specific for ACh. Each postganglionic neuron has short processes and, similar to its preganglionic neuron, re-

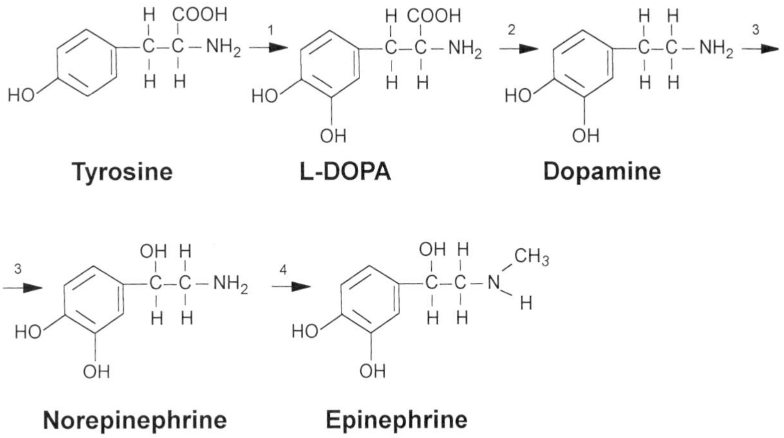

FIGURE 4 The synthesis of several catecholamines from the essential amino acid tyrosine. The key enzymes in the pathway include: 1 (tyrosine hydroxylase), 2 (aromatic amino acid decarboxylase, 3 (dopamine-β-hydroxylase), and 4 (phenylethanolamine-N-methyl transferase).

leases ACh. The receptors found within the targeted visceral organ are muscarinic (Kandel et al., 1991).

IV. CENTRAL COMPONENTS OF THE AUTONOMIC NERVOUS SYSTEM

A. Sympathetic Nervous System

Cell bodies or perikarya of preganglionic neurons of the sympathetic nervous system of mammals form a distinct column within the thoracolumbar region of the spinal cord known as the intermediolateral cell column (IML; Kandel et al., 1991). The avian homolog of this cell column is known as the n. intermediomedialis, pars commissurae dorsalis (Breazile and Hartwig, 1989). Formerly the n. was known as the column of Terni (Terni, 1923; Hosoya et al., 1992). It is the dorsal commissural portion of the intermediate nucleus and is shown as the cell group directly dorsal to the central canal of the spinal cord (Figure 5). Neurons from the brain project to the various levels of the groups of preganglion neurons that are found from the last cervical segment, through all of the thoracic, lumbar, and the first or second synsacral segments of the spinal cord (Macdonald and Cohen, 1973). The major groups of projection neurons from the brain to the nuclei intermediomedialis (nIMM) originate from the central gray as shown in Figure 6. Note that the substantia grisea centralis or central gray (GCt) first appears at the level of the midbrain or mesencephalon (Figure 6A, plane A5.0) and continues through the level of the pons in the chick brain(Figure 6B, plane A1.6), for a total extent of 3.4 mm of neural tissue (Kuenzel and Masson, 1988). It therefore comprises a large number of neurons providing input into the extensive series of preganglionic neurons throughout the thoracolumbar region of the spinal column. Note, however, that retrograde and anterograde tracing studies need to be completed in birds to substantiate the existence of this pathway (GCt to nIMM). In mammals, major projections can be traced to the central gray from the hypothalamus, particularly the ventromedial hypothalamic nucleus (VMN, Krieger et al., 1979; Saper et al., 1976). The findings are significant in that the VMN can then be regarded as forming part of a sympathetic neuroanatomical pathway that has importance regarding homeostasis and energy intake (Luiten et al., 1987). The pathway will be discussed later under the section entitled "Functional Neural Pathways of the Autonomic Nervous System." It should be noted that the projection from the VMN (also referred to as the n. medialis hypothalami posterioris) to the GCt of birds, although present, is not a major projection system compared to that in mammals (Berk and Butler, 1981). Therefore the bird may have a polysynaptic pathway from the VMN to the nIMM rather than a two-neuron pathway (VMN to GCt, GCt to IML) as demonstrated in mammals. There also exists a direct projection from the paraventricular n. to the thoracic spinal cord and therefore presumably to the column of Terni or nIMM (Berk and Finkelstein, 1983).

B. Parasympathetic Nervous System

The effector portion of the parasympathetic nervous system in the brain of birds is illustrated in Figure 7. The accessory oculomotor n. (one of several motor n. of cranial nerve III), considered equivalent to the Edinger-Westphal n. of mammals (Niimi et al., 1958; Heaton and Wayne, 1983; Breazile and Kuenzel, 1993) and the motor n. of the VIIth, IXth, and Xth cranial nerves

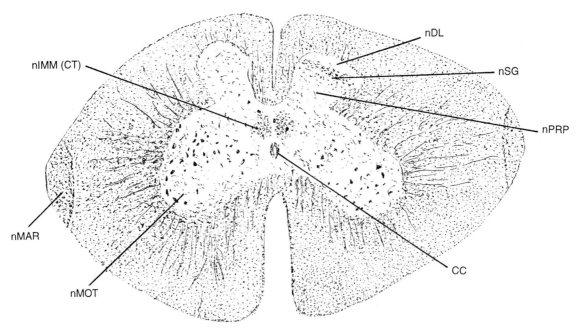

FIGURE 5 A cross section of the spinal cord at the level of thoracic vertebrae. The origin of preganglionic neurons of the sympathetic nervous system occur within the intermediomedial nucleus, also termed the column of Terni. CC, central canal; nDL, dorsolateral nucleus; nIMM, intermediomedial nucleus (column of Terni); nMAR, marginal nucleus; nMOT, motor nucleus; nPRP, nucleus proprius; and nSG, nucleus substantia gelatinosa.

contain preganglionic parasympathetic neurons (Figure 7). In addition, the ventrolateral medulla contains parasympathetic neurons involved in controlling the cardiovascular system (Figure 7). Of the specific brain areas shown in Figure 7, the most important brain components of the parasympathetic nervous system are the dorsal motor nucleus of the vagus (dMnX) and its associated n. tractus solitarius (nTS). The dorsal vagal complex,

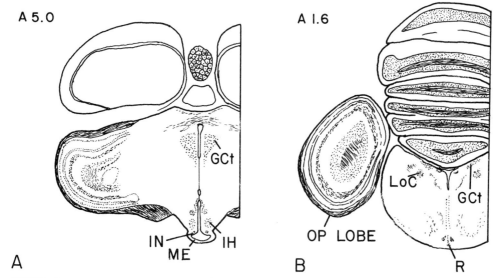

FIGURE 6 The extent of the central gray (GCt) within the brain of the chick. It first appears within the midbrain (A) at the level of plate A5.0 (5.0 mm anterior to zero reference plane; Kuenzel and Masson. *A Stereotaxic Atlas of the Brain of the Chick* (*Gallus domesticus*), © 1988. Johns Hopkins University Press). It then courses over a distance of 3.4 mm and terminates in the pons (B). In mammals (and presumably birds) it contains neurons that are part of the sympathetic nervous system (SNS) which project to preganglionic neurons of the SNS within the thoracolumbar region of spinal cord. IH, inferior hypothalamic nucleus (n); IN, infundibular n.; LoC, locus ceruleus; ME, median eminence; OP LOBE, optic lobe; and R, raphe n.

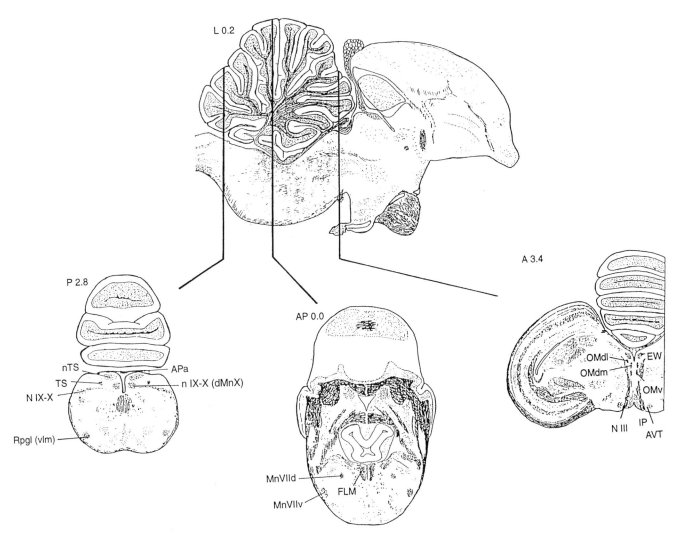

FIGURE 7 Sagittal section of brain near midline showing the origin of preganglionic neurons of the parasympathetic nervous system (PNS). Each vertical line shows the approximate brain region from where the three displayed cross sections were taken. The most rostral section (section A3.4) shows the Edinger Westphal n. (EW), the source of preganglionic neurons of cranial nerve III that innervate the eye and regulate pupil size. The midsection (section AP0.0) shows the origin of preganglionic nerves of cranial nerve VII, which project to the gland of the nictitating membrane, lacrimal gland and salivary glands. The hind-section (section P2.8) shows the most important components of the PNS, (1) the nucleus tractus solitarius (nTS) and (2) the source of preganglionic neurons of cranial nerves IX and X, which project to the heart, lungs, and most portions and organs associated with the digestive tract. In addition, parasympathetic neurons that affect heart and respiratory function are located in the ventrolateral medulla (Rpgl, vlm). APa, area postrema; AVT, area ventralis of Tsai; EW, Edinger-Westphal nucleus; FLM, fasciculus longitudinalis medialis; IP, interpeduncular n.; MnVIId and MnVIIv, dorsal and ventral motor n. of the facial nerve; NIII, third cranial nerve; NIX-X, ninth and tenth cranial nerves; nIX-X, glossopharyngeal and dorsal motor n. of the vagus; nTS, n. tractus solitarius; OMdm, OMdl and OMv, dorsomedial, dorsolateral and ventral oculomotor nerve n.; Rpgl (vlm), n. reticularis paragigantocellularis lateralis (ventrolateral medulla); TS, tractus solitarius.

comprised of two complex nuclear groups, the dMnX and nTS, has been studied extensively in the pigeon *Columba livia* (Figure 7). Katz and Karten (1979, 1983a,b, 1985) subdivided the dMnX and nTS into at least 12 and 19 subnuclei, respectively. Tract-tracing and neuropeptide anatomical studies have been conducted to establish possible functions of subnuclei. For both birds and mammals, the nTS receives primary afferent fibers conveying taste or gustatory and visceral sensory information from the facial, glossopharyngeal, and vagal nerves (Beckstead and Norgren, 1979; Dubbeldam, 1984; Dubbeldam *et al.*, 1976, 1979). Gustatory afferents terminate in rostral portions of the nTS while visceral afferents project to more caudal regions of the nucleus. The complex nTS can also be topographically organized into functional medial and lateral tiers. Gastrointestinal afferents originating from the thoracic and abdominal cavities project upon the medial tier while cardiovascular (heart and aortic arch) and pulmonary (lungs) afferents project upon the lateral tier (Katz and Karten, 1979; 1983a,b). Directly ventral to the nTS is the dMnX, a major source of preganglionic motor neurons of the parasympathetic nervous system (Figure 7). The nTS therefore is strategically located to either relay afferent information to upper brainstem, mid- and forebrain structures (discussed under the section "Functional Neural Pathways of the Autonomic Nervous System") or immediately relay sensory information via vagovagal reflexes directly to the dMnX (Berk and Smith, 1994). Of interest is that within the medial tier of the nTS, there is a rostrocaudal topographic organization of afferents from the gastrointestinal tract that matches the rostrocaudal topography of the gastrointestinal tract. This is paralleled by a partial topographic organization within the dMnX with respect to its series of preganglionic neurons innervating target viscera (Katz and Karten, 1985; Schwaber and Cohen, 1978). Another important anatomical characteristic of the nTS is the presence of the taenia choriodea and the lateralis superficialis of the taenia choroidea which are suggested homologs of the mammalian area postrema (Berk, 1991). The area postrema (APa) is a circumventricular organ (Figures 2, 7, and 9C) and contains modified glial cells and neurons which not only sense constituents within the cerebrospinal fluid but also may release neuromodulators into the ventricular system thereby having a paracrine function (Vigh, 1971; Leonhardt, 1980; Vigh-Teichmann and Vigh, 1983).

The other source of preganglionic parasympathetic neurons that has not been studied rigorously in birds are those responsible for projecting to the pelvic plexus (Figure 1). The pelvic plexus contains both sympathetic and parasympathetic neurons situated within the mesentery supporting the rectum and cloaca. Together with the intestinal nerve (Remak's nerve) the plexus regulates movements in the distal parts of the alimentary canal, ureters, oviduct in the female, and ductus deferens and erectile vascular bodies of the male (Akester, 1979).

C. Hypothalamopituitary and Hypothalamo-ANS Systems

The preoptic–hypothalamic system of birds consists of at least 21 distinct nuclei as described in past stereotaxic atlases (van Tienhoven and Juhász, 1962; Feldman et al., 1973; Youngren and Phillips, 1978; Karten and Hodos, 1967; Zweers, 1971; Stokes et al., 1974; Baylé et al., 1974; Vowles et al., 1975; Kuenzel and van Tienhoven, 1982; Kuenzel and Masson, 1988). Certain nuclei have been shown to contain specific neuropeptides affecting the ANS and have projections to the median eminence, neurohypophysis, dorsal vagal complex, or midbrain central gray and therefore are central components of autonomic pathways. The following is a brief, selected list of nuclei of importance to ANS function.

The medial preoptic n. (POM) within the preoptic area (POA) has been shown to be a sex-related dimorphic nucleus in birds (Viglietti-Panzica et al., 1986; Panzica et al., 1987). It occurs within the central region of the POA and terminates at the cross-sectional plane of the quail brain that shows the presence of the anterior commissure (Figures 8A, 8B). The POM has been demonstrated to be approximately 40% larger in male than female Japanese quail. Further analysis has revealed two cytoarchitectural differences in the POM: dorsolateral and medial subnuclei. Only the former showed hypertrophy of neurons and an overall increase in volume following high levels of testosterone in males but not females (Panzica et al., 1991). Numerous past studies have supported the role of the preoptic area (POA) in sexual behavior and reproductive function. Electrolytic lesions destroying the POA, including the POM, resulted in a decrease in gonadal function (Ralph, 1959; Ralph and Fraps, 1959) and a marked decline in sexual behavior (Barfield, 1965; Meyer and Salzen, 1970; Haynes and Glick, 1974). When androgen was implanted directly into the POA, it activated copulatory behavior in roosters (Barfield, 1969). Autoradiography was used to first show testosterone-binding sites within the POA and other brain areas of the chicken (Meyer, 1973). A polyclonal antibody developed from a synthetic portion of the androgen receptor and immunocytochemistry have been used to confirm the presence of androgen receptors in the POM of Japanese quail (Balthazart et al., 1992). In addition, the conversion of testosterone to estradiol is thought to be critical for the interaction of gonadal steroids with neurons within the brain and activation of copulatory behavior. The enzyme responsible for the conversion of testosterone to estradiol is aromatase. Aromatase-immunoreactive neurons were found within the POM as well as estrogen receptors. Of interest was that few neurons within the POM showed colocalization of aromatase and estrogen receptors (Balthazart et al., 1991). Recently quail aromatase cDNA was utilized to produce a recombinant protein from which a polyclonal antibody was produced. The more specific antibody was utilized in immunocytochemistry and confirmed previous work identifying ir neurons within the POM. More ir neurons for aromatase were found in the quail POA than were found previously utilizing a polyclonal antibody to human aromatase (Foidart et al., 1995). An earlier study (Berk and Butler,

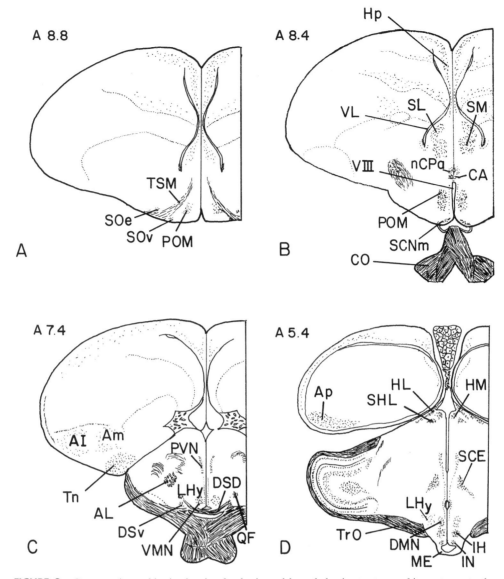

FIGURE 8 Cross sections of brain showing forebrain and hypothalamic structures of importance to the ANS. Plates modified from a stereotaxic atlas of a chick brain (Kuenzel and Masson. *A Stereotaxic Atlas of the Brain of the Chick (Gallus domesticus)*, © 1988. Johns Hopkins University Press). The letter "A" and number in the upper left-hand corner of each brain section shows the distance in millimeters anterior to a zero reference point located in the hindbrain. (A) Depicts the beginning of the preoptic area, medial to the septomesencephalic tract (TSM). (B) The septal area (SM,SL) is shown medial to the ventral portion of the lateral ventricle (VL); the hypothalamus is located ventral to the anterior commissure (CA), dorsal to the optic chiasma (CO) and lateral to the third ventricle (VIII). (C) Section showing the midregion of hypothalamus dorsal to the dorsal supraoptic decussation (DSD). The avian amygdala comprises the n. taeniae (Tn), medial and posterior archistriatum (Am, Ap). (D) Section showing the caudal hypothalamus medial to the optic tract (TrO). Other abbreviations: AL, ansa lenticularis; DMN, dorsomedial hypo. n.; DSv, SCNm, two proposed avian suprachiasmatic n.; HL, HM, SHL, habenular complex; IH, IN, inferior and infundibular hypo., n.; LHy, lateral hypo. area; ME, median eminence; nCPa, bed n. pallial commissure; POM, med. preoptic n.; PVN, paraventricular n.; QF, quintofrontal tract; SCE, stratum cellulare externum; SOe, SOv, supraoptic n.; VMN, ventromedial hypothalamic n.

1981) examining efferent pathways of the POM reported that it contained neurons that projected to the area ventralis of Tsai, n. intercollicularis and central gray (Figure 6) of the midbrain. The latter, in particular, suggests that the POM has input into a structure that is regarded as a central component of the sympathetic nervous system.

The paraventricular n. (PVN) is a complex one and in mammals it comprises 8 to 10 subnuclei (Swanson and Kuypers, 1980). In the chick, it is a very heterogeneous structure consisting of 10 subnuclei (Kuenzel, 1994a). The magnocellular portion of the nucleus extends from a rostral position directly beneath the anterior commissure for approximately 2.0 mm in a caudal direction (Kuenzel and Masson, 1988). The cells project to the neurohypophysis, contain arginine vasotocin and mesotocin, and are involved in water balance and contraction of smooth muscle (Berk and Butler, 1981; Berk and Finkelstein, 1983; Goossens et al., 1977; Bons, 1980). The large cells of the PVN are therefore comparable to those found within the supraoptic n. (SOe, SOv; Figures 8A, and 8C). The PVN of the pigeon, however, also has neurons which project to the external zone of the median eminence and therefore the PVN serves an additional function to regulate anterior pituitary function (Berk and Finkelstein, 1983). In mammals, parvocellular neurons within the PVN contain corticotropin releasing hormone (CRH) and project to the external zone of the median eminence (Lechan et al., 1980). The CRH neurons not only regulate adrenal function, they also affect metabolism and food intake. A detailed account of the PVN and two other nuclei, the ventromedial hypothalamic n. and the lateral hypothalmic area, with respect to their regulation of food intake, will be included under the next major section, "Functional Neural Pathways of the Autonomic Nervous System, Feeding System."

The system of gonadotropin-releasing hormone (GnRH) neurons is discussed in the chapters of the male and female reproductive systems (Chapters 22 and 23). They are included here as they form several groups of neurons in the preoptic, hypothalamic, thalamic, and midbrain regions. A major group of GnRH neurons occurs in the bed n. of the pallial commissure (nCPa; Kuenzel and Blähser, 1991). GnRH neurons found within the nCPa (Figure 8B) and preoptic area project to the median eminence and affect the release of luteinizing hormone and follicle stimulating hormone from the anterior pituitary (Józsa and Mess, 1982; Sterling and Sharp, 1982; Mikami et al., 1988; Foster et al., 1987).

The suprachiasmatic n. (SCN) is a controversial structure within the brain of avian species. Two nuclei have been proposed as the SCN of birds: (1) SCN, pars medialis (SCNm; Crosby and Woodburne, 1940), and (2) n. decussationis supraopticae, pars ventralis (DSv; Figures 8B, and 8C, respectively) or lateral hypothalamic retinorecipient n. (Repérant, 1973). It is not clear which of the two nuclei are homologs of the mammalian SCN as each differs chemoarchitecturally from that of the traditional mammalian SCN (Norgren and Silver, 1990). Due to its involvement with the eyes and the pineal gland for regulating circadian rhythms (see Chapter 21), it is an important hypothalmic n. To date, the afferent and efferent pathways from each of the proposed n. have not been completely described. Perhaps in avian species the functional SCN has been parceled into two or more subnuclei. Certainly past studies involving lesions targeted to the SCNm which disrupted circadian rhythms in birds were large enough such that both the SCNm and DSv were disrupted (Ebihara and Kawamura, 1981; Simpson and Follett, 1981; Takahashi and Menaker, 1982).

The inferior hypothalamic and infundibular n. (IH, IN; Figure 8D), the latter considered equivalent to the arcuate n. of mammals, are two additional groups of neurons that have important interactions with the median eminence and the ANS. The two nuclei, in addition to the caudalmost portion of the VMN, comprise the infundibular or tuberal nuclear complex. Lesions of the infundibular complex, in which the IN and IH have been destroyed, result in termination of gonadal function, including an inability of lesioned birds to show future gonadal growth following photoperiodic stimulation (Graber et al., 1967; Wilson, 1967; Stetson, 1969; Sharp and Follett, 1969; Oliver, 1972; Stetson, 1972; Davies and Follett, 1975). A tuberohypophyseal tract has been described that projects from the infundibular n. to the entire length of the median eminence (Wingstrand, 1951; Oksche and Farner, 1974). Hence, that tract might be the one responsible for maintaining or facilitating gonadal development. Of interest is that neither GnRH neurons nor their fiber tracts have been shown to occur within the IH/IN region (reviewed in Kuenzel, 1993). Two major populations of neurons have been described to occur within the IH/IN region: neuropeptide Y (NPY) neurons (Kuenzel and McMurtry, 1988; Aste et al., 1991; Kuenzel et al., 1994; Kuenzel and Fraley, 1995) and vasoactive intestinal polypeptide (VIP) neurons (Yamada et al., 1982; Silver et al., 1988; Mauro et al., 1989; Sharp et al., 1989; Kuenzel and Blähser, 1994). It has been reported that chicks given a substance, sulfamethazine, that markedly stimulated gonadal development, resulted in a significant increase in darkly immunostained NPY neurons within the IH/IN region (Kuenzel et al., 1994). Concomitantly, plasma LH and FSH were significantly elevated (Kuenzel et al., 1995) in treated chicks compared to controls suggesting that sulfamethazine may facilitate gonadal development via an activation of NPY neurons. Related to the previous immunocytochemical data is the finding that chronic intracerebroventricular injections of NPY resulted in a significant increase in gonadal size in immature chicks (Fraley and Kuenzel, 1993). Therefore past lesion studies within the infundibular nuclear complex of several avian species cited above that significantly reduced gonadal function could have been the result of compromising NPY-like neurons within the IH/IN complex. It is not known, however, if NPY neurons in the IH/IN complex project to the median eminence and therefore are the neurons comprising the tuberohypophyseal tract. In addition, one cannot rule out the significance of VIP neurons found within the IH/IN complex. It has been reported that VIP not only releases prolactin (Opel and Proudman, 1988) but also LH (Macnamee et al., 1986) and a subpopulation of VIP neurons within the mediobasal hypothalamus immunostained with an antibody to opsin, suggesting that some of the VIP neurons may be encephalic photoreceptors (Silver et al., 1988). Future studies are therefore required to establish whether NPY, VIP, or both sets of neurons within the mediobasal hypothalamus are required to function with GnRH neuronal terminals found within the median eminence to effect gonadal development in avian species.

D. Other Systems Involving Circumventricular Organs (CVOs)

Earlier in the chapter it was stated that CVOs (Figure 2) are an important constituent of the ANS of birds in that they contain the appropriate neural and glial components that can secrete as well as internalize neuromodulators via the cerebrospinal fluid. Three CVOs potentially play key roles within the ANS of birds. The first is the lateral septal organ (LSO; Figures 2 and 9A). The LSO has been shown to be a CVO in birds as first proposed and described in the chick (Kuenzel and vanTienhoven, 1982; Kuenzel and Blähser, 1994). It is found at the base of the lateral ventricle and contains highly modified ependymal cells, an incomplete blood–brain barrier, and cerebrospinal fluid-contacting neurons (Figure 9A). A subpopulation of the VIP cerebrospinal fluid-contacting neurons also immunostained with an antibody to opsin suggests that they may be deep encephalic photoreceptors (Silver et al., 1988). The last finding is of significance since avian biologists have been searching for several decades for the location of neurons that respond to photoperiodic stimulation. It has been well documented since the original finding of Rowan (1926) that the reproductive system of birds is activated by increased daylength. The location of the sensory system remains unknown. Compelling evidence has been obtained showing that the eyes and pineal gland are not essential for the progonadal effects of photostimulation. The remaining possibility is that deep encephalic photoreceptors occur within diencephalic tissue (reviewed in Kuenzel, 1993), an hypothesis originally proposed by Karl von Frisch (1911). Two loci have emerged as the most likely candidate(s). The first is the infundibular nuclear complex (described in Chapter 16), while the second is located within the ventral forebrain and is proposed to be the LSO. A subset of the VIP-containing cerebrospinal fluid-contacting neurons

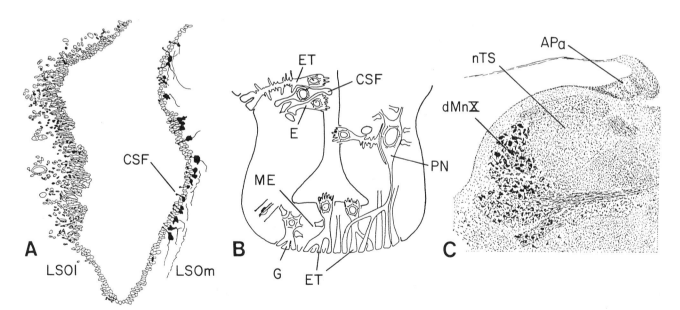

FIGURE 9   Three important circumventricular organs involving the ANS. (A) The lateral septal organ (LSO) consists of two components. The lateral LSO contains VIP receptors and lacks a complete blood–brain barrier. The medial LSO has cerebrospinal fluid contacting neurons that contain VIP. The LSO is proposed to be part of the visceral forebrain system of birds. (B) The median eminence (ME) is an interface between the central nervous system and endocrine system. Diagram modified from Oksche and Farner (1974). CSF, cerebrospinal fluid contacting neuron; E, ependymal cells; ET, ependymal tanycytes; G, glial cells; PN, peptidergic or aminergic neuron. (C) The area postrema (APa) is an interface between peripheral and central components of the ANS. dMnX, dorsal motor n. of the vagus; nTS, n. tractus solitarius.

within the LSO of birds may be the elusive neurons housing deep encephalic photoreceptors. Further study is necessary to provide the critical, direct evidence. Later in the chapter, the LSO will be presented as a component of the visceral forebrain system, a key brain system that directly projects to, and therefore is a part of, the ANS of birds. In summary, the LSO of birds is strategically located to receive as well as integrate information that regulates critical physiological and behavioral functions controlled by the ANS.

The median eminence (ME) is the second CVO and is shown schematically in Figure 9B. It is a most important CVO as it is the interface between the central nervous system and pituitary gland, hence the endocrine system. Directly beneath the ME is the anterior pituitary while the caudal continuation of the ME becomes the posterior pituitary or neurohypophysis (NH; Figure 2). The neurosecretory neurons within the hypothalamus responsible for controlling gonadal (GnRH neurons), thyroid (thyroid-releasing hormone neurons) and adrenal gland (corticotropin-releasing hormone and arginine vasotocin neurons) function, or growth (growth hormone stimulating and somatostatin neurons) and prolactin secretion (VIP neurons) all project to the ME. The terminals release neurohormones or hypophysiotropic factors into a capillary plexus known as the hypophysial portal system. Specific pituicytes within the adenohypophysis respond to the neurohormones by secreting anterior pituitary hormones into the cardiovascular system. Note that the ME contains anatomical constituents similar to the LSO (Figure 9B). The ME consists of modified ependymal cells, aminergic and peptidergic neurons, cerebrospinal fluid contacting neurons, and an incomplete blood–brain barrier (Oksche and Farner, 1974). Please refer to chapters pertaining to the pituitary gland and endocrinology, thyroids, and adrenals for more details on the regulation of the endocrine system. The posterior pituitary is organized very differently. Magnocellular neurons, primarily located within the PVN, SOe, and SOv nuclei (Figures 8A and 8C) project directly into the posterior pituitary. The neurosecretory hormones arginine vasotocin and mesotocin are secreted directly into the posterior pituitary where they can be stored temporarily or released into the circulatory system and transported to either the kidneys or smooth muscle of the reproductive system (see Chapters 11, 22, and 23).

The area postrema (APa) is the third CVO of importance to the ANS (Figures 2 and 9C). It, along with the nucleus tractus solitarius (nTS), serves as an interface between the ANS and the central nervous system. In the literature, the avian APa is referred to as the taenia choroidea. It has been proposed that the taenia choroidea and its adjacent lateral, superficial n. are homologous to the mammalian APa (Berk, 1991). Similar to the LSO and ME, the APa lacks a complete blood–brain barrier. More importantly, afferents from the viscera, particularly from the aortic arch (Katz and Karten, 1979), pulmonary tissue, and gastrointestinal tract (Katz and Karten, 1983a,b) project to the nTS, which is directly ventral and medial to the APa (Figure 9C). The APa and its adjacent nTS comprise an important site within the brainstem for parasympathetic neurons and associated functions of the ANS.

V. FUNCTIONAL NEURAL PATHWAYS OF THE AUTONOMIC NERVOUS SYSTEM

The major functions associated with the ANS include cardiovascular regulation (Jones and West; Chapter 9), respiration (Powell and Mitchell; Chapter 10), osmoregulation (Goldstein and Skadhauge; Chapter 11), thermoregulation (Dawson and Whittow; Chapter 14), circadian rhythms and photoperiodism (Gwinner and Hau; Chapter 21), and the regulation of food intake. The latter will be utilized as an example of how specific groups of cells within the brains of birds project to the motor or efferent components of the ANS and therefore form a functional neural pathway with that system. Thereafter, the visceral forebrain system in birds will be characterized as it is a more global neuroanatomical pathway and therefore subserves more than one function. Recently, it has been proposed to regulate the annual cycles of birds (Kuenzel and Blähser, 1993). Finally, components of the limbic system of birds will be described as it shares several neural structures with the visceral forebrain system and therefore should be considered part of the ANS.

A. An Autonomic Pathway Regulating Food Intake

For many years a dual-center hypothesis was utilized to explain the neural regulation of food intake in mammals (Stellar, 1954). It was based upon the work of Hetherington and Ranson (1940), who demonstrated in rats that bilateral lesions of the ventromedial hypothalamic n. (VMN; Figure 8C) resulted in increased food intake and marked obesity. Further research involving bilateral lesions of the lateral hypothalamic n. (LHy; Figures 8C and 8D) resulted in an opposite effect, complete absence of feeding and drinking (Anand and Brobeck, 1951). The ventromedial hypothalamus, termed a "satiety" center, was proposed to contain a set of neurons that projected to and inhibited a lateral hypothalamic area, the "feeding" center. More recently the past hypothalamic model has been incorporated into a pathway involving the ANS and offers a more complete explanation for regulating food intake. The model, termed the autonomic and endocrine hypothesis, suggests that lesions of the ventromedial hypothalamic n. result in a significant decrease in sympathetic activity. Therefore with or without an increase in food intake, the animal will have a lower metabolic rate and higher release of corticosteriods and insulin from the adrenal cortex and pancreas, respectively (Inoue and Bray, 1979; Bray and York, 1979; Bray, 1991). The result will be an increase in lipogenesis. Neuroanatomical data have supported the "autonomic and endocrine" hypothesis. Tract-tracing studies in the rat have shown that the VMN contains neurons which project to the central gray. In turn, the central gray has neurons that project monsynaptically to the intermediolateral column of the spinal cord (Luiten et al., 1987). As discussed earlier in the chapter, the intermediolateral column houses the preganglionic neurons of the sympathetic nervous system. As suggested by the "autonomic and endocrine" hypothesis, VMN lesions significantly disrupt neuroanatomical connections of the sympathetic nervous system resulting in decreased sympathetic activity. The affected animal gains weight, particularly body lipids. In contrast, the LHy contains neurons that project directly to the nTS and dMnX (Luiten et al., 1987). As mentioned previously, the dMnX comprises preganglionic neurons of the parasympathetic nervous system. Therefore bilateral lesions of the LHy result in an interruption of a neural pathway involving the parasympathetic nervous system. Greater activity thereby occurs in the sympathetic nervous system followed by an increased metabolic rate and loss of body weight. Another hypothalamic n. has been shown to play a key role in food intake, the paraventricular n. (PVN; Figure 8C). Of interest is that the PVN has direct projections to the central gray as well as the nTS and dMnX and therefore has connections to both arms of the ANS. In mammals the PVN has been shown to be one of the most sensitive sites in the brain for stimulating food intake following direct application of norepinephrine, neuropeptide Y, β-endorphin, and galanin (Leibowitz, 1978; Leibowitz and Hor, 1982; Stanley and Leibowitz, 1984; Stanley et al., 1985; Kyrkouli et al., 1990). Further studies have shown that the most sensitive brain site for eliciting food intake in the rat following direct application of NPY occurs in the perifornical hypothalamus which lies next to, but is separate from, the PVN (Stanley et al., 1993).

In avian species, considerably more research is needed to establish the neural pathway responsible for regulating food intake via the ANS. Nonetheless, the following evidence thus far suggests a system similar to that established for mammals. Electrolytic lesions of the VMN of Leghorn chickens (Lepkovsky and Yasuda, 1966; Snapir et al., 1973; Sonoda, 1983) and the VMN and IH of white-throated sparrows (Kuenzel and Helms, 1967, 1970; Kuenzel, 1974) effected hyperphagia and obesity. Bilateral hypothalamic lesions have been performed in chickens (Feldman et al., 1957), white-throated sparrows (Kuenzel, 1972), pigeons (Zeigler, 1976; Zeigler and Karten, 1973; Zeigler et al., 1969) and broiler chicks (Kuenzel, 1982). Lesions targeted in and around the LHy on both sides of the brain reduced food intake. A marked difference between birds and mammals was found; namely, the aphagic response was not permanent. Large, bilateral lesions involving both the ansa lenticularis and quinto frontal tract or the stratum cellulare externum (Figure 8C, 8D) were most effective for reducing food intake, however, intake usually returned to preoperative levels within 4 to 7 days (Kuenzel, 1982). Similar results were obtained when stimulus-bound feeding was attempted in fowl. More than 625 sites were implanted with electrodes for electrical brain stimulation in conscious animals. No positive brain loci were identified for stimulus-bound food intake as has been demonstrated frequently in mammals (Tweeton et al., 1973). Of interest is that neural pathways have been traced from specific hypothalamic areas to the hindbrain in birds. It has been shown that the VMN does project to the central gray, similar to mammals. The beginning and end of the central gray (GCt) are shown in Figure 6. In the chick brain, the GCt continues for at least 3.4 mm and therefore is an extensive structure in the rostrocaudal direction. A difference, however, is that the GCt projection is not as robust as the mammalian one suggesting that important GCt input may be coming from nuclei other than the VMN (Berk and Butler, 1981). Similar to mammals, the LHy has been shown to project directly to the vagal complex. A discrete portion of the LHy, the stratum cellulare externum (SCE), projects to the nTS and dMnX and hence the SCE has been proposed to be homologous to the LHy of mammals (Berk, 1987). The PVN of birds has similar projections to that established for mammals. Monosynaptic connections have been shown to occur from the PVN to the GCt and from the PVN to the thoracic level of the spinal cord. Presumably, therefore, a monosynaptic connection occurs between the PVN and the n. intermediomedialis (the column of Terni) within the spinal cord. The PVN also has monosynaptic connections to the vagal complex (Berk and Finkelstein, 1983). Therefore the PVN has both sympathetic and parasympathetic connections. A compilation of current anatomical connectivity among the VMN, LHy, PVN, and ANS in birds is shown in Figure 10.

Sparce research has been completed regarding the function of the PVN in birds. Bilateral lesions of the PVN have not been completed to date to determine whether birds will become hyperphagic. Norepineph-

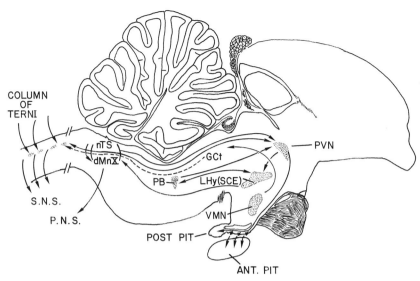

FIGURE 10 Sagittal section of chick brain showing structures important for regulatng food intake. The dashed line shows a pathway that needs to be verified in birds. ANT. PIT., anterior pituitary; dMnX, dorsal motor n. of vagus; GCt, central gray; LHy (SCE), lateral hypo. area (stratum cellulare externum); nTS, n. tractus solitarius; PB, parabrachial n.; PNS, parasympathetic nervous system; POST. PIT. (NH), post. pituitary (neurohypophysis); PVN, paraventricular n.; SNS, sympathetic nervous system; VMN, ventromedial hypo. n.

rine administered to the PVN and anterior medial hypothalamic nuclei resulted in increased food intake in broiler chicks (Denbow and Sheppard, 1993). Direct administration of other biogenic amines and neuropeptides to the PVN have not been attempted.

B. Visceral Forebrain System

A second neural system that has been proposed to exist in mammals is a visceral forebrain system. Similar to the feeding circuit described above, the VFS is intimately involved with the ANS, particularly the parasympathetic nervous system. Its components in birds (displayed as double-underlined structures) and its known anatomical pathways are shown in Figure 11 and Table 1. Two significant structures of the VFS include the nTS and parabrachial n. (PB) as all other nuclei of the mammalian VFS system have reciprocal monosynaptic connections with them (van der Kooy et al., 1984). In birds, all subcortical structures of the VFS have been shown to have single neuron connections with the nTS except the infundibular n. (IN) as illustrated by a dashed line in Figure 11. The major role of the VFS in mammals is to influence cardiovascular, respiratory, and gastrointestinal functions including the possibility of interfering with or overriding brainstem homeostatic mechanisms during periods of stress or emotional activity (van der Kooy et al., 1984). Of interest is that a systematic brain mapping study was conducted in the pigeon where brain loci were determined that significantly altered blood pressure, heart rate, and respiratory rate following electrical stimulation. It was found that all subcortical components of the VFS (refer to Table 1), when electrically stimulated, resulted in cardiovascular and respiratory changes (Macdonald and Cohen, 1973). In addition, a recent immunocytochemical study was completed showing the location of vasoactive intestinal polypeptide (VIP)-immunoreactive neurons and fibers throughout the chick brain. It was noted that all components of the VFS of birds have VIP neurons or fibers (Kuenzel and Blähser, 1993, 1994). When one examines the functions of VIP found in the mammalian literature, a majority directly involve the ANS including vasodilation, increased blood flow and exocrine gland secretion, neuroendocrine release of prolactin, neuroendocrine stimulation or inhibition of luteinizing hormone, stimulation of male sexual behavior and thyroid hormone secretion, alteration of energy metabolism, and modulation of circadian rhythms (reviewed in Kuenzel and Blähser, 1993). Other neural components that could be included in the VFS of birds are the ventrolateral medulla, area ventralis of Tsai, locus ceruleus, and substantia nigra; as each was shown to significantly alter respiratory and cardiovascular function when electrically stimulated (Macdonald and Cohen, 1973), has monosynaptic connections with the nTS (Arends et al., 1988), and contains immunoreactive VIP perikarya and fibers (Kuenzel and Blähser, 1994). To date, the most controversial component of the VFS of birds is the caudal dorsolateral neostriatum (NEOcdl, Figure 11), as it has been proposed

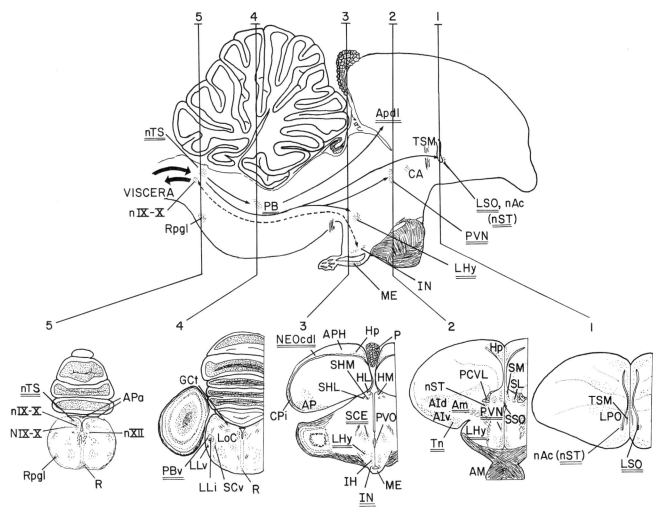

FIGURE 11 The visceral forebrain system of birds. Components are shown as double underline abbreviations on the schematic diagram. See Table 1 for abbreviations.

to be equivalent to the prefrontal cortex of mammals (Mogensen and Divac, 1982; Divac and Mogensen, 1985; Divac *et al.*, 1985; Waldmann and Güntürkün, 1993; Rehkämper and Zilles, 1991); however, in contrast to mammals, the forebrain structure does not have monosynaptic connections with the nTS or PB (Wild *et al.*, 1990). More recently, a larger region of the avian neocortex including the NEOcdl has been termed the pallium externum and it projects to the striatum of the basal ganglia (Veenman *et al.*, 1995). Further studies are therefore needed to determine whether the avian NEOcdl is equivalent to the mammalian prefrontal cortex. Nonetheless, a VFS appears to exist in the avian brain and the complex pathway, with its direct connections to the ANS, as discussed in the last two sections of this chapter, has possible ramifications concerning the current genetic selection programs in the poultry industry as well as with the neuroanatomical system proposed to regulate the annual cycles of migratory birds.

C. Limbic System

The limbic system of mammals includes a number of anatomical substrates involved in emotions. Some of its anatomical structures reside within the cortex and its functions include motivation, memory, and learning. One might question why the system should be associated with the ANS. An overwhelming argument favoring its inclusion is that a primary function of the ANS, particularly when its sympathetic division dominates, is to prepare an organism for a "fight-or-flight" reaction. Such a response involves emotional behavior, memory, and learning to enhance survival of the species. Although it remains an ill-defined neural system in birds, it will be shown that two of its components (hypothalamus

TABLE 1 Components of the Visceral Forebrain System in Mammals and Birds

Mammalian structure	Avian equivalent[a]	Abbreviation
A. Cortical component		
(1) Med. lat. prefontal cortex	Caud. dorsolat. neostriatum[b]	NEOcdl
B. Subcortical components		
Ventrolateral forebrain		
(2) Central n. amygdala	Archistriatum[b,c]	A
	Dorsolat. n. post.	Apdl
	mediale	Am
	n. taeniae	Tn
Septal Area		
(3) Bed n. stria terminalis	n. accumbens	nAc[d]
	Bed. n. stria terminalis	nST
Hypothalamus		
(4) Paraventricular n.	Paraventricular n.	PVN
(5) Posterolateral hypothalamic n.	Lat. hypothalamic area	LHy
	Stratum cellulare externum	SCE
(6) Arcuate n.	Infundibular n.	IN
Pons		
(7) Parabrachial n.	Parabrachial n.	PB
Medulla oblongata		
(8) n. tractus solitarius	n. tractus solitarius	S
C. Proposed additions to VFS of birds[e]		
Midbrain and Pons		
(9) Substantia nigra	Substantia nigra	TPc
(10) Locus ceruleus	Locus ceruleus	LoC
Medulla oblongata		
(11) Ventrolateral medulla	Ventrolateral medulla	vlm

[a]caud. = caudal; dorsolat. = dorsolateral; lat. = lateral; med. = medial; n. = nucleus; post. = posterior.
[b]Controversial equivalents.
[c]Avian equivalent for central n. of amygdala not resolved.
[d]nAc and nST of birds, as shown in past published stereotaxic atlases, proposed to be equivalent to nST of mammals.
[e]Based upon monosynaptic connections to nTS, cardiovascular changes from electrical stimulation and presence of VIP perikarya.

and amygdala) are also part of the VFS discussed previously.

The concept of a limbic lobe within the forebrain of mammals was proposed by James Papez and expanded upon by Paul MacLean, who introduced the term limbic system (Kandel et al., 1991). Specific anatomical components of the mammalian system include the cingulate gyrus and other inner cortical areas that form a ring around the thalamus and hypothalamus. Other constituents include the hippocampal formation, amygdala, septal nuclei, habenula, hypothalamus and the mesolimbic dopamine system (Kandel et al., 1991). In birds, the hippocampus (Hp; Figure 8B) and amygdala (Tn, Am, Ap; Figures 8C and 8D) are relatively large and distinct cortical structures. The hippocampal and parahippocampal areas and their cytochemical composition have been examined in pigeons (Erichsen et al., 1991; Krebs et al., 1991). The equivalent structure of the amygdala in birds includes three proposed nuclear groups: n. taeniae, medial archistriatum, and posterior archistriatum (Zeier and Karten, 1971). The mesolimbic dopamine system, the last structure listed as part of the mammalian limbic system, has been examined in birds (Kitt and Brauth, 1986). The system, originating in a midbrain nucleus, the area ventralis of Tsai (AVT; Figure 7), consists largely of dopaminergic neurons which project to the hypothalamus, habenula (HL, HM, SHL; Figure 8D), preoptic area (anterior portion of hypothalamus ventral to the septomesencephalic tract (TSM; Figure 8A)), septal area (SM, SL; Figure 8B), amygdala, and hippocampal complex (Kitt and Brauth, 1986). All of the structures just listed are considered components of the limbic system.

Functionally, the hippocampus of birds has been shown to be involved with food storage behavior (Healy and Krebs, 1993) and homing (Bingman, 1993). Both behaviors involve memory. The latter has particular significance regarding not only the ability of birds to find home roosts each day, but also the annual feat of migratory birds to journey to their appropriate breeding grounds and wintering grounds each spring and fall, respectively. The amygdala of birds has been shown to function in arousal and to affect emotional behavior. Phillips (1964) reported that mallard ducks placed in cages show considerable fear responses and do not initially breed well in captivity. Bilateral lesions involving the amygdala, particularly its pathway projecting to the hypothalamus, resulted in a marked taming effect. The greatly reduced escape behavior resulted in increased ovarian growth and function in lesioned birds. Additional studies will need to be completed to establish complete, functional neural pathways responsible for emotional and cognitive behaviors expressed by birds.

VI. SHIFTS IN HOMEOSTASIS: AN AUTONOMIC HYPOTHESIS EXPLAINING THE REGULATION OF ANNUAL CYCLES OF BIRDS

Avian species, particularly migratory birds inhabiting the temperate zone of North America, Europe, or Asia, display very pronounced physiological and behavioral events throughout their annual cycles such as increased food intake resulting in large depositions of body fat, molt of feathers, migration, and reproduction. Each of the events occurs in an orderly and predictable sequence and requires a considerable amount of energy. Research efforts over the past few decades have shown that gonadal development (Rowan, 1926), premigratory fattening (King and Farner, 1965), molt, and migratory orientation (Emlen, 1969) can be expressed at different times of the year by manipulating the photoperiod. More interestingly, if one subjects birds to constant environmental conditions over long periods of time, testes development, molt, and nocturnal activity (an expression of a bird's motivation to migrate in a cage environment) persist but in more frequent cycles than would have occurred under a natural photoperiod (Gwinner, 1991). Therefore, circannual cycles behave similarly to circadian rhythms (e.g., sleep–wake cycles and activity and body temperature cycles) of avian or mammalian physiological variables measured under constant environmental conditions. Therefore, it has been proposed that circannual rhythms are long-term manifestations of circadian rhythms and both are the result of an endogenous rhythm that can be synchronized by external factors, particularly photoperiod (Gwinner, 1981).

An unanswered question to date is what constitutes the control system that regulates the annual cycle of birds? It has been argued for some time that prolactin within the pituitary gland is a key hormone, since exogenous injections effect both premigratory fattening (Meier and Farner, 1964) and migratory behavior (Meier et al., 1965). The mediobasal hypothalamus has also been proposed since medial lesions destroying the VMN and part of the IH and IN disrupted all of the ethophysiological events of the annual cycle of the white-throated sparrow, *Zonotrichia albicollis* (Kuenzel, 1974). The pineal gland and suprachiasmatic nucleus have also been implicated (refer to Chapter 21). Recently, the visceral forebrain system (VFS) has been proposed to constitute the neural system that regulates the annual cycle of birds (Kuenzel and Blähser, 1993; 1994; Kuenzel, 1994b). Its components have already been detailed (Figure 11; Table 1). The following is a synopsis of how the VFS functions to control circannual cycles of birds.

Migratory birds that inhabit the temperate zone for both wintering and breeding grounds (short and moderate distant migrants) or breeding grounds only (long distant migrants) begin to display hyperphagia during the early spring in anticipation of migration to the northern breeding grounds. In effect, there occurs a marked shift in homeostasis with respect to maintaining a particular body weight. It is not uncommon for individuals of some species to double their food intake, fat deposition, and display a 50% gain in body weight within a 2- to 4-week period in anticipation of a rigorous migratory flight. The parasympathetic nervous system (PNS) appears to dominate in order to effect the rapid gains in lipid stores. Upon attaining the desired weight gains, birds shift into another phase of their annual cycle as they begin their spring migratory journey. Most passerine species are nocturnal migrants; therefore they change from a diurnal existence to one of nearly complete wakefulness over several days as they migrate until arriving at the breeding grounds. Other drastic changes occur. Prior to and during spring migration, birds feed and migrate in flocks. This social behavior exhibited by many is quickly transformed into a competitive, territorial one upon arrival on the breeding grounds. The agonistic behavior, particularly demonstrated in males, continues until a mate is found and the breeding cycle is initiated. Characteristic of this period is a continuous loss in body weight due to a high energy expenditure involved in securing a mate. Thereafter, yet another dynamic ethophysiological shift occurs and throughout the period of incubating their eggs, there may be a complete cessation of food intake by parents resulting in a significant loss of weight. Finally, a major postnuptial or basic molt occurs where both the flight and body feathers are replaced. A significant increase in daily metabolism has been documented with a nearly complete loss of body lipid reserves (Dolnik and Gavrilov,

1979). In effect, from the initiation of spring migration to the end of postnuptial molt, a shift in the balance of the ANS has occurred such that the sympathetic nervous system (SNS) eventually dominates. Throughout the late spring and early summer, the gradual predominant effects of the SNS are responsible for the continued loss in weight and most body lipid reserves. The physiological processes responsible for molt that occur during the summer are critical. The process of molting not only begins another shifting of the ANS toward a dominance of the PNS, it is responsible for effecting a photorefractory state where birds no longer respond to the remaining long photoperiods of the summer/early fall months. It is therefore proposed that some time between the beginning of postnuptial molt and the beginning of fall migration, a final shift in balance of the ANS occurs with the PSN dominating again. Birds consequently show an increase in food intake and body weight in anticipation of fall migration. The hyperphagic response, body weight gain, and nocturnal activity displayed by migratory birds in the fall have been shown to be significantly less than the same events recorded for spring migration (Kuenzel and Helms, 1974). Data suggest that fall migration may be a more leisurely process for some species of migratory bird. In addition, the data suggest that the PNS is not as dominant during the start of fall migration as it is when birds begin to prepare for the next season's spring migration. The dramatic shifts in homeostasis is orchestrated by the VFS that has direct connections to the PNS and SNS. Vasoactive intestinal peptide (VIP) operating within the VFS may be one of the critical neuromodulators that is responsible for altered balance of ANS. Time will tell whether this simple model may explain the apparent shifts in homeostasis that have been documented in the past literature as a typical migratory bird quickly passes through several distinct phases of an annual cycle to attain optimal conditions for surivival of its own species.

VII. CONSEQUENCES OF GENETIC SELECTION FOR RAPID GROWTH IN BROILERS AND TURKEYS: AN IMBALANCED AUTONOMIC NERVOUS SYSTEM

Since the 1950s geneticists within the commercial poultry industry have been selecting birds for growth rate and feed conversion. In effect, by selecting for rapid growth, broilers and turkeys have also been inadvertently selected for appetite. Broiler breeders are a good example of birds that have a voracious appetite and a tendency to become obese. Their offspring are generally hypoactive and have a significantly lower basal metabolic rate compared to Leghorns (Kuenzel and Kuenzel, 1977). Consequently, broilers show a dominance of the PNS similar to rats receiving lesions to the ventromedial hypothalamic nuclei and in some obese humans (see section "An Autonomic Pathway Regulating Food Intake"). Due to the tendency of broilers to show obesity, avian leg weakness, and diseases associated with the cardiovascular system, it will be important in the future for poultry geneticists to attempt to bring back into balance the SNS and PNS of some commercial poultry strains (Kuenzel, 1994a).

Acknowledgments

The author wishes to acknowledge M. Masson for completing the final ink-and-computer drawings for the figures included in the chapter and to S. Brown for typing the final drafts of the paper. Supported in part by Maryland Agricultural Experiment Station (MAES) Competitive Grants POUL-FY95-43 and FY96–40 and USDA Comptetitive Grant No. 90-37240-5506. Scientific article No. A7884, contribution No. 9218 of the MAES.

REFERENCES

Akester, A. R. (1979). The autonomic nervous system. In "Form and Function in Birds" (A. S. King and J. McLelland, eds.), Vol. 1, pp. 381–441. Academic Press, London.

Anand, B., and Brobeck, J. R. (1951). Hypothalamic control of food intake in rats and cats. *Yale J. Biol. Med.* **24,** 123–146.

Arends, J. J. A., Wild, J. M., and Zeigler, H. P. (1988). Projections of the nucleus of the tractus solitarius in the pigeon (*Columba livia*). *J. Comp. Neurol.* **278,** 405–429.

Aste, N., Viglietti-Panzica, C., Fasolo, A., Andreone, C., Vaudry, H., Pelletier, G., and Panzica, G. C. (1991). Localization of neuropeptide Y-immunoreactive cells and fibres in the brain of the Japanese quail. *Cell Tissue Res.* **265,** 219–230.

Balthazart, J., Foidart, A., Surlemont, C., and Harada, N. (1991). Neuroanatomical specificity in the co-localization of aromatase and estrogen receptors. *J. Neurobiol.* **22,** 143–157.

Balthazart, J., Foidart, A., Wilson, E. M., and Ball, G. F. (1992). Immunocytochemical localization of androgen receptors in the male songbird and quail brain. *J. Comp. Neurol.* **317,** 407–420.

Barfield, R. J. (1965). Effects of preoptic lesions on the sexual behavior of male domestic fowl. *Am. J. Zool.* **5,** 686–687.

Barfield, R. J. 1969. Activation of copulatory behavior by androgen implanted into the preoptic area of the male fowl. *Horm. Behav.* **1,** 37–52.

Baylé, J.-D., Ramade, F., and Oliver, J. (1974). Stereotaxic topography of the brain of the quail (*Coturnix coturnix japonica*). *J. Physiol.* (*Paris*) **68,** 219–241.

Beckstead, R. M., and Norgren, R. (1979). An autoradiographic examination of the central distribution of the trigeminal, facial, glossopharyngeal and vagal nerves in the monkey. *J. Comp. Neurol.* **184,** 455–472.

Bennett, T. (1974). The peripheral and autonomic nervous systems. In "Avian Biology" (D. S. Farner and J. R. King, eds.), Vol. 1, pp. 1–77. Academic Press, New York.

Berk, M. L. (1987). Projections of the lateral hypothalamus and bed nucleus of the stria terminalis to the dorsal vagal complex in the pigeon. *J. Comp. Neurol.* **260,** 140–156.

Berk, M. L. (1991). Distribution and hypothalamic projection of tyrosine hydroxylase containing neurons of the nucleus of the solitary tract in the pigeon. *J. Comp. Neurol.* **312,** 391–403.

Berk, M. L., and Butler, A. B. (1981). Efferent projections of the medial preoptic nucleus and medial hypothalamus in the pigeon. *J. Comp. Neurol.* **203,** 379–399.

Berk, M. L., and Finkelstein, J. A. (1983). Long descending projections of the hypothalamus in the pigeon, *Columba livia. J. Comp. Neurol.* **220,** 127–136.

Berk, M. L., and Smith, S. E. (1994). Local and commissural neuropeptide-containing projections of the nucleus of the solitary tract to the dorsal vagal complex in the pigeon. *J. Comp. Neurol.* **347,** 369–396.

Bingman, V. P. (1993). Vision, cognition, and the avian hippocampus. In "Vision, Brain, and Behavior in Birds" (H. P. Zeigler and H.-J. Bischof, eds.), pp. 391–408. The MIT Press, Cambridge, MA.

Bons, N. (1980). The topography of mesotocin and vasotocin systems in the brain of the domestic mallard and Japanese quail: Immunocytochemical identification. *Cell Tissue Res.* **213,** 37–51.

Bray, G. A. (1991). Obesity, a disorder of nutrient partitioning: The MONA LISA hypothesis. *J. Nutr.* **121,** 1146–1162.

Bray, G. A., and York, D. A. (1979). Hypothalamic and genetic obesity in experimental animals: An autonomic and endocrine hypothesis. *Physiol. Rev.* **59,** 719–809.

Breazile, J. E., and Hartwig, H.-G. (1989). Central nervous system. In "Form and Function in Birds" (A. S. King and J. McLelland, eds.), Vol. 5, pp. 485–566. Academic Press, London.

Breazile, J. E., and Kuenzel, W. J. (1993). Systema nervosum centrale. In "Handbook of Avian Anatomy: Nomina Anatomica Avian" (J. J. Baumel, ed.), pp. 493–554. Nuttall Ornithological Club, Mus. Comp. Zool., Harvard Univ., Cambridge, MA.

Bubien-Waluszewska, A. (1981). The cranial nerves. In "Form and Function in Birds" (A. S. King and J. McLelland, eds.), Vol. 2, pp. 385–438. Academic Press, London.

Cooper, J. R., Bloom, F. E., and Roth, R. H., (1991). "The Biochemical Basis of Neuropharmacology." Oxford Univ. Press, New York.

Crosby, E. C., and Woodburne, R. T. (1940). The comparative anatomy of the preoptic area and the hypothalamus. *Res. Publ. Assoc. Res. Nerv. Ment. Dis.* **20,** 52–169.

Davies, D. T., and Follett, B. K. (1975). The neuroendocrine control of gonadotrophin release in the Japanese quail. I. The role of the tuberal hypothalamus. *Proc. R. Soc. Lond. B.* **191,** 285–301.

Denbow, D. M., and Sheppard, B. J. (1993). Food and water intake responses of the domestic fowl to norepinephrine infusion at circumscribed neural sties. *Brain Res. Bull.* **31,** 121–128.

Divac, I. and Mogensen, J. (1985). The prefrontal "cortex" in the pigeon: Catecholamine histofluorescence. *Neuroscience* **15,** 677–682.

Divac, I., Mogensen, J., and Björklund, A., (1985). The prefrontal "cortex" in the pigeon: Biochemical evidence. *Brain Res.* **332,** 365–368.

Dolnik, V. R., and Gavrilov, V. M. (1979). Bioenergetics of molt in the Chaffinch (*Fringilla coelebs*). *Auk* **96,** 253–264.

Dubbeldam, J. L. (1984). Afferent connections of nervus facialis and nervus glossopharyngeus in the pigeon (*Columba livia*) and their role in feeding behavior. *Brain Behav. Evol.* **24,** 47–57.

Dubbeldam, J. L. (1993). Systema nervosum periphericum. In "Handbook of Avian Anatomy: Nomina Anatomica Avian" pp. 555–584. Nuttall Ornithological Club, Mus. Comp. Zool., Harvard Univ., Cambridge, MA.

Dubbeldam, J. L., Brus, E. R., Menken, S. B. J., and Zeilstra, S. (1979). The central projections of the glossopharyngeal and vagus ganglia in the mallard (*Anas platyrhynchos L.*). *J. Comp. Neurol.* **183,** 149–168.

Dubbeldam, J. L., Karten, H. J., and Menken, S. B. J. (1976). Central projections of the chorda tympani nerve in the mallard (*Anas platyrhynchos L.*). *J. Comp. Neurol.* **170,** 415–420.

Ebihara, S., and Kawamura, H. (1981). The role of the pineal organ and the suprachiasmatic nucleus in the control of circadian locomotor rhythms in the Java sparrow, *Padda oryzivora. J. Comp. Physiol.* **141,** 207–214.

Emlen, S. T. (1969). Bird migration: Influence of physiological state upon celestial orientation. *Science* **165,** 716–718.

Erichsen, J. T., Bingman, V. P., and Krebs, J. R. (1991). The distribution of neuropeptides in the dorsomedial telencephalon of the pigeon (*Columba livia*): A basis for regional subdivisions. *J. Comp. Neurol.* **314,** 478–492.

Feldman, S. E., Larsson, S., Dimick, M. K., and Lepkovsky, S. (1957). Aphagia in chickens. *Am. J. Physiol.* **191,** 259–261.

Feldman, S. E., Snapir, N., Yasuda, M., Treuting, F., and Lepkovsky, S. (1973). Physiological and nutritional consequences of brain lesions: A functional atlas of the chicken hypothalamus. *Hilgardia* **41**(19), 605–629.

Foidart, A., Reid, J., Absil, P., Yoshimura, N., Harada, N., and Balthazart, J. (1995). Critical re-examination of the distribution of aromatase-immunoreactive cells in the quail forebrain using antibodies raised against human placental aromatase and against the recombinant quail, mouse or human enzyme. *J. Chem. Neuroanat.* **8,** 267–282.

Foster, R. G., Plowman, G., Goldsmith, A. R., and Follett, B. K. (1987). Immunocytochemical demonstration of marked changes in the luteinizing hormone-releasing hormone (LH–RH) system of photosensitive and photorefractory European starlings (*Sturnus vulgaris*). *J. Endocrinol.* **115,** 211–220.

Fraley, G. S., and Kuenzel, W. J. (1993). Precocious puberty in chicks (*Gallus domesticus*) induced by central injections of neuropeptide Y. *Life Sci.* **52,** 1649–1656.

Gill, F. B. (1995). "Ornithology." W. H. Freeman, New York.

Goossens, N., Blähser, S., Oksche, A., Vandesande, F. and Dierickx, K. (1977). Immunocytochemical investigation of the hypothalamo-neurohypophysial system in birds. *Cell Tissue Res.* **184,** 1–13.

Graber, J. W., Frankel, A. I., and Nalbandov, A. V. (1967). Hypothalamic center influencing the release of LH in the cockerel. *Gen. Comp. Endocrinol.* **9,** 187–192.

Gwinner, E. (1981). Circannual rhythms: Their dependence on the circadian system. In "Biological Clocks in Seasonal Reproductive Cycles" (B. K. Follett and D. E. Follett, eds.), pp. 153–169. Wright Publishers, Bristol.

Gwinner, E. (1991). Circannual rhythms in tropical and temperate-zone Stonechats: A comparison of properties under constant conditions. *Ökol. Vögel.* **13,** 5–14.

Haynes, R. L. and Glick, B. (1974). Hypothalamic control of sexual behavior in the chicken. *Poult. Sci.* **53,** 27–38.

Healy, S. D., and Krebs, J. R. (1993). Development of hippocampal specialization in a food-storing bird. *Behav. Brain Res.* **53,** 127–131.

Heaton, M. B., and Wayne, D. B. (1983). Patterns of extraocular innervation by the oculomotor complex in the chick. *J. Comp. Neurol.* **216,** 245–254.

Hetherington, A., and Ranson, S. (1940). Hypothalamic lesions and adiposity in the rat. *Anat. Rec.* **78,** 149–172.

Hosoya, Y., Yaginuma, H., Okado, N., and Kohno, K. (1992). Morphology of sympathetic preganglionic neurons innervating the superior cervical ganglion in the chicken: An immunohistochemical study using retrograde labeling of cholera toxin subunit B. *Exp. Brain Res.* **89,** 478–483.

Hsieh, T. M. (1951). "The Sympathetic and Parasympathetic Nervous System of the Fowl." Ph.D. thesis. Edinburgh.

Inoue, S., and Bray, G. A. (1979). An autonomic hypothesis for hypothalamic obesity. *Life Sci.* **25,** 561–566.

Józsa, R., and Mess, B. (1982). Immunohistochemical localization of the luteinizing hormone releasing hormone (LHRH)-containing structures in the central nervous system of the domestic fowl. *Cell Tissue Res.* **227,** 451–458.

Kandel, E.R., Schwartz, J. H., and Jessell, T. M. (1991). "Principles of Neural Science." Appleton and Lange, Norwalk, Connecticut.

Karten, H., and Hodos, W. (1967). "A Stereotaxic Atlas of the Brain of the Pigeon (*Columba livia*)". Johns Hopkins Univ. Press, Baltimore, MD.

Katz, D. M., and Karten, H. J. (1979). The discrete anatomical localization of vagal aortic afferents within a catecholamine-containing cell group in the nucleus solitarius. *Brain Res.* **171,** 187–195.

Katz, D. M., and Karten, H. J. (1983a). Subnuclear organization of the dorsal motor nucleus of the vagus nerve in the pigeon, *Columba livia*. *J. Comp. Neurol.* **217,** 31–46.

Katz, D. M., and Karten, H. J. (1983b). Visceral representation within the nucleus of the tractus solitarius in the pigeon, *Columba livia*. *J. Comp. Neurol.* **218,** 42–73.

Katz, D. M., and Karten, H. J. (1985). Topographic representation of visceral target organs within the dorsal motor nucleus of the vagus nerve of the pigeon *Columba livia*. *J. Comp. Neurol.* **242,** 397–414.

King, J. R., and Farner, D. S. (1965). Studies of fat deposition in migratory birds. *Ann. N. Y. Acad. Sci.* **131,** 422–440.

Kitt, C. A., and Brauth, S. E. (1986). Telencephalic projections from midbrain and isthmal cell groups in the pigeon. II. The nigral complex. *J. Comp. Neurol.* **247,** 92–110.

Krebs, J. R., Erichsen, J. T., and Bingman, V. P. (1991). The distribution of neurotransmitters and neurotransmitter-related enzymes in the dorsomedial telencephalon of the pigeon (*Columba livia*). *J. Comp. Neurol.* **314,** 467–477.

Krieger, M. S., Conrad, L. C. A., and Pfaff, D. W. (1979). An autoradiographic study of the efferent connections of the ventromedial nucleus of the hypothalamus. *J. Comp. Neurol.* **183,** 785–816.

Kuenzel, W. J. (1972). Dual hypothalamic feeding system in a migratory bird, *Zonotrichia albicollis*. *Am. J. Physiol.* **223,** 1138–1142.

Kuenzel, W. J. (1974). Multiple effects of ventromedial hypothalamic lesions in the White-throated Sparrow, *Zonotrichia albicollis*. *J. Comp. Physiol.* **90,** 169–182.

Kuenzel, W. J. (1982). Transient aphagia produced following bilateral destruction of lateral hypothalamic area and quinto-frontal tract of chicks. *Physiol. Behav.* **28,** 237–244.

Kuenzel, W. J. (1993). The search for deep encephalic photoreceptors within the avian brain, using gonadal development as a primary indicator. *Poult. Sci.* **72,** 959–967.

Kuenzel, W. J. (1994a). Central neuroanatomical systems involved in the regulation of food intake in birds and mammals. *J. Nutr.* **124,** 1355S–1370S.

Kuenzel, W. J. (1994b). Proposed function of the visceral forebrain system in avian species. *J. Ornithol.* **135**(3), 418.

Kuenzel, W. J., Abdel-Maksoud, M. M., Macko Walsh, K., Proudman, J. A., and Elsasser, T. (1995). Sulfamethazine advances puberty in chicks via a reduction in thyroid hormone (T_3) followed by a central or pituitary mediation of gonadotropins. *Soc. Neurosci. Abstr.* **21**(1), 102.

Kuenzel, W. J., and Blähser, S. (1991). The distribution of gonadotropin-releasing hormone (GnRH) neurons and fibers throughout the chick brain (*Gallus domesticus*). *Cell Tissue Res.* **264,** 481–495.

Kuenzel, W. J., and Blähser, S. (1993). The visceral forebrain system in birds: Its proposed anatomical components and functions. *Poult. Sci. Rev.* **5,** 29–36.

Kuenzel, W. J., and Blähser, S. (1994). Vasoactive intestinal polypeptide (VIP)-containing neurons: Distribution throughout the brain of the chick (*Gallus domesticus*) with focus upon the lateral septal organ. *Cell Tissue Res.* **275,** 91–107.

Kuenzel, W. J., and Fraley, G. S. (1995). Neuropeptide Y: Its role in the neural regulation of reproductive function and food intake in avian and mammalian species. *Poult. Avian Biol. Rev.* **6**(3), 185–209.

Kuenzel, W. J., and Helms, C. W. (1967). Obesity produced in a migratory bird by hypothalamic lesions. *Bioscience* **17,** 395–396.

Kuenzel, W. J. and Helms, C. W. (1970). Hyperphagia, polydipsia, and other effects of hypothalamic lesions in the white-throated sparrow, *Zonotrichia albicollis*. *Condor* **72,** 66–75.

Kuenzel, W. J., and Helms, C. W. (1974). An annual cycle study of tan-striped and white-striped White-throated Sparrows. *Auk* **91,** 44–53.

Kuenzel, W. J., and Kuenzel, N. T. (1977). Basal metabolic rate in growing chicks *Gallus domesticus*. *Poult. Sci.* **56,** 619–627.

Kuenzel, W. J., Macko Walsh, K., Abdel-Maksoud, M. M., and Advis, J. P. (1994). Increased levels of neuropeptide Y (NPY) and dopamine (DA) in the median eminence of chicks showing early gonadal development. *Soc. Neurosci. Abstr.* **20**(2), 996.

Kuenzel, W. J., and Masson, M. (1988). "A Stereotaxic Atlas of the Brain of the Chick (*Gallus domesticus*)." Johns Hopkins Univ. Press, Baltimore, MD.

Kuenzel, W. J., and McMurtry, J. (1988). Neuropeptide Y: Brain localization and central effects on plasma insulin levels in chicks. *Physiol. Behav.* **44,** 669–678.

Kuenzel, W. J., and van Tienhoven, A. (1982). Nomenclature and location of avian hypothalamic nuclei and associated circumventricular organs. *J. Comp. Neurol.* **206,** 293–313.

Kyrkouli, S. E., Stanley, B. G., Seirafi, R. D., and Leibowitz, S. F. (1990). Stimulation of feeding by galanin: anatomical localization and behavioral specificity of this peptide's effects in the brain. *Peptides* **11,** 995–1001.

Langley, J. N. (1921). "The Autonomic Nervous System, Part I." Heffer and Sons, Cambridge.

Lechan, R. M., Nestler, J. L., Jacobson, S., and Reichlin, S. (1980). The hypothalamic "tubero infundibular" system of the rat as demonstrated by horseradish peroxidase (HRP) microiontophoresis. *Brain Res.* **195,** 13–27.

Leibowitz, S. F. (1978). Paraventricular nucleus: A primary site mediating adrenergic stimulation of feeding and drinking. *Pharmacol. Biochem. Behav.* **8,** 163–175.

Leibowitz, S. F., and Hor, L. (1982). Endorphinergic and α-adrenergic systems in the paraventricular nucleus: Effects on eating behavior. *Peptides* **3,** 421–428.

Leonhardt, H. (1980). Ependym und Circumventriculäre Organe. Hanbuch der mickroskopischen Anatomie des Menschen. Band 4. Nervensystem. Teil 10. *Neuroglia* **I,** 177–666.

Lepkovsky, S., and Yasuda. M. (1966). Hypothalamic lesions, growth and body composition of male chickens. *Poultry Sci.* **45,** 582–588.

Luiten, P. G. M., Ter Horst, G. J., and Steffens, A. B. (1987). The hypothalamus, intrinsic connections and outflow pathways to the endocrine system in relation to the control of feeding and metabolism. *Prog. Neurobiol.* **28,** 1–54.

Macdonald, R. L., and Cohen, D. H. (1973). Heart rate and blood pressure responses to electrical stimulation of the central nervous system in the pigeon (*Columba livia*). *J. Comp. Neurol.* 150:109–136.

Macnamee, M. C., Sharp, P. J., Lea, R. W., Sterling, R. J., and Harvey, S. (1986). Evidence that vasoactive intestinal polypeptide is a physiological prolactin-releasing factor in the Bantam hen. *Gen. Comp. Endocrinol.* **62,** 470–478.

Mauro, L. J., Elde, R. P., Youngren, O. M., Phillips, R. E., and El Halawani, M. E. (1989). Alterations in hypothalamic vasoactive intestinal peptide-like immunoreactivity are associated with reproduction and prolactin release in the female turkey. *Endocrinology* **125,** 1795–1804.

Meier, A. H., and Farner, D. S. (1964). A possible endocrine basis for premigratory fattening in *Zonotrichia leucophrys gambelii. Gen. Comp. Endocrinol.* **4,** 584–595.

Meier, A. H., Farner, D. S., and King, J. R. (1965). A possible endocrine basis for migratory behavior in *Zonotrichia leucophrys gambelii. Anim. Behav.* **13,** 453–465.

Meyer, C. C. (1973). Testosterone concentration in the male chicken brain: an autoradiographic survey. *Science* **180,** *1381–1383.*

Meyer, C. C., and Salzen, E. P. (1970). Hypothalamic lesions and sexual behavior in the domestic chick. *J. Comp. Physiol. Psychol.* **73,** 365–376.

Mikami, S., Yamada, S., Hasegawa, Y., and Miyamoto, K. (1988). Localization of avian LHRH-immunoreactive neurons in the hypothalamus of the domestic fowl, *Gallus domesticus,* and the Japanese quail, *Coturnix coturnix. Cell Tissue Res.* **251,** 51–58.

Mogensen, J., and Divac, I. (1982). The prefrontal "cortex" in the pigeon. Behavioral evidence. *Brain Behav. Evol.* **21,** 60–66.

Niimi, K., Sakai, T., and Takasu, J. (1958). The ontogenetic development of the oculomotor nucleus in the chick. *Tokushima J. Exp. Med.* **5,** 311–325.

Norgren, R. B., and Silver, R. (1990). Distribution of vasoactive intestinal peptide-like and neurophysin-like immunoreactive neurons and acetylcholinesterase staining in the ring dove hypothalamus with emphasis on the question of an avian suprachiasmatic nucleus. *Cell Tissue Res.* **259,** 331–339.

Ohmori, Y., Watanabe, T., and Fujioka, T. (1984). Localization of parasympathetic preganglionic neurons in the sacral spinal cord of domestic fowl. *Jpn. J. Zootech. Sci.* **55,** 792–794.

Oksche, A., and Farner, D. S. (1974). Neurohistological studies of the hypothalamo-hypophysial system of *Zonotrichia leucophrys gambelii. Adv. Anat. Embryol. Cell Biol.* **48,** 1–136.

Oliver, J. (1972). Étude expérimentale des structures hypothalamiques impliquées dans le réflexe photosexuel chez la caille. Thèse Doctorat Spécialité. Montpellier, France.

Opel, H., and Proudman, J. A. (1988). Stimulation of prolactin release in turkeys by vasoactive intestinal peptide. *Proc. Soc. Exp. Biol. Med.* **187,** 455–460.

Panzica, G. C., Viglietti-Panzica, C., Fiori, M. G., Calcagni, M., Anselmetti, G., and Balthazart, J. (1987). Cytoarchitectural analysis of the quail preoptic area: Evidence for a sex-related dimorphism in the medial preoptic nucleus. *Boll. Zool.* **54,** 13–17.

Panzica, G. C., Viglietti-Panzica, C., Sanchez, F., Sante, P., and Balthazart, J. (1991). Effects of testosterone on a selected neuronal population within the preoptic sexually dimorphic nucleus of the Japanese quail. *J. Comp. Neurol.* **303,** 443–456.

Phillips, R. E. (1964). "Wildness" in the mallard duck: Effects of brain lesions and stimulation on "escape behavior" and reproduction. *J. Comp. Neurol.* **122,** 139–155.

Ralph, C. L. (1959). Some effects of hypothalamic lesions on gonadotrophin release in the hen. *Anat. Rec.* **134,** 411–431.

Ralph, C. L., and Fraps, R. M. (1959). Long term effects of diencephalic lesions on the ovary of the hen. *Am. J. Physiol.* **197,** 1279–1283.

Rehkämper, G., and Zilles, K. (1991). Parallel evolution in mammalian and avian brains: Comparative cytoarchitectonic and cytochemical analysis. *Cell Tissue Res.* **263,** 3–28.

Repérant, J. (1973). Nouvelles données sur les projections visuelles chez le pigeon (*Columba livia*). *J. Hirnforsch.* **14,** 151–187.

Rowan, W. (1926). On photoperiodism, reproductive periodicity and the annual migration of birds and certain fishes. *Proc. Boston Soc. Nat. Hist.* **38,** 147–189.

Saper, C. B., Swanson, L. W., and Cowan, W. M. (1976). The efferent connections of the ventromedial nucleus of the hypothalamus of the rat. *J. Comp. Neurol.* **169,** 409–442.

Schwaber, J. S., and Cohen, D. H. (1978). Field potentials and single unit analyses of the avian dorsal motor nucleus of the vagus and criteria for identifying vagal cardiac cells of origin. *Brain Res.* **147,** 79–90.

Scott, M. L., Nesheim, M. C., and Young, R. J. (1969). "Nutrition of the Chicken." M. L. Scott, Ithaca, N.Y.

Sharp, P. J., and Follett, B. K. (1969). The effect of hypothalamic lesions on gonadotrophin release in Japanese quail, *Coturnix coturnix japonica. Neuroendocrinology* **5,** 205–218.

Sharp, P. J., Sterling, R. J., Talbot, R. T., and Huskisson, N. S. (1989). The role of hypothalamic vasoactive intestinal polypeptide in the maintenance of prolactin secretion in incubating bantam hens: Observations using passive immunization, radioimmunoassay and immunohistochemistry. *J. Endocrinol.* **122,** 5–13.

Silver, R. P., Witkovsky, P., Horvath, P., Alones, V., Barnstable, C. J., and Lehman, M. N. (1988). Coexpression of opsin- and VIP-like immunoreactivity in CSF-contacting neurons of the avian brain. *Cell Tissue Res.* **253,** 189–198.

Simpson, S. M., and Follett, B. K. (1981). Pineal and hypothalamic pacemakers: Their role in regulating circadian rhythmicity in Japanese quail. *J. Comp. Physiol.* **144,** 381–389.

Snapir, N., Ravona, H., and Perek, M. (1973). Effect of electrolytic lesions in various regions of the basal hypothalamus in White Leghorn cockerels upon food intake, obesity, blood triglycerides and protein. *Poult. Sci.* **52,** 629–636.

Sonoda, T. (1983). Hyperinsulinemia and its role in maintaining the hypothalamic hyperphagia in chickens. *Physiol. Behav.* **30,** 325–329.

Stanley, B. G., Chin, A. S. and Leibowitz, S. F. (1985). Feeding and drinking elicited by central injection of neuropeptide Y: Evidence for a hypothalamic site(s) of action. *Brain Res. Bull.* **14,** 521–524.

Stanley, B. G., and Leibowitz, S. F. (1984). Neuropeptide Y: Stimulation of feeding and drinking by injection into the paraventricular nucleus. *Life Sci.* **35,** 2635–2642.

Stanley, B. G., Magdalin, W., Seirafi, A., Thomas, W. J., and Leibowitz, S. F. (1993). The perifornical area: The major focus of (a) patchily distributed hypothalamic neuropeptide Y-sensitive feeding system(s). *Brain Res.* **604,** 304–317.

Stellar, E. (1954). The physiology of motivation. *Psychol. Rev.* 61, 5–22.

Sterling, R. J., and Sharp, P. J. (1982). The localisation of LH-RH neurones in the diencephalon of the domestic hen. *Cell Tissue Res.* **222,** 283–298.

Stetson, M. H. (1969). The role of the median eminence in control of photoperiodically induced testicular growth in the white-crowned sparrow, *Zonotrichia leucophrys gambelii. Z. Zellforsch. Mikrosk. Anat.* **93,** 369–394.

Stetson, M. H. (1972). Hypothalamic regulation of testicular function in Japanese quail. *Z. Zellforsch. Mikrosk. Anat.* **130,** 389–410.

Stokes, T. M., Leonard, C. M., and Nottebohm, F. (1974). The telencephalon, diencephalon and mesencephalon in stereotaxic coordinates. *J. Comp. Neurol.* **156,** 337–374.

Takahashi, J. S., and Menaker, M. (1982). Role of the suprachiasmatic nuclei in the circadian system of the house sparrow, *Passer domesticus. J. Neurosci.* **2,** 815–828.

Terni, T. (1923). Richerche anatomiche sul sistema nervosa autonomia degli uccelli. *Archs Ital. Anat. Embriol.* **20,** 433–510.

Tweeton, J. R., Phillips, R. E., and Peek, F. W. (1973). Feeding behavior elicited by electrical stimulation of the brain in chickens, *Gallus domesticus. Poult. Sci.* **52,** 165–172.

van der Kooy, D., Koda, L. Y., McGinty, J. F., Gerfen, C. R., and Bloom, F. E. (1984). The organization of projections from the cortex, amygdala, and hypothalamus to the nucleus of the solitary tract in rat. *J. Comp. Neurol.* **224,** 1–24.

van Tienhoven, A., and Juhász, L. P. (1962). The chicken telencephalon, diencephalon and mesencephalon in stereotaxic coordinates. *J. Comp. Neurol.* **118,** 185–198.

Veenman, C. L., Wild, J. M., and Reiner, A. (1995). Organization of the avian "corticostriatal" projection system: A retrograde and anterograde pathway tracing study in pigeons. *J. Comp. Neurol.* **354,** 87–126.

Vigh, B. (1971). "Das Paraventrikularorgan und das zirkumventrikuläre System des Gehirns," Stud. Biol. Hung, Vol. 10. Akad Kiadó, Budapest.

Vigh-Teichmann, I., and Vigh, B. (1983). The system of cerebrospinal fluid-contacting neurons. *Arch. Histol. Jpn.* **46,** 427–468.

Viglietti-Panzica, C., Panzica, G. C., Fiori, M. G. Calcagni, M., Anselmetti, G. C., and Balthazart, J. (1986). A sexually dimorphic nucleus in the quail preoptic area. *Neurosci. Lett.* **64,** 129–134.

von Frisch, K. (1911). Beiträge zur Physiologie des Pigmentzellen in der Fischhaut. *Pflügers Arch. Gesamte Physiol. Menchen Tiere* **138,** 319–387.

Vowles, D. M., Beazley, L., and Harwood, D. H. (1975). A stereotaxic atlas of the brain of the Barbary dove (*Streptopelia risoria*). *In* "Neural and Endocrine Aspects of Behaviour in Birds" (P. Wright, P. G. Caryl, and D. M. Vowles, eds.), pp. 351–394. Elsevier, Amsterdam.

Waldmann, C., and Güntürkün, O. (1993). The dopaminergic innervation of the pigeon caudolateral forebrain: Immunocytochemical evidence for a "prefrontal cortex" in birds? *Brain Res.* **600,** 225–234.

Wild, J. M., Arends, J. J. A., and Zeigler, H. P. (1990). Projections of the parabrachial nucleus in the pigeon (*Columba livia*). *J. Comp. Neurol.* **293,** 499–523.

Wilson, F. E. (1967). The tubero-infundibular neuron system: A component of the photoperiodic control mechanism of the white-crowned sparrow, *Zonotrichia leucophrys gambelii*. *Z. Zellforsch. Mikrosk. Anat.* **82,** 1–24.

Wingstrand, K. G. (1951). "The Structure and Development of the Avian Pituitary." C. W. K. Gleerup, Lund, The Netherlands.

Yamada, S., Mikami, S., and Yanaihara, N. (1982). Immunohistochemical localization of vasoactive intestinal polypeptide (VIP)-containing neurons in the hypothalamus of the Japanese quail, *Coturnix coturnix*. *Cell Tissue Res.* **226,** 13–26.

Youngren, O. M., and Phillips, R. E. (1978). A stereotaxic atlas of the brain of the three-day-old domestic chick. *J. Comp. Neurol.* **181,** 567–600.

Zeier, H., and Karten, H. J. (1971). The archistriatum of the pigeon: Organization of afferent and efferent connections. *Brain Res.* **31,** 313–326.

Zeigler, H. P. (1976). Feeding behavior of the pigeon. *In* "Advances in the Study of Behavior" (J. S. Rosenblatt, R. A. Hinde, E. Shaw and C. Beer, eds.), pp. 285–389. Academic Press, New York.

Zeigler, H. P., and Karten, H. J. (1973). Brain mechanisms and feeding behavior in the pigeon *Columba livia*. I. Quinto-frontal structures. *J. Comp. Neurol.* **152,** 59–82.

Zeigler, H. P., Karten, H. J., and Green, H. L. (1969). Neural control of feeding in the pigeon. *Psychon. Sci.* **15,** 156–157.

Zweers, G. A. (1971). "A Stereotactic Atlas of the Brainstem of the Mallard (*Anas platyrhynchos L.*)." Van Gorcum, Assen, The Netherlands.

CHAPTER 8

Skeletal Muscle

A. L. HARVEY AND I. G. MARSHALL

Strathclyde Institute for Drug Research
University of Strathclyde
Glasgow G1 1XW, United Kingdom

I. Introduction 123
II. Development of Avian Muscle 123
III. Muscle Fiber Types 126
IV. Innervation 128
V. Electrical Properties of Muscle Fibers 128
VI. Contractile Properties 129
VII. Neuromuscular Transmission 130
VIII. Uses of Avian Muscle in Neuromuscular Pharmacology 133
References 135

I. INTRODUCTION

Avian skeletal muscle and the process of transmission between somatic nerves and skeletal muscle in birds are essentially similar to mammalian skeletal muscle. Figure 1 shows the basic striation pattern of skeletal muscle. Many of the established properties of avian skeletal muscle have been comprehensively reviewed previously (Berger, 1960; George and Berger, 1966; Bowman and Marshall, 1972; van den Berge, 1975). In this chapter we have concentrated on more recent work.

II. DEVELOPMENT OF AVIAN MUSCLE

The chick has been the major species in which morphological studies of avian myogenesis have been made. Muscle development has been studied both *in vivo* and in tissue culture, but owing to the difficulty of defining *in situ* which cells will eventually form skeletal muscle fibers, most of our information on early-stage myogenesis derives from tissue culture studies from single developing muscle cells. Clones, which can be grown in culture from cells capable of forming muscle, can be identified. The change in numbers of muscle precursor cells (myoblasts) has been studied in chick limb muscle (Bonner and Hauschka, 1974); myoblasts amenable to cloning appear on about Day 3 *in ovo*. The numbers increase about six- to sevenfold over the next 6 days. Myogenesis, both *in vivo* and *in vitro*, consists of the fusion of spindle-shaped, uninucleated myoblasts (Figure 2) to form multinucleated myotubes that will eventually grow into mature muscle fibers.

Transmission and scanning electron microscopy has been used to study morphological changes in the early stages of myogenesis of chick embryo muscle cells in culture (Shimada, 1972a,b). Myogenic cells grow very close together, with small projections extending from some myotubes which attach the cell both to adjacent myoblasts and to the surface of the culture dish. Additionally, a cell-surface material is observable that is possibly involved in cellular adhesion. Numerous close junctions with an intercellular distance of 2.5–10 nm and some focal tight junctions with no discernible gap can be detected between pairs of myogenic cells. It is likely that the fusion process is initiated by the formation of close contact between cells, which is followed by the appearance of vesicles and tubules between the adjacent cytoplasms. At the final stage, remnants of broken cell membranes disappear and a common cytoplasm is formed.

Both thick myosin (15–16 nm in diameter) and thin actin (5–6 nm in diameter) filaments are synthesized by clusters of cytoplasmic ribosomes (polysomes). This

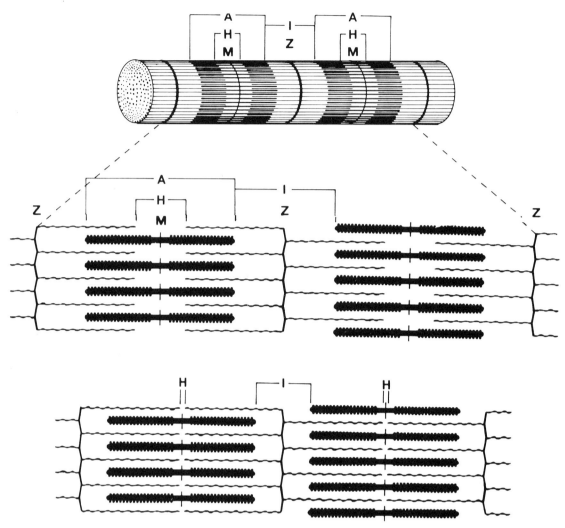

FIGURE 1 Schematic representation of the band pattern of a striated muscle myofibril related to the arrangement of the actin (thin) and myosin (thick) filaments. Two sarcomeres are shown (the area between two adjacent Z-lines is a sarcomere). The myofibrils within the muscle fiber are aligned in a parallel pattern that stretches right across the muscle fiber as a whole. In the contracted state (bottom) the thick and thin filaments slide over one another but neither changes in length. This causes a narrowing of the H and I bands, but no change in the width of the A band which reflects the constant length of the myosin filaments. (Reproduced from Bowman, 1964, with permission.)

contractile protein synthesis greatly increases following myoblast fusion. There is a progressive organization of myofibrils during growth resulting in the mature cell having a cross-striated pattern similar to that of adult skeletal muscle (Shimada et al., 1967; Askanas et al., 1972).

During the earliest myotube stage in embryonic chick muscle, both *in vivo* and *in vitro*, the sarcoplasmic reticulum develops in isolated portions from rough-surfaced endoplasmic reticulum (Ezerman and Ishikawa, 1967; Shimada et al., 1967, Larson et al., 1970, Chan et al., 1990). Subsequently, the isolated portions of sarcoplasmic reticulum join together to create a network around the contractile filaments. The transverse tubular system develops more slowly than the sarcoplasmic reticulum; the surface membrane of the myotubes invaginates to form the T-tubules. Initially the T-tubules consist of shallow vesicles connected to the sarcolemma, but with time they project deeper and deeper into the myotube until contact is made with the sarcoplasmic reticulum (Chan et al., 1990).

FIGURE 2 Phase-contrast micrographs of chick embryo skeletal muscle cells growing in culture. (A) Twenty-four-hour cell culture showing myoblasts (M) and fibroblasts (F). (B) Seven-day culture showing several long myotubes (MT), one of which shows extensive cross-striations (arrows). Scale bar is 50 μm in both A and B.

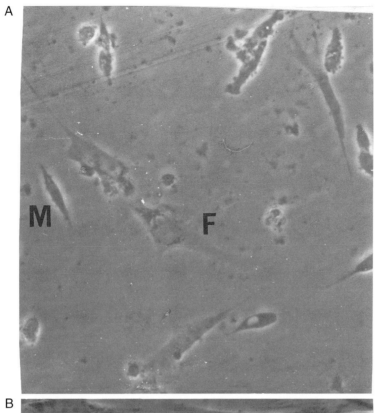
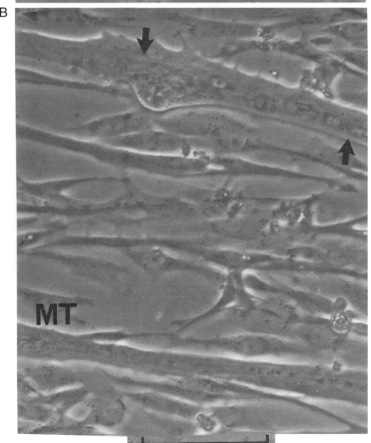

FIGURE 2

The above stages are followed by the main period of growth of the muscle and the differentiation of the fibers into different types. Fibers lengthen by the addition of new sarcomeres and broaden with the increase in myofibrils per fiber (see Burleigh, 1974; Goldspink, 1974; Vrbová et al., 1978 for reviews). An example is the chicken latissimus dorsi muscles where mean fiber diameter increases tenfold from 4 to 6 μm in 18-day-old embryos to 40–60 μm in 8-month-old chickens (Shear and Goldspink, 1971). There are also structural changes in Z-disks during growth in chickens (Ahn et al., 1993). Mechanical strength increases in the weeks after hatching, and the disks in leg muscle become stronger than those in breast muscle. Z-disks are known to be associated with a very large protein (connectin, or titin) that connects them to thick filaments. Titin is also present in chicken muscle (Tan et al., 1993), and it changes its isoform during development (Hattori et al., 1995).

Normal contractile activity is essential for posthatching muscle growth. Immobilization of chicken posterior latissimus dorsi (PLD) muscles for periods up to 11 months results in a dramatic reduction of fiber size (Shear, 1978, 1981). In mature fowls the atrophy is completely reversible, but when the immobilization is initiated immediately posthatching it is only partially reversed.

III. MUSCLE FIBER TYPES

As in amphibians and reptiles, as well as in mammals, some of the avian muscle fibers adapt for rapid, intermittent contraction whereas others adapt for more continuous contraction. These functional differences require differences in the structure and biochemistry of fibers. Muscles are usually described as slow or fast contracting. However, this represents an oversimplification of the situation. The color of the muscles (red or white) does not adequately describe the variety of fiber types that exist either. Indeed, most individual muscles contain a mixture of fiber types, and often more than two types can be distinguished (e.g., Khan, 1976; Toutant et al., 1980; Billeter et al., 1992). Five major fiber types have been recognized on the basis of biochemical and morphological criteria (Table 1) (Barnard et al., 1982). With the important exception of the multiply innervated slow-contracting fibers (which are common in avian, amphibian, and reptilian, but not mammalian muscle), the fiber types in the chicken muscle are closely similar to those

TABLE 1 Comparison of Different Fiber Types in Chicken Muscle[a]

	Twitch fibers			Tonic fibers	
	I	IIA	IIB	IIIA	IIIB
Histochemical criteria					
ATPase (pH 9.4)	No staining	Strong	Strong	Medium	Strong
ATPase (pH 4.6)	Strong	No or weak	Weak	Weak	Medium
ATPase (pH 4.3)	Strong	No staining	No staining	Weak	Medium
NADH-TR[b]	Medium	Weak or medium	No staining	Medium	Medium or strong
Phosphorylase	None or weak	Strong	Srong	Weak	Medium
Fiber innervation	Multiple	Focal	Focal	Multiple	Multiple
Histological characteristics					
Fiber shape	Polygonal	Polygonal	Polygonal	Rounded	Rounded
Fascicle shape	Polygonal	Polygonal	Polygonal	Rounded	Rounded
Mitochondrial density	Very high	High	Low	Very high	Very high
Fiber lipid droplets	No	Yes	No	No	No
Relative fiber size	Small/medium	Medium	Medium	Large	Medium
Myonuclei distribution	Peripheral	Usually peripheral	Usually central	Peripheral	Peripheral
Fiber type composition (%)					
Pectoral	0	<1	>99	0	0
PLD	<3	5–20	80–95	0	0
ALD	0	0	0	65–80	20–35
Sartorius (red)	30–45	35–50	15–25	0	0
Sartorius (white)	0	10–20	80–90	0	0
Plantaris	0	0	0	65–75	25–35

[a]Adapted from Barnard et al. (1982) with permission.
[b]NADH-tetrazolium reductase.

found in mammalian muscle. Very little mammalian muscle is multiply innervated; the extraocular muscles and esophagus are exceptions. Slow-contracting mammalian muscles (e.g., the soleus of the cat) are focally innervated.

In addition to the morphological and enzymatic differences which are shown in Table 1, avian white and red fibers differ in their ultrastructure. Generally, white fibers have a very definite fibrillar appearance (*Fibrillenstruktur*) similar to that of mammalian muscle, whereas red fibers have a more granular and indefinite appearance (*Felderstruktur*).

In *Fibrillenstruktur fibers* (Figure 3A), the myofibrils are polygonal in cross section and uniform in diameter and have a regular arrangement, being separated from each other by a granular sarcoplasm. The cross-striations are evident; the dark A (anisotropic)- and light I (isotropic)-bands. Each I-band is bisected by a smooth Z-line running directly across the fibril (see Figure 1). The H-zone, where only thick filaments are found, can be observed in the midsection of the A-band. The actin and myosin filaments interdigitate on each side of the H-zone causing the greatest optical density in the sarcomere. An M-band can be seen bisecting every H-zone. Taking the chicken posterior latissimus dorsi (PLD) as an example, the myofibrils are 0.5–1 μm in diameter, the A-filaments are 1.55–1.6 μm long, and the I-filaments 1.05 μm long. The I-filaments broaden as they approach the Z-line and overlap 30–40 nm with those of the next sarcomere.

In *Felderstruktur* fibers (Figure 3B), the fibrils are very irregular in both size and distribution. The fibrils connect with each other only at points along the length of the fibrils. In the chicken anterior latissimus dorsi (ALD), the fibrils appear ribbon-shaped in cross section, having diameters of 0.5–1 μm by 2–5 μm. The fibrils are not regularly surrounded by sarcoplasm and granules. The Z-lines take a zigzag course across the width of the fibril. The A-, I-, H-, and M-bands can be distinguished, although the M-line is somewhat less distinct in *Felderstruktur* than in *Fibrillenstruktur* fibers. The A- and I-filaments in the ALD are similar in length to those

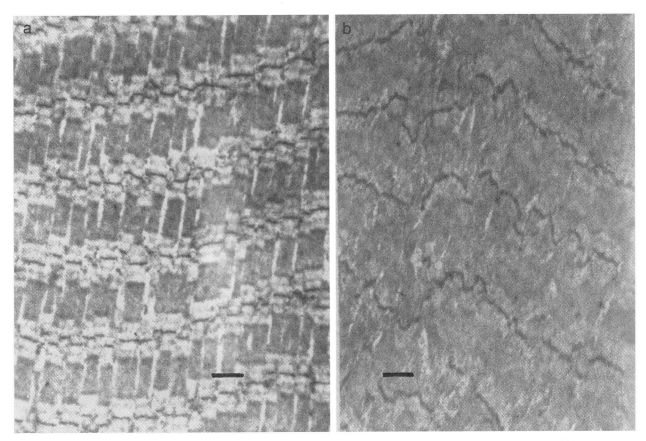

FIGURE 3 Electron micrographs of longitudinal sections of muscles from 2- to 3-month-old chickens. (a) White fibers that have a defined fibrillar appearance ("fibrillenstruktur"). This is a section from the posterior latissimus dorsi. (b) Red fibers that have a more granular and indefinite appearance ("felderstruktur"). This is a section from the anterior latissimus dorsi. Scale bar is 1 micron. Reprinted with permission from Hess (1961). See text for more details.

of the PLD. However, the I-filaments are not regularly arranged at the level of the Z-line, where they can overlap those of the next sarcomere by around 100 nm. One particularly unusual feature is a network of filaments that encircles the fibrils at the Z-line level. Additionally, there is a less well-developed and regular arrangement of the T tubules and sarcoplasmic reticulum in *Felderstruktur* than in *Fibrillenstruktur* fibers. Pigeon ALD fibers appear to have some unique structural features (Khan, 1993).

During development, fibers of posterior latissimus dorsi muscles of the chicken become faster-contracting than ALD fibers at about the same time as the density of triads becomes higher (Takekura et al., 1993). The sarcoplasmic reticulum also begins to differentiate around this stage, but the final fiber-type-specific distribution of T-tubules occurs after hatching. Other studies have correlated the development of the functional Ca^{2+}-channels of the sarcoplasmic reticulum with that of specific forms of foot proteins (Sutko et al., 1991). There are two isoforms of the sarcoplasmic reticulum Ca^{2+}-release channel (or "ryanodine receptor") in chicken skeletal muscles. They have similar conductances but different activation and inactivation properties (Percival et al., 1994). The α-isoform appears to be essential for normal excitation–contraction coupling (Ivanenko et al., 1995).

There are also developmental changes in expression of isoforms of troponin T that correlate with differences in Ca^{2+} sensitivity and contractility in fibers from ALD, PLD, and pectoralis major muscles (Reiser et al., 1992). The fast-contracting PLD and pectoralis major fibers become more sensitive to Ca^{2+} during maturation, which corresponds to changes in isoforms of troponin T, but not of troponin C or I or troponomyosin. There are overall changes in troponin C expression during development, however (Berezowsky and Bag, 1992).

IV. INNERVATION

There is an intimate association between the muscle structure and its innervation. Thus, the final maturation and long-term survival of skeletal muscle is highly dependent on innervation by the motor neurones. The development of the neuromuscular junction is required before individual muscle fibers can fulfill their adult role. The development of nerve–muscle connections has been studied in many species, including in chick muscle (Hirano, 1967; Atsumi, 1971; Landmesser and Morris, 1975; Kikuchi and Ashmore, 1976; Atsumi, 1977; Burrage and Lentz, 1981; Bourgeois and Toutant, 1982; Adachi, 1983) and in pigeon muscle (Torrella et al., 1993). Nerve fibers are observable as early as when the myoblasts are the predominant cell type, but at this early stage there is no evidence of specialization of the nerve ending or of localization of acetylcholinesterase, which is used as a marker of functional transmission. It has been suggested that the cell adhesion molecule, N-cadherin, may be involved in stabilizing neuromuscular contacts (Cifuertis-Diaz et al., 1994). As myotube development proceeds, occasional nerve–muscle contacts can be observed. In the intercostal muscles of chick embryos, a few neuromuscular contacts are observable by Day 6 of incubation. The first development of primitive neuromuscular junctions occurs between Days 7 and 10, and by Days 15–16 mature neuromuscular junctions can be found; these are associated with fully developed muscle fibers. From then on the size of the neuromuscular junction increases with muscle growth, but the basic morphology remains essentially unchanged (Atsumi, 1971). In embryonic chick ALD the morphological development of neuromuscular junctions has been correlated with the onset of transmitter release measured using electrophysiological techniques (Bennett and Pettigrew, 1974). The sequence of innervation has been studied in chick PLD muscles by Bourgeois and Toutant (1982). Additionally, Adachi (1983) has observed that the neuromuscular junctions of different muscles mature at different times, with proximal muscles preceding distal ones.

In the chicken, innervation patterns are related to the different fiber types. White *Fibrillenstruktur* fibers are focally innervated by one or only a few nerve terminals, as in mammalian muscle, whereas the *Felderstruktur*-containing red fibers are multiply innervated by many nerve terminals (Ginsborg and Mackay, 1961; Hess, 1961). The endplate structure differs in focally and multiply innervated fibers: endplates in focally innervated fibers sit on top of the fiber and postjunctional folds like those seen in mammalian muscle are observed; in multiply innervated muscle the endplate is not elevated and postjunctional folds are not seen (Hess, 1967). Muscle spindles (intrafusal fibers) also have different neuromuscular junctions (Maier, 1991). Intracellular microelectrode studies have shown that in multiply innervated fibers the maximum distance between adjacent junctions is of the same order of magnitude as the space constant of the fiber. Ginsborg and Mackay (1961) using electrophysiological techniques to record junctional electrical activity in the form of miniature endplate potentials, measured the average distance between neuromuscular junctions in the ALD of a 2-week-old chick as 225 μm; at 15 weeks this increased to 790 μm. Hess (1961) found a distance of 1000 μm in adult chickens. The distance between endplates appears to be directly proportional to the length of the muscle fiber suggesting that the number of junctions is established early in development and remains constant during growth. Ginsborg and Mackay (1961) estimated that there were about 80 junctions on each of the fibers of the ALD that extend the length of the muscle.

V. ELECTRICAL PROPERTIES OF MUSCLE FIBERS

The resting membrane potential of mature avian muscle fibers is similar to that of other skeletal muscles (i.e., around -70 to -90 mV). In general, adult muscle fiber membranes are much more permeable to K^+ than to Na^+, and this differential permeability develops during growth.

The membrane-passive electrical properties of a muscle fiber determine its response to an electrical stimulus. A high fiber input resistance will result in a large voltage response to a given current pulse; long space and time constants will allow the response to spread over a large area of the membrane. There are very marked differences in membrane-passive properties in avian multiply and focally innervated fibers (Fedde, 1969; Gordon et al., 1977b). The differences are linked to the observations, as in mammalian muscle, that propagating muscle action potentials can be generated in focally innervated fibers, but not always in multiply innervated fibers. In focally innervated chick PLD fibers, propagated muscle action potentials can be elicited by single nerve impulses or by direct electrical stimulation of the muscle fibers (Ginsborg, 1960b; Hník et al., 1967. However, in multiply innervated chick muscle, it is less straightforward. It has been found that *in vivo* the ALD muscle of the chick responds to single-shock nerve stimulation with only local endplate potentials; no action potential is produced. Propagated muscle action potentials can only be elicited *in vivo* by closely spaced twin pulses or by single shocks after a period of high-frequency nerve stimulation (Jirmanová and Vyklický, 1965; Hník et al., 1967). However, *in vitro,* either nerve stimulation or direct muscle stimulation can elicit propagated muscle action potentials in the ALD muscle (Ginsborg, 1960b).

In focally innervated twitch fibers, propagated muscle action potentials emanate from the endplate region as a result of local electrical activity from chemical transmission (see later). These fibers have short space (<1 mm) and time constants (3–4 msec) so that the local potential change evoked by transmitter acetylcholine decays within a very short distance of the endplate (Fedde, 1969; Entrikin and Bryant, 1975; Lebeda and Albuquerque, 1975; Gordon et al., 1977b). However, in the multiply innervated fibers, contraction can be initiated directly by the spread of the acetylcholine-induced potential change. To facilitate this, these fibers have long space (~2 mm) and time constants (~30 msec)
and there are many endplates per muscle fiber. Multiply innervated fibers have high membrane resistances (3000–4000 Ωcm^2) compared to those of focally innervated fibers (500–600 Ωcm^2); this results in a large voltage change in response to the transmitter-induced local current flow through the endplate ion channels. Thus, the combination of electrical properties with the large number of endplates allows simultaneous multiple local responses produced by transmitter acetylcholine to excite the contractile mechanism and produce a muscle contraction.

In terms of the development of the differences between the fiber types, at 14 days *in ovo*, the electrical properties of ALD and PLD fibers are found to be similar. The properties of the PLD change within the first 2 weeks of hatching; some of the changes are associated with the membrane becoming permeable to Cl^- ions (Poznansky and Steele, 1984).

Not surprisingly, given the vastly different mechanism of initiation, the action potentials of the PLD muscles are larger and have a faster maximum rate of rise than those of ALD muscles when they are observed (Cullen et al., 1975). Also, the action potential conduction velocity in the two muscles is markedly different; in the isolated PLD muscle at 31–36°C, the conduction velocity was 2.3–2.8 m/sec, compared to a value of 0.41–0.7 m/sec in isolated ALD muscle at 28–34°C (Ginsborg, 1960b). The difference in conduction velocity is related to the difference in contraction speed of the two muscles.

The different rates of rise of action potentials in the different muscle types presumably relates to different densities of voltage-dependent Na^+-channels. The appearance of such channels in immature chick muscle can be stimulated by the acetylcholine receptor inducing factor ARIA (Corfas and Fischbach, 1993). The numbers of Na^+ channels expressed in cultured chick skeletal muscle may be regulated by intracellular levels of Ca^{2+} (Satoh et al., 1992).

Although the action potentials in mature twitch fibers of avian muscles result from activation of Na^+ channels, less-developed muscle fibers also have voltage-dependent Ca^{2+}-channels (e.g., Yoshida et al., 1990; Satoh et al., 1991). The numbers of such channels can be increased by electrical stimulation (Freud-Silverberg and Shainberg, 1993), although the density of Ca^{2+}-channel binding sites does not appear to be different between T-tubules from normal and dystrophic chickens (Moro et al., 1995).

VI. CONTRACTILE PROPERTIES

Avian skeletal muscle contains actin and myosin filaments, arranged in the classical interdigitated pattern (see above). It is also known to contain the regulatory contractile proteins troponin, tropomyosin, and α-actinin (Allen *et al.*, 1979; Devlin and Emerson, 1978, 1979). It is therefore assumed that the process of excitation–contraction coupling in avian muscle is essentially the same as that in mammalian muscle. Thus, certainly in focally innervated avian muscle fibers, it is assumed that muscle action potentials spread down the T-tubules to activate the contraction mechanism.

The contraction times of multiply innervated muscles with a *Felderstruktur* are 5 to 10 times slower than those of singly innervated muscles with a *Fibrillenstruktur*. This can be observed both *in vivo* (Hnik *et al.*, 1967) and *in vitro* (Ginsborg, 1960b; Gordon and Vrbová, 1975; Gordon *et al.*, 1977b). The time to reach one-half maximum response to a 40-Hz tetanus was about 400–500 msec in the chicken ALD, but only about 50 msec in the chicken PLD. Two other multiply innervated muscles of the chicken, the adductor profundus and the plantaris, have been shown to contract with velocities similar to that of the ALD (Barnard *et al.*, 1982).

Contractile property development has been studied by Gordon *et al.* (1977a,b) in chicken ALD and PLD muscles. After 14–16 days incubation the contraction speeds of both embryo muscles were similar (time to half-maximal tension response to 40 Hz stimulation was ~400–500 msec). Little change in the contraction speed of ALD muscles was observed during subsequent development. In contrast there was a progressive increase in the speed of contraction of PLD muscles

VII. NEUROMUSCULAR TRANSMISSION

The neurotransmitter at avian skeletal muscle neuromuscular junctions is acetylcholine. Evidence for this includes the facts that choline acetyltransferase, the enzyme that synthesizes acetylcholine, is present in chicken ALD and PLD muscles and its activity increases during development (Betz *et al.*, 1980). Drugs such as hemicholinium, which inhibits choline uptake; vesamicol, which blocks synaptic vesicular transport of acetylcholine; and β-bungarotoxin, which blocks acetylcholine release, block neuromuscular transmission in chicken muscle (Marshall, 1969, 1970; Dryden *et al.*, 1974). Thus, it appears from pharmacological inference that prejunctional events at avian junctions share common basic mechanisms with the more widely studied mammalian and amphibian systems. Thus, it is likely that acetylcholine is synthesized from its precursors by choline acetyltransferase in the cytoplasm of the nerve terminal. It is now known that acetylcholine is loaded into synaptic vesicles, their storage structures, by a two-stage concentrative mechanism. In this, protons enter the vesicle by an active transport mechanism involving a V-type ATPase. Intravesicular protons are then exchanged for acetylcholine via the acetylcholine transporter itself (see Parsons *et al.*, 1993 for review). The vesicles are thought to be anchored to the intraterminal cytoskeleton, including actin strands, by a family of synaptic vesicle-associated proteins, the synapsins. Phosphorylation of synapsin by a calcium/calmodulin-dependent protein kinase leads to the freeing of the synaptic vesicle, allowing it to travel toward the active zones or release sites in the prejunctional sarcolemma. A plethora of prejunctional proteins have been associated with the actual process of release, including synaptobrevin, synaptotagmin, syntaxins, synaptophysins, and the α-latrotoxin receptor, rab3A. One possible mechanism is that synaptotagmin, associated with the vesicle, binds to the α-latrotoxin receptor, which is associated with the sarcolemma, forming a fusion pore through which the contents of the vesicle can be released into the synaptic cleft (see Catsicas *et al.*, 1994 for review).

Although the release of acetylcholine from avian motor nerve terminals is dependent on extracellular Ca^{2+}, as in amphibian and mammalian species, the Ca^{2+}-channels are different from those in murine motor nerves. Release of acetylcholine from chicken motor nerves is sensitive to ω-conotoxin GVIA, whereas that from mouse motor nerves is not (De Luca *et al.*, 1991).

On the postjunctional side of the membrane, nicotine itself was long ago demonstrated to cause contracture of avian muscle (Langley, 1905; Gasser and Dale, 1926). As a result of this observation, it was clear that avian skeletal muscle contained the nicotinic type of acetylcholine receptors, a fact that was less clear at that time for mammalian skeletal muscle. Brown and Harvey (1938) showed that close-arterial injection of acetylcholine into the gastrocnemius muscle of anesthetized chickens caused a biphasic response, a rapid contraction of the focally innervated fibers followed by a slower, longer-maintained contracture of the multiply innervated fibers. The presence of acetylcholine receptors in avian muscle has been demonstrated unequivocally by the binding of α-bungarotoxin, an extremely potent and almost irreversible binding component for muscle-type nicotinic receptors from the venom of the Taiwan krait (Chang *et al.*, 1975).

The muscle-type nicotinic receptor, which represents the chemically excitable portion of the muscle membrane, constitutes the recognition site for the chemical transmitter and a cation channel. Thus, it is a ligand-gated cation channel. It is one of a family of such ligand-gated channels, which includes the $5HT_3$, $GABA_A$, glycine, and kainate-type glutamate receptors. It is made up of a protein pentamer inserted into the electrically excitable phospholipid muscle membrane. The receptors are situated on the postjunctional muscle membrane, in close apposition to the sites of release of the immediately available store of acetylcholine; this acetylcholine is stored in synaptic vesicles attached to the specialized release sites, or "active zones."

The muscle-type nicotinic receptor is made up of five glycosylated protein subunits. Each set of subunits spans the muscle membrane to form the wall of an aqueous pore. This is the cation channel through which mainly Na^+ and K^+ flow when the receptor is activated by acetylcholine. The endplate channel, being a relatively nonspecific cation channel, is different from the sodium- and potassium-specific channels involved in the nerve and muscle action potentials.

The five subunits of the mammalian muscle-type nicotinic receptor are designated, and arranged in the order, α-β-α-δ-ε. In *Torpedo* receptors, where much of the original work was performed, and in immature mammalian muscle, the ε-subunit is replaced by a γ-subunit. Acetylcholine interacts with the extracellular N-terminal portions of the two α-subunits, at positions 172–201 of the amino acid sequence; this portion of the receptor is the recognition site for the chemical transmitter.

Each individual subunit of the receptor is made up of alternating hydrophilic and hydrophobic components of the amino acid chain. The hydrophobic sections are associated with the hydrophobic phospholipid muscle membrane. The four hydrophobic sections form the four transmembrane portions of the subunits, designated M1–M4. The five M2 α-helices of the constituent subunits form the inner wall of the channel. The two α-subunits appear identical in amino acid composition. However, their pharmacological binding properties are different. Thus, in *Torpedo* receptor dimers formed from (plus α-or from α-plus-δ have different binding properties. This suggests that the binding sites are close to the boundaries of the α-subunit with these adjoining subunits, rather than to the common β-subunit, which does not appear to play a role in binding.

As indicated earlier, the distribution of endplates and therefore of acetylcholine receptors is different on multiply and focally innervated avian muscle. In the focally innervated fibers, acetylcholine sensitivity appears only at the single neuromuscular junction area of the fiber, whereas there are several peaks of sensitivity in multiply innervated fibers associated with the larger number of endplates per fiber (Fedde, 1969). There is a corresponding difference in the pattern of spontaneous miniature endplate potential activity (Figure 4) (Ginsborg, 1960a). In a focally innervated preparation, miniature endplate potential rise times recorded from any individual fiber do not vary greatly (Figure 4a). In multiply innervated muscles, miniature endplate potentials

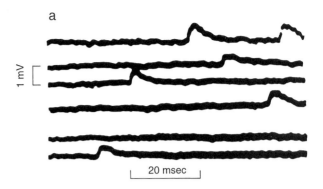

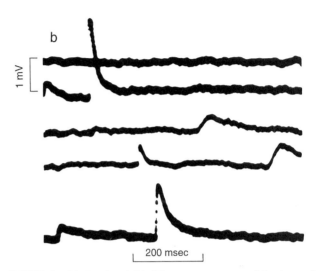

FIGURE 4 (a) Focal and (b) diffuse spontaneous activity (mepps) recorded with intracellular electrodes from two different fibers of the same chick biventer cervicis muscle at 37°C. The relatively constant shape of the potentials in a indicates that they originated at a focal endplate close to which the fiber was impaled. The variations in the shapes of the potentials in b indicate that they originated at junctions at a variety of distances from the point at which the fiber was impaled. Reproduced with permission from Ginsborg (1960a).

have widely varying rise times. The smaller potentials measured have longer rise times than larger potentials, suggesting that the small potentials originate some distance from the site of the recording electrode (Figure 4b). This diffuse pattern of electrical activity has been related to multiple innervation and the more uniform electrical activity to focal (or single) innervation.

Miniature endplate currents from chick ALD and PLD muscles have been recorded using the two-microelectrode voltage clamp technique (Harvey and van Helden, 1981) (Figure 5). In the focally innervated PLD fibers, miniature endplate currents with fast growth times could only be recorded from a single site on each fiber, whereas miniature endplate currents could be recorded from many locations on fibers in ALD muscles. Consistent with Ginsborg's data, obtained using single-

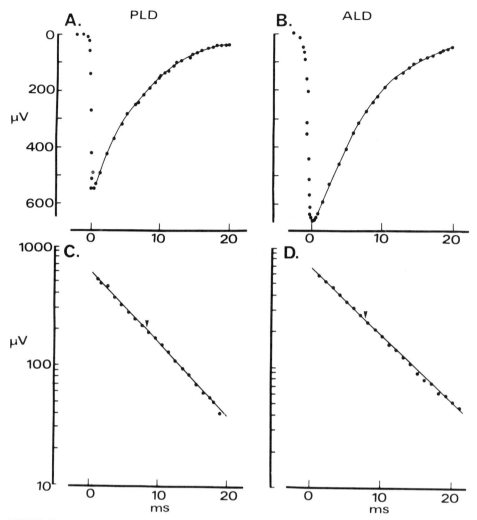

FIGURE 5 The decay of m.e.p.c.s in PLD and ALD muscles is exponential. The averages of 30 extracellularly recorded m.e.p.c.s from a PLD endplate (A) and from and ALD endplate (B) are shown. The amplitude of the decay phase of these averaged m.e.p.c.s is plotted semilogarithmically as a function of time after peak in C (for the PLD) and D (for the ALD). The time constants of decay, τ_D, are shown in the arrows in C and D.

electrode potential recording techniques (Ginsborg, 1960a), the miniature endplate currents in ALD muscles had a much wider range of amplitudes and growth times, consistent with multiple innervation, than those in PLD muscles. In the ALD fibers, very slowly rising miniature endplate currents were frequently observed having amplitudes much greater than predicted from cable theory for the propagation of a transient signal along a muscle fiber. The significance of these giant, slow spontaneous events is unknown, although similar phenomena have been described in mammalian preparations treated with certain drugs (Molgó and Thesleff, 1982).

There is little evidence for significant differences in the functional (Harvey and van Helden, 1981) or biochemical (Sumikawa et al., 1982a,b) properties of receptors from different types of chicken muscle. Thus, in contrast to findings on multiply and singly innervated fibers of the snake (Dionne and Parsons, 1981) and frog (Miledi and Uchitel, 1981), receptors in both types of chicken muscle have similar channel properties (Harvey and van Helden, 1981). Both types of muscle have the same transmitter reversal potential. Similar single-channel conductances (20–40 pS) and single channel lifetimes (4–8 msec at resting potentials of -60 to -80 mV at room temperature) were observed, and similar temperature and voltage sensitivities were measured. One major difference from mammalian or amphibian receptors was that the channel lifetimes of chicken acetylcholine receptors were consistently longer than those of the other species measured under the same conditions. Such long channel lifetimes are also found in cholinoceptors of denervated and in immature mammalian muscle, and there is some evidence for an immunological similarity between such mammalian receptors and those on mature avian ALD muscles (Hall *et al.*, 1985). During development of mammalian muscle it is known that there is a change to short open times (Sakmann and Brenner, 1978). This is associated with the change from a γ- to ε-subunit in the receptor. However, the corresponding change does not occur during the development of chicken ALD or PLD from 16-day-old embryos to 14-week-old animals (Harvey and van Helden, 1981), although there are developmental changes in the molecular forms of acetylcholinesterase at ALD and PLD neuromuscular junctions (Jedrzejczyk *et al.*, 1984).

Acetylcholine receptor turnover has been estimated for *in vivo* developing chicken muscle by measuring the rate of release of bound radioactivity from radiolabelled α-bungarotoxin (Betz *et al.*, 1980; Burden, 1977a,b). Both junctional and extrajunctional receptors had similar half-lives (~30 hr) until about 3 weeks after hatching, when the junctional receptors reached the adult half-life of about 5 days (Burden, 1977b; Betz *et al.*, 1980). The mechanisms responsible for the change in rate of metabolism are not known, but they presumably relate to changes in the cytoskeleton. The change in avian muscle happens much more slowly than in mammalian muscle. Close packing of receptors does not by itself explain the difference, as the half-life of receptors in "hot spots" is similar to that of the diffusely distributed receptors (Schuetze *et al.*, 1978).

The study of the distribution of acetylcholine receptors has been extended to investigations of gene expression and of cytoskeletal proteins that might be associated with clusters of receptors. A glycoprotein has been isolated from chick brain that increases the appearance of receptors in chick myotubes: it has been called ARIA for "acetylcholine receptor-inducing activity." A partial pro-ARIA cDNA has been cloned, and it was shown to produce functionally effective ARIA after transfection into kidney cells (Punn and Tsim, 1995). The glycoprotein ARIA has also been shown to induce the production of Na^+-channels in cultured chick muscle, suggesting that it might regulate several genes important for the development of neuromuscular junctions (Corfas and Fischbach, 1993). Developmental changes in the activity of genes coding for acetylcholine subunits have been studied (e.g., Jia *et al.*, 1992), as has the localization of relevant mRNAs by *in situ* hybridization (Piette *et al.*, 1992). The link between electrical activity of chick muscle (at least in denervated muscle) and inactivation of genes coding for acetylcholine receptors appears to involve the activation of protein kinase C (Huang *et al.*, 1992, 1993), which also regulates other muscle genes (Choi *et al.*, 1991).

Agrin induces the clustering of acetylcholine receptors on membranes of cultured chick myotubes, and agrinlike molecules are present at developing neuromuscular junctions. Agrin also influenced the distribution of some cytoskeletal proteins, including filamin, α-actinin, and vinculin, but had no effect on actin or tropomyosin (Shadiack and Nitkin, 1991). It is probable that proteins like filamin contribute to the stability of the clusters of acetylcholine receptors in mature skeletal muscle.

ATP is released along with acetylcholine from motor nerves because it is contained within synaptic vesicles along with acetylcholine. It is not clear if ATP has a role in neuromuscular transmission in adults, but several ATP-induced responses have been demonstrated in developing chicken muscle and in immature muscle cells in tissue culture. ATP was shown to induce a rapid depolorization by activating a single class of ion channel which was permeable to monovalent cations and anions (Thomas and Hume, 1990). ATP also induced a slower activation of K^+-channels (Hume and Thomas, 1990), which involves two distinct classes of channels with different conductances (23 and 50 pS) (Thomas and Hume, 1993). From pharmacological experiments with a range of ATP agonists and antagonists, it appears that the same class of ATP receptor is linked to both types of response (Thomas *et al.*, 1991).

VIII. USES OF AVIAN MUSCLE IN NEUROMUSCULAR PHARMACOLOGY

It has long been known that avian and amphibian muscles respond quite differently from mammalian muscle to the addition of acetylcholine and other nicotinic agonist drugs such as nicotine and decamethonium which cause endplate depolarization. In early studies of neurotransmission, mimicking of the effects of nerve stimulation by injection of putative chemical transmitters helped to lead to theories of chemical synaptic transmission. However, confirmation of chemical transmission in all types of skeletal muscle was hampered by the fact that innervated mammalian muscle does not respond to intravenously injected acetylcholine or to acetylcholine injected into the solution bathing an isolated muscle. In contrast, nicotine and acetylcholine had long been known to produce contracture responses when injected intravenously into chickens (Langley, 1905; Gasser and Dale, 1926). It was not until after the discovery of the release of acetylcholine from mammalian motor nerves (Dale *et al.*, 1936) that Brown (1938) showed that close intraarterial injection of acetylcholine was required to produce contraction responses in mammals similar to those produced by nerve stimulation.

The difference between the responses of avian and mammalian muscle to endplate depolarizing drugs is related to the previously described innervation and excitation–contraction coupling mechanisms of multiply and focally innervated muscles. Thus, in focally innervated muscles, simultaneous activation of the endplate receptors on many individual muscle fibers is necessary to obtain a synchronized contraction of the muscle. This is difficult to achieve with applied agonists. In multiply innervated muscles, the local endplate depolarizations directly excite the contractile mechanism without the necessity for action potential generation. The muscle will remain in contracture for as long as the depolarizing agent remains activating the receptors. The amplitude of the tension response is related to the number of receptors occupied by the drug, and this phenomenon can be exploited in the study of the action of drugs acting at the neuromuscular junction.

The contracture response of avian multiply innervated muscle has been used to study the actions of nicotinic agonist drugs in the same way as in other multiply innervated muscles, such as the frog rectus abdominis and the leech dorsal muscle. Isolated muscles that have been used for this purpose are the anterior latissimus dorsi (Ginsborg and Mackay, 1960), the semispinalis cervicis (Child and Zaimis, 1960), the biventer cervicis (Ginsborg and Warriner, 1960), and the tibialis anterior (van Reizen, 1968) muscles; the latter two muscles also contain significant numbers of focally innervated fibers. Multiply innervated fibers in the necks and legs of birds has been utilized to differentiate between neuromuscular blocking drugs of the depolarizing and nondepolarizing types (Buttle and Zaimis, 1949; Zaimis, 1953, 1959; Bowman, 1964). These types of drugs are used clinically as skeletal muscle relaxants. However, depolarizing drugs possess several undesirable effects relating to their mechanism of action that necessitate their identification and removal from testing programs designed to discover new agents. Depolarizing agents such as decamethonium, suxamethonium, and carbolonium produce, on intravenous injection in birds, a characteristic spastic paralysis with the neck pulled back and the legs rigidly extended. In contrast, nondepolarizing agents such as tubocurarine produce a flaccid paralysis. Despite the usefulness of this test in unanaesthetized birds, its use is contraindicated on ethical grounds.

It is possible to construct classical concentration–response curves to nicotinic agonists on isolated avian multiply innervated muscle and hence to study the effects of nicotinic antagonists. From these concentration–response curves it is possible to show competitive and noncompetitive blockade of nicotinic receptors by examination of the shape and position of concentration–response lines in the presence of the antagonist.

In our own laboratories, as in others, the isolated chick biventer cervicis nerve–muscle preparation has been widely used as a simple, inexpensive preparation for the initial screening of drugs thought to act at the neuromuscular junction. In this context it has a valuable use both in research and in undergraduate teaching. The muscle can be readily isolated together with its motor nerve, which is encapsulated in its tendon. As the muscle contains a mixture of focally and multiply innervated fibers, stimulation of the motor nerve results in twitch responses mainly of the focally innervated fibers, whereas addition of agonists results in contracture responses of the multiply innervated fibers (Figure 6). Drugs that act to change postjunctional acetylcholine receptor responsiveness affect responses to both nerve stimulation and added agonists. In contrast, drugs that act by changing the release of acetylcholine from the nerve terminals either reduce or increase responses to nerve stimulation, but the responses to added agonists remain unchanged. This is shown with 3,4-diaminopyridine which increases transmitter release (Figure 6). Actions of drugs directly on muscle contractility can be assessed on preparations stimulated directly by addition of KCl or by electrical stimulation after abolition of neuromuscular transmission. We have used this preparation to study the effects of various groups of drugs affecting neuromuscular transmission including competitive neuromuscular blocking agents (Gandiha et al., 1975; Durant et al., 1979; Marshall et al., 1981; Habtemariam et al., 1993; Verma et al., 1994), irreversible postjunctionally acting snake toxins (Dryden et al., 1974; Harvey et al., 1978, 1982; Harvey and Tamiya, 1980), postjunctionally active receptor-associated channel-blocking drugs (Harvey et al., 1984), prejunctionally active blocking drugs interfering with acetylcholine metabolism (Marshall, 1969, 1970a,b), snake toxins and antibiotics that reduce acetylcholine release (Dryden et al., 1974; Singh et al., 1978; Faure et al., 1993), aminopyridines (Bowman et al., 1977; Harvey and Marshall, 1977a,b,c), snake and scorpion toxins that increase transmitter release (Barrett and Harvey, 1979; Harvey and Karlsson, 1980, 1982; Marshall and Harvey, 1989; Rowan et al., 1992; Hollecker et al., 1993), myotoxic and neurotoxic phospholipases A_2 (Rowan et al., 1989a,b, 1991; Mollier et al., 1989; Takasaki et al., 1990; Chwetzoff et al., 1990; Geh et al., 1992; Fatehi et al., 1994), and anticholinesterase agents (Gandiha et al., 1972; Green et al., 1978; Braga et al., 1991). In the case of the last class of drugs, it should be noted that chicken muscle acetylcholinesterase may be insensitive to some types of inhibitors (e.g., Anderson et al., 1985). The preparation has also been used as a model to study the type of drug interaction

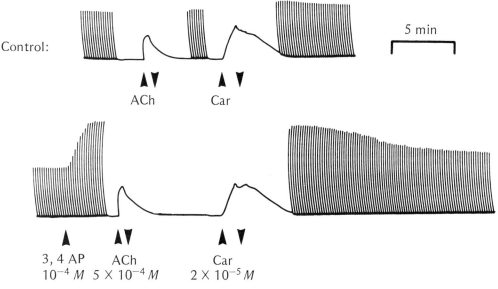

FIGURE 6 Indirectly stimulated chick biventer cervicis nerve–muscle preparation. (A) Responses to single-shock nerve stimulation and to exogenously applied acetylcholine (ACH) and carbachol (car). (B) The augmentation of twitch amplitude produced by 3,4-diaminopyridine (3,4 AP). Note that acetylcholine and carbachol responses are unaffected by the 3,4-diaminopyridine, indicating that the frug is acting to increase the output of transmitter ecetylcholine on nerve stimulation. (Modified from Harvey and Marshall, 1987.)

that causes muscle damage in malignant hyperthermia: a combination of halothane and suxamethonium induced release of creatine kinase, which was prevented by chlorpromazine (McLoughlin et al., 1991).

Because of its simplicity in use, the chick biventer cervicis preparation was recommended as a standard preparation for the screening of snake venoms for neurotoxic and myotoxic effects and for checking that antivenoms can neutralize such effects of venoms (Harvey et al., 1994). This use was demonstrated with six International Reference Antivenoms designated by the World Health Organization (Barfaraz and Harvey, 1994).

References

Adachi, E. (1983). Fluctuation in the development of various skeletal muscles in the chick embryo, with special reference to AChE activity and the formation of neuromuscular junctions. *Dev. Biol.* **95**, 46.

Ahn, D. H., Hattori, A., and Takahashi, K. (1993). Structural changes in z-disks of skeletal-muscle myofibrils during growth of chicken. *J. Biochem.* **113**, 383.

Allen, R. E., Stromer, M. H., Goll, D. E., and Robson, R. M. (1979). Accumulation of myosin, actin, tropomyosin, and α-actinin in cultured muscle cells. *Dev. Biol.* **69**, 655.

Anderson, A. J., Harvey, A. L., and Mbugua, P. M. (1985). Effects of fasciculin 2, an anticholinesterase polypeptide from green mamba venom, on neuromuscular transmission in mouse diaphragm preparations. *Neurosci. Lett.* **54**, 123.

Askanas, V., Shafiq, S. A., and Milhorat, A. T. (1972). Histochemistry of cultured aneural chick muscle. Morphological maturation of fibre types. *Exp. Neurol.* **37**, 218.

Atsumi, S. (1971). The histogenesis of motor neurones with special reference to the correlation of their endplate formation. I. The development of endplates in the intercostal muscle in the chick embryo. *Acta Anat.* **81**, 161.

Atsumi, S. (1977). Development of neuromuscular junctions of fast and slow muscles in chick embryo-light and electron microscopic study. *J. Neurocytol.* **6**, 691.

Barfaraz, A., and Harvey, A. L. (1994). The use of the chick biventer cervicis preparation to assess the protective activity of six international reference antivenoms on the neuromuscular effects of snake venoms *in vitro*. *Toxicon* **32**, 267.

Barnard, E. A., Lyles, J. M., and Pizzey, J. A. (1982). Fibre types in chicken skeletal muscles and their changes in muscular dystrophy. *J. Physiol.* **331**, 333.

Barrett, J. C., and Harvey, A. L. (1979). Effects of venom of the green mamba, *Dendroaspis angusticeps*, on skeletal muscle and neuromuscular transmission. *Br. J. Pharmacol.* **67**, 199.

Berezowsky, C., and Bag, J. (1992). Developmentally regulated troponin-C messenger-RNAs of chicken skeletal muscle. *Biochem. Cell Biol.* **70**, 156.

Berger, A. J. (1960). The Musculature. In "Biology and Comparative Physiology of Birds" (A. J. Marshall, ed.), Vol. 1, p. 301. Academic Press, London/New York.

Billeter, R., Messerli, M., Wey, E., Puntschart, A., Jostarndt, K., Eppenberger, H. M., and Perriard, J. C. (1992). Fast myosin light chain expression in chicken muscles studied by *in situ* hybridization. *J. Histochem. Cytochem.* **40**, 1547.

Bennett, M. R., and Pettigrew, A. G. (1974). The formation of synapses in striated muscle during development. *J. Physiol.* **241**, 515.

Betz, H., Bourgeois, J.-P., and Changeux, J.-P. (1980). Evolution of cholinergic proteins in developing slow and fast skeletal muscles in chick embryo. *J. Physiol.* **302**, 197.

Bonner, P. H., and Hauschka, S. D. (1974). Clonal analysis of vertebrate myogenesis. I. Early developmental events in the chick limb. *Dev. Biol.* **37**, 317.

Bourgeois, J.-P., and Toutant, M. (1982). Innervation of avian latissimus dorsi muscles and axonal outgrowth pattern in the posterior latissimus dorsi motor nerve during embryonic development. *J. Comp. Neurol.* **208,** 1.

Bowman, W. C. (1964). Neuromuscular blocking agents. In "Evaluation of Drug Activities: Pharmacometrics" (D. R. Laurence and A. L. Bacharach, eds.), p. 325. Academic Press, London/New York.

Bowman, W. C., and Marshall, I. G. (1972). Muscle. In "Physiology and Biochemistry of the Domestic Fowl" (D. J. Bell and B. M. Freeman, eds.), Vol. 2, p. 707. Academic Press, London.

Bowman, W. C., Harvey, A. L., and Marshall, I. G. (1977). The actions of aminopyridines on avian muscle. *Naunyn-Schmiedeberg's Arch. Pharmacol.* **297,** 99.

Braga, M. F. M., Harvey, A. L., and Rowan, E. G. (1991). Effects of tacrine, velnacrine (HP029), suronacrine (HP128), and 3,4-diaminopyridine on skeletal neuromuscular transmission *in vitro*. *Br. J. Pharmacol.* **102,** 909.

Brown, G. L. (1938). The preparations of the tibialis anterior (cat) for close-arterial injections. *J. Physiol.* **92,** 22.

Brown, G. L., and Harvey, A. M. (1938). Neuromuscular conduction in the fowl. *J. Physiol.* **93,** 285.

Burden, S. (1977a). Development of the neuromuscular junction in the chick embryo: The number, distribution and stability of acetylcholine receptors. *Dev. Biol.* **57,** 317.

Burden, S. (1977b). Acetylcholine receptors at the neuromuscular junction developmental change in receptor turnover: The number, distribution and stability of acetylcholine receptors. *Dev. Biol.* **61,** 79.

Burleigh, I. G. (1974). On the cellular regulation of growth and development in skeletal muscle. *Biol. Rev. Cambridge Philos. Soc.* **49,** 267.

Burrage, T. G., and Lentz, T. L. (1981). Ultrastructural characterization of surface specializations containing high-density acetylcholine receptors on embryonic chick myotubes *in vivo* and *in vitro*. *Dev. Biol.* **85,** 267.

Buttle, G. A. H., and Zaimis, E. J. (1949). The action of decamethonium iodide in birds. *J. Pharm. Pharmacol.* **1,** 991.

Catsicas, S., Grenningloh, G., and Pich, E. M. (1994). Nerve terminal proteins: To fuse to learn. *Trends Neurosci.* **17,** 368.

Chan, C. Z., Sato, K., and Shimada, Y. (1990). Three dimensional electron microscopy of the sarcoplasmic reticulum and T-system in embryonic chick skeletal muscle cells *in vitro*. *Protoplasma* **154,** 112.

Chang, C. C., Su, M. J., and Lee, M.-C. (1975). A quantification of acetylcholine receptors of the chick biventer cervicis muscle. *J. Pharm. Pharmacol.* **27,** 454.

Child, K. J., and Zaimis, E. (1960). A new biological method for the assay of depolarizing substances using the isolated semispinalis cervicis muscle of the chick. *Br. J. Pharmacol.* **15,** 412.

Choi, J. K., Holtzer, S., Chacko, S. A., Lin, Z. X., Hofmann, R. K., and Holtzer, H. (1991). Phorbol esters selectively and reversibly inhibit a subset of myofibrillar genes responsible for the ongoing differentiation program of chick skeletal myotubes. *Mol. Cell. Biol.* **11,** 4473.

Chwetzoff, S., Mollier, P., Bouet, F., Rowan, E. G., Harvey, A. L., and Menez, A. (1990). On the purification of notexin: Isolation of a single amino acid variant of notexin from the venom of *Notechis scutatus scutatus*. *FEBS Lett.* **261,** 226.

Cifuentes-Diaz, C., Nicolet, M., Goudou, D., Reiger, F., and Mege, R. M. (1994). N-cadherin expression in developing, adult and denervated chicken neuromuscular system—Accumulations at both the neuromuscular junction and the node of Ranvier. *Development* **120,** 1.

Corfas, G., and Fischbach, G. D. (1993). The number of Na^+ channels in cultured chick muscle is increased by ARIA, an acetylcholine receptor inducing activity. *J. Neurosci.* **13,** 2118.

Cullen, M. J., Harris, J. B., Marshall, M. W., and Ward, M. R. (1975). An electrophysiological and morphological study of normal and denervated chicken latissimus dorsi muscles. *J. Physiol.* **245,** 371.

Dale, H. H., Feldberg, W., and Vogt, M., (1936). Release of acetylcholine at voluntary motor nerve endings. *J. Physiol.* **86,** 353.

De Luca, A., Rand, M. J., Reid, J. J., and Story, D. F. (1991). Differential sensitivities of avian and mammalian neuromuscular junctions to inhibition of cholinergic transmission by omega-conotoxin GVIA. *Toxicon* **29,** 311.

Devlin, R. B., and Emerson, C. P. (1978). Coordinate regulation of contractile protein synthesis during myoblast differentiation. *Cell* **13,** 599.

Devlin, R. B., and Emerson, C. P. (1979). Coordinate accumulation of contractile protein mRNAs during myoblast differentiation. *Dev. Biol.* **69,** 202.

Dionne, V. E., and Parsons, R. L. (1981). Characteristics of the acetylcholine-operated channel at twitch and slow fibre neuromuscular junctions of the garter snake. *J. Physiol.* **310,** 145.

Dryden, W. F., Harvey, A. L., and Marshall, I. G. (1974). Pharmacological studies on the bungarotoxins: Separation of the fractions and their neuromuscular activity. *Eur. J. Pharmacol.* **26,** 256.

Durant, N. N., Marshall, I. G., Savage, D. S., Sleigh, T., and Carlyle, I. C. (1979). The neuromuscular and autonomic blocking actions of pancuronium, Prg NC45, and other pancuronium analogues in the cat. *J. Pharm. Pharmacol.* **31,** 831.

Entrikin, R. K., and Bryant, S. H. (1975). Electrophysiological properties of biventer cervicis muscle fibers of normal and Roller pigeons. *J. Neurobiol.* **6,** 201.

Ezerman, E. B., and Ishikawa, H. (1967). Differentiation of the sarcoplasmic reticulum and T system in developing chick skeletal muscle *in vitro*. *J. Cell Biol.* **35,** 405.

Fatehi, M., Rowan, E. G., Harvey, A. L., and Harris, J. B. (1994). The effects of five phospholipase A_2 from the venom of king brown snake, *Pseudechis australis*, on nerve and muscle. *Toxicon* **32,** 1559.

Fedde, M. R. (1969). Electrical properties and acetylcholine sensitivity of singly and multiply innervated avian muscle fibers. *J. Gen. Physiol.* **53,** 624.

Freud-Silverberg, M., and Shainberg, A. (1993). Electric-stimulation regulates the level of Ca channels in chick muscle culture. *Neurosci. Lett.* **151,** 104.

Gandiha, A., Green, A. L., and Marshall, I. G. (1972). Some effects of hexamethonium and tetraethylammonium at a neuromuscular junction of the chicken. *Eur. J. Pharmacol.* **18,** 174.

Gandiha, A., Marshall, I. G., Paul, D., Rodger, I. W., Scott, W., and Singh, H. (1975). Some actions of chandonium iodide, a new short-acting muscle relaxant. *Clin. Exp. Pharmacol. Physiol.* **2,** 159.

Gasser, H. S., and Dale, H. (1926). The pharmacology of denervated muscle. II. Some phenomena of antagonism and the formation of lactic acid in chemical contracture. *J. Pharmacol. Exp. Ther.* **28,** 290.

Geh, S. L., Rowan, E. G., and Harvey, A. L. (1992). Neuromuscular effects of four phospholipases A_2 from the venom of *Pseudechis australis*, the Australian king brown snake. *Toxicon* **30,** 1051.

George, J. C., and Berger, A. J. (1966). "Avian Myology." Academic Press, London/New York.

Ginsborg, B. L. (1960a). Spontaneous activity in muscle fibres of the chick. *J. Physiol.* **150,** 707.

Ginsborg, B. L. (1960b). Some properties of avian skeletal muscle fibers with multiple neuromuscular junctions. *J. Physiol.* **154,** 581.

Ginsborg, B. L., and Mackay, B. (1960). The latissimus dorsi muscles of the chick. *J. Physiol.* **153**, 19.

Ginsborg, B. L., and Mackay, B. (1961). A histochemical demonstration of two types of motor innervation in avian skeletal muscle. *Bibl. Anat.* **2**, 174.

Ginsborg, B. L., and Warriner, J. (1960). The isolated chick biventer cervicis nerve–muscle preparation. *Br. J. Pharmacol.* **15**, 410.

Goldspink, G. (1974). Development of muscle. In "Differentiation and Growth of Cells in Vertebrate Tissues" (G. Goldspink, ed.), p. 69. Chapman and Hall, London.

Gordon, T., and Vrbová, G. (1975). The influence of innervation on the differentiation of contractile speeds of developing chick muscles. *Pflueger's Arch.* **360**, 199.

Gordon, T., Perry, R., Srihari, T., and Vrbová, G. (1977a). Differentiation of slow and fast muscle in chickens. *Cell Tissue Res.* **180**, 211.

Gordon, T., Purves, R. D., and Vrbová, G. (1977b). Differentiation of electrical and contractile properties of slow and fast muscle fibers. *J. Physiol.* **269**, 535.

Green, A. L., Lord, J. A. H., and Marshall, I. G. (1978). The relationship between cholinesterase inhibition in the chick biventer cervicis muscle and its sensitivity to exogenous acetylcholine. *J. Pharm. Pharmacol.* **30**, 426.

Hall, Z. W., Gorin, P. D., Silberstein, L., and Bennett, C. (1985). A postnatal change in the immunological properties of the acetylcholine receptor at rat muscle endplates. *J. Neurosci.* **5**, 730.

Harvey, A. L., and Karlsson, E. (1980). Dendrotoxin from the venom of the green mamba, *Dendroaspis angusticeps:* A neurotoxin that enhances acetylcholine release at neuromuscular junctions. *Naunyn-Schmiedeberg's Arch. Pharmacol.* **312**, 1.

Harvey, A. L., and Karlsson, E. (1982). Protease inhibitor homologues from mamba venoms: Facilitation of acetylcholine release and interactions with prejunctional blocking toxins. *Br. J. Pharmacol.* **77**, 153.

Harvey, A. L., and Marshall, I. G. (1977a). The actions of three diaminopyridines on the chick biventer cervicis muscle. *Eur. J. Pharmacol.* **44**, 303.

Harvey, A. L., and Marshall, I. G. (1977b). The facilitatory actions of aminopyridines and tetraethylammonium on neuromuscular transmission and muscle contractility in avian muscle. *Naunyn-Schmiedeberg's Arch. Pharmacol.* **299**, 53.

Harvey, A. L., and Marshall, I. G. (1977c). A comparison of the effects of amino-pyridines on isolated chicken and rat skeletal muscle preparations. *Comp. Biochem. Physiol. C, Comp. Pharmacol.* **58**, 161.

Harvey, A. L., and Tamiya, N. (1980). Role of phospholipase activity in the neuromuscular paralysis produced by some components isolated from the venom of the sea snake, *Laticauda semifasciata. Toxicon* **18**, 65.

Harvey A. L., and van Helden, D. (1981). Acetylcholine receptors in singly and multiply innervated skeletal muscle fibers of the chicken during development. *J. Physiol.* **317**, 397.

Harvey, A. L., Barfaraz, A., Thomson, E., Faiz, A., Preston, S., and Harris, J. B. (1994). Screening of snake venoms for neurotoxic and myotoxic effects using simple *in vitro* preparations from rodents and chicks. *Toxicon,* **32**, 257.

Harvey, A. L., Jones, S. V. P., and Marshall, I. G. (1984). Disopyramide produces noncompetive, voltage-dependent block at the neuromuscular junction. *Br. J. Pharmacol.* **81**, 169P.

Harvey, A. L., Marshall, R. J., and Karlsson, E. (1982). Effects of purified cardiotoxins from the Thailand cobra (*Naja naja siamensis*) on isolated skeletal and cardiac muscle preparations. *Toxicon,* **20**, 379.

Harvey, A. L., Rodger, I. W., and Tamiya, N. (1978). Neuromuscular blocking activity of two fractions isolated from the venom of the sea snake, *Laticauda semifasciata. Toxicon* **16**, 45.

Hattori, A., Ishii, T., Tatsumi, R., and Takahashi, K. (1995). Changes in the molecular types of connectin and nebulin during development of chicken skeletal-muscle. *Biochim. Biophys. Acta* **1244**, 179.

Hess, A. (1961). Structural differences of fast and slow extrafusal muscle fibres and their nerve endings in chickens. *J. Physiol.* **157**, 221.

Hess, A. (1967). The structure of vertebrate slow and twitch muscle fibres. *Invest. Opthalmol.* **6**, 217.

Hirano, H. (1976). Ultrastructural study on the morphogenesis of the neuromuscular junction in the skeletal muscle of the chick. *Z. Zellforsch. Mikrosk. Anat.* **79**, 198.

Hník, P., Jirmanová, I., Vyklický, L., and Zelená, J. (1967). Fast and slow muscles of the chick after nerve cross-union. *J. Physiol.* **193**, 309.

Hollecker, M., Marshall, D. L., and Harvey, A. L. (1993). Structural features important for the biological activity of the potassium channel blocking dendrotoxins. *Br. J. Pharmacol.* **110**, 790.

Huang, C. F., Neville, C. M., and Schmidt, J. (1993). Control of myogenic factor genes by the membrane depolarization protein kinase C cascade in chick skeletal-muscle. *FEBS Lett.* **319**, 21.

Huang, C. F., Tong, J., and Schmidt, J. (1992). Protein-kinase-C couples membrane excitation to acetylcholine-receptor gene inactivation in chick skeletal-muscle. *Neuron* **9**, 671.

Hume, R. I., and Thomas, S. A. (1990). Activation of potassium channels in chick skeletal muscle by extracellular ATP. *Ann. N. Y. Acad. Sci.* **603**, 486.

Ivanenko, A., McKemy, D. D., Kenyon, J. L., Airey, J. A., and Sutko, J. L. (1995). Embryonic chicken skeletal muscle cells fail to develop normal excitation-contraction coupling in the absence of the alpha ryanodine receptor—implications for a 2-ryanodine receptor system. *J. Biol. Chem.* **270**, 4220.

Jedrzejczyk, J., Silman, I., Lai, J., and Barnard, E. A. (1984). Molecular forms of acetylcholinesterase in synaptic and extrasynaptic regions of avian tonic muscle. *Neurosci. Lett.* **46**, 283.

Jia, H. T., Tsay, H. J., and Schmidt, J. (1992). Analysis of binding and activating functions of the chick muscle acetylcholine receptor gamma-subunit upstream sequence. *Cell. Mol. Neurol.* **12**, 241.

Jirmanova, I., and Vyklický, L. (1965). Post-tetanic potentiation in multiply innervated muscle fibres of the chick (in Czech). *Cslká Fysiol.* **14**, 351.

Khan, M. A. (1976). Histochemical sub-types of three fibre types of avian skeletal muscle. *Histochemistry* **50**, 9.

Khan, M. A. (1993). Unique ultrastructural characteristics of pigeon anterior latissimus dorsi muscle. *Cell. Mol. Biol. Res.* **39**, 65.

Kikuchi, T., and Ashmore, C. R. (1976). Developmental aspects of the innervation of skeletal muscle fibers in the chick embryo. *Cell Tissue Res.* **171**, 233.

Landmesser, L., and Morris, D. G. (1975). The development of functional innervation in the hind limb of the chick embryo. *J. Physiol.* **249**, 301.

Langley, J. N. (1905). On the reaction of cells and nerve endings to certain poisons, chiefly as regards the reaction of striated muscle to nicotine and to curari. *J. Physiol.* **33**, 374.

Larson, P. F., Jenkinson, M., and Hudgson, P. (1970). The morphological development of chick embryo skeletal muscle grown in tissue culture as studied by electron microscopy. *J. Neurol. Sci.* **10**, 385.

Lebeda, F. J., and Alberquerque, E. X. (1975). Membrane cable properties of normal and dystrophic chicken muscle fibers. *Exp. Neurol.* **47**, 544.

Maier, A. (1991). Axon contacts and acetylcholinesterase activity on chicken intrafusal muscle fiber types identified by their myosin heavy chain composition. *Anat. Embryol.* **184**, 497.

Marshall, D. L., and Harvey, A. L. (1989). Block of potassium channels and facilitation of acetylcholine release at the neuromuscular junction by the venom of the scorpion, *Pandinus imperator*. *Toxicon* **27**, 493.

Marshall, I. G. (1969). The effects of some hemicholinium-like compounds on the chick biventer cervicis muscle preparation. *Eur. J. Pharmacol.* **8**, 204.

Marshall, I. G. (1970a). Studies on the blocking action of 2-(4-phenylpiperidino)-cyclohexanol (AH 5183). *Br. J. Pharmacol.* **38**, 503.

Marshall, I. G. (1970b). A comparison between the blocking actions of 2-(4-phenylpiperidino)-cyclohexanol (AH 5183) and its N-methyl quaternary analogue (AH 5954). *Br. J. Pharmacol.* **40**, 68.

Marshall, I. G., Harvey, A. L., Singh, H., Bhardwaj, T. R., and Paul, D. (1981). The neuromuscular and autonomic blocking effects of azasteroids containing choline or acetylcholine fragments. *J. Pharm. Pharmacol.* **33**, 451.

McLoughlin, C., Mirakhur, R. K., Trimble, E. R., and Clarke, R. S. J. (1991). Pathogenesis of suxamethonium-induced muscle damage in the biventer cervicis muscle in the chick. *Br. J. Anaesth.* **67**, 764.

Miledi, R., and Uchitel, O. D. (1982). Properties of post-synaptic channels induced by acetylcholine in different frog muscle fibres. *Nature (London)* **291**, 162.

Mollier, P., Unwetzoff, S., Bouet, F., Harvey, A. L., and Menez, A. (1989). Tryptophanilo, a residue involved in the toxic activity but not in the enzymatic activity of notexin. *Eur. J. Biochem.* **185**, 263.

Moro, G., Saborido, A., Delgado, J., Molano, F., and Megias, A. (1995) Dihydropyridine receptors in transverse tubules from normal and dystrophic chicken skeletal muscle. *J. Muscle Res. Cell Motil.* **16**, 529.

Parsons, S. M., Prior, C., and Marshall, I. G. (1993) The acetylcholine transporter and storage system. *Int. Rev. Neurobiol.* **35**, 279.

Percival, A. L., Williams, A. J., Kenyon, J. L., Grinsell, M. M., Airey, J. A., and Sutko, J. L. (1994). Chicken skeletal muscle ryanodine receptor isoforms—ion channel properties. *Biophys. J.* **67**, 1834.

Piette, J., Huchet, M., Duclert, A., Fujisawasehara, A., and Changeux, J. P. (1992). Localization of messenger-RNAs coding for CMD1, myogenin and the alpha-subunit of the acetylcholine receptor during skeletal muscle development in the chicken. *Mech. Dev.* **37**, 95.

Poznansky, M. J., and Steele, J. A. (1984). Membrane electrical properties of developing fast-twitch and slow-tonic muscle fibres of the chick. *J. Physiol.* **347**, 633.

Punn, S., and Tsim, K. W. K. (1995). Truncated form of pro-acetylcholine receptor-inducing activitiy (ARIA) induces AChR alpha subunit but not AChE transcripts in cultured chick myotubes. *Neurosci. Lett.* **198**, 107.

Reiser, P. J., Greaser, M. L., and Moss, R. L. (1992). Developmental changes in troponin-T isoform expression and tension production in chicken single skeletal muscle fibers. *J. Physiol.* **449**, 573.

Rowan, E. G., Harvey, A. L., and Menez, A. (1991). Neuromuscular effects of nigexine, a basic phospholipase A2 from *Naja nigricollis* venom. *Toxicon* **29**, 371.

Rowan, E. G., Harvey, A. L., Takasaki, C., and Tamiya, N. (1989a). Neuromuscular effects of three phospholipases A_2 from the venom of the Australian king brown snake *Pseudechis australis*. *Toxicon* **27**, 551.

Rowan, E. G., Harvey, A. L., Takasaki, C., and Tamiya, N. (1989b). Neuromuscular effects of a toxic phospholipase A_2 and its nontoxic homologue from the venom of the sea snake, *Laticanda colubrina*. *Toxicon* **27**, 587.

Rowan, E. G., Vatanpour, H., Furman, B. L., Harvey, A. L., Tanira, M. O. M., and Gopalkrishnakone, P. (1992). The effects of Indian red scorpion *Buthus tamulus* venom *in vivo* and in *vitro*. *Toxicon* **30**, 1157.

Sakmann, B., and Brenner, H. R. (1978). Change in synaptic channel gating during neuromuscular development. *Nature (London)* **276**, 401.

Satoh, R., Nakabayashi, Y., and Kani, M. (1991). Pharmacological properties of two types of calcium-channel in embryonic chick skeletal muscle cells in culture. *Neurosci. Lett.* **122**, 233.

Satoh, R., Nakabayashi, Y., and Kani, M. (1992). Chronic treatment with D600 enhances development of sodium channels in cultured chick skeletal muscle cells. *Neurosci. Lett.* **138**, 249.

Schuetze, S. M., Frank, E. F., and Fischbach, G. D. (1978). Channel open time and metabolic stability of synaptic and extrasynaptic acetylcholine receptors on cultured chick myotubes. *Proc. Natl. Acad. Sci. USA* **75**, 520.

Shadiack, A. M., and Nitkin, R. M. (1991). Agrin induces alpha-actinin, filamin, and vinculin to co-localize with AChR clusters on cultured chick myotubes. *J. Neurobiol.* **22**, 617.

Shear, C. R. (1978). Cross-sectional myofibre and myofibril growth in immobilized developing skeletal muscle. *J. Cell. Sci.* **29**, 297.

Shear, C. R. (1981). Effects of disuse on growing and adult chick skeletal muscle. *J. Cell. Sci.* **48**, 35.

Shear, C. R., and Goldspink, G. (1971). Structural and physiological changes associated with the growth of avian fast and slow muscle. *J. Morphol.* **135**, 351.

Shimada, Y. (1972a). Scanning electron microscopy of myogenesis in monolayer culture: A preliminary study. *Dev. Biol.* **29**, 227.

Shimada, Y. (1972b). Early stages in the reorganization of dissociated embryonic chick skeletal muscle cells. *Z. Anat. Entwicklungsgesch.* **138**, 255.

Shimada, Y., Fischman, D. A., and Moscona, A. A. (1967). The fine structure of embryonic chick skeletal muscle cells differentiated *in vitro*. *J. Cell Biol.* **35**, 445.

Singh, Y. N., Marshall, I. G., and Harvey, A. L. (1978). Some effects of the amino-glycoside antibiotic amikacin on neuromuscular and autonomic transmission. *Br. J. Anaesth.* **50**, 109.

Sumikawa, K., Mehraban, F., Dolly, J. O., and Barnard, E. A. (1982a). Similarity of acetylcholine receptors of denervated, innervated and embryonic chicken muscles. 1. Molecular species and their purification. *Eur. J. Biochem.* **126**, 465.

Sumikawa, K., Barnard, E. A., and Dolly, J. O. (1982b). Similarity of acetylcholine receptors of denervated, innervated and embryonic chicken muscles. 2. Subunit compositions. *Eur. J. Biochem.* **126**, 473.

Sutko, J. L., Airey, J. A., Murakami, K., Takeda, M., Beck, C., Deerinck, T., and Ellisman, M. H. (1991). Foot protein isoforms are expressed at different times during embryonic chick skeletal muscle development. *J. Cell Biol.* **113**, 793.

Takekura, H., Shuman, H., and Franzini-Armstrong, C. (1993). Differentiation of membrane systems during development of slow and fast skeletal muscle fibers in chicken. *J. Muscle Res. Cell Motil.* **14**, 633.

Takasaki, C., Sugama, A., Yanagita, A., Tamiya, N., Rowan, E. G., and Harvey, A. L. (1990). Effects of chemical modifications of Pa-11, a phospholipase A2 from the venom of Australian king brown snake (*Pseudechis australis*), on its biological activities. *Toxicon* **28**, 107.

Tan, K. O., Sater, G. R., Myers, A.M., Robson, R. M., and Huiatt, T. W. (1993). Molecular characterization of avian muscle titin. *J. Biol. Chem.* **268**, 22900.

Thomas, S. A., and Hume, R. I. (1990). Permeation of both cations and anions through a single class of ATP-activated ion channels in developing chick skeletal muscle. *J. Gen. Physiol.* **95**, 569.

Thomas, S. A., and Hume, R. I. (1993). Single potassium channel currents activated by extracellular ATP in developing chick skeletal muscle—A role for 2nd messengers. *J. Neurophysiol.* **69**, 1556.

Thomas, S. A., Zawisa, M. J., Lin, X., and Hume, R. I. (1991). A receptor that is highly specific for extracellular ATP in developing chick skeletal muscle *in vitro*. *Br. J. Pharmacol.* **103**, 1963.

Toutant, J. P., Toutant, M. N., Renaud, D., and le Douarin, G. H. (1980). Histochemical differentiation of extrafusal muscle fibres of the anterior latissimus dorsi in the chick. *Cell Differ.* **9**, 305.

Torrella, J. R., Fouces, V., Palomeque, J., and Viscor, G. (1993). Innervation distribution pattern, nerve ending structure, and fiber types in pigeon skeletal muscle. *Anat. Rec.* **237**, 178

Van Den Berge, J. C. (1975). Aves myology. *In* "Sisson and Grossman's The Anatomy of the Domestic Animals," p. 1802. W. B. Saunders, Philadelphia.

Van Reizen, H. (1968). Classification of neuromuscular blocking agents in a new neuromuscular preparation of the chick *in vitro*. *Eur. J. Pharmacol.* **5**, 29.

Vrbrová, G., Gordon, T., and Jones, R. (1978). "Nerve-Muscle Interaction." Chapman and Hall, London.

Yoshida, A., Takahashi, M., Fujimoto, Y., Takisawa, H., and Nakamura, T. (1990). Molecular characterization of 1,4-dihydropyridine sensitive calcium channels of chick heart and skeletal muscle. *J. Biochem.* **107**, 608.

Zaimis, E. J. (1953). Motor endplate differences as a determining factor in the mode of action of neuromuscular blocking substances. *J. Physiol.* **122**, 238.

Zaimis, E. J. (1959). Mechanisms of neuromuscular blockade. *In* "Curare and Curare-like Agents" (D. Bovet, F. Bovet-Nitti, and G. B. Marini-Bettolo, eds.), p. 191. Elsevier, Amsterdam.

CHAPTER 9

The Cardiovascular System

FRANK M. SMITH
Department of Anatomy and Neurobiology
Dalhousie University
Halifax, Nova Scotia
Canada B3H 4H7

NIGEL H. WEST
Department of Physiology
University of Saskatchewan
Saskatoon, Saskatchewan
Canada S7N 5E5

DAVID R. JONES
Department of Zoology
University of British Columbia
Vancouver, British Columbia
Canada V6T 1Z4

I. Introduction 141
II. Heart 142
    A. Gross Structure and Function 142
    B. Cardiac Variables 147
    C. Fine Structure and Cardiac Electrophysiology 148
III. General Circulatory Hemodynamics 154
IV. The Vascular Tree 157
    A. Arterial System 157
    B. Capillary Beds 165
    C. Venous System 173
V. Blood 176
    A. Components 176
    B. Rheology 177
    C. Effects of Altitude 180
    D. Temperature Effects 180
VI. Control of the Cardiovascular System 181
    A. Control Systems 181
    B. Control of Peripheral Blood Flow 181
    C. Control of the Heart 190
    D. Reflexes Controlling the Circulation 215
    E. Integrative Neural Control 221
    References 223

I. INTRODUCTION

Birds have evolved a high-performance cardiovascular system to meet the rigorous demands of running, flying, swimming, or diving in a variety of environments, some of them very harsh. Sustained high levels of activity in these environments place severe demands on the transport capabilities of the cardiovascular system to provide adequate delivery of oxygen to working vascular beds and to provide efficient removal of metabolic products. Furthermore, birds are homeothermic organisms and the cardiovascular system plays a major role in conserving or removing body heat. The descriptions of the component parts of the circulatory system in this chapter illustrate that these transport requirements are met in a variety of ways in birds inhabiting particular environmental niches. This chapter describes the morphology and functional aspects of the avian heart (Section II), hemodynamics of the circulation (Section III), the vascular tree (Section IV), and the physiological properties of the blood (Section V). A common thread running through this discussion is that the component parts of the circulation must cooperate in an integrated fashion in order to ensure that oxygen delivery to the tissues matches demand. The integrative control of the avian circulation by autoregulatory, humoral, and neural mechanisms is described in Section VI.

Modern birds are probably derived from theropod dinosaurs (Chiappe, 1995), while mammals have descended from a group of carnivorous reptiles, the cynodonts. These ancestral lines originated in the Triassic more than 200 million years ago, so in evolutionary terms avian and mammalian stocks have been separated for a substantial period of time. As one might expect, significant differences in cardiovascular structure and function have arisen in the two groups since their separation, yet a number of similarities in their circulatory systems are also evident. Such similarities probably represent both the conservation of characteristics common to organisms ancestral to the two groups and the results of convergent evolution once the stocks had divided. However, our knowledge of cardiovascular structure and function is far more limited in birds than in mammals. In comparing the characteristics of the avian cardiovascular system with those of the mammalian circulation throughout this chapter, we have attempted to clarify the nature of the divergent and convergent features of this system in the two groups.

This review of the avian cardiovascular system draws on a number of excellent previous reviews, including major works by Akester (1971), Jones and Johansen (1972), Bennett (1974), Baumel (1975), Akester (1979), Cabot and Cohen (1980), West *et al.* (1981), Benzo (1986), and the previous edition of this volume (Sturkie, 1986b). This chapter represents an update and extension of these works and summarizes recent contributions in additional areas not previously covered.

II. HEART

A. Gross Structure and Function

1. Functional Anatomy

The avian heart, like that of the mammal, is a four-chambered, muscular fluid pump that intermittently pressurizes the central arteries, inducing blood flow to the capillary beds of both the systemic and pulmonary circulations. Functionally, these circuits lie in series with each other, and blood returns to the heart to be pressurized before entering either circuit. As in mammals, the right ventricle pressurizes the pulmonary circulation, the left ventricle the systemic circulation. In each case, the pressure differential between the central mean arterial pressure and the central venous pressure drives blood flow (the cardiac output) through the resistance to flow offered by the microvessels of the circulation. The left and right atria receive blood at central venous pressure before it enters the ventricles. In common with the atria of mammals, these chambers probably function more as blood reservoirs for their respective ventricles than as important "superchargers" for ventricular pressure. The resistance to blood flow, the peripheral resistance, is lower in the pulmonary than in the systemic circuit, so the right ventricle is required to generate less pressure than the left ventricle to produce the same volume flow rate. This difference in ventricular pressure is reflected in the gross anatomy of the ventricles, the myocardium of the right ventricle being thinner than that of the more powerful left ventricle.

In birds, the heart is located in the cranial part of the common thoracoabdominal cavity, with its long axis slightly to the right of the midline. It is partly enclosed dorsally and laterally by the lobes of the liver. A very thin but tough, fibrous pericardial sac encloses the heart. This sac contains a small volume of serous fluid that provides lubrication for the rhythmic motion of the cardiac contraction cycle. The pericardium is loosely attached to the dorsal surface of the sternum and the surrounding air sacs and more firmly to the liver. It is also attached, via the peritoneum of the hepatic peritoneal cavities, to the vertebral column. These attachments secure the apex of the heart within the median incisura of the liver and in the caudoventral axis of the thoracoabdominal cavity. The outer fibrous layer of the pericardial sac is continuous with the outer adventitial layer of the large central blood vessels. The pericardial membrane is relatively noncompliant and therefore strongly resists large, rapid increases in cardiac size which might be caused by volume overload of a heart chamber. The noncompliant nature of the pericardial sac may result in some degree of mechanical coupling between the ventricles via the contained incompressible lubricating fluid. For example, an increase in diastolic pressure in one ventricle may be transmitted to the other, increasing pressure and decreasing compliance.

2. Heart Size

In birds, heart mass scales in respect to body mass as $M_h = 0.014 M_b^{0.91}$ (Bishop and Butler, 1995). In mammals the relationship is $M_h = 0.0058 M_b^{0.98}$ (Prothero, 1979), where M_h is heart mass and M_b body mass.

When compared with mammals, birds of a given body mass have a significantly heavier heart. This may be due to the high aerobic power input needed to sustain flapping flight. Furthermore, unlike mammals in which heart mass is almost directly proportional to body mass, in birds the exponent denoting proportionality is significantly less than one. This means that larger birds like swans, ducks and geese tend to have proportionally much smaller hearts in relation to their body mass than do smaller birds. Thus, heart mass represents about 1.1% of body mass for a bird like the racing pigeon (421 g), compared with 0.8% for the 2.95 kg Pekin duck; this relationship is shown in Figure 1 (Grubb, 1983; Bishop and Butler 1995). In at least one large species, the barnacle goose (mass 1.6 kg), the heart can hypertrophy before migration to an estimated 1.1% of body mass (Bishop *et al.*, 1995). Therefore, large flying birds may have the genetic potential to increase heart size, and therefore cardiac output, through either seasonal humoral mechanisms or in the long term through natural selection. Hummingbirds have proportionally larger hearts than all other birds (shown separately in Figure 1), probably reflecting the high aerobic demands of hovering flight. For 25 species of hummingbirds: $M_h = 0.025 M_b^{0.95}$ (Hartman, 1961).

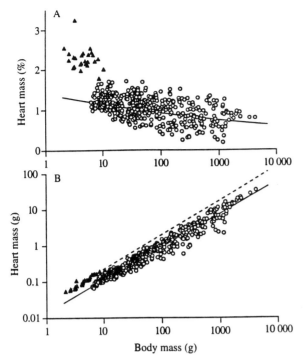

FIGURE 1 (A) Heart mass as a percentage of body mass. (B) Heart mass in grams, plotted against body mass (g) for 488 avian species, including 25 species of hummingbird. Hummingbird data are represented by the filled triangles and dashed line; all other species are represented by open circles and solid line. (After Physiological modelling of oxygen consumption in birds during flight, C. M. Bishop and P. J. Butler, *J. Exp. Biol.* **198,** 2153–2163, 1995, © Springer-Verlag.)

3. Cardiac Chambers

The avian heart has two completely divided atria and ventricles. These chambers are functionally equivalent to those of the mammalian heart, serving to distribute cardiac output both to the systemic circulation and to the lungs. In life, the atria are rounded chambers, distended with blood in atrial diastole. In excised hearts they may collapse, causing auricles to appear. The right atrium tends to be much larger than the left. The wall of the avian atria and ventricles, as in mammals, consists of endocardial, myocardial, and epicardial layers. The atrial walls are generally thin, although atrial muscle is arranged in thick bundles forming muscular arches. The right and left transverse arches are arranged at right angles to the dorsal longitudinal arch and the interatrial septum. The transverse arches branch into smaller bundles which fuse with a circular muscle band (muscularis basianularis atrii) at the ventral limits of the atria. Contraction of atrial muscle nearly empties the atria. In many species the atria lack functional inflow valves, so that the importance of atrial contraction for ventricular filling may be slight.

The muscular architecture of the ventricles is more complex than that of the atria and includes a superficial layer, longitudinal muscle of the right ventricle, and sinuspiral and bulbospiral muscles. The left ventricle is cone-shaped and extends to the apex of the heart. Its right wall forms the interventricular septum. The free wall of the right ventricle is continuous with the outer portion of the wall of the left ventricle and wraps around the right side of the heart to enclose a crescent-shaped cavity which does not reach the apex of the heart. The muscular walls of the two ventricles are differentially developed, the wall of the left ventricle being 2–3 times thicker than that of the right. In addition, the radius of curvature of the wall of the left ventricle is smaller than that of the right (Figure 2). This implies both a greater mechanical advantage for pressure generation in the left than in the right ventricle and, according to LaPlace's law, a smaller wall tension for a given left ventricular pressure increment. Therefore, contraction of the myocardial layers of the thick, small-radius wall of the left ventricle enables it to generate systolic pressures 4–5 times higher than those produced by the right ventricle, without rupturing. The larger radius of curvature and thinner free ventricular wall of the right ventricle reflects the lower systolic pressures generated by this chamber, made possible by the low vascular resistance of the avian lungs. Another consequence of this geometry is that relatively large changes in stroke volume can be made by small changes in the degree of shortening of right ventricular muscle fibers.

4. Valves

Blood entering the left ventricle from the left atrium on atrial systole passes through an orifice guarded by a membranous atrioventricular (AV) valve, similar in general structure to a mammalian AV valve. The valve forms a continuous membrane around the aperture. The valve is tricuspid, not bicuspid as it is in mammals, but in the avian heart the cusps of this valve are poorly defined. The anterior and posterior leaflets are small. The large aortic (medial) leaflet is connected to the bases of the left and noncoronary cusps of the adjacent aortic outflow valve by fibrous tissue. The free margin of the valve is well secured to the left ventricular endocardium by numerous inextensible chordae tendineae. This arrangement prevents valve eversion during ventricular systole.

Blood passing from the right atrium to the right ventricle enters through an orifice guarded by an AV valve that is structurally unique to birds. In pronounced contrast to the fibrous structure characteristic of the mammalian tricuspid valve, in birds the right AV valve consists of a single spiral flap of myocardium attached obliquely to the free wall of the right ventricle (Figure 3; Lu *et al.*, 1993a). This spiral flap is apposed to a downward extension of the free wall of the right atrium.

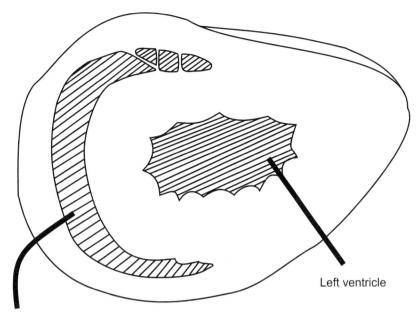

FIGURE 2 Transverse section through the ventricles of the avian heart. The lumen of each ventricle is shaded. (Reprinted from *Form and Function in Birds* **2**, N. H. West, B. L. Langille, and D. R. Jones, Cardiovascular system, pp. 235–339, 1981, by permission of the publisher Academic Press.)

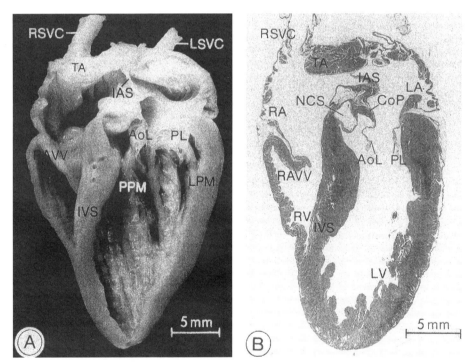

FIGURE 3 Anterior frontal views through the atria and ventricles of a chicken heart, showing both the right and left atrioventricular (AV) valves. (A) Anterior view of heart dissected in the frontal plane. (B) Frontal histological section of 8 μm thickness, Goldner trichrome stain. AOL, aortic leaflet of left AV valve; COP, connecting part of muscle arch; IAS, interatrial septum; IVS, interventricular septum; LA, left atrium; LPM, left papillary muscle; LSVC, left superior vena cava; LV, left ventricle; NCS, noncoronary sinus of the aorta; PL, posterior leaflet of left AV valve; PPM, posterior papillary muscle; RA, right atrium; RAVV, right AV valve; RSVC, right superior vena cava; RV, right ventricle. (After Histological organization of the right and left atrioventricular valves of the chicken heart and their relationship to the atrioventricular Purkinje ring and the middle bundle branch, Y. Lu, T. N. James, M. Bootsma, and F. Terasaki, *Anat. Rec.*, Copyright © 1993 Wiley-Liss, Inc. Reprinted by permission of Wiley-Liss, Inc., a subsidiary of John Wiley & Sons, Inc.)

The atrial component of the valve extends toward the apex of the ventricle for a shorter distance than does the right ventricular flap. Most of the valve is made of ventricular myocardium and is bilaminar only in its upper portion. The mechanism of valve closure at the start of ventricular systole is unknown. It could be active, by contraction of the muscular flaps, or passive, by deflection of the ventricular flap by a brief backflow of blood at the start of ventricular systole.

The idea of valve closure depending at least partially on active muscular contraction is supported by evidence that at the cellular level both atrioventricular valves are closely approached by the electrical conducting system of the myocardium. A complete ring of Purkinje fibers encircles the right AV orifice and connects to the muscular AV valve (Lu *et al.*, 1993a,b). An exception to this anatomical arrangement may be the penguin, in which Adams (1937) did not find a Purkinje ring. The majority of studies, summarized by Lu *et al.* (1993a,b), support the idea that both the atrial and ventricular muscular components of the right AV valve are excited to contract via the Purkinje system. However, Szabo *et al.* (1986) thought that there was an insulating layer of connective tissue between the Purkinje AV ring system and the left ventricular myocardium, suggesting that the muscle flap derived from the left ventricle may work passively. Definitive physiological experiments have not been performed to resolve this issue, but the balance of current evidence suggests that both portions of the valve are electrically activated before the ventricular myocardium to dynamically contract and close the AV orifice at the start of ventricular systole (Lu *et al.*, 1993b). This is clearly very different from the closure mechanism in mammals, in which the leaflets of the tricuspid valve float up into the right AV orifice, moved by the AV pressure differential generated during ventricular systole itself.

The outflow valves from the right and left ventricles are, at first glance, more conventional (mammalian) in nature. The pulmonary outflow valve consists of three semilunar cusps. It prevents regurgitation from the pulmonary artery into the right ventricle, the valvules opening as pressure in the ventricle falls below that in the pulmonary trunk on ventricular diastole. There are also three semilunar cusps in the aortic outflow valve, but they are much more rigid than those of the pulmonary outflow valve and are firmly attached to underlying myocardium. The cusps are linked by a ring of fibrous tissue that lies within a complete ring of underlying, circumferentially arranged, myocardial cells. The ring is completed by an arch of cardiac muscle that lies between the left coronary cusp of the aortic outflow valve and the aortic leaflet of the left AV valve as shown in Figure 4 (Lu *et al.*, 1993a). This anatomical arrangement contrasts with that in the mammalian heart, in which there is only connective tissue, not myocardium, between that part of the muscular ring lying between the aortic wall

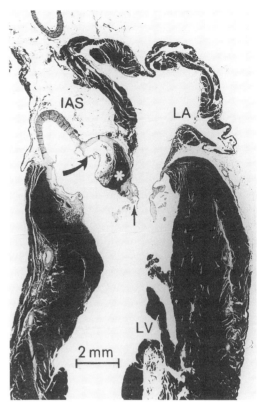

FIGURE 4 Photomicrograph of a histological section of the chicken heart, taken in a modified sagittal plane through the anterior column of the muscle arch, indicated by an asterisk. The muscle arch lies between the left coronary aortic valve cusp (curved arrow) and the aortic leaflet of the left AV valve (straight arrow). IAS, interatrial septum; LA, left atrium; LV, left ventricle. (After Histological organization of the right and left antrioventricular valves of the chicken heart and their relationship to the atrioventricular Purkinje ring and the middle bundle branch, Y. Lu, T. N. James, M. Bootsma, and F. Terasaki, *Anat. Rec.*, Copyright © 1993 Wiley-Liss, Inc. Reprinted by permission of Wiley-Liss, Inc., a subsidiary of John Wiley & Sons, Inc.)

and the adjacent mitral valve; in mammals the myocardial ring is incomplete. In the bird, however, this sphincterlike myocardial cylinder is potentially capable, on contraction, of constricting the left ventricular outflow tract. Lu *et al.* (1993a) propose that the muscular ring could act as a sphincter controlling the rate of left ventricular outflow by modulating outflow resistance. Another attractive possibility is that muscular contraction of the myocardial ring could close the relatively rigid cusps of the aortic outflow valve. The middle bundle branch of the Purkinje system is connected to the arch of muscle, so its contraction may start relatively early in the cardiac cycle. Obviously, physiological studies are urgently needed to determine whether either of these intriguing mechanisms operate in the avian heart.

5. Coronary Circulation

Oxygenated blood destined to supply the avian myocardium via the right and left coronary arteries enters the right ventral and left aortic sinuses, which lie immediately downstream from the cusps of the aortic outflow valve. Most birds have two entrances to the coronary circulation, although there is individual variation so that up to four openings have been observed. In chickens the right ventral sinus leads into the right coronary artery, which then divides immediately into a superficial and a deep branch (Figure 5). The superficial branch follows the groove (coronary sulcus) between the right ventricle and atrium and supplies the cardiac muscle of both chambers. The larger deep branch supplies the ventral wall of the right ventricle, the dorsal walls of both atria, and the muscular right atrioventricular valve. In most species the right coronary artery is dominant and also supplies the ventricular septum, the heart apex, and much of the left ventricular myocardium. The left coronary artery arises from the left aortic sinus and also has a superficial branch which follows the left coronary sulcus. Another superficial branch gives off atrial and ventricular tributaries, and a deep branch supplies the ventral myocardium of the left ventricle. It is not uncommon in chickens for the left coronary artery to be dominant, in which case it supplies almost all of the left ventricular myocardium and the heart apex. There are frequent anastomoses between the branches of the coronary arteries, particularly near the coronary sulcus.

Five groups of cardiac veins, with frequently anastomosing small tributaries, return venous blood from the myocardium into the right atrium via a coronary sinus. Small cardiac veins open directly into the atria and the right ventricle. This basic anatomical pattern of coronary circulation is seen in birds ranging from the chicken, duck, and pigeon (West *et al.*, 1981) to the ostrich (Bezuidenhout, 1984).

The rate of perfusion of the myocardium of the bird heart is high compared with the perfusion rates of most other avian tissues, as shown in Figure 25 (see Section IV,B,3) (Johansen, 1964; Jones *et al.*, 1979). The constantly active cardiac muscle is perfused at a much higher rate than resting skeletal muscle. Presumably, as in mammals, the majority of avian coronary blood flow occurs in diastole, so we might expect coronary flow to increase if the diastolic interval is prolonged, provided arterial driving pressure does not fall. This combination of factors occurs during avian diving, in which the heart slows and the arterial pressure is maintained by an increase in peripheral resistance. In mallards coronary flow is well maintained under these circumstances (Jones *et al.*, 1979).

Generally, a reduction in the oxygen supply, or an increase in myocardial oxygen demand, results in a com-

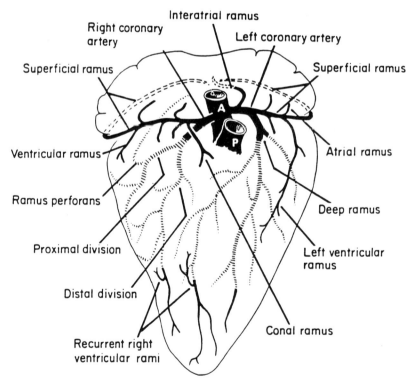

FIGURE 5 Arrangement of the coronary arteries of the chicken, *Gallus*, drawn from the cranioventral aspect. Solid black and dashed lines represent superficial portions of arteries. Cross-hatched lines represent deep arteries embedded in the myocardium of the ventral and right side of the interventricular septum. A, aorta; P, pulmonary trunk. (After West *et al.* (1981) with permission.)

pensatory increase in coronary blood flow. Both of these factors presumably come into play during high-altitude flight in birds. In Pekin ducks and bar-headed geese coronary blood flow (expressed per gram of wet heart weight) was about 3.5 ml min^{-1} g^{-1} at sea level when the inspired partial pressure of oxygen (P_IO_2) was 142 mmHg. Exposure to severe hypoxia (28 mmHg P_IO_2) resulted in increases in coronary blood flow of 5.5 and 2.7 times respectively in ducks and geese (Faraci *et al.*, 1984). These were largely brought about by decreases in coronary vascular resistance in response to hypoxia (Figure 6). Although bar-headed geese are accomplished high-altitude fliers, in contrast to ducks, the geese showed smaller increases in coronary blood flow under the conditions of these experiments. However, these smaller increases in flow still satisfied myocardial oxygen demand because arterial oxygen content (C_aO_2) was higher in the geese than in the ducks.

When migrating at extreme altitude (up to 9000 m) bar-headed geese must hyperventilate to maintain P_aO_2 at adequate levels in the face of reduced P_IO_2. As a consequence of increased ventilation, carbon dioxide is removed from the blood at an increased rate and P_aCO_2 falls. Therefore arterial blood in the geese flying at high altitude is both alkalotic and hypocapnic. Carbon dioxide is a potent coronary vasodilating agent in mammals, coupling increased aerobic metabolism in the myocardium to an increased rate of oxygen delivery, accomplished by decreases in coronary resistance and increases in coronary blood flow. On the other hand, hypocapnia results in increased coronary flow resistance and decreased coronary blood flow. This would obviously be highly undesirable in a bird engaged in active high altitude migratory flight. Interestingly, the relationship between coronary blood flow and P_aCO_2 in bar-headed geese appears to be quite different in the hypocapnic and hypercapnic ranges of PCO_2. Over a hypercapnic range of P_aCO_2 from about 30 to 60 mmHg there is a linear increase in coronary flow with P_aCO_2, as in the mammal. However, in the hypocapnic condition, when P_aCO_2 is 30 mmHg or lower, there appears to be no effect of P_aCO_2 on coronary blood flow or resistance (Faraci and Fedde, 1986). Whether this represents a specific vascular adaptation securing myocardial oxygen delivery during high-altitude flight remains to be determined.

Turkeys sometimes show a congestive cardiomyopathy that is presumed to be of viral origin, the so-called "round heart disease." In this condition, systemic hypotension and low cardiac output, caused by reduced left ventricular myocardial shortening, are probably the result of reduced subendocardial coronary perfusion rate (Einzig *et al.*, 1980). Turkeys with round heart disease also show an altered EKG pattern (see Section II,C,4).

B. Cardiac Variables

The avian cardiovascular system is not merely a replica of the arrangement in mammals, despite similarities in performance between the two systems. Birds have larger hearts, bigger stroke volumes, lower heart rates, and higher cardiac outputs than mammals of corresponding body mass (Grubb, 1983). In addition, in many avian species mean arterial pressure is higher than that found in mammals of comparable body mass (see Smith, 1994).

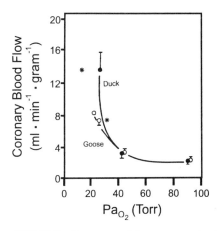
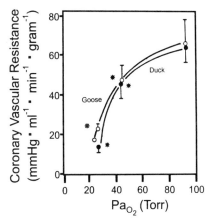

FIGURE 6 Responses of the coronary circulation to hypoxia in Pekin ducks and bar-headed geese. (Left) coronary blood flow (ml min^{-1} g^{-1}); (right) coronary vascular resistance (mmHg ml min^{-1} g^{-1}) as a function of the arterial O_2 partial pressure (P_aO_2). All values are means ± S.E. ($n = 5$) except at 25 Torr, where means for two geese were plotted. An asterisk represents significant difference from normoxia (highest P_aO_2 level) at $P \leq 0.05$. (Reprinted from *Resp. Physiol.* **61**, F. M. Faraci, D. L. Kilgore, and M. R. Fedde, Blood flow distribution during hypocapnic hypoxia in Pekin ducks and barheaded geese, pp. 21–30, Copyright (1985), with permission from Elsevier Science.)

Cardiac output, the product of stroke volume and heart rate, is of particular interest because it is a major determinant of the rate of oxygen delivery to tissues.

In resting birds left ventricular stroke volume (V_s) was found to be almost directly proportional to body mass (M_b) (Grubb, 1983). For nine species of birds ranging in body mass from 0.035 kg (budgerigar) to 37.5 kg (emu) Grubb found that $V_s = 1.72M_b^{0.97}$, where V_s is in milliliters and M_b is in kilograms. Heart rate ($_fH$, beats min^{-1}) at rest was found to be slower in larger birds: $_fH = 178.5M_b^{-0.282}$. Cardiac output (CO, ml kg^{-1} min^{-1}) at rest, the product of stroke volume and heart rate, therefore scaled with the mass of the bird as: CO = $307.0M_b^{0.69}$. The corresponding relationship for mammalian cardiac output (Holt et al., 1968) is CO = $166M_b^{0.79}$. These results show that birds have a proportionally larger cardiac output compared with a mammal of the same body mass. In larger birds resting heart rate is slower than in smaller birds. Recently, Bishop and Butler (1995) found that for 49 species the allometric relationship for heart rate at rest was $_fH = 125M_b^{-0.37}$, while for birds in flight it was $_fH = 480M_b^{-0.19}$. It is interesting that the heart rate–body mass relationship during flight has a more shallow slope than that in resting animals, indicating that larger species show a greater increase in heart rate in absolute terms in the transition from rest to flight (Figure 7). Bishop and Butler (1995) suggest that the body mass exponent of stroke volume in flight should be similar to that at rest ($M_b^{0.96}$), even though the absolute value of stroke volume may increase during flight. Therefore, in flight, it is predicted that cardiac output will scale to body mass as the sum of the exponents for stroke volume and heart rate ($M_b^{0.77}$).

In flight, as at rest, larger species show lower coronary perfusion rates per mass of body tissue. It is likely that this reflects an optimization of the arterial oxygen supply at the tissue level, the body mass exponent for cardiac output being very similar to the exponent for mass-specific oxygen consumption. Thus, natural selection probably acts on cardiac output to maintain the arteriovenous O_2 difference at a similar level across different avian species. Stroke volume is constrained by cardiac geometry such that, on theoretical grounds, V_s should be closely proportional to M_h and M_b (Schmidt-Nielsen, 1984; Astrand and Rodahl, 1986). Therefore, the lower mass–specific cardiac outputs of larger avian species, matching their lower mass–specific VO_2 levels, are reflected in their lower heart rates.

C. Fine Structure and Cardiac Electrophysiology

1. Fine Structure

Histologically the atria and ventricles are quite similar consisting of an external layer, the epicardium, which is separated from an inner endocardium by the mass of heart muscle, the myocardium. Ventricles are much thicker than the atria due to extensive proliferation of the myocardial layer. The epicardial and endocardial layers are morphologically similar, consisting of loose connective tissue and elastic fibers bordered by a single layer of squamous epithelial or endothelial cells, respectively. The atrial and ventricular septa have endocardial layers facing the lumens of their respective cavities with myocardial cells between them. In the sparrow and stork, the atrial septum is very thin and in some regions consists only of two apposed layers of endocardium.

The atrial and ventricular myocardia consist of striated muscle fibers, differing from those of mammals in three notable respects (Sommer and Johnson, 1969, 1970; Hirakow, 1970). (1) Striated muscle bands, which are prominent in mammalian cardiac muscle are also present in the bird heart, except for the M-band. In mammalian cardiac muscle the M-band is a line of protein molecules connecting adjacent myosin filaments. The significance of the lack of an M-band on the contractile properties of avian myocytes is unknown. (2) Avian cardiac muscle fibers are much smaller in diameter than mammalian fibers and hence there are many more of them in similarly sized hearts. Avian myocardial cells are typically 2–7 μm in diameter compared with the 10- to 15-μm diameter of mammalian cells. (3) Avian myocardial cells lack transverse tubules (T-tubules) which are prominent in mammalian cardiac muscle. The membrane surrounding the muscle fibers (sarcolemma) consists of two parts, a cell membrane (plasmalemma)

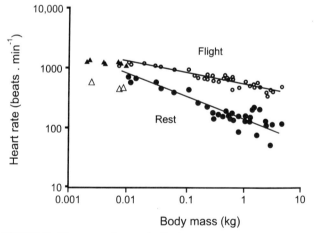

FIGURE 7 Heart rate (beats min^{-1}) plotted against body mass (kg) for 49 species of birds, including 7 species of hummingbirds, at rest and in flight. Hummingbirds, open triangles (rest) and filled triangles (flight). Other species, filled circles (rest) and open circles (flight). (After Bishop and Butler, 1995, J. Exp. Biol., Company of Biologists, Ltd.)

and an external layer interconnected with an interstitial network of collagenous fibers. In mammalian cardiac muscle, T-tubules form as invaginations of the plasmalemma, perpendicular to the long axis of the myofilaments. T-tubules lie next to, and form junctions with, sections of the sarcoplasmic reticulum (diads or triads, Figure 8). In mammals, the T-tubule system increases the surface area of the myocardial cells to the extent that the surface-to-volume ratio of a mouse cardiac cell (15-μm diameter) is the same as that of a finch (8-μm diameter). The finch and the mouse have similar cardiac frequencies (Bossen et al., 1978).

The connection between the sarcoplasmic reticulum and the plasmalemma occurs through "couplings" and in birds, lacking a T-tubule system, these couplings occur at the surface of the cell (Figure 8). Couplings are effected by junctional processes that extend from the cytoplasmic face of the sarcoplasmic reticulum (junctional sarcoplasmic reticulum, JSR) that are very closely apposed to the inner surface of the plasmalemma. Birds also possess an extended junctional sarcoplasmic reticulum (EJSR), which occurs in the region of the Z-bands but it is anomalous in that although it resembles the JSR in most respects, the actual junctional processes may be separated from the plasmalemma by a cleft of several microns. EJSR is much less developed in chicken than in passerines (Sommer et al., 1991). Interestingly, the volume of JSR in mouse hearts and the total volume of JSR (20%) and EJSR (80%) in the finch are virtually identical (Bossen et al., 1978).

2. Excitation–Contraction Coupling

Excitation–contraction coupling describes how an electrical signal, the action potential (AP), traveling along the plasmalemma evokes calcium release from the sarcoplasmic reticulum (SR) in the region of the myofibrils, causing a change in actin–myosin interactions which leads to muscle contraction. The transduction between the electrical signal and Ca^{2+} release from the JSR is effected by a transmitter which is, in fact, calcium itself. In the first step of this process, the AP causes voltage-dependent Ca^{2+} channels to open in the sarcolemmal membrane through a conformational change in the channels. Ca^{2+} then enters the cell and diffuses to receptors on the junctional processes where it acts as a transmitter, opening Ca^{2+}-dependent Ca^{2+} channels which in turn release Ca^{2+} sequestered in the sarcoplasmic reticulum. This Ca^{2+}-induced Ca^{2+} release (CICR) is crucial for the physiological function of the bird heart in which the EJSR, which is the majority of the junctional SR, is separated from the plasmalemma

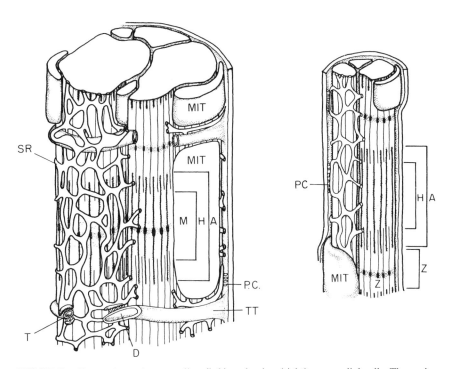

FIGURE 8 Comparison of mammalian (left) and avian (right) myocardial cells. The major distinguishing features of avian fibers are the smaller cell diameters and the presence of M bands and transverse (T)-tubules. SR, sarcoplasmic reticulum; T, triad junction; D, diad junction; PC, peripheral coupling site; TT, transverse tubule; MIT, mitochondria; M, H, A, I, Z, bands of striated muscle (after Sommer and Johnson, 1969; used by permission).

by several microns. Furthermore, CICR allows exquisitely fine regulation of force generation in myocytes. In cardiac muscle, contraction is "all or none" and force modulation must be done at the cellular level by regulating not only the amount of Ca^{2+} entering through the sarcolemma, but also its effect on Ca^{2+} release from the SR.

3. Conduction System

The cardiac conduction system of the avian heart consists of the sinoatrial (SA) node, atrioventricular (AV) node, AV Purkinje ring, His bundle, and three bundle branches (Figure 9). Histologically, three types of cells are associated with the conducting system. (1) Pacemaker cells (P-cells) are small and spherical in shape and are found in both the SA and AV nodes. P-cells have the property of repetitive spontaneous depolarization. (2) Transitional cells (T-cells) are much smaller and have fewer microfibrils than cardiac muscle cells; their structure is intermediate between normal cardiac muscle cells and Purkinje fibers. (3) Purkinje fibers are large, elongated, brick-shaped cells containing few myofibrils. Many Purkinje cells, however, contain longitudinal fibers called intermediate filaments; these are part of the cytoskeleton, serving to maintain cell shape as the myocardium contracts. Purkinje cells may be up to 5 times the diameter of myocardial cells.

The SA node is located close to the opening of the venae cavae into the right atrium although there is considerable species variation in birds. The SA node consists of P- and many T-cells and is enclosed in a loosely organized connective tissue sheath. The T cells transmit impulses from the pacemaker to atrial muscle cells. The SA node is morphologically, and perhaps physiologically, diffuse in birds and the primary electrical pacemaker region appears to change position spontaneously within the node (Hill and Goldberg, 1980).

All the muscle fibers within a given cardiac chamber should contract more or less simultaneously while, for normal cardiac function, it is essential that the atria contract before the ventricles. The wave of excitation which is initiated in the SA node is delayed at the AV node allowing the atria to empty before ventricular contraction begins. The electrical impulse which initiates contraction spreads through the atria and ventricles at

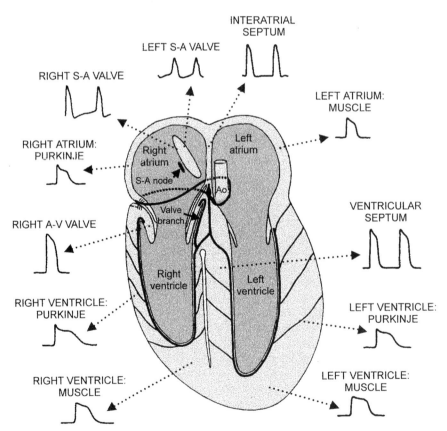

FIGURE 9 The Purkinje system of the bird heart (after Davies, 1930) and transmembrane action potentials recorded from cells at the indicated sites in chicken and turkey (Moore, 1965, with permission of the New York Academy of Sciences; and modified from Jones and Johansen, 1972, with permission).

rates in excess of 1 m sec^{-1}, whereas conduction through the AV node is two to three orders of magnitude slower. Spread of excitation through myocardial muscle occurs from one muscle cell to the next as well as along specialized conducting pathways. Individual muscle cells are discrete entities but they behave electrically as if they were all joined together to form a syncytium. This property results from low electrical resistance in parts of the cell membrane where cell apposition is very close. Junctional complexes, the intercalated disks, commonly join myocytes end-to-end in avian myocardium, occurring at right angles to the long axis of the myofibrils. Intercalated disks consist of two components, desmosomes, which mechanically couple the cells together, and nexuses, which couple cells electrically. A nexus may be viewed as an array of unit electrical resistors with their number being inversely proportional to the electrical resistance between the cells (Sommer, 1983). Interestingly, there are few nexuses along the longitudinal axes of myocardial cells.

The speed at which the wave of electrical excitation propagates through the ventricles is enhanced by a specialized conducting system of Purkinje fibers but whether a specialized conducting system also exists in the atria of birds is controversial. In the atrium, waves of electrical and contractile activity proceed in the same direction, from the SA to AV nodes, which may mean that a specialized conducting system is unnecessary here. However, both atria contain Purkinje cells, and these have been described both morphologically and physiologically as being organized to preferentially direct the wave of activation toward the AV node (Davies, 1930; Hill and Goldberg, 1980). The contrarian view is that since the Purkinje cells in the atria are mixed diffusely among the normal myocardial cells then they may represent the remnants of an embryological anlage left over from the time when that anlage was building the ventricular conducting system (Sommer, 1983).

The atrial wave of excitation crosses to the ventricle through the AV node which, in birds, is a somewhat controversial structure because many morphological investigations have failed to locate it, although its presence has been established in functional studies. In the chicken the AV node is located in the right side of the base of the interatrial septum (Davies, 1930; Lu *et al.*, 1993b; Ying *et al.*, 1993) although in Indian fowl, *Pucnonotus cafer,* house sparrow, *Passer domesticas,* and in *Babo begalensis* it is located in the left AV junction (Mathur, 1973). The His bundle and its three bundle branches of Purkinje cells arise from the AV node. The right and left bundle branches emerge from the septum to form a network in the subendocardium of the right and left ventricles, respectively, penetrating the myocardium along the tracts of the coronary arteries (periarterial Purkinje fibers).

An indication of the theropod ancestry of birds can be inferred from the conducting system of the heart. Birds, unlike mammals, possess an AV ring of Purkinje fibers on the right side of the heart which runs up and around the right AV valve (Figure 10; and see Section II,A,4). The middle bundle branch, after separating from the others, runs around the aorta and connects to the AV ring forming a figure eight (Lu *et al.,* 1993b).

Purkinje cells conduct electrical impulses much faster than cardiac myocytes. In mammals, part of the reason for this high conduction speed is that Purkinje cells lack a T-tubule system. T-tubules increase the surface area of the cell and therefore membrane capacitance (increasing the length of time a given amount of electrical charge will take to alter membrane potential); a high capacitance thus slows conduction velocity. However, in birds, there is no T-tubule system associated with either myocardial or Purkinje cells, yet the latter still conduct impulses at a faster rate. This is because conduction velocity varies directly with cell diameter and avian Purkinje cells are much larger than myocardial cells. Furthermore, a higher conduction velocity may be fostered in Purkinje cells by intermediate filaments which serve to keep the cell round. Also, the electrical resistance between Purkinje cells is lower than that between myocardial cells because nexus size also increases with cell diameter, acting to further increase conduction velocity. Finally, in mammals with extremely large hearts (i.e., ungulates), the Purkinje cells within a bundle are tightly packed together and surrounded by an insulating membrane so that they behave electrically as a single fiber of a diameter equal to that of the whole bundle. Purkinje cells are likewise bundled in the avian heart but whether this enhances conduction velocity is uncertain because the bundles lack a connective tissue sheath and are therefore not insulated from surrounding tissues.

4. Electrophysiology

The Purkinje fibers follow the coronary arteries and therefore take a relatively short course through the thick left myocardium. This accounts for the rapidity of arrival of the wave of excitation at a given point on the surface of the left ventricular wall in the avian heart (Lewis, 1916). The sequence of depolarization is, according to Kisch (1951): right ventricle apex, right ventricle base, left ventricle base, left ventricle apex. Moore (1965) has mapped epicardial activation and suggests that in the turkey the apical third of the right ventricular epicardium is activated earliest, the upper basilar third is intermediate, and the pulmonary outflow tract is the last region activated in the whole heart. The anterior one-

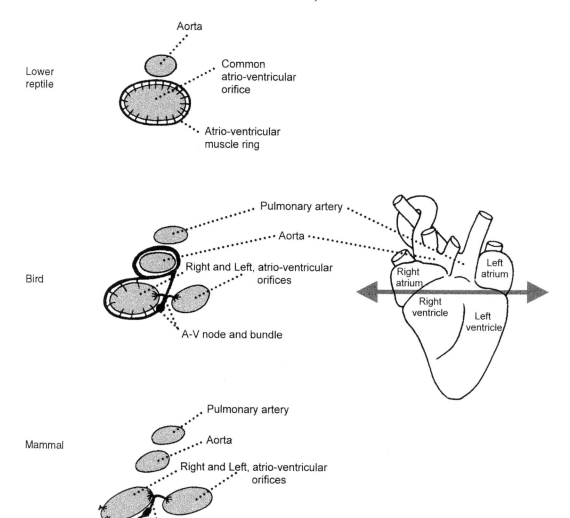

FIGURE 10 Diagram of specialized atrioventricular connections in lower reptiles, birds, and mammals as seen when looking into the ventricles after making a transverse section through the heart, as indicated on the diagram of the chicken heart on the right hand side (after Davies, 1930).

third of the septal region and the middle region of the left ventricle are activated before the basilar regions, the whole left ventricular epicardium being activated in 12.5 ms. Lewis (1916) and Mangold (1919) suggest somewhat different sequences (see Table 1). Kisch's suggestion that the conducting system stimulates heart muscle only at places of direct contact between its terminal fibers and heart muscle and not along the entire course of the conducting system receives support from his own work (Kisch, 1951) showing that subendocardial muscle is activated about 20 ms later than the earliest activated subepicardial muscle, which in turn suggests

TABLE 1 Ventricular Depolarization in Birds[a]

		Moore (1965)	Kisch (1951)	Mangold (1919)	Lewis (1916)
Right ventricle	Apex	1	1	1	1
	Base	4	2	3	4
Left ventricle	Apex	2	4	4	2
	Base	3	3	2	3

[a] Key to table: 1 represents the earliest and 4 the last area to depolarize.

short cuts of the conductive system to subepicardial muscle. However, Davies (1930) and Lu et al. (1993b) both suggest that since the bundle branches lack a fibrous sheath there will be early and widespread propagation of the impulse in the septal region and thence along the bundle to all parts of the ventricles. In fact, Moore (1965) described early activity in the cranial part of the septum which was then followed by contraction of its basal region. Obviously, these two suggestions concerning the conduction network are mutually exclusive, and clarification of the problem must await further experimental work.

The conducting system of the bird heart has been investigated by recording transmembrane potentials from cells in the heart of the chicken and turkey (Moore, 1965; 1967). The pacemaker cells of the SA node, in the absence of any extrinsic influences, set the heart rate. Cells that function as pacemakers show a characteristic slow depolarization during diastole, the steepness of the depolarization being related to the degree of automaticity inherent in the cell (the fastest cells to depolarize drive the slower), whereas cells not spontaneously active show a steady membrane potential during diastole (Figure 9). Action potentials recorded from the junction of the left SA valve with the sinus venosus show diastolic depolarization (prepotentials) with a slow transition to the ascending phase of the actual AP (Figure 9), in contrast to the relatively more rapid rise of the AP recorded in the right SA valve itself, indicating that cells in the right SA valve are triggered by the pacemaker cells. The duration of APs recorded from ventricular muscle cells is longer than those recorded from atrial muscle cells (Figure 9). Purkinje fibers display a prominent sharp peak to their APs, which is followed by a distinct plateau, a feature not seen in APs from atrial or ventricular muscle cells. The duration of depolarization is also much longer in Purkinje fibers although diastolic depolarizations have not been recorded from avian Purkinje fibers. The longer duration of the Purkinje APs as compared with those of ventricular myocytes indicates a long refractory period which would tend to prevent extrasystole and possible fibrillation by assuring a concerted depolarization of the ventricular muscle (Moore, 1965).

The summed electrical activity of the heart (electrocardiogram, ECG) is usually recorded indirectly with electrodes placed on the surface of the body or just under the skin. (A direct ECG recording would be made by dividing the sternum and placing recording electrodes on the surface of the heart). In birds, as in mammals, three standard leads (I, II, and III) are used, following the model first conceived by Einthoven about 100 years ago. The body is a volume conductor of electricity and the waves of depolarization and repolarization that sweep across the heart can be reduced to a single electrical dipole. The dipole has magnitude (volts), direction, and sense (positive or negative), so it is a vector quantity. In Einthoven's concept the cardiac vector is situated at the center of an equilateral triangle (Figure 11) formed by the bipolar lead connections. Lead I connects across the thorax from the right (negative electrode) to the left (positive electrode) wing bases. Lead II is recorded between right wing base (negative electrode) and left thigh (positive electrode). Lead III is connected between the left wing base (negative electrode) and the left thigh (positive electrode). The arrangement of these leads is such that, in humans and many other mammals, the polarity of the recorded signals is positive. In contrast, in birds, the polarity of the major component of the ECG (ventricular contraction) is negative (Figure 12).

There has been some controversy over what the typical avian ECG looks like (Figure 12). For instance, Kisch (1951) reported the presence of P, QRS, and T waves, whereas Mangold (1919) reported that the electrocardiogram of birds has no R component, but instead a deep S wave. Sturkie (1986a) reported the presence of P, a dominant S, and T waves, a small R, but no Q wave. The status of the Q wave has not been clarified by more recent work. There is general agreement that it is absent in the chicken (Goldberg and Bolnick, 1980), small in the turkey (McKenzie et al., 1971), and prominent in some duck ECGs (Figure 12; Cinar et al., 1996). Aside from being useful in monitoring heart rate the ECG can also be used to access the timing of the various phases of the cardiac cycle, since components of the ECG can be identified with atrial (P wave) and ventricular (QRS or RS) depolarization as well as repolarization of the ventricles (T wave). The duration of the P wave is the period of atrial depolarization and repolarization while the P–R duration includes the conduction delay at the AV node. QRS or RS represents ventricular activation and the T wave is the period in which the heart is completely depolarized. QT or RT duration represents the duration of a complete cycle of activation and relaxation of the ventricle (Figure 12). Most of these intervals are fixed, so it is primarily the period between the T and P waves (i.e., interbeat interval), which shortens and lengthens with increases and decreases in heart rate, respectively. In fact, at very high heart rates the T wave of one beat may come to overlie the P wave of the next.

Bipolar recording of the standard limb leads means that the cardiac vector is projected along the line between the two electrodes (Figure 11). When all leads are used then it is possible to reconstruct the orientation of the cardiac vector or mean electrical axis (MEA) for any of the events of the cardiac cycle with respect to

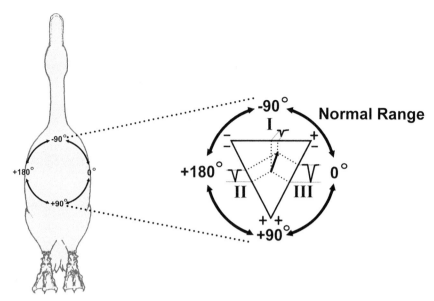

FIGURE 11 Relationship of the electrical axis of the heart to potentials recorded from the three standard leads (I, II, and III) as conceived by Einthoven. The duck is presented in ventral view on the left side of the figure. The electrical axis is upward and close to the midline. The arrow in the middle of Einthoven's triangle shows the direction of the axis, while the length represents its magnitude and degrees its orientation. The voltage changes seen at each lead (I, II, and III) are shown.

the plane of orientation of the leads. Standard leads lie in the frontal plane and, in mammals, the MEA of the QRS is oriented downward (inferiorly) and to the left (+60°). In contrast, the MEA of the QRS wave in the bird heart is close to −90°, being oriented along the long axis of the body, and superior to the frontal plane. Hence, the QRS or RS wave is of negative polarity and barely represented in lead I with the highest voltages being recorded by leads II or III (Figure 11). The mean electrical axes of the P and T waves can be calculated in a similar fashion. By using other cardiac leads in addition to the standard limb lead I, Szabuniewicz and McCrady (1967) were able to determine the MEAs of the QRS or RS, P, and T waves of the heart in the chicken not only in the frontal plane (−77.1°) but also in the horizontal (+72.4°) and sagittal (−55.4°) planes.

As the mean electrical axis of the ventricular depolarization phase is negative while that for the repolarization phase is positive then the QRS or RS component deflects downward or negatively while the T wave is upright or positive (Figure 12). Obviously, the heart does not markedly change its position in the chest with each beat. The waves deflect in opposite directions because depolarization causes the ventricular myocardium to become negative and repolarization drives the myocardium positive. Also, the time courses of these waves are different. Repolarization is slower than depolarization so the T wave is more spread out than the QRS or RS wave. In this context it should be noted that when the ECG is recorded just for purposes of monitoring heart rate, using a single pair of leads (usually II), the RS wave is often presented as deflecting positively. This is achieved by reversing the polarity of the lead II bipolar electrodes (inverting Einthoven's triangle, Figure 12) and is done for artistic reasons.

Further interpretation of the ECG of birds is complicated by variability in electrode recording sites, anatomical differences between species, and the absence of a large bank of data such as has been accumulated for humans and, to a lesser extent, other mammals. In fact, it seems most unlikely that rigorous, detailed investigation of the ECG of birds will ever be used for clinical diagnosis. Boulianne et al. (1992) suggest that only diseases that cause a shift in position of the heart in the torso, and therefore alter the MEA, can be successfully diagnosed by two-dimensional electrocardiography. Round heart disease in chickens and turkeys produces such a shift; the mean RS axis in the frontal plane averages +70° compared with −85° in the normal bird (Hunsaker et al., 1971) (also see Section II,A,5).

III. GENERAL CIRCULATORY HEMODYNAMICS

The three major constituents of the pulmonary and systemic circulations are (1) the arteries or distributing vessels, (2) the capillaries or exchange vessels, and (3), the veins which are storage vessels. Arterioles and venules are muscular vessels located upstream and

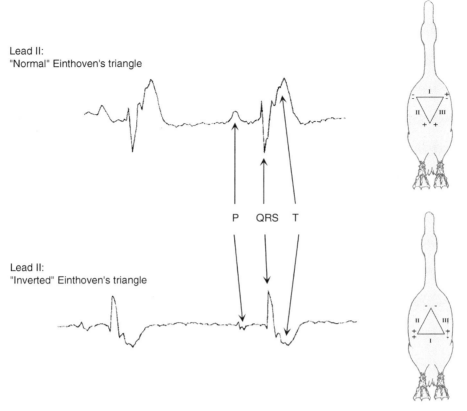

FIGURE 12 Upper ECG trace: voltage recorded from lead II when the recording electrodes are placed in the conventional mammalian manner, on the right shoulder and left leg. Einthoven's triangle is shown on the duck on the upper right in ventral view. Lower ECG trace: voltage recorded from lead II when the recording electrodes are placed on the left shoulder and right leg. Einthoven's triangle is inverted as shown on the duck on the lower right in ventral view. The components of the ECG waveform are shown. P indicates atrial contraction, QRS is ventricular activation, and T represents ventricular depolarization (from L. Liu, D. Syme, and D.R. Jones, unpublished).

downstream of the capillary beds, respectively. They are regulatory vessels, directly controlling blood flow distribution and indirectly controlling exchange of materials across capillary walls by adjustment of capillary pressure.

The major arteries bifurcate many times before the capillary beds and at each bifurcation vascular resistance increases (McDonald, 1974). Volume flow in the parent and daughter vessels remains the same in the steady state but flow velocity in the daughters falls to about 80% of that in the parent vessel. Therefore, the sum of the cross-sectional areas (πr^2, where r is internal radius) of both daughter vessels is greater than that of the parent vessel by about 25%. Hence as the vessels divide flow velocity falls so that in the capillary circulation flow velocity is exceptionally low. This allows adequate time for exchange of blood gases, nutrients, and metabolites with the surrounding cells.

Pressure, generated by cardiac contraction, drives blood flow around the circulation. Poiseuille's law relates volume flow ($\dot{Q}$) to the pressure drop ($P_1 - P_2$) along a tube of radius (r) and length (L) during steady flow, as follows:

$$\dot{Q} = (P_1 - P_2) \times \frac{\pi r^4}{8\mu L}, \quad (1)$$

where μ is blood viscosity. Rearrangement of Eq. 1 hints at a somewhat more familiar form,

$$\frac{8\mu L}{\pi r^4} = \frac{P_1 - P_2}{\dot{Q}} \quad (2)$$

because the term on the left hand side of Eq. 2 is vascular resistance (R). Consequently,

$$R = \frac{P_1 - P_2}{\dot{Q}} \quad (3)$$

or, for the whole body, total peripheral resistance (TPR, kPa sec m^{-3}),

$$\text{TPR} = \frac{\text{MAP} - \text{MVP}}{\text{CO}} \quad (4)$$

where MAP is mean arterial pressure (kPa), MVP is mean venous pressure (kPa), and CO is cardiac output

(m^3 sec^{-1}). In order to compare animals of different sizes it is usual to express cardiac output on a unit weight basis (i.e., m^3 sec^{-1} kg^{-1}).

Since the length (L) of any vascular channel is anatomically fixed while blood viscosity will only vary by 2 to 3 times, then vascular resistance is dominated by the radius of the vessels (Eq. 2). Consequently, with a given pressure drop, halving vessel radius will reduce flow to one sixteenth, as shown in Figure 13, for a change in vessel radius from, for instance, two units (center profile) to one (left profile). This has important implications for the control of blood flow distribution.

Poiseuille's Law (Eq. 1) applies to steady flow but in the major arteries flow is highly pulsatile. In pulsatile flow, due to the inertia of the blood and high heartbeat frequencies, flow amplitude may no longer vary linearly with the pressure gradient. Nevertheless, the extent of the deviation from Poiseuille's law can be assessed from a nondimensional constant α (Womersley, 1957)

$$\alpha = r \sqrt{\frac{2\pi f \rho}{\mu}}, \quad (5)$$

where r is radius, f is heart rate, ρ is blood density, and μ is blood viscosity.

When $\alpha < 0.5$ for the fundamental frequency (i.e., heartbeat frequency), the phase lag is negligible and flow conforms approximately with that predicted by Poiseuille's equation. Calculations for the aorta of a duck give a value for α of 6.0–7.0 for the fundamental frequency (about 3 sec^{-1}) so that estimation of flow by the Poiseuille formula is not reliable. However, it is obvious that for any given heart frequency, the value of α is directly dependent on the size of the vessel. Hence, in the femoral artery, α is certainly below 1, and flow will vary approximately linearly with the pressure gradient in this vessel. The vessels of the capillary circulation are small and α will be likewise, so application of Poiseuille's formula to this vascular bed would appear, superficially, to be most appropriate. Unfortunately blood viscosity, which can be regarded as a constant in larger vessels, may vary unpredictably in vessels of capillary size (Section V,B). Therefore, caution is the watchword when applying steady flow formulations to flow driven by an oscillating pump but, even so, pulsatile flow has a mean, steady flow component and Pouseuille's law can be applied to this component, as in calculations of total peripheral resistance (Eq. 4).

Fourier analysis of pulsatile flow (and pressure) waveforms can be used to isolate the mean from the oscillatory components. The latter are resolved as a series of sinusoidal waves at the harmonics of the original waveform (first harmonic is fundamental frequency (f); second harmonic is 2f, ... n^{th} harmonic is nf). If pressure and flow waveforms are recorded simultaneously then dividing the oscillatory component of pressure by flow, at each harmonic, gives the vascular impedance. Consequently, vascular resistance (mean pressure divided by mean flow) can be considered as the impedance at zero frequency. The vascular impedance of any region of the circulatory system is thus determined by relating the corresponding frequency components of pressure and flow waves recorded simultaneously at that region. If pressure and flow are recorded at the input to the aorta then aortic impedance is an expression not only of the characteristics of the whole systemic circulation but also of the afterload against which the left ventricle must work.

Pressure and flow recorded at the input to the arterial system start synchronously yet peak flow velocity is reached before peak pressure. This rather anomalous

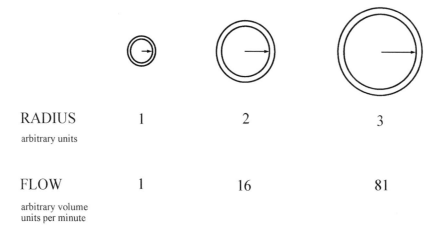

FIGURE 13 Effect of change in radius (length of arrow) of a vessel on fluid flow with a constant driving pressure. Flow decreases in proportion to the fourth power of the decrease in radius. Radius dimensions and relative flow volumes are in arbitrary units.

situation is caused by the fact that the pulses travel through the arterial system. Therefore, for the pressure pulse, a positive gradient between an upstream and downstream point in an artery, established when the crest of the pressure wave traverses the upstream point, will reverse when the crest reaches the downstream point. Hence the pressure gradient oscillates about a mean in all arterial vessels and flow will rise or fall or even reverse with these oscillations in the pressure gradient, although the presence of valves on the outflow tracts of the ventricles may limit the extent of reversal at the root of the aorta. Nevertheless, it is obvious that flow and pressure waves are not "in phase" and the extent of the phase difference, at each harmonic frequency, can be resolved by Fourier analysis.

The input impedance is presented (usually graphically, see Figures 21 and 23 in Section IV) as a set of terms of the values of modulus ($|Z|$) and phase (ϕ) at each frequency obtained by Fourier analysis, thus:

$$(|Z|) = \frac{\text{pulsatile pressure}}{\text{pulsatile flow}} \quad (6)$$

and

$$(\phi) = \text{pressure phase–flow phase.} \quad (7)$$

The phase of the impedance will be negative when the flow leads the pressure and positive when the pressure leads the flow.

IV. THE VASCULAR TREE

A. Arterial System

1. Gross Anatomy

At least six pairs of aortic arches appear in the embryos of all vertebrates, recapitulating their aquatic ancestry. In birds, not all arches are present at one time and some are extremely transitory, such as the fifth pair of aortic arches which make their appearance last. Only three arches persist in the adult, represented by the carotid artery (third arch), the aorta (fourth arch), and the pulmonary artery (sixth arch). In terrestrial vertebrates other than birds and mammals both left and right branches of the fourth aortic arch are retained whereas only the right persists in birds and the left in mammals. In some avian species a remnant of the left aortic arch may remain as a solid core of cells while in a few others, such as the belted kingfisher (*Ceryle alcyon*), the left arch remains patent and functional although it loses its connection with the root of the aorta (Glenny, 1940). In an interesting series of experiments Stéphan (1949) demonstrated that ligation of the right aortic arch in the embryo causes the left to develop, as in mammals.

This finding suggests that the retention or disappearance of aortic arches is dependent upon hemodynamic conditions, and it may be that the persistence of the right arch simply results from the unique development of the ventricular outflow tract in birds.

The first major vascular bed supplied by the aorta is that of the heart. The coronary arteries, supplying the nutritional and respiratory circulation of heart muscle, arise from the ascending aorta close to the heart (Section II,A,5). The ascending aorta then gives rise to two very large brachiocephalic trunks (Figure 14), supplying blood to the head, wings, and flight muscles. Each brachiocephalic vessel is usually larger in diameter than the continuation of the aorta, reflecting the higher blood flow rates in the brachiocephalics.

All the arteries of the head and neck are branches of the carotid arteries. Surprising variation exists in the pattern of the carotid arteries close to the heart. The most common arrangement is two vessels of equal size running side by side (Figure 14). Other patterns are:

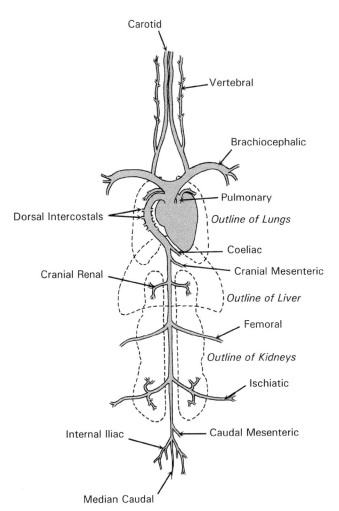

FIGURE 14 The major systemic arteries in the bird.

(1) a single artery formed by fusion of both carotids (herons, bitterns, and kingfishers); (2) a single vessel due to loss of the right (passerines) or left (plovers) carotid; or (3) two arteries of unequal size (flamingos, sulphur-crested cockatoo).

Blood flow to the brain must not be interrupted, or impairment of brain function rapidly ensues. The carotid arteries lie in a groove in the base of the neck vertebrae close to the axis of rotation and are therefore protected from possible obstruction due to compression from neck movements. Other safety measures are provided by anastomoses between the carotid and vertebral arteries and, at the base of the brain, by either an X-, I-, or H-shaped junction between the carotids. This intercarotid anastomosis has been found in all species of bird except those of the suborder Tyranni (Baumel and Gerchman, 1968). Birds do not possess a cerebral arterial circle of Willis comparable to that of mammals but, since the intercarotid anastomosis is relatively large in many cases, it may represent a more effective collateral circulation than the mammalian arterial circle. Blood to the wings and flight muscles is supplied by the subclavian arteries. Each subclavian divides into two branches, the brachial (wing) and pectoral (flight muscles).

The descending aorta runs caudally, ventral to the vertebral column, giving off paired intercostal and lumbar arteries. Blood is supplied to organs within the abdomen and legs by the following vessels, originating from the descending aorta (see Figure 14):

Celiac artery	Liver, spleen, glandular stomach, gizzard, intestine, pancreas
Cranial mesenteric artery	Most of intestine, pancreas
Renal arteries	Kidneys (anterior portion), testes
Femoral arteries	Legs
Ischiatic arteries	Middle and posterior portions of kidney and legs; uterine region of oviduct
Caudal mesenteric artery	Rectum and cloaca
Internal iliac arteries	Walls of pelvis; oviduct
Caudal artery	Tail, terminal branch of aorta

There are in effect three pairs of renal arteries in birds, one pair arising from the aorta and two from the ischiatic arteries. However, in *Ardea cinerea* one pair of renal arteries arises from the femoral arteries instead of the ischiatics.

The ischiatic artery is the major vessel supplying the leg. At the level of the knee it meets and joins the femoral artery to form the popliteal artery. This artery, passing into the lower leg, divides to form the anterior and posterior tibial arteries. In the tarsal region of the leg and in the axillary region of many birds, there are arteriovenous networks of vessels referred to as *rete mirabili* (particularly prominent in wading and aquatic birds). These structures serve as heat exchangers, since warm arterial blood is brought into close proximity to venous blood that has traversed the distal parts of the limbs and is therefore colder. The countercurrent arrangement of blood flow ensures that heat can be transferred from arterial to venous blood all along the length of the artery and vein apposed in the *rete*, thereby reducing heat loss to the environment by reducing the temperature of arterial blood flowing through peripheral thinner sections like the web of the foot or the wing.

2. Functional Morphology of the Arterial Wall

The large arteries have two main functions. First, they serve as low-resistance conduits carrying blood to the arterioles for distribution to the peripheral vascular beds. Second, the whole arterial system serves as a pressure reservoir or *Windkessel*, accepting the volume of blood ejected by the heart and converting the highly pulsatile input into a steady flow of blood through the capillary beds. The *Windkessel* results from wall elasticity, particularly of those vessels close to the heart.

The central arterial vessels are indeed "elastic" while the more peripheral ones, certainly distal to the second order of branching, are "muscular." In elastic arteries, the vast majority of the wall is made up of layers of smooth muscle embedded in elastin fibers alternating with layers of collagen. One layer, composed of a combination of muscular, elastic, and collagen fibers, forms a single lamellar unit within the wall. Large numbers of concentric lamellar units make up the bulk of the wall of elastic arteries of pigeon, chicken, and the mute swan (*Cygnus olor*) as shown in Figure 15 (Bussow, 1973). Interestingly, the lamellar units do not form complete cylinders around the vessel. This is particularly obvious in vessels very close to the heart where an individual lamella may extend around only one quarter, at most, of the vessel circumference.

Wall structure of the muscular arteries is very different. Muscular vessels consist largely of circumferentially arranged smooth muscle cells with elastic fibers distributed, either singly or in bundles, as a wide-meshed plexus between the muscle cells (Hodges, 1974). The collagenous components are transferred to the outer layer of the wall. An interesting embellishment of the normal structure of muscular arteries occurs in the cranial mesenteric artery of the chicken and turkey (Ball et al., 1963). This vessel is invested by longitudinally arranged smooth muscle fibres, the thickness of this layer being approximately the same as that of the circumferentially oriented smooth muscle within the wall proper. The functional significance of the external muscle layer is unclear but it may serve to shorten the vessel to accommodate changes in its position brought about by gut movements (also see Section VI,B,3,a).

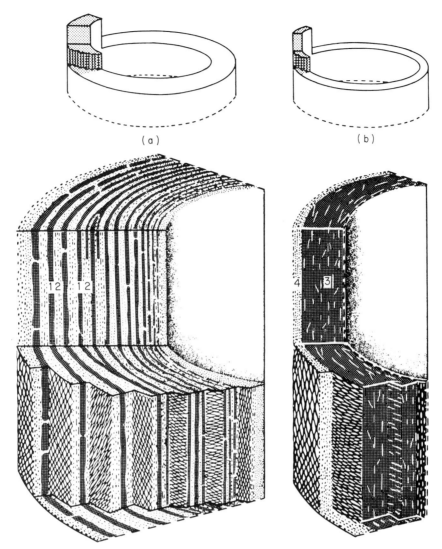

FIGURE 15 Semischematic depiction of the wall of an elastic (a) and muscular (b) artery. (a) Wall structure of an avian elastic artery. The wall is made up of fragmented layers of smooth muscle embedded in a fine network of elastic fibers (1) alternating with layers of collagen (2). (b) Wall structure of an avian muscular artery consisting of a thin intima, a media made up of smooth muscle cells and elastic fibers (3), and a thick adventitia with a well-defined elastica externa (4). Upper diagrams compare wall thickness of elastic and muscular arteries of similar lumen diameter. The wall of the elastic vessel is about three times as thick as that of a muscular vessel (after Bussow, 1973; used by permission).

The elastic arteries include the aortic arch and its major branches, the thoracic aorta up to about the level of the celiac artery, and the extrapulmonary portions of the pulmonary arteries, while all branches of the abdominal aorta as well as the caudal portion of the aorta itself are muscular. In most regions of the arterial tree the change from an elastic to a muscular wall occurs rather abruptly, usually at a branch site. An exception to this is in the aorta itself. In the aorta, the elastic and muscular portions are separated by a segment of the vessel extending from the coeliac artery to the ischiatic arteries, whose wall structure fits neither description very well. Furthermore, in both the pigeon and turkey the wall in this region is transversely asymmetric with a thick, muscular ventral wall and a thin elastic dorsal wall.

The arteries *in vivo* expand and recoil with every heartbeat although this behavior is seldom mimicked in *in vitro* experiments designed to study vessel mechanics. Instead, the static rather than dynamic elastic behavior of the arterial wall is usually investigated. Essentially, a short length of excised blood vessel is inflated from a syringe and the pressure change induced by a given volume change is noted. These "pressure–volume loops" give an immediate and compelling view of how arterial elasticity changes with the degree of inflation. Furthermore, by using blood vessels from different areas of the body, regional variations in elasticity are revealed. Pressure–volume loops are usually J-shaped, showing that the more a vessel is stretched the more resistant it becomes to further stretch (Figure 16; Speckmann and Ringer, 1966). The collagen fibres in the vessel wall inhibit expansion at high pressures, whereas the properties of elastin dominate the lower pressure limb of the curves. It is the compliance of elastin and the stiffness of collagen working in concert that allows uniform and smooth expansion of the vessel wall over a range of distending pressures without formation of aneurysms. In contrast, a wall in which the properties of extension remain constant across the range of distending pressures would be prone to aneurysm formation. Rubber has a straight rather than J-shaped response to distension and aneurysms always occur in the wall of cylindrical balloons when they are inflated.

When a blood vessel is inflated more pressure is required to expand it than is recovered during recoil of the elastic walls. The ratio between the energy recovered in deflation to that expended in inflation is a measure of the "resilience" of that vessel (Figure 16). Surprisingly, "resilience" is similar for both thoracic (elastic) and abdominal (muscular) aortae of the turkey. Over the range of arterial pressures encountered in turkeys "resilience" lies between 85 and 87%, values well above those of most mammals and approaching those obtained for invertebrate blood vessels. In one sense, the higher the "resilience" the better because most of the energy cost of stretching elastic vessels with each cardiac ejection will be returned by elastic recoil on deflation. Unfortunately, if the "resilience" is too high then the vessels may go into uncontrolled oscillations (resonance) particularly at the high repetition frequencies (heart rates) seen in many birds. Obviously, vascular engineering in birds is close to the edge.

Pressure–volume loops reveal characteristics specific to particular segments of whole vessels, whereas the properties of the materials making up a vessel wall are revealed by stress–strain curves (Figure 17A). Stress is the deforming force divided by the area of the vessel wall over which it is applied while strain is the ratio of the stretched radius to the unstretched radius of the vessel. Stress–strain curves for the ascending and descending aortae, the brachiocephalic arteries, and the thoracic aorta of the duck (*A. platyrhynchos*) are shown in Figure 17A. The stress–strain relationship for the abdominal aorta lies to the left of that for the other, more central, vessels indicating that, as in the turkey (Figure 16), the abdominal aorta is much stiffer than the other vessels, a finding which would be expected given that the abdominal portions of the aortae of duck and turkey have more collagen than do more central segments of this vessel.

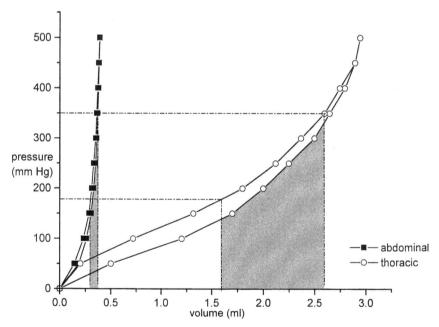

FIGURE 16 Typical pressure–volume loops for the abdominal and thoracic regions of the aorta in a turkey. For each loop, the upper curve is the inflation sequence and the lower curve the deflection sequence. Resilience of the vessel is obtained by dividing the area under the deflation sequence (shaded) by the area under the inflation sequence over the range of blood pressures recorded in turkeys (data from Speckmann and Ringer, 1966).

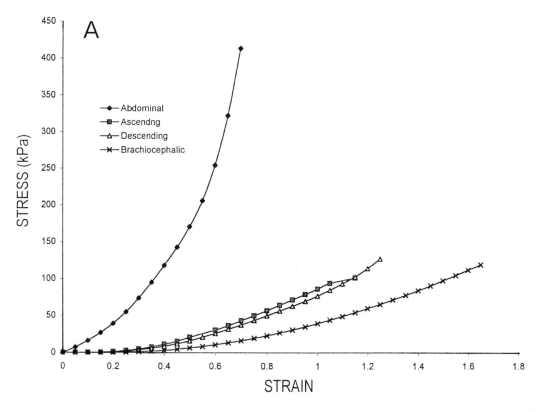

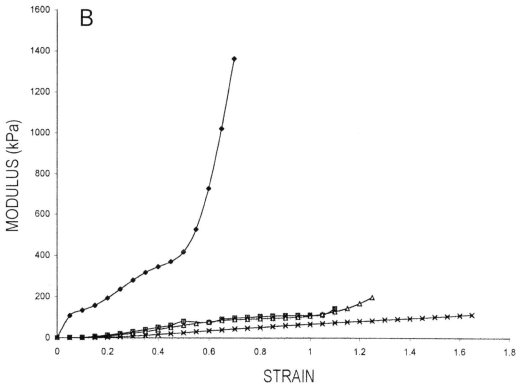

FIGURE 17 (A) Stress–strain curves for the blood vessel wall from four regions of the aorta in a duck (*A. platyrhynchos*). (B) The static elastic modulus of these vessels derived from data in A (M. Braun and D. R. Jones, unpublished).

The slope of the line of a stress–strain curve is the elastic modulus of the material making up the blood vessel wall, but, since the slopes for blood vessels are not linear, the elastic modulus is continually changing. The incremental elastic modulus describes the elastic modulus for a small increment in strain (Bergel, 1961). Incremental elastic moduli for the duck aorta, which are similar to those for the turkey aorta (Speckman and Ringer, 1964), are shown in Figure 17B. The moduli for the abdominal aorta are well above that for the thoracic, confirming that the former is stiffer than the latter. Nevertheless, these moduli are one to two orders of magnitude below those obtained from arteries in corresponding parts of the mammalian vascular tree. In mammals, the lamellar units in the wall form complete cylinders so that those laminae containing primarily collagen must be stretched to the same extent as the more distensible muscular laminae. Consequently, such rigid laminae are more important in determining the degree of extension of the wall. However, in birds, arterial lamellar units do not form complete cylinders so there is some "series" coupling between the rigid and elastic components of the wall which allows more distensible laminae to be extended somewhat independently. Avian blood vessels have much thicker walls than mammalian vessels of the same diameter, thus compensating for their lower elastic modulus (Bussow, 1973).

3. Relationship between Arterial Pressure and Flow

Each heartbeat sends a pulse through the arterial system which arrives later at sites more distal to the heart. The velocity at which the pulse wave travels is lowest in the most distensible vessels and increases in the stiffer peripheral vessels. In ducks, pulse wave velocity increases from 4.4 ± 0.8 m sec^{-1} in the aortic arch to 11.7 ± 1.2 m sec^{-1} in the abdominal aorta, with the major increase in velocity occurring in the thoracic aorta (Langille and Jones, 1975).

When the pulse transit time occupies a considerable proportion of each cardiac cycle then significant phase changes occur between the pressure and flow pulses at different arterial sites. In ducks, the time taken for the pulse to travel from the heart to the distal end of the abdominal aorta is around 20 ms, which is about 5–10% of the cardiac cycle (Langille and Jones, 1975). Consequently, in ducks, there are marked changes in the waveform of the systemic pressure pulse as it travels through the arterial system (Figure 18). Both pulse amplitude and the contour of the pulse waveform are altered, with pulse pressure increasing by about 30%. This peaking of the pressure pulse results from a marked increase in the systolic portion with little change in the diastolic portion (Figure 18).

Peaking of the pressure pulse is due to wave transmission effects, primarily wave reflections. All forms of wave motion can be reflected by physical changes in the system they are traveling through. When such changes occur within the arterial system, incident pressure and flow waves will be reflected back toward the heart. These physical changes can be discrete discontinuities, such as those due to arterial branching (McDonald, 1974) or continuous variations in wall compliance due to an increase in arterial stiffening toward the periphery (Langille and Jones, 1975; 1976). However, the major reflecting site seems to be the terminal vascular bed. From this site pressure and flow pulse waves are reflected back toward the heart to interfere, destructively or constructively, with the incident wave generated by cardiac contraction. This interference means that pressure and flow waves recorded simultaneously at any one site in the arterial system will be quite unlike those recorded at another. In a reflectionless system, pressure and flow pulses sampled at any given site should look similar to those recorded anywhere else in the system.

An essential question concerns the nature of the termination that the peripheral vascular beds present to outgoing pressure and flow waves. Peripheral beds are "closed" terminations if they present a relatively large impedance to pulsatile flow; they are "open" if they present a relatively low impedance. In higher vertebrates, reflections produce large oscillations in peripheral pressures that drive small oscillatory flows through the terminal vascular beds (Figure 18), indicating a high terminal impedance; i.e., of the "closed" type. Hence, the pressure should be reflected at the closed end without a phase shift while, to satisfy the condition that high pressure oscillations are required to drive low oscillatory flows through a high terminal impedance, the reflected flow wave should be inverted. That is, the reflected wave should be 180° out of phase with the incident wave. However, how much of the incident wave reaches the termination (because the incident pulse is attenuated, especially in the smaller vessels, as it propagates through the system) and how much of the reflected wave gets back to the heart (since its amplitude is also reduced by damping) is still a matter of speculation.

The reflection coefficient (that portion of the incident wave reflected by the terminal vascular beds) is extremely sensitive to the state of the peripheral vasculature. Under resting conditions up to 80% of the incident wave may be reflected. Intense vasoconstriction, which occurs in ducks during forced diving or when voluntarily diving birds are trapped underwater and unable to surface, causes 100% of the incident wave to be reflected. In contrast, vasodilatation of peripheral vascular beds, occurring during exercise or hemorrhage, may reduce the reflection coefficient to zero.

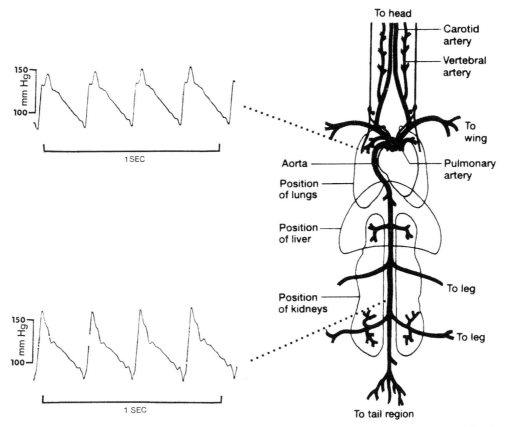

FIGURE 18 Diagrammatic representation of pressure waves recorded simultaneously in the proximal and distal aorta of a duck. Amplification and distortion of the pressure wave occurs during propagation along the aorta. (Enhanced from Langille and Jones (1975), *Am. J. Physiol.* with permission.)

Evaluation of wave propagation through the arterial system is complex, requiring harmonic analyses of pressure and flow waveforms. Fortunately, a simple conceptual analysis of the interaction of incident and reflected waves in the arterial tree is sufficient for the present purpose. For simplicity, consider a pressure wave displaying simple harmonic motion, as illustrated in Figure 19 (first harmonic). At the closed end of the system (terminal vascular beds) both incident and reflected pressure waves will be in phase and the waves will sum giving an antinode, evident as an enlarged pressure oscillation. The reflected wave is shown as 40% of the incident wave in Figure 19 so that the amplitude of the resultant compound wave will be 140% of the incident wave as is shown by the wave envelope. Now consider a point one quarter wavelength back from the termination. The incident wave left here one-quarter of a cycle before it reached the closed end and the reflected wave takes another one-quarter cycle to return to this point, so the incident and reflected waves are now 180° out of phase and destructive interference produces a node, evident as a decrease in pressure. Hence, the amplitude of the resultant wave is some 40% smaller (as shown by the wave envelope in Figure 19) than it would be in the absence of reflections.

The flow wave will also be reflected such that the incident and reflected waves are 180° out of phase at the termination but are in phase one-quarter cycle away from the terminal impedance. Hence, the flow waves will cancel one another at the termination and will sum at the heart. Therefore, highly oscillatory cardiac outflow is generated for less pulsatile pressure than would be the case in a nonreflecting system; the advantage of this is that more external cardiac work is available for any given level of cardiac oxygen consumption (Milnor, 1979).

The benefits of such reflections will be maximized when the heart is one-quarter wavelength upstream of the major reflecting site. In ducks, resting heart rate is 2–3 beats sec^{-1} so even at the lowest pulse wave velocity (4 m sec^{-1}) the major reflecting site would have to be located one third or one half meter from the heart—a most unlikely possibility. During exercise, however, heart frequency may double and the major reflecting site would now be 16–25 cm from the heart. Unfortunately, exercise is associated with vasodilation which will re-

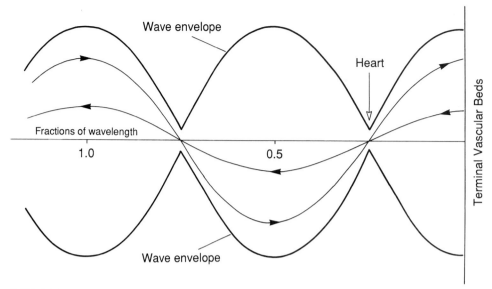

FIGURE 19 Interactions between incident and reflected pressure waves at a closed end (terminal vascular beds). For clarity it is assumed that only 40% of the wave is reflected from the closed end (i.e., reflection coefficient = 0.4). The abscissa is marked in fractions of a wavelength. At the point of reflection both waves are in phase and sum together. With reference to a point one-quarter wavelength away, the incident wave is 90° earlier and the reflected wave 90° later so that they are 180° out of phase and cancel. This point is a node and the only oscillation is the difference between the maximum amplitudes of the incident and reflected waves. For maximum benefits in terms of promoting cardiac efficiency the heart should be located at this point. The total excursion throughout the cycle is represented by the wave envelope (heavy outer lines; figure redrawn and caption modified from McDonald, 1974).

duce the reflection coefficient. In contrast, vasoconstriction during forced or voluntary diving, especially if the bird is trapped underwater, would accentuate reflections but heart frequency is now about 0.33 beats sec^{-1} (or 1 beat every 3 sec).

Consequently, obtaining the necessary balance between heart rate and reflection coefficient to maximize beneficial reflection effects seems unlikely. What then are the consequences of a mismatch between pulse wavelength and distance between the heart and major reflecting sites? Let us assume that Figure 19 describes the second and not the first harmonic of the pressure wave. In this case, the first harmonic will be 90°, not 180°, out of phase at the heart. This represents an antinode for the first harmonic which will add to systolic pressure, causing the heart to expend more energy. The heart will be located at a node for the second harmonic but at an antinode for the third (270° out of phase). As the first three harmonics contribute about 80–90% to the original pulse amplitude then the net effect will be an increase in systolic pressure. In the duck, this is seen as a significant early systolic shoulder in the pressure pulse recorded in the aortic arch (Figure 18).

If the transit time of either pressure or flow waves through the arterial system becomes less than 5% of the cardiac cycle then reflection effects on the shapes of these waves are not obvious. Nevertheless reflections occur but they are diffuse, the pressure wave bouncing back and forth between the heart and periphery, until damped to extinction. In humans, atherosclerosis causes a loss of pressure pulse amplification as the pulse travels through the arterial system (O'Rourke et al., 1968); similar observations have been made in the turkey, Meleagris (Taylor, 1964), where atherosclerosis is very common (Ball et al., 1972; Manning and Middleton, 1972). Loss of pulse amplification in atherosclerosis results from a generalized stiffening of the major arteries, which acts to speed pulse wave propagation and thereby minimize wave transmission phenomena.

The hummingbird is among the smallest homeothermic vertebrates. Is it possible to predict the hemodynamics of the hummingbird from our knowledge of hemodynamics in the duck? Assuming pulse wave velocity is unaltered, then for similar conditions to hold in both hummingbird and duck, heart frequency (f_H) must increase in the same proportion as the linear dimension (L) of the animal decreases with reduction in body mass (M_b). For birds, the allometric equation relating f_H and M_b is $f_H = k_1 \cdot M_b^{-0.282}$ (Grubb, 1983; and see Section II,B), while $L = k_2 \cdot M_b^{-0.33}$, where k_1 and k_2 are constants.

According to this analysis, a hummingbird 400 times smaller than a duck will have a f_H 5.3 times higher but L will decrease nearly seven times. Even at the highest

f_H reported for the giant hummingbird (*Patagona gigas*) of 1020 beats min^{-1} (Lasiewski *et al.*, 1967), pulse transit time as a proportion of the cardiac interval will be considerably less than that in the duck and reflection effects on pulse wave shapes will probably not be obvious (Jones, 1991).

4. Vascular Impedance

While the complex pressure and flow waves recorded in avian arteries (Figure 20) are not directly comparable they, like all periodic signals, can be expressed as a sum of sinusoidal signals of ascending frequency (harmonics). These individual harmonics of pressure and flow are directly comparable (see McDonald, 1974). Comparison can be done most conveniently by determining vascular impedance versus frequency. Impedance modulus (amplitude of a pressure harmonic divided by amplitude of the flow harmonic of the same order) is the analog of vascular resistance that is applicable to pulsatile flows. Impedance phase is simply a measure of the degree to which pressure and flow oscillations are out of synchrony.

Figure 21A illustrates impedance versus frequency curves for the circulation supplied by the descending thoracic aorta of *Anas*, while Fig. 21B illustrates the vascular impedance of the pulmonary circulation. Aortic impedance falls from the value at zero frequency, the peripheral resistance (Z_t), to settle at a steady value at high frequencies; this is an indication of the characteristic impedance (Z_o; Figure 21B). Peripheral resistance and characteristic impedance are measures of arteriolar caliber and aortic distensibility respectively, uninfluenced by wave reflection effects. Often, the modulus and phase are not constant but fluctuate, these fluctuations being caused by wave reflection effects from peripheral vascular sites. A well-defined minimum in impedance modulus of the aortic circulation, and a coincident rise in impedance phase from negative values (pressure lagging flow oscillations) to positive values (pressure leading flow oscillations), are characteristics of a wave reflecting system (Figure 21A). At the frequency of the impedance minimum (10 Hz) the circulation imposes minimal load for pulsatile flow on the heart (Figure 21A).

The reflection coefficient depends on the impedance mismatch between the terminal arteriolar bed and the supply artery and is given by the ratio

$$(Z_t - Z_0)/(Z_t + Z_0) \times 100.$$

For the aortic circulation, the reflection coefficient is high, being over 80%. However, in the pulmonary circulation the absence of clear impedance minima suggests that this low-resistance circulation does not give rise to major reflections of the pulse wave. In fact, the reflection coefficient for the pulmonary circuit is only 25%.

Pulsatile pressure and flow are generated by the first, embryonic cardiac contractions. In Stage 24 chick embryos pressure and flow waveforms resemble those recorded from mature animals despite the absence of the semi-lunar valve apparatus in the heart (Figure 22; Zahka *et al.*, 1989). Both peripheral resistance and vascular impedance modulus decrease with development through stages 18 to 29 (3–6 days, development; Figure 23). Over this period, mean dorsal aortic pressure and flow increase 13 and 10 times, respectively, the increased flow being accommodated by the rapidly expanding arterial bed. The velocity of pulse wave propagation increases with developmental stage, from around 0.5 m sec^{-1} at stage 18 to nearly 1 m sec^{-1} at stage 24 (Yoshigi *et al.*, 1997). Hence, pulse transit time will be a negligible fraction of the cardiac cycle (0.35–0.45 sec) so that reflection effects will be unimportant. Even if this were not the case, reflection effects would be minimized in the later stages of embryonic development because of the marked decline in vascular resistance. Consequently, the fall in vascular impedance during embryonic development is due to the growth of a larger, more distensible, dorsal aorta in which elastic fibers first appear at stage 29 (Hughes, 1942).

B. Capillary Beds

1. Gas Exchange

Systemic capillaries form a vital functional interface between the blood and the systemic tissues of birds. The pathway between erythrocytes in capillary blood and mitochondria in the surrounding tissue represents the last in a series of resistances in the oxygen transport pathway from the lungs. Oxygen and carbon dioxide move between the systemic capillary blood and the surrounding tissue mitochondria by simple diffusion. Therefore, the diffusion distance from erythrocyte to mitochondrion and the partitioning of diffusion resistance along this route is of immense physiological interest. The concept of a capillary domain, a volume of tissue whose oxygen demands could potentially be satisfied by diffusion from one capillary, was first expounded by August Krogh in 1914 and remains a valuable concept in understanding gas exchange in muscle tissue (Krogh, 1919). A simple explanation of the factors important in capillary blood–tissue gas exchange, based on this model, is to be found in West (1995).

Studies on the systemic capillaries of birds, in particular those of flight muscle, have been dominated recently by two interesting themes: (1) the high workload of the avian pectoralis major muscle during flight, reflected in an increase in oxygen consumption of about five times that at rest (Butler *et al.*, 1977), suggests that the func-

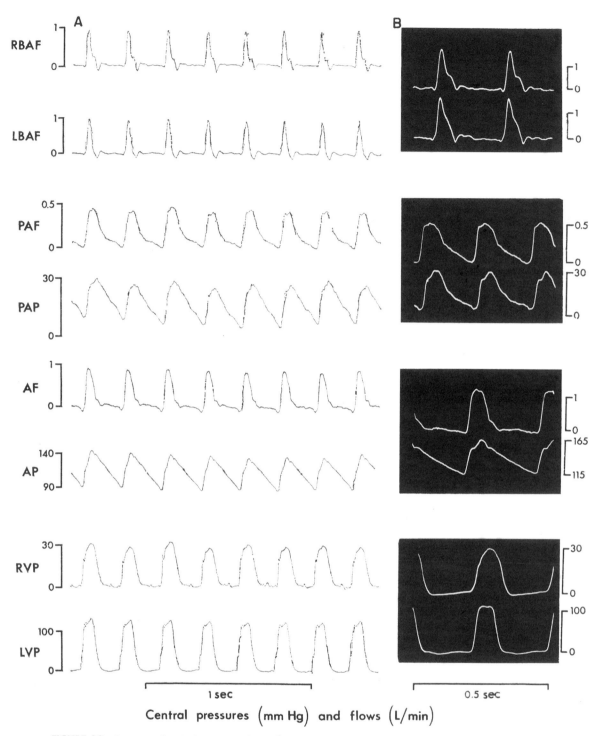

FIGURE 20 Pressures (mmHg) and flows (l min^{-1}) recorded simultaneously in the arterial system of a duck. Traces, from top to bottom: flow in right brachiocephalic artery, RBAF; flow in left brachiocephalic artery, LBAF; pulmonary arterial flow, PAF; pulmonary arterial pressure, PAP; aortic flow, AF; aortic pressure, AP; right ventricular pressure, RVP; left ventricular pressure, LVP. Panels on right represent paired oscilloscope records of central pressures and flows. Traces match those on the left but each pair was recorded from a different animal. (From Langille and Jones (1975), *Am. J. Physiol.* with permission.)

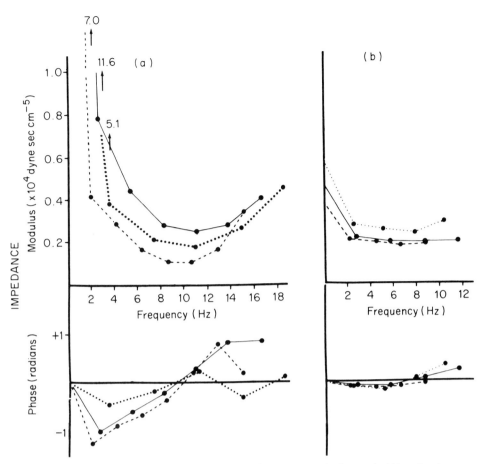

FIGURE 21 Impedance modulus and phase versus frequency graphs for aortic (A) and pulmonary (B) circulations in a duck. (From Langille and Jones (1975), *Am. J. Physiol.*, with permission.)

tional anatomy of the pectoral muscle capillaries may reveal adaptations to both high tissue oxygen demand and mechanical tissue deformation during sarcomere shortening; and (2) some species fly at high altitudes. Therefore, there may be specific adaptations in capillary den-

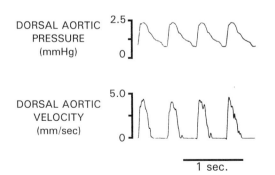

FIGURE 22 Dorsal aortic pressure and flow velocity recorded simultaneously from a stage 24 chick embryo. The pressure and flow waveforms are similar to those recorded from mature animals (see Figure 20) despite the absence of a semilunar valve apparatus. (After K. G. Zahka, N. Hu, K. P. Brin, F. C. Yin, and E. B. Clark, Aortic impedance and hydraulic power in the chick embryo from stages 18 to 29, *Circ. Res.* **64**, 1091–1095.)

sity or geometric arrangement that facilitates the delivery of oxygen to the working pectoralis muscle in the face of a relatively low P_aO_2 (the pressure head for diffusion at the arterial end of systemic capillaries) caused by a reduced atmospheric partial pressure of oxygen.

Three parameters may be considered in determining the capillarity of muscle. These are the number of capillaries per muscle fiber, the cross sectional area of muscle fibers, and the geometrical arrangement of capillaries around each fiber (Snyder, 1990). Gray *et al.* (1983) found that sections of pure slow red muscle fibers from the anterior latissimus dorsi of chicken had 25% more capillaries per square millimeter than did sections of fast white fibers from the posterior part of this muscle. However, in six other species of birds ranging in mass from 11g to 6.2 kg there was no significant correlation between fiber diameter and capillary number per fiber in slow red fibers and fast white fibers of the anterior and posterior latissimus dorsi respectively. Thus, tissue capillary density decreased in muscle with larger fibers. The maximum diffusion distance from capillary to mitochondrion in slow red fibers and fast white fibers of the

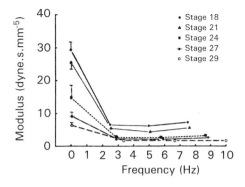
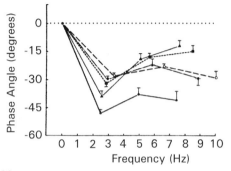

FIGURE 23 Impedance modulus (upper) and phase angle (lower) for chick embryos from stages 18 to 29. The identification key inset in the upper panel also applies to the lower panel. (Modified from K. G. Zahka, N. Hu, K. P. Brin, F. C. Yin, and E. B. Clark, Aortic impedance and hydraulic power in the chick embryo from stages 18 to 29, *Circ. Res.* **64,** 1091–1095.)

gastrocnemius muscle was estimated to average 32.4 and 36.5 μm respectively (Snyder, 1990). This is similar to the values obtained in mammals, suggesting that diffusion distance has been highly conserved in vertebrate evolution. Increasing the number of capillaries per fiber appears to produce diminishing returns such that beyond two capillaries per fiber there do not appear to be further measurable reductions in diffusion distance. In contrast to the situation in neonatal mammals, the capillary-to-fiber ratio in bird muscle appears to be fixed around hatching. Because the fibers hypertrophy during development, diffusion distances are shortest in newly hatched chicks—some 18 μm (Byers and Snyder, 1984).

The benefits conferred by reducing erythrocyte-to-mitochondrion diffusion distance depend on assumptions made about how the resistance to oxygen diffusion is distributed between source (blood) and sink (mitochondrion). It has been argued that the capillary-to-tissue interface represents the major resistance (Gayeski and Honig, 1986), in which case reducing the overall diffusion distance would be of limited effectiveness compared with increasing the area of this interface (see below).

Compared with the geometric arrangement in mammalian hindlimb muscle, there are a larger number of capillary branches running perpendicular to the long axis of muscle fibers in pigeon pectoralis muscle (Mathieu-Costello, 1991, Figure 24). These branch from capillaries running parallel to the long axes of the muscle fibers. The branch points move closer together as the pectoralis muscle shortens on the power stroke, but the perpendicular orientation of the branches to the long axis of the muscle fibers does not change appreciably during contraction. The short segments of capillary parallel to the muscle fiber between these branches bow as the muscle shortens, but do not become particularly tortuous. The branches perpendicular to the long axis of the muscle fibers run around the circumference of the fibers and this arrangement together with their high density ensures that there is an effective envelope of capillary blood surrounding portions of the fibers. This results in very effective blood–tissue O_2 transfer (Ellis et al., 1983). Such an arrangement of capillary branches may compensate for the unfavorable rheological properties of avian blood compared with mammalian blood; these properties include relatively low red cell deformability and a low capillary hematocrit (see Section V).

Flying hummingbirds have the highest mass-specific metabolic rate of any vertebrate and hummingbird flight muscles have the highest oxygen demand of any vertebrate skeletal muscle. It would be expected, therefore, that the adaptations for effective gas exchange at the capillary level would be most obvious in these birds. These could include a reduced diffusion distance from the capillary to the mitochondria or possibly an increase in the capillary-to-fiber contact area. Increases in either of these factors would increase the flux of respiratory gases for a given drop in PO_2 across the capillary–mitochondrial diffusion distance. The ratio of capillary surface area to muscle fiber surface area is about twice as large in hummingbird flight muscle as in the rat soleus muscle, with similar mitochondrial density in the two muscles (Mathieu-Costello *et al.*, 1992). Therefore, according to Fick's law, the rate of O_2 diffusion into the avian fiber would be about double that in the rat, all other factors being equal. This supports the idea that the large area of the capillary–muscle fiber interface in avian flight muscle plays an important role in enabling these muscles to maintain an extremely high oxygen usage during flight. The respiration rates of muscle mitochondria in working hummingbird flight muscle are about double those in the locomotor muscles of mammals. Interestingly the size of this interface in the pectoralis muscle of the only actively flying mammal, the bat, is similar to that in hummingbird flight muscle.

The basic structure of the capillary network in the hummingbird is similar to that in pigeon flight muscle, although capillary density tends to be higher in the hummingbird. This results from the smaller cross-sectional area of the muscle fibers (one-half for aerobic and one-

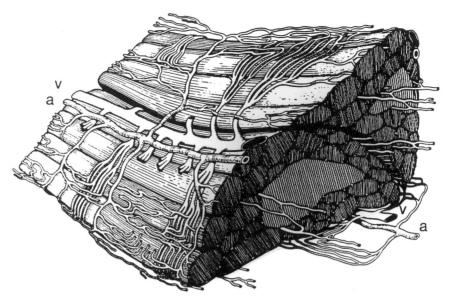

FIGURE 24 Schematic diagram illustrating the microvascular geometry in the pectoralis muscle of the pigeon. For clarity, the width and number of capillary branches running perpendicular to the long axis of the muscle fibers have been reduced. a, artery; b, vein. (Reprinted from *Tissue Cell* **26**, O. Mathieu-Costello, P. J. Agey, R. B. Logemann, M. Florez-Duquett, and M. H. Bernstein, Effect of flying activity on capillary-fiber geometry in pigeon flight muscle, pp. 57–73, (1994), by permission of the publisher Churchill Livingstone.)

tenth for glycolytic) rather than from a greater number of capillaries surrounding each fiber (Mathieu-Costello *et al.*, 1992). The smaller cross sectional area of fibers may be an adaptation to reduce the diffusion distance from capillary blood to the mitochondria. However, some experimental and theoretical evidence shows that the main drop in PO_2 occurs across the resistance offered by the capillary–fiber interface and PO_2 then declines more slowly, largely because of myoglobin-facilitated diffusion (Honig et al., 1991). Thus, a large contact area between capillary and fiber is probably a more important factor in effective oxygen delivery than a short path between capillary and mitochondrion. This is particularly true for flight muscle in both birds and bats where fiber myoglobin content, and therefore the potential for facilitated diffusion, is high. Interestingly, a comparison of actively flying and sedentary pigeons showed little effect of flight conditioning on capillary–fiber relationships. Wild pigeons had a greater aerobic capacity than sedentary birds, achieved by a 30% greater cross-sectional area of aerobic fibers in the pectoralis together with a higher density of mitochondria. However, the capillary–fiber ratio was similar, as was capillary length–fiber volume at a given mitochondrial density (Mathieu-Costello *et al.*, 1994). This finding is consistent with the results of Snyder and Coelho (1989), who caused hypertrophy of the right anterior latissimus dorsi of chickens by taping weights to the humerus. They found that the increased number of capillaries per fiber just matched fiber hypertrophy and concluded that muscle growth was the primary determinant of capillarity.

Birds inhabit niches at a wide range of altitudes and many seasonal migrations occur at high altitude. It is therefore of particular and current interest to know whether we can identify adaptations in the degree of flight muscle capillarity associated with high-altitude residence or flight. In view of the evidence outlined above it might be predicted that any such adaptation would increase the total surface area of contact between capillary and muscle fiber, thereby increasing oxygen conductance. However, the current evidence for adaptation is sparse and contradictory. When wild pigeons actively flying at 3800 m in La Paz, Bolivia were compared with controls at sea level no adaptations in capillary geometry or density could be identified that were attributable to chronic altitude exposure (Mathieu-Costello *et al.*, 1996). On the other hand, Leon-Velarde *et al.* (1993) found that the number of capillaries surrounding a fiber was greater in the pectorals and some limb muscles of Andean coots native to 4200 m compared with controls at sea level. Furthermore, Canada goose goslings hatching from eggs raised under hypoxic conditions have been reported to show an increased capillary-to-fiber ratio (Snyder, 1987). Diffusion distances were also claimed to be shorter in geese incubated in mild hypoxia (94 torr) (Snyder *et al.*, 1984). However, other factors capable of facilitating oxygen delivery, such as changes in P_{50} (partial pressure of oxygen at which hemoglobin is half-saturated) and the oxygen-carrying capacity of blood or increases in the myoglobin content of muscle fibers may ultimately prove to be more significant adaptations. Indeed, increases in carrying capacity and high levels of tissue oxygen extraction have been demonstrated in unexercised pigeons acclimatized to high altitude (Weinstein et al., 1985).

2. Microvascular Fluid Exchange

Capillary fluid balance is maintained by the dynamic interaction between hydrostatic and osmotic forces acting across the capillary wall, as first described by Starling over 100 years ago (Starling, 1896). Starling's original formulation has been modified and refined by Landis (1927) and Kedem and Katchalsky (1958), yielding the following equation to describe microvascular fluid exchange:

$$Jv = K_{FC} [(P_c - P_T) - \sigma_d (\pi_P - \pi_T)],$$

where Jv is net volume flow across the vascular wall, K_{FC} is capillary filtration coefficient, P_c and P_T are capillary and tissue fluid pressures, respectively, $(\pi_P - \pi_T)$ is colloid osmotic pressure difference between plasma ($_P$) and tissue ($_T$), and σ_d is the osmotic reflection coefficient.

In the steady state, the capillary blood pressure opposes the blood colloid osmotic pressure (COP) to maintain tissue fluid balance. Blood pressure exceeds COP at the arteriolar end of the capillary and is usually below COP at the venous end. Fluids are secreted at the arteriolar and absorbed at the venous ends of the capillary. Hence, for adequate fluid exchange, COP pressure must offset capillary pressure, the latter being a reflection of the arterial blood pressure (Landis and Pappenheimer, 1963). The value of the COP is determined by the concentrations and species of blood proteins as well as by cations held in the plasma by the Donnan effect of the proteins (Guyton et al., 1975). However, it is now clear that microvascular fluid exchange is a dynamic process in which extravascular forces such as tissue fluid pressure, tissue colloid osmotic pressure, and the actual flow of lymph can influence transcapillary fluid movement (Taylor and Townsley, 1987). In addition to heterogeneity of Starling forces in different areas of the microvascular beds there is also the possibility that heterogeneity of capillary membrane permeability will also contribute to differences between global and local values of the Starling pressures (Michel, 1997). In the light of recent knowledge, the simplistic steady state view of secretion at the arteriolar end of the capillary and absorption at the venous end of the capillary can only be regarded as a transient phenomenon at best.

Studies on avian species have contributed little to this discussion although microvascular fluid exchange in birds presents some unique and interesting features. For instance, in turkey and duck, the ratio of protein concentration in the interstitial fluid to that in the blood is much lower than the ratio in mammals (Hargens et al., 1974). Hargens et al. (1974) have pointed out that the lower ratio in birds is correlated with a higher arterial blood pressure. Also, birds as a group seem to be highly resistant to hemorrhage, tolerating blood loss much better than mammals. Kovách and Balint (1969) have shown that increased hemorrhage tolerance becomes apparent only during prolonged bleeding because hemodilution continues in the pigeon through the period of blood loss, whereas in the rat no further hemodilution occurs after about 15–20 min of bleeding. Hemodilution is achieved by the inflow of isotonic fluid with a low protein content.

The restoration of blood volume results from absorption of tissue fluid across the capillary walls due to reduced capillary pressure. This fall in capillary pressure could be brought about by an increase in the ratio between pre- and postcapillary resistances as well as by changes in arterial and venous pressures during hemorrhage. Resistance changes across the capillaries seem to be the most important factor in rapid restoration of blood volume in ducks. Blockade of α-adrenergic receptors eliminates vasoconstriction in the skeletal muscle, which forms the major reserve of tissue fluid and leads to a greatly retarded restoration of blood volume (Djojosugito et al., 1968). Djojosugito et al. (1968) attribute the difference in ability to restore blood volume after hemorrhage in ducks and cats to a very pronounced reflex vasoconstriction in duck skeletal musculature and to a capillary surface area in ducks three to five times that in the cat, a condition that increases the rate of absorption of fluid into the vascular system (Folkow et al., 1966).

Many birds have extremely long necks with the head being held a meter or more above the heart. Do these birds have exceptionally high blood pressure to overcome the gravitational effect on the circulation? High blood pressures are not necessary to ensure flow to the head. In a fluid-filled system, the gravitational pressure of blood in the veins will counterbalance the gravitational pressure in the arteries of the neck, much like the loop of a siphon. In other words, it is no more difficult for blood to flow uphill than downhill in a system of closed tubes like the circulation. Overall circulatory flow around the body occurs due to a pressure difference between the aorta and right atrium and it matters little what actual route the blood follows.

If pressure is measured in the cerebral circulation of a long-necked bird, this will be lower than that recorded in the aorta just outside the aortic valves by an amount sufficient to cause the required blood flow along the neck artery (a small difference) and by the gravitational effect due to the height the head is held above the heart. If the head is held one meter above the heart, pressure in the arteries of the head will be about 75 mmHg less than at the heart. Consequently, in the cranial capillary beds it is now possible that the hydrostatic pressure will be lower than the colloid osmotic pressure of the blood and fluid will be continuously removed from the interstitial spaces, with similar consequences as those following excessive alcohol consumption. As drunken ostriches are not a common sight then countermeasures such as markedly increased arterial blood pressure, decreased arteriolar resistance, or reduced blood colloid osmotic pressure must be in effect. It is now possible to obtain a ready supply of long-necked birds (ostriches and emus being bred for food) and it is to be hoped that these countermeasures can be subjected to empirical investigation.

3. Distribution of Blood Flow

a. Distribution of Cardiac Output at Rest

The percentage of cardiac output distributed to different organs is closely related to their aerobic metabolic activity and their size. The distribution of cardiac output is ultimately determined by the relative resistance of systemic vascular beds that are arranged in parallel throughout the body. Vascular resistance, in turn, is determined by a variety of hormonal, autoregulatory, and neural control mechanisms (see Section VI). Values obtained by different investigators for relative blood flow (% cardiac output) to various organs differ widely, probably reflecting differences in experimental technique, species, and "resting" conditions. Nevertheless, it is apparent from the limited data available that the heart, liver, kidneys, and intestines receive relatively large percentages of the total cardiac output (Figure 25). The avian brain appears to receive about 3% of cardiac output, similar to the proportion of cardiac output going to the brain in a small mammal like the rat at rest (Ollenberger and West, 1998).

By far the highest resting blood flows thus far measured in any avian organ (about 16 ml min^{-1} g^{-1} of wet tissue weight) are found in the spleen (Figure 26). This organ receives a disproportionately large percentage of total cardiac output despite its small size (Figure 25). High tissue flow rates also have been found in the mammalian spleen; a flow of some 12 ml min^{-1} g^{-1} has been reported in conscious dogs (Grindlay et al., 1939). Such high rates are almost certainly related to the dual function of this organ: as a filter for aging erythrocytes, which are eliminated by the process of diapedesis; and as an organ of the reticuloendothelial system, in which the blood is cleaned by phagocytic reticuloendothelial cells as it passes through the splenic sinuses and pulp. Obviously, a high flow rate is needed for this dual filtration

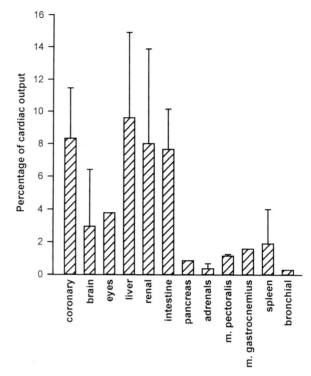

FIGURE 25 Organ blood flow plotted as a percentage of cardiac output in birds at rest. Standard error bars are shown where appropriate. Data are taken from Boelkins et al. (1973) (chicken), Duchamp and Barre (1993) (muscovy duckling), Jones et al. (1979) (Pekin and mallard), Sapirstein and Hartman (1959) (chicken), Stephenson et al. (1994) (Pekin duck) and Wolfenson et al. (1978) (chicken).

role to be effective. In contrast to splenic blood flow, the rate of cerebral blood flow in birds is an order of magnitude less. Blood flow to the whole brain and to individual cerebral regions ranges from 0.43 to 1.6 ml min^{-1} g^{-1} under normoxic conditions, as shown in Figure 26 (Bickler and Julian 1992, Butler et al., 1988; Faraci and Fedde 1986; Faraci et al., 1984; Grubb et al., 1977; Jones et al., 1979; Stephenson et al., 1994; Wolfenson et al., 1982a). Both the heart and kidneys have relatively high rates of mass specific blood flow, reflecting the high oxygen demand of the contracting cardiac muscle and the activity of energy-dependent membrane pumps in renal tissue, respectively. Pectoral and gastrocnemius muscle, on the other hand, show relatively low perfusion rates at rest (Figure 26).

b. Effects of Swimming and Submergence on Tissue Blood Flow

Surface swimming in tufted ducks at close to maximum sustainable speeds resulted in an increase of cardiac output by 70%, from 276 ml min^{-1} to 466 ml min^{-1}. Despite this, there were no increases in blood flow to the brain, liver, adrenals, spleen, or respiratory muscles.

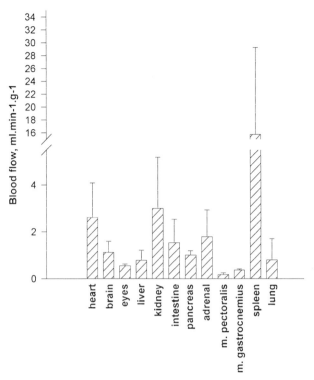

FIGURE 26 Mass-specific organ blood flow (ml min^{-1} g^{-1}) in birds at rest. Standard error bars are shown where appropriate. Data are taken from Butler *et al.* (1988) (tufted duck), Duchamp and Barre (1993) (muscovy duckling), Faraci *et al.* (1985) (Pekin duck), Jones *et al.* (1979) (Pekin and mallard), Stephenson *et al.* (1994) (Pekin duck), and Wolfenson *et al.* (1978) (chicken).

There was about a 30% increase in flow to the myocardium, reflecting an increased oxygen demand of the heart muscle. Blood flow to the active muscles of the legs increased about three times, matching increased oxygen demand. Interestingly, blood flow to the inactive pectoralis muscles, and to some visceral organs like the intestines and kidneys, actually decreased during swimming, despite the increased cardiac output. This suggests that the cardiac output is selectively redistributed during swimming both by vasodilation in active muscle and vasoconstriction in visceral organs and inactive muscle (Figure 27; Butler *et al.*, 1988). Short periods of underwater swimming in redhead ducks resulted in blood flow to the hind legs approximately doubling, suggesting that the pattern of blood flow redistribution is similar to that in surface swimming (Stephenson and Jones, 1992).

During forced submergence aquatic birds show bradycardia, a reduction in cardiac output, and massive peripheral vasoconstriction that in the case of the swimming muscles is not counterpoised by vasodilation; the latter response would normally result from the increased oxygen demand associated with exercise of these muscles under other conditions. Even though cardiac output is reduced during diving, the rate of perfusion of the head and thoracoabdominal areas is kept at or above predive levels, represented by maintained or increased blood flow to the heart and brain (Heieis and Jones, 1988; Jones *et al.*, 1979). In mallards and Pekin ducks myocardial flow was, on average, 0.73 ml min^{-1} g^{-1} presubmersion and 0.88 ml min^{-1} g^{-1} after 144–250 sec of submergence. Cerebral flow increased from 0.43 ml min^{-1} g^{-1} to 3.68 ml min^{-1} g^{-1} over the same time period. In Pekin ducks forcibly submerged until P_aO_2 fell to 50 mmHg, cerebral blood flow increased from 1.58 ml min^{-1} g^{-1} to 3.2 ml min^{-1} g^{-1}. Clearly, regardless of the wide range of absolute values measured, cerebral blood flow increases in forced submersion asphyxia, representing maintained oxygen delivery to brain tissue. A redistribution of blood flow away from more hypoxia-tolerant regions toward more sensitive regions within the brain itself does not seem to occur in the Pekin duck (Stephenson *et al.*, 1994). However such heterogeneous regional changes in cerebral blood flow in response to asphyxia have been proposed to occur in neonatal mammals (Goplerud *et al.*, 1989).

c. Changes in Blood Flow Associated with Flight

The pectoral muscles receive relatively low rates of tissue blood flow at rest (Figure 25), but flow to these muscles during flight has not been measured. During bipedal locomotion in birds, blood flow increases significantly in the locomotory muscles; Bech and Nomoto (1982) reported that sciatic artery blood flow increased 3.7 times in Pekin ducks running on a treadmill. As described above, tissue blood flow to leg muscles in ducks increases some 5-fold in ducks swimming at close to maximum sustainable rates (Butler *et al.*, 1988). Therefore it is probably safe to assume that similar increases in blood flow rates to flight muscle occur during moderate to energetic flapping flight; less strenuous flight may not involve such large changes in muscle flow rates. By analogy, domestic chickens show little change in leg muscle flow when walking on a treadmill, even at metabolic rates over double the resting rate (Brackenbury *et al.*, 1993).

Several studies have addressed the effects of simulated high altitude flight on cerebral blood flow. At altitude the arterial blood is presumably both hypoxemic as a result of the reduced P_IO_2 and hypocapnic as the result of CO_2 washout by increased hypoxemia-driven ventilation (Faraci *et al.*,, 1985). Carbon dioxide is normally a potent vasodilator within the cerebral circulation (Faraci and Fedde, 1986; Grubb *et al.*, 1977), so cerebral vasoconstriction is a distinct possibility under these conditions. However, it turns out that cerebral

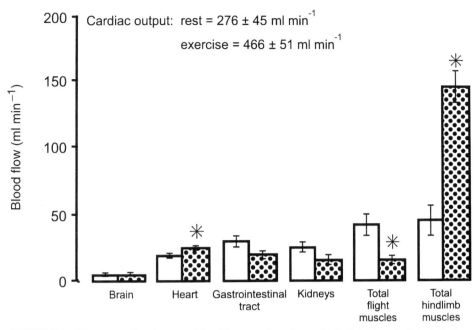

FIGURE 27 Histograms showing mean blood flow to selected vascular beds in six tufted ducks before (open bars) and during swimming at a mean velocity of 0.69 ± 0.01 m sec^{-1} (shaded bars). Asterisks indicate significant differences between preexercise and swimming values ($P < 0.05$). (After Butler et al. (1988), *J. Exp. Biol.*, Company of Biologists Ltd.)

blood flow is fairly insensitive to severe hypocapnia in domestic geese and Pekin ducks and very insensitive in bar-headed geese, a species well adapted to high altitude (Bickler and Julian, 1992, Faraci and Fedde, 1986; Grubb et al., 1977). In domestic geese, cerebral blood flow decreased as P_aCO_2 fell from 50 to 20 mmHg but then reached a plateau. Bickler and Julian (1992), however, found that the PO_2 of cerebral tissue continued to fall as P_aCO_2 fell below 20 mmHg, possibly because brain perfusion plateaued at a level lower than that in normoxia. At altitude arterial blood is both hypocapnic and hypoxemic, and hypoxemia also causes cerebral vasodilation. In this context, Grubb et al. (1977) found that concomitant hypocapnia could shift the onset of hypoxia-driven vasodilation to a lower P_aO_2 range.

d. Changes in Blood Flow Associated with Ovulation

In laying hens there are dynamic changes in regional blood flow to the reproductive organs associated with ovulation (Boelkins et al., 1973). Scanes et al. (1982) found that five major preovulatory follicles received half of the total ovarian blood flow and that blood flow increased during their maturation, but postovulatory follicles received little blood flow. Prostaglandin F-2α administration caused follicular vasoconstriction. Follicular blood flow was also sensitive to the changes in the distribution of cardiac output associated with activity and temperature stress. Treadmill exercise sufficient to cause a 150% increase in metabolic rate significantly increased blood flow to the hindlimb muscles at the expense of flow to the preovulatory follicles, but not to other visceral organs (Brackenbury et al., 1990). Heat stress also affects follicular blood flow. Elevating body temperature by 1–2°C results in blood flow falling to 70–80% of control in the larger ovarian follicles and 58% of control in the uterus, as flow to the skin increases (Wolfenson et al., 1978, 1981). During the passage of an egg down the oviduct, blood flow increased threefold in oviducal segments surrounding the egg. This may have been due to increased oxygen demand associated with contractile activity in the muscular layer of the oviduct wall (Wolfenson et al., 1982b). Blood flow to the shell gland was normally small, but increased fivefold within a few hours after entry of an egg into the gland, paralleling the time course of shell calcification.

C. Venous System

1. Functional Development of Venous System

The embryological development of the avian venous system follows a typical vertebrate pattern. At about the 15-somite stage of the avian embryo, paired cranial and caudal common cardinal veins develop from a vascular plexus in somatic mesoderm (Lillie, 1908; Sabin, 1917). The heart shifts caudally during embryonic development, and the cranial cardinal veins elongate to become the jugular veins. Subclavian veins, returning blood to the right side of the heart from the pectoral region and the wings, arise as tributaries of the caudal cardinal veins (Ede, 1964). The adult avian venous system cranial to the heart differs in detail from the mammalian pattern in that there are two cranial (superior) venae cavae. The right jugular vein is much larger in diameter than the left, and there is an anastomosis between the jugular veins at the base of the head, allowing some blood draining from the left side of the head and neck to return to the heart in the larger right jugular vein.

Development of the venous system caudal to the heart is primarily concerned with the formation of the physiologically important renal and hepatic portal venous circulations (see below). Subcardinal veins, which develop along with the embryonic kidney (mesonephros), initially provide renal drainage into the caudal cardinal veins. Eventually the anterior portions of the caudal cardinal veins disappear and the subcardinals form a connection with the ductus venosus, this connection becoming the caudal vena cava. The liver develops around the ductus venosus, which subdivides into a capillary bed, forming the hepatic portal circulation. The cranial portion of the ductus venosus becomes the hepatic vein and the caudal portion becomes the hepatic portal vein. Finally, the renal portal veins form a junction with the caudal vena cava via the common iliac veins and the caudal subcardinal veins are replaced by the caudal renal veins.

2. Capacitance Function

The walls of the veins in birds are, as in mammals, thinner than those of arteries so that venous distension depends on a positive transmural pressure gradient. The three basic components of blood vessel walls, tunica intima, media, and externa, are present. The tunica media is composed of circumferential smooth muscle fibers. In larger veins elastic laminae appear in the tunica externa, which makes up most of the wall tissue. Veins near the heart are frequently invested with cardiac muscle fibers that are apparently functional. The caudal vena cava of the mallard can occasionally be seen to contract at the same frequency as the sinus venosus.

As in mammals, the avian venous system does not necessarily represent a passive conduit returning blood to the heart. The thin, distensible walls of veins mean that these vessels are relatively compliant compared with arteries. As applied to blood vessels, compliance is the ratio of the change in vessel volume (ΔV) resulting from a change in transmural distending pressure (ΔP):

$$\text{Compliance} = \frac{\Delta V}{\Delta P}.$$

In mammals, the entire vascular system has a compliance of about 3 ml (blood) kg (body mass) mmHg^{-1}. The compliance of the arterial vascular segment is only about 3% of that of the venous segment (Rothe, 1983). Therefore, despite the smaller vascular pressures in the venous side of the circulation, at any one time about 60–80% of blood volume is contained in the veins. They are therefore referred to as capacitance vessels. The capacity of the venous circulation can change either passively by changes in transmural pressure or actively by changes in the contractile state of venous smooth muscle. A reduced transmural pressure and therefore a passive elastic recoil of compliant veins, or a reduction in venous compliance by active contraction of smooth muscle in the venous walls, mediated by α-adrenergic receptors (Section VI,B,3), would both serve to reduce venous capacitance. This transfers blood toward the heart. All other things being equal, this would tend to increase atrial filling and therefore cardiac output. The large veins of the domestic fowl are well innervated with adrenergic motor fibers (Bennett *et al.*, 1974, Bennett and Malmfors, 1975b), and the density of this innervation suggests that there is active control of venous capacitance (Section VI,B,3), although to date there are no physiological studies in birds analogous to those of adrenergic effects on venous capacitance function in mammals (Vanhoutte and Leusen, 1969).

3. Physiological Role of Veins in Exercise and Submersion

Venous pressure in pigeons flying in a low-speed wind tunnel increased to 2.5 mmHg from the resting value of 1.2 mmHg (Butler *et al.*, 1977). Cardiac output increased 4.4 times during flight, mainly accomplished by an increase in heart rate at a constant stroke volume. In flight, the increased pressure gradient from the venous end of capillaries to the right ventricle increases venous return, right heart filling, and cardiac output via the Frank–Starling relationship. Venous pressure is determined by the relationship between venous volume and compliance, which is reduced as venous smooth muscle contracts and the vein walls stiffen. In mammals the venous beds of the liver, spleen, and skin act as reservoirs of blood which can actively reduce their capacitance during exercise, thereby increasing venous pressure and the volume of venous return to the heart (Rothe, 1983). Whether these venous vascular beds also constrict during flight exercise in birds is currently unknown.

Active venoconstriction may also be important in the cardiovascular adjustments to diving in birds. Djojosugito *et al.* (1969) provided indirect evidence that venous pressure increased during diving in ducks and this was confirmed by Langille (1983), whose results suggested that this was due to active venoconstriction. In the latter study, cardiac stroke volume fell if central venous pressure was held constant during diving. This suggests that reduced ventricular contractility, caused by an increase in vagal motor nerve activity to the heart, is normally counteracted during diving by increased venoconstriction-induced filling via the Frank–Starling mechanism (see Section VI,C,2,c).

4. Renal Portal System

In common with most other vertebrate groups, birds possess a renal portal circulation. Venous blood making its way back to the heart from the legs and the lower intestine of birds enters the kidneys through a renal portal system. Within the kidneys this blood mixes with postglomerular efferent arteriolar blood in peritubular sinuses that surround all nonmedullary nephron segments and eventually flows toward the renal veins. The physiological significance of the renal portal system is currently poorly understood. About 50–70% of total renal blood flow is contributed by the renal portal vein. However, this percentage is highly variable between animals and can change rapidly in the same individual for no apparent reason (Odlind, 1978). Part of the variability in flow can be attributed to the status of active, innervated renal portal valves within the iliac veins (Figures 28 and 29; Glahn *et al.*, 1993). The valves receive dense reciprocal motor innervation from the parasympathetic and sympathetic divisions of the autonomic nervous system (see Section VI,B,3). Adrenergic stimulation produces relaxation of the smooth muscle of the valve, and cholinergic stimulation produces contraction. In contrast, smooth muscle of the renal portal vein itself shows a predominantly adrenergic contractile response typical of most vascular smooth muscle (Burrows *et al.*, 1983). The distribution of vascular resistance within the portal system, governed by both the portal valve and the alternate, parallel venous pathways for blood returning to the heart (Figure 30) determine the volume flow of renal portal blood.

There are several options for a "packet" of venous blood returning from the legs in the external iliac veins (Figures 29 and 30): if the renal portal valve is open under the influence of sympathetic nervous system activation, blood can flow through the patent valve into the common iliac vein leading to the vena cava and directly back to the right side of the heart, bypassing the kidney. If the valve is partially closed due to parasympathetic stimulation, and resistance at the valve is high, blood can alternatively enter the renal portal system by flowing into the cranial and caudal portal veins. Thus the portal veins are arranged functionally in parallel with the direct venous route provided by the common iliac vein and the vena cava. Blood entering the cranial and caudal portal systems eventually drains into the internal vertebral venous sinuses and caudal mesenteric veins respectively and makes its way back to the right side of the heart by this route. Anteriorly the caudal mesenteric vein connects with the portal system of the liver, providing another option for blood flow. The cranial and caudal portal veins are arranged in parallel with each other, and a "packet" of blood from the external iliac vein can enter one or the other, but not both. Therefore, variable fractions of venous blood derived from the legs, tail, and lower digestive tract can return directly to the right side of the heart (via the common iliac vein and vena cava) or enter the renal (cranial portal vein) or renal and hepatic portal systems (caudal portal vein) (Figure 30). The overall pattern of venous flow will depend on the distribution of vascular resistance within the portal system.

Several workers have tried to ascribe a functional significance to the renal portal system. This is a difficult task in the face of the possibility of neural, humoral, and local metabolic control all influencing the relative resistances offered by the portal veins and valve. Recent experimental evidence has resulted in the development of the portal compensation hypothesis, in which the parallel interconnecting veins described above are viewed as an anastomosing network, with the peritubular sinuses located in the renal cortex at its center. Therefore, the amount of portal blood flowing to the kidneys will depend on both the resistances and pressures in the parallel shunt pathways described above and resistance and pressure within the peritubular sinuses.

Blood flowing from the renal glomeruli also enters the sinuses and contributes to pressure within them (Wideman *et al.*, 1991), so any reduction in renal arterial pressure below the autoregulatory range should promote inflow to the sinuses from the portal veins. It is known that if the portal system is intact, birds can maintain total renal blood flow in the face of a fall in renal arterial pressure to 40–50 mmHg. Should arterial pressure fall, the glomerular vessels themselves can, by autoregulation, maintain glomerular filtration rate constant but only down to a minimum pressure of 70 mmHg (Wideman *et al.*, 1992). The wider autoregulatory range for total renal blood flow may be due partially to an autoregulatory buffering effect by the portal system. Experimentally reducing portal

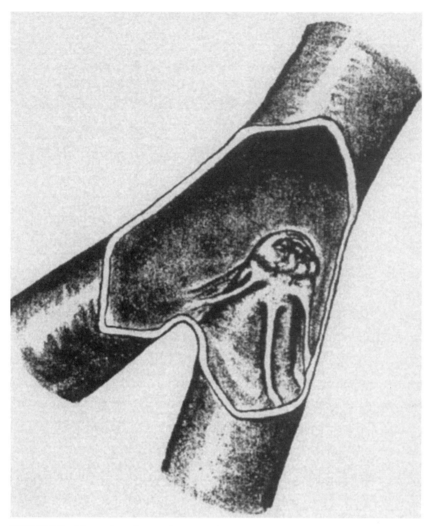

FIGURE 28 Illustration of a renal portal valve. The valve, composed of smooth muscle, is situated in the external iliac vein at the point where the efferent renal vein joins the iliac vein. The valve is anchored to the vein wall by muscular "tethers." (After Burrows *et al.* (1983), *Am. J. Physiol.* with permission.)

blood flow leads to a narrowing of the range of arterial pressures over which total renal blood flow is maintained constant (Wideman *et al.*, 1992). There are regional differences in renal blood flow, with the anterior part of the kidney apparently receiving a greater contribution from the portal veins. This suggests that renal arterial flow is normally lower in the anterior kidney.

The functional significance of the portal system in the environmental physiology of birds may be related to salt loading and dehydration. Dantzler (1989) has proposed that under these conditions the adaptive response of the preglomerular arterial vessels of reptilian-type nephrons in the superficial renal cortex is to constrict, causing sustained cessation of filtration. Renal blood flow could be maintained under these conditions by a compensatory increase in portal flow, maintaining a nutritional blood supply to the cells of cortical nephrons (Wideman and Gregg, 1988).

V. BLOOD

A. Components

Circulating blood is a non-Newtonian fluid, consisting of particles (the cellular components) suspended in plasma. Avian blood characteristically differs from mammalian blood in that erythrocytes are oval and nucleated, thrombocytes are nucleated and are as large as the other leucocytes, blood glucose level is typically double that in mammals, and plasma protein content is substantially lower. A modern account of techniques in avian hematology and cytology is provided by Campbell (1995). Rheologically, the suspension of cells (mainly

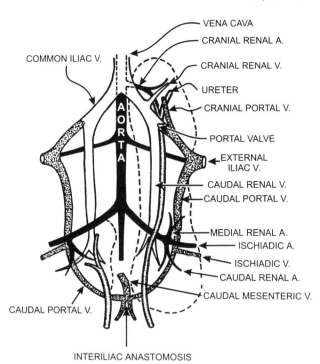

FIGURE 29 Ventral view of the avian kidneys with a simplified representation of the avian renal portal circulation and its connections to the systemic venous system. (After Wideman *et al.* (1992), *Am. J. Physiol.* with permission.)

erythrocytes) in a homogeneous fluid phase (plasma) confers unique properties on the resulting fluid. The apparent viscosity of the blood will vary as a function of

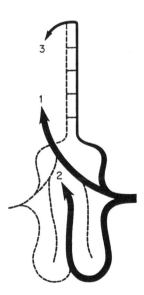

FIGURE 30 Three potential parallel shunt pathways in the renal portal circulation. (1) The renal portal valve is open under influence of its sympathetic motor innervation—venous blood flows from the external iliac vein through the patent valve into the common iliac vein, bypassing the kidney. (2 and 3) The renal portal valve is partially closed and resistance at the valve is high. Blood returning from the legs in the external iliac veins enters the cranial and caudal portal systems. (After Akester (1967), with permission of *J. Anat.*)

its hematocrit, and in the capillary beds and precapillary vessels viscosity is also influenced by the vessel diameter. While little work has been done on these general effects in birds, they have been thoroughly evaluated in mammals; see Berne and Levy (1992) for a lucid discussion.

Nucleation is typical of submammalian erythrocytes, and mature avian erythrocytes have a nucleus positioned centrally in an ovoid cell body. Avian erythrocytes also tend to be larger than those of mammals, although typically smaller than the erythrocytes of reptiles (Hawkey *et al.,* 1991). In birds erythrocyte length ranges from 14.0 to 15.7 μm and width from 7.5 to 7.9 μm, yielding a ratio of length to width of 1.50–2.0 (Palomeque and Planas, 1977). The ovoid shape of the cells accounts for some of the observed differences in rheological properties from those observed in mammalian blood. Nuclei are also ovoid, with a length to width ratio ranging from 1.8 to 3.0. The nuclei therefore have a more elongated shape than do the erythrocytes, and this may be an important factor in determining the extent of erythrocyte deformability as these cells travel through the capillaries.

B. Rheology

The viscosity of a homogeneous (Newtonian) fluid is constant if measured over the range of flow rates and tube diameters that are likely to be found in the cardiovascular system. The properties of the tubing and the fluid are important factors in determining resistance to blood flow, as outlined in Section III. With reference to the tubing, Poiseuille's law (Eq. 1; Section III) states that there is an inverse relationship between blood flow and the fourth power of the radius of a vessel. However, also explicit in Poiseuille's law is a term for fluid viscosity, and resistance to flow is directly proportional to the value of this term. Viscosity is essentially the resistance to bulk flow conferred on a fluid by the frictional, or shearing, interaction of one layer (lamina) of fluid moving over another with a different velocity. In a single-phase (Newtonian) fluid like water, viscosity is constant with changes in tube radius and flow velocity. However, in a two-phase medium like blood, which is essentially a suspension of particles, viscosity depends on (1) the concentration of the particles (essentially the fraction of total volume occupied by erythrocytes, or hematocrit); (2) the velocity of flow; and (3) the radius of the vessel in which viscosity is measured.

1. Effect of Hematocrit

Apparent blood viscosity varies as hematocrit varies. Plasma, devoid of cells and particulate matter, has a viscosity of 1.3 (pure water has a viscosity of 1). Mammalian blood with a normal hematocrit of 45% has a viscosity 2.4 times that of plasma alone. However, there is an exponential relationship between hematocrit and viscosity so that an increase in hematocrit to 70% more than doubles the apparent viscosity and has the same effect on the resistance to blood flow (Berne and Levy, 1992). At low hematocrits, the viscosity of suspensions of turkey erythrocytes in Ringer's solution is comparable to that of mammalian suspensions of the same hematocrit, but as hematocrit approaches 85% viscosity of turkey blood increases disproportionately more than that of mammalian suspensions. This suggests that turkey erythrocytes may be less deformable than mammalian erythrocytes. Avian erythrocytes differ from mammalian erythrocytes in size, shape, presence of a nucleus, and possibly tensile properties of the membrane. The presence of a nucleus with a length to width ratio greater than that of the whole cell may lead to a high internal viscosity, tending to decrease the deformability of the cells (Usami et al., 1970). Another factor that may contribute to a high internal viscosity in avian erythrocytes is a relatively high concentration of free cytoplasmic hemoglobin (Gaehtgens et al., 1981a). In mammalian erythrocytes, velocity gradients in the plasma are transmitted into the cytoplasm by a "tank-tread"-like motion of the membrane, a feature which has been proposed to help reduce the overall viscosity of the blood (Fischer, 1978). This mechanism may not occur in avian erythrocytes if their cell membranes are significantly stiffer than those of mammalian cells.

2. Flow Velocity

As the velocity of blood flow increases in mammals, apparent viscosity decreases. This is called "shear thinning," the term "shear" referring to the shearing of flow laminae past each other in laminar flow. At low flow rates, erythrocytes tend to aggregate, effectively forming large particles which elevate blood viscosity. As flow rates increase these aggregates break up, thus decreasing apparent viscosity. Another flow-related factor responsible for a decrease in viscosity, particularly when hematocrit is high, is the capability for deformation of erythrocytes. At low to medium flow rates erythrocytes assume a biconvex shape, but at higher flow rates the cells deform in the direction of flow and assume a flattened, ellipsoidal morphology (Cohen, 1978). This effect diminishes apparent viscosity. This deformation is important in the capillaries: the diameter of mammalian erythrocytes is around 8 μm, yet cells deformed in the flow stream can pass through a 3 μm lumen.

3. Vessel Radius

If the viscosity of mammalian blood is measured as it flows through small glass tubes, two interesting phenomena are noted. First, hematocrit of flowing blood measured in a small tube connecting a reservoir of blood to a collecting vessel is lower than that measured in blood taken from either vessel. Second, the apparent viscosity of the blood in the tube decreases as a function of tube diameter, if diameters are less than about 0.3 mm. The same effect applies to blood flow through microvessels *in situ*. The hematocrit of blood measured in arterial and venous microvessels in the cat has been found to be lower than that recorded in large systemic vessels (Lipowsky et al., 1980). These hematocrit- and viscosity-related phenomena, discussed in Section V,B,2 above, are linked by the fact that the composition of the blood actually changes as it flows through small tubes of about the same diameter as capillaries. This is because erythrocytes tend toward the center of the flow stream, which has a velocity twice that of the average flow velocity under the laminar flow conditions found in these vessels. This velocity is therefore faster than the average velocity for the plasma component of the blood and results in relatively fewer erythrocytes in the tube or capillary at any one time, compared to the number which would be present in the same volume of blood sampled from a large vessel. This reduction in hematocrit under dynamic conditions is called the Fahreus effect. It is not fully understood why the erythrocytes drift into the center flow stream in small vessels, although it has been shown that flexible particles like red blood cells show this behavior but rigid particles do not.

The viscosity phenomena described above depend on the ability of the erythrocytes to deform at high flow velocities (shear thinning) and their stability of orientation in the axial flow stream in small vessels (Fahreus effect). Mammalian erythrocytes deform readily, and two of the most important prerequisites promoting this phenomenon are the absence of a nucleus and other organelles and a large surface-area-to-volume ratio. While the latter factor is not significantly different from that of duck erythrocytes (Gaehtgens et al., 1981a), it might be predicted that the large, nucleated ovoid avian erythrocytes would not deform as readily as mammalian erythrocytes. Turkey red cells actually show a lower tendency to aggregate in whole blood than do mammalian erythrocytes, although increased concentrations of fibrinogen and globulin can increase viscosity by increasing the tendency to aggregate. Aggregation, however, becomes less important at high shear rates. The viscosity of suspensions of turkey red blood cells in Ringer's solution decreases as shear rates increase, but the suspension still shows higher relative viscosity values than even suspensions of nucleated erythrocytes of other nonmammalian species. This disparity becomes even more apparent at both high hematocrit and high flow rates, as shown in Figure 31 (Chien et al., 1971). As flow rate increases, erythrocyte cell diameter decreases and

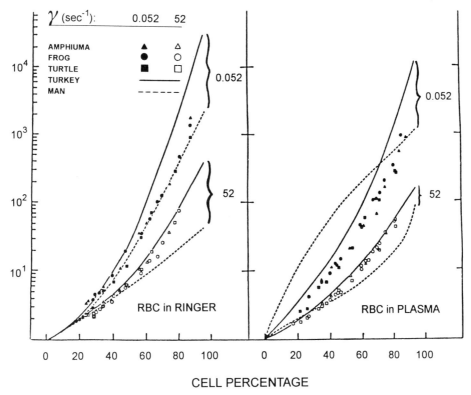

FIGURE 31. Plots of the relative viscosity of turkey red blood cell (rbc) suspensions compared to those of the rbcs of other nonmammalian vertebrates and humans. Left panel: rbcs suspended in Ringer's solution; right panel: rbcs suspended in plasma. Vertical axis, relative viscosity (logarithmic scale), horizontal axis, cell percentage by volume. All nucleated rbc suspensions show higher viscosity than the suspensions of nonnucleated human cells, but turkey cell suspensions exhibit high viscosity compared to those of other nucleated rbcs. (Reprinted from *Biorheology* **8,** S. Chien, S. Usami, R. J. Dellenbeck, and C. A. Bryant, Comparative hemorheology—Hematological implications of species differences in blood viscosity, pp. 35–57, Copyright (1971), with permission from Elsevier Science.)

cell length increases, so that a similar type of deformation occurs in both human and duck erythrocytes; however, human cells are about twice as deformable (Gaehtgens *et al.,* 1981a). Therefore the contribution of erythrocyte deformation to shear thinning is probably less significant in birds than in mammals.

Human red cells orient themselves in a stable fashion in the center of the axial flow stream in narrow capillaries, but duck erythrocytes often tumble in the flow stream and may orient themselves with their long axis at an angle to the tube axis (Gaehtgens *et al.,* 1981a). This is because their relatively low deformability reduces their ability to "go with the flow" and yield to local shear forces. Tumbling and the tendency to travel at an angle to the tube axis both act to reduce the width of the undisturbed plasma layer traveling next to the capillary wall, suggesting that in bird capillaries there may be a smaller Fahreus effect. The velocity of bird red cells in small glass capillary tubes is significantly faster than that of the suspending solution (Gaehtgens *et al.,* 1981b). This results in an effective reduction in

hematocrit and therefore in apparent viscosity of capillary blood by the mechanism described above. The Fahreus effect is thus probably important in reducing resistance and consequently the pressure required to drive flow in bird microvessels. Indeed, there is evidence that this may be the most important mechanism in determining viscosity-related resistance to flow in small capillaries (Gaehtgens *et al.,* 1981b). However, under the same conditions of hematocrit and tube diameter, the relative viscosity of suspensions of duck erythrocytes is 4–6 times greater than that of human cell suspensions, as illustrated in Figure 32 (Gaehtgens *et al.,* 1981b). This difference is probably due to the larger hydrodynamic disturbance produced by a less stable orientation of nucleated erythrocytes in the capillary flow stream. The smaller the viscosity of a flowing suspension, the less the particles will perturb the shearing laminae of flow.

Perhaps in compensation for the relatively high viscous resistance in birds, there is a threefold increase in muscle capillary density compared to cat muscle, and brain capillaries occupy about double the volume of

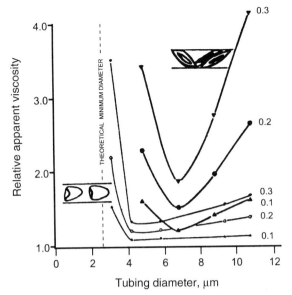

FIGURE 32 Effect of capillary tube diameter and hematocrit on the relative viscosity of duck (filled symbols, heavy lines) and human (open symbols, light lines) red blood cell (rbc) suspensions flowing through glass capillaries. Hematocrit is represented as a fraction of unity (e.g., 0.3 = 30%). The vertical dashed line represents the smallest capillary diameter that can be traversed by rbcs without alteration of their surface area/volume relationship by deformation. Note that the relative viscosity of the duck rbcs is higher at all hematocrits, probably reflecting their less stable orientation in the flow stream. (After Gaehtgens *et al.* (1981b), with permission.)

those in mammals (see Section IV,B). The different red cell morphology in birds may therefore have resulted in compensatory differences in the architecture of the microvasculature compared to that of mammals.

C. Effects of Altitude

A common vertebrate response to the hypoxia produced by exposure to high altitude is to produce more red cells (polycythemia), increasing hematocrit and therefore the amount of circulating hemoglobin. This will in turn improve oxygen conductance by increasing the oxygen carrying capacity of the blood. As we can deduce from the above discussion, development of polycythemia will significantly increase blood viscosity, leading to increased resistance to blood flow through the capillaries. Ultimately this trend will increase the power output needed from the heart, which is already high in the flying bird. In some circumstances, for example, sedentary residence at high altitude, the advantages for oxygen transport might outweigh the disadvantages to cardiac energy output, but in flight at high altitude the balance might be tipped in the other direction. The hematocrit of Pekin ducks increases from 45.4% at sea level to 55.9% after 4 weeks of high-altitude residence (5640 m) with a corresponding increase in hemoglobin concentration in the blood. However, bar-headed geese, which are strong high-altitude fliers, show no changes in hematocrit under similar experimental conditions (Black and Tenney, 1980). In wild quail hematocrit and the concentration of hemoglobin increase with altitude as they migrate to higher altitudes (Prats *et al.*, 1996). These increases were accomplished by the appearance of larger erythrocytes containing more hemoglobin, without an increase in red cell number.

Polycythemia, occurring in response to altitude, seems to be a chronic rather than an acute adaptation in birds, as in mammals. Pigeons acutely exposed to high altitude do not show increases in hematocrit or hemoglobin concentration, although both responses areshown by altitude (7 km)-acclimated pigeons (Weinstein *et al.*, 1985). The variability of the polycythemic response and its interaction with the O_2 affinity of the blood are shown by sharp increases in hematocrit after a week at 3800 m in house finches (native to low altitudes) and parallel changes in rosy finches (native to high altitudes), although these changes occur over a markedly lower range of hematocrits in the latter species (Clemens, 1990). The blood O_2 affinity (of which P_{50} is an index) of rosy finches (P_{50}: 31 Torr) is higher than that of house finches (P_{50}: 37 Torr), suggesting that the adaptation of the high-altitude species to the requirements for O_2 transport at altitude favored increased O_2 affinity over polycythemia, given the attendant increases in blood viscosity associated with the latter response. In pigeons, which are not native to high altitudes, residence at a simulated altitude of 7 km for 6 weeks resulted in both a small reduction in P_{50} (29.0 to 26.5 Torr) and an increase in hematocrit (by 38%; Maginnis *et al.*, 1997). It is clear that the balance of the adaptational changes to altitude between polycythemia (with attendant changes in blood viscosity) and changes in hemoglobin oxygen affinity varies in birds depending on species and the circumstances of altitude exposure.

D. Temperature Effects

If a liquid is cooled, its apparent viscosity increases; as a consequence of this, a "thinner" oil is used in an automobile engine during colder weather to decrease the work that moving parts must perform against internal viscous forces. Blood of Little and Adelie penguins has been shown to have a lower viscosity than chicken blood over a range of temperatures from 0 to 40°C, suggesting that low viscosity may play a similar role in aiding blood flow in small vessels of the flippers when they are immersed in cold water, or come into contact with ice (Clarke and Nicol, 1993). Under these conditions countercurrent heat exchangers conserve body heat, but as a consequence tissue temperature in the extremities approaches ambient temperature, increasing the viscosity of the contained blood. Penguins maintain relatively low blood viscosity despite having very large erythrocytes (1.7 times the volume of chicken erythrocytes) and have significantly higher levels of plasma proteins than chickens. However, penguin red cell count is the lowest recorded in birds (with the exception of the ostrich; Nicol et al., 1988). This may be a trade-off to keep blood viscosity low, but could ultimately limit the scope of aerobic capacity in these animals.

VI. CONTROL OF THE CARDIOVASCULAR SYSTEM

A. Control Systems

Both the output of the heart and the resistance to blood flow in the vascular beds are subject to wide variations, depending on the type of activity in which animals are engaged and the intensity of that activity. As discussed above, cardiac output varies in proportion to total body metabolic requirements while vascular resistance varies on a regional basis according to the blood flow requirements of different parts of the body. Cardiac output and resistance are both under the control of autoregulatory, humoral, hormonal, and neural influences. Arterial blood pressure is the product of cardiac output and total peripheral resistance and it is this pressure which produces the driving force to ensure adequate blood flow to the vascular beds. Blood pressure varies but over a proportionally narrower range than either cardiac output or peripheral resistance. Investigations of the regulation of cardiovascular function have been driven by the broad assumption that blood pressure is maintained within sensible limits to ensure adequate tissue perfusion in the face of the variations in cardiac output and peripheral resistance imposed by changes in the external environment or in activity levels.

The cardiovascular system is controlled by several integrated mechanisms operating over time scales ranging from less than a second to months or longer. The most rapid adjustments in cardiac output and peripheral resistance, which may occur within the span of a few heartbeats, are reflexogenic and primarily function to maintain short-term homeostasis and to effect rapid cardiovascular responses to changes in the internal or external environments. Autoregulatory mechanisms acting within vascular beds to modify blood flow as a result of local changes in metabolites or other influences may operate on a time scale of seconds to minutes. Changes in humoral factors, such as levels of oxygen and carbon dioxide, pH, and metabolic products, can affect cardiovascular receptors or circulatory elements directly, producing changes in cardiac and vascular function directed toward correcting these disturbances on a time scale which may extend over long periods. An example of this is in birds undertaking extended migratory flights at high altitudes where inspired oxygen levels, and therefore blood oxygen levels, are much lower than at sea level. Circulating hormones can also affect both the peripheral circulation and the heart, and levels of many of these hormones in the blood may change depending on the activity state of the animal or on the time of year, as during molting or mating.

B. Control of Peripheral Blood Flow

Contraction of smooth muscle in the walls of arteries provides the means for varying vessel caliber, and the most effective location for altering blood flow by this means will be where the ratio of wall cross-sectional area to lumen area is maximal. The smallest arteries and the arterioles possess the highest wall-to-lumen ratio, and it is here that the contraction of individual muscle fibers, coordinated over the whole cross-section of the wall, will give the greatest change in resistance. The largest component of vascular resistance in the circulation is therefore set by the tone of the muscles in the walls of these vessels. Arterial smooth muscle tone is subject to modulation by several mechanisms: (1) intraluminal pressure changes, leading to mechanical autoregulation of blood flow; (2) humoral factors including oxygen tension, levels of local metabolites, extracellular ion concentrations, locally released vasoactive agents, and circulating hormones and vasoactive agents; as well as (3) transmitters released from autonomic nerve terminals. Smooth muscle fibers in the walls of veins are also influenced by these factors, providing mechanisms for adjusting compliance of the venous walls and thus some control of the rate of return to the heart of blood in the central venous pool. The influences of these regulatory factors on vascular function in birds are discussed in the following sections.

1. Autoregulation

Vascular tone within a region of the circulation can be defined as the average level of contraction of smooth muscle fibers in the blood vessel walls within that region. At a steady intraluminal pressure and in the absence of extrinsic influences, spontaneous contractions of individual smooth muscles occur at an intrinsic rate. In resistance vessels the smooth muscle cells are arranged at a right angle to, or on a shallow helix around, the axis of flow, and the time-averaged tension generated by their contraction, in balance with the intraluminal pressure, will set vessel caliber and thus blood flow. An increase in arterial pressure will distend the vessel wall, increasing the caliber of the vessel and therefore reducing resistance to flow. As a result, blood flow through the vessel will increase. This wall distension also increases the frequency of contraction of the smooth muscles, and this increased vasomotion will then act to reduce the caliber of the vessel, increasing its resistance and restoring blood flow toward its original level. Conversely, a reduction in pressure will reduce wall tension, resulting in a decrease in the rate of smooth muscle spontaneous activity. This leads to vasodilation and an increase in blood flow to offset the effects of reduced perfusion pressure. These myogenic changes in vascular caliber thus provide a mechanism to autoregulate blood flow around a preferred level in the face of variations in tissue perfusion pressure. In most vascular beds, autoregulatory mechanisms are probably limited to the modulation of local blood flow to ensure even blood distribution, as, for example, in the kidney where this mechanism is important in maintaining glomerular flow rate in the face of alterations in arterial blood pressure. Throughout the body local pressure-induced autoregulation will interact with locally released vasoactive agents and with neurogenically mediated vasomotion resulting from activation of autonomic reflexes to provide balanced adjustments in regional peripheral blood flow.

2. Humoral Factors

Three broad classes of humoral factors affect blood flow in the peripheral vasculature. One class includes chemical factors such as PO_2, PCO_2, lactic acid and other metabolic byproducts, electrolyte concentrations, and pH; these factors can act directly on myocytes. The second class of factors consists of vasoactive agents released from the local vascular endothelial cells. This group includes nitric oxide and possibly endothelin; these factors are believed to act on smooth muscle cells by receptor-mediated mechanisms. The third class includes circulating vasoactive agents, also acting via receptor-coupled mechanisms to modify smooth muscle contraction.

a. Chemical Factors

If the metabolic rate of a tissue increases, as, for example, in skeletal muscle during exercise, regional blood flow will increase due partly to local vasodilation of resistance vessels and precapillary sphincters induced by an increase in the concentration of lactic acid and CO_2 and a fall in pH. Vasodilation under these circumstances is apparently produced by direct chemical effects on the contractile apparatus of the vascular myocytes (Mellander and Johansson, 1968). Vasodilation may be further enhanced if local PO_2 falls or the concentrations of extracellular ions such as K^+ increases. The resulting increase in blood flow is called functional hyperemia and serves to accelerate oxygen delivery to the muscle and to increase the clearance rate of tissue metabolites. During exercise, central arterial pressure may also rise (Butler, 1991; Saunders and Fedde, 1994) and local myogenic autoregulation would, as outlined in Section VI,B,1, attempt to limit the rise in blood flow; however in tissue operating at a high metabolic rate, this mechanism is to a large extent overridden by local chemical vasodilatory influences. Vasodilation due to tissue hypoxia and buildup of metabolites also occurs during periods of ischemia, for example, when blood flow to a vascular bed is occluded as shown in Figure 33. Upon release of the occlusion blood flow rises transiently to several times the preocclusion rate, with the increase in flow proportional to the duration of the occlusion. This reactive hyperemia acts to restore tissue oxygen levels and to remove metabolic products accumulating during the period of occlusion. Local hypoxia also reduces the vasoconstrictor effects of norepinephrine applied exogenously or released from sympathetic nerve terminals in avian arteries *in vitro* (Gooden, 1980; and see Section VI,B,2,b below), and might therefore be expected to reduce the *in vivo* effectiveness of vasoconstriction mediated neurogenically or by circulating catecholamines in support of the hyperemic response. Local hypercapnia also produces vasodilation and, in the hind limb vascular bed of the duck, has an even stronger effect than hypoxia in inhibiting neurogenic or catecholamine-induced vasoconstriction (Lacombe and Jones, 1990).

b. Locally Released Vasoactive Agents

Endothelium-derived relaxing factor, now recognized as nitric oxide (NO), is a small, rapidly diffusible molecule released by enzymatic cleavage of L-arginine from vascular endothelial cells of mammals (see Umans and Levi, 1995 for review) and birds (Hasegawa *et al.*, 1993). In both of these vertebrate classes this molecule exerts a powerful vasodilatory effect by relaxing precontracted vascular smooth muscle. The release of NO as a result of acetylcholine (ACh) stimulation of endothelial cells thus provides part of the explanation for the vasodilatory effects of ACh in the circulation, as originally proposed by Furchgott and Zawadzki (1980). In the *in vitro* aorta of the fowl, ACh acting at muscarinic receptors on endothelial cells provokes release of NO which then produces local vasodilation (Hasegawa and Nishimura, 1991). Another factor provoking release of NO from avian vascular endothelial cells is angiotensin II (AII); the vasodilation produced by NO release has been proposed to cause the transient depressor effects on arterial blood pressure observed in some birds imme-

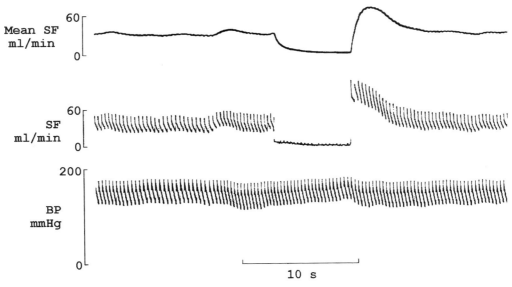

FIGURE 33 Change in pulsatile (SF) and mean (mean SF) blood flow in the sciatic artery of a duck in response to arterial occlusion. The period of occlusion is marked by zero flow. BP, arterial blood pressure. Note the hyperemia after the period of occlusion and the subsequent rapid return to the preocclusion flow rate with minimal change in pressure. (After Jones and Johansen (1972), with permission.)

diately after systemic injection of AII (Stallone et al., 1990; Hasegawa et al., 1993; Takei and Hasegawa, 1990). Mechanical stimuli such as flow-induced shear stress (see Section V,B) may also elicit release of NO from endothelial cells in the vasculature of birds. This effect has been reported for mammalian vasculature (Umans and Levi, 1995) and, given that blood flow rates in birds are generally higher than those in mammals of comparable body size, such a mechanism could provide additional local adjustment of the degree of vasodilation within a vascular bed to match the immediate flow requirements of that bed.

It is likely that other locally released vasoactive agents, such as the peptide endothelin, may also be important in adjusting regional blood flow in birds as in mammals. Endothelin is the most potent vasoconstrictor known in mammals (Inagami et al., 1995), but its vascular effects have not been evaluated in birds, nor has the existence of intravascular endothelin receptors been confirmed in this group of vertebrates. Receptors for this peptide have, however, been demonstrated on avian cardiac myocytes and their activation causes an increase in contractile force (Kohmoto et al., 1993).

c. Circulating Agents

Circulating catecholamines have powerful effects on all elements of the circulation in birds. The catecholamines epinephrine (EPI) and norepinephrine (NE) are released into the circulation from adrenal chromaffin cells and have direct effects on vascular smooth muscle. Significant amounts of NE are also released into the circulation by activation of the sympathetic nervous system (discussed in Section VI,B,3 below). A number of peptides which have vasoactive effects on mammalian vasculature are also present in avian plasma, but the specific actions of most of these peptides on the avian vasculature have not been investigated in detail. Of these peptides, the most extensively studied in birds with respect to vasomotion are AII and avian antidiuretic hormone (ADH).

Circulating NE levels in conscious ducks and fowl at rest are in the range of 3–5 nM (Lacombe and Jones, 1990; Kamimura et al., 1995), while the resting plasma level of EPI is about half the value for NE (Lacombe and Jones, 1990), as illustrated in Figure 34A. EPI appears to be released solely from the adrenal glands since removal of these glands eliminates EPI from the plasma (Lacombe and Jones, 1990). The loss of adrenal glands does not, however, markedly affect the resting level of circulating NE which must therefore be due to spillover into the plasma of NE released from sympathetic nerve terminals by autonomic efferent activity. Circulating levels of both EPI and NE vary under different physiological conditions. For example, plasma catecholamines can increase by factors ranging from 2 to greater than 1000 in ducks during involuntary submersion, with end-dive levels being proportional to dive length as shown in Figure 34B (Huang et al., 1974; Hudson and Jones, 1982; Lacombe and Jones, 1990).

Both NE and EPI produce vasomotion in avian vascular smooth muscle, acting via α- and β-adrenergic receptors (Bolton and Bowman, 1969). NE, injected

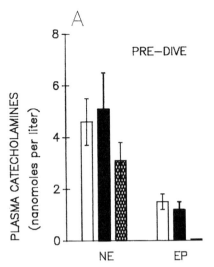 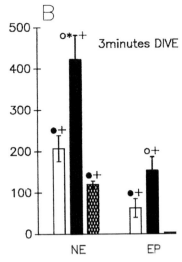

FIGURE 34 Plasma levels of norepinephrine (NE) and epinephrine (EPI) in intact (open bars), sham-operated (filled bars), and adrenalectomized (cross-hatched bars) ducks (A) before and (B) at the 3-min point during forced submergence. The open circles indicate significant differences from intact ducks; closed circles, differences from sham-operated animals; asterisks, differences from adrenalectomized animals; plus signs, differences from predive value. Adrenalectomy eliminated EPI but not NE from the plasma. Note ordinate scale change in B (after Lacombe and Jones, 1990; used by permission).

intravenously into conscious ducks, produces vasoconstriction throughout the body. This vasoconstriction causes an increase in blood pressure produced by the collective effect of increases in resistance to flow in individual vascular beds in the body, illustrated in Figure 35, by increased resistance in the hind limb vascular bed. These vascular responses to NE are primarily mediated by α-adrenergic receptors (Butler et al., 1986; Wilson and West, 1986; Bolton and Bowman, 1969). Activation of α-adrenoceptors on vascular smooth muscle acts through second-messenger systems to mobilize internal calcium stores and to open membrane calcium channels, thus increasing intracellular calcium concentration and activating the actin–myosin contractile apparatus (see Hirst and Edwards, 1989 for review). NE can also have a vasodilatory effect on the vasculature of birds, mediated through β-adrenergic receptors; however, this effect is only apparent systemically after pharmacological blockade of α-adrenoceptors (Butler et al., 1986). β-adrenergic vasodilation in the avian vasculature also works by activating intracellular second messengers, converting the actin–myosin complex to an inactive form to promote relaxation. Combined α- and β-adrenergic blockade appears to eliminate all direct effects of NE on bird vascular smooth muscle. The overall effects of NE on peripheral resistance therefore depend on the relative abundance of α- and β-receptors in individual vascular beds. EPI also acts on adrenoceptors of vascular smooth muscle but binds to α-receptors with a higher affinity than to the β subtype, so exerts a stronger vasoconstrictive effect for the same receptor density than does NE.

Avian AII is similar to the corresponding mammalian peptide in its structure and in the biochemical pathway of its production (see reviews by Wilson, 1989; Henderson and Deacon 1993). Renin, released from the juxtaglomerular cells lining glomerular afferent arterioles in the kidney, produces the peptide angiotensin I by hydrolysis of angiotensinogen, a plasma α-globulin. Angiotensin-converting enzyme, which has been identified in circulating plasma and fixed in the walls of blood vessels in birds (Henderson and Deacon, 1993), then cleaves angiotensin I to produce AII. A number of stimuli such as systemic hypotension, hypovolemia, decreased plasma or distal tubule ion concentrations (particularly Na^+), or the activation of juxtaglomerular β-receptors causes renin to be secreted into the plasma. This promotes an increase in circulating angiotensin I, making it available for conversion to AII. AII affects circulatory function at several levels, evoking responses in both the central nervous system and the peripheral vasculature. These responses are aimed at the conservation of water and electrolytes in order to counter the original renin-secreting stimulus.

Within the central nervous system, AII acts on receptors of some hypothalamic neurons to promote drinking behavior (Evered and Fitzsimmons, 1981). In the periphery exogenous or endogenous AII produces either an increase in systemic arterial blood pressure or a biphasic hypotensive–hypertensive response, depending

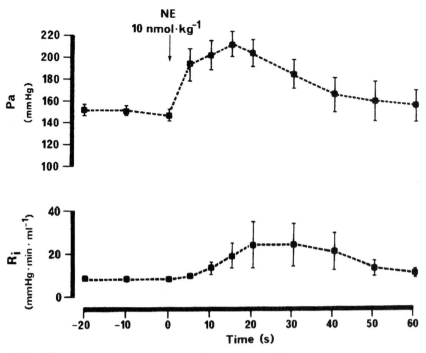

FIGURE 35 Responses of mean arterial blood pressure (P_a) and resistance to blood flow in the ischiatic artery (R_i) of an adult duck to a bolus intravenous injection of norepinephrine (NE, at arrow). The R_i values for each data point were calculated from the corresponding P_a and ischiatic blood flow values (modified from Wilson and West, 1986; used by permission).

on species. In ducks (Wilson and West, 1986; Butler et al., 1986) and pigeons (Evered and Fitzsimmons, 1981) systemic injections of AII produced dose-dependent increases in arterial blood pressure. Butler et al. (1986) and Wilson and West (1986) proposed that this response was due to vasoconstriction resulting from AII-mediated release of NE from sympathetic nerve terminals and enhanced EPI and NE release from the adrenal glands; in their experiments α- and β-adrenergic blockade eliminated the pressor effects of AII injection. Indeed, Moore et al. (1981) maintained that AII has no direct vasoconstrictor effects on arterial smooth muscle in the fowl and Wilson (1989), in a review of the renin–angiotensin system in birds, ascribes AII-induced vasoconstriction entirely to the effects of elevated catecholamine secretion.

In fowl and quail systemic AII injections produce a rapid, transient hypotension followed by a prolonged rise in arterial blood pressure (Nakamura et al., 1982; Takei and Hasegawa, 1990). The hypertensive phase of this response is mediated by adrenergic mechanisms, as in the duck and pigeon, but the transient hypotensive phase appears to be an indirect AII effect on vascular endothelial cells, working via the local release of NO from these cells as described in Section VI,B,2,b above.

Arginine vasotocin (AVT), released from the posterior pituitary into the circulation under conditions of osmotic or hypovolemic challenge, is the avian homolog of the mammalian antidiuretic hormone arginine vasopressin. However, the effects on avian vasculature of increased endogenous AVT levels after salt loading or hemorrhage are not well understood, nor are the vasomotor effects of systemic AVT injections. In mammals arginine vasopressin produces vasoconstriction in systemic arterioles and in glomerular afferent arterioles, both serving to facilitate antidiuresis. In birds AVT also induces vasoconstriction of afferent glomerular arterioles (see Braun, 1982 for review) but its effects on the rest of the circulation are controversial. Several studies have reported no cardiovascular consequences of AVT injection, maintaining that the avian vasculature is not sensitive to this peptide even at doses many times greater than "physiological" levels (Simon-Oppermann et al., 1988; Robinzon et al., 1988). In contrast, Wilson and West (1986) in ducks and chickens and Brummermann and Simon (1990) in ducks found that systemic injections of AVT produced hypotension accompanied by tachycardia. The latter authors proposed that AVT directly relaxes vascular smooth muscle, producing a fall in arterial blood pressure which then evokes a baroreflex-mediated tachycardia. However, Robinzon et al. (1993) found in fowl that the direction of AVT-mediated vascular responses depended on the dose and method of application. Low doses given slowly by intravenous infusion produced hypertension similar to the mammalian response to arginine vasopressin, while bolus intravenous doses produced the hypotensive responses reported in other avian studies. Robinzon et al. (1993) therefore proposed that AVT acts primarily via modulation of vascular caliber but that the degree and direction of vasomotion was not uniform in beds throughout the body. While there appears to be consensus that AVT does have direct vascular effects in birds, the understanding of its specific actions must await further studies on isolated vascular beds *in situ* and on blood vessels *in vitro*.

3. Neural Control

All parts of the systemic and pulmonary vascular trees, except capillary beds, are innervated by the autonomic nervous system. This innervation constitutes the final common pathway for rapid and flexible control by the central nervous system of regional distribution of cardiac output. Autonomic outflow to the vasculature is governed by a variety of reflexogenic inputs from visceral or somatic receptors relayed through brainstem and spinal cord pathways, and is also subject to influences originating at suprabulbar levels of the central nervous system. However, the degree of vasomotion produced in any region of the vasculature for a given intensity of autonomic drive will depend on the densities of effector terminals and postjunctional receptors in that region.

The location of autonomic terminals within the vascular wall is different in arteries and veins. Nerve fibers and terminals in the walls of arteries are, with some exceptions, limited to the tunica adventitia, extending as far as the outer elastic lamina marking the border between the adventitia and the tunica media (Bennett and Malmfors, 1970). In this respect the innervation pattern of avian arteries is similar to that in mammals (see Hirst and Edwards, 1989 for a review of mammalian arterial innervation). In contrast to the arterial innervation pattern, nerve fibers and terminals in avian systemic veins are commonly apposed to smooth muscle in the tunica media, as well as being located in the adventitia (Bennet and Malmfors, 1970) and in this regard also birds are similar to mammals (see Shepherd and Vanhoutte, 1975 for a review of the innervation of mammalian veins).

In addition to intramural nerve fibers and terminals, large and small nerves course over the outer surfaces of both arteries and veins, and some vessels are completely surrounded by plexi of nerve fibers. Furthermore, throughout the body nerves generally accompany blood vessels, running parallel with the vessels to form neurovascular bundles. These close associations between blood vessels and nerves have an important consequence for experimental investigations of neurogenic vasomotion. In assessing the effectiveness of neural control of vascular resistance in particular beds, electrical stimulation of autonomic nerves is commonly employed. However, some neural pathways to these beds may be interrupted if supply vessels to the beds are manipulated or sectioned (for example, to insert a blood flow probe) between the stimulus site and the expected site of vasomotion.

a. Systemic Arterial Innervation

The aorta from its origin to the junction with the celiac artery, the proximal brachiocephalic trunks, and the most proximal parts of the common carotid arteries are elastic vessels (see Section IV,A,2), having relatively little smooth muscle and thus a correspondingly sparse vasomotor innervation (Bennett and Malmfors, 1970; Bennett, 1971). The aorta, especially close to its root and adjacent to the root of the pulmonary trunk, has numerous small cells in the adventitia which contain catecholamines (Bennett, 1971). These cells occur singly or in clusters, appear to be similar to amine-containing cells of the carotid body, and are innervated by branches of the vagus nerve. While these cells may be a source of locally released catecholamines, it has also been suggested that these cells may constitute chemoreceptive "aortic bodies" (Bennett, 1971; Tcheng and Fu, 1962) similar to those found at the homologous site in mammals. In addition the adventitia of the aortic arch is innervated by afferent vagal fibers with nerve terminals transducing wall stretch and thus signaling an index of central arterial blood pressure; these constitute the only systemic arterial baroreceptors in birds (Jones, 1973).

The transition from elastic to muscular wall structure occurs in the arterial tree distal to the branching points of major distribution arteries from the great vessels. This transition also marks an increase in the density of innervation of the arterial wall, as shown in the photograph of a branch of the posterior mesenteric artery of the fowl in Figure 36. In some arteries such as the ischiatic, varicose fibers can be seen in the media as well as at the medial–adventitial border (Bennett and Malmfors, 1970). Of the large muscular arteries, the common carotids appear to be the most heavily innervated (Bennett and Malmfors, 1970). Given that these arteries convey the majority of blood flow to the avian brain, this density of innervation possibly reflects a requirement for greater autonomic control of cephalic blood flow than in other beds. Vasomotor innervation of the smaller branches of arteries within individual vascular beds supplied by the large arteries has not been systematically described but in all beds muscular arterioles, which constitute the resistance vessels responsible for 70 to 80% of total

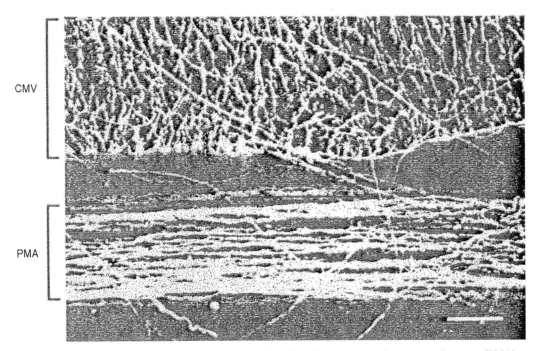

FIGURE 36 Adrenergic innervation patterns of a distal branch of the posterior mesenteric artery (PMA) and the coccygeomesenteric vein (CMV) of the chicken. The vessels run side by side with their long axes aligned from left to right in the photograph. Aldehyde fluorescence histochemistry shows that adrenergic nerve fibers and terminals are present in both vessels; in the artery they run parallel to the vessel axis and are more dense than in the vein. The horizontal bar represents 100 μm (modified from Bennett and Malmfors, 1970; used by permission).

peripheral resistance, are densely innervated (Folkow et al., 1966, Bennett and Malmfors, 1970).

The density of innervation of different regions of the arterial tree is variable among bird species and this variation has important consequences for differences in regional neurogenic control of blood flow among species. Large arteries supplying hindlimb muscles in ducks, for instance, are more densely innervated than those in turkeys, and this difference is correlated functionally with an enhanced capability in the duck to generate and maintain neurogenically mediated intense peripheral vasoconstriction (Folkow et al., 1966). These authors proposed that this was a general physiological adaptation in diving birds, enabling the redistribution of blood flow away from those peripheral vascular beds able to withstand periods of ischemia and toward the central circulation, thus conserving blood oxygen for ischemia-sensitive heart and brain tissue.

Electrical stimulation of the sympathetic innervation of a vascular bed evokes increases in vascular resistance in that bed, mimicking the effect of elevated vasoconstrictor outflow from the central nervous system. As illustrated in Figure 37, graded increases in stimulation frequency evoke proportionally larger increases in peripheral resistance in the hind limb vascular bed of the duck, producing a maximal increase of up to 7 times the prestimulus resistance value at a stimulation frequency of 30 Hz. In contrast, stimulus frequencies of less than 10 Hz are required to produce a maximal increase in resistance in the cat hind limb (Folkow, 1952). The hind limb vascular bed in the duck thus requires a greater degree of sympathetic drive than that of the cat to achieve the same order of increase in resistance to flow. This may reflect differences in the distribution of sympathetic terminals to the arteries in this bed in the two species. In support of this, Bennett and Malmfors (1970) noted that the pattern of adrenergic innervation of small intramuscular arteries in birds was not markedly different from that in mammals, but observed a higher terminal density in larger arteries of avians relative to mammals.

Some regions of the avian arterial vasculature have specialized wall structures and unusual innervation patterns, possibly reflecting enhanced capabilities for regional control of blood flow. The anterior mesenteric artery in several species of birds has, in addition to the normal circular smooth muscle in the media, an outer layer of longitudinal smooth muscle in the adventitia (Bolton, 1969; Bennett and Malmfors, 1970; Bell, 1969; Ball et al., 1963; see Section IV,A,2). Adrenergic fibers and varicose terminals are found at the adventitial–medial border in this as in other arteries, but these

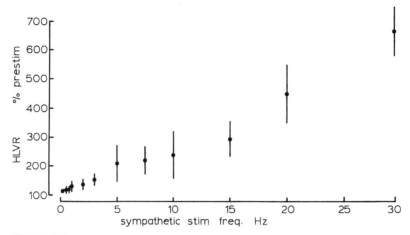

FIGURE 37 Relationship between frequency of electrical stimulation of the sympathetic innervation of the hind limb and the vascular resistance of this bed (HLVR) in a duck. HLVR for each data point was calculated from the corresponding arterial pressure and flow in the ischiatic artery. Increases in HLVR are expressed in percentages relative to the prestimulus value (F. Smith, unpublished).

elements also extend into the longitudinal muscle itself, aligned in the direction of the muscle fibers. Longitudinal muscle of this vessel also receives cholinergic innervation (Bolton, 1967; Bell, 1969; Bennett and Malmfors, 1970), although this does not extend to the inner circular muscle (Bell, 1969). Adrenergic nerve stimulation produces contraction of the circular muscle which is mimicked and blocked by α-adrenergic agonists and antagonists respectively (Bolton, 1969; Bell, 1969; Gooden, 1980). The longitudinal muscle is, however, relaxed by adrenergic nerve stimulation, acting through β-adrenergic receptors (Bolton, 1969; Bell, 1969). Stimulation of cholinergic nerve fibers in an *in vitro* preparation of a segment of the vessel caused the longitudinal muscle to contract, shortening the whole segment. This did not, however, markedly affect resistance to flow through the vessel but in the shortened state adrenergic vasoconstrictor responses of the circular muscle were found to be exaggerated (Bell, 1969). The cholinergic innervation of longitudinal muscle has thus been proposed to potentiate adrenergically mediated control of blood flow, possibly as an adaptation for rapid adjustment of intestinal blood flow during stress (Bell, 1969). The fact that increased sympathetic drive to this vessel has a relaxing effect on the longitudinal muscle as well as a constricting effect on the circular muscle would facilitate the effects of cholinergic input in shortening the vessel segment. Cholinergic input may also help adjust blood flow more precisely when the length of the anterior mesenteric artery changes during gross intestinal movements associated with digestion.

Coronary arteries also possess an outer coat of longitudinal muscle but adrenergic innervation of this muscle layer is very sparse. In these arteries the majority of adrenergic nerve fibers and terminals are found at the adventitiomedial border, as in other arteries (Bennett and Malmfors, 1970). There are no reports of cholinergic innervation of the coronary arteries of birds. Cerebral arteries are, however, dually innervated by adrenergic and cholinergic fibers (Tagawa *et al.*, 1979). Adrenergic varicosities are located in the adventitia near the border with the media throughout the cerebral circulation, in common with most other arteries of the body. Cholinergic fibers as well are located in the tunica adventitia but occur less frequently than adrenergic terminals (Tagawa *et al.*, 1979). It is presumed that, since cerebral arteries have no smooth muscle in the adventitia, neurotransmitters released from both types of terminals diffuse into the media to act on muscle fibers there. Tagawa *et al.* (1979) also reported that, as in mammals, some of the cholinergic innervation of avian cerebral arteries appears to originate from central neurons, especially in the diencephalon, as well as from peripheral neurons. Studies in mammals have shown that the functional significance of reflexogenic vasomotion in overall control of cerebral blood flow is relatively minor compared with neurogenic control of blood flow in other vascular beds in the body; intracerebral blood distribution is influenced mainly by local and possibly circulating humoral factors (Kontos, 1981). This is likely to be true also in birds; Stephenson *et al.* (1994) estimated that neurogenic contributions to changes in cerebral blood flow in ducks during diving were minor.

b. Systemic Venous Innervation

Large veins in birds are more densely innervated than those in mammals. The density of noradrenergic innervation of the chicken caudal vena cava is graded, increasing in portions of the vessel more distal to the heart (Bennett, 1974; Bennett et al., 1974). The caudal vena cava has an outer coat of longitudinal muscle as well as an inner circular layer and both layers receive adrenergic innervation, with the orientation of the terminals and fibers following the direction of the muscle fibers in each layer (Bennett, 1974). There is also a sparse cholinergic innervation of this vessel. The superior venae cavae also have two muscle layers, with a similar adrenergic innervation pattern to the caudal vena cava (Bennett and Malmfors, 1970). The walls of the pectoral, subclavian, celiac, and jugular veins contain both circular and longitudinal smooth muscle, and the innervation of these vessels is through a plexus of varicose fibers and terminals located primarily between the inner and outer muscle layers. Other veins, such as the coccygeomesenteric vein, have primarily circular muscle, innervated in the pattern illustrated in Figure 36.

Functionally, adrenergic vasomotion appears to be preeminent in veins. Vasomotion of the caudal vena cava and other major veins is mediated primarily by α-adrenergic receptors producing vasoconstriction when activated; no functional β-adrenergic receptors are present (Bennett and Malmfors, 1974). Responses of venous smooth muscle to cholinergic nerve stimulation are weak and variable and probably do not contribute directly to neurogenic control of wall compliance; however, ACh released from cholinergic nerve terminals can modulate the release of NE from local adrenergic nerve terminals and thus may provide fine adjustments of adrenergically generated venomotor tone.

Sympathetically mediated contraction of smooth muscle in the major veins serves to decrease wall compliance and vessel diameter, thus providing a reflexogenic mechanism for reducing the volume of the central venous pool and increasing return of blood to the heart, as discussed in Section IV,C,2. Langille (1983) proposed that reflexogenic venoconstriction was responsible for the increased central venous pressure observed in ducks during involuntary submersion. This response may function to aid venous return to the heart to help maintain stroke volume in the face of a reduction in cardiac contractility which may develop during diving (Djojosugito et al., 1969; Langille, 1983).

The renal portal valves, located bilaterally where the external iliac veins join the junction of the caudal vena cava with the caudal renal veins, are unique to birds. These valves (Figure 28) are in the form of sphincters of smooth muscle which can close off the direct route for blood flow from the external iliac veins to the caudal vena cava. Closure increases renal portal blood flow, venous blood from the external iliac vein being then partially redirected into the renal capillaries and thence to the caudal vena cava, as discussed in Section IV,C,4. The renal portal valve is heavily innervated by both adrenergic and cholinergic nerve fibers (Akester and Mann, 1969; Bennett and Malmfors, 1970) with reciprocal effects on the smooth muscle of the valve. Adrenergic nerve stimulation and sympathomimetic agents induce relaxation, mediated by β-adrenergic receptors, while cholinergic nerve stimulation and cholinomimetic agents produce contraction via muscarinic receptors (Bennett and Malmfors, 1975a; Sturkie et al., 1978). The importance of this valve in the control of venous blood distribution in the renal portal system has been the subject of some debate (Akester, 1967; Section IV,C,4). Recent functional studies of the avian kidney in situ by Glahn et al. (1993) have shown that the status of the renal portal valve affects total renal blood flow by altering the amount of flow in the renal portal system. In conditions of lowered arterial blood pressure in which renal arterial perfusion is below the range of autoregulation of glomerular blood flow, closure of the renal portal valve raises renal portal blood flow to compensate total renal blood flow (Glahn et al., 1993). Neural control of the renal portal valve may thus form a component of the suite of reflexogenic responses to hypotension or hypovolemia.

c. Pulmonary Vessel Innervation

Adrenergic innervation of the pulmonary artery in the fowl consists mostly of fibers with few varicose nerve terminals. Fibers and nerve terminals form a plexus at the adventitiomedial border of the artery, with some projections into the circular smooth muscle of the media. There is also smooth muscle oriented longitudinally in the adventitia, but innervation of this is very sparse (Bennett and Malmfors, 1970; Bennett, 1971). Some intrapulmonary arterial branches have dense adrenergic plexi occurring in short segments along their length (Bennett and Malmfors, 1970), which may help to redistribute blood flow to selected gas exchange areas within the lung to optimize local ventilation–perfusion ratios (Hebb, 1969).

The pulmonary veins proximal to the left atrium are very densely innervated with adrenergic nerve fibers and terminals, the density of this innervation being markedly greater than in the pulmonary arteries (Bennett, 1971). Abundant terminal varicosities and nerve fibers are located in a plexus at the adventitiomedial border, with some penetration into the media (Bennett, 1971; Bennett and Malmfors, 1970). There is also longitudinal smooth muscle present in the adventitia, with adrenergic terminals between the muscle fibers (Bennett, 1971; Bennett and Malmfors, 1970). The density of innervation of the longitudinal smooth muscle is reduced close to the left atrium, increasing distally along the vessel until the bifurcation to the lungs. At this junction there is an abrupt decrease in density of innervation of the entire vessel wall, and within the lungs the major branches of the pulmonary veins are very sparsely innervated. Smaller branches of these vessels appear to have no innervation (Bennett, 1971).

d. Autonomic Pathways

The cell bodies of adrenergic postganglionic vasoconstrictor neurons innervating the vasculature are located in paired paravertebral ganglion chains, in prevertebral ganglia and, in some cases, in small ganglia scattered throughout the viscera. Cells in the superior cervical ganglion (representing a fusion of the two most cranial cervical ganglia) innervate blood vessels of the head, including those of the salt and salivary glands, via cephalic extensions anastomosing with several cranial nerves (Bennett, 1974). In birds pairs of sympathetic ganglia are associated with the cervical vertebrae, and cells in these ganglia innervate blood vessels of the neck. The presence of cervical paravertebral ganglia in birds constitutes a major difference in the organization of the sympathetic nervous system between this vertebrate group and mammals. In the caudal part of the neck, in the thorax, and in the wings of birds the vasculature is innervated by neurons in the thoracic paravertebral ganglion chain. Sympathetic fibers reach wing vessels primarily through the brachial plexus. Some thoracic postganglionic neurons also contribute sympathetic fibers to the greater splanchnic nerves innervating anterior abdominal viscera via the celiac plexus (Bennet, 1971).

Vertebrae of the lumbar, sacral, and coccygeal spine are fused to form the synsacrum in birds. The paravertebral ganglion chains from each side in this region are fused in the midline at about the level of the sixth coccygeal segment in avian species so far examined, and this combined sympathetic trunk continues caudally to the pygostyle (Pick, 1970; Akester, 1979; Benzo, 1986). Axons from postganglionic neurons in this part of the sympathetic nervous system innervate abdominal and pelvic viscera via the lesser splanchnic nerves and aortic plexus, the hypogastric plexus, the pelvic plexus, and the cloacal plexus. Some sympathetic postganglionic somata are also located in prevertebral ganglia within these plexi. In addition, lumbosacral sympathetic neurons contribute vasoconstrictor axons to the hind limbs via the lumbosacral plexus (Benzo, 1986; Bennett, 1974).

Sympathetic preganglionic cell bodies synapsing on the postganglionic vasoconstrictor neurons are located in and near a bilateral column of neurons, the column of Terni, in the gray matter near the central canal of the spinal cord (see Section VI,C,2 below). All preganglionic axons innervating postganglionic neurons in the cervical sympathetic chain exit the spinal cord through ventral nerve roots of cranial thoracic segments; there are apparently no connections between the cervical spinal nerve roots and the sympathetic ganglia in the neck (Bennett, 1974; Akester, 1979; Pick, 1970). Although no detailed analysis has been done of the segmental locations of spinal preganglionic neurons projecting to postganglionic vasoconstrictor neurons innervating the thoracic, abdominal, pelvic, and limb regions, it is likely that the locations of these neurons follows the mammalian plan. That is, axons of preganglionic neurons in a particular spinal segment emerge from the cord in the ventral root of that segment to innervate postganglionic neurons in ganglia at that level or within one or two segments rostral or caudal to the exit site (see Gabella, 1976 for review).

Parasympathetic postganglionic neurons innervating the vascular smooth muscle of such organs as salt glands and salivary glands in the head of birds are located in autonomic ganglia near or embedded in the organs (Ash et al., 1969). Cell bodies of preganglionic neurons innervating parasympathetic postganglionic vasodilator neurons associated with structures in the head are located in the oculomotor, facial, glossopharyngeal, and vagal nuclei in the brainstem and course to peripheral ganglia via the respective cranial nerves associated with these nuclei (Akester, 1979). Postganglionic cholinergic nerve fibers innervating arteries and veins of the body originate from the somata of neurons located in plexuses associated with blood vessels themselves or, for vessels of the abdominal viscera, in prevertebral ganglia containing a mixture of adrenergic and cholinergic postganglionic neurons (Bennett and Malmfors, 1970; Bennett, 1974). The location of cells of origin of preganglionic axons innervating the postganglionic neurons producing vasodilation in the viscera and skeletal muscle have not been determined in detail. However, in birds the parasympathetic outflow is organized in cranial and sacral tracts as it is in mammals. It would therefore be expected that preganglionic neuronal somata responsible for vasodilation in thoracic and anterior abdominal viscera and skeletal muscle in the upper part of the body are located in medullary vagal motor nuclei, with their axons running to postganglionic neurons via the vagus nerves. Similarly, vasodilation of posterior abdominal and pelvic viscera and skeletal muscle of the lower part of the body is likely to be mediated by preganglionic neurons with their somata located at sacral levels of the spinal cord and their axons innervating postganglionic neurons via the abdominal and pelvic autonomic nerves.

C. Control of the Heart

1. Catecholamine Effects on the Heart

Both NE and EPI are present in circulating plasma of birds (see Section VI,B,2,c for discussion), and these amines have cardiac effects which include an increase in the rate of pacemaker depolarization (DeSantis *et al.*, 1975; Bolton and Bowman, 1969) and augmented force of myocardial contraction (DeSantis *et al.*, 1975; Bennett and Malmfors, 1974; Bolton, 1967; Bolton and Bowman, 1969). In isolated heart preparations and in myocardial strips *in vitro,* NE has been found to be as effective as, or more potent than, equimolar EPI in producing inotropic and chronotropic augmentation. This is in marked contrast to the condition in the mammalian heart, in which EPI is the more potent stimulant (Gilman *et al.,* 1990). In whole-animal experiments on birds, intravenously injected boluses of NE and EPI augment cardiac output by transiently increasing both rate and force of contraction, contributing along with peripheral vasoconstriction to catecholamine-mediated hypertension (Wilson and Butler, 1983; Wilson and West, 1986; Bolton and Bowman, 1969; Butler *et al.,* 1986). In the avian as in the mammalian heart, catecholamines appear to act primarily via β-adrenergic receptors on myocardial and pacemaker cells (Bolton and Bowman, 1969; Butler *et al.,* 1986; Bolton, 1967).

2. Neural Control

The anatomical organization of the innervation of the heart in birds has been studied for more than 100 years, first by gross dissection and then with a variety of nerve-specific stains (see for example Ábrahám, 1969; Hirsch, 1970; Pick, 1970), but these techniques have not allowed the patterns of cardiac sympathetic and parasympathetic innervation to be anatomicalliy differentiated. This is a factor of major importance in determining the mechanisms involved in neural control of the heart. It is only recently, with the advent of histochemical techniques specific for adrenergic and cholinergic neurotransmitters or enzymes in the catabolic and anabolic pathways of these transmitters, that the patterns of dual, function-specific cardiac innervation have begun to be explored. The most well-established histochemical method for determining the peripheral distribution of adrenergic nerves is that developed by Falck (1962) to render catecholamines brightly fluorescent under ultraviolet illumination in the light microscope. This sensitive technique shows cell bodies, axons, and nerve terminal varicosities, and most of the descriptions of adrenergic innervation of the heart are based on the use of this technique in whole-mounts or sections of atrial and ventricular tissue. Histochemical assays for the presence of acetylcholinesterase (AChE), based on those developed by Koelle and others (reviewed by Koelle, 1963), have been used to determine the distribution of parasympathetic innervation of the heart. There are also more recent immunohistochemical techniques available for identifying enzymes in the biochemical pathways synthesizing neurotransmitters; specifically, antibodies directed against dopamine β-hydroxylase and tyrosine hydroxylase (NE synthesis) and choline acetyltransferase (ACh synthesis) are commercially available. These techniques, however, have yet to be applied to the avian heart.

Interpretation of the results of studies using transmitter-specific histological techniques to determine patterns of autonomic innervation of the cardiovascular system rests on two major assumptions: (1) that the primary neurotransmitter released by the terminals of sympathetic postganglionic neurons is NE, and (2) that released by parasympathetic terminals is ACh. Functional studies in bird hearts largely support this assumption. However, recent work on peripheral autonomic anatomy and function in a wide range of vertebrates has emphasized the diversity of autonomic neurotransmitters and neuromodulators in addition to the classic transmitters utilized by this system (see Nilsson and Holmgren, 1994; Armour and Ardell, 1994; and Furness and Costa, 1987 for reviews), so some caution must be exercised in this area. In the following discussion of the innervation of the avian heart, results from studies using neurotransmitter-specific techniques are emphasized.

a. Sympathetic Innervation

i. Anatomy Postganglionic sympathetic nerve fibers arising from neuronal somata extrinsic to the heart form part of an intracardiac nerve plexus distributed throughout all four cardiac chambers. The adrenergic innervation of the proximal part of the venae cavae appears to be continuous with the intracardiac plexus associated with the right atrium (Bennett and Malmfors, 1970). Sympathetic nerve fibers form a network over the epicardial surface of the right atrium, with some fibers penetrating into the thin atrial wall to lie adjacent to bundles of myocardial cells and others passing through the wall to the subendocardium (Smith, 1971a). The overall appearance of the plexus is that of a three-dimensional latticework of fibers extending from the epicardium through the wall to the subendocardium, with varying concentrations of smooth (nonvaricose) and varicose nerve fibers and nerve endings in different regions of the atrial myocardium.

Bennett and Malmfors (1970) reported that the most densely innervated region of the heart was the external wall of the right atrium; furthermore, within this area the region adjacent to the confluence of the venae cavae with the wall contained the highest density of fibers and terminals. These authors referred to this area as the "sinu-atrial node," presumably by analogy with the corresponding sharply defined sinoatrial node of the mammalian heart. There is, however, some evidence that the cells in the primary pacemaker site in the right atrium of the bird heart may have a functional organization different from that in the mammalian heart, in the area of the junction of the sinus venosus with the sinoatrial valves (Moore, 1965; and see Section II,C,3). These valves are present in birds but are only represented by a vestigial flap in mammals. A large number of the nerve fibers in this area of the avian heart were varicose, running among the myocardial cells and aligned with the longitudinal axes of these cells. Another area with a high density of adrenergic terminals was in the right atrial wall near the atrioventricular border, an area corresponding to the atrioventricular node (Bennett and Malmfors, 1970; Section II,C,3). Many nerve varicosities were also associated with blood vessels in the right atrial wall.

Bogusch (1974, osmium stain) reported in the fowl that the specialized cells of the pacemaker areas and conducting system in the right atrium are well innervated by nerve fibers with frequent varicosities and some bare nerve endings. In contrast, Yousuf (1965, silver stain) found no investment of nerve fibers into the sinoatrial node region and few in the atrioventricular node in the sparrow heart. This disparity could be due to a species difference or to a relative lack of sensitivity of the silver stain; in any case, neither technique is neurotransmitter specific nor as sensitive as the amine fluorescence technique. Full details of the sympathetic innervation of the conducting tissues of the right atrium are still not clear. A combination of immunohistochemical localization of dopamine hydroxylase or tyrosine hydroxylase in nerve terminals and standard histological processing for visualizing pacemaker cells and atrial Purkinje fibers would address this problem.

A small number of strongly fluorescent cell bodies are present in the right atrium and in the walls of some parts of the vasculature after processing for amine-related fluorescence. These cells are, however, not associated with intracardiac ganglia (Bennett and Malmfors, 1970) and probably represent "small, intensely fluorescent" cells (SIF cells) or so-called "paraganglion cells" (Eranko and Eranko, 1977) of as yet uncertain function. There are, however, some fluorescent nerve endings around the cell bodies of nonfluorescent ganglion cells (Bennett and Malmfors, 1970), suggesting some form of sympathetic modulation of the activity of intrinsic cardiac neurons.

The right atrioventricular valve has a central layer of connective tissue between two layers of cardiac muscle (Section II,A,4), and nerve bundles have been observed in association with this connective tissue (Smith, 1971a). In this study several general nerve stains as well as a cholinesterase-specific stain were used, but unfortunately no details are given regarding the specificity of staining of the valvular innervation. However, Bennett and Malmfors (1970) have shown that this valve is innervated by adrenergic nerve fibers with few terminals, arranged in a loose plexus in the leaves of the valve. These observations suggest the existence of some degree of neural control of this valve, possibly by both sympathetic and parasympathetic limbs of the autonomic nervous system.

Adrenergic innervation in the left atrium is less dense than in the right atrium, but more dense than in the ventricles. Adrenergic fibers and terminals of the left atrial cardiac plexus, while distributed in a three-dimensional pattern similar to that in the right atrium, appear to be more evenly spread throughout the left atrial wall with little variation in density. In addition, the interatrial septum also receives adrenergic innervation as an extension of the epicardial plexus (Akester, 1971; Akester et al., 1969). The distal portion of the left atrioventricular valve has a fibroelastic structure, while more proximally cardiac muscle is associated with the fibrous skeleton (Section II,A,4). Smith (1971a) has described a fine network of nerve fibers, continuous with the left atrial subendocardial plexus, investing the fibroelastic portion of this valve; no mention of an innervation of the muscular portion of the valve was made in this study. The staining technique used was again not transmitter specific, so the left atrioventricular valve may be under sympathetic or parasympathetic influence, or both, or the innervation observed may be afferent. However, regarding the latter possibility, Smith (1971a) states that no simple or specialized nerve endings typical of sensory receptors were observed in or near the left atrioventricular valve.

The avian ventricles are relatively sparsely innervated compared with the atria, but even so are more densely innervated than are the ventricles of the hearts of mammals (Smith, 1971b). Akester et al. (1969) and Akester (1971) reported adrenergic innervation of the interventricular septum, and Bennett and Malmfors (1970) observed that innervation of this region was more dense than that in the rest of the ventricle walls. While these authors did not determine the intraseptal targets of innervation, it is possible that some of these axons may innervate intraseptal Purkinje cells since bare nerve endings have been observed close to these cells (Akester, 1971). Regarding the innervation of the pulmonary and aortic valves, Smith (1971a) reported that they were sparsely innervated in comparison with the left atrioventricular valve. Fibers in the pulmonary and aortic valves comprise a plexus arranged to form a widely spaced lattice in the basal parts of the valve leaflets.

In birds the sympathetic cardiac nerves arise from the most rostral ganglia of the thorax and the most caudal cervical ganglia, but there is some variation in detail among different accounts (see Cabot and Cohen,

1980 for summary). Macdonald and Cohen (1970) and Cabot and Cohen (1980) describe the right sympathetic cardiac nerve in the pigeon as a single trunk formed by the anastomosis of postganglionic nerves arising from the three most caudal cervical ganglia. These ganglia are associated with spinal nerves contributing to the brachial plexus; the most caudal ganglion of this group is thus associated with the last spinal segment contributing to the brachial plexus. This description closely follows that of Malinovsky (1962) in the pigeon. In the chicken, however, the origin of the right cardiac sympathetic nerve is limited to the first thoracic paravertebral ganglion (as defined by its location caudal to the head of the first rib; Pick, 1970; Tummons and Sturkie, 1969; see Figure 38).

There has been some controversy over the nomenclature of avian sympathetic ganglia. Birds have a sympathetic ganglion associated with each cervical spinal segment (Gabella, 1976; Pick, 1970) and this observation, as well as variations in opinion on numeration of the ribs in the bird, has made it difficult to be precise about the terminology for the most caudal ganglion contributing to the cardiac nerve. Malinovsky (1962) holds that this is the first thoracic ganglion, since it lies between the heads of the first and second ribs. However, Macdonald and Cohen (1970) have sided with earlier authors in noting that the first rib should not be considered to mark the first thoracic segment since this rib is reduced in size and does not form part of the ribcage proper. As Cabot and Cohen (1980) have pointed out, the issue of terminology is not critical when considering functional aspects of cardiac sympathetic outflow to the bird heart, but may become important when attempting to draw conclusions about the homology of this pathway with the sympathetic innervation of the mammalian heart. Mammals have no cervical paravertebral ganglion chain, and postganglionic neurons efferent to the heart in this vertebrate class are located in the middle cervical ganglion and the stellate ganglion, the latter being formed by the condensation of the caudal cervical and first thoracic ganglia. This arrangement has no parallel in birds.

Regardless of the number of ganglia contributing to the right cardiac sympathetic nerve, the course of this nerve to the heart appears to follow the same general pattern in all avian species so far examined. The nerve courses toward the heart in conjunction with a small vertebral vein arising from the lateral aspect of the vertebral column between the last cervical and first thoracic spinal segments. The vessel and nerve merge with the apical pleura of the lung and run together between the pleural fascia. At the ventral surface of the pleura close to the junction of the vertebral vein with the superior vena cava, the nerve turns caudally along the vena cava toward the heart, there forming two rami. The medial ramus divides further and its fascicles enter the cardiac plexus to distribute within the right atrial wall. The lateral ramus joins the right vagus nerve in the vicinity of the right pulmonary artery (Pick, 1970; Cabot and Cohen, 1980; Tummons and Sturkie, 1969; Cabot and Cohen, 1977a; Macdonald and Cohen, 1970; Malinovsky, 1962). The ganglionic origin of the left sympathetic cardiac nerve is similar to that of the right, arising in the chicken from paravertebral ganglion 14 (first thoracic) and in the pigeon by anastomosis of postganglionic branches from ganglia 12, 13, and 14 (Cabot and Cohen, 1980). In the pigeon, the largest branch contributing to the left sympathetic cardiac nerve arises from ganglion 14, as on the right side. In its path to the heart, the left cardiac nerve in the pigeon divides into two or more fascicles which run in parallel for a short distance then recombine before reaching the superior vena cava. This nerve ramifies again as it runs caudally along the vena cava toward the heart, and the individual rami merge with the cardiac plexus of the left atrium.

The general locations of the cells of origin of the sympathetic postganglionic axons to the avian heart, and the intraspinal locations of the cardiac sympathetic preganglionic neurons have been worked out in greatest detail in the pigeon, by Cabot and Cohen and their co-workers. These workers used a combination of degeneration, neuroanatomical tracing, and electrophysiological stimulation and recording techniques. Macdonald and Cohen (1970) undertook a series of neuronal degeneration studies in order to determine the ganglionic distribution of the somata of postganglionic neurons supplying axons to the heart. After section of the right or

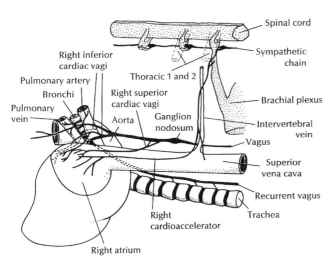

FIGURE 38 Schematic representation of the sympathetic innervation of the chicken heart on the right side. In this species the cardiac sympathetic nerve originates from the first thoracic paravertebral ganglion. (Modified from Tummons and Sturkie (1969), *Am. J. Physiol.* with permission.)

left cardiac sympathetic nerve in the thorax, the greatest number of degenerating neurons was found in ganglion 14, with lesser numbers in ganglia 12 and 13; no degenerating neurons were observed in ganglia rostral or caudal to this level. Postganglionic cells of origin of fibers in the cardiac sympathetic nerves were thus located bilaterally in the same sympathetic ganglia which give rise to the postganglionic nerves constituting the cardiac nerve. In experiments in which the ganglia themselves or the interganglionic sympathetic trunks were stimulated electrically, Macdonald and Cohen (1970) obtained positive cardiac chronotropic responses of short latency from ganglia 12 to 16, occasional longer-latency responses from ganglia 17 and 18, and none from ganglia caudal to 18. Experiments in which stimulation of ganglia was combined with sections of the sympathetic trunk above and below the stimulation site confirmed these results. These data were interpreted to mean that preganglionic axons originating in the spinal cord as caudally as the 16th segment converged on postganglionic cardiac neurons in ganglia 12 to 14. The longer-latency cardiac responses to stimulation of ganglia 17 and 18 were attributed to sympathetic activation of the adrenal glands and consequent release of catecholamines into the bloodstream; nerves arising from ganglia 17 and 18 were observed to join the splanchnic nerves, branches of which innervate the adrenal medulla.

Avian sympathetic preganglionic neurons innervating the heart are located in the same area of the spinal gray matter as those innervating the blood vessels, near the midline dorsal and lateral to the central canal, and extending rostrocaudally throughout segments 14 to 21. Most of the preganglionic neurons are confined to a distinct cell column in the midline, the column of Terni, a nucleus peculiar to the avian spinal cord (Huber, 1936; Cabot and Cohen, 1980). This cell column is the probable homolog of the mammalian intermediolateral cell column. Leonard and Cohen (1975), in a study of the cytoarchitecture of the pigeon spinal gray, reported that the rostral and caudal extents of this nucleus were indistinct due to small clusters of cells which extended into the regions between segments 13 and 14 and caudal to segment 21. Neurotracer studies using retrograde transport of horseradish peroxidase have shown that spinal preganglionic neurons efferent to the postganglionic cells in ganglion 14 of the pigeon are present from the caudal portion of segment 14 to the rostral part of segment 17 (Cabot and Cohen, 1977a). This finding provides strong anatomical support for the earlier conclusion of Macdonald and Cohen (1970), reached on the basis of functional and degeneration studies, that spinal preganglionic inputs to cardiac postganglionic neurons originate from segments 14 to 16, with some inputs possibly coming from segment 17.

The protocol used by Cabot and Cohen (1977a) did not label cardiac preganglionic neurons specifically. In a more recent study of the intraspinal circuitry involved in sympathetic control of the heart, the region of the spinal cord containing cardiac preganglionic motor neurons has been more precisely mapped by Cabot et al. (1991b). In this study, preganglionic neurons associated with ganglion 14 in the pigeon were labeled using fragment C of tetanus toxin (a nontoxic moiety which is retrogradely transported by axons), injected into sympathetic ganglion 14. These experiments confirmed the location of preganglionic neurons in the column of Terni and also demonstrated labeling of neurons lateral to this nucleus (Figure 39) in an area which, in the mammalian spinal cord, is occupied by sympathetic preganglionic neurons of the nucleus intercalatus spinalis (Petras and Cummings, 1972), a term that Cabot et al. (1991b) have applied to the corresponding area in the pigeon spinal cord. Their results show that cardiac neurons in the spinal cord are not confined to the column of Terni. No studies have yet been done to label cardiac postganglionic sympathetic neurons in the bird which innervate specific regions of the heart, but this will be a necessary further step in order to determine the intraspinal locations of groups of neurons controlling cardiac functions such as rate and contractility. A potentially useful ap-

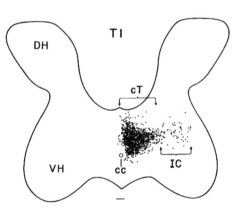

FIGURE 39 Mediolateral distribution of cell bodies of spinal preganglionic neurons labeled in the first thoracic segment (T1) of the spinal cord by an injection of a retrograde neurotracer (fragment C of tetanus toxin) into the right paravertebral ganglion 14 of the pigeon. The largest root of the cardiac sympathetic nerve in the pigeon arises from this ganglion. The diagram shows labeled neurons concentrated in the column of Terni [cT, large cluster of dots closest to the central canal (cc)], along with a lesser concentration of cells located more laterally in the nucleus intercalatus spinalis (IC). Although preganglionic neurons innervating the heart were not labeled specifically in this experiment, their cell bodies will be among the labeled population. Abbreviations: DH, dorsal horn; VH, ventral horn. Horizontal scale bar represents 50 μm. (Reprinted from *Neuroscience* **40,** J. B. Cabot, A. Mennone, N. Bogan, J. Carroll, C. Evinger, and J. T. Erichsen, Retrograde, trans-synaptic and transneuronal transport of fragment C of tetanus toxin by sympathetic preganglionic neurons, pp. 805–823, Copyright (1991), with permission from Elsevier Science.)

proach to this problem would be localized injections of a retrograde neuroanatomical tracer such as fragment C of tetanus toxin or pseudorabies virus into specific regions of the heart. This tracer would then be transported retrogradely and transsynaptically so that both cardiac postganglionic neurons and the spinal preganglionic neurons afferent to them could be visualized. This approach has proven advantageous in studies of cardiac autonomic pathways in mammals (Strack *et al.*, 1988).

The descending projections from higher centers in the central nervous system to sympathetic preganglionic neurons controlling the heart have not in general been delineated in birds. However Cabot *et al.* (1982), in a series of anatomical studies in the pigeon, determined that the overall pattern of projections to spinal preganglionic neurons is very similar to that found in mammals, and this anatomical data corroborates the results of earlier brain stimulation studies in the bird. In particular, Cabot *et al.* (1982) have shown anatomically that avian preganglionic neurons receive direct projections from diencephalic and medullary areas which, when electrically stimulated, provoke cardioaugmentation (Macdonald and Cohen, 1973; Folkow and Rubenstein, 1965; Kotilainen and Putkonen, 1974; Feigl and Folkow, 1963).

ii. Sympathetic Control Electrical activation of intramural adrenergic nerves in the avian heart produces augmented force of contraction and cardioacceleration. Bolton and Raper (1966) and Bolton (1967) first demonstrated sympathetically mediated augmentation of force of contraction of electrically paced strips of *in vitro* left ventricle from the fowl heart. In these studies, field stimulation excited both cholinergic and adrenergic nerves, and sympathetically mediated augmentatory effects were then pharmacologically isolated by the application of atropine to eliminate parasympathetic inhibitory effects. After muscarinic blockade, the increased contractile force produced by field stimulation was attributed to activation of adrenergic nerves since this effect could then be blocked by β-adrenergic antagonists. These data provided the first evidence in birds that the sympathetic nervous system can have a positive inotropic effect directly on ventricular myocardial cells.

Similar field stimulation experiments on left and right atria *in vitro* have shown that activation of intramural sympathetic nerves has powerful effects on these chambers. In the left atrium of the fowl heart, increased force of contraction of myocytes resulted from sympathetic nerve activation. These positive inotropic effects were blocked by β-adrenergic antagonists (Koch-Weser, 1971; Bennett and Malmfors, 1974; Bennett and Malmfors, 1975b). The right atrium *in vitro* responds to field stimulation of intramural sympathetic nerves with an increase in force of contraction and rate of pacemaker discharge, with both effects mediated by β-adrenergic receptors (Pappano and Loffelholz, 1974).

In the isolated chicken heart perfused *in vitro* by Langendorff's method, stimulation of the attached right cardiac sympathetic nerve produces positive chronotropic effects (heart rate increased from 186 to 292 beats per min; Sturkie and Poorvin, 1973). The basal heart rate of this isolated preparation, supplied with (presumably) adequate oxygen in the perfusate at 40°C, might be expected to be similar to that of the *in situ* chicken heart after bilateral section of the vagus and cardiac sympathetic nerves ("decentralized" state), but this was not the case. The mean rate of *in situ* decentralized hearts was in the range of 235–285 beats per min (Tummons and Sturkie, 1968; 1969). Under these conditions stimulation of the peripheral stump of the right cardiac sympathetic nerve increased heart rate to 345 beats per min, an increase of 48%. By comparison, stimulation of the right cardiac nerve stump attached to isolated hearts *in vitro* produced a mean heart rate of 292 beats per min. That is, maximal sympathetic stimulation in this preparation could only raise heart rate to a level equivalent to the basal rate of the decentralized heart *in situ*. However, even though basal heart rate was different in these preparations, the proportional increase in rate during sympathetic stimulation was the same in both cases. Thus sympathetically mediated chronotropic effects in the isolated heart may, in relative terms, reflect the capabilities of this control system in the *in vivo* heart.

In a beating heart *in vivo*, stimulation of the cardiac sympathetic nerves produces cardioacceleration. Tummons and Sturkie (1968), in the unanesthetized chicken, showed that either cardiac sympathetic nerve could produce this effect when stimulated: activation of the right nerve increased heart rate by 48% above the prestimulation value, while activation of the nerve on the left side increased heart rate by 32%. In the pigeon, however, Macdonald and Cohen (1970) found that only the right cardiac nerve mediated cardioacceleration when stimulated (Figure 40, top panel) while stimulation of the left cardiac nerve altered the appearance of the T wave of the electrocardiogram without a chronotropic effect (bottom panel, Figure 40). Such functional asymmetry in cardiac control has also been reported for the mammalian heart (Randall, 1994).

In the right atrium the role of the dense adrenergic innervation of the sinoatrial area (see Section VI,C,2,a) in control of heart rate is obvious and a number of studies have shown that adjustments of heart rate *in vivo* are made by the sympathetic nervous system under a variety of physiological conditions. Furthermore, the anatomical evidence for widespread cardiac innervation and the data cited above for sympathetic influences on myocardial contractility in the atria and ventricles im-

Right cardiac N.

Left cardiac N.

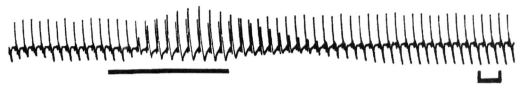

1 Sec.

FIGURE 40 Electrocardiograms showing heart rate responses in the pigeon to electrical stimulation of the right (top) and left (bottom) cardiac sympathetic nerves. The duration of the stimulus train delivered to each nerve is indicated by the length of the solid bars under the traces. (After Cells of origin of sympathetic pre- and postganglionic cardioacceleratory fibers in the pigeon, R. L. Macdonald and D. H. Cohen, *J. Comp. Neurol.*, Copyright © 1970 Wiley-Liss, Inc. Reprinted by permission of Wiley-Liss, Inc., a subsidiary of John Wiley & Sons, Inc.)

plies that the sympathetic nervous system can also produce both global and regional enhancement of contractility. Two important functional consequence of this arrangement are that (1) different patterns of sympathetic outflow from the central nervous system to individual chambers of the heart provide the means for augmenting regional contractility differentially to match the chambers' pumping actions to the hydraulic impedances into which they are working; and (2) the output of each chamber can be controlled independently of pumping rate. However, to understand the function of this control system more thoroughly it is necessary to establish whether different populations of pre- and postganglionic sympathetic neurons do in fact innervate different cardiac regions and whether such subpopulations may be differentially activated through reflexes driven by receptors in specific cardiac or vascular reflexogenic zones. Such an analysis is complicated by the location, remote from the heart, of cells of origin of the postganglionic sympathetic axons innervating the cardiac chambers, in contrast to the intracardiac locations of the postganglionic parasympathetic neurons.

Most studies of sympathetic control of cardiac function have affirmed the role of NE as the transmitter released by avian postganglionic terminals on the myocardium. However, the bird heart contains EPI as well as NE (Sturkie and Poorvin, 1973; De Santis *et al.*, 1975; and data summarized in Holzbauer and Sharman, 1972). It has been suggested that, in other organs such as the rectum of the fowl, adrenergic terminals may release EPI (Komori *et al.*, 1979). In the isolated chicken heart, De Santis *et al.* (1975) proposed that both EPI and NE may act as sympathetic neurotransmitters, on the basis of two main lines of evidence. First, sympathetic nerve stimulation, infusion of tyramine (a compound provoking the release of endogenous amines from sympathetic nerve terminals) or depolarization of intracardiac nerve terminals with a potassium-enriched perfusate all produced elevated NE and EPI efflux from the heart. Second, exogenously applied NE and EPI were equipotent in augmenting cardiac rate and strength of contraction. These authors also treated hearts with 6-hydroxydopamine to destroy sympathetic nerve endings and found that this reduced the intracardiac concentrations of both EPI and NE to very low levels. But De Santis *et al.* (1975) did not perform the critical experiment of determining the effect of chemical sympathectomy on release of catecholamines during cardiac nerve stimulation. Sturkie and Poorvin (1973), on the other hand, concluded that even though they and other workers had identified stores of both EPI and NE in the heart, only NE appeared to be released during sympathetic nerve stimulation. These authors concluded that EPI was sequestered in nonneuronal stores and would not, therefore, be involved in neurogenic control of the myocardium. Currently the most widely accepted view is that NE is the primary sympathetic neurotransmitter in the avian heart, as in the mammalian heart.

Autonomic tone is usually taken to mean the level of spontaneous and ongoing activity in autonomic nerves to the heart under "basal" or "resting" conditions (that is, when the animal is not actively moving or engaged in major physiological responses to its environment). Since monitoring spontaneous nerve activity is technically difficult in any preparation other than the acutely anesthetized animal instrumented for nerve recording, the level of basal heart rate is usually taken as the major indicator of cardiac autonomic tone as rate is easily measured under a variety of conditions in the whole animal. An added complication in analyzing tonic autonomic drive to the heart is that both autonomic limbs can strongly affect rate. Since rate is driven in opposite directions by parasympathetic and sympathetic inputs, the chronotropic effects of tonic activity in one limb can only be accurately assessed in the absence of influences from the other limb. Autonomic inputs to the heart can be selectively ablated by a variety of means including surgical section of the vagus or cardiac sympathetic nerves, chemical sympathectomy (by pretreatment with agents which destroy adrenergic nerve terminals or by adrenergic blockade), blockade of cardiac vagal effects with atropine, or a combination of these methods.

The level of sympathetic tone to the heart can be quantified by determining the change from basal heart rate produced by any of the above methods for functional sympathectomy after vagal influences on the heart are removed; see Sturkie (1986b) for the development of an equation expressing this concept. Widely varying levels of sympathetic tone have been reported among bird species and even among different studies of the same species. A portion of this variability is likely due to the use of anesthetics. Baseline heart rate itself will change as a consequence of general anesthesia, and anesthetics will also have a blunting effect on autonomic control of the heart (Vatner and Braunwald, 1975; Brill and Jones, 1981; Lumb and Jones, 1984). Therefore the most accurate assessment of cardiac sympathetic tone would be made in awake, spontaneously breathing animals in a quiescent state after vagal influences on the heart have been eliminated.

Johansen and Reite (1964) investigated the level of autonomic tone to the heart in both awake ducks and those under general anesthesia; the authors did not differentiate between these states in reporting their data, claiming that this made no difference to the outcome of the experiments. In vagotomized ducks in this study, β-adrenergic blockade produced a large fall in heart rate, implying the existence of strong resting sympathetic tone. Tummons and Sturkie (1969) reported that, in awake chickens at 6 days after recovery from surgical division of the cardiac sympathetic nerves, heart rate was about 16% less than that in intact animals. In anaesthetized ducks Kobinger and Oda (1969), using pharmacological agents to inhibit sympathetic function at central and peripheral levels of the nervous system, also found evidence for significant sympathetic tone to the heart. On the central side, they showed that clonidine-mediated depression of vasopressor areas in the medulla, including the sympathetic cardiomotor area, led to a reduction in heart rate. On the peripheral side, depletion of NE from peripheral sympathetic terminals with reserpine, or prevention of NE release from these terminals, also significantly reduced cardiac rate. None of these experiments was done with accompanying vagal blockade. In contrast to the results of Kobinger and Oda (1969), Folkow et al. (1967) found no significant change in heart rate after β-adrenergic blockade in the same species. However, a complicating factor in the latter study was that the agent used (an experimental β-blocker then under development) was acknowledged by the authors to have a partial β-agonist effect which may have offset any effects of β-blockade on basal heart rate. Butler and Jones (1968; 1971) reported no significant change in heart rate in unanesthetized ducks after β-blockade with propranolol, an agent free of intrinsic β-agonist effects. In awake chickens, Butler (1967) found that β-blockade with the vagi intact produced a significant fall in heart rate to 75% of the control rate. β-blockade after vagotomy produced less of an effect, reducing heart rate to 82% of the level in animals with intact vagi. These results show that significant sympathetic tone is present in the chicken, reinforcing the contention that the level of this tone can only be accurately assessed in the absence of parasympathetic input to the heart. However, the results of Butler (1967) contrast with those of Tummons and Sturkie (1970), who found that sympathetic nerve section produced a fall of about 16% in heart rate from the level before nerve section, while vagotomy produced a rise of about the same proportion. Combined vagotomy and sympathetic nerve section resulted in a heart rate not significantly different from that in intact animals. The authors therefore concluded that, in resting intact animals, the balance between tonic sympathetic and parasympathetic inputs to the heart maintained rate at the same level as the intrinsic rate in animals after cardiac decentralization. It is clear from the foregoing discussion that further studies of tonic sympathetic drive to the heart must be rigorous in taking both the state of anesthesia and the level of concomitant parasympathetic drive into account.

An electrophysiological analysis of the compound action potential of the right cardiac nerve in the pigeon has shown that axons in this nerve can be categorized into two groups separable by conduction velocity, as shown in Figure 41 (Cabot and Cohen, 1977a). Fibers of the more slowly conducting group (range 0.4–2 m sec^{-1}) were shown to mediate sympathetic cardioacceleration. The range of conduction velocities of these axons lies within that of unmyelinated sympathetic postganglionic fibers known to innervate the viscera (Gabella, 1976), and mor-

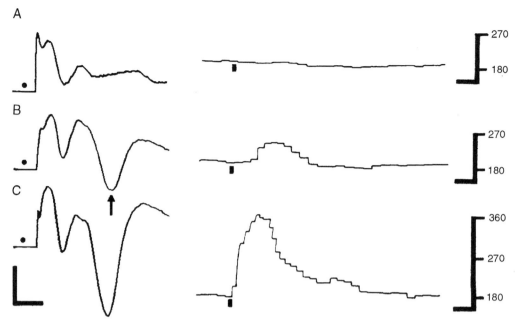

FIGURE 41 Correlation between components of the compound action potentials evoked by graded electrical stimulation of the right cardiac sympathetic nerve and chronotropic responses of the heart in the pigeon. The traces on the left represent the compound action potential at three intensities of stimulation increasing from A to C (round dots associated with each trace indicate the start of the stimulation, which consisted of a 200-ms train of 50 Hz pulses). For these traces the vertical calibration bar represents 100 μV and the horizontal bar represents 5 msec. The traces on the right represent beat-by-beat ratemeter recordings showing heart rate responses produced by the same stimuli (filled squares under each trace) which evoke the nerve responses on the left. The calibration bars to the right of each trace indicate heart rate in beats min^{-1}. For these traces the horizontal bar represents 1 sec. The onset of the component of the compound action potential indicating conduction of cardioaccelerator fibers is indicated by the arrow in trace B; this component strengthens with increased stimulus intensity in trace C, as does the degree of cardioacceleration. (Reprinted from *Brain Res.* **131**, J. B. Cabot and D. H. Cohen, Avian sympathetic cardiac fibers and cells of origin: Anatomical and electrophysiological characteristics, pp. 73–87, Copyright (1977a), with permission from Elsevier Science.)

phological analysis of the pigeon right cardiac nerve confirms that 67% of axons in this nerve are unmyelinated (Macdonald and Cohen, 1970). The faster-conducting group of fibers in this nerve (range 2–5.6 m sec^{-1}) are likely to be myelinated axons. These make up the remaining 33% of the total number of axons and probably represent so-called "sympathetic afferent" fibers (Malliani *et al.*, 1979) with receptor endings in the heart, great vessels or lungs. Afferent fibers in cardiac nerves have been shown to participate in cardiopressor reflexes in the pigeon (Cabot and Cohen, 1977b).

b. Parasympathetic Innervation

i. Anatomy Efferent neurons with their cell bodies in the heart form the final common pathway for parasympathetic control of cardiac function. These postganglionic neurons receive synaptic inputs from terminals of preganglionic neurons with their somata in the brainstem and their axons coursing to the heart in the vagus nerves. It is generally accepted that parasympathetic efferent neurons in the heart are cholinergic, releasing ACh at their myocardial terminals; this neurotransmitter acts to modify myocardial function through postjunctional muscarinic receptors. Indeed, most of the anatomical studies of parasympathetic innervation of the heart have used a histochemical reaction indicating the presence of AChE as a marker to determine the distribution of cholinergic fibers and terminals of the cardiac plexus, as well as the locations of intracardiac neurons. Cabot and Cohen (1980) have extensively reviewed the cholinergic innervation of the heart, so a brief synopsis of that review and the contributions of later workers are combined below.

In all avian species so far examined, all four cardiac chambers receive AChE-positive innervation (Hirsch, 1963; Yousuf, 1965; Smith, 1971a,b; Akester and Akester, 1971; Mathur and Mathur, 1974; Rickenbacher and Müller, 1979; Kirby *et al.*, 1987). Smith (1971a,b), in studies of the innervation pattern in the chicken heart, determined that cholinergic nerves, nerve fibers, and terminals formed a subepicardial ground plexus throughout the atria and ventricles, penetrating into the myocardium. Some fibers were observed to run all the way to the endocardial region where they contributed to a subendocardial plexus which was also distributed throughout the atria and ventricles. In many avian species the overall intracardiac distribution of cholinergic innervation parallels the distribution of adrenergic fibers and terminals (Bennett and Malmfors, 1970). The right atrium has been a particular focus for anatomical studies of cholinergic innervation in view of the location of the primary pacemaker in this chamber and the vagally mediated inhibition of pacemaker node discharge rate. Yousuf (1965) noted in the sparrow heart that the region around the sinoatrial node was the first part of the heart to receive extrinsic cholinergic innervation during embryological development and observed that in later developmental stages this area and the right atrial wall near the atrioventricular node were heavily innervated. However, no nerve fibers were observed to enter the sinoatrial node area proper, and only a few fibers were present among the atrioventricular nodal cells. This pattern has also been observed in the pigeon heart (Mathur and Mather, 1974). However Gossrau (as summarized in Akester, 1971) reported that in pigeon, chaffinch, and canary hearts the pacemaker region had a dense AChE-positive innervation, while in duck and chicken this area was sparsely innervated. Differences among these studies may be partly species dependent, although the contrasting results of Mathur and Mathur (1974) and Gossrau in the pigeon heart may be due to differences in the techniques used. Bogusch (1974) observed a dense cholinergic innervation pattern around subepicardial Purkinje fibers of the fowl right atrium but noted that the density of this innervation decreased as the conducting fibers approached the atrioventricular border. In this study, multiple nerve terminals were only loosely associated with Purkinje fibers, leaving some doubt as to the nature of the neuroeffector–tissue relationship in cholinergic control of conduction in the atrium. Sinoatrial valve remnants have also been reported to be the targets of cholinergic innervation in the bird heart (Akester, 1971; Mathur and Mathur, 1974).

Cholinergic innervation of the left atrium and the ventricles has been less extensively studied than that of the right atrium, but the general pattern of subepicardial and subendocardial plexi with individual nerve fibers and terminals extending into the myocardium, as described by Smith (1971b), appears to hold. Extensive cholinergic innervation has been described in the interatrial and interventricular septa (Akester, 1971), the right atrioventricular valve leaflets (Akester, 1971; Mathur and Mathur, 1974), and the left atrioventricular valve. Here the AChE-positive nerve fibers enter the basal half of the valve from the subendocardial plexus (Smith, 1971a). The cholinergic innervation of the avian chordae tendineae is controversial: Smith (1971b) reported no nerve fibers present in chordae tendineae of the chicken heart, while Mathur and Mathur (1974) found in the pigeon heart that these structures were innervated. The function of such innervation is obscure since the chordae tendineae contain no actively contracting tissue, serving only to anchor the papillary muscles to the aortic valve leaflets. It is most likely that the nerve fibers observed in these structures are efferents *en route* to either the valves or the papillary muscle, but some fibers may subserve an afferent function.

The adventitia of coronary arteries associated with all chambers of the bird heart is innervated by cholinergic nerve fibers (Hirsch, 1963; Akester, 1971; Smith, 1971a) but the origin of this innervation is not known. Cholinergic fibers innervating coronary arteries may originate from postganglionic parasympathetic neurons within the heart or may possibly be extrinsic "sympathetic cholinergic" fibers; in either case this innervation probably functions to increase coronary blood flow by promoting vasodilation.

In development of the chick heart, Rickenbacher and Müller (1979) determined that cell bodies of intracardiac neurons, clustered into ganglia, first developed in the left ventricle, then a large group of ganglia became evident around the coronary sulcus and the ventral surface of the ventricles and, lastly, ganglia developed in association with the dorsal atrial walls.

In the adult avian heart, ganglia are located primarily in the subepicardial plexus, usually in association with plexus nerves and frequently near the branch points of these nerves. The somata of intracardiac neurons have been characterized as multipolar (possessing more than two processes; Smith, 1971b; Yousuf, 1965) but these observations are limited to the hearts of only two species (chicken and sparrow, respectively). No morphological data exists on the variation of somatic dimensions or on the length or specific projection targets of the processes of intracardiac neurons in any avian species.

Histochemical reactions for AChE have been the primary tool used in analyzing the distribution of intracardiac neurons in the bird heart, and there seems little doubt that these techniques allow visualization of most if not all of these neurons. Such anatomical data, along with evidence from the functional studies cited below, supports the contention that the phenotype of avian intracardiac neurons is cholinergic. However, AChE has been detected in some nonneuronal elements associated with the nervous system (see review by Fibiger, 1982) and demonstration of the presence of this enzyme in neuronal somata in the heart, although necessary, may not be a sufficient criterion for designating these cells cholinergic. In the central nervous system the most widely accepted indicator of cholinergic function is the presence of choline acetyltransferase (ChAT), an enzyme in the pathway for ACh synthesis (Fibiger, 1982). Reliable antibodies directed against ChAT are now commercially available and have begun to be applied to the mammalian peripheral autonomic nervous system. A similar application of ChAT immunohistochemical techniques to the avian heart would help verify the assumption that intracardiac neurons in this vertebrate group are in fact of cholinergic phenotype.

In their analysis of the distribution of the intracardiac ganglia during development, Rickenbacher and Müller (1979) found that the right ventricle wall contained about half of the total number of ganglia, the right atrium about one-fifth, the left ventricle about one-sixth, and the left atrium possessed the fewest ganglia. In the adult bird heart no quantitative studies of regional neuron distribution have been done to date, so the absolute number of neurons associated with each chamber is not known. However the general pattern of distribution of intracardiac neurons in the adult heart has been established (see Cabot and Cohen, 1980 for review). Ganglia containing variable numbers of neurons are present on both dorsal and ventral aspects of the left and right atria and ventricles (Yousuf, 1965; Smith, 1971a,b; Rickenbacher and Müller, 1979; Kirby et al., 1987; reviewed by Cabot and Cohen, 1980). Smith (1971b) has reported that a larger proportion of the total number of intracardiac ganglia is found in the ventricles of bird hearts than is the case in mammalian ventricles. Ganglia have been observed near but not within the sinoatrial node region (Yousuf, 1965; Smith, 1971b). Yousuf (1965) reported that some neurons in the sulcus terminalis appeared to send projections in the direction of the nodal tissue. The atrioventricular nodal region was also reported to be devoid of ganglia (Yousuf, 1965; Smith, 1971b; Mathur and Mathur, 1974) but this region and the atrioventricular bundle appeared to be innervated by axons from ganglion neurons in the nearby atrioventricular sulcus (Yousuf, 1975). Smith (1971b) described high concentrations of ganglia within the dorsal right atrial wall near the ostia of the superior and inferior venae cavae, near the roots of the pulmonary veins on the dorsal aspect of the left atrium, within the dorsal portion of the atrioventricular groove, and clustered around the roots of the pulmonary artery and aorta. In the ventricles neurons are located in a scattered pattern reaching from the atrioventricular groove to the apex on the ventral surface (Smith, 1971b; Rickenbacher and Müller, 1979). There are also numerous ganglia associated with nerves accompanying atrial and ventricular coronary arteries (Mathur and Mathur, 1974; Smith, 1971b).

Recent studies in the mammalian heart have shown that a number of neuropeptides are colocalized in axons, terminals, and somata of cholinergic intracardiac neurons, as well as in preganglionic terminals contacting these neurons (Steele et al., 1994; 1996). Peptides constitute an important class of neuromodulators in the peripheral autonomic nervous system, and their presence in specific combinations in some peripheral neurons has been proposed to chemically code subpopulations of these neurons for specific functions such as vasomotion or control of muscle cell contractility. In the bird heart, substance P and vasoactive intestinal peptide have been found in intracardiac neurons and their terminals, while somatostatin is present in intracardiac terminals but not in cell bodies (Corvetti et al., 1988). These neuropeptides have been shown to exert powerful modulatory effects on mammalian intracardiac neuronal activity and cardiodynamics (Armour et al., 1993) and their presence in the bird heart indicates that they may play a prominent role in modulation of ganglionic and neuroeffector transmission in this vertebrate group. This constitutes a promising but as yet unexplored area for the comparative study of mechanisms of neural control of the heart.

The course of the avian vagus nerve and its cardiac branches has been described for a number of species (see Pick, 1970; Jones and Johansen, 1972; Cabot and Cohen, 1980 for reviews). The latter review points out that descriptions of the course and major branches of this nerve are consistent among species so a general summary of the avian vagal cardiac innervation will be given here, based on the comprehensive reports of Malinovsky (1962) and Cohen et al. (1970) in the pigeon and the reviews cited above. Inside the cranium the peripheral trunks of the vagus and glossopharyngeal nerves originate bilaterally from large ganglia composed of a fusion of the proximal ganglion of the glossopharyngeal nerve and the jugular ganglion of the vagus nerve. The trunks of these nerves emerge together from the skull through the jugular foramen, and immediately outside the foramen an anastomosis (of Staderini) connects the vagal trunk to the petrosal ganglion of the glossopharyngeal nerve. The vagal trunk continues caudad in the neck along the dorsomedial aspect of the internal jugular vein, passing over the cervical spinal nerves on their ventral sides. No major vagal branches arise from the trunk along its cervical portion, although Malinovsky (1962) described occasional small anastomoses with the nearby cervical sympathetic trunk. At the level of the thoracic inlet the nodose (alternatively termed distal vagal) ganglia are present as spindle-shaped enlargements of the vagal trunks, as shown in Figure 42. Several afferent nerves carrying the axons of receptors important in the control of cardiovascular and respiratory functions arise from each nodose ganglion. Along the length of this ganglion, branches arise which extend medially to innervate the thyroid, parathyroid and ulti-

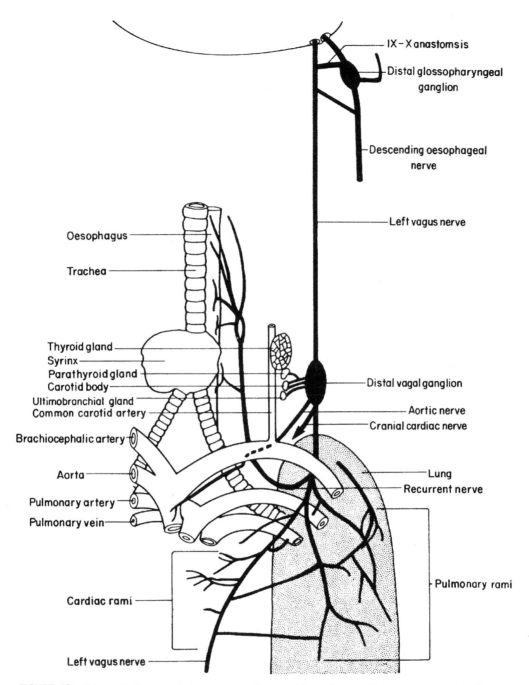

FIGURE 42 Schematic diagram depicting a ventral view of the pathway of the left vagus nerve in the area of the upper thorax of the duck. Details of the vagal innervation of the carotid body, ultimobranchial, thyroid, and parathyroid glands, and the aorta are illustrated. (After Jones and Johansen (1972), with permission.)

mobranchial glands, and the carotid body (Figure 42). The latter structure constitutes the primary locus for peripheral chemoreceptors sensing arterial oxygen and carbon dioxide tensions and pH in the bird (Jones and Purves, 1970; see Section VI,D,1). Branches exiting from the caudal portion of the nodose ganglion course to the root of the aorta. On the right side, Nonidez (1935) reported two such branches in the chick, which he designated "depressor" and "accessory depressor" nerves by analogy with the mammalian condition. Nonidez (1935) reported no equivalent nerve on the left side, but in other species nerves from the nodose ganglion coursing to the aortic root (designated aortic nerves) have been reported to be present bilaterally (Jones and Purves, 1970; Jones, 1973; Cohen et al., 1970; Jones et al., 1983; Smith and Jones, 1990; 1992; summarized by Smith,

1994; see Figure 42). These nerves ramify into a plexus in the aortic wall. Jones (1973) first demonstrated that the aortic nerves carried arterial blood pressure information centripetally and that aortic baroreceptors were involved in regulating and maintaining arterial blood pressure in the duck.

A few millimeters caudal to the nodose ganglion the vagal trunk splits, as shown in Figure 43, to form several major divisions as it approaches the pulmonary artery. Two of these circumscribe the pulmonary artery, rejoin, and continue as the main vagal trunk *en route* to the abdominal cavity while a third forms the recurrent laryngeal nerve, coursing craniad along the trachea; no cardiac vagal branches arise from this nerve. Another major vagal branch courses to the heart to enter the dorsal cardiac plexus (Jones and Johansen, 1972; shown in Figure 42). The remaining vagal branches form part of the pulmonary innervation, running to the lungs along the pulmonary arteries. The main vagal trunk, after reforming caudal to the pulmonary artery, passes ventral to the ipsilateral bronchus and over the dorsal surface of the heart. From this portion of the trunk a variable number of smaller branches arise and enter the cardiac plexus. On the right side, these branches enter the heart near the sinoatrial and atrioventricular nodes and at the caval ostia; on both the left and right sides of the heart vagal branches also enter the cardiac plexus in the vicinity of the atrioventricular groove. That the cardiac nerves described here constitute the major efferent vagal innervation of the heart has been confirmed by functional studies in which electrical stimulation of the vagal trunk was combined with surgical section of the trunk and the various cardiac branches (Cohen *et al.*, 1970). Caudal to the origin of the most inferior cardiac branches, the left and right vagal trunks pass ventral to the pulmonary veins where both turn medially, coming to lie in close approximation as they course to the abdominal viscera (Figure 43).

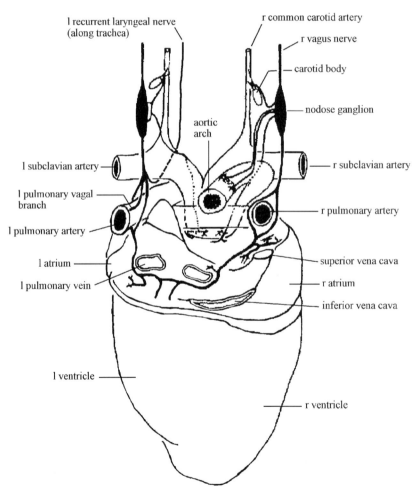

FIGURE 43 Schematic diagram to illustrate a dorsal view of the generalized vagal innervation of the avian heart. Note the pathways of the vagal branches as these nerves split close to the pulmonary arteries. For clarity the right pulmonary and right recurrent laryngeal branches are not shown and the right pulmonary vein, depicted next to the left, is not labeled (r, right; l, left). (Modified from Cabot and Cohen (1980), with permission.)

The location in the central nervous system of cells of origin of vagal cardioinhibitory fibers in birds has been investigated by a variety of anatomical and physiological techniques. The extent of the medullary regions containing vagal preganglionic motor neurons innervating pharyngeal structures and thoracic and abdominal viscera was originally defined anatomically by determining the extent of retrograde degeneration of neuronal somata after section of the cervical vagus nerve (Cohen et al., 1970) and more recently by retrograde labeling of vagal neurons with neurotracers applied to the peripheral vagus nerve (Katz and Karten, 1979; 1983a,b; 1985; Cabot et al., 1991a). Cohen et al. (1970), working in the pigeon, described three major subdivisions of the dorsal motor nucleus of the vagus nerve (DMV) based on cytoarchitectonic and morphological criteria. The principal medullary location of neurons showing signs of degeneration after section of the cardiac vagal branches was in a region extending from the obex to the rostral pole of the DMV, with the highest density of degenerating cells in an area between 0.6 and 0.8 mm rostral to the obex. At this rostrocaudal level, these cells were primarily located in the most ventral region of the DMV. The results of these experiments suggested that the central arrangement of vagal cardioinhibitory neurons in birds was different than the mammalian condition; in mammals the primary locus for these neurons is the nucleus ambiguus, ventrolateral to the DMV (reviewed by Hopkins, 1987).

In the pigeon focal electrical stimulation of the regions of the DMV shown anatomically to contain cell bodies of putative cardioinhibitory neurons produced short-latency decreases in heart rate; this response is shown in Figure 44 (Cohen and Schnall, 1970). The response was rapid, occurring within one or two cardiac cycles after the start of stimulation, suggesting that cardioinhibitory preganglionic cell bodies were being directly stimulated. If stimulation was continued, complete atrioventricular blockade could be produced in some animals; a depressor response invariably occurred secondary to all negative chronotropic responses (Figure 44). There was no lateral asymmetry in this response: stimulating in the DMV on either side produced similar cardioinhibitory responses. The cardiac effects produced by central stimulation could be mimicked by stimulating the vagal trunks in the neck or at the thoracic inlet (Cohen and Schnall, 1970). Field potential and single-unit recordings made in the pigeon DMV during stimulation of the vagal trunk provided further confirmation that the cell bodies of a large number of cardioinhibitory neurons were located in the central zone of the DMV rostral to the obex (Schwaber and Cohen, 1978b).

Schwaber and Cohen (1978a), in an electrophysiological study of the vagus nerves, found that when the cervical vagus was stimulated at progressively greater intensities the onset of bradycardia coincided with the elicitation of a specific component of the compound action potential (B1 wave, Figure 45). This component was generated by the activation of a group of vagal axons conducting in the velocity range of 8 to 14 m sec^{-1}. When the vagus nerve was stimulated with sufficient intensity to evoke both the A and B1 components (trace B of Figure 45) and a polarizing voltage was applied to the nerve to block the A but not the B1 component, the cardioinhibitory response was maintained. This led the authors to conclude that fibers in the vagus nerve responsible for generating the B1 component of the compound action potential were responsible for cardioinhibition. This was confirmed in experiments in which field potential and single-unit activity were mapped in the DMV in correlation with synchronous compound action potentials recorded from the vagus nerve during stimulation of this nerve (Schwaber and Cohen, 1978b). In these experiments, stimulus-evoked activation of the B1 component in the vagus nerve produced the shortest latency, highest amplitude responses in the region of the DMV, which had been shown in previous studies to contain the somata of putative cardioinhibitory neurons. In addition to the evidence from stimulus-evoked potentials, recordings of spontaneously active single units in the DMV rostral to the obex in awake, paralyzed pigeons demonstrated rhythmic discharge patterns phase-locked to mechanical events in the cardiac cycle (Gold and Cohen, 1984). These authors also showed that single-unit neuronal activity in this area was decreased or eliminated by external conditioning stimuli (light flash, foot shock) which caused heart rate to increase (see Section VI,E).

The above anatomical and physiological evidence, taken together, indicates that in the avian brain preganglionic vagal cardioinhibitory neurons are located in the ventrolateral region of the DMV rostral to the obex. However a recent reexamination of the question of the location of these neurons was undertaken by Cabot et al. (1991a), using a new and more sensitive method for retrograde neuroanatomical tracing. These authors injected small volumes of the binding fragment of tetanus toxin into selected regions of the pigeon heart. This neurotracer was taken up by local nerve fibers and terminals at the injection site and transported retrogradely to label the somata of vagal preganglionic cardiac neurons in the medulla by two possible routes, both giving similar end results. The first of these was via direct uptake of neurotracer by fibers or terminals of the preganglionic neurons running through or close to the injection sites in the heart; in this case the tracer would be transported directly back to the cell bodies. The second route was via transsynaptic transport. Fibers and terminals of postganglionic intracardiac neurons took up the neurotracer from the injection sites, and upon traveling to the somata and other processes of these neurons, the

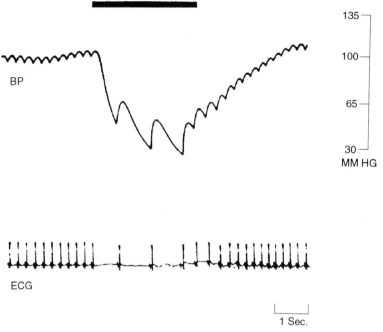

FIGURE 44 Arterial blood pressure (BP) and cardiac chronotropic (ECG) responses to focal electrical stimulation in the area of the dorsal motor nucleus of the vagus nerve in the pigeon. The duration of the stimulus train (50-Hz pulses) is shown by the solid horizontal bar above the traces. (After Medullary cells of origin of vagal cardioinhibitory fibers in the pigeon. II. Electrical stimulation of the dorsal motor nucleus, D. H. Cohen and A. M. Schnall, *J. Comp. Neurol.*, Copyright © 1970 Wiley-Liss, Inc. Reprinted by permission of Wiley-Liss, Inc., a subsidiary of John Wiley & Sons, Inc.)

tracer would then cross synaptic clefts to the preganglionic terminals contacting the postganglionic cells. From these terminals the tracer was transported to the somata of the preganglionic neurons. After intracardiac injection of the tracer, labeling of preganglionic vagal neurons was found in two locations in the medulla (Figure 46). The majority of label was found in neurons located ventrolateral to the DMV, in a site homologous to the nucleus ambiguus of the mammalian brainstem; these neurons were clustered within 0.5 mm of the obex as shown in the bottom panel of Figure 46. A smaller number of labeled neurons were found in more rostral sections in an area bordering the ventrolateral margin of the DMV (top panel of Figure 46), in close proximity to the region delineated in the degeneration studies of Cohen *et al.* (1970) and just ventral to the area described in the functional studies of Schwaber and Cohen (1978a,b). The results of Cabot *et al.* (1991a) have forced a reevaluation of the central organization of neurons controlling the avian heart, indicating that this organization has much more in common with the mammalian condition than was previously believed. However, these new anatomical data have not been confirmed by physiological studies. In addition, it still remains for the membrane and firing properties of vagal cardioinhibitory preganglionic neurons to be investigated to determine if there are functionally discrete subpopulations within this group of neurons and, if so, whether there is any correlation between the functional properties of neurons and their potential roles in controlling specific aspects of cardiodynamic function.

Little is known of the nature and origins of inputs to medullary vagal cardiomotor neurons in birds. Berk and Smith (1994) have shown in the pigeon that peptide-containing projections to these neurons arise from the area of the nucleus of the tractus solitarius (NTS). The NTS is one target for afferent information from visceral receptors carried in the vagus, glossopharyngeal, and other cranial nerves, and such peptidergic projections from the NTS to cardiomotor neurons may represent an important viscerovisceral reflex pathway, as has been described in mammals (Loewy and Spyer, 1990). In addition to inputs of peripheral origin, there is an extensive pattern of projections to the DMV from structures located more rostrally in the brain. Berk and Finkelstein (1983) and Berk (1987) demonstrated projections from the bed nucleus of the stria terminalis, the ventral paleostriatum, and the medial and lateral hypothalamus to the DMV. These forebrain inputs thus represent potential pathways through which central nervous control of cardiac function may be exerted in the interests of homeostasis, as well as providing pathways for neurally

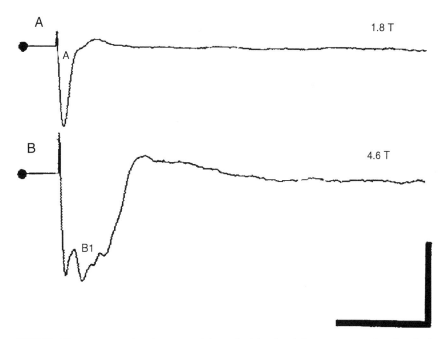

FIGURE 45 Compound action potentials evoked in the right vagus nerve by electrical stimulation at the midcervical level in the pigeon. Responses to stimuli of two intensities are shown. (A) stimulation at 1.8 times the intensity which just evokes a response (threshold intensity, T) produces a short-latency component labeled the A wave, representing the fastest conducting fibers in the vagus (start of stimulus is indicated by the dot at the left side of trace). No change in heart rate is associated with the activation of this group of fibers. (B) stimulation at 4.6 × T evokes responses in an additional, more slowly conducting group of fibers; this component of the compound action potential is labeled the B1 wave. The fibers responsible for this component conduct in the range of 8–14 m sec^{-1} and when activated produce bradycardia. The vertical bar represents 250 μV; the horizontal bar represents 5 msec. (Reprinted from *Brain Res.* **147,** J. S. Schwaber and D. H. Cohen, Electrophysiological and electron microscopic analysis of the vagus nerve of the pigeon, with particular reference to the cardiac innervation, pp. 65–78, Copyright (1978a), with permission from Elsevier Science.)

mediated alterations in cardiac function which may be required during exercise, feeding or other behaviors, or in response to changes in the external environment. In addition, Cohen and coworkers have explored the central anatomical pathways mediating conditioned responses which target medullary cardiomotor neurons in birds (see Section VI,E). These cardiomotor neurons therefore integrate information from visceral and other receptors and from higher levels of the central nervous system to control cardiodynamics, but our knowledge of the integrative mechanisms involved is scant at present.

ii. Parasympathetic Control Acetylcholine acts in the bird heart to depress atrial and ventricular myocyte contractility, rate of discharge of pacemaker tissue, and rate of conduction through the specialized conductive tissues. ACh, released from preganglionic terminals, activates excitatory nicotinic receptors on the membranes of postganglionic neurons in the heart and these neurons in turn release ACh from their effector terminals to inhibit cardiac functions. Intrinsic postganglionic parasympathetic neurons release the majority of ACh which over-

flows from the isolated heart during vagal nerve stimulation. Only a small fraction of the total amount of ACh recovered in these experiments is released from vagal preganglionic terminals, as shown by a large reduction in vagally evoked ACh release after treatment of the isolated heart preparation to prevent release of the neurotransmitter from postganglionic terminals (Loffelholz *et al.,* 1984).

ACh has different effects on the cell membrane conductances and thus on contractile properties of myocytes in the avian atria and ventricles. Inoue *et al.* (1983), using intracellular electrode techniques *in vitro,* investigated the effects of ACh on membrane ion conductances of atrial and ventricular myocytes to determine how these might differ. They found that, in ventricular muscle cells, ACh reduced the force of contraction, diminishing both amplitude and time course of the action potential, but did not change either resting membrane potential or whole-cell resistance. On the other hand, in atrial myocytes, ACh hyperpolarized the membrane and reduced whole-cell resistance (implying an increase in steady-state ionic conductances) as well as causing a reduction in amplitude and time course of the action

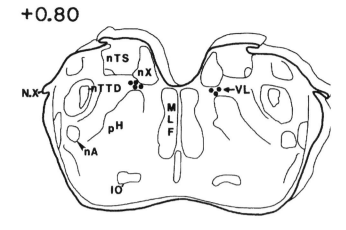

FIGURE 46 Outline drawings of transverse sections of the pigeon medulla in the region of the obex, depicting the location of vagal preganglionic neurons labeled by retrograde transport of tetanus toxin binding fragment C injected into the heart (see text for explanation). The top panel shows a section cut 0.8 mm rostral to the obex, with labeled neurons (filled circles) located just ventral to the dorsal motor nucleus of the vagus (nX) in the nucleus ventrolateralis (VL). The bottom panel, representing a section taken at the level of the obex, shows additional labeled neurons in an area of the medulla which may be the avian homolog of the mammalian nucleus ambiguus (nA), ventrolateral to nX. Abbreviations: IO, inferior olivary nucleus; MLF, medial longitudinal fasciculus; nTS, nucleus and tractus solitarius; nTTD, nucleus and tractus trigemini descendens; NX, vagal nerve rootlet; pH, plexus of Horsley. (Reprinted from *Brain Res.* **544**, J. B. Cabot, J. Carroll, and N. Bogan, Localization of cardiac parasympathetic preganglionic neurons in the medulla oblongata of pigeon, *Columba livia:* A study using fragment C of tetanus toxin, pp. 162–168, Copyright (1991), with permission from Elsevier Science.)

There are several subtypes of muscarinic receptor present in the mammalian heart (see Deighton *et al.*, 1990 for review) and some of these receptor subtypes are also present in the avian heart. The majority of muscarinic receptors on mammalian myocardial cells are of the M_2 subtype (Deighton *et al.*, 1990; Jeck *et al.*, 1988), and this subtype is believed to mediate complex intracellular mechanisms leading to the inhibition of myocyte functions which are ultimately responsible for the parasympathetic control of the heart. In a comparison of muscarinic receptor types in the chicken and guinea pig hearts, Jeck *et al.* (1988) found that receptors of the M_1 subtype predominated in the myocardium of the chicken while the most prevalent type in the guinea pig heart was the M_2 subtype, as found in other studies of the mammalian heart. Detailed analyses of the muscarinic receptor subtypes present in the hearts of other avian species have not been conducted, but if the results of Jeck *et al.* (1988) in the chicken represent the general avian situation, there are likely to be major differences in receptor-mediated intracellular mechanisms of mucarinic inhibition of myocyte function between birds and mammals.

In avian atrial tissue *in vitro,* and in atria in whole *in situ* or isolated hearts, it is widely accepted that ACh has a strong negative inotropic effect (e.g., Jeck *et al.*, 1988 and review by Sturkie, 1986b), but studies of specific cholinergic effects on ventricular inotropy have been few in the bird. Avian ventricular tissue has a higher density of cholinergic innervation than does that of mammals as outlined above, and a higher proportion of the total number of intracardiac neurons is associated with the ventricles of the avian than the mammalian heart (see above, and Smith, 1971b). On the basis of early anatomical evidence suggesting a high density of cholinergic innervation of the ventricular myocytes in birds, Bolton and Raper (1966) compared responses of strips of the right ventricle of fowl and guinea pig hearts *in vitro* to endogenously released and exogenously applied ACh. Field stimulation of electrically paced ventricular strips produced a strong decrease in force of contraction of fowl ventricular tissue, while guinea pig ventricular tissue responded with an increase in force. Atropine, blocking muscarinic receptors, eliminated the negative inotropic response of the fowl ventricle to stimulation, and a strongly positive response was then observed; however, atropine had no effect on the response of the guinea pig ventricle to stimulation. The authors proposed that the fowl heart has a capacity for effective parasympathetic inhibition of ventricular inotropy, mediated by intracardiac release of ACh. Furthermore, blockade of this response unmasked a stimulus-evoked increase in force of contraction which the authors determined was the result of release of NE from sympathetic nerve terminals. They concluded that, in contrast with the mammalian condition, the fowl right ventricle was

potential. ACh binds to muscarinic receptors on myocyte membranes, and the authors found that this induced similar decreases in calcium-dependent sodium currents in the two types of cells. However, atrial myocytes exhibited in addition a muscarinically mediated increase in an outward potassium current which accounted for the hyperpolarization and reduction in resistance induced by ACh; this mechanism was not present in ventricular myocytes. These differences in response to ACh imply that the same neurotransmitter can differentially control atrial and ventricular contractility.

innervated by both the sympathetic and parasympathetic limbs of the autonomic nervous system. Subsequent *in vitro* work has confirmed these observations (Bolton, 1967; Biegon *et al.*, 1980; Biegon and Pappano, 1980).

The first experimental approach used to explore general questions of parasympathetic efferent control of the avian heart, dating from the early part of this century, was to examine the effects of exogenously applied ACh on the activity of pacemaker cells in the right atrium (reviewed by Jones and Johansen, 1972; and Cabot and Cohen, 1980). More detailed characterization of this system has resulted from recent investigations carried out by Loffelholz and co-workers, and others. In the *in vitro* right atrium, ACh release provoked by field stimulation and by stimulation of the attached right vagus nerve produced a fall in pacemaker discharge rate (Pappano and Loffelholz, 1974; Pappano, 1976; Brehm *et al.*, 1992). This chronotropic effect was blocked by atropine but not by the ganglionic blocker hexamethonium, so it must have been mediated by direct release of ACh from postganglionic parasympathetic neurons in the atrial wall (Pappano and Loffelholz, 1974). On the other hand, hexamethonium in the absence of atropine blocked all effects of electrical stimulation of the attached vagal stump. Vagally mediated bradycardia therefore occurred as a result of the synaptic activation of intracardiac postganglionic neurons by preganglionic terminals. As with cholinergic control of the inotropic state of the atrial and ventricular myocardium discussed above, control of pacemaker rate operates via the synaptic relay of impulses from pre- to postganglionic neurons within intracardiac ganglia.

Dromotropic Effects There have been no studies of the direct effects of ACh on the rate of impulse conduction through the specialized conducting cells of the avian heart and few studies of the overall parasympathetic control of this function. There are technical difficulties in identifying the locations of conducting tissues in a viable *in vitro* preparation of cardiac tissue, so work on this intriguing problem in birds has largely been conducted on hearts *in situ*. In the chicken, Goldberg *et al.* (1983) showed that atrioventricular conduction time could be significantly prolonged by stimulation of either vagus nerve; there was no bilateral asymmetry in this response. However, in order to unmask this dromotropic effect the heart was paced through electrodes attached to the sinoatrial node, both to control heart rate and to prevent shifts in pacemaker position during vagal nerve stimulation. Bogusch (1974), using anatomical techniques, identified cholinergic fibers and terminals in the region of conducting cells near the atrioventricular border and proposed a functional role for this innervation but the study of Goldberg *et al.* (1983) appears to be the only physiological confirmation of this role in the bird (also see Section II,C,3).

Chronotropic Effects The study of vagal control of heart rate in birds has a long history going back to the recognition in the last century that activation of the vagus nerves could produce large reductions in heart rate and, in some cases, arrest the heart (see Cabot and Cohen, 1980 for a summary of early work). As these reviewers have pointed out, the nature of vagally mediated bradycardia has been intensively reinvestigated using recently developed techniques in a variety of avian species (Johansen and Reite, 1964; Bopelet, 1974; Peterson and Nightingale, 1976; Langille, 1983; Lindmar *et al.*, 1983; Goldberg *et al.*, 1983; Lang and Levy, 1989; Butler and Jones, 1968; Cohen and Schnall, 1970; Jones and Purves, 1970; and others). The negative chronotropic effects of ACh in the pharmacological studies cited above are paralleled by the effects of electrical stimulation of the vagus nerves. Several preparations have been used to assess the chronotropic consequences of vagal stimulation. These include atrial tissue *in vitro*, isolated beating hearts with attached vagal stumps, open-thorax anesthetized preparations, and anesthetized or awake animals in which only the vagi in the cervical region were exposed. Vagal control of the heart is apparently very robust in all of these preparations and each type of preparation yields results which complement the findings obtained from the others.

Examples of stimulus-induced bradycardia obtained in pigeon are shown in Figure 47. The peripheral cut ends of the right (top panel) and left (bottom panel) cervical vagus nerves were stimulated with trains of pulses at a frequency of 50 Hz. In each case the contralateral vagus nerve was intact. In these examples the intensity of stimulation used was capable of arresting the heart. A close parallel to the negative chronotropic effect of peripheral vagal nerve stimulation can be produced by focal electrical stimulation in the medulla (Cohen and Schnall, 1970) where anatomical studies (see above) have shown that cardiac vagal preganglionic neurons are located.

The effectiveness of the left and right vagi in controlling heart rate has been shown by some workers to be equivalent, while other workers have found strong bilateral asymmetry in this system. Bopelet (1974), Peterson and Nightingale (1976), and Goldberg *et al.* (1983) reported no difference in the chronotropic response of the fowl heart to electrical stimulation of left and right vagus nerves. In the same species, however, Sturkie (1986a) and Lang and Levy (1989) reported that the right vagus was more effective in altering heart rate than was the left. Johansen and Reite (1964), in ducks and seagulls, Jones and Purves (1970) in ducks, and Cohen and Schnall (1970) in pigeons, all reported a similar asymmetrical response to electrical stimulation of the vagus nerves. Furthermore, in a systematic study of vagal control of heart rate in the duck, Butler and Jones (1968) showed by means of cold

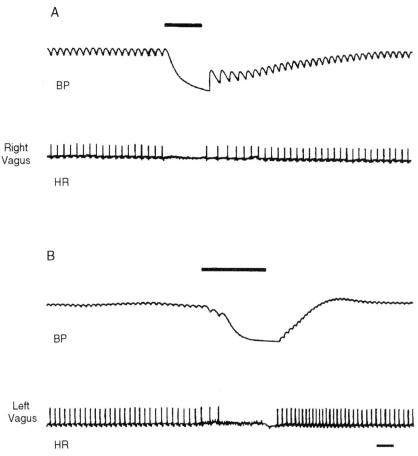

FIGURE 47 Effects of stimulating the right (A) and left (B) vagus nerves on heart rate (HR) and arterial blood pressure (BP) in two pigeons. The nerves were exposed in the neck and stimulated with 50-Hz trains of pulses (stimulus duration indicated by the horizontal bars above the BP traces). Horizontal bar below the bottom trace represents 1 sec. (After Medullary cells of origin of vagal cardioinhibitory fibers in the pigeon. II. Electrical stimulation of the dorsal motor nucleus, D. H. Cohen and A. M. Schnall, *J. Comp. Neurol.*, Copyright © 1970 Wiley-Liss, Inc. Reprinted by permission of Wiley-Liss, Inc., a subsidiary of John Wiley & Sons, Inc.)

blockade of the vagi in the neck and by unilateral and bilateral section of these nerves that vagal dominance could could change sides over time. However, the latter authors did not stimulate the peripheral cut ends of the vagi after nerve section so it is not known whether the origin of this bilateral asymmetry was within the central nervous system or at the heart.

Lindmar *et al.* (1983) have quantified the amounts of ACh released in the isolated chicken heart by stimulation of the attached left and right vagal stumps in an attempt to determine if there were bilateral differences in the intracardiac connections of these nerves. If fibers from each vagus innervated approximately the same number of postganglionic intracardiac neurons, the amount of ACh released by stimulation of either nerve alone would be expected to be similar, and this was in fact found to be the case. If there were no overlap in the populations of intracardiac neurons innervated by each vagus, then the sum of ACh released by unilateral stimulation of each nerve should be the same as that released by simultaneous bilateral nerve stimulation. However, Lindmar *et al.* (1983) found that the sum of ACh released by separate nerve stimulation was significantly greater than that released by bilateral stimulation, indicating that a large proportion of intracardiac neurons must have been bilaterally innervated. In these experiments it was not possible to separate the responses of postganglionic neurons innervating the pacemaker tissue from the responses of neurons subserving other functions, but if it is assumed that the overall pattern of bilateral innervation found by Lindmar *et al.* (1983) can be applied to the specific subpopulation of neurons controlling heart rate, then this pool of neurons may be controlled by preganglionic fibers running in either nerve. There would thus be little reason to expect that side-to-side shifts in vagal dominance result from factors acting to change the relative intracardiac influences of these nerves. This line of reasoning supports the notion that changes in the pattern of activity of vagal preganglionic neurons in the medulla may be responsible for spontaneous shifts in the dominant vagus nerve. Such central nervous factors might be investigated using the approach of Gold and Cohen (1984) to simultaneously record spontaneous activity from neurons in medullary vagal complexes on both sides.

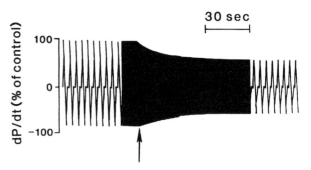

FIGURE 48 Effect of stimulation of the right vagus nerve on the rate of change of left ventricular pressure (dP/dt) in the chicken heart. Vagally induced changes in this variable represent an index of changes in ventricular contractility. The trace depicts the electrically analyzed first derivative of left ventricular chamber pressure; the amplitude of each peak thus represents the maximum rate of change of the corresponding ventricular pressure pulse. The trace was obtained at two speeds; the segment on the left side was recorded at a high chart speed to show the rate of pressure change of individual pulses prior to vagal stimulation. Just before stimulation was started (at the arrow) the chart speed was reduced to display the response to the first part of an 80-sec stimulus train (frequency, 20 Hz); the 30-sec time bar over the trace applies to this segment. On the right the recorder was returned to a high speed to display the rate of change of individual pressure pulses once the response to vagal stimulation had reached a plateau. (After Lang and Levy (1989), *Am. J. Physiol.* with permission.)

Inotropic Effects Despite the evidence cited above for strong negative inotropic effects of endogenously released or exogenously applied ACh in the avian heart *in vitro*, the inotropic effects of vagal stimulation on *in situ* hearts is controversial. Folkow and Yonce (1967) unmasked a strong reduction in one index of left ventricular contractility, that of peak left ventricular chamber pressure, in the duck heart in response to vagal nerve stimulation when heart rate was kept constant by electrical pacing. These authors also found that cardiac output declined with vagal stimulation during pacing. Since vagal stimulation was shown not to affect peripheral resistance in this study, the fall in cardiac output must have been caused by a decrease in stroke volume. While contractile force was not measured directly, the authors attributed both the fall in left ventricular pressure and in cardiac output to a vagally mediated decrease in ventricular contractility. That this effect was neurally mediated was demonstrated by the elimination of the effects of vagal stimulation after atropine was administered intravenously. Furnival *et al.* (1973), using another index of ventricular contractility, that of maximum rate of change of left ventricular pressure, reported results contrary to those of Folkow and Yonce (1967). In their comparative study of the responses of ventricular contractility to vagal stimulation in the dog, duck, and toad, Furnival *et al.* (1973) reported that only the amphibian heart displayed a significant reduction in contractility. These authors proposed that the ventricles of the ducks in the experiments of Folkow and Yonce (1967) had been subject to very high end-diastolic pressures as a consequence of the experimental protocol, were probably in failure as a result of this treatment, and thus responded abnormally to vagal stimulation. Yet Lang and Levy (1989), using the same index of ventricular contractility in the chicken as that employed by Furnival *et al.* (1973) in the duck concluded that vagal stimulation could produce decreases of more than 50% in contractile force (Figure 48). Furthermore, the inotropic responses to vagal stimulation in the chicken heart reported by Lang and Levy (1989) were considerably more powerful than those that could be obtained in mammalian hearts.

The controversy surrounding this issue likely hinges on problems inherent in the methods used to evaluate ventricular contractility. Resolution of this issue will only be possible when more direct indices of contractility, such as direct attachment of Walton-Brodie type force gauges to the ventricular walls, estimation of beat-by-beat ejection fraction, or the measurement of cardiac output in combination with ventricular and systemic pressures for calculating stroke work, are employed.

Tonic Parasympathetic Activity Studies of tonic parasympathetic restraint of heart rate in several bird species have shown that this, like sympathetic tone, varies over a wide range. The factors discussed in Section VI,C,2,a, which affect the evaluation of sympathetic tone (state of anesthesia, presence or absence of tone from the other autonomic limb), apply equally to the analysis of vagal tone, and the protocols used in studies of the extent of vagal restraint will therefore influence the way in which the results of these studies are interpreted. In a study by Johansen and Reite (1964) in awake or anesthetized ducks with intact sympathetic cardiac innervation, section of one vagus nerve (right or left) produced no change in heart rate; the chronotropic response to subsequent section of the remaining vagus was an increase in heart rate up to 65% above that in intact or unilaterally vagotomized animals. These results imply the presence of strong tonic vagal restraint of the heart; however, the authors found that after bilateral vagotomy, β-blockade revealed a high degree of sympathetic tone to the heart. Since the authors did not perform the converse experiment of β-blockade prior to bilateral vagotomy in this study, neither the balance between parasympathetic and sympathetic tone nor the actual degree of parasympathetic restraint on the heart could be ascertained. In unanesthetized ducks Butler and Jones (1968). using a combination of cold block and section of the cervical vagi, showed that mean resting heart rate rose about 180% over the rate prior to these manipulations. In another study by the same authors, pharmacological blockade of the parasympathetic nervous system with atropine caused mean heart rate to increase to about 150% over the control value in awake ducks (Butler and Jones, 1971). In both of these studies the sympathetic nervous system was functional during evaluations of parasympathetic tone. In chickens Butler (1967) found that heart rate increased to 138% over the control value after bilateral vagal nerve section; in contrast to this, Bopelet (1974) reported an increase of only 8%, and Peterson and Nightingale (1974) found no change in rate in chickens after bilateral vagotomy. Sympathetic influence on the heart had not been eliminated in any of these studies. Sturkie and co-workers (reviewed by Sturkie, 1986a) addressed the problem of evaluating parasympathetic tone to the heart by examining the effects of pharmacological or surgical vagotomy in the absence of sympathetic cardiac influences in the chicken using the rationale discussed in section VI,C,2,a. They found that the net restraining effect of tonic parasympathetic activity on the chicken heart was the equivalent of a 20% reduction in heart rate from the rate of the completely decentralized heart. This estimate of vagal tone is substantially lower than the estimates of other workers in chickens (Butler 1967) or in ducks (Johansen and Reite, 1964; Butler and Jones, 1968, 1971) in which the sympathetic cardiac innervation was functional. The degree of parasympathetic restraint on the heart in the latter studies may have been exaggerated by the presence of ongoing sympathetic drive after the lifting of vagal influence.

c. Control of Cardiac Output

The volume of blood pumped by either the left or right ventricle per unit time is termed the cardiac output. The total volume of blood pumped by the heart per unit time is therefore twice the cardiac output, since the outputs of both sides of the heart must be exactly matched over time. Cardiac output, usually expressed in units of milliliters per minute, is the product of the rate of contraction of the heart (beats per minute) and the volume pumped during each beat, or stroke volume (milliliters). Rate and stroke volume are determined by factors which may be intrinsic or extrinsic to the heart. Intrinsic factors include atrial pacemaker cell activity and contractile properties of the cardiac muscle fibers. Extrinsic factors affecting rate and stroke volume include autonomic nervous activity and levels of circulating cardiotropic hormones. Cardiac output over a given time period will thus be determined by the complex interplay of these intrinsic and extrinsic factors. Some of these have been discussed in previous sections, and references will be made to them as necessary.

i. Role of Heart Rate in Control of Cardiac Output

The basal level of heart rate in the absence of external influences is primarily determined by the inherent membrane properties of the pacemaker cells of the sinoatrial node (Section II,C). The rate of depolarization, and thus the rate of discharge of action potentials in these cells, is set by the rate of ion conductances through their membranes (particularly K^+), but this process is itself partly dependent on physical factors such as the concentrations of extracellular ions and temperature. Under most circumstances these physical factors are kept within fairly narrow limits by homeostatic mechanisms and should therefore not affect heart rate substantially.

The interaction of the sympathetic and parasympathetic branches of the autonomic nervous system controls heart rate in a complex, nonlinear manner. Part of this complexity arises from the fact that the full chronotropic effects of vagal stimulation on heart rate occur within a few heart beats whereas it may take up to 30 sec for the full expression of the cardiac response to sympathetic stimulation (Figure 49). Consequently, the most rapid changes in heart rate appear to be dominated by the parasympathetic system. The major reason for the nonlinearity is that the degree of parasympathetic–sympathetic interaction at the heart may be changed by increasing output of one of these limbs.

The relationship of heart rate to bilateral stimulation of the distal cut ends of the vagus and cardiac sympathetic nerves of the duck *A. platyrhynchos* is illustrated in Figure 50 (Furilla and Jones, 1987b). In this figure, 100% represents the frequency of stimulation above which no further changes in heart rate occurred. The heart rate resulting from a given level of vagal and sympathetic stimulation was plotted on "perspective" graph paper and the surface was drawn, by eye, to encompass all heart rates obtained in the stimulation experiments. Area B represents complete cardiac denervation and area E maximal vagosympathetic stimulation. The effects of varying the intensity of sympathetic stimulation at minimal and maximal vagal activity are represented by the lines A–B and D–E, respectively. Similarly, the effects of varying the intensity of vagal stimulation at minimal and maximal sympathetic activ-

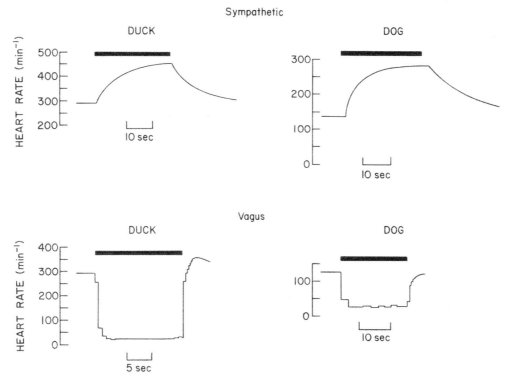

FIGURE 49 Comparison of time courses of changes in heart rate evoked by bilateral stimulation of cardiac sympathetic nerves (top panels) and vagus nerves (bottom panels) in a duck (left-hand traces) and dog (right-hand traces). In each panel the thick horizontal bar represents the duration of the stimulus. (Modified from Furilla and Jones, 1987a, *J. Exp. Biol.*, Company of Biologists, Ltd.)

ity are represented by the lines B–D and A–E, respectively.

Increasing sympathetic activation at zero vagal activity causes heart rate to rise over 200 beats min^{-1} (B to A) while increasing sympathetic drive at maximal vagal activation only increases heart rate by 50 beats min^{-1} (D to E). Similarly, increasing vagal activation causes heart rate to fall by more than 200 beats min^{-1} at zero sympathetic activity (B to D) and by nearly 400 beats min^{-1} at maximal sympathetic activity (A to E). Obviously, the greater the vagal drive, the more this input is able to occlude sympathetic effects on the heart, accounting for the nonlinearity in the interplay between the two branches of the autonomic nervous system at the cardiac pacemaker. This increased parasympathetic effectiveness in cardiac control is termed accentuated antagonism and may be mediated through two mechanisms. First, in response to sympathetic stimulation above a certain threshold, there is a cholinergically mediated reduction in prejunctional release of NE. Second, the magnitude of the postjunctional response to a given level of sympathetic stimulation is attenuated by ACh (Levy, 1971).

Short-term heart rate fluctuations are caused by the continued tug-of-war between the two branches of the autonomic nervous system at the pacemaker. This interplay may overlie long-term changes in heart rate caused, for instance, by changes in levels of circulating hormones. The normal homeostatic processes tend to reduce variability and maintain constancy of internal physiological functions, and short-term fluctuations in heart rate are usually seen as perturbations away from the norm. If a series of cardiac intervals is recorded from an animal, the duration of intervals within the series appears quite irregular with apparently random fluctuations occurring all the time. Whether these fluctuations are truly random or patterned, in the latter case providing evidence for chaotic control of heart rate (Denton *et al.*, 1990; Goldberger *et al.*, 1990), is a matter of some controversy. The strongest evidence for chaotic control of heart rate may lie in the morphology of the nerves innervating the heart. The nerves divide repeatedly, like the branching of a tree which is an intrinsically fractal structure. Hence, if the anatomy is fractal then why should the day to day workings of the system not be fractal as well? (Goldberger, cited in Pool, 1989; Goldberger, 1991).

In any event, it is clear that, rather than maintaining a homeostatic steady state, heart rate fluctuates considerably, even when recorded over short time periods (Figure 51A). The interbeat intervals form a time series which can be transformed into the frequency domain

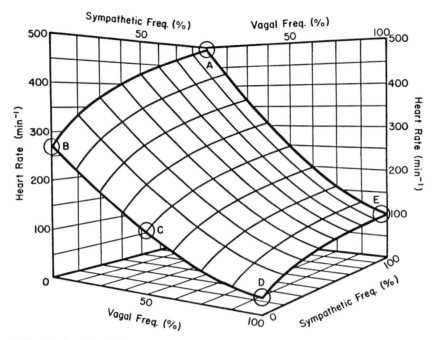

FIGURE 50 The relationship of heart rate to bilateral stimulation of the distal cut ends of the vagus and cardiac sympathetic nerves of the Pekin duck. "One-hundred percent" represents the frequency of stimulation above which no further changes in heart rate occurred with increases in stimulation frequency. The heart rate caused by a given level of vagal and sympathetic stimulation was plotted on "perspective" graph paper and the surface was drawn, by eye, to encompass all heart rates obtained in the stimulation experiments. See the text for an explanation of points A–E. (After Furilla and Jones (1987a), *J. Exp. Biol.*, Company of Biologists, Ltd.)

by Fourier analysis, revealing the presence of periodic components within the series. The square of the absolute value of the Fourier transform yields the power spectrum of the heart rate variability (PS/HRV; Kamath and Fallen, 1993).

The PS/HRV of an intact, resting, duck (*Aythya affinis*) is shown in Figure 51B(i). This plot reveals a single major peak at the respiratory frequency, which is the manifestation of respiratory modulation of cardiac parasympathetic activity (the response to sympathetic heart stimulation is too slow to significantly affect heart rate at a high frequency). Blockade of the sympathetic nervous system with a β-antagonist tends to increase heart rate variability although the amplitude of the high-frequency components are reduced [Figure 53B(ii)], while parasympathetic blockade with atropine gives a regular, unvarying heart rate [Figure 53B(iii)]. This confirms that short-term heart rate control is dominated by the parasympathetic nervous system in *Aythya affinis*.

The PS/HRV is a quantifier of autonomic responsiveness (Saul, 1990) and allows evaluation of cardiovascular regulation in birds over long time courses and during many types of activities. Also, the influence of other periodic functions such as arterial blood pressure and vasomotor fluctuations on PS/HRV can be evaluated using this technique. Telemetric recording of heart rate, combined with frequency analysis of cardiac function, will open new doors for studying control of physiological processes in unrestrained, active birds.

ii. Role of Stroke Volume in Control of Cardiac Output Stroke volume, like heart rate, is dependent upon factors intrinsic and extrinsic to the heart. As all myocytes within the heart contract during each beat, the primary intrinsic factors which determine stroke volume are the inherent contractile properties of each muscle fiber and the resting lengths of all the fibers. The amount of force developed during contraction by a cardiac muscle fiber at a specified precontraction length is properly termed "contractility," but this term has also been used more loosely to describe the collective contractile properties of all of the muscle fibers associated with one chamber of the heart. A major problem in quantifying contractility is that the force developed by a single cardiac muscle fiber is difficult to measure in working hearts. Consequently a number of indirect indices have been developed to estimate this variable. These include measuring cardiac outflow volume over time to calculate stroke volume; recording the ventricular peak systolic pressure developed against a fixed afterload or into a constant arterial pressure; and measuring the rate of rise of ventricular pressure during systole. The major

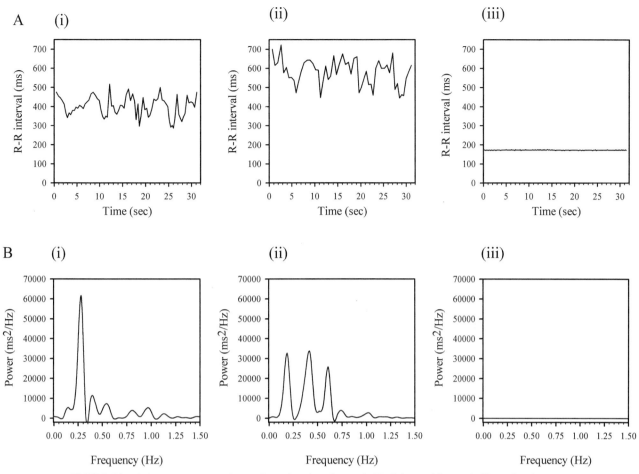

FIGURE 51 (A) 30-sec duration time series of interbeat intervals (R–R intervals) recorded by radiotelemetry from a duck (*Aythya affinis*) while the animal was resting quietly on the water surface. (i) Control; (ii) after blockade of the sympathetic nervous system with Nadolol (a β-adrenergic antagonist); and (iii) after blockade of the parasympathetic nervous system with the muscarinic antagonist atropine. (B) Power spectra of heart rate variability derived from the time series shown in (A). (i) Control; (ii) after Nadolol blockade; and (iii) after atropine blockade. Since the time series were too short to adequately display extremely low frequency components, these components were removed with a high-pass filter from the power spectra (L. McPhail, R. A. Andrews, and D. R. Jones, unpublished).

assumption in all of these methods is that the measured variable reflects the contractility of all muscle fibers integrated over the dimensions of the whole chamber. However, the variety of indices of contractility used by investigators under different experimental conditions has made it difficult to compare estimates across studies. The direct measurement of volume flow from the ventricle would appear to give the most reliable index of cardiac contractility, being independent of the complicating effects of changing arterial or ventricular pressures. This measurement is also among the most difficult to make, requiring highly invasive procedures to place the appropriate instrumentation.

By analogy with the contraction of skeletal muscle, the amount of force developed by a contracting cardiac muscle fiber depends upon its precontraction length ("preload"). This principle was first applied to the heart by Otto Frank (summarized in Rushmer, 1976), who showed that, within limits, the greater the preload on ventricular muscle in diastole, the more tension was developed during the next systole. This length–tension relationship was further investigated by Ernest Starling and co-workers, who demonstrated that the amount of blood ejected by the left ventricle during systole was proportional to the volume of blood in the ventricle at the end of the diastolic filling phase of the cardiac cycle. These concepts have been combined into the Frank–Starling relationship to describe the intrinsic responses of ventricular stroke volume to changes in cardiac venous return, expressed graphically in Figure 52. Elevated contractility of each muscle fiber in the ventricle is evoked by increasing the preload on all of the fibers by increasing the volume of blood filling the ventricle before each beat; this is reflected in an overall increase

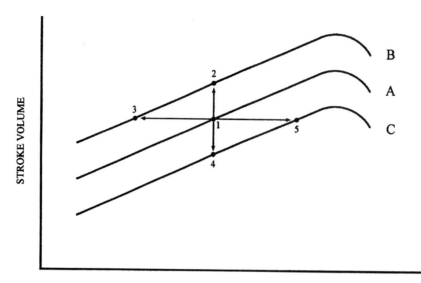

FIGURE 52 Idealized graphical representation of the Frank–Starling relationship for cardiac ventricular muscle. (A) Intrinsic ventricular function curve depicting the relationship between end-diastolic volume (representing degree of stretch of muscle fibers) and stroke volume (index of contractility) in the absence of extrinsic influences. The curve peaks and begins to decline at high end-diastolic volumes because resting sarcomere length is maximal here. (B and C) Factors extrinsic to the heart which alter inotropic function of cardiac muscle reset the ventricular function curve to operate over different ranges of stroke volume, independent of end-diastolic volume or initial fiber length. (B) Elevated cardiac sympathetic drive or circulating catecholamines have positive inotropic effects, resetting the curve toward higher stroke volumes. (C) Elevated vagal drive has negative inotropic effects, resetting the curve toward lower stroke volumes. Points 1 through 5 are the operating points assumed for the text discussion of the effects of extrinsic factors on ventricular function.

in ventricular stroke volume. The curve designated A in Figure 52 is termed a ventricular function curve. The relationship embodied in this curve demonstrates an autoregulatory feature of cardiac function known as heterometric regulation: a change in the resting fiber length (heterometry) results in a change in contractility in the same direction. This regulatory mechanism is an intrinsic property of cardiac muscle. The consequence of this mechanism for the overall function of the heart is that, if all other conditions remain constant, cardiac output will be determined by venous return. An increase in venous return to the left ventricle via the left atrium will result in greater end-diastolic stretch of the ventricle walls and an increase in stroke volume at the next beat; conversely, stroke volume will be reduced if cardiac return falls. In short, the heart "pumps what it gets" if all other factors are unchanging.

Cardiac contractility is influenced by extrinsic factors in addition to the intrinsic Frank–Starling mechanism. Circulating hormones such as EPI and the autonomic neurotransmitters NE and ACh (see Sections VI,C,1 and VI,C,2) directly affect the contractility of cardiac muscle fibers. These extrinsic factors are superimposed on the intrinsic autoregulatory factors governing stroke volume and can shift the whole ventricular function curve (curve A in Figure 52) toward higher (curve B) or lower (curve C) stroke volumes at the same resting muscle fiber length or degree of ventricular filling. This type of regulation of stroke volume is referred to as homeometric regulation, to emphasize the fact that changes in contractility can occur independent of resting fiber length. An increase in sympathetic drive to the heart or an increase in the level of circulating catecholamines will increase ventricular inotropic function homeometrically; thus a greater stroke volume will result from the same degree of cardiac filling, as indicated in Figure 52 by the arrow from point 1 on curve A to point 2 on curve B. Another way to consider this is that after such a shift in the curve a much smaller end-diastolic volume will give the same stroke volume (arrow from point 1 to point 3). On the other hand, elevated vagal drive to the heart of birds can decrease the contractility of ventricular muscle and will shift the ventricular function curve toward a lower stroke volume (curve C in Figure 52). At the new operating point, the same degree of preload will result in a lower stroke volume (arrow from point 1 to point 4); alternatively, a much larger end-diastolic volume will be required to maintain the same stroke volume (arrow from point 1 to point 5).

The arterial pressure against which the ventricle pumps ("afterload") is a major extrinsic factor in determining the magnitude of stroke volume. The pressure generated during the isometric phase of ventricular contraction is a function of the contractility of the muscle fibers, and when chamber pressure exceeds that in the aorta the valves open and blood is ejected from the ventricle during the isotonic phase. If the preload on the ventricle is increased by elevating the arterial blood pressure without a change in contractility or end-diastolic volume, stroke volume of subsequent beats will be reduced because more energy will be required to raise chamber pressure above the new level of arterial pressure. Initially, this will leave a larger fraction of the previous end-diastolic volume still in the chamber at the end of systole, resulting in an increased level of resting tension on the muscle fibers during the next filling phase. This increased tension, according to the Frank–Starling mechanism, will quickly result in increased contractility during subsequent beats, restoring stroke volume by heterometric regulation in the face of the increased arterial pressure.

In many species of birds, cardiac output is adjusted to match perfusion requirements of the tissues in a variety of conditions, such as during exercise, hypoxia, or submersion. These adjustments appear to be made primarily through alterations in heart rate with stroke volume remaining relatively unchanged. Changes in cardiac output during exercise are driven by increased heart rate in ducks (Bech and Nomoto, 1982; Kiley *et al.*, 1985), geese (Fedde *et al.*, 1989), and turkeys (Bouliane *et al.*, 1993a,b). However, in the emu (Grubb *et al.*, 1983) and the chicken (Barnas *et al.*, 1985) stroke volume may increase by up to 100% during exercise, contributing significantly to elevated cardiac output. In the duck, Jones and Holeton (1972a) reported that variations in cardiac output during simulated high-altitude exposure were reflections of changes in heart rate with no significant alteration in stroke volume. In diving ducks, Jones and Holeton (1972b), Lillo and Jones (1982), Jones *et al.* (1983), and Smith and Jones (1992) showed that stroke volume was maintained during the large decreases in cardiac output generated during submersion. Reflex changes in cardiac output mediated by systemic arterial baroreceptor input also appear to operate via alterations in heart rate, leaving stroke volume relatively unchanged (Section VI,D,2). In summary, during exercise, hypoxia, or submersion, birds display significant changes in heart rate, arterial blood pressure, and venous return from the resting condition. In the transition from the resting condition to these altered states, stroke volume also varies. However in most of the species examined so far, stroke volume returns to values close to those at rest after a short period of initial adjustment. This indicates that intrinsic autoregulation of cardiac output has the potential to play an important role in the maintenance of stroke volume in the face of large-scale circulatory adjustments.

D. Reflexes Controlling the Circulation

1. Chemoreflexes

In birds, reflex adjustments of cardiac output and vascular caliber are generated by chemoreflexes in response to changes in levels of oxygen, carbon dioxide, and pH in the cerebrospinal fluid and in arterial blood. Receptors sensitive to CO_2 in cerebrospinal fluid are present in the avian central nervous system (Jones *et al.*, 1982) but no detailed studies of the location or transduction properties of these receptors have been done in birds. However, if these receptors are similar to those found in mammals, they may be located at or near the surface of the ventrolateral medulla (reviewed by Schlaefke, 1981). Arterial chemoreceptors in birds are located primarily in the carotid bodies, bilateral structures lying caudal to the thyroid gland, and close to the ultimobranchial gland and nodose ganglia of the vagus nerves (from which they are innervated) and the carotid artery (from which they are supplied with blood) (Figure 42; Adams, 1958; Jones and Purves, 1970). Putative chemoreceptors have also been reported in "aortic bodies" associated with the roots of the great vessels in several species of birds (Tcheng *et al.*, 1963).

The discharge characteristics of arterial chemoreceptors in response to changes in arterial PCO_2 (P_aCO_2), $PO2$ (P_aO_2) and arterial pH (pH_a) have been studied in the duck and chicken. Receptor discharge rate increases proportionally with P_aCO_2 and in inverse proportion to P_aO_2 (Bouverot and Leitner, 1972; Bamford and Jones, 1976; Nye and Powell, 1984; Hempleman *et al.*, 1992). Discharge sensitivities to changes in P_aCO_2 and P_aO_2 have been quantified by Hempleman *et al.* (1992) for carotid body chemoreceptors in the duck. When P_aO_2 was maintained at a normoxic level (near 100 mmHg) mean chemoreceptor sensitivity to step changes in P_aCO_2 (produced by changing the fraction of CO_2 in the air breathed by the bird) was +0.20 impulses sec^{-1} mmHg $P_aCO_2^{-1}$. Hypoxia (P_aO_2 56 mmHg) potentiated chemoreceptor discharge in response to altered P_aCO_2 (Figure 53). In this condition the same step changes in P_aCO_2 as given in normoxia resulted in a sensitivity of +0.32 impulses sec^{-1} mmHg $P_aCO_2^{-1}$. In the absence of CO_2 in the inspired air, step changes in P_aO_2 from the normoxic level to about half this level resulted in a

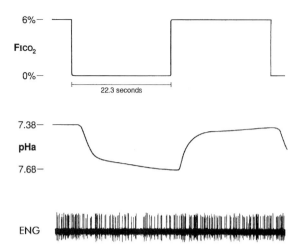

FIGURE 53 Single-fiber arterial chemoreceptor response recorded from vagal slip [bottom trace, electroneurogram (ENG)] to step changes in inspired CO_2 level [top trace, fraction of CO_2 in inspired gas (F_{ICO_2})] during hypoxia (10% O_2 in inspired gas) in a duck. Arterial pH [middle trace (pH_a)] also changes in step with inspired CO_2. Chemoreceptor discharge is sensitive to the level of CO_2 and pH_a and to rate of change of these variables. (Reprinted from *Resp. Physiol.* **90**, S. C. Hempleman, F. L. Powell, and G. K. Prisk, Avian arterial chemoreceptor responses to steps of CO_2 and O_2, pp. 325–340, Copyright (1992), with permission from Elsevier Science.)

mean sensitivity of about -0.10 impulses s^{-1} mmHg $P_aO_2^{-1}$. Carotid body chemoreceptors are thus about twice as sensitive to changes in absolute levels of arterial CO_2 as to O_2. Some of the chemoreceptors sampled in this study were also sensitive to the rate of change of O_2 and CO_2 in the blood, indicating that proportional- and rate-related information on arterial blood gas status are both transmitted to the central nervous system. However the maximum rate of change of discharge of most rate-sensitive chemoreceptors was in the lower frequency part of the range of physiologically occurring blood gas oscillations. Therefore, only relatively low frequencies of blood gas oscillation, such as those occurring at rest or during low-intensity activities, will be faithfully transduced. At higher oscillation frequencies, such as those occurring during panting or high-intensity exercise, chemoreceptor inputs to the central nervous system probably represent "mean" blood gas levels averaged over several oscillatory cycles.

Arterial chemoreceptors are spontaneously active at normoxic and normocapnic blood gas levels in birds and these receptors have been proposed to play an important role in setting the level of eupneic ventilation under these conditions (Bouverot and Leitner, 1972). However, the role of arterial chemoreceptors in reflex control of the avian circulation is less clear. Analyses of the circulatory effects of chemoreceptor activation are complicated by parallel changes in ventilation. During apneic asphyxia in ducks, blood oxygen tension falls, CO_2 tension rises, and carotid body chemoreceptors become progressively more strongly stimulated; this input plays a major role in initiating and maintaining an intense bradycardia (Jones and Purves, 1970; Butler and Taylor, 1973). However, if blood gases in spontaneously breathing ducks are artificially adjusted to mimic the hypoxic and hypercapnic levels achieved during apneic asphyxia, chemoreflex drive acts to increase ventilation, leading to elevated drive from pulmonary receptors. In this state there is little or no change in heart rate (Butler and Taylor, 1973; 1983). When the rise in ventilation (and thus the elevation in pulmonary receptor input) is prevented by controlling tidal volume and respiratory frequency during hypoxic hypercapnia in paralyzed but unanesthetized animals, heart rate falls to a level midway between that in normoxic, normocapnic animals and that obtained at end-dive (Butler and Taylor, 1973; 1983). This effect is illustrated in Figure 54. Subsequent cessation of respiration by stopping the ventilator pump then results in the full expression of chemoreceptor-mediated bradycardia: heart rate falls to the same level as at end-dive (Butler and Taylor, 1973; 1983). These experiments show that in ducks, as in mammals, the cardiovascular responses to strong arterial chemoreceptor stimulation during spontaneous breathing are masked by elevated ventilatory drive. Butler and Taylor (1983) have suggested that pulmonary receptors activated by increased ventilation contribute to occlusion of the cardiac chemoreflex. Chemoreflex-mediated bradycardia contributes to the conservation of blood oxygen stores during submersion in diving birds, and input from arterial chemoreceptors is responsible for about half of the large increase in peripheral resistance

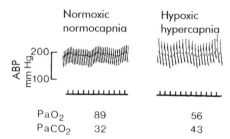

FIGURE 54 Effect of systemic hypoxic hypercapnia on heart rate and blood pressure in mallard duck. Animal was artificially ventilated after spontaneous respiratory movements were suppressed with intravenous pancuronium bromide. (Left panel, upper trace) Arterial blood pressure and heart rate during ventilation with air; systemic arterial blood gas values are shown at bottom. (Right panel) When blood gas values were adjusted to match those at end of 1 min submersion by altering CO_2 and O_2 levels in inspired gas, heart rate decreased as a result of arterial chemoreflex activation. In both panels the ticks on the bars under the pressure traces indicate 1-sec intervals. (Reprinted from *Resp. Physiol.* **19**, P. J. Butler and E. W. Taylor, The effect of hyperoxic hypoxia, accompanied by different levels of lung ventilation, on heart rate in the duck, pp. 176–187, Copyright (1973), with permission from Elsevier Science.)

during submergence in ducks (Jones et al., 1982). However the cardiac chemoreflex and its suppression by ventilation appears to be a general avian phenomenon, since Butler and Taylor (1974) were able to evoke a chemoreflex-driven bradycardia in chickens and pigeons during apnea without submersion.

Input from peripheral arterial chemoreceptors can reflexly alter peripheral vascular resistance but the general role of chemoreceptors in control of the vasculature has not been established. Indirect evidence for chemoreflex effects on the vasculature comes from a study by Bouverot et al. (1979), who showed a trend toward increased peripheral resistance in carotid body-intact ducks subjected to arterial hypoxia while breathing spontaneously. The same stimulus in animals after denervation of the carotid bodies led to a 40% fall in peripheral resistance, indicating a potential role for these chemoreceptors in generating the vascular response to hypoxia. As with cardiac responses to carotid body stimulation, elevated ventilation during hypoxia may mask the full extent of reflexogenic vasoconstriction. In the study of Bouverot et al. (1979) no attempt was made to evaluate this interaction by controlling ventilation.

It is clear that the full role of arterial chemoreflexes in circulatory control has yet to be defined in birds. Elevated carotid body input can result in both ventilatory and cardiac responses, and both of these responses are important in maintaining oxygen delivery to and CO_2 washout from working tissues. Further experiments are necessary to establish the relative importance of these limbs of the chemoreflex in matching ventilation with perfusion to cope with changes in internal and external levels of these gases.

2. Baroreflexes

Arterial blood pressure provides the driving force for perfusion of the systemic vascular beds and must therefore be maintained within limits that ensure optimal tissue blood flow under a variety of physiological conditions. Blood pressure in birds is maintained by the baroreflex, a mechanism employing negative feedback (see reviews by Bagshaw, 1985 and Smith, 1994). Adjustments in blood pressure produced by the baroreflex are driven by afferent signals from arterial baroreceptors. These are mechanoreceptors with their receptor endings embedded in connective tissue of the arterial wall, where they sense changes in arterial pressure as variations in wall tension. An increase in intraarterial pressure results in an increase in circumference of the vessel wall, which in turn stretches baroreceptor endings to increase their frequency of discharge of action potentials. Arterial baroreceptors in birds are located primarily in the walls of the aorta close to the left ventricular valves (Jones, 1973). Their axons course to the brainstem via aortic nerves which arise from the nodose ganglia of the vagus (Nonidez, 1935). In the few studies of avian baroreceptor function done to date, discharge characteristics in response to changes in blood pressure appear to be similar to those of mammalian high-threshold, slowly adapting baroreceptors (Jones, 1969; 1973). Spontaneous baroreceptor impulse generation is phase-locked to mechanical events in systole of the cardiac cycle, as shown in Figure 55. Baroreceptors are also sensitive to the rate of rise of the pressure pulse. It thus appears that baroreceptors are capable of transmitting information on cardiac rate, peak systolic pressure, and the slope of the aortic pressure waveform (which may in turn reflect cardiac contractility) to the central nervous system.

Mean arterial blood pressures of resting birds are higher than those recorded in mammals of equivalent body weight at rest; indeed, mean resting pressures may exceed 150 mmHg in some avian species (Altman and Dittmer, 1971). Avian baroreceptor discharge occurs at resting levels of blood pressure in birds (Jones, 1969; 1973), so it is likely that the baroreflex is tonically active at these pressures. Changes in blood pressure sensed at the receptors are represented to the baroreflex circuitry in the central nervous system by changes in baroreceptor afferent discharge frequency, and the baroreflex acts to adjust pressure in a direction opposite to that of the initial pressure change. This reflex operates through both peripheral vascular and cardiac effectors to return blood pressure toward a set level after disturbances. Pressure is thus maintained within fairly narrow limits over time to ensure a constant head for tissue perfusion. Little is known of the central nervous mechanisms involved in blood pressure regulation in birds but it has been assumed that the basic organization of central components of the baroreflex is similar to the arrangement in mammals.

The primary cardiovascular response to changes in blood pressure is a baroreflex-mediated change in cardiac output, seen in Figure 56A as a rapid fall in heart rate in response to a pharmacologically induced pressure increase. Such reflex responses are completely abolished by bilateral section of the aortic nerves (Figure 56B). The sensitivity of the blood pressure–heart rate relationship ranges from -0.5 to -3.13 beats min^{-1} mmHg^{-1}, depending on species and method of evaluation (chickens, Bagshaw and Cox, 1986; ducks, Smith and Jones, 1990; 1992; and Millard, 1980; see Smith, 1994 for further discussion). The baroreflex-mediated effects of changing pressure at the receptors can be mimicked by electrical stimulation of the central cut end of an aortic nerve in an animal in which both aortic nerves are sectioned. This method was used by Smith and Jones (1992) to explore the dynamic role of barore-

FIGURE 55 Relationship between baroreceptor discharge [upper trace, electoneurogram (ENG)], recorded from peripheral cut end of aortic nerve, and blood pressure [lower trace (ABP)], recorded from a brachiocephalic artery in an anesthetized duck. Baroreceptor discharge is synchronous with systolic peak pressure. ABP scale bar represents a pressure span from 100 to 200 mmHg. (After Jones (1973), with permission.)

ceptors in controlling the circulation. Stimulation of the aortic nerve in barodenervated animals evokes a decrease in arterial blood pressure in proportion to the stimulus frequency when stimulus current is set just above threshold for a response (Figure 57, closed circles). This response works primarily through a fall in cardiac output, mediated by decreased heart rate with no change in stroke volume (Smith and Jones, 1992). This response follows the same pattern as the baroreflex-mediated response to pharmacologically induced pressure changes in baroreceptor-intact birds; that is, in both intact and denervated animals baroreflex activation was expressed primarily through changes in heart rate. However, the baroreflex can engage peripheral vasomotion as well as cardiac responses: in barodenervated animals, stimulation of the aortic nerve with a current intensity several times the threshold level produced decreases in peripheral resistance as well as in heart rate (Smith and Jones, 1992). These data suggest that when relatively small disturbances in pressure occur they will be compensated by adjusting cardiac output, but larger changes in pressure will be corrected by a combination of cardiac and peripheral vascular adjustments.

The effectiveness of baroreflex control of the circulation can be modified by interaction with other reflexes, such as the chemoreflex, which may be concurrently engaged. In an effort to determine the cause of an appar-

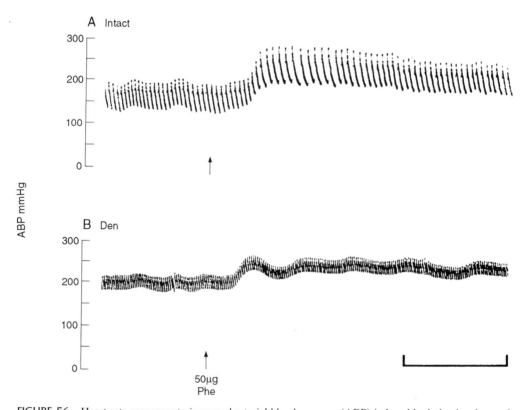

FIGURE 56 Heart rate responses to increased arterial blood pressure (ABP) induced by bolus i.v. doses of phenylephrine (Phe, injected at arrows) in the same duck (A) before and (B) after barodenervation by section of aortic nerves. Baroreflex-mediated bradycardia was eliminated by denervation of baroreceptors. Note increased preinjection blood pressure and heart rate in barodenervated animals, indicating a degree of baroreflex-mediated restraint on cardiovascular system in baroreceptor intact animal (F. Smith, unpublished).

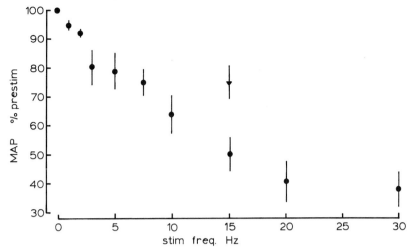

FIGURE 57 Normalized responses of mean arterial blood pressure (MAP) to electrical stimulation of aortic nerve in awake ducks spontaneously breathing air (P_aO_2 88 mmHg, P_aCO_2 26 mmHg; closed circles) or a hypoxic hypercapnic gas mixture (P_aO_2 62 mmHg, P_aCO_2 44 mmHg; triangle). Pressure response at each frequency of nerve stimulation is expressed as a percentage change in MAP relative to MAP just prior to stimulation. Loading the chemoreceptors attenuated the depressor effect of aortic nerve stimulation at 15 Hz. Error bars represent ± 1 S.E.M. for five trials in normoxic conditions and three trials during hypoxic hypercapnia in three animals. Stimulus pulse duration and amplitude were constant within each trial. (F. Smith, unpublished).

ent reduction in baroreflex function observed during submersion in ducks (Jones et al., 1983; Millard, 1980), Smith and Jones (1992) stimulated the aortic nerve in barodenervated ducks before and during periods of elevated chemoreceptor drive. Stimulation of chemoreceptors was accomplished by ventilating animals with a gas mixture which simulated the hypoxic and hypercapnic blood gas values observed at the end of 2 min of submersion; this significantly decreased the capability of aortic nerve stimulation to affect mean arterial pressure. This is shown in Figure 57 by a 50% decrease in the pressure response to aortic nerve stimulation at a frequency of 15 Hz during hypoxic hypercapnia (triangle), compared with the responses during air breathing (closed circles). That this occlusive response was due to chemoreceptor activation was further demonstrated by the attenuation of baroreflex function during perfusion of one vascularly isolated carotid body with venous blood in otherwise normoxic, normocapnic animals which were spontaneously breathing (Smith and Jones, 1992). It therefore appears that the chemoreceptor drive that develops after the first minute of submersion (Jones and Purves, 1970) is at least partly responsible for the attenuation of baroreflex control of cardiovascular variables which occurs during this period.

Baroreceptor input has been shown to have no direct role in generating or maintaining heart rate responses to voluntary submersion in diving ducks, since these animals display the same degree of bradycardia after denervation of arterial baroreceptors as before (Furilla and Jones, 1987a). However, baroreceptors may play a role in control of heart rate in dabbling ducks which have been trained to dive voluntarily. Predive heart rate in dabblers ranged from 100 to about 500 beats min^{-1} but, regardless of the rate preceding any particular dive, rate during the dive tended to a value of approximately 250 beats min^{-1} (Furilla and Jones, 1987b). This response implies that heart rate was being regulated at a set value during voluntary dives. Removal of baroreceptor input by bilateral section of the aortic nerves eliminated the tendency of dive heart rate to approach the "set point" value; after barodenervation dive heart rate varied little from the prevailing predive rate (Furilla and Jones, 1987b). Given that the baroreflex is normally considered to regulate blood pressure by adjusting cardiac output and vascular resistance, the physiological value of a baroreceptor-dependent "set point" for heart rate is uncertain. The behavior of arterial blood pressure during voluntary diving in dabbling ducks has not been established. However there may be some inherent benefit to regulating heart rate under these conditions. If this is true, then strong phasic baroreceptor input to the central nervous system during systole would represent the primary afferent feedback route for heart rate-related information.

3. Reflexes from Cardiac Receptors

Small nerve terminals in the shape of simple knobs, plates or rings as well as straight or spirally wound nerve endings have been observed in anatomical studies of avian atria and ventricles (Ábrahám, 1969; also see review by Jones and Milsom, 1982), but bird hearts do not appear to have the complex and highly developed sensory receptor endings present in large numbers in mammalian hearts. Few functional studies of avian cardiac receptors and their reflexogenic effects have been done so far. Jones (1969) established in the duck that some afferent fibers in the cervical vagus had receptor endings associated with the heart, responding to punctate stimulation near the atrioventricular junction and discharging spontaneously in patterns which were phase-locked to mechanical events in the cardiac cycle. In the chicken, Estavillo and Burger (1973a,b) found that a majority of cardiac receptors with their cell bodies in the nodose ganglia had receptive fields located in the left ventricle near the aortic valves. Discharge patterns of these receptors were either phase-locked to the cardiac cycle or were irregular and apparently unrelated to mechanical events in the cycle. The discharge of receptors of both types could be modulated by varying inspired CO_2 and pH_a independently or together. In this study, the discharge rate of phasically firing cardiac receptors was proportional to arterial blood pressure over a wide pressure range, and this relationship was reset toward lower discharge rates at increased CO_2 levels, as shown in Figure 58. Bilateral section of the middle cardiac nerve, carrying the axons of cardiac receptors in the chicken, produced an immediate rise in arterial blood pressure (Estavillo, 1978; Estavillo et al., 1990) reminiscent of that produced in ducks after section of the aortic nerve (Figure 56).

Avian cardiac receptors have been proposed to contribute to blood pressure regulation and to the control of ventilation, providing sensory feedback on intracardiac pressures and volumes. Such feedback may be modulated by changes in the levels of P_aCO_2 and pH_a (Estavillo and Burger, 1973b; Estavillo et al., 1990). In the latter study, bilateral section of the middle cardiac nerve considerably blunted the increase in ventilation elicited by systemic hypercapnia, in addition to promoting elevated blood pressure.

Birds exhibit a Bezold–Jarisch reflex similar to that in mammals, manifest in ducks as a fall in heart rate and arterial blood pressure when cardiac receptors are stimulated (Blix et al., 1976; Jones et al., 1980). Blix et al. (1976) proposed, on the basis of pharmacological stimulation of cardiac receptors, that this reflex contributed to the generation and maintenance of the cardiac chronotropic response to submersion. In a reexamination of this issue, Jones et al. (1980) loaded and unloaded the cardiac receptors by altering left ventricular pressure to provide more realistic physiological stimulation of these receptors before and during submersion. The results of this study failed to confirm a link between cardiac receptor activation and diving bradycardia.

There is as yet insufficient evidence to determine the overall function of reflexes driven by inputs from cardiac receptors. While these receptors can influence blood pressure and ventilation, they do not appear to have a primary role in pressure or ventilatory regulation. However, they may mediate some of the dynamic interactions between respiratory and circulatory systems, thus helping to correct ventilation–perfusion mismatches which can develop during exercise or in adverse environmental conditions.

4. Reflex Cardiovascular Effects from Skeletal Muscle Afferents

Reflexly mediated changes in arterial blood pressure, heart rate, and other cardiovascular variables accompany exercise, hypoxia, and hypoxic hypercapnia in birds. In several studies designed to deduce the respective roles of peripheral arterial chemoreceptors, chemosensitive areas of the central nervous system and arterial baroreceptors in these cardiovascular responses, it has been suggested that inputs from these receptor groups do not account for all of the changes observed. Thus inputs from some other receptor type must also be involved. In mammals there is a significant reflexogenic increase in mean arterial blood pressure induced by skeletal muscle activity; receptors for this response appear to be intramuscular terminals of group III afferent fibers (small myelinated fibers) and group IV fibers (unmyelinated or C-fibers) coursing in somatic nerves (see Coote, 1975 for review). Kiley et al. (1979) implicated muscle afferents in the cardiovascular responses to exer-

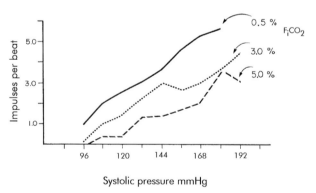

FIGURE 58 Modulatory effect of changes in inspired CO_2 (F_iCO_2) on the relationship between discharge rate of cardiac mechanoreceptors (impulses per heartbeat) and arterial blood pressure (systolic pressure) in chicken. Curves were obtained by plotting receptor discharge rate at a given blood pressure over a range of pressures produced by bolus i.v. injections of mecholylchloride or epinephrine. Progressive increases in F_iCO_2 displaced the receptor discharge–blood pressure relationship to progressively lower discharge rates. (After Estavillo and Burger (1973b), *Am. J. Physiol.* with permission).

cise in ducks, and Lillo and Jones (1983) proposed that somatic muscle afferents in ducks were at least partly responsible for those portions of the cardiac and vascular responses in hypoxic hypercapnia and ischemia which were independent of chemoreceptor activation. Furthermore, Solomon and Adamson (1997) demonstrated that ducks express an "exercise pressor" reflex similar to that in mammals. This effect consists of an increase in mean arterial pressure (largely due to elevation of diastolic pressure) induced by and and sustained during static contraction of a large hind limb muscle, the gastrocnemius (Figure 59A). The authors concluded that this was a reflex effect since section of the sciatic nerve carrying afferent fibers from the muscle to the spinal cord eliminated the pressor response to muscle contraction (Figure 59B). Whether the intramuscular receptors involved in these responses were sensing mechanical events related to muscle contraction or chemical signals resulting from intramuscular metabolite buildup as the primary physiological stimulus for this reflex was not clear from this study. The authors acknowledged that receptors sensing either modality, or both, could be driving reflex changes in pressure. The studies done to date on reflex effects of activating skeletal muscle afferents suggest that direct sensory feedback from exercising or hypoxically stressed muscle to the neural circuitry controlling the circulation could be involved in initiating or intensifying cardiovascular responses to muscle activity. Such a mechanism would serve to increase the perfusion of working muscle, but the full contribution of these reflexes remains to be determined.

E. Integrative Neural Control

As discussed in the previous sections, the major reflexes controlling cardiovascular function are the chemoreflex, the baroreflex, and reflexes driven by receptors within the heart and skeletal muscle. Under some conditions these reflexes may interact, as in the case of chemoreflex occlusion of the baroreflex detailed in Section VI,D,2 above. Furthermore, there are a number of other reflexes which can affect the circulation as part of more global homeostatic control systems, such as those regulating temperature and ventilation. All reflexes influencing the circulation do so through the final common pathways of the autonomic nervous system innervating the heart and vasculature, and several of these reflexes may be engaged at the same time during physiological challenges to the animal. Integrative control of the circulation by the nervous system therefore operates through complex interactions among cardiovascular and other reflexes, but the nature of these interactions is still incompletely understood in any vertebrate group. This large-scale integration makes investigation of the properties of any one reflex a considerable challenge, particularly in unanesthetized animals. One approach to this problem is to ensure that only afferents specific to the reflex in question are stimulated, while activation of afferents associated with other reflexes is prevented. However even this may not be enough to prevent complications in analysis since efferent activity associated with one reflex, and the consequences of this activity, can produce secondary activation of other reflexes. A case in point is the activation of pulmonary reflexes consequent to an increase in ventilation driven by primary stimulation of the arterial chemoreflex. The design of further studies of avian cardiovascular reflexes embedded in complex control systems must therefore take into account the integrative nature of these systems.

A single neurally mediated response, such as a change in heart rate, involving the stimulation of one or more afferent or efferent pathways by a frequently induced or naturally performed behavior, may be subject to habituation or conditioning. Habitation refers to a decrement in the response (not due to sensory adaptation or motor fatigue) resulting from repeated presentations of a single triggering stimulus. In contrast, conditioning involves the animal learning a relationship between two different stimuli. In classical conditioning (Pavlov, 1927) a conditioned and an unconditioned stimulus are pre-

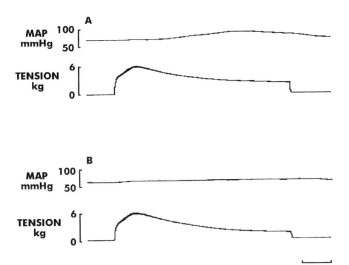

FIGURE 59 Pressor effect [upper traces in each panel, mean arterial blood pressure (MAP)] of static contraction of gastrocnemius muscle [lower traces in each panel (tension)] induced by electrical stimulation of sciatic nerve in anesthetized chicken. (A) MAP increases during static muscle contraction when sciatic nerve–spinal cord connection is intact. (B) No change in MAP during static muscle contraction after section of sciatic nerve. Afferent limb of pressoreflex originates within gastrocnemius muscle. (From Solomon and Adamson (1997), *Am. J. Physiol.* with permission.)

sented with little temporal dissociation between them. The animal learns the relation between these stimuli so that after the initial trials the reflex response can then be evoked by presenting the only the conditioned stimulus. A reflex, once conditioned, will anticipate, augment, and have similar effects as the unconditioned reflex.

Stimulation of nasal receptors in diving ducks and chemoreceptors in dabbling ducks are the proximate causes of the development of diving bradycardia (Furilla and Jones, 1986; 1987a,b). Repeatedly submerging the head of a diving or dabbling duck in a laboratory situation causes the bradycardic response to habituate after 100 to 200 dives (Figure 60; Gabbott and Jones, 1987). In dabbling ducks, however, extending the period of submergence beyond 40 sec virtually eliminates any attenuation of the cardiac response to submergence. Obviously, in dabbling ducks input from the carotid body chemoreceptors is too intense for habitation after 60 sec submergence. Similarly, exposing habituated animals to 10 or 15% oxygen in air before submergence causes prominent bradycardia although the very next trial, after breathing room air, evokes the habituated cardiac response. Interestingly, the heart rate response to diving after breathing air with low levels of oxygen is unaffected by training. Consequently, chemoreceptor input, which will be the same in naive and habituated ducks because blood gas levels are the same after 40 sec submergence, can be habituated. Habituation of the response occurs within the central nervous system, below the thalamic level. Animals with their higher brain centers surgically removed can be trained as easily as intact ducks (Gabbott and Jones, unpublished).

Cohen and his collaborators have developed a model of conditioned learning in the pigeon for exploring cellular neurophysiological mechanisms of long-term associative learning. In this model animals are trained to respond with a transient and quantifiable increase in heart rate (the conditioned response, CR) to a conditioning stimulus (CS, a 6-sec whole-field retinal illumination) by pairing the CS with an unconditional stimulus (US) consisting of a 0.5-sec foot shock. After an initial training period of 30–50 paired-stimulus trials, further CS without US reliably evokes conditioned responses. The properties of this CR in the pigeon are very robust, remaining stable for long periods (up to weeks) without habituating (Cohen, 1980; 1984). Such longevity facilitates electrophysiological investigations of changes in properties of neurons involved in the development of associative learning in this system. Using this model as a platform, Cohen and co-workers have anatomically and physiologically characterized the central visual pathways for the CS, the somatosensory pathways for

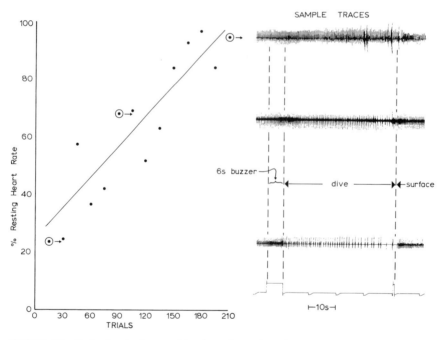

FIGURE 60 Reduction in degree of bradycardia attained in 40-sec dives by a Pekin duck during repeated trials, each consisting of a 40-sec dive immediately preceded by a 6-sec buzzer. The graph is a plot of percentage of resting heart rate achieved at the end of a trial, against the number of trials. Sample ECG traces are taken from the first trial of the series (bottom trace), the ninetieth trial (middle trace), and the final dive (top trace). (After Gabbott et al., (1987), *J. Exp. Biol.*, Company of Biologists, Ltd.)

the US, and the descending tracts converging on the autonomic motor neurons of the final common pathway efferent to the heart in the pigeon. Much of this group's work on the efferent components of this system has already been cited in the description of the neuronal circuitry controlling cardiac function appearing earlier in this chapter.

The central and peripheral nervous pathways involved in the CR have been reviewed in detail by Cohen (1980; 1984) and are briefly summarized here. Ganglion cells throughout the retina respond with a phasic burst of APs at the start of a 6-sec period of whole-field retinal illumination. This phasic wave of excitation is transmitted synchronously along multistage pathways in the central nervous system to preganglionic sympathetic and parasympathetic motor neurons in the spinal cord and brainstem, respectively. These preganglionic neurons in turn control pools of cardiac postganglionic autonomic neurons in a synergistic manner, acting to potentiate a transient sympathetic cardioacceleration and to facilitate concurrent withdrawal of parasympathetic cardioinhibition.

Studies of the development of conditioned responses after lesioning selected components of the central visual pathways and electrophysiological studies during the conditioning process in pharmacologically immobilized pigeons have shown that the CS is conveyed in parallel through multiple pathways to the visual area of the telencephalon. These include (1) a thalamofugal pathway involving the principal optic nucleus (the avian homolog of the mammalian lateral geniculate nucleus) and (2) a tectofugal pathway projecting through the optic tectum and the nucleus rotundus (Cohen, 1980). In addition there may be a third visual input pathway implicated in the conditioned response, projecting through the pretectal area and the thalamus. While retinal responses to repeated CS are not modified during the development of conditioning, responses of second- and higher-order neurons in the CS pathway are facilitated during this process. The time course of these changes parallels the time course of the development of the conditioned heart rate response, so neurons or their impinging synaptic fields in successive stages in the CS pathways are likely sites for modulation of cardiovascular control during associative learning.

The descending pathways involved in this conditioned cardiac response have been well established. Neurons in the medial region of the hypothalamus, along with those located in the ventral brainstem, project to preganglionic sympathetic and parasympathetic cardiomotor neurons. The medial hypothalamus in turn receives inputs from the avian homolog of the mammalian amygdala, an area which, in both groups of animals, evokes marked cardioactive effects when stimulated.

Furthermore, lesions of the avian amygdala or its hypothalamic projection either produce deficits in conditioning or can prevent development of the conditioned response (Cohen, 1980). The telencephalic projections conveying visual information to the avian amygdala have, however, not as yet been established in detail. Convergence of the CS and US has been demonstrated at each of the established stages in the central pathway of this response. It thus appears that training-induced modification of the CS works through long-term heterosynaptic facilitation in this pathway, and such facilitation constitutes an important element in the constellation of neurophysiological mechanisms for associative learning.

The broad outlines of the central and peripheral neural pathways involved in efferent control of the circulation in birds have been established but relatively little is known about details of specific afferent and intermediary connections within the central pathways of any of the cardiovascular reflexes in birds. The working assumption guiding studies of these reflexes is that their pathways are similar to those in mammals, but given the differences in cardiovascular reflexogenic zones between birds and mammals, this is not necessarily a valid assumption. Birds have, for instance, only one major arterial baroreceptor site while mammals have two. In addition, most of the input from avian arterial chemoreceptors originates from the carotid bodies while in mammals both aortic and carotid chemoreceptors contribute to cardiovascular chemoreflexes. The organization of the neural circuitry for cardiovascular control in birds has not been investigated in detail, so the similarity of this circuitry to that of mammals remains an open question.

Acknowledgments

We express our appreciation to the Heart and Stroke Foundation of Canada (FMS, Research Scholarship) and the Natural Sciences and Engineering Research Council of Canada (NHW, DRJ, Operating Grants) for their support during this endeavor. We are also grateful to Agnes Lacombe for invaluable assistance in preparing some of the figures and to Kathy Gorkoff for preparing portions of the manuscript.

References

Ábrahám, A. (1969). "Microscopic Innervation of the Heart and Blood Vessels in Vertebrates including Man." Pergamon Press, Oxford, UK.

Adams, W. E. (1937). A contribution to the anatomy of the avian heart as seen in the kiwi (*Apterix australis*) and the yellow-crested penguin (*Megadypte antipodum*). *J. Zool.* **107,** 417–441.

Adams, W. E. (1958). "Morphology of the Carotid Body and Carotid Sinus." Charles C. Thomas, Springfield, IL.

Akester, A. R. (1967). Renal portal shunts in the kidney of the domestic fowl. *J. Anat.* **101,** 569–594.

Akester, A. R. (1971). The heart. In "Physiology and Biochemistry of the Domestic Fowl" (D. J. Bell and B. M. Freeman, eds.), Vol. 2, pp. 745–781. Academic Press, New York.

Akester, A. R. (1979). The autonomic nervous system. In "Form and Function in Birds" (A. S. King and J. McLelland, eds.), Vol. 1, pp. 381–441. Academic Press, New York,

Akester, A. R. and Akester, B. (1971). Double innervation of the avian cardiovascular system. *J. Anat.* **108**, 618–619.

Akester, A. R., and Mann, S. P. (1969). Adrenergic and cholinergic innervation of the renal portal valve in the domestic fowl. *J. Anat.* **104**, 241–252.

Akester, A. R., Akester, B., and Mann, S. P. (1969). Catecholamines in the avian heart. *J. Anat.* **104**, 591.

Altman, P. L., and Dittmer, D. S. (1971). "Respiration and Circulation," Federation of American Societies for Experimental Biology, Bethesda, MD.

Armour, J. A., and Ardell, J. L. (1994). "Neurocardiology," Oxford Univ. Press, New York.

Armour, J. A., Huang, M. H., and Smith, F. M. (1993). Peptidergic modulation of *in situ* canine intrinsic cardiac neurons. *Peptides* **14**, 191–202.

Ash, R. W., Pearce, J. W., and Silver, A. (1969). An investigation of the nerve supply to the salt gland of the duck. *Q. J. Exp. Physiol.* **54**, 281–295.

Astrand, P. O., and Rodahl, K. (1986). "Textbook of Work Physiology," McGraw–Hill, New York.

Bagshaw, R. J. (1985). Evolution of cardiovascular baroreceptor control. *Biol. Rev.* **60**, 121–162.

Bagshaw, R. J. and Cox, R. H. (1986). Baroreceptor control of heart rate in chickens (*Gallus domesticus*). *Am. J. Vet. Res.* **47**, 293–295.

Ball, R. A., Sautter, J. H., and Katter, M. S. (1963). Morphological characteristics of the anterior mesenteric artery of the fowl. *Anat. Rec.* **146**, 251–256.

Ball, R. A., Sautter, J. H., and Waibel, P. E. (1972). Adaptive features in the turkey aorta which precede plaque formation. *Atherosclerosis* **15**, 241–247.

Bamford, O. S., and Jones, D. R. (1976). The effects of asphyxia on afferent activity recorded from the cervical vagus in the duck. *Pflügers Arch.* **366**, 95–99.

Barnas, G. M., Gleeson, M., and Rautenberg, W. (1985). Respiratory and cardiovascular responses of the exercising chicken to spinal cord cooling at different ambient temperatures. I. Cardiovascular responses and blood gases. *J. Exp. Biol.* **114**, 415–426.

Baumel, J. J. (1975). *Aves* heart and blood vessels. In "Sisson and Grossman's The Anatomy of the Domestic Animals" (R. Getty, ed.). Saunders, Philadelphia.

Baumel, J. J., and Gerchman, L. (1968). The avian intercarotid anastomosis and its homologue in other vertebrates. *Am. J. Anat.* **122**, 1–18.

Bech, C., and Nomoto, S. (1982). Cardiovascular changes associated with treadmill running in the Pekin duck. *J. Exp. Biol.* **97**, 345–358.

Bell, C. (1969). Indirect cholinergic vasomotor control of intestinal blood flow in the domestic chicken. *J. Physiol.* **205**, 317–327.

Bennett, T. (1971). The adrenergic innervation of the pulmonary vasculature, the lung and the thoracic aorta, and on the presence of aortic bodies in the domestic fowl (*Gallus gallus domesticus* L.). *Z. Zellforsch.* **114**, 117–134.

Bennett, T. (1974). Peripheral and autonomic nervous systems. In "Avian Biology Volume 4" (D. S. Farner, J. R. King, and K. C. Parkes, eds.), pp. 1–77. Academic Press, New York.

Bennett, T. and Malmfors, T. (1970). The adrenergic nervous system of the domestic fowl (*Gallus domesticus* (L.)). *Z. Zellforsch.* **106**, 22–50.

Bennett, T., and Malmfors, T. (1974). Regeneration of the noradrenergic innervation of the cardiovascular system of the chick following treatment with 6-hydroxydopamine. *J. Physiol.* **242**, 517–532.

Bennett, T. and Malmfors, T. (1975a). Autonomic control of renal portal blood flow in the domestic fowl. *Experientia* **31**, 1177–1178.

Bennett, T., and Malmfors, T. (1975b). Characteristics of the noradrenergic innervation of the left atrium in the chick (*Gallus gallus domesticus*, L.). *Comp. Biochem. Physiol. C* **52**, 47–49.

Bennett, T., Cobb, J. L. S., and Malmfors, T. (1974). The vasomotor innervation of the inferior vena cava of the domestic fowl (*Gallus gallus domesticus* L.): I. Structural observations. *Cell Tissue Res.* **148**, 521–533.

Benzo, C. A. (1986). Nervous system. In "Avian Physiology" (P. D. Sturkie, ed.), 4th ed. pp. 1–36. Springer-Verlag, Berlin.

Bergel, D. H. (1961). The static properties of the arterial wall. *J. Physiol.* **156**, 445–457.

Berk, M. L. (1987). Projections of the lateral hypothalamus and bed nucleus of the stria terminalis to the dorsal vagal complex in the pigeon. *J. Comp. Neurol.* **260**, 140–156.

Berk, M. L., and Finkelstein, J. A. (1983). Long descending projections of the hypothalamus in the pigeon, *Columba livia*. *J. Comp. Neurol.* **220**, 127–136.

Berk, M. L., and Smith, S. E. (1994). Local and commissural neuropeptide-containing projections of the nucleus of the solitary tract to the dorsal vagal complex in the pigeon. *J. Comp. Neurol.* **347**, 369–396.

Berne, R. M., and Levy, M. N. (1992). "Physiology," Mosby, St. Louis, MO.

Bezuidenhout, A. J. (1984). The coronary circulation of the heart of the ostrich. *J. Anat.* **138**, 385–397.

Bickler, P. E., and Julian, D. (1992). Regional cerebral blood flow and tissue oxygenation during hypocarbia in geese. *Am. J. Physiol.* **263**, R221–R225.

Biegon, R. L., Epstein, P. M., and Pappano, A. J. (1980). Muscarinic antagonism of the effects of phosphodiesterase inhibitor (methylisobutylxanthine) in embryonic chick ventricle. *J. Pharmacol. Exp. Ther.* **215**, 348–356.

Biegon, R. L. and Pappano, A. J. (1980). Dual mechanism for inhibition of calcium-dependent action potentials by acetylcholine in avian ventricular muscle. Relationship to cyclic AMP. *Circ. Res.* **46**, 353–362.

Bishop, C. M. and Butler, P. J. (1995). Physiological modelling of oxygen consumption in birds during flight. *J. Exp. Biol.* **198**, 2153–2163.

Bishop, C. M., Butler, P. J., Egginton, S., el-Haj, A. J., and Gabrielsen, G. W. (1995). Development of metabolic enzyme activity in locomotor and cardiac muscles in the migratory barnacle goose. *Am. J. Physiol.* **269**, R64–R72.

Black, C. P., and Tenney, S. M. (1980). Oxygen transport during progressive hypoxia in high-altitude and sea-level waterfowl. *Resp. Physiol.* **39**, 217–239.

Blix, A. S., Wennergren, G., and Folkow, B. (1976). Cardiac receptors in ducks — A link between vasoconstriction and bradycardia during diving. *Acta Physiol. Scand.* **97**, 13–19.

Boelkins, J. N., Mueller, W. J., and Hall, K. L. (1973). Cardiac output distribution in the laying hen during shell formation. *Comp. Biochem. Physiol. A* **46**, 735–743.

Bogusch, G. (1974). The innervation of Purkinje fibres in the atrium of the avian heart. *Cell Tissue Res.* **150**, 57–66.

Bolton, T. B. (1967). Intramural nerves in the ventricular myocardium of the domestic fowl and other animals. *Br. J. Pharmacol. Chemother.* **31**, 253–268.

Bolton, T. B. (1969). Spontaneous and evoked release of neurotransmitter substances in the longitudinal muscle of the anterior mesenteric artery of the domestic fowl. *Br. J. Pharmacol.* **35**, 112–120.

Bolton, T. B. and Bowman, W. C. (1969). Adrenoreceptors in the cardiovascular system of the domestic fowl. *Eur. J. Pharmacol.* **5,** 121–132.

Bolton, T. B., and Raper, C. (1966). Innervation of domestic fowl and guinea-pig ventricles. *J. Pharm. Pharmacol.* **18,** 192–193.

Bopelet, M. (1974). Normal electrocardiogram of the chicken: its variations during vagal stimulation and following vagotomies. *Comp. Biochem. Physiol. A* **47,** 361–369.

Bossen, E., Sommer, J. R., and Waugh, R. A. (1978). Comparative stereology of the mouse and finch left ventricle. *Tissue Cell* **10,** 773–784.

Boulianne, M., Hunter, D. B., Julian, R. J., O'Grady, M. R., and Physick Sheard, P. W. (1992). Cardiac muscle mass distribution in domestic turkey and relationship to electrocardiogram. *Avian Dis.* **36,** 582–589.

Boulianne, M., Hunter, D. B., Physick-Sheard, P. W., Viel, L., and Julian, R. J. (1993a). Effect of exercise on cardiac output and other cardiovascular parameters of heavy turkeys and relevance to the sudden death syndrome. *Avian Dis.* **37,** 98–106.

Boulianne, M., Hunter, D. B., Viel, L., Physick-Sheard, P. W., and Julian, R. J. (1993b). Effect of exercise on the cardiovascular and respiratory systems of heavy turkeys and relevance to sudden death syndrome. *Avian Dis.* **37,** 83–97.

Bouverot, P., and Leitner, L. M. (1972). Arterial chemoreceptors in the domestic fowl. *Resp. Physiol.* **15,** 310–320.

Bouverot, P., Douguet, D., and Sébert, P. (1979). Role of the arterial chemoreceptors in ventilatory and circulatory adjustments to hypoxia in awake Pekin ducks. *J. Comp. Physiol. B* **133,** 177–186.

Brackenbury, J. H., el-Sayed, M. S., and Darby, C. (1990). Effects of treadmill exercise on the distribution of blood flow between the hindlimb muscles and abdominal viscera of the laying fowl. *Br. Poult. Sci.* **31,** 207–214.

Brackenbury, J. H., el-Sayed, M. S., and Jacques, A. L. (1993). Blood flow distribution during graded treadmill exercise in domestic cockerels. *Br. Poult. Sci.* **34,** 758–792.

Braun, E. J. (1982). Glomerular filtration in birds — its control. *Fed. Proc.* **41,** 2377–2381.

Brehm, G., Lindmar, R., and Loffelholz, K. (1992). Inhibitory and excitatory muscarinic receptors modulating the release of acetylcholine from the postganglionic parasympathetic neuron of the chicken heart. *Naunyn-Schmiedeberg's Arch. Pharmacol.* **346,** 375–382.

Brill, R. W., and Jones, D. R. (1981). On the suitability of Innovar, a neuroleptic analgesic, for cardiovascular experiments. *Can. J. Physiol. Pharmacol.* **59,** 1184–1189.

Brummermann, M., and Simon, E. (1990). Arterial hypotension in ducks adapted to high salt intake. *J. Comp. Physiol. B* **160,** 127–136.

Burrows, M. E., Braun, E. J., and Duckles, S. P. (1983). Avian renal portal valve: A reexamination of its innervation. *Am. J. Physiol.* **245,** H628–H634.

Bussow, H. (1973). Zar wandstruktur der großen arterien der vogel. *Z. Zellforsch.* **142,** 263–288.

Butler, D. G., Wilson, J. X., and Graves, L. E. (1986). α- and β-adrenergic mechanisms mediate blood pressure control by norepinephrine and angiotensin in ducks. *Gen. Comp. Endocrinol.* **61,** 323–329.

Butler, P. J. (1967). The effect of progressive hypoxia on the respiratory and cardiovascular systems of the chicken. *J. Physiol.* **191,** 309–324.

Butler, P. J. (1991). Exercise in birds. *J. Exp. Biol.* **160,** 233–262.

Butler, P. J., and Jones, D. R. (1968). Onset of and recovery from diving bradycardia in ducks. *J. Physiol.* **196,** 255–272.

Butler, P. J., and Jones, D. R. (1971). The effect of variations in heart rate and regional distribution of blood flow on the normal pressor response to diving in ducks. *J. Physiol.* **214,** 457–479.

Butler, P. J., and Taylor, E. W. (1973). The effect of hyperoxic hypoxia, accompanied by different levels of lung ventilation, on heart rate in the duck. *Resp. Physiol.* **19,** 176–187.

Butler, P. J., and Taylor, E. W. (1974). Responses of the respiratory and cardiovascular systems of chickens and pigeons to changes in PaO_2 and $PaCO_2$. *Resp. Physiol.* **21,** 351–363.

Butler, P. J., and Taylor, E. W. (1983). Factors affecting the respiratory and cardiovascular responses to hypercapnic hypoxia, in Mallard ducks. *Resp. Physiol.* **53,** 109–127.

Butler, P. J., Turner, D. L., Al-Wassia, A., and Bevan, R. M. (1988). Regional distribution of blood flow during swimming in the tufted duck (*Aythya fuligula*). *J. Exp. Biol.* **135,** 461–472.

Butler, P. J., West, N. H., and Jones, D. R. (1977). Respiratory and cardiovascular responses of the pigeon to sustained, level flight in a wind-tunnel. *J. Exp. Biol.* **71,** 7–26.

Byers, R. L., and Snyder, G. K. (1984). Effects of maturation on tissue capillarity in chickens. *Resp. Physiol.* **58,** 137–150.

Cabot, J. B., and Cohen, D. H. (1977a). Avian sympathetic cardiac fibers and their cells of origin: Anatomical and electrophysiological characteristics. *Brain Res.* **131,** 73–87.

Cabot, J. B., and Cohen, D. H. (1977b). Anatomical and physiological characterization of avian sympathetic cardiac afferents. *Brain Res.* **131,** 89–101.

Cabot, J. B., and Cohen, D. H. (1980). Neural control of the avian heart. In "Hearts and Heart-like Organs" (G. B. Bourne, ed.), Vol. 1, pp. 199–258. Academic Press, NY.

Cabot, J. B., Carroll, J., and Bogan, N. (1991a). Localization of cardiac parasympathetic preganglionic neurons in the medulla oblongata of pigeon, *Columba livia*: A study using fragment C of tetanus toxin. *Brain Res.* **544,** 162–168.

Cabot, J. B., Mennone, A., Bogan, N., Carroll, J., Evinger, C., and Erichsen, J. T. (1991b). Retrograde, trans-synaptic and transneuronal transport of fragment C of tetanus toxin by sympathetic preganglionic neurons. *Neuroscience* **40,** 805–823.

Cabot, J. B., Reiner, A., and Bogan, N. (1982). Avian bulbospinal pathways: anterograde and retrograde studies of cells of origin, funicular trajectories and laminar terminations. *Prog. Brain Res.* **57,** 79–108.

Campbell, T. W. (1995). "Avian Hematology and Cytology," 2nd ed. Iowa State Univ. Press, Ames.

Chiappe, L. M. (1995). The first 85 million years of avian evolution. *Nature* **378,** 349–355.

Chien, S., Usami, S., Dellenback, R. J., and Bryant, C. A. (1971). Comparative hemorheology—Hematological implications of species differences in blood viscosity. *Biorheology* **8,** 35–57.

Cinar, A., Bagci, C., Belge, F., and Uzun, M. (1996). The electrocardiogram of the Pekin duck. *Avian Dis.* **40,** 919–923.

Clarke, J. and Nicol, S. (1993). Blood viscosity of the little penguin, *Eudyptula minor* and the Adelie penguin, *Pygoscelis adeliae*: Effects of temperature and shear rate. *Physiol. Zool.* **66,** 720–731.

Clemens, D. T. (1990). Interspecific variation and effects of altitude on blood properties of rosy finches *Leucosticte arctoa* and house finches *Carpodacus mexicanus*. *Physiol. Zool.* **63,** 288–307.

Cohen, D. H. (1980). The functional neuroanatomy of a conditioned response. In "Neural Mechanisms of Goal-Directed Behavior and Learning" (R. F. Thompson, L. H. Hicks, and V. B. Shvyrkov, eds.), pp. 283–302. Academic Press, New York.

Cohen, D. H. (1984). Identification of vertebrate neurons modified during learning: Analysis of sensory pathways. In "Primary Neural Substrates of Learning and Behavioral Change" (D. L. Alkon and J. Farley, eds.), pp. 129–154. Cambridge Univ. Press, Cambridge, UK.

Cohen, D. H., and Schnall, A. M. (1970). Medullary cells of origin of vagal cardioinhibitory fibers in the pigeon. II. Electrical stimulation of the dorsal motor nucleus. *J. Comp. Neurol.* **140,** 321–342.

Cohen, D. H., Schnall, A. M., Macdonald, R. L., and Pitts, L. H. (1970). Medullary cells of origin of vagal cardioinhibitory fibers in the pigeon. I. Anatomical studies of peripheral vagus nerve and the dorsal motor nucleus. *J. Comp. Neurol.* **140**, 299–320.

Cohen, W. D. (1978). On erythrocyte morphology. *Blood Cells* **4**, 449–451.

Coote, J. H. (1975). Physiological significance of somatic afferent pathways from skeletal muscle and joints with reflex effects on the heart and circulation. *Brain Res.* **87**, 139–144.

Corvetti, G., Andreotti, L., and Sisto-Daneo, L. (1988). Chick heart peptidergic innervation: localization and development. *Basic Appl. Histochem.* **32**, 485–493.

Dantzler, W. H. (1989). "Comparative Physiology of the Vertebrate Kidney," Springer-Verlag, New York.

Davies, F. (1930). The conducting system of the bird's heart. *J. Anat.* **64**, 129–146.

Deighton, N. M., Motomura, S., Borquez, D., Zerkowski, H. R., Doetsch, N., and Brodde, O. E. (1990). Muscarinic cholinoceptors in the human heart: demonstration, subclassification, and distribution. *Naunyn-Schmiedebergs Arch. Pharmacol.* **341**, 14–21.

Denton, T., Diamond, G. A., Helfant, R. H., Khan, S., and Karagueuzian, H. (1990). Fascinating rhythm: A primer on chaos theory and its application to cardiology. *Am. Heart J.* **120**, 1419–1440.

DeSantis, V. P., Lindmar, L. R., and Loffelholz, K. (1975). Evidence for noradrenaline and adrenaline as sympathetic transmitters in the chicken. *Br. J. Pharmacol.* **55**, 343–350.

Djojosugito, A. M., Folkow, B., and Kovách, A. G. B. (1968). The mechanisms behind the rapid blood volume restoration after hemorrhage in birds. *Acta Physiol. Scand.* **74**, 114–122.

Djojosugito, A. M., Folkow, B., and Yonce, L. R. (1969). Neurogenic adjustments of muscle blood flow, cutaneous a-v shunt flow and of venous tone during "diving" in ducks. *Acta Physiol. Scand.* **75**, 377–386.

Duchamp, C., and Barre, H. (1993). Skeletal muscle as the major site of non-shivering thermogenesis in cold-acclimated ducklings. *Am. J. Physiol.* **265**, R1076–R1083.

Ede, D. A. (1964). "Bird Structure," Hutchinson Educational, London.

Einzig, S., Staley, N. A., Mettler, E., Nicoloff, D. M., and Noren, G. R. (1980). Regional myocardial blood flow and cardiac function in a naturally occurring congestive cardiomyopathy of turkeys. *Cardiovas. Res.* **14**, 396–407.

Ellis, C. G., Potter, R. F., and Groom, A. C. (1983). The krogh cylinder geometry is not appropriate for modelling O_2 transport in contracted skeletal muscle. *Adv. Exp. Med. Biol.* **159**, 253–268.

Eranko, O., and Eranko, L. (1977). Morphological indications of SIF cell functions. *Adv. Biochem. Psychopharmacol.* **16**, 525–531.

Estavillo, J. A. (1978). Fiber size and sensory endings of the middle cardiac nerve of the domestic fowl (*Gallus domesticus*). *Acta Anat.* **101**, 104–109.

Estavillo, J., and Burger, R. E. (1973a). Cardiac afferent activity in depressor nerve of the chicken. *Am. J. Physiol.* **225**, 1063–1066.

Estavillo, J., and Burger, R. E. (1973b). Avian cardiac receptors: Activity changes by blood pressure, carbon dioxide, and pH. *Am. J. Physiol.* **225**, 1067–1071.

Estavillo, J. A., Adamson, T. P., and Burger, R. E. (1990). Middle cardiac nerve section alters ventilatory response to PaCO2 in the cockerel. *Resp. Physiol.* **81**, 349–358.

Evered, M. D., and Fitzsimons, J. T. (1981). Drinking and changes in blood pressure in response to angiotensin II in the pigeon *Columba livia*. *J. Physiol.* **310**, 337–352.

Falck, B. (1962). Observations on the possibilities of the cellular localization of monoamines by a fluorescence method. *Acta Physiol. Scand.* **197** (Suppl.).

Faraci, F. M. and Fedde, M. R. (1986). Regional circulatory responses to hypocapnia and hypercapnia in bar-headed geese. *Am. J. Physiol.* **250**, R499–R504.

Faraci, F. M., Kilgore, D. L., and Fedde, M. R. (1984). Oxygen delivery to the heart and brain during hypoxia: Pekin duck vs. bar-headed goose. *Am. J. Physiol.* **247**, R69–R75.

Faraci, F. M., Kilgore, D. L., and Fedde, M. R. (1985). Blood flow distribution during hypocapnic hypoxia in Pekin ducks and bar-headed geese. *Resp. Physiol.* **61**, 21–30.

Fedde, M. R., Orr, J. A., Shams, S., and Scheid, P. (1989). Cardiopulmonary function in exercising bar-headed geese during normoxia and hypoxia. *Resp. Physiol.* **77**, 239–252.

Feigl, E., and Folkow, B. (1963). Cardiovascular responses in "diving" and during brain stimulation in ducks. *Acta Physiol. Scand.* **57**, 99–110.

Fibiger, H. C. (1982). The organization and some projections of cholinergic neurons of the mammalian forebrain. *Brain Res. Rev.* **4**, 327–388.

Fischer, T. M. (1978). On erythrocyte morphology, A commentary: A comparison of the flow behavior of disc shaped versus elliptic red blood cells (RBC). *Blood Cells* **4**, 453–461.

Folkow, B. (1952). Impulse frequency in sympathetic vasomotor fibres correlated to the release and elimination of the transmitter. *Acta Physiol. Scand.* **25**, 49–76.

Folkow, B. and Rubenstein, E. H. (1965). Effect of brain stimulation on "diving" in ducks. *Hvalradets Skr.* **48**, 30–41.

Folkow, B., and Yonce, L. R. (1967). The negative inotropic effect of vagal stimulation on the heart ventricles of the duck. *Acta Physiol. Scand.* **71**, 77–84.

Folkow, B., Fuxe, K., and Sonnenschein, R. R. (1966). Responses of skeletal musculature and its vasculature during "diving" in the duck: Peculiarities of the adrenergic vasoconstrictor innervation. *Acta Physiol. Scand.* **67**, 327–342.

Folkow, B., Nilsson, N. J., and Yonce, L. R. (1967). Effects of "diving" on cardiac output in ducks. *Acta Physiol. Scand.* **70**, 347–361.

Furchgott, R. F., and Zawadzki, J. V. (1980). The obligatory role of endothelial cells in the relaxation of arterial smooth muscle by acetylcholine. *Nature* **288**, 373–376.

Furilla, R. A., and Jones, D. R. (1986). The contribution of nasal receptors to the cardiac response to diving in restrained and unrestrained redhead ducks (*Aythya americana*). *J. Exp. Biol.* **121**, 227–238.

Furilla, R. A., and Jones, D. R. (1987a). The relationship between dive and predive heart rates in restrained and free dives by diving ducks. *J. Exp. Biol.* **127**, 333–348.

Furilla, R. A., and Jones, D. R. (1987b). Cardiac responses to dabbling and diving the mallard, *Anas platyrhynchos*. *Physiol. Zool.* **60**, 406–412.

Furness, J. B., and Costa, M. (1987). "The Enteric Nervous System," Churchill Livingston, Edinburgh.

Furnival, C. M., Linden, R. J., and Snow, H. M. (1973). The inotropic effect on the heart of stimulating the vagus in the dog, duck and toad. *J. Physiol.* **230**, 155–170.

Gabbott, G. R. J., and Jones, D. R. (1987). Habituation of the cardiac response to involuntary diving in diving and dabbling ducks. *J. Exp. Biol.* **131**, 403–415.

Gabella, G. (1976). "Structure of the Autonomic Nervous System." J Wiley, New York.

Gaehtgens, P., Schmidt, F., and Will, G. (1981a). Comparative rheology of nucleated and non-nucleated red blood cells. I. Microrheology of avian erythrocytes during capillary flow. *Pflügers Arch.* **390**, 278–282.

Gaehtgens, P., Schmidt, F., and Will, G. (1981b). Comparative rheology of nucleated and non-nucleated red blood cells. II. Rheological properties of avian red cell suspensions in narrow capillaries. *Pflügers Arch.* **390,** 283–287.

Gayeski, T. E. J., and Honig, C. R. (1986). O_2 gradients from sarcolemma to cell interior in red muscle at maximal VO_2. *Am. J. Physiol.* **251,** H789–H799.

Gilman, A. G., Rall, T. W., Nies, A. S., and Taylor, P. (1990). "The Pharmacological Basis of Therapeutics," 8th ed. Pergamon, Oxford, UK.

Glahn, R. P., Bottje, W. G., Maynard, P., and Wideman, R. F. (1993). Response of the avian kidney to acute changes in arterial perfusion pressure and portal blood supply. *Am. J. Physiol.* **264,** R428–R434.

Glenny, F. H. (1940). A systematic study of the main arteries in the region of the heart — Aves. *Anat. Rec.* **76,** 371–380.

Gold, M. R., and Cohen, D. H. (1984). The discharge characteristics of vagal cardiac neurons during classically conditioned heart rate change. *J. Neurosci.* **4,** 2963–2971.

Goldberg, T. M., and Bolnick, D. A. (1980). Electrocardiograms from the chicken, emu, red-tailed hawk and Chilean tinamou. *Comp. Biochem. Physiol. A* **67,** 15–19.

Goldberg, J. M., Johnson, M. H., and Whitelaw, K. D. (1983). Effect of cervical vagal stimulation on chicken heart rate and atrioventricular conduction. *Am. J. Physiol.* **244,** R235–R243.

Goldberger, A. L. (1991). Is the normal heartbeat chaotic or homeostatic? *NIPS* **6,** 87–91.

Goldberger, A. L., Rigney, D. R., and West, B. J. (1990). Chaos and fractals in human physiology. *Sci. Am.* **262,** 42–49.

Gooden, B. A. (1980). The effect of hypoxia on vasoconstrictor responses of isolated mesenteric arterial vasculature from chicken and duckling. *Comp. Biochem. Physiol. C* **67,** 219–222.

Goplerud, J. M., Wagerle, L. C., and Delivoria-Papadopoulos, M. (1989). Regional cerebral blood flow response during and after acute asphyxia in newborn piglets. *J. Appl. Physiol.* **66,** 2827–2832.

Gray, S. D., McDonagh, P. F., and Gore, R. W. (1983). Comparison of functional and total capillary densities in fast and slow muscles of the chicken. *Pflügers Arch.* **397,** 209–213.

Grindlay, J. H., Herrick, J. F., and Mann, F. C. (1939). Measurement of the blood flow of the spleen. *Am. J. Physiol.* **127,** 106–118.

Grubb, B. R. (1983). Allometric relations of cardiovascular function in birds. *Am. J. Physiol.* **245,** H567–H572.

Grubb, B. R., Jorgensen, D. D., and Conner, M. (1983). Cardiovascular changes in the exercising emu. *J. Exp. Biol.* **104,** 193–201.

Grubb, B. R., Mills, C. D., Colacino, J. M., and Schmidt-Neilsen, K. (1977). Effect of arterial carbon dioxide on cerebral blood flow in ducks. *Am. J. Physiol.* **232,** H596–H601.

Guyton, A. C., Young, D. B., DeClue, J. W., Trippodo, N., and Hall, J. E. (1975). Fluid balance, renal function and blood pressure. *Clin. Nephrol.* **4,** 122–126.

Hargens, A. R., Millard, R. W., and Johansen, K. (1974). High capillary permeability in fishes. *Comp. Biochem. Physiol. A* **48,** 675–680.

Hartman, F. A. (1961). "*Smithsonian Miscellaneous Collections,*" Vol. 143, pp. 1–91.

Hasegawa, K., and Nishimura, H. (1991). Humoral factor mediates acetylcholine-induced endothelium-dependent relaxation of chicken aorta. *Gen. Comp. Endocrinol.* **84,** 164–169.

Hasegawa, K., Nishimura, H., and Khosla, M. C. (1993). Angiotensin II-induced endothelium-dependent relaxation of fowl aorta. *Am. J. Physiol.* **264,** R903–R911.

Hawkey, C. M., Bennett, P. M., Gascoyne, S. C., Hart, M. G., and Kirkwood, J. K. (1991). Erythrocyte size, number and haemoglobin content in vertebrates. *Br. J. Haematol.* **77,** 392–397.

Hebb, C. (1969). Motor innervation of pulmonary blood vessels. In "The Pulmonary Circulation and Interstitial Space" (A. P. Fishman and H. H. Hecht, eds.), Part 3. Univ. of Chicago Press, Chicago.

Heieis, M., and Jones, D. R. (1988). Blood flow and volume distribution during forced submergence in Pekin ducks (*Anas platyrhynchos*). *Can. J. Zool.* **66,** 1589–1596.

Hempleman, S. C., Powell, F. L., and Prisk, G. K. (1992). Avian arterial chemoreceptor responses to steps of CO_2 and O_2. *Resp. Physiol.* **90,** 325–340.

Henderson, I. W., and Deacon, C. F. (1993). Phylogeny and comparative physiology of the renin-angiotensin system. In "The Renin-Angiotensin System: Biochemistry and Physiology" (J. I. S. Robertson and M. G. Nicholls, eds.), Vol. 1, pp. 2.1–2.28. Mosby, New York.

Hill, J. R., and Goldberg, J. M. (1980). P-wave morphology and atrial activation in the domestic fowl. *Am. J. Physiol.* **239,** R483–R488.

Hirakow, R. (1970). Ultrastructural characteristics of the mammalian and sauropsidian heart. *Am. J. Cardiol.* **25,** 195.

Hirsch, E. F. (1963). The innervation of the human heart. V. A comparative study of the intrinsic innervation of the heart in vertebrates. *Exp. Mol. Pathol.* **2,** 384–401.

Hirsch, E. F. (1970). "The Innervation of the Vertebrate Heart," Charles C. Thomas, Springfield, IL.

Hirst, G. D. S. and Edwards, F. R. (1989). Sympathetic neuroeffector transmission in arteries and arterioles. *Physiol. Rev.* **69,** 546–604.

Hodges, R. D. (1974). "The Histology of the Domestic Fowl," Academic Press, London.

Holt, J. P., Rhode, E. A., and Kines, H. (1968). Ventricular volumes and body weight in mammals. *Am. J. Physiol.* **215,** 704–715.

Holzbauer, M., and Sharman, D. F. (1972). The distribution of catecholamines in vertebrates. In "Handbook of Experimental Pharmacology: Catecholamines" (H. Blaschko and E. Muscholl, eds.), Vol. 33, pp. 110–185. Springer-Verlag, Berlin.

Honig, C. R., Gayeski, T. E. J., and Groebe, K. (1991). Myoglobin and oxygen gradients. In "The Lung: Scientific Foundations" (R. G. Crystal, J. B. West, B. J. Barnes, N. S. Cherniak, and E. R. Weibel, eds.), Raven, New York.

Hopkins, D. A. (1987). The dorsal motor nucleus of the vagus nerve and the nucleus ambiguus: Structure and connections. In "Cardiogenic Reflexes" (R. Hainsworth, P. N. McWilliam, and D. A. S. G. Mary, eds.), pp. 185–203. Oxford Univ. Press, Oxford, UK.

Huang, H. C., Sung, P. K., and Huang, T. F. (1974). Blood volume, lactic acid and catecholamines in diving response in ducks. *Taiwan I. Hsueh Hui Tsa Chih.* **73,** 203–210.

Huber, J. F. (1936). Nerve roots and nuclear groups in the spinal cord of the pigeon. *J. Comp. Neurol.* **65,** 43–91.

Hudson, D. M., and Jones, D. R. (1982). Remarkable blood catecholamine levels in forced dived ducks. *J. Exp. Zool.* **224,** 451–456.

Hughes, A. F. W. (1942). The histogenesis of the arteries of the chick embryo. *J. Anat.* **77,** 266–287.

Hunsaker, W. G., Robertson, A., and Magwood, S. E. (1971). The effect of round heart disease on the electrocardiogram and heart weight of turkey poults. *Poult. Sci.* **50,** 1712–1720.

Inagami, R., Naruse, M., and Hoover, R. (1995). Endothelium as an endocrine organ. *Annu. Rev. Physiol.* **57,** 171–189.

Inoue, D., Hachisu, M., and Pappano, A. J. (1983). Acetylcholine increases resting membrane potassium conductance in atrial but not in ventricular muscle during muscarinic inhibition of Ca^{++}-dependent action potentials in chick heart. *Circ. Res.* **53,** 158–167.

Jeck, D., Lindmar, R., Löffelholz, K., and Wanke, M. (1988). Subtypes of muscarinic receptor on cholinergic nerves and atrial cells of chicken and guinea-pig hearts. *Br. J. Pharmacol.* **93,** 357–366.

Johansen, K. (1964). Regional distribution of circulating blood during submersion asphyxia in the duck. *Acta Physiol. Scand.* **62,** 1–9.

Johansen, K., and Reite, O. B. (1964). Cardiovascular responses to vagal stimulation and cardioaccelerator nerve blockade in birds. *Comp. Biochem. Physiol.* **12,** 479–487.

Jones, D. R. (1969). Avian afferent vagal activity related to respiratory and cardiac cycles. *Comp. Biochem. Physiol. A* **28,** 961–965.

Jones, D. R. (1973). Systemic arterial baroreceptors in ducks and the consequences of their denervation on some cardiovascular responses to diving. *J. Physiol.* **234,** 499–518.

Jones, D. R. (1991). Cardiac energetics and design of arterial systems. In "Efficiency and Economy in Animal Physiology" (R. W. Blake, ed.), pp. 159–168. Cambridge Univ. Press, Cambridge, UK.

Jones, D. R. and Holeton, G. F. (1972a). Cardiovascular and respiratory responses of ducks to progressive hypocapnic hypoxia. *J. Exp. Biol.* **56,** 657–666.

Jones, D. R. and Holeton, G. F. (1972b). Cardiac output of ducks during diving. *Comp. Biochem. Physiol. A* **41,** 639–645.

Jones, D. R. and Johansen, K. (1972). The blood vascular system of birds. In "Avian Biology" (D. S. Farner and J. E. King, eds.), Vol. 2, pp. 157–285. Academic Press, New York.

Jones, D. R., and Milsom, W. K. (1982). Peripheral receptors affecting breathing and cardiovascular function in non-mammalian vertebrates. *J. Exp. Biol.* **100,** 59–91.

Jones, D. R., Milsom, W. K., and Gabbott, G. R. J. (1982). Role of central and peripheral chemoreceptors in diving responses of ducks. *Am. J. Physiol.* **243,** R537–R545.

Jones, D. R., Milsom, W. K., and West, N. H. (1980). Cardiac receptors in ducks: The effect of their stimulation and blockade on diving bradycardia. *Am. J. Physiol.* **238,** R50–R56.

Jones, D. R., and Purves, M. J. (1970). The carotid body in the duck and the consequences of its denervation upon the cardiac responses to immersion. *J. Physiol.* **211,** 279–294.

Jones, D. R., Bryan, R. M., West, N. H., Lord, R. H., and Clark, B. (1979). Regional distribution of blood flow during diving in the duck(*Anas platyrhynchos*). *Can. J. Zool.* **57,** 995–1002.

Jones, D. R., Milsom, W. K., Smith, F. M., West, N. H., and Bamford, O. S. (1983). Diving responses in ducks after acute barodenervation. *Am. J. Physiol.* **245,** R222–R229.

Kamath, M. V., and Fallen, E. L. (1993). Power spectral analysis of heart rate variability: A noninvasive signature of cardiac autonomic function. *Crit. Rev. Biomed. Eng.* **21,** 245–311.

Kamimura, K., Nishimura, H., and Bailey, J. R. (1995). Blockade of beta-adrenoceptor in control of blood pressure in fowl. *Am. J. Physiol.* **269,** R914–R922.

Katz, D. M., and Karten, H. J. (1979). The discrete anatomical localization of vagal aortic afferents within a catecholamine-containing cell group in the nucleus solitarius. *Brain Res.* **171,** 187–195.

Katz, D. M., and Karten, H. J. (1983a). Subnuclear organization of the dorsal motor nucleus of the vagus nerve in the pigeon, *Columba livia. J. Comp. Neurol.* **217,** 31–46.

Katz, D. M., and Karten, H. J. (1983b). Visceral representation within the nucleus of the tractus solitarius in the pigeon, *Columba livia. J. Comp. Neurol.* **218,** 42–73.

Katz, D. M., and Karten, H. J. (1985). Topographic representation of visceral target organs within the dorsal motor nucleus of the vagus nerve of the pigeon *Columba livia. J. Comp. Neurol.* **242,** 397–414.

Kedem, O., and Katchalsky, A. (1958). Thermodynamic analysis of the permeabililty of biological membanes to non-electrolytes. *Biochim. Biophys. Acta* **27,** 229–246.

Kiley, J. P., Faraci, F. M., and Fedde, M. R. (1985). Gas exchange during exercise in hypoxic ducks. *Resp. Physiol.* **59,** 105–115.

Kiley, J. P., Kuhlmann, W. D., and Fedde, M. R. (1979). Respiratory and cardiovascular responses to exercise in the duck. *J. Appl. Physiol.* **47,** 827–833.

Kirby, M. L., Conrad, D. C., and Stewart, D. E. (1987). Increase in the cholinergic cardiac plexus in sympathetically aneural chick hearts. *Cell Tissue Res.* **247,** 489–496.

Kisch, B. (1951). The electrocardiogram of birds: chicken, duck, pigeon. *Exp. Med. Surg.* **9,** 103–124.

Kobinger, W., and Oda, M. (1969). Effects of sympathetic blocking substances on the diving reflex of ducks. *Eur. J. Pharmacol.* **7,** 289–295.

Koch-Weser, J. (1971). Beta-receptor blockade and myocardial effects of cardiac glycosides. *Circ. Res.* **28,** 109–118.

Koelle, G. B. (1963). Cytological distributions and physiological functions of cholinesterases. In "Handbook of Experimental Pharmacology" (O. Eichler and A. Farah, eds.). Springer-Verlag, Heidelberg.

Kohmoto, O., Ikenouchi, H., Hirata, Y., Momomura, S., Serizawa, T., and Barry, W. H. (1993). Variable effects of endothelin-1 on $[Ca2+]i$ transients, pHi, and contraction in ventricular myocytes. *Am. J. Physiol.* **265,** H793–H800.

Komori, S., Ohashi, H., Okada, T., and Takewaki, T. (1979). Evidence that adrenaline is released from adrenergic neurones in the rectum of the fowl. *Br. J. Pharmacol.* **65,** 261–269.

Kontos, H. A. (1981). Regulation of the cerebral circulation. *Annu. Rev. Physiol.* **43,** 397–407.

Kotilainen, P. V., and Putkonen, P. T. S. (1974). Respiratory and cardiovascular responses to electrical stimulation of the avian brain with emphasis on inhibitory mechanisms. *Acta Physiol. Scand.* **90,** 358–369.

Kovách, A. G. B., and Balint, T. (1969). Comparative study of haemodilation after haemorrhage in the pigeon and the rat. *Acta Physiol. Acad. Sci. Hung.* **35,** 231–243.

Krogh, A. (1919). The number and distribution of capillaries in muscles with calculations of the oxygen pressure head necessary for supplying the tissue. *J. Physiol.* **52,** 409–415.

Lacombe, A. M. A., and Jones, D. R. (1990). The source of circulating catecholamines in forced dived ducks. *Gen. Comp. Endocrinol.* **80,** 41–47.

Landis, E. M. (1927). Micro-injection studies of capillary permeability. II. The relation between capillary pressure and the rate at which fluid passes through the walls of single capillaries. *Am. J. Physiol.* **82,** 217–238.

Landis, E. M., and Pappenheimer, J. R. (1963). Exchange of substances through the capillary walls. In "Handbook of Physiology, Section 2: Circulation" (W. F. Hamilton and P. Dow, eds.), Vol. II, pp. 961–1034. American Physiological Society, Washington, DC.

Lang, S. A., and Levy, M. N. (1989). Effects of vagus nerve on heart rate and ventricular contractility in chicken. *Am. J. Physiol.* **256,** H1295–H1302.

Langille, B. L. (1983). Role of venoconstriction in the cardiovascular responses of ducks to head immersion. *Am. J. Physiol.* **244,** R292–R298.

Langille, B. L., and Jones, D. R. (1975). Central cardiovascular dynamics of ducks. *Am. J. Physiol.* **228,** 1856–1861.

Langille, B. L., and Jones, D. R. (1976). Examination of elastic non-uniformity in the arterial system using a hydraulic model. *J. Biomechan.* **9,** 755–761.

Lasiewski, R. C., Weathers, W. W., and Bernstein, M. H. (1967). Physiological responses of the giant hummingbird, *Patagona gigas. Comp. Biochem. Physiol.* **23,** 797–813.

Leon-Velarde, F., Sanchez, J., Bigard, A. X., Brunet, A., Lesty, C., and Monge, C. C. (1993). High altitude tissue adaptation in Andean coots: Capillarity, fibre area, fibre type and enzymatic activities of skeletal muscle. *J. Comp. Physiol.B* **163,** 52–58.

Leonard, R. B., and Cohen, D. H. (1975). Responses of sympathetic postganglionic neurons to peripheral nerve stimulation in the pigeon (*Columba livia*). *Exp. Neurol.* **49,** 466–486.

Levy, M. N. (1971). Sympathetic-parasympathetic interactions in the heart. *Circ. Res.* **29,** 437–445.

Lewis, T. (1916). The spread of the excitatory process in the vertebrate heart. V. The bird's heart. *Phil. Trans. Roy. Soc. (Lond., Ser. B)* **207,** 298–311.

Lillie, F. R. (1908). "Development of the Chick," Holt, New York.

Lillo, R. S., and Jones, D. R. (1982). Effect of cardiovascular variables on hyperpnea during recovery from diving in ducks. *J. Appl. Physiol.* **52,** 206–215.

Lillo, R. S., and Jones, D. R. (1983). Influence of ischemia and hyperoxia on breathing in ducks. *J. Appl. Physiol.* **55,** 400–408.

Lindmar, R., Loffelholz, K., Weide, W., and Weis, S. (1983). Evidence for bilateral vagal innervation of postganglionic parasympathetic neurons in chicken heart. *J. Neur. Transmiss.* **56,** 239–247.

Lipowsky, H. H., Usami, S., and Chien, S. (1980). In vivo measurements of "apparent viscosity" and microvessel hematocrit in the mesentary of the cat. *Microvasc. Res.* **19,** 297.

Loewy, A. D., and Spyer, K. M. (1990). "Central Regulation of Autonomic Functions", Oxford Univ. Press, Oxford, UK.

Loffelholz, K., Brehm, R., and Lindmar, R. (1984). Hydrolysis, synthesis, and release of acetylcholine in the isolated heart. *Fed. Proc.* **43,** 2603–2606.

Lu, Y., James, T. N., Bootsma, M., and Terasaki, F. (1993a). Histological organization of the right and left atrioventricular valves of the chicken heart and their relationship to the atrioventricular Purkinje ring and the middle bundle branch. *Anat. Rec.* **235,** 74–86.

Lu, Y., James, T. N., Yamamoto, S., and Terasaki, F. (1993b). Cardiac conduction in the chicken: Gross anatomy plus light and electron microscopy. *Anat. Rec.* **236,** 493–510.

Lumb, W. V., and Jones, E. W. (1984). "Veterinary Anesthesia," 2nd ed. Lea and Febiger, Philadelphia, PA.

Macdonald, R. L., and Cohen, D. H. (1970). Cells of origin of sympathetic pre- and postganglionic cardioacceleratory fibers in the pigeon. *J. Comp. Neurol.* **140,** 343–358.

Macdonald, R. L., and Cohen, D. H. (1973). Heart rate and blood pressure responses to electrical stimulation of the central nervous system in the pigeon (*Columba livia*). *J. Comp. Neurol.* **150,** 109–136.

Maginnis, L. A., Bernstein, M. H., Deitch, M. A., and Pinshow, B. (1997). Effects of chronic hypobaric hypoxia on blood oxygen binding in pigeons. *J. Exp. Zool.* **277,** 293–300.

Malinovsky, L. (1962). Contribution to the anatomy of the vegetative nervous system in the neck and thorax of the domestic pigeon. *Acta Anat.* **50,** 326–347.

Malliani, A., Pagani, M., and Bergamaschi, M. (1979). Positive feedback sympathetic reflexes and hypertension. *Am. J. Cardiol.* **44,** 860–865.

Mangold, E. (1919). Elektrographischel Untersuchungen del Erregungsverlaufes im Vogelherzen. *Arch. Ges. Physiol. (Pflügers).* **175,** 327–354.

Manning, P. J., and Middleton, C. C. (1972). Atherosclerosis in wild turkeys: Morphological features of lesions and lipids in serum and aorta. *Am. J. Vet. Res.* **33,** 1237–1246.

Mathieu-Costello, O. (1991). Morphometric analysis of capillary geometry in pigeon pectoralis muscle. *Am. J. Anat.* **191,** 74–84.

Mathieu-Costello, O., Agey, P. J., and Normand, H. (1996). Fiber capillarization in flight muscle of pigeons native and flying at high altitude. *Resp. Physiol.* **103,** 187–194.

Mathieu-Costello, O., Suarez, R. K., and Hochachka, P. W. (1992). Capillary-to-fiber geometry and mitochondrial density in hummingbird flight muscle. *Resp. Physiol.* **89,** 113–132.

Mathieu-Costello, O., Agey, P. J., Logemann, R. B., Florez-Duquett, M., and Bernstein, M. H. (1994). Effect of flying activity on capillary-fiber geometry in pigeon flight muscle. *Tissue Cell* **26,** 57–73.

Mathur, P. N. (1973). Distribution of the specialized conducting tissue in the avian heart. *Indian J. Zool.* **1,** 17–27.

Mathur, R., and Mathur, A. (1974). Nerves and nerve terminations in the heart of *Columba livia*. *Anat. Anz.* **136,** 40–47.

McDonald, D. A. (1974). "Blood Flow in Arteries." Williams and Wilkins, Baltimore, MD.

McKenzie, B. E., Will, J. A., and Hardie, A. (1971). The electrocardiogram of the turkey. *Avian Dis.* **15,** 737–744.

Mellander, S., and Johansson, B. (1968). Control of resistance, exchange, and capacitance functions in the peripheral circulation. *Pharmacol. Rev.* **20,** 117–196.

Michel, C. C. (1997). Starling: The formulation of his hypothesis of microvascular fluid exchange and its significance after 100 years. *Exp. Physiol.* **82,** 1–30.

Millard, R. W. (1980). Depressed baroreceptor-cardiac reflex sensitivity during simulated diving in ducks. *Comp. Biochem. Physiol. A* **65,** 247–249.

Milnor, W. R. (1979). Aortic wavelength as a determinant of the relation between heart rate and body size in mammals. *Am. J. Physiol.* **237,** R3–R6.

Moore, A. F., Strong, J. H., and Buckley, J. P. (1981). Cardiovascular actions of angiotensin in the fowl (*Gallus domesticus*). I. Analysis. *Res. Commun. Chem. Path. Pharmacol.* **32,** 423–445.

Moore, E. N. (1965). Experimental electrophysiological studies on avian hearts. *Ann. N.Y. Acad. Sci.* **127,** 127–144.

Moore, E. N. (1967). Phylogenetic observations on specialized cardiac tissues. *Bull. N.Y. Acad. Med.* **43,** 1138–1159.

Nakamura, Y., Nishimura, H., and Khosla, M. C. (1982). Vasodepressor action of angiotensin in conscious chickens. *Am. J. Physiol.* **243,** H456–H462.

Nicol, S. C., Melrose, W., and Stahel, C. D. (1988). Hematology and metabolism of the blood of the little penguin, *Eudyptula minor*. *Comp. Biochem. Physiol. A* **89,** 383–386.

Nilsson, S., and Holmgren, S. (1994). "Comparative Physiology and Evolution of the Autonomic Nervous System," Harwood Academic Publishers, London, UK.

Nonidez, J. F. (1935). The presence of depressor nerves in the aorta and carotid of birds. *Anat. Rec.* **62,** 47–73.

Nye, P. C. G., and Powell, F. L. (1984). Steady-state discharge and bursting of arterial chemoreceptors in the duck. *Resp. Physiol.* **56,** 369–384.

O'Rourke, M. F., Blazek, J. V., Morrells, C. L., and Krovetz, L. J. (1968). Pressure wave transmission along the human aorta. *Circ. Res.* **23,** 567–579.

Odlind, B. (1978). Blood flow distribution in the renal portal system of the intact hen: A study of a venous system using microspheres. *Acta Physiol. Scand.* **102,** 342–356.

Ollenberger, G. P., and West, N. H. (1998). Distribution of regional cerebral blood flow in voluntarily diving rats. *J. Exp. Biol.* **201,** 549–558.

Palomeque, J., and Planas, J. (1977). Dimensions of the erythrocytes of birds. *Ibis* **119,** 533–535.

Pappano, A. J. (1976). Onset of chronotropic effects of nicotinic drugs and tyramine on the sino-atrial pacemaker in chick embryo heart: Relationship to the development of autonomic neuroeffector transmission. *J. Pharmacol. Exp. Therapeut.* **196,** 676–684.

Pappano, A. J., and Loffelholz, K. (1974). Ontogenesis of adrenergic and cholinergic neuroeffector transmission in chick embryo heart. *J. Pharmacol. Exp. Therapeut.* **191,** 468–478.

Pavlov, I. P. (1927). "Conditioned Reflexes: An Investigation of the Physiological Activity of the Cerebral Cortex" (Translated by G. V. Anrep), Oxford Univ. Press, Oxford, UK.

Peterson, D. F., and Nightingale, T. E., (1976). Functional significance of thoracic vagal branches in the chicken. *Resp. Physiol.* **27,** 267–275.

Petras, J. M., and Cummings, J. F. (1972). Autonomic neurons in the spinal cord of the rhesus monkey: A correlation of the findings of cytoarchitectonics and sympathectomy with fiber degeneration following dorsal rhizotomy. *J. Comp. Neurol.* **146,** 189–218.

Pick, J. (1970). "The Autonomic Nervous System," Lippincott, Philadelphia, PA.

Pool, R. (1989). Is it healthy to be chaotic? *Science* **243,** 604–607.

Prats, M. T., Palacios, L., Gallego, S., and Riera, M. (1996). Blood oxygen transport properties during migration to higher altitude of wild quail, *Coturnix coturnix coturnix. Physiol. Zool.* **69,** 912–929.

Prothero, J. (1979). Heart weight as a function of body weight in mammals. *Growth* **43,** 139–150.

Randall, W. C. (1994). Efferent sympathetic innervation of the heart. In "Neurocardiology" (J. A. Armour and J. L. Ardell, eds.), pp. 77–94. Oxford Univ. Press, Oxford, UK.

Rickenbacher, J., and Müller, E. (1979). The development of cholinergic ganglia in the chick embryo heart. *Anat. Embryol.* **155,** 253–258.

Robinzon, B., Koike, T. I., and Marks, P. A. (1993). At low dose, arginine vasotocin has vasopressor rather than vosodepressor effect in chickens. *Gen. Comp. Endocrinol.* **91,** 105–112.

Robinzon, B., Koike, T. I., Neldon, H. L., Hendry, I. R., and el-Halawani, M. E. (1988). Physiological effects of arginine vasotocin and mesotocin in cockerels. *Br. Poult. Sci.* **29,** 639–652.

Rothe, C. F. (1983). Venous system: physiology of the capacitance vessels. In "Handbook of Physiology: The Cardiovascular System, Volume III. Circulation and Organ Blood Flow, Part 1" (J. T. Shepherd and F. M. Abboud, eds.), American Physiological Society, Bethesda, MD.

Rushmer, R. F. (1976). "Cardiovascular Dynamics," 4th ed. W. B. Saunders, Philadelphia.

Sabin, F. R. (1917). Origin and development of the primitive vessels of the chick and of the pig. *Contr. Embryol.* **6,** 63.

Sapirstein, L. A., and Hartman, F. A. (1959). Cardiac output and its distribution in the chicken. *Am. J. Physiol.* **196,** 751.

Saul, P. J. (1990). Beat-to-beat variations of heart rate reflect modulation of cardiac autonomic outflow. *NIPS* **5,** 32–37.

Saunders, D. K., and Fedde, M. R. (1994). Exercise performance of birds. In "Advances in Veterinary Science and Comparative Medicine: Comparative Vertebrate Exercise Physiology; Phyletic Adaptations" (J. H. Jones, ed.), Vol. 38B, pp. 139–190. Academic Press, New York.

Scanes, C. G., Mozelic, H., Kavanagh, E., Merrill, G., and Rabii, J. (1982). Distribution of blood flow in the ovary of the domestic fowl (*Gallus domesticus*) and changes after prostaglandin F-2 alpha treatment. *J. Reprod. Fertil.* **64,** 227–231.

Schlaefke, M. E. (1981). Central chemosensitivity: A respiratory drive. *Rev. Physiol. Biochem. Pharmacol.* **90,** 171–244.

Schmidt-Neilsen, K. (1984). "Scaling: Why is Animal Size so Important?" Cambridge Univ. Press, Cambridge, UK.

Schwaber, J. S., and Cohen, D. H. (1978a). Electrophysiological and electron microscopic analysis of the vagus nerve of the pigeon, with particular reference to the cardiac innervation. *Brain Res.* **147,** 65–78.

Schwaber, J. S., and Cohen, D. H. (1978b). Field potential and single unit analyses of the avian dorsal motor nucleus of the vagus and criteria for identifying vagal cardiac cells of origin. *Brain Res.* **147,** 79–90.

Shepherd, J. T., and Vanhoutte, P. M. (1975). "Veins and Their Control," W. B. Saunders, London.

Simon-Oppermann, C., Simon, E., and Gray, D. A. (1988). Central and systemic antidiuretic hormone and angiotensin II in salt and fluid balance of birds as compared to mammals. *Comp. Biochem. Physiol. A* **90,** 789–803.

Smith, F. M. (1994). Blood pressure regulation by aortic baroreceptors in birds. *Physiol. Zool.* **67,** 1402–1425.

Smith, F. M., and Jones, D. R. (1990). Effects of acute and chronic baroreceptor denervation on diving responses in ducks. *Am. J. Physiol.* **258,** R895–R902.

Smith, F. M., and Jones, D. R. (1992). Baroreflex control of arterial blood pressure during involuntary diving in ducks (*Anas platyrhynchos* var.). *Am. J. Physiol.* **263,** R693–R702.

Smith, R. B. (1971a). Intrinsic innervation of the avian heart. *Acta Anat.* **79,** 112–119.

Smith, R. B. (1971b). Observations on nerve cells in human, mammalian and avian cardiac ventricles. *Anat. Anz.* **129,** 436–444.

Snyder, G. K. (1987). Muscle capillarity in chicks following hypoxia. *Comp. Biochem. Physiol. A* **87,** 819–822.

Snyder, G. K. (1990). Capillarity and diffusion distances in skeletal muscles in birds. *J. Comp. Physiol. B* **160,** 583–591.

Snyder, G. K., and Coelho, J. R. (1989). Microvascular development in chick anterior latissimus dorsi following hypertrophy. *J. Anat.* **162,** 215–224.

Snyder, G. K., Byers, R. L., and Kayar, S. R. (1984). Effects of hypoxia on tissue capillarity in geese. *Resp. Physiol.* **58,** 151–160.

Solomon, I. C., and Adamson, T. P. (1997). Static muscular contraction elicits a pressor reflex in the chicken. *Am. J. Physiol.* **272,** R759–R765.

Sommer, J. R. (1983). The implications of structure and geometry on cardiac electrical activity. *Ann. Biomed. Eng.* **11,** 149–157.

Sommer, J. R., and Johnson, E. A. (1969). Cardiac muscle: A comparative ultrastructural study with special reference to frog and chicken hearts. *Z. Zellforsch. mikrosk. anat.* **98,** 437–468.

Sommer, J. R., and Johnson, E. A. (1970). Comparative ultrastructure of cardiac cell membrane specialization: A review. *Am. J. Cardiol.* **25,** 184–194.

Sommer, J. R., Bossen, E., Dalen, H., Dolber, P., High, T., Jewett, P., Johnson, E. A., Junker, J., Leonard, S., Nassar, R., Scherer, B., Spach, M., Spray, T., Taylor, I., Wallace, N. R., and Waugh, R. (1991). To excite a heart: A bird's view. *Acta Physiol. Scand.* **S599,** 5–21.

Speckmann, E. W., and Ringer, R. K. (1964). Static elastic modulus of the turkey aorta. *Can. J. Physiol. Pharmacol.* **42,** 553–561.

Speckmann, E. W., and Ringer, R. K. (1966). Volume pressure relationships in the turkey aorta. *Can. J. Physiol. Pharmacol.* **44,** 901–907.

Stallone, J. N., Nishimura, H., and Nasjletti, A. (1990). Angiotensin II binding sites in aortic endothelium of domestic fowl. *Am. J. Physiol.* **258,** E777–E782.

Starling, E. H. (1896). On the absorption of fluids from connective tissue spaces. *J. Physiol.* **19,** 312–326.

Steele, P. A., Gibbins, I. L., and Morris, J. L. (1996). Projections of intrinsic cardiac neurons to different targets in the guinea-pig heart. *J. Auton. Nerv. Syst.* **47,** 177–187.

Steele, P. A., Gibbins, I. L., Morris, J. L., and Mayer, B. (1994). Multiple populations of neuropoptide-containing intrinsic neurons in the guinea-pig heart. *Neuroscience* **62,** 241–250.

Stéphan, F. (1949). Les suppliances obtenues experimentalement dans le systeme des arcs aortiques de l'embryon d'oiseau. *C. R. Soc. Anat.* **36,** 647.

Stephenson, R., and Jones, D. R. (1992). Blood flow distribution in submerged and surface-swimming ducks. *J. Exp. Biol.* **166,** 285–296.

Stephenson, R., Jones, D. R., and Bryan, R. M. (1994). Regional cerebral blood flow during submergence asphyxia in Pekin duck. *Am. J. Physiol.* **266,** R1162–R1168.

Strack, A. M., Sawyer, W. B., Marubio, L. M., and Loewy, A. D. (1988). Spinal origin of sympathetic preganglionic neurons in the rat. *Brain Res.* **455,** 187–191.

Sturkie, P. D. (1986a). Heart: Contraction, conduction, and electrocardiography. In "Avian Physiology" (P. D. Sturkie, ed.), pp. 167–190. Springer-Verlag, Berlin.

Sturkie, P. D. (1986b). "Avian Physiology," Springer-Verlag, Berlin.

Sturkie, P. D. and Poorvin, D. W. (1973). The avian neurotransmitter. *Proc. Soc. Exp. Biol. Med.* **143**, 644–646.

Sturkie, P. D., Dirner, G., and Gister, R. (1978). Role of renal portal valve in the shunting of blood flow in renal and hepatic circulations of chickens. *Comp. Biochem. Physiol. C* **59**, 95–96.

Szabo, E., Viragh, S., and Challice, C. E. (1986). The structure of the atrioventricular conducting system in the avian heart. *Anat. Rec.* **215**, 1–9.

Szabuniewicz, M., and McCrady, J. D. (1967). The electrocardiogram of the chicken. *Southwest Vet.* **20**, 287–294.

Tagawa, T., Ando, K., and Wasano, T. (1979). A histochemical study of the innervation of the cerebral blood vessels in the domestic fowl. *Cell Tissue Res.* **198**, 43–51.

Takei, Y., and Hasegawa, Y. (1990). Vasopressor and depressor effects of native angiotensins and inhibition of these effects in the Japanese quail. *Gen. Comp. Endocrinol.* **79**, 12–22.

Taylor, A. E., and Townsley, M. I. (1987). Evaluation of the Starling fluid flux equation. *NIPS* **2**, 48–52.

Taylor, M. G. (1964). Wave travel in arteries and the design of the cardiovascular system. In "Pulsatile Arterial Blood Flow" (E. O. Attinger, ed.), McGraw–Hill, New York.

Tcheng, K. T., and Fu, S. K. (1962). The structure and innervation of the aortic body of the yellow-breasted bunting. *Sci. Sin.* **11**, 221–232.

Tcheng, K. T., Fu, S. K., and Chen, T. Y. (1963). Supracardial encapsulated receptors of the aorta and the pulmonary artery in birds. *Sci. Sin.* **12**, 73–81.

Tummons, J., and Sturkie, P. D. (1968). Cardio-accelerator nerve stimulation in chickens. *Life Sci.* **7**, 377–380.

Tummons, J. L., and Sturkie, P. D. (1969). Nervous control of heart rate during excitement in the adult White Leghorn cock. *Am. J. Physiol.* **216**, 1437–1440.

Tummons, J. L., and Sturkie, P. D. (1970). Beta adrenergic and cholinergic stimulants from the cardioaccelerator nerve of the domestic fowl. *Z. Vergl. Physiol.* **68**, 268–271.

Umans, J. G., and Levi, R. (1995). Nitric oxide in the regulation of blood flow and arterial pressure. *Ann. Rev. Physiol.* **57**, 771–790.

Usami, S., Magazinovic, V., Chien, S., and Gregersen, M. I. (1970). Viscosity of turkey blood: Rheology of nucleated erythrocytes. *Microvasc. Res.* **2**, 489–499.

Vanhoutte, P. M. and Leusen, I. (1969). The reactivity of isolated venous preparations to electrical stimulation. *Pflügers Arch.* **306**, 341–353.

Vatner, S. F., and Braunwald, E. (1975). Cardiovascular control mechanisms in the conscious state. *New Engl. J. Med.* **293**, 970–976.

Weinstein, Y., Bernstein, M. H., Bickler, P. E., Gonzalez, D. V., Samaniego, F. C., and Escobedo, M. A. (1985). Blood respiratory properties in pigeons at high altitudes: Effects of acclimation. *Am. J. Physiol.* **249**, R765–R775.

West, J. B. (1995). "Respiratory Physiology—The Essentials," 5th ed. Williams and Wilkins, Baltimore, MD.

West, N. H., Langille, B. L., and Jones, D. R. (1981). Cardiovascular system. In "Form and Function in Birds" (A. S. King and J. McLelland, eds.), Vol. 2, pp. 235–339. Academic Press, New York.

Wideman, R. F., and Gregg, C. M. (1988). Model for evaluating avian renal hemodynamics and glomerular filtration rate autoregulation. *Am. J. Physiol.* **254**, R925–R932.

Wideman, R. F., Braun, E. J., and Anderson, G. L. (1991). Microanatomy of the renal cortex in the domestic fowl. *J. Morphol.* **168**, 249–267.

Wideman, R. F., Glahn, R. P., Bottje, W. G., and Holmes, K. R. (1992). Use of a thermal pulse decay system to assess avian renal blood flow during reduced renal arterial perfusion pressure. *Am. J. Physiol.* **262**, R90–R98.

Wilson, J. X. (1989). The renin-angiotensin system in birds. In "Progress in Avian Osmoregulation" (M. Hughes and A. Chadwick, eds.), pp. 61–79. Leeds Philosophical and Literary Society, Leeds, UK.

Wilson, J. X., and Butler, D. G. (1983). Catecholamine-mediated pressor responses to angiotensin II in the Pekin duck, *Anas platyrhynchos*. *Gen. Comp. Endocrinol.* **51**, 477–489.

Wilson, J. X., and West, N. H. (1986). Cardiovascular responses to neurohormones in conscious chickens and ducks. *Gen. Comp. Endocrinol.* **62**, 268–280.

Wolfenson, D., Frei, Y. F., and Berman, A. (1982a). Blood flow distribution during artificially induced respiratory alkalosis in the fowl. *Resp. Physiol.* **50**, 87–92.

Wolfenson, D., Frei, Y. F., and Berman, A. (1982b). Responses of the reproductive vascular system during the egg–formation cycle of unanaesthetized laying hens. *Br. Poult. Sci.* **23**, 425–431.

Wolfenson, D., Frei, Y. F., Snapir, N., and Berman, A. (1978). Measurement of blood flow distribution by radioactive microspheres in the laying hen. *Comp. Biochem. Physiol. A* **61A**, 549.

Wolfenson, D., Frei, Y. F., Snapir, N., and Berman, A. (1981). Heat stress effects on capillary blood flow and its redistribution in the laying hen. *Pflügers Arch.* **390**, 86–93.

Womersley, J. R. (1957). The mathematical analysis of the arterial circulation in a state of oscillatory motion. *Tech. Rep. Wright AFB Dev. Ctr.* **WADC-TR-56-614.**

Ying, L., James, T. N., Yamamoto, S., and Terasaki, F. (1993). Cardiac conduction system in the chicken: gross anatomy plus light and electron microscopy. *Anat. Rec.* **236**, 493–510.

Yoshigi, M., Ettel, J. M., and Keller, B. B. (1997). Developmental changes in flow-wave propagation velocity in embryonic chick vascular system. *Am. J. Physiol.* **273**, H1523–H1529.

Yousuf, N. (1965). The conducting system of the heart of the house sparrow, *Passer domesticus indicus*. *Anat. Rec.* **152**, 235–250.

Zahka, K. G., Hu, N., Brin, K. P., Yin, F. C., and Clark, E. B. (1989). Aortic impedance and hydraulic power in the chick embryo from stages 18 to 29. *Circ. Res.* **64**, 1091–1095.

CHAPTER 10

Respiration

F. L. POWELL
Division of Physiology
Department of Medicine
School of Medicine
University of California, San Diego
La Jolla, California 92093-0623

I. Overview 233
  A. Oxygen Cascade 234
  B. Symbols and Units 234
II. Anatomy of the Avian Respiratory System 234
  A. Upper Airways 235
  B. Lungs 235
  C. Air Sacs 236
  D. Respiratory System Volumes 238
III. Ventilation and Respiratory Mechanics 238
  A. Respiratory Muscles 238
  B. Mechanical Properties 239
  C. Ventilatory Flow Patterns 240
IV. Pulmonary Circulation 243
  A. Anatomy of the Pulmonary Circulation 243
  B. Pulmonary Capillary Volume 243
  C. Pulmonary Vascular Resistance and the Distribution of Blood Flow 244
  D. Fluid Balance 244
V. Gas Transport by Blood 244
  A. Oxygen 245
  B. Carbon Dioxide 247
  C. Acid–Base 248
  D. Blood Gas Measurements 249
VI. Pulmonary Gas Exchange 249
  A. Basic Principles of Oxygen Transport 250
  B. Cross-Current Gas Exchange 250
  C. Lung Diffusing Capacity 251
  D. Heterogeneity in the Lung 252
  E. Frontiers: Gas Exchange during High-Altitude Flight 254
  F. Summary of O_2 and CO_2 Exchange in Avian Lungs 254
VII. Tissue Gas Exchange 254
  A. Microcirculation 255
  B. Myoglobin 255
  C. Effects of Hypoxia and Exercise 255
VIII. Control of Breathing 256
  A. Respiratory Rhythm Generation 256
  B. Sensory Inputs 256
  C. Ventilatory Reflexes 258
  References 259

I. OVERVIEW

The primary function of the respiratory system in air-breathing vertebrates is gas exchange—delivering oxygen from the environment to the tissues and removing carbon dioxide from the tissues. The respiratory system is also critical for thermoregulation (by evaporative water loss) in birds and nonrespiratory functions such as vocalization, but these are not covered in this chapter. Generally, the respiratory system acts as a servant to the rest of the organism by delivering enough oxygen and removing sufficient carbon dioxide for metabolic demands. As oxygen demand increases, a variety of respiratory responses insure an adequate supply of oxygen. This involves the lung, respiratory mechanics, the pulmonary circulation, transport of oxygen and carbon dioxide in blood, pulmonary and tissue gas exchange, and the coordination of all these mechanisms by the respiratory control system.

Individual sections in this chapter focus on each of these physiological mechanisms. References are made to mammalian respiratory physiology, so the reader can consult the extensive literature available for more details on fundamentals concepts. In addition, important unanswered questions and topics ripe for future research are highlighted in sections labeled "Frontiers."

A. Oxygen Cascade

Figure 1 shows how these physiological transport steps function in series to transport oxygen from the environment to the cells. This is often referred to as the "oxygen cascade" because the oxygen level (quantified as partial pressure, or P_{O_2}) decreases at each step in the model. Breathing movements bring fresh air into the lungs, and the heart pumps oxygen-poor blood to the lungs. Oxygen diffuses from gas to blood in the lungs, and this oxygen-rich blood returns to the heart via the pulmonary circulation. Arterialized blood is pumped to the various organs and tissues of the body via the systemic circulation. Finally, oxygen diffuses out of the systemic capillaries to metabolizing tissues and ultimately to the mitochondria inside cells. Carbon dioxide moves out of the cells to the environment through these same steps in the opposite direction from oxygen. Each of these steps is covered in the sections below, with an emphasis on respiratory structure–function relationships that are unique to birds, especially in comparison with mammals. Mammals are the only other vertebrates which achieve avian levels of oxygen demand during thermoregulation and activity.

B. Symbols and Units

Table 1 provides a list of abbreviations used in this chapter, which are based on a few simple conventions (Macklem, 1987). Primary variables are symbolized with a capital letter, and a dot over the variable indicates the first derivative with respect to time. Modifiers are small capitals for the gas phase and lowercase letters for liquid or tissues. Finally, a specific gas species is indicated with a subscript. Respiratory gas volumes (e.g., ventilation) are reported for physiological conditions (BTPS, body temperature and pressure, saturated) unless otherwise noted. Amounts of oxygen (O_2) or carbon dioxide (CO_2) are expressed in millimoles (e.g., O_2 concentration in mmol/liter = mM). Pressure is expressed in Torr (7.5 Torr = 1 kPa).

II. ANATOMY OF THE AVIAN RESPIRATORY SYSTEM

The structure of the avian respiratory system is unique among the vertebrates, with small lungs which do not change volume during breathing and nine large

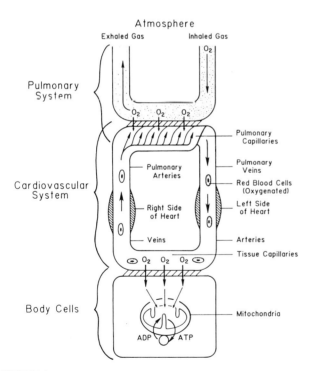

FIGURE 1 General model of the oxygen transport in birds. (Reprinted from *Respir. Physiol.* **44**, C. R. Taylor and E. R. Weibel, Design of the mammalian respiratory system. I. Problem and strategy, pp. 1–10, Copyright (1981), with permission from Elsevier Science.)

TABLE 1 Symbols in Respiratory Physiology

Primary variables (and units)
  C, Concentration or content (mM, mmmol/liter)
  D, Diffusing capacity (mmol$_{O_2}$/(min · Torr)
  P, Partial pressure or hydrostatic pressure (Torr or cm H_2O)
  V, Gas volume (liters or ml)
  $\dot{V}$, Ventilation (liters/min)
  $\dot{Q}$, Blood flow or perfusion (liters/min)
  $\dot{M}$, gas flux (mmol/min)

Modifying symbols
  D, Dead space gas
  E, Expired gas
  Ē, Mixed expired gas
  I, Inspired gas
  T, Tidal gas
  a, Arterial blood
  c, Capillary blood
  m, Membrane
  t, Tissue
  v, Venous blood
  v̄, Mixed venous blood

Examples
  $P_{I_{O_2}}$, Partial pressure of O_2 in inspired gas
  $P_{a_{O_2}}$, Partial pressure of O_2 in arterial blood
  $P_{\bar{v}_{O_2}}$, Partial pressure of O_2 in mixed venous blood
  $\dot{M}_{O_2}$, O_2 consumption per unit time
  $\dot{V}_P$, Ventilation of the parabronchi per unit time

air sacs which act as bellows to ventilate the lung but do not participate directly in gas exchange (Figure 2). The total volume of the respiratory system in a bird (i.e., lungs and air sacs) is larger than that in a comparably sized mammals (ca., 15 vs 7% of body volume), but the avian lung itself is smaller (ca., 1 to 3% of body volume). Apparently during evolution, birds segregated the functions of gas exchange and ventilation as the respiratory organ was subdivided into smaller functional units to increase gas exchange surface area. Such heterogeneous partitioning of the respiratory organ contrasts with the homogenous partitioning in mammals (Duncker, 1978). Alveoli in mammalian lungs perform both respiratory functions of ventilation and gas exchange.

Also in contrast to mammals, the avian thoracic cavity is essentially at atmospheric pressure (versus subatmospheric), and there is no diaphragm to functionally separate it from the abdominal cavity (see Section II,B). This section covers basic respiratory system anatomy necessary to understand respiratory function, but the reader is referred to several excellent monographs and reviews for more details (Duncker, 1971; King and Molony, 1971; McLelland, 1989b). Terminology according to the *Nomina Anatomica Avium* is used here (King, 1979). The anatomy of the pulmonary circulation is covered in a separate section (III,A).

A. Upper Airways

Birds can breathe through the nares or mouth. Oronasal structures tend to heat and humidify inspired gas and filter out large particles which could potentially damage the delicate respiratory surfaces. The oronasal cavity is separated from the trachea by the larynx, which opens into the trachea through the slit-like glottis. The laryngeal muscles contract with breathing, to open the glottis during inspiration and decrease the resistance to inspiratory air flow. This rhythmic opening of the glottis is useful when attempting to intubate a bird. The trachea has complete cartilaginous rings in most avian species and plentiful smooth muscle. Interesting exceptions include the "double trachea" of penguins, with a medial septum dividing the trachea into two tubes, and the slit-like opening on the ventral surface of the trachea in emus, which is responsible for their characteristic booming call (McLelland, 1989a). The anatomy and physiology of the larynx and trachea have been reviewed in detail (McLelland, 1989a).

Tracheal volume is an important determinant of "dead space" ventilation and therefore gas exchange. Hinds and Calder (1971) measured tracheal volume in 27 species of birds and found *in situ* volume (V in milliliters) was related to body mass (M_B in kilograms) as: $V = 3.7 \, M_B^{1.09}$. This equation underestimates tracheal volume in Pekin ducks (Bech *et al.*, 1984; Hastings and Powell, 1986b) and pigeons (Powell, 1983b) and overestimates the value for male chickens (Kuhlmann and Fedde, 1976), but the error is less than 25%. Tracheal volume is 4.5 times larger in birds than in comparably sized mammals (Hinds and Calder, 1971) and birds generally compensate for this increased dead space with a deep and slow breathing pattern (Bouverot, 1978). Several species of birds possess tracheal elongations, which loop inside the neck, but their function is unknown (McLelland, 1989a).

The trachea bifurcates into two primary bronchi at the syrinx. In many species (e.g., chickens, ducks), but not all, this bifurcation occurs inside the thoracic cavity where the trachea runs through the clavicular air sac. The syrinx is responsible for vocalization in birds but relatively little is known about the precise mechanisms. Readers are referred to comprehensive reviews about this structure for details (King, 1989). Similar to the avian trachea, the syrinx shows considerable variation between species, and males in some species exhibit large bullae with unknown functions.

B. Lungs

The avian lung is located dorsally in the thoracoabdominal cavity of birds (Figure 2), with invaginations from spinal processes typically on the dorsal surface of the lung. Figure 3 shows the pulmonary bronchial branching pattern in a representative species.

1. Conducting Airways

In most cases, the extrapulmonary primary bronchi, between the syrinx and the lungs, are relatively short. The intrapulmonary primary bronchus travels through the entire length of the lung, entering on the medioventral aspect and exiting at the caudal border of the lung

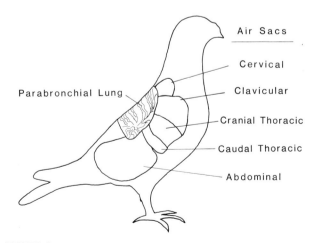

FIGURE 2 Respiratory system of a pigeon consisting of the parabronchial lung and air sacs.

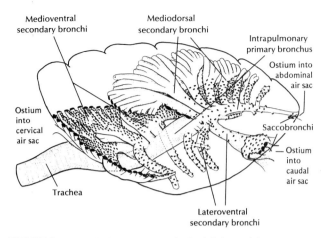

FIGURE 3 Bronchial arrangement in the left lung of the mute swan (*Cygnus olor*). (After Duncker, 1971).

into the ostium of the abdominal air sac (see Section I,C). The secondary bronchi can be considered in two functional groups based on their origin from the primary bronchus. The cranial group consists of 4 or 5 medioventral secondary bronchi, originating from the medioventral intrapulmonary primary bronchus. These cranial secondary bronchi branch further to form a fan covering the medioventral surface of the lung. The caudal group consists of 6 to 10 mediodorsal secondary bronchi, which also branch to form a fan over the mediodorsal surface of the lung.

A third group of secondary bronchi includes a variable number of lateroventral bronchi in most species, which also branch off caudal parts of the primary bronchus. The first or second laterobronchus forms a short connection to the posterior thoracic air sac (see below). Other lateroventral bronchi may penetrate lateroventral parts of the lung to variable degrees in different species, but they do not form a regular branching fan like the other secondary bronchi.

The primary and secondary bronchi are conducting airways because they do not participate in gas exchange. Cartilaginous semirings and smooth muscle support the primary bronchi, but the walls of the secondary bronchi are flaccid and require on adhesion to the surrounding lung or pleura to remain open. The respiratory epithelium is ciliated with variable amounts of goblet cells in different species in the trachea and primary and secondary bronchi (Duncker, 1974).

2. Parabronchi

Parabronchi are the functional unit of gas exchange in the avian lung. They are also called "tertiary bronchi" because they can originate from the secondary bronchi, but parabronchus is the preferred terminology because they also originate from further branches of secondary bronchi (Figure 3). Most of the parabronchi are organized as a parallel series of several hundred tubes connecting the medioventral and mediodorsal secondary bronchi (Figure 3). Such parabronchi are called paleopulmonic parabronchi and together with the primary and cranial and caudal groups of secondary bronchi they compose the simplest scheme of bronchial branching in the avian lung (Duncker, 1972, 1974).

In all birds except some penguins there are additional parabronchi called neopulmonic parabronchi (Duncker, 1972, 1974). These parabronchi are not organized as regular parallel stacks of tubes, but may exhibit irregular branching patterns. Neopulmonic parabronchi may connect another set of caudal laterodorsal secondary bronchi to caudal air sacs (see Section I,C) or other parabronchi. Neopulmonic parabronchi never compose more than 25% of the parabronchi, and there are large species variations. However, it has not been possible to demonstrate any phylogenetic or evolutionary significance to neopulmonic versus paleopulmonic parabronchi, despite the implication of the terms (Maina, 1989). The functional significance of these different kinds of parabronchi is considered below (see Section III,C,2.).

Figure 4 shows the detailed structure of a parabronchus. These gas-exchanging tubes can be several millimeters long and 0.5 to 2.0 mm in diameter, depending on the size of the bird (Duncker, 1971; Maina, 1989). The parabronchi are separated from each other along their length by a boundary of connective tissue and larger pulmonary blood vessels. The parabronchial lumen is lined by a meshwork of smooth muscle, which outlines the entrances to atria radiating from the parabronchial lumen. The atria lead to infundibula and, ultimately, the air capillaries, which are 2 to 10 μm in diameter and as long as about one-quarter of the total parabronchial diameter (Figure 4). The air capillaries intertwine with a similar network of pulmonary blood capillaries in the parabronchial mantle, where the air–blood capillary interface is the site of gas exchange. Details about the air–blood capillary interface are considered later (see Section V,C).

C. Air Sacs

The air sacs are thin membranous structures connected to the primary or secondary bronchi via ostia and they comprise most of the volume of the respiratory system (Figure 2). Air sacs are poorly vascularized by the systemic circulation and do not directly participate in significant gas exchange but act as a bellows to ventilate the lungs. In most species, there are nine air sacs which can be considered in cranial and caudal functional groups (Maina, 1989). Air sac diverticulae may also penetrate the skeleton, but there are large species differ-

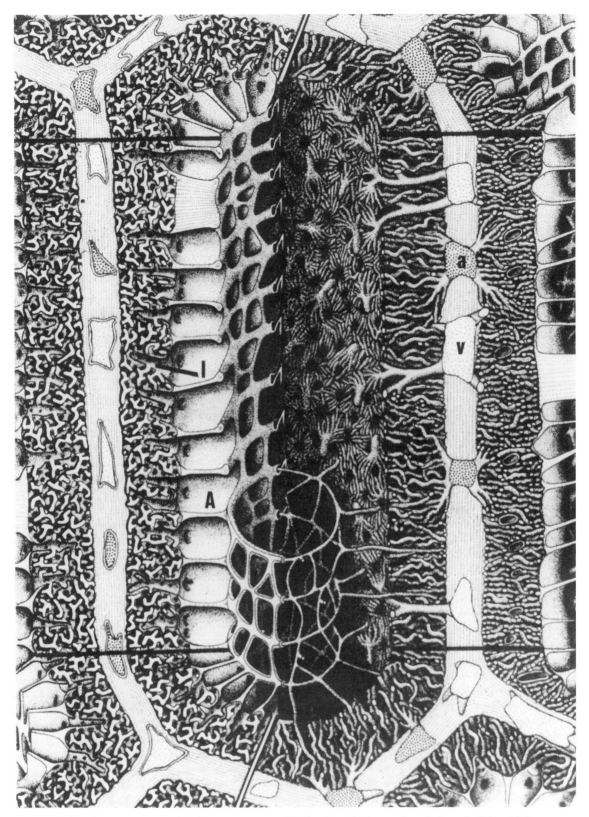

FIGURE 4 Parabronchus. (Left) Pathway for gas diffusion from the lumen through the atria (A) and infundibula (l) to the air capillaries. (Right) Pathway for blood flow from interparabronchial arteries (a) and intraparabronchial arterioles to the blood capillaries. Blood is collected in venules at the base of the atria and carried back to interparabronchial veins (v) which drain into large pulmonary vein. (Reprinted from *Respir. Physiol.* **22,** H. R. Duncker, Structure of the avian respiratory tract, pp. 1–19, Copyright (1974), with permission from Elsevier Science.)

ences and the functional significance of such connections for respiration has not been established (Maina, 1989).

The cranial group consists of the paired cervical air sacs, the unpaired clavicular air sac, and the paired cranial thoracic air sacs. The cervical sacs directly connect to the first medioventral secondary bronchus. The clavicular air sac directly connects to the third medioventral secondary bronchus and may also have indirect connections via parabronchi to other cranial (medioventral) secondary bronchi in some species (e.g., chickens). The cranial thoracic air sacs generally connect to the third medioventral secondary bronchi and also to parabronchi originating from other cranial secondary bronchi in some species.

The caudal group consists of the paired caudal thoracic air sacs and paired abdominal air sacs. The caudal thoracic air sac is directly connected to the lateroventral secondary bronchus and may have indirect connections to other lateroventral or even cranial (medioventral) secondary bronchi in species with large amounts of neopulmonic parabronchi (e.g., chickens). The abdominal air sacs connect to the caudal end of the intrapulmonary primary bronchus and may have more indirect connections to parabronchi from laterodorsal secondary bronchi and the last mediodorsal secondary bronchi. Air sac connections with parabronchi are frequently grouped into a funnellike structure called the saccobronchus.

D. Respiratory System Volumes

The upper airways and bronchial branches proximal to the parabronchi compose anatomical dead space as described above (see Section I,A.). Such conducting airways are called "dead space" because they do not participate directly in gas exchange. The intrapulmonary conducting airways make a relatively small contribution to total dead space, but total dead space volume is generally larger than most mammals, consistent with the generally long neck of birds. Physiological measures of dead space are considered below (see Section III,C).

The actual volume of air in the avian lung involved in gas exchange at any moment is in the air capillaries. This is considerably less than the volume in gas exchanging portions of alveolar lungs from comparably sized mammals (Powell and Mazzone, 1983). However, the unique pattern of air flow in the open-ended parabronchi renews this gas exchanging volume more frequently than does tidal ventilation in alveolar lungs (see Section II,C,2). Therefore, birds do not need as large of a functional residual capacity (FRC) in the lungs as mammals to smooth out variations in gas exchange and O_2 and CO_2 levels that could occur during the breathing cycle.

Most of the respiratory system volume in birds is in the air sacs, and there is no comparable volume in the mammalian lung. Unlike mammalian alveoli, which change volume during ventilation, the air sacs are not important sites of gas exchange. There is tremendous variation in the volumes of air sacs reported in the literature, because the value is very sensitive to the method of measurement. For example, the volume of plastic casting material that can be instilled under a pressure head into the air sacs of a dead bird may be much greater than the gas volume in live birds with muscle tone in the thorax and abdominal wall. Also, air sac volume *in vivo* can vary with posture and digestive and reproductive states as different structures in the body (e.g., eggs) displace volume. Casting under controlled pressure conditions (Duncker, 1971) and gas dilution *in vivo* (Scheid *et al.*, 1974) are probably the most accurate methods available for determining air sac volume.

III. VENTILATION AND RESPIRATORY MECHANICS

Respiratory muscles generate the forces (pressures) to move air in and out of the air sacs and through the parabronchial lung. The air sacs follow changes in body volume with respiratory muscle activity and act as a bellows to ventilate the parabronchial lung which is essentially constant in volume (Jones *et al.*, 1985; Macklem *et al.*, 1979). In contrast to mammals, the avian lung volume is maintained by attachments to the body wall and not by a subatmospheric pressure in an intrapleural space surrounding it. Also in contrast to mammals, birds have no diaphragm separating the body cavity into separate thoracic and abdominal compartments. Hence, pressures are relatively uniform in the avian thoracoabdominal cavity, which behaves mechanically as a single compartment (Scheid and Piiper, 1989).

Ventilation ($\dot{V}$) is the product of the volume per breath, or tidal volume (V_T) and the respiratory frequency (f_R), so $\dot{V}$ can be increased by breathing faster or deeper. The distribution of gas flow in the avian respiratory system depends upon the magnitude and pattern of respiratory muscle activity, as well as the mechanical properties of the body wall, lungs, and air sacs as described below.

A. Respiratory Muscles

Figure 5 shows the changes in thoracic skeleton between normal inspiration and expiration in a bird (King and Molony, 1971; Zimmer, 1935). During inspiration, the sternum rocks cranially and ventrally with the coracoids and furcula rotating at the shoulder. Simultane-

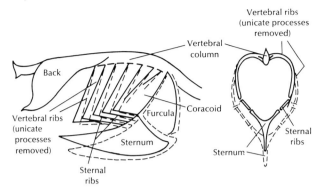

FIGURE 5 Changes in the position of the thoracic skeleton during breathing in a standing bird. Solid lines show thoracic position at the end of expiration and dotted lines show the end of inspiration. (After Zimmer, 1935.)

ously, the vertebral ribs move cranially to expand the sternal ribs and thoracoabdominal cavity laterally.

In small birds (e.g., starlings) during flight, the furcula (wishbone) and sternum are mechanically coupled such that the wing beat assists ventilation (Jenkins *et al.*, 1988). During rest, both inspiration and expiration require active contraction of the respiratory muscles as listed in Table 2. The innervation for these muscles is summarized in deWet *et al.* (1967). Increases in ventilatory volumes are achieved by recruiting more motor units in active muscles and additional respiratory muscles and during expiration the opposite occurs (Fedde *et al.*, 1964b; Fedde *et al.*, 1963, 1969; Kadono and Okada, 1962; Kadono *et al.*, 1963). Therefore, the relaxed resting volume of the avian respiratory system is midway between inspiratory and expiratory volumes (Seifert, 1896) in contrast to mammals, which relax to functional residual capacity (FRC) at end-expiratory volume. The costoseptal muscles control the tension of the horizontal septum covering the ventral surface of the lung, but unlike the mammalian diaphragm, these are not effective at changing lung volume (Fedde *et al.*, 1964a).

B. Mechanical Properties

1. Compliance

Compliance (C) defines the effectiveness of small pressure changes (ΔP) at inducing volume changes (ΔV):

$$C = \Delta V / \Delta P.$$

Because the pressure changes with breathing are essentially uniform throughout the coelom in birds, they can be measured as the difference between pressure in an air sac (P_{AS}) and atmospheric pressure outside the bird. Changes in respiratory system volume (V_{RS}) can be measured with a plethysomograph, which measures changes in whole body volume during breathing, or a pneumotachograph, which quantifies the amount of air inhaled or exhaled at the mouth or trachea. Compliance, measured as the slope of the steepest part of a graph plotting V_{RS} versus P_{AS} in an artificially ventilated bird, ranges from 10 ml/cm H_2O in chickens (Scheid and Piiper, 1969) to 30 ml/cm H_2O in ducks (Gillespie *et al.*, 1982b). These values are similar to compliance in mammals when correcting for body size.

Different results are obtained when compliance is measured by applying small oscillations in volume on spontaneously breathing birds. Compliance measured with this forced oscillation technique is only 7.7 ml/cm H_2O in ducks (Gillespie *et al.*, 1982b), or much less than in mammals. In contrast, compliance measured with this technique in pigeons (2.8 ml/cm H_2O, Kampe and Crawford, 1973), is 3.7 times greater than the value predicted for a similar-sized mammal (Powell, 1983b). The reasons for these differences are not clear but they suggest that compliance is exquisitely sensitivity to posture and muscular tone. Compliance primarily depends on the viscoelastic properties of the body wall and air sacs in birds, in contrast to the elastic properties of the lung in mammals (Macklem *et al.*, 1979). Therefore, compliance in birds is high when the volume changes occur by "unfolding" air sacs and stretching the abdominal wall, but it is low when volume changes are opposed by muscle tone in the body wall.

2. Resistance

Ohm's law defines the relationship between pressure, flow, and resistance (R) for the respiratory system as:

$$R = \Delta P / \dot{V},$$

where ΔP is the pressure gradient between the atmosphere and air sacs driving ventilatory air flow ($\dot{V}$).

TABLE 2 Respiratory Muscles of the Chicken

Inspiratory	Expiratory
M. scalenus	Mm. intercostales externi of fifth and sixth spaces
Mm. intercostales externi (except in fifth and sixth spaces)	Mm. intercostales interni of third to sixth spaces
intercostalis interni in second space	M. costosternalis pars minor
M. costosternalis pars major	M. obliquus externus abdominis
Mm. levatores costarum	M. obliquus internus abdominis
M. serratus profundus	M. transversus abdominis
	M. rectus abdominis
	serratus superficialis, pars cranialis and caudalis
	M. costoseptalis

Expiration decreases air sac volume and creates a small positive pressure which drives air flow out of the sac across small airway resistances; the opposite occurs during inspiration. Air sac pressure changes during breathing are small and similar in all of the air sacs (± 1 cm H_2O), so resistance can be measured by measuring pressure and volume changes during artificial or spontaneous breathing as described above for compliance measurements (Scheid and Piiper, 1989). Airway resistance ranges from 4.8 cm H_2O/(L/sec) in ducks (Gillespie et al., 1982a) to 41 cm H_2O/(L/sec) in pigeons at the resonant frequency of their respiratory system (Kampe and Crawford, 1973).

Airway geometry is an important determinant of resistance. Poiseuille's law predicts resistance to laminar air flow is directly proportional to the length of an airway, and inversely proportional to the fourth power of the airway radius. Therefore, resistance can vary between the different anatomical pathways possible for air flow in the avian respiratory system (see Section II,C,2). Different pathways presumably explain why airway resistance is generally greater during inspiration than during expiration in birds (Brackenbury, 1971, 1972; Cohn and Shannon, 1968). However, measurements using the forced oscillation technique on awake ducks find similar resistance during inspiration and expiration (Gillespie et al., 1982a). Resistance measured with this technique does not include any contributions from the body wall (Scheid and Piiper, 1989), so some differences between inspiratory and expiratory resistance may reflect muscle tone.

Physiological factors also affect airway resistance. Decreased lung P_{CO_2} increases resistance by a local effect on the openings of the mediodorsal secondary bronchi into the primary bronchus (Molony et al., 1976). However, P_{CO_2} does not affect smooth muscle contraction in the parabronchi (Barnas and Mather, 1978). Turbulent flow, which may occur in large airways at high ventilation rates or at bronchial bifurcations, can increase resistance (Brackenbury, 1972; Molony et al., 1976). Finally, resistance can change with breathing frequency. The pathway for airflow may be different with ventilation at high frequencies and small volumes (Banzett and Lehr, 1982; Hastings and Powell, 1987) and therefore affect resistance as described above. Also, the respiratory system has a resonant frequency at which the overall impedance to breathing is minimized (e.g., 9.4 breaths/sec in pigeons, Kampe and Crawford, 1973).

3. Air Capillary Surface Forces

Surface tension at the gas–liquid interface of the respiratory exchange surface will tend to collapse the air capillaries. This tendency to collapse also decreases interstitial pressure between the air and blood capillaries, which increases capillary filtration and can lead to thickening of the blood–gas barrier by edema (see Section III,D). However, the air capillaries are lined by a surfactant, which lowers surface tension and counteracts these potentially deleterious effects (McLelland, 1989b). Lamellated osmophilic bodies in the parabronchial atria presumably secrete the surfactant, which spreads as a trilaminar substance over the air capillary surface (King and Molony, 1971; Pattle, 1978). This trilaminar substance is unique to birds (McLelland, 1989b). The law of LaPlace predicts that the tendency for a bubble to collapse is inversely proportional to its radius. Hence, surfactant should be even more important for maintaining the patency and proper fluid balance in birds than in mammals, because the air capillary radius (ca. 3 μm) is so much smaller than alveolar radius (ca. 300 μm). Although compliance of the avian lung is relatively low, the parabronchi are not perfectly rigid and trilaminar substance may be important for preventing air capillary collapse with compression of the parabronchi (Macklem et al., 1979).

C. Ventilatory Flow Patterns

1. Air Sac Ventilation

The air sacs are ventilated roughly in proportion to their volume, such that the cranial group (clavicular and cranial thoracic air sacs) and caudal group (caudal thoracic and abdominal air sacs) each receive about 50% of the inspired volume (Scheid et al., 1974). The ventilation/volume ratio affects the O_2 and CO_2 levels in the air sacs (see Section II,C,3). Hence, the increase in air sac volume with changes in body wall muscle tone that may accompany anesthesia can alter air sac composition (Scheid and Piiper, 1969). There is no evidence for air flow between the sacs during normal breathing (Scheid and Piiper, 1989), although this has been postulated as a mechanism to enhance pulmonary gas exchange during breath-hold diving in birds (Boggs et al., 1996).

2. Pulmonary Ventilation

Figure 6 shows the general pattern of ventilatory air flow during inspiration and expiration in the avian lung. The unidirectional flow in a caudal-to-cranial direction through the paleopulmonic parabronchi during both phases of ventilation has been established by a variety of methods (Scheid and Piiper, 1989). Early researchers noted that soot deposited primarily in the caudal regions of the lungs of pigeons collected in train stations, suggesting that inspired gas entered the lungs from a caudal

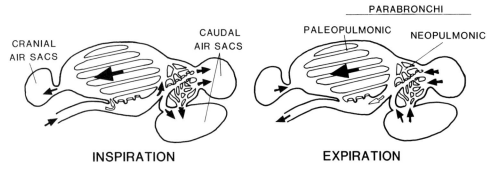

FIGURE 6 Pathway of airflow in the avian respiratory system during inspiration and expiration. Flow in paleopulmonic parabronchi is always caudal-to-cranial during both phases of breathing (large solid arrows) but neopulmonic flow is bidirectional. Open arrows show possible ventilatory shunts.

direction (Dotterweich, 1930). More recently, researchers have confirmed this pattern with direct measurements of air flow (Brackenbury, 1971; Bretz and Schmidt-Nielsen, 1971; Scheid et al., 1972) and respiratory gases (Powell et al., 1981) in different parts of the lung.

Upon inspiration, about half of the tidal volume goes into the caudal air sacs and half goes into the cranial air sacs. Figure 6 (filled arrows) shows how inspiratory flow bypasses the cranial secondary bronchial openings in the primary bronchus and flows directly into the caudal air sacs and caudal secondary bronchi. When gas enters the caudal secondary bronchi, it continues through the paleopulmonic parabronchi in a caudal-to-cranial direction and enters the cranial air sacs via the cranial secondary bronchi. If tidal volume is large enough, some of the inspired gas may reach the cranial air sacs via the paleopulmonic parabronchi during the same breath.

Upon expiration, the air sacs expel gas which eventually leaves the bird via the primary bronchi and trachea. Figure 6 (filled arrows) shows how expiratory flow from the cranial air sacs leaves the lung via the cranial secondary bronchi emptying into the primary bronchi. Expiratory flow from the caudal air sacs is routed through the paleopulmonic parabronchi in a caudal-to-cranial direction via caudal secondary bronchi. If the expired volume is large enough, gas from the caudal air sacs will also leave the lung through the cranial secondary bronchi and mix with the gas emptying from the cranial air sacs.

Hence, airflow in the paleopulmonic parabronchi is in a caudal-to-cranial direction during both inspiration and expiration. "Aerodynamic valving" (see Section II,C,6.) is responsible for rectifying air flow in the paleopulmonic parabronchi and determining the distribution of ventilation in the avian lung. There is no evidence for anatomical valves, for example, closing the primary bronchial openings of the cranial secondary bronchi during inspiration (reviewed by Scheid and Piiper, 1989). In contrast to the paleopulmonic parabronchi, flow is bidirectional in the neopulmonic parabronchi, which are functionally in series with the caudal air sacs (Figure 6). Inspiratory airflow is cranial-to-caudal through neopulmonic parabronchi and into caudal air sacs, and in the caudal-to-cranial direction during expiration (Scheid and Piiper, 1989). The implications of these flow patterns for gas exchange are discussed in later (see Section V,D,4).

3. Air Sac P_{O_2} and P_{CO_2}

This pattern of air flow is an important determinant of P_{O_2} and P_{CO_2} in the air sacs (Table 3). Cranial air sacs only receive gas from the parabronchi, so their P_{O_2} and P_{CO_2} levels are very near end-expired values. However, caudal air sacs contain a mixture of reinhaled dead space gas (also end-expired P_{O_2} and P_{CO_2} levels) and fresh air, which raises their P_{O_2} and lowers their P_{CO_2}. Other factors which decrease P_{O_2} and increase P_{CO_2} in the air sacs include "stratification," gas exchange across the air sac wall, and gas exchange in neopulmonic parabronchi in series with the air sacs (Geiser et al., 1984).

Gas exchange across the air sac walls is less than 5% of the total respiratory gas exchange and a minor factor in determining air sac P_{O_2} and P_{CO_2} (Magnussen et al., 1976). Stratification (i.e., incomplete mixing of freshly inspired gas and resident gas in the air sacs) has been observed in ducks (Torre-Bueno et al., 1980), but its effect on air sac gas concentrations is not clear (Powell and Hempleman, 1985). Gas exchange in neopulmonic parabronchi in series with caudal air sacs seems to be the most important factor causing differences in measured P_{O_2} and P_{CO_2} values and those predicted from reinhaled dead space (Piiper, 1978). Bidirectional flow may also occur in a small fraction of parabronchi in the purely paleopulmonic lungs of penguins and affect P_{O_2} and

TABLE 3 Partial Pressure of O_2 and CO_2 in Air Sacs and in End-Expiratory Gas of Awake Birds

	Goose[a]	Goose[b]	Chicken[c]	Duck[d]	Pigeon[e]
Clavicular					
P_{CO_2} (Torr)	35	39	44.0	39.2	32
P_{O_2} (Torr)	100	92	83.9	99.4	109
Cranial thoracic					
P_{CO_2} (Torr)	35	38	41.6	35.7	34
P_{O_2} (Torr)	100	95	99.1	104.3	105
Caudal thoracic					
P_{CO_2} (Torr)	28	20	24.2	18.9	29
P_{O_2} (Torr)	115	124	120.3	123.9	111
Abdominal					
P_{CO_2} (Torr)	28	18	14.7	17.5	27
P_{O_2} (Torr)	115	128	130.0	126.7	110
End expiratory					
P_{CO_2} (Torr)	35	39	36.7	35.7	—
P_{O_2} (Torr)	100	100	94.3	100.1	—

[a] Cohn and Shannon (1968), Figure 6.
[b] Scheid et al. (1991).
[c] Piiper et al. (1970).
[d] Vos (1934), calculated assuming 700 Torr dry pressure.
[e] Scharnke (1938), calculated assuming 700 Torr dry pressure.

P_{CO_2} in the caudal lungs of these birds too (Powell and Hempleman, 1985).

4. Effective Parabronchial Ventilation

A quantitative description of gas exchange requires a measure of the effective ventilation of the lung (see Section V,A). By analogy with alveolar ventilation in mammals, this is defined as parabronchial ventilation ($\dot{V}_P$) birds and it differs from inspired ventilation ($\dot{V}_I$) because of dead space ventilation ($\dot{V}_D$). $\dot{V}_D$ in birds includes not only anatomic dead space in the upper airways, but it may result from ventilatory shunts in which airflow bypasses the parabronchi. Figure 6 (open arrows) shows how gas might bypass the parabronchi during inspiration by directly entering the cranial secondary bronchi and air sacs or during expiration by flowing back out the primary bronchus (Powell, 1988).

P_{CO_2} measurements in the cranial secondary bronchi show that an "inspiratory shunt," with inspired gas directly entering cranial air sacs from the primary bronchi does not occur (Powell et al., 1981). Some inspiratory flow may enter the fourth medioventral (cranial secondary) bronchus, however, and flow in a cranial-to-caudal direction through some paleopulmonic parabronchi (Powell and Hempleman, 1985), but this would not be a "shunt." In contrast, P_{CO_2} measurements indicate that an "expiratory shunt" with 10 to 25% of expired gas from the caudal air sacs gas flowing out through the primary bronchus (Powell, 1988; Powell et al., 1981).

Other experiments indicate that the relative proportion of expiratory flow in a mesobronchial shunt may vary from 100 to 75% at the beginning of expiration, to 0% near the midpoint of expiration (Hastings and Powell, 1986a). The magnitude of expiratory mesobronchial shunting changes with tidal volume, the pattern of expiratory volume changes (Hastings and Powell, 1986a) and with physiological changes such as thermal panting (Bretz and Schmidt-Nielsen, 1971).

5. Artificial Ventilation

The flow-through design of the rigid parabronchial lung allows a unique form of artificial ventilation called unidirectional ventilation (Burger and Lorenz, 1960). Fresh humidified gas can be insufflated through a cannula in the trachea or an air sac, so it flows through the parabronchi before leaving the body through another cannula. This technique can be used clinically to support gas exchange during surgery which opens one of the air sacs (preventing effective spontaneous ventilation), to administer anesthetic gas or nebulized drugs (Fedde, 1978; Whittow et al., 1970), or for experimental studies (e.g., Burger et al., 1979; Fedde et al., 1974a). Artificial ventilation can also be performed manually by alternately compressing and lifting the sternum, for example, in a bird that may be anesthetized too deeply. It is also important to note that the sternum should not be compressed when holding a bird because this may lead to suffocation.

6. Frontiers: Aerodynamic Valving

"Aerodynamic valving" refers to fluid mechanical properties of gas flow in the avian lung that determine the distribution of ventilation. Differences in resistances between various groups of secondary bronchi or parabronchi could influence the flow pattern (Scheid and Piiper, 1989). However, local pressure measurements show that branch points, such as the openings of the cranial secondary bronchi into the primary bronchi, are critical sites for determining flow (Kuethe, 1988; Molony et al., 1976). As early as 1943, fluid dynamic models of the avian lung were used to show that these branches were critical for unidirectional caudal-to-cranial air flow in the parabronchi (Hazelhoff, 1951). More recently, a theory of aerodynamic valving has been used to predict decreased effectiveness of valving with decreased gas density and velocity, and experiments show valve failure and inspiratory shunts with reduced gas density (Banzett et al., 1987, 1991; Butler and Turner, 1988; Wang et al., 1988; Want et al., 1992). $\dot{V}_P$ changes with barometric pressure in some experiments on resting ducks (Shams et al., 1990) but not others (Shams and Scheid, 1993), so shunting may occur at altitude with decreased gas density. The physiological significance of these findings, for example, during flight at altitude, when gas velocities are high but gas density is low, remain to be determined. In contrast, expiratory aerodynamic valving is not sensitive to gas density and future experiments are necessary to determine its mechanism (Brown et al., 1995).

IV. PULMONARY CIRCULATION

Chapter 5 covers the basic physiology of the pulmonary circulation, but this section highlights some details necessary to understand pulmonary gas exchange. The pulmonary circulation is unique because the lung is the only organ to receive the entire cardiac output. The "in series" arrangement of the systemic and pulmonary circulations in birds and mammals means that the lungs receive the same amount of blood flow as the whole rest of the body. However, the resistance to blood flow in the lungs is lower, and this allows lower perfusion pressure than in the systemic circulation, with the complete separation of the left and right ventricles. If pressures are too high, then pulmonary capillaries can suffer "stress failure," which can allow blood to leak into air spaces of the lung and impair gas exchange (West and Mathieu-Costello, 1992). This section describes some structural and functional factors which determine pressures and volumes and flows in the pulmonary capillaries, which are important determinants of gas exchange.

A. Anatomy of the Pulmonary Circulation

The functional anatomy of the pulmonary circulation has been studied in detail for the domestic fowl (Abdalla and King, 1975). Interparabronchial arteries arise from main rami of the pulmonary arteries and run between the parabronchi and may perfuse more than one parabronchus. These vessels give rise to intraparabronchial arteries which perfuse the parabronchial mantle at several points along a parabronchus. The intraparabronchial arteries branch into the pulmonary blood capillaries near the outside edge of the parabronchial mantle which form a meshwork with air capillaries as described above (Section I,B,2). Pulmonary capillary blood is collected in intraparabronchial veins near the parabronchial lumen. These veins deliver blood flow to interparabronchial veins located near the outside edges of the parabronchus. The interparabronchial veins run between the parabronchi and collect blood from several points along a parabronchus and from several parabronchi. There are no anastomoses between the arterioles and veins in the lung (Abdalla, 1989).

This anatomy has two important consequences for respiratory gas exchange. First, all of the parabronchi are perfused along their entire length by oxygen-poor mixed venous blood, and the oxygenated blood returning to the heart in the pulmonary vein (i.e., systemic arterial blood) is a mixture of blood draining the entire length of all the parabronchi. This allows cross-current gas exchange to occur, which is more efficient than alveolar gas exchange as described below (Section V,B,1). Second, it means that blood flow within the parabronchial mantle is directed from the periphery toward the lumen, which also affects the efficiency of gas exchange in the air capillaries (see Section IV,C,1).

B. Pulmonary Capillary Volume

Pulmonary capillary blood volume in the parabronchial lungs of birds is essentially constant under all conditions. This contrasts with the alveolar lungs of mammals, which can increase pulmonary capillary volume by recruitment and distention, when perfusion pressure increases. This is important, for example during exercise, because it increases the capillary surface area for diffusion and the capillary transit time, which allows more time for oxygen to diffuse into blood (see Section V,C). However, recruitment or distention of blood capillary volume in the parabronchial lung would collapse the adjacent air capillaries and reduce gas exchange efficiency by causing a shunt (see Section V,D,2).

C. Pulmonary Vascular Resistance and the Distribution of Blood Flow

1. Pulmonary Vascular Resistance (PVR)

By analogy with Ohm's law:

$$PVR = \Delta P/\dot{Q},$$

where $\dot{Q}$ is cardiac output and ΔP is the difference between mean pulmonary artery and left atrial pressure. Pulmonary vascular pressures, cardiac output, and therefore PVR are similar in resting birds and mammals. However, PVR increases more with increases in cardiac output in birds compared to mammals. Recruitment and distention in alveolar lungs increases vascular cross-sectional area and decreases PVR, but this cannot occur in the constant-volume parabronchial lung. For example, doubling the blood flow through one lung of a domestic duck almost doubles mean pulmonary artery pressure but causes no change in the resistance calculated for that lung and no change in the capillary dimensions (Powell et al., 1985). A similar mechanism may also explain the increase in pulmonary artery pressure with hypoxia observed in birds (Black and Tenney, 1980a; Burton et al., 1968). Hypoxic stimulation of cardiac output (see Chapter 5) will cause pulmonary hypertension if vascular resistance is constant (Powell, 1983a). The direct effects of hypoxic pulmonary vasoconstriction (see Section III,C,2) on PVR are uncertain.

2. Distribution of Blood Flow

Local and regional changes in vascular resistance are more important than overall PVR for respiratory gas exchange. For example, regional control of blood flow between parabronchi has important effects on the efficiency of gas exchange (see Section V,D,4). Hypoxia has been shown to decrease local parabronchial blood flow, and gradients in P_{O_2} can explain differences in perfusion along parabronchi (Holle et al., 1978; Parry and Yates, 1979). The physiological mechanism of this response is not known. It may be similar to hypoxic pulmonary vasoconstriction in mammals, which is a direct effect of alveolar hypoxia on small vessels, although increases in O_2 or CO_2 do not affect pulmonary blood flow in birds like they do in mammals (Parry and Yates, 1979). Smooth muscle capable of controlling local blood flow has been described for interparabronchial arteries and veins in chickens, and this could be responsible for control of blood flow between and along the lengths of parabronchi (Abdalla, 1989).

D. Fluid Balance

Fluid balance in the lungs, as in all organs, depends on the balance of hydrostatic and colloid osmotic pressures across the capillaries and capillary permeability (see Chapter 7). Although capillary pressures are similar in birds and mammals, plasma colloid osmotic pressure in birds can be less than half the value in mammals. Consequently, the effect of volume loading on the accumulation of extravascular water in the lung interstitium is greater in birds than in mammals (Weidner et al., 1993). This suggests that birds should have a particularly well-developed lymphatic system to protect the constant volume parabronchial lung from interstitial and air capillary edema. However, the normal pathway and rates pulmonary lymph drainage are still unknown, so this remains speculation.

1. Frontiers: Pulmonary Circulation and Ascites

The pulmonary circulation is also involved in an important clinical problem involving systemic fluid balance, namely ascites in fast-growing chickens bred for meat production (Julian, 1993). Ascites (i.e., fluid accumulation in the peritoneum) also occurs frequently in chickens raised at high altitude, so researchers hypothesized that abnormal oxygen sensitivity may explain ascites in fast-growing chickens at sea level (Peacock et al., 1990). However, now it appears that ascites in both cases are the result of increased cardiac output, which increases pulmonary artery pressure as described above (Section III,C,1) and ultimately leads to right heart failure. This increases venous and capillary pressures, which cause fluid loss from the vascular system. At high altitude, hypoxia stimulates cardiac output (Bouverot, 1985). In fast-growing birds at sea level, the increased growth rate requires increased O_2 consumption, which stimulates cardiac output, presumably by the same mechanism that operates during exercise (Julian, 1993).

The chronic increase in cardiac output of fast-growing birds causes pulmonary hypertension and ascites because the vascular capacity of the lungs has not increased in proportion to the metabolic demands of the rest of the body. The pulmonary capillary volume of normal chickens is significantly smaller than that predicted for a bird of comparable size (Maina, 1989), probably because domestic chickens have been bred for other traits for thousands of years. Understanding the genetics of respiratory traits such as lung or capillary volume could greatly enhance the commercial potential of breeding programs for fast growth.

V. GAS TRANSPORT BY BLOOD

Equilibrium curves, also called dissociation curves, quantify the amount of O_2 and CO_2 in blood as functions of partial pressure. It is necessary to consider both partial pressure and concentration because partial pressure gradients drive diffusive gas transport in lungs and tissues, but concentration differences determine convective gas transport rates in lungs and the circulation (see Section V,A). The concentration of a physically dissolved gas in a liquid is directly and linearly proportional to its partial pressure according to Henry's law: $C = \alpha P$, where α = solubility in millimolar per Torr. This means that inert gases such as nitrogen, and even anesthetic gases, increase in blood in direct proportion to their partial pressure. However, O_2 and CO_2 also enter into chemical reactions with blood. These reactions result in more complex relationships between concentration and partial pressure, but they serve to (1) increase O_2 and CO_2 concentrations in blood, (2) allow physiological modulation of O_2 and CO_2 transport in blood, and (3) make respiratory CO_2 exchange an important mechanism of acid–base balance in the body.

A. Oxygen

Oxygen concentration in normal arterial blood (Ca_{O_2}) of, for example, a pigeon is about 8.3 mM. However, the physical solubility of O_2 in blood (α_{O_2}) is only 0.00124 mM/Torr at 41°C, so only 0.117 mM of arterial O_2 content is dissolved gas with a normal arterial P_{O_2} of 95 Torr in pigeons (Powell, 1983b). Most of the O_2 in blood is chemically bound to hemoglobin.

1. Hemoglobin

Hemoglobin is a large molecule consisting of four individual polypeptide chains, each with a heme (iron-containing) protein that can bind O_2 when iron is in the *ferrous* (Fe^{++}) form. Methemoglobin occurs when the iron is in the *ferric* form (Fe^{+++}), and it cannot bind O_2. Small amounts of methemoglobin, which occur under normal conditions, slightly reduce the amount of O_2 that can be bound to hemoglobin. One gram of pure mammalian hemoglobin can bind 0.060 mmol of O_2 when fully saturated, and this value appears similar in birds (Powell, 1983b). Hemoglobin is concentrated inside red blood cells, or erythrocytes. This cellular packaging is important for the biophysics of the microcirculation, and it provides physiological control of O_2 binding through cellular changes in the hemoglobin microenvironment (see Section IV,A,3).

2. O_2–Blood Equilibrium Curves

Figure 7 shows the O_2 equilibrium curve for duck blood as saturation (S_{O_2}) versus P_{O_2}, where S_{O_2} is defined as the percentage of the total hemoglobin sites available for binding O_2 which are occupied by O_2. Therefore, the maximum S_{O_2} is 100% and independent of hemoglo-

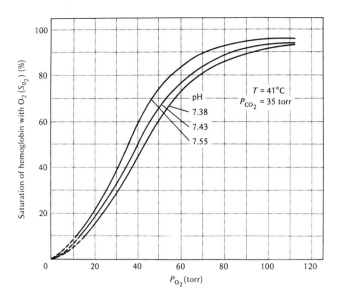

FIGURE 7 O_2–blood equilibrium curves for duck. The Bohr effect is demonstrated by shifts of the curve as pH changes. (Reprinted from *Respir. Physiol.* **24,** G. Scheipers, T. Kawashiro, and P. Scheid, Oxygen and carbon dioxide dissociation of duck blood, pp. 1–13, Copyright (1975), with permission from Elsevier Science.)

bin concentration in blood. In contrast, O_2–hemoglobin equilibrium curves plotting concentration versus P_{O_2} quantify the absolute amount of O_2 in blood at a given P_{O_2}, and the maximum O_2 concentration depends on the amount of hemoglobin available. O_2 capacity defines the maximum O_2 concentration in blood when hemoglobin is 100% saturated with O_2. Total O_2 concentration in blood (C_{O_2}), including chemically bound and dissolved O_2, can be calculated as:

$$C_{O_2} = (O_2 \text{ capacity } [S_{O_2} / 100]) + (\alpha_{O_2} P_{O_2})$$

The sigmoidal (or "S") shape of the blood–O_2 equilibrium curve results from cooperative, allosteric interactions between the four subunits of hemoglobin, which determine the three dimensional shape of the molecule. O_2 equilibrium curves for individual hemoglobin subunits are not sigmoidal, but simple convex curves like the O_2 equilibrium curve for myoglobin (see Section VI). The shape of the blood–O_2 hemoglobin equilibrium curve facilitates O_2 loading on blood in the lungs and O_2 unloading from blood in the tissues.

The cooperativity between functional subunits of hemoglobin is quantified with Hill's coefficient, n (Powell and Scheid, 1989). High values of n, exceeding the theoretical limit of 4, have been observed in bird blood and may reflect increased cooperativity between aggregates of multiple hemoglobin molecules or interactions between different isoforms of hemoglobin within a blood sample (Black and Tenney, 1980b; Lapennas and Reeves, 1983).

3. Factors Affecting O_2–Hemoglobin Affinity

P_{50}, defined as the P_{O_2} at 50% saturation, is used to quantify changes in the affinity of hemoglobin for O_2. For example, a decrease in P_{50}, or a "left shift," indicates an increase in O_2 affinity because S_{O_2} is greater for a given P_{O_2}. The most important physiological factors that affect the P_{50} in a given species are (1) organic phosphate levels, (2) pH, and (3) temperature. Evolutionary differences in the amino acid sequence of the four hemoglobin subunits explain the differences in P_{50} between species and during development.

Myinositol 1,3,4,5,6-pentophosphate (IPP) is the primary organic phosphate affecting P_{50} in birds (Weber and Wells, 1989). The effect of IPP binding with hemoglobin inside erythrocytes is considerable; for example, increasing P_{50} from less than 3 Torr in stripped hemoglobin from chickens to over 40 Torr under *in vivo* conditions (Weingarten *et al.*, 1978). Physiological changes in organic phosphates have not been studied extensively in birds (Maginniss and Kilgore, 1989; Weber and Wells, 1989). In mammals, acclimatization to conditions such as altitude can modulate P_{50} by altering organic phosphates (2,3 diphosphoglycerate, or 2,3-DPG). However, it is important to note that differences in P_{50} between birds adapted over generations to low or high altitude cannot be explained by different organic phosphate concentrations. Erythrocyte inorganic phosphate levels are similar in the greylag and Canada goose, natives to low altitude (with P_{50} = 39 and 42 Torr, respectively), and the bar-headed goose, which is native to high altitude (with P_{50} = 29 Torr) (Petschow *et al.*, 1977). Differences in the binding of IPP to hemoglobin from different species explains such differences in P_{50} (Rollema and Bauer, 1979).

Figure 7 shows the effect of pH on the O_2 affinity, which is known as the Bohr effect. Increases in pH cause decreases in P_{50} (i.e., increase O_2–hemoglobin affinity) and vice versa. Hydrogen ion binds to histidine residues in hemoglobin, which changes the molecular conformation and ability to heme sites to bind O_2. The physiological advantage of the Bohr effect is that it facilitates O_2 loading in the lungs, where CO_2 is low and pH is high (see Section IV,D). In muscles, the opposite occurs and decreased pH facilitates O_2 unloading to the tissues. The Bohr effect is independent of saturation in most birds (Lapennas and Reeves, 1983; Maginniss, 1985; Meyer *et al.*, 1978) and similar to values reported for mammals (Table 4).

In most birds, and in contrast to mammals, there is no independent effect of CO_2 on P_{50} (Meyer *et al.*, 1978). Carbon dioxide forms carbamino compounds with hemoglobin in mammals, and these cause small increases in P_{50}. In some birds, such as sparrows and burrowing owls, the Bohr effect is greater when pH is changed with CO_2 compared to fixed acid (Maginniss, 1985; Maginniss and Kilgore, 1989). Therefore, carbamino formation does occur and can decrease O_2 affinity in stripped avian hemoglobin. However, strong binding of IPP to hemoglobin in most birds prevents an independent CO_2 Bohr effect (Lapennas and Reeves, 1983; Weingarten *et al.*, 1978).

The *in vivo* physiological O_2 dissociation curve is steeper than the individual *in vitro* curves in Figure 7 because P_{CO_2} increases and pH decreases between arterial and venous blood. This is an advantage for gas exchange because it increases the change in O_2 concentration for a given change in P_{O_2}, thereby increasing O_2 uptake or delivery. The slope of the physiological O_2–blood equilibrium curve, in terms of O_2 concentration, is called βb_{O_2} (mM/Torr) and is used for quantitative descriptions of gas exchange (see Section V,A).

TABLE 4 Respiratory Parameters in Avian Whole Blood[a]

Reference[b]	Burrowing owl (*Athene cunicularia*) (15)	Pigeon (12)	Female domestic fowl (2–4, 6–10, 14)	Pekin duck (1, 5)	Muscovy duck (*Cairina moschata*) (8, 11, 13)
Hematocrit (%)	33.7 ± 2.1	48.7	26 to 30	45.4	37.3 ± 1.3
Hemoglobin (g%)	10.7 ± 0.4	14.3	8.6 to 9.3	15.5	—
O_2 capacity (mmol l^{-1} P_{50} Torr)	—	8.6	8.6 to 12.3	≥8.9	7.3 ± 0.51
	42.3 ± 0.8	40.8 ± 1.4	47.7 ± 4.2	42.7 ± 45.0	40.1 ± 3.7
Hill's n (−)	2.60 to 3.42	2.75	3.4 ± 0.1	4.3	2.9
Bohr coefficient ($\Delta \log P_{50}/\Delta pH$)	0.42 to 0.46	0.42 to 0.53	0.50 ± 0.08	0.40 to 0.44	0.44 to 0.53
Temperature coefficient ($\Delta \log P_{50}/\Delta T$)	—	0.015 to 0.026	0.014 to 0.015	—	—
Haldane effect ($\Delta C_{CO_2}/\Delta C_{O_2}$)	—	—	0.42	—	0.30

[a] Where three or more measurements are available; SD is given; for less than three; the range is indicated.

[b] References: (1) Andersen and Lovo (1967); (2) Bartels *et al.* (1966); (3) Bauer *et al.* (1978); (4) Baumann and Baumann (1977); (5) Black and Tenney (1980b); (6) Henning *et al.* (1971); (7) Hirsowitz *et al.* (1977); (8) Holle *et al.* (1977); (9) Lapennas and Reeves (1983); (10) Meyer *et al.* (1978); (11) Morgan and Chichester (1935); (12) Powell (1983); (13) Scheipers *et al.* (1975); (14) Wells (1976); (15) Maginnis and Kilgore (1989).

Because the combination of O_2 with hemoglobin is a chemical reaction which releases heat, increased temperature reduces the affinity of hemoglobin for O_2. This facilitates O_2 off-loading at exercising muscle with relatively high temperatures. It has also been hypothesized to facilitate O_2 on-loading in the lungs of birds flying at high altitude, when high rates of ventilation with extremely cold air might cool the respiratory exchange surfaces (Faraci, 1991, 1986). However, experiments to date have not been able to demonstrate decreased temperature of blood in the lungs. The effect of temperature on P_{50} in birds is similar to the effect in mammals (Table 4).

In general, P_{50} in avian blood is greater than in mammalian blood. Table 4 shows P_{50} values for several common avian species in the 40-Torr range, while P_{50} in comparably sized mammals are nearer 30 Torr. Some studies have used erythrocyte suspensions (Lutz, 1980), and this might explain the low P_{50} values they have found. However, determinations of P_{50} in avian hemoglobin solutions agree well with other published values for whole blood; for example, in pigeons (Powell, 1983b). Because the efficiency of O_2 uptake in the avian lung is greater than mammals, birds may have evolved blood with low O_2 affinity to maximize O_2 delivery in tissue.

Developmental changes in hemoglobin–O_2 affinity are explained by differences in the type of hemoglobin expressed in an individual. For example, the dramatic decrease of P_{50} in chickens, from 75 Torr at 8 days to 35 Torr at 14 to 16 days of development, is explained by the replacement fetal hemoglobin with adult hemoglobin expressed in the erythrocytes (Baumann and Baumann, 1978).

4. Factors Affecting O_2 Capacity

Changes in hemoglobin concentration [Hb] in blood will change the O_2 capacity and therefore the O_2 concentration at any P_{O_2} as described above. Hemoglobin concentration depends on both the mean corpuscular hemoglobin concentration (MCHC) and the hematocrit. Typical hematocrit and [Hb] values are given in Table 4; [Hb] is expressed as g/100 ml of blood instead of mM because the molecular weight was not known for all hemoglobins when they were originally measured. Typical values for MCHC in birds are 30 to 40 g Hb/100 ml of erythrocytes (Palomeque et al., 1979), similar to the MCHC in mammals.

If [Hb] decreases (for example, with decreased hematocrit in anemia) then O_2 capacity and concentration decreases at any given P_{O_2}. The O_2 capacity increases when [Hb] increases, for example, by the stimulation of red blood cell production in bone marrow by the hormone erythropoietin (EPO). EPO is released from cells in the kidneys in response to decreases in arterial O_2 levels (Sturkie, 1986). Significant gender differences in [Hb], for example, chickens are explained by the effects of sex hormones on hematocrit (Sturkie, 1986).

5. Frontiers: Hemoglobin Molecular Engineering

Hemoglobin–O_2 affinity is generally greater in birds native to high altitude compared to those living at low altitudes (Black and Tenney, 1980b; Jessen et al., 1991; Petschow et al., 1977). For example, P_{50} in the bar-headed goose, which lives at high altitude in Tibet and migrates across the Himalayas to India, is 10 Torr less than P_{50} in the closely related greylag goose, which lives on the Indian planes year round (Petschow et al., 1977). Low P_{50} should enhance O_2 loading in the lungs in extreme hypoxia (Bencowitz et al., 1982). Amino acid sequences from the bar-headed goose and greylag goose revealed a single substitution on one of the four hemoglobin subunits, where two α- and two β-chains compose a hemoglobin molecule (Allen, 1983). This led molecular biologists to hypothesize that the substitution creates a two-carbon gap at the contact point between α_1- and β_1-chains and increases O_2-affinity by relaxing tension in the hemoglobin molecule in its deoxygenated state.

To test this idea, the researchers introduced the amino acid substitution into human globin synthesized in the bacteria *Escherichia coli* and studied the reconstituted hemoglobins. The engineered hemoglobins combined in the normal cooperative fashion with O_2, but their P_{50} differed even more than the difference found between bar-headed and greylag geese. Structural studies of the crystallized hemoglobins confirmed the gap between the individual chains predicted from the amino acid sequence data. Such discoveries raise interesting possibilities for improving respiratory function through genetic engineering.

B. Carbon Dioxide

Carbon dioxide–blood equilibrium (or dissociation) curves are nonlinear, but they have a different shape and position than O_2–blood equilibrium curves (Figure 8). Carbon dioxide is carried in three forms by blood, so CO_2 concentrations in blood are generally much higher than O_2 concentrations. This results in a smaller range of P_{CO_2} values in the body, compared to the range P_{O_2} values, although the differences between arterial and venous concentrations are similar for CO_2 and O_2. The resulting physiological CO_2 dissociation curve between the arterial and venous points is much more linear than the physiological O_2 dissociation curve (Figure 8).

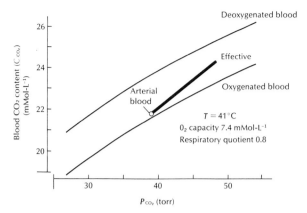

FIGURE 8 CO$_2$–blood equilibrium curves from the duck. Upper (deoxygenated blood) and lower (oxygenated blood) curves are derived from *in vitro* equilibration of blood samples. Heavy line is the physiological or *in vivo* dissociation curve from unanesthetized undisturbed birds. It illustrates the changes in CO$_2$ content (C_{CO_2}) in blood as it changes from arterial blood to venous blood in the tissue capillaries. (Reprinted from *Respir. Physiol.* **24**, G. Scheipers, T. Kawashiro, and P. Scheid, Oxygen and carbon dioxide dissociation of duck blood, Copyright (1975), with permission from Elsevier Science.)

1. Forms of CO$_2$ in Blood

Carbon dioxide solubility in water or plasma is 0.0278 mM/Torr, or about 20 times more soluble than O$_2$. Still, dissolved CO$_2$ only contributes about 5% of total CO$_2$ concentration in arterial blood. Carbon dioxide can also combine with terminal amine groups in hemoglobin to form carbamino compounds (see Section IV,A,3). Bicarbonate ion (HCO$_3^-$) is the most important form of CO$_2$ carriage in blood. Carbon dioxide combines with water to form carbonic acid and this dissociates to HCO$_3^-$ and H$^+$:

$$CO_2 + H_2O \leftrightarrow H_2CO_3 \leftrightarrow HCO_3^- + H^+$$

Carbonic anhydrase is the enzyme that catalyzes this reaction and it occurs mainly in red blood cells (Maren, 1967). The reaction is almost instantaneous with carbonic anhydrase, but the uncatalyzed reaction will occur much more slowly in any aqueous medium (requiring over 4 min for equilibrium). The rapid conversion of CO$_2$ to bicarbonate results in about 90% of the CO$_2$ in arterial blood being carried in that form. The H$^+$ produced from CO$_2$ reacts with hemoglobin and affects both the O$_2$ dissociation curve (Bohr effect) and CO$_2$ dissociation curve as described next.

2. Factors Affecting Blood–CO$_2$ Equilibrium Curves

Hemoglobin–O$_2$ Saturation is the major factor affecting the position of the CO$_2$ equilibrium curve. The Haldane effect increases CO$_2$ concentration when blood is deoxygenated, or decreases CO$_2$ concentration when blood is oxygenated, at any given P_{CO_2} (Figure 8). The Haldane effect is actually another view of the same molecular mechanism causing the Bohr effect on the O$_2$ equilibrium curve (see Section IV,A,3). Hydrogen ions and O$_2$ can be thought of as competing for hemoglobin binding, so increasing O$_2$ decreases the affinity of hemoglobin for H$^+$ (Haldane effect), and increased [H$^+$] decreases the affinity of hemoglobin for O$_2$ (Bohr effect).

The physiological advantages of the Haldane effect are to promote unloading of CO$_2$ in the lungs when blood is oxygenated, and CO$_2$ loading in the blood when O$_2$ is released to tissues. The Haldane effect also results in a steeper physiological CO$_2$–blood equilibrium curve (Figure 8), which has the physiological advantage of increasing CO$_2$ concentration differences for a given P_{CO_2} difference. Finally, the Haldane effect can cause apparent negative blood-gas CO$_2$ gradients when it is amplified by cross-current gas exchange in avian lungs (see Section V,B,2).

C. Acid–Base

The chemical equilibrium between CO$_2$ and H$^+$/HCO$_3^-$ ions has tremendous implications for acid–base physiology. Every mole of metabolic CO$_2$ produced results in one mole of acid, and over 95% of this acid is excreted by the lungs (Skadhauge, 1983). The ability to change blood P_{CO_2} levels rapidly by changing ventilation has a powerful effect on blood pH, so acid–base balance depends on the integrated function of respiratory and renal systems.

1. Henderson-Hasselbalch Equation

This equation describes the relationship between P_{CO_2}, pH, and [HCO$_3^-$] in blood as:

$$pH = pK_a + \log([HCO_3^-]/\alpha P_{CO_2})$$

where pK_a is -log$_{10}$ of K_a, the dissociation constant for carbonic acid; [HCO$_3^-$] is bicarbonate concentration in mEq/liter or mM; and α is the physical solubility for CO$_2$ in water. A normal value for arterial pH (pHa) in chickens is 7.52 (Table 5), which can be calculated from pK_a = 6.09 and α_{CO_2} = 0.03 mM/Torr in chicken plasma at 41°C (Helbacka *et al.*, 1963), arterial P_{CO_2} = 33 Torr (Table 5), and arterial [HCO$_3^-$] = 27.2 mM. The buffer value for plasma in ducks is similar to the value for humans (Scheipers *et al.*, 1975) if corrections are made for differences in [Hb]. At pH = 7.5, the [H$^+$] is only 30 nM, or significantly less than many other important ions in the body, such as Na$^+$, Cl$_3^-$, HCO$_3^-$, which occur in the millimolar range. Small changes in pH, corresponding to very small changes in [H$^+$] (see Chapter 7), can lead to dramatic changes in physiological function.

The Henderson-Hasselbalch equation shows how the physiological control of pH depends on the ratio of [HCO$_3^-$] to [αP_{CO_2}]. Notice that a normal pH can occur

TABLE 5 Gas Exchange Variables in Awake Resting Birds[a]

Reference	Pigeon (1)	Female domestic fowl (2)	Pekin duck (3)	(4)	Muscovy duck (*Cairina moschata*) (4)
M_B	0.38	1.6	2.37	2.4	2.16
$\dot{M}_{O_2}$ (mmol/min)	0.35	1.09	1.67	—	—
f_R (min^{-1})	27.3	23	15.6	8.2	10.5
V_T (ml)	7.5	33	58.5	98	69
$\dot{V}_E$ (L/min)	0.204	0.760	0.910	0.807	0.700
$\dot{Q}$ (L/min)	0.127	0.430	0.423	0.973	0.844
$P_{E_{O_2}}$ (Torr)	—	101.8	—	100.1	96.6
Pa_{O_2} (Torr)	95	87	100	93.1	96.1
$P\bar{v}_{O_2}$ (Torr)	50	40.8	69.6	63.3	55.9
$P_{E_{CO_2}}$ (Torr)	—	33.0	—	34.2	34.2
Pa_{CO_2} (Torr)	34	29.2	33.8	36.3	35.9
$P\bar{v}_{CO_2}$ (Torr)	—	39.3	—	37.3	42.6

[a] Data collected from birds in body plethysmographs except references (2) and (4), which used endotracheal tubes in lightly restrained upright birds. $\dot{M}_{O_2}$ not given for reference (4) because mixed-expired gases were not measured.

[b] References: (1) Bouverot *et al.* (1976); (2) Piiper *et al.* (1970); (3) Bouverot *et al.* (1979); (4) Jones and Holeton (1972).

with a variety of [HCO_3^-] and P_{CO_2} values; so, for example, pH = 7.52 in a chicken does not necessarily indicate normal acid–base status. The primary cause of a chronic acid–base disturbance cannot be determined from P_{CO_2}, pH, and [HCO_3^-] data alone. Other details of the disease history, pulmonary function, or blood chemistry must be obtained for a proper diagnosis.

The respiratory system controls pH primarily by changing arterial P_{CO_2} (Pa_{CO_2}). Pa_{CO_2} is determined by parabronchial ventilation at any given metabolic rate (see Section V,B). Increasing ventilation will decrease Pa_{CO_2} and increase pHa, while decreasing ventilation will have the opposite effects. Therefore, ventilation is an extremely effective mechanism for changing pHa quickly, and ventilatory reflex responses to pH are the most important physiological mechanisms for rapid control of pH. The kidneys can also control pH by changing [HCO_3^-] independent of CO_2 changes, as described in Chapter 7, but renal changes in pH generally take longer than respiratory changes in pH.

D. Blood Gas Measurements

Avian blood presents special challenges to the accurate measurement of P_{O_2}, P_{CO_2}, and pH with traditional equipment designed for humans. In contrast to mammals, avian erythrocytes are nucleated (as are most other vertebrates) and this may be the reason for high rates of O_2 consumption compared to mammals. Therefore, care must be taken to analyze arterial blood gases in birds as soon as possible after the sample is drawn, and to correct for any decreases in P_{O_2} with time if necessary. Storing the samples in ice water may help if immediate analysis is not possible, but the analyzed value can still differ from the *in vivo* value if any delay occurs, especially in normoxia where the blood–O_2 equilibrium curve is relatively flat and small O_2 content changes cause large changes in P_{O_2} (Scheid and Kawashiro, 1975). Sampling delays may also explain reports of 0 Torr P_{O_2} values in mixed venous blood from pigeons at simulated high altitude (Weinstein *et al.*, 1985).

In addition to sampling delays, care must be taken to perform the analysis at body temperature, which is usually greater than the human value of 37°C, or to apply temperature correction values established for avian blood (Kiley *et al.*, 1979). Also, a blood gas correction factor needs to be established with a tonometer to account for differences in P_{O_2} measured in the liquid phase after calibrating electrodes with a gas phase (Nightingale *et al.*, 1968). These, and other factors which may affect O_2–hemoglobin saturation measurements, have been discussed in other reviews (e.g., Powell and Scheid, 1989).

Finally, the most important determinant of arterial blood gas values is the physiological state of the bird. Tables 5 and 6 present arterial blood gases for several species breathing room air, but these may not be "normal" if the bird was excited by the sampling procedure. Remote-controlled sampling devices have been used for resting ducks (Scheid and Slama, 1975), and this technology could presumably be extended to other interesting conditions such as diving birds, which have been studied while carrying instruments to measure the depth of a dive and heart rate (Kooyman, 1989).

VI. PULMONARY GAS EXCHANGE

The unique anatomy of the avian respiratory system results, theoretically, in a model of gas exchange which is more efficient than the mammalian model (Piiper and Scheid, 1975). For a given level of ventilation to the gas exchange surfaces ($\dot{V}_P$), cardiac output ($\dot{Q}$), and lung diffusing capacity ($D_{L_{O_2}}$), arterial O_2 loading and CO_2 elimination are predicted to be better in a parabronchial lung, compared to an alveolar lung with the same inspired gases and metabolic demands (Powell and Scheid, 1989). This section describes the structural and functional basis for this model and how it actually behaves in nature under physiological conditions.

A. Basic Principles of Oxygen Transport

1. Convection

Convection, or bulk flow of gas, is used to transport oxygen into the lungs by ventilation and to the tissues by blood flow. The Fick principle, which is simply conservation of mass applied to respiratory gas transport, can be used to quantify O_2 uptake as:

$$\dot{M}_{O_2} = \dot{V} \beta g_{O_2} (P_{I_{O_2}} - P_{\bar{E}_{O_2}}),$$

where $\dot{M}_{O_2}$ is O_2 uptake, $\dot{V}$ is ventilation, βg_{O_2} is the capacitance coefficient for O_2 in the gas phase (0.512 mM/Torr at 41°C), and ($P_{I_{O_2}} - P_{\bar{E}_{O_2}}$) is the difference between inspired and mixed-expired P_{O_2} (Piiper et al., 1971; Powell and Scheid, 1989). Inspired and expired ventilation are assumed equal in this formulation of the Fick principle, so the amount O_2 consumed is the difference between the amount of O_2 inspired and the amount expired. Mixed-expired P_{O_2} is used when total ventilation is measured; end-expired P_{O_2} is used if parabronchial ventilation ($\dot{V}_P$) is available (see Section V,D,1). The same principles described for O_2, also apply to CO_2 exchange.

The Fick principle can be written for the cardiovascular transport of O_2 out of the lungs and to the tissues also:

$$\dot{M}_{O_2} = \dot{Q} \beta b_{O_2} (P_{a_{O_2}} - P_{\bar{v}_{O_2}}),$$

where $\dot{Q}$ is cardiac output and βb_{O_2} is the physiological slope of the blood–O_2 equilibrium curve (see Section IV,A,3). Hence, the amount of O_2 taken up by blood in the lungs is the difference between the amount of O_2 leaving the lungs in arterial blood and the amount that entered the lungs in mixed venous blood. In a steady state, $\dot{M}_{O_2}$ is equal at each step of the O_2 cascade, so the Fick principle can be rearranged to calculate $\dot{M}_{O_2}$, $\dot{Q}$, $\dot{V}$, or P_{O_2} from measurements of other variables in the equation. Table 5 lists the important variables for quantifying gas exchange in several birds under resting conditions. Changes in these variables under different physiological conditions such as exercise, hypoxia, and thermal stress have been summarized in several excellent reviews (Brackenbury, 1984; Butler, 1991; Faraci, 1991; Powell and Scheid, 1989).

2. Diffusion

O_2 movement over the very short distances across the blood–gas barrier occurs effectively by the "passive" mechanism of diffusion; active transport of O_2 does not occur in the body. Fick's law of diffusion describes O_2 transport from the air capillaries to the blood capillaries as:

$$\dot{M}_{O_2} = \Delta P_{O_2} \cdot D_{L_{O_2}},$$

where ΔP_{O_2} is the average P_{O_2} gradient between the air capillary and blood in the pulmonary capillary and $D_{L_{O_2}}$ is the diffusing capacity of the lung for O_2. The equation shows that a larger $D_{L_{O_2}}$ can transport more O_2 for a given P_{O_2} gradient. Determinants of $D_{L_{O_2}}$ are described in the section on lung diffusing capacity below (see Section V,C).

B. Cross-Current Gas Exchange

1. O_2 Exchange

Figure 9 shows how air flow and blood flow in the parabronchus can be viewed as occurring perpendicular to one another, and this is the basis for describing avian gas exchange with a "cross-current" model (Piiper and Scheid, 1975, 1972). In this idealized model of cross-current exchange, air flow is assumed continuous through the parabronchus which is uniformly perfused along its length by mixed venous blood. At the inspiratory end of the parabronchus, there is a large P_{O_2} gradient driving diffusion of O_2 into the capillary blood,

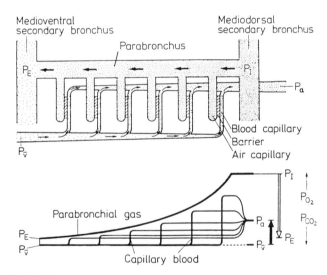

FIGURE 9 Cross-current model of gas exchange in the parabronchus as described in the text. (Bottom) How Pa_{O_2} results from a mixture of blood from capillaries all along the parabronchus, where P_{O_2} ranges from P_I to P_E. Overlap of P_{O_2} in gas (open arrows) and blood (filled arrows) shows how Pa_{O_2} can exceed $P_{E_{O_2}}$ in birds. (From Scheid, 1990, with permission).

which raises capillary P_{O_2} and drops parabronchial P_{O_2}. As air flows along the parabronchus, the P_{O_2} gradient driving O_2 diffusion decreases with parabronchial P_{O_2}, while $P\bar{v}_{O_2}$ is constant. At the expiratory end of the parabronchus, P_{O_2} has decreased to end-expired levels, and P_{O_2} in the capillary blood leaving this part of the parabronchus is correspondingly low. However, notice that arterialized blood returning to the heart is a mixture of capillary blood draining the entire length of the parabronchus. Therefore, arterial P_{O_2} is greater than end-expired P_{O_2} in ideal cross-current gas exchange.

A negative-expired arterial P_{O_2} difference, shown in Figure 9 as the overlap of P_{O_2} arrows, is not possible in alveolar gas exchange. The best situation that can be achieved in ideal alveolar gas lungs is equilibrium between expired and arterial P_{O_2}, so $(P_{E_{O_2}} - P_{a_{O_2}}) = 0$. However, negative-expired arterial P_{O_2} differences are not always observed in birds either (Table 5), because of limitations to gas exchange as discussed below (see Section V,D).

Notice that the efficiency of cross-current exchange should not depend on the direction of ventilatory flow in the parabronchus. This is supported by experiments which reversed the direction of parabronchial flow in ducks and chickens and found no difference in the expired arterial P_{O_2} differences for O_2 or CO_2 (Powell, 1982; Scheid and Piiper, 1972). It is physiologically significant because air flow is presumably bidirectional in neopulmonic parabronchi (see Section II,C).

2. Cross-Current CO_2 Exchange

Figure 9 shows how expired P_{CO_2} can exceed arterial P_{CO_2} in a parabronchus, which is also impossible in alveolar gas exchange. This occurs similar to the "overlap" of arterial and expired P_{O_2} as explained above. However, experimental observations of expired P_{CO_2} exceeding mixed venous P_{CO_2} in birds can only be explained by an interesting interaction of O_2 and CO_2 exchange in the parabronchus. Because of the shape of the O_2 and CO_2 blood equilibrium curves, the ratio of CO_2 elimination to O_2 uptake (R = respiratory exchange ratio) decreases near the expiratory ends of the parabronchus (Meyer et al., 1976). As originally postulated by Zeuthen (1942), this can lead to oxygenation of mixed venous blood in expiratory ends of the parabronchi and increased P_{CO_2} in capillaries by the Haldane effect. Because P_{CO_2} of oxygenated mixed venous blood is greater than $P\bar{v}_{CO_2}$ (Figure 8), P_{CO_2} in gas equilibrating with such blood at the expiratory end of the parabronchus (i.e., $P_{E_{CO_2}}$) can exceed true $P\bar{v}_{CO_2}$.

Overlap between arterial and expired P_{CO_2} is more commonly observed than overlap for O_2 in birds (Table 5). This is because of differences in the O_2 and CO_2 blood equilibrium curves, and the fact that CO_2 is less sensitive than O_2 to some of the factors limiting the efficacy of cross-current gas exchange (Powell and Scheid, 1989).

C. Lung Diffusing Capacity

The diffusing capacity of the lung for O_2 ($D_{L_{O_2}}$) is a complex variable that depends on several physiological processes (Powell, 1982; Powell and Scheid, 1989). These include gas-phase diffusion in the air capillaries, diffusion across the blood–gas barrier, and the chemical reaction between O_2 and hemoglobin.

1. Air Capillary Diffusion

Ventilatory flow in the parabronchi prevents significant P_{O_2} gradients from developing in the lumen of the parabronchus. However, P_{O_2} must decrease along the length of the air capillaries, and this P_{O_2} gradient drives diffusion of O_2 into the air capillaries (Figure 9). This gradient is predicted to be extremely small and does not present a significant limitation to gas exchange under any physiological conditions (Burger et al., 1979; Crank and Gallagher, 1978; Scheid, 1978). In contrast to diffusion in tissue, gas phase diffusion is an extremely effective transport mechanism over distances as long as the air capillaries. Furthermore, air–blood capillary exchange is enhanced by an arrangement similar to countercurrent exchange (Scheid, 1978). Oxygen diffuses from the lumen toward the periphery in air capillaries, which is opposite to the direction of flow in the blood capillaries (Figures 4 and 9). Therefore, blood at the end of a capillary is equilibrating with the relatively high P_{O_2} levels in air capillaries near the parabronchial lumen.

2. Blood–Gas Barrier Diffusion

The effect of the blood–gas barrier on diffusion in the lung can be evaluated by the membrane diffusing capacity (Dm_{O_2}), which can be estimated from morphometric measurements of the lung (Maina, 1989; Powell and Scheid, 1989). The membrane diffusing capacity is directly proportional to the surface area and inversely proportional to the thickness of the blood–gas barrier. These variables have been measured in several species of birds now, using perfusion fixation and rapid freezing to preserve tissue for electron microscopy and stereological analysis (Dubach, 1981; Maina et al., 1982, 1989; Powell and Mazzone, 1983). The values depend on body size (which determines metabolic levels; see Chapter 9), but in general they show that Dm_{O_2} is larger in birds than in comparably sized terrestrial mammals (Maina, 1989). For example, Dm_{O_2} in canada geese is 1.7 times greater than a comparably sized mammal (Powell and Mazzone, 1983). However, blood–gas barrier surface area and thickness are similar in birds and bats of the same body size, indicating the importance of favorable diffusion for high O_2 consumption levels during flight (Maina, 1989).

3. O_2–Hemoglobin Reaction Rates

Finite reaction rates between O_2 and hemoglobin behave as an additional resistance to O_2 uptake across the blood–gas barrier. Consequently, D_{LO_2} increases if there is more hemoglobin available with increased capillary volume. Pulmonary capillary blood volume is similar in birds and terrestrial mammals of similar body sizes, except for chickens which have a relatively low capillary volume (Maina, 1989). Also, estimates of the reaction rates between O_2 and avian hemoglobin in birds and mammals are similar (Phu et al., 1986). Therefore, O_2–hemoglobin reaction rates probably contribute a similar amount to diffusion resistances in avian and mammalian lungs. In mammals, this is estimated to comprise about half of the diffusive resistance to O_2 diffusion.

4. Physiological Estimates of D_{LO_2}

The diffusing capacity of the lung for O_2 can be estimated from experimental measurements on birds if certain conditions are met to satisfy the assumptions necessary for an ideal cross-current analysis of the data. For example, measurements should be made in hypoxia, where the O_2–blood equilibrium curve is linear, to satisfy the assumption of constant βb_{O_2}. Physiological measurements of D_{LO_2} which satisfy these assumptions have been made in ducks and chickens (Burger et al., 1979; Scheid and Piiper, 1970), and they generally agree with morphometric estimates (Maina and King, 1982). Physiological D_{LO_2} in ducks ranges from 38 to 68 μmol/(min Torr kg) (Hempleman and Powell, 1986). Other methods for estimating potential diffusion limitations indicate complete diffusion equilibrium for O_2 in resting birds (Powell and Scheid, 1989).

Exercise and pharmacological stimulation of metabolic rate increase D_{LO_2} in ducks (Geiser et al., 1984; Hempleman and Powell, 1986; Kiley et al., 1985), and the change is correlated closely with increased cardiac output (Hempleman and Powell, 1986). In alveolar lungs, D_{LO_2} increases with cardiac output by recruitment and distension of the pulmonary capillaries, which increases surface area and capillary volume. However, these enhancements can be offset by shorter transit times in the pulmonary capillaries allowing less time for diffusion equilibrium. Recruitment and distension are not expected in avian lungs (see Section III,B), so mechanisms of increasing D_{LO_2} in birds during exercise are not known.

D. Heterogeneity in the Lung

The avian lung is a complex structure consisting of hundreds of parabronchi. Mismatching of ventilation, blood flow, or diffusing capacity between these functional units can reduce the efficacy of gas exchange. Temporal variations in flow rates and inspired gas composition can also impair gas exchange. Under normal resting conditions at sea level, such heterogeneity in lung function is the most important factor reducing gas exchange efficacy from ideal levels in birds and in mammals.

1. Physiological Dead Space

Physiological dead space is defined as the difference between total inspired or expired ventilation ($\dot{V}_E$ or $\dot{V}_I$) and effective parabronchial ventilation ($\dot{V}_P$):

$$\dot{V}_{Dphys} = \dot{V}_I - \dot{V}_P.$$

$\dot{V}_{Dphys}$ includes anatomic dead space ventilation plus any heterogeneity such as inspiratory or expiratory mesobronchial shunts (see Section III,A) or ventilation to regions of the lung with high $\dot{V}/\dot{Q}$ ratios (see Section V,D,3). Therefore, $\dot{V}_{Dphys}$ considers ventilation "as if" total ventilation was partitioned between a single ideal parabronchus with $\dot{V}_P$ and anatomic dead space (Scheid and Piiper, 1989).

Many of the techniques used to estimate $\dot{V}_{Dphys}$ in alveolar lungs are not applicable to birds (Powell, 1988). However, with a computer model of O_2 and CO_2 dissociation curves and cross-current gas exchange, $\dot{V}_P$ can be calculated from measured ventilation, mixed venous blood, and mixed expired gas (Hastings and Powell, 1986b). $\dot{V}_{Dphys}$ in artificially ventilated ducks was almost 10 ml greater than anatomic plus instrument dead space and two-thirds of the 15 ml anatomic dead space (Hastings and Powell, 1986b). This large amount of physiological dead space is consistent with relatively large amounts of total ventilation going to high $\dot{V}/\dot{Q}$ regions of the lung (see Section V,D,3) and indicates that ventilatory heterogeneity within the lung has a significant impact on $\dot{V}_P$ and gas exchange.

2. Shunt

Shunting of pulmonary blood flow past effective gas exchange areas is very small in birds. As described above (see Section III,A), there are no arteriovenous anastomoses in the pulmonary circulation. Shunt ranges from less than 1 to 2.7% of cardiac output in anesthetized artificially ventilated geese and ducks, respectively, using inert gas methods to quantify true intrapulmonary shunt (Burger et al., 1979; Powell and Wagner, 1982). Oxygen can be used to quantify intrapulmonary plus extrapulmonary shunts, such as drainage of systemic venous blood from bronchial or Thebesian veins into pulmonary venous blood. Oxygen shunts average 6.3 to 8% of cardiac output in ducks, which is much greater than would be predicted given the magnitude of Thebesian and bronchial circulations in mammals (Bickler et al., 1986). One possible explanation for this large shunt is the connections between the vertebral venous and pulmonary circulations described for chickens (Burger and Estavillo, 1977), which also occur in ducks (Bickler et al., 1986). The sensitivity of shunts to various physiological conditions have not been measured.

3. $\dot{V}/\dot{Q}$ Mismatching

Differences in the ventilation/perfusion ratio ($\dot{V}/\dot{Q}$) between individual parabronchi is the main factor reducing arterial P_{O_2} from ideal cross-current levels in birds under resting conditions at sea level (Powell and Scheid, 1989). Ventilation and blood flow can differ between parabronchi depending on small differences in resistance or pressure gradients that can occur along the multiple parallel pathways through the lung. Physiological mechanisms, such as smooth muscle tone in the bronchi and parabronchi (see Section II,C) or interparabronchial arteries (see Section III,A), may act to reduce such heterogeneity, but matching is never perfect.

It is important to point out that the effect of such spatial $\dot{V}/\dot{Q}$ heterogeneity on gas exchange is different than the effect of changes in the *overall* $\dot{V}/\dot{Q}$ ratio. The overall $\dot{V}/\dot{Q}$ ratio can affect Pa_{O_2}, so, for example, decreases in ventilation at constant cardiac output cause decreases in Pa_{O_2}. Also, the overall $\dot{V}/\dot{Q}$) ratio affects the magnitude of the arterial-expired P_{O_2} difference in a perfectly homogeneous cross-current gas exchanger (Powell and Scheid, 1989). However, spatial mismatching of $\dot{V}/\dot{Q}$ ratios between parabronchi will decrease Pa_{O_2} further and make the arterial-expired P_{O_2} difference more positive than predicted for a homogeneous cross-current exchanger with the same overall $\dot{V}/\dot{Q}$ ratio; this is analogous to increasing the ideal alveolar–arterial P_{O_2} difference in mammals.

Several techniques have been used to measure distributions of $\dot{V}/\dot{Q}$ ratios in the avian lung, but they are relatively complicated and have not been applied yet to awake birds under many physiological conditions (Burger et al., 1979; Hempleman and Powell, 1986; Powell, 1988; Powell and Wagner, 1982). Significant amounts of ventilation go to high $\dot{V}/\dot{Q}$ regions of the lung in some cases (Powell and Wagner, 1982), and this contributes to physiological dead space. Heterogeneity in $\dot{V}/\dot{Q}$ ratios near the overall parabronchial $\dot{V}/\dot{Q}$ ratio has a large impact on O_2 exchange in birds. Compared to mammals, $\dot{V}/\dot{Q}$ heterogeneity is slightly greater (Powell and Wagner, 1982) and cross-current gas exchange is more sensitive to $\dot{V}/\dot{Q}$ mismatching than alveolar gas exchange (Powell and Hempleman, 1988; Powell and Scheid, 1989). In normoxic artificially ventilated geese, Pa_{O_2} is 25 Torr less than the ideal level predicted for homogeneous cross-current gas exchange, and 15 Torr of this difference can be explained by $\dot{V}/\dot{Q}$ heterogeneity (Powell, 1993). The remaining 10 Torr difference between measured and ideal Pa_{O_2} is hypothesized to be caused by postpulmonary shunt. Pulmonary shunt is small (Bickler et al., 1986; Powell and Wagner, 1982) and there is no evidence for a diffusion limitation in normoxia at rest (Powell and Scheid, 1989).

There is no difference between $\dot{V}/\dot{Q}$ distributions measured in anesthetized ducks ventilated with hypoxic or normoxic gas mixtures (Powell and Hastings, 1983). However, the difference between measured and predicted Pa_{O_2} is only a few Torr in hypoxia (Powell, 1993). $\dot{V}/\dot{Q}$ heterogeneity only decreases Pa_{O_2} 1 Torr from ideal cross-current levels in hypoxia because exchange occurs on the steep linear portion of the O_2–blood equilibrium curve (Powell and Scheid, 1989). The remaining difference between ideal and measured Pa_{O_2} can be explained by a small diffusion limitation that may occur in hypoxia, and postpulmonary shunts must not be occurring in hypoxia (Powell, 1993).

In addition to parallel $\dot{V}/\dot{Q}$ mismatching between parabronchi, serial $\dot{V}/\dot{Q}$ mismatch can occur along a single parabronchus if the longitudinal distribution of blood flow is not even. Several studies have shown that blood flow is greater at the inspiratory ends of the parabronchi (Holle et al., 1978; Jones, 1982; Parry and Yates, 1979). However, such serial heterogeneity does not affect gas exchange unless there is a diffusion limitation (Holle et al., 1978; Powell and Scheid, 1989), and this does not occur under most conditions (see Section V,C,4).

4. Temporal Heterogeneity

Changes in instantaneous ventilation of the parabronchi during normal breathing could result in temporal variations in the $\dot{V}/\dot{Q}$ ratio. In theory, this could significantly decrease Pa_{O_2} relative to the ideal level predicted for continuous ventilation (Powell, 1988; Powell and Scheid, 1989). For example, a ventilatory pause could act like a breath-hold and rapidly decrease P_{O_2} in the small gas volume of the avian lung (Powell and

Scheid, 1989). However, experiments indicate that this does not occur during normal breathing because the effective parabronchial gas volume is increased by mixing with larger bronchi (Scheid et al., 1977). Temporal changes in P_{O_2} entering the parabronchi during breathing could also impact Pa_{O_2}, but the effects are predicted to be relatively small (Powell, 1988). A time-averaged PI_{O_2} can be calculated for the parabronchi, and it is similar to caudal air sac P_{O_2} (Scheid et al., 1978).

E. Frontiers: Gas Exchange during High-Altitude Flight

The question of how birds manage to sustain adequate levels of O_2 consumption for flight in extreme hypoxia at very high altitudes has stimulated the minds of countless respiratory physiologists. However, a smaller number of scientists have actually researched the problem. The only study of respiratory physiology in a bird actually flying under conditions simulating altitude is on hummingbirds hovering in a hypobaric chamber (Berger, 1974). $\dot{M}_{O_2}$ during hovering at sea level, which is less than the maximal $\dot{M}_{O_2}$ for a hummingbird (Wells, 1993), was 32 mmol/(kg min). This is greater than the maximal $\dot{M}_{O_2}$ for a comparably sized mammal. Furthermore, the hummingbird was able to maintain this high level of $\dot{M}_{O_2}$ at 6000 m simulated altitude, while maximal $\dot{M}_{O_2}$ is reduced to half the sea level value in mammals that have been studied at this altitude (cf. Powell, 1993).

Experiments to explain the physiological basis for such avian–mammalian differences have been limited to studies on larger birds exercising at low levels; for example, ducks running on a treadmill (Kiley et al., 1985) or resting in hypoxia (Black and Tenney, 1980b). Some experiments on resting birds indicate that high levels of ventilation at extreme altitude would eliminate the advantage of cross-current gas exchange in comparison with alveolar exchange (Shams and Scheid, 1989). However, these conclusions are extremely sensitive to cardiac output, which varies greatly in the same species studied during hypoxic rest in different laboratories (Black and Tenney, 1980b; Shams and Scheid, 1989). Depending on the cardiac output, the advantage of cross-current compared to alveolar gas exchange could increase Pa_{O_2} a couple of Torr in birds at 11 km altitude. This is significant when the maximum P_{O_2} gradient between inspired gas and mixed venous blood is only 20 Torr (Powell, 1993). Experiments measuring all of the variables necessary for a quantitative analysis of gas exchange during hard exercise in hypoxia will be required to define the advantage of the avian lung at altitude. Other factors which may allow birds to exercise at extreme altitudes have been reviewed elsewhere (Faraci, 1991; Fedde, 1990).

F. Summary of O_2 and CO_2 Exchange in Avian Lungs

Pa_{O_2} is relatively low in birds compared to mammals in resting conditions at sea level, primarily because of heterogeneity. Diffusion limitations are not predicted for birds or mammals under these conditions. However, spatial $\dot{V}/\dot{Q}$ mismatching between functional units of gas exchange is larger in avian lungs than in mammalian lungs. Although, cross-current gas exchange in parabronchial lungs is more efficient than alveolar gas exchange, $\dot{V}/\dot{Q}$ mismatching impairs cross-current exchange more. Postpulmonary shunts also decrease Pa_{O_2} in anesthetized birds, but the role of such shunts remains to be investigated under more physiological conditions. Hypoxic conditions "unmask" the inherent efficiency of cross-current gas exchange and the parabronchial lung may provide an advantage for birds exercising at altitude, compared to mammals.

CO_2 exchange can be affected by the same factors, but Pa_{CO_2} is most sensitive to changes in ($\dot{V}_{Dphys}$) and effective $\dot{V}_P$. If O_2 is not diffusion limited, then CO_2 should not be diffusion limited either because O_2 and CO_2 require similar times for diffusion equilibrium (Wagner, 1977). $\dot{V}/\dot{Q}$ mismatch and shunt increases Pa_{CO_2} less than they decrease Pa_{O_2} because of differences between the O_2– and CO_2–blood equilibrium curves (Powell and Scheid, 1989).

VII. TISSUE GAS EXCHANGE

Oxygen moves out of systemic capillaries to the mitochondria in cells by diffusion. Therefore, O_2 transport in tissues is described by Fick's first law of diffusion, similar to diffusion across the blood–gas barrier in the lung:

$$\dot{M}_{O_2} = \Delta P_{O_2} \cdot Dt_{O_2},$$

where ΔP_{O_2} is the average P_{O_2} gradient between capillary blood and the mitochondria, and Dt_{O_2} is a tissue diffusing capacity for O_2, analogous to the lung diffusing capacity (see Section V,A,2). The main difference between O_2 diffusion in tissue and in the lung is that diffusion pathways are much greater in tissue. Tissue capillaries may be 50 μm apart, so the distance from a capillary surface to mitochondria can be 50 times longer than the thickness of the blood–gas barrier (<0.5 μm).

A. Microcirculation

1. Skeletal Muscle

Long diffusion distances can lead to significant P_{O_2} gradients in muscle. Also, the P_{O_2} gradient varies along the length of a capillary as O_2 leaves the blood and capillary P_{O_2} decreases from arterial to venous levels. However, birds have some unique structural features in the skeletal muscle microcirculation to minimize diffusion distances and enhance tissue gas exchange. For example, the number of capillaries per cross-sectional area of flight muscle fiber in hummingbirds is six times greater than the value for rat soleus muscle, and the value in pigeon flight muscle is 3 times greater (Mathieu-Costello, 1993). This clearly decreases radial diffusion distances for O_2 leaving capillaries in birds.

In addition, skeletal muscle capillaries in birds are very tortuous and have extensive manifolds connecting capillaries running along adjacent muscle fibers (Mathieu-Costello, 1991; Mathieu-Costello et al., 1992). This geometry increases the exchange surface area so the muscle fibers is functionally surrounded by "sheet" of capillary blood. This provides better tissue oxygenation than the traditional mammalian model of straight capillaries running along a muscle fiber (i.e., the Krogh cylinder); Krogh's model predicts that P_{O_2} at the venous end of the capillary may be zero when O_2 supply decreases or demand increases (Mathieu-Costello, 1991). Chronic hypoxia further increases the capillary–fiber surface area contact for aerobic flight muscles in pigeons (Mathieu-Costello and Agey, 1997).

The effects of hypoxia and exercise on skeletal muscle tissue gas exchange are considered in a separate section below (see Section VI,C).

2. Cerebral Circulation

Most evidence indicates cerebral blood flow increases with hypoxia, but does not change with CO_2 in birds (Faraci, 1991). This results in significant improvements in tissue O_2 delivery during hypoxia (Faraci et al., 1984; Grubb et al., 1977). In mammals, hypoxia also increases cerebral blood flow, but this is partially offset by a vasoconstrictor effect of the decrease in Pa_{CO_2} that accompanies the reflex increase in ventilation during hypoxia (see Section VII,C,2). This difference in cerebral vascular control may help explain how some birds are able to tolerate severe hypoxia better than some mammals (Faraci, 1986).

B. Myoglobin

Myoglobin is an O_2-binding protein, similar to a single polypeptide chain of the hemoglobin molecule (see Section IV,A,1), which has an extremely high affinity for O_2. For example, the P_{50} for hummingbirds myoglobin is 2.5 Torr (Johansen et al., 1987), so myoglobin can readily accept O_2 from capillary blood. Consequently, myoglobin is thought to be important for facilitating O_2 diffusion in muscle, by shuttling O_2 to sites far away from capillaries or toward the venous end of capillaries. High levels of myoglobin are present in the heart and skeletal muscles of the diving birds (Giardina et al., 1985; Weber et al., 1974), birds native to high altitudes (Fedde, 1990), and in hummingbirds with extremely high metabolism (Johansen et al., 1987). Increases in myoglobin with physical training in birds provides further evidence for the physiological significance of myoglobin in tissue gas exchange (Butler and Turner, 1988).

C. Effects of Hypoxia and Exercise

Generally, decreases in O_2 supply (e.g., hypoxia) or increases in O_2 demand (e.g., exercise) are satisfied at the tissue level by increased O_2 extraction from blood or increasing blood flow. This is illustrated by the Fick equation (see Section V,A,1) applied to the cardiovascular system:

$$\dot{M}_{O_2} = \dot{Q}\,(Ca_{O_2} - C\bar{v}_{O_2}).$$

Increased O_2 extraction from the blood decreases $C\bar{v}_{O_2}$, but $P\bar{v}_{O_2}$ does not decrease much because the slope of the blood–O_2 equilibrium curve is steep around the venous point (Figure 7). High $P\bar{v}_{O_2}$ levels are advantageous by keeping average capillary P_{O_2} levels high to drive O_2 diffusion into tissue. Oxygen consumption is maintained in duck skeletal muscle during hypoxia without any change in blood flow (Grubb, 1981). During severe hypoxia in resting ducks and bar-headed geese, Cv_{O_2} can be less than 0.5 mM (Black and Tenney, 1980b). This suggests nearly complete O_2 extraction, although such low venous O_2 values could also result from measurement error (see Section IV,E). Oxygen extraction in the cerebral and coronary circulations is not known for birds.

In exercising birds, extraction and blood flow increase to satisfy metabolic demands (Faraci, 1986; Faraci et al., 1984; Fedde, 1990). Increasing blood flow helps maintain high average capillary P_{O_2} because it raises mixed venous O_2 concentration for any given arterial concentration and O_2 consumption (see equation above). Increases in blood flow are also observed in the ventilatory muscles of resting birds during hypoxia, which presumably reflects increased work in these muscles during increased breathing (Faraci, 1986). Under some extreme conditions of hypoxic exercise, muscle blood flow or tissue O_2 diffusion may actually limit maximal $\dot{M}_{O_2}$ in birds (Fedde et al., 1989).

VIII. CONTROL OF BREATHING

Breathing originates as rhythmic motor output from the central nervous system. This basic respiratory rhythm is modulated by several reflexes, in response to changes in activity and the environment. These reflexes are examples of negative feedback control and tend to maintain normal arterial blood gases and pH (Table 6). For example, if dead space increases, then Pa_{CO_2} will increase if ventilation is constant. However, increased Pa_{CO_2} stimulates an increase in tidal volume, which compensates for the increased dead space and returns Pa_{CO_2} toward the control value. Like all reflexes, the ventilatory reflexes include (1) a sensory or afferent component, (2) an integrating component in the central nervous system (CNS), and (3) a motor or efferent component.

This topic bridges respiratory physiology and neuroscience, but the emphasis here is on respiratory aspects and ventilatory reflex responses to changes in blood gases, often called the "chemical control of ventilation." Several excellent reviews cover more details about the neuroscience of ventilatory control and control of breathing under different physiologic conditions (Bouverot, 1978; Davey and Seller, 1987; Gleeson and Molony, 1989; Jones and Milsom, 1982; Powell, 1983b; Scheid and Piiper, 1986).

A. Respiratory Rhythm Generation

The basic respiratory rhythm is generated by a "central pattern generator," composed of networks of neurons in the brainstem of the (CNS). Respiratory rhythm can be measured in neural outputs from isolated hindbrain of chicks (Fortin et al., 1994), and transecting the brainstem between the XIth and XIIth cranial nerve roots results in apnea and eventual death in pigeons (von Saalfeld, 1936). Reciprocal inhibition between medullary inspiratory and expiratory neurons are a common feature in birds (Peek et al., 1975), but recent evidence in mammals indicates that respiratory rhythm generation can occur without inhibitory synaptic interactions and pacemaker cells may be involved (Smith et al., 1991). Other CNS structures important for sending motor outputs to ventilatory muscles in birds (Table 2) have been identified with stimulation experiments to study vocalization (e.g., Peek et al., 1975) and panting (e.g., Richards, 1971) in chickens or pigeons (Davey and Seller, 1987).

B. Sensory Inputs

1. Central Chemoreceptors

In mammals, relatively discrete regions on the ventrolateral surface of the medulla in mammals show chemosensitivity to changes in Pa_{CO_2} and local pH. These so-called central chemoreceptors can explain most of the reflex increase in ventilation when Pa_{CO_2} increases in mammals (Bouverot, 1978). Central chemoreceptors have not been identified by neurophysiological or anatomic methods in birds. However, conscious ducks increase ventilation when Pa_{CO_2} is increased in blood perfusing only the head, indicating an important physiological role for central chemoreceptors in birds (Milsom et al., 1981; Sèbert, 1979).

2. Arterial Chemoreceptors

Arterial chemoreceptors are sensitive to changes in Pa_{O_2}, Pa_{CO_2}, and pH, and they explain all of the ventilatory response to hypoxia in birds and mammals (Bouv-

TABLE 6 Arterial Blood Gases and pH in Unanesthetized Birds Breathing Air

Bird	P_{O_2} (Torr)	P_{CO_2} (Torr)	pH
Female Black Bantam chicken[a]	—	29.9	7.48
Female White Leghorn chicken[d]	82	33.0	7.52
Male White Rock chicken[a]	—	29.2	7.53
Mallard duck[f]	81	30.8	7.46
Muscovy duck[d]	82	38.0	7.49
Muscovy duck[e]	96.1	35.9	7.46
Pekin duck[c]	93.5	28.0	7.46
Pekin duck[g]	100	33.8	7.48
Emu[l]	99.7	33.5	7.45
Bar-headed goose[c]	92.5	31.6	7.47
Domestic goose[b]	97	32	7.52
Herring gull[a]	—	27.2	7.56
Red-tailed hawk[k]	108	27.0	7.49
Burrowing owl[m]	97.6	32.6	7.46
White pelican[a]	—	28.5	7.50
Adelie penguin[i]	83.8	36.9	7.51
Chinstrap penguin[i]	89.1	37.1	7.52
Gentoo penguin[i]	77.1	40.9	7.49
Pigeon[n]	95	30	7.503
Roadrunner[a]	—	24.5	7.58
Abdim stork[j]	—	27.9	7.56
Mute swan[h]	91.3	27.1	7.50
Turkey vulture[a]	—	27.5	7.51

[a] Calder and Schmidt-Nielsen (1968).
[b] Scheid et al. (1991).
[c] Black and Tenney (1980).
[d] Kawashiro and Scheid (1975).
[e] Jones and Holeton (1972).
[f] Butler and Taylor (1983).
[g] Bouverot et al. (1979).
[h] Bech and Johansen (1980).
[i] Murrish (1982).
[j] Marder and Arad (1975).
[k] Kollias and McLeish (1978).
[l] Jones et al. (1983).
[m] Kilgore, D. L., F. M. Faraci, and M. R. Fedde (unpublished data).
[n] Powell (1983).

erot, 1978). They are also important for the ventilatory response to CO_2 and pH. The carotid bodies are very small (<1-mm diameter) organs located bilaterally in the thorax between the carotid artery and the nodose ganglion of the vagus nerve (Adamson, 1958). They are richly perfused by a branch of the carotid artery, and they are innervated by a branch of the vagus. Carotid bodies are near the parathyroid and ultimobranchial glands in birds and are enveloped within the parathyroid gland in some species (Kobayashi, 1969; Yamatsu and Komeda 1995); the physiological significance of this association is unknown. Arterial chemoreceptors have also been identified in other regions of the neck, along the carotid artery and aorta by anatomical and physiological methods (reviewed by Gleeson and Molony, 1989).

Figure 10 shows the response of carotid body arterial chemoreceptors to changes in Pa_{O_2} and Pa_{CO_2} in a duck. Afferent information about hypoxia or hypercapnia is transmitted to the CNS via the vagus nerve as increased frequency of action potentials from the carotid body. The pattern of action potential firing can differ in single chemoreceptors depending on the stimulus modality (O_2 versus CO_2) but the physiological significance of this is not known (Nye and Powell, 1984; Powell and Hempleman, 1990). Avian arterial chemoreceptors can also respond to oscillations in Pa_{O_2}, Pa_{CO_2}, and pH which can occur during breathing (Hempleman et al., 1992).

The cell types and ultrastructure of the avian and mammalian carotid bodies are similar (reviewed by Gleeson and Molony, 1989), and similar chemoreceptor mechanisms on glomus cells in the carotid body probably explain P_{O_2} and P_{CO_2}/H^+ sensitivity in both classes. However, cellular mechanism of arterial chemosensitivity are not completely understood (Gonzales et al., 1995).

3. Intrapulmonary Chemoreceptors (IPC)

In contrast to mammals, the lungs of birds (and reptiles) contain intrapulmonary chemoreceptors (IPC) which respond to physiological changes in P_{CO_2} (Burger et al., 1974; Fedde et al., 1974a; Peterson and Fedde, 1968). The sensory endings of IPC have not been identified (reviewed by Gleeson and Molony, 1989), but physiological evidence indicates that these vagal afferents have multiple endings in the parabronchial mantle at several points along the length of one or more parabronchi (Hempleman and Burger, 1984). IPC can also respond to changes in CO_2 delivery to the lung by the pulmonary arteries (Banzett and Burger, 1977). Figure 11 shows how IPC are stimulated by decreases in P_{CO_2}, in contrast to arterial chemoreceptors, which are inhibited by hypocapnia. However, increases in IPC activity cause ventilation to decrease, so the reflex response to CO_2 is similar in direction for central, arterial, and intrapulmonary chemoreceptor reflexes (see Section VII, C). The mechanism of chemoreception in IPC is unknown, but it may involve intracellular pH changes (Scheid et al., 1978). IPC respond to changes in extracellular pH in the blood (Powell et al., 1978b).

IPC are extremely sensitive to changes in P_{CO_2}, and show large overshoots or undershoots in action potential

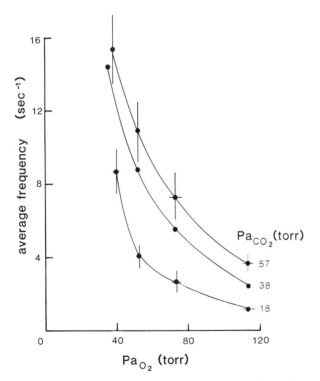

FIGURE 10 Average frequency of action potentials (± SEM) for arterial chemoreceptors ($n = \times 14$) in domestic ducks exposed to different combinations of Pa_{O_2} and Pa_{CO_2}. (S. C. Hempelman and F. L. Powell, unpublished.)

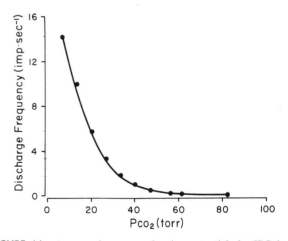

FIGURE 11 Average frequency of action potentials for IPC ($n = 54$) exposed to CO_2 levels ranging from 7 to 82 Torr. (Reprinted from Respir. Physiol. **33**, P. C. G. Nye and R. E. Burger, Chicken intrapulmonary chemoreceptors: Discharge at static levels of intrapulmonary carbon dioxide and their location, pp. 299–322, Copyright (1978), with permission from Elsevier Science.)

frequency with the kinds of periodic changes in P_{CO_2} that occur in the lung during normal breathing (Fedde et al., 1976; Scheid et al., 1978). This makes IPC well suited for fine tuning the pattern of ventilation, similar to the role of vagal pulmonary stretch receptors in mammals (see Section VII,C). However, in contrast to mammalian pulmonary stretch receptors, which are sensitive to mechanical stimuli and P_{CO_2}, avian IPC are not sensitive mechanical stretching of the lung (Bouverot, 1978; Fedde et al., 1974b).

4. Other Receptors Affecting Breathing

Ventilation also respond to changes in activity from air sac mechanoreceptors, thermal receptors in the spinal cord, proprioceptors in the skin and maybe skeletal muscle, upper airway receptors sensitive to irritants, cold and water, and, perhaps, arterial baroreceptors (reviewed by Gleeson and Molony, 1989).

C. Ventilatory Reflexes

1. CO₂ Response

Most birds are not normally exposed to increases in ambient CO_2 levels, except perhaps in specialized nests or burrows. However, ventilatory responses to common stimuli, such as exercise, hot or cold temperatures, and altitude, are influenced by CO_2, so it is important to understand the response to CO_2. Figure 12 shows how ventilation increases with increasing inspired CO_2 in conscious ducks. Most species show increased tidal volume but the frequency response can vary (Bouverot, 1978). At low levels of inspired CO_2, ventilation increases sufficiently to maintain Pa_{CO_2} at normal levels (Osborne and Mitchell, 1978). Decreases in Pa_{CO_2} or intrapulmonary P_{CO_2} can decrease ventilation also. Hence, when ventilation is increased by another stimulus, such as hypoxia or hyperthermia, the decrease in Pa_{CO_2} will act to inhibit ventilatory drive, and ventilation will be the net result of stimulation and inhibition.

Central chemoreceptors contribute to the ventilatory response to Pa_{CO_2} as described above (see Section VII,B,1). Arterial chemoreceptors play an important role in the response to dynamic changes in Pa_{CO_2} (Fedde et al., 1982; Jones and Purves, 1970; Seifert, 1986), such as Pa_{CO_2} oscillations which may occur during breathing and perhaps also in the response to static changes in Pa_{CO_2} (Gleeson and Molony, 1989). IPC may also contribute to the ventilatory response to changes in Pa_{CO_2}, although their role in this response is controversial (Bouverot, 1978; Gleeson and Molony, 1989). Experimental evidence seems to favor a role for IPC in determining the pattern of breathing, but not the overall level of ventilation in conditions where Pa_{CO_2} increases (reviewed by Gleeson and Molony, 1989). IPC are well

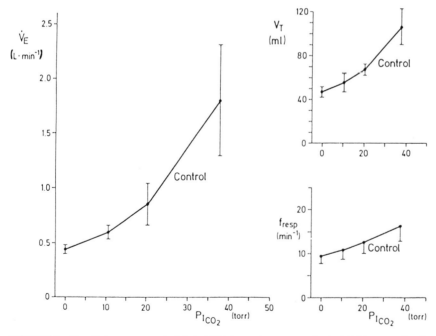

FIGURE 12 Ventilatory response to inhaled CO_2 in awake muscovy ducks (*Cairina moschata*). $\dot{V}_E$, expired ventilation; V_T, tidal volume; f_{resp}, frequency. (Reprinted from *Respir. Physiol.* **35**, F. L. Powell, M. R. Fedde, R. K. Gratz, and P. Scheid, Ventilatory responses to CO_2 in birds. I. Measurements in the unanesthetized duck, pp. 349–359, Copyright (1978), with permission from Elsevier Science.)

suited to sense breath-by-breath changes in ventilation as P_{CO_2} changes in the rigid avian lung. Therefore, IPC may play a similar role in the control of breathing to pulmonary stretch receptors in the alveolar lungs, which also sense instantaneous changes in ventilation as changes in lung volume.

2. Hypoxic Ventilatory Response

The ventilatory response to acute decreases in Pa_{O_2} is absent without intact arterial chemoreceptors Pa_{CO_2} (Jones and Purves, 1970; Seifert, 1896). The hypoxic ventilatory response (ventilation versus Pa_{O_2}) is similar in birds and mammals (Black and Tenney, 1980b; Bouverot, 1978), and the response curve has a similar shape to the arterial chemoreceptor stimulus response (Figure 10). Increasing Pa_{O_2} above normal levels does not cause large decreases in ventilation, indicating that normoxic ventilatory drive from arterial chemoreceptors is relatively small (Gleeson and Molony, 1989). Also, ventilation does not change much until Pa_{O_2} decreases below about 60 Torr in normal conditions, because small increases in ventilation will decrease Pa_{CO_2} and ventilatory drive. Figure 10 shows isocapnic responses to Pa_{O_2}, in which Pa_{CO_2} was held constant by experimental manipulations when ventilation increased. Under natural conditions of high altitude, for example, the ventilation would increase less as the a bird moves to progressively lower Pa_{CO_2} curves on Figure 10 (e.g., from Pa_{CO_2} = 38 Torr at Pa_{O_2} = 110 Torr to Pa_{CO_2} = 18 Torr at Pa_{O_2} = 60 Torr).

Chronic hypoxia causes a time-dependent increase in ventilation above the acute response level, which is called ventilatory acclimatization to hypoxia (Black and Tenney, 1980b; Bouverot, 1985; Bouverot *et al.,* 1979, 1976; Powell, 1990). Such physiological acclimatization helps return Pa_{O_2} toward normoxic values. In species genetically adapted to high altitude, such as the bar-headed goose, the hypoxic ventilatory response is blunted compared to low-altitude species (Black and Tenney, 1980b). This suggests that other components in the oxygen cascade can evolve to solve the problem of adequate O_2 delivery.

3. Ventilatory Response to Exercise

Exercise is the most common cause of increased ventilation, but the exact physiological mechanism for this is still unknown. The best evidence to date indicates that ventilation increases during exercise through a combination of "feed forward" mechanisms and feedback from chemoreceptors (Dempsey *et al.,* 1995). Feed forward mechanisms, also called central command, are neural signals from higher centers in the CNS which may stimulate respiratory centers directly. For example, neural signals to locomotor muscles may also stimulate ventilatory muscles and could contribute to some of the phase locking between wing beats and respiration (Funk *et al.,* 1992a,b). Feedback from chemoreceptors prevents ventilation from increasing too much. Pa_{CO_2} usually decreases in exercising birds (Kiley *et al.,* 1979), and this hypocapnia would be even worse if ventilatory chemoreflexes did not inhibit ventilation. Other ventilatory stimuli such as body temperature and hypoxia at altitude can modify the response to exercise (reviewed by Gleeson and Molony, 1989).

References

Abdalla, M. A. (1989). The blood supply to the lung. *In* "Form and Function in Birds" (A. S. King and J. McLelland, eds.), pp. 281–306. Academic Press, London.

Abdalla, M. A., and King, A. S. (1975). The functional anatomy of the pulmonary circulation of the domestic fowl. *Respir. Physiol.* **23,** 267–290.

Adamson, T. P. (1958). "The Comparative Morphology of the Carotid Body and Carotid Sinus." Chas. C. Thomas, Springfield.

Allen, R. L. (1983). Haemoglobins. *In* "Physiology and Biochemistry of the Domestic Fowl" (B. M. Freeman, ed.), pp. 313–319, Academic Press, London.

Andersen, H. T., and Lövö, A. (1967). Indirect estimation of partial pressure of oxygen in arterial blood of diving ducks. *Respir. Physiol.* **2,** 163–167.

Banzett, R. B., and Burger, R. E. (1977). Response of avian intrapulmonary chemoreceptors to venous CO_2 and ventilatory gas flow. *Respir. Physiol.* **29,** 63–72. [Abstract]

Banzett, R. B., and Lehr, J. L. (1982). Gas exchange during high-frequency ventilation of the chicken. *J. Appl. Physiol.* **53**(6), 1418–1422.

Banzett, R. B., Butler, J. P., Nations, C. S., Barnas, G. M., Lehr, J. L., and Jones, J. H. (1987). Inspiratory aerodynamic valving in goose lungs depends on gas density and velocity. *Respir. Physiol.* **70,** 287–300.

Banzett, R. B., Nations, C. S., Wang, N., Fredberg, J. J., and Butler, J. P. (1991). Pressure profiles show features essential to aerodynamic valving in geese. *Respir. Physiol.* **84,** 295–309.

Barnas, G. M., and Mather, F. B. (1978). Response of avian intrapulmonary smooth muscle to changes in carbon dioxide concentration. *Poult. Sci.* **57,** 1400–1407.

Bartels, H., Hiller, G., and Reinhardt, W. (1996). Oxygen affinity of chicken blood before and after hatching. *Respir. Physiol.* **1,** 345–356.

Bauer, C., Jelkmann, W., and Rollema, H. S. (1978). Mechanisms controlling the oxygen affinity of bird and reptile blood: A comparison between the functional properties of chicken and crocodile haemoglobin. *In* "Respiratory Function in Birds, Adult and Embryonic" (J. Piiper, ed.), pp. 61–66. Springer-Verlag, Berlin.

Baumann, F. H., and Baumann, R. (1977). A comparative study of the respiratory properties of bird blood. *Respir. Physiol.* **31,** 333–343.

Baumann, R., and Baumann, F. H. (1978). Respiratory function of embryonic chicken hemoglobin. *In* "Respiratory Function in Birds, Adult and Embryonic" (J. Piiper, ed.), pp. 292–297. Springer-Verlag, Berlin.

Bech, C., Johansen, K., Brent, R., and Nicol, S. (1984). Ventilatory and circulatory changes during cold exposure in the Pekin duck *Anas platyrhynchos. Respir. Physiol.* **57,** 103–112.

Bencowitz, H. Z., Wagner, P. D., and West, J. B. (1982). Effect of change in P50 on exercise tolerance at high altitude: A theoretical study. *J. Appl. Physiol.* **53**, 1487–1495.

Berger, M. (1974). Energiewechsel von Kolibris beim Schwirrflug unter Höhenbedingungen. *J. Ornithol.* **115**, 273–288.

Bickler, P. E., Maginniss, L. A., and Powell, F. L. (1986). Intrapulmonary and extrapulmonary shunt in ducks. *Respir. Physiol.* **63**, 151–160.

Black, C. P., and Tenney, S. M. (1980a). Pulmonary hemodynamic responses to acute and chronic hypoxia in two waterfowl species. *Comp. Biochem. Physiol. A* **67**, 291–293.

Black, C. P., and Tenney, S. M. (1980b). Oxygen transport during progressive hypoxia in high-altitude and sea-level waterfowl. *Respir. Physiol.* **39**, 217–239.

Boggs, D. F., Butler, P. J., and Warner, M. (1996). Fluctuations in differential pressure between the anterior and posterior air sac of Tufted ducks, *Aythya fuligula*, during breath-hold dives. *Physiologist*, **39**, A27. [Abstract]

Bouverot, P. (1978). Control of breathing in birds compared with mammals. *Physiol. Rev.* **58**(3), 604–655.

Bouverot, P. (1985). "Adaptation to Altitude-Hypoxia in Vertebrates." Springer-Verlag, Berlin.

Bouverot, P., Douguest, D., and Sèbert, P. (1979). Role of the arterial chemoreceptors in ventilatory and circulatory adjustments to hypoxia in awake Pekin ducks. *J. Comp. Physiol.* **133**, 177–186.

Bouverot, P., Hildwein, G., and Oulhen, P. (1976). Ventilatory and circulatory O_2 convection at 4000 m in pigeon at neutral or cold temperature. *Respir. Physiol.* **28**, 371–385.

Brackenbury, J. (1984). Physiological responses to birds to flight and running. *Biol. Rev.* **59**, 559–575.

Brackenbury, J. H. (1971). Airflow dynamics in the avian lung as determined by direct and indirect methods. *Respir. Physiol.* **13**, 319–329.

Brackenbury, J. H. (1972). Physical determinants of air flow pattern within the avian lung. *Respir. Physiol.* **15**, 384.

Bretz, W. L., and Schmidt-Nielsen, K. (1971). Bird respiration: Flow patterns in the duck lung. *J. Exp. Biol.* **54**, 103–118.

Brown, R. E., Kovacs, C. E., Butler, J. P., Wang, N., Lehr, J. L., and Banzett, R. B. (1995). The avian lung: Is there an aerodynamic expiratory valve? *J. Exp. Biol.* **198**, 2349–2357.

Burger, R. E., and Estavillo, J. A. (1977). Pulmonary circulation—Vertebral venous interconnections in the chicken. *Anat. Rec.* **188**(1), 39–44.

Burger, R. E., and Lorenz, F. W. (1960). Artificial respiration in birds by unidirectional air flow. *Poult. Sci.* **39**(1), 236–237.

Burger, R. E., Meyer, M., Graf, W., and Scheid, P. (1979). Gas exchange in the parabronchial lung of birds: Experiments in unidirectionally ventilated ducks. *Respir. Physiol.* **36**, 19–37.

Burger, R. E., Osborne, J. L., and Banzett, R. B. (1974). Intrapulmonary chemoreceptors in *Gallus domesticus*: Adequate stimulus and functional localization. *Respir. Physiol.* **22**, 87–97.

Burton, R. R., Besch, E. L., and Smith, A. H. (1968). Effect of chronic hypoxia on the pulmonary arterial blood pressure of the chicken. *Am. J. Physiol.* **214**(6), 1438–1442.

Butler, P. J. (1991). Exercise in birds. *J. Exp. Biol.* **160**, 233–262.

Butler, P. J., and Taylor, E. W. (1983). Factors affecting the respiratory and cardiovascular responses to hypercapnic hypoxia in mallard ducks. *Respir. Physiol.* **53**, 109.

Butler, P. J., and Turner, D. L. (1988). Effect of taining on maximal oxygen uptake and aerobic capacity of locomotory muscles in tufted ducks. *Aythya fuligula. J. Physiol.* **401**, 347–359.

Butler, P. J., Banzett, R. B., and Fredberg, J. J. (1988). Inspiratory valving in avian bronchi: Aerodynamic considerations. *Respir. Physiol.* **72**, 241–256.

Calder, W. A., and Schmidt-Nielsen, K. (1968). Panting and blood carbon dioxide in birds. *Am. J. Physiol.* **215**, 477–482.

Cohn, J. E., and Shannon, R. (1968). Respiration in unanesthetized geese. *Respir. Physiol.* **5**, 259–268.

Crank, W. D., and Gallagher, R. R. (1978). Theory of gas exchange in the avian parabronchus. *Respir. Physiol.* **35**, 9–25.

Davey, N. J., and Seller, T. J. (1987). Brain mechanisms for respiratory control. In "Bird Respiration" (T. J. Seller, ed.), pp. 169–188. CRC Press, Boca Raton, FL.

Dempsey, J. A., Forster, H. V., and Ainsworth, D. M. (1995). Regulation of hyperpnea, hyperventilation, and respiratory muscle recruitment during exercise. In "Regulation of Breathing" (J. A. Dempsey and A. I. Pack, eds.), pp. 1065–1134. Marcel Dekker, New York.

deWet, P. D., Fedde, M. R., and Kitchell, R. L. (1967). Innervation of the respiratory muscles of *Gallus domesticus. J. Morphol.* **123**(1), 17–34.

Dotterweich, H. (1930). Versuche Über den weg der atemluft in der vogellunge. *Z. Vergleich. Physiol.* **11**, 271–284.

Dubach, M. (1981). Quantitative analysis of the respiratory system of the house-sparrow, budgerigar and violet-eared hummingbird. *Respir. Physiol.* **46**, 43–60.

Duncker, H. R. (1971). The lung air sac system of birds. *Adv. Anat. Embryol. Cell Biol.* **45**, 7–171.

Duncker, H. R. (1972). Structure of avian lungs. *Respir. Physiol.* **14**, 44–63.

Duncker, H. R. (1974). Structure of the avian respiratory tract. *Respir. Physiol.* **22**, 1–19.

Duncker, H. R. (1978). General morphological principles of amniotic lungs. In "Respiratory Function in Birds, Adult and Embryonic" (J. Piiper, ed.), pp. 1–18. Springer-Verlag, Berlin.

Faraci, F. M., Kilgore, D. L., and Fedde, M. R. (1984). Oxygen delivery to the heart and brain during hypoxia: Pekin duck vs. bar-headed goose. *Am. J. Physiol.* **16**, R69–R75.

Faraci, F. M. (1986). Circulation during hypoxia in birds. *Comp. Biochem. Physiol. A* **85**(4), 613–620.

Faraci, F. M. (1991). Adaptations to hypoxia in birds: How to fly high. *Annu. Rev. Physiol.* **53**, 59–70.

Fedde, M. R. (1978). Drugs used for avian anesthesia. *Poult. Sci.* **57**, 1376–1399.

Fedde, M. R. (1990). High-altitude bird flight: Exercise in a hostile environment. *NIPS* **5**, 191–193.

Fedde, M. R., and Scheid, P. (1976). Intrapulmonary CO_2 receptors in the duck. IV. Discharge pattern of the population during a respiratory cycle. *Respir. Physiol.* **26**, 223–227.

Fedde, M. R., Burger, R. E., and Kitchell, R. L. (1963). Electromyographic studies of the effects of bodily position and anesthesia on the activity of the respiratory muscles of the domestic cock. *Poult. Sci.* **43**, 839–846.

Fedde, M. R., Burger, R. E., and Kitchell, R. L. (1964a). Anatomic and electromyographic studies of the costo-pulmonary muscles in the cock. *Poult. Sci.* **43**, 1177–1184.

Fedde, M. R., Burger, R. E., and Kitchell, R. L. (1964b). Electromyographic studies of the effects of bilateral, cervical vagotomy on the action of the respiratory muscles of the domestic cock. *Poult. Sci.* **43**, 1119–1125.

Fedde, M. R., deWet, P. D., and Kitchell, R. L. (1969). Motor unit recruitment pattern and tonic activity in respiratory muscles of *Gallus domesticus. J. Neurophysiol.* **32**, 995–1004.

Fedde, M. R., Gatz, R. N., Slama, H., and Scheid, P. (1974a). Intrapulmonary CO_2 receptors in the duck. I. Stimulus specificity. *Respir. Physiol.* **22**, 99–114.

Fedde, M. R., Gatz, R. N., Slama, H., and Scheid, P. (1974b). Intrapulmonary CO_2 receptors in the duck. II. Comparison with mechanoreceptors. *Respir. Physiol.* **22**, 115–121.

Fedde, M. R., Kiley, J. P., Powell, F. L., and Scheid, P. (1982). Intrapulmonary CO_2 receptors and control of breathing in ducks: Effects of prolonged circulation time to carotid bodies and brain. *Respir. Physiol.* **47**, 121–140.

Fedde, M. R., Orr, J. A., Shams, H., and Scheid, P. (1989). Cardiopulmonary function in exercising bar-headed geese during normoxia and hypoxia. *Respir. Physiol.* **77**, 239–262.

Fortin, G., Champagnat, J., and Lumdsen, A. (1994). Onset and maturation of branchio-motor activities in the chick hindbrain. *Neuro Report* **5**, 1149–1152.

Funk, G. D., Milsom, W. K., and Steeves, J. D. (1992a). Coordination of wingbeat and respiration in the Canada goose. I. Passive wing flapping. *J. Appl. Physiol.* **73**, 1014–1024.

Funk, G. D., Steeves, J. D., and Milsom, W. K. (1992b). Coordination of wingbeat and respiration in birds. II. "Fictive" flight. *J. Appl. Physiol.* **73**, 1025–1033.

Geiser, J., Gratz, R. K., Hiramoto, T., and Scheid, P. (1984). Effects of increasing metabolism by 2,4-dinitrophenol on respiration and pulmonary gas exchange in the duck. *Respir. Physiol.* **57**, 1–14.

Giardina, B., Corda, M., Pellegrini, M. G., Condo, S. G., and Brunori, M. (1985). Functional properties of the hemoglobin system of two diving birds (*Podiceps nigricollis* and *Phalacrocorax carbo sinensis*). *Molec. Physiol.* **7**, 281–292.

Gillespie, J. R., Gendner, J. P., Sagot, J. C., and Bouverot, P. (1982a). Impedance of the lower respiratory system in ducks measured by forced oscillations during normal breathing. *Respir. Physiol.* **47**, 51–68.

Gillespie, J. R., Sagot, J. C., Gendner, J. P., and Bouverot, P. (1982b). Respiratory mechanics of Pekin ducks under four conditions: Pressure breathing, anesthesia, paralysis or breathing CO_2-enriched gas. *Respir. Physiol.* **47**, 177–191.

Gleeson, M., and Molony, V. (1989). Control of breathing. In "Form and Function in Birds" (A. S. King and J. McLelland, eds.), pp. 439–484. Academic Press, London.

Gonzales, C., Dinger, B. G., and Fidone, S. J. (1995). Mechanisms of carotid body chemoreception. In "Regulation of Breathing" (J. A. Dempsey and A. I. Pack, eds.), pp. 391–472. Marcel Dekker, New York.

Grubb, B. R. (1981). Blood flow and oxygen consumption in avian skeletal muscle during hypoxia. *J. Appl. Physiol.* **50**, 450–455.

Grubb, B., Mills, C. D., Colacino, J. M., and Schmidt-Nielsen, K. (1977). Effect of arterial carbon dioxide on cerebral blood flow in ducks. *Am. J. Physiol.* **232**(6), H596–H601.

Hastings, R. H., and Powell, F. L. (1986a). Single breath CO_2 measurements of dead space in ducks. *Respir. Physiol.* **63**, 139–149.

Hastings, R. H., and Powell, F. L. (1986b). Physiological dead space and effective parabronchial ventilation in ducks. *J. Appl. Physiol.* **60**(1), 85–91.

Hastings, R. H., and Powell, F. L. (1987). High-frequency ventilation of ducks and geese. *J. Appl. Physiol.* **63**(1), 413–417.

Hazelhoff, E. H. (1951). Structure and function of the lung of birds (reprinted from 1943). *Poult. Sci.* **30**, 3–10.

Helbacka, N. V. L., Casterline, J. L., Jr., Smith, C. J., and Shaffner, C. S. (1963). Investigation of plasma carbonic acid pK' of the chicken. *Poult. Sci.* **43**, 138–144.

Hempleman, S. C., and Burger, R. E. (1984). Receptive fields of intrapulmonary chemoreceptors in the Pekin duck. *Respir. Physiol.* **57**, 317–330.

Hempleman, S. C., and Powell, F. L. (1986). Influence of pulmonary blood flow and O_2 flux on D_{O_2} in avian lungs. *Respir. Physiol.* **63**, 285–292.

Hempleman, S. C., Powell, F. L., and Prisk, G. K. (1992). Avian arterial chemoreceptor responses to steps of CO_2 and O_2. *Respir. Physiol.* **90**, 325–340.

Henning, B., Scheid, P., and Piiper, J. (1971). Determination of the Haldane effect in chicken blood. *Respir. Physiol.* **11**, 279–284.

Hinds, D. S., and Calde, W. A. (1971). Tracheal dead space in the respiration of birds. *Evolution* **25**, 429–440.

Hirsowitz, L. A., Fell, K., and Torrance, J. D. (1977). Oxygen affinity of avian blood. *Respir. Physiol.* **31**, 51–62.

Holle, J. P., Heisler, N., and Scheid, P. (1978). Blood flow distribution in the duck lung and its control by respiratory gases. *Am. J. Physiol.* **234**(3), R146–R154.

Holle, J. P., Meyer, M., and Scheid, P. (1977). Oxygen affinity of duck blood determined by *in vivo* and *in vitro* technique. *Respir. Physiol.* **29**, 355–361.

Jenkins, F. A., Dial, K. P., and Goslow, G. E. (1988). A cineradiographic analysis of bird flight: The wishbone in starlings is a spring. *Science* **241**, 1495–1498.

Jessen, T. H., Weber, R. E., Fermi, G., Tame, J., and Braunitzer, G. (1991). Adaptation of bird hemoglobins to high altitudes: Demonstration of molecular mechanism by protein engineering. *Proc. Natl. Acad. Sci. USA* **88**, 6519–6522.

Johansen, K., Berger, M., Bicudo, J. E. P., Ruschi, A., and De Almeida, P. J. (1987). Respiratory properties of blood and myoglobin in hummingbirds. *Physiol. Zool.* **60**(2), 269–278.

Jones, J. H. (1982). Pulmonary blood flow distribution in panting ostriches. *J. Appl. Physiol.* **53**, 1411–1417.

Jones, D. R., and Holeton, G. F. (1972). Cardiovascular and respiratory responses of ducks to progressive hypocapnic hypoxia. *J. Exp. Biol.* **56**, 657–666.

Jones, D. R., and Milsom, W. K. (1982). Peripheral receptors affecting breathing and cardiovascular function in non-mammalian vertebrates. *J. Exp. Biol.* **100**, 59.

Jones, D. R., and Purves, M. J. (1970). The effect of carotid body denervation upon the respiratory response to hypoxia and hypercapnia in the duck. *J. Physiol.* **211**, 295–308.

Jones, J. H., Effmann, E. L., and Schmidt-Nielsen, K. (1985). Lung volume changes during respiration in ducks. *Respir. Physiol.* **59**, 15–25.

Jones, J. H., Grubb, B., and Schmidt-Nielsen, K. (1983). Panting in the emu causes arterial hypoxemia. *Respir. Physiol.* **54**, 189.

Julian, R. J. (1993). Ascites in poultry. *Avian Pathol.* **23**, 419–454.

Kadono, H., and Okada, T. (1962). Electromyographic studies on the respiratory muscles of the domestic fowl. *Jpn. J. Vet. Sci.* **24**(4), 215–223.

Kadono, H., Okada, T., and Ono, K. (1963). Electromyographic studies on the respiratory muscles of the chicken. *Poult. Sci.* **42**(1), 121–128.

Kampe, G., and Crawford, E. C. (1973). Oscillatory mechanics of the respiratory system of pigeons. *Respir. Physiol.* **18**, 188–193.

Kiley, J. P., Faraci, F. M., and Fedde, M. R. (1985). Gas exchange during exercise in hypoxic ducks. *Respir. Physiol.* **59**, 105–115.

Kiley, J. P., Kuhlmann, W. D., and Fedde, M. R. (1979). Respiratory and cardiovascular responses to exercise in the duck. *J. Appl. Physiol.* **47**, 827–833.

King, A. S. (1979). Systema respiratorium. In "Nomina Anatomica Avium" (J. J. Baumel, A. M. King, A. M. Lucas, J. E. Breazile, and H. E. Evans, eds.), pp. 227–265. Academic Press, London.

King, A. S. (1989). Functional anatomy of the syrinx. In "Form and Function in Birds" (A. S. King and J. McLelland, eds.), pp. 105–192. Academic Press, London.

King, A. S., and Molony, V. (1971). The anatomy of respiration. In "Physiology and Biochemistry of the Domestic Fowl" (O. J. Bell and B. M. Freeman, eds.), p. 227. Academic Press, New York.

Kobayashi, S. (1969). Catecholamines in the avian carotid body. *Specilia* **25**, 1075–1076.

Kolias, G. V., and McLeish, I. (1978). Effects of ketamine hydrochloride in red-tailed hawks (*Buteo jamaicensis*). I. Arterial blood gas and acid base. *Comp. Biochem. Physiol. C* **60,** 57.

Kooyman, G. L. (1989). "Diverse Divers: Physiology and Behavior." Springer-Verlag, Berlin.

Kuethe, D. O. (1988). Fluid mechanical valving of air flow in bird lungs. *J. Exp. Biol.* **136,** 1–12.

Kuhlmann, W. D., and Fedde, M. R. (1976). Upper respiratory dead space in the chicken: Its fraction of the tidal volume. *Comp. Biochem. Physiol. A* **54,** 409–411.

Lapennas, G. N., and Reeves, R. B. (1983). Oxygen affinity of blood of adult domestic chicken and red jungle fowl. *Respir. Physiol.* **52,** 27–39.

Lutz, P. L. (1980). On the oxygen affinity of bird blood. *Am. Zool.* **20,** 187–198.

Macklem, P. T. (1987). Symbols and abbreviations. In "Handbood of Physiology: The Respiratory System—Gas Exchange" (L. E. Farhi and S. M. Tenney, eds.), p. ix. American Physiological Society, Bethesda, MD.

Macklem, P. T., Bouverot, P., and Scheid, P. (1979). Measurement of the distensibility of the parabronchi in duck lungs. *Respir. Physiol.* **38,** 23–35.

Maginniss, L. A. (1985). Red cell organic phosphates and Bohr effects in house sparrow blood. *Respir. Physiol.* **59,** 93–103.

Maginniss, L. A., and Kilgore, D. L. (1989). Blood oxygen binding properties for the burrowing owl, *Athene cunicularia*. *Respir. Physiol.* **76,** 205–214.

Magnussen, H., Willmer, H., and Scheid, P. (1976). Gas exchange in air sacs: Contribution to respiratory gas exchange in ducks. *Respir. Physiol.* **26,** 129–146.

Maina, J. N. (1989). The morphometry of the avian lung. In "Form and Function in Birds" (A. S. King and J. McLelland, eds.), pp. 307–368. Academic Press, London.

Maina, J. N., and King. A. S. (1982). Morphometrics of the avian lung. 2. The wild mallard (*Anas platyrhynchos*) and Graying goose (*Anser anser*). *Respir. Physiol.* **50,** 299–310.

Maina, J. N., Abdalla, A., and King. A. S. (1982). Light microscopic morphometry of the lung of 19 avian species. *Acta Anat.* **112,** 264–270.

Maina, J. N., King, A. S., and Settle, G. (1989). An allometric study of pulmonary morphometric parameters in birds, with mammalian comparisons. *Phil. Trans. R. Soc. London B* **326,** 1–57.

Marder, J., and Arad, Z. (1975). The acid–base balance of abdim's stork (*Sphenorhynchus abdimii*) during thermal panting. *Comp. Biochem. Physiol. A* **51,** 887–889.

Maren, T. H. (1967). Carbonic anhydrase: Chemistry, physiology and inhibition. *Physiol. Rev.* **47,** 595–781.

Mathieu-Costello, O. (1991). Morphometric analysis of capillary geometry in pigeon pectoralis muscle. *Am. J. Anat.* **191,** 74–84.

Mathieu-Costello, O. (1993). Comparative aspects of muscle capillary supply. *Annu. Rev. Physiol.* **55,** 503–525.

Mathieu-Costello, O., and Agey, P. J. (1997). Chronic hypoxia affects capillary density and geometry in pigeon pectoralis muscle. *Respir. Physiol.* **109,** 39–52.

Mathieu-Costello, O., Suarez, R. K., and Hochachka, P. W. (1992). Capillary-to-fiber geometry and mitochondrial density in hummingbird flight muscle. *Respir. Physiol.* **89,** 113–132.

McLelland, J. (1989a). Larynx and trachea. In "Form and Function in Birds" (A. S. King, and J. McLelland, eds.), pp. 69–104. Academic Press, London.

McLelland, J. (1989b). Anatomy of the lungs and air sacs. In "Form and Function in Birds" (A. S. King and J. McLelland, eds.), pp. 221–280. Academic Press, London.

Meyer, M., Holle, J. P., and Scheid, P. (1978). Bohr effect induced by CO_2 and fixed acid at various levels of O_2 saturation in duck blood. *Pflügers Arch. Ges. Physiol.* **376,** 237–240.

Meyer, M., Worth, H., and Scheid, P. (1976). Gas–blood CO_2 equilibrium in parabronchial lings of birds. *J. Appl. Physiol.* **41,** 302.

Milsom, W. K., Jones, D. R., and Gabbott, G. R. J. (1981). On chemoreceptor control of ventilatory responses to CO_2 in unanesthetized ducks. *J. Appl. Physiol.* **50,** 1121–1128.

Molony, V., Graf, W., and Scheid, P. (1976). Effects of CO_2 on pulmonary air flow resistance in the duck. *Respir. Physiol.* **26,** 333–349.

Morgan, V. E., and Chichester, D. F. (1935). Properties of the blood of the domestic fowl. *J. Biol. Chem.* **110,** 285–298.

Murrish, D. E. (1982). Acid–base balance in three species of antarctic penguins exposed to thermal stress. *Physiol. Zool.* **55**(2), 137–143.

Nightingale, T. E., Boster, R. A., and Fedde, M. R. (1968). Use of the oxygen electrode in recording PO_2 in avian blood. *J. Appl. Physiol.* **25**(4), 371–375.

Nye, P. C. G., and Burger, R. E. (1978). Chicken intrapulmonary chemoreceptors: Discharge at static levels of intrapulmonary carbon dioxide and their location. *Respir. Physiol.* **33,** 299–322.

Nye, P. C. G., and Powell, F. L. (1984). Steady-state discharge and bursting of arterial chemoreceptors in the duck. *Respir. Physiol.* **56,** 369–384.

Osborne, J. L., and Mitchell, G. S. (1978). Intrapulmonary and systemic CO_2-chemoreceptor interaction in the control of avian respiration. *Respir. Physiol.* **33,** 349–357.

Palomeque, J., Palacios, L., and Planas, J. (1979). Comparative respiratory functions of blood in some passeriform birds. *Comp. Biochem. Physiol. A* **66,** 619–624.

Parry, K., and Yates, M. S. (1979). Observations on the avian pulmonary and bronchial circulation using labelled microspheres. *Respir. Physiol.* **38,** 131–140.

Pattle, R. E. (1978). Lung surfactant and lung lining in birds. In "Respiratory Function in Birds, Adult and Embryonic" (J. Piiper, ed.), p. 23. Springer-Verlag, New York.

Peacock, A. J., Pickett, C., Morris, K., and Reeves, J. T. (1990). Spontaneous hypoxaemia and right ventricular hypertrophy in fast growing broiler chickens reared at sea level. *Comp. Biochem. Physiol. A* **97,** 537–541.

Peek, F. W., Youngren, O. M., and Phillips, R. E. (1975). Repetitive vocalizations evoked by electrical stimulation of avian brains. *Brain Behav. Evol.* **12,** 1–41.

Peterson, D. F., and Fedde, M. R. (1968). Receptors sensitive to carbon dioxide in lungs of chicken. *Science* **162,** 1499–1501.

Petschow, D., Wurdinger, I., Baumann, R., Duhm, J., Braunitzer, G., and Bauer, C. (1977). Causes of high blood O_2 affinity of animals living at high altitude. *J. Appl. Physiol.* **42**(2), 139–143.

Phu, D. N., Yamaguchi, K., Scheid, P., and Piiper, J. (1986). Kinetics of oxygen uptake and release by red blood cells of chicken and duck. *J. Exp. Biol.* **125,** 15–27.

Piiper, J. (1978). Origin of carbon dioxide in caudal airsacs of birds. In "Respiratory Function in Birds, Adult and Embryonic" (J. Piiper, ed.), pp. 148–153. Springer-Verlag, Berlin.

Piiper, J., and Scheid, P. (1972). Maximum gas transfer efficacy of models for fish gills, avian lungs and mammalian lungs. *Respir. Physiol.* **14,** 115–124.

Piiper, J., and Scheid, P. (1975). Gas transport efficacy of gills, lungs and skin: Theory and experimental data. *Respir. Physiol.* **23,** 209–221.

Piiper, J., Dejours, P., Haab, P., and Rahn, H. (1971). Concepts and basic quantities in gas exchange physiology. *Respir. Physiol.* **13,** 292–304.

Piiper, J., Drees, F., and Scheid, P. (1970). Gas exchange in the domestic fowl during spontaneous breathing and artificial ventilation. *Respir. Physiol.* **9,** 234–245.

Powell, F. L. (1982). Diffusion in avian lungs. *Fed. Proc. Am. Sci.* **41,** 2131–2133.

Powell, F. L. (1983a). Circulation. *In* "Physiology and Behaviour of the Pigeon" (M. Abs, ed.), pp. 97–116. Academic Press, New York.

Powell, F. L. (1983b). Respiration. *In* "Physiology and Behavior of the Pigeon" (M. Abs, ed.), pp. 73–95. Academic Press, New York.

Powell, F. L. (1988). Lung structure and function. *In* "Comparative Pulmonary Physiology: Current Concepts" (S. C. Wood, ed.), pp. 237–255. Marcel Dekker, New York.

Powell, F. L. (1990). Acclimatization to high altitude. *In* "Hypoxia: The Adaptations" (J. R. Sutton, G. Coates, and J. E. Remmers, eds.), pp. 41–44. B. C. Decker, Toronto.

Powell, F. L. (1993). Birds at altitude. *In* "Respiration in Health and Disease" (P. Scheid, ed.), pp. 352–358. G. Fisher, Stuttgart/New York.

Powell, F. L., Fedde, M. R., Gratz, R. K., and Scheid, P. (1978a). Ventilatory responses to CO_2 in birds. I. Measurements in the unanesthetized duck. *Respir. Physiol.* **35,** 349–359.

Powell, F. L., Gratz, R. K., and Scheid, P. (1978b). Response to intrapulmonary chemoreceptors in the duck to changes in P_{CO_2} and pH. *Respir. Physiol.* **35,** 65–77.

Powell, F. L., Geiser, J., Gratz, R. K., and Scheid, P. (1981). Airflow in the avian respiratory tract: Variations of O_2 and CO_2 concentrations in the bronchi of the duck. *Respir. Physiol.* **44,** 195–213.

Powell, F. L., and Hastings, R. H. (1983). Effects of hypoxia on ventilation-perfusion matching in birds. *Physiologist* **26,** A50. [Abstract]

Powell, F. L., and Mazzone, R. W. (1983). Morphometrics of rapidly frozen goose lungs. *Respir. Physiol.* **51,** 319–332.

Powell, F. L., and Hempleman, S. C. (1988). Comparative physiology of oxygen transfer in lungs. *In* "Oxygen Transfer from Atmosphere to Tissues" (N. C. Gonzalez and M. R. Fedde, eds.), pp. 53–65. Plenum, New York.

Powell, F. L., and Hempleman, S. C. (1990). Information content of arterial chemoreceptor discharge pattern. *In* "Arterial Chemoreception" (C. Eyzaguirre, S. J. Fidone, R. S., Fitzgerald, S. Lahiri, and D. M. McDonald, eds.), pp. 247–253. Springer-Verlag, New York.

Powell, F. L., Hastings, R. H., and Mazzone, R. W. (1985). Pulmonary vascular resistance during unilateral pulmonary arterial occlusion in ducks. *Am. J. Physiol.* **249**(18), R39–R43.

Powell, F. L., and Hempleman, S. C. (1985). Sources of carbon dioxide in penguin air sacs. *Am. J. Physiol.* **248,** R748–R752.

Powell, F. L., and Scheid, P. (1989). Physiology of gas exchange in the avian respiratory system. *In* "Form and Function in Birds" (A. S. King and J. McLelland, eds.), Vol. 4, pp. 393–437. Academic Press, London.

Powell, F. L., and Wagner, P. D. (1982). Ventilation-perfusion inequality in avian lungs. *Respir. Physiol.* **48,** 233–241.

Richards, S. A. (1971). Brain stem control of polypnoea in the chicken and pigeon. *Respir. Physiol.* **11,** 315–326.

Rollema, H. S., and Bauer, C. (1979). The interaction of inositol pentaphosphate with the hemoglobin of the highland and the lowland geese. *J. Biol. Chem.* **254,** 12038–12043.

Scharnke, H. (1938). Experimentelle Beiträge zur Kenntnis der Vogelatmung. *Z. vergl. Physiologie* **25,** 548–583.

Scheid, P. (1978). Analysis of gas exchange between air capillaries and blood capillaries in avian lungs. *Respir. Physiol.* **32,** 27–49.

Scheid, P. (1990). Avian respiratory system and gas exchange. *In* "Hypoxia: The Adaptations" (J. R. Sutton, G. Coates, and J. E. Remmers, eds.), pp. 4–7. B. C. Decker, Toronto.

Scheid, P., and Kawashiro, T. (1975). Metabolic changes in avian blood and their effects on determination of blood gases and pH. *Respir. Physiol.* **23,** 291–300.

Scheid, P., and Piiper, J. (1969). Volume, ventilation and compliance of the respiratory system in the domestic fowl. *Respir. Physiol.* **6,** 298–308.

Scheid, P., and Piiper, J. (1970). Analysis of gas exchange in the avian lung: Theory and experiments in the domestic fowl. *Respir. Physiol.* **9,** 246–262.

Scheid, P., and Piiper, J. (1972). Cross-current gas exchange in avian lungs: Effects of reversed parabronchial air flow in ducks. *Respir. Physiol.* **16,** 304–312.

Scheid, P., Slama, H., and Piiper, J. (1972). Mechanisms of unidirectional flow in parabronchi of avian lungs: Measurements in duck lung preparations. *Respir. Physiol.* **14,** 83–95.

Scheid, P., and Piiper, J. (1986). Control of breathing in birds. *In* "Handbood of Physiology: The Respiratory System—Control of Breathing" (N. S. Cherniack and J. G. Widdicombe, eds.), pp. 815–832. American Physiological Society, Bethesda, MD.

Scheid, P., and Piiper, J. (1989). Respiratory mechanics and air flow in birds. *In* "Form and Function in Birds" (A. S. King and J. McLelland, eds.), pp. 369–391. Academic Press, London.

Scheid, P., and Slama, H. (1975). Remote-controlled device for sampling arterial blood in unrestrained animals. *Pflügers Arch.* **356,** 373–376.

Scheid, P., Fedde, M. R., and Piiper, J. (1991). Gas exchange and airsac composition in the unanaesthetized, spontaneously breathing goose. *J. Exp. Biol.* **142,** 373–385.

Sceid, P., Gratz, R. K., Powell, F. L., and Fedde, M. R. (1978). Ventilatory response to CO_2 in birds. II. Contribution by intrapulmonary CO_2 receptors. *Respir. Physiol.* **35,** 361–372.

Scheid, P., Slama, H., and Willmer, H. (1974). Volume and ventilation of air sacs in ducks studied by inert gas wash-out. *Respir. Physiol.* **21,** 19–36.

Scheid, P., Worth, H., Holle, J. P., and Meyer, M. (1977). Effects of oscillating and intermittent ventilatory flow on efficacy of pulmonary O_2 transfer in the duck. *Respir. Physiol.* **31,** 251–258.

Scheipers, G., Kawashiro, T., and Scheid, P. (1975). Oxygen and carbon dioxide dissociation of duck blood. *Respir. Physiol.* **24,** 1–13.

Sèbert, P. (1979). Mise en evidence de l'action centrale du stimulus $CO_2[H^+]$ de la ventilation chez le Canard Pekin. *J. Physiol. Paris* **75,** 901–909.

Seifert, E. (1896). Über die Atmung der Reptilien und Vögel. *Pflügers Arch. Ges. Physiol.* **64,** 321–506.

Shams, H., and Scheid, P. (1989). Efficiency of parabronchial gas exchange in deep hypoxia: Measurements in the resting duck. *Respir. Physiol.* **77,** 135–146.

Shams, H., and Scheid, P. (1993). Effects of hypobaria on parabronchial gas exchange in normoxic and hypoxic ducks. *Respir. Physiol.* **91,** 155–163.

Shams, H., Powell, F. L., and Hempleman, S. C. (1990). Effects of normobaric and hypobaric hypoxia on ventiltion and arterial blood gases in ducks. *Respir. Physiol.* **80,** 163–170.

Skadhauge, E. (1983). Formation and composition of urine. *In* "Physiology and Biochemistry of the Domestic Fowl" (B. M. Freeman, ed.), pp. 108–135. Academic Press, London.

Smith, J. C., Ellenberger, H. H., Ballanyi, K., Richter, D. W., and Feldman, J. L. (1991). Pre-Bötzinger complex: A brainstem region that may generate respiratory rhythm in mammals. *Science* **254,** 726–729.

Sturkie, P. D. (1986). Body fluids: Blood. *In* "Avian Physiology" (P. D. Sturkie, ed.), pp. 102–139. Springer-Verlag, New York.

Taylor, C. R., and Weibel, E. R. (1981). Design of the mammalian respiratory system. I. Problem and strategy. *Respir. Physiol.* **44,** 1–10.

Torre-Bueno, J. R., Geiser, J., and Scheid, P. (1980). Incomplete gas mixing in air sacs of the duck. *Respir. Physiol.* **42,** 109–122.

von Saalfeld, E. (1936). Untersuchungen Über das hacheln bei tauben. *Z. vergl. Physiologie* **23,** 727–743.

Vos, H. F. (1935). Über den Weg der Atemluft in der Entenlunge. *Z. vergl. Physiologie* **20,** 552–578.

Wagner, P. D. (1977). Diffusion and chemical reaction in pulmonary gas exchange. *Physiol. Rev.* **57,** 257–312.

Wang, N., Banzett, R. B., Butler, J. P., and Fredberg, J. J. (1988). Bird lung models show that convective inertia effects inspiratory aerodynamic valving. *Respir. Physiol.* **73,** 111–124.

Wang, N., Banzett, R. B., Nations, C. S., and Jenkins, F. A. (1992). An aerodynamic valve in the avian primary bronchus. *J. Exp. Biol.* **262,** 441–445.

Weber, R. E., and Wells, R. M. G. (1989). Hemoglobin structure and function. *In* "Comparative Pulmonary Physiology: Current Concepts" (S. C. Wood, ed.), pp. 279–310. Marcel Dekker, New York.

Weber, R. E., Hemmingsen, E. A., and Johansen, K. (1974). Functional and biochemical studies of penguin myoglobin. *Comp. Biochem. Physiol. B* **49,** 197–214.

Weidner, W. J., Selna, L. A., McClure, D. E., and DeFouw, D. O. (1993). Effect of extracellular fluid volume expansion on avian lung fluid balance. *Respir. Physiol.* **91,** 125–136.

Weingarten, J. P., Rollema, H. S., Bauer, C., and Scheid, P. (1978). Effects of inositol hexaphosphate on the Bohr effect induced by CO_2 and fixed acids in chicken hemoglobin. *Pflügers Arch.* **377,** 135–141.

Weinstein, Y., Bernstein, M. H., Bickler, P. E., Gonzales, D. V., Samaniego, F. C., and Escobedo, M. A. (1985). Blood respiratory properties in pigeons at high altitudes: Effects of acclimation. *Am. J. Physiol.* **249,** R765–R775.

Wells, D. J. (1993). Ecological correlates of hovering flight of hummingbirds. *J. Exp. Biol.* **178,** 59–70.

Wells, R. M. G. (1976). The oxygen affinity of chicken hemoglobin in whole blood and erythrocyte suspensions. *Respir. Physiol.* **27,** 21.

West, J. B., and Mathieu-Costello, O. (1992). Strength of the blood-gas barrier. *Respir. Physiol.* **88,** 141–148.

Whittow, G. C., and Ossorio, N. (1970). A new technique for anesthetizing birds. *Lab Anim. Care* **20,** 651–656.

Yamatsu, Y., and Kameda, Y. (1995). Accessory carotid body within the parathyroid gland III of the chicken. *Histochemistry* **103,** 197–204.

Zeuthen, E. (1942). The ventilation of the respiratory tract in birds. *K. danske Vidensk. Selsk. Skr.* **17,** 1–50.

Zimmer, K. (1935). Beitrage zur Mechanik der Atmung be den Vogeln in Stand und Flug. *Zoologica* **33,** 1.

CHAPTER 11

Renal and Extrarenal Regulation of Body Fluid Composition

DAVID L. GOLDSTEIN
Department of Biological Sciences
Wright State University
Dayton, Ohio 45435

ERIK SKADHAUGE
Department of Anatomy and Physiology
Royal Veterinary and Agricultural University
DK-Frederiksberg C, DK-1870 Denmark

I. A Review of Reviews 265
II. Introduction 265
III. Intake of Water and Solutes 266
   A. Drinking 266
   B. Solute Intake 267
IV. The Kidneys 267
   A. Anatomy 267
   B. Physiology 270
V. Extrarenal Organs of
   Osmoregulation: Introduction 282
VI. The Lower Intestine 283
   A. Introduction 283
   B. Transport Properties of Coprodeum, Colon, and Cecum 283
   C. Postrenal Modification of Ureteral Urine 287
VII. Salt Glands 288
   A. Anatomy 288
   B. Function 289
   C. Contribution of the Salt Glands to Osmoregulation 290
VIII. Evaporative Water Loss 291
   References 291

I. A REVIEW OF REVIEWS

Recent reviews of topics in avian osmoregulation can be found in Skadhauge (1981), Wideman (1988), Hughes and Chadwick (1989), Braun and Duke (1989), Dantzler (1989), Gerstberger and Gray (1993), Skadhauge (1993), Elbrønd et al. (1993), and Brown et al. (1993). These reviews are often referenced in the present chapter for the sake of conciseness; the reviews themselves should be consulted for additional references to the primary literature.

II. INTRODUCTION

Avian osmoregulation—regulation of the balance of water and electrolytes—involves the interacting contributions of a number of organs and organ systems, including the kidneys, intestinal tract, salt glands (when present), and skin and respiratory tracts (as routes of evaporative water loss). Among these, the kidneys are usually considered the primary organs of regulation. However, the kidneys empty their output into the lower intestine, after which the urine may reside in and be modified by the coprodeum, colon, and ceca; function of the kidney must therefore be integrated with these latter organs. Furthermore, under some circumstances (and in some species) the salt glands subsume a primary osmoregulatory role, and the majority of total water loss may in fact be evaporative, not excretory. The present chapter will focus most heavily on anatomical and functional organization of the kidneys and cloaca, but will also examine the physiology of these other organs of osmoregulation.

Before discussing the physiological mechanisms of regulation, it is worth considering the normal physio-

TABLE 1 Typical Normal Values of Some Regulated Osmoregulatory Variables[a]

Total body water	60–70 ml/100 g body mass
Extracellular fluid volume	20–25 ml/100 g body mass
Plasma volume	3.5–6.5 ml/100 g body mass
Plasma osmolality	320–370 mosmol/kg water
Plasma Na^+	150–170 meq/liter
Plasma K^+	2–5 meq/liter
Plasma uric acid	0.1–1 mmol/liter

[a]Values are typical for adult birds. Data are taken primarily from Skadhauge (1981), which should be consulted for references to original literature; the ranges presented encompass most of the variability among data compiled in that review.

logic state—what is it that is protected by the osmoregulatory systems. Table 1 provides normal values for several of these variables.

III. INTAKE OF WATER AND SOLUTES

A. Drinking

Many birds (especially carnivores and frugivores) routinely acquire all of the water they require through their food; a few species, particularly small, xerophilic birds, can survive on the metabolic water produced from a dry diet even in the absence of drinking water (Bartholomew, 1972). For most birds, though, drinking remains an important route of water intake.

The rate of drinking by birds in the laboratory follows a regular relation with body mass: birds weighing 100 g or more drink approximately 5% of body mass per day, but at progressively smaller body masses drinking rate rises, to about 50% per day for birds 10–20 g (Bartholomew and Cade, 1963; Skadhauge, 1981, Chapter 2). Birds with salt glands, drinking saline water, may drink substantially more than those without salt glands, using some of the water to excrete (via the salt glands) the ingested salt and thereby retaining a volume of pure water (Hughes et al., 1989). Except for those species which are able to survive without any drinking water (see Krag and Skadhauge, 1972; Skadhauge and Bradshaw, 1974), minimum water requirements are typically one-third to one-half of the *ad libitum* drinking rate (Skadhauge, 1981).

The physiological stimuli which initiate drinking are threefold: cellular dehydration, extracellular dehydration, and angiotensin II. The osmoreceptor cells which respond to cellular dehydration (i.e., to an osmotic loss of cellular water induced by a rise in extracellular fluid osmolality) are localized in periventricular regions of the hypothalamus (Thornton, 1986a; Kanosue et al., 1990). Their response may be specific to the Na^+ concentration in the cerebrospinal fluid, rather than to the total osmolarity (Thornton, 1986b; Kanosue et al., 1990).

The stimulation of drinking by extracellular dehydration (loss of extracellular fluid volume) can be demonstrated through the effects of hemorrhage or injection into various fluid compartments of a nonabsorbed, osmotically active compound such as polyethylene glycol. Both of these manipulations elicit responses consistent with the interpretation that the receptors (presumably mechanoreceptors sensitive to stretch or tension) are in the extravascular (interstitial), rather than the vascular, compartment of the extracellular fluid (Kaufman and Peters, 1980; Takei et al., 1989). The diposgenic effects of extracellular volume depletion may be mediated by angiotensin II (Takei et al., 1989).

Angiotensin II, administered either centrally (into the cerebrospinal fluid) or peripherally, elicits drinking in a wide variety of birds (Takei et al., 1989). The primary sites of action of this hormone appear to be the preoptic area and subfornical organ (Takei, 1977; Gerstberger et al., 1987, Simon et al. 1992). The angiotensin II reaching these circumventricular regions may cross their incomplete blood–brain barrier or may arise from a local, independent brain renin–angiotensin system.

In addition to angiotensin II, other compounds within the brain may be involved within the brain in eliciting drinking. At least some pathways between dipsogenic centers in the brain (e.g., connecting preoptic area and subfornical organ) appear to use adrenergic nerve fibers (Takei et al., 1989, Denbow and Sheppard 1993). A variety of peptides can either stimulate or inhibit drinking in birds (Takei et al., 1989), though the physiological significance of these compounds is not clear.

When birds are dehydrated, all of the stimuli for drinking (intracellular and extracellular dehydration and angiotensin II) are elicited simultaneously. When dehydrated birds regain access to water, they may drink substantially more water than required to restore their cellular and extracellular water deficits (Takei et al., 1988; Goldstein, 1995). However, much of this water is excreted in a dilute urine (Goldstein, 1995) or is absorbed slowly from the intestinal tract, and so aspects of fluid balance may remain in deficit despite the copious intake. In these circumstances, restoration of cellular fluid balance (osmolality) and distension of the intestinal tract with water appear to terminate drinking, even while blood volume remains depressed and angiotensin

II levels remain elevated (Takei *et al.*, 1988). With drinking as with release of antidiuretic hormone (see below), the osmolarity of the extracellular fluid is a more potent stimulus (is more closely regulated) than its volume.

B. Solute Intake

At least one bird, the pigeon (*Columba livia*), expresses a salt appetite in response to sodium depletion (Epstein and Massi, 1987). This response appears to depend on a synergistic action of angiotensin II and aldosterone (Massi and Epstein, 1990). In a variety of other birds, drinking rates may increase with increasing salinity of the drinking water, with maximal drinking rates typically at 0.15–0.3 M NaCl (see review in Skadhauge, 1981, Chapter 2). The maximum saline concentration that is drunk voluntarily is likely set by the bird's ability to extract free water, which results from the balance between renal concentrating ability and other routes of water gain and loss.

IV. THE KIDNEYS

A. Anatomy

1. Gross Anatomy

The avian urinary organs consist of a pair of kidneys and the ureters, which transport urine to the urodeum of the cloaca. There is no urinary bladder, though the cloaca may serve as its functional equivalent in some species and circumstances (see below).

The avian kidney lies within a cavity formed by the ventral surface of the synsacrum. The mass of the two kidneys is proportional to (body mass)$^{0.9}$ and represents approximately 0.8% of body mass in birds without salt glands, 1.4% in birds with salt glands (Hughes, 1970a).

The external appearance of the avian kidney is elongate and trilobed, with anterior, middle, and posterior divisions (Johnson, 1968). Within each division, the kidney is divided into numerous subunits. In the cortex, nephrons are arranged around branches (central veins) of the efferent venous system. These cortical units (sometimes called cortical lobules) contribute nephron elements (collecting ducts and loops of Henle) to several medullary subunits; conversely, each medullary subunit receives nephron elements from several cortical units (Johnson *et al.*, 1972). The medullary elements are wrapped in a sheath of connective tissue to form structures known as medullary cones. A kidney lobule (Figure 1) is defined as a medullary cone and the region of cortex that it drains. The number of kidney lobules, both absolutely and per gram of kidney, varies substantially among species (Johnson and Mugaas, 1970; Goldstein and Braun, 1989).

2. Nephron Types and Numbers

The structure of avian nephrons is highly heterogeneous. The smallest nephrons are located toward the surface of the lobule (Figure 1) and have simple glomeruli (Figure 2a). Because they resemble nephrons of reptilian kidneys, particularly in that they lack loops of Henle, these nephrons have been termed reptilian-type (RT) nephrons (Huber, 1917). Nephron size increases progressively with depth from the kidney surface. Those located most deeply have larger, more complex glomeruli (Figure 2b) and do include a loop of Henle, leading to the name mammalian-type (MT) nephron. Between definitive RT and MT nephrons is a continuous gradation of nephrons (Goldstein and Braun, 1989; Roush and Spotts, 1988), including those with elongated, looping intermediate segments that are not bound into the medullary cones (Boykin and Braun, 1993). Braun has proposed abandoning the RT/MT terminology and replacing it with the terms loopless (LLN), transitional (TN), and looped (LN) nephrons (Braun, 1993; Boykin and Braun 1993).

Both nephrons without loops of Henle and those with loops empty in a highly regular pattern into common collecting ducts (Boykin and Braun, 1993). Collecting ducts conjoin as they descend through the medullary cone, and the terminus of the medullary cone is a single large collecting duct that empties directly into the ureter. The avian kidney has no renal pelvis.

Of the total nephron population, between 10 and 30% possess loops of Henle (i.e., are of the MT variety) (Goldstein and Braun, 1989). Several studies suggest habitat-related patterns in kidney structure, such as smaller kidneys, larger volume of medulla, or smaller volume of cortex in desert species (Thomas and Robin, 1977; Warui, 1989; Casotti and Richardson, 1992); it is not yet clear how these variations relate to the different proportions of RT and MT nephrons.

3. Blood Flow

The avian kidney receives both arterial and afferent venous (portal) blood supplies and is drained by efferent venous flow (Figure 3). The overall pattern of flow is complex; the reader is referred to Wideman (1988) for a detailed review.

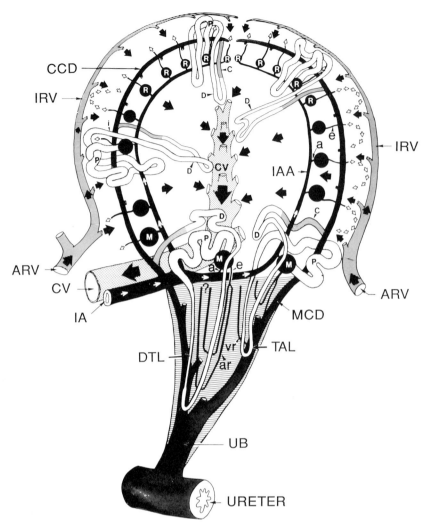

FIGURE 1 A typical kidney lobule from the domestic fowl. The lobule consists of a medullary cone and all cortical tissue that feeds into it. Abbreviations (in alphabetical order) are: a, afferent arteriole; ar, arteriole rectae (descending vasa recta); ARV, afferent renal (portal) vein; C, connecting tubule; CCD, cortical collecting duct; CV, central (intralobular renal) vein; D, distal tubule; DTL, descending thin limb of Henle's loop; e, efferent arteriole; i, intermediate segment; IA, intralobar artery; IAA, intralobular artery; IRV, interlobular renal vein; M, mammalian-type (MT) glomerulus; MCD, medullary collecting duct; P, proximal tubule; R, reptilian type (RT) glomerulus; TAL, thick ascending limb of Henle's loop; UB, ureteral branch; vr, venule rectae (ascending vasa recta). (Figure is from R. F. Wideman, Jr., Avian kidney anatomy and physiology, *CRC Critical Reviews in Poultry Biology,* 1988, pp. 133–176. Reprinted by permission of CRC Press, Boca Raton, Florida.)

a. Arterial Supply

Each division of the kidney is supplied by a separate artery (Figure 3). The arteries branch upon entering the kidney to form intralobar and then numerous intralobular arteries, from which branch the short afferent arterioles supplying the glomeruli.

Efferent arterioles from glomeruli of RT nephrons empty into sinuses surrounding the cortical tubular elements, thereby forming the peritubular blood supply. In contrast, the efferent arterioles leaving glomeruli of MT nephrons enter the medullary cones to form a meshlike network of vessels, the vasa recta (Figure 4). Whether the outflow from the vasa recta enters peritubular sinuses surrounding cortical components of the MT nephrons or, instead, whether it enters directly into the intralobular veins is as yet unclear.

b. Renal Portal System

The second afferent blood supply to the kidney, the renal portal supply, receives blood flow both from the ischiadic and the external iliac veins. A valve, the renal portal valve, is located between the renal portal vein

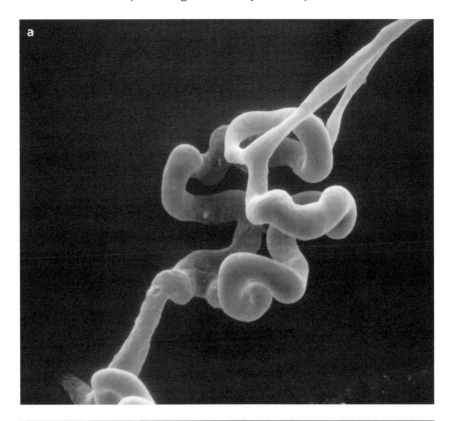

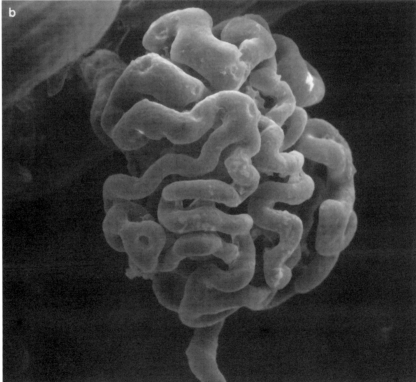

FIGURE 2 (a) Glomerulus from a reptilian-type nephron. Note the simple looping pattern and lack of cross-branching in these capillaries. (b) Glomerulus from a mammalian-type nephron with a longer, more complex capillary network. Photographs courtesy of G. Casotti and E. J. Braun.

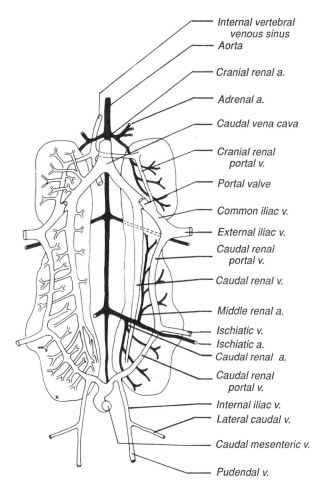

FIGURE 3 Major blood vessels in the avian kidney. The kidneys are drawn as though transparent, and in the left kidney only the major venous trunks are shown to permit an indication of finer levels of vessel branching. The internal vertebral venous sinus has been displaced to one side so that it can be seen. Reproduced from King (1975) with permission.

and the common iliac vein (leading to the posterior vena cava) (Figure 3). This smooth muscle sphincter receives both adrenergic (stimulatory, causing valve closure) and cholinergic (inhibitory, allowing valve opening) innervation (Burrows et al., 1983). Blood from the external iliac vein may flow directly into the vena cava when the renal portal valve is open or, in contrast, be forced into the renal portal vein when the valve is closed.

Once in the renal portal vein, blood may enter the low-pressure peritubular sinuses formed from the efferent arterioles; the mixed portal and postglomerular blood then flows out of the kidney through efferent veins. Alternatively, blood may flow out of the kidney into vertebral sinuses or, via the caudal mesenteric vein, to the liver. The pattern of flow through these various pathways may be highly variable both among animals and within an individual animal over time (Akester, 1967; Odlind, 1978; Oelofsen, 1973; Wideman, 1988), presumably because of varying states of regulation of vascular resistances and of valve status. The functional significance of the portal system and its valve may relate in part to regulating systemic hemodynamics, particularly during times of leg muscle activity (see Wideman, 1988), and in part to regulating renal hemodynamics, particularly when arterial blood pressure or flow is reduced (Wideman, 1991; see below).

Since Sperber's (1948) convincing demonstration of portal circulation to the renal cortex, many investigators have used the so-called "Sperber technique" in studies of avian kidney function. In this approach, a compound is infused into one leg vein (and hence into the renal portal circulation), and function of the ipsilateral kidney is compared with that of the contralateral kidney that does not receive the portal infusate. The Sperber technique has been used the evaluate both the mechanisms and regulation of transport of organic and inorganic compounds by cortical tubular elements; portal blood does not perfuse the glomeruli or the renal medulla.

c. Venous Drainage

Blood from the peritubular sinuses drains into central (intralobular) veins, which coalesce into interlobar veins. These join to form the efferent renal vein, which empties into the posterior vena cava.

B. Physiology

1. Overview

The primary function of the kidneys is eliminative—that is, they rid the body of wastes and excess water and solutes. However, they accomplish this largely in reverse. The kidneys receive a large fraction of the cardiac output, and from this the entire body fluid volume is filtered through the glomeruli several times each day. Most of this volume and its solutes are then reclaimed from the urine back to the blood, a process known as tubular reabsorption. Some substances, though, may additionally be added to the urine by the renal epithelia in the process of secretion. Finally, the end product of filtration, reabsorption, and secretion enters the ureters as urine.

2. Renal Blood Flow

For any substance, renal clearance is defined as the volume of blood from which that substance is removed (cleared) in the kidneys per unit time. Hence, if a substance is completely removed from the plasma (by filtration and secretion) during a single pass through the renal vasculature, renal clearance and renal plasma flow are the same. (Renal plasma flow can be corrected to renal blood flow by dividing by [1 − the hematocrit].)

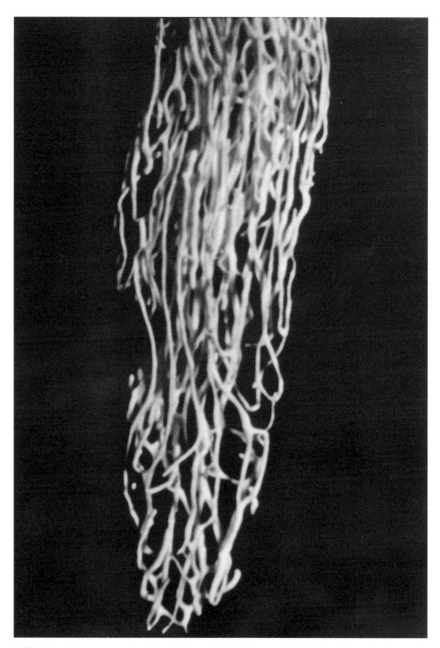

FIGURE 4 The vasa recta of Gambel's quail. Note the mesh-like (rather than simple ascending and descending) appearance of the vessel network. Photograph courtesy of E. J. Braun.

The clearance (C, ml/min) of any substance, X, is calculated from the formula: $C = U_X V/P_X$, where U_X and P_X are the urine and plasma concentrations of the substance and V is the urine flow rate (ml/min). In mammals, renal plasma flow has typically been measured as the clearance of paraamino hippuric acid (PAH), which is cleared with an efficiency of approximately 90%. This substance has been used in several studies of birds as well. However, the complexity and inaccessibility of the avian renal vasculature has made it difficult to determine the extent to which PAH is actually removed from the blood in the kidneys. Recent measurements in a reduced avian kidney model (with a single arterial supply and accessible renal vein) indicated PAH extraction efficiency between 50 and 75% (Wideman and Gregg, 1988), suggesting that PAH clearance is likely to underestimate the true renal plasma flow. Nevertheless, the values of RPF measured by Wideman (and corrected for PAH extraction) are on the lower end of values measured in other studies, which average approximately 40 ml/kg body mass/min (e.g., Nechay and Nechay, 1959; Sperber, 1960; Holmes *et al.*, 1968; Wideman, 1988; Roberts, 1992).

Measurement of RPF as C_{PAH} includes both arterial and portal contributions to RPF. Alternatively, RPF has been measured from the distribution of radioactive microspheres infused into the heart, thereby providing a measure of renal arterial blood flow (without the portal contribution). Results from these studies suggest that avian kidneys receive 10–15% of cardiac output.

Along with mathematical models of avian renal blood flow (Shideman *et al.*, 1981), these studies collectively suggest that approximately 50% of avian renal blood flow might derive from the renal portal system, the remainder from arterial flow. However, as described earlier, renal portal flow patterns can be highly variable.

In the domestic fowl, renal blood flow can be autoregulated (maintained constant) over a range of arterial pressures from 100 to below 40 mm Hg (Wideman and Gregg, 1988). An important contributor to this autoregulation is the renal portal flow; as arterial pressure falls to the lower end of the autoregulatory range, arterial flow diminishes, but portal flow increases complementarily (Wideman, 1991). The mechanism for this coordination is unknown, but it presumably functions to maintain perfusion of the cortical tubule elements during times of reduced glomerular (and postglomerular) blood flow (see below).

Despite the ability to autoregulate, avian renal blood flow varies substantially in some circumstances. For example, in feral chickens salt loading increased renal blood flow by more than 100% (Roberts, 1992). The mechanism of such changes, including the relative contribution of changing portal flow, is undescribed.

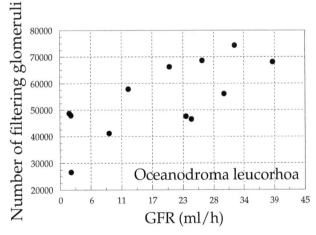

FIGURE 5   The number of filtering glomeruli (evaluated by infusion of the stain Alcian blue, which binds to the negatively charged filtration barrier as it is filtered) in Leach's storm petrels (*Oceanodroma leucorhoa*) whose GFR varies over a wide range in response to intravenous infusion of NaCl. Experimental conditions are described in Goldstein (1993).

3. Glomerular Filtration

Filtration of plasma through the glomerular capillaries is the primary step in formation of avian urine. The whole kidney glomerular filtration rate (GFR) is the sum of the single nephron glomerular filtration rates (SNGFR). SNGFR is determined by the difference between net hydrostatic pressure favoring filtration and oncotic pressure opposing it, by the hydraulic conductivity (permeability to water) of the glomerular filtration barrier, and by the surface area for filtration. Primarily because of the simplicity of avian (as compared with mammalian) glomeruli, and hence their small surface area, SNGFR in birds is low. However, because birds have more glomeruli per kidney, whole kidney GFRs are similar in birds and mammals (Yokota *et al.*, 1985).

GFR (whole-kidney or single-nephron) is measured as a special case of clearance. If a substance is freely filtered and is neither secreted nor reabsorbed by the renal tubules, then the amounts filtered from the plasma and excreted in the urine are the same. The clearance of such a substance provides a measure of the GFR. Avian GFR has been measured as the clearance of inulin, polyethylene glycol, sodium ferrocyanide, iothalamate, and others.

SNGFR during mannitol diuresis has been measured in the kidneys of two species, Gambel's quail (*Callipepla gambelii*) and the European starling (*Sturnus vulgaris*), using the ferrocyanide precipitation technique (Braun and Dantzler, 1972; Braun, 1978), which permits simultaneous evaluation of filtration rates in both superficial and deep nephrons. SNGFR on average was 6–7 nl/min in RT nephrons of both species, about 11 nl/min in "short-looped" MT (transitional) nephrons of the quail, and about 15 nl/min in "long-looped" MT nephrons. SNGFR has also been measured in the most superficial, and hence smallest, RT nephrons by micropuncture, in which filtered ferrocyanide is sampled directly from the renal tubules using a fine glass micropipette (Laverty and Dantzler, 1982; Roberts and Dantzler, 1990). The micropuncture technique yielded much lower values of SNGFR, from 0.25 to 0.5 nl/min. It is likely that nephrons exist with a continuum of filtration rates ranging from these very low values to the highest MT values.

Whole-kidney GFR (ml/hr) in normally hydrated birds varies with body mass (M, in grams) as GFR = $1.24 \times M^{0.69}$ (Yokota *et al.*, 1985). This equation predicts a GFR of 6 ml/hr for a 10-g bird, 30 ml/hr for a 100-g bird, and 145 ml/hr for a 1000-g bird. Within this general relationship exists substantial variability both among and within species. For example, the GFRs of Gambel's quail and coturnix quail (*Coturnix coturnix*), both approximately 150 g, are about 15 and 36 ml/hr, respectively (Braun and Dantzler, 1972; Clark and Sasayama, 1981); female Pekin ducks (*Anas platyrhynchos*) have higher GFR than males (Hughes *et al.*, 1989), perhaps associated with different kidney sizes (Hughes *et al.*, 1992). It is tempting to speculate on adaptive variation in filtration rates; the low GFR of Gambel's quail, for example, would presumably promote a lesser urine flow in this desert bird. However, the data are yet insufficient to generate any robust generalizations about habitat- or diet-related variation in GFR.

Like renal plasma flow, GFR of birds is autoregulated over a broad range of blood pressures (from 110 down to 60 mm Hg in the domestic fowl; Wideman and Gregg, 1988). This autoregulation was not affected by dietary sodium intake, as might be expected if tubuloglomerular feedback (changing renal vascular constriction in response to varying NaCl delivery to the distal tubule) were the responsible mechanism (Vena *et al.*, 1990); these authors therefore concluded that a myogenic mechanism, in which renal arteriolar smooth muscle responds directly to stretch (pressure), is a likely mechanism of autoregulation.

Also like renal plasma flow, despite the potential for autoregulation, GFR in birds is variable and changes as part of the renal mechanism for varying urinary output of water and sodium (see below).

4. Regulation of Water Excretion

The avian kidney filters a large volume of fluid (approximately 11 times the entire body water each day for a 100-g bird) and then reclaims most filtered water by tubular reabsorption. In normally hydrated birds the percentage of filtered water that is reabsorbed is typically greater than 95%. Variation from this "normal" situation can be achieved through regulation of either the rate of filtration or the rate of reabsorption, both of which may be influenced by the avian antidiuretic hormone arginine vasotocin (AVT).

a. Arginine Vasotocin

AVT is a small (8-amino-acid) peptide hormone released from the neurohypophysis. Circulating concentrations of AVT are about 10 pg/ml in normally hydrated birds (e.g., Möhring *et al.*, 1980; Rice *et al.*, 1985; Stallone and Braun, 1985; Gray and Erasmus, 1988; Gray *et al.*, 1988). The primary stimulus for additional AVT release is a rise in extracellular fluid osmolality. In several species, the circulating concentration of AVT increases by 0.25–2 pg/ml for each millimole per kilogram rise in plasma osmolality (Möhring *et al.*, 1980; Arad *et al.*, 1986; Stallone and Braun, 1986; Gray *et al.*, 1988). A decrease in extracellular fluid volume (experimental hemorrhage) may also stimulate AVT release, but with less sensitivity than the response to osmolarity (Simon-Oppermann *et al.*, 1984; Stallone and Braun, 1986). Concentrations of AVT in dehydrated birds are typically 30–60 pg/ml (e.g., Stallone and Braun, 1986; Gray *et al.*, 1988; Gray and Erasmus, 1988), though some higher values have been reported (Goldstein and Braun, 1988; Hughes *et al.*, 1993).

b. Regulation of GFR

Reduction in GFR is a consistent component of the avian response to dehydration (Table 2) and to infusion of AVT (e.g., Ames *et al.*, 1971, Gerstberger *et al.*, 1985; Figure 6). This could reduce urine flow both directly (by a reduced throughput of water) and indirectly (by allowing enhanced tubular reabsorption and urine concentration). The diminished GFR may be brought about by an AVT-induced constriction of afferent arterioles (Braun and Dantzler, 1974), leading to a reduction or complete cessation (intermittency) of filtration in some or all nephrons (Figure 5). The RT nephrons appear most sensitive to AVT, and AVT therefore reduces the contribution of RT nephrons (which lack loops of Henle and the ability to produce a concentrated urine) to urine production.

Birds are also able to increase GFR in response to extracellular fluid expansion induced by oral or intravenous water loading (Korr, 1939; Krag and Skadhauge, 1972; Roberts, 1992). The mechanism of this increase is not known.

c. Tubular Water Reabsorption and the Urinary Concentrating Mechanism

Tubular reabsorption of water in birds can range from less than 70% to more than 99% of the filtered volume. As a result, the final concentration of avian urine can vary from dilute (approximately 40 mmol/kg) to hyperosmotic (2–3 times the plasma osmolality).

Both micropuncture studies of superficial (RT) proximal tubules (Laverty and Dantzler, 1982) and studies of isolated, perfused proximal tubules from deeper (transitional) nephrons (Brokl *et al.*, 1994) suggest that, as in mammals, the avian proximal tubule absorbs about 70% of the filtered volume of water. The reabsorption depends on active sodium reabsorption but not on the presence or reabsorption of bicarbonate (Brokl *et al.*, 1994).

Regulation of water reabsorption occurs not in the proximal tubule, though, but in more distal nephron segments. As noted above, birds, like mammals, are able to produce a urine more concentrated than the plasma. The fundamental mechanism driving this ability is thought to be the same in birds and mammals—that

TABLE 2 Representative Responses of Avian GFR to Osmoregulatory Challenge[a]

Species	Condition	GFR (% of control)	Reference
House sparrow (*Passer domesticus*)	30 hr dehydration	46	1
Starling (*Sturnus vulgaris*)	1 day dehydration	42	2
Stubble quail (*Coturnix pectoralis*)	21–120 days dehydration	66	2
Emu (*Dromaius novaehollandiae*)	7 days dehydration	70	2
Mallard duck (*Anas platyrhynchos*)[b]	Saline acclimation	100	3
Glaucous-winged gull (*Larus glaucescens*)[b]	Saline acclimation	210	3
Canada goose (*Branta canadensis*)[b]	Saline acclimation	183	3
Starling (*Sturnus vulgaris*)	24 meq/kg NaCl infusion	200	4
Gambel's quail (*Callipepla gambellii*)[c]	20 meq/kg NaCl infusion	90	3
Chicken (*Gallus domesticus*) Domestic[c]	15 meq/kg NaCl infusion	100	3
Feral	12 meq/kg NaCl infusion	160	5
Leach's storm petrel (*Oceanodroma leucorhoa*)[b]	24 meq/kg NaCl infusion	15	7
Pekin duck (*Anas platyrhynchos*)[b]	20 meq/kg NaCl infusion	100	6

[a]References: (1) Goldstein and Braun (1988); (2) From summary table in Roberts and Dantzler (1989); (3) From summary table in Roberts *et al.* (1985); (4) Laverty and Wideman (1989); (5) Roberts (1992); (6) Gerstberger *et al.* (1985); (7) Goldstein (1993).
[b]Species with salt glands.
[c]At higher salt loads (35–50 meq/kg NaCl infusion) GFR decreased to approximately 40% of control.

is, a countercurrent multiplier system producing an osmolarity gradient in the renal medulla.

Earlier studies of the avian renal medulla revealed a number of the basic features associated with a countercurrent multiplier mechanism of urine concentration. The MT nephrons possess loops of Henle that extend into the renal medulla in parallel with collecting ducts. ^{22}Na autoradiography suggested active Na$^+$ reabsorption within the medullary cone (Emery *et al.*, 1972), and this was consistent with an osmotic gradient composed primarily of NaCl along the length of the medulla (Skadhauge and Schmidt-Nielsen, 1967b; Emery *et al.*, 1972). However, for the avian descending and ascending limbs of Henle to act as single-effect countercurrent multipliers (i.e., countercurrent multipliers whose energy source is a single active transport process, as proposed for the mammalian kidney (see e.g., Jamison and Kriz, 1982)), they must possess three characteristics: (1) ability to generate energy sufficient to develop a transverse difference in osmotic pressure, (2) differences in permeabilities, and (3) countercurrent flow. The last of these requirements is presumed from the anatomy. Thus, recent studies have focused on evaluating the transport and permeability properties of the nephron.

i. Descending Thin Limb (DTL) The transition from pars recta (straight descending limb of the proximal tubule) to DTL occurs with an abrupt transition in cell morphology near the base of the medullary cone (Braun and Reimer, 1988). The physiological properties of avian DTL have been studied for just the upper segment of a single species (*Coturnix coturnix*; Nishimura *et al.*, 1989). *Coturnix* DTL segments fail to generate an electrical potential both when bathed in isosmotic media and when exposed to a transepithelial ion gradient; these results derive from an absence of electrogenic ion transport, combined with high but nearly equal permeabilities to Na$^+$ and Cl$^-$ (Nishimura *et al.*, 1989). In contrast to the high ion permeabilities, *Coturnix* DTL has a low permeability to water (Nishimura *et al.*, 1989).

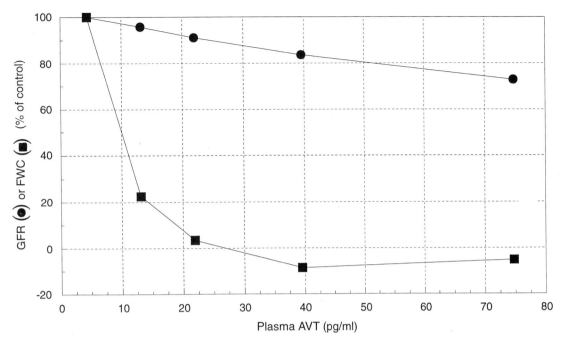

FIGURE 6 The effect of arginine vasotocin (AVT, the avian antidiuretic hormone) on free water clearance (FWC) and glomerular filtration rate (GFR). Note that FWC decreases markedly in response to low concentrations of AVT while GFR is essentially unchanged. (Data taken from Stallone and Braun (1985), *Am. J. Physiol.* with permission.)

ii. Thick Ascending Limb (TAL) The avian loop of Henle always makes the transition from DTL to TAL before the hairpin turn (Figures 1 and 7). In cellular morphology the avian TAL resembles other vertebrate-diluting segments (Nishimura, 1993; Braun and Reimer, 1988). Functional properties of this segment include a lumen-positive voltage that requires Na^+ and Cl^- and is inhibited by the loop diuretics bumetanide and furosemide, the ability to transport NaCl against transepithelial electrochemical gradients, and a low permeability to water and net water transport (Miwa and Nishimura, 1986; Nishimura et al., 1986). Ion reabsorption by the TAL is sensitive to solute delivery (enhanced either by increased flow or concentration; Osono and Nishimura, 1994), but is not affected by AVT (Miwa and Nishimura, 1986), forskolin, or isoproterenol (Osono and Nishimura, 1994). The combination of active salt reabsorption and low water permeability implies that the urine is diluted as it passes through the TAL.

iii. Collecting Duct The avian collecting duct receives dilute urine from both the TAL and from RT nephrons, and so final concentration of the urine (extraction of water into the hyperosmotic interstitium) requires that this nephron segment is, or can become, permeable to water. A variety of whole animal experiments suggest that, as in mammals, an increased permeability of the distal nephron to water may be induced by AVT. For example, infusion of AVT into the renal portal system increased urine osmolarity (though the portal flow should not have perfused medullary nephron elements and urine osmolarity never rose above plasma levels; Skadhauge 1964). Moreover, infusion of AVT into the systemic circulation, at least at low doses, may increase urine osmolarity and reduce urine flow with little or no change in GFR (Ames et al., 1971; Stallone and Braun, 1985; Figure 6).

Despite these findings, direct assessment AVT's actions on isolated, perfused avian collecting ducts has failed to reveal a significant effect of the hormone (Nishimura, 1993). The *Coturnix* quail collecting duct had a moderate baseline permeability to water even in the absence of AVT. Although forskolin ($10^{-4}\,M$), which activates cyclic AMP production, induced a substantial increase in this permeability, AVT did so only slightly.

iv. Urinary Concentrating Mechanism Nishimura (Nishimura, 1993; Osono and Nishimura, 1994) has proposed a model of the avian urinary concentrating mechanism that incorporates the evidence described above (Figure 7). The single effect that drives urine concentration is active ion reabsorption by the TAL. These reabsorbed ions enter the DTL by passive diffusion without accompanying water flux. The low water permeability of the DTL maintains volume flow through the loop, which should favor ion reabsorption by the flow-

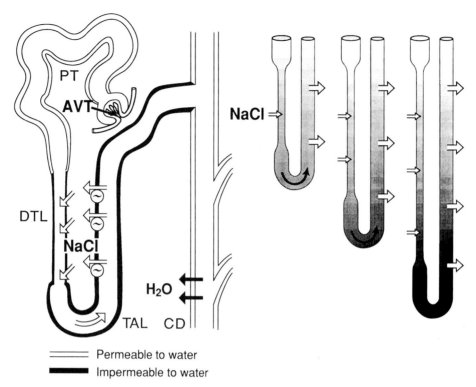

FIGURE 7 A proposed model for the avian urine concentrating mechanism. Tubule segments illustrated with heavy lines have low permeability to water. Possible sites of regulation include AVT-induced reduction of the filtered load (shown), flow-sensitivity of the thick ascending limb, and cAMP-mediated changes in water permeability of the collecting duct. To the right is pictured the hypothetical cascade effect of the concentrating mechanism that results from the presence of loops of Henle of varying length. (From H. Nishimura (1993) *New Insights in Vertebrate Kidney Function*, pp. 189–208, © Cambridge University Press 1993. Reprinted with the permission of Cambridge University Press.)

sensitive TAL. The result is that urine is concentrated as it descends through the DTL, then diluted as it ascends the TAL. Simultaneously, the interstitium surrounding the loop accumulates ions and becomes concentrated. The sensitivity of the TAL to solute delivery means that the tip of the loop (at the start of the TAL epithelium) should be the site of greatest ion reabsorption, and this should contribute to generating a concentration gradient in the medulla. These effects are amplified by the countercurrent flow and may be further enhanced by interaction among loops of Henle which turn at different depths in the medullary cone (Layton, 1986). As the urine subsequently descends through the collecting duct, it may equilibrate osmotically with the interstitium via the baseline permeability to water of this nephron segment. The exact role of AVT in this mechanism remains to be clarified. However, even if AVT acts primarily by reducing fitration rates of RT nephrons, not by a tubular effect, then the decreased flow through the collecting ducts would permit greater equilibration of water between tubule and interstitial fluid. It also remains for the model to be harmonized with emerging details of avian renal anatomy. For example, in at least some groups of birds the descending and ascending limbs of Henle are separated from each other by a ring of collecting ducts throughout the depth of the medullary cone (Cassoti and Richardson, 1993). Lastly, it is notable that NaCl may account for just a small proportion of total urine osmolality (Skadhauge, 1977; Hughes *et al.*, 1992; Goldstein 1993b); the roles of ions other than NaCl in urine concentration deserve examination.

v. Net Effect: Avian Urinary Concentrating Ability The net effect of the medullary concentrating mechanism is that birds are typically able to concentrate urine to an osmolarity 2–3 times that of plasma. (An exception to this is a report that salt-marsh savannah sparrows, *Passerculus sandwichensis beldingi,* when drinking seawater, excreted cloacal fluid with an osmolarity 4.5 times that of plasma (Poulson and Bartholomew, 1962).) The urinary concentrating ability generally varies inversely with body mass, so that small birds (10–25 g) typically concentrate to about 1000 mmol/kg, larger birds (>500–1000 g) mostly to about 600–700 mmol/kg (Goldstein and Braun, 1989). This pattern may result from the dependence of urine concentration on metabolically active processes (active ion reabsorption), along with the decreasing rate of cellular metabolism that accompanies increasing body mass (Greenwald, 1989). It remains unclear the extent to which other features of kidney anatomy, such as the proportions or RT and MT nephrons, the lengths of medullary cones, or the proportions of kidney mass composed of medullary tissue, relate to urinary concentrating ability independent of the body mass effect (Johnson, 1974; Goldstein and Braun, 1989; Warui, 1989). At least within closely related species, urine concentration does seem to be related to the proportion of medullary tissue in the kidneys (Skadhauge, 1981).

5. Regulation of Sodium Excretion

a. Patterns of Response

One of the important functions of the avian kidney is the regulation of sodium excretion; sodium is the primary extracellular cation, and its content in the extracellular fluid largely determines the extracellular fluid volume and hence contributes to blood pressure regulation. Regulation of sodium excretion by the avian kidney has been examined under two general circumstances. In the first, dietary sodium intake is varied; typically these animals are able to regulate the plasma sodium concentration within the normal range. In the second, birds are infused with saline, typically at a high concentration that challenges the regulatory abilities and ultimately leads to a rise in plasma sodium concentration and osmolarity.

Acclimation to increasing salt intake results, as expected, in an enhanced renal salt excretion. This may involve an increase (Hughes, 1980), decrease (Hughes et al., 1989; Dawson et al., 1991), or no change (Holmes et al., 1968; Hughes, 1980; Goldstein, 1990) in GFR (Table 2). The renal response depends at least in part on the level of NaCl intake. For example, chukars (*Alectoris chukar*) responded to an increase from 0.25 to 2 meq Na^+/d primarily by reducing intestinal sodium absorption, with little renal response, whereas a further increase to 10 meq/d induced a reduction in tubular sodium reabsorption (Goldstein, 1990). NaCl drinking solutions at concentrations beyond the kidneys' regulatory ability may induce dehydration and an accompanying reduction in GFR (Roberts and Hughes, 1983).

Infusion of hyperosmotic salt loads induces different responses in different species (Table 2). Some birds respond with large increases in GFR and reduced fractional Na^+ reabsorption (Laverty and Wideman, 1989; Roberts, 1992). In others, the response is no change (Skadhauge and Schmidt-Nielsen 1967a; Gerstberger et al., 1985; Hughes et al., 1993) or reduction in GFR (Braun and Dantzler, 1972; Dantzler, 1966; Goldstein, 1993b). In birds with salt glands, the lack of change in GFR in response to saline drinking or infusion may result from the ability of the salt glands to excrete the NaCl and thereby preserve pure water. Among other species, the different responses may relate to different strategies of osmoregulation—whether to preserve extracellular volume at the expense of a rising osmolarity (i.e., to retain the infused sodium, but thereby retain water) or to preserve osmolarity at the expense of reduced volume (i.e., to excrete the solute load, but lose water along with it).

b. Mechanisms of Regulation

A number of hormones may be involved in producing the varied patterns of sodium excretion.

i. Arginine Vasotocin A rising extracellular osmolarity would be expected to stimulate release of AVT and, indeed, saline infusion can result in elevated AVT concentrations (Koike et al., 1979; Arad et al., 1986). Conversely, saline infusion may induce an expansion of the extracellular fluid, and the consequent decrease in circulating AVT levels may allow GFR to rise. Yet other studies have measured stable or substantially increased GFR even while AVT was increased (Roberts, 1992; Hughes et al., 1993). This may relate to counteracting influences of expanded extracellular fluid volume, perhaps mediated by atrial natriuretic peptide (see below). The precise role of AVT in the response to saline infusion remains uncertain.

ii. Renin/Angiotensin The evidence for a complete renin/angiotensin system in birds is now well accepted. Avian nephrons have a juxtaglomerular apparatus (Morild et al., 1985), and their plasma contains renin activity, angiotensin I (ANGI), and angiotensin II (ANGII) (reviewed by Wilson, 1989). In ducks (*A. platyrhynchos*), circulating levels of angiotensin are elevated during extracellular fluid volume contraction (Gray and Simon, 1985), as may occur at high saline concentrations of drinking water (Zenteno-Savin, 1991). Seemingly in contrast with this, infusion of angiotensin II into ducks (Gray et al., 1986) or into the renal portal vein of chickens (Langford and Fallis, 1966; Stallone and Nishimura, 1985) stimulates natriuresis and diuresis without any change in GFR, suggesting that ANGII inhibits sodium reabsorption through a direct tubular action. It now seems that the physiological effects of angiotensin vary with the osmoregulatory status of the animal. With salt and volume depletion, angiotensin stimulates reduction in GFR, urine flow, and sodium excretion (Gray and Erasmus, 1989); salt and volume loading, on the other hand, changes the response so that angiotensin becomes natriuretic and diuretic. The mediator of this change has not been determined, though Gray and Erasmus (1989) suggest that prostaglandins are likely candidates.

iii. Aldosterone Circulating concentrations of aldosterone may or may not be elevated by a diet low in NaCl or during dehydration (Skadhauge et al., 1983; Klingbeil, 1985; Arnason et al., 1986; Goldstein, 1993a). The hormone is presumed to have an action on the avian kidney similar to that seen in mammals—that is, promoting sodium conservation.

iv. Atrial Natriuretic Peptide Birds, like other vertebrates, secrete a peptide hormone (29 amino acids in the chicken) from the heart atria (Toshimori et al., 1990). Infusion of chicken ANP increases renal sodium and water excretion both in the domestic duck (Gray et al., 1991; Schütz et al., 1992) and chicken (Gray, 1993). Although receptors for ANP have been localized on both the glomeruli and the renal tubules (Schütz et al., 1992), the mechanism of action of the hormone, including the relative contributions of glomerular versus tubular effects, remains unresolved.

6. Regulation of Calcium and Phosphate Excretion

a. Calcium

Normally the avian kidney reabsorbs more than 98% of filtered calcium (Wideman, 1987) by a pathway that is apparently saturable and normally operates near its maximum capacity. Procedures that increase the filtered load of calcium, such as intravenous infusion of calcium or injection of exogenous parathyroid hormone (PTH, which increases GFR and plasma calcium levels), result in enhanced renal calcium excretion (Clark et al., 1976). On the other hand, in most birds (but not all; Clark and Mok, 1986) the normal pattern of reabsorption depends of parathyroid hormone (PTH); calcium excretion is markedly increased by parathyroidectomy and is restored to control levels by PTH administration (Clark and Wideman, 1977; Clark and Sasayama, 1981). The single clearance study of the effects of calcitonin failed to demonstrate any significant effect on renal calcium excretion (Clark and Wideman, 1980).

A cellular mechanism of calcium reabsorption and stimulation by PTH, which may occur in the thick ascending limb (Braun and Dantzler, 1987), has been proposed by Clark and Wideman (1989). Calcium likely moves across the apical cell membrane down an electrochemical gradient. Reabsorption across the basolateral membrane would then be accomplished by active transport, perhaps involving a calcium ATPase. Parathyroid hormone, which is known to activate cellular cyclic AMP production (e.g., Dousa, 1974; Pines et al., 1983), could promote this reabsorption via actions either on the apical permeability or the basolateral mechanism of reabsorption.

b. Phosphate

In normal chickens, approximately 60% of the filtered load of phosphate is excreted in the urine. This balance results from the simultaneous operation of a reabsorptive pathway, which reclaims about half of the filtered phosphate, and a low level of secretion (Wideman 1987). PTH both inhibits phosphate reabsorption (Cole, 1985; in Clark and Wideman, 1989) and stimulates its secretion (Wideman and Braun, 1981). As a consequence, in the absence of PTH (i.e., following parathyroidectomy) urinary phosphate excretion falls to near zero (Clark and Wideman, 1977). In contrast, administration of PTH to intact birds can result in renal excretion of more than twice the filtered load of phosphate (Wideman and Braun, 1981). The single clearance study of the effects of calcitonin failed to demonstrate any significant effect on renal phosphate excretion (Clark and Wideman, 1980).

In starlings, individual proximal tubules of superficial RT nephrons may exhibit either net secretion or net reabsorption of phosphate (Laverty and Dantzler, 1982), and this may be the site of regulation of phosphate transport. The cellular mechanism of phosphate reabsorption involves a saturable, sodium-dependent transporter located in the apical membrane (Grahn and Butterworth, 1982; Renfro and Clark, 1984). Inhibition of this transport is mediated by cyclic AMP (Cole, 1985). The mechanism of activation of phosphate secretion remains unknown. Phosphate infused only into blood supplying the renal tubules (via the renal portal vasculature) accounted for only a small fraction of net phosphate transport into the tubule lumen during times of net secretion (Wideman and Braun, 1981). The secreted phosphate may derive primarily from a pool that is sequestered within the tubule cell (Wideman, 1984).

In addition to its effects on calcium and phosphate transport, PTH may have other actions on the kidney, including most consistently increases in GFR, sodium excretion, and urine flow (Clark and Wideman, 1989).

7. Nitrogen Excretion

The end products of nitrogen metabolism excreted in the urine of birds include urates, ammonia, urea, creatinine, amino acids, and others (Table 3). Of these, urates are the predominant compounds under all circumstances, though ammonia may account for as much as 25% of total nitrogen. The present discussion will

TABLE 3 Patterns of Nitrogen Excretion in Avian Urine[a]

Species	Condition	Total N (g/liter)	Urate (% of total)	NH$_4$ (% of total)	Urea (% of total)
Domestic fowl	Fed	4.4	84.1	6.8	5.2
(*Gallus domesticus*)	Fasted	2.4	57.8	23.0	2.9
	Low protein	11	54.7	17.3	7.7
	High protein	13	72.1	10.8	9.7
Turkey vulture	Fed	61	87	9	4
(*Cathartes aurea*)	Fasted	13	76	17	7

[a]Data extracted from summary table in Skadhauge (1981). Other nitrogenous components of the urine, including creatinine, amino acids, and purines, always accounted for <10% of total nitrogen.

focus on the excretion of urates; urinary ammonia will be discussed in the following section, as it relates to acid/base regulation.

Urate in avian plasma is thought to be all or mostly unbound to proteins and hence filterable through the glomeruli (see discussion in Dantzler, 1978). Nevertheless, with a plasma urate concentration of 0.1–0.7 mM, glomerular filtration can only account for 10–20% of the total urate excreted by the kidneys. Shannon (1938) was the first to demonstrate conclusively that the clearance of urate exceeds that of inulin and hence that tubular secretion provides a significant fraction of urinary urates. It is now accepted that most (≥90%) urinary urate derives from tubular secretion. The source of the secreted urates is primarily the liver, though renal synthesis may account for 3–20% of the total (Martindale, 1976; Chin and Quebbemann, 1978).

Slices from avian renal cortex (Platts and Mudge, 1961; Dantzler, 1969), but not of medullary cones (Dantzler, 1969), accumulate urate *in vitro,* suggesting that urate secretion occurs in cortical nephron elements. Micropuncture studies of superficial proximal tubules of reptilian-type nephrons, as well as studies of urate transport by isolated proximal tubules from deeper (transitional) nephrons, have confirmed that these nephron segments secrete urate (Laverty and Dantzler, 1983; Brokl *et al.,* 1994). However, direct evidence for the roles of more distal segments or of mammalian-type nephrons is lacking. Recent studies of isolated proximal tubules (Brokl *et al.,* 1994) indicate that urate secretion involves transport across the basolateral membrane against an electrochemical gradient; this transport can be inhibited by PAH and probenecid, though it is not clear that PAH and urate necessarily share a transporter. The flux from cell to tubule lumen is down an electrochemical gradient. Net urate transport in proximal tubules is always secretory, but perfused tubules show a substantial simultaneous reabsorptive flux that apparently occurs through a paracellular route. *In vivo,* though, there is no evidence for such a reabsorptive pathway (Nechay and Nechay, 1959).

Once secreted, the chemistry of urates in urinary solutions is complex. The acid form of urate (uric acid) has a low aqueous solubility (0.38 mmol/liter), but with a pK of 5.6 and a typical urinary pH of 6–7, most molecules will exist as monobasic urate. The urate could form salts with a variety of cations in the urine, but sodium and potassium urate are most likely because of the abundance of these ions. The solubilities of sodium and potassium urate (6.8 and 12.1 mmol/liter, respectively) substantially exceed those of uric acid; nevertheless, the high concentrations of urates in the liquid fraction of the urine exceed values accounted for by these solubilities (McNabb *et al.,* 1973; McNabb, 1974).

Urates are apparently able to exist in solution at supersaturated concentrations because they form stable colloidal suspensions. However, the concentrations of urates in avian urine exceed even the stability of these colloids, and this stability is increased through the interaction of the urate colloid with mucoid materials in urine (Porter, 1963). These mucopolysaccharides and glycoproteins are most abundant in distal nephron segments where the urine concentration is likely to be highest (McNabb *et al.,* 1973; Nicholson, 1982). The concentrated aqueous colloids probably minimize formation of urate crystals and thereby facilitate passage of urates through the nephron.

Nevertheless, urates do also exist in precipitated form in avian urine. This is not in the form of crystalline masses, but rather as spheres a few microns in diameter (Figure 8; Folk, 1969; Lonsdale and Sutor, 1971). The spheres may form as alternating layers of urate and mucoid materials with associated water (Minnich, 1976).

The trapped water and mucopolysaccharides in turn may sequester cations within the urate spheres, perhaps because of the charge configurations existing between urate layers (McNabb and McNabb, 1975). Various studies have measured that anywhere from less than 5% to more than 75% of urinary Na$^+$ and K$^+$ are associated with the precipitated urate rather than existing in free solution (Hughes, 1972; McNabb *et al.,* 1973; Braun,

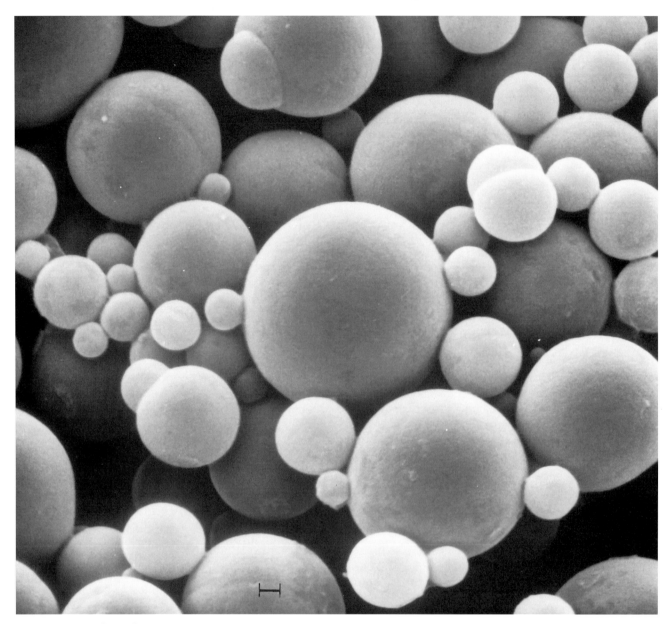

FIGURE 8 Scanning electron micrograph of urate spheres from the urine of Gambel's quail (*Callipepla gambelii*). The scale bar represents 1 μm. Photograph courtesy of G. Casotti and E. J. Braun.

1978; Long and Skadhauge, 1980; Long and Skadhauge, 1983b; Laverty and Wideman, 1989; Dawson *et al.*, 1991). The trapped ions would not contribute to urinary osmolarity and may present a means of enhancing ion excretion despite the limited concentrating ability of the avian kidney.

8. Renal Contribution to Acid/Base Regulation

Birds maintain an alkaline arterial pH of approximately 7.5, despite a constant metabolic production of acid. Most of this acid, resulting from hydration of CO_2, is excreted as respiratory CO_2. However, nonvolatile acids (e.g., H_2SO_4 and H_3PO_4) are also a threat to acid/base homeostasis, and the kidneys must excrete the protons equivalent to these metabolic end-products. Thus, though avian urine pH may range under experimental conditions from less than 5 to about 8 (Sykes, 1971; Long and Skadhauge, 1983a), avian urine is typically acidic, with a pH in the range of 5.5–7.5. The renal defense of arterial alkaline pH is thus presumed to consist of two components, conservation of base (bicarbonate) and excretion of acid (H^+, largely buffered).

Avian renal conservation of bicarbonate is thought to be accomplished by mechanisms similar to those existing in the mammalian kidney. That is, secreted H^+

combines in the tubule lumen with filtered HCO_3^- to form CO_2 that diffuses back into the tubule cells. There, catalyzed by carbonic anhydrase, it rehydrates and redissociates into H^+ and HCO_3^-. The HCO_3^- is then reabsorbed into the blood by carrier-mediated transport across the basolateral membrane. Overall, birds are indeed as efficient at conserving bicarbonate as are mammals, with virtual complete reabsorption of filtered bicarbonate during periods of acidosis (Anderson, 1967). It is likely that bicarbonate reabsorption in the avian kidney occurs in the distal nephron, where carbonic anhydrase activity is most prominent (see Laverty, 1989). Consistent with this, micropuncture studies of starling proximal tubules found no evidence for preferential reabsorption of bicarbonate in this nephron segment (Laverty and Alberici, 1987).

The lack of acidification of proximal tubule fluid, even in acidotic birds, suggests that H^+ secretion, too, may take place in distal nephron segments (Laverty and Alberici, 1987). The mechanism of avian renal H^+ secretion is little studied, though it is likely to be accomplished at least in part by Na^+/H^+ exchange (Laverty, 1989). The minimum urinary pH achievable by the avian nephron is 4.5–5, and so adequate acid excretion requires that most urinary hydrogen ions be excreted in buffered form. Three compounds—ammonia, phosphate, and urate—serve as the primary urinary buffers in birds.

As noted above, ammonia may constitute up to one-quarter of the total nitrogen excreted in avian urine. The urinary excretion of ammonia is inversely related to the pH of the urine (Wolbach, 1955; Long and Skadhauge, 1983a). This ammonia is presumed to derive from cellular metabolism (deamination reactions) and secretion in the distal nephron. The production of ammonia can be enhanced by infusion of the amino acids D,L-alanine, L-leucine, and glycine (Wolbach, 1955), and chronic metabolic acidosis in the chicken leads to increased activity of glutaminase in the kidney (Craan et al., 1982). The standard mechanism for urinary buffering by ammonia has been nonionic diffusion. The uncharged, lipid-soluble NH_3 is synthesized in the tubule cell and diffuses through cell membranes to equilibrate in the various fluid compartments (including the tubule lumen). The abundance of ammonium, NH_4^+, in the various compartments then depends on their H^+ concentrations. Because the pK for ammonium is 9, this ion is more abundant in the more acidic urine than in the cells or plasma. With its positive charge, NH_4^+ permeates cell membranes more slowly than NH_3 and so is effectively trapped in the urine, from where it may be excreted. In the mammalian nephron this mechanism may operate in the proximal tubule, where most ammonia is initially secreted, but other processes, including carrier-mediated transport and countercurrent multiplication of NH_4^+, occur more distally (Knepper, 1991; Good, 1994). It is not known whether such mechanisms operate in the avian nephron.

The phosphate buffer system (HPO_4/H_2PO_4), with a pK of 6.8, also plays an important role in urinary buffering. H_2PO_4 may account for approximately 20–30% of total acid excretion at typical urine pH, but may increase substantially when phosphate is infused during times of production of acidic urine (Wolbach, 1955; Sykes, 1971). The contribution of phosphate to acid excretion declines as urinary pH rises, perhaps because of a preferential reabsorption of the basic form of the buffer (HPO_4), which would be more abundant at the higher pH (Long and Skadhauge, 1983a).

The final major contributor to buffering of avian urine is urate/uric acid (pK 5.8). The contribution of this system to acid excretion is undoubtedly substantial (Wolbach, 1955) but is difficult to quantify accurately because of uncertainty about the chemical form of the colloidal precipitate. The proportions of urate and uric acid in the supernatant fraction of the urine can be calculated using the Henderson-Hasselbalch equation, but this can not be done for the precipitate. Calculations based on the assumptions either that all precipitated U/UA exists as UA (Lonsdale and Sutor, 1971) or that the precipitated U/UA is in equilibrium with the supernatant, suggest that U/UA constitutes 25–50% of net acid excretion (Long and Skadhauge, 1980; 1983a).

9. The Final Urine—Composition and Flow

An analysis of kidney function usually involves an examination of the urine and its response to changing circumstance. However, collection of urine from conscious birds is difficult. Urine flows from the ureters into the coprodeum, colon, and ceca, where it may be modified. A variety of approaches have been used to circumvent these difficulties. In some studies, the possibility of urine flow into the hindgut is removed, either by colostomy (in which the lower intestine is exteriorized and the ureters empty into a blind cloacal pouch; Hart and Essex, 1942; Holmes et al., 1982; Karasawa and Maeda, 1992; Belay et al., 1993) or by exteriorizing the ureters (Hart and Essex, 1942; Dicker and Haslam, 1972). Besides the problems associated with general maintenance and recovery after surgery, these techniques suffer from the possibility that urine flow is influenced by the very fact that the urine can no longer interact with the lower intestine (and that normal osmotic regulation is therefore disrupted). Alternatively, urine can be collected directly from the cloaca. This may be done over an extended time via collecting devices sewn over the ureteral openings (Anderson and

Braun, 1985), though this again may disrupt normal osmotic regulation. Alternatively, a filtration marker can be infused at a constant rate (such as by an implanted micropump); the "instantaneous" urine flow rate can then be calculated from a brief urine collection made by inserting a cannula over the ureteral openings (Goldstein and Braun 1988, Goldstein and Rothschild, 1993). This approach has the benefit of minimally disturbing normal function, but suffers the drawback of being unable to provide flow rates integrated over a longer time span.

In unanesthetized birds with free access to water, urine flow rate is in the range of 0.5–3 ml/kg^{-7}hr (Goldstein, 1990; using the same body mass scaling factor, kg$^{0.7}$, as pertains to GFR) and urine osmolality is typically close to isosmotic with plasma. During dehydration, urine flow diminishes and osmotic concentration increases to two to three times that of plasma. The composition of this urine has been analyzed in detail in only a small number of studies. In chickens eating grain (30 meq Na$^+$/kg), Na$^+$ and Cl$^-$ together accounted for just over 10% of the total osmolarity of 582 mM/kg (Skadhauge, 1977). The most abundant solutes were NH$_4^+$, phosphate, and K$^+$, with small concentrations of other measured solutes (Table 4). In this study, 41% of the osmotic space could not be accounted for by measured solutes. In chickens eating a diet with three times the sodium content of the grain, Na$^+$ and Cl$^-$ in the urine rose about threefold, whereas NH$_4^+$ was substantially diminished; other measured solutes changed by just small amounts (Table 4; Skadhauge, 1977). These findings—relatively high concentrations of NH$_4^+$, phosphate, and K$^+$, with Na$^+$ and Cl$^-$ together accounting for less than 50% of the osmotic space—are representative of other studies as well. However, with salt loading (in petrels, with salt glands; and partridges, without) Na$^+$ concentrations can rise to as much as 30–50% of the total osmotic space (Goldstein, 1990, 1993b).

10. Function of the Ureters

The ureters of birds are surrounded by smooth muscle which normally possesses tone and is capable of peristalsis (Gibbs, 1929). The ureters act to "milk" the urine from the kidneys, but they also offer significant resistance to the flow of urine. Because the ureters are direct extensions of the collecting ducts, the ureteral resistance may influence the rate of fluid flow through the nephron. Consistent with this, Hughes (1987) found that severing the ureter just caudal to the kidney increased urine flow and reduced urine osmolarity, suggesting that a more rapid transit through the nephron impaired ion and/or water reabsorption. The state of ureteral tone and extent of peristalsis may be under the influence of the sympathetic nervous system (Gibbs 1929).

V. EXTRARENAL ORGANS OF OSMOREGULATION: INTRODUCTION

Despite their limited urine concentrating ability, birds are as effective as mammals at water conservation. In both groups, small, desert-dwelling representatives are able to survive on seed diets without drinking water (Krag and Skadhauge, 1972; Skadhauge and Bradshaw, 1974). The means by which birds accomplish this, including the specific features of desert birds which differ from their nondesert counterparts, are not well defined. Water conservation in birds does result at least in part from renal mechanisms, including the ability of the avian kidney to reduce GFR in addition to increasing tubular reabsorption of water. Uricotelism, too, provides a means of excreting nitrogenous waste with a minimum of water loss (because of the insolubility of the urates and hence their lack of contribution to the urinary osmotic concentration) (Smith, 1953). Beyond this, avian

TABLE 4 Composition of Supernatant of Avian Urine[a]

Species	Condition	Na[b]	Cl	K	Mg	Ca	NH$_4$	PO$_4$	Osm[c]
Domestic fowl	Wheat and barley	41	36	73	13	7	120	130	582
(*Gallus domesticus*)	Commercial food	141	121	62	11	11	50	132	594
Galah									
(*Cacatua roseicapilla*)	Dehydrated	78	51	30	—	—	124	118	618
Ostrich									
(*Struthio camelus*)	Control	85	158	266	—	1	50	3	—

[a]Data extracted from summary table in Skadhauge (1981).
[b]All ion concentrations in mmol/liter.
[c]Osmolality in mosmol/kg water.

osmoregulation depends additionally on several other organs that regulate salt and water losses.

VI. THE LOWER INTESTINE

A. Introduction

In most birds ureteral urine flows backward by retrograde peristalsis from the urodeum into both coprodeum and colon (rectum) (Figure 9). This orally directed flow brings the absorbing epithelia of the lower gut into contact with a mixture of urine and chymus coming from the ileum. Both coprodeum and colon have a single layer of columnar epithelial cells which permit absorption/secretion of dissolved particles (mainly ions) and water. No urine (or feces) is stored in urodeum or proctodeum, which have nontransporting, multilayered squamous epithelia. Some urine/chymus may enter the paired caeca, which are present in a majority of birds. Not only ions and water, but nutrients, particularly short-chain fatty acids (SCFA), the products of fermentation, are absorbed here. Overall, urinary refluxing may serve a variety of functions, including possibly reclamation of water, electrolytes, nitrogen, or energy that would otherwise be lost.

B. Transport Properties of Coprodeum, Colon, and Cecum

1. Basic Transport Mechanisms in Coprodeum and Colon

Although coprodeum and colon, in most birds, form a common storage chamber for ureteral urine and chyme, their transport properties differ markedly.

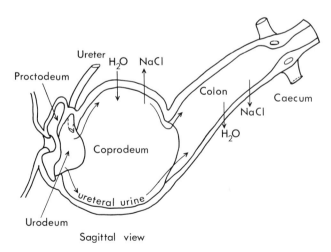

FIGURE 9 Lower intestine of the domestic fowl. Arrows indicate the retrograde flow of urine from urodeum to coprodeum, colon, and ceca, as well as possible directions for net fluxes of water and NaCl in coprodeum and colon. (Modified from Choshniak *et al.* (1977), with permission.)

This section is largely based on studies on the domestic fowl, as surveyed by Skadhauge (1973, 1981, 1993), Goldstein (1989), and Elbrønd *et al.* (1993), but it should be emphasized that the mechanisms observed in the fowl are generally observed in other avian species as different as the galah (*Cacatua roseicapilla*) and the emu (*Dromaius novaehollandiae*).

a. Transport of NaCl and Water

Both coprodeum and colon have electrolyte transport parameters which are heavily influenced by the NaCl intake (Figure 10). Particularly, the rate of Na^+ absorption is activated by NaCl depletion and suppressed by NaCl loading. The transport patterns must therefore be described and characterized in relation to the NaCl content of the diet.

Coprodeum is, as observed *in vitro*, a "medium-tight" epithelium (3–500 ohm $\times$ cm^2). On a low-NaCl diet coprodeum absorbs Na^+ at a very high rate, 10-20 μeq/cm^2 hour *in vitro* under short-circuit conditions (Figure 10), and about 100 μeq/kg bodyweight $\times$ hr *in vivo*. Under open-circuit conditions a transepithelial lumen-negative electrical potential difference (PD) of 40–60 mV develops *in vitro*, and similar differences from lumen to plasma can be measured *in vivo*. The PD drives passive Cl^- transport *in vivo*, so the net result is absorption of near-neutral NaCl, whereas there is no net Cl^- transport under short-circuit conditions *in vitro*. The Na^+ transport is completely suppressed by NaCl loading; the PD may be reversed due to a small, persisting K^+ and H^+ secretion. The cellular mechanism behind the huge change in sodium absorption is a large increase in the Na^+ permeability of the apical membrane of the enterocytes (Bindslev, 1979; Clauss *et al.*, 1987). The Na^+ permeability is characterized by a single channel conductance of 4 pico-Siemens (Christensen and Bindslev, 1982). The Na^+ transport occurs through a uniform layer of cells with near-identical transport pattern (Holtug *et al.*, 1991). There seems to be little adaptation of the Na-pump located at the basolateral membrane, as the concentration of ouabain-inhibitable Na/K-stimulated ATPase or the K-paranitrophenylphosphatase (Mayhew *et al.*, 1992) did not increase after NaCl depletion. Neither did the histochemical location of the latter differ (Mayhew *et al.*, 1992). In coprodeum no transport of nutrients (glucose/amino acids) is measurable and neither can a solute-linked water flow be detected *in vivo* (Rice and Skadhauge, 1982a) or *in vitro* (Bindslev, 1981).

Colon from birds (domestic fowl and galah) on a low-NaCl diet has a Na-absorbing capacity and other transport characteristics just like coprodeum, particularly no stimulation of Na transport induced by hexoses or amino acids. The colonic epithelium is more "leaky"

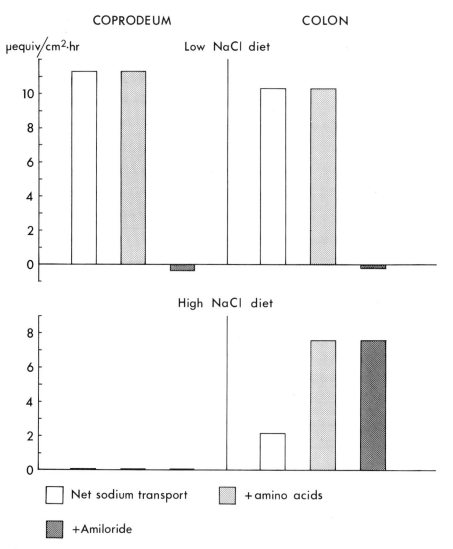

FIGURE 10 Effect of dietary NaCl on transport properties of coprodeum and colon in domestic fowl. Illustrated are Na^+ transport rates in the absence of luminal organic substrates (open bars), in the presence of luminal leucine and lysine (light stippling), and in the presence of the sodium channel blocker amiloride (dark stippling).

(80–100 ohm $\times$ cm^2) than coprodeum and consequently has a transmural PD of only 15–20 mV *in vitro* and *in vivo*. Concomitant with this, a solute-linked water absorption can be measured of 150 μl H$_2$O/kg bodyweight $\times$ hr *in vivo* (Rice and Skadhauge, 1982a) and 16 μl/cm^2 hr *in vitro* (Bindslev, 1981).

On a high-NaCl intake the colonic transport parameters are remarkably different from those of coprodeum (Figure 9): rather than being suppressed, Na^+ absorption continues at 5–10 microeq/cm^2 hr, but only when glucose and amino acids are present on the luminal side (Lind *et al.*, 1980a; Clauss *et al.*, 1991; Rice and Skadhauge, 1982b). These nonelectrolytes are now cotransported with Na^+ (Lind *et al.*, 1980b). This Na^+ transport is not affected by the blocker of apical Na^+ channels amiloride, which totally suppresses Na^+ absorption in both coprodeum and colon from NaCl-depleted birds. NaCl-loading thus induces in the avian colon a switchover from a Na^+ channel to Na^+/nutrient cotransport. The latter is half-saturated by glucose and amino acid (lysine or leucine) concentrations less than 2 mM; the Na^+ absorption will therefore be activated whenever there is chyme in the colon.

b. Transport of Other Ions

K^+ is secreted into coprodeum and colon, as observed in the domestic fowl and the emu *in vivo* and in the domestic fowl *in vitro* (see Skadhauge, 1981). Only at high intraluminal concentrations of K^+ will this ion be absorbed. At concentrations of K^+ typically found in ureteral urine the net secretion of K$^+$ from coprodeum and colon is around 10 μmol/kg body weight × hr (Skadhauge and Thomas, 1979). As measured in the Ussing chamber (see survey by Skadhauge, 1993, pp.73–74), the net secretion of K$^+$ in coprodeum and colon is usually around 0.5 mmol/cm^2 hr, less than 5% of the concomitant Na$^+$ absorption.

Both ammonium and phosphate ions are absorbed from the coprodeum and colon *in vivo* in the domestic fowl (Skadhauge and Thomas, 1979) and in the emu (Dawson *et al.*, 1985). In both species higher absorption rates were observed at higher luminal concentrations. H$^+$ is secreted, presumably by an apical H$^+$/Na$^+$ exchanger (see Skadhauge, 1993). This secretion leads to an acid microclimate on the mucosal surface (Holtug *et al.*, 1992), though this is also influenced by products of enterocyte metabolism (Laverty *et al.*, 1994). The acid microclimate aids in colonic absorption of SCFAs, particularly propionate (Holtug *et al.*, 1992).

c. Quantitative Role of Coprodeum versus Colon

When the absorptive areas and the local transport rates of the two organs are compared, the colon has in all species examined a much higher total capacity. The colon therefore functions as the "workhorse" (Thomas, 1982) for recovery of salt and water from ureteral urine and chyme, whereas coprodeum functions as a final "tuner" of total excretion.

2. Dietary and Hormonal Regulation of Coprodeal and Colonic Transport

Both the amount of NaCl in the diet, which turns on the low/high switch-over of Na$^+$ transport, and the time sequences of the change-over have been elucidated in the domestic fowl. No other dietary components have been identified which have a measurable influence on NaCl transport.

Árnason and Skadhauge (1991) exposed domestic fowls to six levels of NaCl intake ranging from 0.25 to 25 mmol/kg bodyweight day, and observed the plasma levels of aldosterone, corticosterone, prolactin, and arginine vasotocin, and the sodium absorption and induced chloride secretion of colon and coprodeum. The low levels of NaCl intake resulted in increasing levels of plasma aldosterone, and colon and coprodeum absorbed sodium by an amiloride-suppressible mechanism. The highest levels of NaCl intake increased the concentration of prolactin and arginine vasotocin in plasma. The intake range of 3–6 mmol NaCl/kg body mass/day presented the least osmoregulatory stress (i.e., being sensed neither as NaCl depletion nor as a NaCl load). The study points clearly to aldosterone as the main inducer of the high rate of Na$^+$ transport in coprodeum.

Studies following the temporal development of Na$^+$ depletion have shown Na$^+$ transport and aldosterone levels to go up to a new steady state over a week (Thomas and Skadhauge, 1982; Árnason *et al.*, 1986). Following acute resalination (NaCl loading) plasma aldosterone is suppressed in 8 hr, and net Na$^+$ absorption in 24 hr (Skadhauge *et al.*, 1983; Árnason *et al.*, 1986). The immediate conclusion from these findings is that absence of aldosterone brings about rapid closure of the functional apical Na$^+$ channels.

Studies with external aldosterone injections reveal, however, that the hormonal (and dietary) regulation of coprodeal Na$^+$ transport is more complicated. Supramaximal injections of aldosterone over 24 hr, the time normally needed to cause a full switch-over of an epithelium to maximal Na$^+$ transport rate, only induced 10–30% of the net Na$^+$ transport of that caused by full adaptation to a low-NaCl diet (Clauss *et al.*, 1984, 1987). If, however, the aldosterone was injected during the first 24 hr of acute resalination, the sodium absorption was maintained at the high rate characteristic of low-NaCl adapted birds. This effect declined over 3–5 days, when the 24-hr aldosterone injection was postponed later and later, to the moderate level of Na$^+$ transport induced by aldosterone in birds on a chronic high-NaCl intake (Clauss *et al.*, 1984). The reason for this apparent change in sensitivity to aldosterone is a structural change induced by the hormone: increase in number of both enterocytes and apical microvilli on the individual cell (see next section). This enlarges the apical area by 400%. The apparent change in aldosterone sensitivity was not caused by recruitment of aldosterone receptors (Sandor *et al.*, 1989).

The colon of the domestic fowl does not exhibit these delayed transport changes (Clauss *et al.*, 1984, 1991), but this organ does not change apical microvilli with diet. On the other hand, the difference in histologic appearance of colon from salt-marsh versus upland savannah sparrows (Goldstein *et al.*, 1990) is indeed similar to that of coprodeum from high- versus low-salt chickens.

3. Ultrastructural Adaptation and Molecular Induction

Continued NaCl depletion induces in white Plymouth Rock chickens a very late effect: a structural change of the individual epithelial cells (Figure 11) as well as of the entire organ. This was investigated by tandem measurements of coprodeal ultrastructure and sodium transport (amiloride-suppressible short-circuit current) *in vivo* (Mayhew *et al.*, 1990). The high aldosterone/low NaCl diet involves modulations of the level of mRNA that codes either for the Na-channel or a posttranscrip-

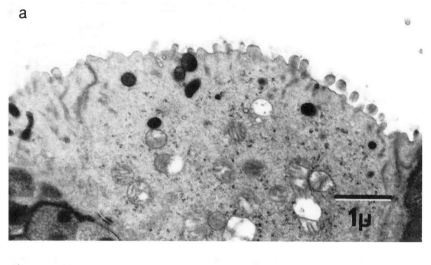

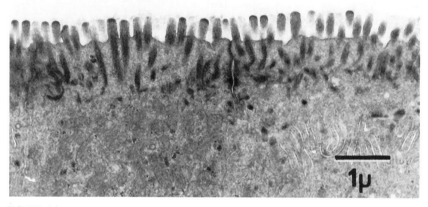

FIGURE 11 Tranmission electron micrographs of the apical region of coprodeal epithelial cells from chickens eating either (a) a high-NaCl diet or (b) a low-NaCl diet. Note the differences in height and density of apical microvilli, providing the low-NaCl birds with a substantially enhanced absorptive surface area. Photograph courtesy of Dr. Vibeke Elbrønd.

tional regulator of the channel, since the mRNA from low-NaCl but not high-NaCl diet coprodeum expresses an amiloride-suppressible Na^+ uptake after injection into *Xenopus* oocytes (Asher *et al.*, 1992). The microvillous surface area increased in birds chronically exposed to the low-NaCl diet, from 32 to 49 μm^2/cell without changing cell size (Figure 11), and the total cell number per organ increased from 270 to 420 million. The total effect was a 135% rise in apical area (Elbrønd *et al.*, 1991; Mayhew *et al.*, 1992). Supramaximal doses of aldosterone injected over 2–6 days also induced a substantial increase of the microvillous surface area (Elbrønd *et al.*, 1993). This enlargement of apical area occurred more rapidly than the rate of cell turnover and hence involved remodeling of the existing cell population (Elbrønd *et al.*, 1993). The late effect of aldosterone to increase cell number then leads to synthesis of cells with near-identical sodium transport capacity, as demonstrated by uniform histochemical localization of Na/K ATPase activity (Mayhew *et al.*, 1992), by immunochemical localization of the amiloride-sensitive Na^+ channel (Smith *et al.*, 1993) and by measurements of the local Na^+ transport rates by the vibrating microprobe technique (Holtug *et al.*, 1991). The basolateral membrane area did not change significantly per individual cell following chronic NaCl depletion (Mayhew *et al.*, 1992).

4. Salt and Water Transport in the Caeca

The paired caeca are of variable size; large caeca are typically associated with herbivory and granivory (McLelland, 1989). *In vivo* perfusion studies in the domestic fowl demonstrate substantial absorption of NaCl and water and secretion of K^+. The Na^+ absorption is stimulated by acetate; NaCl absorption is enhanced by luminal glucose and stimulated by aldosterone (Rice and Skadhauge, 1982c; Thomas and Skadhauge, 1989a). Anatomically the caecum in the chicken has a large end (the bulbus) and a more narrow neck region (see Dantzer, 1989). The bulbus had higher transport rates of both Na^+ and amino acids than the neck region (Thomas and Skadhauge, 1989b).

C. Postrenal Modification of Ureteral Urine

1. Basic Patterns: Hydration/NaCl Loading

The interplay of kidney and cloaca depends on flow rate and ionic and osmotic concentration of ureteral urine in relation to the transport capacities of the epithelia of the lower gut. The osmotic permeability coefficient and the sodium transport rate are most important, since these parameters determine the net rates of absorption/secretion in the parts of the lower gut in which urine is stored (or flows through). One would expect the fractional effect of cloacal storage to be limited if the flow rate of ureteral urine is high. This would apply both to water reabsorption in the hydrated state and for NaCl absorption during NaCl loading. In the domestic fowl (see Skadhauge, 1973), consideration of all relevant parameters, including the resulting osmolality and Na^+ concentration of the contents of the lower gut, permits calculation that only about 2% of ureterally excreted water and NaCl will be reabsorbed in the hydrated and salt-loaded state, respectively. Cloacal storage little modifies the excretory function of the kidney in these functional states. In contrast, substantial modification of urine (particularly reabsorption of NaCl) occurred in the starling even during brisk urine flow induced by mannitol diuresis (Laverty and Wideman 1989); the transport properties of the lower intestine of this species are unstudied.

2. Basic Patterns: Dehydration/NaCl Depletion

Quantitative studies of flow rate and composition of ureteral urine of dehydrated birds, and the transport parameters of coprodeum and colon, have been carried out in three species of seed-eating land birds, the domestic fowl, the galah, and the house sparrow (Goldstein and Braun, 1986, 1988). The values obtained on the fowl and the galah have been treated by a computer simulation of the retrograde flow into the cloaca and the integrated effect on salt and water transport (Skadhauge and Kristensen, 1972; Skadhauge, 1973). The result obtained is that 10–20% of ureterally excreted water and 70% of Na^+ is absorbed in the cloaca of both species. This occurs in the galah despite its higher urine osmolality because of a counteracting higher rate of solute-linked water flow and lower osmotic permeability coefficient (Skadhauge, 1981). The house sparrow (Goldstein and Braun, 1988) and the budgerigar (Krag and Skadhauge, 1972) seem to have a similar interaction of cloacal recovery of NaCl and water, though in the house sparrow the transport properties of the colon more closely parallel those of hen coprodeum (Goldstein, 1993a).

The relevance of the computer calculations has received independent support from two sources: (1) slow perfusion experiments of coprodeum/colon in the domestic fowl with fluids simulating ureteral urine in the dehydrated state have demonstrated net water absorption from liquids up to 200 mosm higher osmolality than that of plasma, (2) the computer simulation (Skadhauge, 1973) resulted in near-identical absorption rates as calculated from a detailed study with and without exteriorization of the ureters (Dicker and Haslam, 1972).

In only one study (Skadhauge and Thomas, 1979) has the cloacal fractional absorption of K^+, NH_4^+, and PO_4 been quantitatively assessed. In the domestic fowl it was shown that the cloaca changes the ureteral output by an additional secretion of 20% of K^+, and absorption of 8% of NH_4^+ and 2% of PO_4. Thus the cloaca seems to assist significantly in secretion of K^+, but the storage affects NH_4^+ and PO_4 transport only little.

3. Special Case: Birds with Salt Glands

Schmidt-Nielsen *et al.* (1963) suggested that birds with salt glands might maintain high rates of intestinal salt and water reabsorption during times of substantial NaCl intake, routing the NaCl to the salt glands for excretion and thereby retrieving free water. Examinations of this problem have focused on two species, the domestic duck and the glaucous-winged gull. In both species, Na^+ uptake in the small intestine remained undiminished during saline acclimation (Hughes and Roberts, 1988). In the hindguts of the two species, the colon carries the majority of NaCl absorption and solute-linked water flow, with coprodeum being of minor size and with lower transport capacity per square centimeter of serosal area. Even during salt loading, continued urine flow is necessary to maintain nitrogen excretion, and at least in ducks urine apparently continues to enter the rectum and even the ceca in saline acclimated birds (Hughes and Raveendran, 1994). Colons of both species uphold high NaCl absorption rates in the salt-loaded state (Skadhauge *et al.*, 1984a; Goldstein *et al.*, 1986); any NaCl that is excreted by the kidney may therefore be reabsorbed with water postrenally and finally secreted by the salt gland, thereby leaving a volume of "free water." This system operates in the duck to save approximately 0.5 ml water/kg body mass/hr (Skadhauge *et al.,* 1984a) but, because of the gull's small lower intestine and relativley low urinary NaCl concentrations, its contribution is likely minor in this species (Goldstein *et al.,* 1986).

4. Special Case: The Ratites

In the large, flightless ratite birds the weight of the gut and its contents is not as critical as in flying species, and special adaptation of the lower gut to act as fermentation chamber is possible. This is the case in the ostrich (Skadhauge *et al.,* 1984b) in which the more than 10 meter long colon of the adult male has SCFA concentrations up to 120 mmol. To a minor extent this is parallelled by the *Rhea americana* (the nandu). As both ostrich (Skadhauge *et al.,* 1984b) and nandu (Skadhauge *et al.,* 1993) have renal concentrating ability of approximately 800 mmol/kg, the question of how far this ureteral urine is passed retrograde into the coprodeum/colon is of importance. The ostrich has a strong sphincter between coprodeum and terminal colon, and no influx of urine into colon has been observed (Skadhauge, 1983). No feces are stored in the coprodeum, which therefore functions as a real bladder in this species. The transport properties of the coprodeal wall have not been measured in the ostrich. The dorsal diverticulum from the proctodeum of the nandu does contain ureteral urine and is lined by a single layer of cuboidal cells thus possibly having resorptive capacity (Skadhauge *et al.,* 1993). However, the small size of this diverticulum relegates its possible functional role to minor importance.

In contrast to these species, the emu has a conventional, small coprodeum/colon where little fermentation occurs. The dehydrated emu has a low urine-to-plasma ratio of osmolality (1.4) and a fairly high urine flow rate. To compensate, the absorption capacity of coprodeum/colon is high, resulting from extensive mucosal folding (Skadhauge *et al.,* 1991). The cassowary (*Casuarius casuarius*) has a macroscopic anatomy of coprodeum/colon simlilar to that of the emu (Skadhauge *et al.,* 1993).

5. Quantitative Role of the Caeca in Osmoregulation

The main function of caeca in avian physiology is nutritional, as these organs, when present and large, take part in the digestion of fine particulate matter, food fiber, and the production of SCFA (see reviews in Braun and Duke, 1989). Furthermore, saving of nitrogen by bacterial breakdown of uric acid of ureteral origin is possible (Braun, 1993). Large fractions of intestinal and ureteral water and solutes move together with solids of small particle size into the caeca by peristalsis and antiperistalsis in the ileum and colon, respectively (see Skadhauge, 1981; Chapter 3). Quantitative studies of caecal function surveyed by Skadhauge (1981) based on inflow studies of unabsorbable water markers from either ureteral urine or after oral ingestion suggest inflow rates about 20% from both sources. In the domestic fowl the simple (serosal) surface area of the ceca is of equal magnitude to that of coprodeum and colon, but the Na^+ and water transport rates per unit body mass are approximately threefold higher (Thomas and Skadhauge, 1989b). The transport capacity for NaCl and water reabsorption from chymus and urine is therefore large, perhaps larger than colon plus coprodeum. The quantitative rate of caecal water absorption has been estimated from 5 to 38% of body mass per day. Despite this, ligation of the ceca may have little overall effect on osmoregulation (Skadhauge, 1981, p. 38, and others (see Braun and Duke 1989)), probably because of compensation by other organs (Hughes *et al.,* 1992). The contribution of the ceca may become essential only during the combined stress of poor feeding and salt and water depletion (Thomas, 1982; Thomas and Skadhauge, 1989b).

VII. SALT GLANDS

Although the presence of supraorbital glands in marine birds was known for many years, it was not until 1958 that their salt-secreting function was defined (Schmidt-Nielsen *et al.,* 1958). Functional salt glands exist in approximately 10 orders of birds, including all those with marine representatives and at least two orders of terrestrial birds (several falconiforms (Cade and Greenwald, 1966) and the roadrunner (*Geococcyx californianus,* order Cuculiformes; Ohmart, 1972)). Reports of salt gland function in two other terrestrial orders have recently been disputed (Thomas *et al.,* 1982; Brown and Gray, 1994). Within related groups, glands are larger in species exposed to hypersaline food and water; thus, marine species have larger glands than coastal or terrestrial birds (Staaland, 1967).

A. Anatomy

The salt glands, which constitute from 0.1 to 2% of body mass, are located in depressions usually in or above the orbits (e.g., Schmidt-Nielsen, 1959; Siegel-Causey, 1990; Figure 12). They are distinct from lacrimal or Harderian glands. The basic structural unit of the salt gland, the lobe, consists of blind-ended secretory acini (tubules) which drain into central ducts (canals). Toward the blind end of the tubule are small, relatively undifferentiated peripheral cells, whereas most of the length of the secretory tubules is composed of principal

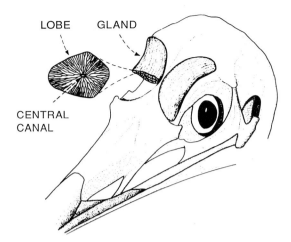

FIGURE 12 The supraorbital salt gland of the herring gull. Branching secretory tubules are arranged radially around the central canal. (From K. Schmidt-Nielsen, The salt-secreting gland of marine birds, *Circ.* (1960), **XXI,** 962. Reproduced with permission. *Circulation.* Copyright (1960) American Heart Association.)

secretory cells (Ernst and Ellis, 1969). Capillary blood flow runs counter to flow in the secretory tubules. Each gland contains numerous lobes whose final common ducts drain into the nasal cavity. Salt gland secretion either drips from or is shaken from the bird's beak.

Acclimation of a bird to increasing salt intake induces growth of the salt gland, accompanied by an enhanced secretory capacity. Growth of the secretory tubules includes cell hypertrophy and hyperplasia, increased membrane folding and mitochondrial density, and enhanced activity of several cellular enzymes, including Na/K ATPase, the enzyme responsible for establishing the ion gradients that drive secretion (Holmes and Phillips, 1985).

The main innervation to the salt glands is from the VIIth cranial nerve, a branch of which enters the secretory nerve ganglion near the orbit (Ash *et al.,* 1969). Postganglionic parasympathetic fibers, which may release both acetylcholine and vasoactive intestinal peptide (VIP), then ramify through the gland, with terminals on both secretory cells and blood vessels. The blood vessels and tubules also receive adrenergic input from neurons likely running both in the walls of arteries and through the secretory nerve ganglion (Peaker and Linzell, 1975; Gerstberger, 1991).

B. Function

1. Stimulus for Secretion

Secretion by the salt glands may be initiated by stimulation of either central or peripheral receptors. The central receptors appear to be osmoreceptors in the region of the third cerebral ventricle, with a responsiveness to the Na^+ (or other cation) concentration, not just to total osmotic concentration (Gerstberger *et al.,* 1984).

The peripheral receptors, which are located in the vicinity of the heart and large arteries nearby, communicate with the central nervous system via the vagus nerve, the severing of which eliminates salt gland responsiveness to systemic salt-loading (Hanwell *et al.,* 1972). These receptors act as osmoreceptors, as indicated by the inability of isosmotic expansion of the extracellular fluid (ECF) to initiate salt gland secretion, at least in some experiments (see Hughes, 1989b). There are also reports of salt gland secretion during dehydration (e.g., Stewart, 1972), when ECF volume would have been reduced but osmolarity increased.

Numerous experiments also point to an important role for systemic volume receptors. In some species (mallard ducks, but not glaucous-winged gulls; Hughes, 1989a) isosmotic ECF expansion is a sufficient stimulus for salt gland secretion. Moreover, a rise in extracellular Na^+ concentration without a rise in ECFV may be insufficient to initiate secretion (Ruch and Hughes, 1975; Hughes, 1989b); volume and osmolarity act in concert to jointly achieve a threshold for secretion (Hammel, 1989). Some evidence suggests that the ECF receptors occur not in the vasculature, but in the interstitial fluid (Hammel *et al.,* 1980).

It is likely that variation in experimental results, including differing importance of volume and osmolarity as stimuli for salt gland secretion, relates to variation among species. The most detailed experimental work in this field has been done using domestic ducks, descendants of wild mallards. These birds are never exposed to seawater in their natural habitat and they are quite intolerant of saline (Schmidt-Nielsen and Kim, 1964). Studies of birds more highly adapted to saline may yield different results (Hughes, 1989b).

2. Secretion Mechanism and Fluid Composition

The fluid produced by avian salt glands is typically nearly pure NaCl, with trace concentrations of K^+, HCO_3^-, Ca^{2+}, and Mg^{2+} (though K^+ concentrations up to 100 mM and Ca^{2+} concentrations up to 50 mM may occur in some situations; Hughes, 1970b; Cade and Greenwald, 1966). The concentration of NaCl in the secretion varies among species from approximately 500 to 1000 mM (see Skadhauge, 1981; Chapter 7), being more concentrated in species with higher salt intakes (such as those which eat marine invertebrates).

The cellular mechanism of secretion by salt gland principal cells is difficult to study *in situ* because of the complex anatomy of the tubules and their encasement in connective tissue. Knowledge of this process has therefore come from a variety of *in vitro* approaches, including studies of tissue slices, isolated dispersed cells (Shuttleworth and Thompson, 1989), and, more recently, cell culture (Lowy et al., 1989). These studies collectively have suggested a model of secondary active Cl⁻ transport. In this model (Figure 13), a basolateral ouabain-sensitive Na/K ATPase establishes an inwardly directed Na$^+$ gradient. This gradient provides the energy for furosemide-sensitive inward transport of Na$^+$/K$^+$/2Cl$^-$, also across the basolateral membrane. Chloride ions then pass down their electrochemical gradient through (putative) apical Cl⁻ channels into the tubule lumen. The resulting transtubule electrical potential (inwardly negative) provides the driving force for passive peritubular Na$^+$ transport. Activation of secretion also entails a brief intracellular acidification which is compensated for by a HCO$_3^-$-dependent alkalinization as secretion is sustained (Shuttleworth and Wood, 1992). Activation of secretion can be stimulated via either of two intracellular second messenger systems, phosphatidyl inositol/Ca$^+$ (linked to stimulation of muscarinic acetylcholine receptors (Snider et al., 1986; Hildebrandt and Shuttleworth, 1993)) or cyclic AMP (linked to β-adrenergic stimulation or vasoactive intestinal peptide (Lowy and Ernst, 1987; Lowy et al., 1987)).

The concentration of the solution as elaborated by the cells lining the secretory tubules is unknown, and a variety of mechanisms have been proposed for attaining the final concentration of secreted fluid. Marshall et al. (1985) proposed that tubular cells secrete an isotonic fluid that is concentrated by the subsequent withdrawal of water along the ducts. With a final secretion rate as high as 10% of gland blood flow rate, secretion of virtually the entire plasma would be initially required. Other theories (see Holmes and Phillips, 1985; Butler et al., 1989, Gerstberger and Gray, 1993) suggest instead that the hypertonicity is generated within the tubules themselves. In various studies, the concentration of secreted fluid either increases, decreases, or does not change with the rate of secretion, and so this relationship does little to help clarify the concentrating mechanism.

Adequate secretion by the salt glands requires adequate blood flow, and this increases more than 10-fold during active secretion (Hanwell et al., 1971; Kaul et al., 1983). Over a wide range of blood flow and secretion rates the salt glands extract a relatively constant fraction (between 15 and 20% in the Pekin duck) of the NaCl passing through the gland (Kaul et al., 1983).

3. Regulatory Mediators

Stimulation of salt gland secretion is activated by the parasympathetic nervous system (Fänge et al., 1958) acting through muscarinic receptors for acetylcholine and independent receptors for VIP (Gerstberger, 1988). These mediators activate the principle cell secretory mechanism as well as vasodilation and enhanced blood flow (see Butler et al., 1989; Gerstberger, 1991). Secretion is inhibited by anaesthesia or disturbance, at least in part via sympathetically mediated vasoconstriction.

The only hormone that directly stimulates secretion is atrial natriuretic peptide, which activates the gland via high affinity receptors (Schütz and Gerstberger, 1990; Gerstberger, 1991). A substantial literature did suggest a role for corticosterone. However, though adrenalectomy may reduce salt gland secretion, this apparently results from the cardiovascular effects of NaCl and water depletion. Maintenance of normal fluid balance through saline drinking water and adequate feeding restores salt gland function even in the absence of corticosterone (Butler, 1987).

Similarly, angiotensin II has been thought to have a direct influence on salt gland secretion, though in this case inhibitory. More recently, Butler et al. (1989) have demonstrated that the effect is not a peripheral one but instead acts by stimulating centrally mediated vasoconstriction of salt gland blood vessels, probably via sympathetic release of norepinephrine. Consistent with this, secretion is inhibited by low doses of angiotensin II administered to the cerebrospinal fluid of the third ventricle (Gerstberger et al., 1984). Peripherally produced angiotensin may act by crossing the blood–brain barrier and activating these central responses.

C. Contribution of the Salt Glands to Osmoregulation

The quantitative contribution of the salt glands relative to the kidneys has been evaluated in a few species, and Skadhauge (1981, Chapter 7) has summarized the

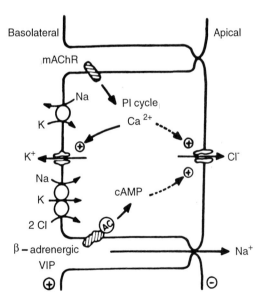

FIGURE 13 Proposed mechanisms of secretion by principal secretory cells in the avian salt gland. (From Lowy et al. (1989), Am. J. Physiol. with permission.)

major points that derive from these studies. First, during conditions that stimulate the salt glands (salt loading or dehydration), typically ≥75% of excreted Na^+ is lost via the salt glands. Second, the division of water loss between salt glands and kidneys depends of the water load; during a copious saline load the two organs may be equally responsible for water excretion, whereas during dehydration the renal losses of water may be much more reduced. Last, despite the relatively low concentration of K^+ in the salt gland fluid, the salt glands may eliminate more than one-third of excreted K^+ following a salt load. It is worth noting that these generalizations derive from relatively few studies conducted over short experimental time spans; the relative roles of the organs under natural conditions undoubtedly reflect dietary intakes of salts and water and must vary considerably over time. For example, the spontaneous salt gland secretions of Pekin ducks varied with, and nearly matched, the NaCl concentration of their drinking water (Hughes *et al.*, 1992).

VIII. EVAPORATIVE WATER LOSS

Evaporative losses of water from the respiratory tract and the skin constitute an important component of avian water balance. Respiratory water loss is an inevitable consequence of the high rates of ventilation demanded by endothermy. It has long been appreciated that birds modulate respiratory evaporation (e.g., by panting) as a means of defending body temperature in the heat, and most research into regulation of rates of evaporation has been conducted within this context. In contrast, there is little evidence as to whether respiratory evaporation is regulated as a component of osmoregulation (e.g, whether respiratory frequencies or volumes are altered in dehydrated birds (Crawford and Schmidt-Nielsen, 1967)). For a discussion of the role of evaporative water loss in thermoregulation, please see Chapter 14.

Despite birds' lack of sweat glands, a substantial fraction (often more than 50%, and sometimes above 80%; summarized in Dawson, 1982) of total evaporation occurs across the skin. It is now clear that the rate of cutaneous evaporation is regulated. In a number of species, the cutaneous resistance to diffusion of water vapor decreases markedly at elevated ambient temperatures, thereby enhancing evaporative heat loss and contributing to thermoregulation (e.g., Marder and Ben-Asher, 1983; Webster and Bernstein, 1987; Withers and Williams, 1990). Decreased cutaneous resistance is partly due to adrenergically increased cutaneous blood flow (Marder and Raber, 1989); other mechanisms, such as changing structure of skin lipids, may be involved, but their contributions are unresolved. Osmoregulatory status may influence regulation of cutaneous evaporation, since cutaneous resistance was decreased in hydrated but not in dehydrated heat-stressed pigeons (Arad *et al.*, 1987).

Evaporation typically accounts for approximately 50% of total water losses in a variety of species given unlimited water in the laboratory (summarized in Skadhauge, 1981; Chapter 4). During dehydration, birds greatly decrease excretory water loss, thus evaporation constitutes a greater fraction (up to 80%) of total water loss (Skadhauge, 1981).

Acknowledgments

Thanks to Maryanne Hughes and Eldon Braun for reading and commenting on portions of this manuscript and to Eldon Braun and Vibeke Elbrond for providing unpublished photographs. E. Skadhauge's contribution to this chapter was supported by the Desirée and Niels Yde Foundation.

References

Akester, A. R. (1967). Renal portal shunts in the kidney of the domestic fowl. *J. Anat.* **101,** 569–594.

Ames, E., Steven, K., and Skadhauge, E. (1971). Effects of arginine vasotocin on renal excretion of Na^+, K^+, Cl^-, and urea in the hydrated chicken. *Am. J. Physiol.* **221,** 1223–1228.

Anderson, G. L., and Braun, E. J. (1985). Postrenal modification of urine in birds. *Am. J. Physiol.* **248,** R93–R98.

Anderson, R. S. (1967). Acid–base changes in the excreta of the laying hen. *Vet. Rec.* **30,** 314–315.

Arad, Z., Chadwick, A., Rice, G. E., and Skadhauge, E. (1986). Osmotic stimuli and NaCl-intake in the fowl: Release of arginine vasotocin and prolactin. *J. Comp. Physiol. B* **156,** 399–406.

Arad, Z., Gavrieli-Levin, I., Eylath, U., and Marder, J. (1987). Effect of dehydration on cutaneous water evaporation in heat-exposed pigeons (*Columba livia*). *Physiol. Zool.* **60**(6), 623–630.

Árnason, S. S., and Skadhauge, E. (1991). Steady-state sodium absorption and chloride secretion of colon and coprodeum, and plasma levels of osmoregulatory hormones in hens in relation to sodium intake. *J.Comp.Physiol. B* **161,** 1–14.

Árnason, S. S., Rice, G. E., Chadwick, A., and Skadhauge, E. (1986). Plasma levels of arginine vasotocin, prolactin, aldosterone and corticosterone during prolonged dehydration in the domestic fowl: Effect of dietary NaCl. *J.Comp.Physiol. B* **156,** 383–397.

Ash, R. W., Pearce, J. W., and Silver, A. (1969). An investigation of the nerve supply to the salt gland of the duck. *Quart. J. Exp. Physiol.* **54,** 281–295.

Asher, C., Singer, D., Eren, R., Yeger, O., Dascal, N., and Garty, H. (1992). NaCl-dependent expression of amiloride-blockable Na^+ channel in *Xenopus* oocytes. *Am. J. Physiol.* **262,** G244–G248.

Bartholomew, G. A. (1972). The water economy of seed-eating birds that survive without drinking. *Proc. XVth Int. Ornithol. Congr.* 237–254.

Bartholomew, G. A., and Cade, T. J. (1963). The water economy of land birds. *Auk* **80,** 504–539.

Belay, T., Bartels, K. E., Wiernusz, C. J., and Teeter, R. G. (1993). A detailed colostomy procedure and its application to quantify water and nitrogen balance and urine contribution to thermobalance in broilers exposed to thermoneutral and heat-distressed environments. *Poult. Sci.* **72,** 106–115.

Bindslev, N. (1979). Sodium transport in the hen lower intestine. Induction of sodium sites in the brush border by a low sodium diet. *J. Physiol.* **288,** 449–466.

Bindslev, N. (1981). Water and NaCl transport in the hen lower intestine during dehydration. *In* "Water Transport Across Epithelia" (H. H. Ussing, N. Bindslev, N. A. Lassen, and O. Sten-Knudsen, eds.), Alfred Benzon Symposium 15, pp. 468–481. Munkdsgaard, Copenhagen.

Boykin, S. L. B., and Braun, E. J. (1993). Entry of nephrons into the collecting duct network of the avian kidney: A comparison of chickens and desert quail. *J. Morphol.* **216,** 259–269.

Braun, E. J. (1978). Renal response of the starling (*Sturnus vulgaris*) to an intravenous salt load. *Am. J. Physiol.* **231,** 1111–1118.

Braun, E. J. (1993). Renal function in birds. *In* "New Insights in Vertebrate Kidney Function" (J. A. Brown, R. J. Balment, and J. C. Rankin, eds.), pp. 167–188. Cambridge Univ. Press, Cambridge.

Braun, E. J., and Dantzler, W. H. (1972). Function of mammalian-type and reptilian type nephrons in kidney of desert quail. *Am. J. Physiol.* **222**(3), 617–629.

Braun, E. J., and Dantzler, W. H. (1974). Effects of ADH on single nephron glomerular filtration rates in the avian kidney. *Am. J. Physiol.* **226,** 1–12.

Braun, E. J., and Dantzler, W. H. (1987). Mechanisms of hormone actions on renal function. *In* "Vertebrate Endocrinology: Fundamentals and Biomedical Implications" (P. K. T. Pang and M. P. Schreibman, eds.), Vol. 2, pp. 189–210. Academic Press, San Diego, CA.

Braun, E. J., and G. E. Duke, eds. (1989). Function of the avian cecum. *J. Exp. Biol.* **3,**(Suppl.), 1–130.

Braun, E. J., and Reimer, P. R. (1988). Structure of avian loop of Henle as related to countercurrent multiplier system. *Am. J. Physiol.* **255,** F500–F512.

Brokl, O. H., Braun, E. J., and Dantzler, W. H. (1994). Transport of PAH, urate, TEA, and fluid by isolated perfused and nonperfused avian renal proximal tubules. *Am. J. Physiol.* **266,** R1085–R1094.

Brown, J. A., Balment, R. J., and Rankin, J. C. (1993). "New Insights in Vertebrate Kidney Function." Cambridge Unviersity Press, Cambridge.

Brown, C. R., and Gray, D. A. (1994). Ostrich salt glands are not stimulated by increase in plasma osmolality. *J. Ornithol.* **135,** 76.

Burrows, M. E., Braun, E. J., and Duckles, S. P. (1983). Avian renal portal valve: A reexamination of its innervation. *Am. J. Physiol.* **245,** H629–H634.

Butler, D. G. (1987). Adrenalectomy fails to block salt gland secretion in Pekin ducks (*Anas playrhynchos*) preadapted to 0.9% saline drinking water. *Gen. Comp. Endocrinol.* **16,** 171–181.

Butler, D. G., Siwanowicz, H., and Puskas, D. (1989). *In* "Progress in Avian Osmoregulation" (M. R. Hughes and A. Chadwick, eds.), pp. 127–141. Leeds Philosophical and Litararary Society, Leeds, United Kingdom.

Cade, T. J., and Greenwald, L. (1966). Nasal salt secretion in falconiform birds. *Condor* **68,** 338–350.

Casotti, G., and Richardson, K. C. (1992). A stereological analysis of kidney structure of honeyeater birds (*Meliphagidae*) inhabiting either arid or wet environments. *J. Anat.* **180,** 281–288.

Casotti, G., and Richardson, K. C. (1993). A qualitative analysis of the kidney structure of Meliphagid honeyeaters from wet and arid environments. *J. Anat.* **182,** 239–247.

Chin, T. Y., and Quebbemann, A. J. (1978). Quantitation of renal uric acid synthesis in the chicken. *Am. J. Physiol.* **234,** F446.

Choshniak, I., Munck, B. G., and Skadhauge, E. (1977). Sodium chloride transport across the chicken coprodeum: Basic characteristics and dependence on sodium chloride intake. *J. Physiol.* **271,** 489–504.

Christensen, O., and Bindslev, N. (1982). Fluctuation analysis of short-circuit current in a warm-blooded sodium-retaining epithelium: Site current, density, and interaction with triamterene. *J. Membrane Biol.* **65,** 19–30.

Clark, N. B., and Mok, L. L. S. (1986). Renal excretion in gull chicks: Effect of parathyroid hormone and calcium loading. *Am. J. Physiol.* **250,** R41–R50.

Clark, N. B., and Sasayama, Y. (1981). The role of parathyroid hormone on renal excretion of calcium and phosphate in the Japanese quail. *Gen. Comp. Endocrinol.* **45,** 234–241.

Clark, N. B., and Wideman, R. F. (1977). Renal excretion of phosphate and calcium in parathyroidectomized starlings. *Am. J. Physiol.* **233,** F138–F144.

Clark, N. B., and Wideman, R. F., Jr. (1980). Calcitonin stimulation of urine flow and sodium excretion in the starling. *Am. J. Physiol.* **238,** R406–R412.

Clark, N. B., and Wideman, R. F., Jr. (1989). Actions of parathyroid hormone and calcitonin in avian osmoregulation. *In* "Progress in Avian Osmoregulation" (M. R. Hughes and A. Chadwick, eds.). Leeds Philosophical and Literary Society, Leeds, United Kingdom.

Clark, N. B., Braun, E. J., and Wideman, R. F. (1976). Parathyroid hormone and renal excretion of phosphate and calcium in normal starlings. *Am. J. Physiol.* **231,** 1152–1158.

Clauss, W., Árnason, S. S., Munck, B. G., and Skadhauge, E. (1984). Aldosterone-induced sodium transport in lower intestine: Effects of varying NaCl intake. *Pflügers Arch.* **401,** 354–360.

Clauss, W., Dürr, J. E., Guth, D., and Skadhauge, E. (1987). Effects of adrenal steroids on Na transport in the lower intestine (coprodeum) of the hen. *J. Membrane Biol.* **96,** 141–152.

Clauss, W., Dantzer, V., and Skadhauge, E. (1991). Aldosterone modulates electrogenic Cl secretion in the colon of the hen (*Gallus domesticus*). *Am. J. Physiol.* **261,** R1533–R1541.

Cole, J. A. (1985). The role of parathyroid hormone in the seasonal regulation of calcium and phosphate metabolism in the European starling *Sturnus vulgaris*. Doctoral dissertation, University of Connecticut.

Craan, A. G., Lemieux, G., Vinay, P., Gougoux, A., and Quenneville, A. (1982). The kidney of chicken adapts to chronic metabolic acidosis: *In vivo* and *in vitro* studies. *Kid. Int.* **22,** 103–111.

Crawford, E. C., and Schmidt-Nielsen, K. (1967). Temperature regulation and evaportative cooling in the ostrich. *Am. J. Physiol.* **212,** 347–353.

Dantzer, V. (1989). Ultrastructural differences between the two major components of chicken ceca. *J. Exp. Zool.* **3**(Suppl.), 21–31.

Dantzler, W. H. (1966). Renal response of chickens to infusion of hyperosmotic sodium chloride solution. *Am. J. Physiol.* **210,** 640–646.

Dantzler, W. H. (1969). Effects of K, Na, and ouabain on urate and PAH uptake by snake and chicken kidney slices. *Am. J. Physiol.* **217,** 1510.

Dantzler, W. H. (1978). Urate excretion in nonmammalian vertebrates. *In:* "Uric Acid" (W. N. Kelley and I. M. Weiner, eds.), pp. 185–210. Springer-Verlag, Berlin/Heidelberg/New York.

Dantzler, W. H. (1989). "Comparative Physiology of the Vertebrate Kidney." Springer-Verlag, New York.

Dawson, T. J., Herd, R. M., and Skadhauge, E. (1985). Osmotic and ionic regulation during dehydration in a large bird, the emu (*Dromaius novaehollandiae*): An important role for the cloaca-rectum. *Quart. J. Exp. Physiol.* **70**, 423–436.

Dawson, T. J., Maloney, S. K., and Skadhauge, E. (1991). The role of the kidney in electrolyte and nitrogen excretion in a large flightless bird, the emu, during different osmotic regimes, including dehydration and nesting. *J. Comp. Physiol. B* **161**, 165–171.

Dawson, W. R. (1982). Evaporative losses of water by birds. *Comp. Biochem. Physiol. A* **71**, 495–509.

Denbow, D. M., and Sheppard, B. J. (1993). Food and water intake responses of the domestic fowl to norepinephrine infusion at circumscribed neural sites. *Brain Res. Bull.* **31**, 121–128.

Dicker, S. E., and Haslam, J. (1972). Effects of exteriorization of the ureters on the water metabolism of the domestic fowl. *J. Physiol. (London)* **224**, 515–520.

Dousa, T. P. (1974). Effects of hormones on cyclic AMP formation in kidneys of nonmammalian vertebrates. *Am. J. Physiol.* **226**, 1193–1197.

Elbrønd, V. S., Dantzer, V., Mayhew, T. M., and Skadhauge, E. (1991). Avian lower intestine adapts to dietary salt (NaCl) depletion by increasing transepithelial sodium transport and microvillous membrane surface area. *Exp. Physiol.* **76**, 733–744.

Elbrønd, V. S., Dantzer, V., Mayhew, T. M., and Skadhauge, E. (1993). Dietary and aldosterone effects on the morphology and electrophysiology of the chicken coprodeum. In "Avian Endocrinology" (P. J. Sharp, ed.), pp. 217–226. Journal of Endocrinology Ltd., Bristol.

Emery, N., Poulson, T. L., and Kinter, W. B. (1972). Production of concentrated urine by avian kidneys. *Am. J. Physiol.* **223**, 180–187.

Epstein, A. N., and Massi, M. (1987). Salt appetite in the pigeon in response to pharmacological treatments. *J. Phys.(London)* **393**, 555–568.

Ernst, S. A., and Ellis, R. A. (1969). The development of surface specialization in the secretory epithelium of the avian salt gland in response to osmotic stress. *J. Cell. Biol.* **40**, 305–321.

Fänge, R., Schmidt-Nielsen, K., and Robinson, M. (1958). Control of secretion from the avian salt gland. *Am. J. Physiol.* **195**, 321–326.

Folk, R. L. (1969). Spherical urine in birds: Petrography. *Science* **166**, 1516–1518.

Gerstberger, R. (1988). Functional vasocative intestinal polypeptide (VIP)-system in salt gands of the Pekin Duck. *Cell Tissue Res.* **252**, 39–48.

Gerstberger, R. (1991). Regulation of salt gland function. *Acta XX Congr. Int. Ornithol. Symp.* **38**, 2114–2121.

Gerstberger, R., and Gray, D. A. (1993). Fine structure, innervation, and functional control of avian salt glands. *Int. Rev. Cytol.* **144**, 129–215.

Gerstberger, R., Gray, D., and Simon, E. (1984). Circulatory and osmoregulatory effects of angiostensin II perfusion of the third ventricle in a bird with salt glands. *J. Physiol.* **349**, 167–182.

Gerstberger, R., Kaul, R., Gray, D. A., and Simon, E. (1985). Arginine vasotocin and glomerular filtration rate in saltwater-acclimated ducks. *Am. J. Physiol.* **248**, F663.

Gerstberger, R., Healy, D. P., Hammel, H. T., and Simon, E. (1987) Autoradiographic localization and characterization of circumventricular angiotensin II receptors in duck brain. *Brain Res.* **400**, 165–170.

Gibbs, O. S. (1929). The function of the fowl's ureters. *Am. J. Physiol.* **88**, 87–100.

Goldstein, D. L. (1989). Transport of water and electrolytes by the lower intestine and its contribution to avian osmoregulation. In "Progress in Avian Osmoregulation" (M. R. Hughes and A. Chadwick, eds.), pp. 271–294. Leeds Philosophical and Literarary Society, Leeds, United Kingdom.

Goldstein, D. L. (1990). Effects of different sodium intakes on renal and cloacal sodium excretion in Chukars (*Aves: Phasianidae*). *Physiol. Zool.* **63**(2), 408–419.

Goldstein, D. L. (1993a). Influence of dietary sodium and other factors on plasma aldosterone concentrations and *in vitro* properties of the lower intestine in house sparrows (*Passer domesticus*). *J. Exp. Biol.* **176**, 159–174.

Goldstein, D. L. (1993b). Renal response to saline infusion in chicks of Leach's storm petrel (*Oceanodroma leucorhoa*). *J. Comp. Physiol. B* **163**, 167–173.

Goldstein, D. L. (1995). Effects of water restriction during growth and adulthood on renal function in bobwhite quail. *J. Comp. Physiol. B* **164**, 663–670.

Goldstein, D. L., and Braun, E. J. (1986). Lower intestinal modification of ureteral urine in hydrated house sparrows. *Am. J. Physiol.* **250**, R89–R95.

Goldstein, D. L., and Braun, E. J. (1988). Contributions of the kidneys and intestines to water conservation, and plasma levels of antidiuretic hormone, during dehydration in house sparrows (*Passer domesticus*). *J. Comp. Physiol. B* **158**, 353–361.

Goldstein, D. L., and Braun, E. J. (1989). Structure and concentrating ability in the avian kidney. *Am. J. Physiol.* **256**, R501–R509.

Goldstein, D. L., and Rothschild, E. L. (1993). Daily rhythms in rates of glomerular filtration and cloacal excretion in captive and wild song sparrows (*Melospiza melodia*). *Physiol. Zool.* **66**(5), 708–719.

Goldstein, D. L., Hughes, M. R., and Braun, E. J. (1986). Role of the lower intestine in the adaptation of gulls (*Larus glaucescens*) to sea water. *J. Exp. Biol.* **123**, 345–357.

Goldstein, D. L., Williams, J. B., and Braun, E. J. (1990). Osmoregulation in the field by salt-marsh savannah sparrows *Passerculus sandwichensis beldingi*. *Physiol. Zool.* **63**, 669–682.

Good, D. W. (1994). Ammonium transport by the thick ascending limb of Henle's loop. *Annu. Rev. Physiol.* **56**, 623–647.

Grahn, M. F, and Butterworth, P. J. (1982). Phosphate uptake by proximal tubule cells isolated from chick kidney. *Biochem. Soc. Trans.* **9**, 465–466.

Gray, D. A. (1993). Plasma atrial natriuretic factor concentrations and renal actions in the domestic fowl. *J. Comp. Physiol. B* **163**, 519–523.

Gray, D. A., and Erasmus, T. (1988). Plasma arginine vasotocin and angiotensin II in the water deprived kelp gull (*Larus dominicanus*), Cape gannet (*Sula capensis*) and jackass penguin (*Spheniscus demersus*). *Comp. Biochem. Physiol. A* **91**, 727–732.

Gray, D. A., and Erasmus, T. (1989). Control of renal and extrarenal salt and water excretion by plasma angiotensin II in the kelp gull (*Larus dominicanus*). *J. Comp. Physiol. B* **158**, 651–660.

Gray, D. A., and Simon, E. (1985). Control of plasma angiotensin II in a bird with salt glands (*Anas platyrhynchos*). *Gen. Comp. Endocrinol.* **160**, 1–13.

Gray, D. A., Hammel, H. T., and Simon, E. (1986). Osmoregulatory effects of angiotensin II in a bird with salt glands (*Anas platyrhynchos*). *J. Comp. Physiol. B* **156**, 315–321.

Gray, D. A., Naude, R. J., and Erasmus, T. (1988). Plasma arginine vasotocin and angiotensin II in the water deprived ostrich (*Struthio camelus*). *Comp. Biochem. Physiol. A* **89**(2), 251–256.

Gray, D. A., Schütz, H., and Gerstberger, R. (1991). Interaction of atrial natriuretic factor and osmoregulatory hormones in the Pekin duck. *Gen. Comp. Endocrinol.* **81**, 246–255.

Greenwald, L. (1989). The significance of renal relative medullary thickness. *Physiol. Zool.* **62**(5), 1005–1014.

Hammel, H. T. (1989). Neural control of salt gland excretion: Enhanced and sustained by autofacilitation. In "Progress in Avian Osmoregulation" (M. R. Hughes and A. Chadwick, eds.), pp. 163–181. Leeds Philosophical and Literarary Society, Leeds, United Kingdom.

Hammel, H. T., Simon-Oppermann, C., and Simon, E. (1980). Properties of body fluids influencing salt gland secretion in Pekin ducks. *Am. J. Physiol.* **239,** R489–R496.

Hanwell, A., Linzell, J. L., and Peaker, M. (1971). Salt-gland secretion and blood flow in the goose. *J. Physiol. (London)* **213,** 373–387.

Hanwell, A., Linzell, J. L., and Peaker, M. (1972). Nature and location of the receptors for salt-gland secretion in the goose. *J. Physiol. (London)* **226,** 453–472.

Hart, W. M., and Essex, H. E. (1942). Water metabolism of the chicken (*Gallus domesticus*) with special reference to the role of the cloaca. *Am. J. Physiol.* **136,** 657–668.

Hildebrandt, P., and Shuttleworth, T. J. (1993). G_q-type G protein couples muscarinic receptors to inositol phosphate and calcium signaling in exocrine cells from the avian salt gland. *J. Membr. Biol.* **133,** 183–190.

Holmes, W. N., and Phillips, J. G. (1985). The avian salt gland. *Biol. Rev.* **60,** 213–256.

Holmes, W. N., Fletcher, G. L., and Stewart, D. J. (1968). The patterns of renal electrolyte excretion in the duck (*Anas platyrhynchos*) maintained on freshwater and on hypertonic saline. *J. Exp. Biol.* **48,** 487–508.

Holmes, W. N., Gorsline, J., and Wright, A. (1982). Patterns of cloacal water and electrolyte excretion in constantly-loaded intact and colostomized ducks (*Anas platyrhynchos*). *Comp. Biochem. Physiol. A* **3,** 675–677.

Holtug, K., McEwan, G. T. A., and Skadhauge, E. (1992). Effects of propionate on the acid microclimate of hen (*Gallus domesticus*) colonic mucosa. *Comp. Biochem. Physiol. A* **103,** 649–652.

Holtug, K., Shipley, A., Dantzer, V., Sten-Knudsen, O., and Skadhauge, E. (1991). Localization of sodium absorption and chloride secretion in an intestinal epithelium. *J. Membr. Biol.* **122,** 215–229.

Huber, G. C. (1917). On the morphology of the renal tubulus of vertebrates. *Anat. Res.* **13,** 305–339.

Hughes, M. R. (1970a). Relative kidney size in nonpasserine birds with functional salt glands. *Condor* **72,** 164–168.

Hughes, M. R. (1970b). Flow rate and cation concentration in salt gland secretion of the glaucous-winged gull, *Larus glaucescens*. *Comp. Biochem. Physiol.* **32,** 807–812.

Hughes, M. R. (1972). The effect of salt gland removal on cloacal ion and water excretion in the growing kittiwake, *Rissa tridactyla*. *Can. J. Zool.* **50,** 603–610.

Hughes, M. R. (1980). Glomerular filtration rate in saline acclimated ducks, gulls, and geese. *Comp. Biochem. Physiol. A* **65,** 211.

Hughes, M.R. (1987). The effects of ureteral resistance on gull urine composition and flow rate. *Can. J. Zool.* **65,** 2669–2771.

Hughes, M. R. (1989a). Extracellular fluid volume and the initiation of salt gland secretion in ducks and gulls. *Can. J. Zool.* **67,** 194–197.

Hughes, M. R. (1989b). Stimulus for avian salt gland secretion. In "Progress in Avian Osmoregulation" (M. R. Hughes and A. Chadwick, eds.), pp. 143–161. Leeds Philosophical and Literarary Society, Leeds, United Kingdom.

Hughes, M. R., and Chadwick, A. (eds.) (1989). "Progress in Avian Osmoregulation." Leeds Philosophical and Literarary Society, Leeds, United Kingdom.

Hughes, M. R., and Raveendran, L. (1994). Ion and luminal marker concentrations in the gut of saline-acclimated ducks. *Condor* **96,** 295–299.

Hughes, M. R., and Roberts, J. R. (1988). Sodium uptake from the gut of freshwater- and seawater-acclimated ducks and gulls. *Can. J. Zool.* **66,** 1365–1370.

Hughes, M. R., Roberts, J. R., and Thomas, B. R. (1989). Renal function in freshwater and chronically saline-stressed male and female Pekin ducks. *Poult. Sci.* **68,** 408–416.

Hughes, M. R., Kojwang, D., and Zenteno-Savin, T. (1992). Effects of caecal ligation and saline acclimation on plasma concentration and organ mass in male and female Pekin ducks, *Anas platyrhynchos*. *J. Comp. Physiol.* **162,** 625–631.

Hughes, M. R., Goldstein, D. L., and Raveendran, L. (1993). Osmoregulatory responses of glaucous-winged gulls (*Larus glaucescens*) to dehydration and hemorrhage. *J. Comp. Physiol. B* **163,** 524–531.

Jamison, R. L., and Kriz, W. (1982). "Urinary Concentration Mechanism: Structure and Function" Oxford University Press, New York.

Johnson, O. W. (1968). Some morphological features of avian kidneys. *Auk* **85,** 216–228.

Johnson, O. W. (1974). Relative thickness of the renal medulla in birds. *J. Morphol.* **142**(3), 277–284.

Johnson, O. W., and Mugaas, J. N. (1970). Quantitative and organizational features of the avian renal medulla. *Condor* **72,** 288–292.

Johnson, O. W., Phipps, G. L., and Mugaas, J. N. (1972). Injection studies of cortical and medullary organization in the avian kidney. *J. Morphol.* **136,** 181–190.

Kanosue, K., Schmid, H., and Simon, E. (1990). Differential osmoresponsiveness of periventricular neurons in duck hypothalamus. *Am. J. Physiol.* **258,** R973–R981.

Karasawa, Y., and Maeda, M. (1992). Effect of colostomy on the utilisation of dietary nitrogen in the fowl fed on a low protein diet. *Br. Poult. Sci.* **33,** 815–820.

Kaufman, S., and Peters, G. (1980). Regulatory drinking in the pigeon, *Columba livia*. *Am. J. Physiol.* **301,** 91–99.

Kaul, R., Gerstberger, R., Meyer, U., and Simon, E. (1983). Salt gland blood flow in saltwater-adapted Pekin ducks: Microsphere measurement of the proportionality to secretion rate and investigation of controlling mechanisms. *J. Comp. Physiol. B* **149,** 457–462.

Klingbeil, C. K. (1985). Effects of chronic changes in dietary electrolytes and acute stress on plasma levels of corticosterone and aldosterone in the duck (*Anas platyrhynchos*). *Gen. Comp. Endocrinol.* **58,** 10–19.

Knepper, M. A. (1991). NH_4^+ transport in the kidney. *Kidney Int.* **40,** S95–S102.

Koike, T. I., Pryor, L. R., and Neldon, H. L. (1979). Effect of saline infusion on plasma immunoreactive vasotocin in conscious chickens. *Gen. Comp. Endocrinol.* **37,** 451–458.

Korr, I. M. (1939). The osmotic function of the chicken kidney. *J. Cell Comp. Physiol.* **13,** 175–194.

Krag, B., and Skadhauge, E. (1972). Renal salt and water excretion in the budgerygah (*Melopsittacus undulatus*). *Comp. Biochem. Physiol. A* **41,** 667–683.

Langford, H. G., and Fallis, N. (1966). Diuretic effect of angiotensin in the chicken. *Proc. Soc. Exp. Biol. Med.* **123,** 317–321.

Laverty, G. (1989). Renal tubular transport in the avian kidney. In "Progress in Avian Osmoregulation" (M. R. Hughes and A. Chadwick, eds.). Leeds Philosophical and Literary Society, Leeds, United Kingdom.

Laverty, G., and Alberici, M. (1987). Micropuncture study of proximal tubule pH in avian kidney. *Am. J. Physiol.* **253,** R587–R591.

Laverty, G., and Dantzler, W. H. (1982). Micropuncture of superficial nephrons in avian (*Sturnus vulgaris*) kidney. *Am. J. Physiol.* **243,** F561.

Laverty, G., and Dantzler, W. H. (1983). Micropuncture study of urate transport by superficial nephrons in avian (*Sturnus vulgaris*) kidney. *Pflugers Arch.* **397,** 232–236.

Laverty, G., and Wideman, R. F., Jr. (1989). Sodium excretion rates and renal responses to acute salt loading in the European starling. *J. Comp. Physiol. B* **159**, 401–408.

Laverty, G., Holtug, K., Elbrønd, V. S., Ridderstråle, Y., and Skadhauge, E. (1994). Mucosal acidification and an acid microclimate in the hen colon in vitro. *J. Comp. Physiol. B* **163**, 633–641.

Layton, H. E. (1986). Distribution of Henle's loops may enhance urine concentrating capability. *Biophys. J.* **49**, 1033–1040.

Lind, J., Munck, B. G., Olsen, O., and Skadhauge, E. (1980a). Effects of sugars, amino acids and inhibitors on electrolyte transport across hen colon at different sodium chloride intakes. *J. Physiol. (London)* **305**, 315–325.

Lind, J., Munck, B. G., and Olsen, O. (1980b). Effects of dietary intake of sodium chloride on sugar and amino acid transport across isolated hen colon. *J. Physiol. (London)* **305**, 327–336.

Long, S., and Skadhauge, E. (1980). Renal reabsorption of Na and K in *Gallus*: Role of urinary precipitates. *Acta. Physiol. Scand.* **109**, 31A.

Long, S., and Skadhauge, E. (1983a). Renal acid excretion in the domestic fowl. *J. Exp. Biol.* **104**, 51.

Long, S., and Skadhauge, E. (1983b). The role of urinary precipitates in the excretion of electrolytes and urate in the domestic fowl. *J. Exp. Biol.* **104**, 41–50.

Lonsdale, K., and Sutor, D. J. (1971). Uric acid dihydrate in bird urine. *Science* **172**, 958–959.

Lowy, R. J., and Ernst, S. A. (1987). β-adrenergic stimulation of ion transport in primary cultures of avian salt gland. *Am. J. Physiol.* **252** (21), C670–C676.

Lowy, R. J., Schreiber, J. H., and Ernst, S. A. (1987). Vasoactive intestinal peptide stimulates ion transport in avian salt gland. *Am. J. Physiol.* **253**, R801–R803.

Lowy, R. J., Dawson, D. C., and Ernst, S. A. (1989). Mechanisms of ion transport by avian salt gland primary cell cultures. *Am. J. Physiol.* **256**, R1184–R1191.

Marder, J., and Ben-Asher, J. (1983). Cutaneous water evaporation. I. Its significance in heat-stressed birds. *Comp. Biochem. Physiol. A* **75**(3), 425–431.

Marder, J., and Raber, P. (1989). Beta-adrenergic control of transcutaneous evaporative cooling mechanisms in birds. *J. Comp. Physiol. B* **159**, 97–103.

Marshall, A. T., Hyatt, A. D., Phillips, J. G., and Condron, R. J. (1985). Isosmotic secretion in the avian nasal salt gland: X-ray microanalysis of luminal and intracellular ion distributions. *J. Comp. Physiol. B.* **156**, 213–227.

Martindale, L. (1976). Renal urate synthesis in the fowl (*Gallus domesticus*). *Comp. Biochem. Physiol. A* **53**, 389–391.

Massi, M., and Epstein, A. N. (1990). Angiotensin/aldosterone synergy governs the salt appetite of the pigeon. *Appetite* **14**, 181–192.

Mayhew, T. M., Dantzer, V., Elbrønd, V. S., and Skadhauge, E. (1990). A sampling scheme intended for tandem measurements of sodium transport and microvillous surface area in the coprodaeal epithelium of hens on high- and low-salt diets. *J. Anat.* **173**, 19–31.

Mayhew, T. M., Elbrønd, V. S., Dantzer, V., Skadhauge, E., and Møller, O. (1992). Structural and enzymatic studies on the plasma membrane domains and sodium pump enzymes of absorptive epithelial cells in the avian lower intestine. *Cell Tissue Res.* **270**, 577–585.

McLelland, J. (1989). Anatomy of the avian cecum. *J. Exp. Zool.* **3**(Suppl.), 2–9.

McNabb, F. M. A., McNabb, R. A., and Steeves, H. R. (1973). Renal mucoid materials in pigeons fed high and low protein diets. *Auk* **90**, 14–18.

McNabb, R. A. (1974). Urate and cation interactions in the liquid and precipitated fractions of avian urine, and speculations on their physico-chemical state. *Comp. Biochem. Physiol. A* **48**, 45–54.

McNabb, R. A., and McNabb, F. M. A. (1975). Urate excretion by the avian kidney. *Comp. Biochem. Physiol. A* **51**, 253–258.

Minnich, J. E. (1976). Adaptations in the reptilian excretory system for excreting insoluble urates. *Israel J. Med. Sci.* **12**, 854–861.

Miwa, T., and Nishimura, H. (1986). Diluting segment in avian kidney. II. Water and chloride transport. *Am. J. Physiol.* **250**, R341.

Möhring, J., Schoun, J., Simon-Oppermann, C., and Simon, E. (1980). Radioimmunoassay for arginine-vasotocin (AVT) in serum of Pekin ducks: AVT concentrations after adaptation to freshwater and salt water. *Pflügers Arch.* **387**, 91–97.

Morild, I., Monwinckel, R., Bohle, A., and Christensen, J. A. (1985). The juxtaglomerular apparatus in the avian kidney. *Cell Tissue Res.* **240**, 209–214.

Nechay, B. R., and Nechay, L. (1959). Effects of probenecid sodium, salicylate, 2,4-dinitrophenol and pyrazinamide or renal secretin of uric acid in chickens. *J. Pharmacol. Exp. Therapeut.* **26**, 291–295.

Nicholson, J. K. (1982). The microanatomy of the distal tubules, collecting tubules and collecting ducts of the starling kidney. *J. Anat.* **134**, 11.

Nishimura, H. (1993). Countercurrent urine concentration in birds. In "New Insights in Vertebrate Kidney Function" (J. A. Brown, R. J. Balment, and J. C. Rankin, eds.), pp. 189–208. Cambridge Univ. Press, Cambridge.

Nishimura, H., Imai, M., and Ogawa, J. (1986). Diluting segment in avian kidney. I. Characterization of trans-epithelial voltages. *Am. J. Physiol.* **250**, R333.

Nishimura, H., Koseki, C., Imai, M., and Braun, E. J. (1989). Sodium chloride and water transport in the thin descending limb of Henle of the quail. *Am. J. Physiol.* **257**, F994–F1002.

Odlind, B. (1978a). Blood flow distribution in the renal portal system of the intact hen: A study of a venous system using microspheres. *Acta Physiol. Scand.* **102**, 342.

Oelofsen, B. W. (1973). Renal function in the penguin (*Spheniscus demersus*) with special reference to the role of the renal portal valves. *Zool. Afr.* **8**, 41.

Ohmart, R. D. (1972). Physiological and ecological observations concerning the salt-secreting nasal glands of the roadrunner. *Comp. Biochem. Physiol. A* **43**, 311–316.

Osono, E., and Nishimura, H. (1994). Control of sodium and chloride transport in the thick ascending limb in the avian nephron. *Am. J. Physiol.* **267**, R455–R462.

Peaker, M., and Linzell, J. L. (1975). "Salt Glands in Birds and Reptiles." Cambridge Univ. Press, Cambridge.

Pines, M., Polin, D., and Hurwitz, S. (1983). Urinary cyclic AMP excretion in birds: Dependence on parathyroid hormone activity. *Gen. Comp. Endocrinol.* **49**, 90–96.

Platts, M. M., and Mudge, G. H. (1961). Accumulation of uric acid by slices of kidney cortex. *Am. J. Physiol.* **200**, 387–391.

Porter, P. (1963). Physico-chemical factors involved in urate calculus formation. II. Colloidal flocculation. *Res. Vet. Sci.* **4**, 592–602.

Poulson, T. L., and Bartholomew, G. A. (1962). Salt balance in the savannah sparrow. *Physiol. Zool.* **35**, 109–119.

Renfro, J. L., and Clark, N. B. (1984). Parathyroid hormone induced phosphate excretion following preequilibration with ^{32}P. *Am. J. Physiol.* **246**, F373–F378.

Rice, G. E., and Skadhauge, E. (1982a). The in vivo dissociation of colonic and coprodeal transepithelial transport in NaCl depleted domestic fowl. *J. Comp. Physiol. B* **146**, 51–56.

Rice, G. E., and Skadhauge, E. (1982b). Colonic and coprodeal transepithelial transport parameters in NaCl-loaded domestic fowl. *J. Comp. Physiol. B* **147**, 65–69.

Rice, G. E., and Skadhauge, E. (1982c). Caecal water and electrolyte absorption and the effects of acetate and glucose, in dehydrated, low-NaCl diet hens. *J. Comp. Physiol. B* **147**, 61–64.

Rice, G. E., Arnason, S. S., Arad, S., and Skadhauge, E. (1985). Plasma concentrations of arginine vasotocin, prolactin, aldosterone and corticosterone in relation to oviposition and dietary NaCl in the domestic fowl. *Comp. Biochem. Physiol. A* **81,** 769–777.

Ridderstrale, Y. (1980). Ultrastructure and localization of carbonic anhydrase in the avian nephron. *Swed. J. Agric. Res.* **10,** 41.

Roberts, J. R. (1992). Renal function and plasma arginine vasotocin during an acute salt load in feral chickens. *J. Comp. Physiol. B* **162,** 54–58.

Roberts, J. R., and Dantzler, W. H. (1990). Micropuncture study of the avian kidney: infusion of mannitol or sodium chloride. *Am. J. Physiol.* **258,** R869–R874.

Roberts, J. R., and Hughes, M. R. (1983). Glomerular filtration rate and drinking rate in Japanese quail, *Coturnix coturnix japonica*, in response to acclimation to saline drinking water. *Can. J. Zool.* **61,** 2394.

Roberts, J. R., Baudinette, R. V., and Wheldrake, J. F. (1985). Renal clearance studies in stubble quail *Coturnix pectoralis* and king quail *Coturnix chinensis* under conditions of hydration, dehydration, and salt loading. *Physiol. Zool.* **58,** 340.

Roush, W. B., and Spotts, C. B. (1988). Strain differences in the number and size of glomeruli in domestic fowl. *Br. Poult. Sci.* **29,** 113–117.

Ruch, F. E., Jr., and Hughes, M. R. (1975). The effects of hypertonic sodium chloride injection on body water distribution in ducks (*Anas platyrhynchos*), gulls (*Larus glaucescens*), and roosters (*Gallus domesticus*). *Comp. Biochem. Physiol. A* **52,** 21–28.

Sandor, T., Skadhauge, E., Dibattista, J. A., and Mehdi, A. Z. (1989). Interrelations of the intestinal glucocorticoid and mineralocorticoid receptor systems with salt homeostasis. *In* "Progress in Avian Osmoregulation" (M. R. Hughes and A. Chadwick, eds.), pp. 305–332. Leeds Philosophical and Literarary Society, Leeds, United Kingdom.

Schmidt-Nielsen, K. (1959). Salt glands. *Sci. Am.* **200,** 109–116.

Schmidt-Nielsen, K., and Kim, Y. T. (1964). The effect of salt intake on the size and function of the salt glands of ducks. *Auk* **81,** 160–172.

Schmidt-Nielsen, K., Jorgensen, C. B., and Osaki, H. (1958). Extrarenal salt excretion in birds. *Am. J. Physiol.* **193,** 101–107.

Schmidt-Nielsen, K., Borut, A., Lee, P., and Crawford, E. (1963). Nasal salt excretion and the possible function of the cloaca in water conservation. *Science* **142,** 1300–1301.

Schütz, H., and Gerstberger, R. (1990). Atrial natriuretic factor controls salt gland secretion in the Pekin duck (*Anas platyrhynchos*) through interaction with high affinity receptors. *Endocrinology* **127,** 1718–1726.

Schütz, H., Gray, D. A., and Gerstberger, R. (1992). Modulation of kidney function in conscious Pekin ducks by atrial natriuretic factor. *Endocrinology* **130,** 678–684.

Shannon, J. A. (1938a). The excretion of exogenous creatinine by the chicken. *J. Cell. Comp. Physiol.* **11,** 123–134.

Shideman, J. R., Evans, R. L., Bierer, D. W., and Quebbemann, A. J. (1981). Renal venous portal contribution to PAH and uric acid clearance in the chicken. *Am. J. Physiol.* **240,** F46–F53.

Shuttleworth, T. J., and Thompson, J. L. (1989). Intracellular [Ca^{2+}] and inositol phosphates in avian nasal gland cells. *Am. J. Physiol.* **257,** C1020–C1029.

Shuttleworth, T. J., and Wood, C. M. (1992). Changes in pH_i associated with activation of ion secretion in avian nasal salt gland cells. *Am. J. Physiol.* **262,** C221–C228.

Siegel-Causey, D. (1990). Phylogenetic patterns of size and shape of the nasal gland depression in phalacrocoracidae. *Auk* **107,** 110–118.

Simon, E., Gerstberger, R., and Gray, D. A. (1992). Central nervous angiotensin II responsiveness in birds. *Prog. Neurobiol.* **39,** 179–207.

Simon-Oppermann, C, Szczepanska-Sadowska, E., Gray, D. A., and Simon, E. (1984). Blood volume changes and arginine vasotocin (AVT) blood concentration in conscious fresh water and salt water adapted ducks. *Pflugers Arch.* **400,** 151–159.

Skadhauge, E. (1964). Effects of unilateral infusion of arginine-vasotocin into the portal circulation of the avian kidney. *Acta Endocrinol.* **47,** 321–330.

Skadhauge, E. (1973). Renal and cloacal salt and water transport in the fowl (*Gallus domesticus*). *Danish Med. Bull.* **20**(Suppl. 1), 1–82.

Skadhauge, E. (1977). Solute compositon of the osmotic space of ureteral urine in dehydrated chickens (*Gallus domesticus*). *Comp. Biochem. Physiol. A* **56,** 271–274.

Skadhauge, E. (1981). "Osmoregulation in Birds". Springer-Verlag, New York.

Skadhauge, E. (1983). Ionic and osmotic regulation in birds. *Verh. Dtsch. Zool. Ges.* 69–81.

Skadhauge, E. (1993). Basic characteristics and hormonal regulation of ion transport in avian hindguts. *Adv. Comp. Environ. Physiol.* **16,** 67–93.

Skadhauge, E., and Bradshaw, S. D. (1974.) Saline drinking and cloacal excretion of salt and water in the zebra finch. *Am. J. Physiol.* **227,** 1263–1267.

Skadhauge, E., and Kristensen, K. (1972). An analogue computer simulation of cloacal resorption of salt and water from ureteral urine in birds. *J. Theor. Biol.* **35,** 473–387.

Skadhauge, E., and Schmidt-Nielsen, B. (1967a). Renal function in domestic fowl. *Am. J. Physiol.* **212,** 793–798.

Skadhauge, E., and Schmidt-Nielsen, B. (1967b). Renal medullary electrolyte and urea gradient in chickens and turkeys. *Am. J. Physiol.* **212,** 1313–1318.

Skadhauge, E., and Thomas, D. H. (1979). Transepithelial transport of K^+, NH_4^+, inorganic phosphate and water by hen (*Gallus domesticus*) lower intestine (colon and coprodeum) perfused luminally in vivo. *Pflügers Arch.* **379,** 237–243.

Skadhauge, E., Thomas, D. H., Chadwick, A., and Jallageas, M. (1983). Time course of adaptation to low and high NaCl diets in the domestic fowl: Effects on electrolyte excretion and on plasma hormone levels (aldosterone, corticosterone and prolactin). *Pflügers Arch. Eur. J. Physiol.* **396,** 301–307.

Skadhauge, E., Munck, B. G., and Rice, G. E. (1984a). Regulation of NaCl and water absorption in duck intestine. *In* "Osmoregulation in Estuarine and Marine Animals: Lecture Notes on Coastal and Estuarine Studies" (A. Pequeux, R. Gilles, and L. Bolis, eds.), 118–133. Springer-Verlag. Berlin.

Skadhauge, E., Warüi, C. N., Kamau, J. M. Z., and Maloiy, G. M. O. (1984b). Function of the lower intestine and osmoregulation in the ostrich: preliminary anatomical and physiological observations. *Quart. J. Exp. Physiol.* **69,** 809–818.

Skadhauge, E., Maloney, S. K., and Dawson, T. J. (1991). Osmotic adaptation of the emu (*Dromaius novaehollandiae*). *J. Comp. Physiol. B* **161,** 173–178.

Skadhauge, E., Dantzer, V., McLean, A., and Dawson, T. J. (1993). Comparative aspects of kidney-cloaca interaction in ratite birds. *Proc. XXXII Cong. Int. Union. Phys. Sci.* 124.

Smith, H. W. (1953). "From Fish to Philosopher." Little, Brown, and Company, New York.

Smith, P. R., Bradford, A. L., Dantzer, V., Benos, D. J., and Skadhauge, E. (1993). Immunocytochemical localization of amiloride-sensitive sodium channels in the lower intestine of the hen. *Cell Tissue Res.* **272,** 129–136.

Snider, R. M., Roland, R. M., Low, R. J., Agranoff, B. A., and Ernst, S. A. (1986). Muscarinic receptor-stimulated Ca^{2+} signaling and inositol lipid metabolism in avian salt gland cells. *Biochim. Biophys. Acta* **889,** 216–224.

Sperber, I. (1948). Investigations on the circulatory system of the avian kidney. *Zool. Bidrag.* **27,** 429–448.
Sperber, I. (1960). Excretion. *In* "Biology and Physiology of Birds" (A. J. Marshall, ed.), Vol. 1, pp. 469–492. Academic Press, London/New York.
Staaland, H. (1967). Anatomical and physiological adaptations of the nasal glands in charadriiformes birds. *Comp. Biochem. Physiol.* **23,** 933–944.
Stallone, J. N., and Braun, E. J. (1985). Contributions of glomerular and tubular mechanisms to antidiuresis in conscious domestic fowl. *Am. J. Physiol.* **249,** F842.
Stallone, J. N., and Braun, E. J. (1986). Regulation of plasma arginine vasotocin in conscious water-deprived domestic fowl. *Am. J. Physiol.* **250,** R658–R664.
Stallone, J. N., and Nishimura, H. (1985). Angiotensin II (AII)-induced natriuresis in anesthetized domestic fowl. *Fed. Proc. Natl. Acad. Sci. USA* **44,** 1364.
Stewart, D. J. (1972). Secretion by salt gland during water deprivation in the duck. *Am. J. Physiol.* **223,** 384–386.
Sykes, A. H. (1971). Formation and composition of urine. *In* "Physiology and Biochemistry of the Domestic Fowl" (D. J. Bell and B. M. Freeman, eds.), Vol. 1, Chap. 9, 233–278. Academic Press, London/New York.
Takei, Y. (1977). The role of the subfornical organ in drinking induced by angiotensin in the Japanese quail *Coturnix coturnix japonica*. *Cell Tiss. Res.* **185,** 175–181.
Takei, Y., Okawara, Y., and Kobayashi, H. (1988). Water intake induced by water deprivation in the quail, *Coturnix coturnix japonica*. *J. Comp. Physiol. B* **158,** 519–525.
Takei, Y., Okawara, Y., and Kobayashi, H. (1989). Control of drinking in birds. *In* "Progress in Avian Osmoregulation" (M. R. Hughes and A. Chadwick, eds.), 1–12. Leeds Philosophical and Literarary Society, Leeds, United Kingdom.
Thomas, D. H. (1982). Salt and water excretion by birds: the lower intestine as an integrator of renal and intestinal excretion. *Comp. Biochem. Physiol. A* **71,** 527–535.
Thomas, D. H., and Robin, A. P. (1977). Comparative studies of thermoregulatory and osmoregulatory behaviour and physiology of five species of sandgrouse (*Aves:Pterocliidae*) in Morocco. *J. Zool. (London)* **183,** 229–249.
Thomas, D. H., and Skadhauge, E. (1982). Time course of adaptation to low and high NaCl diets in the domestic fowl: Effects on electrical behaviour of isolated epithelia from the lower intestine. *Pflügers Arch. Eur. J. Physiol.* **395,** 165–170.
Thomas, D. H., and Skadhauge, E. (1989a). Water and electrolyte transport by the avian ceca. *J. Exp. Zool.* **3**(Suppl.), 95–102.
Thomas, D. H., and Skadhauge, E. (1989b). Function and regulation of the avian caecal bulb: Influence of dietary NaCl and aldosterone on water and electrolyte fluxes in the hen (*Gallus domesticus*) perfused *in vivo*. *J. Comp. Physiol. B* **159,** 51–60.
Thomas, D. H., Degen, A., and Pinshow, B. (1982). Do phasianid birds really have functional salt glands? Absence of nasal salt secretion in salt-loaded sand partridges and chukars *Ammoperdix heyi* and *Alectoris chukar sinaica*. *Physiol. Zool.* **55**(3), 323–326.
Thornton, S. N. (1986a). Osmoreceptor localization in the brain of the pigeon (*Columba livia*). *Brain Res.* **377,** 96–104.
Thornton, S. N. (1986b). The influence of central infusions on drinking due to peripheral osmotic stimuli in the pigeon (*Columba livia*). *Physiol. Behav.* **36,** 229–233.
Toshimori, H., Toshimori, K., Minamino, N., Kangawa, K., Oura, C., Matsukura, S., and Matsuo, H. (1990). Chicken atrial natriuretic peptice (chANP) and its secretion. *Cell Tissue Res.* **259,** 293–298.
Vena, V. E., Lac, T. H., and Wideman, R. F. (1990). Dietary sodium, glomerular filtration rate autoregulation, and glomerular size distribution profiles in domestic fowl (*Gallus gallus*). *J. Comp. Physiol. B* **160,** 7–16.
Warui, C. N. (1989). Light microscopic morphometry of the kidneys of fourteen avian species. *J. Anat.* **162,** 19–31.
Webster, M. D., and Bernstein, M. H. (1987). Ventilated capsule measurements of cutaneous evaporation in mourning doves. *Condor* **89,** 863–868.
Wideman, R. F., Jr. (1984). Organic phosphate probes of the avian renal phosphate secretory mechanism. *Comp. Biochem. Physiol. B* **78,** 315.
Wideman, R. F., Jr. (1987). Renal regulation of avian calcium and phosphorus metabolism. *J. Nutr.* **117,** 808–814.
Wideman, R. F., Jr. (1988). Avian kidney anatomy and physiology. *In* "CRC Critical Reviews in Poultry Biology," Vol. 1, pp. 133–176. CRC Press, Boca Raton, FL.
Wideman, R. F., Jr. (1991). Autoregulation of avian renal plasma flow: Contribution of the renal portal system. *J. Comp. Physiol. B* **160,** 663–669.
Wideman, R. F., Jr., and Braun, E. J. (1981). Stimulation of avian renal phosphate secretion by parathryoid hormone. *Am. J. Physiol.* **241,** F263.
Wideman, R. F., and Gregg, C. M. (1988). Model for evaluating avian hemodynamics and glomerular filtration rate autoregulation. *Am. J. Physiol.* **254,** R925–R932.
Wilson, J. X. (1989). The renin-angiotensin system in birds. *In* "Progress in Avian Osmoregulation" (M. R. Hughes and A. Chadwick, eds.), pp. 61–79. Leeds Philosophical and Literarary Society, Leeds, United Kingdom.
Withers, P. C., and Williams, J. B. (1990). Metabolic and respiratory physiology of an arid- adapted Australian bird, the spinifex pigeon. *Condor* **92,** 961–969.
Wolbach, R. A. (1955). Renal regulation of acid-base balance in the chicken. *Am. J. Physiol.* **181,** 149–158.
Yokota, S. D., Benyajati, S., and Dantzler, W. H. (1985). Comparative aspects of glomerular filtration in vertebrates. *Renal Physiol.* **8,** 193–221.
Zenteno-Savin, T. (1991). Plasma arginine vasotocin and angiotensin II concentrations during saline acclimation in birds with salt glands. M.Sc. thesis, University of British Columbia, Canada.

CHAPTER 12

Gastrointestinal Anatomy and Physiology

D. MICHAEL DENBOW
Department of Animal and Poultry Sciences
Virginia Tech
Blacksburg, Virginia 24061-0306

I. Anatomy of the Digestive Tract 299
    A. Beak, Mouth, and Pharynx 299
    B. Esophagus and Crop 302
    C. Stomach 303
    D. Small Intestine 303
    E. Ceca 304
    F. Rectum and Cloaca 304
II. Anatomy of the Accessory Organs 305
    A. Pancreas 305
    B. Liver 305
III. Motility 305
    A. Esophagus 305
    B. Gastrointestinal Cycle 306
    C. Small Intestine 308
    D. Ceca 309
    E. Rectum 309
    F. Other Influences on Motility 310
IV. Neural and Hormonal Control of Motility 310
    A. Rate of Passage 313
V. Secretions and Digestion 313
    A. Mouth 313
    B. Esophagus and Crop 313
    C. Stomach 314
    D. Intestines 315
    E. Colon 316
    F. Pancreas 316
    G. Bile 317
VI. Absorption 317
    A. Carbohydrates 317
    B. Amino Acids and Peptides 318
    C. Fatty Acids and Bile Acids 318
    D. Volatile Fatty Acids 318
    E. Electrolytes 319
    F. Water, Sodium, and Chloride 319
    G. Vitamins 320
VII. Age-Related Effects on Gastrointestinal Function 320
VIII. Food Intake Regulation 320
    References 321

I. ANATOMY OF THE DIGESTIVE TRACT

The digestive system (Figure 1) has adaptations designed to facilitate flight. The length of the intestinal tract is shorter in birds relative to mammals (Table 1). Also, birds lack teeth and heavy jaw muscles which have been replaced with a light-weight bill or beak. Food particles are swallowed whole and then reduced in size by the ventriculus or gizzard located within the body cavity. This chapter will not attempt to describe the many species variations in detail but will instead describe differences between birds and mammals. The reader is referred to the excellent reviews of McLelland (1975, 1979) for specific information on various species.

A. Beak, Mouth, and Pharynx

Birds have a bill or beak. This structure shows extensive anatomical variations among species thus allowing for adaptations for different feeding styles. The upper bill is usually covered with a hard keratin. However, in some types of birds the whole bill is relatively soft (i.e., Charadrii or shorebirds) whereas in others only the tip

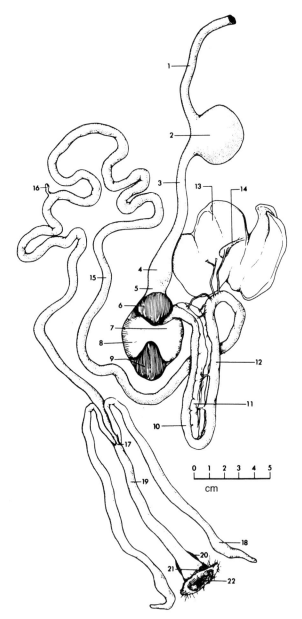

FIGURE 1 Digestive tract of a 12-week-old turkey. 1, precrop esophagus; 2, crop; 3, postcrop esophagus; 4, proventriculus; 5, isthmus; 6, thin craniodorsal muscle; 7, thick cranioventral muscle; 8, thick caudodorsal muscle; 9, thin caudoventral muscle (6–9, gizzard); 10, proximal duodenum; 11, pancreas; 12, distal duodenum; 13, liver; 14, gallbladder; 15, ileum; 16, Meckel's diverticulum; 17, ileocecocolic junction; 18, ceca; 19, rectum; 20, bursa of Fabricius; 21, cloaca; 22, vent. Scale is in centimeters. (Reprinted from Gary E. Duke, "Avian Digestion," in *Duke's Physiology of Domestic Animals,* Tenth Edition, edited by Melvin J. Swenson, p. 360. Copyright 1984 by Cornell University Press. Used by permission of Cornell University Press.)

is hard (i.e., Anatidae or waterfowl). The culmen, the medial dorsal area of the upper beak, has a pointed protuberance in the embryo, the egg tooth, which drops off after hatching.

Birds, unlike mammals, have no sharp distinction between the pharynx and mouth. Birds lack a soft palate and a pharyngeal isthmus; the combined oral and pharyngeal cavities are referred to as the oropharynx. The palate contains a longitudinal fissure, the choana, which connects the oral and nasal cavities. Caudal to the choana is the infundibular cleft, which is medially located and is the common opening to the auditory tubes (Figure 2). The palate generally also has ridges which aid in opening the shell of seeds.

The tongue shows adaptations for collecting, manipulating, and swallowing food (Harrison, 1964). In birds where the tongue is used for collecting food, it can be extended from the mouth during the collection process. Such tongues typically have lateral barbs and may be coated with mucous secreted by the mandibular salivary gland. The tongue may act as either a brush, a spear, or as a capillary tube, sometimes extending four times the length of the beak.

Tongues used to manipulate food, such as in piscivorous species, are nonprotruding and covered with stiff, sharp, caudally directed papillae. In birds of prey, the tongue is a rasp-like structure with the rostral portion frequently being very hard and rough (McLelland, 1979). On the tongue of birds that typically strain food particles (e.g. ducks), the rostral portion forms a scoop-like structure with the lateral borders having a double row of overlapping bristles. The bristles work in conjunction with the lamellae of the bill to filter particles.

In birds where the tongue is utilized to aid in the swallowing of food, caudally directed papillae tend to be located near the root of the tongue. These papillae function to propel food caudally. Tongues specialized for swallowing are nonprotruding.

Salivary glands also show considerable species variation. While salivary glands are generally well developed in granivorous species, they are less developed in birds of prey, poorly developed in piscivores, and absent in the Anhinga and Great Cormorant (Antony, 1920). As a generalization, the maxillary, palatine, and sphenopterygoid glands are located in the roof of the mouth. The buccal gland is in the cheeks whereas the mandibular, lingual, and cricoarytenoid glands are in the floor of the mouth. While the salivary glands of *Gallus* and *Meleagris* are reported to secrete little amylase, the house sparrow secretes considerable amounts (Jerrett and Goodge, 1973).

Taste buds are variably located on the upper beak epithelium, in the anterior mandible, and the mandibular epithelium posterior to the tongue. There are a small number of taste buds also located ventrolaterally on the anterior tongue. It is believed that chickens have as many as 300 taste buds (Saito, 1966; Ganchrow and Ganchrow, 1985).

In some species of birds, the floor of the mouth contains sac-like diverticuli called oral sacs. These can act

TABLE 1 Dimensions of the Digestive Tract of Various Species of Birds[a]

Species	Body wt (kg)	Esophagus Length (mm)	Esophagus Total %	Esophagus Wt (g)	Proventriculus and gizzard Length (mm)	Proventriculus and gizzard Total %	Proventriculus and gizzard Wt (g)	Small intestine Length (mm)	Small intestine Total %	Small intestine Wt (g)	Cecum Length (mm)	Cecum Wt (g)	Rectum Length (mm)	Rectum Total %	Rectum Wt (g)	Total Length (mm)	Total Length/BW
Chicken																	
Leghorn	1.2	136	9.9	8.2	86	6.3	26.7	1082	78.9	29.5	127	5.2	68	5.0	2.3	1372	1.14
Broiler	3.0	140	6.4	16.8	101	4.7	43.5	1796	82.7	73.6	188	10.7	134	6.2	5.1	2171	0.72
Turkey	3.0	123	5.7	8.5	110	5.1	52.9	1853	85.7	85.3	278	20.1	75	3.5	4.4	2161	0.72
Japanese quail[b]	NA	75	11.5	—	38	5.8	—	510	78.1	—	100	—	30	4.6	—	653	—
Domestic duck[c]	2.2	310	11.7	—	130	4.9	—	2110	79.9	—	140	—	90	3.4	—	2640	1.20
Emu[c]	53.0	790	12.1	—	260	4.0	—	5200	79.4	—	120	—	300	4.6	—	6550	0.12
Rhea[d]	25.0	—	NA	—	310	—	—	1400	—	—	480	—	400	—	—	—	—
Ostrich[d]	122.0	—	NA	—	480	—	—	6400	—	—	940	—	8000	18.1	—	—	—
Cedar waxwing[e]	NA	51	16.2	—	36	11.4	—	171	54.3	—	0	0	57	18.1	—	315	—

[a] The length and weight of the gastrointestinal tract can change depending on the environment in which the birds are raised (see Deaton et al., 1985). NA, not available.
[b] From Fitzgerald (1969).
[c] From Herd (1985).
[d] From Fowler (1991).
[e] From Levey and Duke (1992).

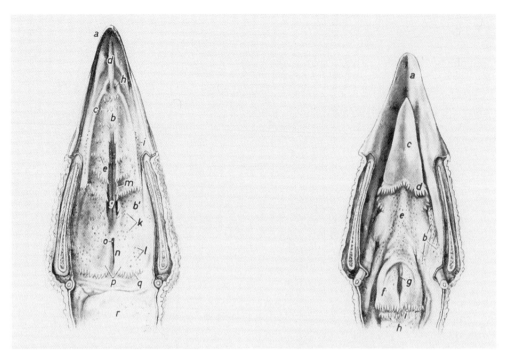

FIGURE 2 (Left) Roof of the mouth cavity and pharynx of the fowl. a, upper beak; b, b', palate; c, lateral palatine ridges; d, median swelling and e, papillae of the palate; g, palatine cleft (choanal cleft); h, orifice of the gland. maxillaris; i, orifice of the glandd. palatinae latt.; k, orifice of the glandd. palatinae medd.; l, orifice of the glandd. pterygoideae and glandd. tubariae; m, orbital folds; n, pharyngeal folds; o, infundibular cleft; p, pharynx; q, rows of pharyngeal papillae; r, oesophagus. (Right) Floor of the mouth and pharyngeal cavities of the fowl. a, lower beak; b, orifice of the glandd. mandibulares postt.; c, tongue; d, row of lingual papillae; e, orifice of the glandd. linguales postt.; f, larynx (larynx cran.); g, laryngeal cleft; h, oesophagus; i, rows of pharyngeal papillae. (Reprinted from R. Nickel, A. Schummer, E. Seiferle, W. G. Siller, and P. A. L. Wight, "Anatomy of the Domestic Birds" p. 43. Copyright 1977 by Springer-Verlag, New York. Used by permission of the publisher.)

to either carry food or as a display apparatus during the breeding season.

Immediately behind the tongue is the laryngeal mound. It contains a narrow slit-like opening into the glottis of the larynx. The laryngeal mound generally contains rows of caudally directed cornified papillae which aid in moving food toward the esophagus during swallowing.

B. Esophagus and Crop

The esophagus is a thin-walled, distensible tube which transports food from the pharynx to the stomach allowing birds to swallow their food whole. Thus, it contains a number of longitudinal folds which provide distensibility. The avian esophageal wall consists of four layers: mucosal, submucosal, muscle tunic, and the serosal layer and generally contains only smooth muscle cells, with a circular muscle layer predominating (McLelland, 1979).

Unlike mammals, the avian esophagus is divided into a cervical and a thoracic region. In addition, the esophagus of birds lacks both upper and lower esophageal sphincters which are present in mammals (see Mule, 1991).

In many, but not all (e.g., gulls, penguins, ostriches), species of birds the cervical esophagus is expanded to form a crop. The crop functions to store food and may be spindle-shaped, bilobed, or unilobed. In the chicken, the crop is a ventral diverticulum of the esophagus and contains longitudinal folds on the inner surface thus making it distensible. Beyond the crop, the esophagus continues as the thoracic esophagus to connect with the proventriculus.

A small number of species have a diverticulum or bilaterally symmetrical expansion of the cervical esophagus, the esophageal sac. In most species which have such a structure, it occurs only in the male and functions as a display during the breeding season and for the production of mating calls.

The esophagus and crop are lined with incompletely keratinized stratified squamous epithelia into which open numerous mucous glands. These glands are generally more numerous in the thoracic esophagus and may even be absent in the cervical region. Mucous glands are located in the crop only near the junction with the esophagus.

The cervical esophagus is innervated by parasympathetic nerves. The thoracic esophagus is innervated by the vagus and the coeliac plexus. The esophagus is innervated by a few adrenergic fibers which synapse with the myenteric plexus rather than the muscles (McLelland, 1975; Mule, 1991).

C. Stomach

In mammals, the stomach consists of a single chamber. However, in birds, the stomach consists of two chambers, the proventriculus (pars glandularis) and the gizzard (pars muscularis), with the former being the mammalian counterpart. The proventriculus or glandular stomach is located orad to the gizzard or muscular stomach. The proventriculus is a fusiform organ varying in size and shape among species, being relatively large and distensible in aquatic carnivores while being relatively small in granivorous species. In ostriches, which lack a crop, the proventriculus is especially large. The proventriculus is the site of acid secretion whereas the gizzard functions in mechanical digestion and is the site of gastric proteolysis. The pyloric region connects the gizzard to the duodenum.

There are two extremes in gastric anatomy (McLelland, 1979). The first type, characteristic of carnivorous and piscivorous species, is adapted for storage and the digestion of a relatively soft diet. The two stomach chambers contain little separation, although one chamber may be more developed than the other, depending on the species. The second stomach type, characteristic of omnivores, insectivores, herbivores, and granivores, is adapted for very hard diets. The proventriculus and gizzard are separated by an isthmus termed the zona intermedia gastric. The proventriculus is relatively small, while the gizzard is large and powerful. The gizzard consists of two pairs of opposing muscles termed thick and thin pairs (Figure 3), which are composed of circular muscle. These pairs of muscles are lacking in some species (e.g., herons, hawks, owls).

The proventriculus, which generally lacks ridges except in fish- and meat-eating species, is lined with a mucous membrane. Projecting into the lumen are papillae on the surface of which can be seen the openings of the compound glands which secrete gastric juices. These glands generally contain only oxynticopeptic cells and secrete hydrochloric acid, pepsin, and mucous.

The interior surface of the gizzard is lined with a cuticle, sometimes called koilin, which is produced by the mucosal glands. The cuticle protects the gizzard from acid and proteolytic enzymes secreted by the pro-

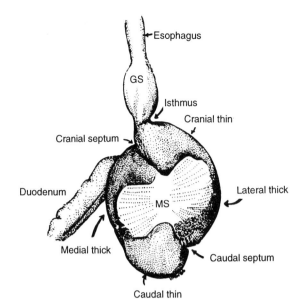

FIGURE 3 Anatomical features of muscular stomach of domestic turkeys. GS, glandular stomach; MS, muscular stomach. Note that cranial thin muscle of MS is continuous with lateral thick muscle and separated from caudal thin muscle at caudal septum. Similarly, the caudal thin muscle is continuous with the medial thick muscle and separated from cranial thin muscle by cranial septum. (From Chaplin and Duke (1990), *Am. J. Physiol.* with permission.)

ventriculus and from injury during grinding of hard food items. The greenish or brownish color of the cuticle is due to the reflux of bile pigments from the duodenum. While the cuticle is continuously worn away and replaced in most birds, it is shed in some species.

The pyloric portion of the stomach (pars pylorica gastric) shows a considerable range in development. In *Gallus,* it is only 0.5 cm long. Its mucosal glands secrete mucous instead of a cuticle. In other species, the pyloric portion is enlarged and contains a cuticle. While the function of this region is unknown, it is believed that it may slow the movement of large particles into the duodenum (Vergara *et al.,* 1989).

The stomach is innervated by the vagus and perivascular nerve fibers from the coeliac and mesenteric plexii. The muscle cells are innervated by cholinergic fibers while noradrenergic fibers innervate mainly blood vessels. The myenteric nerve plexus lies just under the serosa, while the submucosal plexus is lacking. Since the longitudinal muscle layer is absent, the myenteric nerve plexus is normally visible through the transparent serosa.

D. Small Intestine

The small intestine is sometimes divided into the duodenum, jejunum, and ileum, although these are not distinguishable based on histology or gross observation.

While there is a distinct duodenal loop, the yolk stalk (diverticulum vitellinum; formally called Meckel's diverticulum) is often used as a landmark to separate the jejunum and ileum. Intestinal length varies considerably between species, being relatively shorter in frugivores, carnivores, and insectivores and long in granivores, herbivores, and piscivores. However, the length appears to be relatively shorter than in mammals.

The intestinal wall can contain either folds or villi depending on the species. The type of mucosal projections are not necessarily consistent between the small and large intestine. *Gallus* species have villi which decrease in length from 1.5 mm in the duodenum to 0.4–0.6 mm in the ilium and rectum. The number of villi decreases from 1 to 10 days of age, but thereafter remains constant.

Genetic selection for growth has altered villi morphology (Yamauchi and Isshiki, 1991). Compared to White Leghorns, the villi of broilers are larger and show more epithelial cell protrusions from the apical surface of the duodenal villi. Nevertheless, the villi from both types of chickens form a zig-zag arrangement which is thought to slow ingesta flow.

The intestinal wall contains the same four layers as the remainder of the tract including the mucosal, submucosal, muscle tunic, and the serosal layer. The mucosal layer consists of the muscularis mucosa, lamina propria, and epithelium. However, the muscularis mucosa and lamina propria are poorly developed in birds, possibly because the villi lack a central lacteal. Although Brunner's glands, common to mammals, are absent (Calhoun, 1954), tubular glands possibly homologous to Brunner's glands, are present in some species (Ziswiler and Farner, 1972). The epithelium contains chief cells, goblet cells, and endocrine cells. The crypts of Lieberkuhn are the source of epithelial cells lining the villi. The crypts contain undifferentiated cells, goblet cells, endocrine cells, and lymphocytes. Globular leukocytes and Paneth cells appear near the base of the crypts.

The intestines contain extensive innervation from both the sympathetic and parasympathetic nervous system. As described by Bennett (1974), innervation is both cholinergic and adrenergic. Contraction of the rectum appears to be mediated by noncholinergic, nonadrenergic nerves (Bartlet, 1974; Takewaki *et al.*, 1977). The enteric nerve plexuses were described by Ali and McLelland (1978). Except in the rectum, the longitudinal muscle is sparsely innervated.

The intestinal nerve (Nerve of Remak), which runs the length of the small and large intestine, is unique to birds. While it has no mammalian homolog, it may be analogous to the prevertebral ganglion (Hodgkiss, 1984a). This nerve is thought to be a mixed nerve containing both sympathetic and parasympathetic autonomic fibers (Hodgkiss, 1986).

E. Ceca

Arising at the junction of the ileum and rectum are the ceca. In some species, the ceca may be absent (e.g., Psittaciformes, Apodiforms, and Piciforms) or rudimentary (e.g., Columbiformes and Piciformes). In other species they are either paired (e.g., herbivores, most granivores, and owls), singular (Ardeidae), or consist of a double pair (secretary bird). McLelland (1979) has grouped ceca into four types based on morphology: (1) intestinal, which resemble the remainder of the intestine; (2) glandular, which are long and contain many actively secreting crypts; (3) lymphoid, which are reduced in size and contain many lymphocytes and occasional nonsecreting crypts; and (4) vestigial, which are small with a reduced lumen. A correlation between diet and cecal development and between the size of the ceca and the length and width of the rectum has not been discovered (McLelland, 1989).

In chickens, a cecum can be morphologically divided into three regions (Ferrer *et al.*, 1991). Near the ileocecal junction is the basis ceci where the villi are well developed. The medial cecal region (corpus ceci) has longitudinal folds with small villi while the distal cecal region (apex ceci) similarly has small villi and contains both longitudinal and transverse folds. The combination of villi and musculature near the ileocecal junction effectively prevents even very small particles from entering the ceca (Ferrando *et al.*, 1987), although fluid contents do enter.

There is growing understanding in the importance of the ceca. Cecectomy results in lower metabolizability of food, greater loss of amino acids, and lower digestibility of crude fiber (Chaplin, 1989). The role of the ceca in absorption is discussed later.

F. Rectum and Cloaca

The rectum, sometimes called colon, is relatively short, linking the ileum with the coprodeal compartment of the cloaca. Whereas the colon of mammals has no villi and many goblet cells, the rectum of birds has numerous flat villi and relatively few goblet cells (Clauss *et al.*, 1991). In addition, the avian rectum has relatively few crypts, and they are shorter than in mammals. As discussed under "Absorption," the cloaca and rectum have an important role in water reabsorption.

The cloaca serves as a common pathway for excretory, reproductive, and digestive wastes. It contains three chambers: coprodeum, urodeum, and proctodeum. The coprodeum is the most cranial portion into which empties the rectum. The urodeum is the middle and smallest compartment of the cloaca separated from the coprodeum and proctodeum by the coprourodeal fold and the uroproctodeal fold, respectively. The urinary and reproductive tracts empty into the urodeum. The final chamber, the proctodeum, opens externally through the anus. The bursa of Fabricius, involved in immune function, projects dorsally into the proctodeum. Also projecting into the dorsal portion of the proctodeum is the dorsal proctodeal gland, sometimes called the foam gland, which secretes a white, frothy fluid.

II. ANATOMY OF THE ACCESSORY ORGANS

A. Pancreas

The pancreas is a pale yellow or red organ located within the duodenal loop, although in some species, such as the budgerigar, part of it may be found outside the loop. The gland has both an endocrine, which will be discussed later, and an exocrine function. It is relatively small in carnivores and granivores but large in piscivores and insectivores. The pancreas is generally divided into three lobes, dorsal, ventral, and splenic, but their function in unknown (Paik *et al.*, 1974).

The exocrine pancreas consists of compound tubuloacinar glands which are divided into lobules. The number of pancreatic ducts varies from one to three (three in the domestic fowl). The pancreatic ducts generally drain into the distal part of the ascending duodenum and rarely drain into the descending loop of the duodenum. In domestic birds, the pancreatic and bile ducts drain into the ascending loop of the duodenum by a common papilla (Figure 1). The pancreas secretes amylase, lipase, proteolytic enzymes, and sodium bicarbonate.

B. Liver

The liver also functions as both an endocrine and an exocrine gland. It is divided into right and left lobes which are joined cranially at the midline. The right lobe is larger and, in the domestic fowl and turkey, the left lobe is subdivided into the dorsal and ventral parts. The bile canaliculi drain into the interlobular ducts. The lobular ducts then combine to form the right and left hepatic ducts. In birds, unlike in mammals, bile is transported to the duodenum via two ducts. The right and left hepatic duct combine to form the common hepatoenteric duct which then goes to the duodenum. However, a hepatocystic duct branches from the right hepatic duct and connects to the gall bladder which, in turn, is drained by the cysticoenteric duct into the duodenum. In species without a gall bladder (e.g., pigeons, some parrots, and ostriches), the branch of the right hepatic duct, called the right hepatoenteric duct, drains directly into the duodenum. The bile ducts generally drain into the duodenum at a site very near the pancreatic ducts. This generally occurs on the ascending loop of the duodenum. However, in some species including the ostrich and *Columba,* the ducts empty into the descending loop of the duodenum.

III. MOTILITY

A. Esophagus

Deglutition has been studied in *Gallus* by White (1968, 1970) and Suzuki and Nomura (1975). For prehension, the head is first lowered after which food is grasped with the beak and then moved towards the oropharynx with the tongue. The choana reflexively closes. The oral phase of swallowing involves the rapid rostrocaudal movements of the tongue for 1–3 sec which moves the food particles caudally (Suzuki and Nomura, 1975). This movement is assisted by the caudally directed papillae.

During the next phase, the pharyngeal phase, the infundibular mound and glottis close, the hyoid apparatus becomes concave, the tongue is moved backward, and the esophagus is moved forward thus decreasing the distance between the oral cavity and the esophagus. The head is raised and further tongue movements, assisted by the rostrocaudal movements of the infundibular mound, propel the food particles from the tongue to the esophagus.

Primary peristalsis within the esophagus moves the bolus toward the stomach (esophageal phase). Contractions are more rapid in the cranial esophagus than in the thoracic esophagus. In a fasted bird, the longitudinal muscle layer obliterates the esophagoingluvial fissura, thus preventing a bolus from entering the crop (Ashcraft, 1930). After partial gizzard filling, the esophagoingluvial fissura is relaxed and food can either enter the crop or stomach depending on the contractile state of the gizzard. The crop acts as a temporary food storage site (Hill, 1971). The destination of food appears to be controlled by the contractile state of the gizzard with food entering or bypassing the crop when the gizzard is contracting or relaxing, respectively. In 6- to 10-week-old turkeys fasted overnight, almost no ingested food enters the crop during the first 4–6 hr after dawn. During each meal the gastrointestinal tract fills above the upper one-third of ileum. In late afternoon meals, the crop also fills (G. Duke, personal communication).

Food is evacuated from a crop as a result of contractions by the crop wall. Such contractions last about 6 sec with a force of approximately 20 cm H$_2$O (Fileccia et al., 1984). During primary peristalsis, there is a cessation of spontaneous electrical activity which is associated with relaxation of the muscular wall. This is followed by a propagated, long-lasting spike burst of high amplitude. As the peristaltic wave moves aborad, it is preceded by inhibition of the muscles directly in front of the wave. The rate of emptying of the crop is not influenced by particle size or whether the compounds are soluble or insoluble (Vergara et al., 1989).

Unlike mammals, spontaneous electrical activity and contractions have been recorded in the esophagi of birds (see Mule, 1991). This activity is myogenic in origin. It occurs independent of slow waves which are absent in the esophagi of birds. While the function of these spontaneous contractions is unknown, they may act to clear contents from the esophagus.

Primary peristalsis is mediated entirely by the extrinsic nervous system (Mule, 1991). The enteric nervous system of birds is not responsible for propagating peristalsis, at least beyond short distances. Sectioning the glossopharyngeal or vagus nerve can disrupt peristalsis in the cervical and thoracic esophagus, respectively. Denervation of the esophagus prevents the propagation of electrical wave activity, indicating that the muscle cells act similar to multiunit smooth muscle cells.

B. Gastrointestinal Cycle

Motility of the mammalian stomach is regulated by slow waves thought to arise from the interstitial cells of Cajal which lie adjacent to the myenteric plexus. These waves propagate through the circular smooth muscle. The stomachs of most birds lack longitudinal smooth muscle and do not display slow waves. As a result, gastric motility is more complex in birds than in mammals (Dziuk and Duke, 1972).

During the gastrointestinal cycle of birds, the thin muscles of the muscular stomach contract and the isthmus closes, after which the pylorus opens and gastric contents flow into the duodenum (Figure 4). Next, the duodenum contracts followed by relaxation of the isthmus and a contraction of the thick muscles of the muscular stomach. This precipitates a refluxing of the contents of the muscular stomach into the glandular stomach. The cycle is completed with contraction of the glandular stomach. A gastroduodenal cycle occurs with a frequency of 3.3 cycles/min in turkey (Duke, 1982). As might be expected, the largest change in intraluminal pressure is associated with contraction of the thick muscles (Figure 5).

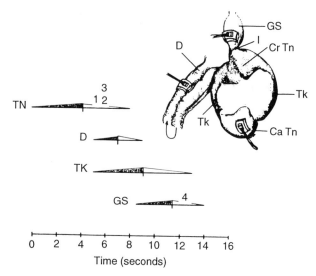

FIGURE 4 (Top right) Illustration of gastroduodenal apparatus of turkeys showing anatomical relationships of organs and placement of strain gauge transducers. Letters refer to organs as follows: GS, glandular stomach; I, isthmus; CrTN, cranial thin muscle of muscular stomach; TK, thick muscle pair (dorsal and ventral) of muscular stomach; CaTN, caudal thin muscle of muscular stomach; D, duodenum. (Bottom left) Triangles graphically represent sequence and duration of events in gastroduodenal cycle. Stippled areas indicate duration of contraction and clear areas indicate duration of relaxation of thin muscle pair (TN) and thick muscle pair (TK) of the muscular stomach, duodenum (D), and glandular stomach (GS). Numbers refer to noncontractile events in gastroduodenal cycle; 1, pylorus open; 2, isthmus open; 3, pylorus closed; 4, isthmus closed. (From Chaplin and Duke (1988), Am. J. Physiol. with permission.)

As stated earlier, the pyloric region appears to control the movement of materials from the gizzard to the duodenum. Whereas soluble material is readily transported from the gizzard to the duodenum, larger particles are retained longer within the muscular stomach (Vergara et al., 1989).

Initiation of the gastrointestinal cycle is not dependent on extrinsic innervation in sated birds, suggesting that there is an intrinsic pacemaker controlling the cycle (Chaplin and Duke, 1988). Denervation of the stomach slows the rate of the gastrointestinal cycle in fasted birds and disrupts its normal synchronization (Figure 6), indicating that the gastrointestinal cycle is not independent of external neural input.

The pacemaker for the gastrointestinal cycle appears to reside in the isthmus (Chaplin and Duke, 1990). Destruction of the myenteric plexus in this area reduced contractions of the muscular stomach and duodenum by 50%, while simultaneously abolishing glandular stomach contractions.

The muscular stomach of raptors lack the two pairs of opposing muscles characteristic of other types of birds and, as a result, the gastroduodenal cycle is simplified. It is described as having three phases including mechani-

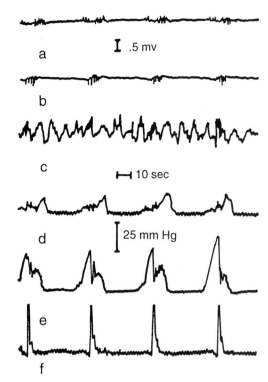

FIGURE 5 Tracings of typical records of electrical potential and intraluminal pressure changes from the stomach and duodenum of turkeys. (a, b, and c) Tracings of electrical potential changes recorded from the proventriculus, thick cranioventral muscle of the gizzard, and proximal duodenum, respectively. Slow waves with spikes are evident in the duodenum; only spikes associated with contractions are evident in the proventriculus and gizzard. (d, e, and f) Tracings of the corresponding intraluminal pressure changes recorded from these same organs, respectively. Time constant for electrical recordings was 1.1 sec. (From Duke et al., 1975.)

cal, chemical, and pellet formation and egestion (Kostuch and Duke, 1975). A peristaltic wave originating in the proventriculus moves aborad through the gizzard and into the small intestine. While the sequence of motility in the stomach of raptors resembles that in mammals, slow waves have not been recorded in these species.

Egestion is another gastrointestinal function unique to birds (Rea, 1973). This process which involves the oral expulsion of nondigestible material is more common in carnivores. Whenever bone, fur, or feathers are ingested, these materials are compacted and orally egested. This process is unlike regurgitation in ruminants or vomiting in mammals (Duke et al., 1976a). Beginning approximately 12 min prior to egestion, gizzard contractions increase in frequency and amplitude. This process both compacts the material into a pellet and moves it into the lower esophagus. Beginning 8 to 10 sec prior to egestion, the pellet is moved orad by esophageal antiperistalsis. Neither the abdominal nor duodenal muscles are involved.

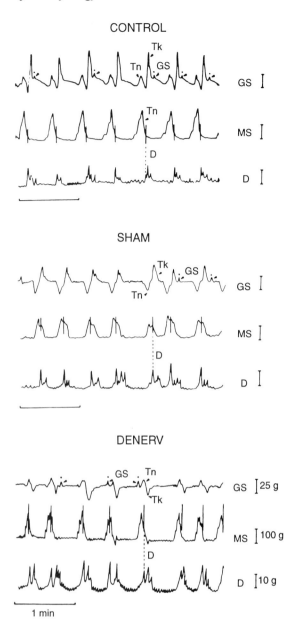

FIGURE 6 Tracings of recordings of glandular stomach (GS), muscular stomach (MS), and duodenal (D) contractions detected by implanted strain gauge transducers. Recordings were obtained on day 3 after surgery in three treatment groups used. Arrows and lettering identify contractions of each organ during one gastroduodenal cycle. Glandular stomach contractions are marked with an arrow and a closed circle; duodenal synchronization is identified with a dashed line. Note normal synchronization of contractions in the control tracings. (From Chaplin and Duke (1988), Am. J. Physiol. with permission.)

Enterogastric reflexes control gastric emptying. Simply increasing intraduodenal pressure, or administering intraduodenal injections of HCl, hypertonic NaCl, or amino acid solutions inhibits gastric motility (Duke and Evanson, 1972; Duke et al., 1972). Inhibition was related to dose and volume, it generally occurred within 3 to

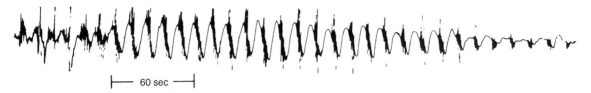

FIGURE 7 Myoelectric recording of action potentials on slow waves (*Gallus*). (From Clench *et al.* (1989), *Am. J. Physiol.* with permission.)

30 sec, and it persisted 2 to 35 min. Following an intraduodenal injection of a lipid solution, gastric motility decreased after 4 to 6 min and remained inhibited for 24 to 45 min. This latter response appears to involve hormonal regulation, presumably by an enterogastrone.

C. Small Intestine

The migrating myoelectric complex (MMC) is characterized by electrical potential changes known as slow waves which travel aborad and are associated with periodic spike potentials and smooth muscle contraction (Figure 7). While the MMC has not been extensively studied in birds, available data suggest that the MMC is similar in birds and mammals (Clench *et al.*, 1989). The MMC has three phases: quiescence (phase 1); irregularly spaced spike activity superimposed on slow waves (phase 2); and high-amplitude, regular spike activity superimposed on slow waves (phase 3). In chickens, the MMC has a periodicity of 77 to 122 min, while in owls it is only 21 min. The duration of phase 3 lasts approximately 5 to 8 min (Figure 8). These values are similar to those in mammals. Nevertheless, the propagation speed in birds is relatively slow, ranging from .48 to .62 cm/min. Whereas the MMC has been observed in the duodenum of chickens, quail, and owls, it has only been observed aborad of the duodenum in turkeys (Mueller *et al.*, 1990).

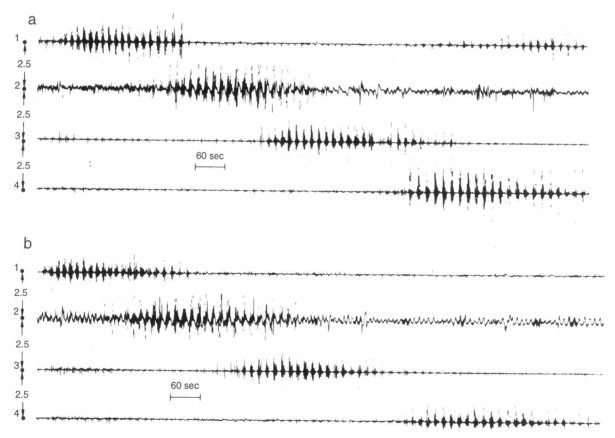

FIGURE 8 Representative myoelectric recording of a migrating myoelectric complex (MMC) in *Strix*. The electrodes were placed on the proximal ileum 2.5 cm apart. (a) Bird in fed state; (b) bird in fasted state. High frequency of MMCs in this owl species is indicated in a, with a second regular spike activity beginning on *lead 1* before previous complex has propagated through *lead 4*. (From Clench *et al.* (1989), *Am. J. Physiol.* with permission.)

Intestinal refluxes appear unique to birds (Duke et al., 1972). Increases in intraluminal pressure of the duodenum normally follow increased pressure in the muscular stomach. However, approximately every 15 to 20 min in the turkey, large pressure changes were observed in the duodenum which were associated with an inhibition of gastric motility and a reflux of intestinal contents (Figure 9). This has been observed in a number of other species as well and appears more frequently as dietary fat levels increase (Duke et al., 1989).

D. Ceca

Motility in the ceca is not well understood. The ceca are filled as a result of the convergence of rectal antiperistaltic waves and ileal peristaltic waves. Due to the morphology of the ileocecal junction, only fluid or very small particles are allowed entrance to the ceca. In fact, 87–97% of cecal fluid originates from the urine (Bjornhag and Sperber, 1977). The importance of the movement of urine to the ceca is discussed below.

While MMC-like bursts of electrical activity have been observed within the cecum, this activity does not migrate and, thus, does not constitute a MMC (Clench et al., 1989). Two types of contractions have been recorded in the ceca of turkeys (Duke et al., 1980). One type has a low amplitude occurring at a rate of 2.6/min, while the other has a higher amplitude and occurs at a rate of 1.2/min. Low-amplitude contractions are associated with mixing whereas high-amplitude contractions are propulsive. While aboral contractions were more common than orad contractions, the latter contractions had a greater amplitude, and, thus, prevented the collection of digesta in the distal ceca. Peristalsis in a cecum appears to be myogenically mediated, with inhibitory neural input apparently able to suppress such contraction (Hodgkiss, 1984b). In mammals, distention causes an ascending stimulation and a descending inhibition which is neurogenically mediated by the enteric nervous system and can be blocked by tetrodotoxin. By contrast, in birds, distention causes contraction of the circular muscle which is unaffected by tetrodotoxin and, thus, apparently not controlled by the enteric nervous system.

The contents of the ceca are much different in consistency than that of the rectum and can, therefore, be easily distinguished from rectal feces. The ceca evacuate relatively infrequently compared to the rectum, with rectal evacuations preceding cecal evacuations (Duke et al., 1980). Cecal evacuations in young turkeys typically occur within 1 to 5 min of lights-on and again in late afternoon. There is an increase in the frequency of high-amplitude contractions, with four to seven such contractions occurring during the 2 min preceding cecal evacuation. These contractions are associated with high-amplitude electrical spiking. One high-amplitude contraction also occurs in the ileum and rectum at the time of cecal evacuation. The ratio of cecal to rectal evacuations is also influenced by diet, ranging from 1:7.3 when feeding barley to 1:11.5 when feeding corn (Roseler, 1929). As discussed below under "Absorption," the extended time that digesta spends in the ceca provides a unique role for this organ.

E. Rectum

The rectum displays almost continuous antiperistalsis. Such motility is responsible for carrying urine from the urodeum into the colon and ceca (Akester et al., 1967; Polin et al., 1967). Two types of colonic slow waves have been recorded (Lai and Duke, 1978). These include small, short-duration (sSW) and large, long-duration (lSW) slow waves (Figure 10). The sSW are associated with small contractions and, as shown with radiography, are antiperistaltic. The lSW are associated with large contractions but, radiographically, could not be associated with movements of rectal contents.

The frequencies of sSW and lSW are shown in Table 2. The frequency of sSW is highest in the distal colon. In contrast, the frequency of lSW is highest in the proximal colon and cannot be recorded from the distal colon. This pattern of slow waves, when compared to the motility pattern, suggests that sSW arise at the distal colon and are responsible for antiperistaltic movement, whereas

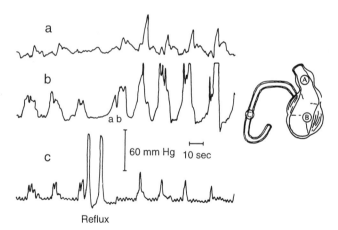

FIGURE 9 Tracings of typical records of pressure changes obtained from the proventriculus (a), gizzard (b), and upper proximal duodenum (c) of a turkey, showing pressure events occurring during a duodenal reflux. Positions of open-tipped tubes within GI tract are indicated by letters A, B, and C (circled) on the diagram of a sagittal section of the stomach. The biphasic patter (b) of the tracing representing the contraction of the gizzard is quite variable; two phases of one cycle are (a) pressure wave due to contraction of thin muscle pair; and (b) of thick muscle pair. (From Duke et al., 1972, with permission from The American Physiological Society.)

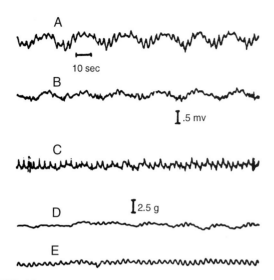

FIGURE 10 Electrical potential changes and contractile forces recorded from three bipolar electrodes (A, B, C) and two strain gauges (D, E) in the colon of a turkey. Electrodes and strain gauges were implanted at 10, 8, 6, 3, and 1 cm from the cloaca, respectively. Both large and small slow waves are evident in A and B; only small slow waves are evident in C. Small contractions (antiperistalsis are evident in D and E but large contractions (peristalsis) can be seen only in D. (From Lai and Duke (1978) *Dig. Dis.* **23**, 673–681. Reproduced with permission of S. Karger AG, Basel.)

the lSW begin in the proximal colon and are responsible for peristaltic movement of rectal contents.

The nearly continuous antiperistaltic activity of the colon is interrupted only near the time of defecation (Duke, 1982). Beginning approximately 10 min prior to defecation, the amplitude of the sSW begins to decrease while the frequency gradient of lSW increases. These conditions favor inhibition of antiperistalsis and stimulation of peristalsis. Defecation is associated with a strong contraction beginning at the proximal colon, moving aborally, and carrying the contents the length of the colon in less than 4 sec.

F. Other Influences on Motility

Motility of the gastrointestinal tract shows diurnal variations. In turkeys, the frequency and amplitude of muscular stomach contractions increased during light periods relative to dark periods (Duke and Evanson, 1976). The increased and decreased gastric contractions coincided with just prior to "lights-on" and "lights-off," respectively. Although less pronounced, this diurnal rhythm of gastric activity continued when birds were fasted.

Birds also display cephalic, gastric, and intestinal phases of motility. In 24-hr-fasted great horned owls, turkeys, and red-tailed hawks, the sight of food stimulated gastric contractions (Duke *et al.*, 1976b). When allowed to eat, gastric activity increased further, indicative of the gastric phase. During the intestinal phase, entrance of food into the duodenum slows the frequency of gastric contractions to allow time for digestion (Duke *et al.*, 1973). While the cephalic phase is mediated largely by the nervous system, there appears to also be an as yet unidentified endocrine component which increases motility.

Motility is influenced by many factors. For example, anesthetics, including nembutal and methoxyflurane, reduce gastroduodenal motility as well as decrease gastric secretion (Kessler *et al.*, 1972; Duke *et al.*, 1977). High environmental temperatures decrease gastrointestinal motility whereas cold temperatures differentially affect motility in various parts of the digestive system but appear to have an overall effect of decreasing transit time (Tur and Rial, 1985).

IV. NEURAL AND HORMONAL CONTROL OF MOTILITY

Contractions of the esophagus are increased by acetylcholine (ACh) and vagal stimulation and unaffected by sympathetic nerve stimulation (Bowman and Ever-

TABLE 2 Mean Frequency and Amplitude of Small and Large Contractions and of Small and Large Slow Waves in the Rectums of Turkeys[a]

	Contractions				Slow waves			
	Small		Large		Small		Large	
	Frequency (cycles/min)	Amplitude (g)	Frequency (cycles/min)	Amplitude (g)	Frequency (per min)	Amplitude (mV)	Frequency (per min)	Amplitude (mV)
Proximal[b]	14.6 ± 0.85	0.45 ± 0.24	2.66 ± 0.26	0.54 ± 0.20	15.4 ± 1.07	0.17 ± 0.08	2.83 ± 0.26	0.21 ± 0.09
Middle	—	—	—	—	15.8 ± 1.12	0.16 ± 0.09	2.76 ± 0.24	0.12 ± 0.06
Distal	15.4 ± 0.69	<0.70 ± 0.33	—	—	16.4 ± 2.16	0.25 ± 0.12	—	—

[a] From Lai and Duke (1978).
[b] Proximal, middle, and distal refer to electrode implant sites on the colon 10, 6, and 1 cm from the cloaca, respectively, or to strain gauge transducer implants at 8 (proximal) and 3.5 (distal) cm from the cloaca. (Large contractions were not recorded from the distal strain gauge nor were large slow waves recorded from the distal electrode.)

ett, 1964; Ohashi and Ohga, 1967; Taneike et al., 1988). However, there is evidence for nonadrenergic, noncholinergic (NANC) inhibition of esophageal smooth muscles (Sato et al., 1970; Postorino et al, 1985). Serotonin also causes contraction of the esophagus (Mule et al., 1987), and this affect appears to be mediated indirectly by activation of both cholinergic and NANC excitatory neurons (Fileccia et al., 1987).

As shown in Table 3 many peptides have been identified within the gastrointestinal tract of birds. The function of these peptides remains to be elucidated. Although the identity of the NANC excitatory neurotransmitter was not identified, it has been shown that neurotensin can induce contraction of crop smooth muscle (Denac and Scharrer, 1987). This latter effect is a postjunctional response not mediated by acetylcholine, prostaglandins, or opioids.

Histamine interacts with cholinergic neurotransmission to control esophageal contractions (Taneike et al., 1988). Histamine dose-dependently induces esophageal contractions, but this effect is blocked by tetrodotoxin suggesting that its effect is mediated by the release of ACh. Contractions induced by vagal stimulation are increased by the presence of histamine, whereas histamine-induced contractions were enhanced by acetylcholinesterase inhibitors. It appears, therefore, that histamine modulates, via H_1 receptors, the release of ACh to control the contraction of esophageal smooth muscle.

As indicated in Figure 11, enkephalins also cause contraction of the esophagus, but their effect is mediated by serotonergic neurons.

The crop appears to be controlled similarly to the esophagus. Electrical stimulation of the crop causes contraction which is largely, but not completely, due to ACh release since it can be significantly blocked by atropine (Denac et al., 1990). While it is unclear as to which neurotransmitters are responsible for the contraction not blocked by atropine (i.e., NANC-induced), neurotensin, bombesin, and substance P have been shown

TABLE 3 Distribution and Frequency of the Endocrine Cells in the Avian Gastrointestinal Tract[a]

	Proventriculus	Gizzard	Pyloric region	Duodenum	Jejunum	Ileum	Cecum	Rectum
Gastrin								
Chicken	—	—	■	╫	+	—	—	—
Quail	—	—	■	+	±	+	—	—
Duck	—	╫	■	+	+	+	—	—
Somatostatin								
Chicken	■	—	■	+	—	—	—	—
Quail	□	+	■	╫	+	+	—	±
Glucagon								
Chicken	╫	—	—	—	—	—	—	—
Quail	╫	—	—	—	—	—	—	—
Gilcentin								
Quail	—	—	—	+	+	□	±	±
Neurotensin								
Chicken	—	—	□	+	+	+	±	╫
Quail	—	—	—	╫	+	□	±	╫
Motilin								
Quail	—	—	—	╫	+	+	—	—
Pancreatic polypeptide								
Chicken	±	—	—	—	±	±	—	—
Gastrin releasing peptide (bombesin)								
Chicken	□	—	—	—	—	—	—	—
Secretin								
Chicken	—	—	—	+	+	+	—	—
Quail	—	—	—	+	+	+	—	—
Substance P								
Chicken	—	—	—	+	+	+	+	+
Vasoactive intestinal peptide								
Chicken	╫	+	NA	╫	╫	╫	╫	□

Note. (−) none; (±) very few, irregularly observed; (+) few, regularly observed; (╫) moderate; (□) numerous; (■) very numerous; NA, not analyzed.
[a] Adapted from Yamada et al. (1983).

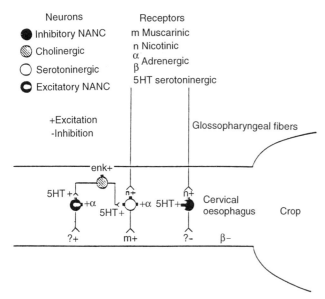

FIGURE 11 Schematic diagram illustrating the putative pattern of the intrinsic innervation of the pigeon cervical esophagus. Different neuron types are present in the intramural plexuses; excitatory cholinergic, excitatory NANC, inhibitory NANC and serotoninergic neurons. (Reprinted from *Comp. Biochem. Physiol.* 99A, Mule, F., 491–498 © (1991), with kind permission from Elsevier Science Ltd, The Boulevard, Langford Lane, Kidlington OX5 1GB, UK.)

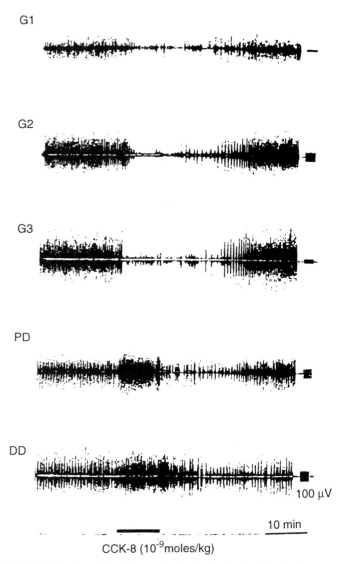

FIGURE 12 Recording of gastroduodenal electrical activity showing effect of cholecystokinin octapeptide (CCK-8; 10^{-9} mol·kg^{-1}) infusion. Studied gastric areas are as follows: G1, proventriculus; G2, craniodorsal thin muscle; G3, caudodorsal thick muscle; PD, proximal duodenum; DD, distal duodenum. Similar responses were observed in all animals. (From Martinez *et al.* (1993), *Am. J. Physiol.* with permission.)

to cause atropine-resistant contractions following electrical stimulation of the crop (Denac and Scharrer, 1987, 1988). Norepinephrine caused crop muscle relaxation which was mediated by β-adrenoceptors. It appears that the crop and esophagus are controlled by three types of nerves: stimulatory cholinergic neurons, stimulatory NANC neurons (probably peptides), and inhibitory noradrenergic nerves.

Cholecystokinin octapeptide (CCK-8) is the most studied regulator of intestinal motility. Intravenous infusion of CCK-8 inhibited gastric and duodenal motility (Savory *et al.*, 1981). CCK-8 and CCK-tetrapeptide (CCK-4) inhibited gastric electrical activity (Martinez *et al.*, 1992). Whereas CCK-4 inhibited duodenal electrical activity, CCK-8 stimulated such activity (Figure 12). CCK-A and CCK-B receptor antagonists were unable to block the effects of CCK. It appears that the action of CCK on gastric and duodenal motility is similar in birds and mammals. CCK inhibits stomach motility, and the increase in duodenal activity is suggested to cause segmental contractions and, thus, also delay gastric emptying (Martinez *et al.*, 1992).

Vagotomy and hexamethonium blocked the response to CCK-8 in the chicken stomach (i.e., proventriculus and gizzard), but had no effect on the duodenal response (Martinez *et al.*, 1993). Furthermore, the action of CCK-8 in the duodenum was not altered by atropine or methysergide. Phentolamine and propranolol had no effect on the gastric or duodenal response to CCK-8. It appears, therefore, that the action of CCK-8 on the stomach is mediated via the vagus, whereas the action in the duodenum probably involved a direct effect on smooth muscle cells. The action of CCK-8 on the stomach was blocked by N^G-nitro-L-arginine methyl ester (L-NAME), suggesting that the inhibitory effect of CCK involves the release of nitric oxide. Since L-NAME did not completely block the CCK effect on the caudodorsal thick muscle, it is likely that another neurotransmitter, possibly vasoactive intestinal peptide, is involved in this latter response. Interestingly, CCK-8 caused an excitatory action in the stomach of vagotomized chickens indicating that CCK-8 may, in addition to acting via the vagus, have a direct action on gastric muscles. The increase in electrical activity caused by CCK-8 in the duodenum was enhanced by L-NAME, indicating that nitric oxide may be a tonic inhibitor of duodenal electrical activity.

Chicken gastrin (cG), which belongs to the gastrin/CCK family, has been isolated from the chicken antrum (Dimaline et al., 1986). Intravenous infusions of cG caused effects similar to CCK-4 (Martinez et al., 1992). This suggests the existence of one receptor subtype in the stomach recognizing both CCK and cG, but possibly two receptors in the duodenum, one recognizing CCK-8 and the other CCK-4 and cG.

Opioid peptides appear to be involved in the MMC. Infusion of met-enkephalin, morphine, and β-casomorphin (5×10^{-7} moles/kg) induced intense electrical activity, similar to phase 3 of the MMC, in the distal duodenum, which migrated through the small intestine (Jiménez et al., 1992). The effect of morphine was blocked by naloxone (5×10^{-7} moles/kg), while higher doses of naloxone reduced gastroduodenal motility. Simultaneous to increasing duodenal electrical activity, activity in the stomach is inhibited.

A. Rate of Passage

The rate of passage of material through the digestive tract has been measured in many ways. Since digesta consists of both solid and liquid components, different types of markers have been used. Insoluble markers such as chromium-mordanted rice, cerium-mordanted rice, Cr_2O_3 or radiopaque plastic pellets (Branch and Cummings, 1978; Uden et al., 1980; Ferrando et al., 1987) have been used as indicators of solid transit time whereas a soluble marker such as Cr-EDTA (Vergara et al., 1989) or phenol red (Goñalons et al., 1982) has been used to measure liquid transit time. In general, it was found that larger particles are retained longer in the digestive tract.

In chickens, insoluble markers first appear in the excreta 1.6 to 2.6 hr after ingestion. However, mean retention time is a better indicator of transit time than is time to initial appearance of the marker. Mean retention time for insoluble markers can vary form 5 to 9 hr depending on the nature of the ingesta and its size.

Transit time of digesta is influenced by genetics. When comparing broiler and Leghorn-type chickens, the overall mean retention time is not different, but the time food spent in various parts of the digestive tract is different (Table 4).

The rate of food passage is affected by many factors. Feed transit time through the small and large intestine increases with age (Shires et al., 1987). This may account

TABLE 4 Mean Retention Time (min) of Solid Phase Markers in Various Segments of the Digestive System of Broilers and Leghorns

Gastrointestinal tract segment	Broilers[a]	Leghorns[a]	Broilers[b]
Crop	31	48	41
Proventriculus + gizzard	39	71	33
Duodenum	10	7	5
Jejunum	84	85	71
Ileum	97	84	90
Ceca	119	112	—
Rectum	56	51	26

[a]From Shires et al. (1987). The values have been adjusted for birds weighing 1800 g.
[b]From Van Der Klis et al. (1990).

for increases in metabolizable energy values of feedstuffs noted in older birds. Adding lipid (Sell et al., 1983) or protein (Sibbald, 1979) to the diet can increase passage time. Increases in environmental temperature also slows transit time.

V. SECRETIONS AND DIGESTION

A. Mouth

The salivary glands secrete mucous and, depending on the species, amylase. While amylase is not present in the saliva of *Gallus* and *Meleagris*, it is found in the saliva of the house sparrow (Jerrett and Goodge, 1973) and other species (Bhattacharya and Ghose, 1971). The volume of daily salivary secretion in *Gallus* ranges from 7 to 25 ml (Leasure and Link, 1940).

Mucous functions to lubricate food and allow it to move down the esophagus. However, in some species, mucous also functions as an adhesive coating on the tongue to aid in capturing insects or as a material which cements components during the construction of nests.

B. Esophagus and Crop

The esophagus is not important in chemical digestion, as its major secretion is mucous. The secretion of esophageal mucous is, nevertheless, important, for it is necessary to supplement the limited secretion of saliva. However, in some species, including the greater flamingo and male Emperor penguin, a nutritive merocrine-type secretion is produced by the wall of the esophagus which is fed to the young.

Some carbohydrate digestion may occur in the crop due to the presence of amylase activity (Philips and Fuller, 1983). Amylase activity at this site comes from either salivary secretions, intestinal reflux, or plant and/

or bacterial sources. Bolton (1965) reported that starch is hydrolyzed within the crop where it can either be absorbed; converted to alcohol, lactic or other acids; or transported down the gastrointestinal tract. Pinchasov and Noy (1994) showed that substantial amylolysis occurs in the crop. Sucrose is also hydrolyzed within the crop. While absorption of sugars from the crop appears possible, it is probably minimal.

The crop is not essential for normal growth when access to food is sufficient. Cropectomy has no effect on growth rate of *ad libitum*-fed chickens, but it does decrease growth rate when food intake is limited. This supports the view that the primary function of the crop is food storage, and it is not essential for digestion.

In pigeons and doves, "crop-milk" is produced during the breeding season under the influence of prolactin. Crop milk contains 12.4% protein, 8.6% lipids, 1.37% ash, and 74% water (Vandeputte-Poma, 1968). Therefore, while rich in protein and essential fatty acids (Desmeth, 1980), it is devoid of carbohydrates and calcium.

C. Stomach

The oxynticopeptic cells found in birds secrete both HCl and pepsinogen. Pepsinogen, under the influence of acid or pepsin that is already present, is converted to pepsin. While lipase has been found in gastric secretion, this is probably due to reflux from the duodenum. The basal gastric secretory rate is 15.4 ml/hr and contains 93 mEq/liter of acid and 247 Pu/ml of pepsin (Long, 1967) with a pH of 2.6 (Joyner and Kokas, 1971). The pH of gastric contents, however, is normally above 2.6 due to the presence of ingesta. The pH has been determined in several species immediately after sacrifice (Table 5). Higher pH values have been reported when measurements were made on live birds. For example, Winget *et al.* (1962) reported the following values for chickens: mouth, 6.7; crop, 6.4; ileum, 6.7; rectum, 7.1. Age has no effect on pH of the digestive tract (Herpol, 1966). Acid secretion of chickens is high relative to mammals, possibly because of the rapid digestive transit time (Table 6).

While amylolysis occurs in the crop, it is not evident in the ventriculus. This is the result of the low pH of the stomach, which is unfavorable for amylase activity (Pinchasov and Noy, 1994).

There are three phases to gastric secretion: the cephalic phase, the gastric phase, and the intestinal phase. All three phases are present in birds (Burhol, 1982). The cephalic phase entails an increase in hydrogen ion (H^+) and pepsin secretion caused by the sight, smell, or expectation of food. This phase is under vagal control.

TABLE 5 The Ph of Contents of the Digestive Tract of Avian Species[a]

	Chicken	Pigeon	Pheasant	Duck	Turkey
Crop	4.51	6.3[b] 4.28	5.8	4.9	6.0
Proventriculus	4.8	1.4[b] 4.8	4.7	3.4	4.7
Gizzard	4.74[c] 2.50	2.0	2.0	2.3	2.2
Duodenum	5.7–6.0 6.4[c]	6.4[b] 5.2–5.4	5.6–6.0	6.0–6.2	5.8–6.5
Jejunum	5.8–5.9 6.6[c]	5.3–5.9	6.2–6.8	6.1–6.7	6.7–6.9
Ileum	6.3–6.4 7.2[c]	6.8[b] 5.6	6.8	6.9	6.8
Rectum	6.3	5.4 6.6[b]	6.6	6.7	6.5
Ceca	5.7 6.9[c] 5.5–7.0[d]		5.4	5.9	5.9
Bile	7.7[e] 6.6[c] 5.9		6.2	6.1	6.0

[a]From Sturkie (1976), based on work of Farner (1942).
[b]From Herpol (1966).
[c]From Herpol and van Grembergen (1967).
[d]Sudo and Duke (1980).
[e]From Lin *et al.* (1974).

As summarized by Duke (1986), vagal stimulation increases both gastric secretion rate and pepsin secretion. Vagal stimulation causes a greater increase in gastric secretion than do cholinergic agents (Gibson *et al.*, 1974). This suggests, as discussed below, that gastric secretion is stimulated by other neurotransmitters acting together with ACh.

In birds, vagal stimulation causes greater pepsin than H^+ accumulation (Burhol, 1982). In contrast, insulin injection inhibits gastric H^+ secretion without affecting pepsin secretion. Therefore, H^+ and pepsin secretion may be under different control.

Pepsin from chicken and duck has been well characterized (see Pichova and Kostka, 1990). Duck pepsino-

TABLE 6 Basal Acid Secretion in Various Species[a]

Species	Body weight (kg)	Acid output (mEq/kg/hr)	Pepsin output (PU/kg/hr)
Man	70	0.03	862
Dog	15	0–0.004	0–62
Rat	0.35	0.25	2230
Monkey	2.5	0.12	730
Chicken	1.75	0.78	2430

[a]From Long (1967).

TABLE 7 Gastrointestinal Hormones in the Domestic Fowl

Hormone	Site of origin	Biological actions
Gastrin	Proventriculus	Stimulates gastric acid and pepsin secretion
Cholecystokinin	Duodenum, jejunum	Stimulates gall-bladder contraction and pancreatic enzyme secretion and gastric acid secretion; inhibits gastric emptying; potentiates secretin-induced stimulation of pancreatic electrolyte secretion
Secretin	Duodenum, jejunum	Stimulates bicarbonate secretion by pancreas
Vasoactive intestinal peptide	Duodenum, jejunum	May be a more potent stimulator of pancreatic electrolyte secretion than secretin; inhibits smooth muscle contraction
Pancreatic polypeptide	Pancreas, proventriculus, duodenum	Stimulates gastric acid and pepsin secretion
Gastrin-releasing peptide (bombesin)	Proventriculus	Stimulates pancreatic enzyme secretion; stimulates crop contraction
Somatostatin	Pancreas, gizzard, proventriculus, duodenum, ileum	Inhibits secretion of other gut hormones

gen and pepsin have 374 and 324 amino acids, respectively. Duck pepsin has a pH optimum of 4, is stable up to pH 7.5, and is inactivated at pH 9.6.

Many hormones are involved in gastric secretion (Table 7). Gastrin plays a role in the gastric phase of secretion. Chicken gastrin has been isolated from the chicken pylorus, which is equivalent to the mammalian antrum (Dimaline et al., 1986). While structurally similar to cholecystokinin (CCK), cG has markedly different secretory functions (Dimaline and Lee, 1990). Infusion of cG increases both acid and pepsin secretion. Unlike CCK, gastrin has no effect on gall bladder contractions or pancreatic secretion. Another peptide, gastrin-releasing peptide, also induces acid secretion, but it is not known whether it acts via gastrin release (Campbell et al., 1991).

The intestinal phase of gastric secretion is controlled by several hormones including CCK, secretin, and avian pancreatic polypeptide (APP). APP, originally discovered in chickens (Kimmel et al., 1968; Larsson et al., 1974), is released from the pancreas postprandially in response to amino acids and HCl (Hazelwood et al., 1973; Duke et al., 1982). APP does not appear to act during the cephalic phase of gastric secretion since it is not released during sham feeding (Kimmel and Pollock, 1975). APP increases gastric acid and pepsin secretion, and this effect is independent of the vagus nerve (Hazelwood et al., 1973).

CCK (Dockray, 1977) and secretin (Nilsson, 1974) have been isolated from the duodenum and jejunum of birds. CCK stimulates gastric acid secretion while having no effect on pepsin secretion. Contrary to its effects in mammals in which secretin inhibits acid and stimulates pepsin secretion, in chickens secretin stimulates both acid and pepsin secretion (Burhol, 1974).

As in mammals, histamine is involved in gastric acid release. Injection of cimetidine, an H_2-receptor blocker, raised the pH of the proventriculus and duodenal contents (Ward et al., 1984). The increase in gastric acid secretion induced by the iv injection of 2-deoxy-D-glucose was also blocked by metiamide, an H_2-receptor blocker (Nakagawa et al., 1983).

D. Intestines

Amylase is produced by both the pancreas and intestine (Osman, 1982). While found in all parts of the small intestine, it is found in particularly high concentrations in the jejunum with 80% of the activity found there. The high levels found in the jejunum are presumable due to the fact that the openings of the pancreatic ducts discharge near the anterior jejunum. Amylase is found in only trace amounts in the ceca. The pH optimum of the pancreatic and intestinal amylase is 7.5 and 6.9, respectively.

Intestinal enzymes provide the last step in digestion. These secretions are responsible for digesting starch, sucrose, fats, and protein (Table 8). The small intestine of birds contains maltase, sucrase, and palatinase, but

TABLE 8 Enzymes Secreted by the Intestines

Enzyme	Substrate	Product or function
Maltase	Maltose	Glucose
Isomaltase	Dextrins	Glucose
Sucrase	Sucrose	Glucose, fructose
Enterokinase	Trypsinogen	Trypsin
Lipase	Monoglycerides	Glycerol, fatty acids
Peptidases	Di- and tripeptides	Amino acids

does not contain trehalase (Siddons, 1969). Whether lactase is present appears to be debatable. However, it has been reported that lactase is not present in germ-free chicks and that the rate of mortality of germ-free chicks fed lactose as the sole energy source is very high (Siddons and Coates, 1972). Enzyme activity is highest in the jejunum and decreases both proximally and distally. These enzymes are located in the epithelial cells of the villi. The maltase, sucrase, and palatinase activity found in the large intestine comes from the small intestine whereas the lactase activity found in the large intestine is probably of a cecal bacterial origin.

Relatively little is known about the control of intestinal secretions in birds. Intestinal secretion is increased by duodenal distention, vagal stimulation, and secretin. Vagal stimulation has a greater effect on stimulating mucous secretion than enzyme secretion.

E. Colon

Chloride ions (Cl^-) are secreted in the rectum, ceca, and coprodeum. This is discussed below under "Absorption."

F. Pancreas

As mentioned above, pancreatic and bile secretions enter the gastrointestinal tract near the anterior jejunum. Pancreatic secretions have a pH of 6.4–6.8 in chickens (Hulan et al., 1972) and 7.4–7.8 in turkeys (Duke, 1986). Secretions include an aqueous phase containing water and bicarbonate ions and an enzymatic phase.

Digestive enzymes found in the pancreas of broiler-type chickens are listed in Table 9. Although not shown, the pancreas is also reported to secrete ribonuclease and deoxyribonuclease (Dal Borgo et al., 1968). Amylase is found in the duodenum, jejunum, ileum, and colon. Both trypsin and amylase are found in highest concentrations in the jejunum (Bird, 1971; Osman, 1982), presumably because the pancreatic ducts enter near the end of the duodenum. Both pancreatic and intestinal amylase have a requirement for chloride ion. Characterization of these enzymes suggests that pancreatic amylase is similar to mammalian α-amylase, while intestinal amylase is similar to glucoamylase.

Pancreatic secretion is controlled by both nervous and hormonal mechanisms. The secretion rate is higher in birds than in mammals (Table 10). Secretion has both a cephalic and intestinal phase. When fasted birds are allowed to eat, pancreatic secretion increases immediately (Kokue and Hayama, 1972). This response is blocked by vagotomy or atropine and can be augmented by cholinergic agents (Hokin and Hokin, 1953).

Secretinlike activity is released in response to dilute HCl placed in the duodenum (Nilsson, 1974). Secretin, when injected iv, increases the aqueous component of pancreatic secretion. However, in contrast to mammals, vasoactive intestinal peptide (VIP) more potently stimulates the secretion of pancreatic juice (Vaillant et al., 1980). VIP is found in neurons both in the gastrointestinal tract and pancreas. It is believed that VIP, rather than secretin, is the primary regulator of pancreatic juice secretion and that this response may be either neuronally or hormonally mediated (Dockray, 1988). VIP does not stimulate pancreatic enzyme secretion.

CCK is released in response to lipids and amino acids. The administration of CCK has been shown to increase pancreatic secretion in pigeons (Sahba et al., 1970) and to increase the flow rate and protein secretion rate in turkeys (Dockray, 1975).

Two gastrin-releasing peptides (GRP), which are structurally related to bombesin, have been isolated from the proventriculus (Campbell et al., 1991). These peptides are found in endocrine cells and contain either 27 or 6 amino acids. Distension of the proventriculus with peptone stimulates pancreatic juice and enzyme secretion. This effect appears to be mediated by GRP-27, with GRP-6 being ineffective. Distension with saline is less effective.

Diet can influence the secretory rate of pancreatic enzymes. Increasing the carbohydrate and fat content

TABLE 9 Pancreatic Digestive Enzymes[a]

Enzyme	Percentage of total
Trypsinogen	10
Chymotrypsinogen (A, B, and C)	20
Trypsin inhibitor	11.3
Amylase	28.9
Procarboxypeptidase (A and B)	29.8

[a]From Pubols (1991).

TABLE 10 Pancreatic Secretory Rate and Influence of Fasting in Chicken, Dog, Rat, and Sheep[a]

Species	Starvation time (hr)	Pancreatic secretory volume (ml/kg/hr)
Chicken	24	0.70
	48	0.68
	72	0.65
Dog	24	0.1–0.3
	48	Negligible
Rat	24	0.6–0.7
Sheep	24	0.13
	48	0.07

[a]Adapted from Kokue and Hayama (1972).

of the diet increases amylase and lipase activity in pancreatic secretions (Hulan and Bird, 1972). Increasing the protein content of the diet increases chymotrypsin activity in the duodenum and jejunum (Dal Borgo *et al.*, 1968; Bird and Moreau, 1978).

G. Bile

Bile, produced and secreted by the liver, is essential for fat digestion. It acts to emulsify lipid so that it can be more efficiently digested by lipase. In addition, amylase begins to appear in chicken bile at 4–8 weeks of age (Farner, 1943). Therefore, it is also involved in carbohydrate digestion.

Relatively little is known about biliary secretion in birds possibly because of the complex anatomy in which bile enters the small intestine via both the hepatoenteric duct and the cysticoenteric duct. The biliary secretion rate is 24.2 μl/min in fasted broilers (Lisbona *et al.*, 1981). Chenodeoxycholyltaurine and cholyltaurine are the predominant bile acids in chickens and turkeys, while chenodeoxycholyltaurine and phocaecholyltaurine predominate in ducks (Elkin *et al.*, 1990). These are secreted by an active transport system.

The bile salts glycocholate and taurocholate are readily absorbed through the intestinal wall. The absorption rate is higher near the distal end of the small intestine (Lindsay and March, 1967). This allows for recirculation of bile acids, thus allowing for their reuse in lipid digestion. It is estimated that 90% of bile salts are reabsorbed in the jejunum and ileum (Hurwitz *et al.*, 1973).

Since chickens have low levels of liver glucuronyl transferase and little or no biliverdin reductase, the secretion rate of biliverdin is high relative to that of bilirubin (14.7 vs. 0.9 μg/kg/min). The green color observed in coprodeal droppings is likely due to biliverdin. The reason for the brown color associated with cecal droppings is unknown, but may be due to bacterial reduction of biliverdin to bilirubin and a subsequent dehydrogenation (Hill, 1983).

VI. ABSORPTION

A. Carbohydrates

Absorption of carbohydrates in birds occurs by mechanisms similar to those found in mammals. Absorption occurs more rapidly from the small intestine compared to the cecum. Carbohydrates are absorbed by both active and passive mechanisms. Those sugars containing a six-membered ring with the hydroxyl group in the number 3 position oriented similar to glucose are actively transported. Active transport accounts for at least 80% of glucose absorption. D-Glucose and D-galactose are absorbed faster than D-xylose, D-fructose, and D-arabinose. These sugars are absorbed faster than L-arabinose, L-xylose, D-ribose, D-mannose, and D-cellobiose.

The absorption of sugars against a concentration gradient involves an apically located sodium-dependent, phloridzin-sensitive transport. This system is coupled to the Na^+-K^+-ATPase located on the basolateral membrane. Contrary to what is believed to occur in mammals, the active transport of one molecule of carbohydrate in birds is coupled to the movement of 2 molecules of Na (Kimmich and Randles, 1984). Sugars leave the enterocytes on their way to the bloodstream crossing the basolateral membrane by either simple diffusion or facilitated by a Na-independent mechanism.

Within the small intestine, the greatest absorption of glucose occurs in the duodenum (Figure 13). Cumulatively, up to 65, 85, and 97% of ingested starch is digested through the duodenum, jejunum, and the terminal ileum, respectively (Riesenfeld *et al.*, 1980). Virtually all of the glucose released from starch digestion is absorbed within the small intestine.

Significant absorption of glucose also occurs in the cecum (Savory and Mitchell, 1991). While the entire cecum is able to accumulate sugars at hatch, this ability is soon limited to the proximal region (Planas *et al.*, 1986). The ability of the cecum to actively absorb sugars at low concentrations appears higher than that of the jejunum (Vinardell and Lopera, 1987).

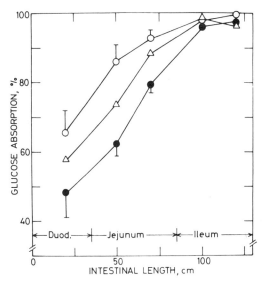

FIGURE 13 Cumulative percentage of glucose and starch disappearance from the intestine of 7-week-old chicks. Glucose absorption from glucose monohydrate as the sole dietary carbohydrate source ($\bigcirc$), starch digestion ($\triangle$) and absorption ($\bullet$) in chicks fed starch as the sole dietary carbohydrate source. Values given in the figure are means of six chickens $\pm$ SE. (From Riesenfeld *et al.* (1980). © *J. Nutr.*: (**110**, 117–121), American Society for Nutritional Sciences.)

There is a greater affinity for active transport of glucose in the ileum (Levin et al., 1983). The ileum appears well suited for transporting glucose which may not be absorbed within the jejunum.

The gastrointestinal tract also has a role in glucose homeostasis (Riesenfeld et al., 1982). Plasma glucose levels are maintained relatively constant in chickens when fed semipurified diets in which glucose is replaced with either fructose, soybean oil, or cellulose. While this is partially the result of varying glucose turnover rates, it is largely a result of varying rates of conversion of glucose to lactate by the intestine.

B. Amino Acids and Peptides

Amino acid absorption in birds is similar to that of mammals (Levin, 1984). Amino acid transport occurs via a secondary active transport system as described for sugars. This process is saturable, coupled to Na^+ transport, and utilizes ATP for energy. The amino acid transport systems can be classified in four groups: (1) neutral amino acids; (2) proline, β-alanine, and related amino acids; (3) basic amino acids; and (4) acidic amino acids. Such a classification, however, is not rigid since many amino acids share transport with more than one group. For example, leucine, a neutral amino acid, can inhibit the uptake of proline and arginine, a basic amino acid. Glycine transport is partially inhibited by both proline and β-alanine.

The primary site for amino acid absorption is the small intestine, although absorption is also reported to occur in the crop, gizzard, and proventriculus. It is still unclear as to which section of the small intestine has the greatest capacity for absorption since there is a lack of agreement among studies. The rectum of hens is also capable of absorbing methionine via a saturable process.

As with glucose transport, the K_m for various amino acids is lower in the ileum than in the duodenum (Table 11).

TABLE 11 Read K_m (mM) and J_{max} (pmol cm^{-2}d^{-1}) for Saturable Amino Acid Absorption in the Chicken Small Intestine[a]

Amino acid	Jejunum		Ileum	
	K_m	J_{max}	K_m	J_{max}
Amino-isobutyric acid	4.6 ± 0.9	46 ± 7	2.5 ± 0.2	56 ± 6
Glycine	4.2 ± 0.4	37 ± 5	2.7 ± 0.2	55 ± 6
Histidine	3.4 ± 0.7	132 ± 12	0.8 ± 0.2	129 ± 4
Methionine	4.9 ± 0.6	147 ± 14	1.9 ± 0.6	148 ± 5
Valine	3.2 ± 0.7	38 ± 7	1.5 ± 0.2	82 ± 13

[a] From Levin (1984).

While not extensively investigated in birds, there is evidence that, in addition to amino acids, peptides are also absorbed. Peptides appear to be absorbed more rapidly than amino acids. Absorption of peptides acts to eliminate composition between amino acids for uptake and thus helps increase speed of absorption.

The ceca are also important in amino acid absorption. Proline is absorbed in the cecum via a sodium-dependent carrier-mediated transport system with transport rates being higher in the proximal than distal cecum (Obst and Diamond, 1989). In addition, leucine, phenylalanine, proline, and glycylsarcosine are absorbed against a concentration gradient (Gasaway, 1976; Calonge et al., 1990; Moretó et al., 1991). Proline and methionine have also been shown to be transported by a Na-independent system (Moretó et al., 1991).

The ceca has a greater ability to transport amino acids than sugars (Moretó et al., 1991). This may be functionally important when considering the following: (1) uric acid, which is retrogradely carried to the ceca from the coprodeum, may be microbially converted to amino acids; (2) proteases are in high concentration within the ceca and may release amino acids from such proteins. Therefore, the ceca may be important in amino acid absorption.

C. Fatty Acids and Bile Acids

Fatty acid absorption occurs in the distal half of the jejunum and, to a lesser extent, in the ileum. Since the bile duct enter the gastrointestinal tract of birds near the distal duodenum, emulsification of fats is delayed in birds relative to mammals.

In mammals, fatty acids enter the enterocytes where they are reesterified to triglycerides and packaged into chylomicrons which enter the lymphatic system. However, in birds, after being reesterified, lipids are packaged into portomicrons which pass directly into the hepatic portal blood supply.

D. Volatile Fatty Acids

The concentration of volatile fatty acid (VFA), mainly acetate, but some propionate and butyrate, in the ceca is high. It can reach 125 mM in the chicken (Annison et al., 1968) and 70 mM in the goose (Clemens et al., 1975). VFAs are one of the byproducts of microbial decomposition of uric acid (see Braun and Campbell, 1989). Levels of VFAs are higher in the portal blood of conventionally raised compared to germ-free birds, suggesting that VFAs formed by bacteria are absorbed from the gastrointestinal tract. VFAs are absorbed from both the small intestine and the ceca by passive transport (Sudo and Duke, 1980). Since the concentration of VFAs is higher inside the lumen, absorption occurs by the passive movement of these compounds down their concentration gradient. While the rate of absorption of propionate and butyrate is equal between the small intestine and ceca, acetate is absorbed faster within the ceca. Microbial fermentation within the ceca of the ptarmigan can provide enough VFAs to meet 11–18% of the energy needs for basal metabolic rate (Gasaway, 1976).

E. Electrolytes

The major site of calcium and phosphate absorption, which are both actively absorbed, is the upper jejunum (Levin, 1984). When fed diets adequate in phosphorus and calcium, there is a net secretion of phosphate in the duodenum. However, when fed a phosphorous-deficient diet, there is a net absorption within the duodenum.

Calcium absorption occurs via an active transport system which is influenced by the vitamin D hormone 1,25-dihydroxyvitamin D_3 ($1,25(OH)_2D_3$). While calcium enters the mucosal membrane via a diffusion-type process, it is then secreted into the blood by an active transport system located on the basolateral membrane.

The hormonal byproduct of vitamin D, $1,25(OH)_2D_3$, induces the synthesis of calbindin-D_{28k} in birds (Wasserman, 1992) and calcium pump units in the basolateral membrane of chicken enterocytes (Wasserman et al., 1992). Feeding chickens a calcium- or phosphorus-deficient diet, or repleting vitamin D-deficient chickens with vitamin D or injecting $1,25(OH)_2D_3$ causes an increase in plasma membrane calcium pump mRNA in the duodenum, jejunum, ileum, and rectum (Cai et al., 1993).

The absorption and secretion of Na, potassium (K), calcium (Ca), and magnesium (Mg) in the gastrointestinal tract of broilers has recently been characterized (Van Der Klis et al., 1990). Sodium transport is discussed below. Potassium, Ca, and Mg are secreted in the duodenum. In addition, there is some secretion of Mg in the ileum and rectum. The major site of absorption for these minerals is the proximal jejunum with some absorption of Ca and Mg occurring in the gizzard. This coincides with the fact that dry matter spends approximately 25% of its time in the jejunum. There is little absorption of these minerals beyond the jejunum.

F. Water, Sodium, and Chloride

Water is absorbed throughout the small and large intestine and the ceca. Absorption of water occurs as a secondary response to the active absorption of other compounds such as glucose, Na, and amino acids.

Birds secrete a very dilute urine. Urine can travel into the coprodeum and, via antiperistalsis, continue orad into the ceca. The excreta of cecectomized, compared to intact chickens, is higher in water indicating that the lower part of the intestinal tract has an important role in water and salt balance.

Sodium is secreted in the proximal portion of the intestinal tract (Van Der Klis et al., 1990). The major sites for Na absorption are the proximal jejunum followed by the rectum. The ileum also has a small capacity for Na absorption.

The coprodeum, rectum, and ceca serve an important role in Na, chloride (Cl), and water balance (Thomas and Skadhauge, 1988; Skadhauge, 1989). When fed a low-NaCl diet, there is a net absorption of Na from the coprodeum, rectum, and ceca. Sodium absorption is a secondary active transport system dependent on the Na-K-ATPase located on the basolateral membrane. The absorption of Na is linked with Cl and water absorption and K secretion. This uptake is blocked by amiloride, a competitive blocker of Na absorption. When fed a high-NaCl diet, absorption of Na from the coprodeum nearly ceases, while that in the colon remains high provided a low concentration of hexoses and amino acids is maintained on the mucosal side.

In addition, there is also a net secretion of Cl from the coprodeum and rectum, but the secretory ability of the coprodeum disappears in birds fed a high-NaCl diet (Skadhauge, 1989). The secretion of Cl is induced by serotonin (Hansen and Bindslev, 1989a). Receptor antagonists for $5-HT_{1A}$, $5-HT_{1B}$, $5-HT_{1C}$, $5-HT_{1D}$, $5-HT_2$, $5-HT_3$, adrenergic, cholinergic, and histaminergic receptors were unable to block the response. However, the second messenger for serotonin-induced chloride secretion is cAMP (Hansen and Bindslev, 1989b).

The absorption of Na and secretion of Cl within the lower gut is regulated by plasma aldosterone (Clauss et al., 1991). Increases in plasma aldosterone stimulate amiloride-sensitive Na uptake in the coprodeum, rectum, and ceca. Feeding high-NaCl diets causes a decrease in plasma aldosterone and a decrease in amiloride-dependent Na absorption. Furthermore, there is a switch to amiloride-independent hexose/amino acid stimulated Na transport within the rectum and cecum, but a cessation of Na transport in the coprodeum. The hexose/amino acid stimulated Na transport in the rectum functions to counteract the osmotic water loss due to feeding a high-NaCl diet (Skadhauge et al., 1985), and is stimulated by increases in plasma arginine vasotocin and/or prolactin (Arnason and Skadhauge, 1991).

Increased Na transport results from an increase in the number of Na channels in the apical membrane of enterocytes. Unlike the rectum which always has microvilli, increased Na uptake is associated with an increase in the number of apical microvilli within the coprodeum; these villi disappear in birds fed a high-NaCl diet.

G. Vitamins

Absorption of vitamin B6 occurs throughout the small intestine, but primarily in the duodenum (Heard and Annison, 1986). Absorption can also occur in the crop and cecum, but this is minimal. The mechanism of absorption is passive diffusion. Although microbial synthesis of vitamin B6 does occur, this is not important in meeting the birds vitamin requirement.

VII. AGE-RELATED EFFECTS ON GASTROINTESTINAL FUNCTION

Growth rate has increased due to genetic selection in poultry. There is increasing interest about whether the physiological capacity of the gastrointestinal tract may limit continued advances in growth rate. It was hypothesized that postnatal growth rate is at least partially determined by the differential growth of various organs (Lilja, 1983).

With regard to pancreatic enzyme activity within the intestines, trypsin, protease, and amylase activity increased rapidly during the first 21 days post-hatch in turkey poults (Krogdahl and Sell, 1989). However, lipase activity did not begin to increase until after 21 days of age. Feeding a high-fat diet did not significantly increase lipase activity until after 21 days of age. At least during the first few weeks post-hatch, it appeared that lipase activity may be a limiting factor in digestion.

Several notable changes have been found in the development of nutrient transport systems during ontogeny in broilers (Obst and Diamond, 1992). During the first week post-hatch, proline uptake by the small intestine is high relative to glucose uptake. Since the relative growth rate of broilers is greatest during the first week, it appears that amino acid uptake may parallel this growth pattern. The uptake of glucose displays a transitory increase during week 2. This increase is hypothesized to be a result of a switch from lipid metabolism to carbohydrate metabolism due to depletion of the yolk reserves. Because of the allometric growth of the intestine, during the second week post-hatch the intestine displays a decrease in size relative to body weight. This may be a second reason for the increase in uptake of glucose occurring at this time. There is a transitory increase in proline uptake during week 6. This increase parallels the first postjuvenile molt and a rise in absolute growth rate.

Interestingly, the intestinal uptake capacity of broilers closely matches the birds nutrient needs. This contrasts with results in mammals where uptake far exceeds the animals needs. It remains to be determined whether this may indicate that nutrient uptake represents a potential constraint for increased growth in broilers, or whether it means that broilers are better at allocating resources.

VIII. FOOD INTAKE REGULATION

Food intake regulation in birds has recently been reviewed (Denbow, 1994a,b; Kuenzel, 1994). Studies in which birds have been given a choice between high-energy and high-protein diets have confirmed that birds can regulate both their protein and energy intake. Interestingly, genetic selection has altered the protein and energy intakes resulting in strain differences.

Regulation occurs both within and outside the central nervous system. Outside the central nervous system, food intake is regulated by both the gastrointestinal tract and the liver. Infusion of hyperosmotic solutions into the duodenum reduces food intake suggesting the existence of osmoreceptors. While it has been suggested that infusion of glucose into the gastrointestinal tract decreases food intake via a chemosensitive effect, this effect appears more likely an osmotic response. The mechanism whereby intraduodenal infusions of hyperosmotic solutions decrease food intake may involve a decrease in gastrointestinal motility (Chaplin et al., 1992).

The crop does not appear to play a major role in food intake regulation. Filling of the crop is generally not associated with meal termination. Instead, meal termination is more closely associated with the filling of the stomach, duodenum, and proximal one-third of the ileum. The crop only appears to affect growth when the availability of food is limited.

The liver is an important site for food intake control in birds. Intrahepatic infusion of glucose, lysine, or lipids decrease food intake. Interestingly, these effects vary with the strain of bird. Intrahepatic infusion of glucose decreased food intake in Leghorns, but not broilers (Lacy et al., 1985). Similarly, infusions of lipids decreased food intake in free-feeding Leghorns, but not broilers (Lacy et al., 1986).

Although the liver appears able to alter food intake in response to changes in hepatic portal blood glucose concentration, there is little evidence that plasma or brain concentrations of glucose influences food intake. Savory (1987a) reported no correlation between plasma glucose concentrations and food intake. Furthermore, injections of glucose, or its antimetabolites, into the lateral cerebroventricle have no effect on food intake. Whereas insulin injections increase food intake in mammals, similar injections either have no effect or decrease food intake in birds.

The hypothalamus is a major site of food intake regulation in birds. As in mammals, lesioning the medial hypothalamus of avian species increases food intake whereas lesioning the lateral hypothalamic area decreases food intake. While these sites have traditionally been considered as satiety and feeding centers, respectively, it is currently believed that they are better considered as parts of neural circuits involved in food intake regulation. Kuenzel (1989) recently identified five neural pathways concerned with mandibulation, taste, smell, sight, and the autonomic nervous system. Alterations in hypothalamic function which result in changes in food intake likely alter the autonomic nervous system. Increased evidence suggests that obesity results from decreased sympathetic nervous system activity (Bray, 1991). Interestingly, lines of chickens genetically selected for high body weight did not increase food intake in response to ventromedial hypothalamic lesions whereas low-weight birds increased food intake.

The neurochemical control of food intake is complex. Within the central nervous sytem many neurochemicals have been shown to function in food intake control. As previously summarized (Denbow, 1994a), food intake is decreased by intracerebroventricular (ICV) injections of serotonin, cholecystokinin, and bombesin. Food intake is increased by similar injections of epinephrine, neuropeptide Y, neuropeptide YY, avian pancreatic polypeptide, β-endorphin, and [Met^5]-enkephalin. Endogenous opioids increase food intake in the central nervous system via interaction with delta receptors. It has recently been shown that nitric oxide also functions both inside and outside the blood–brain barrier to increase food intake (Choi et al., 1994).

Selection for body weight appears to have altered the response to at least some of these compounds. Whereas ICV injections of epinephrine increased food intake in broilers, it was without effect in Leghorns. While the ICV injection of serotonin decreased food intake in fasted Leghorns, it had no effect in fasted broilers.

These neurochemicals also work outside the central nervous system to alter food intake. Peripheral administration of opioids stimulates feeding presumably by interacting with mu and delta receptors. Intrahepatic infusion of epinephrine decreases food intake in chickens. Similarly, intravenous injections of CCK or bombesin decrease food intake. However, it has been questioned whether the effect of CCK and bombesin is specific to food intake. Savory (1987) showed that intravenous injections of these peptides induced abnormal GIT motility with resumption of normal feeding occurring after recovery of normal motility. Therefore, the peptides may be causing malaise or nausea rather than specifically controlling food intake.

References

Akester, A. R., Anderson, R. S., Hill, K. J., and Osbaldiston, G. W. (1967). A radiographic study of urine flow in the domestic fowl. *Br. Poult. Sci.* **8,** 209–212.

Ali, H. A., and McLelland, J. (1978). Avian enteric nerve plexuses: A histochemical study. *Cell Tissue Res.* **189,** 537–548.

Annison, E. F., Hill, K. J., and Kenworthy, R. (1968). Volatile fatty acids in the digestive tract of the fowl. *Br. J. Nutr.* **22,** 207–216.

Antony, M. (1920). Uber die Speicheldrusen der Vogel. *Zool. Jb. Abt. Anat. Ontog. Tiere* **41,** 547–660.

Arnason, S. S., and Skadhauge, E. (1991). Steady-state sodium absorption and chloride secretion of colon and coprodeum, and plasma levels of osmoregulatory hormones in hens in relation to sodium intake. *J. Comp. Physiol. B* **161,** 1–14.

Ashcraft, D. W. (1930). The correlative activities of the alimentary canal of the fowl. *Am. J. Physiol.* **93,** 105–110.

Bartlet, A. L. (1974). Actions of putative transmitters in the chicken vagus nerve/oesophagus and Remak nerve/rectum preparations. *Br. J. Pharmacol. Chemother.* **51,** 549–558.

Bennett, T. (1974). Peripheral and autonomic nervous systems. In "Avian Biology" (D. S. Farner and J. R. King, eds.), pp 1–77. Academic Press, New York.

Bhattacharya, S., and Ghose, K. C. (1971). Influence of food on amylase system in birds. *Comp. Biochem. Physiol. B* **40,** 317–320.

Bird, F. H. (1971). Distribution of trypsin and amylase activities in the duodenum of the domestic fowl. *Br. Poult. Sci.* **12,** 373–378.

Bird, F. H., and Moreau, G. E. (1978). The effect of dietary protein levels in isocaloric diets on the composition of avian pancreatic juice. *Poult. Sci.* **57,** 1622–1628.

Bjornhag, D., and Sperber, I. (1977). Transport of various food components through the digestive tract of turkeys, geese and guinea fowl. *Swedish J. Agric. Sci.* **7,** 57–66.

Bolton, W. (1965). Digestion in crop. *Br. Poult. Sci.* **6,** 97–102.

Bowman, W. C., and Everett, S. D. (1964). An isolated parasympathetically-innervated oesophagus preparation from the chick. *Br. J. Pharmacol.* **16,** 72T–79T.

Branch, J., and Cummings, J. H. (1978). Comparison of radio-opaque pellets and chromium sesquioxide as inert markers in studies requiring accurate fecal collections. *Gut* **19,** 371–376.

Braun, E. J., and Campbell, C. E. (1989). Uric acid decomposition in the lower gastrointestinal tract. *J. Exp. Zool.* **3**(Suppl.), 70–74.

Bray, G. A. (1991). Obesity, a disorder of nutrient partitioning: The MONO LISA hypothesis. *J. Nutr.* **1212,** 1146–1162.

Burhol, P. G. (1974). Gastric stimulation by intravenous injection of cholecystokinin and secretin in fistula chickens. *Scand. J. Gastroenterol.* **9,** 49–53.

Burhol, P. G. (1982). Regulation of gastric secretion in the chicken. *Scand. J. Gastroenterol.* **17,** 321–323.

Cai, Q., Chandler, J. S., Wasserman, R. H., Kumar, R., and Penniston, J. T. (1993). Vitamin D and adaptation to dietary calcium and phosphate deficiencies increase intestinal plasma membrane calcium pump gene expression. *Proc. Natl. Acad. Sci. USA* **90,** 1345–1349.

Calhoun, M. (1954). "Microscopic Anatomy of the Digestive System." Iowa State College Press, Ames, Iowa.

Calonge, M. L., Ilundain, A., and Bolufer, J. (1990). Glycylsarcosine transport by epithelial cells isolated from chicken proximal cecum and rectum. *Am. J. Physiol.* **258,** G660–G664.

Campbell, B., Garner, A., Dimaline, R., and Dockray, G. J. (1991). Hormonal control of avian pancreas. *Am. J. Physiol.* **261,** G16–G21.

Chaplin, S. B. (1989). Effect of cecectomy on water and nutrient absorption of birds. *J. Exp. Zool.* **3**(Suppl.) 81–86.

Chaplin, S. B., and Duke, G. E. (1988). Effect of denervation on initiation and coordination of gastroduodenal motility in turkeys. *Am. J. Physiol.* **255**, G1–G6.

Chaplin, S. B., and Duke, G. E. (1990). Effect of denervation of the myenteric plexus on gastroduodenal motility in turkeys. *Am. J. Physiol.* **259**, G481–G489.

Chaplin, S. B., Raven, J., and Duke, G. E. (1992). The influence of the stomach on crop function and feeding behavior in domestic turkeys. *Physiol. Behav.* **52**, 261–266.

Choi, Y.-H., Furuse, M., Okumura, J.-I., and Denbow, D. M. (1994). Nitric oxide controls feeding behavior in the chicken. *Brain Res.* **654**, 163–166.

Clauss, W., Dantzer, V., and Skadhauge, E. (1991). Aldosterone modulates Cl secretion in the colon of the hen (*Gallus domesticus*). *Am. J. Physiol.* **261**, R1533–R1541.

Clemens, E. T., Stevens, C. E., and Southworth, M. (1975). Sites of organic acid production and pattern of digesta movement in the gastrointestinal tract of geese. *J. Nutr.* **105**, 1341–1350.

Clench, M. H., Pineiro-Carrero, V. M., and Mathias, J. R. (1989). Migrating myoelectric complex demonstrated in four avian species. *Am. J. Physiol.* **256**, G598–G603.

Dal Borgo, G. A., Salman, J., Pubols, M. H., and McGinnis, J. (1968). Exocrine function of the chick pancreas as affected by dietary soybean meal and carbohydrate. *Proc. Soc. Exp. Biol. Med.* **129**, 877–881.

Deaton, J. W., Branton, J. L., and Lott, B. D. (1985). Noted difference in the digestive system in cages and floor-reared commercial egg-type pullets. *Poult. Sci.* **64**, 1035–1037.

Denac, M., and Scharrer, E. (1987). Effect of neurotensin on the smooth muscle of the chicken crop. *Comp. Biochem. Physiol. C* **87**, 325–327.

Denac, M., and Scharrer, E. (1988). Effect of bombesins and substance P on the smooth muscle of the chicken crop. *Vet. Res. Commun.* **12**, 447–452.

Denac, M., Kumin, G., and Scharrer, E. (1990). Effect of electrical field stimulation on muscle strips from chicken crop. *Exp. Physiol.* **75**, 69–73.

Denbow, D. (1994a). Appetite and its control. *Poult. Sci. Rev.* **5**, 209–219.

Denbow, D. (1994b). Peripheral regulation of food intake in poultry. *J. Nutr.* **124**, 1349S–1354S.

Desmeth, M. (1980). Lipid composition of pigeon cropmilk. II. Fatty acids. *Comp. Biochem. Physiol. B* **66**, 135–138.

Dimaline, R., and Lee, C. M. (1990). Chicken gastrin: A member of the gastrin/CCK family with novel structure–activity relationships. *Am. J. Physiol.* **259**, G882–G888.

Dimaline, R., Young, J., and Gregory, H. (1986). Isolation from chicken antrum, and primary amino acid sequence of a novel 36-residue peptide of the gastrin/CCK family. *FEBS Lett.* **205**, 318–322.

Dockray, G. J. (1975). Comparison of the actions of porcine secretin and extracts of chicken duodenum on pancreatic exocrine secretion in the cat and turkey. *J. Physiol.* **244**, 625–637.

Dockray, G. J. (1977). Molecular evolution of gut hormones: Application of comparative studies on the regulaton of digestion. *Gastroenterology* **72**, 344–358.

Dockray, G. J. (1988). Evolutionary aspects of gastrointestinal hormones. *In* "Advances in Metabolic Diseases," Vol. 11, p. 85. Academic Press, San Diego.

Duke, G. E. (1982). Gastrointestinal motility and its regulation. *Poult. Sci.* **61**, 1245–1256.

Duke, G. E. (1986). Alimentary canal: Secretion and digestion, special digestive functions, and absorption. *In* "Avian Physiology" (P. D. Sturkie, ed.), 4th ed. pp. 269–288. Springer-Verlag, New York.

Duke, G. E., and Evanson, O. A. (1972). Inhibition of gastric motility by duodenal contents in turkeys. *Poult. Sci.* **51**, 1625–1636.

Duke, G. E., and Evanson, O. A. (1976). Diurnal cycles of gastric motility in normal and fasted turkeys. *Poult. Sci.* **55**, 1802–1807.

Duke, G. E., Dzuik, H. E., and Evanson, O. A. (1972). Gastric pressure and smooth muscle electrical potential changes in turkeys. *Am. J. Physiol.* **222**, 167.

Duke, G. E., Evanson, O. A., Ciganek, J. G., Miskowiec, J. F., and Kostuch, T. E. (1973). Inhibition of gastric motility in turkeys by intraduodenal injections of amino acid solutions. *Poult. Sci.* **52**, 1749–1756.

Duke, G. E., Dziuk, H. E., Evanson, O. A., and Miller, J. E. (1977). Studies of methods for *in situ* observations of gastric motility in domestic turkeys. *Poult. Sci.* **56**, 1575–1578.

Duke, G. E., Evanson, O. A., Redig, P. T., and Rhoades, D. D. (1976a). Mechanism of pellet egestion in great horned owls (*Bubo virginianus*). *Am. J. Physiol.* **213**, 1824–1829.

Duke, G. E., Evanson, O. A., and Redig, P. T. (1976b). A cephalic influence on gastric motility upon seeing food in domestic turkeys, great horned owls (*Bubo virginianus*) and red-tailed hawks (*Buteo janaicensis*). *Poult. Sci.* **55**, 2155–2165.

Duke, G. E., Evanson, O. A., and Huberty, B. J. (1980). Electrical potential changes and contractile activity of the distal cecum of turkeys. *Poult. Sci.* **59**, 1925–1934.

Duke, G. E., Kimmel, J. R., Durham, K., Pollock, H. G., Bertoy, R., and Rains-Epstein, D. (1982). Release of avian pancreatic polypeptide by various intraluminal contents in the stomach, duodenum or ileum of turkeys. *Dig. Dis. Sci.* **27**, 782–786.

Duke, G. E., Place, A. R., and Jones, B. (1989). Gastric emptying and gastointestinal motility in Leach's Storm-Petrels chicks (*Oceanodroma leuchorhoa*). *Auk* **106**, 80–85.

Dziuk, H. E., and Duke, G. E. (1972). Cineradiographic studies of the gastric motility in turkeys. *Am. J. Physiol.* **222**, 159–166.

Elkin, R. G., Wood, K. V., and Hagey, L. R. (1990). Biliary bile acid profiles of domestic fowl as determined by high performance liquid chromatography and fast atom bombardment mass spectrometry. *Comp. Biochem. Physiol. B* **96**, 157–161.

Farner, D. S. (1942). The hydrogen ion concentration in avian digestive tracts. *Poult. Sci.* **21**, 445–450.

Farner, D. S. (1943). Biliary amylase in the domestic fowl. *Biol. Bull.* **84**, 240–243.

Ferrando, C., Vergara, P., Jiménez, M., and Goñalons, E. (1987). Study of the rate of passage of food with chromium-mordanted plant cells in chickens (*Gallus gallus*). *Quart. J. Exp. Physiol.* **72**, 251–259.

Ferrer, R., Planas, J. M., Durfort, M., and Moretó, M. (1991). Morphological study of the caecal epithelium of the chicken (*Gallus Gallus domesticus l.*). *Br. Poult. Sci.* **32**, 679–691.

Fileccia, R., Mule, F., Postorino, A., Serio, R., and Abbadessa, U. S. (1987). 5-Hydroxytryptamine involvement in the intrinsic control of oesophageal EMG activity. *Arch. Int. Physiol. Biochim.* **95**, 281–288.

Fileccia, R., Postorino, A., Serio, R., Mule, F., and Abbadessa, U. S. (1984). Primary peristalsis in pigeon cervical oesophagus: two EMG patterns. *Arch. Int. Physiol. Biochim.* **92**, 185–192.

Fitzgerald, T. C. (1969). "The Coturnix Quail Anatomy and Histology," pp. 207–260. Iowa State Univ. Press, Ames Iowa.

Fowler, M. E. (1991). Comparative clinical anatomy of ratites. *J. Zoo. Wildlife Med.* **22**, 204.

Ganchrow, D., and Ganchrow, J. R. (1985). Number and distribution of taste buds in the oral cavity of hatchling chicks. *Physiol. Behav.* **34**, 889–894.

Gasaway, W. C. (1976). Seasonal variation in diet, volatile fatty acid production and size of the cecum of rock ptarmiga. *Comp. Biochem. Physiol. A* **53**, 109–114.

Gibson, R. G., Colvin, H. W., Jr., and Hirschowitz, B. I. (1974). Kinetics for gastric response in chickens to graded electrical vagal stimulation. *Proc. Soc. Exp. Biol. Med.* **145**, 1058–1060.

Goñalons, E., Rial, R., and Turk, J. A. (1982). Phenol red as indicator of digestive tract motility in chickens. *Poult. Sci.* **61**, 581–583.

Hansen, M. B., and Bindslev, N. (1989a). Serotonin receptors for chloride secretion in hen colon. *Comp. Biochem. Physiol. A* **94**, 189–197.

Hansen, M. B., and Bindslev, N. (1989b). Serotonin-induced chloride secretion in hen colon, possible second messengers. *Comp. Biochem. Physiol. A* **94**, 315–321.

Harrison, J. G. (1964). Tongue. In "A New Dictionary of Birds" (A. L. Thomson, ed.), pp. 825–827. Nelson, London.

Hazelwood, R. L., Turner, S. D., Kimmel, J. R., and Pollock, H. G. (1973). Spectrum effects of a new polypeptide (third hormone?) isolated from the chicken pancreas. *Gen. Comp. Endocrinol.* **21**, 485–497.

Heard, G. S., and Annison, E. F. (1986). Gastrointestinal absorption of vitamin B-6 in the chicken (*Gallus domesticus*). *J. Nutr.* **116**, 107–120.

Herd, R. M. (1985). Anatomy and histology of the gut of the emu (*Dromaius novaehollandiae*). *Emu* **85**, 43–46.

Herpol, C. (1966). Influence de l'ago sur le pH dans le tube digestif de gallus domesticus. *Ann. Biol. Anim. Biochim. Biophys.* **4**, 239–244.

Herpol, C., and van Grembergen, G. (1967). La signification du pH dans le tube digestif de gallus domesticus. *Ann. Biol. Anim. Biochim. Biophys.* **7**, 33–38.

Hill, K. (1971). Physiology of digestion. In "Physiology and Biochemistry of the Domestic Fowl" (D. J. Bell and B. M. Freeman, eds.), Vol. 1, pp. 25–49. Academic Press, New York.

Hill, K. J. (1983). Physiology of the digestive tract. In "Physiology and Biochemistry of the Domestic Fowl" (B. M. Freeman, ed.), Vol. 4, pp. 31–47. Academic Press, London.

Hodgkiss, J. P. (1984a). Evidence that enteric cholinergic neurones project orally in the intestinal nerve of the chicken. *Quart. J. Exp. Physiol.* **69**, 797–807.

Hodgkiss, J. P. (1984b). Peristalsis and antiperistalsis in the chicken caecum are myogenic. *Quart. J. Exp. Physiol.* **69**, 161–170.

Hodgkiss, J. P. (1986). The unmyelinated fibre spectrum of the main trunk and side branches of the intestinal nerve in the chicken (*Gallus gallus* var. *domesticus*). *J. Anat.* **148**, 99–110.

Hokin, L. R., and Hokin, M. R. (1953). Enzyme secretion and the incorporation of ^{32}P into phospholipids of pancreas slices. *J. Biol. Chem.* **203**, 967–977.

Hulan, H. W., and Bird, F. H. (1972). Effect of fat level in isonitrogenous diets on composition of avian pancreatic juice. *J. Nutr.* **102**, 459–468.

Hulan, H. W., Moreau, G., and Bird, F. H. (1972). Effect of fat level in isonitrogenous diets on composition of avian pancreatic juice. *J. Nutr.* **102**, 459–468.

Hurwitz, S., Bar, A., Katz, M., Sklan, D., and Budowski, P. (1973). Absorption and secretion of fatty acids and bile acids in the intestine of the laying fowl. *J. Nutr.* **103**, 543–547.

Jerrett, S. A., and Goodge, W. R. (1973). Evidence for amylase in avian salivary glands. *J. Morphol.* **139**, 27–46.

Jiménez, M., Martinez, V., Goñalons, E., and Vergara, P. (1992). Opioid-induction of migrating motor activity in chickens. *Life Sci.* **50**, 465–468.

Joyner, W. L., and Kokas, E. (1971). Action of serotonin on gastric (proventriculus) secretion in chickens. *Comp. Gen. Pharmacol.* **2**, 145–150.

Kessler, C. A., Hirschowitz, B. I., Burhol, P. G., and Sachs, G. (1972). Methoxyflurane (penthrane) anesthesia effect on histamine stimulated gastric secretion in the chickens. *Proc. Soc. Exp. Biol. Med.* **139**, 1340–1343.

Kimmich, G., and Randles, J. (1984). Sodium-sugar coupling stoichiometry in chick intestinal cells. *Am. J. Physiol.* **247**, C74–C82.

Kimmel, J. R., and Pollock, H. G. (1975). Factors affecting blood levels of avian pancreatic polypeptide (APP), a new pancreatic hormone. *Fed. Proc. Fed. Am. Soc. Exp. Biol.* **34**, 454.

Kimmel, J. R., Pollock, H. G., and Hazelwood, R. L. (1968). Isolation and characterization of chicken insulin. *Endocrinology* **83**, 1323–1330.

Kokue, E., and Hayama, T. (1972). Effects of starvation and feeding in the endocrine pancreas of chicken. *Poult. Sci.* **51**, 1366–1370.

Kostuch, T. E., and Duke, G. E. (1975). Gastric motility in great horned owls. *Comp. Biochem. Physiol. A* **51**, 201–205.

Krogdahl, A., and Sell, J. L. (1989). Influence of age on lipase, amylase, and protease activities in pancreatic tissue and intestinal contents of young turkeys. *Poult. Sci.* **68**, 1561–1568.

Kuenzel, W. J. (1989). Neuroanatomical substrates involved in the control of food intake. *Poult. Sci.* **68**, 926–937.

Kuenzel, W. J. (1994). Central neuroanatomical systems involved in the regulation of food intake in birds and mammals. *J. Nutr.* **124**, 1355S–1370S.

Lacy, M. P., Van Krey, H. P., Skewes, P. A., and Denbow, D. M. (1985). Effect of intrahepatic glucose infusion on feeding in heavy and light breed chicks. *Poult. Sci.* **64**, 751–756.

Lacy, M. P., Van Krey, H. P., Skewes, P. A., and Denbow, D. M. (1986). Food intake in the domestic fowl: Effect of intrahepatic lipid and amino acid infusion. *Physiol. Behav.* **36**, 533–538.

Lai, H. C., and Duke, G. E. (1978). Colonic motility in domestic turkeys. *Dig. Dis.* **23**, 673–681.

Larsson, L. I., Sundler, F., Hakanson, R., Pollock, H. G., and Kimmel, J. R. (1974). Localization of APP, a postulated new hormone to a pancreatic cell type. *Histochemistry* **42**, 377–382.

Leasure, E. E., and Link, R. P. (1940). Studies on the saliva of the hen. *Poult. Sci.* **19**, 131–134.

Lepkovski, S., and Furuta, F. (1970). Lipase in pancreas and intestinal contents of chickens fed heated and raw soybean diets. *Poult. Sci.* **49**, 192–198.

Levin, R. J. (1984). Absorption from the alimentary tract. In "Physiology and Biochemistry of the Domestic Fowl" (B. M. Freeman, ed.), Vol. 5, pp. 1–22. Academic Press, London.

Levin, R., Mitchell, M. A., and Barber, D. C. (1983). Comparison of jejunal and ileal absorptive functions for glucose and valine *in vivo*—A technique for estimating real K_m and J_{max} in the domestic fowl. *Comp. Biochem. Physiol. A* **74**, 961–966.

Levy, D. J., and Duke, G. E. (1992). How do frugivores process fruit? Gastrointestinal transit and glucose absorption in cedar waxwing (*Bambycilla cedorum*). *Auk* **109**, 722–730.

Lilja, C. (1983). A comparative study of postnatal growth and organ development in some species of birds. *Growth* **47**, 317–339.

Lin, G. L., Himes, J. A., and Cornelius, C. E. (1974). Bilirubin and biliverdin excretion by the chicken. *Am. J. Physiol.* **226**, 881–885.

Lindsay, O. B., and March, B. E. (1967). Intestinal absorption of bile salts in the cockerel. *Poult. Sci.* **46**, 164–168.

Lisbona, F., Jiménez, M., Esteller, A., and Lopez, M. A. (1981). Basal biliary secretion in conscious chicken and role of enterohepatic circulation. *Comp. Biochem. Physiol. A* **69**, 341–344.

Long, J. F. (1967). Gastric secretion in unanesthetized chickens. *Am. J. Physiol.* **212**, 1303–1307.

Martinez, V., Jiménez, M., Goñalons, E., and Vergara, P. (1992). Effects of cholecystokinin and gastrin on gastroduodenal motility and coordination in chickens. *Life Sci.* **52**, 191–198.

Martinez, V., Jiménez, M., Goñalons, E., and Vergara, P. (1993). Mechanism of action of CCK in avian gastroduodenal motility: Evidence for nitric oxide involvement. *Am. J. Physiol.* **265**, G842–G850.

Matsushita, S. (1983). Purification and partial characterization of chick intestinal sucrase. *Comp. Biochem. Physiol. B* **76**, 465–470.

McLelland, J. (1975). Aves digestive system. *In* "Sisson and Grossman's The Anatomy of the Domestic Animals" (R. Getty, ed.), Vol. 2, pp. 1857–1882. Saunders, Philadelphia.

McLelland, J. (1979). Digestive system. *In* "Form and Function in Birds" (A. S. King and J. McLelland, eds.), pp. 69–181. Academic Press, London.

McLelland, J. (1989). Anatomy of the avian cecum. *J. Exp. Zool.* **3**(Suppl.), 2–9.

Moretó, M., Amat, C., Puchal, A., Buddington, R. K., and Planas, J. M. (1991). Transport of L-proline and α-methyl-D-glucoside by chicken proximal cecum during development. *Am. J. Physiol.* **260**, G457–G463.

Mueller, L. R., Duke, G. E., and Evanson, O. A. (1990). Investigations of the migrating motor complex in domestic turkeys. *Am. J. Physiol.* **259**, G329–G333.

Mule, F. (1991). The avian oesophageal motor function and its nervous control: Some physiological, pharmacological and comparative aspects. *Comp. Biochem. Physiol. A* **99**, 491–498.

Mule, F., Fileccia, R., Postorino, A., Serio, R., Abbadessa, U. S., and La Grutta, G. (1987). Pigeon oesophageal EMG activity: Analysis of intramural neural control. *Arch. Int. Physiol. Biochim.* **95**, 269–280.

Nakagawa, H., Nishimura, M., and Urakawa, N. (1983). 2-Deoxy-D-glucose stimulates acid secretion from chicken proventriculus. *Jpn. J. Vet. Sci.* **45**, 721–726.

Nilsson, A. (1974). Isolation, amino acid composition and terminal amino acid residue of the vasoactive octacosapeptide from chicken intestine: Partial purification of chicken secretin. *FEBS Lett.* **47**, 284–289.

Obst, B. S., and Diamond, J. M. (1989). Interspecific variation in sugar and amino acid transport by the avian cecum. *J. Exp. Zool.* **3**(Suppl.), 117–126.

Obst, B. S., and Diamond, J. (1992). Ontogenesis of intestinal nutrient transport in domestic chickens (*Gallus gallus*) and its relation to growth. *Auk* **109**, 451–464.

Ohashi, H., and Ohgua, A. (1967). Transmission of excitation from the parasympathetic nerve to the smooth muscle. *Nature* **216**, 291–292.

Osman, A. M. (1982). Amylase in chicken intestine and pancreas. *Comp. Biochem. Physiol. B* **73**, 571–574.

Paik, Y. K., Fujioka, T., and Yasuda, M. (1974). Comparative and topographical anatomy of the fowl. LXXVIII. Division of pancreatic lobes and distribution of pancreatic ducts. *Jap. J. Vet. Sci.* **36**, 213–229.

Philips, S. M., and Fuller, R. (1983). The activities of amylase and a trypsin-like protease in the gut contents of germ-free and conventional chickens. *Br. Poult. Sci.* **24**, 115–121.

Pichova, I., and Kostka, V. (1990). Molecular characteristics of pepsinogen and pepsin from duck glandular stomach. *Comp. Biochem. Physiol. B* **97**, 89–94.

Pinchasov, Y., and Noy, Y. (1994). Early postnatal amylolysis in the gastrointestinal tract of turkey poults *Meleagris gallopavo*. *Comp. Biochem. Physiol. A* **107**, 221–226.

Planas, J. M., Villá, M. C., Ferrer, R., and Moretó, M. (1986). Hexose transport by chicken cecum during development. *Pfluegers Arch.* **407**, 216–220.

Polin, D., Wynosky, E. R., Loukides, M., and Porter, C. C. (1967). A possible urinary back flow to ceca revealed by studies on chicks with artificial anus and fed amprolium-C^{14} or thiamine-C^{14}. *Poult. Sci.* **46**, 89–93.

Postorino, A., Serio, R., Fileccia, R., Mule, F., and Abbadessa, U. S. (1985). Electrical stimulation of glossopharyngeal nerve and oesophageal EMG response in the pigeon. *Arch. Int. Physiol. Biochim.* **93**, 321–330.

Pubols, M. H. (1991). Ratio of digestive enzymes in the chick pancreas. *Poult. Sci.* **70**, 337–342.

Rea, A. M. (1973). Turkey vultures casting pellets. *Auk* **90**, 209–210.

Riesenfeld, G., Geva, A., and Hurwitz, S. (1982). Glucose homeostasis in the chicken. *J. Nutr.* **112**, 2261–2266.

Riesenfeld, G., Sklan, D., Barr, A., Eisner, U., and Hurwitz, S. (1980). Glucose absorption and starch digestion in the intestine of the chicken. *J. Nutr.* **110**, 117–121.

Roseler, M. (1929). Die Bedeutung der Blinddarme des Haushuhnes fur die Resorption der Nahrung und Verdauung der Rohfaser. *Z. Tierz. Zuechtungsbid* **13**, 281–310.

Sahba, M. M., Morisset, J. A., and Webster, P. D. (1970). Synthetic and secretory effects of cholecystokinin-pancreozymin on the pigeon pancreas. *Proc. Soc. Exp. Biol. Med.* **134**, 728–732.

Saito, I. (1966). Comparative anatomical studies of the oral organs of the poultry. V. Structures and distribution of taste buds of the fowl. *Bull. Fac. Agric., Miyazaki Univ.* **13**, 95–102.

Sato, H., Ohga, A., and Nakazato, Y. (1970). The excitatory and inhibitory innervation of the stomachs of the domestic fowl. *Jpn. J. Pharmacol.* **20**, 382–397.

Savory, C. J. (1987a). How closely do circulating blood glucose levels reflect feeding state in fowls? *Comp. Biochem. Physiol. A* **88**, 101–106.

Savory, C. J. (1987b). An alternative explanation for apparent satiating properties of peripherally administered bombesin and cholecystokinin in domestic fowls. *Physiol. Behav.* **39**, 191–202.

Savory, C. J., and Mitchell, M. A. (1991). Absorption of hexose and pentose sugars *in vivo* in perfused intestinal segments in the fowl. *Comp. Biochem. Physiol. A* **100**, 969–974.

Savory, C. J., Duke, G. E., and Bertoy, R. W. (1981). Influence of intravenous injections of cholecystokinin on gastrointestinal motility in turkeys and domestic fowls. *Comp. Biochem. Physiol. A* **70**, 179–189.

Sell, J. L., Eastwood, J. A., and Mateos, G. G. (1983). Influence of supplemental fat on diet metabolizable energy and ingesta transit time in laying hens. *Nutr. Rep. Intern.* **28**, 487–495.

Shires, A., Thompson, J. R., Turner, B. V., Kennedy, P. M., and Goh, Y. K. (1987). Rate of passage of corn-canola meal and corn-soybean meal diets through the gastrointestinal tract of broiler and White Leghorn chickens. *Poult. Sci.* **66**, 289–298.

Sibbald, I. R. (1979). Passage of feed through the adult rooster. *Poult. Sci.* **58**, 446–459.

Siddons, R. C., (1969). Intestinal disaccharidase activities in the chick. *Biochem. J.* **112**, 51–59.

Siddons, R. C. and Coates, M. E. (1972). Effect of diet on disaccharidase activity in the chick. *Br. J. Nutr.* **27**, 343–352.

Skadhauge, E. (1989). Hormonal regulation of sodium absorption and chloride secretion across the lower intestines of birds. *Zool. Sci.* **6**, 437–444.

Skadhauge, E., Clauss, W., Arnason, S. S., and Thomas, D. H. (1985). Mineralocorticoid regulation of lower intestinal ion transport. *In* "Transport processes, Iono- and Osmoregulation" (R. Gilles and M. Gilles-Baillien, eds.), pp. 118–133. Springer-Verlag, Berlin/Heidelberg/New York.

Sturkie, P. D. (1965). "Avian Physiology" (P. D. Sturkie, ed.), 2nd Ed., pp. 300–306. Cornell Univ. Press, Ithaca, NY.

Sturkie, P. D. (1976). "Avian Physiology" (P. D. Stukie, ed.), 3rd Ed., pp 270, 280. Springer-Verlag, New York.

Sudo, S. Z., and Duke, G. E. (1980). Kinetics of absorption of volatile fatty acids from the ceca of domestic turkeys. *Comp. Biochem. Physiol. A* **67,** 231–237.

Suzuki, M., and Nomura, S. (1975). Electromyographic studies on the deglutition movements in the fowl. *Jap. J. Vet. Sci.* **37,** 289–293.

Takewaki, T., Ohashi, H., and Okada, T. (1977). Non-cholinergic and non-adrenergic mechanism in the contraction and relaxation of the chicken rectum. *Jap. J. Pharmac.* **27,** 105–115.

Taneike, T., Miyazaki, H., and Ohga, A. (1988). Histamine-induced potentiation of cholinergic transmission in chick oesophagus (*Gallus gallus*). *Comp. Biochem. Physiol. C* **89,** 271–276.

Thomas, D. H., and Skadhauge, E. (1988). Transport function and control in bird caeca. *Comp. Biochem. Physiol. A* **90,** 591–596.

Tur, J. A., and Rial, R. V. (1985). The effect of temperature and relative humidity on the gastrointestinal motility of young broilers. *Comp. Biochem. Physiol. A* **80,** 481–486.

Uden, P., Colucci, P. E., and Van Soest, P. J. (1980). Investigation of chromium, cerium and cobalt as markers in digesta. Rate of passage studies. *J. Sci. Food Agric.* **31,** 625–632.

Vaillant, C., Dimaline, R., and Dockray, G. J. (1980). The distribution and cellular origin of vasoactive intestinal polypeptide in the avian gastrointestinal tract and pancreas. *Cell. Tissue Res.* **211,** 511–523.

Van Der Klis, J. G., Verstegen, M. W. A., and De Wit, W. (1990). Absorption of minerals and retention time of dry matter in the gastrointestinal tract of broilers. *Poult. Sci.* **69,** 2185–2194.

Vandeputte-Poma, J. (1968). Quelques donnees sur la composition du "Lait de Pigeon." *Z. Verg. Physiol.* **58,** 356–363.

Vergara, P., Ferrando, C., Jiménez, M., Fernández, E., and Goñalons, E. (1989). Factors determining gastrointestinal transit time of several markers in the domestic fowl. *Quart. J. Exp. Physiol.* **74,** 867–874.

Vinardell, M. P., and Lopera, M. T. (1987). Jejunal and cecal 3-oxymethyl-D-glucose absorption in chicken using a perfusion system *in vivo. Comp. Biochem. Physiol. A* **86,** 625–627.

Ward, N. E., Jones, J. E., and Maurice, D. V. (1984). Increase in intestinal pH of chickens due to cimetidine injection. *Fed. Proc.* **43,** 856.

Wasserman, R. H. (1992). "Extra- and Intracellular Calcium and Phosphate Regulation" (F. Bronner and M. Peterlik, eds). CRC, Boca Raton, FL.

Wasserman, R. J., Smith, C. A., Brindak, M. E., de Talamoni, N., Fullmer, C. S., Penniston, J. T., and Kumar, R. (1992). Vitamin D and mineral deficiencies increases the plasma membrane calcium pump of chicken intestine. *Gastroenterology* **102,** 886–894.

White, S. S. (1968). Mechanisms involved in deglutition in *Gallus domesticus. J. Anat.* **104,** 177.

White, S. S. (1970). "The larynx of *Gallus domesticus.*" Ph.D. Thesis, University of Liverpool, Liverpool.

Winget, C. M., Ashton, G. C., and Cawley, A. J. (1962). Changes in gastrointestinal pH associated with fasting in laying hen. *Poult. Sci.* **41,** 1115–1120.

Yamada, J., Kitamura, N., and Yamashita, T. (1983). Avian gastrointestinal endocrine cells. *In* "Avian Endocrinology: Environmental and Ecological Perspectives" (S-i. Mikami, K. Homma, and M. Wada, eds.), pp. 67–79. Japan Scientific Societies Press, Tokyo.

Yamauchi, K.-E., and Isshiki, Y. (1991). Scanning electron microscopic observations on the intestinal villi in growing White Leghorn and broiler chickens from 1 to 30 days of age. *Br. Poult. Sci.* **32,** 67–78.

Ziswiler, V., and Farner, D. S. (1972). Digestion and digestive system. *In* "Avian Biology" (D. S. Farner and James R. King, eds.), Vol. II, pp. 343–430. Academic Press, London.

CHAPTER 13

Energy Balance

CHARLES R. BLEM
Department of Biology
Virginia Commonwealth University
Richmond, Virginia 23284-2012

I. Introduction 327
II. The Measurement of Energy Exchange 327
   A. Elements of Energy Demand in Birds 327
   B. Basal Metabolism 328
   C. Resting Metabolism 329
   D. Existence Energy 329
   E. Metabolized Energy 330
   F. Metabolizable Energy 330
   G. Body Size and Energetics 331
III. Energy Costs of Activity 331
   A. Locomotion 332
   B. Foraging 333
   C. Reproductive Activities 333
IV. Energy Storage and Production 334
   A. Starvation 334
   B. Premigratory Fattening 334
   C. Overwinter Fattening 334
   D. Gonad Growth and Egg Production 335
   E. Molt 335
   F. Growth of Embryos and Newly Hatched Young 336
V. Daily Energy Budgets and Energetics of Free-Living Birds 337
VI. Geographic Variation in Energy Balance 338
VII. Energy Requirements of Populations and Communities 338
VIII. Summary 338
   References 339

I. INTRODUCTION

Allocation of ingested energy to the many demands of avian existence is a fundamental process linked inextricably to the life histories of birds. Walsberg (1983a) correctly suggested that the most basic question in avian energy balance is: "How do thermal relations affect . . . survival, reproduction, and consequent contribution to succeeding generations?" I further suggest that many adaptations of avian energy balance minimize energy expenditure in such as way as to maximize production of DNA and its conveyances while minimizing exposure to predators, inclement weather, and injury that result from excessive foraging activity. Additionally, genetic diversity promoting these adaptations provides the basis for development of productive strains of domestic birds.

There are hundreds of papers dealing with energy balance of domestic and wild birds. A number of authors previously have reviewed aspects of the subject, and anyone new to the topic should read accounts by Gessaman (1973, 1987), Calder and King (1974), Kendeigh *et al.* (1977), Walsberg (1983b), Whittow (1986), and Blem (1990). Many general statements made in the present account are more thoroughly documented in those sources.

II. THE MEASUREMENT OF ENERGY EXCHANGE

A. Elements of Energy Demand in Birds

Superficially, energy use by birds can be summarized as: energy intake = metabolized energy + energy in egesta (feces, urine) (Figure 1) and is expressed as kcal or kJ per unit time or watts (W). In practice, ingested energy is measured as *gross energy intake* (GEI), the product of the mass of food ingested multiplied by the heat of combustion of the food materials (see below).

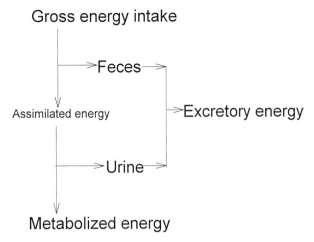

FIGURE 1 General avian energy balance.

Excretory energy (EXE) is the energy content of urine and fecal materials. *Metabolized energy* (ME) is the total cost of basal metabolic processes, thermoregulation, specific dynamic action, physical activity, and production and is obtained by subtracting EXE from GEI. *Production* (P) is the allocation of energy to growth, reproduction, lipid deposition, and synthesis of feathers and is the amount of energy expended above resting metabolism or existence energy levels. A comprehensive model of avian energy balance must include consideration of rates of energy intake minus several additive layers of expenditures including basal metabolism and the costs of thermoregulation, physical activity, molt, growth, lipid deposition, reproduction, and energy lost in digestion and assimilation (Figure 2). Furthermore, these measurements may be affected by body composition, sex, and ambient temperature. Previous thermal history of the bird also is influential; summer-acclimatized birds usually have higher metabolic rates than winter-acclimatized birds (Blem, 1973b; Kendeigh *et al.*, 1977).

B. Basal Metabolism

Basal metabolism (BM) is the rate of energy use by birds at rest within the zone of thermoneutrality (i.e., not stimulated by low ambient temperature). For measurements of BM, birds must be postabsorptive (not fed) and therefore not affected by heat produced during assimilation of food (specific dynamic action). BM generally is measured by respirometric determination of the volumetric rate of oxygen consumption (see Withers, 1977, for details on technique) or (rarely) carbon dioxide production. To convert these measurements to energy terms, the volumes of gases must be multiplied by appropriate caloric equivalents of the gas in question (see Gessaman, 1973, 1987). There is no natural corre-

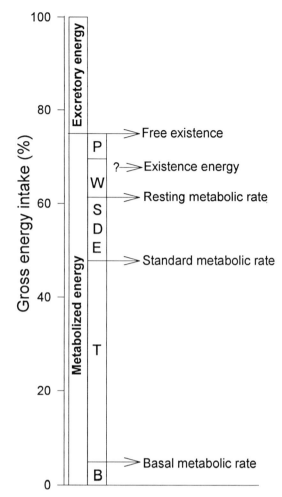

FIGURE 2 Physiological levels of energy expenditure in birds (B, energy expended in basal physiological processes; T, heat of thermoregulation; SDE, specific dynamic action (effect); W, energy expended in activities involving work; P, production).

late of BM under most conditions. Furthermore, many authors (e.g., Lasiewski and Dawson, 1967; Calder and King, 1974) have used the term *standard metabolism* (SM) to refer to basal levels of energy utilization because true basal measurements are difficult to make. Here I follow others (e.g., Kendeigh *et al.*, 1977) who use standard metabolism to refer to energy expenditure by postabsorptive (unfed) birds at rest at ambient temperatures below the lower critical temperature (i.e., below the zone of thermal neutrality). There are additional sources of variation in metabolism. For example, basal/standard measurements of unfed birds at rest in the zone of thermal neutrality differ between the inactive and active phases of the daily cycle (usually night/day differences; see Aschoff and Pohl, 1970). Reasons for these differences are not clear.

Basal metabolism can be predicted with accuracy from the body mass of the individual bird, although one should take care to account for mass variations arising from differences in lipid stores, water balance, and/or reproductive condition, when possible. Basal measurements are useful for comparison of basic differences in metabolic capacity, but usually are not useful for studies of costs of natural activity.

C. Resting Metabolism

Resting metabolism (RM) represents the energy used by birds who are eating and behaving normally, except that costs of activity are minimal. RM is measured by respirometric determinations of oxygen consumption or carbon dioxide production and may include the costs of thermorcgulation as it often is measured at ambient temperatures below the zone of thermoneutrality. Temperature regulation is a major energy expenditure below the zone of thermal neutrality and at extreme lower limits of temperature tolerance and may amount to more than 70% of energy expenditure.

Small birds have relatively large surface areas, therefore rates of heat loss at low ambient temperatures may become so stressful that the bird undergoes moderate to severe hypothermia. In large birds such decreases are small or nonexistent. Extreme hypothermia may result in moderate to severe loss of activity as the bird enters torpor. Few birds have evolved sufficient thermal tolerance to survive deep torpor. Hummingbirds (Trochilidae), goatsuckers (Caprimulgidae), and swifts (Apodidae) may have temperature decreases of as much as 30°C and may go into deep torpor for hours (hummingbirds) or days (Poor-will). A less-extreme form of hypothermia (i.e., body temperature decreases of 4–12°C) may occur in tits (*Parus;* Haftorn, 1972; Chaplin, 1974), kinglets (*Regulus;* Blem and Pagels, 1984), and manakins (*Manacus, Pipra;* Bartholomew *et al.,* 1983), although there are tit, kinglet, and manakin species that do not seem to go into hypothermia at all. The energy savings in the modest torpor of small passerines may be 10–75% of BM of normothermic birds (Reinertsen, 1983; Reinertsen and Haftorn, 1986), depending on the degree and duration of hypothermia. Hummingbirds going into prolonged, deeper hypothermia can save energy equal to 75% of BM (Calder and King, 1974).

Resting metabolism also includes specific dynamic action (SDA), also known as the calorigenic effect or heat increment. SDA is the elevation of the metabolic rate that occurs after a food item is absorbed from the digestive tract. It is not a result of increased gut motility or assimilation, as injection of food constituents into the blood stream can cause SDA (Whittow, 1986). SDA is a function of food composition and roughly amounts to 15% of basal rates for fat, 29% for carbohydrates, and 45% for protein (Ricklefs, 1974). Heat from SDA can substitute for heat used in thermoregulation, but its effect is influenced by other variables, including thermal history and nutrition of the animal (Calder and King, 1974; King, 1974). The degree of substitution may be highly variable or, in some instances, absent (Mugaas and King, 1981).

D. Existence Energy

Existence energy (EE) is the amount of energy expended by caged birds fed *ad lib,* whose movements are restricted (Figure 3). Such measurements include costs of thermoregulation, specific dynamic action, and minimal activity. Molt, lipid deposition, reproduction, or growth must not occur during these measurements. In most situations the magnitude of EE is similar to RM. The technique involves quantification of the energy both in food eaten and excreta voided (EXE). The heat of combustion of food and waste materials in kJ or kcal per unit mass is determined by bomb calorimetry. The bird's weight must remain constant during such measurements; that is, mass change should be less than 2% (Gessaman, 1987). Methods of measurement of heats of combustion by bomb calorimetry are summarized in Grodzinski *et al.* (1975) and in manuals available from manufacturers of calorimeters. Bomb calorimetry involves combustion of dry materials in an atmosphere of pure oxygen in a heavy metal cylinder. The heat is captured in a surrounding water jacket containing a known amount of water. The heat of combustion in kJ

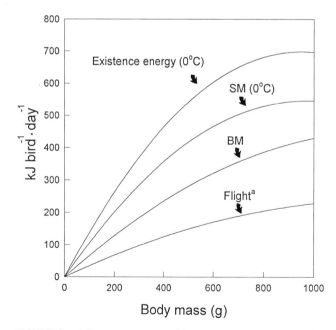

FIGURE 3 Existence energy at 0°C, standard metabolism at 0°C (SM), basal metabolism (BM), and metabolism in flight for passerine birds weighing 0–1000 g (flight costs are kJ·bird^{-1} hr^{-1}).

or kcal per unit mass is computed from the temperature increase of the water (see Table 1 for examples). All of the energy of the heat of combustion is available to the organism digesting the food. Aspects of the techniques needed to assess avian energy balance are comprehensively covered by Kleiber (1975) and Gessaman (1973, 1987; also see Blem, 1968).

E. Metabolized Energy

Metabolized energy is the energy expended by caged birds undergoing activities such as fattening, egg-laying, or molt. Energy mobilized in excess of existence metabolism (productive energy) can be included. It is determined in the same manner as existence energy and includes the same basic components, except that body mass can change. The method is useful in a general way in comparisons of metabolic capacity of caged birds, particularly in studies of controlled production and assessment of digestibility of food items. Since activity cannot be included to any great degree in such measurements, the technique should not be applied to estimates of power consumption of birds in nature. Conversely, the concept of metabolized energy as the sum of all power-consuming avian activities is a useful one. Several models have been presented that suggest that avian energy expenditure reaches a maximum sustainable level determined by the capacity of the bird to ingest and assimilate food (e.g., Kendeigh, 1969b; Goldstein,

1988). Studies of ME indicate the model is correct. For example, under caged conditions, Japanese quail (*Coturnix coturnix*) cease laying eggs at ambient temperatures 1–2°C above lower lethal temperatures because costs of thermoregulation are near the maximum sustainable threshold of energy mobilization and therefore no surplus energy is available for development of ova (Blem, unpublished data).

F. Metabolizable Energy

Although the term has been confused with metabolized energy, *metabolizable energy*, strictly defined, is the amount of energy assimilated from food and is equal to gross energy intake minus fecal energy. Energy in food absorbed from the gastrointestinal tract is *digestible energy*. Energy in feces and urine is *excretory energy*. In birds, true assimilation is difficult to measure since feces and urine are voided together and the energy content of urine represents material that has been assimilated while feces are unassimilated materials (Figure 1). Assimilation is affected by food composition; simple sugars and lipids are used with great efficiency (see Blem, 1976b; Castro *et al.*, 1989; Karasov, 1990); for example, hummingbirds extract more than 98% of the energy in nectar (Table 2). Proteins cannot be fully catabolized and significant energy is lost in nitrogenous wastes in the urine. As a result, metabolizable energy largely is a function of food composition (Table 2; also see Karasov, 1990). However, some birds have evolved the ability to ingest, assimilate, and catabolize complex materials including wax esters and beeswax (Place and Roby, 1986) and coarse herbaceous materials including twigs, pine needles, and grass (Table 2).

The efficiency with which birds utilize ingested energy has been called *energetic efficiency* (Blem, 1976b), *metabolizable energy coefficient* (see Kendeigh *et al.*, 1977), and *assimilation efficiency* or *energy assimilation efficiency* (Handrich et al., 1993). Metabolizable energy coefficient (MEC) is the oldest and most consistently used term. It is computed as:

$$MEC = 100(ME/GEI).$$

MEC values may differ among species eating the same food (Table 2), among individuals for unknown causes, and may change with ambient temperature. For example, in some species it is lowest at low ambient temperatures when food intake rates are maximal (Blem, 1973b). Food selection and survival of cold stress may be a function of the bird's ability to extract usable energy (Sprenkle and Blem, 1984). It has been hypothesized that MECs may increase with physiological adjustment of rates of assimilation, playing a role in accumulation of surplus energy for activities such as premigratory fattening (Bairlein, 1985).

TABLE 1 Energy Contents of Avian Materials

Material	kJ/g	Source
Feathers		
Keratin	22.5	King and Farner (1961)
Tissue		
Adult birds, wet mass	5.6–11.1	Blem (unpublished data)
Young birds, wet mass	2.7–9.4	Blem (1975, 1978, unpublished data)
Eggs		
Altricial birds; wet mass	4.3–5.2	Kendeigh *et al.* (1977)
Semi-altricial birds; wet mass	4.8	Carey *et al.* (1980)
Semi-precocial birds; wet mass	6.8	Carey *et al.* (1980)
Precocial birds; wet mass	6.4–10.2	Kendeigh *et al.* (1977)
Seabirds; dry mass	25.9–29.7	Pettit *et al.* (1984)
Seabirds; wet mass	3.8–7.1	Walsberg (1983b)
Lipid		
Extracted dry depot	37.7–39.8	Blem (1990)
Petrel stomach oil	40.6	Obst and Nagy (1993)
Food		
Seeds (dry mass)	18.0	Kendeigh *et al.* (1977)
Insects (dry mass)	23.0	Kendeigh, *et al.* (1977)
Vertebrate prey (dry mass)	23.1	Goldstein (1990)

TABLE 2 Metabolized Energy Coefficients (MEC)[a]

Food items	Taxa of birds	MEC	References
	Adult birds		
Pine needles	Spruce Grouse (*Dendragapus canadensis*)	29.8	Pendergast and Boag (1971)
Willow twigs	Ptarmigan (*Lagopus sp.*)	32.0	Moss (1973)
Grass	Brant (*Branta bernicla*)	30.1–51.2	Buchsbaum et al. (1986)
Nectar	Hummingbirds (Trochilidae)	98.1–98.2	Hainsworth (1974)
Seeds	Twelve species of passerines	76–91	Karasov (1990)
Seeds	Bobwhite (*Colinus virginianus*)	31.2–84.6	Robel and Arruda (1986)
Chick mash	Twelve species of granivorous birds	62–89	Kendeigh et al. (1977)
Arthropods	Insectivorous birds	75.9–91.6	Woods (1982)
Squid	King penguin	79.8–84.3	Adams (1984)
Fish	Double-crested cormorant (*Phalacrocorax auritus*)	77.9–89.2	Brugger (1993)
Fish	Bald eagle (*Haliaeetus leucocephalus*)	75.0	Stalmaster and Gessaman (1982)
Mammals	Five species of owls (Strigidae)	77.5–94.2	Woods (1982)
Ducks	Bald eagle	85.2	Stalmaster and Gessaman (1982)
	Nestlings		
Soft bill mix	Several species of passerines	75.3	Blem (1973b)
Chick mash	Bobwhite	30.2–77.5	Blem and Zara (1980)
Chick mash	Japanese quail	44.8–64.0	Blem (1978)
Chick mash	Black-bellied whistling-duck	69.3–86.5	Cain (1976)
Fish	Double-crested cormorant	79.9–88.1	Dunn (1975)

[a] 100× (GEI-EE)/GEI (see section on metabolized energy).

G. Body Size and Energetics

All measurements of energy balance are positive curvilinear functions of body mass (Figure 3). Total metabolism (M) of the bird increases exponentially with body mass (W), therefore this relationship usually is presented as equations of the form: $M = aW^b$, or $\log M = b \log W + \log a$. When metabolism is expressed as energy expended per unit body mass, the rate rises so sharply at low body weight that minimum size may be energetically constrained. Difficulties in maintaining a positive energy balance may limit minimum size in passerine birds at about 5 g and in nonpasserines at 3 g (Kendeigh, 1970, 1972). Equations for avian basal metabolism, standard metabolism, and existence energy are available in numerous sources, including Aschoff and Pohl (1970), Kendeigh (1970), Calder (1974, 1984), Ricklefs (1974), and Kendeigh et al. (1977). Photoperiod, thermal history, season, and ambient temperature affect the slopes of these equations, and there are significant phylogenetic differences among groups of birds as well (see Calder, 1974, 1984; Kendeigh et al., 1977).

III. ENERGY COSTS OF ACTIVITY

Measurements of metabolized energy, while useful for evaluation of the processes of digestion and SDA, should not be considered to be accurate assessments of energy use by free-living birds under natural conditions. However, if existence energy, resting metabolism, and/or metabolized energy are known, the expenditure of energy in productive processes in nest cavities or by caged birds can be computed. For example, Blem (1975) calculated costs of activity in young birds by subtracting energy involved in growth (measured from carcass analyses) plus resting metabolism (from respirometry) from metabolized energy (determined by energy balance methods). Using such techniques one can compare costs of processes such as egg production as amount of ME or RM in excess of SM and compare this with the energy content of the eggs themselves.

Goldstein (1988) reviews the ways in which daily energy expenditure can be measured in birds: (1) time-energy budgets and (2) doubly labeled water techniques. Time-energy budgets initially were the most common way of estimating daily energy expenditures (Goldstein, 1988). Such budgets are computed from measurements of the duration of a bird's daily activities multiplied by the respective costs of each activity. In many ways the more satisfactory method for measuring the total energy expenditure of free-living birds is the doubly labeled water technique. In this method the bird's body water is labeled with the isotopes of hydrogen and oxygen (^2H, ^3H, and ^{18}O). Both deuterium and tritium have been used as the radioactive label (Nagy, 1983, 1989; Powers and Conley, 1994). Body water then is sampled periodically to determine changes in the ratios of the two isotopes. Oxygen is lost both in expired carbon dioxide and through water lost in evaporation/expiration, but hydrogen is lost only through water. The difference in the two rates of loss is a measure of carbon dioxide production which can be converted to energy terms (although the respiratory quotient must be estimated). The energy budget of breeding acorn woodpeckers (*Melanerpes formicivorus*) constructed by Weathers et al. (1990) is an outstanding example of the use of multiple techniques to assess metabolic rate of a free-living bird.

The costs of foraging or reproductive behavior also have been evaluated from measurements of time birds spent in activity multiplied by the energy expended per unit time (time–energy budget). To construct time–energy budgets one must determine the time the bird spends in all significant activities, assign energy costs to each of them, and total the result for overall estimates. These computations can quickly become cumbersome and of uncertain accuracy. However, Williams and Nagy (1984) and Mock (1991) compared time–energy budgets of western bluebirds (*Sialia mexicana*) with measurements made using doubly labeled water methods; they obtained similar values from the two techniques (also see Weathers and Nagy, 1980; Weathers et al., 1984).

A. Locomotion

Energetic costs of avian locomotion vary with body mass, type of locomotion, velocity of travel, and, in some cases, ambient temperature. In all forms of locomotion the energy required to transport a unit mass of bird varies inversely with the mass of the bird. Costs of locomotion are expressed as energy expended per unit body mass moving through one unit of distance; the correct units are $kJ \cdot kg^{-1} \ km^{-1}$.

1. Flight

The energy requirement has been measured in more than a dozen species of birds. Per unit time, flight is the most expensive avian activity. Cost of flight varies with forward velocity (Figure 4). Low velocities are expensive to maintain because of lack of momentum and high velocities are costly because of various forms of friction on the surface of the bird. Minimum costs of flight are attained at intermediate velocities determined to some extent by the mass of the bird. As a result, large birds fly more cheaply at higher velocities than do small birds. Costs range from 2.7× to 23× BM, depending on the style of flight and the taxonomic group of birds involved (Tatner and Bryant, 1986; Goldstein, 1988). Natural migrants appear to fly at velocities slightly higher than birds flying in wind tunnels. This may be because this results in cheaper travel per unit distance; flight costs increase less rapidly than velocity in this part of the

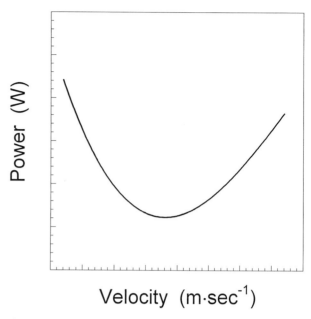

FIGURE 4 Power costs of flight in a bird as a function of velocity.

energy–velocity curve (see Blem, 1980, for a review). During migration, costs of flight increase as lipid stores decrease; loss of mass results in slower optimum flight speeds and higher costs per unit mass. Migrants such as the blackpoll warbler (*Dendroica striata*), a 10-g bird that deposits about 10 g of lipid in preparation for a transoceanic flight of 80–90 hr, are believed to fly more slowly and at great expense at the end of migration when body reserves are nearly depleted.

2. Walking

When body mass is considered, the energetic costs of walking or hopping generally are less than those of flight. The cost of walking/hopping by birds is a linear function of velocity of travel and varies among species although it has been measured few times. Energy expended in hopping in European robins (*Erithacus rubecula*) is a function of ambient temperature and is 2.0–2.4× nighttime resting metabolic rates (Tatner and Bryant, 1986). Most birds spend very little time walking, but in some penguins walking to and from colonies may require significant amounts of energy.

3. Swimming

Swimming is the least expensive form of locomotion. It tends to differ between birds that swim underwater, using wings and feet (e.g., penguins) and those that swim on the surface using only feet (ducks). Energy demand during swimming is relatively low; costs range from 2.2 to 2.8× resting metabolic rates.

B. Foraging

It is difficult to quantify the energy required in foraging. Some authors have been able to estimate, from counts of captures of specific food items and their caloric equivalents, the gross or net energy acquired during relatively brief bouts of foraging. However, comprehensive models must include costs of defense of food resources, searching for food, the costs of chase, capture, and/or collection of prey, and processing of food items. Most of these have been studied very little, and the cost of terrestrial foraging in Gambel's quail (*Callipepla gambelii*; Goldstein, 1990) were 3.5× resting metabolic rates (Goldstein and Nagy, 1985). Costs of foraging in the Laysan Albatross (*Diomedea immutabilis*) were 2.6× resting levels (Petit *et al.*, 1988).

C. Reproductive Activities

1. Incubation

Costs of incubation vary with ambient temperature and with artificial manipulation of egg temperatures. For example, Vleck (1981) found that adult zebra finches could double their metabolic rate when egg temperatures were lowered artificially. The calorigenic effect of food can supply some of the heat used in incubation, but apparently only at very low ambient temperatures (Biebach, 1984). Several authors have used biophysical models to calculate potential energy expenditure of incubating birds (see reviews by Walsberg, 1983a,b). Their estimates predict that incubating birds should have metabolic rates 15–40% above BMR. Others have directly measured the oxygen consumption of cavity-nesting species, using the nest cavity as a respiratory chamber in an open-flow respirometry system (e.g., Biebach, 1977, 1979; Gessaman and Findell, 1979; Vleck, 1981). In such measurements, incubating birds consumed 25–43% more energy than resting birds not involved in incubation, and costs of incubation increase with clutch size (Biebach, 1981, 1984). Haftorn and Reinertsen (1985) found that energy costs of incubation in blue tits (*Parus caerulescens*) increased by 6–7% RM per egg in the clutch. El-Wailly (1966) measured metabolized energy in feeding trials and concluded that incubating zebra finches (*Taeniopygia castanotis*) expended up to 34% more energy than control birds.

However, it should be noted that incubation has not been shown to be costly in all studies. Pettit *et al.* (1988) found that the energy expenditure of incubating Laysan albatrosses was similar to that of resting birds not incubating. Walsberg (1983a) provides important criticism of most studies of the energetics of incubation. He correctly points out several difficulties in determining how much energy expenditure is directly attributable to incubation and in deciding the proper ecological framework in which to compare elevated expenditures. He notes (Walsberg and King, 1978; Walsberg, 1983a) that, in some cases, a bird incubating eggs and attending the nest may actually involve a 15–20% *decrease* of energy expenditure in comparison to nonbreeding adults outside of the nest.

2. Territorial Defense

Territorial defense may be relatively expensive during brief periods of the year and may amount to 0.9–33.1% of daily energy expenditures (Walsberg, 1983b). Bryant and Westerterp (1980a), using the doubly labeled water technique, found that territorial male housemartins (*Delichon urbica*) used 2.44× BM. Territorial behavior may be quite expensive; captive male Carolina wrens (*Thyrothorus ludovicianus*) used 2.7–8.7× BM during singing and this exceeded the costs of all other activities except flight (Eberhardt, 1994).

3. Rearing Young

Walsberg (1983a) indicated that care and feeding of young was one of the "most conspicuously neglected areas in avian energetics." Newly hatched birds survive mostly on contents of their yolk sacs which may provide energy for several days. Altricial young are fed and incubated by their parents throughout the developmental period, although the frequency and duration decreases as the young mature. Nestling food is mostly energy-rich materials, either collected by the parents or secreted by them. Petrels, for example, regurgitate oils with high heats of combustion (Table 1) into the throats of their young. Pigeons secrete a milky, energy-rich substance from their crops which similarly is delivered to their young.

Bryant and Westerterp (1983) documented the complexity of the energetics of rearing young. They found that 13 of 39 environmental and biological factors tested were correlated with daily metabolic rates of housemartins rearing broods and the overall expenditure was about 3.6× BM over extended periods of measurement (Bryant and Westerterp, 1980a). However, it appears that feeding of young is not very demanding. For example, Williams and Nagy (1985) found that female savannah sparrows (*Passerculus sandwichensis*) used 2.2–3.4× BM, only about 11% more than caged existence levels. Hails and Bryant (1979) calculated that housemartins expended 2.2–5.3× SM rates while feeding young. Weathers and Sullivan (1989), using concurrent time–energy budgets and doubly labeled water techniques, constructed extensive energy budgets for yellow-eyed juncos (*Junco phaeonotos*) and their young. Adult juncos feeding young were not food limited and were not working maximally. Adults obtained energy for their young and themselves by foraging for 75% or less of daylight hours. Dykstra and Karasov (1993) found that brood size of housewrens (*Troglodytes aedon*) was not limited by maximal metabolic rates of adults as constrained by ability to assimilate energy. Williams and Nagy (1985) found that energy use by savannah sparrows feeding broods of two did not differ significantly from sparrows feeding broods of three.

IV. ENERGY STORAGE AND PRODUCTION

A. Starvation

Birds deprived of food eventually use food in their digestive tracts, become postabsorptive, and deplete glycogen reserves after a short period (usually less than 24–48 hr, depending upon the size of the bird). They then begin to metabolize fat reserves. The respiratory quotient drops as the substrate being used changes from carbohydrate to lipid. Basal metabolic rate also may drop, perhaps as much as 23.4% (Shapiro and Weathers, 1981). If starvation persists, the bird may begin to use protein for glucose synthesis and the RQ will rise. The time required for depletion of all carbohydrate and lipid reserves varies with size of the bird and amount of initial reserve. In large birds the process may require days; emperor penguins (*Aptenodytes foresteri*) fast for more than 100 days while incubating. Hummingbirds may be severely stressed by food deprivation after only a few hours. During recovery from starvation, efficiency of assimilation in common barn-owls (*Tyto alba*) increased slightly and apparently increased the rate of recovery of body mass (Handrich et al., 1993).

B. Premigratory Fattening

Deposition of lipid reserves in preparation for migration represents a major seasonal energy expenditure (King, 1974; Blem, 1976a, 1980). Although the energy involved in weight gain or loss has been estimated in a few cases (Kendeigh et al., 1977), the costs of lipid deposition have been quantified only roughly. This is because techniques for making precise estimates of lipid depots in living birds do not, as yet, exist. Domestic fowl require about 65.5 kJ to deposit 1 g of fat (Whittow, 1986).

Avian energy reserve mostly is stored as lipid, specifically triglycerides, whose caloric density is very high (Table 1) and, unlike carbohydrate, whose storage requires little, if any, addition of water (see Blem, 1980, 1990). There are fewer metabolic intermediates in lipid oxidation than in the breakdown of carbohydrates (see Blem, 1990), and glycolysis can be avoided in lipid oxidation.

Daily energy demands have not been directly measured during fattening and in most instances the daily magnitude of lipid deposition is not known with precision. However, Norstrom et al. (1986) estimated that fat deposition in herring gulls (*Larus argentatus*) was less than 1% of an adult gull's annual energy budget. It has been estimated that 1 week of fat deposition is required for one night of migration in some small birds (Kendeigh et al, 1977). Using equations in Kendeigh et al. (1977) and assuming that the costs of lipid deposition are equal only to the average heat of combustion of lipid (i.e., about 38 kJ; Table 1), then the 10 g blackpoll warbler would use 88% of a week's existence metabolism at 0°C to deposit the 10 g of lipid needed for migration. At 30°C this expenditure would be about 184%.

Although spring fattening for migration occurs more precisely within a restricted time frame than that for fall migration (Figure 5), it still requires more than a few days to attain. Lipid deposition for fall migration appears generally to be less intensive and may be spread over even longer periods.

C. Overwinter Fattening

Although the time course and magnitude of avian fat deposition during the winter (Figure 5) has been documented for several species (King, 1972; Blem, 1976a, 1990), very little is known about the energetics of the process. Some birds have small lipid reserves in the winter, either because of their dependence on immediate food supplies or because they are able to conserve energy by hypothermia. Small birds that reduce the energy demands of overnight survival at low ambient temperatures by nocturnal hypothermia may have relatively small daily cycles of lipid deposition (Blem and Pagels, 1984; Blem, 1990). Tropical birds usually do not exhibit seasonal variation in lipid deposition, but temperate-zone permanent-resident species generally show substantial midwinter fattening (Figure 5). The magnitude of lipid depots in small passerine birds have been shown to be more closely related to long-term average temperature than to temperatures of the past several days (Blem and Shelor, 1986; Dawson and Marsh, 1986) and, in fact, may not be very responsive to immediate conditions. In most passerine birds, maximum fat depots are accumulated by sundown (Blem and Shelor, 1986) since they do not feed after dark and must obtain sufficient reserves to survive the nightly fast. Late afternoon lipid depots are a function of overnight demands and are greatest in midwinter when nighttime fasts are longest and ambient tempera-

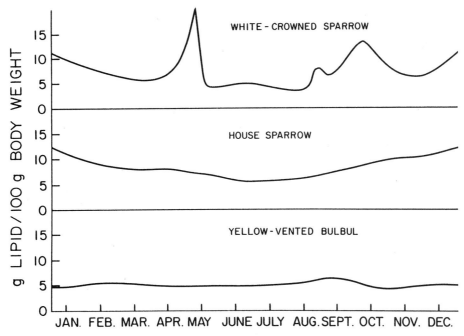

FIGURE 5 Annual cycles of lipid deposition in three passerine birds (from Blem, 1976a, with permission). The white-crowned sparrow (*Zonotrichia albicollis*) is a temperate-zone migrant, the house sparrow (*Passer domesticus*) is a temperate-zone nonmigrant, and the yellow-vented bulbul (*Pycnonotus goiavier*) is a tropical nonmigrant.

ture lowest. In species of small birds the amount of lipid deposited per unit body mass is large (Blem, 1976a, 1980), while the relative amount of lipid is less in larger bird species. However, in similar birds of different mass, the ratio of lipid store/metabolism is greater in larger birds (Blem, 1990). The rate and degree of fattening apparently are functions of ambient temperatures, but the nature of the response is poorly understood (Blem, 1981). Domestic fowl fed high-energy foods *ad lib* often have greater body mass and fat than those on normal or restricted diets. However, most species of passerine birds held in cages lose mass and have lower lipid reserves even when provided unlimited amounts of high-energy food.

D. Gonad Growth and Egg Production

Seasonal growth of avian testes (recrudescence) is relatively extensive in comparison with that of other vertebrates. The testes may increase in mass by 100× from early winter to spring, yet the cost only amounts to about 2% or less of daily BMR over a 10- to 30-day period of growth (Walsberg, 1983b). On the other hand, ovarian growth may amount to 2–9% of BMR over a similar time span. The part of gametogenesis of major energetic significance is egg production (Walsberg, 1983a,b), and energy availability can affect clutch size and time of breeding (Yom-Tov and Hilborn, 1981).

Larger birds such as ducks, geese, and penguins meet the demands of ovogenesis by using body fat, and sometimes protein, reserves. Most small birds rely entirely on their diet for the energy to produce eggs, but females of many small passerines have reduced body mass and virtually no lipid reserves by the time of fledging of young. Only a few attempts have been made to make direct measurements of the costs of egg production in wild birds (King, 1973; Ricklefs, 1974; Krementz and Ankney, 1986). The energy demands of egg formation include: (1) the energy contained in the egg materials and (2) costs of biosynthesis of the materials. The former is relatively easy to measure, and several attempts have been made to do so (King, 1973; Ricklefs, 1974; Walsberg, 1983b; Bancroft, 1985). The latter is more difficult but it appears that peak daily costs of ovogenesis range from 37–55% of BM of passerine species to 125–216% of BM in precocial birds, including the amazing brown kiwi (*Apteryx australis*), a 2200-g bird that lays a 400-g egg (Ricklefs, 1974; Walsberg, 1983a,b)!

E. Molt

Molt is a period of intensive energy demand; nearly 25% of the bird's lean dry mass is lost and regenerated in feathers over a period of a few days (King, 1981). Oxygen consumption of molting birds is 9–46% greater than that of nonmolting birds (Walsberg, 1983b). The total costs of molt include several elements: (1) energy content of feathers, (2) costs of biosynthesis of feather materials, (3) heat loss due to decreased insulation during the molting period, (4) changes in activity, and (5) energy intake required to supply the sulfur-containing amino acids necessary for feather synthesis (King, 1981). Separation of these costs has not been done, but the total energy requirement for feather production ranges from 420 to 505 kJ (Dolnik and Gavrilov, 1979; King, 1981). This is 18.7–22.4× the energy content of the feathers (heat of combustion of keratin = 22.5 kJ/g; King and Farner, 1961). BMR increases during molt by 9–46%, depending on the size and species involved (Walsberg, 1983b).

F. Growth of Embryos and Newly Hatched Young

Metabolism of avian embryos increases throughout incubation in patterns that reflect variations in growth rate and mode of development for that species. Metabolic rates of precocial birds such as chickens, quail, and ducks increase until about 80% of incubation is complete. Metabolism then either declines, remains constant, or increases slowly (Vleck *et al.*, 1980). Embryonic metabolism of altricial species such as small passerine birds increases at an accelerating rate throughout incubation (Figure 6). The energy cost of growth averages about 1.23 kJ/g of embryo (Vleck *et al.*, 1980). Maintenance costs are 0.17 kJ/g, appear to be independent of the mass of the embryo and account for much of the high development costs of precocial young or other birds that require long incubation periods (Vleck *et al.*, 1980).

Newly hatched birds grow relatively quickly and synthesize new tissues rapidly. The energy content of lean dry body materials decreases with age, as more skeletal mass is accumulated during growth. Energy contained in dry material increases because of greater lipid deposition in older nestlings. Energy content of fresh tissue (live mass) increases with growth mainly because the relative amount of water in tissues declines with maturation (Ricklefs, 1974; Blem, 1975, 1978). The energy balance of young altricial birds is greatly influenced by costs of thermoregulation, particularly endothermy. Early growth is rapid but requires little energy because the nestling is largely poikilothermic and body heat is supplied by the adult through incubation. Later, as endothermy begins and lipid depositions increases, the energy demand of nestlings becomes maximal. Nestling energy budgets vary in accordance with adult reproductive strategy (Dunn, 1980) and with the degree of development at hatching (Blem, 1975, 1978; Blem and Zara, 1980; Figure 7).

Altricial young, born naked, blind, and defenseless, may progress from the egg to adult size in 10–14 days. The metabolic rate of newly hatched altricial birds is low because thermoregulation is poor and there is little ability to generate heat. After the midpoint of growth, shivering begins, insulation becomes at least partly ef-

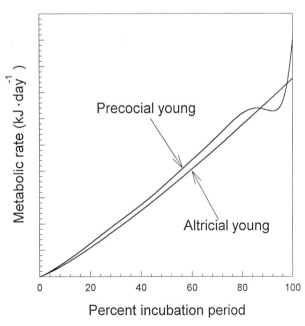

FIGURE 6 Metabolic rates of avian embryos (after Vleck *et al.*, 1980, by permission of American Society of Zoologists).

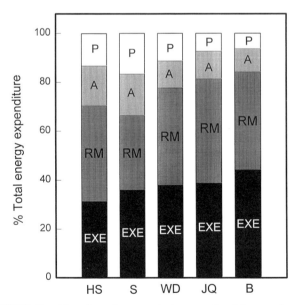

FIGURE 7 Allocation of energy during growth of five species of birds. HS, house sparrow (Blem, 1975); S, European starling (*Sturnus vulgaris*; O'Connor, 1983); WD, black-bellied whistling duck (*Dendrocygna autumnalis*; Cain, 1976); JQ, Japanese quail (*Coturnix coturnix*; Blem and Zara, 1980). EXE, excretory energy; RM, resting metabolism; A, energy costs of activity; and P, production.

fective, and the energy expended in temperature regulation becomes substantial. It appears that surplus energy resulting from lower basal rates and low costs of thermoregulation may be directed to growth in some species.

The Japanese quail, a precocial bird able to run and feed shortly after hatching, reaches adult size and sexual maturity in about 35 days. Most other precocial species require a longer period of development. The major limits of growth rate may be physiological ability to process food (Dunn, 1980) and/or limits of the ability of parents to deliver energy (Ricklefs, 1974; but see Weathers and Sullivan, 1989; Dykstra and Karasov, 1993). Assimilation efficiency (MEC) starts low in newly hatched young, particularly precocial chicks, possibly because of the conversion of yolk remaining after hatching. MEC slowly increases through development of most species (Blem, 1973a, 1978). The energetics of growth of young birds largely involves production of protein in muscle tissue, plus some storage of carbohydrate and assembly of bones and connective tissue. Late stages of development, shortly before fledging, may involve relatively expensive deposition of lipid for a postfledging period of food deprivation. In altricial young, peak energy demands usually are reached sometime past the midpoint of development, but before fledging. In this period endothermy has begun and costs of thermoregulation overlap with fairly large costs of tissue synthesis and activity (Figure 8). Precocial birds tend to allocate a smaller percentage of energy to growth than do altricial species (Blem, 1978; Dunn, 1980).

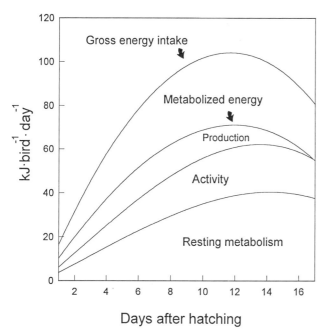

FIGURE 8 Generalized model of the energetics of growth of a small passerine bird (from data in Blem, 1973a, 1975).

V. DAILY ENERGY BUDGETS AND ENERGETICS OF FREE-LIVING BIRDS

Computation or measurement of daily energy budgets is complex. Approaches to the problem have been of two types: (1) measurement of total energy demand accompanied by attempts to partition the total among various activities or (2) measurement of costs of activities summed to calculate total energy demand. The first approach usually involves doubly labeled water (see above) or more theoretical approaches which focus on physical measurements of the environment. Environmental factors that influence heat loss, including wind and solar radiation, play a major role in determining metabolic rate of birds (Hayes and Gessaman, 1980). Some investigators have used operative and standard operative temperatures to evaluate the total effect of avenues of heat balance and to relate laboratory energetics data to field conditions (see Bakken, 1980; Bakken et al., 1985). These are pseudotemperatures which define the effective ambient temperature resulting from combined environmental conditions such as evaporative cooling, wind, convection, thermal radiation, and the like. They are either calculated from measurements of physical environment or are measured from heat loss through heated taxidermic mounts. The power required to maintain simulated body temperature is an estimate of minimal energy requirement of the inactive bird.

There have been few comprehensive studies of annual energy budgets of free-living birds, although smaller time/activity intervals have been examined frequently (e.g., Bryant and Gardiner, 1979; Bryant and Westerterp, 1980a,b; Stalmaster and Gessaman, 1984). Three of the more notable attempts are those by West (1960), Mugaas and King (1981), and Walsberg (1983b). Construction of such budgets are difficult not only because of the many avenues of energy loss, but because free-living birds have numerous options for conservation of energy, including huddling (Case, 1973), use of nest cavities for nesting and roosting (Kendeigh, 1961), and selection of favorable microclimate (Lustick et al., 1982). Expensive activities do not overlap much, if at all. Molt, reproduction, migration, and major costs of thermoregulation occur at different times and energy expenditure does not vary a great deal as a result (Figure 9).

It is axiomatic that energy models should differ greatly among birds of different latitudes, between sexes, and with variations in timing of molt, fattening, migration, and reproduction. In general, it appears that basal and thermoregulatory costs are the major expenditures in birds outside of tropical habitats (Figure 9). This suggests that energy use would be greatest in winter in high-latitude species and, in fact, can amount to more

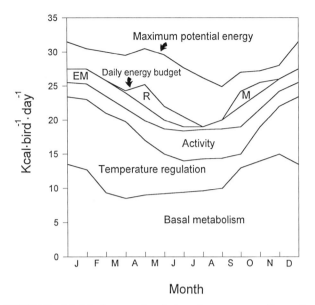

FIGURE 9 Model of seasonal variation of energy expenditure of a small temperate-zone bird (after West, 1960; Kendeigh et al., 1977). Maximum potential energy is defined as the greatest rate of energy utilization possible under the present conditions. (After Kendeigh et al. (1977), Avian energetics, in *Granivorous Birds in Ecosystems*, pp. 129–204. Reprinted with the permission of Cambridge University Press.)

than 90% of daily energy expenditure over much of the year.

VI. GEOGRAPHIC VARIATION IN ENERGY BALANCE

Geographic differences in avian energetics are difficult to quantify because of the many sources of variation and the diversity of processes involved and seldom have been studied. The trend, both intraspecifically and interspecifically, seems to be that BM and EE is greater than expected in northern latitudes on the basis of mass—metabolism relationships (Blem, 1977; Weathers, 1979). Kendeigh's (1976) analysis of latitudinal trends in the metabolic adjustments of the housesparrow (*Passer domesticus*), based largely on measurements by Blem (1973b), demonstrated that expenditures for temperature regulation, existence metabolism, reproduction, and daily energy budgets all increased northward. Housesparrows also demonstrate significant geographic adaptation to environmental extremes including greater lipid reserves in colder climates (Blem, 1973b) and greater capacity for heat production at higher latitudes (Hudson and Kimzey, 1966; Blem, 1973b; Kendeigh and Blem, 1974). Much of this variation was due to latitudinal changes in conductance (Blem, 1974) and metabolic rate (Kendeigh and Blem, 1974; Blem, 1977; Weathers, 1979). Latitudinal intraspecific increases in body mass (Bergmann's rule) and decreases in leg length (Allen's rule) likely are energetically adaptive (Kendeigh, 1972, 1976). Increases in mass may allow greater capacity for heat production with only slightly increased surface area through which heat is lost. Appendage length probably plays an important role in heat conservation or dissipation.

VII. ENERGY REQUIREMENTS OF POPULATIONS AND COMMUNITIES

Computation of the energy requirements of assemblages of birds is of practical interest to students of agricultural productivity (e.g., Wiens and Dyer, 1977; White et al., 1985). Total energy demand of populations or communities is calculated from: (1) counts of the number of birds and (2) measurements of the energy expended by free-living birds. Coarse attempts have been made to calculate comprehensive energy budgets for bird communities by using assumed values of energy expenditure for individual birds, usually resting metabolism or existence energy levels multiplied by some correction factor (e.g., Blem and Blem, 1976). The accuracy of such estimates is questionable although the general magnitude may be indicated by the results. Such models have been improved by increased partitioning of activity and refined estimates of the energy involved in each of them (see Wiens, 1984; White et al., 1985). More elegant models have involved two approaches: (1) summation of the costs of individual activities or (2) tracing the fate of ingested energy as it is dissipated in individual activities (Wiens, 1984). Both involve amassing considerable information about environmental conditions, metabolism of birds, and bird abundance and activity (Wiens and Dyer, 1977; Furness, 1978; Wiens, 1984). In some cases the avian community consumes a significant amount of the energy in the ecosystem. For example, Wiens and Scott (1975) and Furness (1978) found that seabird communities used 22–29% of the local annual fish production. White et al. (1985) estimated that a major winter flock of more than 1 million European starlings (*Sturnus vulgaris*), common grackles (*Quiscalus quiscula*), and red-winged blackbirds (*Agelaius phoeniceus*) consumed 1200–6406 metric tons of corn from the foraging area each winter. This, however, amounted to less than 2% of corn fed to livestock and a tiny portion of the total agricultural production of the study area. The significance to farm production was minor.

VIII. SUMMARY

More than a decade ago Walsberg (1983b) summarized the questions that remained to be answered in avian energetics. At that time he suggested that more research was needed regarding the thermal stresses imposed upon birds and the relationship between thermoregulatory metabolism, production, and locomotor activity. While progress has been made in those areas, much still remains to be done. In recent years avian energy balance studies have progressed from descriptive studies and baseline measurements to more ecological and experimental analyses. Physiological measurements now usually are made only in the context of behavior and adaptation. It always has been clear that energetic phenomena constrain and/or shape behavior, and simultaneous investigations of behavior and energetics will continue to provide insight into the role of thermal stress on avian behavior and energy balance.

Acknowledgment

I am indebted to Leann B. Blem for critically reading this manuscript and for making my prose more readable as she has done for 30 years.

References

Adams, N. J. (1984). Utilization efficiency of a squid diet by adult King Penguins. *Auk* **101**, 884–886.

Aschoff, J., and Pohl, H. (1970). Der Ruheumsatz von Vögeln als Funktion der Tageszeit und der Körpergrosse. *J. Ornithol.* **111**, 38–47.

Bairlein, F. (1985). Efficiency of food utilization during fat deposition in the long-distance migratory garden warbler, *Sylvia borin*. *Oecologia (Berlin)* **68**, 118–125.

Bakken, G. S. (1980). The use of standard operative temperature in the study of the thermal energetics of birds. *Physiol. Zool.* **53**, 108–119.

Bakken, G. S., Santee, W. R., and Erskine, D. J. (1985). Operative and standard operative temperature: Tools for thermal energetics studies. *Am. Zool.* **25**, 933–943.

Bancroft, G. T. (1985). Nutrient content of eggs and the energetics of clutch formation in the Boat-tailed Grackle. *Auk* **102**, 43–48.

Bartholomew, G. A., Vleck, C. M., and Bucher, T. L. (1983). Energy metabolism and nocturnal hypothermia in two tropical passerine frugivores, *Manacus vitellinus* and *Pipra mentalis*. *Physiol. Zool.* **56**, 370–379.

Biebach, H. (1977). Der Energieaufwand für das Brüten beim Star. *Naturwissenschaften* **64**, 343.

Biebach, H. (1979). Energetik des Brütens beim Star (*Sturnus vulgaris*). *J. Ornithol.* **120**, 121–138.

Biebach, H. (1981). Energetic costs of incubation on different clutch sizes in Starlings (*Sturnus vulgaris*). *Ardea* **69**, 141–142.

Biebach, H. (1984). Effect of clutch size and time of day on the energy expenditure of incubating starlings (*Sturnus vulgaris*). *Physiol. Zool.* **57**, 26–31.

Blem, C. R. (1968). Determination of caloric and nitrogen content of excreta voided by birds. *Poult. Sci.* **47**, 1205–1208.

Blem, C. R. (1973a). Laboratory measurements of metabolized energy in some passerine nestlings. *Auk* **90**, 895–897.

Blem, C. R. (1973b). Geographic variation in the bioenergetics of the house sparrow. *Ornithol. Monogr.* **14**, 96–121.

Blem, C. R. (1974). Geographic variation of thermal conductance in the house sparrow. *Comp. Biochem. Physiol. A* **47**, 101–108.

Blem, C. R. (1975). Energetics of nestling house sparrows, *Passer domesticus*. *Comp. Biochem. Physiol. A*, **52**, 305–312.

Blem, C. R. (1976a). Patterns of lipid storage and utilization in birds. *Am. Zool.* **16**, 671–684.

Blem, C. R. (1976b). Efficiency of energy utilization of the House Sparrow, *Passer domesticus*. *Oecologia* **25**, 257–264.

Blem, C. R. (1977). Reanalysis of geographic variation of House Sparrow energetics. *Auk* **94**, 358–359.

Blem, C. R. (1978). The energetics of young Japanese Quail, *Coturnix japonica*. *Comp. Biochem. Physiol. A* **59**, 219–223.

Blem, C. R. (1980). The energetics of migration. *In* "Animal Migration, Orientation and Navigation" (S. A. Gauthreaux, ed.), pp. 175–224. Academic Press, Orlando. Florida.

Blem, C. R. (1981). Geographic variation in mid-winter body composition of starlings. *Condor* **83**, 370–376.

Blem, C. R. (1990). Avian energy storage. *In* "Current Ornithology" (D. M. Power, ed.), Vol. 7, pp. 59–113. Plenum, New York.

Blem, C. R., and Blem, L. B. (1976). Density, biomass and energetics of the bird and small mammal populations of an Illinois deciduous forest. *Transact. Ill. Acad. Sci.* **68**, 156–164.

Blem, C. R., and Pagels, J. F. (1984). Mid-winter lipid reserves of the Golden-crowned Kinglet. *Condor* **86**, 491–492.

Blem, C. R., and Shelor, M. H. (1986). Multiple regression analyses of midwinter fattening of the white-throated sparrow. *Can. J. Zool.* **64**, 2405–2411.

Blem, C. R., and Zara, J. (1980). The energetics of young Bobwhite (*Colinus virginianus*). *Comp. Biochem. Physiol. A* **67**, 611–615.

Brugger, K. E. (1993). Digestibility of three fish species by Double-crested Cormorants. *Condor* **95**, 25–32.

Bryant, D. M., and Gardiner, A. (1979). Energetics of growth in House Martins (*Delichon urbica*). *J. Zool. (London)* **189**, 275–304.

Bryant, D. M., and Westerterp, K. R. (1980a). The energy budget of the house martin (*Delichon urbica*). *Ardea* **68**, 91–102.

Bryant, D. M., and Westerterp, K. R. (1980b). Energetics of foraging and free existence in birds. *Proc. Int. Ornithol. Congr. Berlin.* **1**, 292–299.

Bryant, D. M., and Westerterp, K. R. (1983). Short-term variability in energy turnover by breeding House Martins *Delichon urbica*: A study using doubly-labelled water ($D_2^{18}O$). *J. Anim. Ecol.* **52**, 535–543.

Buchsbaum, P., Wilson, J., and Valiela, I. (1986). Digestibility of plant constituents by Canada Geese and Atlantic Brant. *Ecology* **67**, 386–393.

Cain, B. W. (1976). Energetics of growth for Black-bellied Tree Ducks. *Condor* **78**, 124–128.

Calder, W. A., III. (1974). Consequences of body size for avian energetics. *In* "Avian Energetics" (R. A. Paynter, Jr., ed.). Publication of the Nuttall Ornithological Club 15, Cambridge, Massachusetts.

Calder, W. A., III. (1984). "Size, Function, and Life History." Harvard Univ. Press, Cambridge, Massachusetts.

Calder, W. A., III, and King, J. R. (1974). Thermal and caloric relations of birds. *In* "Avian Biology" (D. S. Farner and J. R. King, eds.), Vol. IV, pp. 259–413. Academic Press, New York.

Carey, C., Rahn, H., and Parisi, P. (1980). Calories, water, lipid, and yolk in avian eggs. *Condor* **82**, 335–343.

Case, R. M. (1973). Bioenergetics of a covey of Bobwhites. *Wilson Bull.* **85**, 52–59.

Castro, G., Stoyan, N., and Myers, J. P. (1989). Assimilation efficiency in birds: a function of taxon or food type? *Comp. Biochem. Physiol. A* **92**, 217–278.

Chaplin, S. B. (1974). Daily energetics of the black-capped chickadee, *Parus atricapillus*, in winter. *J. Comp. Physiol.* **89**, 321–330.

Dawson, W. R., and Marsh, R. L. (1986). Winter fattening in the American Goldfinch and the possible role of temperature in its regulation. *Physiol. Zool.* **59**, 357–368.

Dolnik, V. R., and Gavrilov, V. M. (1979). Bioenergetics of molt in the Chaffinch (*Fringilla coelebs*). *Auk* **96**, 253–264.

Dunn, E. H. (1975). Caloric intake of nestling Double-crested Cormorants. *Auk* **92**, 553–565.

Dunn, E. H. (1980). On the variability in energy allocation of nestling birds. *Auk* **97**, 19–27.

Dykstra, C. R., and Karasov, W. H. (1993). Nesting energetics of House Wrens (*Troglodytes aedon*) in relation to maximal rates of energy flow. *Auk* **110**, 481–491.

Eberhardt, L. S. (1994). Oxygen consumption during singing by male Carolina Wrens (*Thryothorus ludovicianus*). *Auk* **111**, 124–130.

El-Wailly, J. (1966). Energy requirements for egg-laying and incubation in the zebra finch, *Taeniopygia castanotis*. *Condor* **68**, 582–594.

Furness, R. W. (1978). Energy requirements of seabird communities: a bioenergetics model. *J. Anim. Ecol.* **47**, 39–53.

Gessaman, J. A. (ed.) (1973). "Ecological Energetics of Homeotherms." Utah State University Monograph Series 20, Logan, Utah.

Gessaman, J. A. (1987). Energetics. *In* "Raptor Management Techniques Manual" (B. A. G. Pendleton, B. A. Millsap, Keith W. Cline, and D. M. Bird, eds.), pp. 289–320. National Wildlife Federation Scientific and Technical Series No. 10. Washington, D.C.

Gessaman, J. A., and Findell, P. R. (1979). Energy cost of incubation in the American Kestrel. *Comp. Biochem. Physiol. A* **65**, 273–289.

Goldstein, D. L. (1988). Estimates of daily energy expenditure in birds: The time-energy budget as an integrator of laboratory and field studies. *Am. Zool.* **28**, 829–844.

Goldstein, D. L. (1990). Energetics of activity and free living in birds. *Studies Avian Biol.* **13**, 423–426.

Goldstein, D. L., and Nagy, K. A. (1985). Resource utilization by desert quail: time and energy, food and water. *Ecology* **66**, 378–387.

Grodzinski, W., Klekowski, R. Z., and Duncan, A. (eds.) (1975). Methods for ecological bioenergetics. *In* "International Biological Program Handbook," Vol. 24. Blackwell, Oxford, U.K.

Haftorn, S. (1972). Hypothermia in tits in the Arctic winter. *Ornis Scand.* **3**, 153–166.

Haftorn, S., and Reinertsen, R. E. (1985). The effect of temperature and clutch size on the energetic cost of incubation in a free-living Blue Tit (*Parus caeruleus*). *Auk* **102**, 470–478.

Hails, C. J., and Bryant, D. M. (1979). Reproductive energetics of a free-living bird. *J. Anim. Ecol.* **48**, 471–482.

Hainsworth, F. R. (1974). Food quality and foraging efficiency of sugar assimilation by hummingbirds. *J. Comp. Physiol.* **88**, 425–431.

Handrich, Y., Nicolas, L., and Le Maho, Y. (1993). Winter starvation in captive Common Barn-owls: bioenergetics during refeeding. *Auk* **110**, 470–480.

Hayes, S. R., and Gessaman, J. A. (1980). The combined effects of air temperature, wind and radiation on the resting metabolism of avian raptors. *J. Thermal Biol.* **5**, 119–125.

Hudson, J. W., and Kimzey, S. L. (1966). Temperature regulation and metabolic rhythms in populations of the House Sparrow, *Passer domesticus*. *Comp. Biochem. Physiol.* **17**, 203–217.

Karasov, W. H. (1990). Digestion in birds: Chemical and physiological determinants and ecological implications. *Studies Avian Biol.* **13**, 391–415.

Kendeigh, S. C. (1961). Energy of birds conserved by roosting in cavities. *Wilson Bull.* **73**, 140–147.

Kendeigh, S. C. (1969b). Energy responses of birds to their thermal environment. *Wilson Bull.* **81**, 441–449.

Kendeigh, S. C. (1970). Energy requirements for existence in relation to size of bird. *Condor* **72**, 60–65.

Kendeigh, S. C. (1972). Energy control of size limits in birds. *American Naturalist* **106**, 79–88.

Kendeigh, S. C. (1976). Latitudinal trends in the metabolic adjustments of the house sparrow. *Ecology* **57**, 509–519.

Kendeigh, S. C., and Blem, C. R. (1974). Metabolic adaptation to local climate in birds. *Comp. Biochem. Physiol. A* **48**, 175–187.

Kendeigh, S. C., Dolnik, V. R., and Gavrilov, V. M. (1977). Avian energetics. *In* "Granivorous Birds in Ecosystems" (J. Pinowski and S. C. Kendeigh, eds.), pp. 129–204. Cambridge Univ. Press, Cambridge, U.K.

King, J. R. (1972). Adaptive periodic fat storage by birds. *In* "Proc. 15th Int. Ornithol. Congr." (K. Voous, ed.), pp. 200–217. E. J. Brill, Leiden, The Netherlands.

King, J. R. (1973). Energetics of reproduction in birds. *In* "Breeding biology of Birds" (D. S. Farner, ed.). National Academy of Sciences, Washington, D.C.

King, J. R. (1974). Seasonal allocation of time and energy resources in birds. *In* "Avian energetics" (R. A. Paynter, ed.), Vol. 15, pp. 4–85. Publications of the Nuttall Ornithological Club, Cambridge, Massachusetts.

King, J. R. (1981). Energetics of avian molt. *Proc. 17th Int. Ornithol. Congr.* **1**, 312–317.

King, J. R., and Farner, D. S. (1961). Energy metabolism, thermoregulation, and body temperature. *In* "Biology and Comparative Physiology of Birds" (A. J. Marshall, ed.), pp. 215–288. Academic Press, New York.

Kleiber, M. (1975). "The Fire of Life: An Introduction to Animal Energetics." Krieger Pub. Co., Huntington, New York.

Krementz, D. G., and Ankney, C. D. (1986). Bioenergetics of egg production by female House Sparrows. *Auk* **103**, 299–305.

Lasiewski, R. C., and Dawson, W. R. (1967). A re-examination of the relation between standard metabolism and body weight in birds. *Condor* **69**, 13–23.

Lustick, S., Battersby, B., and Mayer, L. (1982). Energy exchange in the winter acclimatized American Goldfinch, *Carduelis* (*spinus*) *tristis*. *Comp. Biochem. Physiol. A* **72**, 715–719.

Mock, P. J. (1991). Daily allocation of time and energy of Western Bluebirds feeding nestlings. *Condor* **93**, 598–611.

Moss, R. (1973). The digestion and intake of winter foods by wild Ptarmigan in Alaska. *Condor* **75**, 293–300.

Mugaas, J. N., and King, J. R. (1981). Annual variation of daily energy expenditure by the black-billed magpie: A study of thermal and behavioral energetics. *Studies Avian Biol.* **5**, 1–75.

Nagy, K. A. (1983). The doubly-labeled water ($^3HH^{18}O$) method: a guide to its use. *University of California Publication* **12**, 1417.

Nagy, K. A. (1989). Field bioenergetics: Accuracy of models and methods. *Physiol. Zool.* **62**, 237–252.

Norstrom, R. J., Clark, T. P., Kearney, J. P., and Gilman, A. P. (1986). Herring Gull energy requirements and body constituents in the Great Lakes. *Ardea* **74**, 1–23.

Obst, B. S., and Nagy, K. A. (1993). Stomach oil and the energy budget of Wilson's Storm-Petrel nestlings. *Condor* **95**, 792–805.

O'Connor, R. J. (1984). "The Growth and Development of Birds." John Wiley & Sons, New York.

Pendergast, B. A., and Boag, D. A. (1971). Nutritional aspects of the diet of spruce grouse in central Alberta. *Condor* **73**, 437–443.

Pettit, T. N., Nagy, K. A., Ellis, H. I. and G. C. Whittow. (1988). Incubation energetics of the Laysan Albatross. *Oecologia* **74**, 546–550.

Pettit, T. N., Whittow, G. C., and G. S. Grant. (1984). Caloric content and energetic budget of tropical seabird eggs. *In* "Seabird Energetics" (G. C. Whittow and H. Rahn, eds.). Plenum, New York.

Place, A. R., and Roby, D. D. (1986). Assimilation and deposition of dietary fatty alcohol in Leach's Storm Petrel, *Oceanodroma leucorhoa*. *J. Exp. Zool.* **240**, 149–161.

Powers, D. R., and Conley, T. M. (1994). Field metabolic rate and food consumption of two sympatric hummingbird species in southeastern Arizona. *Condor* **96**, 141–150.

Reinertsen, R. E. (1983). Nocturnal hypothermia and its energetic significance for small birds living in the arctic and subarctic regions: A review. *Polar Res.* **1**, 269–284.

Reinertsen, R. E., and Haftorn, S. (1986). Different metabolic strategies of northern birds for nocturnal survival. *J. Comp. Physiol.* **156**, 655–663.

Ricklefs, R. E. (1974). Energetics of reproduction in birds. *In* "Avian Energetics" (R. A. Paynter, Jr., ed.), Vol. 15, pp. 152–297. Publications of the Nuttall Ornithological Club, Cambridge, Massachusetts.

Robel, R. J., and Arruda, S. M. (1986). Energetics and weight changes of Northern Bobwhites fed 6 different foods. *J. Wildl. Manag.* **50**, 236–238.

Shapiro, C. J., and Weathers, W. W. (1981). Metabolic and behavioral responses of American Kestrels to food deprivation. *Comp. Biochem. Physiol. A* **68**, 111–114.

Sprenkle, J. M., and Blem, C. R. (1984). Metabolism and food selection of eastern House Finches. *Wilson Bull.* **96**, 184–195.

Stalmaster, M. V., and Gessaman, J. A. (1982). Food consumption and energy requirements of captive Bald Eagles. *J. Wildl. Manag.* **46**, 646–654.

Stalmaster, M. V., and Gessaman, J. A. (1984). Ecological energetics and foraging behavior of overwintering bald eagles. *Ecol. Monogr.* **54**, 407–428.

Tatner, P., and Bryant, D. M. (1986). Flight cost of a small passerine measured using doubly labeled water: Implication for energetic studies. *Auk* **103**, 169–180.

Vleck, C. M. (1981). Energetic cost of incubation in the Zebra Finch. *Condor* **83**, 229–237.

Vleck, C. M., Vleck, D., and Hoyt, D. F. (1980). Patterns of metabolism and growth in avian embryos. *Am. Zool.* **20**, 405–416.

Walsberg, G. E. (1983a). Ecological energetics: What are the questions? *In* "Perspectives in Ornithology" (A. H. Brush and G. A. Clark, Jr., eds.), pp. 135–158. Cambridge Univ. Press, New York.

Walsberg, G. E. (1983b). Avian ecological energetics. *In* "Avian Biology" (D. S. Farner and J. R. King, eds.), Vol. 7, pp. 161–220. Academic Press, New York.

Walsberg, G. E., and King, J. R. (1978). The energetic consequences of incubation for two passerine species. *Auk* **95**, 644–655.

Weathers, W. W. (1979). Climatic adaptation in avian standard metabolic rate. *Oecologia* **42**, 81–89.

Weathers, W. W., and Nagy, K. A. (1980). Simultaneous doubly labeled water ($^{3}HH^{18}O$) and time budget estimates of daily energy expenditure in *Phainopepla nitens*. *Auk* **97**, 861–867.

Weathers, W. W., and Sullivan, K. A. (1989). Juvenile foraging proficiency, parental effort, and avian reproductive success. *Ecol. Monogr.* **59**, 223–246.

Weathers, W. W., Buttemer, W. A., Hayworth, A. M., and Nagy, K. A. (1984). An evaluation of time-budget estimates of daily energy expenditure in birds. *Auk* **101**, 459–472.

Weathers, W. W., Koenig, W. D., and Stanback, M. T. (1990). Breeding energetics and thermal ecology of the Acorn Woodpecker in central coastal California. *Condor* **92**, 341–359.

West, G. C. (1960). Seasonal variation in the energy balance of the tree sparrow in relation to migration. *Auk* **77**, 306–329.

White, S. B. Dolbeer, R. A., and Bookhout, T. A. (1985). Ecology, bioenergetics, and agricultural impacts of a winter-roosting population of blackbirds and starlings. *Wildl. Monogr.* **93**, 1–42.

Whittow, G. C. (1986). Energy metabolism. *In* "Avian Physiology" (P. D. Sturkie, ed.), 4th edition, Springer-Verlag, New York.

Wiens, J. A. (1984). Modelling the energy requirements of seabird populations. *In* "Seabird Energetics" (G. C. Whittow, ed.). Plenum, New York.

Wiens, J. A., and Dyer, M. J. (1977). Assessing the potential impact of granivorous birds. *In* "Granivorous Birds in Ecosystems" (J. Pinowski and S. C. Kendeigh, eds.). Cambridge Univ. Press, London.

Wiens, J. A., and Scott, J. M. (1975). Model estimation of energy flow in Oregon coastal seabird populations. *Condor* **77**, 439–452.

Williams, J. B., and Nagy, K. A. (1984). Validation of the doubly labeled water technique for measuring energy metabolism in Savannah Sparrows. *Physiol. Zool.* **57**, 325–328.

Williams, J. B., and Nagy, K. A. (1985). Daily energy expenditure by female Savannah Sparrows feeding nestlings. *Auk* **102**, 187–190.

Withers, P. C. (1977). Measurements of V_{O_2}, V_{CO_2} and evaporative water loss with a flow-through mask. *J. Appl. Physiol.* **42**, 120–123.

Woods, P. E. (1982). Vertebrate digestive and assimilation efficiencies: taxonomic and trophic comparisons. *The Biologist* **64**, 58–77.

Yom-Tov, Y., and Hilborn, R. (1981). Energetic constraints on clutch size and time of breeding in temperate zone birds. *Oecologia* **48**, 234–243.

CHAPTER 14

Regulation of Body Temperature

W. R. DAWSON
Department of Biology
The University of Michigan at Ann Arbor
Ann Arbor, Michigan 48109

G. C. WHITTOW
Department of Physiology
John A. Burns School of Medicine
University of Hawaii at Manoa
Honolulu, Hawaii 96822

I. Introduction 344
II. Body Temperature 344
    A. Deep Body ("Core") Temperature 344
    B. Measurement 344
    C. Interspecific Variation 345
    D. Circadian Rhythm 345
    E. Ambient Temperature 347
    F. Acclimatization and Acclimation 347
    G. Dehydration 347
    H. Food Deprivation 347
III. Heat Balance 348
    A. Units 348
IV. Changes in Bodily Heat Content (S) 348
V. Heat Production (H) 349
    A. Measurement 349
    B. Basal, Standard, and Resting Metabolic Rates 349
    C. Circadian Rhythm 350
    D. Interspecific Variation in Standard Metabolic Rate 350
    E. Influence of Ambient Temperature on Avian Metabolic Rates 350
    F. Intraspecific Variation in Avian Metabolic Rate: Acclimatization, Acclimation, and Geographical Patterns 354
VI. Heat Transfer within the Body 355
    A. Vascular Heat Exchange 355
    B. Cold Vasodilatation 355
    C. Brood Patch 355
    D. Thermal Conductance of the Tissues 355
    E. Acclimatization and Acclimation 355
VII. Heat Loss 356
    A. Nonevaporative Heat Loss; Total Thermal Conductance 356
    B. Evaporative Heat Loss 356
    C. Partition of Heat Loss 363
VIII. Heat Exchange under Natural Conditions 364
IX. Behavioral Thermoregulation 364
    A. Introduction 364
    B. Behavior Reducing Heat Loss or Facilitating Heat Gain 364
    C. Behavior Facilitating Heat Loss or Reducing Heat Gain 366
    D. Behavioral Augmentation of Evaporative Cooling 367
X. Control Mechanisms 367
    A. Thermosensitivity and Control of Physiological Thermoregulatory Responses 367
    B. Control of Behavioral Thermoregulatory Responses 371
    C. Afferent Inputs Contributing to Thermal Tachypnea 371
    D. Miscellaneous Observations on Thermoregulatory Control 372
XI. Thermoregulation at Reduced Body Temperatures 372
    A. Patterns of Avian Hypothermia 372
    B. The Hypothermic State 373
    C. Arousal from Hypothermia 375
XII. Development of Thermoregulation 375
    A. Embryo 375
    B. Hatchling 375
    C. Nestling 377
XIII. Summary 379
References 379

I. INTRODUCTION

Birds are among a minority of animals capable of effective regulatory thermogenesis. Such endothermic capacities are also evident either bodywide or, in some cases, within particular tissues in various insects; some large, fast swimming fish; a few large reptiles; and, of course, mammals. Like the last group, birds show inpressive homeothermic capacities involving the balancing of rates of thermolysis and thermogenesis. These homeotherms are thus able to regulate body temperature within a narrow range over a wide span of thermal conditions.

There is evidence that the high metabolic rates permitting endothermy did not exist in the earliest known bird *Archeopteryx lithographica* and may not have been completely established in the Class Aves until midlate Cretaceous times. The elevated metabolic levels characterizing modern members of this class may have originally evolved in response to selection for an increased capacity for long-distance flight, rather than as a requisite for flight itself. Endothermic temperature regulation would have been a by-product of this development (Ruben, 1996).

Despite the fact that homeothermy probably arose independently in the avian and mammalian lines (Ruben, 1995), the general features of thermoregulation in the two groups appear quite similar. Nonetheless there are some behavioral and functional differences between them that influence the manner in which this regulation proceeds. Birds, with some important exceptions, appear in winter to make far less use than mammals of the thermal protection provided by underground or subnivian burrows, nests, and cavities in trees. Significantly, seasonal dormancy (hibernation) is rare in birds. However, many of these animals surpass mammals in evasion of adverse thermal conditions by long-distance migration (Dawson *et al.*, 1983a). The relation between hypothalamic and spinal thermosensitivity differs between these groups (Simon, 1989). Furthermore, nonshivering thermogenesis appears absent or less prominent in birds than in the array of mammals possessing brown adipose tissue (Marsh and Dawson, 1989; Marsh, 1993). There are also differing considerations affecting insulation in the two groups. One relates to the fact that the avian plumage, unlike the mammalian pelage, generally must meet aerodynamic as well as insulative needs. Additionally, distribution of subcutaneous fat tends to be more localized in many birds than in mammals, which affects tissue insulation (Marsh and Dawson, 1989). Sweat glands have not been found in the former group and avian cutaneous water loss, which can be substantial (Marder and Gavrieli-Levin, 1987), must therefore proceed by a different mechanism than that employed by sweating mammals. Additionally, virtually all birds appear able to enhance respiratory evaporation by panting, whereas this ability is more restricted among mammals. Finally, the contrasting developmental patterns of the two groups (oviparity versus viviparity, excepting the prototherians) have important thermoregulatory implications. The development of the avian egg outside the body of the hen ties one or both parents to the nest and, except for mound builders and brood parasites, necessitates effective incubation patterns serving to maintain embryonic temperature within a narrow range suitable for development. These patterns can expose the incubating parent to extremely demanding thermal conditions (see, for example, Morton, 1978; Grant, 1982).

This chapter examines the general features of avian physiological and behavioral temperature regulation. A special effort has been made to include information on wild as well as domesticated species. Whittow (1965, 1976, 1986) should be consulted for some older references and for further background.

II. BODY TEMPERATURE

A. Deep Body ("Core") Temperature

As noted above, avian thermoregulatory processes serve to maintain the temperature of interior tissues and organs within fairly narrow limits. It is this deep-body or "core" temperature (T_b here) that is commonly used in interspecific comparisons (Table 1) and to assess the effectiveness of these processes. On the other hand, temperature of more superficial areas comprising the "shell" may vary substantially with the bird's activity states and external conditions and also may differ from one anatomical region to another. Maintenance of appropriate thermal gradients between the core and the shell is a crucial element in the bird's control of heat exchange with its environment.

B. Measurement

Deep-body temperature of birds is commonly measured with a thermistor or thermocouple inserted into the proventriculus or large intestine (rectum). However, temperature of the brain or spinal cord is also frequently measured under experimental conditions, owing to the importance of these structures in thermoregulatory control. Telemetry techniques involving implantation of thermal sensitive transmitters permit long-term recording of avian body temperature in undisturbed birds in the laboratory or field (e.g., Whittow *et al.*, 1978).

C. Interspecific Variation

Although ranges of T_b for birds and mammals overlap, normothermic values for many of the former, including songbirds, lie somewhat above the mammalian range (Table 1). Core temperatures of large flightless birds (ostrich, emu, rhea) and certain aquatic species (e.g., penguins, albatrosses, shearwaters, and petrels) are somewhat lower, falling within this mammalian range. Also, T_b of some insectivorous birds that naturally enter torpor (e.g., white-throated swift, *Aëronautes saxatalis*, and common poor-will *Phalaenoptilus nuttallii*) can fall to relatively low levels (35–37.5°C) when the birds are normally alert at moderate ambient temperatures (T_a; Bartholomew et al., 1957, 1962).

D. Circadian Rhythm

The T_b of birds varies over the course of the day (Table 1), being highest during the respective active phases of the daily cycle in both diurnal and nocturnal species. Because the thermal maxima for the latter occur at night when the coolest T_a prevail, it is tempting to conclude the rhythm of T_b is independent of external temperature. However, this is not the case, for the amplitude of the temperature cycle in birds subjected to constant warm T_a was reduced (Dawson, 1954). Moreover, that in individuals exposed to cold was increased (Graf, 1980). The amplitude also tends to vary inversely with body size (as well as with certain other factors including nutritional plane [Ostheim, 1992]), ranging from as much as 8°C in hummingbirds to less than 1°C in the ostrich (*Struthio camelus*).

The daily cycles of avian T_b fundamentally reflect entrained circadian rhythms for which photic cues seem to serve as a primary *Zeitgeber* or entrainment factor. Such temperature rhythms have a free-running period of approximately 24 hr and persist in intact birds in continuous darkness (see, for example, Binkley et al., 1971). The pineal body and/or the eyes have a pacemaking function for these rhythms (Underwood and Siopes, 1984). Additionally, experimental studies of avian activity rhythms have indicated that the hypothalamic suprachiasmatic nuclei (SCN) are involved in the avian circadian control system. (Takahashi and Menaker, 1982). The prominence of the various elements in this system varies interspecifically. For example, pinealectomy abolished the temperature cycle in house sparrows (*Passer domesticus*) when they were maintained in the dark (Binkley et al., 1971). On the other hand, this treatment had little or no effect on the free-running T_b rhythm in Japanese quail (*Coturnix coturnix japonicum*). However, removal of the eyes (bilateral enucleation) resulted in the birds becoming arrhythmic in constant darkness (Underwood, 1994). Additionally, the pineal as well as the eyes appeared involved in control of thermal rhythms in the rock pigeon (the familar domestic pigeon, *Columba livia*), for both pinealectomy and bilateral enucleation were required to abolish these rhythms in constant dim light (Oshima et al., 1989).

The contributions of the pineal or the eyes to pacemaking appear to involve their rhythmical synthesis and release of melatonin. In the rock pigeon, this hormone normally is produced principally in the pineal body and the eyes (Foà and Menaker, 1988). Exposure of rock pigeons to constant bright light suppressed nocturnal synthesis of this hormone and *N*-acetyl transferase, a key enzyme in melatonin production. This treatment also suppressed sleep and rhythms involving T_b, activity, and other variables (Berger and Phillips, 1994). Pinealectomy and blinding by bilateral enucleation reduced plasma melatonin to the minimal detectable level in rock pigeons maintained in constant dim light. This effect was accompanied by the disappearance of rhythms of T_b and activity. Daily administration of melatonin to these animals restored circadian rhythms of locomotor activity, which entrained to the injections (Oshima et al., 1989). A robust daily rhythm in melatonin level was evident in Japanese quail and this again has been attributed to the secretion of this hormone into the vascular system by both the pineal and the eyes (Underwood and Siopes, 1984).

In most studies, cycles of avian T_b and locomotor activity appear closely synchronized, suggesting their reliance on the same circadian oscillators(s) (Oshima et al., 1989; Refinetti and Menaker, 1992). However, the thermal rhythms appear to be something more than a mere byproduct of thermogenic and thermolytic changes associated with locomotor activity, even though changes in activity level can produce changes in T_b (Refinetti and Menaker, 1992). An indication of the discreteness of the two rhythms is provided by the fact that the circadian rhythm of T_b in a given animal was frequently phase advanced relative to that for activity (Refinetti and Menaker, 1992). Such discreteness is further suggested by results of returning pinealectomized house sparrows to a light–dark (LD) cycle from constant darkness, where they had been arrhythmic. An activity rhythm was apparent from day 1 of LD, whereas a temperature rhythm did not appear until at least day 4 (Binkley et al., 1971). The proximate basis of circadian rhythms of T_b in birds is incompletely understood. They appear to depend more on modulations of heat loss than of heat production. Diurnal variation in thermoregulatory set point also might be involved, but evidence on this point is inconsistent for both birds and mammals (Refinetti and Menaker, 1992). Control of the circadian

TABLE 1 Deep Body Temperatures of Selected Birds at Rest under Thermoneutral Conditions[a]

Species	Body mass (kg)	Deep body temperature[b] (°C)	Reference[c]
Ostrich (*Struthio camelus*)	100.0	38.3	*
Emu ♂♂ (*Dromaius novaehollandiae*)	40.7	37.7S	1
Emu ♀♀	45.4	38.3S	1
Emu ♂♂	39.7	37.7	1
Emu ♀♀	37.0	38.2	1
Rhea (*Rhea americana*)	21.7	39.7	*
Mute swan (*Cygnus olor*)	8.3	39.5	*
Domestic goose (*Anser anser*)	5.0	41.0	*
Gentoo penguin (*Pygoscelis papua*)	4.9	38.3	*
Giant petrel (*Macronectes giganteus*)	3.9	39.2	2
Peruvian penguin (*Spheniscus humboldti*)	3.9	39.0	*
Domestic turkey (*Meleagris gallopavo*)	3.7	41.2	*
Adélie penguin (*Pygoscelis adeliae*)	3.5	38.5	*
Chinstrap penguin (*Pygoscelis antarctica*)	3.1	39.4	*
Great spotted kiwi (*Apteryx haastii*)	2.5	38.4	3
Domestic fowl (*Gallus gallus*)	2.4	41.5	*
Domestic duck (*Anas platyrhynchos*)	1.9	42.1	*
Double-crested cormorant (*Phalacrocorax auritus*)	1.33	41.2D, 40.2N	*
South polar skua (*Catharcta maccormicki*)	1.250	40.9	2
Black grouse (*Lyrurus tetrix*)	1.079	41.3S	*
	0.931	40.2W	
Kelp gull (*Larus dominicanus*)	0.98	41.0	2
Anhinga (*Anhinga anhinga*)	1.33	39.9D	*
		39.1N	*
Great horned owl (*Bubo virginianus*)	1.00	39.9	4
Little penguin (*Eudyptula minor*)	0.9	38.4	*
Brünnich's guillemot (*Uria lomvia*)	0.819	39.6	5
Fulmar (*Fulmarus glacialis*)	0.651	38.7	5
Brown-necked raven (*Corvus corax ruficollis*)	0.610	39.9	*
Willow ptarmigan (*Lagopus lagopus*)	0.573	39.9	*
Mexican spotted owl (*Strix occidentalis lucida*)	0.571	39.1	4
Tawny frogmouth (*Podargus striatus*)	0.420	38.6[d]	6
European coot (*Fulica atra*)	0.387	39.6[d]	7
Black-legged kittiwake (*Rissa tridactyla*)	0.365	40.2	5
Black guillemot (*Cepphus grylle*)	0.342	39.9	5
Papuan frogmouth (*Podargus papuensis*)	0.315	38.8[d]	6
Rock pigeon (*Columbia livia*)	0.3	42.2	*
Bobwhite (*Colinus virginianus*)	0.210	38.9DS[d]	8
		37.0NS[d]	8
	0.228	37.7DW[d]	8
		37.4NW[d]	8
Brown noddy (*Anous stolidus*)	0.142	40.3	9
California quail (*Callipepla californica*)	0.139	41.3	*
Tengmalm's owl (*Aegolius funereus*)	0.127	39.4	10
American kestrel (*Falco sparverius*)	0.119	39.3	*
Acorn woodpecker (*Melanerpes formicivorus*)	0.082	42.4[d]	11
Green woodhoopoe (*Phoeniculus purpureus*)			
♂♂	0.080	39.6N	12
♀♀	0.072	39.7N	12
Evening grosbeak (*Coccothraustes vespertinus*)	0.060	41.0	*
Barred button quail (*Turnix suscitator*)	0.058	39.5	13
Blue-breasted quail (*Coturnix chinensis*)	0.053	39.0	13
Speckled mousebird (*Colius striatus*)	0.053	39.0	*
Wilson's storm petrel (*Oceanites oceanicus*)	0.034	39.2	2
Common redpoll (*Carduelis flammea*)	0.015	40.1	*
Sunbirds[e]	0.007–0.017	42.5D, 38.9N	14
Zebra finch (*Poephila guttata*)	0.012	40.3	*
Anna hummingbird (*Calypte anna*)	0.005	42.0 (Median T_b)	15

[a]Modified and expanded from Whittow (1976, 1986), which should be consulted for references marked by an asterisk. Thermoneutral temperatures are those requiring neither regulatory thermogenesis nor active evaporative cooling (see Section V,E). The vast majority of the

(*continues*)

rhythm of T_b in the rock pigeon persisted in febrile individuals (Nomoto, 1996; see Section X,D).

E. Ambient Temperature

Avian T_b generally fall within a narrow band over a wide range of T_a. For example, mean values for the spinifex pigeon (*Geophaps plumifera*) only varied from 41.8 ± 0.1°C (SE) at T_a of 40 to 42°C to 40.5 ± 0.1°C at −4 to −9°C (Withers and Williams, 1990). The actual lower limit of the T_a range depends on the size of the bird, density of plumage, and acclimation/acclimatization state (see Section V,F), among other things. At high T_a, T_b tends to increase in a manner influenced by the nature and extent of evaporative cooling responses and the spinifex pigeon's T_b rose to 43.4 ± 0.2°C at 47 to 51°C (Withers and Williams, 1990). Controlled hyperthermia can serve to maintain a favorable temperature difference between body and environment, which enhances the opportunity for nonevaporative heat dissipation by birds under heat challenges. It ranked among the variable seedeater's (*Sporophila aurita*) principal means of dealing with the high operative temperatures and humidity characterizing its lowland tropical environment (Fig. 2; Weathers, 1997).

F. Acclimatization and Acclimation

The terms acclimatization and acclimation are often used interchangeably in physiology, referring to phenotypic modifications that serve to reduce the stress imposed by naturally occurring or experimentally induced changes in particular climatic factors. However, in thermophysiology the former term preferably should refer to a modification in response to stressful changes (e.g., seasonal or geographical) imposed by one or more climatic factors in nature, whereas acclimation is reserved for modifications in physiological response induced in the laboratory by controlled changes in specific climatic factors (e.g., ambient temperature; IUPS Commission for Thermal Physiology, 1987). A presumed effect of acclimatization is seen in T_b, which was significantly higher in black grouse (*Lyrurus tetrix*) during summer than in winter (Table 1), suggesting some type of compensatory adjustment (Rintamäki *et al.*, 1983). Diurnal, but not nocturnal T_b, was significantly higher in bobwhite (*Colinus virginianus*) in summer than in winter (Table 1; Swanson and Weinacht, 1997). On the other hand, temperature appeared seasonally stable in such divergent species as the emu (Table 1) and the dark-eyed junco *Junco hyemalis* (Swanson, 1991). Studies of domestic fowl and rock pigeons document acclimation effects on T_b. Temperature reached a maximum in laying hens on the first days of a 3-day period of heat exposure and then declined (Fujita *et al.*, 1990). Male broilers acclimated to heat for 5 days and unacclimated controls, both given access to food, had similar T_b during heat exposure. However, the former had significantly lower temperatures than the latter at the end of such exposure without food (Lott, 1991). Additionally, heat-acclimated rock pigeons regulated T_b between 41.2° and 42.0°C, at T_a between 30° and 60°, whereas T_b of unacclimated individuals rose 0.11° per 1°C increase in T_a above 35°C (Marder and Arieli, 1988).

G. Dehydration

Dehydration affects avian heat defense, resulting in significantly greater hyperthermia at high T_a in, for example, the ostrich, domestic fowl, rock pigeon, and Japanese quail (Crawford and Schmidt-Nielsen, 1967; Arad *et al.*, 1985, 1987; Itsakiglucklich and Arad, 1992). This effect generally results from a decrease in the extent of evaporative cooling. A change in the mode of this activity was evident with dehydration in the rock pigeon, which shifted from cutaneous to respiratory evaporation as the primary cooling mechanism (Arad *et al.*, 1987).

H. Food Deprivation

The T_b of American kestrels (*Falco sparverius*) decreased 0.2-0.4°C/day over a 79-hr fast (Shapiro and Weathers, 1981). Food scarcity or deprivation is particularly effective in increasing the depth of nocturnal hypo-

TABLE 1 (*Continued*)

values for T_b were obtained during the daytime, the active phase of the daily cycle for most species with the exception of such birds as owls and frogmouths. In cases where measurements were made during both day and night the particular period is specified.

[b] D, N, S, and W refer to daytime, nightime, summer, and winter measurements, respectively.

[c] References: 1, Maloney and Dawson (1993); 2, Morgan *et al.* (1992); 3, McNab (1996); 4, Ganey *et al.* (1993); 5, Gabrielsen *et al.* (1988); 6, McNab and Bonaccorso (1995); 7, Brent *et al.* (1985); 8, Swanson and Weinacht (1997); 9, Ellis *et al.* (1995); 10, Hohtola *et al.* (1994); 11, Weathers *et al.* (1990); 12, Williams *et al.* (1991); 13, Prinzinger *et al.* (1993); 14, Prinzinger *et al.* (1989); 15, Powers (1992).

[d] T_b independent of T_a over a range of temperatures that includes or closely approaches zone of thermal neutrality.

[e] Values for T_b at 26.5°C T_a calculated from equations relating T_b to T_a based on data for five species of sunbirds: *Aethopyga siparaja, Anthreptes collaris, Nectarinia cuprea, Nectarinia tacazze,* and *Nectarinia klimensis.*

III. HEAT BALANCE

Organismal gain or loss of heat has several components and these are indicated for an animal not performing external work in the familiar heat balance equation

$$S = H - E \pm R \pm C \pm K,$$

where S = the gain or loss of bodily heat; H = metabolic heat production; E = evaporative heat loss; R = radiative heat gain or loss; C = convective heat gain or loss; and K = conductive heat gain or loss. Body temperature will, of course, remain unchanged when S is zero (i.e., when heat gain matches heat loss). If more heat is produced and gained than lost (S is positive), T_b will rise. Conversely, if heat loss exceeds heat gain (S is negative), it will fall.

A. Units

The Système International (SI) units such as watts or joules (J) are used wherever possible in this chapter in keeping with the effort in recent decades to standardize units. Some metabolically relevant conversions from the older convention of using calories (cal) or kilocalories (kcal) for energy expressions are as follows:

$$1 \text{ kcal} = 4187 \text{ J} = 4.187 \text{ kJ}$$

$$1 \text{ kcal/hr} = 1.163 \text{ W}.$$

IV. CHANGES IN BODILY HEAT CONTENT (S)

The mean specific heat of avian tissues is approximately 3.5 kJ (kg · °C)$^{-1}$ and the change in bodily heat content for a particular rise or fall of mean T_b is given by the following equation

$$S = (3.5) \, (M) \, (\Delta \overline{T}_b),$$

where S is the change in bodily heat content in kilojoules, M is the body mass in kilograms, and $\Delta \overline{T}_b$ is the change in mean T_b in degrees Centigrade.

The $\overline{T}_b$ is determined from a large number of temperatures measured both superficially and deeply within the body of an animal. The procedure is tedious and it is usually approximated by considering the animal to consist of an inner "core" at a relatively high temperature and an outer "shell" at a lower and more variable temperature. Following measurement of the core temperature, an overall shell temperature is determined from measurements taken at several different points on the skin. These measurements are weighted, the weighting factors reflecting the respective surface areas represented by the various points, and then averaged to represent a shell temperature. The $\overline{T}_b$ is then computed by averaging the core temperature and the mean skin temperature, each of which has been weighted by a factor appropriate to the amount of mass it represents. Insufficient information on these factors is available for birds.

Changes in the heat content of the body can contribute significantly to birds' successfully meeting thermal challenges. As noted in Section II,E, they may store some heat in hot situations (Dawson, 1984a). The resultant hyperthermia serves either to increase nonevaporative heat loss to the environment or to decrease heat gain from it, depending on the relation between T_b and the standard operative temperature (T_{es}) of this environment (see Bakken, 1992, for a discussion of T_{es}). These effects on heat transfer should be especially important in hot desert regions through their lessening of the need to expend water in evaporative cooling (Dawson, 1984a). As noted above, many birds show greater hyperthermia when they are dehydrated (see Section II,G). However, there is a limit to avian tolerance of hyperthermia so that the amount of heat stored cannot be such that T_b rises to lethal levels, which in many birds are between 46° and 47°C, only a few degrees above normothermic levels (Table 1). Heat-induced hyperthermia can be controlled by the mechanisms of evaporative heat dissipation (see Section VII,B), but at the cost of increased water loss. These matters seem especially serious for small birds with their low thermal inertia.

Brain tissues appear to be the most susceptible to heat damage. It is therefore of interest that various species of birds can maintain the temperature of this organ up to 1°C or so cooler than the general body core during hyperthermia resulting from vigorous activity or heat stress (Pinshow et al., 1982). This depends on a transfer of heat from the branches of the external ophthalmic arteries, which supply a major portion of the blood to the brain, to the venous drainage of the orbit and buccopharyngeal cavity. These small arterial and venous juxtapositions on each side are organized into a temporal *rete mirabile ophthalmicum* that serves as a countercurrent heat exchanger in such species as rock pigeons (Kilgore et al., 1979); mallard ducks (*Anas platyrhynchos;* Arad et al., 1984); domestic fowl (Midtgård et al., 1983); various seabirds (Pettit et al., 1981; Grant, 1985); and turkey vultures (*Cathartes aura;* Arad et al., 1989). This *rete* is not well developed in zebra finches (*Poephila guttatta*), which maintained brain temperature only 0.2°C below body temperature during heat stress (Bech and Midtgård, 1981). However, the body-to-brain temperature differential in calliope hummingbirds (*Stellula calliope*), which lack the *rete mirabile ophthalmicum*, averaged 0.73°C at high T_b (Burgoon *et al.*, 1987), presumably due to the use of other sites for arteriovenous heat exchange.

Just as controlled hyperthermia can benefit birds contending with heat, reduced T_b can lead to energetic savings in the cold (see Section XI). Although only one bird, the common poor-will, is known to hibernate, a number of species, principally smaller ones, can show nocturnal hypothermia to varying extents when energy challenged (Reinertsen, 1996; see Section XI). This latter response, which slows metabolism and reduces heat loss and thus the requirement for expenditure of energy in regulatory thermogenesis, allows birds to husband their energy reserves during nocturnal fasts. More localized effects also are important and vasoconstriction leading to peripheral cooling reduced thermoregulatory costs in cold water for sea-acclimated Gentoo penguins (*Pygoscelis papua*; Figure 1; see Section VI,E). Peripheral heterothermia involving cooling of the limbs and other peripheral body tissues curtails heat loss of many birds in cold environments. It is especially important for aquatic species and it often depends on vascular arrangements in the limbs facilitating transfer heat from warmer arterial blood leaving the core to cooler venous blood returning to it. In many instances these involve vascular retia that are especially effective in facilitating the countercurrent exchange of heat (Johansen and Bech, 1983; Midtgård, 1989a; see Section VI,A).

V. HEAT PRODUCTION (H)

A. Measurement

Heat production is usually determined from measurements of oxygen consumption ($\dot{V}_{O_2}$). If the bird is at rest, and if it is not growing or depositing fat (energy storage; see Chapter 13), then virtually all of the energy released in the tissues is in the form of heat. This release comes about through oxidative reactions and a direct relation exists between oxygen consumed and heat produced.

In measurements of ($\dot{V}_{O_2}$), a bird is housed in a chamber or, less commonly, fitted with a mask through which air is passed at a known rate. The product of the flow rate and the difference in fractional contents of oxygen in the incurrent and excurrent air (corrected for the effect of respiratory exchange ratios differing from 1.0) equals ($\dot{V}_{O_2}$). The dry volume of this gas consumed per unit time is corrected to standard conditions of temperature (0°C) and pressure (760 mmHg = 101.3 kPa), i.e., to STPD, and then multiplied by 20.1 kJ/liter O_2 to convert it into energy units. This conversion factor varies slightly with the nature of the energy substrates being metabolized. However, the errors are small and they can be determined with reasonable precision (Withers, 1977).

Other methods for measuring or estimating the heat production of birds are available. One depending on measurements of temperatures of the skin and plumage surface, plus calculations of heat transfer through the plumage, compared very favorably with direct measurements of ($\dot{V}_{O_2}$) (Hayes and Gessaman, 1982).

B. Basal, Standard, and Resting Metabolic Rates

The minimal heat production of a normothermic bird that is awake while resting and fasting (i.e., has reached a postabsorptive state) at a T_a requiring neither regulatory thermogenesis nor expenditure of energy in active evaporative cooling, represents the animal's basal metabolic rate (BMR). This is commonly measured in the inactive

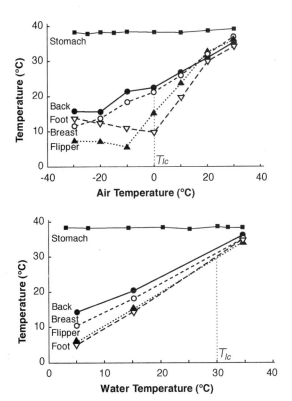

FIGURE 1 Body and skin temperatures of adult Gentoo penguins (*Pygoscelis papua*) in air (upper panel) and water (lower panel). Each point represents the mean of at least six measurements. Stomach temperature was taken as T_b (■). Skin temperatures were measured on back (●), breast (○), foot (▽), and flipper (▲). The vertical dashed line marked by T_{lc} in each panel locates the lower critical temperature for this species in the particular medium. (Redrawn from Dumonteil *et al.*, 1994, *Am. J. Physiol.* with permission.)

phase of the daily cycle (see next section). The range of T_a within which BMR occurs is the zone of thermal neutrality (TNZ; see Section V,E). Mission (1974) reported that several training sessions were required to accustom domestic fowl to the experimental procedures before truly basal rates were obtained. Moreover, 24–48 hr, depending on the size of the fowl, were required for it to reach a postabsorptive state. On the other hand, small birds become postabsorptive in just a few hours (see, for example, Weathers et al., 1983). Representatives of some species, especially freshly captured ones, may not reach a truly basal state during metabolic measurements. Results of metabolic measurements for such animals resting in a postabsorptive state in the TNZ are referred to as standard metabolic rates (SMR) when the conditions are clearly specified (IUPS Commission for Thermal Physiology, 1987). Measurements of birds resting in the TNZ, but not in a postabsorptive state, represent resting metabolic rates (RMR), according to the IUPS Commission for Thermal Physiology (1987). Ingestion and assimilation of food tend to raise metabolic rate (this heat increment of feeding, or specific dynamic action, is greatest with protein). Unfortunately, there is variation among investigators in the definitions of BMR, SMR, and RMR. Consequently, great care must be exercised in interpreting measured values of heat production, particularly concerning the conditions under which the measurements were made. We have chosen the conservative course of referring to metabolic rates of postabsorptive animals resting in their respective TNZ as standard metabolic rates throughout the remainder of this chapter. As documented in the next section, it is important to specify the phase of a bird's daily activity cycle in which SMR is determined.

C. Circadian Rhythm

Avian metabolic rate varies over the day (Aschoff and Pohl, 1970). The pattern is not completely independent of the effects of feeding. For example, the amplitude of metabolic rhythms varied inversely with the amount of food received in rock pigeons, primarily due to effects on dark-phase metabolism. Timing of food consumption mainly affected the light-phase segment of metabolic rate; when feeding was delayed until late in the light-phase, metabolic rate was greatly depressed early in that phase, then rose substantially near the scheduled time of feeding. Rashotte et al. (1995) have raised the possibility that this distinctive light phase pattern, which developed quickly, reflects the influence of a circadian, food-entrainable oscillator. The SMR of domestic fowl is highest in the forenoon and lowest near 8:00 PM. The nocturnal reduction amounts to 18–30% in this species, but can be as much as 49% in the house sparrow. The difference reflects a general tendency for the relative amplitude of such rhythms to be greater in small than in large birds (King and Farner, 1961). As with rhythms of T_b, activity contributes to circadian oscillations in metabolism, but the circadian rhythm in fasting fowls is not exclusively the result of changes in muscular exertion. Photic cues also play a part, for reversal of daily photoperiod was followed by an inversion of the phases of the daily oscillation in metabolic rate. Such oscillations persisted under conditions of constant light and temperature, but the rhythm was lost in dim light, suggesting that more intense illumination is necessary for the generation of circadian metabolic rhythms (Berman and Meltzer, 1978). An underlying circadian rhythm in thyroid activity in the domestic fowl (Klandorf et al., 1982) might also contribute to metabolic rhythms.

D. Interspecific Variation in Standard Metabolic Rate

Considerable information is available on avian SMR and this has led to a profusion of allometric equations linking these rates to body mass (for examples, see Table 2). Claims of a difference between metabolic levels of passerine and nonpasserine birds indicated in Table 2 have been questioned by Prinzinger and Hänssler (1980) and Prinzinger et al. (1981), based on an analysis of their measurements of birds of nine nonpasserine orders. This analysis yielded a regression similar to that for passerines. Bennett and Harvey (1987) confirmed the existence of a significant difference between the regressions for empirical SMR of passerines and nonpasserines using metabolic values reported in the literature. As they noted, "nonpasserines" comprise an artificial group. Importantly, Reynolds and Lee (1996) found no significant difference in SMR between passerines and nonpasserines, once procedures designed to eliminate phylogenetic effects were applied. Their general equations for birds are given in Table 2. Reynolds and Lee (1996) pointed out that empirically based allometric relations are critically dependent on the underlying assumptions of the statistical model employed. Consequently, so-called adaptive explanations of observed departures of such variables as SMR from expectation (see Dawson and O'Connor, 1996, for examples) should be corroborated by experimental data and use of alternative evolutionary models.

E. Influence of Ambient Temperature on Avian Metabolic Rates

Avian thermoregulatory requirements result in a complex relation between metabolic rate and T_a, which resembles that in mammals. The general form of this

TABLE 2 Equations for Calculating Standard Metabolic Rates (SMR in W)[a] and Total Evaporative Water Loss at 25°C (TEWL in Milliliters of H_2O/Day) of Birds in Relation to Body Mass (M, kg)

Equation	Group/state	Reference[b]
Metabolic Rate		
log SMR = 0.833 + 0.704 (log M)	Passerine birds/active phase[c]	1
log SMR = 0.740 + 0.726 (log M)	Passerine birds/rest phase[c]	1
log SMR = 0.643 + 0.729 (log M)	Other birds/active phase[c]	1
log SMR = 0.544 + 0.734 (log M)	Other birds/rest phase[c]	1
log SMR = 0.599 + 0.670 (log M)	All birds[c]	2
log SMR = 0.591 + 0.635 (log M)	All birds[d]	2
Total evaporative water loss at 25°C		
log TEWL = 1.511 + 0.678 (log M)	All birds[c]	3
log TEWL = 1.534 + 0.789 (log M)	All birds[d]	3

[a] SMR (W) here refers to metabolic rates of postabsorptive birds resting in their respective zones of thermal neutrality.

[b] References: 1, Aschoff and Pohl (1970); 2, the two equations for the SMR–body mass relationship for all birds presented in this table are based on the corresponding equations accompanying Fig. 3 in Reynolds and Lee (1996)—the versions of these equations presented in the text (p. 743) appear to be incorrect; 3, Williams (1996).

[c] Standard least-squares regression; phylogenetic effects not considered.

[d] Based on statistical procedures designed to eliminate phylogenetic effects.

relation in short-term tests is illustrated in Figure 2 for variable seedeaters resting in the dark during the day. Over an intermediate range of T_a, metabolism remains constant at standard or resting rate, depending on the nutritional state of the bird (see Section V,B for definitions of SMR and RMR). The resting rate tends to be higher due to the heat increment of feeding. The temperature range over which heat production is independent of T_a is, of course, the TNZ (see Section V,B), which may vary in width from a degree or two in small birds to more than 30°C in larger ones such as the emperor penguin *Aptenodytes foresteri* (Pinshow et al., 1976), winter-acclimatized rock ptarmigan *Lagopus mutus*, and willow ptarmigan *Lagopus lagopus* (West, 1972). In the TNZ, the bird achieves thermal balance by matching the rate of heat loss to the SMR or RMR. Below the lower limit of thermal neutrality (the lower critical temperature, T_{lc}), where insulation is at or near a maximum, heat production varies inversely with T_a. It can reach a maximum of ~3–8× SMR, in the cold, depending on the species (see, for example, Saarela et al., 1989a; Marsh and Dawson, 1989; Sutter and MacArthur, 1989; Hinds et al., 1993; Saarela et al., 1995; Dutenhoffer and Swanson, 1996). Often, these multiples for cold-induced summit metabolism are lower than the maxima seen in birds during peak activity (Marsh and Dawson, 1989). Avian terrestrial locomotion or flight can involve metabolic rates as high as 12× (running) or 14× (flying) resting rates (Brackenbury, 1984), though factors of 5–10× are more common. A question has existed as to whether the heat increment of feeding and/or activity metabolism of birds in the cold can substitute in some part for regulatory thermogenesis. Evidence on this point for the former is conflicting (see Dawson and O'Connor, 1996, for discussion), though Meienberger and Dauberschmidt (1992) concluded that granivorous songbirds could use the heat increment of feeding in thermoregulation. This is also the case in nestlings of the house wren *Troglodytes aedon* (Chappell et al., 1997; see XII,B,1). In the dark phase of the rock pigeon's daily cycle, T_b falls to a nocturnal plateau that is directly correlated with the amount of food consumed in the preceding light phase and with the temporal pattern and quantity of cloacal droppings produced in the night (Rashotte et al., 1997). As food consumption rises, the plateau, dark-phase T_b is elevated while dark-phase shivering is suppressed (Geran and Rashotte, 1997; Rashotte and Chambers, 1998). Shivering and digestion thus appear related in a reciprocal fashion in this situation, providing a further suggestion of substitution of thermogenesis resulting from food processing for muscular heat production. In a previous review, Marsh and Dawson (1989) found indications of substitution of activity metabolism for regulatory thermogenesis in some studies (Tucker, 1968; Berger and Hart, 1972; Nomoto et al., 1983; Rothe et al., 1987), but not in others (Pohl, 1969; Schuchmann, 1979). The extent to which such substitution occurs may not only vary among species, but also with T_a. For instance, Pohl and West (1973) found no substitution of activity metabolism for regulatory thermogenesis in the common redpoll (*Carduelis flammea*) between 0° and −30°C, but nearly complete substitution at −45°C. Substitution in hopping white-crowned sparrows (*Zonotrichia leucophrys gambelii*) increased with declining temperature down to −10°C, at which point it was complete (Paladino and King, 1984). Thus a sparrow

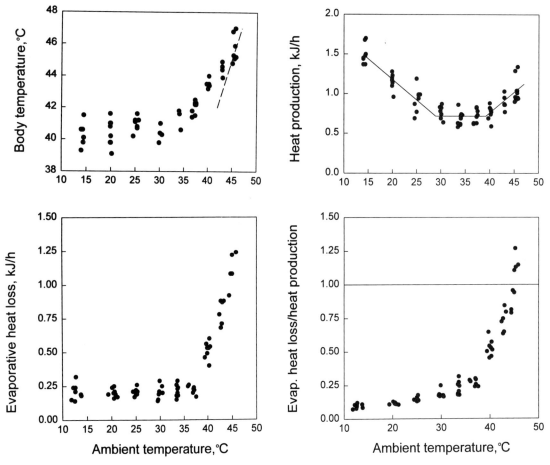

FIGURE 2 Relationships of various thermoregulatory variables to ambient temperature (T_a) in the variable seedeater (*Sporophila aurita*), a 10-g passerine bird occurring from southern Mexico through Central America to western Colombia, western Ecuador, and northwestern Peru. Measurements were made on postabsorptive individuals resting in the dark during the active phase of their daily cycle. Upper panels: (left) body temperature (°C), the dashed line marking equivalence between T_b and T_a; (right) heat production (kJ/hr). Lower panels: (left) evaporative heat loss (kJ/hr); (right) ratio of evaporative heat loss to heat production. (Modified from Weathers, 1997.)

hopping at −10°C would expend no more energy than a quiescent one engaged in regulatory thermogenesis at this T_a.

A problem in the determination of the extent of substitution of activity metabolism for regulatory thermogenesis at any T_a is that the active and quiescent animals must be examined under environmental conditions that impose similar thermal demands. Zerba and Walsberg (1992) determined the heat production of Gambel's quail (*Callipepla gambelii*) at T_a of 10° and 20°C while the birds were resting or running at 1.5 m·s^{-1}. These values were compared with previous data (Goldstein, 1983) on the effects of windspeed on the thermogenesis of resting Gambel's quail. The thermogenesis of exercising birds exceeded that of resting ones in still air, but was similar to that of the resting individuals exposed to wind at 1.5 m·s^{-1}. The latter findings indicated essentially complete substitution of activity metabolism for regulatory thermogenesis at 10° and 20°C. Studies combining time–activity budget (TAB) and doubly labeled water (DLW) estimates of field metabolic rate (FMR) by verdins, *Auriparus flaviceps* (Webster and Weathers, 1990), yellow-eyed juncos, *Junco phaeonotus*, and dark-eyed juncos (Weathers and Sullivan, 1993) under cool conditions have also suggested that heat produced in activity can substitute for regulatory thermogenesis.

In consideration of the form of avian metabolism–T_a relations, it is important to note that the transition between the "physical" thermoregulation (here principally involving adjustment of insulation) occurring in the TNZ and the "chemical" thermoregulation (involving regulatory thermogenesis) evident at lower T_a is abrupt in some species, occurring at the lower boundary of the TNZ, the lower critical temperature (T_{lc}). In this case, if T_b and insulation remain stable below T_{lc}, the regression line describing heat production at T_a below the TNZ should extrapolate to zero metabolic rate at a temperature matching T_b, as illustrated in the model provided by Scholander *et al.* (1950). This is simply an expression of Fourier's Law, which predicts that heat loss under these conditions below the T_{lc} will be proportional to the difference in temperature between the bird and its environment. Not all birds conform to this model, due principally to their changing insulation at T_a somewhat below thermal neutrality. Where insulation does change below thermoneutrality, it generally continues to increase (see, for example, West, 1962). However, relatively high shell temperatures indicated that it was reduced below the T_{lc} in tests conducted in air with juvenile Gentoo penguins before their initial immersion in cold water, and with adults of this species (Dumonteil *et al.*, 1994; see Section VI,E). Management of insulation below the T_{lc} may differ between closely related species. In one Hawaiian honeycreeper (*Loxops virens*) it continued to increase with decreasing T_a below the TNZ, whereas in a congener (*Loxops parva*), it remained constant but T_b declined (MacMillen, 1974).

The regulatory thermogenesis evident in birds below the TNZ seems to result primarily from shivering (see Marsh and Dawson, 1989; Dawson and O'Connor, 1996), which appears largely supported by oxidation of fatty acids (Carey *et al.*, 1978). The prominence of the flight muscles in most adult birds has led to emphasis of their importance in the shivering process (see, for example, Marsh and Dawson, 1989). However, the leg musculature also plays an important role in the shivering of house finches (*Carpodacus mexicanus*) in severe cold (Carey *et al.*, 1989). Furthermore, it is the major contributor in young turkeys (*Meleagris gallopavo*) and guinea fowl (*Numida meleagris*; Dietz *et al.*, 1997). Shivering thermogenesis was sufficient to allow 14-g winter-acclimatized American goldfinches (*Carduelis tristis*) to withstand ambient temperatures of −60° to −70°C for at least 3 hr (Dawson and Carey, 1976). Attempts have been made to demonstrate nonshivering thermogenesis (NST) in birds (see Dawson and O'Connor, 1996, for review) and some indirect evidence for such a process has been obtained (see, for example, Saarela and Heldmaier, 1987). However, conclusive direct results under ecologically appropriate circumstances (see Connolly *et al.*, 1989) have not thus far been reported. Birds appear to lack brown adipose tissue and uncoupling protein (see Dawson and O'Connor, 1996), which are central to mammalian NST. Moreover, Marsh (1993) has noted that the site of any avian NST has not been convincingly demonstrated; no mechanism for activating it during cold exposure has been elucidated; and tests with reputed calorigenic substances such as glucagon may represent pharmacological rather than physiological effects. In cold-acclimated Muscovy ducklings (*Cairina moschata*), lack of electromyographic activity in the gastrocnemius muscle coincident with increased regulatory thermogenesis has led to suggestions of NST (Barré *et al.*, 1985). Vittoria and Marsh (1996) also found that shivering was often absent in the gastrocnemius muscle of such birds. However, they also determined that shivering in two thigh muscles (iliofibularis and flexor cruris lateralis) increased coincident with increases in metabolic rate. These thigh muscles were among those that showed large and sustained increases in citrate synthase activity during cold acclimation. Vittoria and Marsh (1996) concluded that shivering is probably the major source of thermogenesis in cold-acclimated ducklings.

Deep body temperature of some birds has been shown to undergo a minor rise with increasing T_a in the TNZ (Weathers, 1981; Ellis *et al.*, 1995), even though heat production, by definition, remained constant. The rise in metabolic rate with increasing T_a above the upper critical temperature (the upper boundary of the TNZ, T_{uc}) probably results from the further hyperthermia evident in most birds at high T_a, plus the effort of effecting evaporative cooling (Figure 2). The impact of high T_a on metabolism of resting birds at low humidities generally is smaller than that resulting from severe cold challenges (see above). With exposure to such humidities, which facilitate efficient evaporative cooling, resting metabolic rates at T_a of ca. 45°C seldom exceed 1.5–2× SMR (Dawson and O'Connor, 1996). Indeed, under these conditions, the rise in metabolic rate is negligible in a number of species (see Table 4.2 in Dawson and O'Connor, 1996).

The metabolism–T_a relation evident in short-term tests for birds resting in the inactive phase of their daily cycle may not be evident under other conditions. For example, the 55-g evening grosbeak (*Coccothraustes vespertinus*) had a TNZ that was more than 15°C wide at night (Dawson and Tordoff, 1959), but not apparent during daytime measurements, which yielded higher metabolic rates (Hart, 1962). Such a zone was also lacking in longer-term tests where metabolism was estimated from food consumption and the birds have had an opportunity to become at least partially acclimated to each test temperature (West and Hart, 1966). Metabolism–T_a curves resulting from such tests typically have shallower slopes than those evident below the TNZ in short-term tests. Prinzinger (1982) has noted another source of variability in the metabolic response of birds to temperature. Sinusoidal fluctuations of temperature around a mean value elicited higher heat production in Japanese quail than did continued exposure to the T_a matching that value. Evidently, the manner in which temperatures are presented can affect the nature of the metabolic response.

F. Intraspecific Variation in Avian Metabolic Rate: Acclimatization, Acclimation, and Geographical Patterns

Seasonal variation of SMR is sometimes apparent in free-living birds at particular localities. For example, the European goldfinch *Carduelis carduelis* (Gelineo, 1969), common redpoll (Pohl and West, 1973), mute swan *Cygnus olor* (Bech, 1980), dark-eyed junco (Swanson, 1990), and black-capped chickadee *Parus atricapillus* (Cooper and Swanson, 1994) have higher rates in winter than in summer on a total (per bird) and/or mass-specific (per gram) basis. Mass-specific SMR of house sparrows in Iowa was higher in November–February than in April–June (Miller, 1939), but such birds at Ottawa, Canada, where cold winters are also the norm, were found to have a seasonally stable daytime SMR on either a total or mass-specific basis (Hart, 1962). Other passerines show similar stability of SMR (see Dawson and Marsh, 1989). These include the house finch, in which SMR did not differ between winter and late spring within populations from southern California, Colorado, and Michagan/Ohio (Dawson *et al.*, 1985; Root *et al.*, 1991). This seasonal stability contrasts with the geographic variation in winter SMR noted below between the first and latter two populations of house finches. The Eurasian kestrel *Falco tinnunculus* also had an SMR that was independent of season (Masman *et al.*, 1988). This was also the case for the bobwhite between summer and winter (Swanson and Weinacht, 1997).

Metabolic levels of individual birds also may vary with thermal acclimation. A rise of 10–85% occurred in SMR within 1 to 4 weeks after transfer of birds from ca. 25–30°C to 12–22°C or −14° to +10°C (see Gelineo, 1964; Arieli *et al.*, 1979). A drop in SMR developed within the same intervals for transfers in the reverse direction.

Some geographic variation in metabolic characteristics also is evident in birds. For example, North American populations of the house sparrow from warmer climates tended to have lower SMR than their counterparts from cooler areas (Hudson and Kimzey, 1966). Furthermore, winter SMR of house finches from Colorado and Michigan/Ohio, where they encounter relatively severe cold seasonally, were significantly higher than those of southern Californian birds living under milder conditions (Root *et al.*, 1991). At the interspecific level, Weathers (1979) has reported that SMR tends to be higher in birds from cold climates and lower in tropical forms than anticipated from their respective body masses. On the average, SMR rose 1% per degree increase in latitude. However, penguins, which undergo prolonged fasts during the breeding season, don't show elevated rates typical of high-latitude birds.

The above variations in SMR in response to experimental, seasonal, or geographical conditions may be functionally significant. A relatively low SMR could be advantageous in hot weather by reducing a bird's endogenous heat burden and the need for evaporative cooling. It might also increase survival of xerophilic populations in the periods of low productivity characterizing deserts. The case of the white-browed scrubwren (*Sericornis frontalis*) is instructive. Scrubwrens living in a dry part of Western Australia had SMR averaging 19% lower in summer than in winter, a difference viewed as an adjustment reducing energy expenditure, water loss, and thermoregulatory problems in the hot, dry portion of the year (Ambrose and Bradshaw, 1988). Interestingly, white-browed scrubwrens did not show such a metabolic change in more mesic regions.

The importance of a high SMR for existence in the cold is less obvious. At first glance, a high rate would seem to represent a needless encumbrance, given the capacities of birds for regulatory thermogenesis. Perhaps, in the case of birds experimentally exposed to cold, it represents an emergency response linked with protection of peripheral tissues from cold injury. It also might lower the T_a threshold for shivering. Additionally, cold-induced increases in avian SMR can be accompanied by improved thermogenic capacity and increased cold resistance (Gelineo, 1955, 1964; Swanson, 1990; Liknes and Swanson, 1996). Whether the higher SMR is a contributing factor to these improvements, a byproduct of them, or a separate response is unclear. As noted above, not all free-living birds show higher SMR in winter than in warmer parts of the year (see Dawson and Marsh, 1989, for discussion), indicating that shifts in these rates are not a mandatory part of avian seasonal acclimatization.

Seasonal acclimatization or acclimation may also affect the positioning of avian metabolism–temperature curves of the type illustrated in Figure 2. For example, the T_{lc} of some species was displaced downward in winter relative to the summer value (Kendeigh *et al.*, 1977; Rintamäki *et al.*, 1983) and it was 4.2°C lower in cold-acclimated American coots (*Fulica americana*) than in warm-acclimated ones (Sutter and MacArthur, 1989). A shallower slope for the metabolism–temperature relationship below thermal neutrality also may be evident (Hissa and Palokangas, 1970; Kendeigh *et al.*, 1977). These differences primarily reflect an enhanced insulation during the winter season. Such changes were apparent in the 0.5-kg willow ptarmigan, an arctic species in which the T_{lc} declines from 7.7°C to −6.3°C from summer to winter, with a corresponding lowering of the slope of the metabolism–temperature relation below thermal neutrality (West, 1972). Dry mass of the contour plumage in dark-eyed juncos (body mass ca. 18–19 g) was 32% heavier in winter then in summer (Swanson, 1991). Nevertheless, insulative changes apparently play only a secondary role in the winter adjustments of smaller birds (Marsh and Dawson, 1989; Cooper and Swanson, 1994), which instead increase their cold resistance in this season primarily by metabolic means (Dawson et al., 1983a; Marsh and Dawson, 1989; Swanson, 1991; Cooper and Swanson, 1994; O'Connor, 1996). This metabolic acclimatization to winter conditions affects regulatory thermogenesis, increasing thermogenic endurance (the ability to sustain elevated rates of heat production in the cold) and, sometimes, thermogenic capacity (the ability to increase heat production). The latter was increased by melatonin treatment in the Japanese quail. Such treatment, cold, and short days additionally improved cold resistance and thermal insulation relative to quail on long days (Saarela and Heldmaier, 1987). The extent and form of metabolic acclimatization can vary geographically within species. For example, house finches from southern California showed no seasonal variation in cold resistance, whereas those from Michigan and Colorado had significantly greater thermogenic endurance during winter than in late spring (Dawson et al., 1983b; O'Connor, 1995). Additionally, thermogenic capacity was greater at the former season in Michigan birds, but not in Colorado birds (O'Connor, 1996).

VI. HEAT TRANSFER WITHIN THE BODY

Heat produced in deep-seated organs such as the liver must be transported to the skin surface or the mucosa lining the upper respiratory tract before it can be lost to the environment. As long as there is some blood flow to the skin or mucosal surface, some of this heat is transported by way of the blood stream. Under cold conditions, the blood vessels in the skin may be fully constricted and heat transfer then has to occur by conduction through the tissues (Whittow, 1986).

A. Vascular Heat Exchange

In the bare extremities, blood flow to the tissues can be maintained without a concomitantly high heat loss, by directing the blood through special vascular structures in which the heat in the arterial blood going to an extremity is transferred to the venous blood returning to the heart, rather than being lost from the extremity to the enviroment. These structures take the form of *venae comitantes* or *retia* (Midtgård, 1989b). The former consist of veins grouped around an artery; they permit the direct transfer of heat from the arterial blood supplying the limb to the veins returning blood from the limbs to the heart. In this way, heat that would otherwise be lost, is retained in the body. Retia (networks of arteries and veins) serve the same purpose. Arteriovenous (A-V) anastomoses in the skin are a more ubiquitous vascular arrangement. They permit a larger volume of warm blood to flow to the skin than if the circulatory pathway was only through the capillaries. In this way, they facilitate heat loss. A-V anastomosis blood flow occurs mostly in unfeathered skin in the domestic fowl, representing 17, 53, and 83% of total blood flow in cold, thermoneutral, and hot conditions, respectively (Wolfenson, 1983).

B. Cold Vasodilatation

The possibility that tissues in the cold extremities might freeze is circumvented by periodic increases in blood flow to the extremities (cold vasodilatation). This occurs at the expense of increased heat loss from the bird.

C. Brood Patch

The brood patches of birds are special, well-vascularized areas of thoracic skin that facilitate transfer of body heat to the egg or hatchling. In broody hens of the domestic fowl, cooling their thoracic skin (brood patch), as might occur when returning to a cool egg, led to an increase in the metabolic heat production of the bird and compensatory vasoconstriction in the feet, but no constriction of the brood patch vasculature. These changes ensure that heat transfer to the egg or hatchling is maintained (Brummermann and Reinertsen, 1991).

D. Thermal Conductance of the Tissues

Changes in heat transfer through the tissues can be inferred from changes in the calculated tissue thermal conductance ($C_{tissues}$ in $W[°C]^{-1}$ or $W[kg·°C]^{-1}$; see Whittow, 1986)

$$C_{tissues} = (H - E_{ex})/(T_b - \overline{T}_{sk})$$

where H is metabolic heat production (W or $W[kg]^{-1}$) computed from oxygen consumption (see Section V,A), E_{ex} is the respiratory evaporative heat loss (W or $W[kg]^{-1}$; see Section VII,B,1), and T_b and $\overline{T}_{sk}$ (°C) are the core and mean skin temperatures, respectively.

E. Acclimatization and Acclimation

Limited, inconsistent information exists on changes in thermal conductance of tissues during the adjustment of individual birds to cold. In species such as monk parakeets (*Myiopsitta monarchus*) and black-capped chickadees, overall thermal conductance did not vary between winter and summer (Weathers and Caccamise, 1978; Cooper and Swanson, 1994), suggesting that tissue insulation was seasonally constant. On the other hand, skin temperature of the trunk and extremities was higher in cold- than in warm-acclimated rock pigeons. This reflected a higher thermal conductance in the tissues of the former (i.e., lower tissue insulation), presumably due to increased peripheral blood flow. However, fat deposits were greater in rock pigeons acclimated to a T_a of 10°C than in those acclimated to 29°C (Rautenberg, 1969a). Acclimation to cold water by juvenal Gentoo penguins involved an opposite effect. With successive immersions, they switched from maintaining a relatively high shell temperature to a greater peripheral vasoconstriction serving to increase body insulation (Dumonteil *et al.*, 1994). Tissue insulation of juvenal king penguins (*Aptenodytes patagonicus*) in cold water also increased over the first several immersions, probably due to an increasing vasoconstriction reaction (Barré and Roussel, 1986). A decreased thermal conductance of the tissues in the legs of cold-acclimatized ring-necked pheasants (*Phasianus colchicus*) appeared to result from the operation of the countercurrent heat exchange mechanism (Ederstrom and Brumleve, 1964).

Increased subcutaneous fat deposits in aquatic birds during the colder parts of the year also must serve to increase body insulation. However, as noted in Section I, the winter fattening of many land birds appears to involve only localized deposits with apparently limited effects on insulation (Newton, 1969; see Marsh and Dawson, 1989, for discussion). Nevertheless, the higher tissue insulation noted in gray jays (*Perisoreus canadensis*) in winter was attributed to greater amounts of fat (Veghte, 1964). Moreover, house sparrows from northern populations in North America have more insulation than their southern counterparts and this has been attributed to increased subcutaneous lipid deposition (Blem, 1974).

VII. HEAT LOSS

Heat transported to the surface of the skin may be lost by evaporation or by nonevaporative means. As most of the bird's surface is covered with feathers, this heat and any water evaporated on the skin must traverse the plumage before being lost to the environment (see Section VII,B,2 and Whittow, 1986). The depth of the plumage is not fixed; it varies inversely with T_a, the elevation of the feathers being effected by the pinnamotor muscles (Marsh and Dawson, 1989).

The effect of immersion in water on plumage insulation of aquatic birds has been analyzed. In the common eider (*Somateria mollissima*) this insulation was less in water than in air, but this was compensated for, in part, by greater tissue insulation in the former than in the latter medium (Jenssen *et al.*, 1989).

A. Nonevaporative Heat Loss; Total Thermal Conductance

Nonevaporative heat loss is relatively difficult to measure because of the multiple pathways (conduction, convection, and radiation—see Whittow, 1986) by which it can proceed. Under natural conditions, the determination of such heat loss can be especially complex. Convective heat loss is a significant component in the heat balance equation for flying birds because of their movement through the air (see Chapter 15). However, it is exposure to solar radiation that contributes most to the variability in nonevaporative heat loss in nature.

In practice, many investigators have used measurements of T_b, T_a, metabolic heat production (H in W or W[kg]$^{-1}$, determined from oxygen consumption), and evaporative heat loss (E in W or W[kg]$^{-1}$, determined from evaporative water loss) to estimate the total "dry" (evaporative heat loss taken into account) thermal conductance (C_{total} in W[°C]$^{-1}$ or W[kg·°C]$^{-1}$; other terms and units as in equation for $C_{tissues}$):

$$C_{total} = (H - E)/(T_b - T_a).$$

Total thermal conductance provides an indication of the facility with which heat is transferred through the tissues and plumage and lost (at $T_b > T_a$) to the surrounding medium at a given difference between T_b and T_a. It includes the $C_{tissues}$ as well as a coefficient describing the heat loss from the bird to its surroundings.

Figure 3 presents data for the wedge-tailed shearwater (*Puffinus pacificus*). The C_{total} for this bird was low between T_a of 5 and 29°C. It rose considerably at higher T_a, probably reflecting an increased cutaneous blood flow, particularly in the extremities (Whittow *et al.*, 1987). Other work has shown that the minimal total thermal conductance of the American coot resting in water was 1.6–1.7 times that recorded in air (Sutter and MacArthur, 1989).

B. Evaporative Heat Loss

1. Total Evaporative Heat Loss (E)

Williams (1996) recently analyzed the total evaporative water loss (TEWL)–body mass relationship at 25°C for birds ranging in size from hummingbird to ostrich. Both least-squares regression and phylogenetically independent contrasts yield somewhat higher values for the slope of this relationship than previously established. These analytical methods also indicate that birds from arid environments have a statistically lower TEWL than those from mesic ones. Moreover, small birds have

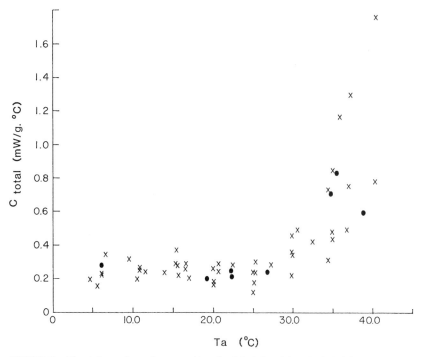

FIGURE 3 Total thermal conductance (C_{total}) of fledgling (x) and adult (o) wedge-tailed shearwaters (*Puffinus pacificus*) at different air temperatures (T_a). (From *J. Comp. Physiol. B*, Temperature regulation in a burrow-nesting seabird, the wedge-tailed shearwater (*Puffinus pacificus*), G. C. Whittow, T. N. Pettit, R. A. Ackerman, and C. V. Paganelli, **157**, 607–614, 1987, © Springer-Verlag.)

ratios of TEWL to oxygen consumed that are similar to those of heavier species, which negates a generalization that small desert birds replenish a lower fraction of their evaporative losses through oxidative production of water than larger ones. Williams' (1996) general equations for TEWL–body mass relationships of birds are presented in Table 2.

Total evaporative heat loss (E) is the product of the rate of total evaporative water loss ($\dot{m}_e$) and the latent heat of vaporization of water (λ, approximately 2.4 J/mg H_2O):

$$E = \dot{m}_e \lambda.$$

Evaporation of water in birds involves both respiratory (including the buccopharyngeal cavity) and cutaneous (including corneal surfaces) components, so total evaporative heat loss may be represented as follows:

$$E = E_{ex} + E_{sw},$$

where E_{ex} is evaporative heat loss from the respiratory tract and E_{sw} is evaporative heat loss from the body surface. As with E, both E_{ex} and E_{sw} represent the products of the rate of evaporation and λ.

Total evaporative water loss of a bird can be measured in an open circuit metabolism system by passing the excurrent air through a desiccant column and determining the gain in mass of this column over a precisely timed interval. Correction of the gain for the water content of the incurrent air yields the evaporative water loss for the interval. If the rates of air flow into and out of the metabolism chamber are known, evaporative water loss also can be determined from the difference in the water vapor densities in the incurrent and excurrent air. Additionally, it can be measured most simply by recording the decrease in mass of the animal over a defined period, making appropriate corrections for any change in mass due to respiratory gas exchange (oxygen, carbon dioxide) or defecation.

The heat that birds can lose in hot environments through total evaporation can be substantial, matching or exceeding that produced in metabolism in many species (Table 3). The respective contributions of respiratory and cutaneous evaporation to this performance appear to vary from group to group, as discussed below. It is also important to note that effective evaporative cooling is widespread but not universal among birds. Two exceptions are the Hawaiian honeycreepers *Loxops virens* and *Loxops parva*, which are relatively intolerant of heat (MacMillen, 1974).

2. Cutaneous Evaporative Heat Loss (E_{sw})

Marder and Ben-Asher (1983) have determined cutaneous evaporation from measurements of the resistance of avian skin to the diffusion of water. Ventilated capsules can also be used to determine water loss at various

TABLE 3 Total Evaporative Heat Loss (E) of Birds, Expressed as a Percentage of Metabolic Heat Production (H) at High Ambient Temperatures[a]

Species	Body mass (kg)	T_a (°C)	E/H (%)
Ostrich (*Struthio camelus*)	100.0	44.5	100
Bedouin fowl (*Gallus gallus*)	1.427	48	159
Brown-necked raven (*Corvus corax ruficollis*)	0.610	50	167
Rock pigeon (*Columba livia*)	0.315	44.5	118
	0.239	60	304[b]
Roadrunner (*Geococcyx californianus*)	0.285	44.5	137
Galah (*Cacatua roseicapilla*)	0.271	>45	170
Burrowing owl (*Speotyto cunicularia*)	0.143	44.1	95
Japanese quail (*Coturnix coturnix japonica*)	0.100	43	144
Spinifex pigeon (*Geophaps plumifera*)	0.089	45–50	>241[b]
Spotted nightjar (*Eurostopodus guttatus*)	0.088	52.8	ca. 300[b]
Common poor-will (*Phalaenoptilus nuttallii*)	0.050	47	352[b]
Speckled coly (*Colius striatus*)	0.044	44	99
Blue-breasted quail (*Coturnix chinensis*)	0.043	43.5	116
Inca dove (*Scardafella inca*)	0.042	43.5	108
House sparrow (*Passer domesticus*)	0.025	44.5	106
House finch (*Carpodacus mexicanus*)	0.020	44.5	130
Gouldian finch (*Poephila gouldiae*)	0.014	44.5	105
Zebra finch (*Poephila guttata*)	0.012	43.5	123
Verdin (*Auriparus flaviceps*)	0.007	50	179[b]
Costa's hummingbird (*Calypte costae*)	0.003	40	66

[a]Based on Table 9-3 in Whittow (1986).
[b]Additional data from Marder and Arieli (1988), Withers and Williams (1990), Dawson and Fisher (1969), Lasiewski (1969), and Wolf and Walsberg (1996a) for the rock pigeon, spinifex pigeon, spotted nightjar, common poor-will, and verdin, respectively.

points on the skin (Webster and Bernstein, 1987). Additionally, cutaneous evaporative heat loss can also be measured using a chamber fitted with an elastic membrane containing a small aperture through which the bird's head is inserted. This membrane, which encloses the neck, serves as a partition between two ventilated compartments, one allowing measurement of respiratory and cephalic cutaneous water loss and the other cutaneous loss from the remainder of the body (e.g., Wolf and Walsberg, 1996a). The arrangement, after suitable corrections for cutaneous evaporation in the chamber enclosing the head, largely separates cutaneous from respiratory water evaporation. Cutaneous water loss also has been determined indirectly as the difference between TEWL and the respiratory water loss determined from respiratory minute volume and the saturated aqueous vapor density of air for the temperature at which it leaves the bird's respiratory tract (see, for example, Withers and Williams, 1990).

The rate of evaporative heat transfer from the skin to the air (E_{sw} in W/m^2) is described by the following equation:

$$E_{sw} = \frac{h_e (\phi_{sk} P_{ws} - \phi_a P_{wa}) \lambda}{100},$$

where h_e is the evaporative heat transfer coefficient, a function of air movement, viscosity, density, the diffusivity of water vapor, and the dimensions of the body ($g[sec \cdot m^2 \cdot kPa]^{-1}$); ϕ_{sk} and ϕ_a are the relative humidities at the skin surface and of the air, respectively (%); P_{ws} is the saturated aqueous vapor pressure of skin at temperature T_{sk} (kPa); P_{wa} is the saturated aqueous vapor pressure of air at T_a (kPa); and λ is the aqueous latent heat of vaporization at T_{sk} (J/g; see Section VII,B,1).

The partition of evaporative heat loss into its respiratory and cutaneous components has been accomplished for birds of several orders (see Wolf and Walsberg, 1996a). In most species studied, cutaneous evaporation represented half or more of the total evaporation at low humidities and moderate to warm T_a (25 to 35°C or 37.5°C; see Dawson, 1982, for summary of earlier work; Webster and King, 1987; Withers and Williams, 1990; Wolf and Walsberg, 1996a). It constituted 40–44% in the ostrich (Withers, 1977). The proportion drops in exercising birds, judging by Taylor et al.'s (1971) observations on the rhea (*Rhea americana*).

The importance of cutaneous evaporation for birds in very hot environments seems, on the basis of limited information, to vary among groups (Wolf and Walsberg, 1996a). It appears secondary to respiratory evaporation in ducks and geese (Order Anseriformes); fowl, pheasants, and quail (Order Galliformes); and perching or passerine birds (Order Passeriformes). For example, cutaneous water loss accounted for only 25% of total evaporation at 40°C (T_a) in the domestic fowl studied during winter in Britain (Richards, 1976) and but 14% in the verdin at 50°C (Wolf and Walsberg, 1996a). The emu *Dromaius novaehollandiae* (Order Struthioniformes) effected only 30% of its evaporative heat dissipation cutaneously at 45°C and so should be included in this group as well (Maloney and Dawson, 1994). In contrast, cutaneous water loss had a much more prominent role in heat dissipation at high T_a and low humidities in pigeons, doves, and sandgrouse (Order Columbiformes; see Marder and Ben-Asher, 1983; Marder *et al.*, 1986). Under low humidity conditions, cutaneous evaporation was sufficient in heat-acclimated rock pigeons to allow effective temperature regulation at T_a of 48–60°C without the birds' having to resort to either panting or gular fluttering (Marder and Gavrieli-Levin, 1987; see Section VII,B,3). Marder and Arieli (1988) estimated that cutaneous evaporation accounted for 75% of the heat dissipation under these conditions, with the remainder resulting from respiratory evaporation. The resistance to diffusion of water vapor through the feather coat and associated boundary layer has been estimated to represent only 6.2–25% of the total vapor resistance in rock pigeons (Webster *et al.*, 1985) and the skin therefore seemed to be the primary barrier to cutaneous water loss. The mechanisms serving to increase cutaneous evaporation at high T_a are unclear (Withers and Williams, 1990; Wolf and Walsberg, 1996a), but could involve vasomotor adjustments, stearic changes in skin lipids, or increased hydration of the epidermal stratum corneum (Webster *et al.*, 1985). The epidermis of heat-acclimated rock pigeons differs in several respects from that of non- or cold-acclimated members of this species. In heat-acclimated individuals, both the dorsal and abdominal skin include modified areas characterized by increased vascularization, epidermis of greater thickness, and changes in intracellular structures all suggestive of a high rate of cutaneous evaporation (Peltonen *et al.*, 1998). β-Adrenergic blockade decreased skin resistance and increased cutaneous water loss in rock pigeons at a T_a of 30°C (Marder and Raber, 1989). Marder *et al.* (1989) concluded that increased cutaneous water loss in acclimated rock pigeons is probably elicited by inputs generated by dermal warm receptors in response to increased T_a. Warming the brain to 42.5° was also noted to increase such loss during exposure of these birds to a T_a of 26°C.

Cutaneous water loss is apparently affected by the hydration state of the bird. In contrast to hydrated individuals (see above), exposure of dehydrated rock pigeons (ca. 16% loss of body mass after 48 hr of water deprivation) to 45–50°C did not increase cutaneous evaporation beyond minimal levels and panting seemed to be the major evaporative cooling mechanism (Arad *et al.*, 1987). Water restriction of zebra finches (*Poephila guttata*) at 25°C was followed by a reduction in evaporative water loss, evidently due to lower cutaneous loss (Lee and Schmidt-Nielsen, 1971).

3. Respiratory Evaporative Heat Loss (E_{ex})

Respiratory evaporative heat loss (in W) can be represented as follows:

$$E_{ex} = \dot{V}(\rho_{ex} - \phi_a \rho_{in} [10^{-2}])\lambda,$$

where $\dot{V}$ is respiratory minute volume (liter/min), ρ_{ex} is water content (g/liter) of expired air saturated at the expiration temperature, ϕ_a is the relative humidity (%) of inspired (ambient) air, ρ_{in} is the water content (g/liter) of the inspired air saturated at ambient air temperature, and λ is the latent heat of vaporization of water in the expired air (J/g). Respiratory evaporative cooling is a primary means of heat defense in most avian orders (see Section VII,B,2) and birds almost without exception use it under sufficiently challenging circumstances. For example, it was evident in rock pigeons at higher humidities and T_a, despite the impressive capacities of properly acclimated individuals for cutaneous water loss (see Section VII,B,2; Marder and Arieli, 1988). The means by which respiratory evaporative cooling is effected are discussed in the following sections. However, before proceeding, it is worth mentioning the apparent importance of wind-assisted mouth cooling in flying birds. Ram ventilation of the buccal cavity of rock pigeons during simulated fast flight produced evaporative cooling equivalent to more than 3.5× resting heat production (St.-Laurent and Larochelle, 1994; heat balance in flying birds is considered in more detail in Chapter 15). Even slight breezes could increase evaporation in a resting pigeon if it were to open its mouth.

a. Thermal Tachypnea

Under heat challenges, birds characteristically augment respiratory evaporation by increasing respiratory frequency (f_{resp}), culminating in thermal tachypnea (thermal polypnea). This commonly involves open-mouth respiration, or thermal panting. In the verdin, which relies primarily on such cooling, panting increased respiratory water loss by 30.5× between 30° and 50°C (Wolf and Walsberg, 1996a). The increased ventilation responsible for this is achieved by augmenting f_{resp} sufficiently to overcome the effect of a concurrent reduction in tidal volume (V_T) characteristic of type I panting (see below). Typically, in a hyperthermic, panting bird respiratory minute volume may increase sixfold (Richards, 1970). However, as the T_b of the bird increases to very high levels, f_{resp} reaches a maximum and subsequently declines. These developments are accompanied by an increased V_T so that respiratory minute volume actually increases further (type II panting). However, this minute volume also eventually declines as the limit of the animal's thermal tolerance is approached.

The reduction in V_T accompanying thermal tachypnea in various birds under conditions of mild or moderate heat stress largely tends to confine the increased ventilation to the respiratory dead space (see, for example, Bech and Johansen, 1980), thus permitting a substantial increase in respiratory minute volume and respiratory evaporative cooling, without substantially affecting blood CO_2 and acid–base balance (summarized in Tables 1 and 2 in Marder and Arad, 1989). Nevertheless, significant hypocapnia and alkalosis (mean values for arterial CO_2 tension and pH, 14.2 torr [1.89 kPa] and 7.70, respectively, vs values of 27.1 torr [3.61 kPa] and 7.51 for birds at thermoneutral T_a) have been observed in a number of other species during heat challenges (data, primarily from Calder and Schmidt-Nielsen, 1968, summarized in Tables 1 and 2 in Marder and Arad, 1989). Marder and Arad (1989) attributed the hypocapnia and alkalosis to undue stress imposed by experimental procedures and insufficient acclimation to heat (see Section VII,B,4 concerning extent of conditioning required to produce sufficient heat acclimation in the rock pigeon). Additionally, Bech and Johansen (1980) suggested that induction of type II panting by exposure of birds to severe heat stress could be a factor in inducing hypocapnia and alkalosis in certain cases.

Bech and Johansen (1980) identified three separate ventilatory responses in panting birds, all of which involved reduced parabronchial ventilation but increased respiratory minute volume. In many species (the mute swan *Cygnus olor* studied by Bech and Johansen is an example), the V_T decreases to near that of the respiratory dead space. In other birds such as the rock pigeon, the panting is superimposed on slower, deeper breathing, with the combination being referred to as "compound ventilation" by Ramirez and Bernstein (1976). The shallow, rapid breathing primarily ventilating the respiratory dead space increased evaporative heat loss while gas exchange was achieved by the slower, deeper parabronchial component, which made respiratory alkalosis less likely than in simple panting. The greater flamingo (*Phoenicopterus ruber*) illustrates the third pattern, which involves shallow breathing at V_T less than respiratory dead space. This rhythm was interrupted at regular intervals by a brief series of deeper breaths ("flushouts;" Figure 4), which provided parabronchial ventilation (Bech *et al.*, 1979). Domestic fowl also displayed this pattern (Brackenbury *et al.*, 1981a; Arad and Marder, 1983). A different strategy for protecting acid–base balance was found in Adélie penguin chicks (*Pygoscelis adeliae*), in which panting led to a reduced carbon dioxide tension in the blood, but blood pH remained unchanged due to an increase in lactic acid concentration (Murrish, 1983). It is of interest that the ostrich could pant at 50°C (T_a) for 8 hr without suffering alkalosis. Air sac P_{CO_2} fell markedly, suggesting a functional shunt system regulating lung ventilation in this species (Schmidt-Nielsen *et al.*, 1969). Jones' (1982) observations of pulmonary blood flow distribution appear consistent with such a view.

The specific manner in which thermal tachypnea is initiated varies interspecifically. In some birds (e.g., cormorants [*Phalacrocorax*], nightjars, frogmouths [*Podargus*], mousebirds [*Colius*], passerines), respiratory frequency increased steadily with increasing heat load, whereas in others (e.g., the roadrunner *Geococcyx californianus*, pigeons, doves, owls, and ostrich) only a narrow band of frequencies was used under such loads (Lasiewski, 1972). It has been suggested for some members of the latter group that the frequency employed matches the resonant frequency of the respiratory system (see, for example, Bartholomew *et al.*, 1968), making it possible to effect panting movements by expending only enough energy to keep the thoracoabdominal structure oscillating at its natural frequency. Crawford and Kampe (1971) provided evidence for such a relation in the rock pigeon, though Weathers (1972) reported contradictory observations. In neither the ostrich (Schmidt-Nielsen *et al.*, 1969) nor the domestic fowl did panting frequencies appear to exploit the resonant properties of the respiratory system (Lacey and Burger, 1972). Use of a narrow band of respiratory frequencies in panting means, of course, that substantial changes in evaporative cooling can only be effected by changing the proportion of time spent panting and/or modulating the V_T. Despite the possibility of some species using resonant frequencies in panting, it can be a vigorous activity, especially at higher humidities. It and, possibly, the hyperthermia with which it is often associated were found to increase metabolic rate in brown-necked ravens (*Corvus ruficollis*) by factors of 1.68 and 2.14 at T_a of 45° and 50°C, respectively (Marder, 1973a). However, panting individuals of most of the species surveyed by Dawson and O'Connor (1996) had metabolic rates at 44° or 45°C that were less than 1.5X BMR. The factor was only 1.04 at 45°C for panting rock pigeons unacclimated to heat (Marder and Arieli, 1988).

b. Gular Flutter

Many birds (e.g., cormorants, pelicans, boobies [*Sula*], turkey vultures, quail [e.g., *Callipepla, Colinus*], goatsuckers [nightjars, nighthawks, poor-wills], roadrunners, pigeons, doves, owls, and mousebirds) supplement panting with rapid fluttering of the gular area (Lasiewski, 1972; Dawson, 1982; Arad *et al.*, 1989). Rhythmic inflation or pulsation of the well-vascularized

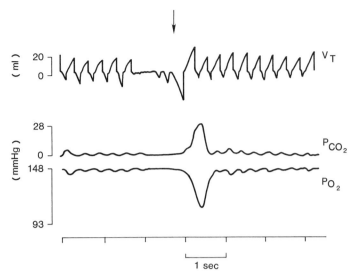

FIGURE 4 Tidal volume (V_T) and the partial pressures of carbon dioxide (P_{CO_2}) and oxygen (P_{O_2}) in the air from a flow-through respirometer in the flamingo (*Phoenicopterus ruber*) during exposure to heat. The abscissa is a time scale. Rapid, shallow panting is interrupted by a single "flush-out" at the arrow. (Redrawn from C. Bech, K. Johansen, and G. M. O. Maloiy (1979), *Physiol. Zool.* **52**, 313–328 by permission of The University of Chicago Press. © 1979 by The University of Chicago.)

esophagus contributes to the effectiveness of this activity in pigeons and doves (Gaunt, 1980; Baumel *et al.*, 1983). Gular flutter was absent in some species studied (e.g., gulls [*Larus*], frogmouths, and passerines [Lasiewski, 1972; W. R. Dawson, unpublished observations]). The flutter, which is produced by alternate flexure and relaxation of the hyoid apparatus, appears to have two advantages over thermal tachypnea. First, the air movement only affects surfaces that do not participate in gas exchange, so problems with hypocapnia and alkalosis are avoided. Second, the energetic cost of fluttering the gular area is probably in most cases considerably less than that of moving the larger thoracoabdominal structure, leading to greater cooling efficiency.

Goatsuckers appeared especially proficient at gular fluttering, which contributed to the ability of such birds as the common poor-will and the spotted nightjar (*Eurostopodus guttatus*) at very high T_a to attain rates of evaporative cooling that were more than 3× their concurrent rates of heat production (Lasiewski, 1969; Dawson and Fisher, 1969). Relatively low metabolic rates at high T_a assisted in this performance. During heat challenges, the estimated proportions of total evaporative water loss accounted for by gular flutter in Japanese quail and by buccopharyngeal ventilation in the domestic fowl were 20% and up to 35%, respectively (Weathers and Schoenbechler, 1976; Brackenbury *et al.*, 1981b). Results for the common poor-will suggested that gular flutter may contribute more than half of total evaporative water loss at T_a above 39.5°C (Lasiewski, 1972).

Frequencies of gular flutter range from 176 to over 1000 cycles/min, depending on the species (cf. Table VI in Dawson and Hudson, 1970). In some species (e.g., quail), the frequency rose with increasing heat load, but in others (e.g., pelicans, cormorants, goatsuckers, roadrunner, pigeons, doves, owls, and mousebirds) only a narrow band of temperature-independent frequencies was used (Lasiewski, 1972). Bartholomew *et al.* (1968) suggested that these frequencies match the resonant frequencies of the birds' respective hyoid apparati. However, direct evidence for this appears lacking. Such an arrangement would allow operation of this structure at a low energetic cost, thereby contributing to cooling efficiency. Whatever the case, restriction of flutter frequencies to a narrow range means that adjustment of the contribution of gular flutter to evaporative cooling must be achieved primarily by altering the duration of bouts of fluttering, the amplitude of the gular movements, and/or the area of the buccopharyngeal region involved.

In birds that respond to heat loads with both gular flutter and thermal tachypnea, the frequencies of the two movements can be very dissimilar (Lasiewski, 1972). For example, in cormorants, cattle egret (*Bubulcus ibis*), and common poor-will, the frequency of gular flutter exceeded that of thermal tachypnea (e.g., one 12-day-old cattle egret panted and fluttered at ≤81 and 895–965 cycles/min, respectively, Hudson *et al.*, 1974). Perhaps the difference in frequencies employed reflects in part the substantial difference in mass between the thoracic cage and the gular–hyoid structure. However, panting and flutter frequencies are similar in some birds (e.g., roadrunners, pigeons, doves, owls, and blue-breasted quail [*Coturnix chinensis*]), and it has been suggested that they are linked with the resonant properties of the thoracoabdominal region (Bartholomew *et al.*, 1968). Again, direct experimental evidence concerning such an association appears lacking.

In the response of several species to heat challenges, gular flutter commenced before thermal tachypnea. However, the sequence was reversed in the rock pigeon and barn owl (*Tyto alba*). This was also the case for the 12-day-old cattle egret studied by Hudson *et al.* (1974) in which gular flutter began at slightly higher cloacal temperatures than did thermal tachypnea.

c. Site of Evaporative Cooling

The principal sites of the evaporation produced by thermal tachypnea are the nasal, buccopharyngeal, and upper tracheal regions, with some possibility of participation of the walls of the air sacs in the ostrich (Lasiewski, 1972). The principal site for that resulting from thermal tachypnea in the domestic fowl is also well anterior, in the upper respiratory tract (Figure 5). Blood flow to the upper respiratory tract increased during thermal panting by the Pekin duck *Anas platyrhynchos* (Bech and Johansen, 1980). Gular flutter permits evaporation from the moist surfaces of the buccal regions and the upper digestive tract (portions of the pharynx and anterior esophagus [Lasiewski, 1972]). As noted in Section VII,B,3,b, rhythmical inflation of the esophagus contributed significantly to the evaporative cooling of doves (Gaunt, 1980; Baumel *et al.*, 1983). In these birds, a collar plexus of subcutaneous veins that appeared to facilitate heat exchange is associated with this portion of the digestive tract (Baumel *et al.*, 1983).

d. Osmotic State and Respiratory Evaporative Cooling

In addition to thermal factors, the onset and extent of thermal tachypnea are affected by the osmoregulatory state of the animal, with dehydrated individuals tending to delay its appearance until higher T_b during heat challenges (see, for example, Bartholomew and Dawson, 1954). Trained rock pigeons abandoned thermal tachypnea in favor of instrumental cooling, which provided them with cool air on demand, following water deprivation or intravenous or intracarotid infusion of hyperosmotic salt loads (Brummermann and Rautenberg, 1989). Also, water deprivation resulting in a 16% loss of body mass after 48 hr produced a shift from cutaneous to respiratory evaporative cooling in rock pigeons. The greater energy cost and lower output of water with the latter activity led to a lowered ratio of heat lost by evaporation to heat produced (Arad *et al.*, 1987). During exposure to heat f_{resp} and evaporation were reduced in the dehydrated ostrich relative to values for normal individuals (Crawford and Schmidt-Nielsen, 1967). This was also the case in domestic fowl (Arad, 1983) with an ~15% mass loss due to dehydration. These birds showed a relative hyperthermia, but acid–base balance was not impaired. Interestingly, water deficits from heat stress or flight involving mass losses of 13–15% did not reduce plasma volume appreciably in white-necked ravens (*Corvus cryptoleucus*), rock pigeons, or Japanese quail. Carmi *et al.* (1983) speculated that such conservation of plasma volume is an adaptation originally involved in dealing with the high heat loads incurred by birds during vigorous flight.

e. Respiratory Water and Heat Conservation

As birds inspire, the air is warmed nearly to T_b and saturated with water vapor while still in the nasal passages. Heat is removed from the nasal mucosa in the conversion of water from a liquid to vapor state and the mucosal temperature falls, in some cases to a level cooler than the initial temperature of the inhaled air. Expired air, essentially saturated with water vapor, leaves the lungs and air sacs at T_b. Passage over the cool nasal surfaces during exhalation cools this air and a portion of its water content condenses on the mucosa.

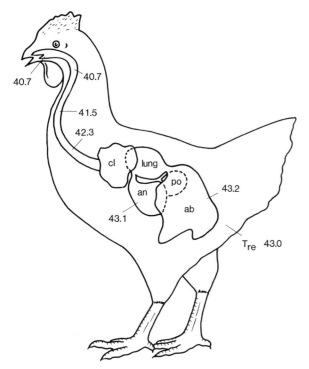

FIGURE 5 Respiratory surface temperatures (°C) of the domestic fowl during thermal panting. The air temperature was 43.0°C and the respiratory frequency 141 breaths/min. Abbreviations: cl, clavicular; an, anterior thoracic; po, posterior thoracic; ab, abdominal air sacs; T_{re}, rectal temperature. (Reprinted from *Respir. Physiol.* **25**, B. Menaum and S. A. Richards, Observations on the sites of respiratory evaporation in the fowl during thermal panting, pp. 39–52, Copyright 1975, with permission from Elsevier Science.)

For example, at 15°C and 25% relative humidity, cactus wrens (*Campylorhynchus brunneicapillum*) recovered three-quarters of the water and heat added to the inspired air. The heat conserved equaled 16% of the heat production of this bird at 15°C (Schmidt-Nielsen *et al.*, 1970). The mechanism also operates in aquatic birds, with penguins (*Pygoscelis* spp.) reclaiming on exhalation more than four-fifths of the water and heat added to inspired air (Murrish, 1973). The total amount of heat (evaporative and nonevaporative) recovered in the nasal passages equaled 17% of the metabolic heat production of these animals. Brent *et al.* (1984) reported that the respiratory heat loss from the European coot (*Fulica atra*) at a T_a of −25°C was only 9.6% of the theoretical maximum, largely as a result of cooling of the expired air. Such cooling also leads to substantial caloric savings in the prairie falcon *Falco mexicanus* (Kaiser and Bucher, 1985). The nasal passages of all these birds are regarded as countercurrent heat exchangers in which flow is separated temporally rather than spatially (Schmidt-Nielsen *et al.*, 1970). This action of the nasal passages is particularly effective below the lower critical temperature, where avian evaporative water loss tends to decline with decreasing T_a despite the opposite trend in metabolism (see Dawson, 1982). Overriding of this arrangement is necessary at higher T_a where respiratory evaporative cooling is required, and this is accomplished by a shift to the open-mouth respiration characterizing thermal panting.

Some other factors act to minimize heat loss from the respiratory tract during exposure of birds to cold (see Johansen and Bech, 1984, and Marsh and Dawson, 1989, for discussion). Their distinctive respiratory tract allows given levels of oxygen extraction with substantially lower levels of ventilation than in mammals of comparable size (Bernstein and Schmidt-Nielsen, 1974; Bech and Johansen, 1980) and this should place them in a more favorable situation regarding restriction of respiratory heat loss. Additionally, certain birds can increase the efficiency of such extraction at cooler T_a, allowing a lower respiratory minute volume than would otherwise be required (Bucher 1981, 1985; Brent *et al.*, 1983, 1984; Bech *et al.*, 1984). This is the case in the giant petrel (*Macronectes giganteus*) in which increasing oxygen demand in the cold was accommodated mainly by increasing extraction efficiency, with an increase in V_T being of secondary importance (Morgan *et al.*, 1992). The black-legged kittiwake (*Rissa tridactyla*) also altered its breathing pattern in the cold in a manner that contributed to heat conservation (Brent *et al.*, 1983). However, not all birds display these capacities (Kaiser and Bucher, 1985; Chappell and Bucher, 1987; Morgan *et al.*, 1992). For example, oxygen extraction in the rock pigeon was independent of both T_a and spinal cooling (Bouverot *et al.*, 1976; Barnas and Rautenberg, 1984; Bech *et al.*, 1985).

4. Acclimation and Acclimatization

Acclimation and acclimatization have a number of effects on birds affecting their responses to heat stress. Secretory unit densities in the lateral nasal glands and arteriovenous anastomoses in the nasal mucosa were significantly greater in the rostral nasal conchae of domestic fowl exposed to heat for 4 hr/day over 2 months than in control birds. Midtgård (1989a) suggested that these differences reflected an increased capacity for evaporative cooling from the nasal mucosa of the former birds. The rate of total evaporative water loss at high T_a by heat-acclimated rock pigeons was significantly lower than that by unacclimated individuals (Marder and Arieli, 1988). Fully heat-acclimated pigeons relied on cutaneous evaporative cooling even at T_a of >60°C. In a subsequent study, Marder (1990) noted that some heat-acclimated individuals panted at 55–60°C. At these higher T_a both nonpanting and panting individuals regulated blood pH at normal levels (7.544 ± 0.011 [SD] and 7.531 ± 0.022, respectively) accompanied by a slight hypocapnia (Pa_{CO_2} = 24.8 ± 4.0 and 23.8 ± 2.49 torr [3.31 ± 0.53 and 3.17 ± 0.33 kPa], respectively). Birds adjusted to lower T_a panted vigorously on exposure to 50°C and underwent a severe hypocapnia (Pa_{CO_2} = 9.1 ± 2.52 torr [1.21 ± 0.34 kPa]) and alkalosis (pH = 7.702 ± 0.048). Thirteen, 4- to 6-hr/day exposures to 50°C significantly improved the capacity of these panting individuals to maintain an almost normal acid–base balance. In reporting these results, Marder (1990) suggested that acclimation to high T_a (50–60°C) is needed for fine adjustment of the competing needs for heat dissipation, pulmonary gas exchange, and acid–base regulation in heat-challenged rock pigeons.

Heat treatment of neonatal domestic fowl has persistent effects on heat resistance, which may involve different mechanisms than those operating in the acclimation/acclimatization of adults (Arjona *et al.*, 1990). At 43 days after hatching, broiler cockerels that had been exposed to 35–37.8°C for 24 hr at 5 days of age showed significantly lower mortality at high T_a than controls that had not received neonatal heat exposure.

C. Partition of Heat Loss

Several investigators have measured the heat loss from domestic fowl using a gradient-layer calorimeter, a device that partitions the heat loss from an animal among convection, radiation, and evaporation (Roller and Dale, 1962; Deshazer, 1967). Most of the heat loss at low T_a occurs by nonevaporative means; convection or radiation is the major pathway, depending on the rate of air movement. The importance of evaporation increases as T_a rises and at or above 40°C it may account for virtually all of the heat loss (see Table 3).

VIII. HEAT EXCHANGE UNDER NATURAL CONDITIONS

In recent years, thermal physiology has expanded from being primarily concerned with responses of organisms under controlled conditions in the laboratory to include attempts at analyzing their heat exchange in natural environments. Natural environments present very complicated situations involving fluctuations in a number of thermally relevant physical variables. Although air temperature is the principal one considered in laboratory experiments, it cannot adequately describe the situation outdoors, where complex radiational and convective conditions exert substantial effects on heat exchange. For example, in a broad-tailed hummingbird (*Selasphorus platycercus*) sitting on its nest at T_a of 0–4.6°C, radiation comprised an estimated 9–35% of total heat loss while convection accounted for 44–46% (Calder, 1973).

Recently, the "standard operative temperature" (T_{es}; see Bakken, 1992, for discussion) has been used as an index of heat flow between the bird and its surroundings in nature (see Dawson and O'Connor, 1996, for examples). The T_{es} of a particular natural environment equals the air temperature in a black body environment under standardized convective conditions that produces the same heat flux to or from the animal. Two approaches are used in determining this index. One is based on measurements of thermally relevant environmental variables (e.g., air temperature, wind speed, and short- and long-wave radiation) as well as pertinent animal properties (e.g., body size, shape, orientation, and surface reflectivity to short- and long-wave radiation). The practical difficulties with this approach, which relate primarily to the complexity of radiative heat transfer through animal coats have been discussed by Walsberg and Wolf (1996). The other technique for determination of T_{es} involves use of taxidermic mounts. For birds or mammals, these are typically hollow metal casts of an animal's body covered by its integument. Operative temperature (T_e), which represents the sum of T_a and a temperature increment or decrement produced by convective and radiative factors, is obtained from the temperature of the mounts following thermal equilibration in a particular environment. It is used in calculating standard operative temperature employing mathematical models incorporating information on factors such as wind speed. In some cases, more elaborate taxidermic mounts are used, which are not only covered with the integument of the bird or mammal of interest, but also fitted with internal electrical heaters that are thermostatically controlled to maintain temperature near that of the live animal. The advantage of these more elaborate devices is that they can, with suitable calibration, be used to determine T_{es} directly (Bakken *et al.*, 1981, 1983). They, in effect, serve as T_{es} thermometers that greatly simplify requirements for micrometeorological data. However, Walsberg and Wolf (1996) have cautioned that the adequacy of taxidermic mounts for predicting physiological responses of animals varied widely in their study with the species, type of mount, and environmental conditions. They noted further that individual variation in the characteristics of the mounts required that multiple ones should be used and that they require careful calibration under the range of conditions, including appropriate radiant fluxes actually encountered in the field by the animal under study. In one noteworthy study, Buttemer (1985) showed by use of heated taxidermic mounts that American goldfinches (*Carduelis tristis*) reduced their nocturnal thermoregulatory costs as much as 19% in winter by roosting in sheltered locations in spruce trees (*Picea pungens*). Dawson and O'Connor (1996) have discussed other examples involving the use of heated taxidermic mounts.

IX. BEHAVIORAL THERMOREGULATION

A. Introduction

Behavior plays an important role in the thermal relations of birds, often allowing them to save energy, conserve water, and generally reduce thermal stress. It typically involves some form of movement leading to a change in position, posture, or orientation to wind and/or sun, though in some instances remaining quiescent is an effective means of minimizing thermal stress. The most extreme example of thermally significant movement is long-distance migration, which allows many species to divide their time between seasonally productive middle- and high-latitude environments for breeding and areas with milder winters.

B. Behavior Reducing Heat Loss or Facilitating Heat Gain

In cold environments, postural adjustments contribute to heat conservation by birds. The domestic fowl reduced its surface area and hence its heat loss by "hunching." The head, especially in small birds and near the eyes, tends to be a region of substantial heat loss (Veghte and Herreid, 1965; Hill *et al.*, 1980). Birds sleeping in the cold often tuck the head beneath the scapular feathers on the back for protection. Heat loss from unfeathered feet and tarsi can also be reduced by as much as 20–50% by the bird's squatting and enclosing them within its ventral contour feathers. Emperor penguins (*Aptenodytes foresteri*), in addition to squatting, minimized conductive heat loss to the ice by supporting themselves on their tarsometatarsal joints at rest (LeMaho *et al.*, 1976). Huddling is an effective means of reducing heat loss and it was extremely important in helping the penguins, fasting on their nesting grounds, to cope with the Antarctic winter (LeMaho *et al.*, 1976; LeMaho, 1983). A variety of smaller birds pass cold winter nights in more or less tight clusters (Löhrl, 1955; MacKenzie, 1959; Robertson and Schnapf, 1987; Thaler, 1991). Such huddling produced significant energy savings in small passerines (Brenner, 1965; Chaplin, 1982). However, despite its energetic advantages, this activity actually contributed to increased mortality among swallows, presumably due to their jamming themselves together in old nests during cold, inclement weather (Weatherhead *et al.*, 1985).

During cold, windy periods, birds may seek shelter (for example, see Watson, 1972; Grubb, 1975, 1978) or avoid ruffling their feathers by facing into the wind (see Kessel 1976). Willow tits (*Passer montanus*) moved downward and inward in foliage during windstress, an option denied them in the presence of crested tits (*Parus cristatus*), which were dominant (Lens, 1996). Both small birds such as black-capped chickadees (*Parus atricapillus*) and larger ones such as partridge (*Perdix perdix*) used protected roost sites at night during very cold weather, and they sometimes remained in these roosts well into the next day (Delane and Hayward, 1975; Kessel, 1976). Nocturnal roost sites include dense foliage, cavities in trees, and other structures such as nests. The large communal nests of sociable weavers (*Philetairus socius*) in the deserts of southern Africa provide a striking example of the last (White *et al.*, 1975). Verdins construct winter nests which substantially reduced their overnight energy expenditures (Buttemer *et al.*, 1987). Walsberg's (1990) observations of 15 black-tailed gnatcatchers (*Polioptila melanura*, ~6 g) roosting in one of these nests has provided a further illustration of use of shelter and huddling in dealing with cold. Under particularly severe northern winter conditions, small songbirds such as goldcrests (*Regulus regulus*), tits (*Parus* spp.) redpolls (*Carduelis* spp.), and bullfinches (*Pyrrhula pyrrhula*) may spend the night in cavities within the snow (references in Marsh and Dawson, 1989). The entrances to these cavities were typically left open (Sulkava, 1969), in contrast to the pattern in grouse, which closed their snow burrows (for example, see Irving, 1972; Korhonen, 1980). The use of subnivian shelter provides birds with somewhat more moderate T_a and protection from wind and the cold night sky, thereby allowing them to save energy that would otherwise be expended in thermogenesis. Small birds probably must still endure T_a below thermal neutrality (Korhonen, 1981), but the T_a in the snow burrows of grouse such as the willow ptarmigan, capercaillie, and black grouse appeared to be near or even above their respective T_{lc} (Korhonen, 1980; Marjakangas *et al.*, 1984).

Under sunny conditions, birds may reduce requirements for regulatory thermogenesis by "sunbathing." Such behavior typically involves maximizing the surface area of the body exposed to the sun. Sunbathing roadrunners oriented the back perpendicularly to the incoming sunlight, drooped and held their wings slightly away from the body, and erected the cervical plumage, consequently exposing the black skin of the interscapular apterium and the soft black plumage of the dorsal spinal tract. This allowed the birds under simulated solar radiation to maintain SMR at T_a as low as 9°C, well below their normal T_{lc}. Sunbathing also allowed hypothermic roadrunners to rewarm themselves at a reduced metabolic cost (Ohmart and Lasiewski, 1971). In another example, the heat production of white-crowned sparrows at low T_a in the laboratory was reduced in birds exposed to simulated solar radiation (Figure 6; De Jong, 1976), further indicating the energy savings that can be effected by the absorption of solar heat. Finally, anhingas (*Anhinga anhinga*) spread their wings much more frequently in cool than in warm weather (Henneman,

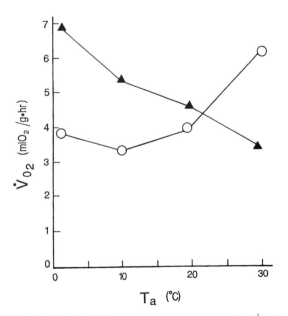

FIGURE 6 Relationship between oxygen consumption ($\dot{V}_{O_2}$) and air temperature (T_a) in the white-crowned sparrow (*Zonotrichia leucophrys*) with (○) and without (▲) solar radiation. (Redrawn from De Jong, 1976.)

1982). In view of the low metabolic heat production and high thermal conductance of these birds, the absorption of solar radiation by the spread wings may be an important element in their heat balance equation. Unfortunately, much confusion exists in the avian literature between birds that are sunbathing and those that are manifesting heat-stress behavior (see following section).

C. Behavior Facilitating Heat Loss or Reducing Heat Gain

Birds in hot environments frequently minimize their heat loads and requirements for evaporative cooling by seeking shade and remaining inactive in the shelter of rocks or vegetation during the most stressful periods (Dawson, 1984a). These behaviors are well illustrated by a verdin and several black-tailed gnatcatchers, which persistently selected cooler shaded microsites during the heat of summer days in the Arizona desert (Wolf *et al.,* 1996). The importance of such behavior in the verdin has been confirmed by laboratory analysis (Wolf and Walsberg, 1996b). Behavioral thermoregulation in the heat is also evident in the Bedouin fowl (*Gallus gallus*), which sought shade during the middle of hot days in the Negev Desert and became prone in contact with cooler soil (Marder, 1973b). Pursuit of cooler surroundings doesn't always proceed at ground level. Some large raptors dealt with intense heat by thermal soaring to heights where the air was substantially cooler owing to adiabatic expansion (Madsen, 1930). Many tropical seabirds face extremely demanding conditions by nesting in exposed situations where the need to protect eggs and young from intense solar radiation can impose substantial heat burdens. The birds augment their physiological responses by several behavior adjustments. For example, the masked booby (*Sula dactylatra*) faced away from the sun, thereby placing its feet and gular area in the shade of its body. Additionally, the wings were held away from the body and the scapular feathers elevated so that convective heat loss was facilitated (Bartholomew, 1966). This posture was reminiscent of that of sunbathing birds (see Section IX,B). The incidence of a similar thermal posture in great frigatebirds (*Fregata minor*) correlated well with the equivalent temperature

FIGURE 7 A juvenile Laysan albatross (*Diomedea immutabilis*) on Tern Island, French Frigate Shoals in the Northwestern Hawaiian Islands. The bird still has vestiges of its juvenile plumage around the head and neck. Note that it has its bill agape, is panting, and that the scapular feathers are raised and ruffled by the breeze. The juvenile's back is to the sun so that the feet, which are raised off the ground, are in the shade of its own body. (Photo: G. C. Whittow.)

of the environment, an index that takes into account the size, shape, and color of the bird (Mahoney et al., 1985). The albatrosses *Diomedea nigripes* and *Diomedea immutabilis*, which breed at low latitudes, also increased convective heat loss on their nesting grounds by pivoting on their heels and raising their large webbed feet off the ground in the shade of their bodies (Figure 7; Howell and Bartholomew, 1961; Whittow, 1980). Incubating Heermann's gulls (*Larus heermanni*) displayed an elaborate suite of postural adjustments that contributed either to heat conservation or dissipation, depending on environmental conditions. These adjustments were closely correlated with soil temperature, which apparently served as a metaphor for the intensity of solar radiation (Bartholomew and Dawson, 1979).

D. Behavioral Augmentation of Evaporative Cooling

Birds may behaviorally augment evaporative cooling in a variety of ways. Heat-challenged wood storks (*Mycteria americana*) and New World vultures resorted to the gross expedient of directing liquid excrement onto the bare portions of their legs ("urohidrosis"), which were cooled by the evaporation of the water in this material (Kahl, 1963; Hatch, 1970). The domestic fowl achieved a similar result in a more genteel way by splashing water over its comb and wattles before initiating physiological responses to heat (Horowitz et al., 1978). Some waders, e.g., the black-necked stilt (*Himantopus mexicanus*), nesting under the extremely hot conditions around the Salton Sea, California, cooled not only themselves but their eggs by wetting the abdominal feathers, i.e., "belly soaking" (Grant, 1982). The Egyptian plover (*Pluvianus aegyptius*) buries its eggs—and its chicks—in the sand, achieving both concealment and cooler temperatures. The parents also wet their abdominal feathers, which in turn moistened the sand at the nest site, cooling the eggs or chicks in the process (Howell, 1979). Bathing also helps a number of other birds to remain cool in hot environments (Dawson and Bartholomew, 1968).

Some plasticity is evident in avian thermoregulatory behavior. Ring doves under laboratory conditions could be trained to press a switch that turned on a heater. By this means, the birds controlled their environmental temperature, maintaining it at an average 32.4°C, which is within their TNZ (Budgell, 1971). Chukars (*Alectoris chukar*) increased their average time in a warm or cold chamber as chamber temperature approached a neutral or preferred T_a between 25.9° and 31.9°C (Laudenslager and Hammel, 1977). Though they did not avail themselves of a source of heat in a cold laboratory environment, domestic fowl could be trained to actuate a source of cold air in a hot environment (Horowitz et al., 1978).

Rock pigeons also were capable of this behavior and, when osmotically stressed, preferred such instrumental cooling over thermal tachypnea during heat challenges (Rautenberg et al., 1980). Body temperature actually fell in certain heat-stressed individuals as they increased their reliance on such cooling following salt loading or water deprivation (Brummermann and Rautenberg, 1989; see Section VII,B,3,d).

X. CONTROL MECHANISMS

A. Thermosensitivity and Control of Physiological Thermoregulatory Responses

Evidence exists for the presence in birds of both peripheral temperature receptors and temperature-sensitive elements deeper within the body. Within the central nervous system, indications exist that thermoresponsiveness of both birds and mammals may involve temperature-dependent synaptic transmission as well as intrinsically thermoresponsive central neurons (Simon, 1989). In attempts to establish functional roles of thermally sensitive central components, it is important to distinguish between nonspecific temperature sensitivity and responsive capabilities that actually contribute to thermoregulation (Simon et al., 1977).

1. Peripheral Thermoreception

Shivering could be detected in domestic fowl exposed to a cold environment before any change occurred in deep T_b (Randall, 1943). This result is consistent with the involvement of peripheral thermoreceptors in the control of heat production. Birds appear well provided with such receptors in the skin, tongue, and beak. Additionally, Necker (1977) has reported a high degree of responsiveness to heating and cooling of feathered skin areas, especially the upperparts, of the rock pigeon. On the other hand, the beak and unfeathered areas of the legs and feet were relatively insensitive. However, Bech et al. (1988) found that selectively varying the temperature of air inspired by rock pigeons in 5° and 25°C environments elicited appropriate changes in heat production and, especially at the higher temperature, in vasomotor state. The results suggested that peripheral thermoreceptors important for the initiation of thermoregulatory effector mechanisms existed in the head region. Thermosensory functions in peripheral areas have been documented by neural recording and confirmation of appropriate thermoeffector responses to local heating and cooling. Neural recording has identified characteristics similar to those of mammalian thermoreceptors in various warm- and cold-sensitive peripheral units of birds (Kitchell *et al.*, 1959; Leitner and Roumy, 1974; Necker, 1972, 1973). These units occurred in the proportion of 3:1 for cold- to warm-sensitive units, respectively (Necker, 1973). Some of these thermosensors, e.g., those in the brood patch and beak, may be more important in mediating behavioral and physiological responses affecting egg and nest temperature than in control of T_b in the incubating parent (Rautenberg, 1986). The response to stimulation by deep body thermoreceptors can be modified in some species by local heating or cooling of the skin (Rautenberg, 1971; Inomoto and Simon, 1981). Furthermore, Eissel and Simon (1980) have cited results for Pekin ducks showing changes in discharge frequency of certain thermally sensitive hypothalamic neurons when the deep body temperature and/or the temperature of the web of the foot were altered, but the hypothalamic temperature remained constant. The importance of peripheral inputs is further suggested by the fact the heat production of the goose (*Anser anser*) increased when T_b was lowered, but that the threshold T_b for this was reduced at higher T_a (Helfmann *et al.*, 1981). Nevertheless, integumentary receptors are usually regarded as playing only a subsidiary role in the physiological regulation of deep T_b in birds (Rautenberg, 1986; Simon *et al.*, 1986).

2. Role of the Hypothalamus in Physiological Thermoregulatory Control

The situation regarding central control of physiological thermoregulation is more complex. Cooling the preoptic/anterior hypothalamic area (PO/AH) elicited appropriate, though relatively weak, thermogenic responses in emus (Jessen *et al.*, 1982). This was also the case in willow ptarmigan, *Lagopus lagopus*, subjected to moderate hypothalamic cooling at a T_a of $-10°C$ (Mercer and Simon, 1987). However, reduction of hypothalamic temperatures below 34°C caused a fall in T_b due to an inhibition of shivering responses. A somewhat stronger increase in heat production occurred in the house sparrow with hypothalamic cooling (Mills and Heath (1972a), but the possibility that midbrain, spinal, or other deep body receptors were affected by the treatment in this relatively small bird could not be excluded.

The preceding indications of some appropriate hypothalamic thermoresponsiveness in birds appear to be the exception rather than the rule. In contrast to the case for mammals, hypothalamic cooling in the rock pigeon (Rautenberg *et al.*, 1972); California quail *Callipepla californica* (Snapp *et al.*, 1977); Pekin duck (Simon-Oppermann *et al.*, 1978; Martin *et al.*, 1981; Bech *et al.*, 1982b); Adélie penguin (Simon *et al.*, 1976); and domestic fowl (Avery and Richards, 1983) was either ineffective in eliciting shivering, or, paradoxically, inhib-

ited this process. In some cases at cold or neutral T_a, hypothalamic cooling produced no or inappropriate vasomotor responses (Simon *et al.*, 1976; Simon-Oppermann *et al.*, 1978; Mercer and Simon, 1987). Hypothalamic cooling at neutral and higher T_a also elicited inappropriate effects on panting and other thermolytic activities in the Adélie penguin and Pekin duck (Simon *et al.*, 1976; Simon-Opperman *et al.*, 1978; Simon *et al.*, 1981). Additionally, cooling of the rostral brain stem in rock pigeons engaged in intermittent panting at warm T_a resulted in transient increases in f_{resp} while hypothalamic heating led to a similarly inappropriate temporary suppression of this function (Schmidt, 1976a). On the other hand, weak appropriate heat defensive responses to warming the rostral brain stem at warmer T_a have been reported in the Adélie penguin, Pekin duck, California quail, and domestic fowl (Simon *et al.*, 1976; Simon-Opperman *et al.*, 1978; Snapp *et al.*, 1977; Avery and Richards, 1983). Rautenberg *et al.* (1972) also found that cooling the brainstem of the rock pigeon led to appropriate ptiloerection and vasoconstriction (see also Section X,A,4). Furthermore, the extent of ptiloerection by ring doves appeared to be under the proportional control of hypothalamic temperature, with the slope of the inverse relation reduced by water deprivation and increased by inanition (McFarland and Budgell, 1970). In contrast to such proportional arrangements, a small number of what are referred to as "temperature guardian neurons" have recently been identified in the preoptic area of the hypothalamus of the Muscovy duck (Basta *et al.*, 1997). These are exclusively sensitive to temperatures of ~ 36.1 or ~42.3°C. Perhaps they serve as "limit switches" that contribute to maintaining brain temperature within this range.

That certain birds differ from mammals in their response to changes in hypothalamic temperature is not explained by discrepancies in thermal sensitivity within the rostral brain stem. Some of the warm-responsive neurons in this region in the Pekin duck did not differ with regard to frequency range, temperature coefficient, or Q_{10} from so-called "high Q_{10} units" in the mammalian rostral brainstem to which hypothalamic thermosensory function has been ascribed (Eissel and Simon, 1980). Moreover, warm- and cold-responsive units in the PO/AH region of Pekin ducks were in similar proportions to those in mammals, as shown both *in vivo* (cats, rabbits, and dogs; Simon *et al.*, 1977) and *in vitro* (rats; Nakashima *et al.*, 1987). *In vitro* comparison of PO/AH neuronal thermoresponsiveness in rats and ducks revealed no differences at the single unit level which might account for the divergent contributions of the avian and mammalian hypothalami to deep body temperature perception. A tentative explanation for the difference (Eissel and Simon, 1980; Schmidt and Simon,

1982; Lin and Simon, 1982) attributes many of the paradoxical effects of local temperature change in the avian hypothalamus to differential Q_{10} effects on hypothalamic neurons (cold-sensitive units with higher Q_{10} than warm-sensitive ones) participating in receiving cold and warm thermosensory inputs from other regions of the body and in controlling thermoregulatory activities. Simon (1989) has provided the overview that birds share with mammals the general temperature dependence of central signal transmission and the existence of specific temperature receptors at multiple sites of the skin, deep body tissues, and the central nervous system. These two classes of animals appear to differ only quantitatively in the distribution of thermoreceptive elements along the central neural axis. In mammals, receptor density at the hypothalamic level seems high and coupling of their signals to efferent control of effectors generally tight. In birds, on the other hand, experimental evidence suggests a low density of thermoreceptor elements, with less tight and variable coupling with thermoregulatory effectors. These avian arrangements are thought to produce the inappropriate physiological thermoregulatory effects observed, particularly with hypothalamic cooling.

Whatever the extent of its direct role in physiological thermoreception, the avian hypothalamus appears prominently involved in the control of behavioral thermoregulation (see Section X,B). Moreover, it contributes to physiological thermoregulation in a variety of ways, judging by both physiological and pharmacological observations. Thermoregulation is disrupted by lesions in the anterior hypothalamus in the domestic fowl (Rogers and Lackey, 1923; Feldman *et al.*, 1957; Lepkovsky *et al.*, 1968), house sparrow (Mills and Heath, 1972b), and Pekin duck (Hagan and Heath, 1980). Additionally, despite the sensitivity of the rock pigeon's lower cervical and thoracic spinal cord to heating and cooling (see Section X,A,4), appropriate thermoeffector responses require that neural information ascend to higher centers. Cold blockage of neural transmission in the cervical spinal cord prevented shivering in response to cooling of more caudal segments of the cord (Rautenberg *et al.*, 1972). However, some degree of postcranial control of effector output does occur in spinal rock pigeons, which responded to spinal cooling by shivering, though this activity was weak and relatively uncoordinated in comparison to that of intact birds (Görke and Pierau, 1979). Intracellular recording in the cervical spinal cord of the rock pigeon has revealed the presence of temperature-sensitive ascending neurons (Necker, 1975), whose destinations may include the hypothalamus. Also, as noted previously, direct neural recording in the hypothalamus of this species and the Pekin duck has detected neurons that respond to changes of temperature in other parts of the body including the spinal cord (Rosner, 1977; Eissel and Simon, 1980; Lin and Simon, 1982).

As to further indications of the involvement of the rostral brainstem in avian thermoregulation, it is important to note that thermoeffector activity and discharge rates of warm- and cold-responsive neurons in the PO/AH area of birds can also be affected by application of various neuroactive compounds such as amines, prostaglandins, and other substances, some of which are putative avian neurotransmitters. Compounds such as norepinephrine (NE) and epinephrine (E) have been found in significant concentrations in the avian brain (Juorio and Vogt, 1967). Hissa (1988) provided a summary of the thermoregulatory effects of central application of several of these, concluding that injection of such substances as NE, 5-hydroxytryptamine (5-HT), dopamine (DA), and acetylcholine (ACh) into the preoptic/anterior hypothalamic area of the rock pigeon brain overall causes hypothermia at T_a below thermoneutrality. More specifically, injections of DA, E, NE, 5-HT, and ACh into the anterior hypothalamus of this bird produced an inhibition of shivering, a decrease in heat production, and peripheral vasodilation leading to a reduction in T_b in cold-exposed individuals (Hissa and Rautenberg, 1974, 1975; Hohtola *et al.*, 1989). Regarding the suppression of shivering in the rock pigeon, the order of potency of anterior hypothalamic injections of putative neurotransmitters was NE > DA > ACh > 5−HT (Hohtola *et al.*, 1989). Injection of ACh, 5-HT, and NE into the third ventricle near the anterior hypothalamus produced mostly hypothermic effects in domestic fowl (Hillman *et al.*, 1980). Some of the neural actions of these compounds were investigated by Sato and Simon (1988). Their microiontophoretic application of ACh, NE, and 5-HT singly or in combination to the PO/AH area of conscious Pekin ducks led to a variety of effects on the activity of warm- and cold-responsive units, which represented 17 and 20%, respectively, of 355 neurons evaluated. Acetylcholine more consistently stimulated cold units, whereas NE inhibited and 5-HT stimulated the majority of warm units (Sato and Simon 1988).

Norepinephrine, 5-HT, and ACh generally produce hypothermia when injected into the avian posterior hypothalamus (Hissa, 1988). On the other hand, injection of 5-HT into this region in cold-challenged rock pigeons led to a brief bout of hyperthermia (Pyörnilä and Hissa, 1979), accompanied in most cases by a fall in foot temperature indicative of vasoconstriction.

Prostaglandins (PG) are also known from the avian brain (Horton and Main, 1967). Hyperthermia in rock pigeons and domestic fowl resulted from experimental introduction of PGA_1, PGE_1, or PGE_2, into the third ventricle or the hypothalamus (Nisticò and Marley, 1973, 1976; Hissa *et al.*, 1980), with the effectiveness of a given dose being greater at the latter site (Nisticò and Marley, 1976). Macari *et al.* (1993) also reported a fever response to PGE_2 in domestic fowl. Prostaglandin F_{2a} lowered T_b in the fowl (Nisticò and Marley, 1976), but raised it in the rock pigeon (Hissa *et al.*, 1980). It has been suspected that PGE_1 and E_2 might be involved as mediators of pyrogen fever (Nisticò and Marley, 1976). However, administration of *Escherichia coli* lipopolysaccharide produced a rise in T_b but no increase in PGE_2 production *ex vivo* in domestic fowl (Fraifeld *et al.*, 1995).

Injection of arginine vasotocin (AVT) and angiotensin II (ANG II) into the preoptic/anterior hypothalamic area or intravenously has been shown to have thermoregulatory effects in rock pigeons. The AVT reduced shivering, $\dot{V}_{O_2}$, and T_b at 2°C, but was without effect at 32°C. On the other hand, these variables, and foot skin temperature were increased by injections of ANG II (Hassinen *et al.*, 1994).

Taken together, the various observations on hypothalamic function in birds support the view that, in terms of temperature regulation, this region of the brain should primarily be considered an integrator of inputs from the spinal cord, brain, and periphery, rather than the sole controller of thermoregulatory response (see, for example, Barnas and Rautenberg, 1987).

3. Contributions of Other Areas of the Brain to Physiological Thermoregulatory Control

Functionally relevant thermoeffector responses to changes in temperature of more posterior regions of the avian brain are also evident. Appropriate thermogenic and thermolytic responses were produced by respectively cooling and heating the lower mid brain/pontine (or rostral rhombencephalic) region in the Pekin duck (Simon-Oppermann and Martin, 1979; Martin *et al.*, 1981; Simon *et al.*, 1981; Bech *et al.*, 1982b). Furthermore, the domestic goose showed a strong thermogenic response when its entire brain was cooled (Helfmann *et al.*, 1981).

As previously noted, the avian hypothalamus appears to have only a minimal role in control of panting, which Richards (1975) described as "facilitative" (Barnas and Rautenberg, 1987). Inputs to the respiratory center from other regions, especially from the midbrain, spinal cord, deep body sites outside the central nervous system, and periphery seem more effective in evoking panting, and this activity can occur with a sufficient rise in T_b in birds lacking an intact hypothalamus (Richards, 1971; Richards and Avery, 1978; Barnas and Rautenberg, 1987). The dorsal portion of this region in the rock pigeon and domestic fowl contains a "panting center" (Saalfeld, 1936; Richards, 1971). Electrical stimulation of this locus, which may be a counterpart of the pneumotaxic center in mammals, elicited a breathing pattern closely resembling that in thermally induced panting (summarized in Richards, 1975).

4. Role of the Spinal Cord in Thermoregulatory Control

Selective thermal stimulation of posterior levels of the avian spinal cord has not been possible and the extent of their contribution to thermosensitivity is unknown. The cervical and thoracic levels of the spinal cord appear conspicuously involved in thermoreception in various birds, particularly as it relates to heat production and thermal tachypnea. This pattern is not unique to birds and the sensitivity of the proportional thermogenic response (W/°C) appears similar to that of mammals (see Table 3 in Simon *et al.*, 1986). The rock pigeon showed an especially pronounced thermogenic response to spinal cooling (Rautenberg *et al.*, 1972), with the thermosensitivity of the thoracic region of the spinal cord surpassing that of the cervical region (Østnes and Bech, 1992). Spinal cooling in this species was similar in effectiveness to brainstem cooling in evoking appropriate vasomotor and ptilomotor responses (Rautenberg *et al.*, 1972; see Section X,A,2). Significant spinal thermoresponsiveness to cooling was also evident in such species as the Adélie penguin (Hammel *et al.*, 1976; Simon *et al.*, 1976), Pekin duck (Inomoto and Simon, 1981; Bech *et al.*, 1982b), California quail (Snapp *et al.*, 1977), and domestic fowl (Avery and Richards, 1983). Heating the spinal cord alone caused immediate thermal tachypnea in rock pigeons and fowl, whereas cooling this structure quickly inhibited bouts of this activity induced by ambient heating (Rautenberg *et al.*, 1972; Avery and Richards, 1983). In the former species, the f_{resp} raised by spinal heating was not lowered by brain stem cooling (Rautenberg *et al.*, 1978). On the other hand, spinal thermosensitivity was found to be low in the domestic goose (Helfmann *et al.*, 1981).

5. Role of Deep Body Receptors outside the Central Nervous System in Thermoregulatory Control

Despite the apparent significance of thermosensory elements in the avian brain and spinal cord, it is important to note that a major portion of the thermogenic response of birds to cold appears driven by input from deep body thermosensors lying outside the central nervous system (Mercer and Simon, 1984; Necker, 1984). A particularly clear example of this is provided by the Pekin duck (Simon *et al.*, 1981), in which the thermosensitivity to whole body cooling was approximately 14-fold greater than to spinal cooling (-4.02 vs

-0.28 W[kg·°C]$^{-1}$). Also, only a small fraction of the whole-body thermosensitivity of the domestic goose resides in the CNS. Deep body thermosensitivity in this bird was a remarkably high -11.7 W[kg·°C]$^{-1}$ (Helfmann et al., 1981). A somewhat greater share of whole-body thermosensitivity was associated with the CNS in the rock pigeon with its higher sensitivity to spinal cooling (Rautenberg et al., 1972). Snapp et al. (1977), among others, have speculated that locating primary monitoring of T_b in the spinal cord and other deep body sites rather than in the brain may have been an evolutionary solution to the lability of cephalic temperature in birds, particularly during flight.

B. Control of Behavioral Thermoregulatory Responses

Behavioral thermoregulatory responses in birds appear regulated by the PO/AH region of the brain and by peripheral receptors. For example, drooping of the wings by the domestic fowl was abolished by lesions in the hypothalamus (Lepkovsky et al., 1968). Hypothalamic rather than peripheral temperature controlled thermoregulatory behavior in ring doves and the control mechanism appeared to operate as a simple "on–off" system (Budgell, 1971). In rock pigeons trained to work for cooling reinforcement under an ambient heat load (51°C), warming the spinal cord and/or rostral brain stem led to an increase in response rate, whereas cooling these structures produced a decrease (Schmidt, 1976b). Schmidt (1978a) also noted that appropriate behavioral thermoregulatory responses could be elicited from the hypothalamus, whereas physiological thermoregulatory responses were thought to arise largely in the spinal cord (Schmidt, 1978a). Additionally, in this species behavioral responses appeared related to the rate of change of skin temperature on the exposed areas of the body (Schmidt, 1978b). Further analysis of the effects of changing temperatures of the spinal cord, hypothalamic, and facial skin temperature of the rock pigeon (Schmidt, 1983) has bolstered the conclusion that displacement of the former primarily evokes physiological thermoregulatory responses, whereas localized changes in the temperature of the facial region exert a more powerful effect on behavioral thermoregulation. Relative effects of localized changes in hypothalamic temperature on behavioral thermoregulation differed between heating and cooling.

A thermoregulatory role for higher levels of CNS involved in conditioned reflexes has been suggested by work on rock pigeons (Sieland et al., 1981), which became conditioned to change heart rate to interrupt an artificially induced aversive temperature in their spinal cord.

C. Afferent Inputs Contributing to Thermal Tachypnea

As noted above, central neural elements involved in control of panting include a "panting center" in the dorsal midbrain (Saalfeld, 1936; Richards, 1971) and thermosensitive elements in the spinal cord (Rautenberg, 1969b, 1983). Though input from central elements appears primary, other sources also may affect thermal tachypnea or gular flutter. There is evidence of the initiation of these activities with exposure to solar radiation before any increase in T_b has taken place (Shallenberger et al., 1974). Also, though an increase in skin temperature per se probably does not evoke an increased f_{resp}, there is an indication that it can influence thermal tachypnea once established (Woods and Whittow, 1974).

Afferent neural information, particularly that conveyed by the vagal afferent system, is important in determining breathing patterns during hyperthermia. The vagi of rock pigeons contain fibers identified respectively as intrapulmonary chemoreceptors (IPC) and mechanoreceptors (Mückenhoff et al., 1989). The IPC include both phasic and tonic elements which increase their activity at low P_{CO_2}. Information provided by phasic IPC and the mechanoreceptors is thought to contribute to setting the respiratory pattern during hyperthermic tachypnea. Increased firing of the tonic IPC may be a factor in producing the high f_{resp} and reduced V_T with which birds reliant on respiratory evaporative cooling deal with manageable heat challenges (Mückenhoff et al., 1989). Sensitivity of IPC to CO_2 is reduced at higher T_b and this was interpreted as contributing to the increased tidal volumes that develop in birds under extreme heat stress (Barnas et al., 1983). Extrapulmonary chemoreceptors appeared to have an approximately equal influence to that of IPC on ventilation of awake, hyperthermic domestic fowl and they may also contribute to the diminution in V_T (Barnas and Burger, 1983).

Sectioning of one vagus nerve during hyperthermic panting reduced f_{resp} slightly in various birds and, except for the rock pigeon, bilateral vagotomy abolished rapid breathing during hyperthermia (Hiestand and Randall, 1942; Richards, 1968). In the bilaterally vagotomized rock pigeon, f_{resp} rose less, Pa_{CO_2} decreased more, and pH_a increased more for a given rise in T_b than in intact individuals (Barnas et al., 1986). All three of these functions were more variable in vagotomized individuals than in controls. Evidently, central chemoreception alone is insufficient to maintain stability of breathing pattern during thermal panting. Electrical stimulation of the cut central ends of the vagi reinstated panting in the bilaterally vagotomized domestic fowl (Hiestand and Randall, 1942; Richards, 1968).

D. Miscellaneous Observations on Thermoregulatory Control

Exposure of rock pigeons to 7% O_2 in N_2 at a T_a of 5°C inhibited electromyographic (EMG) activity indicative of shivering (Gleeson *et al.*, 1986a; Barnas and Rautenberg, 1990). The effect was reduced if the birds were allowed to breathe a similar hypoxic mixture enriched with 4.5% CO_2, which resulted in higher arterial P_{O_2} and oxygen content. Birds subjected to bilateral cervical vagotomy under normoxic conditions could still rapidly effect cardiorespiratory adjustments to shivering, despite the loss of inputs from IPC, carotid bodies, and other receptors (Gleeson *et al.*, 1986b). However, the shivering process in these individuals was somewhat more sensitive to hypoxia than in controls (Gleeson *et al.*, 1986b). Direct effects of low CO_2 and/or high pH in the blood and, ultimately, the cerebrospinal fluid of hypoxic birds could not be definitely ruled out in the inhibition of shivering. Whether hypoxic inhibition of shivering involves thermoregulatory neurons or those from areas that coordinate motor activity was not resolved (Barnas and Rautenberg, 1990).

Unlike mammals, Pekin ducks do not suffer disturbances of physiological heat and cold defense in response to capsaicin (Geisthövel *et al.*, 1986). Birds do develop a fever in response to some bacterial infections, and antipyretic drugs such as acetylsalicylic acid produce an attentuation of fever (Kluger, 1979; see also Section X,A,2). The response pattern to injection of bacterial endotoxin (lipopolysaccharide) in rock pigeons varied diurnally. The febrile state developed sooner and was of greater extent but of shorter duration with nocturnal than with daytime injection. From these observations, Nomoto (1996) has suggested that control of the circadian rhythmicity of T_b is not disturbed during fever.

XI. THERMOREGULATION AT REDUCED BODY TEMPERATURES

A. Patterns of Avian Hypothermia

Under certain circumstances, representatives of at least nine avian orders allow their T_b to fall from euthermic levels (Reinertsen, 1996). Such reductions are common in small birds that depend on insects or nectar, the supplies of which can be seasonally or temporarily disrupted. In fact, Brown *et al.* (1978) suggested that the evolution of small nectar-feeding hummingbirds has been related to their ability to become dormant at night. Without such an ability, a 2-g hummingbird might exhaust its energy reserves during an overnight fast.

Reinertsen (1996) has recognized two major categories of short-term response involving cooling from euthermic levels of T_b. Both usually occur during the inactive phase of the animals' circadian cycle. *Nocturnal hypothermia* involves a relatively modest decline in T_b to no lower than approximately 30°C. This has been observed in such birds as turkey vultures, several doves and pigeons, greater roadrunner, smooth-billed ani (*Crotophaga ani*), Japanese quail, and species in more than nine families of passerines including manakins, sunbirds, honeyeaters, tits, and certain finches. *Torpor* involves more substantial reductions in T_b, sufficient to render the animal comatose over a few hours. Short-term bouts of torpor have been documented in goatsuckers, especially the common poor-will, many species of hummingbirds, swifts, and mousebirds (*Colius* spp.). Reinertsen (1996) has called attention to a third, more-prolonged hypothermic response, *hibernation* (natural multiday bouts of torpor during winter). This appears to depend on the same basic processes as nocturnal hypothermia and torpor. However, Geiser and Ruf's (1995) analysis of 104 species of heterothermic birds and mammals indicated that no overlap exists between hibernators and daily heterotherms in maximum duration of torpor bouts. Moreover, the minimum T_b and minimum metabolic rate reached during these bouts tend to be lower in the former than in the latter. However, in considering the application of these findings to birds, it is important to note that hibernation has been documented in only one species of bird, the common poor-will (Jaeger, 1949; French, 1993). French's (1993) observations suggest that multiday periods of dormancy in this species are interrupted by brief episodes of euthermia, as in many hibernating mammals.

Instances of lowered T_b in birds have most commonly, but not exclusively, been linked with reduced availability of food or depletion of energy reserves, whatever the season (Reinertsen, 1986; Wang, 1989). Water restriction appears to be a factor in some species (MacMillen and Trost, 1967; see below). With respect to energetic considerations, rock pigeons and Japanese quail developed shallow nocturnal hypothermia with food restriction (Graf *et al.*, 1989; Hohtola *et al.*, 1991). Increased filling of the digestive tract with either nutrients or cellulose pellets tended to reverse this effect in rock pigeons (Reinertsen and Bech, 1994; Geran and Rashotte, 1997), the effective stimulus apparently being gastrointestinal distension as an indicator of the animals' energy reserves. Reinertsen (1996) concluded that an inverse relation exists between the length and depth of bouts of nightly torpor or hypothermia and body mass at roosting. Lowered T_b produce an indisputable saving of energy (see Calder and King, 1974). Nevertheless, simple relations between an individual's energy reserves and hypothermia or torpor do not appear to be the whole story. The willow tit resorted to shallow nocturnal hypothermia on a regular basis during the Norwegian winter, even when ample food supplies were available (Reinertsen and Haftorn, 1983). Additionally, some common poor-wills found torpid in nature by Jaeger (1949) and French (1993) were quite heavy, suggesting ample energy reserves. Perhaps cases such as these represent use of torpor or hypothermia as insurance against future energy problems. Hiebert (1993) provided evidence concerning such a possibility in the rufous hummingbird (*Selasphorus rufus*). The use of nocturnal torpor by other birds of this species in the process of accumulating fat prior to migration (Carpenter and Hixon, 1988; Carpenter *et al.*, 1993) appears to represent a further pattern in which depletion of energy reserves or reduced food availability was not a consideration. Entry into a torpid state evidently facilitated the fattening process by reducing nocturnal existence costs during the premigratory period. The use of torpor in connection with this process implies some shift in the threshold value of these energy reserves for the induction of a hypothermic state, relative to the situation where simple overnight survival is the challenge (Calder, 1994). These and similar observations suggest, for at least some birds, that reductions of T_b leading to nocturnal hypothermia or torpor must result from a complex integration of information on such matters as food availability, anticipated foraging success, nocturnal T_a, and other factors relating to future energy demands, as well as on the state of physiological energy reserves (see, for example, Bartholomew *et al.*, 1983).

Tendencies toward hypothermia and torpor may vary seasonally or with changes in circumstances. The response is not a regular event in the broad-tailed hummingbird, having been found only on infrequent occasions of inadequate daily energy storage occurring early on the wintering grounds, in a few breeding territories, and after summer storms that precluded feeding (Calder, 1994). Common poor-wills fitted with temperature-sensitive radiotransmitters only rarely became torpid during the breeding season in British Columbia during late spring and early summer and no individuals involved in active nesting attempts did so (Brigham, 1992; Csada and Brigham, 1994). However, they regularly entered this state in April, May, and September, even though their body masses indicated no reduction in energy reserves (Brigham, 1992). Torpor during these 3 months was most closely correlated with T_a at dusk. Responses of the rufous hummingbird to food restriction also varied seasonally, with birds exposed to such restriction spending less time torpid in summer, when initial body mass was higher, than in spring (Hiebert, 1991). Seasonal effects were also evident in the nocturnal hypothermia shown by the black-capped chickadee (*Parus atricapillus*) and the willow tit (*Parus montanus*). The greatest reduction in T_b occurred in winter-acclimatized individuals of both these species (Chaplin, 1976; Reinertsen and Haftorn, 1983).

B. The Hypothermic State

Perhaps the relatively shallow depressions of T_b characterizing bouts of nocturnal hypothermia in birds such as passerines reflect some phylogenetic constraint limiting tolerance of cooling. For example, American goldfinches, which under certain circumstances may allow body temperature to fall a few degrees under nocturnal cold challenges (Lustick *et al.*, 1982), succumb at T_b below approximately 20–22°C (W. R. Dawson, unpublished observations), whereas the common poor-will and the Anna hummingbird (*Calypte anna*), which can enter deep torpor, survive T_b below 10°C (Bartholomew *et al.*, 1957; Withers, 1977). Even a reduction of a few degrees during bouts of hypothermia should help conserve energy during the lengthy nocturnal fasts imposed by cold winter nights at middle and high latitudes. Reinertsen (1986) estimated that willow tits could reduce their overnight energy costs by 10 and 15% with a 6°C reduction in T_b at T_a of $-20°$ and 0°C, respectively. Nocturnal reductions in T_b of up to 5.5°C in the Australian silvereye (*Zosterops lateralis*) led to energetic savings of 15, 21, and 24% (compared with daytime RMR) at T_a of 7, 16, and 25°C, respectively, under an 11-hr scotophase (Maddocks and Geiser, 1997). A saving of 58% would occur if the small tropical manakin *Manacus vitellinus* maintained a T_b of 27°C for 10 hr in a T_a of 22°C (Bartholomew *et al.*, 1983). Nightly energetic savings at T_a of 19–23°C by the hummingbirds *Amazilia versicolor*, *Melanotrochilus fuscus*, and *Eupetomena macroura* torpid with T_b near 24°C ranged from 49 to 61% (Bech *et al.*, 1997). These savings could be quite important for all these small birds and Maddocks and Geiser (1997) suggested that they contribute to the ability of the Australian silvereye to occupy a wide array of habitats and climates.

The shallow nocturnal hypothermia observed in some birds is not confined to cold-challenged species. For example, Inca doves (*Scardafella inca*) underwent a pronounced nocturnal hypothermia at moderate T_a during either food or water deprivation (MacMillen and Trost, 1967). This state also developed in some individuals of the two tropical manakins *Pipra mentalis* and *Manacus vitellinus* during tests at T_a of 14.6° to ~29°C. In fact, one individual of the latter species cooled to 26.8°C at the former T_a (Bartholomew *et al.*, 1983).

The proximate physiological events leading to avian hypothermia and torpor are not well understood. Evidence suggests that hypothermia and sleep in birds are closely related phenomena (Heller et al., 1983; Walker et al., 1983). Declining T_b in ring doves (*Streptopelia risoria*) was characterized by a progressive reduction in rapid eye movement sleep, with the periods of hypothermia involving almost continuous slow-wave sleep (Walker et al., 1983). Generally, temperature regulation is not abandoned by birds at reduced T_b; rather, there appears to be a lowering of the thermal set point. This is readily appreciated in hypothermic birds such as passerines and pigeons in which substantial differences are maintained between T_b and T_a. The lowered set point is not necessarily fixed, for the threshold of spinal temperature for the shivering response in hypothermic rock pigeons fell to successively lower levels over 5 nights during an episode of food deprivation (Graf et al., 1989). A lowering of the thermal setpoint also occurs in species entering torpor, where the temperature difference between body and environment can become relatively small. For example, the West Indian hummingbird *Eulampis jugularis* maintained its T_b between 18° and 20°C at T_a of 2.5–18°C. Metabolic rate increased linearly with decreasing temperature below 18°C (Hainsworth and Wolf, 1970). The corresponding figures for the Brazilian hummingbirds *Melanotrochilus fuscus* and *Eupetomena macroura* were 19–23°C and 13–16°C, respectively (unpublished data cited in Bech et al., 1997). Body temperature of the highland tropical hummingbirds *Panterpe insignis* and *Eugenes fulgens* was not allowed to fall below 10–12°C in cool environments (Wolf and Hainsworth, 1972). Torpid *Selasphorus* hummingbirds (*S. rufus* and *S. platycercus*) likewise defended T_b near 12°C (Bucher and Chappell, 1992). Strikingly, the Andean hillstar hummingbird (*Oreotrochilus estella*) generally did not allow T_b to fall below 6.5°C during nights spent roosting in a cool montane cave (Carpenter, 1974). Hummingbirds that engage in torpor at moderate T_a (19–23°C) may not actively regulate T_b, initiating such activity only if cooler conditions are imposed (Bech et al., 1997). At moderate T_a, bouts of torpor were sometimes interrupted, possibly by external stimuli, in *Amazilia versicolor, Melanotrochilus fuscus,* and *Eupetomena macroura,* with the animals arousing before once again entering torpor (Bech et al., 1997). Relatively brief bouts of nocturnal hypothermia (≤2.5–4 hr) also have been noted in other hummingbirds (Hainsworth et al., 1977; Hiebert, 1990).

Entry into torpor by common poor-wills and Anna and Allen (*Selasphorus sasin*) hummingbirds was preceded by pronounced oscillations in $\dot{V}_{O_2}$ involving "test drops" (Withers, 1977). On the other hand, this process usually was characterized by a smooth decline of $\dot{V}_{O_2}$ in *Selasphorus rufus* and *S. platycercus*. This was associated with a coincident decline in respiratory frequency (f_{resp}), but the proportional changes in the two variables were not identical and their relationship varied at different T_a. The decline in f_{resp} involved a lengthened time of active inspiration that was ultimately interrupted by a nonventilatory pause. A slight increase in time of expiration also developed (Bucher and Chappell, 1992). Electromyographic activity was reduced or absent during entry of the giant hummingbird (*Patagona gigas*) and the common poor-will into torpor, indicating a curtailment of shivering (Bartholomew et al., 1962; Lasiewski et al., 1967). Additionally, Withers (1977) found that thermal conductance of individuals of the latter species was increased during entrance into torpor. However, it did not differ between normothermic and torpid individuals of the hummingbird *Eulampis jugularis* (Hainsworth and Wolf, 1970). Consistent with the inverse relations between avian body mass and both mass-specific surface area and mass-specific thermal conductance, the rate of entry of birds into hypothermia is inversely related to body mass. Lasiewski and Lasiewski's (1967) analysis indicated that lengthy nocturnal bouts of deep hypothermia are only practicable for relatively small birds (see Bicudo, 1996, for discussion). The Lasiewskis estimated that a bird weighing between 80 and 100 g would require 12 hr merely to lower its T_b to 20°C.

At the lowest f_{resp}'s observed in the *Selasphorus* hummingbirds during torpor (1–2 breaths/min vs normothermic values of 300–400 breaths/min), ventilation occurred in bursts of breaths separated by nonventilatory pauses that lengthened into apneas of up to 5 min. Withers (1977) found that V_T and respiratory minute volume as well as f_{resp} were reduced during torpor in the common poor-will and Anna and Allen hummingbirds. Heart rate was also depressed, with values in the common poor-will falling below 30 beats/min at a T_b of 8–9°C, compared with rates of >500 beats/min in normothermic birds at cool T_a (Bartholomew et al., 1962). The metabolic rates of torpid birds are also greatly reduced, representing only 5 or 6% of normothermic values at a T_a of 20°C in common poor-wills and Anna and Allen's hummingbirds (Withers, 1977). Even lower percentages are evident during torpidity at still lower T_b (Dawson and Hudson, 1970). The respiratory exchange ratio (R) of well-fed swallow-tailed hummingbirds (*Eupetomena macroura*) declined with fasting and could reach unusual values as low 0.2–0.3 in torpid individuals (Chaui-Berlinck, and Bicudo, 1995; Bicudo, 1996). However, these remarkable Rs were not exclusively linked with torpor, also occurring during fasting with normothermic T_b and metabolic rates. They did not seem to be associated with a change in breathing pattern, instead possibly reflecting usage of some unidentified metabolic pathway (Bicudo, 1996). On the other hand, nocturnal Rs for both normothermic and torpid broad-tailed and rufous hummingbirds declined from about 0.86 only to approximately 0.71 (Bucher and Chappell, 1997), a more conventional value.

Analysis of Q_{10}'s for the relation between SMR and metabolic rates at reduced T_b (Geiser, 1988) suggested that the latter rates are primarily a product of direct effects of temperature on rate processes in daily heterotherms and large hibernators. Physiological inhibition of metabolism at reduced body temperatures, reflected in relatively high Q_{10}'s, may also be involved in small hibernators. For birds, the small number of species included in this analysis, Q_{10}'s >3 in several of them, and the rarity of hibernation in the Class Aves make the applicability of these generalizations concerning direct effects of temperatures on rate processes uncertain. However, Bucher and Chappell (1997) recently concluded from rufous and broad-tailed hummingbirds, which were not actively defending T_b in torpor, that a normal Q_{10} effect (Q_{10} = 1.9–2.2) was sufficient to account for the birds' reduced metabolic rates.

C. Arousal from Hypothermia

Arousal from nocturnal hypothermia or torpor typically occurs well before the beginning of the daily photophase (see, for example, Hiebert, 1990). This apparent persistence of circadian organization at reduced T_b should serve to reduce predation risk. (It appears relevant in attempting to account for this persistence to note that pineal cells from domestic fowl chicks show temperature compensation over a range of 34–41°C in their circadian rhythm of melatonin secretion *in vitro* [Barrett and Takahashi, 1995].) The minimal T_b from which birds can spontaneously arouse tend to be higher than in many hibernating mammals, ranging for most species from approximately 10–12°C to ≥20°C (Bartholomew *et al.,* 1957; Dawson and Hudson, 1970; Reinertsen, 1996; Bicudo, 1996), though the minimal T_b for the Andean hillstar hummingbird (*Oreotrochilus estella*) was 6.5°C (Carpenter, 1974). Rewarming rates are inversely related to body size (Lasiewski and Lasiewski, 1967) and this, like the relation between entry rate and size, may hinder the use of short-term torpor by larger birds (see Bicudo, 1996, for discussion). Arousing Anna hummingbirds could raise their T_b 1.0–1.5°C/min at a T_a of 23°C (Bartholomew *et al.,* 1957). Perhaps the relatively high T_b threshold for arousal in most birds is related to the apparent absence of any specialized thermogenic elements such as brown adipose tissue (Olson *et al.,* 1988; Saarela *et al.,* 1989b, 1991; Brigham and Trayhurn, 1994) and a consequent primary reliance on shivering thermogenesis, which may not be readily activated at very low T_b. However, it is of interest that a common poor-will actively aroused from T_b as low as 5 or 6°C, though its rewarming required 12 hr (Ligon, 1970). The Andean hillstar hummingbird apparently approached this performance (Carpenter, 1974). Thermal conductance of arousing common poor-wills was reduced, facilitating the retention of heat. For these birds, values for f_{resp}, V_T, respiratory minute volume, and $\dot{V}_{O_2}$ during arousal greatly exceeded those at similar T_b during entry into torpor. In fact, $\dot{V}_{O_2}$ of rewarming individuals reached 6.3–8.9× resting metabolic rate (Withers, 1977). Spontaneous rewarming is a vigorous process in hummingbirds as well. In the two *Selasphorus* species studied by Bucher and Chappell (1992), f_{resp} usually increased faster than $\dot{V}_{O_2}$ during the early stages of arousal. In cases where the initial T_b exceeded 20°C, conspicuous decoupling of f_{resp} and $\dot{V}_{O_2}$ was evident, with f_{resp} often increasing 3- to 10-fold within 30 sec of the initiation of arousal. Bucher and Chappell (1992) speculated that this initial rapid increase in respiratory frequency might be a hyperventilatory mechanism for eliminating a torpor-related metabolic acidosis. However, they later found in *Selasphorus* hummingbirds that, while transient CO_2 storage often occurred during entrance into torpor, it was not a prerequisite for this process. Similarly, CO_2 release was not a requirement for arousal by these birds (Bucher and Chappell, 1997). Regarding acidosis at reduced T_b, it is of interest that P_{CO_2} in the abdominal and interclavicular air sacs of rock pigeons increased during a fasting-induced bout of shallow nocturnal hypothermia, indicating an effective parabronchial hypoventilation (Jensen and Bech, 1992).

XII. DEVELOPMENT OF THERMOREGULATION

A. Embryo

The development of thermoregulation in the embryo is discussed in Chapter 24.

B. Hatchling

Birds vary enormously in their thermoregulatory capacities immediately after hatching, and they have been classified according to these and other capabilities by Nice (1962) and others. Precocial species, e.g., the hatchling of the domestic fowl, are covered with down and they are able to respond effectively to heat and to cold. In contrast, altricial birds, e.g., members of the Passeriformes, are naked when hatched and have little ability to regulate T_b at T_a below 35°C or above 40°C. Other birds (semi-precocial, semi-altricial) are intermediate between precocial and altricial species in their thermoregulatory capacities.

1. Physiological Responses to Cold

Some precocial hatchlings are able to increase their metabolic heat production as much as five times in response to exposure to cold (Eppley, 1984). Deep body cold sensitivity similar to that of adult Pekin ducks is evident in hatchlings of this species. Nevertheless, peripheral receptors seem of primary importance under natural conditions in providing cold signal inputs in Pekin ducklings during the first 12 days after hatching (Østnes and Bech, 1997a).

A recent series of studies compared the thermoregulatory responses of the hatchlings of three species of semi-precocial tropical seabirds (Matthiu et al., 1991, 1992, 1994). In all three species, there were well-defined responses of the hatchling to cold and to heat. The responses of one of the species, the wedge-tailed shearwater (*Puffinus pacificus*), are shown in Figure 8. Comparison of the responses of the hatchling with those of the adult reveal that T_b is lower within the TNZ in the former and that the boundaries (in terms of T_a) of the thermoneutral zone are higher. The RMR of the hatchling within the TNZ is almost at the adult level when corrected for differences in body mass ("mass-independent metabolism"; Bucher, 1986). However, hatchling shearwaters are surpassed by adults in their thermogenic capacity (ability to increase in heat production) in the cold and in the effectiveness of their evaporative cooling in the heat. Hatchling shearwaters are very much smaller than adults with a correspondingly larger surface area relative to their body mass from which to lose heat. This may be an advantage in the warm environment in which the wedge-tailed shearwater lives.

In contrast to the responses of precocial and semi-precocial hatchlings to cold, altricial birds at this stage fail to increase their heat production during cold exposure (Whittow and Tazawa, 1991). Altricial hatchlings are therefore entirely dependent on heat derived from the parent birds in order to prevent hypothermia. This is achieved by brooding. In the semi-precocial arctic tern (*Sterna paradisaea*), brooding spared the hatchling from expending a considerable amount of its own energy (Klaassen et al., 1989).

2. Brooding Responses

Many hatchlings can influence their own T_b by eliciting appropriate brooding responses from their parents. The American white pelican (*Pelecanus erythrorhynchos*) neonate was observed to do this by calling (squawking). The call rate increased as T_b of chicks declined precipitously during exposure to cool T_a (20–23°C). The vocalizations, which were quite distinct from

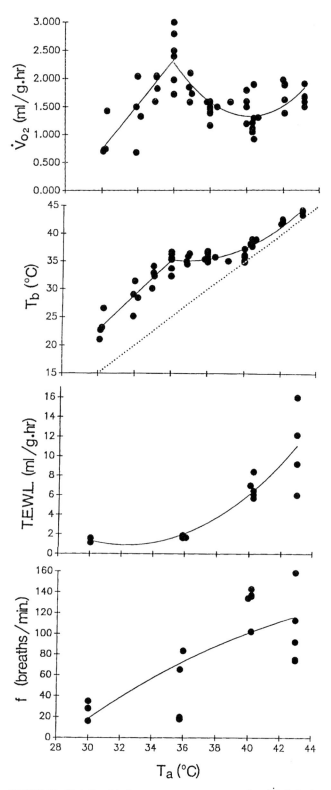

FIGURE 8 Relationship between oxygen consumption ($\dot{V}_{O_2}$), body temperature (T_b), total evaporative water loss (TEWL), respiratory frequency (f), and air temperature (T_a) in hatchling wedge-tailed shearwaters, *Puffinus pacificus*. (Redrawn from P. M. Mathiu, W. R. Dawson, and G. C. Whittow (1992), *J. Therm. Biol.* **16**, 317–325, by permission of The University of Chicago Press. © 1992 by The University of Chicago.)

food-begging calls (Evans, 1992), could elicit sufficient brooding to reverse the decline in T_b (Evans, 1994). Semi-precocial newly hatched ring-billed gulls (*Larus delawarensis*) and herring gulls (*L. argentatus*) also responded to moderate chilling (20°C) by calling. When it was arranged that the calls would provide brief periods of instrumental heating (34°C), cold-induced vocalizations allowed the hatchling gulls to maintain T_b near normothermic levels (Wiebe and Evans, 1994). Even the precocial American coot (*Fulica americana*) neonate was observed to call in response to cooling (Bugden and Evans 1991).

3. Responses to Heat

The differences between precocial and altricial hatchlings are less obvious when they are exposed to heat. Thus, the precocial neonate mallee fowl (*Leipoa ocellata*) could dissipate heat in excess of metabolic heat production when subjected to heating (Booth, 1984) and altricial hatchlings also have well developed responses to heat (see Whittow, 1986; Montevecchi and Vaughan, 1989).

C. Nestling

1. Growth Rates

The overall growth of the nestling has important implications for thermoregulation: the surface area/body mass ratio of the young bird diminishes as the nestling gets larger, so that the surface area from which heat might be lost diminishes relative to the mass of the tissue available to produce heat and in which to store it. The faster the growth of the nestling, the greater the impact of this change. The growth rates of altricial birds are considerably greater than those of precocial young, and there is evidence in the red-winged blackbird (*Agelaius phoeniceus*) that the high post-natal growth rate of altricial nestlings precedes the development of thermoregulation (Hill and Beaver, 1982). In this way, the maximal amount of energy is diverted to growth rather than being dissipated as heat for thermoregulatory purposes. It is not only the overall growth rate that is important in thermoregulation; the relative growth rates of individual tissues and organs also profoundly affect the capacity of the nestling for thermogenesis. Skeletal muscle is a tissue of paramount importance for thermogenesis in the nestling (Ricklefs, 1983); in precocial species, the leg muscles are used for locomotion, which may predispose them toward their use, also, in shivering (Eppley, 1996). In altricial birds such as the red-winged blackbird, the onset of endothermy correlates with the detection of shivering in the pectoral muscles (Olson, 1994).

2. Responses of Altricial and Precocial Species

The study of the development of thermoregulation in the field (*Spizella pusilla*) and chipping (*S. passerina*) sparrows by Dawson and Evans (1957) documented the transition from poikilothermy to homeothermy in two very small (hatchling mass 1.7 g) altricial birds. During the first few days of life, T_b and oxygen consumption declined *pari passu* with T_a, but, by the end of the first week, the nestlings could increase their heat production threefold and maintain their body temperature during exposure to low air temperatures. Figure 9 presents data for the vesper sparrow (*Pooecetes gramineus*), which weighed approximately 2 g at hatching. The development of the plumage clearly was of considerable importance in providing the young bird with a means of restricting the loss of heat produced by metabolism. However, as Dawson and Hudson (1970) put it, referring to the feathers, ". . . these structures appear more important for the perfection of temperature regulation than for its initiation"—the sparrows studied by Dawson and Evans (1957, 1960) all displayed fairly effective control of body temperature at moderate and low temperatures before the eruption of the contour feathers from their sheaths. As Odum (1942) proposed over 56 years ago, shivering precedes body temperatures regulation which, in turn, precedes feather development in the altricial house wren. The increase in body mass (and of capacity for thermogenesis) thus seems more important than the development of the plumage. Many nestlings are members of a brood, however, and, as such, can modulate their collective insulation. This occurred as early as 3 days after hatching in altricial red-winged blackbirds, whereas thermoregulatory thermogenesis began at 4–5 days (Hill and Beaver, 1982).

As pointed out by Dawson (1984b), insights concerning the underlying mechanisms of the development of cold thermogenesis in altricial nestlings have come from tissue enzyme studies. Marsh and Wickler (1982) could correlate the attainment of maximal metabolic rates in the cold with myofibrillar ATPase activity in the pectoral muscles in bank swallows (*Riparia riparia*). Activity of this enzyme in skeletal muscle is a measure of muscle contractile function (Marsh and Wickler, 1982). In an earlier study on the larger, albeit altricial, cattle egret (*Bubulcus ibis*), Hudson et al. (1974) obtained electromyographical evidence of a progressive increase in shivering activity with age in nestlings. In the altricial European starling (*Sturnus vulgaris*), citrate synthase, an enzyme indicative of aerobic capacity, increased with growth of the pectoral muscles. This did not occur in two species of precocial quail (Choi et al., 1993). However, increase in citrate synthase in the starling was not accompanied by development of thermogenic capacity.

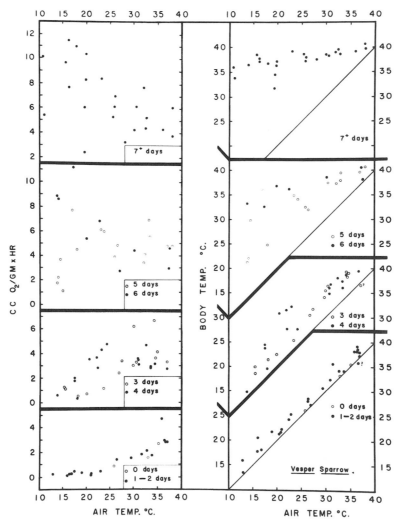

FIGURE 9 The relationship between oxygen consumption (left panel) and body temperature (right panel) and air temperature (abscissas) in the nestling vesper sparrows *Pooecetes gramineus*. (From Dawson and Evans (1960) *Condor* **62,** 329–340.)

Nonshivering thermogenesis is controversial in birds (see Section V,E) with much of the evidence suggesting its occurrence in this class coming from young individuals in which skeletal muscle is implicated as the site (Barré *et al.,* 1989; Duchamp *et al.,* 1991). Information on Muscovy ducklings has been mentioned previously (Section V,E). Measurements of electromyograms from the gastrocnemius muscle of winter-acclimatized king penguin chicks showing that the threshold for shivering lay ~9°C below their T_{lc} have provided indirect evidence for NST in these birds (Duchamp *et al.,* 1989; but see comments in Section V,E about similar information for Muscovy ducklings). This process is thought to be mediated by the hormone glucagon in these penguin chicks (Duchamp *et al.,* 1989; Jansky, 1995). Regarding heat derived from other sources than shivering, larger endothermic chicks of the house wren (>8 g) could substitute the heat increment of feeding for thermoregulatory heat at lower T_a (Chappell *et al.,* 1997).

The metabolic responses of young altricial nestlings to lowered T_a are in striking contrast to those of the precocial nestling of the domestic fowl, elegantly portrayed by Barrott and Pringle (1946) over 50 years ago. The increase in heat production was greatest in the hatchling, and the magnitude of the increase diminished progressively with age. In the semi-precocial black-headed gull (*Larus ridibundus;* Palokangas and Hissa, 1971) and Manx shearwater (*Puffinus puffinus*; Bech *et al.,* 1982a), the increase in heat production in the cold diminished with the age of the nestling. In the black-headed gull, as in the domestic fowl, there is clearly a shift in the TNZ to a lower range of air temperature with increasing age. Minimal thermal conductance in the cold declined with increasing age and body size in the three petrels investigated by Ricklefs and Roby (1983).

3. Heat Loss

As far as heat loss from the nestling is concerned, in the white-crowned sparrow, feather development is the major factor reducing heat loss from the nestling after 2 days of age (Webb and King, 1983). The nest substantially reduced wind speed near the nestling, and, therefore, heat loss from it. Net long-wave radiant heat exchange was small, short-wave radiation gain from the sun being more important.

4. Solar Heat Gain

Exposure to solar radiation is sufficient in some nestlings to impose a heat stress on them. The thermal consequences of such exposure depend largely on the reflectance of the plumage. In the herring gull, the reflectance by the nestling's down of solar radiation was greater than that of the adult plumage (Dunn, 1976). This would result in less heat gain by the bird from the environment. Under cool conditions, this would be to the nestling's disadvantage, but, at high T_a, it would benefit from the resultant reduced heat stress.

5. Brooding

Brooding by the parent may continue as the nestling grows (Klaassen *et al.*, 1989), as prompted in the American white pelican by the young bird's vocalization. In the 1-week-old nestling, the decline of T_b with cooling was less rapid than in the hatchling, but vocalizations still increased in response to cold, despite the presence of shivering. In contrast, at 2 weeks, vocalization rate did not increase as T_b fell (Evans, 1994). The behavior of 1- to 9-day-old shag (*Phalacrocorax aristotelis*) nestlings resembles that of the younger pelicans and Østnes and Bech (1997b) suggest that the thermally significant vocalizations of these young shags are mediated by fast-acting cutaneous thermoreceptors.

Huddling of nestlings is also effective (Hill and Beaver, 1982). For example, a brood of white-crowned sparrows proves to have more insulation than does an adult (Webb and King, 1983).

6. Hypothermia

The absence of a thermoregulatory response in the cold should not be confused with elective hypothermia, which seems to occur in certain nestlings when they are not fed. The criteria for elective hypothermia should be that the nestling responds to a low T_a with regulatory thermogenesis when fed, whereas, in the absence of food, shivering is held in abeyance and heat production and T_b fall. Unfortunately, such rigorous criteria have not been observed in many studies, but, nevertheless, evidence of elective hypothermia has been obtained for the young of several petrels (Ricklefs *et al.*, 1980; Pettit *et al.*, 1982; Simons and Whittow, 1984; Boersma, 1986). The T_b of the fork-tailed storm petrel (*Oceanodroma furcata*) decreased to as low as 10.6°C during periods of inanition (Boersma, 1986).

Under natural conditions, nests and nesting burrows may provide considerable protection from wind, rain, insolation, and extreme T_a for eggs, nestlings, and incubating adults, thereby affecting both heat loss and solar heat gain. Thermal insulation varies enormously among nests. It can be of the same order of magnitude as that of the tissues and plumage of the bird.

XIII. SUMMARY

Birds have well-developed powers of temperature regulation that depend on generally high metabolic levels, substantial capacities for regulatory thermogenesis, effective physiological and behavioral mechanisms affecting heat transfer between these animals and their environment, and the ability to dissipate considerable quantities of heat evaporatively. Their endothermic homeothermy appears to have evolved independently from that characterizing mammals and some differences are evident between the two groups involving such things as level of body temperature, mechanisms for evaporative cooling, incidence and mechanisms of nonshivering thermogenesis, and extent of reliance on hibernation. Some differences in the neural organization of thermoregulatory control are also evident between birds and mammals, but these appear more quantitative than qualitative. Differences in detail should not obscure the overall similarity of thermoregulation by animals in the two classes, nor deflect attention from the importance of this regulatory activity in shaping their biology.

See Note Added in Proof following the References on p. 390.

References

Ambrose, S. J., and Bradshaw, S. D. (1988). Seasonal changes in standard metabolic rates in the white-browed scrubwren *Sericornis frontalis* (Acanthizidae) from arid, semi-arid, and mesic environments. *Comp. Biochem. Physiol. A* **89,** 79–83.

Arad, Z. (1983). Thermoregulation and acid–base status in the panting dehydrated fowl. *J. Appl. Physiol.* **54,** 234–243.

Arad, Z., and Marder, J. (1983). Acid–base regulation during thermal panting in the fowl (*Gallus domesticus*): Comparison between breeds. *Comp. Biochem. Physiol. A* **74**, 125–130.

Arad, Z., Arnason, S. S., Chadwick, A., and Skadhauge, E. (1985). Osmotic and hormonal responses to heat and dehydration in the fowl. *J. Comp. Physiol. B* **155**, 227–234.

Arad, Z., Gavrieli-Levin, I., Eylath, U., and Marder, J. (1987). Effect of dehydration on cutaneous water evaporation in heat-exposed pigeons (*Columba livia*). *Physiol. Zool.* **60**, 623–630.

Arad, Z., Midtgård, U., and Bernstein, M. H. (1989). Thermoregulation in turkey vultures: Vascular anatomy, arteriovenous heat exchange, and behavior. *Condor* **91**, 505–514.

Arad, Z., Toledo, C. S., and Bernstein, M. H. (1984). Development of brain temperature regulation in the hatchling mallard duck *Anas platyrhynchos*. *Physiol. Zool.* **57**, 493–499.

Arieli, A., Berman, A., and Meltzer, A. (1979). Cold thermogenesis in summer-acclimatized and cold-acclimated domestic fowl. *Comp. Biochem. Physiol. C* **63**, 7–12.

Arjona, A. A., Denbow, D. M., and Weaver, W. D., Jr. (1990). Neonatally-induced thermotolerance: Physiological responses. *Comp. Biochem. Physiol. A* **95**, 393–399.

Aschoff, J., and Pohl, H. (1970). Rhythmic variations in energy metabolism. *Fed. Proc.* **29**, 1541–1552.

Avery, P., and Richards, S. A. (1983). Thermosensitivity of the hypothalamus and spinal cord in the domestic fowl. *J. Therm. Biol.* **8**, 237–239.

Bakken, G. S. (1992). Measurement and application of operative and standard operative temperatures in ecology. *Am. Zool.* **32**, 194–216.

Bakken, G. S., Buttemer, W. A., Dawson, W. R., and Gates, D. M. (1981). Heated taxidermic mounts: A means of measuring the standard operative temperature affecting small animals. *Ecology* **62**, 311–318.

Bakken, G. S., Erskine, D. J., and Santee, W. R. (1983). Construction and operation of heated taxidermic mounts used to measure standard operative temperature. *Ecology* **64**, 1658–1662.

Barnas, G. M., and Burger, R. E. (1983). Interaction of temperature with extra- and intrapulmonary chemoreceptor control of ventilatory movements in the awake chicken. *Respir. Physiol.* **54**, 223–232.

Barnas, G. M., and Rautenberg, W. (1984). Respiratory responses to shivering produced by external and central cooling in the pigeon. *Pflügers Arch.* **401**, 228–232.

Barnas, G. M., and Rautenberg, W. (1987). Temperature control. *In* "Bird Respiration" (T. J. Seller, ed.), Vol., 1, pp. 131–153. CRC Press, Boca Raton, FL.

Barnas, G. M., and Rautenberg, W. (1980). Shivering and cardiorespiratory responses during normocapnic hypoxia in the pigeon. *J. Appl. Physiol.* **68**, 84–87.

Barnas, G. M., Hempleman, S. C., and Burger, R. E. (1983). Effect of temperature on the CO_2 sensitivity of avian intrapulmonary chemoreceptors. *Respir. Physiol.* **54**, 233–240.

Barnas, G. M., Gleeson, M., and Rautenberg, W. (1986). Effect of bilateral vagotomy on arterial acid–base balance stability during panting in the pigeon. *Respir. Physiol.* **66**, 293–302.

Barré, H., and Roussel, B. (1986). Thermal and metabolic adaptation to first cold-water immersion in juvenile penguins. *Am. J. Physiol.* **251**, R456–R462.

Barré, H., Duchamp, C., Rouanet, J.-L., Dittmar, A., and Delhomme, G. (1989). Muscular nonshivering thermogenesis in cold-acclimated ducklings. *In* "Physiology of Cold Adaptation in Birds" (C. Bech and R. E. Reinertsen, eds.), pp. 49–57. Plenum, New York.

Barré, H., Geloen, A. Chatonnet, J., Dittmar, A., and Rouanet, J.-L. (1985). Potentiated muscular thermogenesis in cold-acclimated muscovy duckling. *Am. J. Physiol.* **249**, R533–R538.

Barrett, R. K., and Takahashi, J. S. (1995). Temperature compensation and temperature entrainment of the chick pineal cell circadian clock. *J. Neurosci.* **15**, 5681–5692.

Barrott, H. G., and Pringle, E. M. (1946). Energy and gaseous metabolism of the chicken from hatch to maturity as affected by temperature. *J. Nutr.* **31**, 35–50.

Bartholomew, G. A. (1966). The role of behavior in the temperature regulation of the masked booby. *Condor* **68**, 523–535.

Bartholomew, G. A., Jr., and Dawson, W. R. (1954). Body temperature and water requirements of the western mourning dove, *Zenaidura macroura marginella*. *Ecology* **35**, 181–187.

Bartholomew, G. A., and Dawson, W. R. (1979). Thermoregulatory behavior during incubation in Heermann's gulls. *Physiol. Zool.* **52**, 422–437.

Bartholomew, G. A., Howell, T. R., and Cade, T. J. (1957). Torpidity in the white-throated swift, Anna hummingbird, and poor-will. *Condor* **59**, 145–155.

Bartholomew, G. A., Hudson, J. W., and Howell, T. R. (1962). Body temperature, oxygen consumption, evaporative water loss, and heart rate in the poor-will. *Condor* **64**, 117–125.

Bartholomew, G. A., Lasiewski, R. C., Crawford, E. C., Jr. (1968). Patterns of panting and gular flutter in cormorants, pelicans, owls, and doves. *Condor* **70**, 31–34.

Bartholomew, G. A., Vleck, C. M., and Bucher, T. L. (1983). Energy metabolism and nocturnal hypothermia in two tropical passerine frugivores, *Manacus vitellinus* and *Pipra mentalis*. *Physiol. Zool.* **56**, 370–379.

Basta, D., Tzschentke, B., and Nichelmann, M. (1997). Temperature guardian neurons in the preoptic area of the hypothalamus. *Brain Res.* **767**, 361–362.

Baumel, J. J., Dalley, A. F., and Quinn, T. H. (1983). The collar plexus of subcutaneous thermoregulatory veins in the pigeon, *Columba livia*; its association with esophageal pulsation and gular flutter. *Zoomorphology* **102**, 215–239.

Bech, C. (1980). Body temperature, metabolic rate, and insulation in winter and summer acclimatized mute swans (*Cygnus olor*). *J. Comp. Physiol.* **136**, 61–66.

Bech, C., and Johansen, K. (1980). Ventilatory and circulatory responses to hyperthermia in the mute swan (*Cygnus olor*). *J. Exp. Biol.* **88**, 195–204.

Bech, C., and Midtgård, U. (1981). Brain temperature and the *rete mirabile ophthalmicum* in the zebra finch (*Poephila guttata*). *J. Comp. Physiol.* **145**, 89–93.

Bech, C., Abe, A. S., Steffensen, J. F., Berger, M., and Bicudo, J. E. P. W. (1997). Torpor in three species of Brazilian hummingbirds under semi-natural conditions. *Condor* **99**, 780–788.

Bech, C., Brent, R., Pedersen, P. F., Rasmussen, J. G., and Johansen, K. (1982a). Temperature regulation in chicks of the Manx Shearwater *Puffinus puffinus*. *Ornis Scand.* **13**, 206–210.

Bech, C., Johansen, K., Brent, R., and Nicol, S. (1984). Ventilatory and circulatory changes during cold exposure in the Pekin duck (*Anas platyrhynchos*). *Respir. Physiol.* **57**, 103–112.

Bech, C., Johansen, K., and Maloiy, G. M. O. (1979). Ventilation and expired gas composition in the flamingo, *Phoenicopterus ruber*, during normal respiration and panting. *Physiol. Zool.* **52**, 313–328.

Bech, C., Rautenberg, W., and May-Rautenberg, B. (1988). Thermoregulatory responses of the pigeon (*Columba livia*) to selective changes in inspired air temperature. *J. Comp. Physiol. B* **157**, 747–752.

Bech, C., Rautenberg, W., and May, B. (1985). Ventilatory oxygen extraction during cold exposure in the pigeon. *J. Exp. Biol.* **116**, 499–502.

Bech, C., Rautenberg, W., May, B., and Johansen, K. (1982b). Regional blood flow changes in response to thermal stimulation of the brain and spinal cord in the Pekin duck. *J. Comp. Physiol.* **147,** 71–77.

Bennett, P. M., and Harvey, P. H. (1987). Active and resting metabolism in birds: Allometry, phylogeny, and ecology. *J. Zool. (London)* **213,** 327–363.

Berger, M., and Hart, J. S. (1972). Die Atmung beim Kolibri *Amazilia fimbriata* während des Schwirrfluges bei verscheidenen Umgebungstemperaturen. *J. Comp. Physiol.* **81,** 363–380.

Berger, R. J., and Phillips, N. H. (1994). Light suppresses sleep and circadian rhythms in pigeons without consequent sleep rebound in darkness. *Am. J. Physiol.* **267,** R945–R952.

Berman, A., and Meltzer, A. (1978). Metabolic rate: Its circadian rhythmicity in the female domestic fowl. *J. Physiol. (London)* **282,** 419–427.

Bernstein, M. H. (1971). Cutaneous water loss in small birds. *Condor* **73,** 468–469.

Bernstein, M. H., and Schmidt-Nielsen, K. (1974). Ventilation and oxygen extraction in the crow. *Respir. Physiol.* **21,** 393–401.

Bicudo, J. E. P. W. (1996). Physiological correlates of daily torpor in hummingbirds. In "Animals and Temperature: Phenotypic and Evolutionary Adaptation" (I. A. Johnston and A. F. Bennett, eds.), pp. 293–311. *Soc. Exp. Biol. Sem., Ser.* 59. Cambridge Univ. Press, Cambridge, UK.

Binkley, S., Kluth, E., and Menaker, M. (1971). Pineal function in sparrows: Circadian rhythms and body temperature. *Science* **174,** 311–314.

Blem, C. R. (1974). Geographic variation of thermal conductance in the house sparrow *Passer domesticus*. *Comp. Biochem. Physiol. A* **47,** 101–108.

Boersma, P. D. (1986). Body temperature, torpor, and growth in chicks of fork-tailed storm-petrels (*Oceanodroma furcata*). *Physiol. Zool.* **59,** 10–19.

Booth, D. T. (1984). Thermoregulation in neonate mallee fowl *Leipoa ocellata*. *Physiol. Zool.* **57,** 251–260.

Bouverot, O., Hildwein, G., and Oulhen, P. (1976). Ventilatory and circulatory O_2 convection at 4000 m in pigeons at neutral or cold temperature. *Respir. Physiol.* **28,** 371–385.

Brackenbury, J. (1984). Physiological responses of birds to flight and running. *Biol. Rev.* **59,** 559–575.

Brackenbury, J. H., Avery, P., and Gleeson, M. (1981a). Air sac gases and ventilation during panting in fowl, *Gallus gallus*. *J. Exp. Biol.* **90,** 343–345.

Brackenbury, J. H., Avery, P., and Gleeson, M. (1981b). Respiratory evaporation in panting fowl: Partition between the respiratory and buccopharyngeal pumps. *J. Comp. Physiol.* **145,** 63–66.

Brenner, F. J. (1965). Metabolism and survival time of grouped starlings at various temperatures. *Wilson Bull.* **77,** 388–395.

Brent, R., Pedersen, P. F., Bech, C., and Johansen, K. (1984). Lung ventilation and temperature regulation in the European coot *Fulica atra*. *Physiol. Zool.* **57,** 19–25.

Brent, R., Pedersen, P. F., Bech, C., and Johansen, K. (1985). Thermal balance in the European coot *Fulica atra* exposed to temperatures from −28°C to 40°C. *Ornis Scand.* **16,** 145–150.

Brent, R., Rasmussen, J. G., Bech, C., and Martin, S. (1983). Temperature dependence of ventilation and O_2 extraction in the kittiwake, *Rissa tridactyla*. *Experientia (Basel)* **39,** 1092–1093.

Brigham, R. M. (1992). Daily torpor in a free-ranging goatsucker, the common poorwill (*Phalaenoptilus nuttallii*). *Physiol. Zool.* **65,** 457–472.

Brigham, R. M., and Trayhurn, P. (1994). Brown fat in birds? A test for the mammalian BAT-specific mitochondrial uncoupling protein in common poor-wills. *Condor* **96,** 208–211.

Brown, J. H., Calder, W. A., III, and Kodric-Brown, A. (1978). Correlates and consequences of body size in nectar-feeding birds. *Am. Zool.* **18,** 687–700.

Brummermann, M., and Rautenberg, W. (1989). Interaction of autonomic and behavioral thermoregulation in osmotically stressed pigeons (*Columba livia*). *Physiol. Zool.* **62,** 1102–1116.

Brummermann, M., and Reinertsen, R. E. (1991). Adaptation of homeostatic thermoregulation: Comparison of incubating and non-incubating bantam hens. *J. Comp. Physiol. B* **161,** 133–140.

Bucher, T. L. (1981). Oxygen consumption, ventilation and respiratory heat loss in a parrot *Bolborhynchus lineola*, in relation to air temperature. *J. Comp. Physiol.* **142,** 479–488.

Bucher, T. L. (1985). Ventilation and oxygen consumption in *Amazona viridigenalis*: A reappraisal of "resting" respiratory parameters in birds. *J. Comp. Physiol. B* **155,** 269–276.

Bucher, T. L. (1986). Ratios of hatchling and adult mass-independent metabolism: A physiological index to the altricial–precocial continuum. *Respir. Physiol.* **65,** 69–83.

Bucher, T. L., and Chappell, M. A. (1992). Ventilatory and metabolic dynamics during entry into and arousal from torpor in *Selasphorus* hummingbirds. *Physiol. Zool.* **65,** 978–993.

Bucher, T. L., and Chappell, M. A. (1997). Respiratory exchange and ventilation during nocturnal torpor in hummingbirds. *Physiol. Zool.* **70,** 45–52.

Budgell, P. (1971). Behavioural thermoregulation in the Barbary dove (*Streptopelia risoria*). *Anim. Behav.* **19,** 524–531.

Bugden, S. C., and Evans, R. M. (1991). Vocal responses to chilling in embryonic and neonatal American Coots. *Wilson Bull.* **103,** 712–717.

Burgoon, D. A., Kilgore, D. L., Jr., and Motta, P. J. (1987). Brain temperature in the calliope hummingbird (*Stellula calliope*): A species lacking a *rete mirabile ophthalmicum*. *J. Comp. Physiol. B* **157,** 583–588.

Buttemer, W. A. (1985). Energy relations of winter roost-site utilization by American goldfinches (*Carduelis tristis*). *Oecologia (Berlin)* **68,** 126–132.

Buttemer, W. A., Astheimer, L. B., Weathers, W. W., and Hayworth, A. M. (1987). Energy savings attending winter-nest use by verdins (*Auriparus flaviceps*). *Auk* **104,** 531–535.

Calder, W. A. (1973). An estimate of the heat balance of a nesting hummingbird in a chilling climate. *Comp. Biochem. Physiol. A* **46,** 291–300.

Calder, W. A. (1994). When do hummingbirds use torpor in nature? *Physiol. Zool.* **67,** 1051–1076.

Calder, W. A., and King J. R. (1974). Thermal and caloric relations of birds. In "Avian Biology" (D. S. Farner and J. R. King, eds.), Vol. IV, pp. 259–413. Academic Press, New York.

Calder, W. A., and Schmidt-Nielsen, K. (1968). Panting and blood carbon dioxide in birds. *Am. J. Physiol.* **215,** 477–482.

Carey, C., Dawson, W. R., Maxwell, L. C., and Faulkner, J. A. (1978). Seasonal acclimatization to temperature in cardueline finches: II. Changes in body composition and mass in relation to season and acute cold stress. *J. Comp. Physiol.* **125,** 101–113.

Carey, C., Marsh, R. L., Bekoff, A., Johnston, R., and Olin, A. M. (1989). Enzyme activities in muscles of seasonally acclimatized house finches. In "Physiology of Cold Adaptation in Birds" (C. Bech and R. E. Reinertsen, eds.), pp. 95–104. Plenum, New York.

Carmi, N., Pinshow, B., and Horowitz, M. (1993). Birds conserve plasma volume during thermal and flight-incurred dehydration. *Physiol. Zool.* **66,** 829–846.

Carpenter, F. L. (1974). Torpor in an Andean hummingbird: Its ecological significance. *Science* **183,** 545–547.

Carpenter, F. L., and Hixon, M. A. (1988). A new function for torpor: Fat conservation in a wild migrant hummingbird. *Condor* **90,** 373–378.

Carpenter, F. L., Hixon, M. A., Beuchat, C. A., Russell, R. W., and Paton, D. C. (1993). Biphasic mass gain in migrant hummingbirds:

Body composition changes, torpor, and ecological significance. *Ecology* **74**, 1173–1182.

Chaplin, S. B. (1976). The physiology of hypothermia in the black-capped chickadee, *Parus atricapillus. J. Comp. Physiol.* **112**, 335–344.

Chaplin, S. B. (1982). The energetic significance of huddling in common bushtits (*Psaltiparus minimus*). *Auk* **99**, 424–430.

Chappell, M. A., and Bucher, T. L. (1987). Effects of temperature and altitude on ventilation and gas exchange in chukars (*Alectoris chukar*). *J. Comp. Physiol. B* **157**, 126–136.

Chappell, M. A., Bachman, G. C., and Hammond, K. A. (1997). The heat increment of feeding in house wren chicks: Magnitude, duration, and substitution for thermostatic costs. *J. Comp. Physiol. B* **167**, 313–318.

Chaui-Berlinck, J. G., and Bicudo, J. E. P. W. (1995). Unusual metabolic shifts in fasting hummingbirds. *Auk* **112**, 774–778.

Choi, I.-H., Ricklefs, R. E., and Shea, R. E. (1993). Skeletal muscle growth, enzyme activities, and the development of thermogenesis: A comparison between altricial and precocial birds. *Physiol. Zool.* **66**, 455–473.

Connolly, E., Nedergaard, J., and Cannon, B. (1989). Shivering and nonshivering thermogenesis in birds: a mammalian view. *In* "Physiology of Cold Adaptation in Birds" (C. Bech and R. E. Reinertsen, eds.), pp. 37–48. Plenum, New York.

Cooper, S. J., and Swanson, D. L. (1994). Seasonal acclimatization of thermoregulation in the black-capped chickadee. *Condor* **96**, 638–646.

Crawford, E. C., Jr., and Kampe, G. (1971). Resonant panting in pigeons. *Comp. Biochem. Physiol.* **40**, 549–552.

Crawford, E. C., Jr., and Schmidt-Nielsen, K. (1967). Temperature regulation and evaporative cooling in the ostrich. *Am. J. Physiol.* **212**, 347–353.

Csada, R. D., and Brigham, R. M. (1994). Reproduction constrains the use of daily torpor by free-ranging common poorwills (*Phalaenoptilus nuttallii*) (Aves: Caprimulgidae). *J. Zool. (London)* **234**, 209–216.

Dawson, W. R. (1954). Temperature regulation and water requirements of the brown and Abert towhees, *Pipilo fuscus* and *Pipilo aberti*. *Univ. Calif. Publ. Zool.* **59**, 81–124.

Dawson, W. R. (1982). Evaporative losses of water by birds. *Comp. Biochem. Physiol. A* **71**, 495–509.

Dawson, W. R. (1984a). Physiological studies of desert birds: Present and future consideration. *J. Arid. Envir.* **7**, 133–155.

Dawson, W. R. (1984b). Metabolic responses of embryonic seabirds to temperature. *In* "Seabird Energetics" (G. C. Whittow and H. Rahn, eds.), pp. 139–157. Plenum, New York.

Dawson, W. R., and Bartholomew, G. A. (1968). Temperature regulation and water economy of desert birds. *In* "Desert Biology" (G. W. Brown, Jr., ed.), Vol. I, pp. 357–394. Academic Press, New York.

Dawson, W. R., and Carey, C. (1976). Seasonal acclimatization to temperature in cardueline finches: I. Insulative and metabolic adjustments. *J. Comp. Physiol.* **112**, 317–333.

Dawson, W. R., and Evans, F. C. (1957). Relation of growth and development to temperature regulation in nestling field and chipping Sparrows. *Physiol. Zool.* **30**, 315–327.

Dawson, W. R., and Evans, F. C. (1960). Relation of growth and development to temperature regulation in nestling vesper sparrows. *Condor* **62**, 329–340.

Dawson, W. R., and Fisher, C. D. (1969). Responses to temperature by the spotted nightjar. *Condor* **71**, 49–53.

Dawson, W. R., and Hudson, J. W. (1970). Birds. *In* "Comparative Physiology of Thermoregulation" (G. C. Whittow, ed.), Vol. I, pp. 223–310. Academic Press, New York.

Dawson, W. R., and Marsh, R. L. (1989). Metabolic acclimatization to cold and season in birds. *In* "Physiology of Cold Adaptation in Birds" (C. Bech and R. E. Reinertsen, eds.), pp. 83–94. Plenum, New York.

Dawson, W. R., and O'Connor, T. P. (1996). Energetic features of avian thermoregulatory responses. *In* "Avian Energetics and Nutritional Ecology" (C. Carey, ed.), pp. 85–124. Chapman & Hall, New York.

Dawson, W. R., and Tordoff, H. B. (1959). Relation of oxygen consumption to temperature in the evening grosbeak. *Condor* **61**, 388–396.

Dawson, W. R., Marsh, R. L., and Yacoe, M. E. (1983a). Metabolic adjustments of small passerine birds for migration and cold. *Am. J. Physiol.* **245**, R755–R767.

Dawson, W. R., Marsh, R. L., Buttemer, W. A., and Carey, C. (1983b). Seasonal and geographic variation of cold resistance in house finches *Carpodacus mexicanus. Physiol. Zool.* **56**, 353–369.

Dawson, W. R., Buttemer, W. A., and Carey, C. (1985). A reexamination of the metabolic response of house finches to temperature. *Condor* **87**, 424–427.

De Jong, A. A. (1976). The influence of simulated solar radiation on the metabolic rate of white-crowned sparrows. *Condor* **78**, 174–179.

Delane, T. M., and Hayward, J. S. (1975). Acclimatization to temperature in pheasants (*Phasianus colchicus*) and partridge (*Perdix perdix*). *Comp. Biochem. Physiol. A* **51**, 531–536.

Deshazer, J. A. (1967). Heat loss variations of the laying hen. PhD. thesis, North Carolina State University, Raleigh, NC.

Dietz, M. W., Mourik, S., Tøien, Ø., Koolmees, P. A., and Tersteeg-Zijderveld, M. H. G. (1997). Participation of breast and leg muscles in shivering thermogenesis in young turkeys and guinea fowl. *J. Comp. Physiol. B* **167**, 451–460.

Duchamp, C., Barré, H., Delage, D., Berne, G., Brebion, P., and Rouanet, J.-L. (1989). Non-shivering thermogenesis in winter-acclimatized king penguin chicks. *In* "Physiology of Cold Adaptation in Birds" (C. Bech and R. E. Reinertsen, eds.), pp. 59–67. Plenum, New York.

Duchamp, C., Barré, H., Rouanet, J.-L., Lanni, A., Cohen-Adad, F., Berne, G., and Brebion, P. (1991). Non-shivering thermogenesis in king penguin chicks: 1. Role of skeletal muscle. *Am. J. Physiol.* **261**, R1438–R1445.

Dumonteil, E., Barré, H., Rouanet, J.-L., Diarra, M., and Bouvier, J. (1994). Dual core and shell temperature regulation during sea acclimatization in Gentoo penguins (*Pygoscelis papua*). *Am. J. Physiol.* **266**, R1319–R1326.

Dunn, E. H. (1976). The development of endothermy and existence energy expenditure in herring gull chicks. *Condor* **78**, 493–498.

Dutenhoffer, M. S., and D. L. Swanson. (1996). Relationship of basal to summit metabolic rate in passerine birds and the aerobic capacity model for the evolution of endothermy. *Physiol. Zool.* **69**, 1232–1254.

Ederstrom, H. E., and Brumleve, S. J. (1964). Temperature gradients in the legs of cold-acclimated pheasants. *Am. J. Physiol.* **207**, 457–459.

Eissel, K., and Simon, E. (1980). How are neuronal thermosensitivity and lack of thermoreception related in the duck's hypothalamus? A tentative answer. *J. Therm. Biol.* **5**, 219–223.

Ellis, H. I., Maskrey, M., Pettit, T. N., and Whittow, G. C. (1995). Thermoregulation in the brown noddy (*Anous stolidus*). *J. Therm. Biol.* **20**, 307–313.

Eppley, Z. A. (1984). Development of thermoregulatory abilities in Xantus' murrelet chicks *Synthliboramphus hypoleucus. Physiol. Zool.* **57**, 307–317.

Eppley, Z. A. (1996). Development of thermoregulation in birds: Physiology, interspecific variation and adaptation to climate. In "Animals and Temperature" (I. A. Johnston and A. F. Bennett, eds.), pp. 313–419. Soc. Exp. Biol. Sem. Ser. 59. Cambridge Univ. Press, Cambridge, UK.

Evans, R. M. (1992). Embryonic and neonatal vocal elicitation of parental brooding and feeding responses in American white pelicans. *Anim. Behav.* **44,** 667–675.

Evans, R. M. (1994). Cold-induced calling and shivering in young American white pelicans—Honest signalling of offspring need for warmth in a functionally integrated thermoregulatory system. *Behavior* **129,** 13–34.

Feldman, S. E., Larsson, S., Dimick, M. K., and Lepkovsky, S. (1957). Aphagia in chickens. *Am. J. Physiol.* **191,** 259–261.

Foà, A., and Menaker, M. (1988). Contribution of pineal and retinae to the circadian rhythms of circulating melatonin in pigeons. *J. Comp. Physiol. A* **164,** 25–30.

Fraifeld, V., Blaicher-Kulick, R., Degen, A. A., and Kaplanski, J. (1995). Is hypothalamic prostaglandin E_2 involved in avian fever? *Life Sci.* **56,** 1343–1346.

French, A. R. (1993). Hibernation in birds: Comparisons with mammals. In "Life in the Cold" (C. Carey, G. L. Florant, B. A. Wunder, and B. Horwitz, eds.), pp. 43–53. Westview Press, Boulder, CO.

Fujita, M., Shimizu, M., and Yamamoto, S. (1990). Effects of short term heat exposure on physiological response and plasma substrate concentration in laying hens. *Jap. J. Zootech. Sci.* **61,** 707–713.

Gabrielsen, G. W., Mehlum, F., and Karlsen, H. E. (1988). Thermoregulation in four species of arctic seabirds. *J. Comp. Physiol. B* **157,** 703–708.

Ganey, J. L., Balda, R. P., and King, R. M. (1993). Metabolic rate and evaporative water loss of Mexican spotted and great horned owls. *Wilson Bull.* **105,** 645–656.

Gaunt, S. L. L. (1980). Thermoregulation in doves (Columbidae): A novel esophageal heat exchanger. *Science* **210,** 445–447.

Geiser, F. (1988). Reduction of metabolism during hibernation and daily torpor in mammals and birds: Temperature effect or physiological inhibition? *J. Comp. Physiol. B* **158,** 25–37.

Geiser, F., and Ruf, T. (1995). Hibernation versus daily torpor in mammals and birds: Physiological variables and classification of torpor patterns. *Physiol. Zool.* **68,** 935–966.

Geisthövel, E., Ludwig, O., and Simon, E. (1986). Capsaicin fails to produce disturbances of autonomic heat and cold defence in an avian species (*Anas platyrhynchos*). *Pflügers Arch.* **406,** 343–350.

Gelineo, S. (1955). Température d'adaptation et production de chaleur chez les oiseaux de petite tailles. *Arch. Sci. Physiol.* **9,** 225–243.

Gelineo, S. (1964). Organ systems in adaptation: The temperature regulating system. In "Handbook of Physiology, Section 4: Adaptation to the Environment" (D. B. Dill, ed.), pp. 259–282. American Physiological Society, Washington, DC.

Gelineo, S. (1969). Heat production in birds in summer and winter. *Srpska Akad. Nauka I Umetnosti Belgrad, Bull. Classe Sci. Math. Sci. Nat.* **12**(n.s.), pp. 99–105.

Geran, L. C., and Rashotte, M. E. (1997). Participation of gastrointestinal-load volume in "setting" the pigeon's nocturnal body temperature. *Naturwissenschaften* **84,** 350–353.

Gleeson, M., Barnas, G. M., and Rautenberg, W. (1986a). The effects of hypoxia on the metabolic and cardiorespiratory responses to shivering produced by external and central cooling in the pigeon. *Pflügers Arch.* **407,** 312–319.

Gleeson, M., Barnas, G. M., and Rautenberg, W. (1986b). Cardiorespiratory responses to shivering in vagotomized pigeons during normoxia and hypoxia. *Pflügers Arch.* **407,** 664–669.

Goldstein, D. L. (1983). Effect of wind on avian metabolic rate with particular reference to Gambel's quail. *Physiol. Zool.* **56,** 485–492.

Görke, K., and Pierau, F.-K. (1979). Initiation of muscle activity in spinalized pigeons during spinal cord cooling and warming. *Pflügers Arch.* **381,** 47–52.

Graf, R. (1980). Diurnal changes of thermoregulatory functions in pigeons: I. Effector mechanisms. *Pflügers Arch.* **386,** 173–179.

Graf, R. Krishna, S., and Heller, H. C. (1989). Regulated nocturnal hypothermia induced in pigeons by food deprivation. *Am. J. Physiol.* **256,** R733–R738.

Grant, G. S. (1982). Avian incubation: Egg temperature, nest humidity, and behavioral thermoregulation in a hot environment. *Ornithol. Monogr.* **30,** ix + 75 pp.

Grant, G. S. (1985). Rete mirabile ophthalmicum and intercarotid anastomosis in Procellariiformes taken off the North Carolina coast. *Brimleyana* **11,** 81–86.

Grubb, T. C., Jr. (1975). Weather-dependent foraging behavior of some birds wintering in a deciduous woodland. *Condor* **77,** 175–182.

Grubb, T. C., Jr. (1978). Weather-dependent foraging rates of wintering woodland birds. *Auk* **95,** 370–376.

Hagan, A. A., and Heath, J. E. (1980). Effects of preoptic lesions on thermoregulation in ducks. *J. Therm. Biol.* **5,** 141–150.

Hainsworth, F. R., and Wolf, L. L. (1970). Regulation of oxygen consumption and body temperature during torpor in a hummingbird, *Eulampis jugularis*. *Science* **168,** 368–369.

Hainsworth, F. R., Collins, B. G., and Wolf, L. L. (1977). The function of torpor in hummingbirds. *Physiol. Zool.* **50,** 215–222.

Hammel, H. T., Maggert, J., Kaul, R., Simon, E., and Simon-Oppermann, C. (1976). Effects of altering spinal cord temperature on temperature regulation in the Adelie penguin, *Pygoscelis adeliae*. *Pflügers Arch.* **362,** 1–6.

Hart, J. S. (1962). Seasonal acclimatization in four species of small wild birds. *Physiol. Zool.* **35,** 224–236.

Hassinen, E., Pyörnilä, A., and Hissa, R. (1994). Vasotocin and angiotensin II affect thermoregulation in the pigeon, *Columba livia*. *Comp. Biochem. Physiol. A* **107,** 545–551.

Hatch, D. E. (1970). Energy conserving and heat dissipating mechanisms of the turkey vulture. *Auk* **87,** 111–124.

Hayes, S. R., and Gessaman, J. A. (1982). Prediction of raptor resting metabolism: Comparison of measured values with statistical and biophysical estimates. *J. Therm. Biol.* **7,** 45–50.

Helfmann, W., Jannes, P., and Jessen, C. (1981). Total body thermosensitivity and its spinal and supraspinal fractions in the conscious goose. *Pflügers Arch.* **391,** 60–67.

Heller, H. C., Graf, R., and Rautenberg, W. (1983). Circadian and arousal state influences on thermoregulation in the pigeon. *Am. J. Physiol.* **245,** R321–R328.

Hennemann, W. W. (1982). Energetics and spread-winged behavior of anhingas in Florida. *Condor* **84,** 91–96.

Hiebert, S. M. (1990). Energy costs and temporal organization of torpor in the rufous hummingbird (*Selasphorus rufus*). *Physiol. Zool.* **63,** 1082–1097.

Hiebert, S. M. (1991). Seasonal differences in the response of rufous hummingbirds to food restriction: Body mass and the use of torpor. *Condor* **93,** 526–537.

Hiebert, S. M. (1993). Seasonality of daily torpor in a migratory hummingbird. In "Life in the Cold" (C. Carey, G. L. Florant, B. A. Wunder, and B. Horwitz, eds.), pp. 25–32. Westview Press, Boulder, CO.

Hiestand, W. A., and Randall, W. C. (1942). Influence of proprioceptive vagal afferents on panting and accessory panting movements in mammals and birds. *Am. J. Physiol.* **138,** 12–15.

Hill, R. W., and Beaver, D. L. (1982). Inertial thermostability and thermoregulation in broods of red-winged blackbirds. *Physiol. Zool.* **55,** 250–266.

Hill, R. W., Beaver, D. L., and Veghte, J. H. (1980). Body surface temperatures and thermoregulation in the black-capped chickadee (*Parus atricapillus*). *Physiol. Zool.* **53**, 305–321.

Hillman, P. E., Scott, N. R., and van Tienhoven, A. (1980). Effect of 5-hydroxytryptamine and acetylcholine on the energy budget of chickens. *Am. J. Physiol.* **239**, R57–R61.

Hinds, D. S., Baudinette, R. V., MacMillen, R. E., and Halpern, E. A. (1993). Maximum metabolism and the aerobic factorial scope of endotherms. *J. Exp. Biol.* **182**, 41–56.

Hissa, R. (1988). Controlling mechanisms in avian temperature regulation: A review. *Acta Physiol. Scand.* **132**, 567 (Suppl.).

Hissa, R., and Palokangas, R. (1970). Thermoregulation in the titmouse (*Parus major* L.). *Comp. Biochem. Physiol.* **33**, 941–953.

Hissa, R., and Rautenberg, W. (1974). The influence of centrally applied noradrenaline on shivering and body temperature in the pigeon. *J. Physiol. (London)* **238**, 427–435.

Hissa, R., and Rautenberg, W. (1975). Thermoregulatory effects of intrahypothalamic injections of neurotransmitters and their inhibitors in the pigeon. *Comp. Biochem. Physiol. A* **51**, 319–326.

Hissa, R., Pyörnilä, A., and George, J. C. (1980). The influence of intrahypothalamic injections of prostaglandins E_1 and F_{2a} and ambient temperature on thermoregulation in the pigeon. *J. Therm. Biol.* **5**, 163–167.

Hohtola, E., Hissa, R., Pyörnilä, A., Rintamäki, H., and Saarela, S. (1991). Hypothermia in fasting Japanese quail: The effect of ambient temperature *Physiol. Behav.* **49**, 563–567.

Hohtola, E., Pyörnilä, A., and Rintamäki, H. (1994). Fasting endurance and cold resistance without hypothermia in a small predatory bird: The metabolic strategy of Tengmalm's owl, *Aegolius funereus*. *J. Comp. Physiol. B* **164**, 430–437.

Hohtola, E., Saarela, S., Harjula, R., and Hissa, R. (1989). Cardiovascular and thermoregulatory responses to intrahypothalamically injected neurotransmitters in the pigeon. *J. Therm. Biol.* **14**, 41–45.

Horowitz, K. A., Scott, N. R., Hillman, P. E., and van Tienhoven, A. (1978). Effects of feathers on instrumental thermoregulatory behavior in chickens. *Physiol. Behav.* **21**, 233–238.

Horton, E. W., and Main, I. H. M. (1967). Identification of prostaglandins in central nervous tissues of the cat and chicken. *Br. J. Pharmacol. Chemother.* **30**, 582–602.

Howell, T. R. (1979). Breeding biology of the Egyptian plover, *Pluvianus aegyptius*, *Univ. Calif. Publ. Zool.* **113**, 1–76.

Howell, T. R., and Bartholomew, G. A. (1961). Temperature regulation and Laysan and black-footed albatrosses. *Condor* **63**, 185–197.

Hudson, J. W., and Kimzey S. L. (1966). Temperature regulation and metabolic rhythms in populations of the house sparrow, *Passer domesticus*. *Comp. Biochem. Physiol.* **17**, 203–217.

Hudson, J. W., Dawson, W. R., and Hill, R. W. (1974). Growth and development of temperature regulation in nestling cattle egrets. *Comp. Biochem. Physiol. A* **49**, 717–741.

Inomoto, T., and Simon, E. (1981). Extracerebral deep-body cold sensitivity in the Pekin duck. *Am. J. Physiol.* **241**, R136–R145.

International Union of Physiological Sciences Commission for Thermal Physiology (1987). Glossary of terms for thermal physiology, second edition. *Pflügers Arch.* **410**, 567–587.

Irving, L. (1972). "Zoophysiology and Ecology, Vol. 2: Arctic Life of Birds and Mammals Including Man" (D. S. Farner, ed.). Springer-Verlag, Berlin.

Itsakigluckich, S., and Arad, Z. (1992). The effect of dehydration on brain temperature regulation in Japanese quail (*Coturnix coturnix japonica*). *Comp. Biochem. Physiol. A* **101**, 583–588.

Jaeger, E. C. (1949). Further observations on hibernation of the poorwill. *Condor* **51**, 105–109.

Jansky, L. (1995). Humoral thermogenesis and its role in maintaining thermal balance. *Physiol. Rev.* **75**, 237–259.

Jensen, C., and Bech, C. (1992). Ventilation and gas exchange during shallow hypothermia in pigeons. *J. Exp. Biol.* **165**, 111–120.

Jenssen, B. M., Ekker, M., and Bech, C. (1989). Thermoregulation in winter-acclimatized common eiders (*Somateria mollissima*) in air and water. *Can. J. Zool.* **67**, 669–673.

Jessen, C., Hales, J. R. S., and Molyneux, G. S. (1982). Hypothalamic thermosensitivity in an emu, *Dromiceius novae-hollandiae*. *Pflügers Arch.* **393**, 278–280.

Johansen, K., and Bech, C. (1983). Heat conservation during cold exposure in birds (vasomotor and respiratory implications). *Polar Res.* **1** (n.s.), 259–268.

Johansen, K., and Bech, C. (1984). Breathing and thermoregulation. *In* "Thermal Physiology" (J. R. S. Hales, ed.), pp. 341–346. Raven, New York.

Jones, J. H. (1982). Pulmonary blood flow distribution in panting ostriches. *J. Appl. Physiol.* **53**, 1411–1417.

Juorio, A. V., and Vogt, M. (1967). Monoamines and their metabolites in the avian brain. *J. Physiol. (London)* **189**, 489–518.

Kahl, M. P., Jr. (1963). Thermoregulation in the wood stork, with special reference to the role of the legs. *Physiol. Zool.* **36**, 141–151.

Kaiser, T. J., and Bucher, T. L. (1985). The consequences of reverse sexual size dimorphism for oxygen consumption, ventilation, and water loss in relation to ambient temperature in the prairie falcon, *Falco mexicanus*. *Physiol. Zool.* **58**, 748–758.

Kendeigh, S. C., Dol'nik, V. R., and Gavrilov, V. M. (1977). Avian energetics. *In* "Granivorous Birds in Ecosystems" (J. Pinowski and S. C. Kendeigh, eds.), pp. 127–204. Cambridge University Press, Cambridge, UK.

Kessel, B. (1976). Winter activity patterns of black-capped chickadees in interior Alaska. *Wilson Bull.* **88**, 36–61.

Kilgore, D. L., Jr., Boggs, D. F., and Birchard, G. F. (1979). Role of *Rete mirabile ophthalmicum* in maintaining the body-to-brain temperature difference in pigeons. *J. Comp. Physiol.* **129**, 119–122.

King, J. R., and Farner, D. S. (1961). Energy metabolism, thermoregulation and body temperature. *In* "Biology and Comparative Physiology of Birds" (A. J. Marshall, ed.), Vol. II, pp. 215–288. Academic Press, New York.

Kitchell, R. L., Ström, L., and Zotterman, Y. (1959). Electrophysiological studies of thermal and taste reception in chickens and pigeons. *Acta Physiol. Scand.* **46**, 133–151.

Klaassen, M., Bech, C., Masman, D., and Slagsvold, G. (1989). Energy partitioning in arctic tern chicks (*Sterna paradisaea*) and possible metabolic adaptations in high latitude chicks. *In* "Physiology of Cold Adaptation in Birds" (C. Bech and R. E. Reinertsen, eds.), pp. 339–347. Plenum, New York.

Klandorf, H., Lea, R. W., and Sharp, P. J. (1982). Thyroid function in laying, incubating, and broody bantam hens. *Gen. Comp. Endocrinol.* **47**, 492–496.

Kluger, M. J. (1979). Phylogeny of fever. *Fed Proc.* **38**, 30–34.

Korhonen, K. (1980). Microclimate in the snow burrows of willow grouse (*Lagopus lagopus*). *Ann. Zool. Fenn.* **17**, 5–9.

Korhonen, K. (1981). Temperature in the nocturnal shelters of the redpoll (*Acanthis flammea* L.) and the Siberian tit (*Parus cinctus* Budd.) in winter. *Ann. Zool. Fenn.* **18**, 165–168.

Lacey, R. A., Jr., and Burger, R. E. (1972). Personal communication cited in Lasiewski (1972).

Lasiewski, R. C. (1969). Physiological responses to heat stress in the poorwill *Am. J. Physiol.* **217**, 1504–1509.

Lasiewski, R. C. (1972). Respiratory function in birds. *In* "Avian Biology" (D. S. Farner and J. R. King, eds.), Vol. II, pp. 287–342. Academic Press, New York.

Lasiewski, R. C., and Lasiewski, R. J. (1967). Physiological responses of the blue-throated and Rivoli's hummingbirds. *Auk* **84**, 34–48.

Lasiewski, R. C., Weathers, W. W., and Bernstein, M. H. (1967). Physiological responses of the giant hummingbird, *Patagona gigas*. *Comp. Biochem. Physiol.* **23,** 797–813.

Laudenslager, M. L., and Hammel, H. T. (1977). Environmental temperature selection by the Chukar partridge. *Physiol. Behav.* **19,** 543–548.

Lee, P., and Schmidt-Nielsen, K. (1971). Respiratory and cutaneous evaporation in the zebra finch: Effect on water balance. *Am. J. Physiol.* **220,** 1598–1605.

Leitner, L.-M., and Roumy, M. (1974). Thermosensitive units in the tongue and in the skin of the duck's bill. *Pflügers Arch.* **346,** 151–155.

LeMaho, Y. (1983). Metabolic adaptations to long-term fasting in antarctic penguins and domestic geese. *J. Therm. Biol.* **8,** 91–96.

LeMaho, Y., Delclitte, P., and Chatonnet, J. (1976). Thermoregulation in fasting emperor penguins under natural conditions. *Am. J. Physiol.* **231,** 913–922.

Lens, L. (1996). Wind stress affects foraging site competition between crested tits and willow tits. *J. Avian Biol.* **27,** 41–46.

Lepkovsky, S., Snapir, N., and Furuta, F. (1968). Temperature regulation and appetitive behavior in chickens with hypothalamic lesions. *Physiol. Behav.* **3,** 911–915.

Ligon, J. D. (1970). Still more responses of the poor-will to low temperatures. *Condor* **72,** 496–497.

Liknes, E. T., and Swanson, D. L. (1996). Seasonal variation in cold tolerance, basal metabolic rate, and maximal capacity for thermogenesis in white-breasted nuthatches *Sitta carolinensis* and downy woodpeckers *Picoides pubescens*, two unrelated arboreal temperate residents. *J. Avian Biol.* **27,** 279–288.

Lin, M. T., and Simon, E. (1982). Properties of high Q_{10} units in the conscious duck's hypothalamus responsive to changes of core temperature. *J. Physiol.* (*London*) **322,** 127–137.

Löhrl, H. (1955). Schlafgewohnheitender Baumläufer (*Certhia brachydactyla; C. familiaris*) und anderer Kleinvögel in kalten Winternächten. *Vogelwarte* **18,** 71–77.

Lott, B. D. (1991). The effect of feed intake on body temperature and water consumption of male broilers during heat exposure. *Poultry Sci.* **70,** 756–759.

Lustick, S., Battersby, B., and Mayer, L. (1982). Energy exchange in the winter acclimatized American goldfinch, *Carduelis* (*Spinus*) *tristis*. *Comp. Biochem. Physiol. A* **72,** 715–719.

Macari, M., Furlan, R. L., Gregorut, F. P., Secato, E. R., and Guerreiro, J. R. (1993). Effects of endotoxin, interleukin-1-beta and prostaglandin injections on fever response in broilers. *Br. Poultr. Sci.* **34,** 1035–1042.

McFarland, D., and Budgell, P. (1970). The thermoregulatory role of feather movements in the barbary dove (*Streptopelia risoria*). *Physiol. Behav.* **5,** 763–771.

MacKenzie, J. M. D. (1959). Roosting of treecreepers. *Bird Study* **6,** 8–14.

MacMillen, R. E. (1974). Bioenergetics of Hawaiian honeycreepers: The amakihi (*Loxops virens*) and the anianiau (*L. parva*). *Condor* **76,** 62–69.

MacMillen, R. E., and Trost, C. H. (1967). Nocturnal hypothermia in the Inca dove, *Scardafella inca*. *Comp. Biochem. Physiol.* **23,** 243–253.

McNab, B. K. (1996). Metabolism and temperature regulation of kiwis (Apterygidae). *Auk* **113,** 687–692.

McNab, B. K., and Bonaccorso, F. J. (1995). The energetics of Australasian swifts, frogmouths, and nightjars. *Physiol. Zool.* **68,** 245–261.

Maddocks, T. A., and Geiser, F. (1997). Energetics, thermoregulation and nocturnal hypothermia in Australian silvereyes. *Condor* **99,** 104–112.

Madsen, H. (1930). Quelques remarques sur la cause pourquoi les grande oiseaux au Soudan planent si haut au milieu de la journée. *Vidensk. Medd. Dansk Naturhist. Forening, Copenhagen* **8,** 301–303.

Mahoney, S. A., Fairchild, L., and Shea, R. E. (1985). Temperature regulation in great frigatebirds *Fregata minor*. *Physiol. Zool.* **58,** 138–148.

Maloney, S. K., and Dawson, T. J. (1993). Sexual dimorphism in basal metabolism and body temperature of a large bird, the emu. *Condor* **95,** 1034–1037.

Maloney, S. K., and Dawson, T. J. (1994). Thermoregulation in a large bird, the emu (*Dromaius novaehollandiae*). *J. Comp. Physiol. B* **164,** 464–472.

Marder, J. (1973a). Body temperature regulation in the brown-necked raven (*Corvus corax ruficollis*): I. Metabolic rate, evaporative water loss and body temperature of the raven exposed to heat stress. *Comp. Biochem. Physiol. A* **45,** 421–430.

Marder, J. (1973b). Temperature regulation in the Bedouin fowl (*Gallus domesticus*). *Physiol. Zool.* **46,** 208–217.

Marder, J. (1990). The effect of acclimation on the acid–base status of pigeons exposed to high ambient temperatures. *J. Therm. Biol.* **15,** 217–221.

Marder, J., and Arad, Z. (1989). Panting and acid–base regulation in heat stressed birds. *Comp. Biochem. Physiol. A* **94,** 395–400.

Marder, J., and Arieli, Y. (1988). Heat balance of acclimated pigeons (*Columba livia*) exposed to temperatures up to 60°C Ta. *Comp. Biochem. Physiol. A* **91,** 165–170.

Marder, J., and Ben-Asher, J. (1983). Cutaneous water evaporation: I. Its significance in heat-stressed birds. *Comp. Biochem. Physiol. A* **75,** 425–431.

Marder, J., and Gavrieli-Levin, I. (1987). The heat-acclimated pigeon: An ideal physiological model for a desert bird. *J. Appl. Physiol.* **62,** 952–958.

Marder, J., and Raber, P. (1989). Beta-adrenergic control of transcutaneous evaporative cooling mechanisms in birds. *J. Comp. Physiol. B* **159,** 97–103.

Marder, J., Gavrieli-Levin, I., and Raber, P. (1986). Cutaneous water evaporation in heat-stressed spotted sandgrouse. *Condor* **88,** 99–100.

Marder, J., Peltonen, L. M., Raber, P., and Arieli, Y. (1989). CNS thermosensitivity and control of latent heat dissipation in the pigeon. *J. Therm. Biol.* **14,** 243–247.

Marjakangas, A., Rintamäki, H., and Hissa, R. (1984). Thermal responses in the capercaillie *Tetrao urogallus* and the black grouse *Lyrurus tetrix* roosting in the snow. *Physiol. Zool.* **57,** 99–104.

Marsh, R. L. (1993). Does regulatory nonshivering thermogenesis exist in birds? *In* "Life in the Cold" (C. Carey, G. L. Florant, B. A. Wunder, and B. Horwitz, eds.), pp. 535–538. Westview Press, Boulder, CO.

Marsh, R. L., and Dawson, W. R. (1989). Avian adjustments to cold. *In* "Advances in Comparative and Environmental Physiology" (L. C. H. Wang, ed.), Vol. 4, pp. 205–253. Springer-Verlag, Berlin.

Marsh, R. L., and Wickler, S. J. (1982). The role of muscle development in the transition to endothermy in nestling bank swallows, *Riparia riparia*. *J. Comp. Physiol.* **149,** 99–105.

Martin R., Simon, E., and Simon-Oppermann, C. (1981). Brain stem sites mediating specific and non-specific effects on thermoregulation in the Pekin duck. *J. Physiol.* (*London*) **314,** 161–174.

Masman, D. M., Daan, S., and Beldhuis, H. J. A. (1988). Ecological energetics of the kestrel: Daily energy expenditure throughout the year based on time-energy budget, food intake and doubly labeled water methods. *Ardea* **76,** 64–81.

Mathiu, P. M., Dawson, W. R., and Whittow, G. C. (1991). Development of thermoregulation in Hawaiian brown noddies (*Anous stolidus*). *J. Therm. Biol.* **16,** 317–325.

Mathiu, P. M., Dawson, W. R., and Whittow, G. C. (1994). Thermal responses of late embryos and hatchlings of the sooty tern. *Condor* **96,** 280–294.

Mathiu, P. M., Whittow, G. C., and Dawson, W. R. (1992). Hatching and the establishment of thermoregulation in the wedge-tailed shearwater. *Physiol. Zool.* **65,** 583–603.

Meienberger, C., and Dauberschmidt, C. (1992). Kann die "spezifisch dynamische Wirkung" einen Beitrage zur Thermoregulation körnerfressender Singvögel leisten? *J. f. Ornithol.* **133,** 33–41.

Menuam, B., and Richards, S. A. (1975). Observations on the sites of respiratory evaporation in the fowl during thermal panting. *Respir. Physiol.* **25,** 39–52.

Mercer, J. B., and Simon, E. (1984). A comparison between total body thermosensitivity and local thermosensitivity in mammals and birds. *Pflügers Arch.* **400,** 228–234.

Mercer, J. B., and Simon, E. (1987). Appropriate and inappropriate hypothalamic cold sensitivity of the willow ptarmigan. *Acta Physiol. Scand.* **131,** 73–80.

Midtgård, U. (1989a). The effect of heat and cold on the density of arteriovenous anastomoses and tissue composition in the avian nasal mucosa. *J. Therm. Biol.* **14,** 99–102.

Midtgård, U. (1989b). Circulatory adaptations to cold in birds. In "Physiology of Cold Adaptation in Birds" (C. Bech, and R. E. Reinertsen, eds.), pp. 211–222. Plenum, New York.

Midtgård, U., Arad, Z., and Skadhauge, E. (1983). The *rete ophthalmicum* and the relation of its size to the body-to-brain temperature difference in the fowl (*Gallus domesticus*). *J. Comp. Physiol.* **153,** 241–246.

Miller, D. S. (1939). A study of the physiology of the sparrow thyroid. *J. Exp. Zool.* **80,** 259–281.

Mills, S. H., and Heath, J. E. (1972a). Responses to thermal stimulation of the preoptic area in the house sparrow, *Passer domesticus*. *Am. J. Physiol.* **222,** 914–919.

Mills, S. H., and Heath, J. E. (1972b). Anterior hypothalamus/preoptic lesions impair normal thermoregulation in house sparrows. *Comp. Biochem. Physiol. A* **43,** 125–129.

Misson, B. H. (1974). An open circuit respirometer for metabolic studies on the domestic fowl: Establishment of standard operating conditions. *Br. Poult. Sci.* **15,** 287–297.

Montevecchi, W. A., and Vaughan, R. (1989). The ontogeny of thermal independence in nestling gannets. *Ornis Scand.* **20,** 161–168.

Morgan, K. R., Chappell, M. A., and Bucher, T. L. (1992). Ventilatory oxygen extraction in relation to ambient temperature in four antarctic seabirds. *Physiol. Zool.* **65,** 1092–1113.

Morton, M. L. (1978). Snow conditions and the onset of breeding in the mountain white-crowned sparrow. *Condor* **80,** 285–289.

Mückenhoff, K., Barnas, G., and Scheid, P. (1989). Afferent vagal activity during hyperthermic polypnea in the pigeon. *Respir. Physiol.* **75,** 267–278.

Murrish, D. E. (1973). Respiratory heat and water exchange in penguins. *Respir. Physiol.* **19,** 262–270.

Murrish, D. E. (1983). Acid–base balance in penguin chicks exposed to thermal stress. *Physiol. Zool.* **56,** 335–339.

Nakashima, T., Pierau, F.-K., Simon, E., and Hori, T. (1987). Comparison between hypothalamic thermoresponsive neurons from duck and rat slices. *Pflügers Arch.* **409,** 236–243.

Necker, R. (1972). Response of trigeminal ganglion neurons to thermal stimulation of the beak in pigeons. *J. Comp. Physiol.* **78,** 307–314.

Necker, R. (1973). Temperature sensitivity of thermoreceptors and mechanoreceptors on the beak of pigeons. *J. Comp. Physiol.* **87,** 379–391.

Necker, R. (1975). Temperature-sensitive ascending neurons in the spinal cord of pigeons. *Pflügers Arch.* **353,** 275–286.

Necker, R. (1977). Thermal sensitivity of different skin areas in pigeons. *J. Comp. Physiol.* **116,** 239–246.

Necker, R. (1984). Central thermosensitivity: CNS and extra-CNS. In "Thermal Physiology" (J. R. S. Hales, ed.), pp. 53–61. Raven, New York.

Newton, I. (1969). Winter fattening in the bullfinch. *Physiol. Zool.* **42,** 96–107.

Nice, M. M. (1962). Development of behavior in precocial birds. *Trans. Linnean Soc. N.Y.* **8,** 1–211.

Nisticò, G., and Marley, E. (1973). Central effects of prostaglandin E_1 in adult fowls. *Neuropharmacology* **12,** 1009–1016.

Nisticò, G., and Marley, E. (1976). Central effects of prostaglandins E_2, A_1, and F_{2a} in adult fowls. *Neuropharmacology* **15,** 737–741.

Nomoto, S. (1996). Diurnal variations in fever induced by intravenous LPS injection in pigeons. *Pflügers Arch.* **431,** 987–989.

Nomoto, S., Rautenberg, W., and Iriki, M. (1983). Temperature regulation during exercise in the Japanese quail (*Coturnix coturnix japonica*). *J. Comp. Physiol. B* **149,** 519–525.

O'Connor, T. P. (1995). Metabolic characteristics and body composition in house finches: Effects of seasonal acclimatization. *J. Comp. Physiol. B* **165,** 298–305.

O'Connor, T. P. (1996). Geographic variation in metabolic seasonal acclimatization in house finches. *Condor* **98,** 371–381.

Odum, E. P. (1942). Muscle tremors and the development of temperature regulation in birds. *Am. J. Physiol.* **136,** 618–622.

Ohmart, R. D., and Lasiewski, R. C. (1971). Roadrunners: Energy conservation by hypothermia and absorption of sunlight. *Science* **172,** 67–69.

Olson, J. M. (1992). Growth, the development of endothermy, and the allocation of energy in red-winged blackbirds (*Agelaius phoeniceus*) during the nestling period. *Physiol. Zool.* **65,** 124–152.

Olson, J. M. (1994). The ontogeny of shivering thermogenesis in the red-winged blackbird (*Agelaius phoeniceus*). *J. Exp. Biol.* **191,** 59–88.

Olson, J. M., Dawson, W. R., and Camilliere, J. J. (1988). Fat from black-capped chickadees: Avian brown adipose tissue? *Condor* **90,** 529–537.

Oshima, I., Yamada, H., Goto, M., Sato, K., and Ebihara, S. (1989). Pineal and retinal melatonin is involved in the control of circadian locomotor and body temperature rhythms in the pigeon. *J. Comp. Physiol. A.* **166,** 217–226.

Ostheim, J. (1992). Coping with food-limited conditions: Feeding behavior, temperature preference, and nocturnal hypothermia in pigeons. *Physiol. Behav.* **51,** 353–361.

Østnes, J. E., and Bech, C. (1992). Thermosensitivity of different parts of the spinal cord of the pigeon (*Columba livia*). *J. Exp. Biol.* **162,** 185–196.

Østnes, J. E., and Bech, C. (1997a). Ontogeny of deep-body cold sensitivity in Pekin ducklings (*Anas platyrhynchos*). *J. Comp. Physiol. B* **167,** 241–248.

Østnes, J. E., and Bech, C. (1997b). The early emergence of cold sensation in shag nestlings *Phalacrocorax aristotelis*. *J. Avian Biol.* **28,** 24–30.

Paladino, F. V., and King, J. R. (1984). Thermoregulation and oxygen consumption during terrestrial locomotion by white-crowned sparrows *Zonotrichia leucophrys gambelii*. *Physiol. Zool.* **57,** 226–236.

Palokangas, R., and Hissa, R. (1971). Thermoregulation in young black-headed gulls (*Larus ridibundus* L.). *Comp. Biochem. Physiol. A* **38,** 743–750.

Peltonen, L., Arieli, Y., Pyörnilä, A., and Marder, J. (1998). Adaptive changes in the epidermal structure of the heat-acclimated rock pigeon (*Columba livia*): A comparative electron microscope study. *J. Morphol.* **235,** 17–29.

Pettit, T. N., Grant, G. S., and Whittow, G. C. (1982). Body temperature and growth of Bonin petrel chicks. *Wilson Bull.* **94**, 358–361.

Pettit, T. N., Whittow, G. C., and Grant, G. S. (1981). Rete mirabile ophthalmicum in Hawaiian seabirds. *Auk* **98**, 844–846.

Pinshow, B., Bernstein, M. H., Lopez, G. E., and Kleinhaus, S. (1982). Regulation of brain temperature in pigeons: effects of corneal convection. *Am. J. Physiol.* **242**, R577–R581.

Pinshow, B., Fedak, M. A., Battles, D. R., and Schmidt-Nielsen, K. (1976). Energy expenditure for thermoregulation and locomotion in emperor penguins. *Am. J. Physiol.* **231**, 903–912.

Pohl, H. (1969). Factors influencing the metabolic response to cold in birds. *Fed. Proc.* **28**, 1059–1064.

Pohl, H., and West, G. C. (1973). Daily and seasonal variation in metabolic response to cold during rest and forced exercise in the common redpoll. *Comp. Biochem. Physiol. A* **45**, 851–867.

Powers, D. R. (1992). Effect of temperature and humidity on evaporative water loss in Anna's hummingbird (*Calypte anna*). *J. Comp. Physiol. B* **162**, 74–84.

Prinzinger, R. (1982). The energy costs of temperature regulation in birds: The influence of quick sinusoidal temperature fluctuations on the gaseous metabolism of the Japanese quail (*Coturnix coturnix japonica*). *Comp. Biochem. Physiol. A* **71**, 469–472.

Prinzinger, R., and Hänssler, I. (1980). Metabolism-weight relationship in some small non-passerine birds. *Experientia* (*Basel*) **36**, 1299–1300.

Prinzinger, R., Krüger, K., and Schuchmann, K.-L. (1981). Metabolism-weight relationship in 17 humming-bird species at different temperatures during day and night. *Experientia* (*Basel*) **37**, 1307–1309.

Prinzinger, R., Lübben, I., and Schuchmann, K.-L. (1989). Energy metabolism and body temperature in 13 sunbird species (Nectariniidae). *Comp. Biochem. Physiol. A* **92**, 393–402.

Prinzinger, R., Misovic, A., and Schleucher, E. (1993). Energieumsatz und Körpertemperatur bei der Zwergwachtel *Coturnix chinensis* und beim Bindenlaufhühnchen *Turnix suscitator*. *J. f. Ornithol.* **134**, 79–84.

Pyörnilä, A., and Hissa, R. (1979). Opposing temperature responses to intrahypothalamic injections of 5-hydroxytryptamine in the pigeon exposed to cold. *Experientia* (*Basel*) **35**, 59–60.

Ramirez, J. M., and Bernstein, M. H. (1976). Compound ventilation during thermal panting in pigeons: A possible mechanism for minimizing hypocapnic alkalosis. *Fed. Proc.* **35**, 2562–2565.

Randall, W. C. (1943). Factors influencing the temperature regulation of birds. *Am. J. Physiol.* **139**, 56–63.

Rashotte, M. E., and Chambers, J. B. (1998). Circadian variation in the pigeon's body temperature: Role of digestion, shivering, and locomotor activity. *In* "Programme and Abstracts. Symposium on Avian Thermal Physiology and Energetics" (E. Hohtola and S. Saarela, eds.), p. 28, Oulu, Finland, 9–14 August, 1998.

Rashotte, M. E., Basco, P. S., and Henderson, R. P. (1995). Daily cycles in body temperature, metabolic rate, and substrate utilization in pigeons—Influence of amount and timing of food consumption. *Physiol. Behav.* **57**, 731–746.

Rashotte, M. E., Phillips, D. L., and Henderson, R. P. (1997). Nocturnal digestion, cloacal excretion, and digestion-related thermogenesis in pigeons (*Columba livia*). *Physiol. Behav.* **61**, 83–92.

Rautenberg, W. (1969a). Untersuchungen zur Temperaturregulation wärme- und kälteakklimatisierter Tauben. *Z. vergl. Physiol.* **62**, 221–234.

Rautenberg, W. (1969b). Die Bedeutung der zentralnervösen Thermosensitivität für die Temperregulation der Taube. *Z. vergl. Physiol.* **62**, 235–266.

Rautenberg, W. (1971). The influence of skin temperature on the thermoregulatory system of pigeons. *J. Physiol.* (*Paris*) **63**, 396–398.

Rautenberg, W. (1983). Thermoregulation. *In* "Physiology and Behaviour of the Pigeon" (M. Abs, ed.), pp. 131–148. Academic Press, London.

Rautenberg, W. (1986). Neural control of cold defense mechanisms in the avian thermoregulation system. *In* "Living in the Cold: Physiological and Biochemical Adaptations" (H. C. Heller, X. J. Musacchia, and L. C. H. Wang, eds.), pp. 151–166. Elsevier, Amsterdam.

Rautenberg, W., May, B., and Arabin, G. (1980). Behavioral and autonomic temperature regulation in competition with food intake and water balance of pigeons. *Pflügers Arch.* **384**, 253–260.

Rautenberg, W., May, B., Necker, R., and Rosner, G. (1978). Control of panting by thermosensitive neurons in birds. *In* "Respiratory Function in Birds, Adult and Embryonic" (J. Piiper, ed.), pp. 204–210. Springer-Verlag, Berlin.

Rautenberg, W., Necker, R., and May, B. (1972). Thermoregulatory responses of the pigeon to changes of the brain and spinal cord temperatures. *Pflügers Arch.* **338**, 31–42.

Refinetti, R., and Menaker, M. (1992). The circadian rhythm of body temperature. *Physiol. Behav.* **51**, 613–637.

Reinertsen, R. E. (1986). Hypothermia in northern passerine birds. *In* "Living in the Cold" (H. C. Heller, X. J. Musacchia, and L. C. H. Wang, eds.), pp. 419–426. Elsevier, New York.

Reinertsen R. E. (1996). Physiological and ecological aspects of hypothermia. *In* "Avian Energetics and Nutritional Ecology" (C. Carey, ed.), pp. 125–157. Chapman & Hall, New York.

Reinertsen, R. E., and Bech, C. (1994). Hypothermia in pigeons; relating body temperature regulation to the gastrointestinal system. *Naturwissenschaften* **81**, 133–136.

Reinertsen, R. E., and Haftorn, S. (1983). Nocturnal hypothermia and metabolism in the willow tit, *Parus montanus*, at 63°N. *J. Comp. Physiol.* **151**, 109–118.

Reynolds, P. S., and Lee, R. N., III. (1996). Phylogenetic analysis of avian energetics: Passerines and non-passerines do not differ. *Am. Nat.* **147**, 735–759.

Richards, S. A. (1968). Vagal control of thermal panting in mammals and birds. *J. Physiol.* (*London*) **199**, 89–101.

Richards, S. A. (1970). The biology and comparative physiology of thermal panting. *Biol. Rev.* **45**, 223–264.

Richards, S. A. (1971). Brain stem control of polypnoea in the chicken and pigeon. *Respir. Physiol.* **11**, 315–326.

Richards, S. A., (1975). Thermal homeostasis in birds. *In* "Avian Physiology" (M. Peaker, ed.), pp. 65–96. Academic Press, New York.

Richards, S. A. (1976). Evaporative water loss in domestic fowls and its partition in relation to ambient temperature. *J. Agric. Sci.* **87**, 527–532.

Richards, S. A., and Avery, P. (1978). Central nervous mechanisms regulating thermal panting. *In* "Respiratory Function in Birds, Adult and Embryonic" (J. Piiper, ed.), pp. 196–203. Springer-Verlag, Berlin.

Ricklefs, R. E. (1983). Avian postnatal development. *In* "Avian Biology" (D. S. Farner, J. R. King, and K. C. Parkes, eds.), Vol. 7., pp. 1–83. Academic Press, New York.

Ricklefs, R. E., and Roby, D. D. (1983). Development of homeothermy in the diving petrels *Pelecanoides urinatrix exul* and *P. georgicus*, and the antarctic prion *Pachyptila desolata*. *Comp. Biochem. Physiol. A* **75**, 307–311.

Ricklefs, R. E., White, S. C., and Cullen, J. (1980). Energetics of postnatal growth in Leach's storm-petrel. *Auk* **97**, 566–575.

Rintamäki, H., Saarela, S., Marjakangas, A., and Hissa, R. (1983). Summer and winter temperature regulation in the black grouse *Lyrurus tetrix*. *Physiol. Zool.* **56**, 152–159.

Robertson, P. B., and Schnapf, A. F. (1987). Pyramiding behavior in the Inca dove: adaptive aspects of day-night differences. *Condor* **89**, 185–187.

Rogers, F. T., and Lackey, R. W. (1923). Studies on the brain stem: VII. The respiratory exchange and heat production after destruction of the body temperature-regulating centers of the hypothalamus. *Am. J. Physiol.* **66**, 453–460.

Roller, W. L., and Dale, A. C. (1962). Heat losses from Leghorn layers at warm temperatures. Paper No. 62-428, Annual Meeting of the American Society of Agriculture Engineers, Washington, DC.

Root, T. L., O'Connor, T. P., and Dawson, W. R. (1991). Standard metabolic level and insulative characteristics of eastern house finches, *Carpodacus mexicanus* (Müller). *Physiol. Zool.* **64**, 1279–1295.

Rosner, G. (1977). Response of hypothalamic neurons to thermal stimulation of spinal cord and skin in pigeons. *Pflügers Arch.* **368**, (Suppl.), R29.

Rothe, H.-J., Biesel, W., and Nachtigall, W. (1987). Pigeon flight in a wind tunnel. II. Gas exchange and power requirements. *J. Comp. Physiol. B* **157**, 99–109.

Ruben, J. (1995). The evolution of endothermy in mammals and birds: From physiology to fossils. *Annu. Rev. Physiol.* **57**, 69–95.

Ruben, J. (1996). Evolution of endothermy in mammals, birds and their ancestors. *In* "Animals and Temperature: Phenotypic and Evolutionary Adaptation" (I. A. Johnston and A. F. Bennett, eds.), pp. 347–376. *Soc. Exp. Biol. Sem., Ser.* 59. Cambridge Univ. Press, Cambridge, UK.

von Saalfeld, E. (1936). Untersuchungen über das Hacheln bei Tauben: I. Mitteilung. *Z. vergl. Physiol.* **23**, 727–743.

Saarela, S., and Heldmaier, G. (1987). Effect of photoperiod and melatonin on cold resistance, thermoregulation and shivering/nonshivering thermogenesis in Japanese quail. *J. Comp. Physiol. B* **157**, 625–633.

Saarela, S., Klapper B., and Heldmaier, G. (1989a). Thermogenic capacity of greenfinches and siskins in winter and summer. *In* "Physiology of Cold Adaptation in Birds" (C. Bech and R. E. Reinertsen, eds.), pp. 115–122. Plenum, New York.

Saarela, S., Hissa, R., Pyörnilä, A., Harjula, R., Ojanen, M., and Orell, M. (1989b). Do birds possess brown adipose tissue? *Comp. Biochem. Physiol. A* **92**, 219–228.

Saarela, S., Keith, J. S., Hohtola, E., and Trayhurn, P. (1991). Is the "mammalian" brown fat-specific mitochondrial uncoupling protein present in adipose tissues of birds? *Comp. Biochem. Physiol. B.* **100**, 45–49.

Saarela, S., Klapper, B., and Heldmaier, G. (1995). Daily rhythm of oxygen consumption and thermoregulatory responses in some European winter- or summer-acclimatized finches at different ambient temperatures. *J. Comp. Physiol. B* **165**, 366–376.

St.-Laurent, R., and Larochelle, J. (1994). The cooling power of the pigeon head. *J. Exp. Biol.* **194**, 329–339.

Sato, H., and Simon, E. (1988). Thermal characterization and transmitter analysis of single units in the preoptic and anterior hypothalamus of conscious ducks. *Pflügers Arch.* **411**, 34–41.

Schmidt, I. (1976a). Paradoxical changes of respiratory rate elicited by altering rostral brain stem temperature in the pigeon. *Pflügers Arch.* **367**, 111–113.

Schmidt, I. (1976b). Effect of central thermal stimulation on the thermoregulatory behavior of the pigeon. *Pflügers Arch.* **363**, 271–272.

Schmidt, I. (1978a). Behavioral and autonomic thermoregulation in heat stressed pigeons modified by central thermal stimulation. *J. Comp. Physiol.* **127**, 75–87.

Schmidt, I. (1978b). Interactions of behavioral and autonomic thermoregulation in heat stressed pigeons. *Pflügers Arch.* **374**, 47–55.

Schmidt, I. (1983). Weighting regional thermal inputs to explain autonomic and behavioural thermoregulation in the pigeon. *J. Therm. Biol.* **8**, 47–48.

Schmidt, I., and Simon, E. (1982). Negative and positive feedback of central nervous system temperature in thermoregulation of pigeons. *Am. J. Physiol.* **243**. R363–R372.

Schmidt-Nielsen, K., Hainsworth, F. R., and Murrish, D. E. (1970). Counter-current heat exchange in the respiratory passages: Effect on water and heat balance. *Respir. Physiol.* **9**, 263–276.

Schmidt-Nielsen, K., Kanwisher, J., Lasiewski, R. C., Cohn, J. E., and Bretz, W. L. (1969). Temperature regulation and respiration in the ostrich. *Condor* **71**, 341–352.

Scholander, P. F., Hock, R., Walters, V., Johnson, F., and Irving, L. (1950). Heat regulation in some arctic and tropical mammals and birds. *Biol. Bull.* **99**, 237–258.

Schuchmann, K.-L. (1979). Metabolism of flying hummingbirds. *Ibis* **121**, 85–86.

Shallenberger, R. J., Whittow, G. C., and Smith, R. M. (1974). Body temperature of the nesting red-footed booby (*Sula sula*). *Condor* **76**, 476–478.

Shapiro, C. J., and W. W. Weathers. (1981). Metabolic and behavioral responses of American kestrels to food deprivation. *Comp. Biochem. Physiol. A* **68**, 111–114.

Sieland, M., Delius, J. D., Rautenberg, W., and May, B. (1981). Thermoregulation mediated by conditioned heart-rate changes in pigeons. *J. Comp. Physiol.* **144**, 375–379.

Simon, E. (1981). Effects of CNS temperature on generation and transmission of temperature signals in homeotherms: A common concept for mammalian and avian thermoregulation. *Pflügers Arch.* **392**, 79–88.

Simon, E. (1989). Nervous control of cold defence in birds. *In* "Physiology of Cold Adaptation in Birds" (C. Bech and R. E. Reinertsen, eds.), pp. 1–15. Plenum, New York.

Simon, E., Hammel, H. T., and Oksche, A. (1977). Thermosensitivity of single units in the hypothalamus of the conscious Pekin duck. *J. Neurobiol.* **8**, 523–535.

Simon, E., Martin, R., and Simon-Oppermann, C. (1981). Central nervous versus total body thermosensitivity of the duck. *Internat. J. Biometeorol.* **25**, 249–256.

Simon, E., Pierau, F.-K., and Taylor, D. C. M. (1986). Central and peripheral thermal control of effectors in homeothermic temperature regulation. *Physiol. Rev.* **66**, 235–300.

Simon, E., Simon-Oppermann, C., Hammel, H. T., Kaul, R., and Maggert, J. (1976). Effects of altering rostral brain stem temperature on temperature regulation in the Adélie penguin (*Pygoscelis adeliae*). *Pflügers Arch.* **362**, 7–13.

Simon-Oppermann, C., and Martin, R. (1979). Mammalian-like thermosensitivity in the lower brainstem of the Pekin duck. *Pflügers Arch.* **379**, 291–293.

Simon-Oppermann, C., Simon, E., Jessen, C., and Hammel, H. T. (1978). Hypothalamic thermosensitivity in conscious Pekin ducks. *Am. J. Physiol.* **235**, R130–R140.

Simons, T. R., and Whittow, G. C. (1984). Energetics of breeding dark-rumped petrels. *In* "Seabird Energetics" (G. C. Whittow and H. Rahn, eds.), pp. 159–181. Plenum, New York.

Snapp, B. D., Heller, H. C., and Gospe, S. M., Jr. (1977). Hypothalamic thermosensitivity in California quail (*Lophortyx californicus*). *J. Comp. Physiol.* **117**, 345–357.

Sulkava, S. (1969). On small birds spending the night in the snow. *Aquilo Ser. Zool.* **7**, 33–37.

Sutter, G. C., and MacArthur, R. A. (1989). Thermoregulatory performance of fledgling American coots (*Fulica americana*) in air and water. *Can. J. Zool.* **67**, 1339–1346.

Swanson, D. L. (1990). Seasonal variation in cold hardiness and peak rates of cold-induced thermogenesis in the dark-eyed junco (*Junco hyemalis*). *Auk* **107,** 561–566.

Swanson, D. L. (1991). Seasonal adjustments in metabolism and insulation in the dark-eyed junco. *Condor* **93,** 538–545.

Swanson, D. L., and Weinacht, D. P. (1997). Seasonal effects on metabolism and thermoregulation in northern bobwhite. *Condor* **99,** 478–489.

Takahashi, J. S., and Menaker, M. (1982). Role of the suprachiasmatic nuclei in the circadian system of the house sparrow, *Passer domesticus. J. Neurosci.* **2,** 815–828.

Taylor, C. R., Dmi'el, R., Fedak, M., and Schmidt-Nielsen, K. (1971). Energetic cost of running and heat balance in a large bird, the rhea. *Am. J. Physiol.* **221,** 597–601.

Thaler, E. (1991). Survival strategies in goldcrest and firecrest (*Regulus regulus, R. ignicapillus*) during winter. In "Acta XX Congressus Internationalis Ornithologici" (B. D. Bell, R. O. Cossee, J. E. C. Flux, B. D. Heather, R. A. Hitchmough, C. J. R. Robertson, and M. J. Williams, eds.), Vol. III, pp. 1791–1798. New Zealand Ornithological Congress Trust Board, Wellington.

Tucker, V. A. (1968). Respiratory exchange and evaporative water loss in the flying budgerigar. *J. Exp. Biol.* **48,** 67–87.

Underwood, H. (1994). The circadian rhythm of thermoregulation in Japanese quail: I. Role of the eyes and pineal. *J. Comp. Physiol. A* **175,** 639–653.

Underwood, H., and Siopes, T. (1984). Circadian organization in Japanese quail. *J. Exp. Zool.* **232,** 557–566.

Veghte, J. H. (1964). Thermal and metabolic responses of the gray jay to cold stress. *Physiol. Zool.* **37,** 316–328.

Veghte, J. H., and Herreid, C. F. (1965). Radiometric determination of feather insulation and metabolism of arctic birds. *Physiol. Zool.* **38,** 267–275.

Vittoria, J. C., and Marsh, R. L. (1996). Cold-acclimated ducklings shiver when exposed to cold. *Am. Zool.* **36,** 66A.

Walker, L. E., Walker, J. M., Palca, J. W., and Berger, R. J. (1983). A continuum of sleep and shallow torpor in fasting doves. *Science* **221,** 194–195.

Walsberg, G. E. (1990). Communal roosting in a very small bird: Consequences for the thermal and respiratory gas environments. *Condor* **92,** 795–798.

Walsberg, G. E., and Wolf, B. O. (1996). An appraisal of operative temperature mounts as tools for studies of ecological energetics. *Physiol. Zool.* **69,** 658–681.

Wang, L. C. H., and Wolowyk, M. W. (1988). Torpor in mammals and birds. *Can. J. Zool.* **66,** 133–137.

Watson, A. (1972). The behaviour of the ptarmigan. *Br. Birds* **65,** 6–26.

Weatherhead, P. J., Sealy, S. G., and Barclay, R. M. R. (1985). Risks of clustering in thermally-stressed swallows. *Condor* **87,** 443–444.

Weathers, W. W. (1972). Thermal panting in domestic pigeons, *Columba livia,* and the barn owl, *Tyto alba. J. Comp. Physiol.* **79,** 79–84.

Weathers, W. W. (1979), Climatic adaptation in avian standard metabolic rate. *Oecologia* (*Berlin*) **42,** 81–89.

Weathers, W. W. (1981). Physiological thermoregulation in heat stressed birds: Consequences of body size. *Physiol. Zool.* **64,** 345–361.

Weathers, W. W. (1997). Energetics and thermoregulation by small passerines of the humid, lowland tropics. *Auk* **114,** 341–353.

Weathers, W. W., and Caccamise, D. F. (1978). Seasonal acclimatization to temperature in monk parakeets. *Oecologia* (*Berlin*) **35,** 173–183.

Weathers, W. W., and Schoenbaechler, D. C. (1976). Contribution of gular flutter to evaporative cooling in Japanese quail. *J. Appl. Physiol.* **40,** 521–524.

Weathers, W. W., and Sullivan, K. A. (1993). Seasonal patterns of time and energy allocation by birds. *Physiol. Zool.* **66,** 511–536.

Weathers, W. W., Koenig, W. D., and Stanback, M. T. (1990). Breeding energetics and thermal ecology of the acorn woodpecker in central coastal California. *Condor* **92,** 341–359.

Weathers, W. W., Weathers, D. L., and van Riper, C., III. (1983). Basal metabolism of the apapane: Comparison of freshly caught birds with long-term captives. *Auk* **100,** 977–978.

Webb, D. R., and King, J. R. (1983). Heat transfer relations of avian nestlings. *J. Therm. Biol.* **8,** 301–310.

Webster, M. D., and Bernstein, M. H. (1987). Ventilated capsule measurements of cutaneous evaporation in mourning doves. *Condor* **89,** 863–868.

Webster, M. D., and King, J. R. (1987). Temperature and humidity dynamics of cutaneous and respiratory evaporation in pigeons, *Columba livia. J. Comp. Physiol. B* **157,** 253–260.

Webster, M. D., and Weathers, W. W. (1990). Heat produced as a by-product of foraging activity contributes to thermoregulation by verdins, *Auriparus flaviceps. Physiol. Zool.* **63,** 777–794.

Webster, M. D., Campbell, G. S., and King, J. R. (1985). Cutaneous resistance to water-vapor diffusion in pigeons and the role of plumage. *Physiol. Zool.* **58,** 58–70.

West, G. C. (1962). Responses and adaptations of wild birds to environmental temperature. In "Comparative Physiology of Temperature Regulation" (J. P. Hannon and E. Viereck, eds.), Part 3, pp. 291–333. Arctic Aeromedical Laboratory, Fort Wainwright, Alaska.

West. G. C. (1972). Seasonal differences in resting metabolic rate of Alaskan ptarmigan. *Comp. Biochem. Physiol. A.* **42,** 867–876.

West, G. C., and Hart, J. S. (1966). Metabolic responses of evening grosbeaks to constant and to fluctuating temperatures. *Physiol. Zool.* **39,** 171–184.

White, F. N., Bartholomew, G. A., and Howell, T. R. (1975). The thermal significance of the nest of the sociable weaver. *Ibis* **117,** 171–179.

Whittow, G. C. (1965). Regulation of body temperature. In "Avian Physiology" (P. D. Sturkie, ed.), 2nd Ed., pp. 186–238. Cornell Univ. Press, Ithaca, NY.

Whittow, G. C. (1976). Regulation of body temperatures. In "Avian Physiology" (P. D. Sturkie, ed.), 3rd Ed., pp. 146–173. Springer-Verlag, New York.

Whittow, G. C. (1980). Thermoregulatory behavior of the Laysan and black-footed albatross. *Elepaio* **40,** 97–98.

Whittow, G. C. (1986). Regulation of body temperature. In "Avian Physiology" (P. D. Sturkie, ed.), 4th Ed., pp. 221–252. Springer-Verlag, New York.

Whittow, G. C., and Tazawa, H. (1991). The early development of thermoregulation in birds. *Physiol. Zool.* **64,** 1371–1390.

Whittow, G. C., Araki, C. T., and Pepper, R. L. (1978). Body temperature of the great frigatebird *Fregata minor. Ibis* **120,** 358–360.

Whittow, G. C., Pettit, T. N., Ackerman, R. A., and Paganelli, C. V. (1987). Temperature regulation in a burrow-nesting seabird, the wedge-tailed shearwater (*Puffinus pacificus*). *J. Comp. Physiol. B* **157,** 607–614.

Wiebe, M. O., and Evans, R. M. (1994). Development of temperature regulation in young birds—Evidence for a vocal regulatory mechanism in 2 species of gulls (Laridae). *Can. J. Zool.* **72,** 427–432.

Williams, J. B. (1996). A phylogenetic perspective of evaporative water loss in birds. *Auk* **113,** 457–472.

Williams, J. B., du Plessis, M. A., and Siegfried, W. R. (1991). Green woodhoopoes (*Phoeniculus purpureus*) and obligate cavity roosting provide a test of the thermoregulatory insufficiency hypothesis. *Auk* **108,** 285–293.

Withers, P. C. (1977). Respiration, metabolism, and heat exchange of euthermic and torpid poor-wills and hummingbirds. *Physiol. Zool.* **50,** 43–52.

Withers, P. C., and Williams, J. B. (1990). Metabolic and respiratory physiology of an arid-adapted Australian bird, the spinifex pigeon. *Condor* **92,** 961–969.

Wolf, B. O., and Walsberg, G. E. (1996a). Respiratory and cutaneous evaporative water loss at high environmental temperatures in a small bird. *J. Exp. Biol.* **199,** 451–457.

Wolf, B. O., and Walsberg, G. E. (1996b). Thermal effects of radiation and wind on a small bird and implications for microsite selection. *Ecology* **77,** 2228–2236.

Wolf, B. O., Wooden, K. M., and Walsberg, G. E. (1996). The use of thermal refugia by two small desert birds. *Condor* **98,** 424–428.

Wolf, L. L., and Hainsworth, F. R. (1972). Environmental influence on regulated body temperature in torpid hummingbirds. *Comp. Biochem. Physiol. A* **41,** 167–173.

Wolfenson, D. (1983). Blood flow through arteriovenous anastomoses and its thermal function in the laying hen. *J. Physiol. (London)* **334,** 395–407.

Woods, J. J., and Whittow, G. C. (1974). The role of central and peripheral temperature changes in the regulation of thermal polypnea in the chicken. *Life Sci.* **14,** 199–206.

Zerba, E., and Walsberg, G. E. (1992). Exercise-generated heat contributes to thermoregulation by Gambel's quail in the cold. *J. Exp. Biol.* **171,** 409–422.

Note Added in Proof

Ostnes and Bech (1998) have shown that, in the pigeon, an increase in heat production during acute exposure to cold is triggered by peripheral thermosensors. Over a longer period (Circadian Cycle), internal thermosensors are also involved; they may be stimulated by the flushing of cold blood through the body during cold vasodilatation in the periphery. [Ostnes, J. E. and Bech, C. (1998). Thermal control of metabolic cold defence in pigeons *Columbia livia*. *J. Exp. Biol.* **201,** 793–803.]

CHAPTER 15

Flight

P. J. BUTLER
School of Biological Sciences
The University of Birmingham
Birmingham B15 2TT, United Kingdom

C. M. BISHOP
School of Biological Sciences
University of Wales–Bangor
Gwynedd LL57 2UW, United Kingdom

I. Introduction 391
II. Scaling 392
III. Energetics of Bird Flight 393
    A. Techniques Used to Study the Mechanical Power Output Required for Flight 394
    B. Techniques Used to Measure the Power Input Required for Flight 396
    C. Empirical Data Concerning the Power Input during Flight 399
IV. The Flight Muscles of Birds 406
    A. Flight Muscle Morphology and Fiber Types 406
    B. Biochemistry of the Flight Muscles 408
    C. Neurophysiology and Muscle Function 410
V. Development of Locomotor Muscles and Preparation for Flight 413
VI. Metabolic Substrates and Fuel Deposits 415
VII. The Cardiovascular System 416
    A. Cardiovascular Adjustments during Flight 417
    B. The Cardiac Muscles 417
VIII. The Respiratory System 418
    A. Ventilatory Adjustments during Flight and Ventilatory/Locomotor Coupling 418
    B. Temperature Control 423
    C. Respiratory Water Loss 424
IX. Migration and Long-Distance Flight Performance 425
    A. Preparation for Migration 425
    B. Migratory Behavior 427
X. Flight at High Altitude 428
    References 430

I. INTRODUCTION

The vast majority of birds have two independent locomotor systems, the legs and the wings. There are, however, a number of species of birds which rely exclusively or predominantly on one of these systems. Flightlessness, although not common among birds, is not only a feature of the ratites (e.g., ostriches and emus) and penguins, but has evolved on a number of occasions in rails, geese, ducks, grebes, and ibises (McNab, 1994). As this author points out, flightlessness in terrestrial birds is always associated with a reduction in the mass of the pectoral muscles with a consequent reduction in basal energy consumption, which may be related to overall energy conservation. Conversely, a complete reliance on flight as a means of locomotion is seen in only a few families, such as the swifts, swallows, hummingbirds, and sunbirds. Hummingbirds exhibit an exceptional form of flight in that they are the only birds known routinely to use prolonged periods of "stationary" or hovering flight and are even capable of flying backwards due to their unique wing kinematics. However, most birds have the best of both worlds, the ability to walk, run or swim using the legs and the ability to fly (in air or water) using the wings and tail.

Forward flapping flight is, for birds of similar mass, energetically more costly per unit of time than running or swimming at the surface (Butler, 1991). However, during flight over long distances, birds are able to maintain relatively high velocities, which means that the energy required for a given mass to travel a given distance (i.e., the cost of transport) is considerably less than the energy cost of transport during walking, running, or surface swimming (Tucker, 1970; Schmidt-Nielsen, 1972). Thus, a number of species of birds are able to migrate over long distances in a relatively short time period. Flight also enables birds to travel relatively long distances on a daily basis, to and from their nests or roosting areas.

The high rate of energy consumption required for flapping flight to be sustained over a prolonged period of time can only be maintained by "oxidative" or "aerobic" metabolism of fuel substrates and places large demands on the respiratory, cardiovascular, and muscular systems of birds. Migration over long distances has the added problems of the birds initially carrying a relatively large amount of fuel (see Alerstam and Lindström, 1990; Witter and Cuthill, 1993), while at the same time controlling body temperature and evaporative water loss in the face of the large amount of heat that is generated by the intense muscular activity. The energy cost of spring migration may be so high as to influence the subsequent energy dynamics of reproduction (e.g., Bromley and Jarvis, 1993). The postbreeding autumn migration follows a period of relatively sedentary behaviours such as incubation, brooding, and molt for the adults, during which there is evidence, for some species at least, that the main flight muscles atrophy (Mainguy and Thomas, 1985; Piersma, 1988; Bishop et al., 1996). Thus, for postbreeding adults, as well as for the season's fledglings, the flight muscles and supporting systems have to achieve an adequate level of aerobic fitness in preparation for the forthcoming migration. Those species of birds that fly at high altitudes during their migration also have the combined problems of engaging in this energetically costly activity in a severely hypoxic and cold environment and at a reduced air density.

At the other extreme from sustained flight is the explosive "burst" flapping flight of many species of galliform birds, which they utilize to escape terrestrial predators or to roost in trees. These birds rely on "anaerobic" metabolism and are unable to sustain such activity for more than a few minutes and may be rendered incapable of flying at all after repeated bursts of activity (Marden, 1994). In between these two extremes are birds such as vultures, storks, and albatrosses which are able to remain airborne by soaring and gliding and expend a fraction of the energy required for forward flapping flight during foraging trips and migration. In addition, small species of birds appear to use different "gaits" or flight modes, such as flap-bounding and flap-gliding (Tobalske and Dial, 1994), in order to maintain the wing beat frequency at the most energy efficient rate (Rayner, 1985a).

This chapter will focus primarily on: the energetics of bird flight, including long distance migration; the function, physiology, and biochemistry of the flight muscles, including "fitness" in preparation for migration; deposition of fuel stores in preparation for migration; respiratory and cardiovascular adjustments associated with flight, including temperature control and water loss; and flight at high altitudes. However, it is not easy to obtain experimental data from a flying animal, and it should be remembered that it is doubtful if any one experimental method can adequately give all of the necessary information required for a full analysis of the physiological responses to flight. Thus, a number of techniques used to study the flight of birds will be discussed. To begin with, however, there is a short introduction to the significance of animal size and scale.

II. SCALING

It is important to be able to assess the significance of a particular characteristic or adaptation and compare it with those of other individuals and/or species, even thought these other animals may vary greatly in size or mass. The study of the consequences of changes in shape or mass of different animals is called scaling (Schmidt-Nielsen, 1984). If two bodies are the same shape or "geometrically similar" but of different volume (V) and mass (M) (i.e., their various characteristics are directly proportional to their difference in mass), then these bodies are said to scale isometrically with regard to each other. If we consider two cubes of different size, then each side will have a length (l), and the area will be proportional to the length squared ($A \propto l^2$) while the volume or mass (assuming a constant density) will be proportional to the length cubed ($V \propto l^3$). Thus, area will be proportional to volume or mass to the two-thirds power ($A \propto V^{2/3}$) and any length will be proportional to volume or mass to the one-third power ($l \propto V^{1/3}$). Because body mass (M_b) is a relatively easy variable to measure in most animals and one which represents the large size range seen in animals, it is most frequently used as the independent variable in scaling. Thus:

$$l \propto M_b^{1/3} \text{ or } M_b^{0.33} \text{ and } A \propto M_b^{2/3} \text{ or } M_b^{0.67}.$$

Pennycuick (1982) studied the flight and morphometric parameters of 11 species of Procellariiformes (e.g., petrels and albatrosses), ranging from 0.03 to 9 kg, and found that wing span (b) scaled in proportion to $M_b^{0.37}$ (Figure 1a) and wing area (S) in proportion to $M_b^{0.627}$ (Figure 1b). These results indicate that there is a slight tendency for larger birds to have relatively longer wings and smaller wing areas, but that overall they are geometrically very similar. However, Figure 1d shows that, if the outlines of each species are traced and then displayed so that they all have the same wing span, as the birds become smaller the shape of their wings changes in a systematic way (Pennycuick, 1992). This can be assessed by calculating the dimensionless variable called aspect ratio (Λ), which is b^2/S. Figure 1c shows that Λ scales in proportion to $M_b^{0.116}$ and is highly significantly different from the expected isometric value of $M_b^{0.0}$. Thus, this feature of birds scales nonisometrically or allometrically and the study of such relationships is often called "allometry."

Such analyses can also be used to test predictions regarding the scaling of animal energetics. For example,

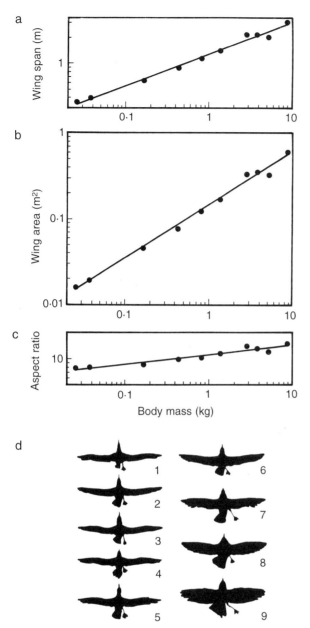

FIGURE 1 (a) Wing span, S; (b) wing area, A; (c) aspect ratio, Λ, of 11 species of Procellariiform birds, plotted against body mass, M_b (From C. J. Pennycuick, "Newton Rules Biology: A Physical Approach to Biological Problems," 1992, by permission of Oxford University Press.) (d) Wing tracings from nine of these species, with the smaller species enlarged such that all have the same wing span. 1, wandering albatross *Diomedia exulans*; 2, black-browed albatross *D. melanophrys*; 3, grey-headed albatross *D. chysostoma*; 4, light-mantled sooty albatross, *Phoebetria palpebrata*; 5, giant petrel *Macronectes* sp.; 6, white-chinned petrel *Procellaria aequinoctialis*; 7, cape pigeon *Daption capensis*; 8, dove prion *Pachyptila desolata*; 9, Wilson's storm petrel *Oceanites oceanicus*. The wing span of the largest species, *D. exulans*, is 3.03 m and that of the smallest, *O. oceanicus*, is 0.393 m. (From C. J. Pennycuick, The flight of petrels and albatrosses (Procellariiformes), observed in South Georgia and its vicinity, *Phil. Trans. R. Soc. Lond. B* **300,** 75–106/Figure 4, 1982, The Royal Society.)

as "geometrically similar" animals increase in mass, their surface area should decline proportional to $M_b^{0.67}$ and so should the rate at which energy will consequently be lost to the environment. Thus, it might be expected that basal metabolic rate (BMR) should also decline in proportion to $M_b^{0.67}$. Most studies of vertebrates have shown that BMR actually scales allometrically, at around $M_b^{0.72-0.75}$, i.e., slightly greater than predicted from the surface area law. However, some authors have suggested that $M_b^{0.67}$ is probably the correct term and that the calculations are biased by taxonomic (Bennett and Harvey, 1987) or statistical (Heusner, 1991) artifacts. Allometric analysis can be used to address a wide variety of issues in biology (Schmidt-Nielsen, 1984), and the scaling of metabolic and biomechanical parameters associated with bird flight will be considered throughout this chapter.

III. ENERGETICS OF BIRD FLIGHT

One of the most useful physiological variables is the rate at which a bird expends energy (i.e., power input (P_i) during flight). The P_i of resting and exercising animals are important considerations in many ecological and physiological studies of animal behaviour and evolution. Therefore, some of the various techniques used to study the biology of flight will be discussed, primarily with respect to the estimation of P_i during flight. However, the mechanical power available for flight comes from the relevant locomotor muscles. The chemical energy available from the anaerobic or aerobic catabolism of the various fuel molecules is transformed by the myofibrillar proteins of the flight muscles into mechanical power output (P_o) and heat. The exact quantity of P_o produced from a given P_i will depend on the efficiency with which the locomotor muscles can convert the chemical energy into mechanical power. The heat produced can be used to regulate body temperature, but is almost invariably "lost" to the environment, while the biomechanical P_o results in various forms of locomotory behaviors.

The SI unit of power is the watt (W) which is equivalent to 1 joule per second (J sec^{-1}). An estimate of P_i can be obtained by converting the rate of oxygen consumption ($\dot{V}O_2$, ml sec^{-1}) to W. The conversion of $\dot{V}O_2$ to P_i depends on the metabolic substrate; for metabolism of pure carbohydrate [i.e., with a respiratory quotient (RQ) of 1], 1 ml O_2 sec^{-1} = 21.1 W whereas for pure fat metabolism (i.e., with an RQ of 0.71), 1 ml O_2 sec^{-1} = 19.6 W (Lusk, 1919; Brobeck and Dubois, 1980). Pure protein metabolism in birds would yield an RQ of 0.74 and 1 ml O_2 sec^{-1} = 18.4 W (Schmidt-Nielsen, 1997). During short-duration exercise, it is likely that carbohydrate oxidation will dominate; for example, short hovering flights during foraging in hummingbirds (Suarez et al., 1990) and soon after take-off in pigeons (Butler et al., 1977), but during longer bouts of exercise it is likely that fat oxidation will become dominant (Rothe et al., 1987). However, this may be influenced by the feeding regime and by the season: well-fed and summer birds start at a higher RQ and take longer to reach an RQ of approximately 0.7 (Nachtigall, 1995). Thus, if RQ is unknown, a compromise figure for an RQ of 0.8, of 1 ml O_2 sec^{-1} = 20.1 W is often used for calculations of P_i during aerobic activity (Schmidt-Nielsen, 1997).

A. Techniques Used to Study the Mechanical Power Output Required for Flight

1. Aerodynamic and Biomechanical Models

The use of aerodynamic models provides a theoretical and biomechanical framework for understanding the physiological adaptations of birds and the likely constraints acting on different species. In addition, it is theoretically possible to estimate the P_i of a particular bird, using estimates for P_o based on aerodynamic models (Rayner, 1979; Pennycuick, 1989), provided the efficiency by which the flight muscles are able to transduce chemical energy into mechanical energy is known. The P_o required for different modes of flight is primarily determined by the overall mass, wing kinematics and detailed morphology of the bird under investigation (Pennycuick, 1968), but it is also modified by the prevailing environmental conditions such as air density, strength of gravity, local air thermals and deflected currents from nearby surfaces (Pennycuick, 1989).

During forward flapping flight, a bird has to generate sufficient mechanical power to overcome the downward acceleration due to gravity acting on its own mass (induced power) and the frictional drag of the airflow over its body (parasite power) and wings (profile power), as well as inertial power to accelerate and decelerate the wings during each beat. Inertial power requirements are usually ignored for medium to fast flight speeds, but may be important during hovering and slow flight (Norberg, 1990). A number of aerodynamic models (Pennycuick 1969; Tucker, 1973; Greenewalt, 1975; Rayner, 1979) have indicated that these components of the overall P_o of the flight muscles should vary with flight speed (U). Calculations indicate that parasite and profile powers should both increase with speed, whereas induced power is high at low speeds and declines at higher speeds.

Thus, overall P_o should vary with speed in a U-shaped fashion (Figure 2). This means that there is a speed at which P_o (and P_i) are at a minimum (U_{min}) and at which the bird should fly in order to maximize its flying time. However, this is not the speed at which the bird should fly in order to maximize the distance flown for a given amount of fuel consumption (i.e., the speed at which the energy cost of transport is lowest, E_{min}). This is the maximum range speed (U_{mr}, Figure 2) and is, theoretically, the speed at which a long distance migrant should fly (Pennycuick, 1969), particularly if the supply of energy is limited for any reason (Alerstam and Lindström, 1990). Heden-

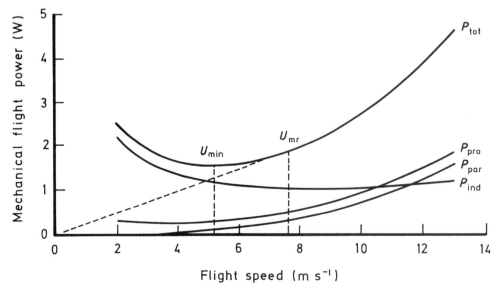

FIGURE 2 Calculated mechanical power output (P_{tot}) and its separate components (induced power, P_{ind}; parasite power, P_{par}; profile power, P_{pro}) for the European kestrel *Falco tinnunculus* (mass 0.21 kg) flying at different speeds. Also shown are the minimum power speed (U_{min}) and maximum range speed (U_{mr}). (From Rayner, *Contemp. Math.* **141** (1993), 351–399 by the American Mathematical Society.)

ström and Alerstam (1995) contend that birds should adapt their flight speed differently when migrating or transporting food compared with when they are foraging.

Measurement of drag components and the overall body drag coefficient of living birds is relatively difficult (Pennycuick *et al.*, 1988). Recent data from a teal (*Anas crecca*) and a thrush nightingale (*Luscinia luscinia*) flying in a wind tunnel indicate that the drag coefficients for these species may be some 20% of what was previously assumed (Pennycuick *et al.*, 1996). A phenomenon that could reduce induced drag is "ground effect", which is the result of a bird flying close above a plane surface, be it water (e.g., petrels) or land (e.g., certain raptors). It may also be important during take-off for heavy birds such as swans and vultures. Rayner (1991) has developed a theory for ground effect under fixed wing conditions. While acknowledging the limitations of the model, particularly with respect to its neglect of flapping wings, Rayner (1991) concludes that there are improvements in performance for flight in ground effect, provided the flight speed is not too low.

The inter-species scaling of the biomechanical power required to hold the wings out and to support the mass of the body during gliding flight is discussed by Pennycuick (1989). Given the limited amount of data available, he tentatively concludes that the power required should scale with respect to body mass between $M_b^{0.67}$ and $M_b^{0.83}$, i.e., that the power required is approximately a fixed multiple of the aerobic basal metabolic rate in birds of different mass. This is consistent with the limited amount of data on the cost of gliding in birds (see below).

Pennycuick (1968, 1969, 1989) has also discussed the factors which may limit the flapping flight performance of different species of birds. He hypothesizes that the P_o required for flight should scale proportional to $M_b^{1.17}$ and predicts that the P_o available from the flight muscles should scale proportional to $M_b^{0.67}$. Therefore, in order to maintain an equivalent flight performance, larger birds should develop relatively larger flight muscles than smaller birds. In fact, the mass of the flight muscles appears to scale in direct proportion to body mass (Greenewalt, 1962; Marden, 1987; Rayner, 1988; Bishop and Butler, 1995), which means that larger birds should have reduced flight performance in terms of rate of climb, speed of take-off, ability to hover, and so on. Thus, because the mass exponents are so different, there should be a clear upper limit to the size of birds which can perform flapping flight (Figure 3), and this would appear to be at approximately 12 kg (Pennycuick, 1968).

Pennycuick (1968, 1969, 1989) suggests that the range of flight performance of any particular bird will depend on the relation between the P_o/U curve and the maximum P_o available from the flight muscles. He distinguishes two categories of maximum power: absolute

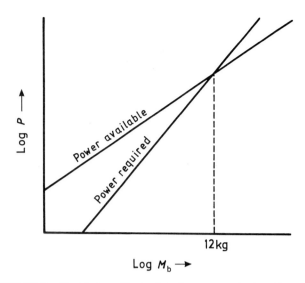

FIGURE 3 Double logarithmic plot of power required and power available (P) against body mass (M_b) for geometrically similar flying animals of different size. (From Pennycuick (1968), *J. Exp. Biol.*, Company of Biologists, Ltd.)

maximum power (P_{max}), which includes the aerobic and anaerobic capacity of the flight muscles, and maximum sustainable power (P_{ms}), which only includes the aerobic capacity of the muscles and is limited by the rate of delivery of oxygen and metabolites by the cardiorespiratory system. In hummingbirds, both P_{ms} and P_{max} lie above the power required for hovering (Figure 4) as these birds can hover aerobically. In contrast, for the white-backed vulture (*Pseudogyps africanus*) both P_{ms} and P_{max} lie between the minimum power required for

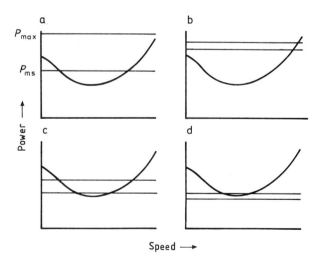

FIGURE 4 Power curves (see Figure 1) for (a) pigeon, (b) hummingbird, (c) white-backed vulture, and (d) California condor (not to scale). The upper horizontal line represents maximum power available (P_{max}), the lower line represents the maximum sustainable power (P_{ms}). (Modified from Pennycuick (1968), *J. Exp. Biol.*, Company of Biologists, Ltd.)

flight and that required for hovering, as they cannot hover, whereas they can fly horizontally. Birds such as the pigeon are intermediate between these two, possessing a P_{max} sufficient to hover for brief periods using anaerobic metabolism.

2. Air Flow Visualization and Direct Force Measurements

Although the aerodynamic models for estimating P_o have been available for many years, very few experiments have attempted to test their accuracy by obtaining direct measurements of the forces produced by the wings. Some ingenious experiments performed by Spedding *et al.* (1984) and Spedding (1986) enabled the wake of pigeons (*Columba livia*) and jackdaws (*Corvus monedula*) flying in a 4-m long cage to be visualized as they flew slowly through a cloud of neutrally buoyant helium bubbles. Although the data obtained supported the hypothesis (Rayner, 1979) that the wake is composed of a chain of small vortex rings (Figure 5a), indicating that the upstroke is inactive, the energy (induced power) calculated from the rings was only 60 and 35%, respectively, of those predicted from the model. The authors were unable to explain this "deficit", other than the possibility that the flight was of too short a duration and that, therefore, the birds were not flying at a constant speed as they flew through the cloud of bubbles. In a further study, Spedding (1987) flew a kestrel (*Falco tinnunculus*) through helium bubbles along a 36-m long corridor at a "medium" speed of approximately 7 m sec^{-1}. In this case, the wake consisted of a pair of continuous, undulating trailing vortices (Figure 5b), indicating that the upstroke of the wing was aerodynamically active. As a result, a more simple wake model can be used to calculate induced power, which is very close to that calculated from the model of Pennycuick (1975), based on classical aerodynamic theory. It would appear that the wake geometry of the kestrel flying at medium speed may be characteristic of a range of birds flying at their cruising speeds (Spedding, 1987; Rayner, 1990). Using wing-beat kinematics to infer lift production, Tobalske and Dial (1996) concluded that black-billed magpies (*Pica pica*) use a vortex-ring gait at all speeds, whereas pigeons use a vortex-ring gait at 6–8 m sec^{-1}, a transitional gait at 10 m sec^{-1}, and a continuous-vortex gait at higher speeds.

Recently, experimentalists have made a technological advance by using calibrated strain recordings from the humerus of flying birds to measure the force and estimate the P_o generated by the pectoralis muscle (Biewener *et al.*, 1992; Dial and Biewener 1993; Dial *et al.*, 1997). In general, their results give values for the mechanical power output required during flight that are lower than those generated by recent aerodynamic models (Rayner 1979; Pennycuick 1989), and the shape of the power curve with regards to the speed of flight is flatter (Dial *et al.*, 1997). Dial and Biewener (1993) indicate the possible sources of error in their studies. They acknowledge that there were uncertainties about the airspeed and flow conditions experienced by the birds in the wind tunnel, and that there were limitations in the calibration of the strain gauges, which may have led to an underestimation of the maximal force and P_o.

B. Techniques Used to Measure the Power Input Required for Flight

1. Mass Loss

Some of the earliest measurements of P_i were estimated from the mass loss recorded during a long, nonstop flight, with the assumption that fat constituted by far the major part of this loss and that the net loss of water was negligible (Nisbet *et al.*, 1963). Berger and Hart (1974) criticized this method, as they believe that total water loss may exceed the production of metabolic water, particularly at high environmental temperatures (cf., Nachtigall, 1995). Even so, Jehl (1994) has used this method to estimate the energy expenditure of black-necked grebes (*Podiceps nigricollis*) forced to land during migration as a result of bad weather. The data were inconclusive, mainly because of uncertainty over the time the birds were flying. Another problem with this method is that there is evidence to indicate that migrat-

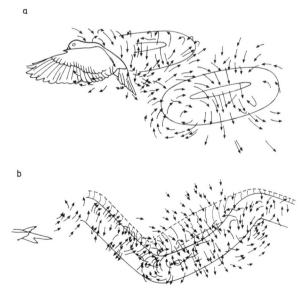

FIGURE 5 Reconstruction of the air movements and the position of the vortex core in the wake of (a) a pigeon, *Columba livia*, and (b) a kestrel, *Falco tinnunculus*, flying through a cloud of neutrally buoyant, helium-filled soap bubbles. (From Rayner (1988), "Current Orinthology," Vol. 5, pp. 1–66, Plenum Publishing Company.)

ing birds also catabolize protein (e.g., muscle and intestine) during flight (Piersma and Jukema, 1990; Jenni-Eiermann and Jenni, 1991) and, whereas fat has an energy density of 39.3 kJ g^{-1}, the value for muscle (which is composed of approximately 70% water) is only 5.4 kJ g^{-1} (cf., Schmidt-Nielsen, 1997).

Masman and Klaassen (1987) developed a more complex method of using changes in body mass to estimate flight costs. They studied the energy budgets of trained kestrels (*Falco tinnunculus*) performing directional flights in the laboratory. Energy expenditure during flight was calculated by monitoring daily metabolizable energy intake, oxygen consumption and carbon dioxide production during rest, and time spent flying per day. A similar approach has recently been applied to the study of a single thrush nightingale (*Luscinia luscinia*) flying for very long periods (seven different 12 h flight sessions) in a wind tunnel (Klaassen *et al.*, 1998). However, in general, estimates of energy consumption based on changes in body mass may be subject to considerable error as they are very sensitive to the values of protein and/or carbohydrate utilized in the calculation.

2. Doubly Labeled Water

The most commonly used method to determine the field metabolic rate (FMR) of air-breathing vertebrates is the doubly labeled water (DLW) method, and it has been used extensively with free-ranging birds (Hails, 1979; Ellis, 1984; Tatner and Bryant, 1989). This method was first proposed as a means of determining CO_2 production by Lifson and his co-workers over 40 years ago (Lifson *et al.*, 1949, 1955). It involves the injection of a mixture of ^2H_2O (or ^3H_2O) and $H_2^{18}O$ into the animal, allowing these isotopes to equilibrate with the body water pool(s), taking a body fluid sample after injection and releasing the animal into the field. After a number of days, the animal is recaptured and another body fluid sample is taken. The difference between the rate of loss of the different isotopes between the two samples is assumed to be equivalent to the rate of CO_2 production ($\dot{V}CO_2$), from which the metabolic rate and P_i of the animal can be estimated. The maximum time between these sampling points (the maximum duration of the experiment) varies according to the initial enrichment (the greater this is, the longer the time), the mass of the animal (for the same relative enrichment, the greater the mass, the longer the time) and, of course, the level of activity of the animal. The many assumptions that are made when using this technique have been thoroughly discussed by Nagy (1980) and Speakman (1990).

Although this is a relatively simple method to use in the field, it has two major limitations. Perhaps most importantly for field work is the fact that the duration of an experiment is limited by the turnover rate of the ^{18}O, thus it is necessary to recapture the animal after a sufficient delay following injection of the isotopes to enable a reasonable decline in the enrichment of ^{18}O, but before the enrichment is too low (Nagy, 1983). The other problem is that the technique only gives an average value for energy expenditure between the two sampling points and this is usually over a 24-hr daily cycle of activity (Speakman and Racey, 1988). When DLW is used to determine the metabolic cost of a specific activity (e.g., flying) then, as with the body mass method, it is necessary to have accurate information on the duration of the flying period(s) between the initial and final sampling times. Ideally, the respiratory quotient (RQ) also needs to be known to convert the calculated $\dot{V}CO_2$ to an estimate of $\dot{V}O_2$ and/or P_i (see above).

Recent validation studies with captive birds exercising at different levels of activity (but not flying) over a period of 72 hr have indicated that, provided mean data are used from a number of birds, the DLW method can give estimates of metabolic rate that are within a few percent of those measured by respirometry (Nolet *et al.*, 1992; Bevan *et al.*, 1994, 1995a). As far as the authors are aware, no such validations have been performed with flying birds, although the data from the first use of the technique with a flying bird (LeFebvre, 1964) compare well with those obtained from estimated fat loss by the same author and with direct measurements during flight of the same species (pigeon) in a wind tunnel (Butler *et al.*, 1977).

3. Telemetry and Data Logging

In recent years, much progress has been made in the use of biotelemetry and data logging devices for studying the physiology of freely flying birds. For example, heartbeat frequency (f_H) can be measured and used as an indicator of energy consumption. Some of the early studies involved attaching relatively large radiotransmitters to the back of birds and recording data during flights of only a few seconds duration (Hart and Roy, 1966; Berger *et al.*, 1970a,b). Of course, any externally mounted object is likely to influence the flight performance of a bird, not only as a result of the added mass, but also by increasing the aerodynamic drag of the body. These effects may be minimal for a relatively small transmitter that the bird can preen under its contour feathers (Obrecht *et al.*, 1988). However, in some circumstances, the effects can be quite substantial, particularly for high-performance birds like homing pigeons (Gessaman and Nagy, 1988). Butler and Woakes (1980) overcame the problem of increasing body drag by implanting relatively small (<10 g) radiotransmitters into the abdominal cavity of barnacle geese (*Branta leucopsis*) and were able to record heart rate (mean, 512 beats min^{-1}) and respiratory frequency (mean, 99 breaths min^{-1}) from two birds (mean mass, 1.6 kg) flying behind an open-topped vehicle. The problem with this technique is that it is necessary to be sufficiently close to the subject in order to receive and record the transmitted data. It is now possible, however, to implant data loggers (weighing less than 25 g) that will store basic physiological information (heart rate, body temperature, etc.) over periods of many days. The data can then be downloaded after the animal is recaptured (Woakes *et al.*, 1995; Bevan *et al.*, 1995b).

Thus, it has recently been demonstrated that, if properly calibrated and if mean data are used from a number of animals, f_H may be used as an indicator of oxygen uptake in free-living birds (Nolet *et al.*, 1992; Bevan *et al.*, 1994, 1995a,b). By storing f_H in implantable data loggers, Bevan *et al.* (1995b) estimated the energy cost of gliding flight, while having minimal effect on the behavior of the birds. However, the relationships between f_H and $\dot{V}O_2$ that are obtained during walking and running, or during swimming, may not be identical to those obtained when the birds are flying (Gessaman, 1980; Nolet *et al.*, 1992). In addition, it is recommended that the f_H values recorded in the field should not exceed the range of f_H measured during calibration. Thus, before this technique can be used reliably for birds during flapping flight, it should be calibrated under those conditions, perhaps in a wind tunnel. As the technology improves to enable more variables to be monitored, and for the storage of the data not to rely on the continued operation of the battery, this could be a very useful method for determining the physiological responses of birds flying under completely natural conditions.

4. Wind Tunnel

P_i is most easily estimated indirectly, and under laboratory conditions, by directly measuring the $\dot{V}O_2$ of birds flying in a wind tunnel and then converting that to P_i (see above). The increased use of wind tunnels during the past 30 years has greatly added to our knowledge and understanding of the physiology of birds during flight (Butler and Woakes, 1990; Norberg, 1990). Perhaps the main advantage of being able to use a wind tunnel is that it is possible to make physiological measurements that would be virtually impossible by any other means (e.g., repeated blood samples, blood pressure and flow measurements, direct measurements of force production, etc.). However, it must also be borne in mind that, as with almost any other method for studying a living animal, the wind tunnel itself can influence the behaviour of the animal flying in it and hence affect the data obtained when compared with those from free-flying birds (Butler *et al.*, 1977; Rayner, 1994). In order to obtain physiological data from a bird flying in a wind tunnel, it is often necessary to attach recording equipment to them, which will have some affect on the measurements. For example, gas exchange ($\dot{V}O_2$ and $\dot{V}CO_2$) is usually determined by means of a lightweight mask which covers the beak and nose and through which air is drawn via an attached tube (Tucker, 1968b; Tucker, 1972; Bernstein *et al.*, 1973; Butler *et al.*, 1977; Gessaman, 1980; Hudson and Bernstein, 1983; Rothe *et al.*, 1987). The mask and tubing will influence the recorded data because of their mass, but more so because of the extra drag that they impose. The only studies in which gas exchange has been determined in a wind tunnel without the use of a mask, are those by Tucker (1966) and Torre-Bueno and Larochelle (1978). These authors used air-tight, closed-circuit wind tunnels and measured the changing concentration of gases in the tunnel during the flight periods.

An aerodynamic analysis by Rayner (1994) suggests that the P_o required by a bird to fly is reduced when it flies in a wind tunnel with a closed section (unless the tunnel is at least 2.5× the wingspan of the bird and it flies at or near the center of the tunnel) and increased when it flies in a tunnel with an open section (again, the effect is related to the relative size of the tunnel), compared with the situation during free flight. However, at the moment there are few reliable experimental data to support these conclusions. In this context, it is interesting to note that the value for $\dot{V}O_2$ obtained by Butler *et al.* (1977) for pigeons (mean mass 0.442 kg) flying in a relatively large wind tunnel at 10 m sec^{-1} (182 ml O$_2$ min^{-1} kg^{-1}, adjusted for 10% effect of drag of mask and tubing; Tucker, 1972) is similar to that obtained by LeFebvre (1964) using the DLW method for free-flying pigeons (199 ml O$_2$ min^{-1} kg^{-1}, mean mass 0.384 kg). However, it is somewhat less than the mean minimum value obtained by Rothe *et al.* (1987) for pigeons (mean mass 0.330 kg) flying in a relatively small wind tunnel (248.5 ml O$_2$ min^{-1} kg^{-1}, also adjusted for drag of mask and tubing, but for a 22% effect; Rothe *et al.*, 1987). According to Rayner's analysis, aerodynamic factors alone would lead to $\dot{V}O_2$ being lower for the pigeons flown by Rothe *et al.* (1987). Rayner (1994) also concludes that, in closed wind tunnels, the P_o-versus-speed curve (see above) will be flatter than that for free-flying birds.

Contrary to the above, other studies have suggested that data on the energy cost of flight obtained from birds flying in wind tunnels are 30–50% higher than those obtained from some of the other methods discussed above, such as mass loss and DLW (Masman and Klaassen, 1987; Rayner, 1990). However, the former authors included in their wind tunnel data measurements from hovering hummingbirds and from other experiments that most definitely did not involve the use of a wind tunnel (e.g., Teal, 1969 and Berger *et al.*, 1970a). Rayner (1990) used the same data set of 71 birds as Masman and Klaassen (1987), but supplemented it with 40 more values from more recent publications. Unfortunately, he did not give the source of these additional data, so it is not possible to check and if necessary reanalyze them. However, taking the 9 values obtained by the DLW technique and used by Masman and Klaassen (1987) and reanalyzing them using the method of reduced major axis (RMA; Sokal and Rohlf, 1981; Rayner, 1985b), P_i (W) during forward flapping flight is calculated as: $P_i = 69.5 M_b^{0.87}$, $r^2 = 0.83$, $n = 9$ M_b = body mass in kg). Taking the 7 values for birds that did actually fly in wind tunnels from the list of Masman and Klaassen (1987) (i.e., numbers 44, 54, 57, 63, 64, 67, 68; note, P_i for bird 44 should be 4.08 W and not 40.8 W) gives $P_i = 58.8 M_b^{0.76}$, $r^2 = 0.98$. Neither the slopes nor the elevations of these regressions are significantly different from one another. If the minimum values are taken from the 7 studies of birds flying in wind tunnels, and if the drag of the mask and tubing is taken into account, then the equation is $P_i = 52.6 M_b^{0.74}$, $r^2 = 0.95$. Thus, there appears to be no justification for the conclusions made by Masman and Klaassen (1987) and Rayner (1990). It should be noted that Norberg (1996) made similar erroneous inclusions in her "wind tunnel" data (see her Table 7.3) to those of Masman and Klaassen (1987).

5. Modeling of Cardiovascular Function

Some of the shortcomings of using the aerodynamic models for estimating P_i (by assuming a value for efficiency) could possibly be avoided by devising a model to estimate total P_i directly. This has recently been attempted by Bishop and Butler (1995), who assumed that the mass of an animal's heart could be used as a basis for estimating cardiac stroke volume and calculating the animal's aerobic capacity. The role of the various components of the cardiovascular system in presenting O_2 to (and removing CO_2 from) the the exercising muscles can best be described by a form of Fick's formula for convection:

$$\dot{V}O_2 = f_H \times V_S \times (C_aO_2 - C_{\bar{v}}O_2)$$

where f_H is heart beat frequency (beats min^{-1}); V_S is cardiac stroke volume (ml); and $(C_aO_2 - C_{\bar{v}}O_2)$ is the difference between the oxygen content in arterial and mixed venous blood (milliliter of O_2 per milliliter of blood). In theory, each of these variables can scale independently with respect to body mass such that:

$$(\dot{V}O_2)M_b^z = (f_H)M_b^w \times (V_s)M_b^x \times (C_aO_2 - C_{\bar{v}}O_2)M_b^y.$$

There has only been one study in which all four variables of the Fick equation have been measured in a bird during forward flapping flight (Butler *et al.*, 1977), so Bishop and Butler (1995) substituted each of the values from this study into the above equation, based on a series of general allometric equations ($y = a x^b$) for the scaling of f_H, V_s, and $(C_aO_2 - C_{\bar{v}}O_2)$, respectively. For birds flying close to their minimum power speeds, the individual components of the Fick equation were estimated to scale as follows: $f_H = 574 M_b^{-0.19}$, $V_s = 3.48 M_b^{0.96}$, $(C_aO_2 - C_{\bar{v}}O_2) = 0.083 M_b^{0.00}$. The estimated scaling of V_s during flight was based on the assumption that V_s should be almost directly proportional to the mass of the heart (M_h), or $V_s = 0.3 M_h^{1.05}$, while f_H during flight and M_h may be readily determined experimentally.

The determination of $(C_aO_2 - C_{\bar{v}}O_2)$ may be more problematic, but using the value of 0.083 milliliter of O_2 per milliliter of blood for a 1.6-kg barnacle goose with a heart rate of 512 beats min^{-1}, Bishop and Butler (1995) give an estimate of $\dot{V}O_2$ (264 ml min^{-1} = 165 ml min^{-1} kg^{-1}) which is not very different from the directly measured value (Figure 6). Bishop and Butler (1995) concluded that, for birds with a high aerobic capacity (such as pigeons), $\dot{V}O_2$ (ml min^{-1}) = 166 $M_b^{0.77}$, for birds flying close to their minimum power speeds. This compares favorably with the relationship ($\dot{V}O_2 = 160 M_b^{0.74}$) calculated by Butler (1991) and reanalyzed using RMA for seven species of birds during forward flapping flight in a wind tunnel. Bishop (1997) used a similar approach to estimate the maximum $\dot{V}O_2$ of birds during flight, again using M_h as the basis of the calculation. Although there is a need to obtain more data on the cardiovascular changes that occur in birds of different body mass and during different modes of flight, it would appear that the use of M_h to estimate the aerobic capacity of different species of birds may have more general applicability than the use of the allometric relationship obtained from a few species of birds flying in a wind tunnel.

C. Empirical Data Concerning the Power Input during Flight

An individual flight by a free-ranging bird will include at least one take-off and a period of climbing, probably several maneuvers (turns, short periods of "burst" flying, changes of gait) as well as various periods of gliding or soaring and flapping flight. The power requirements associated with all of these various types of flight will be different and may be provided by either anaerobic or aerobic metabolic pathways or by a combination of the two.

Species belonging to the Galliformes (e.g., pheasants, grouse, and quail) routinely engage in a powerful "burst" type of flight during take-off and short-duration

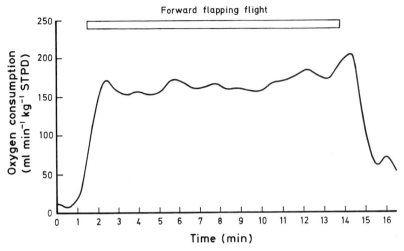

FIGURE 6 Direct measurement of the rate of oxygen consumption (STPD) of a 1.59 kg barnacle goose, *Branta leucopsis*, during forward flapping flight in a wind generator at a speed of 12 m sec^{-1}. Note that toward the end of the flight, as the air speed decreased, the bird progressively began to hover and the rate of oxygen consumption increased accordingly. (Redrawn from P. J. Butler, A. J. Woakes, R. M. Bevan, and R. Stephenson, submitted.)

flights. Their muscles use intracellular stores of muscle glycogen utilizing predominantly anaerobic metabolism, which leads to the accumulation of lactic acid and to a metabolic acidosis. Fatigue soon develops and, subsequently, the lactate is either oxidized as fuel by tissues such as the heart or reconverted to glycogen by the liver. Thus, a large proportion of the total cost of these explosive flights is "paid" while the bird is resting on the ground following the flight (Bishop and Butler, 1995) and an assessment of the true cost of such flights, if based on $\dot{V}O_2$, would have to include the postflight period of recovery. In other words, measurements of $\dot{V}O_2$ can only be used for accurate estimates of P_i if the activity is sustainable for relatively long periods of time (i.e., if the bird is metabolizing aerobically).

1. Gliding and Soaring Flight

In windy conditions, albatrosses exclusively use soaring and gliding flight and even in near calm conditions, they only beat their wings occasionally (Alerstam *et al.*, 1993). Using the DLW method, Adams *et al.* (1986), studying the wandering albatross (*Diomedia exulans*), and Costa and Prince (1987), studying the grey-headed albatross (*D. chrysostoma*), estimated that the metabolic cost of flight is 3.0× (49.7 W for birds of 8.4 kg mean mass) and 3.2× (36.3 W for birds of 3.7 kg mean mass) predicted basal metabolic rate (BMR), respectively. These estimates were based on the assumption that, when on water, the metabolic rates of the birds were close to those when they were at rest on land and their mean activity budgets were similar to those obtained from other studies on the same species. However, for herring gulls (*Larus argentatus*) gliding in a wind tunnel, P_i (calculated from measured $\dot{V}O_2$) is, at 13.9 W for birds of 0.91 kg mean mass, only 2.1× the resting value (Baudinette & Schmidt-Nielsen, 1974).

In a study on black-browed albatrosses (*D. melanophrys*), Bevan *et al.* (1995b) used the f_H method to determine FMR and salt water-switches to determine when the birds were on the water. It was, therefore, possible for these authors to estimate the metabolic rate of their birds when on water and this was 2.6× that recorded from the birds during incubation and 1.9× the predicted BMR; that is, somewhat higher than was assumed by Adams *et al.* (1986) and Costa and Prince (1987). Interestingly, the average metabolic cost of flying was, at 21.7 W for birds of 3.5 kg mean mass, 60% of that estimated for the similar sized grey-headed albatross by Costa and Prince (1987) and not significantly different from the value obtained when the birds were on the water. Other studies on sea birds, in which the DLW method has been used to estimate FMR, have used salt water switches in order to determine when the bird is airborne, but the species studied (Sulidae) use a combination of gliding and flapping flight, so it was not possible to determine the energetic cost of either mode (Birt-Friesen *et al.*, 1989; Ballance, 1995). Similar limitations apply to other studies in which the assumption is made that, when not on the nest, the birds are continuously flying (e.g., Flint and Nagy, 1984; Obst *et al.*, 1987).

The study by Bevan *et al.* (1995b) illustrates the importance of accurately determining the behavior of the animals under study and the advantages of using a method for determining FMR that allows the metabolic cost of the separate behaviors to be estimated. By using satellite transmitters and the logging of f_H data, Bevan et al. (1995b) were able to determine the metabolic costs of different sections of a foraging trip of an individual black-browed albatross (Figure 7). During section 2 of the trip, it covered a distance of 80.1 km in 1.7 hr, a mean ground speed of 13.1 m sec^{-1}. This is very similar to the mean speed of 12.9 m sec^{-1} that Alerstam et al. (1993) report for a visually tracked black-browed albatross. This suggests that the bird was flying continually during section 2. Metabolic rate was low during this period (2.4 W kg^{-1}) and barely greater than that during incubation (2.2 W kg^{-1}). The bird was probably searching for patches of prey and being carried by the wind (Alerstam et al., 1993). On the other hand, during section 4 of the trip, the straight-line distance traveled was only 6.3 km at a mean ground speed of 1.6 m sec^{-1}. This suggests that the bird was relatively stationary on the water and/or was flying within a restricted area. It exhibited its highest metabolic rate (8.8 W kg^{-1}) during this period, indicating that the bird was engaged in some form of relatively strenuous activity, such as more active flight and/or feeding.

2. Forward Flapping Flight

Some of the most comprehensive, and probably the most reliable (see above), data we have on the energy consumed (P_i) during forward flapping flight are those obtained from birds flying in wind tunnels. The use of wind tunnels has made it possible to estimate P_i at different flight speeds (Figure 8), and this relationship appears to vary between species. Three species have a J-shaped curve, one has a flat "curve," and only two, the budgerigar (*Melopsittacus undulatus*) and the pigeon, have the U-shaped curve predicted by aerodynamic theory for mechanical power output (Pennycuick, 1969; Tucker, 1973; Greenewalt, 1975; Rayner, 1979). Arguably, the J-shaped P_i curves may not be so different from the predicted U-shaped P_o curve, as very low air speeds were not obtained with these species. However, partly on the basis of data obtained from hummingbirds while hovering and during forward flapping flight (Berger, 1985), Ellington (1991) has argued that there may, in fact, be "little variation in mechanical power from hov-

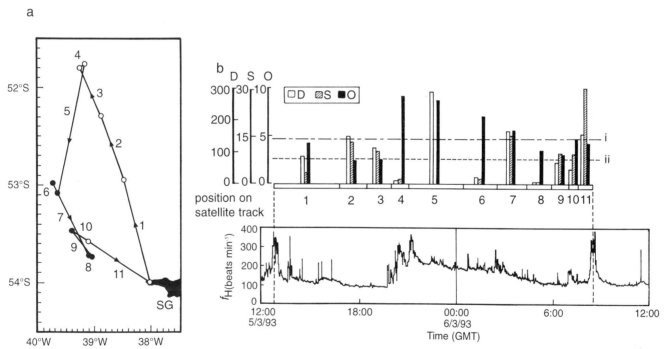

FIGURE 7 (a) Flight track of a single foraging trip from South Georgia (SG) by a black-browed albatross, *Diomedea melanophrys*, equipped with a satellite transmitter and a heart rate data logger. Circles represent the position of each satellite uplink, with the solid circles showing positions during the hours of darkness. The arrows show the direction traveled between each successive position, each section being numbered consecutively. (b) Top graph: distance traveled (D, km), ground speed (S, m sec^{-1}), and estimated energy expenditure (O, W kg^{-1}) for each section of the trip depicted in a. The horizontal bars represent the time between each satellite uplink. Mean ground speed (horizontal broken line, i) is from Alerstam et al. (1993) and the incubating metabolic rate (horizontal broken line, ii) is that measured while birds were on the nest incubating (W kg^{-1}). Bottom graph: heart rate (f_H) trace over the same period. The vertical dashed lines joining the two graphs show when the bird left and returned to the colony. (Modified from R. M. Bevan, P. J. Butler, A. J. Woakes, and P. A. Prince, The energy expenditure of free-ranging back-browed albatrosses, *Phil. Trans. R. Soc. Lond. B* **350**, 119–131/Figure 9, 1995b, The Royal Society.)

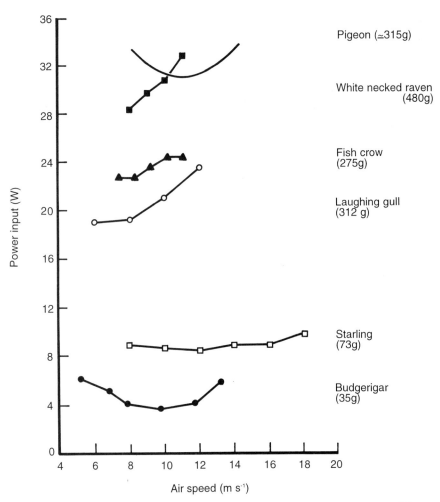

FIGURE 8 Power input (W) at different air speeds for six different species of birds of different mass during horizontal flapping flight in a wind tunnel. (From *Bird Migration: Physiology and Ecophysiology,* The physiology of bird flight, P. J. Butler and A. J. Woakes, pp. 300–318, 1990, © Springer-Verlag.)

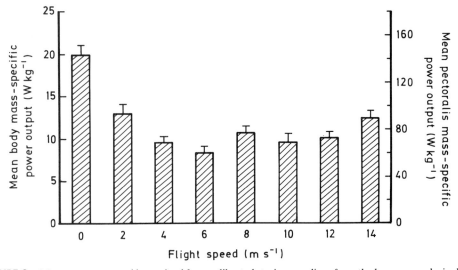

FIGURE 9 Mean power output (determined from calibrated strain recordings from the humerus and wingbeat frequency) of three black-billed magpies, *Pica pica,* flying at different speeds in a wind tunnel. Power output during hovering was significantly higher ($P < 0.001$) than at all other speeds, but no significant differences in power output were observed between 4 and 14 m sec^{-1} (reprinted with permission from *Nature,* Dial et al., 1997, copyright 1997 Macmillan Magazines Limited).

ering to intermediate flight speeds and then a sharp increase at higher speeds." In other words, the J-shaped P_o curve may be more realistic for these species. Thus, an increase in both P_i and P_o with increasing speed is widely accepted and this would indicate a near constant mechanochemical efficiency with increasing speed. In constrast, the first ever direct determinations of the forces produced, and the P_o of the pectoral muscles of a bird (the black-billed magpie) flying at a range of speeds, indicates that there is an L-shaped (i.e., a mirror image of the J-shape) relationship between the two, with P_o being relatively constant at speeds above 4 m sec^{-1} (Figure 9).

Figure 10 illustrates an allometric plot of the available data for the minimum P_i of seven species of birds during forward flapping flight in wind tunnels (Tucker, 1968b, 1972; Bernstein *et al.*, 1973; Butler *et al.*, 1977; Torre Bueno and Larochelle, 1978; Gessaman, 1980; Hudson

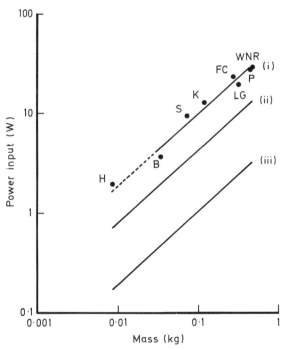

FIGURE 10 The relationships between power input (P_i) and body mass, M_b, using least-squares regression analysis for: (i) seven species of birds during forward flapping flight in a wind tunnel; B, budgerigar *Melopsittacus undulatus;* S, starling *Sturnus vulgaris;* K, American kestrel *Falco sparverius;* LG, laughing gull *Larus atricilla;* FC, fish crow *Corvus ossifragus;* P, pigeon *Columba livia;* WNR, white-necked raven *Corvus cryptoleucus.* Minimum values of power input (P_{im}) were used to construct this curve and $P_{im} = 50.7 M_b^{0.72}$; (ii) small mammals during maximum sustainable exercise where $P_i = 22.6 M_b^{0.73}$ (data from Pasquis *et al.*, 1970); (iii) resting nonpasserine birds where $P_i = 5.5 M_b^{0.73}$ (data from Prinzinger and Hanssler, 1980). (From *Circulation, Respiration, and Metabolism: Current Comparative Approaches,* Exercise in normally ventilating and apnoeic birds, P. J. Butler and A. J. Woakes, pp. 39–55, 1985, © Springer-Verlag.) Also included in (i), the minimum P_i of the hummingbird *Colibri coruscans* during forward flapping flight in a wind tunnel (H) (from Berger, 1985).

and Bernstein, 1983) and it demonstrates two interesting features. First, on average, the minimum P_i (after adjusting for the drag of the mask and tubing, where necessary) of the seven species (mass range, 35 to 480 g) is 9.2× the BMR calculated for nonpasserine birds (Prinzinger and Hanssler, 1980) and second, it is 2.2× the P_i of mammals of similar mass (up to 900 g) running at their maximum sustainable speed (Pasquis *et al.*, 1970). There has been one study in which metabolic rate was recorded from a species of hummingbird (*Colibri coruscans*) during forward flapping flight in a wind tunnel (Berger, 1985). It can be seen from Figure 10, that the minimum metabolic rate during flight is close (within 20%) to that predicted from the extrapolated regression line determined previously for the other seven species. If the value from the hummingbird data is included in the regression analysis, the equation becomes: 48.5 $M_b^{0.69}$, $r^2 = 0.98$, using least-squares regression analysis.

Data from four wild barnacle geese migrating from their breeding grounds in Svalbard (79°N) to their wintering grounds in southern Scotland (55°N), indicate that the average heart rate during flight at the beginning of migration was 317 beats min^{-1}, but that it decreased during the migratory period, reaching its lowest value (226 beats min^{-1}) toward the end. This was, no doubt, related to an estimated 26% reduction in body mass during the migration (Butler *et al.*, 1998). Average heart rate during all of the migratory flights was only 253 beats min^{-1} (i.e., 50% of that obtained from geese flying behind a truck, see above). The mean mass of four premigratory geese was 2.3 kg and, by estimating cardiac stroke volume (Bishop and Butler, 1995) and assuming a $(C_aO_2 - C_{\bar{v}}O_2)$ of 0.123 milliliter of O_2 per milliliter of blood, Butler *et al.*, (1998) estimated average minimum $\dot{V}O_2$ at the beginning of migration to be 302 ml O_2 min^{-1}, or 99 W. This is remarkably similar to the value (296 ml O_2 min^{-1} for a 2.3-kg goose) obtained from an extrapolation of the allometric equation derived from seven species of birds flying in a wind tunnel (Butler, 1991) recalculated using RMA regression (see above).

3. Hovering Flight

Hovering flight is, in theory, relatively costly (see Figure 2); however, the only study in which $\dot{V}O_2$ for both hovering and forward flapping flight have been determined for the same bird (Berger, 1985) indicates that there is, in fact, no difference in P_i between the two forms of flight (Figure 11). This is consistent with the fact that the values for $\dot{V}O_2$ from six species of hovering hummingbirds lie close to the extrapolated line obtained from larger birds during forward flapping flight, even though the allometric relationships for the two groups are substantially different [$\dot{V}O_2 = 160 M_b^{0.74}$

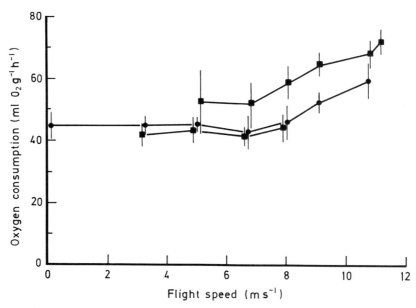

FIGURE 11 Mean (± SD) of direct measurements of oxygen consumption of two species of hummingbird, *Colibri coruscans*, 8.5 g (●) and *C. thalassinus*, 5.5 g (■) flying at different speeds in a windtunnel. [From Berger (1985), with permission from Gustav Fischer, Stuttgart.]

for birds during forward flapping flight and 463 $M_b^{0.91}$ for hovering hummingbirds; Butler (1991), and reanalyzed using RMA]. If data from three more species of hummingbirds are added to those used by Butler (1991) in his analysis (those from Wolf and Hainsworth (1971) and Wells (1993a), the relationship for the nine species of hummingbirds, using RMA analysis, becomes: $\dot{V}O_2 = 449 \, M_b^{0.90}$, $r^2 = 0.94$. It is interesting to note that the analysis by Bishop and Butler (1995) indicates that minimum $\dot{V}O_2$ during forward flapping flight of hummingbirds should scale $314 \, M_b^{0.90}$.

4. Scaling of Flight Muscle Efficiency and Elastic Energy Storage

The efficiency with which the flight muscles can produce mechanical power from chemical energy is of great physiological significance as it determines the overall energy required by the animal during exercise. Excess heat must be lost to the environment in order to prevent hyperthermia. However, in order to estimate efficiency it is necessary to measure both the P_i and the P_o of the flight muscle.

By using detailed kinematic and morphological analyses, Wells (1993a) calculated an estimate for P_o (100 W kg^{-1}) during hovering of the broad-tailed hummingbird (*Selasphorus platycercus*) based on the aerodynamic model of Ellington (1984). $\dot{V}O_2$ was measured as 50 ml g^{-1} h^{-1}, which gave an estimated efficiency of 9–11% (assuming perfect elastic storage of energy). By adding weights to the birds, Wells (1993b) determined that maximum P_o of the pectoral muscles for this species during hovering was around 117 W kg^{-1} (assuming perfect elastic storage of energy) and that the efficiency remained approximately 10%. Using the novel and elegant method of flying ruby-throated hummingbirds (*Archilochus colubris*) in an increasing atmosphere of He/O$_2$, Chai and Dudley (1995) were able to reduce the density of the medium from the normal air value of 1.23 kg m^{-3} to the point (on average, 0.54 kg m^{-3}) at which the birds were no longer able to hover for more than a few seconds. At this density, the P_o of the highly oxidative pectoral muscles was calculated to be around 133 W kg^{-1} (assuming perfect elastic storage of energy), $\dot{V}O_2$ was measured as 62 ml g^{-1} h^{-1}, and the efficiency was estimated to be around 10% regardless of the intensity of the flight "effort." The above calculations include an allowance of 10% for the cost of cardiorespiratory function (Tucker, 1973; Pennycuick, 1989), but there is no allowance for BMR. However, taking this into account (Krüger *et al.*, 1982) only increases efficiency by approximately 1% in both studies.

Biewener *et al.* (1992) and Dial and Biewener (1993) used calibrated strain recordings from the humerus of flying birds to measure the force and estimate the power generated by the pectoralis muscles. In the first of these studies, mean P_o for three starlings (*Sturnus vulgaris*) between 70–73 g mass and flying in a wind tunnel (for unstated duration) at an estimated speed of 13.7 m sec^{-1} was 1.12 W. Torre-Bueno and Larochelle (1978) reported mean P_i to be 8.9 W for three starlings of mean mass 73 g and flying in a wind tunnel for 30 min over a range of speeds from 8–18 m sec^{-1}. Thus, after subtracting BMR (Prinzinger and Hanssler, 1980) and allowing 10% for the cost of cardiorespiratory function, the estimated efficiency of the flight muscles for this species is 15.4%. In the second study, mean P_o for four pigeons between 301–314 g and flying at approximately 8 m sec^{-1} along a 47 m long corridor, was 10.5 W kg^{-1}. Dial and Biewener (1993) used a P_i value of 106 W kg^{-1} (cf., Rothe et al., 1987). Thus, estimated flight muscle efficiency is approximately 11.2%. However, Rothe et al. (1987), estimated that their measurements of P_i were elevated by between 15–30% by the presence of the mask and tubing. If a mean overestimate of 22% is assumed, the actual value of P_i would be 87 W kg^{-1}, which would give an efficiency of 13.8%. If the value of P_i, after adjustment for the drag on the mask and tube, is taken from Butler et al. (1977) for pigeons of mean mass 442 g and flying at 10 m sec^{-1} (61 W kg^{-1}), a flight muscle efficiency of 19.9% is obtained. Thus taken together, these studies indicate that there may be an increase in overall flight efficiency with an increase in body mass.

It must be emphasized that the above calculations may not give the "true" efficiencies at which the flight muscles operate. It is only the "apparent" efficiency calculated when specifically combining these estimates for aerobic P_i and P_o of the flight muscles. Errors in either measurement/calculation will lead to an error in the calculation for efficiency. Theoretically, it should not be possible for muscle fibers to operate with an overall efficiency greater than 28% (Rall, 1985) and, in practice, muscle fibers operate with maximal efficiencies of between 20 and 25% (Taylor, 1994). Currently, there are insufficient data to differentiate completely between the hypotheses that the "true" conversion efficiency is either a constant or that it scales with a slightly positive exponent with respect to M_b. As the maximum P_i available for flight is calculated to scale in proportion to approximately $M_b^{0.82}$ (Bishop, 1997), while P_o required scales approximately to $M_b^{1.17}$ (see above), this would indicate that overall efficiency (P_o/P_i) is proportional to approximately $M_b^{0.35}$. In other words, the overall efficiency is theoretically greater in the larger flying birds (cf., Rayner, 1990). Bishop and Butler (1995) estimated a lower scaling of overall efficiency of $M_b^{0.21}$ and even this may be an overestimate. However, this is a controversial subject and may involve issues such as elastic storage of wing kinetic energy (see below).

Maximum mass-specific P_o of the flight muscles would be expected to be much greater in birds such as hummingbirds compared to that of larger species of birds due to their higher wingbeat frequencies (Pennycuick and Rezende, 1984). Dial and Biewener (1993) calculated a figure of 119 W kg^{-1} of flight muscle for the pigeon, which is similar to that calculated for hummingbirds hovering for a maximum of a few seconds (Wells, 1993a,b; Chai and Dudley, 1995), despite the fact that the pigeon has a wingbeat frequency of around 15 Hz compared to that of 50 Hz for the hummingbird. However, studies by Chai and Millard (1997) and Chai et al. (1997) on five species of hummingbirds hovering while lifting maximum loads for less than 1 sec, show that maximum anaerobic output of the flight muscle varies between 206 and 327 W kg^{-1} (assuming perfect elastic storage of inertial energy). Thus, mass-specific P_o of the pectoral muscle of hummingbirds may, in fact, be considerably greater than that of pigeons.

Wells (1993a) and Chai and Dudley (1995) concluded that hummingbirds are able to store the "inertial energy" required to accelerate the wing through each beat in an elastic component of their muscles. This assumption greatly reduces the value for the calculation of the power required to fly, especially for small species of birds that have high wingbeat frequencies. Bird flight muscles lack any associated specialized elastic components, such as long tendons, so any elastic component is likely to be within the cross-bridges of the muscle (Alexander and Bennett-Clark 1977). Dial et al. (1987) measured the electrical activity of the pectoralis muscles of the pigeon during flight and demonstrated that the pectoralis muscles are active before the end of the upstroke of the wings. They conclude that the cross-bridges of the flight muscles are active during the final phase of the wing stroke and that the muscle could act like a spring (Goslow and Dial 1990).

As pointed out by Spedding (1994), if zero elastic storage is assumed (cf., Weis-Fogh, 1972) the calculation of efficiency during hummingbird flight is around 25% and that this is still within the theoretical possibilities for vertebrate muscle (Weis-Fogh and Alexander 1977; Pennycuick, 1992). The aerodynamic models of Pennycuick (1975; 1989) and Rayner (1979) also do not include the costs of accelerating and decelerating the wing during each cycle, as it is argued that the inertial costs are thought to be small during forward flight at medium to fast speeds. However, Van den Berg and Rayner (1995) suggest that it may be necessary to consider this cost. Van den Berg and Rayner (1995) estimated that the inertial power requirement should scale as approximately 5.8 $M_b^{0.80}$ (allowing for a 25% reduction due to wing flexion on the upstroke). Thus, in the light of the reduced estimates of flight costs of Pennycuick et al. (1996), the relative importance of the inertial power requirements may have to be increased. For a 10-kg bird, inertial power could be approximately an additional 15% of the P_o required, while for a 0.1-kg bird inertial power could be approximately 34% of the P_o required.

Whatever the merits of each side of the debate, it seems possible that perfect elastic storage may prove to be unrealistic. Thus, any increment of the inertial costs that must be met by the flight muscles will increase the estimate of the "apparent" efficiency of the flight muscles. If "near" zero elastic storage is correct, then the inertial costs would dominate the total cost of hovering flight in hummingbirds (Spedding, 1994).

IV. THE FLIGHT MUSCLES OF BIRDS

A. Flight Muscle Morphology and Fiber Types

The flight muscles of birds are structurally similar to the striated muscles of other vertebrates and consist of large numbers of long fibers, or cells, aligned essentially in parallel (Figure 12a). Each fiber can have a distinct biochemical character. At its simplest, each fiber can be specialised either for aerobic or anaerobic energy consumption, they can vary in diameter and cross-sectional area and are supplied with various metabolites and oxygen via a network of capillaries (Figure 12b). Different species of birds have different amounts of flight muscle with respect to body mass, termed the flight muscle:mass ratio by Marden (1987), and these muscles are composed of four main cell or fiber types (cf., Rosser and George, 1986a), so called slow oxidative fibers (SO), fast oxidative glycolytic fibers (FOG), fast glycolytic fibers (FG) and fibers which are intermediate (I) in their oxidative ability between the FOG and FG fibers.

Table 1 lists the major characteristics of these different fiber types. Essentially, the SO fibers have relatively slow rates of shortening, but are relatively efficient in producing force during isometric contractions, utilize oxidative metabolic pathways in the biosynthesis of energy-rich adenosine triphosphate (ATP), and are resistant to fatigue. FOG fibers are also capable of oxidative metabolism and resistance to fatigue, but can also use carbohydrates as a fuel during anaerobic metabolism, have relatively faster rates of shortening, and are relatively efficient in producing force while shortening in length. FG fibers are susceptible to fatigue and are restricted to anaerobic metabolism of carbohydrates, but produce more power per unit mass of muscle than SO and FOG fibers. One reason for the latter, is that they contain a lower volume fraction of mitochondria and associated structures and, therefore, have room for more myofibrilar proteins.

The function of SO fibers is considered primarily to be of advantage in postural muscles (e.g., in the back and neck muscles) and also in the legs during rest, due to their economic production of force during isometric contractions. So far, SO fibers have only been found in the deep layers of the pectoralis muscles of a limited number of species of birds which are especially adapted for gliding and soaring modes of flight, that is, species belonging to the avian orders of the Procellariformes, Pelecaniformes, Ciconiiformes, and the Gruiformes (Rosser and George, 1986a; Rosser et al., 1994). Thus, Rosser et al. (1994) found that the deep "belly" of the pectoralis muscle of the American white pelican (*Pelecanus erythrorhynchos*) is composed exclusively of SO fibers, and other authors have reported SO fibers in the deep layers of the pectoralis muscles of the turkey vulture *Cathartes aura* (Rosser and George, 1986b), red-tailed hawk *Buteo jamaicensis* (Rosser and George, 1986a), double-crested cormorant *Phalacrocorax auritus* (Rosser and George, 1986a), and the domestic chicken *Gallus gallus* (Matsuda et al., 1983; Rosser and George 1986a). However, in general, the occurrence of SO fibers is rare in birds, and they make up only a small percentage of the pectoralis muscle when they do occur. Thus, while SO fibers may have some role in the thermal soaring flight of the pelican and the red-tailed hawk, or perhaps in the "outstretched" wing posture of the cormorant while drying its feathers, it would appear that most species of birds can manage without this specialized fiber type in their flight muscles (cf., Talesara and Goldspink, 1978; Rosser and George, 1986a).

The most abundant type of fiber to be found in the flight muscles of birds capable of prolonged flapping flight is the FOG (George and Berger 1966; Rosser and George, 1986a). In general, the flight muscles of small birds are entirely composed of FOG fibers while larger birds have a mixture of different fiber types. However, there is a large degree of interspecies variation. George and Berger (1966) classified the pectoralis muscle of birds into six groups, with each fiber type listed in order of relative numerical abundance: (1) fowl type (FG, I, FOG); (2) duck type (FOG, FG, I); (3) pigeon type (FOG, FG); (4) kite type (mainly I); (5) starling type (FOG, I); and (6) sparrow type (FOG). While most species can be broadly assigned to one of these groups, there are many exceptions. Clearly, each individual species is adapted to cope with its own unique requirements based on its behavioral ecology, its detailed morphology and the environmental conditions in which it lives.

Even more importantly, the histochemical classification of muscle fiber types can be misleading from a functional perspective. In reality, the fast-twitch fiber types are arbitrary classifications within a continuum of oxidative to glycolytic potentials. In addition, the term "fast" is also arbitrary, as there is likely to be a continuum for the rates at which the flight muscles are optimized to contract, due to the scaling of wingbeat fre-

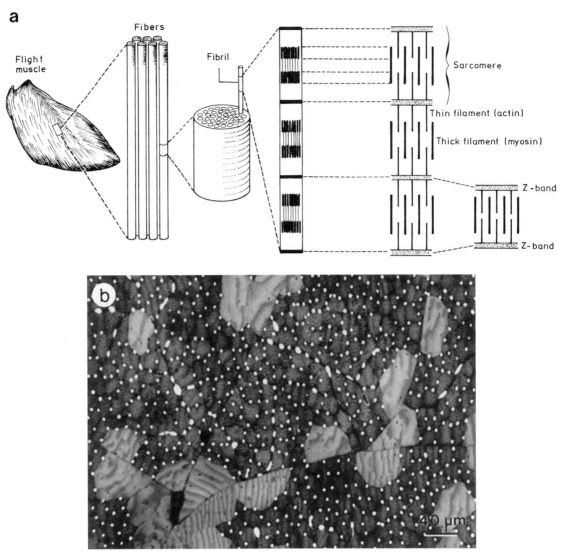

FIGURE 12 (a) Schematic illustration of the structure of a bird's pectoral muscle. The whole muscle is composed of many different fibers or cells, each of which consists of many fibrils and each fibril contains many thin actin filaments interspersed between thick myosin filaments. These filaments are oganized into sarcomeres, which shorten during contraction as the actin and myosin filaments slide past each other. (Modified from K. Schmidt-Nielsen, "Animal Physiology: Adaptation and Environment," 5th ed., 1997, Cambridge University Press.) (b) Light micrograph of a 1-μm-thick transverse section of pectoralis muscle of the pigeon *Columba livia*. Fast oxidative glycolytic fibers (dark and small) and fast glycolytic fibers (pale and large) can be clearly identified. Capillaries are empty following vascular perfusion fixation. Bar = 40 μm. [From Mathieu-Costello, 1991, *Am. J. Anat.* **191**, 74–84. Copyright © 1991 Wiley-Liss Inc., a division of John Wiley & Sons, Inc.]

TABLE 1 Simplified Characteristics of Muscle Fiber Types

Property	Slow oxidative SO (type I)	Fast oxidative glycolytic FOG (type IIA)	Fast glycolytic FG (type IIB)
Speed of contraction	Slow	Medium–fast	Medium–very fast
Aerobic Capacity	High	High	Low
Anaerobic capacity	Low	Moderate	High
Capillary supply	Good	Good	Poor
Triglyceride stores	High	Medium–high	Low
Glycogen stores	Medium	Medium–high	Medium–high
Fatigue resistance	High	High	Low
Cross-sectional area	Small	Small–medium	Medium–large

quency with body mass (Pennycuick, 1990, 1996). A small hummingbird may beat its wings at up to 50 Hz, while a large swan may only have a wingbeat frequency of around 2.4 Hz. Therefore, there are likely to be many associated differences in the molecular characteristics of the contractile proteins from different species of birds. As a consequence, the FOG fibers of a hummingbird are not biochemically identical to the FOG fibers of a swan. It is probable that, even within the muscle of an individual bird, the FOG fibers of the deep layers may have slightly different characteristics to the FOG fibers from the more peripheral layers. Rosser and George (1986a) suggest that, given sufficient detailed knowledge about the protein isoforms of muscle, it is conceivable that almost all muscle fibers could be described as a unique subtype.

Despite the difficulties outlined above, it is still possible to draw some interesting conclusions about the functioning of avian flight muscles by quantifying their fiber type compositions. The Galliformes, such as the domestic chicken and the ruffed grouse (*Bonasa umbellus*), have a majority of FG fibers in their pectoralis muscles (Rosser and George 1986a). This is matched by their relatively small heart masses and a low capacity to supply the muscles with oxygen (Bishop and Butler, 1995; Bishop, 1997). These birds are restricted to anaerobic flights of short duration. Typically, galliform birds live close to the edges of forests, have relatively short wings and relatively large flight muscles, and escape predation by being adapted for rapid take-off followed by short flights into dense vegetation or into the branches of trees. The South American ecological equivalent of the Old World Galliformes, the tinamous, also have very small hearts and large flight muscles (Hartman, 1961) and live on the forest floor.

The Galliformes contrast with their closest relatives (Sibley and Alquist, 1990), the Anseriformes (the ducks, geese, and swans). Although of similar body size, the Anseriformes have relatively longer wings and live in open, uncluttered habitats. Their pectoralis muscles contain a high proportion of FOG fibers interspersed among the FG fibers, and their relative heart mass is also much larger (Magnan, 1922; Hartman, 1961). Thus, their wing morphology determines that they have a slightly more efficient form of flight (Rayner, 1988; Pennycuick, 1989), and their muscle physiology and cardiovascular adaptations determine that they are able to prolong these flights over relatively long distances. Similar adaptations are seen in the flight muscles of the doves and pigeons (Columbiformes).

The aerobic nature of the flight muscles of some relatively large species of birds, such as the bustards and swans, has an important consequence for the interpretation of the size limitations to flight performance in birds. It is necessary to distinguish between "burst" flight performance and "prolonged" or aerobic flight performance (Marden, 1994) and to compare flight muscles which are adapted for similar types of flight. The data of Marden (1987) and Pennycuick *et al.* (1989) on the maximum lifting ability of flying animals during take-off refer only to "burst" flight performance, and bird species that have predominantly anaerobic FG flight muscle fibers will be at an advantage during take-off compared to species that have predominantly aerobic FOG muscle fibers.

Caldow and Furness (1993) studied the histochemical adaptations of two ecologically similar seabirds, the great skua (*Catharacta skua*) and the herring gull (*Larus argentatus*). They found that both species had FOG fibers exclusively in their pectoralis and supracoracoideus flight muscles, but that the mass-specific activity of both the oxidative and glycolytic enzymes (see below) were higher in the muscles of the great skua and concluded that this adaptation enabled the skua (Stercorariidae) to be a more effective aerial kleptoparasite than the herring gull (Lariidae). The mass of the pectoralis muscles is relatively small in the herring gull (12% of body mass), but it has a moderately large heart mass (0.9% of body mass), which is consistent with the fact that its flight muscles consist entirely of FOG fibers.

B. Biochemistry of the Flight Muscles

As discussed by Marsh (1981), the sustained, long-distance flights performed by many migrating birds require two interacting adaptions: the ability to store sufficient fuel reserves for the flight and the ability to maintain supplies of oxygen and fuel to the working tissues. Due to the very high energy density of lipids (39.3 kJ g^{-1}), compared to pure carbohydrates (17.6 kJ g^{-1}) and proteins (17.8 kJ g^{-1}), the energy for migratory flights is predominantly stored as fat in adipocytes. Thus, adaptations for the efficient oxidation of fatty acids and the potential for a high aerobic capacity are some of the most important features of most avian flight muscles. The relatively high mass-specific energy output required from the flight muscles necessitates high capillary and mitochondrial densities and high levels of mass-specific activities for the associated mitochondrial enzymes.

Hummingbirds are able to sustain the highest mass-specific metabolic rates of any vertebrate (Lasiewski, 1963), and Table 2 shows data from Suarez *et al.*, (1986) for the maximum activities of various catabolic enzymes in hummingbird flight muscle. The following enzymes are of particular interest: citrate synthase (CS), as an indicator of general aerobic capacity; 3-hydroxyacyl-CoA dehydrogenase (HAD) and carnitine palmitoyltransferase (CPT), as indicators of fatty acid utilization;

TABLE 2 Maximum Enzyme Activities in the Flight Muscle and Heart of the Rufous Hummingbird *Selophorus rufus*[a]

Enzyme	Flight muscle	n	Heart	n
Glycogen phosphorylase	31.22 ± 2.5	6	Not measured	
Hexokinase	9.18 ± 0.31	4	10.08 ± 1.9	4
Phosphofructokinase	109.8 ± 13	6	Unstable	
Pyruvate kinase	672.4 ± 27	6	507.3 ± 25	6
Lactate dehydrogenase	230.3 ± 23	6	357.4 ± 30	6
Carnitine palmitoyltransferase	4.42 ± 0.46	6	2.83 ± 0.72	4
3-Hydroxyacyl-CoA dehydrogenase	97.10 ± 13	6	68.51 ± 10	4
Glutamate-oxaloacetate transaminase	1,388 ± 70	5	576.4 ± 29	5
Glutamate-pyruvate transaminase	75.97 ± 6.0	5	16.31 ± 2.2	5
Citrate synthase	343.3 ± 8.8	6	190.3 ± 4.8	4
Creatine kinase	2,848 ± 337	5	348.9 ± 59	5
Malate dehydrogenase	3,525 ± 331	6	2,024 ± 191	6
α-Glycerophosphate dehydrogenase	9.37 ± 2.1	6	8.10 ± 2.2	6

[a] Values are expressed in μmol · min^{-1} · g wet wt^{-1} and presented as means ± SE; n, number of birds (from Suarez et al., 1986).

phosphofructokinase (PFK), as an indicator of general glycolytic flux; pyruvate kinase (PK), as an indicator of anaerobic glycolysis; and hexokinase (HEX) and glycogen phosphorylase (GPHOS), as indicators of plasma glucose and intracellular glycogen utilization, respectively. As expected, the flight muscles of hummingbirds have some of the highest catabolic enzyme activities ever measured. In particular, mass-specific CS and HEX activities are high and even higher (at 448.4 and 18.4 μmol g^{-1} min^{-1}, respectively) in a subsequent study with a modified extraction procedure (Suarez et al., 1990).

These two studies (Suarez et al., 1986, 1990) indicate that while it is possible for hummingbirds to meet almost all their flight requirements by oxidizing lipid fuel, it is also possible for them to meet the requirements of hovering flight utilizing carbohydrate fuels. The latter possibility makes sense given that hummingbirds feed mainly on sucrose-rich nectar and so during the day they would have a good supply of pure carbohydrate direct from the environment. However, for long-term storage, carbohydrates are stored as glycogen which consists of at least 75% water, thus giving it a relatively poor energy density (4.4 kJ g^{-1}). It would not be possible, therefore, to store enough carbohydrates to fuel long-distance flight. In fact, it is unlikely that stored carbohydrate could support hovering flight for more than around 5 min (Suarez et al., 1990). However in the short-term, it is more efficient to oxidize plasma glucose directly to support flight, due to the 16% lower net yield of ATP resulting from the cost of synthesizing fatty acids from glucose for long-term storage. This led these authors to suggest that premigratory hummingbirds should behave as "carbohydrate maximizers" by keeping their foraging flights well under 5 min duration. This would provide efficient foraging while allowing excess glucose to be stored as fat. However, while the availability of plasma glucose may not be limiting for hummingbirds, the majority of bird species may place a higher emphasis on carbohydrate "sparing." Bishop et al. (1995) found that HEX activity was relatively low in the pectoralis muscles of adult premigratory barnacle geese, and similar results have been reported for the pigeon and domestic chicken (Crabtree and Newsholme, 1972) and for the migratory semipalmated sandpiper *Calidris pusilla* (Driedzic et al., 1993).

Many studies have shown that prior to long-distance flights, such as those performed during seasonal migrations, birds increase in body mass due primarily to the laying down of fat stores (Blem, 1976). This is usually correlated with hypertrophy of the pectoralis muscle in order to increase the power available for take-off and forward flapping flight and species-specific changes in catabolic enzyme activities. While Marsh (1981) found no change in the premigratory activity of CS in the pectoralis muscle of the catbird (*Dumatella carolinensis*), he reported a significant increase in the level of HAD (indicating an increase in the potential for fatty acid oxidation). A similar increase in the ability to oxidize fatty acids was found in the semipalmated sandpiper, but the level of CS actually decreased (Driedzic et al. 1993). However, some species of small migratory passerine birds do appear to show an increase in the premigratory activity of both CS and HAD (Lundgren and Kiessling, 1985, 1986), while CS activity in the premigratory barnacle goose was around 30% greater than that of postmoulting birds approximately 4 weeks earlier in the season (Bishop et al., 1995).

Avian muscle capillarity and diffusion distances have been studied in a variety of different tissues, but no significant allometric trends with respect to different body masses were found for any of the variables measured (Snyder, 1990). However, Snyder (1990) did find a small but significant difference in the capillary to fiber ratio (C/F) between the aerobic fiber types (1.8 caps fiber^{-1}) and the glycolytic fibers (1.4 caps fiber^{-1}). Capillary density and diffusion distances are primarily determined by the cross-sectional area of the individual fibers, and the author argues that there may be little selective advantage in having a C/F greater than 2.0. Studies on avian pectoralis muscles have generally resulted in similar conclusions. Mathieu-Costello *et al.* (1992) found that the C/F of the rufous hummingbird (*Selaphorus rufus*) is only 1.55 caps fiber^{-1} compared to that of the pigeon, which is 2.0 caps fiber^{-1} (Mathieu-Costello, 1991). However, fiber diameter and the cross-sectional area are much smaller in the hummingbirds than in pigeons and, therefore, the capillary density per unit area is much higher in hummingbirds. Mathieu-Costello *et al.* (1992) consider that one of the most important determinants of the rate of oxygen flux in aerobic fibers is the size of the capillary-to-fiber interface (i.e., the capillary-surface-per-fiber-surface area). However, in order to measure this feature accurately, it is necessary to standardize with respect to both the sarcomere length and the degree of tortuosity of the capillaries.

C. Neurophysiology and Muscle Function

Birds are capable of synchronizing the extension of the elbow and wrist joints of the wing during flapping flight, using automated coordinating mechanisms involving skeletal and muscular adaptations (Vazquez, 1994). However, the power and thrust generated during the wingbeat cycle, and the complex kinematics of the wing during the flapping flight of different species, require an active neuromuscular control mechanism and specialized adaptations of the flight muscles.

The detailed movements of the wing during flight actually involve a large number of different muscles (Dial *et al.*, 1991; Dial, 1992a) and their differential (Dial, 1992a; Tobalske and Dial, 1994; Tobalske, 1995; Tobalske *et al.*, 1997) or regional (Boggs and Dial, 1993) activation. For most studies, however, it is sufficient to focus on the functioning of the largest flight muscles, the pectoralis and supracoracoideus. Dial (1992b) showed that birds were unable to take-off or perform controlled landings without the use of the muscles of the forearm, but that they could sustain level flapping flight following an assisted take-off.

The largest flight muscle, the pectoralis major, can be divided into two anatomical parts (Figure 13), the sternobrachialis (SB: which is superficial and lies along the sternum) and the thoracobrachialis (TB: which forms a deep layer lying along the sternum), and these are separated by the membrana intermuscularis (Dial *et al.*, 1987). The SB is primarily innervated by the rostral nerve branch and the TB by the caudal nerve branch of the brachial ventral cord. In the pigeon, the fibers within the SB and TB have different orientations (Dial *et al.*, 1987), while the SB has a lower percentage of FOG fibers and relatively more FG fibers, and the TB is primarily made up of FOG fibers (Kaplan and Goslow, 1989). Thus, the histochemical analysis and the neuroanatomy support the view that the pectoralis major is made up of at least two functional subunits, each with a potential for independent action on the wing during flight (Dial *et al.*, 1987, 1988; Kaplan and Goslow, 1989; Goslow and Dial, 1990). Dial *et al.* (1987) showed that both the SB and TB muscles of the pigeon were maximally activated during take-off or during large-amplitude wingbeats. However, during slow flight, the SB was relatively more important and during low-amplitude wingbeats the activity of the TB could be almost zero. Interestingly, Dial *et al.* (1987) also indicate that there is a gradual reduction of large-amplitude electromyogram (EMG) spikes during the transition from take-off to level flight and that they reappear during landing. This supports the hypothesis that FG fibers are derecruited during level flight.

Tobalske *et al.* (1997) studied the neuromuscular control and wing kinematics of the black-billed magpie, both in the wild and while flying in a wind tunnel, and showed that this species has a particularly complex flight style, consisting of alternating high- and low-amplitude wingbeats, and occasional brief glides. This wide-ranging study found that the pectoralis consisted of FOG and I fiber types, as in the woodpecker (see below), and that the I fiber types were probably only recruited during the high-amplitude wingbeats. EMG recordings were made from six different wing muscles, and all showed that the relative intensity of the EMG signal exhibited a U-shaped output with respect to flight speed. Minimum intensity was recorded around 4.4 m sec^{-1} and the highest were recorded during hovering and the top speed of 13.4 m sec^{-1}. The pectoralis and the biceps brachii muscles showed strong activity during the end of the upstroke and the beginning of the downstroke, while the supracoracoideus, the humerotriceps, the scapulotriceps, and the scapulohumeralis caudalis were important during the end of the downstroke and the beginning of the upstroke.

A more detailed study of the EMG activity of 17 different muscles in the shoulder and forelimb of the pigeon during five different modes of flight was conducted by Dial (1992a). He concluded that the temporal pattern of activity did not vary much between different modes of flight, but that the intensity of the EMG signal

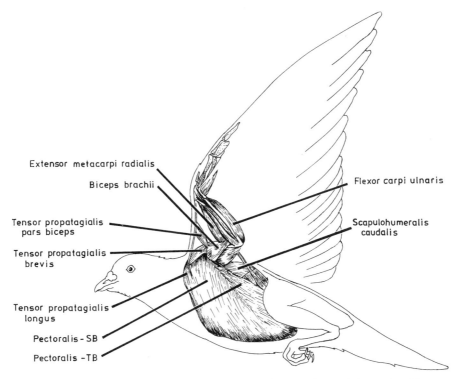

FIGURE 13 Illustration of the superficial flight muscles of the wings of a pigeon, *Columba livia*. The supracoracoideus muscles (not shown) are primarily used to power the upstroke and lie directly under the larger pectoralis muscles, which primarily power the downstroke. [Modified from Activity patterns of the wing muscles of the pigeon (*Columba livia*) during different modes of flight, K. P. Dial, *J. Exp. Zool.*, Copyright © 1992 John Wiley & Sons, Inc. Reprinted by permission of Wiley-Liss, Inc., a subsidiary of John Wiley & Sons, Inc.]

was very important in determining the role of the different muscles during different modes of flight (Figure 14).

Tobalske (1996) studied the muscle composition and wing morphology of six different species of woodpecker, ranging in mass from 27 to 263 g, and related their morphology to the scaling of intermittent flight behavior. Intermittent flight consists of regular alternation between periods of flapping and nonflapping. During the nonflapping phases, the bird's wings are either held folded at its side (flap-bounding flight) or held fully extended (flap-gliding flight). Biomechanical and physiological analyses suggest that intermittent flight is energetically efficient relative to continuous flapping (Rayner, 1985a), particularly for small species that have high wingbeat frequencies which are relatively "fixed" (Rayner, 1985a). Only FOG and I fiber types were found in these six species of woodpecker, with small species having almost exclusively FOG fibers in their pectoralis muscles, while species >100 g tended to have significant amounts of I fibers. All six species were capable of exhibiting flap-bounding behavior, although the percentage of time spent flapping increased with increasing body mass. The empirical observations of Tobalske (1996) greatly exceed the theoretical considerations (Rayner, 1985a) that the upper limit for flap-bounding flight should be around 100 g of body mass.

However, the heterogeneous scaling of the fiber type composition of the pectoralis is likely to be one reason why the theoretical figures underestimate bird flight performance (Tobalske, 1996). FG and I fibers have the potential to produce more force per unit area than FOG fibers and, therefore, can produce more power for a given degree of fiber shortening and frequency. Thus, Tobalske (1996) argued that the positive scaling of the percentage and cross-sectional diameter of I fibers in the pectoralis of woodpeckers may be a direct result of the positive scaling of the power required for flight compared to the power available from the flight muscles (cf., Pennycuick, 1989; Marden, 1987; Ellington, 1991). Interestingly, among woodpeckers, the scaling of relative heart muscle mass is negative relative to increasing body mass (Hartman, 1961), which matches the decline in FOG content of the flight muscles. Thus, while large woodpecker species may be able to meet the power required to perform flap-bounding flight, it is likely that they will not be able to prolong these types of flight as easily as the smaller species.

Lewis' woodpecker (*Melanerpes lewis*) was found to be able to perform both flap-bounding and flap-gliding flight, contrary to theoretical predictions that it should not perform the former (Tobalske, 1996), and this behavior was not well correlated with the morphology of

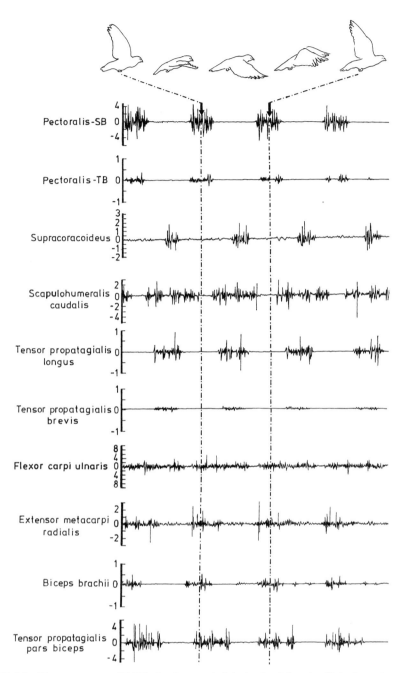

FIGURE 14 Electromyogram (EMG) signals recorded during level flapping flight from the various muscles of a pigeon, *Columba livia,* illustrated in Figure 13. EMG activity is presented with reference to the phase of the wingbeat cycle, with each downstroke indicated by the vertical dashed line. [Modified from Activity patterns of the wing muscles of the pigeon (*Columba livia*) during different modes of flight, K. P. Dial, *J. Exp. Zool.,* Copyright © 1992 John Wiley & Sons, Inc. Reprinted by permission of Wiley-Liss, Inc., a subsidiary of John Wiley & Sons, Inc.]

the wing. In addition, Tobalske and Dial (1994) found that the 35-g budgerigar was also capable of performing flap-gliding behavior when flying slowly, but switched to flap-bounding behavior during fast flight, despite the observation that they only have a single fiber type (FOG) in their pectoralis muscles (Rosser and George, 1986a). Thus, the hypothesis that the FOG fibers of small birds should not be used for gliding as it is inefficient appears to be undermined by the budgerigars, with the pectoralis being active during the gliding phases, but inactive during the bound. Tobalske and Dial (1994) predicted that all other species that show intermittent flight should also be facultative flap-gliders and flap-bounders and that the choice of flight mode is simply dependent on flight speed.

V. DEVELOPMENT OF LOCOMOTOR MUSCLES AND PREPARATION FOR FLIGHT

The development of the locomotor muscles and heart, together with other morphological features, have been studied in a Svalbard population of migratory barnacle geese, from hatch until it migrates to its wintering grounds in southern Scotland at the age of approximately 12 weeks (Bishop *et al.*, 1996). Up until the age of 7 weeks, the goslings are unable to fly and spend their time walking and foraging. At the age of 5 weeks, the relative mass of the leg muscles is approximately 13% of body mass, while that of the pectoral muscles is only 3.5% of body mass. During the first 5 weeks of their life, there is a strong relationship between mass of the ventricles and mass of the leg muscles (Figure 15b). During the following 2 weeks, the flight muscles continue to grow in an exponential manner and by the time the goslings are 7 weeks old, these muscles are an impressive 14% of body mass, whereas the relative mass of the leg muscles has declined to approximately 9% of body mass. From the age of 7 weeks, the mass of the ventricles is proportional to the mass of the pectoral muscles (Figure 15a). At 5 weeks of age, ventricular mass is 5.2% of total mass of the leg muscles and at 7 weeks of age it is 5.5% of mass of the pectoral muscles. Thus, as might be expected, the output capacity of the heart is closely matched to the oxygen requirements of the dominant set of locomotory muscles.

Choi *et al.* (1993) studied the skeletal muscle growth and development of thermogenesis in the European starling (*Sturnus vulgaris*), the northern bobwhite (*Colinus virginianus*), and the Japanese quail (*Coturnix japonica*). The body mass of the altricial starling developed more rapidly than that of the precocial quail species, although the percentage of the body mass that was accounted for by muscle mass was quite similar. The main differences between these groups was in the expression of muscle mass-specific enzyme activities. In particular, the activity of CS did not exceed 50 and 30 μmol min^{-1} g^{-1} in the pectoralis muscles of *Coturnix* and *Colinus* quail, respectively, throughout development. In the starling pectoralis muscle, CS activity increased rapidly, and linearly, during development and reached 142 μmol min^{-1} g^{-1} around fledging at 16 days of age and approximately 230 μmol min^{-1} g^{-1} in adult starlings. In the bank swallow (*Riparia riparia*) CS activity increased from around 20 μmol min^{-1} g^{-1} at 2 days

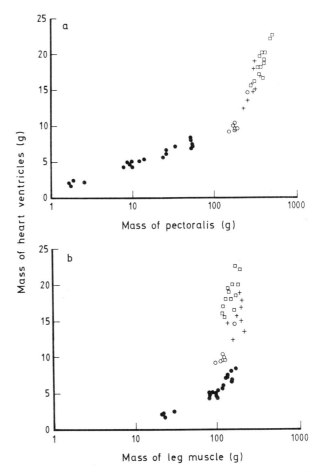

FIGURE 15 Scatter diagram of the masses (g) of the heart ventricles plotted aginst (a) the mass of the pectoral muscles and (b) the mass of the total leg muscles of wild barnacle geese, *Branta leucopsis*. Goslings of 1–5 weeks of age (●), fledgling goslings of 7 weeks of age (○), adult geese captured 7 weeks after the population hatch date (+), and premigratory adult geese (□) (modified from Bishop *et al.*, 1996, with permission).

of age and reached around 150 μmol min^{-1} g^{-1} after 10 days (Marsh and Wickler, 1982).

A similar study of the barnacle goose found an even more exaggerated developmental pattern (Bishop *et al.*, 1995), with CS expression showing almost no change in activity from hatch (12 μmol min^{-1} g^{-1}) through to 5 weeks of age (20 μmol min^{-1} g^{-1}). Subsequently, the activity increased rapidly to 75 μmol min^{-1} g^{-1} at 7 weeks of age and reached around 100 μmol min^{-1} g^{-1} in the premigratory goslings and adults (see Figure 24). The mass-specific activity of four different mitochondrial enzymes showed broadly similar patterns of activity to each other during development (Bishop *et al.*, 1995) and are typified by the results obtained for HAD activity in pectoralis, cardiac, and semimembranosus leg muscle (Figure 16a). Throughout most of development,

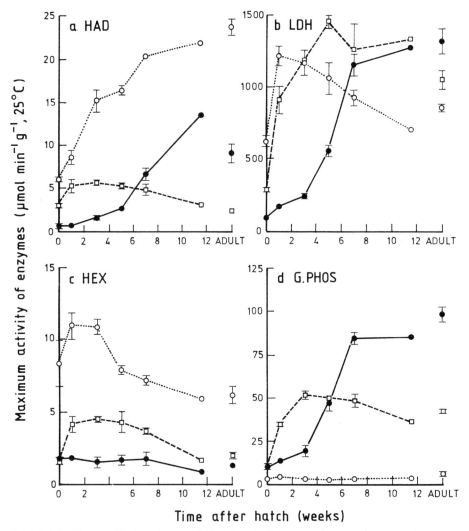

FIGURE 16 Mean ± SE of maximum activities of catabolic enzymes (mol min^{-1} g wet wt^{-1} at 25°C) in pectoral muscle (●——●), heart muscle (○······○) and semimembranosus muscle (□------□) plotted against age of wild barnacle geese, *Branta leucopsis*. (a) 3-hydroxyacyl-CoA dehydrogenase (HAD), (b) lactate dehydrogenase (LDH), (c) hexokinase (HEX) and (d) glycogen phosphorylase (G.PHOS). n = 2 − 8, except for a single individual at 11.5 weeks of age. Adult geese and 11.5 week old gosling are pre-migratory (see text). (Modified from Bishop *et al.* (1995), *Am. J. Physiol.* with permission.)

the cardiac muscle had the highest relative capacity for aerobic metabolic flux and fatty acid oxidation. HAD activity in the semimembranosus leg muscle was intermediate in value prior to fledging at 7 weeks (when the birds first begin to fly), but subsequently declined, while CS activity in the pectoralis flight muscle was initially very low, but reached a peak in the premigratory birds (see Figure 24). The developmental profiles of various glycolytic enzymes, lactate dehydrogenase (LDH, Figure 16b), hexokinase (HEX, Figure 16c), and glycogen phosphorylase (GPHOS, Figure 16d), were more tissue-specific compared to those of the mitochondrial enzymes. Development of LDH and GPHOS activities in the pectoralis muscle show a similar exponential increase as that for CS, but values for HEX remain very low at all ages.

Overall, the results suggest that the relative use of circulating plasma glucose by the pectoralis muscle during flight is low, while intra muscular stores of glycogen are of primary importance during short-burst activity and of intermediate importance during longer-term activity and that fatty acid oxidation is the main fuel during long-distance flight (Bishop *et al.*, 1995). In general, moderate activity levels were detected in the semimembranosus muscle for all the enzymes that were measured, suggesting that the leg muscles are capable of utilizing all fuel types in almost equal amounts. However, the cardiac muscle had extremely low levels of GPHOS, indicating that it can only utilize aerobic pathways and that metabolic fuel must be primarily supplied by the blood. LDH values were relatively high in the heart, while pyruvate kinase levels were very low (data not shown), indicating that the heart is capable of oxidizing circulating lactate as a fuel.

VI. METABOLIC SUBSTRATES AND FUEL DEPOSITS

The use of fatty acids as the primary fuel during long-distance flights is an important adaptation, as the energy density of fat is much greater than that of other stored fuels. As pointed out earlier, the energy density of pure carbohydrate is 17.6 kJ g^{-1}; however, the storage of glycogen in cells involves between 3 and 5 g of water for each gram of glycogen, giving an effective energy density of approximately 4.4 kJ g^{-1} compared with that for lipids of 39.3 kJ g^{-1} (Schmidt-Nielsen, 1997). Protein may also be used to provide energy, but tissues such as muscle also consist of 70–80% water, so that the energy density of stored protein *in vivo* is around 5.4 kJ g^{-1}. Thus, during a long flight, it is likely that the majority of energy is provided by fatty acids. However, fatty acid transport in the plasma (from the adipocytes to the working muscles or the liver) may be a rate-limiting step during intense exercise (Newsholme and Leech, 1983; Butler, 1991) as it must be transported by albumin and, until recently, it has not been clear how small birds could be capable of delivering the high rates of fatty acid required by the flight muscles. Jenni-Eiermann and Jenni (1992) measured various metabolite concentrations in the plasma of three species of small night-migrating passerines (Figure 17a), the robin (*Erithacus rubecula*), garden warbler (*Sylvia borin*), and pied flycatcher (*Ficedula hypoleuca*). Their results indicated that during long-distance flights, plasma levels of triglyceride, glycerol, free-fatty acids (FFA), and very low density lipoproteins (VLDL) all increased significantly above values at rest.

The rise in triglycerides and VLDL was unexpected, but could provide a mechanism for increasing the FFA supply to the flight muscles. FFA is reesterified to triglyceride in the liver and subsequently delivered to the blood as VLDL. Thus, the removal of FFA from the blood stream by the flight muscles and the liver will increase the uptake of FFA from the adipocytes. Secondly, the VLDL can be hydrolyzed by lipoprotein lipase in the endothelium of the flight muscle capillaries, thus allowing the uptake of the released FFA (Figure 17b). Apart from increasing the availability of FFAs to the flight muscles, this mechanism has the additional advantage that the overall protein (albumin) concentration of the blood is kept relatively low and prevents an unacceptable rise in blood viscosity. In addition, the rate of uptake of the triglycerides via the VLDL can be controlled by the rate of lipoprotein lipase activity at the site of demand. During the migratory period, these birds accumulate fat in both their flight muscles and the adipocytes. Therefore, overnight-fasted birds were able preferentially to metabolize intramuscular stores of triglycerides, allowing a lower rate of utilization of adipocyte fatty acids and efficient sparing of protein and carbohydrate stores (Jenni-Eiermann and Jenni, 1996). It appears that this process is not as important in pigeons (Schwilch *et al.*, 1996) and, in the barnacle goose, the maximum concentration of fatty acid-binding protein (considered to act as an intracellular carrier for fatty acids) in the pectoral muscle occurs at the same time as the maximum activity of HAD, just before the autumn migration (Pelsers *et al.*, 1999).

Although protein stores are less energy dense than fatty acid adipocyte stores (see above), in reality it may not be possible entirely to exclude their use during periods of starvation and/or during prolonged periods of exercise. The brain and other nervous tissues are extremely dependent on glucose metabolism for the provision of energy (Newsholme and Leech, 1983). As carbohydrate stores are very limited in most animals, during starvation the major source of glucose is via protein (amino acid) degradation to various metabolic intermediates (called oxoacids), most of which can then enter the gluconeogenesis pathway in the liver. In addition, glycerol that is produced following the hydrolysis of triglycerides can also be converted to glucose. A small number of amino acids (leucine and lysine) are degraded directly to acetyl-CoA and so can contribute to ATP synthesis via the citric acid cycle (CAC) or be converted to fat, as acetyl-CoA cannot enter the gluconeogenesis pathway. Alternatively, excess amino acid production can provide intermediates (e.g., 2-oxalogluterate, succinyl-CoA, fumarate, and oxaloacetate) for the CAC if these should be in short supply or be converted to pyruvate and then to acetyl-CoA either to contribute to ATP synthesis or to be converted to fat. Certainly, there are a number of studies which show that protein, in the form of muscle and the intestinal tract, is catabolized during long periods of flight (Jehl, 1997) or during "simulated" flights by starvation (Biebach, 1998).

Associated with the premigratory increase in body mass is, not surprisingly, an increase in food intake (hyperphagia). A number of hormones have been implicated in these processes (Wingfield *et al.*, 1990), but recent data do not support a major role for prolactin in the regulation of premigratory deposition of fat in the European quail *Coturnix coturnix* (Boswell *et al.*, 1995). On the other

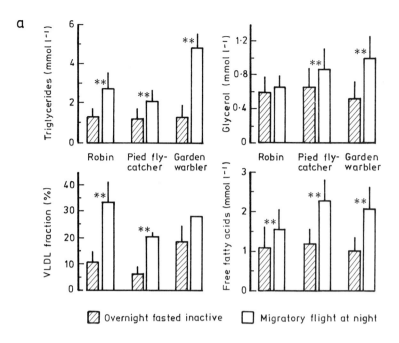

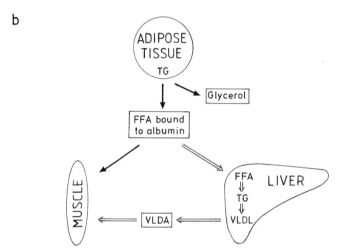

FIGURE 17 (a) Means ± SD of various fat metabolite concentrations and very low-density lipoproteins (VLDL) in the plasma of three species of birds during overnight fasting (cross-hatched) or during night time migratory flights (clear). **$P < 0.01$, Mann-Whitney U-test. (b) Diagram illustrating the proposed pathway for fatty acid transport to the flight muscles of small birds. Free fatty acids (FFA) are released from the triglyceride (TG) stores of the adipose tissue and either taken up directly by the muscles or converted to TG by the liver and transported to the muscles as VLDL. The activity of the latter pathway (double arrows) is suggested to increase in small birds during flight. (Modified from *Physiol. Zool.*, S. Jenni-Eiermann and L. Jenni, with permission from The University of Chicago Press. © University of Chicago Press 1992.)

hand, neuropeptide Y (NPY), a member of the pancreatic family of polypeptides, does cause white-crowned sparrows (*Zonotrichia leucophrys gambelii*) to consume more food when administered into the third ventricle of the brain and may be a natural stimulator of feeding in this species (Richardson *et al.*, 1995).

VII. THE CARDIOVASCULAR SYSTEM

A major role of the cardiovascular system is to deliver metabolic substrates and oxygen to the tissues undergoing aerobic metabolism and to remove the waste products of that metabolism. Thus, the cardiovascular system in general, and the heart in particular, must be capable of meeting the demands of the flight muscles during sustained, flapping flight.

A. Cardiovascular Adjustments during Flight

There is only one study (Butler et al., 1977, working on pigeons) in which all but one of the variables of the Fick equation (see above) were measured (thus allowing the absent one (V_s) to be calculated). When flying in a wind tunnel at 10 m sec^{-1}, $\dot{V}O_2$ in these birds was 10× the resting value (Table 3). The respiratory system maintained C_aO_2 at slightly below the resting value, but $C_{\bar{v}}O_2$ was halved, giving a 1.8-fold increase in ($C_aO_2 - C_{\bar{v}}O_2$). There was no significant change in V_s, so the major factor in transporting the extra oxygen to the muscles was the 6-fold increase in f_H. Unfortunately, all of these variables were not measured simultaneously on the same birds and Bishop and Butler (1995) have argued that the original calculation of V_s at rest may have been inappropriate. This being the case, there would have been a 1.4-fold increase in V_s during flight. As mean arterial blood pressure did not change, total peripheral resistance must have declined by the same proportion as cardiac output ($\dot{V}_b$, which is $f_H \times V_s$) increased.

In general, birds have larger hearts and lower resting heart rates than those of mammals of similar body mass (Lasiewski and Calder, 1971; Grubb, 1983). The higher $\dot{V}_b$ in birds is an important factor in their attaining a higher maximum $\dot{V}O_2$ ($\dot{V}O_{2,\text{max}}$) during flight than similar-sized mammals do when running. Bats have larger hearts and a higher blood oxygen-carrying capacity than other mammals of similar size (Jurgens et al., 1981) and their $\dot{V}O_2$ during flight is similar to that of birds of similar mass (Butler, 1991). Among birds, those that are not capable of sustained flight, such as the Galliformes, tend to have slightly lower hematocrit (Hct, packed cell volume) and hemoglobin concentration ([Hb]) than those of good fliers (Balasch et al., 1974) and weaker fliers have a slightly lower Hct than stronger fliers (Carpenter, 1975). Wild barnacle geese migrating from their breeding grounds in the high Arctic to their wintering grounds in southern Scotland have surprisingly low heart rates during flight (see earlier), which means either that the energy cost of migratory flight is lower than was previously thought (maybe as a result of formation flight, see later) and/or that ($C_aO_2 - C_{\bar{v}}O_2$) is greater in migrating geese than was assumed from the data obtained from pigeons flying in a wind tunnel (see Butler et al., 1998).

B. The Cardiac Muscles

Studies of exercising animals have indicated that both the locomotor and cardiac musculature are dynamic structures that can vary in mass seasonally and in direct response to demand (Hickson et al., 1983; Marsh, 1984; Dreidzic et al., 1993; Bishop et al., 1996). Thus, it might be anticipated that where additional energetic "costs" occur seasonally (e.g., due to migratory fattening or the development of secondary sexual characteristics) then the relevant cardiac and locomotor musculature might also be regulated seasonally. Bishop et al., (1996) showed that both cardiac and pectoralis muscles hypertrophy during the premigratory fattening period of the barnacle goose.

Bishop and Butler (1995) suggest that heart mass (M_h) can be used to model the availability of oxygen to the flight muscles of birds. Assuming that the adaptations of the flight muscles are appropriately matched to the cardiovascular system, then the relative M_h of birds of a similar body mass should be a good indicator of the relative aerobic power input available to the flight muscles. Bishop (1997) calculated that maximum $\dot{V}O_2$ ($\dot{V}O_{2,\text{max}}$) of birds should scale with respect to body mass as

TABLE 3 Mean Values of Oxygen Uptake and Cardiovascular Variables Measured in the Pigeon *Columba livia* and Emu *Dromaius novaehollandiae*[a]

	Pigeon (0.442 kg)		Emu (37.5 kg)	
	Rest	Flying	Rest	Running
Oxygen uptake (ml min^{-1} STPD)[b]	9.0	88.4	156.7	1807
Heart rate (beats min^{-1})	115	670	45.8	180
Cardiac stroke volume (ml)	1.44 (1.14)	1.58	57	102.7
Oxygen content of arterial blood (vol%)	15.1	13.7	15.2	15.2
Oxygen content of mixed venous blood (vol%)	10.5	5.4	9.0	4.6

[a] Measurements taken at rest and after 6 min of steady level flight in a wind tunnel at a speed of 10 m sec^{-1} for the pigeon and after 20 min running on a treadmill at a 6° incline and a speed of 1.33 m sec^{-1} for the emu (from Butler, 1991). Value for cardiac stroke volume in parentheses is recalculated from Butler et al. (1977) as discussed by Bishop and Butler (1995).

[b] STPD, Standard temperature and pressure, dry.

230 $M_b^{0.82}$ (ml min^{-1}), that maximum cardiac output of birds should scale with respect to M_h as 213 $M_h^{0.88}$ (ml min^{-1}), and that maximum aerobic power input (aerobic $P_{i,max}$) should scale as 11 $M_h^{0.88}$ (W). Thus, for studies of the prolonged aerobic flight performance of birds, the estimates of $\dot{V}O_{2,max}$ or $P_{i,max}$ based on relative M_h should be of more practical value than the use of general scaling equations based on body mass alone.

The mean values for M_h from selected avian families are plotted in Figure 18a (data from Hartman, 1961) and Figure 18b (data from Magnan, 1922) and indicate the interfamily adaptive diversity in aerobic capacity. The very small M_h of all three species of tinamou indicate that the members of this family have the lowest aerobic ability of all bird species. Other relatively sedentary forest birds with small M_h include the guans, motmots, puffbirds, and antbirds. This contrasts with the relatively large M_h of other bird families that live predominantly in forest, such as the parrots, trogons, kingfishers, and hummingbirds. Interestingly, Figure 18b shows that the M_h values of the bustard family (Otidae) appear to range above the general allometric trend. The relative M_h of the 0.83-kg little bustard and the 8.95-kg great bustard are 1.8 and 1.4%, respectively. These relative M_h values are even larger than those of the Anatidae which typically range from 0.8 to 1.1%. Two other observations appear to substantiate this finding. Crile and Quiring (1940) gave the following values for the relative heart masses of the African Kori bustard; a 5.54-kg female Kori bustard (1.1%) and a 10-kg male Kori bustard (1.0%). In addition, Stickland (1977) shows that while 100% of the area of the pectoralis muscle of the African helmeted guinea-fowl is made up of "white" anaerobic muscle fibers (i.e., FG fibers), 82% of the area of the pectoralis muscle of the similarly sized white-bellied bustard is made up of "red" aerobic fibers (i.e., FOG fibers). This author also comments on the fact that the European species of bustards undergo local migrations, which is a further indication that they are capable of a "prolonged" mode of flight.

There are also some interesting intrafamily differences in aerobic capacity indicated in the M_h data of Hartman (1961) and Magnan (1922). The different genera of the Columbidae (pigeons and doves) have relative M_h means ranging from 1.29 to 0.57%, while among the genera of the Falconidae (falcons) these range from 1.14 to 0.6%. In both these families, the genera with the relatively small hearts occur predominantly in tropical forests and are likely to be relatively sedentary. As intrafamily pectoralis muscle mass is fairly constant in these examples, it is reasonable to assume that the relatively sedentary species are more dependent on anaerobic metabolism to support their flight activity and that their flight muscles consist of a relatively greater proportion of FG fiber types.

VIII. THE RESPIRATORY SYSTEM

The respiratory system is not only concerned with gas exchange (i.e., the supply of oxygen to and the removal of carbon dioxide from the circulatory system and the metabolizing tissues), but also with the control of body temperature and evaporative water loss.

A. Ventilatory Adjustments during Flight and Ventilatory/Locomotor Coupling

As for the cardiovascular system, there is a variation of Fick's formula for convection that describes the relationship between $\dot{V}O_2$ and the various components of the respiratory system:

$$\dot{V}O_2 = f_{resp} \times V_T \times (C_IO_2 - C_EO_2),$$

where f_{resp} is respiratory frequency (breaths min^{-1}), V_T is respiratory tidal volume (ml), and $(C_IO_2 - C_EO_2)$ is the difference in the oxygen content in the inspired and expired gas (milliliter of O_2 per milliliter of gas). Data on ventilation are given in Table 4 for 5 species of birds at rest and during forward flapping flight in a wind tunnel. At relatively low ambient temperatures (<23°C), minute ventilation volume $\dot{V}_I$ (= $f_{resp} \times V_T$) increases by a similar proportion as $\dot{V}O_2$ and the proportion of oxygen extracted from the inspired gas ($O_{2,ext}$, effectively equal to $C_IO_2 - C_EO_2/C_IO_2$) during flight is similar to that in resting birds. The relative contributions of f_{resp} and V_T to the increase in $\dot{V}_I$ during flight vary between species. In white-necked ravens (*Corvus leucocephalus*), V_T does not change at all, whereas in fish crows and black-billed magpies it doubles and in starlings there is a fourfold increase. Thus, in the first three species, f_{resp} makes the greater contribution, but in the fourth, volume predominates. Respiratory frequency and V_T in the fish crow (Bernstein, 1976) and f_{resp} in the barnacle goose (Butler and Woakes, 1980) are independent of flight speed. This could mean that, like the starling and the fish crow, $\dot{V}O_2$ of the barnacle goose is also largely independent of flight speed. However, in the budgerigar, both $\dot{V}O_2$ and f_{resp} change with speed in a U-shaped fashion (Tucker, 1968b).

Despite the apparent matching between the increases in $\dot{V}_I$ and $\dot{V}O_2$ during flight in the four species listed in Table 4, there does appear to be an increase in effective lung ventilation above that required by metabolic rate (hyperventilation) during flight in starlings, as indicated by a decrease in the partial pressure of CO_2 (P_{CO2}, hypocapnia) in the anterior and posterior airsacs (Torre-

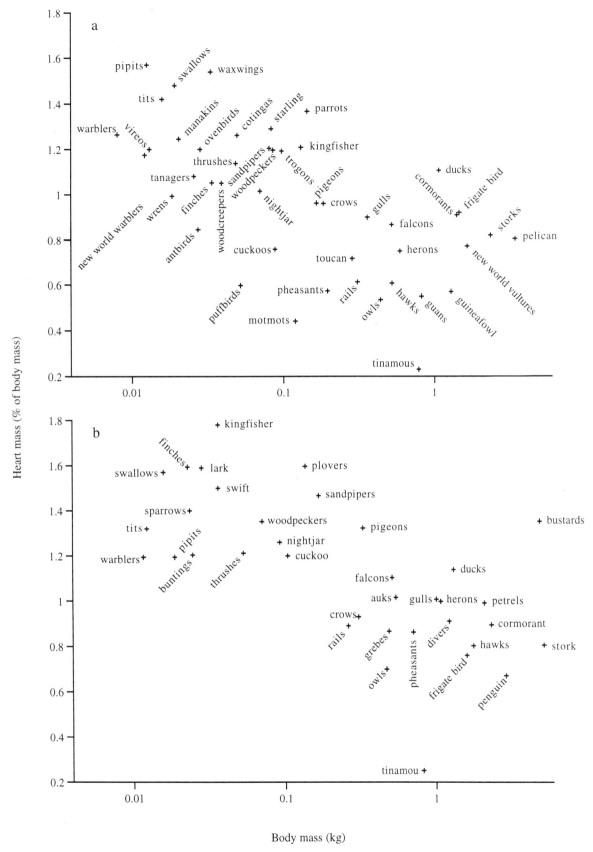

FIGURE 18 Mean values for cardiac muscle mass as a percentage of body mass (%), plotted against body mass (kg) for different families of birds. (a) Data from Hartman (1961); (b) data from Magnan (1922).

TABLE 4 Mean Values of Respiratory Frequency, Tidal Volume, Minute Ventilation Volume, Oxygen Extraction, and Oxygen Uptake[a]

	Mass (kg)	Rest					Flight				
		f_{resp}	V_T	$\dot{V}_I$	$O_{2\,ext}$	$\dot{V}O_2$	f_{resp}	V_T	$\dot{V}_I$	$O_{2\,ext}$	$\dot{V}O_{2\,min}$
Hummingbird (>20°C)[b] (*Amazilia fimbriata fluviatilis*)	0.006	—	—	—	—	—	280	0.63	0.18	0.135	4.1
Hummingbird (36°C)[c] (*Colibri coruscans*)	0.008	—	—	—	—	—	330	0.38	0.12	0.24	5.0
Budgerigar (18–20°C)[d] (*Melopsittacus undulatus*)	0.035	—	—	0.047	0.27	2.62	199	1.15	0.232	0.26	10.9
Starling (10–14°C)[e] (*Sturnus vulgaris*)	0.073	92	0.67	0.061	0.28	3.16	180	2.8	0.504	0.31	28.1
Black-billed magpie[f] (*Pica pica*)	0.165	52.4	2.95	0.154	—	—	162	6.1	0.953	—	—
Fish crow (12–22°C)[g] (*Corvus ossifragus*)	0.275	27.3	8.2	0.223	0.19	8.5	120	14.9	1.79	0.19	68.0
White-necked raven (14–22°C)[h] (*Corvus cryptoleucus*)	0.48	32.5	10.5	0.34	0.24	17.0	140	10.7	1.40	0.29	84.9

[a] Abbreviations: f_{resp}, respiratory frequency (min^{-1}); V_T, tidal volume, (ml); $\dot{V}_I$, minute ventilation volume (1 min^{-1} BTPS, except for fish crow where it is 1 min^{-1} STPD); $O_{2\,ext}$, oxygen extraction; $\dot{V}O_2$, rate of oxygen uptake (ml O_2 min^{-1} STPD) during rest and while hovering in two species of hummingbirds and while flying in a wind tunnel for five other species of birds. The values for $\dot{V}O_2$ during flight are the minima that have been recorded and have been corrected for the drag and mass of mask etc., where necessary [see Butler et al. (1977) for further details]. BTPS, body temperature and pressure, saturated. STPD, standard temperature and pressure, dry.
[b] Berger and Hart (1972).
[c] Berger (1978).
[d,e,g,h] Butler and Woakes (1990).
[f] Boggs et al (1997a).

Bueno, 1978a). It can also be seen from Table 4 that $O_{2,\,ext}$ for the hovering hummingbird *Colibri corusans* is similar to those for the four birds during forward flapping flight, whereas that for *Amazilia fimbriata fluviatilis* is 50% or less.

Ever since Marey's (1890) pioneering studies on bird flight, it has been known that f_{resp} may be coordinated with the beating of the wings and Marey himself suggested that the flapping of the wings during flight might have some impact on the airsacs. In crows (*Corvus brachyrhynchos*) and pigeons there is a 1:1 correspondence between f_{resp} and the frequency of wing beating (f_{wb}), whereas ratios as high as 5:1 (f_{wb}:f_{resp}) have been reported for the black duck *Anas rubripes*, quail *C. coturnix*, and ring-necked pheasant *Phasianus colchicus* (Hart and Roy, 1966; Berger et al., 1970b). However, as mentioned above, the flights reported by these authors were of only a few-seconds duration and it was concluded that coordination was not obligatory. During flights of up to 10-min duration, pigeons in a wind tunnel showed a very close relationship between these two activities. Wingbeats occurred in bursts and, although f_{resp} was often either slower or faster than the wing beat frequency (f_{wb}) between bursts of wing beating, there was always close coordination between the two when the wings were flapping (Figure 19a). Almost immediately upon landing, the pigeons panted at a frequency identical to the mean resonant frequency of the respiratory system (i.e., 10 Hz, Kampe & Crawford, 1973). During flapping flight, however, f_{resp} was slower, at approximately 7 Hz (i.e., the same as f_{wb}).

With a 1:1 correspondence, it is possible to imagine how contractions of the flight muscles could assist respiratory air flow, but with higher ratios this is not always so obvious. Nevertheless, a correspondence of 3:1 has been found in free-flying barnacle (1.6 kg) and Canada (3.8 kg) geese (Butler and Woakes, 1980; Funk et al., 1993) and it is clear that the wing beat is tightly locked to fixed phases of the respiratory cycle during flights of relatively long duration (Figure 19b). These phase relationships can be mantained, even following transient changes in one of the activities (Butler and Woakes, 1980). As Tucker (1968b) comments, "It is hard to believe that the contractions of the flight muscles have no influence in ventilation. . . ." Recent studies have attempted to investigate this intriguing possibility.

Benzett et al. (1992) recorded respiratory airflow and the timing of each wingbeat for three starlings flying at 11 m sec^{-1} for up to 5 min in a wind tunnel. Triggering on wingbeat and using the technique of ensemble averaging, these authors found that there was usually a 3:1 ratio between f_{wb} and f_{resp}, but that the effect of wingbeat

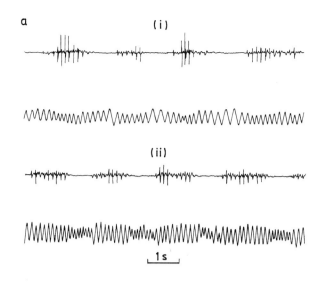

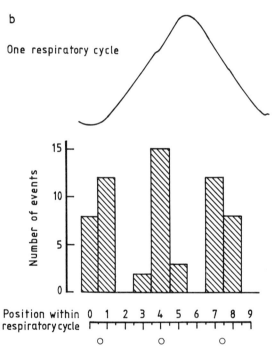

FIGURE 19 (a) Traces from a pigeon, *Columba livia* (0.45 kg), flying at a speed of 10 m sec^{-1} in a wind tunnel, showing changes in respiratory frequency associated with periods of wing beating and with periods of gliding (i) at the beginning of a flight (1 min after take-off) when respiratory frequency decreased during the gliding period and (ii) later in the flight (6 min after take-off) when respiratory frequency increased during the gliding period. In each series, the traces from top to bottom are: Electromyogram from the pectoralis muscle, respiratory movements (up to trace, inspiration). (From Butler *et al.* (1977), *J. Exp. Biol.*, Company of Biologists, Ltd.) (b) Histogram showing the positions during the respiratory cycle at which the wings were fully elevated (called "events") during the flight of a barnacle goose, *Branta leucopsis*, trained to fly behind a truck. Data are from 20 respiratory cycles, which were divided into 10 equal parts. Below the histogram is plotted the mean position of each group of events (○). Above the histogram is a trace of one of the respiratory cycles (inspiration, up). (From Butler and Woakes (1980), *J. Exp. Biol.*, Company of Biologists, Ltd.)

on V_T ranged from 3 to 11%. They concluded, therefore, that wingbeat and breathing in starlings are essentially mechanically independent. A problem with ensemble averaging is that it examines the systems when they are out of synchrony; that is, when there would be a low influence of locomotion on ventilation. On the other hand, Jenkins *et al.* (1988) took high-speed X-ray cine films of starlings flying in a wind tunnel and suggested that the lateral movement of the furcula during the downstroke of the wing and the recoil during the upstroke may facilitate inflation and deflation, respectively, of the clavicular airsac. They also found, however, that the sternum is moved in an expiratory direction during the downstroke and in an inspiratory direction during the upstroke. As the action of the sternum will influence the more posterior airsacs, these authors proposed that the combined action of the movements of the furcula and of the sternum, which are caused by the beating wings, is to produce a secondary respiratory cycling mechanism between the airsacs and the lungs that performs independently of (the slower) inhalation and exhalation.

Boggs *et al.* (1997a,b) recorded pressures in, and simultaneous cineradiographic images of, the anterior and posterior airsacs, airflow in the trachea, and EMGs of the pectoral muscles in black-billed magpies during short (10–20 sec) flights in a wind tunnel (Figure 20a). Although they found similar patterns of movements of the furcula and sternum in the magpie during flight as Jenkins *et al.* (1988) described for the starling, they did not find that the pressure changes in the anterior and posterior air sacs are consistent with the internal, secondary cycling between the airsacs and the lungs, as postulated by Jenkins *et al.* (1988). However, the downstrokes and upstrokes of the wings do have compressive and expansive effects, respectively, on the thoracoabdominal cavity, most probably by way of inertial mechanisms.

It was only possible to quantify the effect of these influences when the movements of the wings were opposing the action of the respiratory muscles (Figure 20b). Thus, the average change in airsac pressure caused when a downstroke occurred during inspiration was 94%, whereas the average change caused by an upstroke occurring during expiration was 41%. The corresponding average changes in flow and volume were 75 and 23% and 35 and 11%, respectively. The conclusion is, therefore, that when the effects of the wings and the respiratory muscles are acting together (i.e., when downstroke occurs during expiration and upstroke during inspiration), ventilation of the lung is substantially enhanced. With the normal 3:1 ratio between f_{wb} and f_{resp}, the pattern of phasic coordination means that there are two upstrokes during inspiration (cf. Figure 19b) and two downstrokes during expiration, thus giving a

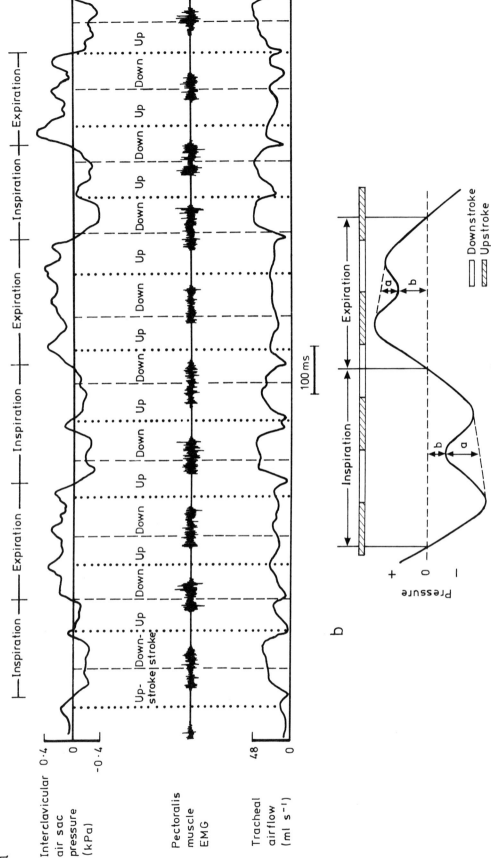

FIGURE 20 (a) The relationships between interclavicular airsac pressure, tracheal airflow, and electromyogram (EMG) of a pectoral muscle during flight of a black-billed magpie, *Pica pica*, in a wind tunnel. As the flow signal is not directional, inspiration is taken as that period during which pressure is subatmospheric (below 0) and expiration is that period during which pressure is above atmospheric (above 0). The EMG of the pectoral muscle indicates the upstrokes and downstrokes of the wings. When a downstroke occurs during inspiration, the subatmospheric pressure is driven up toward or above zero and when upstroke occurs during expiration, the supraatmospheric pressure is reduced toward zero (from Boggs, 1997, with permission). (b) Diagram to indicate how the average changes in airsac pressure shown in (a) were quantified in terms of the percentage change in pressure caused during inspiration or expiration by downstroke or upstroke of the wings, respectively. [From Boggs *et al.* (1997a), *J. Exp. Biol.*, Company of Biologists, Ltd.]

net assistance to inspiration during the former and a net assistance to expiration during the latter. This phasic coordination is not disrupted when the birds breathe 5% CO_2 during flight. When they spontaneously switch to a 2:1 ratio, they shorten inspiratory time to ensure that upstroke occurs during most of inspiration and that downstroke corresponds with the transition to expiration (Boggs *et al.*, 1997b). Both of these phenomena indicate that such coordination is probably of fundamental functional significance.

Studies with decerebrate Canada geese (Funk *et al.*, 1992a,b) indicate that, in the absence of peripheral feedback from the flapping wings, there is predominantly a 1:1 ratio between f_{resp} and f_{wb} and that peripheral feedback is required to create the patterns of coordination seen in free-flying birds.

B. Temperature Control

At least 70% of the total energy expended during forward flapping flight is wasted as heat, and this heat load has to be dissipated in some way. One possible route is by evaporative heat loss via the respiratory system. The hyperventilation that occurs in starlings during forward flapping flight (Torre Bueno, 1978a) could be the result of the increase in body temperature (T_b), by at least 2°C, that occurs in all birds so far studied, even at low (0°C) ambient temperatures (T_a) (Torre-Bueno, 1976; Hudson and Bernstein, 1981; Hirth *et al.*, 1987). However, when the usual elevation in T_b during running is prevented in ducks and domestic fowl, there are still signs of hyperventilation (Kiley *et al.*, 1982; Brackenbury and Gleeson, 1983). In ducks, there was an increase in lactic and hence a slight acidosis which could have stimulated ventilation and in the fowl, hyperventilation seemed only to occur at higher work loads. Thus, it appears as if factors, other than hyperthermia, contribute to the overall hyperventilation in birds during exercise.

At relatively low T_a (<23°C), T_b does not change with varying T_a at constant speed in starlings, white-necked ravens, and pigeons flying in a wind tunnel (Torre-Bueno, 1976; Hudson and Bernstein, 1981; Hirth *et al.*, 1987). Above a T_a of 23°C, however, T_b does increase with increasing T_a in the same three species. In the raven, $\dot{V}_I$ increases with increasing T_b so that, in this species and in the fish crow flying at constant speed, $\dot{V}_I$ progressively rises above that required by the metabolic demands of the bird as T_a increases above about 23°C. Thus, $O_{2, ext}$ falls from a value of 0.19 at a T_a of 20°C to 0.13 at a T_a of 25°C in the flying fish crow (Figure 21). The increased hyperventilation at higher T_a would tend to cause further hypocapnia (see Torre-Bueno, 1978a) and thus alkalosis. It has been demonstrated that the white-necked raven resorts to so-called compound ventilation when flying at

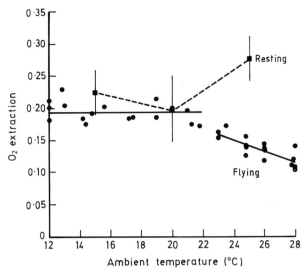

FIGURE 21 The relationship between ambient temperature and steady-state oxygen extraction of fish crows, *Corvus ossifragus*, while at rest and while flying in a wind tunnel. (Reprinted from *Resp. Physiol.* **26**, M. H. Bernstein, Ventilation and respiratory evaporation in the flying crow, *Corvus ossifragus*, pp. 371–382, Copyright 1976 with permission from Elsevier Science–NL, Sara Burgerhartstraat 25, 1005 KV Amsterdam, The Netherlands.)

high T_a. A high-frequency, shallow ventilatory component is superimposed upon a deeper, lower-frequency component and may serve to reduce hypocapnia during thermal panting (Hudson and Bernstein, 1978).

The physiological significance of the hyperthermia during flapping flight, even at low T_a, is uncertain, although Torre-Bueno (1976) concluded that birds adjust their insulation to allow a certain increase in T_b during flight in order to improve muscle efficiency and to increase maximal work output. Hudson and Bernstein (1981) point out that, assuming an overall efficiency of 25%, the almost 3°C increase in T_b in white-necked ravens while flying at a speed of 10 m sec^{-1} reflects the storage of up to half of the metabolic heat produced during a flight of 5 min duration. At lower overall efficiencies, more heat would be produced, so the proportion stored under the above conditions would be less. A similar situation exists for the calculations of the proportion of metabolic heat lost by respiratory evaporation (see below).

Whatever the significance of the increase in core temperature during flapping flight, brain temperature is maintained at a lower level. In fact, even in birds under resting, thermoneutral conditions, there is approximately a 1°C difference between brain and body temperatures (Bernstein *et al.*, 1979a). This difference is at least maintained as kestrels (*Falco sparverius*) become hyperthermic during flapping flight and may even increase (Figure 22). The structure that appears to be largely responsible for this phenomenon is the *rete mirabile ophthalmicum* (RMO) (Kilgore *et al.*, 1979; Bern-

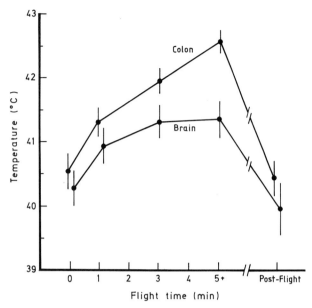

FIGURE 22 Mean (± 2 SE) temperatures in the colon and brain of American kestrels, *Falco sparverius*, before, during and after flying at a speed of 10 m sec^{-1} and at an ambient temperature of 23°C. Data at 5+ min were obtained as mean values over the period between 5 min and 15 min after the onset of flight and represent steady-state values. Postflight values are those obtained after reestablishment of steady-state following each flight. (From Bernstein *et al.* (1979a), *Am. J. Physiol.* with permission.)

stein *et al.*, 1979a,b). It is closely associated with the circulatory system of the eye and warm arterial blood from the body is thought to be cooled by the countercurrent exchange with venous blood returning from the relatively cool beak, evaporative surfaces of the upper respiratory tract, and eye (Midtgård, 1983).

Hyperventilation during flight no doubt serves a thermoregulatory function and as T_a increases, a greater proportion of total heat loss during flight is by respiratory evaporation. However, even at a T_a of 30°C, this proportion is only approximately 20% of total heat loss for the budgerigar and fish crow (Tucker, 1968b; Bernstein, 1976) and 30% for the white-necked raven and pigeon (Hudson and Bernstein, 1981; Biesel and Nachtigall, 1987). In the hummingbird *Amazilia fimbriata*, maximum respiratory heat loss of 40% of the total occurred at a T_a of 35°C (Berger and Hart, 1972). So, most metabolic heat must be dissipated by means other than respiratory evaporation. Indeed, herring gulls can lose up to 80% of total heat production through their webbed feet during flight (Baudinette *et al.*, 1976). In the pigeon, the value is probably less, but nonetheless significant, at 50–65% (Martineau and Larochelle, 1988).

C. Respiratory Water Loss

During flapping flight at T_a above 18°C for the budgerigar and 7.5°C for the pigeon, water is lost at a greater rate than it is produced metabolically (i.e., the birds are dehydrating; Tucker, 1968b; Biesel and Nachtigall, 1987). The starling is in water balance only at T_a below 7°C (Figure 23; Torre-Bueno, 1978b). However, because of the uncertainty regarding the rate of metabolic water production during flight, the author points out that the critical temperature lies between 0° and 12°C. Only below a T_a of 0°C was the hovering hummingbird *Amazilia fimbriata* in water balance (Berger and Hart, 1972). Torre-Bueno (1978b) suggested that during migrations, birds ascend to altitudes where the air is cool enough to enable a greater proportion of heat to be dissipated by nonevaporative means, thus keeping them in water balance.

Some of the larger species of birds, do indeed, appear to migrate at high altitudes, in excess of 8000 m (Swan, 1970; Stewart, 1978). However, Carmi *et al.* (1992) used program 1 from Pennycuick (1989) to calculate P_i and concluded that smaller species might be better off staying below 1000 m because of the greater ventilation (and hence increased water loss) required by the reduced air density and partial pressure of oxygen (PO_2) associated with higher altitudes. These authors also emphasize the importance of $O_{2,\,ext}$ and the temperature of the exhaled air (T_{exp}) in water conservation during flight. The greater the $O_{2,\,ext}$, the lower the $\dot{V}_I$ for a given $\dot{V}O_2$ and the lower the T_{exp}, the less water will be lost in expired air, provided the birds have an effective countercurrent heat exchanger in their respiratory passages (Schmidt-Nielsen *et al.*, 1970). Thus, it should be advantageous for small migrants crossing the Sahara to fly at low altitude at night and to rest during the day. However, a study of the behavior of nocturnal migrants in a desert area indicates that only tailwind speed is closely related to the altitude at which the birds fly (Bruderer and Liechti, 1995). Thus, in autumn, the birds make use of the northerly winds at relatively low altitudes, despite the high temperatures, while in the spring they fly at higher altitudes in order to take advantage of the southwesterly winds. That there is an overall advantage of flying at a given altitude with a helpful tailwind, despite the implications for water balance, may not be too surprising, but how nocturnal migrants are able to detect wind direction is a fascinating question.

Carmi *et al.* (1993) also demonstrated that when at rest or during flight, birds have an ability to maintain plasma volume, despite significant reductions in body mass which are largely the result of water loss. This is of obvious significance during flight, particularly during long distance migratory flights, when a reduction in plasma volume would result in an increase in viscosity of the blood which, in turn, would increase the work load of the heart resulting in an inadequate cardiac output and supply of oxygen to the metabolizing tissues.

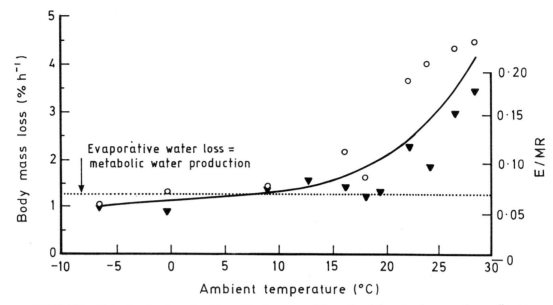

FIGURE 23 Data showing the rate of loss of body mass and the ratio of the rate of evaporative cooling to the metabolic rate (E/MR) as a function of ambient temperature in two starlings, *Sturnus vulgaris,* flying in a wind tunnel at between 9 and 14 m sec^{-1}. (From Torre-Bueno (1978b), *J. Exp. Biol.,* Company of Biologists, Ltd.)

IX. MIGRATION AND LONG-DISTANCE FLIGHT PERFORMANCE

A. Preparation for Migration

As discussed by Fry *et al.* (1972), it is a reasonable expectation that there will be correlated physiological adaptations associated with the extra power required to fly due to the laying down of fat as fuel for long-distance flights. Their work indicated that early in the fattening process of the yellow wagtail (*Motacilla flava*) the flight muscles showed a small hypertrophy. Other studies have also found muscle hypertrophy during fattening for both the pectoralis muscles (Marsh, 1984; Driedzic *et al.,* 1993; Bishop *et al.,* 1996; Jehl, 1997) and the cardiac muscles of birds (Driedzic *et al.,* 1993; Bishop *et al.,* 1996; Jehl, 1997). However, the mechanisms underlying these processes are currently obscure.

Bishop *et al.* (1997) compared the development of a captive population of the barnacle goose with that of the wild migratory population in order to investigate to what extent some of the migratory specializations of the cardiac and locomotory muscles might be determined by developmental processes and to what extent they might be modulated by differences in relative levels of activity. Postflight increases in the masses of the pectoralis muscles, of both wild and captive geese, tend to show an appropriate amount of hypertrophy in response to changes in body mass (Figures 24a and 24b). Regression equations were calculated for both the wild adult premigratory geese (pectoralis mass = 0.5 $M_b^{0.86}$, r^2 = 0.93) and captive adult geese (pectoralis mass = 0.3 $M_b^{0.92}$, r^2 = 0.90). The body mass exponents (or slopes) of the two regression equations are not significantly different between the two populations, although the coefficients are slightly smaller in the captive geese). Thus, approximately, 92% of the pectoralis muscle mass of premigratory barnacle geese appears to be almost independent of the experience of flight per se. This is in contrast to the conclusion of Marsh and Storer (1981) who suggested that the correlation between flight muscle mass and body mass in Cooper's hawk (*Accipiter cooperii*) is a natural analog of "power" training during flight.

The results of the barnacle goose study suggest that either the pectoralis muscle mass is determined by endogenous processes during premigratory fattening or that the flight muscle is able to detect and respond appropriately to the consequent increase in wing loading during this period. Perhaps the relatively infrequent bouts of wing-flapping and occasional take-offs exhibited by captive birds provided a sufficient mechanistic link by which the birds are able to maintain the flight muscle to body mass relationship. However, Gaunt *et al.* (1990) studied the flight muscle changes of the eared grebe (*Podiceps nigricollis*) during a staging period at Mono Lake, California, where the food resources were not limiting. Following a postbreeding moult at the lake, the birds become flightless and their flight muscles show a slight atrophy, but due to an abundance of food the birds put on a large amount of fat and body mass greatly increases. Just prior to departure for the winter grounds, the birds metabolize much of the fat and it is at this

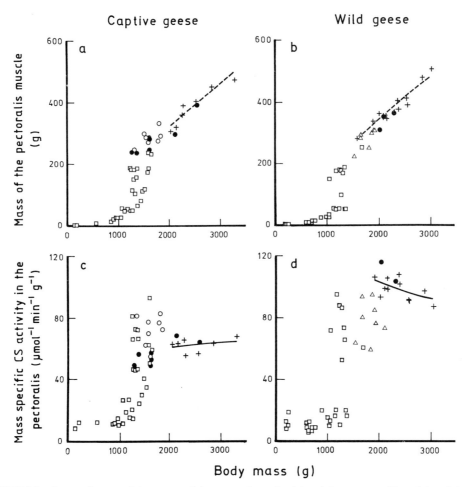

FIGURE 24 Scatter diagram of the masses of the pectoral muscles (g) and the mass-specific activity of citrate synthase (CS) in the pectoral muscles (μmol min^{-1} g wet wt^{-1}) of captive (a and c) and wild (b and d) barnacle geese, *Branta leucopsis*, plotted against body mass (g). Goslings of 1–7 weeks of age ($\square$), goslings of 12–20 weeks of age ($\bullet$), adults at 7 weeks after the population hatch date ($\triangle$) and premigratory adults >10 weeks post-hatch (+). (From *Physiol. Zool.*, C. M. Bishop, P. J. Butler, A. J. El Haj, S. Egginton, with permission from The University of Chicago Press. © University of Chicago Press 1998.)

time that the flight muscles hypertrophy. In this example, flight muscle atrophy occurs when the birds have plenty of food, while subsequent hypertrophy occurs while the birds are actually reducing in body mass. Thus, there is no obvious mechanistic link between hypertrophy of the flight muscle and body mass, although the authors did associate increased wing flapping with the flight muscle hypertrophy.

Figures 24c and 24d show the change in maximum mass-specific CS activity in the same barnacle geese as for the data on pectoralis muscle mass (Figure 24a and 24b). There is little difference between the two populations up to 7 weeks of age. Specific activities of CS in the pectoralis muscles of the two wild 11.5-week-old juveniles are, at 103 and 116 μmol of substrate min^{-1} g^{-1}, outside the range of that for the 12-week-old captive birds (55–59 μmol min^{-1} g^{-1}) and the captive adult birds (56–58 μmol min^{-1} g^{-1}), but similar to the range of wild adult birds (92–106 μmol min^{-1} g^{-1}). The main conclusion from these results is that the captive adults have mass-specific values of activity for CS (mean = 62.1 ± 1.7 μmol min^{-1} g^{-1}) which do not scale with body mass (CS = 20 $M_b^{0.14}$, $r^2 = 0.11$) and are substantially below those of the wild premigratory birds (mean = 98.6 ± 1.9 μmol min^{-1} g^{-1}, $P < 0.0001$). Results from captive goslings aged between 12 and 20 weeks also appear to be closely associated with the regression line resulting from the data from captive adult geese.

Peak values for the activity of CS in the pectoralis muscles of the wild geese are found in the premigratory birds, although there is a tendency for the mass-specific activity to decline with increasing body mass (CS = 862 $M_b^{-0.28}$, $r^2 = 0.33$). Thus, the activity of CS in the pectoralis of adult captive geese, and captive goslings over 11 weeks of age, is only around 60% of that measured in wild geese. In addition, CS activity in the pectoralis muscles of a group of postmoulting, wild, adult geese was similar to that of long-term captive birds. Thus, it is suggested that the rise in the activity of CS in the premigratory birds may be a reaction to the increase in flight activity *per se*, while the rise in CS activity during development of the goslings up to fledging could be primarily under endogenous control. In addition, heart ventricular mass is reduced in captive and postmolt wild adults compared to that in wild premigratory birds and is qualitatively similar to the reduction shown in the aerobic capacity of the pectoralis muscle. It is suggested that this reduction in ventricular mass is also likely to be a direct result of the lower activity levels experienced by captive birds (Bishop *et al.*, 1998).

One of the factors which may be important in the development of the aerobic capability of the pectoral muscles of barnacle geese is thyroxine (T_4). Circulating levels of T_4 show a similar developmental profile as mass and mass-specific activity of CS of the pectoral muscles (Bishop *et al.*, 1997). While artificially accelerating the increase in circulating T_4 did not have a significant effect on either relative mass or on the mass-specific CS activity of these muscles (Deaton *et al.*, 1997), hypothyroidism during development, induced by administration of the drug methimazole, did result in the retardation of their growth, mass-specific CS activity, fractional volume of mitochondria, and capillarity (Deaton *et al.*, 1998). There was no such effect on the mass of the muscles of the leg.

If the mass-specific CS values for the pectoralis muscles of premigratory geese are generalized to the whole muscle, then the largest birds will have the lowest body mass-specific aerobic capacity despite requiring the highest relative power outputs (Pennycuick, 1989). Consequently, during sustained flight, the larger birds may have to fly nearer to their minimum power speeds than the smaller geese. This conclusion is also supported by the data on the heart ventricular mass (Bishop *et al.*, 1996), which indicates that there is a tendency for relative heart mass to decline slightly with increasing body mass. Thus, given that maximum oxygen consumption is likely to be closely correlated with heart mass (Bishop and Butler 1995; Bishop, 1997), it is suggested that during premigratory fattening, P_i available is likely to scale relative to body mass with an exponent value considerably lower than that of the P_o exponent required ($M_b^{1.59}$) for flight performance to be maintained at a similar level according to aerodynamic theory (Rayner, 1990).

B. Migratory Behavior

When migrating, it might be more appropriate for the bird to minimize the time of migration (T_{min}), rather than the energy consumed, particularly if it is important for it to arrive at its destination before most of its competitors (Alerstam and Lindström, 1990). In this case, the optimal flight speed will be greater than U_{mr}. In a recent analysis of data in the literature for 48 species during migration, Welham (1994) concluded that lighter species do, indeed, fly faster than U_{mr}, whereas heavier species tend to fly slower. However, recent modifications to the aerodynamic model of Pennycuick (cf., Pennycuick 1997) indicate that previous estimates of U_{mr} may be underestimates. Both U_{mr} and the speed for T_{min} are similarly affected by winds; optimal speed increases with headwinds and decreases with tailwinds (Pennycuick, 1978).

It has been suggested that some species of birds, such as geese, may be able to reduce induced drag by flying in V-formation (Lissaman and Scholenberger, 1970; Hummel, 1995). This would entail their maintaining positions so that the tips of the wings on the inside of the V are close to the centers of the trailing vortices from the wings on the outside of the bird ahead (Hainsworth, 1987). Theoretically, an "infinite" number of birds may save up to 71% of their induced power in this way and measurements of wing tip spacing (WTS) have been used to indicate whether or not such savings are possible. Assuming a fixed wing at maximum span, Hainsworth (1987) found considerable variability in WTS (−128 and 289 cm) both between and within formations of Canada geese (*Branta canadensis*) but found that median WTS for 55 geese (−19.8 cm) was close to the theoretical optimum (−16 cm) and corresponded to an estimated saving in induced power of 36%, compared with the maximum of 51% for 9 geese flying in V-formation. Even so, in order to achieve such savings with beating wings, there would have to be a high degree of synchrony in wingbeat frequencies, as there is little variation in depth (the distance between adjacent birds' body centers, parallel to the flight path) between individuals (Hainsworth, 1989). This author found close synchrony (a difference of <0.1 beat sec^{-1}) to be present in 48% of the birds he studied and, because of the calm conditions at the time, concluded that this probably represents an upper limit. Indeed, even when not taking wing beat frequency into account, Cutts and Speakman (1994) calculated mean saving in induced power for 54 skeins of pink-footed geese (*Anser brachyrhynchus*) to be only 14% and concluded that the relationship between depth and WTS supported the communication hypothesis (Gould and Heppner, 1974). This contends that the birds position themselves so as to avoid collisions and to maintain flock unity.

Gudmundsson *et al.* (1995) studied the migration of brent geese (*Branta bernicla*) from Iceland to Greenland and tracked them by satellite as they flew up and across the Greenland ice cap. These authors found that the brent geese (fitted with 60-g satellite transmitters) were not able to fly continuously over the Greenland ice cap and had to make frequent and long rests during the climb. Gundmundsson *et al.* (1995) concluded that this evidence supported the biomechanical predictions of Pennycuick (1989) that these birds were up against the limit of the biomechanical power available from the flight muscles due to negative scaling of wing beat frequency with respect to body mass. However, the analysis of Bishop and Butler (1995) and Bishop (1997) suggest that the limit to the sustained climbing performance of birds, such as brent geese, is more likely to be related to the biomechanical performance of the heart. As geese have both FG and FOG fiber types in their flight muscles (Rosser and George, 1986a), the net effect is that during "prolonged" activity the mass of the flight muscle is effectively reduced to that of the "aerobic" FOG fibers and the P_o available becomes proportional to the ability of the cardiovascular and respiratory system to supply the working (aerobic) muscle tissue. This appears to be a reasonable notion, as the idea that these birds were capable of flying for more than a few minutes while using a relatively large amount of anaerobic metabolism is difficult to accept. The additional contribution from the FG fibers would have meant that even with the satellite transmitter attached there should have been plenty of potential to fly up the slope using anaerobic power, but the birds would require constant stops (as was observed). The empirical data of Gudmundsson *et al.* (1995) actually appear to indicate that it is not the biomechanical power available from the whole flight muscles *per se* that is limiting, but the power available from the mass of the FOG fibers, which is likely to be closely matched to the ability of the cardiovascular system to perfuse the flight muscle with blood.

X. FLIGHT AT HIGH ALTITUDE

Two aspects of flying at high altitude (air temperature and evaporative water loss) have already been mentioned. Other aspects are the effect of the decreases in air density and in the partial pressure of oxygen with increasing altitude. Pennycuick (1975) argues that, as the effective lift:drag ratio increases with increased height, birds, particularly small ones, should fly at the maximum height at which they are able to maintain their maximum range speed (U_{mr}). However, as this speed increases with height, so does the power required to maintain it. Also, because the rate of gas exchange is affected by PO_2, there will be a particular altitude at which the bird can only just obtain O_2 at a sufficient rate to maintain U_{mr} and this is the optimum cruising height. As the bird uses fuel and thus loses mass during a long flight, it would be expected to climb progressively in order to maintain U_{mr} (see also, Pennycuick, 1978). However, the evidence for this being the case is rather weak (Hedenström and Alerstam, 1992).

Early studies indicated that most passerines migrating at night fly below 2 km (Lack, 1960; Nisbet, 1963), although it has been demonstrated more recently (Bruderer and Liechti, 1995) that wind direction may cause the majority of birds crossing the south of Israel to fly above 1.8 km in spring. Even so, 90% were below 3.5 km. Some birds have been observed at extremely high altitudes. A flock of 30 swans (probably whooper, *Cygnus cygnus*) was located by radar off the west coast of Scotland at an altitude of 8–8.5 km, where the temperature was −48°C (Stewart, 1978; Elkins, 1979). Bar-headed geese, *Anser indicus*, have been observed flying at altitudes of approximately 9 km (where PO_2 is, at 6.9 kPa, approximately one-third of the sea-level value) during their migration across the Himalayas (Swan, 1961). Thus, birds can fly at altitudes at which nonacclimatized mammals can barely survive (Tucker, 1968a). Unfortunately, there have been few physiological studies on birds flying at high altitude, real or simulated, although Tucker (1968a) did train a budgerigar wearing a $\dot{V}O_2$ mask to fly in a hypobaric wind tunnel for 30 sec at the equivalent of 6.1 km. Work on captive animals has indicated some of the various adaptations that may be of great importance in those species that fly at high altitudes.

For obvious reasons, the bar-headed goose has been the subject of many studies concerned with the physiological adaptations for flying at high altitude. Unlike Pekin ducks, bar-headed geese do not increase their Hct and [Hb] when exposed to simulated high altitude (Black and Tenney, 1980). This means that there is no increase in the viscosity of the blood, thus preventing a possible reduction in its circulation. This is more than counterbalanced by the higher affinity for oxygen (low P_{50}) of the Hb of the goose (P_{50} for blood of the bar-headed goose at pH 7.5 is approximately 5 kPa compared with 7.5 kPa for the duck), which allows the maintenance of a higher C_aO_2 (and hence $C_aO_2 - C_{\bar{v}}O_2$) at high altitudes than in the duck.

A comparison between the bar-headed goose and the plain-dwelling greylag goose, *Anser anser*, indicates that their Hb differs by only four amino acid substitutions, one of which (α^{119}-Pro→Ala) may, by altering the contact between the α_1 and β_1 chains, confer a small increase in the O_2 affinity of the Hb from the bar-headed goose (Perutz, 1983), and this is then amplified by the interaction with inositol pentaphosphate (Rollema and Bauer, 1979). The complementary substitution (β^{55}-Leu→Ser) at the same position may have a similar effect on the P_{50} of the Hb of the Andean goose *Cloephaga melanoptera* (Weber *et al.*, 1993).

Temporary movements to high altitude seem to have different effects. The natural annual movements of wild quail from 200 to 1200 m in Catalonia is accompanied by an increase in [Hb] of 24% but a decrease in the affinity of Hb for O_2 (P_{50} from 3.6 to 4.2 kPa, Prats *et al.*, 1996). Thus, there is an increase in O_2 carrying capacity of the blood and an enhancement of its ability to release O_2 to the tissues.

It is clear from the data of Black and Tenney (1980) that the respiratory system of bar-headed geese is able to maintain a very small difference between the PO_2 of inspired air (P_IO_2) and that in the arterial blood (P_aO_2) when at high altitude. For example, when at sea level, $P_IO_2 - P_aO_2$ in bar-headed geese is approximately 7 kPa, whereas at a simulated altitude of 10.67 km it is only 0.5 kPa. The large increase in $\dot{V}_I$ that is required to maintain such a small difference between P_IO_2 and P_aO_2 also causes a decline in $PaCO_2$ and an associated increase in arterial pH (pHa); the bird becomes hypocapnic and alkalotic.

In a number of mammals (dog, monkey, rat, man) hypocapnia causes a reduction in cerebral blood flow. This is not the case in ducks and bar-headed geese (Grubb *et al.*, 1977; Faraci and Fedde, 1986). Also, hypoxia causes a greater increase in cerebral blood flow in ducks than it does in dogs, rats, and man (Grubb *et al.*, 1978). At high altitude, of course, both of these factors occur together (hypocapnic hypoxia) and under these conditions hypocapnia appears to attenuate the increase in cerebral blood flow caused by hypoxia (Grubb *et al.*, 1979). However, these authors found that blood flow is similar at a given O_2 content in both normocapnic and hypocapnic ducks. This is because during hypocapnia (and alkalosis), there is a leftward shift of the O_2 equilibrium curve, as a result of the Bohr effect, so that a given O_2 content is achieved at a lower PO_2. In fact, in bar-headed geese, the alkalosis during severe hypoxia is greater than that in Pekin ducks which, together with the higher affinity of their Hb for O_2, means that at a given (low) PaO_2, CaO_2 is much (at least two times) greater in the geese (Faraci *et al.*, 1984a). Thus, similar (or even greater) O_2 deliveries to the brain and heart can be achieved in hypoxic bar-headed geese at lower cerebral and coronary blood flows than in hypoxic ducks.

Unlike the duck, this species of goose is able to maintain, or even to increase, its perfusion of all tissues during severe hypocapnic hypoxia (Faraci *et al.*, 1985). Also, capillary density and capillary : fiber ratio are greater in the skeletal muscle (gastrocnemius) of bar-headed geese than in that of Canada geese (Snyder *et al.*, 1984) and the pulmonary vasoconstrictor response to hypoxia is smaller in the bar-headed goose than in other birds and mammals (Faraci *et al.*, 1984b). All of these factors are, no doubt, very important features when the geese are flying over the Himalayas.

It has been suggested that the more effective lungs of birds, compared with those of mammals, may contribute significantly to the ability of birds better to tolerate high altitude (Scheid, 1985). However, Shams and Scheid (1989) concluded that, although the parabronchial lung of birds does confer an advantage at a PO_2 equivalent to that at the top of Mt. Everest, at the highest simulated altitude (11.58 km) tolerated by their ducks, there was no advantage at all. The authors concluded that the major difference between birds and mammals at these extreme altitudes is the ability of the former to tolerate lower $PaCO_2$, thus enabling the respiratory system to maintain PaO_2 at as high a level as possible. An intriguing aspect of this for birds actually flying at these extremely high altitudes is how they protect the upper airways from cold damage potentially caused by hyperventilating air at $-40°C$ to $-50°C$.

In a subsequent paper, the same authors (Shams and Scheid, 1993) reported that when hypoxia is accompanied by the appropriate hypobaria (reduced atmospheric pressure), $\dot{V}_I$ and PaO_2 are slightly, but significantly greater. Although the authors have no explanation for these differences, they would allow a bird at the elevation of Mt. Everest to gain another 700 m in height before the "increase" in PaO_2 was eliminated. Gas exchange across the *rete mirabile ophthalmicum*, thus enhancing the supply of oxygen to the brain (Bernstein *et al.*, 1984) and improvement in gas transport as a result of acclimation to high altitude (Weinstein *et al.*, 1985), may be other ways in which birds are adapted for survival and activity at high altitudes.

Despite all of these apparent adaptations to life at high altitude, experiments on bar-headed geese running on a treadmill under hypoxic conditions, similar to those at the top of Mt. Everest, indicate severe limitations to oxygen uptake (Fedde *et al.*, 1989). Under normoxic conditions, running at 0.6 m sec^{-1} at a 2° incline caused a doubling of $\dot{V}O_2$ and of $\dot{V}b$. Under hypoxic conditions, there was a decline in resting $\dot{V}O_2$ and this did not change significantly during 6 min of exercise. $\dot{V}b$ also remained unchanged during exercise, although there was a significant reduction in V_s. The authors argue that diffusion of O_2 from the muscle capillaries to the mitochondria may limit the aerobic capacity of the leg muscles during hypoxia, but that this would not be the case for the pectoral muscles during flight at high altitude. The question is, how does the heart manage to cope with the extra demands of flight at high altitude when it appears to be at its limit during running under hypoxic conditions similar to those at the top of Mt. Everest? The authors suggest that there may be a hypoxic depression of cardiac contractility, but again this cannot occur when the birds are flying at high altitude. It may be that the birds used in this study were stressed by the surgical techniques associated with the cannulations and so on (cf., Woakes and Butler, 1986). Certainly, their resting heart rate was almost 3× higher than that of resting barnacle geese (Butler and Woakes, 1980). Whether or not this is the explanation remains to be seen. There is no doubt that understanding how birds are able to perform such a high level of exercise as flight at such high altitudes is yet another challenge to students of avian flight.

Acknowledgments

The authors wish to thank NERC, BBSRC, The Royal Society, and the Leverhulme Trust for financial support for their recent and current work on this topic. We are grateful to Mrs. Pauline Hill and Mrs. Jackie Harris for their assistance in the preparation of this chapter.

References

Adams, N. J., Brown, C. R., and Nagy, K. A. (1986). Energy expenditure of free-ranging wandering albatrosses *Diomedea exulans*. *Physiol. Zool.* **59**, 583–591.

Alerstam, T., and Lindström, Å. (1990). Optimal bird migration: The relative importance of time, energy, and safety. In "Bird Migration" (Gwinner, E. ed.), pp. 331–351. Springer-Verlag, Berlin/Heidelberg/New York.

Alerstam, T., Gudmundsson, G. A., and Larsson, B. (1993). Flight tracks and speeds of Antarctic and Atlantic seabirds: Radar and optical measurements. *Phil. Trans. R. Soc. Lond. B* **340**, 55–67.

Alexander, R. McN., and Bennett-Clark, H. C. (1977). Storage of elastic strain energy in muscle and other tissues. *Nature* **265**, 114–117.

Balasch, J., Palomeque, J., Palacios, L., Musquera, S., and Jimenez, M. (1973). Hematological values of some great flying and aquatic-diving birds. *Comp. Biochem. Physiol. A* **49**, 137–145.

Ballance, L. T. (1995). Flight energetics of free-ranging red-footed boobies (*Sula sula*). *Physiol. Zool.* **68**, 887–914.

Banzett, R. B., Nations, C. S., Wang, N., Butler, J. P., and Lehr, J. L. (1992). Mechanical independence of wingbeat and breathing in starlings. *Resp. Physiol.* **89**, 27–36.

Baudinette, R. V., and Schmidt-Nielsen, K. (1974). Energy cost of gliding flight in herring gulls. *Nature* **248**, 83–84.

Baudinette, R. V., Loveridge, J. P., Wilson, K. J., Mills, C. D., and Schmidt-Nielsen, K. (1976). Heat loss from feet of herring gulls at rest and during flight. *Am. J. Physiol.* **230**, 920–924.

Bennett, P. M., and Harvey, P. H. (1987). Active and resting metabolism in birds—Allometry, phylogeny and ecology. *J. Zool. London* **213**, 327–363.

Berger, M. (1978). Ventilation in the humming birds *Colibri coruscans* during altitude hovering. In "Respiratory Function in Birds, Adult and Embryonic" (J. Piiper, ed.), pp 85–88. Springer-Verlag, Berlin/Heidelberg/New York.

Berger, M. (1985). Sauerstoffverbrauch von Kolibris (*Colibri coruscans* und *C. thalassinus*) beim Horizontalflug. In "BIONA report 3" (W. Nachtigall, ed.), pp. 307–314. Gustav Fischer, Stuttgart.

Berger, M., and Hart, J. S. (1972). Die Atmung beim Kolibri *Amazilia fimbriata* während des Schwirrfluges bei verschiedenen Umgebungstemperaturen. *J. Comp. Physiol.* **81**, 363–380.

Berger, M., and Hart, J. S. (1974). Physiology and energetics of flight. In "Avian Biology" (D. S. Farner and J. R. King, eds.), Vol. IV, pp. 415–477. Academic Press, New York and London.

Berger, M., Hart, J. S., and Roy, O. Z. (1970a). Respiration, oxygen consumption and heart rate in some birds during rest and flight. *Z. vergl. Physiol.* **66**, 201–214.

Berger, M., Roy, O. Z., and Hart, J. S. (1970b). The co-ordination between respiration and wing beats. *Z. vergl. Physiol.* **66**, 190–200.

Bernstein, M. H. (1976). Ventilation and respiratory evaporation in the flying crow, *Corvus ossifragus*. *Resp. Physiol.* **26**, 371–382.

Bernstein, M. H., Curtis, M. B., and Hudson, D. M. (1979a). Independence of brain and body temperatures in flying American kestrels, *Falco sparverius*. *Am. J. Physiol.* **237**, 58–62.

Bernstein, M. H., Duran, H. L., and Pinshow, B. (1984). Extrapulmonary gas exchange enhances brain oxygen in pigeons. *Science* **226**, 564–566.

Bernstein, M. H., Sandoval, I., Curtis, M. B., and Hudson, D. M. (1979b). Brain temperature in pigeons: effects of anterior respiratory bypass. *J. Comp. Physiol.* **129**, 115–118.

Bernstein, M. H., Thomas, S. P., and Schmidt-Nielsen, K. (1973). Power input during flight of the fish crow, *Corvus ossifragus*. *J. Exp. Biol.* **58**, 401–410.

Bevan, R. M., Butler, P. J., Woakes, A. J., and Prince, P. A. (1995b). The energy expenditure of free-ranging back-browed albatrosses. *Phil. Trans. R. Soc. Lond. B* **350**, 119–131.

Bevan, R. M., Woakes, A. J., and Butler, P. J. (1994). The use of heart rate to estimate oxygen consumption of free-ranging black-browed albatrosses *Diomedea melanophrys*. *J. Exp. Biol.* **193**, 119–137.

Bevan, R. M., Woakes, A. J., Butler, P. J., and Croxall, J. P. (1995a). Heart rate and oxygen consumption of exercising gentoo penguins. *Physiol. Zoo.* **68**, 855–877.

Biebach, H. (1998). Phenotypic plasticity in the digestive system and the breast muscle in trans-Sahara migrating passerines. *J. Avian Biol.*, in press.

Biesel, W., and Nachtigall, W. (1987). Pigeon flight in a wind tunnel. IV. Thermoregulation and water homeostasis. *J. Comp. Physiol B* **157**, 117–128.

Biewener, A. A., Dial, K. P., and Goslow, G. E., Jr. (1992). Pectoralis muscle force and power output during flight in the starling. *J. Exp. Biol.* **164**, 1–18.

Birt-Friesen, V. L., Montevecchi, W. A., Cairns, D. K.,. and Macko, S. A. (1989). Activity-specific metabolic rates of free-living northern gannets and other seabirds. *Ecology* **70**, 357–367.

Bishop, C. M. (1997). Heart mass and the maximum cardiac output of birds and mammals: Implications for estimating the maximum aerobic power input of flying animals. *Phil. Trans. R. Soc. Lond. B* **352**, 447–456.

Bishop, C. M., and Butler, P. J. (1995). Physiological modelling of oxygen consumption in birds during flight. *J. Exp. Biol.* **198**, 2153–2163.

Bishop, C. M., Butler, P. J., Egginton, S., El Haj, A. J., and Gabrielsen, G. W. (1995). Development of metabolic enzyme activity in locomotor and cardiac muscles of the migratory barnacle goose. *Am. J. Physiol.* **269**, R64–R72.

Bishop, C. M., Butler, P. J., El Haj, A. J., Egginton, S., and Loonen, M. J. J. E. (1996). The morphological development of the locomotor and cardiac muscles of the migratory barnacle goose (*Branta leucopsis*). *J. Zool. London* **239**, 1–15.

Bishop, C. M., Butler, P. J., El Haj, A. J., and Egginton, S. (1998). Comparative development of captive and migratory populations of the barnacle goose. *Physiol. Zool.*, **71**, 198–207.

Black, C. P., and Tenney, S. M. (1980). Oxygen transport during progressive hypoxia in high-altitude and sea-level waterfowl. *Resp. Physiol.* **39**, 217–239.

Blem, C. R. (1976). Patterns of lipid storage and utilization in birds. *Amer. Zool.* **16**, 671–684.

Boggs, D. F. (1997). Coordinated control of respiratory pattern during locomotion in birds. *Am. Zool.* **37**, 41–53.

Boggs, D. F., and Dial, K. P. (1993). Neuromuscular organization and regional EMG activity of the pectoralis in the pigeon. *J. Morphol.* **218**, 43–57.

Boggs, D. F., Jenkins, F. A., Jr. and Dial, K. P. (1997a). The effects of wingbeat cycle on respiration in black-billed magpies *Pica pica*. *J. Exp. Biol.* **200**, 1403–1412.

Boggs, D. F., Seveyka, J. J., Kilgore, D. L., Jr., and Dial, K. P. (1997b). Coordination of respiratory cycles with wingbeat cycles in the black-billed magpie (*Pica pica*). *J. Exp. Biol.* **200**, 1413–1420.

Boswell, T., Sharp, P. J., Hall, M. R., and Goldsmith, A. R. (1995). Migratory fat deposition in European quail: A role for prolactin? *J. Endocrinol.* **146**, 71–79.

Brackenbury, J. H., and Gleeson, M. (1983). Effects of P_{CO_2} on respiratory pattern during thermal and exercise hyperventilation in domestic fowl. *Resp. Physiol.* **54**, 109–119.

Brobeck, J. R., and DuBois, A. B. (1980). Energy exchange. *In* "Medical Physiology" (V. B. Mountcastle, ed.), Vol. 2, pp. 1351–1365. St Louis.

Bromley, R. G., and Jarvis, R. L. (1993). The energetics of migration and reproduction of dusky Canada geese. *Condor* **95**, 193–210.

Bruderer, B., and Liechti, F. (1995). Variation in density and height distribution of nocturnal migration in the south of Israel. *Israel J. Zool.* **41**, 477–487.

Butler, P. J. (1991). Exercise in birds. *J. Exp. Biol.* **160**, 233–262.

Butler, P. J., and Woakes, A. J. (1980). Heart rate, respiratory frequency and wing beat frequency of free flying barnacle geese *Branta leucopsis*. *J. Exp. Biol.* **85**, 213–226.

Butler, P. J., and Woakes, A. J. (1985). Exercise in normally ventilating and apnoeic birds. *In* "Circulation, Respiration and Metabolism: Current Comparative Approaches" (R. Gilles, ed.), pp. 39–55. Springer-Verlag, Berlin.

Butler, P. J., and Woakes, A. J. (1990). The physiology of bird flight. *In* "Bird Migration: Physiology and Ecophysiology" (E. Gwinner, ed.), pp. 300–318. Springer-Verlag, Berlin/Heidelberg.

Butler, P. J., West, N. H., and Jones, D. R. (1977). Respiratory and cardiovascular responses of the pigeon to sustained, level flight in a wind-tunnel. *J. Exp. Biol.* **71**, 7–26.

Butler, P. J., Woakes, A. J., and Bishop, C. M. (1998). Behaviour and physiology of Svalbard barnacle geese, *Branta leucopsis*, during their autumn migration. *J. Avian Biol.*

Butler, P. J., Woakes, A. J., Bevan, R. M., and Stephenson, R. Heart rate and rate of oxygen consumption during forward flapping flight of the barnacle goose, *Branta leucopsis*. *J. Avian Biol.*, submitted.

Caldow, R. W. G., and Furness, R. W. (1993). A histochemical comparison of fibre types in the M. pectoralis and M. supracoracoideus of the great skua *Catharacta skua* and the herring gull *Larus argentatus* with reference to kleptoparasitic capabilities. *J. Zool. London* **229**, 91–103.

Carmi, N., Pinshow, B., Horowitz, M., and Bernstein, M. H. (1993). Birds conserve plasma volume during thermal and flight-incurred dehydration. *Physiol. Zool.* **66**, 829–846.

Carmi, N., Pinshow, B., Porter, W. P., and Jaeger, J. (1992). Water and energy limitations on flight duration in small migrating birds. *Auk* **109**, 268–276.

Carpenter, F. L. (1975). Bird hematocrits: Effects of high altitude and strength of flight. *Comp. Biochem. Physiol A.* **50**, 415–417.

Chai, P., and Dudley, R. (1995). Limits to vertebrate locomotor energetics suggested by hummingbirds hovering in heliox. *Nature* **377**, 722–725.

Chai, P., and Millard, D. (1997). Flight and size constraints: Hovering performance of large hummingbirds under maximal loading. *J. Exp. Biol.* **200**, 2757–2763.

Chai, P., Chen, J. S. C., and Dudley, R. (1997). Transient hovering performance of hummingbirds under conditions of maximal loading. *J. Exp. Biol.* **200**, 921–929.

Choi, I., Ricklefs, R. E., and Shea, R. E. (1993). Skeletal muscle growth, enzyme activities and the development of thermogenesis: A comparison between altricial and precocial birds. *Physiol. Zool.* **66**, 455–473.

Costa, D. P., and Prince, P. A. (1987). Foraging energetics of grey-headed albatrosses *Diomedea chrysostoma* at Bird Island, South Georgia. *Ibis* **129**, 149–158.

Crabtree, B., and Newsholme, E. A. (1972). The activities of phosphorylase, hexokinase, phosphofructokinase, lactate dehydrogenase and glycerol 3-phosphate dehydrogenase in muscles from vertebrate and invertebrates. *Biochem J* **126**, 49–58.

Crile, G., and Quiring, D. P. (1940). A record of the body weight and certain organ and gland-weights of 3690 animals. *Ohio J. Sci.* **XL**, 219–259.

Cutts, C. J., and Speakman, J. R. (1994). Energy savings in formation flight of pink-footed geese. *J. Exp. Biol.* **189**, 251–261

Deaton, K. E., Bishop, C. M., and Butler, P. J. (1997). The effect of thyroid hormones on the aerobic development of locomotor and cardiac muscles in the barnacle goose. *J. Comp. Physiol.* **167**, 319–327.

Deaton, K. E., Bishop, C. M., and Butler, P. J. (1998). Tissue-specific effects of hypothyroidism on postnatal muscle development in the barnacle goose. *J. Exp. Biol.*, **201**, 827–836.

Dial, K. P. (1992a). Activity patterns of the wing muscles of the pigeon (*Columba livia*) during different modes of flight. *J. Exp. Zool.* **262**, 357–373.

Dial, K. P. (1992b). Avian forelimb muscles and nonsteady flight: Can birds fly without using the muscles in their wings? *Auk* **109**, 874–885.

Dial, K. P., and Biewener, A. A. (1993). Pectoralis muscle force and power output during different modes of flight in pigeons (*Columba livia*). *J. Exp. Biol* **176**, 31–54.

Dial, K. P., Biewener, A. A., Tobalske, B. W., and Warrick, D. R. (1997). Mechanical power output of bird flight. *Nature* **390**, 67–70.

Dial, K. P., Kaplan, S. R., Goslow, G. E. J., and Jenkins, F. A, J, (1987). Structure and neural control of the pectoralis in pigeons: Implications for flight mechanics. *Anat. Rec.* **218**, 284–287.

Dial, K. P., Kaplan, S. R., and Goslow, G. E. J. (1988). A functional analysis of the primary upstroke and downstroke muscles in the domestic pigeon (*Columba livia*) during flight. *J. Exp. Biol.* **134**, 1–16.

Dial, K. P., Goslow, G. E. J., and Jenkins, F. A. J. (1991). The functional anatomy of the shoulder in the european starling (*Sturnus vulgaris*). *J. Morphol* **207**, 327–344.

Driedzic, W. R., Crowe, H. L., Hicklin, P. W., and Sephton, D. H. (1993). Adaptations in pectoralis muscle, heart mass, and energy metabolism during premigratory fattening in semipalmated sandpipers (*Calidris pusilla*). *Can. J. Zool.* **71**, 1602–1608.

Elkins, N. (1979). High altitude flight by swans. *Br. Brids* **72**, 238–239.

Ellington, C. P. (1984). The aerodynamics of hovering insect flight. VI. Lift and power requirements. *Phil. Trans. Soc. Lond. B* **305**, 145–181.

Ellington, C. P. (1991). Limitations on animal flight performance. *J. Exp. Biol.* **160,** 71–91.

Ellis, H. I. (1984). Energetics of free-ranging seabirds *In* "Seabird Energetics" (C. G. Whittow and H. Rahn, eds.), pp. 203–234. Plenum, New York.

Faraci, F. M., and Fedde, M. R. (1986). Regional circulatory responses to hypocapnia and hypercapnia in bar-headed geese. *Am. J. Physiol.* **250,** R499–R504.

Faraci, F. M., Kilgore, D. L., Jr., and Fedde, M. R. (1984a). Oxygen delivery to the heart and brain during hypoxia: Pekin duck vs. bar-headed goose. *Am. J. Physiol.* **247,** R69–R75.

Faraci, F. M., Kilgore, D. L., Jr., and Fedde, M. R. (1984b). Attenuated pulmonary pressor response to hypoxia in bar-headed geese. *Am. J. Physiol.* **247,** R402–R403.

Faraci, F. M., Kilgore, D. L., Jr., and Fedde, M. R. (1985). Blood flow distribution during hypocapnic hypoxia in Pekin ducks and bar-headed geese. *Resp. Physiol.* **61,** 21–30.

Fedde, M. R., Orr, J. A., Shams, H., and Scheid, P. (1989). Cardiopulmonary function in exercising bar-headed geese during normoxia and hypoxia. *Resp. Physiol.* **77,** 239–262.

Flint, E. N., and Nagy, K. A. (1984). Flight energetics of free-living sooty terns. *Auk* **101,** 288–294.

Fry, C. H., and Ferguson-Lees, I. F. (1972). Flight muscle hypertrophy and ecophysiological variation of yellow wagtail (*Motacilla flava*) races at Lake Chad. *J. Zool. London* **167,** 293–306.

Funk, G. D., Milsom, W. K., and Steeves, J. D. (1992a). Coordination of wingbeat and respiration in the Canada goose. I. Passive wing flapping. *J. Appl. Physiol.* **73,** 1014–1024.

Funk, G. D., Sholomenko, G. N., Valenzuela, I. J., Steeves, J. D., and Milsom, W. K. (1993). Coordination of wing beat and respiration in Canada geese during free flight. *J. Exp. Biol.* **175,** 317–323.

Funk, G. D., Steeves, J. D., and Milsom, W. K. (1992b). Coordination of wingbeat and respiration in birds. II. "Fictive" flight. *J. Appl. Physiol.* **73,** 1025–1033.

Gaunt, A. S., Hikida, R. S., Jehl, J. R., Jr., and Fenbert, L. (1990). Rapid atrophy and hypertrophy of an avian flight muscle. *Auk.* **107,** 649–659.

George, J. C., and Berger, A. J. (1966). "Avian Myology." Academic Press, London/New York.

Gessaman, J. A. (1980). An evaluation of heart rate as an indirect measure of daily energy metabolism of the American kestrel. *Comp. Biochem. Physiol. A* **65,** 273–289.

Gessaman, J. A., and Nagy, K. A. (1988). Transmitter loads affect the flight speed and metabolism of homing pigeons. *Condor* **90,** 662–668.

Goslow, G. E. J., and Dial, K. P. (1990). Active stretch-shorten contractions of the M. Pectoralis in the European starling (*Sturnus vulgaris*): Evidence from electromyography and contractile properties. *Neth. J. Zool.* **40,** 106–114.

Gould, L. L., and Heppner, F. (1974). The vee formation of Canada geese. *Auk* **91,** 494–506.

Greenewalt, C. H. (1962). Dimensional relationships for flying animals. *Smithson. Misc. Collect.* **144,** 1–46.

Greenewalt, C. H. (1975). The flight of birds. *Trans. Am. Phil.* **65,** 1–67.

Grubb, B. R. (1983). Allometric relations of cardiovascular function in birds. *Am. J. Physiol.* **245,** H567–H572.

Grubb, B., Mills, C. B., Colacino, J. M., and Schmidt-Nielsen, K. (1977). Effect of arterial carbon dioxide on cerebral blood flow in ducks. *Am. J. Physiol.* **232,** H596–H601.

Grubb, B., Colacino, J. M., and Schmidt-Nielsen, K. (1978). Cerebral blood flow in birds: Effect of hypoxia. *Am J. Physiol.* **234,** H230–H234.

Grubb, B., Jones, J. H., and Schmidt-Nielsen, K. (1979). Avian cerebral blood flow: Influence of the Bohr effect on oxygen supply. *Am. J. Physiol.* **236,** H744–H749.

Gudmundsson, G. A., Benvenuti, S., Alterstam, T., Papi, F., Lilliendahl, K., and Åkesson, S. (1995). Examining the limits of flight and orientation performance: satellite tracking of brent geese migrating across the Greenland ice-cap. *Proc. R. Soc. Lond. B.* **261,** 73–79.

Hails, C. J. (1979). A comparison of flight energetics in hirundines and other birds. *Comp. Biochem. Physiol. A* **63,** 581–585.

Hainsworth, F. R. (1987). Precision and dynamics of positioning by Canada geese flying on formation. *J. Exp. Biol.* **128,** 445–462.

Hainsworth, F. R. (1988). Wing movements and positioning for aerodynamic benefit by Canada geese flying in formation. *Can. J. Zool.* **67,** 585–589.

Hart, J. S., and Roy, O. Z., (1966). Respiratory and cardiac responses to flight in pigeons. *Physiol. Zool.* **39,** 291–306.

Hartman, F. A. (1961). Locomotor mechanisms of birds. *Smithson. Misc. Collect.* **143,** 1–91.

Hedenström, A., and Alerstam, T. (1992). Climbing performance of migrating birds as a basis for estimating limits for fuel-carrying capacity and muscle work. *J. Exp. Biol.* **164,** 19–38.

Hedenström, A., and Alerstam, T. (1995). Optimal flight speed of birds. *Phil. Trans. R. Soc. Lond. B* **348,** 471–487.

Heusner, A. A. (1991). Size and power in mammals. *J. Exp. Biol.* **160,** 25–54.

Hickson, R. C., Galassi, T. M., and Dougherty, K. A. (1983). Repeated development and regression of exercise-induced cardiac hypertrophy in rats. *J. Appl. Physiol.* **54,** 794–797.

Hirth, K. D., Biesel, W., and Nachtigall, W. (1987). Pigeon flight in a wind tunnel. III. Regulation of body temperature. *J. Comp. Physiol. B* **157,** 111–116.

Hudson, D. M., and Bernstein, M. H. (1978). Respiratory ventilation during steady-state flight in the white-necked raven, *Corvus cryptoleucus*. *Fed. Proc.* **37,** 472.

Hudson D. M., and Bernstein, M. H. (1981). Temperature regulation and heat balance in flying white-necked ravens, *Corvus cryptoleucus*. *J. Exp. Biol.* **90,** 267–281.

Hudson, D. M., and Bernstein, M. H. (1983). Gas exchange and energy cost of flight in the white-necked raven, *Corvus cryptoleucus*. *J. Exp. Biol.* **103,** 121–130.

Hummel, D. (1995). Formation flight as an energy-saving mechanism. *Israel J. Zool.* **41,** 261–278.

Jehl, J. R., Jr. (1994). Field estimates of energetics in migrating and downed black-necked grebes. *J. Avian Biol.* **25,** 63–68.

Jehl, J. R., Jr. (1997). Cyclical changes in body composition in the annual cycle and migration of the Eared Grebe, *Podiceps nigricollis*. *J. Avian Biol.* **28,** 132–142.

Jenkins, F. A., Jr., Dial, K. P., and Goslow, G. E., Jr. (1988). A cineradiographic analysis of bird flight: The wishbone in starlings is a spring. *Science* **241,** 1495–1498.

Jenni-Eiermann, S., and Jenni, L. (1991). Metabolic responses to flight and fasting in night-migrating passerines. *J. Comp. Physiol. B* **161,** 465–474.

Jenni-Eiermann, S., and Jenni, L. (1992). High plasma triglyceride levels in small birds during migratory flight: A new pathway for fuel supply during endurance locomotion at very high mass-specific metabolic rates? *Physiol. Zool.* **65,** 112–123.

Jenni-Eiermann, S., and Jenni, L. (1996). Metabolic differences between the postbreeding, moulting and migratory periods in feeding and fasting passerine birds. *Funct Ecol.* **10,** 62–72.

Jurgens, K. D., Bartels, H., and Bartels, R. (1981). Blood oxygen transport and organ weights of small bats and small nonflying mammals. *Resp. Physiol.* **45,** 243–260.

Kampe, G., and Crawford, E. C., Jr. (1993). Oscillatory mechanics of the respiratory system of pigeons. *Resp. Physiol.* **18,** 188–193.

Kaplan, S. R., and Goslow, G. E. J. (1989). Neuromuscular organization of the pectoralis (*Pars thoracicus*) of the pigeon (*Columbia livia*): Implications for motor control. *Anat. Rec.* **224,** 426–430.

Kiley, J. P., Kuhlmann, W. D., and Fedde, M. R. (1982). Ventilatory and blood gas adjustments in exercising isothermic ducks. *J. Comp. Physiol.* **147,** 107–112.

Kilgore, D. L., Jr., Boggs, D. F., and Birchard, G. F. (1979). Role of the *rete mirable ophthalmicum* in maintaining the body-to-brain temperature difference in pigeons. *J. Comp. Physiol.* **129,** 119–122.

Klaassen, M., Kvist, A., and Lindström, Å. (1998). Energy and protein metabolism in a bird migrating in a wind tunnel. *J. Avian Biol.,* in press.

Krüger, K., Prinzinger, R., and Schuchmann, K.-L. (1982). Torpor and metabolism in hummingbirds. *Comp. Biochem. Physiol. A* **73,** 679–689.

Lack, D. (1960). Migration across the north sea studied by radar. Part 2. The spring departure 1956–59. *Ibis* **102,** 27–59.

Lasiewski, R. C. (1963). Oxygen consumption of torpid, resting, active and flying hummingbirds. *Physiol. Zool.* **36,** 122–140.

Lasiewski, R. C., and Calder, W. A., Jr. (1971). A preliminary allometric analysis of respiratory variables in resting birds. *Resp. Physiol.* **11,** 152–166.

LeFebvre, E. A. (1964). The use of D_2O^{18} for measuring energy metabolism in *Columba livia* at rest and in flight. *Auk* **81,** 403–416.

Lifson, N., Gordon, G. B., and McClintock, R. (1955). Measurement of total carbon dioxide production by means of D_2O^{18}. *J. Appl. Physiol.* **7,** 704–710.

Lifson, N., Gordon, G. B., Visscher, M. B., and Nier, A. O. (1949). The fate of utilized molecular oxygen and the source of the oxygen of respiratory carbon dioxide, studied with the aid of heavy oxygen. *J. Biol. Chem.* **180,** 803–811.

Lissaman, P. B. S., and Shollenberger, C. A. (1970). Formation flight of birds. *Science* **168,** 1003–1005.

Lundgren, B. O., and Kiessling, K.-H. (1985). Seasonal variation in catabolic enzyme activities in breast muscle of some migratory birds. *Oecologia* **66,** 468–471.

Lundgren, B. O., and Kiessling, K.-H. (1986). Catabolic enzyme activities in the pectoralis muscle of premigratory and migratory juvenile reed warblers *Acrocephalus scirpaceus*. *Oecologia* **68,** 529–532.

Lusk, G. (1919). "The Elements of the Science of Nutrition". W. B. Saunders, Philadelphia/London.

Magnan, A. (1922). Les caracteristiques des oiseaux suivant le mode de vol. *Ann. Sci. Nat.* **5,** 125–334.

Mainguy, S. K., and Thomas, V. G. (1985). Comparisons of body reserve buildup and use in several groups of Canada geese. *Can. J. Zool.* **63,** 1765–1772.

Marden, J. H. (1987). Maximum lift production during takeoff in flying animals. *J. Exp. Biol.* **130,** 235–258.

Marden, J. H. (1994). From damselflies to pterosaurs: how burst and sustainable flight performance scale with size. *Am. J. Physiol.* **266,** R1077–R1084.

Marey, E. J. (1890). "Le Vol des Oiseaux." Masson, Paris.

Martineau, L., and Larochelle, J. (1988). The cooling power of pigeon legs. *J. Exp. Biol.* **136,** 193–208.

Marsh, R. L. (1981). Catabolic enzyme activities in relation to premigratory fattening and muscle hypertrophy in the gray catbird *Dumetella carolinensis*. *J. Comp. Physiol.* **141,** 417–423.

Marsh, R. L. (1984). Adaptations of the gray catbird *Dumatella carolinensis* to long-distance migration: Flight muscle hypertrophy associated with elevated body mass. *Physiol. Zool.* **57,** 105–117.

Marsh, R. L., and Storer, R. W. (1981). Correlation of flight muscle size and body mass in Cooper's Hawks: A natural analogue of power training. *J. Exp. Biol.* **91,** 363–368.

Marsh, R. L., and Wickler, S. J. (1982). The role of muscle development in the transition to endothermy in nestling bank swallows *Riparia riparia*. *J. Comp. Physiol.* **149,** 99–105.

Masman, D., and Klaassen, M. (1987). Energy expenditure during free flight in trained and free-living Eurasian kestrels (*Falco tinnunculus*). *Auk* **104,** 603–616.

Mathieu-Costello, O. (1991). Morphometric analysis of capillary geometry in pigeon pectoralis muscle. *Am. J. Anat.* **191,** 74–84.

Mathieu-Costello, O., Suarez, R. K., and Hochachka, P. W. (1992). Capillary-to-fiber geometry and mitochondrial density in hummingbird flight muscle. *Respir. Physiol.* **89,** 113–132.

Matsuda, R., Bandman, E., and Strohman, R. C. (1983). Regional differences in the expression of myosin light-chains and tropomyosin subunits during development of chicken breast muscle. *Dev. Biol.* **95,** 484–491.

McNab, B. K. (1994). Energy conservation and the evolution of flightlessness in birds. *Am. Nat.* **144,** 628–642.

Midtgård, U. (1983). Scaling of the brain and the eye cooling system in birds: A morphometric analysis of the *Rete ophthalmicum*. *J. Exp. Zool.* **225,** 197–207.

Nachtigall, W. (1995). Impositions of energy balance in prolonged flight: Wind tunnel measurements with "model birds." *Israel J. Zool.* **41,** 279–295.

Nagy, K. A. (1980). CO_2 Production in animals: analysis of potential errors in the doubly labeled water method. *Am. J. Physiol.* **238,** R466–R473.

Nagy, K. A. (1983). The doubly labeled water ($^3HH^{18}O$) method: A guide to its use. *University of California at Los Angeles,* Publication No. 12-1417, 45 pages.

Newsholme, E. A., and Leech, A. R. (1983). "Biochemistry for the Medical Sciences." Wiley, Chichester/New York.

Nisbet, I. C. T. (1963). Measurements with radar of the height of nocturnal migration over Cape Cod, Massachusetts. *Bird Band.* **34,** 57–67.

Nisbet, I. C. T., Drury, W. H., Jr. and Baird, J. (1963). Weight-loss during migration, Part I: Deposition and consumption of fat by the blackpoll warbler *Dendroica striata*. *Bird Band.* **34,** 107–159.

Nolet, B. A., Butler, P. J., Masman, D., and Woakes, A. J. (1992). Estimation of daily energy expenditure from heart rate and doubly labeled water in exercising geese. *Physiol. Zool.* **65,** 1188–1216.

Norberg, U. M. (1990). "Vertebrate Flight. Mechanics, Physiology, Morphology, Ecology and Evolution". Springer-Verlag, Berlin/Heidelberg.

Norberg, U. M. (1996). Energetics of flight. *In* "Avian Energetics and Nutritional Ecology" (C. Carey, ed.), pp. 199–249. Chapman & Hall, New York.

Obrecht, H. H., Pennycuick, C. J., and Fuller, M. R. (1988). Wind tunnel experiments to assess the effect of back-mounted radio transmitters on bird body drag. *J. Exp. Biol.* **135,** 265–273.

Obst, B. S., Nagy, K. A., and Ricklefs, R. E. (1987). Energy utilization by Wilson's storm-petrel (*Oceanites oceanicus*). *Physiol. Zool.* **60,** 200–210.

Pasquis, P., Lacaisse, A., and Dejours, P. (1970). Maximal oxygen uptake in four species of small mammals. *Resp. Physiol.* **9,** 298–309.

Pelsers, M. A. L., Bishop, C. M., Butler, P. J., and Glatz, J. F. C. (1999). Fatty acid-binding protein in heart and skeletal muscles of the migratory barnacle goose throughout development. *Am. J. Physiol.,* in press.

Pennycuick, C. J. (1968). Power requirements for horizontal flight in the pigeon *Columba livia*. *J. Exp. Biol.* **49,** 527–555.

Pennycuick, C. J. (1969). The mechanics of bird migration. *Ibis* **111,** 525–556.

Pennycuick, C. J. (1975). Mechanics of flight. *In* "Avian Biology" (D. S. Farner, J. R. King, and K. C. Parkes, eds.), Vol. V, pp. 1–75. Academic Press, New York.

Pennycuick, C. J. (1978). Fifteen testable predictions about bird flight. *Oikos* **30**, 165–176.

Pennycuick, C. J. (1982). The flight of petrels and albatrosses (Procellariiformes), observed in South Georgia and its vicinity. *Phil. Trans. R. Soc. Lond. B* **300**, 75–106.

Pennycuick, C. J. (1988). Empirical estimates of body drag of large waterfowl and raptors. *J. Exp. Biol.* **135**, 253–264.

Pennycuick, C. J. (1989). "Bird Flight Performance: A Practical Calculation Manual". Oxford Univ. Press, Oxford.

Pennycuick, C. J. (1990). Predicting wingbeat frequency and wavelength of birds. *J. Exp. Biol.* **150**, 171–185.

Pennycuick, C. J. (1992). "Newton Rules Biology: A Physical Approach to Biological Problems." Oxford Univ. Press, Oxford/New York/Tokyo.

Pennycuick, C. J. (1996). Stress and strain in the flight muscles as constraints on the evolution of flying animals. *J. Biomech.* **29**, 577–581.

Pennycuick, C. J. (1997). Actual and "optimum" flight speeds: Field data reassessed. *J. Exp. Biol.* **200**, 2355–2361.

Pennycuick, C. J., and Rezende, M. A. (1984). The specific power output of aerobic muscle, related to the power density of mitochondria. *J. Exp. Biol.* **108**, 377–392.

Pennycuick, C. J., Fuller, M. R., and McAllister, L. (1989). Climbing performance of Harris hawks (*Parabuteo unicinctus*) with added load: Implications for muscle mechanics and for radiotracking. *J. Exp. Biol.* **142**, 17–29.

Pennycuick, C. J., Klaassen, M., Dvist, A., and Lindström, Å. (1996). Wingbeat frequency and the body drag anomaly: Wind-tunnel observations on a thrush nightingale (*Luscinia luscinia*) and a teal (*Anas crecca*). *J. Exp. Biol.* **199**, 2757–2765.

Perutz, M. F. (1983). Species adaptation in a protein molecule. *Mol. Biol. Evol.* **1**, 1–28.

Piersma, T. (1988). Breast muscle atrophy and constraints on foraging during the flightless period of wing moulting great crested grebes. *Ardea* **76**, 96–106.

Piersma, T., and Jukema, J. (1990). Budgeting the flight of a long-distance migrant: Changes in nutrient reserve levels of bar-tailed godwits at successive spring staging sites. *Ardea* **78**, 315–337.

Prats, M.-T., Palacios, L., Gallego, S. and Riera, M. (1996). Blood oxygen transport properties during migration to higher altitude of wild quail, *Coturnix coturnix coturnix*. *Physiol. Zool.* **69**, 912–929.

Prinzinger, R., and Hänssler, I. (1980). Metabolism–weight relationship in some small nonpasserine birds. *Experientia* **36**, 1299–1300.

Rall, J. A. (1985). Energetic aspects of skeletal muscle contraction: Implications of fiber types. *Exercise Sport Sci. Rev.* **13**, 33–74.

Rayner, J. M. V. (1979). A new approach to animal flight mechanics. *J. Exp. Biol.* **80**, 17–54.

Rayner, J. M. V. (1985a). Bounding and undulating flight in birds. *J. Theor. Biol.* **117**, 47–77.

Rayner, J. M. V. (1985b). Linear relations in biomechanics: the statistics of scaling functions. *J. Zool. London* **206**, 415–439.

Rayner, J. M. V. (1988). Form and function in avian flight. *In* "Current Ornithology" (R. F. Johnston, ed.), Vol. 5. pp. 1–66. Plenum, New York.

Rayner, J. M. V. (1990). The mechanics of flight and bird migration performance. *In* "Bird Migration" (E. Gwinner, ed.), pp. 283–299. Springer-Verlag, Berlin/Heidelberg.

Rayner, J. M. V. (1991). On the aerodynamics of animal flight in ground effect. *Phil. Trans. R. Soc. Lond. B* **334**, 119–128.

Rayner, J. M. V. (1993). On aerodynamics and the energetics of vertebrate flapping flight. *Contemp. Math.* **141**, 351–399.

Rayner, J. M. V. (1994). Aerodynamic corrections for the flight of birds and bats in wind tunnels. *J. Zool. London* **234**, 537–563.

Richardson, R. D., Boswell, T., Raffety, B. D., Seeley, R. J., Wingfield, J. C., and Woods, S. C. (1995). NPY increases food intake in white-crowned sparrows: effect in short and long photoperiods. *Am. J. Physiol.* **268**, 1418–1422.

Rollema, H. S., and Bauer, C. (1979). The interaction of inositol pentaphosphate with the hemoglobins of highland and lowland geese. *J. Biol. Chem.* **254**, 12038–12043.

Rosser, B. W. C., and George, J. C. (1986a). The avian pectoralis: histochemical characterization and distribution of muscle fiber types. *Can. J. Zool.* **64**, 1174–1185.

Rosser, B. W. C., and George, J. C. (1986b). Slow muscle fibres in the pectoralis of the turkey vulture (*Cathartes aura*): An adaptation for soaring flight. *Zool. Anz.* **217**, 252–258.

Rosser, B. W. C., Waldbillig, D. M., Wick, M. and Bandman, E. (1994). Muscle fiber types in the pectoralis of the white pelican, a soaring bird. *Acta Zool.* **75**, 329–336.

Rothe, H. J., Biesel, W., and Nachtigall, W. (1987). Pigeon flight in a wind tunnel. II. Gas exchange and power requirements. *J. Comp. Physiol. B* **157**, 99–109.

Scheid, P. (1985). Significance of lung structure for performance at high altitude. *In* "ACTA XVII, Congressus Internationalis Ornithologici, Moscow" (V. D. Ilyichev and V. M. Gavrilov, eds.), pp. 976–977. Nauka.

Schmidt-Nielsen, K. (1972). Locomotion: Energy cost of swimming, flying, and running. *Science* **177**, 222–228.

Schmidt-Nielsen, K. (1984). "Scaling: Why Is Animal Size so Important?" Cambridge Univ. Press.

Schmidt-Nielsen, K. (1997). "Animal Physiology: Adaptation and Environment," 5th ed. Cambridge Univ. Press.

Schmidt-Nielsen, K., Hainsworth, F. R., and Murrish, D. E. (1970). Counter-current heat exchange in the respiration passages: Effect on water and heat balance. *Resp. Physiol.* **9**, 263–276.

Schwilch, R., Jenni, L., and Jenni-Eiermann, S. (1996). Metabolic responses of homing pigeons to flight and subsequent recovery. *J. Comp. Physiol.* **166**, 77–87.

Shams, H., and Scheid, P. (1989). Efficiency of parabronchial gas exchange in deep hypoxia: measurements in the resting duck. *Resp. Physiol.* **77**, 135–146.

Shams, H., and Scheid, P. (1993). Effects of hypobaria on parabronchial gas exchange in normoxic and hypoxic ducks. *Resp. Physiol.* **91**, 155–163.

Sibley, C. G., and Ahlquist, J. E. (1990). "Phylogeny and Classification of Birds: A Study in Molecular Evolution." Yale Univ. Press, New Haven/London.

Snyder, G. K. (1990). Capillarity and diffusion distances in skeletal-muscles in birds. *J. Comp. Physiol. B* **160**, 583–591.

Snyder, G. K., Byers, R. L., and Kayar, S. R. (1984). Effects of hypoxia on tissue capillarity in geese. *Resp. Physiol.* **58**, 151–160.

Sokal, R. R., and Rohlf, F. J. (1981). "Biometry: The Principles and Practice of Statistics in Biological Research." W. H. Freeman, New York.

Speakman, J. R. (1990). Principles, problems and a paradox with the measurement of energy expenditure of free-living subjects using double-labelled water. *Stats. Med.* **9**, 1365–1380.

Speakman, J. R., and Racey, P. A. (1988). Consequences of non steady-state CO_2 production for accuracy of the doubly labelled water technique: The importance of recapture interval. *Comp. Biochem. Physiol. A* **90**, 337–340.

Spedding, G. R. (1986). The wake of a jackdaw (*Corvus monedula*) in slow flight. *J. Exp. Biol.* **125**, 287–307.

Spedding, G. R. (1987). The wake of a kestrel (*Falco tinnunculus*) in flapping flight. *J. Exp. Biol.* **127,** 59–78.

Spedding, G. R. (1994). On the significance of unsteady effects in the aerodynamic performance of flying animals. *Contemp. Math.* **141,** 401–419.

Spedding, G. R., Rayner, J. M. V., and Pennycuick, C. J. (1984). Momentum and energy in the wake of a pigeon (*Columba livia*) in slow flight. *J. Exp. Biol.* **111,** 81–102.

Stewart, A. G. (1978). Swans flying at 8,000 metres. *Brit. Birds* **71,** 459–460.

Stickland, N. C. (1977). Succinic dehydrogenase distribution in the pectoralis muscle of several East African birds. *Acta Zool.* **58,** 41–44.

Suarez, R. K., Brown, G. S., and Hochachka, P. W. (1986). Metabolic sources of energy for hummingbird flight. *Am. J. Physiol.* **251,** 537–542.

Suarez, R. K., Lighton, J. R. B., Moyes, C. D., Brown, G. S., Gass, C. L., and Hochachka, P. W. (1990). Fuel selection in rufous hummingbirds: Ecological implications of metabolic biochemistry. *Proc. Natl. Acad. Sci. USA.* **87,** 9207–9210.

Swan, L. W. (1961). The ecology of the high himalayas. *Sci. Am.* **205,** 68–78.

Swan, L. W. (1970). Goose of the Himalayas. *Nat. Hist.* **79,** 68–74.

Talesara, G. L., and Goldspink, G. (1978). A combined histochemical and biochemical study of myofibrillar ATPase in pectoral, leg and cardiac muscle of several species of bird. *Histochem. J.* **10,** 695–710.

Tatner, P., and Bryant, D. M. (1989). Doubly-labelled water technique for measuring energy expenditure. *In* "Techniques in Comparative Respiratory Physiology: An Experimental Approach" (C. R. Bridges and P. J. Butler, eds.). Cambridge Univ. Press.

Taylor, C. R. (1994). Relating mechanics and energetics during exercise. *Adv Vet Sci Comp Med. A* **38,** 181–215.

Teal, J. M. (1969). Direct measurement of CO_2 production during flight in small birds. *Zoologica* **54,** 17–24.

Tobalske, B. W. (1995). Neuromuscular control and kinematics of intermittent flight in the European starling (*Sturnus vulgaris*). *J. Exp. Biol.* **198,** 1259–1273.

Tobalske, B. W. (1996). Scaling of muscle composition, wing morphology, and intermittent flight behavior in woodpeckers. *Auk.* **113,** 151–177.

Tobalske, B. W., and Dial, K. P. (1994). Neuromuscular control and kinematics of intermittent flight in budgerigars (*Melopsittacus undulatus*). *J. Exp. Biol.* **187,** 1–18.

Tobalske, B. W., and Dial, K. P. (1996). Flight kinematics of black-billed magpies and pigeons over a wide range of speeds. *J. Exp. Biol.* **199,** 263–280.

Tobalske, B. W., Olson, N. E., and Dial, K. P. (1997). Flight style of the black-billed magpie: Variation in wing kinematics, neuromuscular control and muscle composition. *J. Exp. Zool.* **279,** 313–329.

Torre-Bueno, J. R. (1976). Temperature regulation and heat dissipation during flight in birds *J. Exp. Biol.* **65,** 471–482.

Torre-Bueno, J. R. (1978a). Respiration during flight in birds. *In* "Respiratory Function in Birds, Adult and Embryonic" (J. Piiper, ed.). pp. 89–94. Springer-Verlag, Berlin/Heidelberg/New York.

Torre-Bueno, J. R. (1978b). Evaporative cooling and water balance during flight in birds. *J. Exp. Biol.* **75,** 231–236.

Torre-Bueno, J. R., and Larochelle, J. (1978). The metabolic cost of flight in unrestrained birds. *J. Exp. Biol.* **75,** 223–229.

Tucker, V. A. (1966). Oxygen consumption of a flying bird. *Science* **154,** 150–151.

Tucker, V. A. (1968a). Respiratory physiology of house sparrows in relation to high-altitude flight. *J. Exp. Biol.* **48,** 55–66.

Tucker, V. A. (1968b). Respiratory exchange and evaporative water loss in the flying budgerigar. *J. Exp. Biol.* **48,** 67–87.

Tucker, V. A. (1970). Energetic cost of locomotion in animals. *Comp. Biochem. Physiol.* **34,** 841–846.

Tucker, V. A. (1972). Metabolism during flight in the laughing gull, *Larus atricilla. Am. J. Physiol* **222,** 237–245.

Tucker, V. A. (1973). Bird metabolism during flight: Evaluation of a theory. *J. Exp. Biol.* **58,** 689–709.

Van den Berg, C., and Rayner, J. M. V. (1995). The moment of inertia of bird wings and the inertial power requirement for flapping flight. *J. Exp. Biol.* **198,** 1655–1664.

Vazquez, R. J. (1994). The automating skeletal and muscular mechanisms of the avian wing (Aves). *Zoomorphol.* **114,** 59–71.

Weber, R. E., Jessen, T.-H., Malte, H., and Tame, J. (1993). Mutant hemoglobins (α^{119}-Ala and β^{55}-Ser): Functions related to high-altitude respiration in geese. *J. Appl. Physiol.* **75**(6), 2646–2655.

Weinstein, Y., Bernstein, M. H., Bickler, P. E., Gonzales, D. V., Samaniego, F. C., and Escobedo, M. A. (1985). Blood respiratory properties in pigeons at high altitudes: effects of acclimation. *Am. J. Physiol.* **249,** R765–R775.

Weis-Fogh, T. (1972). Energetics of hovering flight in hummingbirds and in *Drosophila. J. Exp. Biol.* **56,** 79–104.

Weis-Fogh, T., and Alexander, R. McN. (1995). The moment of inertia of bird wings and the inertial power requirement for flapping flight. *J. Exp. Biol.* **198,** 1655–1664.

Welham, C. V. (1992). Flight speeds of migrating birds: A test of maximum range speed predictions from three aerodynamic equations. *Behav. Ecol.* **5,** 1–8.

Wells, D. J. (1993a). Muscle performance in hovering hummingbirds. *J. Exp. Biol.* **178,** 39–57.

Wells, D. J. (1993b). Ecological correlates of hovering flight of hummingbirds. *J. Exp. Biol.* **178,** 59–70.

Wingfield, J. C., Schwabl, H., and Mattocks, P. W. J. (1990). Endocrine mechanisms of migration. *In* "Bird Migration" (E. Gwinner, ed.), pp. 232–256. Springer-Verlag, Berlin/Heidelberg.

Witter, M. S., and Cuthill, I. C. (1993). The ecological costs of avian fat storage. *Phil. Trans. R. Soc. Lond. B.* **340,** 73–92.

Woakes, A. J., and Butler, P. J. (1986). Respiratory, circulatory and metabolic adjustments during swimming in the tufted duck, *Aythya fuligula. J. Exp. Biol.* **120,** 215–231.

Woakes, A. J., Butler, P. J., and Bevan, R. M. (1995). Implantable data logging system for heart rate and body temperature: Its application to the estimation of field metabolic rates in Antarctic predators. *Med. Biol. Eng. Comput.* **33,** 145–151.

Wolf, L. L., and Hainsworth, F. R. (1971). Time and energy budgets of territorial hummingbirds. *Ecology* **52,** 980–988.

CHAPTER 16

Introduction to Endocrinology: Pituitary Gland

COLIN G. SCANES

Department of Animal Science
Cook College
Rutgers—The State University
New Brunswick, New Jersey 08903

I. Introduction 437
II. Anatomy of the Hypothalamic–Hypophyseal Complex 438
III. Gonadotropins 438
   A. Structure 438
   B. Action of Gonadotropins in the Female 438
   C. Action of Gonadotropins in the Male 439
   D. Control of Gonadotropin Release 441
IV. Thyrotropin 443
   A. Structure 443
   B. Assay 443
   C. Control of Thyrotropin Release 443
   D. Pituitary Origin of Thyrotropin 443
   E. Ontogeny of Thyrotropin 444
   F. Role of Thyrotropin 444
   G. Mechanism of Thyrotropin Action 444
V. Growth Hormone 444
   A. Chemistry 444
   B. Growth Hormone Variants 444
   C. Assay 444
   D. Actions of Growth Hormone 444
   E. GH Receptor 445
   F. Control of GH Secretion 446
   G. Pituitary Origin of Growth Hormone 447
VI. Prolactin 447
   A. Chemistry 447
   B. Pituitary Origin of Prolactin 447
   C. Assay 447
   D. Actions of Prolactin 448
   E. Receptor and Mechanism of Action 448
   F. Control of Prolactin Release 449
VII. Adrenocorticotropic Hormone 449
   A. Chemistry 449
   B. Pituitary Origin of ACTH 449
   C. Control of ACTH Release 450
   D. Actions of ACTH 450
VIII. Other Adenohypophyseal Peptides 450
IX. Neurohypophysis 451
   A. Introduction 451
X. Arginine Vasotocin and Mesotocin 451
   A. Chemistry 451
   B. Content 451
   C. Actions of AVT 451
   D. Actions of MT 452
   E. Control of AVT and MT Release 452
   F. Clearance of AVT 452
References 452

I. INTRODUCTION

Birds have a complement of endocrine organs similar to those of mammals; these are the pituitary–hypothalamus complex, gonads, pancreatic islets, adrenal glands, thyroid glands, parathyroid glands, ultimobranchial glands (producing calcitonin) and the endocrine cells of the gut. These ductless or endocrine glands release hormones into the bloodstream. The hormone(s) then acts on specific tissues, cells, or organs (referred to as the target organ). Each hormone acts by interacting with receptors on the surface of the cells (proteins and polypeptides) or within the cytoplasm and nucleus (steroids, thyroid hormones). In addition to the "classical" endocrine glands, hormones are produced by other organs including the pineal (melatonin), liver, (insulinlike growth factor I), and kidney (renin, 1,25-dihydroxy vitamin D_3, erythropoietin).

II. ANATOMY OF THE HYPOTHALAMIC–HYPOPHYSEAL COMPLEX

The pituitary gland (hypophysis) is intimately connected to the hypothalamus at the base of the brain. The pituitary gland has a complex structure and an interesting embryonic development. Primary tissue can be classified as either adenohypophysis or neurohypophysis; each with a distinct embryonic origin. The adenohypophysis is derived from Rathke's pouch (probably ectoderm from the roof of the mouth) and the neurohypophysis is derived from the infundibulum (an outgrowth of the brain). The adenohypophysis, in mammals, goes on to form the pars distalis (anterior pituitary gland), the pars intermedia, and the pars tuberalis. However, in birds there is no pars intermedia, and hence the adenohypophysis forms the anterior pituitary gland or pars distalis. The neurohypophysis forms the pars nervosa (which is equivalent of the posterior pituitary gland in birds), the infundibular stalk, and the median eminence. The structure of the hypothalamic–hypophyseal complex is shown in Figure 1. The avian pars distalis produces a full complement of hormones: the gonadotropins (luteinizing hormone, LH; follicle stimulating hormone, FSH), thyrotropin (TSH), prolactin, growth hormone (GH), and adrenocorticotropic hormone (ACTH). The induction of specific cells type involves specific transcription factors. In mammals, Pit-1 is thought to play a role in the initial transcription of both GH and prolactin genes. The avian homolog of the pituitary-specific transcription factor Pit-I has been characterized based on the cDNA sequence (Wong et al., 1992).

The anterior lobe of the pituitary gland is well supplied with blood vessels, including the hypophyseal portal vessels. These latter provide a route from the neurosecretory nerve terminals in the median eminence to the anterior pituitary gland. Indeed it is by way of the portal blood vessels that the anterior pituitary gland is controlled, by *releasing factors* (or hypophysiotropic factors) from the median eminence (Table 1, Figure 1).

The posterior pituitary gland consists of the neurosecretory terminals, which release either mesotocin or arginine vasotocin. These hormones are synthesized in cell bodies in nuclei in the hypothalamus and are transported to the posterior pituitary gland through modified axons (Table 1, Figure 1).

III. GONADOTROPINS

A. Structure

Avian LH and FSH (molecular weight ~30 kDa) are glycoproteins (Burke et al., 1979a; Papkoff et al., 1982). Avian LH and FSH consist of two glycoprotein subunits; an α-subunit (common to LH, FSH, and TSH) and a β-subunit (hormone specific). Both subunits are required for biological activity (Burke et al., 1979b). The nucleotide sequence of the cDNA for chicken and turkey α-subunit and the β-subunit for both chicken and Japanese quail have been established and the amino acid sequence deduced (Foster et al., 1991; reviewed Ishii, 1993).

B. Action of Gonadotropins in the Female

The roles of LH and FSH in the control of ovarian function in birds are becoming increasingly well established.

1. Luteinizing Hormone

A major role for LH is to induce ovulation. Premature ovulation is provoked following injection of LH (Imai, 1973). Administration of LH is followed by a series of effects on the most mature follicle: breakdown of the germinal vesicle, dissociation of the junctions between the granulosa projections and the oocyte surface, development of a perivitelline space, and the formation of the first and then second maturation spindle (Yoshimura et al., 1993).

Other effects of LH include stimulation of steroidogenesis and reduction of plasminogen activator activity. *In vivo*, mammalian LH increases progesterone and testosterone production by the hen ovary (Shahabi et al., 1975). Not only can LH stimulate the production of progesterone and androstenedione *in vitro* but also decreases both secreted and cell associated plasminogen activator by granulosa cells for the largest follicle (F1) (Tilly and Johnson, 1987; 1988a,b). LH also stimulates production of androstenedione by theca cells from the next largest yellow follicle (F2) (Tilly and Johnson, 1989) and of progesterone, DHEA, androsterone, and estradiol from theca cells from small follicles (6–8 mm) (Kowalski et al., 1991; Tilly et al., 1991a). Based on studies in the Japanese quail, ^{125}I-LH can bind specifically to both theca and granulosa cells with F_1 theca having much reduced numbers of binding sites (Kikuchi and Ishii, 1989, 1992).

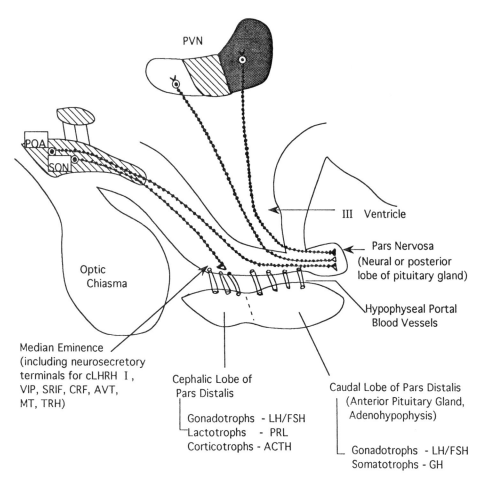

FIGURE 1 Schematic diagram of the avian hypothalamus and anterior pituitary gland. Location of AVT (sold white shading) and MT (solid black shading) cell bodies based on Bons et al. (1980); (>○●●●●●●●●●◀) neurosecretory neuron; (○━━━━○) blood vessel; POA, Preoptic area; PVN, paraventricular nucleus; SON, supraoptic nucleus.

The mechanism by which LH exerts its effect on theca cells is via adenylate cyclase and cAMP (Tilly and Johnson, 1989). While on granulosa cells, LH acts by adenylate cyclase and cAMP (Johnson and Tilly, 1988) and subsequently calcium mobilization (Asem et al., 1987; Tilly and Johnson, 1988a,b).

2. Follicle Stimulating Hormone

Based on *in vitro* studies, FSH influences the functioning of less mature large yolky follicles and small follicles but not the large preovulatory follicles. FSH stimulates both cAMP formation and progesterone secretion by the granulosa of intermediate stage follicles (e.g., F5 or the fifth-largest follicle) (Calvo and Bahr, 1983). In addition, FSH acutely increases cAMP, progesterone, androstenedione, and estradiol production by theca cells from small follicles (6–8 and 9–12 mm) (Kowalski et al., 1991). In addition, FSH stimulates maturation of the granulosa cells of small follicles such that they attain the ability to respond to LH, FSH, or cAMP with increased progesterone production (Tilly et al., 1991b). This is due, at least in part, to the induction of the P_{450} side-chain cleavage enzyme mRNA (Li and Johnson, 1993).

C. Action of Gonadotropins in the Male

Both LH and FSH are essential to controlling testicular functioning in the male bird.

1. Luteinizing Hormone

As in mammals, LH acts primarily to stimulate the Leydig cells to differentiate and produce testosterone. The administration of chicken LH to hypophysectomized Japanese quail greatly increases the number of mature Leydig cells and decreases the number of fibroblasts and transitional cell types in the interstitium (Brown et al., 1975). The injection of preparations of

TABLE 1 Hypothalamic and Extrahypothalamic Locus of Neuropeptides

Peptide	Location[a] Hypothalamic	Extrahypothalamic	Reference
Arginine vasotocin (AVT)/mesotocin (MT)	(F) Median eminence (C) Supraoptic Nucleus (C) Preoptic Area (C) Paraventricular nucleus	—	Bons et al. (1980); Tennyson et al. (1985)
Corticotropin releasing factor (CRF)	(F) Median Eminence (C) Paraventricular (C) Preoptic and (C) Mammillary nuclei	(C) Dorsomedial and dorsolateral thalamic nuclei (C) Nucleus accumbens septi, (C) lobus parolfactorus	Jozsa et al. (1984); reviewed Jozsa and Mess (1993a)
Enkephalin	(C) Paraventicular	(C) Paleostriatal complex in telencephalon, telencephalon lateral	Reiner et al. (1984)
Galanin	(C) Periventricular, (C) Paraventricular, (C) Dorsomedial and (C) Tuberal hypothalamic nuclei	(C) Septal nucleus hippocampus, (C) Nucleus of solitary tract of the medulla oblongata	Jozsa and Mess (1993b)
cLHRH-I	(F) Median Eminence, (C) Lateral hypothalamic area, (C) Ventriculus lateralis, (C) Nucleus accumbens, (F) Nucleus ventromedialis hypothalami	(F) Hippocampus, (F) Nucleus subhabenularis	Millam et al. (1993)
cLHRH-II	(C) Lateral hypothalamus	(C) Magnocellular area, (F) Limbic structures olfactory area	
Somatostatin (SRIF)	(C) Paraventricular nuclei, (C) Supraoptic nucleus (F) Median eminence	(C) Telencephalon-hippocampus, (C) Olfactory lobe, (C,F) Mesencephalon, (F) Rhombencephalon, (F) Raphe nucleus	Blahser et al. (1978); Jozsa and Mess (1993a)
Thyrotropin releasing hormone (TRH)	(C) Paraventricular nucleus, (C) Preoptic area Median eminence	(C) Mesencephalon (C) Band of Broca	Jozsa and Mess (1993a)
Vasoactive intestinal peptide (VIP)	(F) Median eminence, (C) Supraoptic, Magnocellular Preoptic, Anterior medial, Ventromedial hypothalamic and Paraventricular nuclei	(C) Telencephalon-septal area, (C) Mesencephalon-optic lobe, dorsal midbrain, central midbrain, ventral midbrain, (C) Medula oblongata	Norgen and Silver (1990); Esposito et al. (1993); Kuenzel and Blahser (1994)

[a] Abbreviations: C, cell body; F, fibers.

avian or mammalian LH into Japanese quail elevates the plasma concentration of testosterone (Maung and Follett, 1978). Similarly, the *in vitro* synthesis of testosterone is stimulated by preparations of avian or mammalian LH in avian testicular tissue (e.g., Maung and Follett, 1977).

2. Follicle Stimulating Hormone

The major role for FSH is stimulation of spermatogenesis. Avian FSH, when injected into hypophysectomized quail, increases testicular size (8.8-fold) and seminiferous tubule diameter (2.1-fold) and promotes Sertoli cell differentiation and spermatogenesis (Brown *et al.*, 1975). FSH has little if any effect on testosterone production by bird testes (e.g. Maung and Follett, 1978). It would appear that FSH can upregulate the FSH receptors (Tsuisui and Ishii, 1980).

D. Control of Gonadotropin Release

1. Luteinizing Hormone Releasing Hormone (LHRH)

As in mammals, LH release is under the control of LHRH. As the ability of this peptide to influence FSH release in birds is not fully established, the term "gonadotropin releasing hormone" (GnRH), will not be employed.

a. Chemistry of LHRH

Two distinct forms of LHRH have been isolated from chicken hypothalamus:

Chicken LHRH-I (cLHRH-I)
(from King and Millar, 1982)
pGlu-His-Trp-Ser-Tyr-Gly-Leu-Gln-Pro-Gly-NH$_2$

Chicken LHRH-II (cLHRH-II)
(from Miyamoto *et al.*, 1984)
pGlu-His-Trp-Ser-His-Gly-Trp-Tyr-Pro-Gly-NH$_2$

An identical situation exists in the phylogenetically distant ostrich (Powell *et al.*, 1987). This would suggest that cLHRH-I and cLHRH-II are found throughout the class Aves.

b. Action of LHRH

Secretion of LH in birds is stimulated by cLHRH-I, with cAMP being involved in signal transduction (Bonney and Cunningham, 1977). It is unclear whether the release of FSH is influenced by LHRH. Administration of LHRH evokes either less release of FSH compared to that with LH in birds (Hattori *et al.*, 1986) or is completely ineffective (Scanes *et al.*, 1977; Krishnan *et al.*, 1993). However, strong evidence that both LH and FSH are under hypothalamic control comes from deafferentation and lesioning studies (Davies and Follett, 1980).

Chicken LHRH-II is considerably more (4.7-fold) potent than cLHRH-I in stimulating LH release from chicken pituitary cells *in vitro* (Chou *et al.*, 1985; Millar *et al.*, 1986) but there is little difference in *in vivo* activity (Chou *et al.*, 1985; Sharp *et al.*, 1987). It is likely that cLHRH-I exerts a more important role than cLHRH-II in the control of LH secretion. This is due to their respective anatomical location, lack of release of cLHRH-II from the median eminence, and the failure of active immunization agonist cLHRH-II to influence LH release (Katz *et al.*, 1990; Sharp *et al.*, 1990). However, cLHRH-II levels do change with reproductive state (see below).

c. Control of cLHRH-I Release

The cLHRH-I neurosecretory terminals in median eminence release cLHRH-I in response to various stimuli. These include progesterone during the preovulatory LH surge and photoperiodic induction of reproduction.

The neurotransmitters involved directly or indirectly in influencing cLHRH-I release include stimulation by norepinephrine (or epinephrine) via α_1-adrenergic receptors (Knight *et al.*, 1982; Millam *et al.*, 1984) and also a positive effect by neuropeptide Y (Fraley and Kuenzel, 1992; Contijoch *et al.*, 1993a). Inhibitory neurotransmitters include dopamine (Knight *et al.*, 1982; Contijoch *et al.*, 1992) and β endorphin (Contijoch *et al.*, 1993b).

d. Changes in Hypothalamic LHRH Content

The hypothalamic LHRH content appears to be related to reproductive state. For instance, the hypothalamic content of LHRH rises in photosensitive starlings on transfer to long photoperiods but declines with the onset of photorefractoriness (Dawson *et al.*, 1985). Subsequently, on short photoperiods, when photorefractoriness is overcome, hypothalamic LHRH levels rise with a particularly large increase in median eminence immunoreactive (IR)-LHRH (Goldsmith *et al.*, 1989). Similarly, the density of LHRH nerve fibers and the intensity of LHRH immunostaining in the median eminence is greater in Japanese quail on long daylength than short daylength (Foster *et al.*, 1988).

There are changes with physiological state in the hypothalamic contents of both cLHRH-I and cLHRH-II in turkeys (Rozenboim *et al.*, 1993). The LHRH-I levels are higher in laying hens compared to nonphotostimulated birds (with photostimulated birds intermediate). Unexpectedly if it is accepted that LHRH-II is not involved in reproduction, the levels of LHRH-II are also increased by photostimulation, higher in laying hens, but reduced in incubating and photorefractory turkeys

(Rozenboim et al., 1993). During embryonic development of the chicken, there are changes in the brain concentrations of cLHRH-I and cLHRH-II but these are not in a parallel manner (Millam et al., 1993). In addition to the presence of IR-LHRH in hypothalamic neurons, IR-LHRH-containing mast cells have been reported in the dove medial habenular regions with the number and distribution depending acutely on reproductive behavior (e.g., immediately after courtship behavior (Silver et al., 1992; Zhung et al., 1993)).

2. Negative and Positive Feedback

Sex steroids influence gonadotropin secretion in birds, primarily by influencing the release of LHRH from the median eminence in the portal blood vessels. This is predominantly an inhibitory effect, as gonadectomy leads to large increases in the circulating concentration of both LH and FSH. Studies in the castrated Japanese quail by Davies and colleagues (1980) demonstrated that LH secretion in the male is inhibited by estrogens (estradiol) and androgens. The effectiveness of the androgens was similar irrespective of whether they could be aromatized to estrogens (such as testosterone or androstenedione) or were nonaromatizable (5-α-dihydrotesterone, DHT). The secretion of FSH in quail is similarly depressed by testosterone and estradiol, but is not affected by DHT. In young male chickens, it would appear that testosterone exerts a negative feedback following aromatization to estrogens. The high postcastration levels of LH are reduced by implanting either testosterone or the readily aromatizable 19-nortesterone with the nonaromatizable DHT having only a small inhibiting effect (Fennell et al., 1990). Secretion of FSH in male chick embryos is thought to be controlled by endogenous estrogens or androgens following aromatization; the day 13 peak in FSH being prevented by estrogen treatment and accentuated by an aromatase inhibitor (Rombauts et al., 1993).

In female birds, while estradiol exerts a *negative feedback* effect on LH secretion, progesterone can either stimulate or inhibit LH release. Indeed, the preovulatory LH surge in the hen is induced by progesterone. In the ovariectomized domestic hen, plasma concentrations of LH are decreased by a single injection of either progesterone or estradiol (Wilson and Sharp, 1976a,b). Progesterone, acting via the hypothalamus, has a *positive feedback* effect on LH release in the intact or in the ovariectomized hen that has been primed with progesterone and estradiol (Wilson and Sharp, 1976a,b).

3. Female Reproductive Cycle

The ovulation cycle is considered elsewhere in this volume. However, it is pertinent to outline briefly the secretion of gonadotropins during the ovulation cycle of the domestic chicken. Preovulatory surge in plasma LH concentrations occurs approximately 4–6 hr before ovulation (Furr et al., 1973). Similar patterns of plasma LH concentrations occur during the ovulation cycles of other birds. Little (major) changes in the circulatory concentrations of FSH are observed by radioimmunoassay during the ovulatory cycle of the hen (Scanes et al., 1977; Krishnan et al., 1993).

4. Seasonal Breeding

Reproduction is restricted to a specific season in many avian species with this being predominantly in the spring in temperate zone birds. The selective advantage of this seasonal breeding is to maximize the chance of survival of the chicks due particularly to feed availability. In many birds from outside the equatorial zone, reproductive activity is induced by increasing daylength (photoperiodism). This is followed by a period of reproductive quiescence despite the ambient long photoperiods. This is known as photorefractoriness or failure to respond to what would otherwise be a stimulatory photoperiod. For instance, plasma concentrations of FSH increase in temperate zone birds on exposure to long photoperiods but subsequently the birds become refractory to the stimulation by light (photorefractory) (e.g., starlings; Dawson et al., 1985). This is discussed in more detail below.

a. Photoperiodism

Increasing daylengths are thought to stimulate reproduction via increasing LHRH release and hence LH and FSH secretion (e.g., Japanese quail; Follett and Maung, 1978). For instance, as little as 1 long day has been shown to stimulate the release of both LH and FSH in the Japanese quail (Follett et al., 1977). These mechanisms by which long daylength evokes this effect involves light being detected during the photosensitive phase of the rhythm. There is evidence that this rhythm is circadian in the white crowned sparrow (*Zonotrichia leucophys gambelii*) (Follett et al., 1974) but is a rapidly damping oscillator in the Japanese quail (Follett and Pearce-Kelly, 1991).

b. Photorefractoriness

Prolonged exposure of birds to long daylengths does not continue to be stimulatory, the birds becoming photorefractory. A period on a short daylength is required to overcome this photorefractoriness. Photorefractoriness is not immediately terminated by exposure to short daylengths. Rather prolonged exposure (at least 5 weeks) to short photoperiods is required to dissipate photorefractoriness (Japanese quail: Follett and Pearse-Kelly, 1990).

Thyroid hormones and sensitivity to negative feedback appear to be involved in refractoriness. Castration does not affect circulating concentrations of FSH in photorefractory starlings on a long daylength. Thyroidectomy overcomes the photorefractoriness as indicated by the ability of castration to elevate plasma concentrations of FSH in starlings held for a prolonged period on long daylengths (Goldsmith and Nicholls, 1984a).

c. Other Factors Influence Seasonal Breeding

For instance, temperature influences plasma concentrations of LH. The decline in plasma concentrations of LH following transfer from long daylength to short or intermediate photoperiods is accelerated in Japanese quail subjected to low nocturnal temperature (Tsuyoshi and Wada, 1992).

IV. THYROTROPIN

A. Structure

Assuming avian TSH is similar to its mammalian homolog, it is heterodimer consisting of α- and a β-glycoprotein subunits; the α subunit being common to LH, FSH, and TSH. An ostrich TSH has been purified which is a glycoprotein (Papkoff *et al.*, 1982).

B. Assay

There are no homologous radioimmunoassays for avian TSH. Almeida and Thomas (1981), however, reported that a human TSH radioimmunoassay detects immunoreactive(IR)-TSH in the plasma of Japanese quail. As might be expected if the assay were measuring TSH, plasma concentrations of IR-TSH were found to be elevated by the administration of TRH or a goitrogen and depressed by injection of either T_4 or T_3.

An alternative strategy to measure avian TSH has been adopted by Berghman and colleagues (1993). This entails estimating TSH by subtracting the LH and FSH concentrations from a determination of concentration of the α-subunit (common to LH, FSH, and TSH). As might be expected if the system was indeed measuring TSH, plasma concentrations of this IR-TSH is elevated in hypothyroid chicks, depressed by hyperthyroid chicks, and its release is stimulated by TRH *in vitro* (Berghman *et al.*, 1993).

C. Control of Thyrotropin Release

The release of TSH from the hypothalamus is under predominantly stimulatory control. This is supported by the failure of cold exposure to increase circulating concentrations of T_4 following hypothalamic deafferentation in quail (Herbute *et al.*, 1984). In mammals, TSH release is under stimulatory control by thyrotropin-releasing hormone (TRH) (L-pyro-glutamyl-L-histidyl-L-proline amide).

A similar situation exists in birds. TRH has been found in the chicken hypothalamus (Jackson and Reichlin, 1974; Thommes *et al.*, 1985). Moreover, synthetic TRH stimulates the release of bioassayable TSH from chicken pituitary glands incubated *in vitro* (Scanes, 1974). Similarly *in vivo*, TRH provokes TSH release in chickens as indicated by increases in plasma concentrations of T_4 (Klandorf *et al.*, 1978; Kuhn and Nouwen, 1978). The release of TSH may also be under inhibitory control. This is based on the ability of somatostatin to depress circulating concentrations of T_4 in young chickens (Lam *et al.*, 1986).

As in mammals, T_3 exerts a *negative feedback* effect on TSH release in birds. This is supported by the ability of goitrogens which depress circulating concentrations of thyroid hormones to increase thyrotroph number and pituitary TSH content (Sharp *et al.*, 1979b; Thommes *et al.*, 1983).

Environmental factors probably influence TSH release. For instance, cold evokes a rapid increase in TSH release as indicated by circulating concentrations of T_4 (Herbute *et al.*, 1984). However, prolonged exposure to low environmental temperatures do not consistently elevate plasma concentrations of T_4 (e.g., Cogburn and Freeman, 1987). Fasting results in elevated plasma concentrations of T_4 in young chickens (Decuypere *et al.*, 1988). In view of the possible effects of fasting on T_4 clearance/monodeiodination, it is difficult to conclude with confidence whether chronic exposure to cold or dietary restriction increases TSH release in birds. Physiological state (fasting) influences the change in plasma concentrations of T_3 (resulting from monodeiodination) following T_4 administration (Sharp and Klandorf, 1985).

D. Pituitary Origin of Thyrotropin

The thyrotrophs, the cells producing TSH, are located almost entirely in the cephalic (or rostral) lobe of the avian anterior pituitary gland (Figure 1). This is based on immunocytochemistry with antisera against mammalian TSH β-subunit (Sharp *et al.*, 1979). It is not surprising, based on the stimulatory control of TSH release by TRH, that TRH binding sites have been reported for membrane preparations from cephalic lobes of chicken pituitary gland (Harvey and Baidwan, 1989). The proportion of adenohypophyseal cells which are thyrotrophs as been determined to range from 2% at hatching to 6% in adult ring doves (Reichardt, 1993).

E. Ontogeny of Thyrotropin

Both TSH (pituitary) and TRH (hypothalamus) are detectible very early in the development of the chicken embryo (TSH at day 6.5, Thommes et al., 1983; TRH 4.5, Thommes et al., 1985). Obviously changes in cell number and TSH/TRH content occur during development. There is evidence that the hypothalamo–pituitary–thyroid axis is functional by midembryonic development. A marked increase in pituitary (to TRH) and/or thyroid (to TSH) responsiveness occurs between 10.5 and 12.5 days of embryonic development of the chicken, as evidenced by the change in plasma T_4 concentrations following TRH challenge (Thommes et al., 1984). A second phase is the maturation of the axis occurs at about the time of hatching when the machinery to 5' monodeiodinate T_4 to T_3 is established (e.g., Decuypere et al., 1988; McNichols and McNabb, 1988).

F. Role of Thyrotropin

In addition to stimulating avian thyroid growth (Robinson et al., 1976), TSH acts to increase the release of T_4 from the thyroid glands of birds. Injection of TSH is followed by increases in the circulating concentrations of T_3 and T_4 in chickens (e.g., MacKenzie, 1981; Williamson and Davison, 1985; McNichols and McNabb, 1988). In view of the extremely low level of T_3 in the avian thyroid gland (McNichols and McNabb, 1988) together with the ready conversion of T_4 to T_3 in vivo (Sharp and Klandorf, 1985) and in vitro with liver homogenates (Decuypere et al., 1988; McNichols and McNabb, 1988), it is likely that TSH does not directly increase T_3 release from the thyroid. Rather TSH stimulates T_4 release which is rapidly converted to T_3.

G. Mechanism of Thyrotropin Action

Based on analogy to the situation in mammals, TSH acts by binding to the TSH receptors which are linked by G_s proteins to adenylate cyclase. Activation of adenylate cyclase results in increased intracellular cAMP and activation of protein kinase A. In birds, TSH increases cAMP release from thyroid tissue in vitro (chicken; Tonoue and Kitch, 1978) and elevates the thyroid cAMP content in vivo (quail: McNichols and McNabb, 1988).

V. GROWTH HORMONE

A. Chemistry

Somatotropin GH has been isolated from avian pituitary tissue (e.g., Harvey and Scanes, 1977; Papkoff et al., 1982). Moreover, the cDNA for chicken GH has been sequenced (chickens; Lamb et al., 1988). Recombinant chicken GH has been produced (Souza et al., 1984; Burke et al., 1987) and as with pituitary-derived native avian GH preparation, this is biologically active in rat in vivo biological assays.

B. Growth Hormone Variants

Although, chicken GH is reported to have an isoelectric point of 7.5 and a molecular weight of approximately 23,000 (Harvey and Scanes, 1977), there is evidence for a series of charge and size variants of avian GH. Unlike the situation with human GH, where there are both multiple genes and where different splicing patterns of mRNA occur, GH variants in birds appear to be due entirely to posttranslation modification. There are reports of a glycosylated GH (Berghman et al., 1987), a phosphorylated GH (Aramburo et al., 1990, 1992), dimeric, and other oligomeric forms (Houston and Goddard, 1988) in the chicken pituitary gland.

C. Assay

With the availability of chicken and turkey GH (either native or biosynthetic), it has been possible to establish radioimmunoassays for avian GH (e.g., Harvey and Scanes, 1977). These systems appear very similar in terms of sensitivity, specificity, and precision and have been extensively employed to examine the physiology of GH secretion.

Radioreceptor assays for avian GH have also been established (e.g., Leung et al., 1984; Krishnan et al., 1989). In addition, effects of physiological state on chicken GH mRNA levels have been quantitated by Northern blot analysis (e.g., McCann-Levorse et al., 1993).

D. Actions of Growth Hormone

1. GH and Growth

GH is required for normal posthatching growth. Lack of avian GH resulting from hypophysectomy is followed by a dramatic decrease in the growth rate of chickens (e.g., King and Scanes, 1986). The effect of pituitary ablation on growth rate is partially overcome by replacement therapy with either mammalian or avian GH (King and Scanes, 1986; Scanes et al., 1986). In contrast chicken GH does not stimulate growth, more than at most transitorily, in young intact chicks (Leung et al., 1986; Burke et al., 1987; Cogburn et al., 1989). Stimulation of growth in midgrowth phase chickens is reported when GH is administered in pulsatile manner (Vasilatos-Younken et al., 1988).

GH is thought to exert its effect on somatic growth by increasing circulating concentrations of insulinlike growth factor-I (IGF-I) via stimulation of IGF-I production from the liver. Plasma concentrations of IGF-I in chicks are reduced following hypophysectomy (Huybrechts *et al.*, 1985) but increased by chronic GH treatment in either hypophysectomized chicks (Scanes *et al.*, 1986) or in young pullets receiving pulsatile GH administrations and in which growth is enhanced (Vasilatos-Younken *et al.*, 1988). *In vitro*, GH increases IGF-I release from chicken hepatocytes (Houston and O'Neill, 1991).

2. GH and Lipid Metabolism

GH exerts short-term effects on metabolism in birds, stimulating lipolysis, exerting an insulinlike/antilipolytic effect in inhibiting glucagon-stimulated lipolysis, and also reducing lipogenesis. Administration of GH can influence adiposity of chickens with the direction of the effect affected by the mode of administration. For instance, body fat is increased by daily injection (Cogburn *et al.*, 1989) but decreased by pulsatile administration of GH (Vasilatos-Younken *et al.*, 1988).

a. Lipolytic Effect of GH

Chicken GH stimulates lipolysis *in vitro* by adipose explants of chickens, turkeys, and hypophysectomized pigeons (Harvey *et al.*, 1977). Moreover, this lipolytic effect of increased glycerol release is observed with either native or recombinant GH and of either mammalian or avian origins (Campbell and Scanes, 1985; Campbell *et al.*, 1990). Further evidence for the specificity of this *in vitro* lipolytic effect comes from ability of a mutant form of mammalian GH with GH antagonist activity to block the lipolytic effect of GH on chicken adipose tissue explants (Campbell *et al.*, 1993). Moreover *in vivo*, GH can increase circulating concentrations of free fatty acids (FFA) acutely in conscious or anesthetized chickens (Hall *et al.*, 1987; Scanes, 1992). Similarly, the administration of mammalian GH chronically elevates circulating concentrations of FFA in hypophysectomized ducks (Foltzer and Mialhe, 1976).

b. Insulinlike/Antilipolytic Effect of GH

There is evidence that GH exerts an effect on avian adipose like that observed with insulin in mammals. *In vitro*, GH reduces the release of glycerol from adipose tissue explants from adult chicks observed in the presence of either glucagon or cAMP (Campbell and Scanes, 1987; Campbell *et al.*, 1990).

c. GH and Lipogenesis

An inhibitory effect of GH on lipogenesis in birds is a distinct possibility. Bovine or chicken GH prevented insulin-stimulated synthesis of fatty acids from ^{14}C-labeled acetate by chick hepatocytes *in vitro* (Harvey *et al.*, 1977). Pulsatile administration of GH to young pullets *in vivo* also reduced hepatic lipogenesis (Rosebrough *et al.*, 1991).

3. GH and Thyroid Hormones

It is not established whether GH stimulates T_4 release from the avian thyroid. Circulating concentrations of T_4 in chickens have been reported to be elevated (MacKenzie, 1981) or decreased (Kuhn, 1985) following GH administration. In young chicks, GH increases circulating concentrations of T_3 (e.g. Kuhn *et al.*, 1985). It may be due to GH inducing hepatic 5'-monodeiodinase which metabolizes T_4 to T_3 (e.g., Darras *et al.*, 1990) and/or reducing the catabolic activity of hepatic type III monodeiodinase which metabolizes T_3 to T_2 (Darras *et al.*, 1992). This latter appears to be a physiological effect of GH as type III monodeiodinase is elevated following hypophysectomy and partially restored to normal by GH replacement therapy (Darras *et al.*, 1993).

4. GH and Adrenocortical Hormones

There is good evidence that GH can influence the avian adrenal cortex. Plasma concentrations of corticosterone are increased acutely following the administration of recombinant chicken GH (Cheung *et al.*, 1988) or chronically following the continuous infusion of native chicken GH (Rosebrough *et al.*, 1991). There are also effects of hypophysectomy, in the presence or absence of GH replacement therapy, on corticosteroidogenesis *in vitro* (Carsia *et al.*, 1985c).

5. GH and Immune Functioning

GH is thought to exert stimulatory effects on the avian immune system. For instance, GH administration increases thymus weights in hypophysectomized chicks (King and Scanes, 1986; Scanes *et al.*, 1986), bursal weight if T_3 is also present in hypophysectomized chicks (Scanes *et al.*, 1986a) and influences peripheral white blood cells (Johnson *et al.*, 1993).

E. GH Receptor

With the sequencing of chicken GH receptor (GHR) cDNA (Burnside *et al.*, 1991), it is possible to predict the structure of the receptor itself. As with the mammalian GHR, the chicken GHR consists of three domains: extracellular, a single transmembrane domain, and an intracellular domain. While the GHR is not a protein kinase, it is likely that GH evokes its effects by protein kinase(s) associated with the GH receptor. Sex-linked dwarf chickens have mutations such that the GHR is not expressed normally (Burnside *et al.*, 1992). In chickens, GH "down-regulates" the GHR; with hypophysectomy increasing and GH administration decreasing ^{125}I-GH binding to chicken liver membranes (Vanderpooten *et al.*, 1991a). It is not surprising, therefore, that both hepatic ^{125}I-GH binding (chicken; Vanderpooten *et al.*, 1991b) and GHR mRNA expression (Burnside and Cogburn, 1992) are low in young birds when plasma concentrations of GH are high.

F. Control of GH Secretion

The release of GH from the avian pituitary gland is under hypothalamic control. There are at least three hypophysiotropic factors in birds, of which two are stimulatory. There are (1) TRH, (2) a GH releasing factor (GHRF), and (3) somatostatin, which inhibits GH secretion.

1. Thyrotropin Releasing Hormone

A major physiological stimulus for GH release in birds is the tripeptide TRH. This is found in the avian hypothalamus (see above) while TRH receptors are found in the cells of caudal lobe of chicken pituitary gland (Harvey and Baidwan, 1989) where GH is produced. In the domestic fowl, TRH stimulates GH secretion from the pituitary gland, both *in vivo* and *in vitro* (Harvey *et al.*, 1978b; Scanes and Harvey, 1984). The *in vivo* effect of TRH is most apparent in the young chick, but is also present in the anesthetized adult chicken (Scanes and Harvey, 1984). In addition, TRH potentiates GHRF induced GH release *in vivo* (Harvey and Scanes, 1985) and *in vitro* (Perez *et al.*, 1987).

Marked increases in GH release with TRH are observed with chicken pituitary cells perfused *in vitro* (Scanes *et al.*, 1986b) with a smaller effect in static incubations (Harvey *et al.*, 1978b; Perez *et al.*, 1987) perhaps due to TRH down-regulating its receptors. Evidence that TRH is, in fact, a physiological regulator of GH secretion comes from the ability of antiserum to TRH to suppress circulating concentrations of GH in young chicks (Klandorf *et al.*, 1985).

2. Growth Hormone Releasing Factor

As yet, avian GHRF has not been isolated. However, mammalian GHRF has been found to increase GH release in chickens *in vivo* and *in vitro* (e.g., Scanes *et al.*, 1984). This GHRF effect is not due to the TRH receptor. Evidence for this comes from the following: (1) the GHRF effect is potentiated by TRH *in vitro* (Perez *et al.*, 1987) and (2) that repeated GHRF injections becomes progressively less effective but if challenged with TRH, a large increase in GH is observed (Scanes and Harvey, 1985). In addition, GHRF increases GH mRNA *in vivo* (Vasilatos-Younken *et al.*, 1992) and *in vitro* (Radecki *et al.*, 1994).

3. Somatostatin (SRIF)

The structure of avian $SRIF_{14}$ is identical to the mammalian homolog H-Ala-Gly-Cys-Lys-Asn-Phe-Phe-Trp-Lys-Thr-Phe-Thr-Ser-Cys-OH (pigeon, Spiess *et al.*, 1979; chicken, Hasegawa *et al.*, 1984).

Somatostatin plays a inhibitory role in the hypothalamic control of GH release in birds. Not only are there somatostatin neurons present in the hypothalamus of the duck (Blahser *et al.*, 1978), but also SRIF binding sites have been demonstrated in the chicken anterior pituitary gland (Harvey *et al.*, 1990). Synthetic SRIF depresses both the *in vivo* and *in vitro* GH response to TRH in the domestic fowl (e.g., Harvey *et al.*, 1978b; Scanes and Harvey, 1989a).

4. Hypothalamic Control of the Release of Hypophysiotropic Peptides

Pharmacological studies indicate that (nor-)adrenergic neurons may be playing a stimulatory role in maintaining endogenous GH release (Buonomo *et al.*, 1984). This may be envisaged as GHRF release being under (nor-)adrenergic control.

5. Other Hormones

Growth hormone secretion in birds is inhibited by peripheral hormones. If GH is acting via IGF-I, then negative feedback by IGF-I on GH release would seem to be likely. IGF-I has been found to inhibit GH release *in vitro* and *in vivo* in chicks (Perez *et al.*, 1985; Buonomo *et al.*, 1987). Similarly as GH increases the circulating concentration of the active thyroid hormone, T_3, then it is probable that T_3 would inhibit GH secretion. This is the case. The presence of elevated levels of T_3 reduces GH release from the chicken pituitary gland *in vivo* (e.g. Scanes and Harvey, 1987b) and *in vitro* (Donoghue and Scanes, 1991).

6. Nutrition and GH Secretion

Nutritional deprivation increases the plasma concentration of GH in birds. Plasma concentrations of GH are elevated by a short period (1–2 days) of starvation in chickens (Harvey *et al.*, 1978a) and turkeys (Anthony *et al.*, 1990); the latter reflecting an increase in the basal GH level in the plasma and not changes in pulsatile GH release (amplitude or frequency). Plasma concentrations of GH are also increased by calorie-protein deprivation in chronically feed restricted young chickens and turkeys (Engster *et al.*, 1979). Similarly, long-term protein deficiency is accompanied by elevated plasma concentrations of GH in young chickens (Scanes *et al.*, 1981).

7. Ontogeny of GH Secretion

A common pattern of circulating concentrations of GH has been observed in all species of birds examined. This consists of progressively rising circulating concentrations of GH in late embryonic development and early posthatch development, high plasma concentrations of GH during the period of rapid posthatching growth, and low GH concentrations are seen in older and adult birds (e.g., Harvey *et al.*, 1979a;b). The mechanistic bases for this ontogenic profile includes changes in GH synthesis, somatotroph numbers (see below), and in pituitary sensitivity to secretagogues. The decline in plasma concentrations of GH in the middevelopment/growth does not appear to be due to gonadal steroid(s) as castration does not prevent the decrease (Scanes and Johnson, 1984). The ontogenic profile for GH mRNA is very similar to that of plasma concentrations of GH (McCann-Levorse *et al.*, 1993).

G. Pituitary Origin of Growth Hormone

The somatotrophs are located in the caudal lobe of the anterior pituitary gland (Mikami, 1980; Reichardt *et al.*, 1993) (Figure 1). During late embryological development and early posthatch life, the number of somatotrophs increases rapidly until a plateau level is achieved (S. Malamed and C.G. Scanes, unpublished observations; Reichardt *et al.*, 1993). There are also ontogenic changes in the structure of the somatotroph with increases in secretory granule number during late embryonic development (Malamed *et al.*, 1993) and a further increase between young and adult chickens (Malamed *et al.*, 1985, 1988).

VI. PROLACTIN

A. Chemistry

Avian prolactin has been isolated from pituitary tissue (e.g. Papkoff *et al.*, 1982). In addition, recombinant chicken prolactin has been purified following expression in *Escherichia coli* (Hanks *et al.*, 1989a). In addition, the nucleotide sequence of avian prolactin cDNA has been established and the amino acid residue sequence deduced (chicken: Hank *et al.*, 1989b; Watahiki *et al.*, 1989).

As in mammals, there is evidence that posttranslationally modified variants of prolactin exist in birds. In the turkey, two glycosylated prolactin variants (one O- and one N-glycosylated) have been identified (with reduced radioreceptor activity) in addition to a nonglycosylated form (Corcoran and Proudman, 1991). Similarly, a glycosylated prolactin is reported in the chicken (Berghman *et al.*, 1992). There is also evidence that avian prolactin or some variants are phosphorylated (Aramburo *et al.*, 1992).

B. Pituitary Origin of Prolactin

The prolactin-producing cells (lactotrophs) are largely restricted to the cephalic or rostral lobe of the anterior pituitary gland (Figure 1) (Burke and Papkoff, 1980; Mikami, 1980). The relative proportion of lactotrophs changes with physiological state. For instance, the proportion drops from 12% in newly hatched pigeons to less than 2% in adult pigeons (Reichardt, 1993).

C. Assay

Avian prolactin levels have been determined by biological assay, radioimmunoassay, and radioreceptor assay, while prolactin mRNA has been quantitated from Northern blotting. The response of the pigeon crop sac to prolactin (Nicholl, 1967) has been widely employed to determine prolactin biological activity. The precision of the assay can be improved by using the ratio of the concentrations of two crop proteins; one induced by prolactin and one decreased by prolactin (Lebovic and Nicoll, 1992).

Homologous radioimmunoassays have been developed for avian prolactin (e.g., Scanes *et al.*, 1976; Proudman and Opel, 1981). In addition, a heterologous prolactin (McNeilly *et al.*, 1978) has been widely employed to measure prolactin in a number of avian species including starlings (Goldsmith and Nicholls, 1984b) and doves (Goldsmith *et al.*, 1987).

Avian prolactin radioreceptor assays have been developed using pigeon crop sac gland membranes (Kledyck *et al.*, 1975) or kidney membrane preparations from either chickens or turkeys (Krishnan *et al.*, 1991). An *in vitro* bioassay for prolactin proliferation of rat Nb2 lymphoma cells has been employed to measure avian prolactin; the circulating concentrations of prolactin in incubating turkeys being increased by 2.8-fold as determined by bioassay and 2.4-fold by radioimmunoassay (Soarses and Proudman, 1991). Changes in prolactin mRNA expression with reproductive state have been reported which parallel those seen with plasma concentrations of prolactin (Talbot *et al.*, 1991; Wong *et al.*, 1991).

D. Actions of Prolactin

Prolactin stimulates the production of "crop milk" and the proliferation of mucosal cells of the crop sac gland in pigeons and doves. This unique feature of the *Columbidae* (pigeons and doves) has been used as the basis of the biologic assay for prolactin. During the latter part of the period when both male and female doves are incubating eggs, the crop increases in weight, and there is a concomitant increase in the plasma prolactin concentrations (Goldsmith *et al.*, 1981). It is also possible that prolactin also directly influences behavior in pigeons and doves as discrete prolactin binding sites (receptors) have been reported in dove brains (Buntin *et al.*, 1993).

The effects of prolactin on the crop sac gland are thought to be enhanced by synlactin which is a prolactin induced liver factor (Nicoll *et al.*, 1985). The identify of synlactin is not established but it may be related to an IGF or IGF-binding protein. It might be noted that relaxin, a peptide which has homologies to insulin, IGF-I, and IGF-II, can stimulate the pigeon crop sac (Bani *et al.*, 1990).

In birds, prolactin is involved in incubation behavior and broodiness (reviewed e.g., El Halawani and Rozenboim, 1993). Prolactin can act to reduce circulating concentrations of gonadotropins and to induce ovarian regression in turkeys (reviewed El Halawani *et al.*, 1988). In turkeys, at least, prolactin appears to act centrally to increase nesting activity and together with sex steroids to induce incubating behavior (El Halawani *et al.*, 1986, 1993; Youngren *et al.*, 1991). Similarly in bantam chickens, prolactin increases incubation behavior as indicated by the increase in nesting behavior in previously nest-deprived birds (Sharp *et al.*, 1988). A positive feedback loop between prolactin and incubation can be envisaged below.

Circulating concentrations of prolactin are elevated during the period when the birds are incubating their eggs (reviewed El Halawani *et al.*, 1988, 1993; Sharp *et al.*, 1988). Furthermore, the plasma concentration of prolactin falls rapidly if incubation behavior or nesting is interrupted (reviewed El Halawani *et al.*, 1983).

Prolactin may also be involved in photorefractoriness (i.e., the lack of ability to respond (or continue to respond to) a stimulating photoperiod). Plasma concentrations of prolactin rise gradually in starlings transferred to long photoperiods with the peak temporally linked to the onset of photorefractoriness (Goldsmith and Nicholls, 1984b). Moreover, if the rise in prolactin can be abolished so is photorefractoriness. For instance, neither the increase in prolactin not photorefractoriness is not observed following thyroidectomy (Goldsmith and Nicholls, 1984c). In a complimentary manner, thyroxine administration can induce photorefractoriness and enhance prolactin secretion in starlings (Goldsmith and Nicholls, 1984c).

Prolactin may also have an osmoregulatory role in birds. The evidence for this comes from studies in which chick embryos are treated with mammalian prolactin. Allantoic concentrations of both sodium and chloride ions are depressed by ovine or bovine prolactin while metanephric kidney Na^+–K^+-ATPase activity is increased (Doneen and Smith, 1982; Murphy *et al.*, 1986).

E. Receptor and Mechanism of Action

The amino acid sequence of the avian prolactin receptor has been established based on the nucleotide sequence of prolactin receptor cDNA in the chicken (Tanaka *et al.*, 1992) and pigeon (Chan and Horseman, 1994). The avian receptor, like its mammalian homolog consists of extracellular, transmembrane, and intracellular domains. However, the avian prolactin receptor has a unique feature with a double-antenna structure of two putative prolactin binding sites in the extracellular domain. Although the prolactin receptor does not appear to be a protein kinase, it is likely that a receptor-associated protein kinase(s) is a signal transducers which activates enzyme and induces specific genes. For instance, prolactin evokes marked changes in the level of

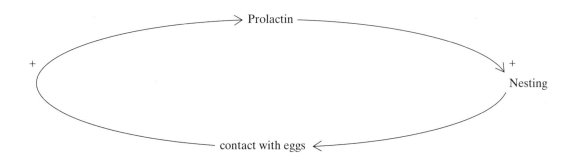

(and also synthesis of) of specific proteins in the crop epithelium. Prolactin increases the levels of proteins with molecular weights of 25 and 154 kDa and decreases the levels of proteins with weights of 16 and 30 kDa (Pukac and Horseman, 1984; Lebovic and Nicoll, 1992). Moreover, prolactin has been found to induce a 70-fold increase on the expression of specific mRNA a 35.5-kDa protein (Pukac and Horseman, 1986).

F. Control of Prolactin Release

Prolactin secretion is under predominantly stimulatory hypothalamic control, although some inhibitory influences are also probable.

1. Vasoactive Intestinal Peptide (VIP)

The major hypothalamo-hypophysiotropic factor stimulating prolactin release in vasoactive intestinal peptide (VIP). The amino acid sequence of chicken VIP is HSDAVFTDNYSRFRKQMAVKKLYNSVLT (from Nilsson, 1975; see Figure 2 for key).

Prolactin release is stimulated by VIP in chickens (*in vivo*, Macnamee et al., 1986) and turkeys (*in vivo*, Opel and Proudman, 1988; *in vitro*, Proudman and Opel, 1988; El Halawani et al., 1990). Moreover, VIP has been identified in the hypothalamus, particularly in the median eminence (Macnamee et al., 1986). Immunoneutralization of VIP decreases circulating concentrations of prolactin in chicken (Sharp et al., 1989) and abolishes the prolactin-releasing activity of turkey hypothalamic extracts (El Halawani et al., 1990). It might be noted that the hypothalamic VIP content varies with physiological state in a manner which is consistent with VIP being the prolactin-releasing factor (Mauro et al., 1989, 1992; Rozenboim et al., 1993). For instance, hypothalamic content of VIP has been found to be elevated in photostimulated compared to nonphotostimulated turkeys and further increased in incubating hens (Rozenboim et al., 1993).

2. Thyrotropin Releasing Hormone (TRH)

Another putative prolactin-releasing factor in birds is TRH. Synthetic TRH increases prolactin tissue *in vitro* (Hall et al., 1975; Harvey et al., 1978) and in the turkey *in vivo* (Saeed and El Halawini, 1986). However, the TRH content of avian hypothalamic extracts does not account for a major proportion of the hypothalamic prolactin-releasing activity.

3. Arginine Vasotocin (AVT)

Posterior pituitary extracts can provoke prolactin release from turkey pituitary cells; this being due to both AVT and VIP. Not only can AVT mimic this effect and also act as an antisera to either AVT or VIP (but not to MT), but it can also partially immunoneutralize the posterior pituitary prolactin releasing principle(s) (El Halawani et al., 1992).

4. Other Stimulatory Factors

Other peptides also stimulate prolactin secretion in birds. For instance, the peptide histidine isoleucine (PHI) can stimulate prolactin release in turkeys (Proudman and Opel, 1990).

5. Dopamine

Dopamine can inhibit prolactin secretion in birds (reviewed Hall et al., 1986). Further evidence that dopamine exerts a physiological effect on prolactin secretion in birds comes from the observation that there are dopamine receptors on cephalic lobe of chicken anterior pituitary gland and that dopamine binding sites decrease in broody bantam chickens (MacNamee and Sharp, 1989).

VII. ADRENOCORTICOTROPIC HORMONE

A. Chemistry

Adrenococorticotropic hormone (ACTH) is synthesized as part of a large protein (proopiomelanocortin). Proopiomelanocortin also contains the sequences of β-endorphin (β-EP), which in itself is part of β-lipotropin (β-LPH), together with α- and β-melanophore-stimulating hormone (MSH).

ACTH, β-EP, and β-LPH have been purified from adenohypophyseal tissue from several avian species (see Figure 2). Avian ACTH, like its mammalian homolog, is a simple polypeptide containing 39 amino acids (Naude and Oelofson, 1977; Li et al., 1978; Chang et al., 1980, Hayashi et al., 1991). In the mammalian pars intermedia, proteolytic cleavage of proopiomelanocortin results in the production of αMSH. Birds lack a pars intermedia and there is little production of αMSH. However, the presence of some αMSH has been reported in the avian pars distalis (Hayashi et al., 1991). β-LPH and β-EP have been purified from ostrich and turkey adenohypophyseal tissue (Chang et al., 1980; Naude et al., 1980; 1981a,b). As in mammals, the 31-amino-acid sequence of ostrich β-EP is identical to the C-terminal of ostrich β-LPH (see Figure 2).

B. Pituitary Origin of ACTH

ACTH is produced by acidophilic cells (corticotrophs) in the cephalic lobe of the anterior pituitary gland (Hayashi et al., 1991) (Figure 1). ACTH and

ACTH (α MSH sequence italics)

```
                          10             20
     Ostrich     SYSMEHFRWGKPVGRKRRPV
     Turkey      SYSMEHFRWGKPVGRRKRPI
     Chicken     SYSMEHFRWGKPVGRKRRPI
                 21       30        39
     Ostrich     KVYPNGVQEETSEGFPLEF        Li et al. (1978)
     Turkey      KVYPNGSVDEEQASTPVEF        Chang et al. (1980);
                                            Yamashiro et al. (1984)
     Chicken     KVYPNGVDEESAESYPMEF        Hayashi et al. (1991)
```

βLPH [βEP sequence (Naude et al. (1981b) as italics]

```
                       10             20
     Ostrich     ALPPAAMLPAAAEEEEGEEE       Naude et al. (1981a)
                       30          40
     Ostrich     EEGEAEKEDGGSYRMRHFRW
                       50            60
     Ostrich     QAPLKDKRYGGFMSSERGRA
     Turkey               YGGFMTSEHSQM      Chang et al. (1980)
     Ostrich     PLVTLFKNAIVKSAYKKGO
     Turkey*     PLLTLFKNAIVKSAYKKGO
```

FIGURE 2 Structure of avian ACTH and βLPH. Abbreviations: A, Ala; C, Cys; D, Asp; E, Glu; F, Phe; G, Gly; H, His; I, Ile; K, Lys; L, Leu; M, Met; N, Asn; P, Pro; Q, Gln; R, Arg; S, Ser; T, Thr; V, Val; W, Trp; and Y, Tyr. An asterisk refers to the sequence of turkey βEP only.

αMSH coexist in the same cells (Iturriza et al., 1980; Hayashi et al., 1991). The chicken anterior pituitary gland has a high content of ACTH (1.6 μg) but low levels of αMSH (10 ng) (Hayashi et al., 1991). The corticotrophs are first observed on day 7 in embryonic development in the domestic fowl (Jozsa et al., 1979).

C. Control of ACTH Release

The secretion of ACTH is under the control of hypothalamic–hypophysiotropic factors and glucocorticoids. The releasing factor(s) is released from the median eminence and travels to the anterior pituitary gland via the portal blood vessels (Figure 1). Avian corticotrophin-releasing factor (CRF) has not yet been characterized. However, an IR-CRF has been observed in the avian hypothalamus, particularly the median eminence (Jozsa et al., 1984). Moreover, mammalian CRF (a 44-amino-acid residue peptide) stimulates ACTH release in vitro from chicken and duck pituitary cells (Carsia et al., 1986; Castro et al., 1986). Both posterior pituitary hormones, arginine vasotocin, and mesotocin also stimulate ACTH release and also potentiate the effect of CRF in elevating ACTH release (Castro et al., 1986).

Glucocorticoids inhibit ACTH release in vivo (Herold et al., 1992) and in vitro (Carsia et al., 1986). It might be noted that protein-deprived chicks which have elevated circulating concentrations of corticosterone also have reduced circulating concentrations of ACTH (Carsia et al., 1988) presumably due to increased glucocorticoid negative feedback.

D. Actions of ACTH

ACTH acts to stimulate avian adrenal cortical cells to produce corticosterone (the major glucocorticoid), aldosterone (the major mineralocorticoid), and deoxycorticosterone (e.g., Carsia et al., 1985a,b; Collie et al., 1992). The effect of ACTH is mediated by cAMP (Carsia et al., 1985a). It might be noted that the adrenal response to ACTH varies with physiological state (Carsia et al., 1985b). For instance, the sensitivity to ACTH declines during growth. In addition ACTH evokes marked changes in the morphology of adrenocortical cells due to changes in the cytoskeleton (Cronshaw et al., 1992). Central nervous and direct effects of ACTH on avian behavior have been reported (Pamksepp and Normansell, 1990).

Despite the lack of substantial production of α-MSH, there is evidence that, in birds, α-MSH can exert biological effects distinct from those evoked by ACTH. For instance, α-MSH but not ACTH stimulates sodium excretion from the nasal salt glands of ducks (Ituzziza et al., 1992).

VIII. OTHER ADENOHYPOPHYSEAL PEPTIDES

Chromogranin A is an acidic peptide which can be secreted with protein hormones. A peptide with close (~80%) homologies to mammalian chromogranin A has been purified and sequenced for ostrich adenohypophyseal tissue (Lazure et al., 1990). While its function is not known, fragments of chromogranin A may have biological effects.

IX. NEUROHYPOPHYSIS

A. Introduction

The anatomy of the hypothalamic neurosecretory tracts (neurons) leading to their secretory terminals in the pars nervosa is shown in Figure 1. It is apparent that the two hormones of the neurohypophysis, arginine vasotocin (AVT) and mesotocin (MT), are produced by and secreted from separate neurosecretory neurons. However, the AVT and MT cell bodies are located in both separate and overlapping areas within the hypothalamus; the hormones of the avian pars nervosa being synthesized in the hypothalamic cell bodies (Goosens et al., 1977; Bons, 1980; Tennyson et al., 1985). The embryonic development of the AVT- and MT-containing cell bodies and axons occurs during embryonic development between days 6–17 in the chicken (Tennyson et al., 1986).

AVT and MT are transported bound to carrier proteins (neurophysins) by axoplasmic transport. The hormones are then stored in pars nervosa prior to release. Two avian neurophysins have been characterized and sequenced in the ostrich (Lazure et al., 1987, 1989).

In view of the similarities of their chemical structures and also, (to some extent) of their actions, the hormones of the pars nervosa will be considered together.

X. ARGININE VASOTOCIN AND MESOTOCIN

A. Chemistry

The hormones of the avian neurohypophysis have been characterized. The avian antidiuretic hormone is AVT and that the oxytocic principle is mesotocin (Acher et al., 1970). arginine vasotocin (8-arginine oxytocin): CyS-Tyr-Ile-Glu(NH_2)-Asp(NH_2)-CyS-Pro-Arg-Gly(NH_2); mesotocin (8-isoleucine oxytocin): CyS-Tyr-Ile-Glu(NH_2)-Asp(NH_2)-CyS-Pro-Ile-Gly(NH_2).

AVT differs from arginine vasopressin, the mammalian antidiuretic hormone, by one amino acid residue (isoleucine at position 3 in place of phenylalanine), while MT differs from the mammalian homolog, oxytocin, by the substitution of isoleucine for leucine (position 8).

B. Content

The posterior pituitary gland contains high levels of AVT and MT (chicken 4.0 μg AVT; 0.9 μg MT) (Robinzon et al., 1988). In addition, significant levels of these neuropeptides, albeit <0.5% that of the neurohypophysis, are found in other body tissues including throughout the brain (Robinzon et al., 1988b) and in the ovary (Saito et al., 1990).

C. Actions of AVT

The major roles of AVT are on renal functioning and reproduction.

1. Renal Functioning

There is no doubt that AVT is the major antidiuretic hormone in birds. The absence of AVT, following either surgical removal of the pars nervosa or in AVT-deficient chicken, is accompanied by very large increases in the volume of urine produced (Shirley and Nalbandov, 1956; Dunson et al., 1972). The administration of AVT to birds has an antidiuretic effect. In hydrated (hypotonic glucose solution-infused) chickens, AVT depresses urine flow while increasing urine osmolality (Ames et al., 1971). At low doses, AVT exerts its principal effect on tubule function, reducing renal free water clearance. At higher concentrations, AVT reduces the glomerular filtration rate (GFR) (Gerstberger et al., 1985; Stallone and Braun, 1985). Based on studies in the desert quail, AVT reduces GFR by decreasing both the number of reptilian-type nephrons filtering and the single-nephron GFR of mammalian-type nephrons (Braun and Dantzler, 1974). Marked but contradictory cardiovascular effects of AVT have been reported. For instance, AVT can have both vasodepressor and vasopressor activity in birds, perhaps depending on how it is administered. Bolus injections of AVT to conscious adult chickens or either adult or young ducks results in a marked drop in mean arterial pressure (Wilson and West, 1986; Robinzon et al., 1988a). In contrast, infusion of AVT can increase arterial blood pressure in conscious chickens (Robinzon et al., 1993). In an analogous manner, bolus injection of AVT has been reported to elevate heart rate while infusion of AVT depressed cardiac frequency (Wilson and West, 1986; Robinzon et al., 1988a, 1993).

2. Oviposition

Oviposition, the physical process of laying eggs in birds, is controlled by AVT. The hen uterus contracts in response to AVT but not to MT (e.g., Koike et al., 1988). The mechanism by which AVT provoked premature oviposition may involve AVT stimulation of local production of prostaglandins (probably E_1) by the uterus which, in turn, causes uterine contractions (Rzasa, 1978, 1984).

At the time of oviposition, circulating concentrations of AVT are greatly increased, as indicated by bioassay

TABLE 2 Factors Affecting Circulating Concentrations of AVT and MT in Birds

Bird	AVT	MT	Reference
Chickens			
Oviposition	↑↑	→	Nouven et al. (1984); Koike et al. (1988)
Water deprivation	↑	↑	Koike et al. (1977); Nouven et al. (1984)
Hemorrhage (anesthesized)	↑	↓	Bottje et al. (1989)
Anesthesia (pentobarbitone)	↓	↑	Bottje et al. (1989)
NaCl infusion	↑		Koike et al. (1979)
NaCl injection	↑		Klempt et al. (1992)
Chronic high NaCl	↑		Arnason et al. (1986)
Angiotensin II	↑		Goto et al. (1986)
Ducks			
Atrial naturetic factor	→		Gray et al. (1991)
Water deprivation	↑		Gray and Simon (1987)
Chronic high salt (NaCl) intake	↑		Gray and Simon (1987)
Hypotonic NaCl infusion	↑		Simon-Oppermann et al. (1980)
Cerebral hypothalamic NaCl	↑		Simon-Oppermann et al. (1980)
Hypothalamic cooling	↑		Simon-Oppermann et al. (1980)
Hemorrhage	↑		Simon-Oppermann et al. (1984)
Angiotensin II	→		Gray et al. (1986)
Sparrows			
Water deprivation	↑		Goldstein and Braun (1988)

(Sturkie and Lin, 1966) or radioimmunoassay (Nouwen et al., 1984; Koike et al., 1988). No changes in MT level are observed, however (Nouwen et al., 1984; Koike et al., 1988). The pars nervosa is the likely source of AVT at the time of oviposition but ovarian AVT may contribute significantly; follicular AVT levels decline immediately prior to oviposition (Saito et al., 1990). Not only do circulating concentrations of AVT rise markedly at oviposition but also the sensitivity of uterus to AVT is increased (Rzasa, 1978).

3. Other Roles

Neurohypophyseal peptides may control secretion of other hormones in birds. For instance, AVT stimulates prolactin release *in vitro* (El Halawani et al., 1992) and ACTH secretion (Castro et al., 1986).

D. Actions of MT

There is relatively little information on the physiological role of MT in birds. Unlike its mammalian homolog, MT does not appear to affect the uterus (e.g., Koike et al., 1988). Infusion of MT to chickens does not influence the major cardiovascular indices of arterial blood pressure or heart rate. MT does influence blood flow to some organs (as indicated by reduced temperature of the shank and comb) (Robinzon et al., 1988). In addition, MT infusion reduces circulating concentrations of aldosterone (Robinzon et al., 1988).

E. Control of AVT and MT Release

Table 2 summarizes the effects of factors known to influence circulating concentrations of AVT and MT. Despite the relative paucity of information on circulating concentrations of MT, the available evidence suggests that the release of AVT and MT are under independent control.

AVT is released when blood osmolality is high in either chicken or ducks (see Table 2) and consequently water loss should be prevented. Frequently, conditions where blood osmolality is increased also reduce blood volume. The release of AVT is under osmotic rather than volemic control (e.g., Stallone and Braun, 1986).

F. Clearance of AVT

A bolus injection of AVT is followed by a monophasic exponential decline in circulating concentrations of AVT with a half-life of 6.3 min in chickens (Arad et al., 1986) and 7 min in ducks (Simon-Oppermann and Simon, 1982). Based on infusion studies the plasma metabolic clearance rate for AVT in ducks is 25 ml · min^{-1} kg^{-1} (Gray and Simon, 1983).

References

Acher, R., Chauvet, J., and Chauvet, M. T. (1970). Phylogeny of the neurohypophyseal hormones: The avian active peptides. *Eur. J. Biochem.* **17**, 509–513.

Almeida, O. F. X., and Thomas, D. H . (1981). Effects of feeding pattern on the pituitary thyroid axis in the Japanese quail. *Gen. Comp. Endocrinol.* **44,** 508–513.

Ames, E., Steven, K., and Skadhauge, E. (1971). Effect of arginine vasotocin on renal excretion of Na^+, K^+, Cl^-, and urea in the hydrated chicken. *Am. J. Physiol.* **221,** 1223–1228.

Anthony, N. B., Vasilatos-Younken, R., Bacon, W. L., and Lilburn, M. S . (1990). Secretory pattern of growth hormone, insulin and related metabolites in growing male turkeys: Effect of overnight fasting and refeeding. *Poult. Sci.* **69,** 801–811.

Arad, Z., Chadwick, A., Rice, G. E., and Skadhauge. E. (1986). Osmotic stimuli and NaCl-intake in the fowl: Release of arginine vasotocin and prolactin. *J. Comp. Physiol. B* **156,** 399–406.

Aramburo, C., Donoghue, D., Montiel, J. L., Berghman, L. R., and Scanes, C. G. (1990). Phosphorylation of chicken growth hormone. *Life Sci.* **47,** 947–952.

Aramburo, C., Montiel, J. L. Proudman, J. A., Berghman, L. R., and Scanes, C. G. (1992). Phosphorylation of prolactin and growth hormone. *J. Mol. Endocrinol.* **8,** 183–191.

Arnason, S. S., Rice, G. F., Chadwick, A., and Skadhauge, E. (1986). Plasma levels of arginine vasotocin, prolactin, aldosterone and corticosterone during prolonged dehydration in the domestic fowl: Reflect of dietary NaCl. *J. Comp. Physiol. B* **156,** 383–397.

Asem, E. K., Molnar, M., and Hertelendy, F. (1987). Luteinizing hormone-induced intracellular calcium mobilization in granulosa cells: Comparison with forskolin and 8-bromo-adenosine $3',5'$ monophosphate. *Endocrinology* **120,** 853–859.

Bani, G., Sacchi, T. B., and Bigazzi, M. (1990). Response of the pigeon crop sac to mammotrophic hormones: Comparison between relaxin and prolactin. *Gen. Comp. Endocrinol.* **80,** 16–23.

Berghman, L. R., Darras, V. M., Chiasson, R. B., Decuypere, E., Kuhn, E. R., Buyse, J., and Vandesande, F. (1993). Immunocytochemical demonstration of chicken hypophyseal thyrotropes and development of a radioimmunological indicator for chicken TSH using monoclonal and polyclonal hemologous antibodies in a subtractive strategy. *Gen. Comp. Endocrinol.* **92,** 189–200.

Berghman, L. R., Grauwels, L., Vanhamme, L., Proudman, J. A., Foidart, A., Balthazart, J., and Vandesande, F. (1992). Immunocytochemistry and immunoblotting of avian prolactins using polyclonal and monoclonal antibodies toward a synthetic fragment of chicken prolactin. *Gen. Comp. Endocrinol.* **85,** 346–357.

Berghman, L. R., Lens, P., Decuypere, E., Kuhn, E. R., and Vandesande, F. (1987). Glycosylated chicken growth hormone. *Gen. Comp. Endocrinol.* **68,** 408–414.

Blahser, S., Fellman, D., and Bugnon, C. (1978). Immunocytochemical demonstration of somatostatin-containing neurons in the hypothalamus of the domestic mallard. *Cell Tissue Res.* **195,** 183–187.

Bonney, R. C., and Cunningham, F. J. (1977). A role for cyclic AMP as a mediator for the action of LH-RH on chicken anterior pituitary cells. *Mol. Cell. Endocrinol.* **7,** 233–244.

Bons, N. (1980). The topography of mesotocin and vasotocin systems in the brain of the domestic mallard and Japanese quail: Immunocytochemical identification. *Cell Tissue Res.* **213,** 37–51.

Bottje, W. G., Holmes, K. R., Neldon, H. L., and Koike, T. K. (1989). Relationships between renal hemodynamic and plasma levels of arginine vasotocin and mesotocin during hemorrhage in the domestic fowl (*Gallus domesticus*). *Comp. Biochem. Physiol. A* **92,** 423–427.

Braun, E. J., and Dantzler, W. D. (1974). Effects of ADH on single-nephron glomerular filtration rates in the avian kidney. *Am. J. Physiol.* **226,** 1–8.

Brown, N. L., Bayle, J. D., Scanes, C. G., and Follett, B. K. (1975). The actions of avian LH and FSH on the testes of hypophysectomized quail. *Cell Tissue Res.* **156,** 499–520.

Buntin, J. D., Ruzychi, E., and Witebsky, J. (1993). Prolactin receptors in dove brain: Autoradiographic analysis of binding characteristics in discrete brain regions and accessibility to blood-borne prolactin. *Neuroendocrinology* **57,** 738–750.

Buonomo, F. C., Lauterio, T. J., Baile, C. A., and Daughaday, W. H. (1987). Effects of insulinlike growth factor I (IGF-I) on growth hormone-releasing hormone (GRF) and thyrotropin-releasing hormone (TRH) stimulation of growth hormone (GH) secretion in the domestic fowl (*Gallus domesticus*). *Gen. Comp. Endocrinol.* **66,** 274–279.

Buonomo, F. C., Zimmerman, N. G., Lauterio, T. J., and Scanes, C. G. (1984). Catecholamine involvement in the control of growth hormone secretion. *Gen. Comp. Endocrinol.* **54,** 360–371.

Burke, W. H., Licht, P., Papkoff, H., and Bona Gallo, A. (1979a). Isolation and characterization of luteinizing hormone and follicle-stimulating hormone for pituitary glands of the turkey (*Meleagris gallopavo*). *Gen. Comp. Endocrinol.* **37,** 508–520.

Burke, W. H., Moore, J. A., Ogez, J. R., and Builder, S. E. (1987). The properties of recombinant chicken growth hormone and its effects on growth, body composition, feed efficiency and other factors in broiler chickens. *Endocrinology* **120,** 651–658.

Burke, W. H., Papkoff, H., Licht, P., and Bona Gallo, A. (1979b). Preparation and properties of luteinizing hormone (LH) subunits from the turkey (*Meleagris gallopavo*) and their recombination with subunits of ovine LH. *Gen. Comp. Endocrinol.* **37,** 501–507.

Burnside, J., and Cogburn, L. A. (1992). Developmental expression of hepatic growth hormone receptor and insulin-like growth factor-I mRNA in the chicken. *Mol. Endocrinol.* **89,** 91–96

Burnside, J., Liou, S. S., Zhong, C., and Cogburn, L. A. (1992). Abnormal growth hormone receptor gene expression in the sex-linked dwarf chicken. *Gen. Comp. Endocrinol.* **88,** 20–28.

Burnside, J., Liou, S. S., and Cogburn, L. A. (1991). Molecular cloning of the chicken growth hormone receptor complimentary DNA : mutation of the gene in sex-linked dwarf chickens. *Endocrinology* **128,** 3183–3192.

Calvo, F. O., and Bahr, J. M. (1983). Adenylyl cyclase system of the small preovulatory follicles of the domestic hen: Responsiveness of follicle-stimulating hormone and luteinizing hormone. *Biol. Reprod.* **29,** 542–547.

Campbell, R. M., and Scanes, C. G. (1985). Lipolytic activity of purified pituitary and bacterially derived growth hormone on chicken adipose tissue *in vitro*. *Proc. Soc. Exp. Biol. Med.* **180,** 513–517.

Campbell, R. M., and Scanes, C. G. (1987). Growth hormone inhibition of glucagon- and cAMP-induced lipolysis by chicken adipose tissue *in vitro*. *Proc. Soc. Exp. Biol. Med.* **184,** 456–460.

Campbell, R. M., Chen, W. Y., Wiehl, P., Kelder, B., Kopchick, J. J., and Scanes, C. G. (1993). A growth hormone (GH) analog that antagonizes the lipolytic effect but retains full insulin-like (antilipolytic) activity of GH. *Proc. Soc. Exp. Biol. Med.* **203,** 311–316.

Campbell, R. M., Kostyo, J. L., and Scanes, C. G. (1990). Lipolytic and antilipolytic effects of human growth hormone, its 20 kilodalton-variant, a reduced and carboxymethylated derivative and human placental lactogen on chicken adipose tissue *in vitro*. *Proc. Soc. Exp. Biol. Med.* **193,** 269–273.

Carsia, R. V., Scanes, C. G., and Malamed, S. (1985a). Isolated adrenocortical cells of the domestic fowl (*Gallus domesticus*): Steroidogenic and ultrastructural properties. *J. Steroid Biochem.* **22,** 273–279.

Carsia, R. V., Scanes, C. G., and Malamed, S. (1985b). Loss of sensitivity to ACTH of adrenocortical cells isolated from maturing domestic fowl. *Proc. Soc. Exp. Biol. Med.* **179,** 279–282.

Carsia, R. V., Weber, H., King, D. B., and Scanes, C. G. (1985c). Adrenocortical cell function in the hypophysectomized domestic fowl: Effects of growth hormone and 3,5,3′-triiodothyronine. *Endocrinology* **117**, 928–933.

Castro, M. G., Estivariz, F. E., and Iturriza, F. C. (1986). The regulation of the corticomelanotropic cell activity in aves. II. Effect of various peptides on the release of ACTH from dispensed, perfused duck pituitary cells. *Comp. Biochem. Physiol. A* **83**, 71–75.

Carsia, R. V., Weber, H., and Perez, F. M. (1986). Corticotropin-releasing factor stimulates the release of adrenocorticotropin from domestic fowl pituitary cells. *Endocrinology* **118**, 143–148.

Carsia, R. V., Weber, H., and Lauterio, T. J. (1988). Protein malnutrition in the domestic fowl induced alterations in adrenocortical function. *Endocrinology* **122**, 673–680.

Chang, W. C., Chung, D., and Li, C. H. (1980). Isolation and characterization of β–lipotropin and adrenocorticotropin from turkey pituitary glands. *Int. J. Pept. Protein Res.* **15**, 261–270.

Chen, X., and Horseman, N. D. (1994). Cloning, expression, and mutational analysis of the pigeon prolactin receptor. *Endocrinology* **135**, 269–276.

Cheung, A., Hall, T. R., and Harvey. S. (1988). Stimulation of corticosterone release in the fowl by recombinant DNA-derived chicken growth hormone. *Gen. Comp. Endocrinol.* **69**, 128–132.

Chou, H.-F., Johnson, A. L., and Williams, J. B. (1985). Luteinizing hormone releasing activity of [Gln^8]-LHRH and [His^5, Trp^7, Tyr^8]-LHRH in the cockerel, *in vivo* and *in vitro*. *Life Sci.* **37**, 2459–2465.

Cogburn, L. A., and Freeman, R. M. (1987). Response surface of daily thyroid hormone rhythms in young chickens exposed to constant ambient temperatures. *Gen. Comp. Endocrinol.* **68**, 113–123.

Collie, M. A., Holmes, W. N., and Cronshaw, J. (1992). A comparison of the response of dispersed steroidogenic cells derived from embryonic adrenal tissue from the domestic chicken (*Gallus domesticus*), the domestic Pekin duck and the wild Mallard duck (*Anas platyrhynchos*), and the domestic Muscovy duck (*Cairina moschata*). *Gen. Comp. Endocrinol.* **88**, 375–387.

Contijoch, A. M., Gonzolez, C., Singh, H. M., Malamed, S., Trancoso, S., and Advis, J. P. (1992). Dopaminergic regulation of luteinizing hormone-releasing hormone release at the median eminence level: Immunocytochemical and physiological evidence in hens. *Neuroendocrinology* **55**, 290–300.

Contijoch, A. M., Malamed, S., McDonald, J. K., and Advis, J. P. (1993a). Neuropeptide Y regulation of LHRH release in the median eminence: Immunocytochemical and physiological evidence in hens. *Neuroendocrinology* **57**, 135–145.

Contijoch, A. M., Malamed, S., Sarkar, D. K., and Advis, J. P. (1993b). β-endorphin regulation of LHRH release at the median eminence level: Immunocytochemical and physiological evidence in hens. *Neuroendocrinology* **57**, 365–373.

Corcaran, D. H., and Proudman, J. A. (1991). Isoforms of turkey prolactin: Evidence for differences in glycosylation and in tryptic peptide mapping. *Comp. Biochem. Physiol. B* **99**, 563–570.

Cronshaw, J., Reese, B. K., Collie, M. A., and Holmes, W. N. (1992). Cytoskeletal changes accompanying ACTH-induced steroidogenesis in cultured embryonic adrenal gland cells from the Pekin duck. *Cell Tissue Res.* **268**, 157–165.

Darras, V. M., Berghman, L. R., Vanderpooten, A., and Kuhn, E. R. (1992). Growth hormone acutely decreases type III iodothyronine deiodinase in chicken liver. *FEBS Lett.* **310**, 5–8.

Darras, V. M., Huybrechts, L. M., Berghman, L., Kuhn, E. R., and Decuypere, E. (1990). Ontogeny of the effect of purified chicken growth hormone on the liver 5′monodeiodination activity in the chicken: Reversal of the activity after hatching. *Gen. Comp. Endocrinol.* **77**, 212–220.

Darras, V. M., Rudas, P., Visser, T. J., Hall, T. R., Huybrechts, L. M., Vanderpoolen, A., Berghman, R., Decuypere, E., and Kuhn, E. R. (1993). Endogenous growth hormone controls high plasma levels of 3,3′,5-triiodothyronine (T_3) in growing chickens by decreasing the T_3-degrading type III deiodinase activity. *Dom. Anim. Endocrinol.* **10**, 55–65.

Davies, D. T., and Follett, B. K. (1980). Neuroendocrine regulation of gonadotrophin-releasing hormone secretion in the Japanese quail. *Gen. Comp. Endocrinol.* **40**, 220–225.

Davies, D. T., Massa, R., and James, R. (1980). Role of testosterone and of its metabolites in regulating gonadotrophin secretion in the Japanese quail. *J. Endocrinol.* **84**, 211–222.

Dawson, A., Follett, B. K., Goldsmith, A. R., and Nicholls, T. J. (1985). Hypothalamic gonadotrophin-releasing hormone and pituitary and plasma FSH and prolactin during photostimulation and photorefractoriness in intact and thyroidectomized starlings (*Sturnus vulgaris*). *J. Endocrinol.* **105**, 71–77.

Decuypere, E., Igbal, A., Michels, H., Kuhn, E. R., Scheider, R., and El Azeem, A. A. (1988). Thyroid hormone response to thyrotropin releasing factor after cold treatment during pre- and post-natal development in the domestic fowl. *Horm. Metab. Res.* **20**, 484–489.

Doneen, B. A., and Smith, T. E. (1982). Ontogeny of endocrine control of osmoregulation in chick embryo. II. Actions of prolactin, arginine, vasopression, and aldosterone. *Gen. Comp. Endocrinol.* **48**, 310–318.

Donoghue, D. J., and Scanes, C. G. (1991). Triiodothyronine (T_3) inhibition of growth hormone secretion by chicken pituitary cells *in vitro*. *Gen. Comp. Endocrinol.* **84**, 344–354.

Dunson, W. A., Buss, E. G., Sawyer, W. H., and Sokol, H. (1972). Hereditary polydipsia and polyuria in chickens. *Am. J. Physiol.* **222**, 1167–1176.

El Halawani, M. E., and Rozenboim, I. (1993). Ontogeny and control of incubation behavior in turkeys. *Poult. Sci.* **72**, 906–911.

El Halawani, M. E., Burke, W. H., and Dennison, P. T. (1980). Effect of nest deprivation on serum prolactin level in resting female turkeys. *Biol. Reprod.* **23**, 118–123.

El Halawani, M. E., Fehrer, S., Hargis, B., and Porter, T. (1988). Incubation behavior in the domestic turkey: Physiological correlations. *CRC Crit. Rev. Poult. Biol.* **1**, 285–314.

El Halawani, M. E., Silsby, J. L., Koike, T. I., and Robinzon, B. (1992). Evidence of a role for the turkey posterior pituitary in prolactin release. *Gen. Comp. Endocrinol.* **87**, 436–442.

Engster, H. M., Carew, L. B., Harvey, S., and Scanes. C. G. (1979). Growth hormone metabolism in essential fatty acid-deficient and pair-fed non-deficient chicks. *J. Nutr.* **109**, 330.

Esposito, V., De Girolamo, P., and Gargiulo, G. (1993). Immunoreactivity to vasoactive intestinal polypeptide (VIP) in the hypothalamus of the domestic fowl, *Gallus domesticus*. *Neuropeptides* **25**, 83–90.

Fennell, M. J., Johnson A. L., and Scanes, C. G. (1990). Influence of androgens on plasma concentrations of growth hormone in growing castrated and intact chickens. *Gen. Comp. Endocrinol.* **77**, 466–475.

Follett, B. K., and Maung, S. L. (1978). Rate of testicular maturation, in relation to gonadotrophin and testosterone levels in quail exposed to various artificial photoperiods and to natural day lengths. *J. Endocrinol.* **78**, 267–280.

Follett, B. K., and Pearce-Kelly, A. S. (1990). Photoperiodic control of the termination of reproduction in Japanese quail (*Coturnix coturnix Japonica*). *Proc. R. Soc. London B* **242**, 225–230.

Follett, B. K., and Pearce-Kelly, A. S. (1991). Photoperiodic induction in quail as a function of the period of the light-dark cycle: Implications for models of time measurement. *J. Biol. Rhythms* **6**, 331–341.

Follett, B. K., Davies, D. T., and Gledhill, B. (1977). Photoperiodic control of reproduction in Japanese quail: Changes in gonadotrophin secretion on the first day of induction and their pharmacological blockade. *J. Endocrinol.* **74**, 449–460.

Follett, B. K., Mattocks, P. W., and Farner, D. S. (1974). Circadian function in the photoperiodic induction of gonadotropin secretion in the White-crowned sparrow, *Zonotrichia leucophrys gambelii*. *Proc. Natl. Acad. Sci. USA* **71,** 1666–1669.

Foltzer, C., and Mialhe, P. (1976). Pituitary and adrenal control of pancreatic endocrine function in the duck. II. Plasma free fatty acids and insulin variations following hypophysectomy and replacement therapy with growth hormone and corticosterone. *Diabet. Metab.* **2,** 101–105.

Foster, D. N., Galehouse, D., Giordano, T., Min, B., Lamb, I. C., Porter, D. A., Intehar, K. J., and Bacon, W. L. (1991). Nucleotide sequence of the cDNA encoding the common α subunit of the chicken pituitary glycoprotein hormones. *J. Mol. Endocrinol.* **8,** 21–27.

Foster, R. G., Panzica, G. C., Parry, D. M., and Viglietti-Panzica, C. (1988). Immunocytochemical studies on the LHRH system of the Japanese quail: Influence by photoperiod and aspects of sexual differentiation. *Cell Tissue Res.* **253,** 327–335.

Fraley, G. S., and Kuenzel, W. J. (1992). Precocious puberty in chicks (*Gallus domesticus*) induced by central injections of neuropeptide Y. *Life Sci.* **52,** 1649–1956.

Furr, B. J. A., Bonney, R. C., England, R. J., and Cunningham, F. J. (1973). Luteinizing hormone and progesterone in peripheral blood during the ovulatory cycle of the hen (*Gallus domesticus*). *J. Endocrinol.* **57,** 159–169.

Gerstberger, R., Kaul, R., Gray, D. A., and Simon, E. (1985). Arginine vasotocin and glomerular filtration rate in saltwater-acclimated ducks. *Am. J. Physiol.* **248,** F663–F667.

Goldsmith, A. R., and Nicholls, T. J. (1984a). Thyroidectomy prevents the development of photorefractoriness and the associate rise in plasma prolactin in starlings. *Gen. Comp. Endocrinol.* **54,** 256–263.

Goldsmith, A. R., and Nicholls, T. J. (1984b). Prolactin is associated with the development of photorefractoriness in intact, castrated and testosterone-implanted starlings. *Gen. Comp. Endocrinol.* **54,** 247–255.

Goldsmith, A. R., and Nicholls, T. J. (1984c). Thyroxine induces refractoriness and stimulates prolactin secretion in European starlings (*Sturnus vulgaris*). *J. Endocrinol.* **101,** RI–R3.

Goldsmith, A. R., Edwards, C., Koprucu, M., and Silver, R. (1981). Concentrations of prolactin and luteinizing hormone in plasma of doves in relation to incubation and development of the crop. *J. Endocrinol.* **90,** 437–443.

Goldsmith, A. R., Ivings, W. E., Pearce-Kelly, A. S., Parry, D. M., Plowman, G., Nicholls, T. J., and Follett, B. K. (1989). Photoperiodic control of the development of the LHRH neurosecretory system of European starlings (*Sturnus vulgaris*) during puberty and the onset of photorefractoriness. *J. Endocrinol.* **122,** 255–268.

Goldstein, D. L., and Braun, E. J. (1988). Contributions of the kidneys and intestines to water conservation, and plasma levels of antidiuretic hormone, during dehydration in house sparrows (*Passer domesticus*). *J. Comp. Physiol. B* **158,** 353–366.

Goosens, N. S., Blahser, S., Oksche, A., Vandesande, F., and Dierickx, K. (1977). Immunocytochemical investigation of the hypothalo-neurohypophyseal system in birds. *Cell Tissue Res.* **184,** 1.

Goto, K., Koike, T. I., Neldon, H. L., and McKay, D. W. (1986). Peripheral angiotensin II stimulates release of vasotocin in conscious chickens. *Am. J. Physiol.* **251,** R333–R339.

Gray, D. A., and Simon, E. (1983). Mammalian and avian antidiuretic hormone: Studies related to possible species variation in osmoregulatory systems. *J. Comp. Physiol.* **151,** 241–246.

Gray, D. A., and Simon, E. (1987). Dehydration and arginine vasotocin and angiotensin II in CSF and plasma of Pekin ducks. *Am. J. Physiol.* **253,** R285–R291.

Gray, D. A., Hammel, H. T., and Simon, E. (1986). Osmoregulatory effect of angiotensin II in a bird with salt glands (*Anas platyrhynchos*). *J. Comp. Physiol.* **156,** 315–321.

Gray, D. A., Schutz, H., and Gerstberger, R. (1991). Interaction of atrial natriuretic factor and osmoregulatory hormones in the Pekin duck. *Gen. Comp. Endocrinol.* **81,** 246–255.

Hall, T. R., Harvey, S., and Chadwick, A. (1986). Control of prolactin secretion in birds: a review. *Gen. Comp. Endocrinol.* **62,** 171–184.

Hall, T. R., Cheung, A., and Harvey, S. (1987). Some biological activities of recombinant DNA-derived growth hormone on plasma metabolite concentrations in domestic fowl. *Comp. Biochem. Physiol. A* **86,** 29–32.

Hanks, M. C., Talbot, R. T., and Sang, H. M. (1989a). Expression of biologically active recombinant-derived chicken prolactin in *Escherichia coli*. *J. Mol. Endocrinol.* **3,** 15–21.

Hanks, M. C., Alonzi, J. A., Sharp, P. J., and Sang, H. M. (1989b). Molecular cloning and sequence analysis of putative chicken prolactin cDNA. *J. Mol. Endocrinol.* **2,** 21–30.

Harvey, S., and Baidwan, J. S. (1989). Thyrotrophin-releasing hormone (TRH)-induced growth hormone secretion in fowl: Binding of TRH to pituitary membranes. *J. Mol. Endocrinol.* **3,** 23–32.

Harvey, S., and Scanes, C. G. (1977). Purification and radioimmunoassay of chicken growth hormone. *J. Endocrinol.* **73,** 321–329.

Harvey, S., and Scanes, C. G. (1978). Effect of adrenaline and adrenergic active drugs on growth hormone secretion in immature cockerels. *Experientia* **34,** 1096–1097.

Harvey, S., and Scanes, C. G. (1981). Inhibition of growth hormone release by prostaglandins in immature domestic fowl (*Gallus domesticus*). *J. Endocrinol.* **91,** 69–73.

Harvey, S., and Scanes, C. G. (1985). Interaction between pancreatic growth hormone releasing factor (hpGRF) and thyrotropin releasing hormone (TRH) on growth hormone secretion in the domestic fowl. *Horm. Metab. Res.* **17,** 113–114.

Harvey, S., Scanes, C. G., and Howe, T. (1977). Growth hormone effects on *in vitro* metabolism of avian adipose and liver tissue. *Gen. Comp. Endocrinol.* **33,** 322–328.

Harvey, S., Scanes, C. G., Chadwick, A., and Bolton, N. J. (1978a). Influence of fasting, glucose and insulin on the levels of growth hormone and prolactin in the plasma of the domestic fowl (*Gallus domesticus*). *J. Endocrinol.* **78,** 501–506.

Harvey, S., Scanes, C. G., Chadwick, A., and Bolton, N. J. (1978b). The effect of thyrotropin-releasing hormone (TRH) and somatostatin (GHRIH) on growth hormone and prolactin secretion *in vitro* and *in vivo* in the domestic fowl (*Gallus domesticus*). *Neuroendocrinology* **26,** 249–260.

Harvey, S., Davison, T. F., and Chadwick, A. (1979a). Ontogeny of growth hormone and prolactin secretion in the domestic fowl (*Gallus domesticus*). *Gen. Comp. Endocrinol.* **39,** 270–273.

Harvey, S., Scanes, C. G., Chadwick, A., and Bolton, N. J. (1979b). Growth hormone and prolactin secretion growing domestic fowl; influence of sex and breed. *Br. Poult. Sci.* **20,** 9–17.

Harvey, S., Baidwan, J. S., and Attardo, D. (1990). Homologous and heterologous regulation of somatostatin-binding sites on chicken adenohypophysial membranes. *J. Endocrinol.* **127,** 417–425.

Hasegawa, Y., Miyamoto, K., Nomura, M., Igarashi, M., Kangawa, K., and Matsuo, H. (1984). Isolation and amino acid compositions of four somatostatin-like substances in chicken hypothalamic extract. *Endocrinology* **115,** 433–435.

Hattori, A., Ishii, S., and Wada, M. (1986). Effects of two kinds of chicken luteinizing hormone-releasing hormone (LHRH), mammalian LH-RH and its analogs on the release of LH and FSH in Japanese quail and chicken. *Gen. Comp. Endocrinol.* **64,** 446–455.

Hayashi, H., Imai, K., and Imai, K. (1991). Characterization of chicken ACTH and α-MSH: The primary sequence of chicken ACTH is more similar to xenopus ACTH than to other avian ACTH. *Gen. Comp. Endocrinol.* **82,** 434–443.

Herbute, S., Pintat, R., Ramade, F., and Bayle, J. D. (1984). Effect of short exposure to cold on plasma thyroxine in *Coturnix* quail: Role of the infundibular complex and its neural afferents. *Gen. Comp. Endocrinol.* **56,** 1–8.

Herold, M., Brezinschek, H. P., Gruschwitz, M., Dietrich, H., and Wick, G. (1992). Investigation of ACTH responses of chickens with autoimmune disease. *Gen. Comp. Endocrinol.* **88,** 188–198.

Houston, B., and Goddard, C. (1988). Molecular forms of growth hormone in the chicken pituitary gland. *J. Endocrinol.* **116,** 35–41.

Houston, B., and O'Neill, I. E. (1990). Insulin and growth hormone act synergistically to stimulate insulin-like growth factor-I production by cultured chicken hepatocytes. *J. Endocrinol.* **128,** 389–393.

Huybrechts, L. M., King, D. B., Lauterio, T. J., Marsh, J., and Scanes, C. G. (1985). Plasma concentrations of somatomedin-C in hypophysectomized, dwarf and intact growing domestic fowl as determined by heterologous radioimmunoassay. *J. Endocrinol.* **104,** 233–239.

Imai, K. (1973). Effects of avian and mammalian pituitary preparations on induction of ovulation in the domestic fowl, *Gallus domesticus*. *J. Reprod. Fertil.* **33,** 91–98.

Ishii, S. (1993). The molecular biology of avian gonadotropin. *Poult. Sci.* **72,** 856–866.

Iturriza, F. C., Estivariz, F. E., and Levitin, H. P. (1980). Coexistence of α-melanocyte stimulating hormone and adrenocorticotrophin in all cells containing either of the two hormones in the duck pituitary. *Gen. Comp. Endocrinol.* **42,** 110–115.

Iturriza, F. C., Venosa, R. A., Pijol, M. G., and Quintas, N. B. (1992). α-Melanocyte-stimulating hormone stimulates sodium excretion in the salt gland of the ducks. *Gen. Comp. Endocrinol.* **87,** 369–374.

Jackson, I. M. D., and Reichlin, S. (1974). Thyrotropin-releasing hormone (TRH); distribution in hypothalamic and extrahypothalamic brain tissues of mammalian and submammalian chordates. *Endocrinology* **95,** 854–862.

Johnson, A. L., and Tilly, J. L. (1988). Effects of vasoactive intestinal peptide on steroid secretion and plasminogen activator activity in granulosa cells of the hen. *Biol. Reprod.* **38,** 296–303.

Johnson, B. E., Scanes, C. G., King, D. B., and Marsh, J. A. (1993). Effect of hypophysectomy and growth hormone on immune development in the domestic fowl. *Dev. Comp. Immunol.* **17,** 331–339.

Jozsa, R., and Mess, B. (1993a). Localization and ontogeny of peptidergic neurons in birds. *Akademia Kiado, Budapest, Hungary*.

Jozsa, R., and Mess, B. (1993b). Galanin-like immunoreactivity in the chicken brain. *Cell Tiss. Res.* **273,** 391–399.

Jozsa, R., Scanes, C. G., Vigh, S., and Mess, B. (1979). Functional differentiation of the embryonic chicken pituitary gland studied by immunohistological approach. *Gen. Comp. Endocrinol.* **39,** 158–163.

Jozsa, R., Vigh, S., Schally, A. V., and Mess, B. (1984). Localization of corticotropin-releasing factor-containing neurons in the brain of the domestic fowl. *Cell Tiss. Res.* **236,** 245–248.

Katz, I. A., Millar, R. P., and King, J. A. (1990). Differential regional distribution and release of two forms of gonadotropin-releasing hormone in the chicken brain. *Peptides* **11,** 443–450.

Kikuchi, M., and Ishii, S. (1992). Changes in luteinizing hormone receptors in the granulosa and theca layers of the ovarian follicle during follicular maturation in the Japanese quail. *Gen. Comp. Endocrinol.* **85,** 124–137.

King, D. B., and Scanes, C. G. (1986). Effect of mammalian growth hormone in the growth of hypophysectomized chickens. *Proc. Soc. Exp. Biol. Med.* **182,** 201–207.

King, J. A., and R. P. Millar. (1982). Structure of chicken hypothalamic luteinizing hormone-releasing hormone: II isolation and characterization. *J. Biol. Chem.* **257,** 10729–10732.

Klandorf, H., Sharp, P. J., and Sterling, R. (1978). Induction of thyroxine and triiodothyronine release by thyrotrophin-releasing hormone in the hen. *Gen. Comp. Endocrinol.* **34,** 377–379.

Klandorf, H., Harvey, S., and Fraser, H. M. (1985). Physiological control of thyrotrophin-releasing hormone in the domestic fowl. *J. Endocrinol.* **105,** 351–355.

Kledzik, G., Marshall, S., Gelato, M., Campbell, G., and Meites, J. (1975). Prolactin binding activity in the crop sacs of juvenile, mature, parent and prolactin-injection pigeons. *Endocr. Res. Commun.* **2,** 345–355.

Klempt, M., Ellendorff, F., and Grossmann, R. (1992). Functional maturation of arginine vasotocin secretory responses to osmotic stimulation in the chick embryo and the newborn chicken. *J. Endocrinol.* **133,** 269–274.

Knight, P. G., Wilson, S. C., Gladwell, R. T., and Cunningham, F. J. (1982). Evidence for the involvement of control catecholaminergic mechanisms in mediating the preovulatory surge of luteinizing hormone in the domestic hen. *J. Endocrinol.* **94,** 295–304.

Koike, T. K., Pryor, L. R., Neldon, H. L., and Venable, R. S. (1977). Effect of water deprivation on plasma radioimmunoassayable arginine vasotocin in conscious chickens (*Gallus domesticus*). *Gen. Comp. Endocrinol.* **33,** 359–364.

Koike, T. I., Pryor, L. R., and Neldon, H. L. (1979). Effect of saline infusion on plasma immunoreactive vasotocin in conscious chickens (*Gallus domesticus*). *Gen. Comp. Endocrinol.* **37,** 451–458.

Koike, T. I., Shimada, K., and Cornett, L. E. (1988). Plasma levels of immunoreactive mesotocin and vasotocin during oviposition in chickens: Relationship to oxytocic action of the peptides *in vitro* and peptide interaction with myometrial membrane binding sites. *Gen. Comp. Endocrinol.* **70,** 119–126.

Kowalski, K. I., Tilly, J. L., and Johnson, A. L. (1991). Cytochrome P450 side-chain cleavage (P450scc) in the hen ovary. 1. Regulation of P450scc messenger RNA levels and steriodogenesis in theca cells of developing follicles. *Biol. Reprod.* **45,** 955–966.

Krishnan, K. A., Proudman, J. A., and Bahr, J. M. (1989). Avian growth hormone receptor assay: Use of chicken and turkey liver membranes. *Molec. Cell. Endocrinol.* **66,** 125–134.

Krishnan, K. A., Proudman, J. A., and Bahr, J. M. (1991). Radioligand receptor assay for prolactin using chicken and turkey kidney membranes. *Comp. Biochem. Physiol. B* **100,** 769–774.

Krishnan, K. A., Proudman, J. A., Bolt, D. J., and Bahr, J. M. (1993). Development of an homologous radioimmunoassay for chicken follicle-stimulating hormone and measurement of plasma FSH during the ovulatory cycle. *Comp. Biochem. Physiol. A* **105,** 729–734.

Kuenzel, W. J., and Blahser, S. (1994). Vasoactive intestinal polypeptide (VIP)-containing neurons: distribution throughout the brain of the chick (*Gallus domesticus*) with focus upon the lateral septal organ. *Cell Tissue Res.* **275,** 91–107.

Kuhn, E. R., and Nouwen, E. J. (1978). Serum levels of triiodothyronine and thyroxine in the domestic fowl following mild cold exposure and injection of synthetic thyrotropin-releasing hormone. *Gen. Comp. Endocrinol.* **34,** 336–342.

Kuhn, E. R., Verheyer, G., Chiasson, R. B., Huts, C., and Decuypere, E. (1985). Ovine growth hormone reverses the fasting induced decrease of plasma T_3 in adult chickens. *IRCS Med. Sci.* **13,** 451–452.

Lam, S. K., Harvey, S., and Scanes, C. G. (1986). Somatostatin inhibits thyroid function in fowl. *Gen. Comp. Endocrinol.* **63,** 134–138.

Lamb, I. C., Galehouse, D. M., and Foster, D. N. (1988). Chicken growth hormone cDNA sequence. *Nucl. Acids* **16,** 9339.

Lazure, C., Paquet, L., Litthauer, D., Naude, R. J., Oelofsen, W., and Chretien, M. (1990). The ostrich pituitary contains a major peptide homologous to mammalian chromogranin A(1-76). *Peptides* **11**, 79–87.

Lazure, C., Saayman, H. S., Naude, R. J., Oelofsen, W., and Chretien, M. (1987). Complete amino acid sequence of a VLDV-type neurophysin from ostrich differs markedly from known mammalian neurophysins. *Int. J. Peptide Prot. Res.* **30**, 634–645.

Lazure, C., Saayman, H. S., Naude, R. J., Oelofsen, W., and Chretien, M. (1989). Ostrich MSEL-neurophysin belongs to the class of two-domain "big" neurophysin as indicated by complete amino acid sequence of the neurophysin/copeptin. *Int. J. Peptide Prot. Res.* **33**, 46–58.

Lebovic, D. I., and Nicoll, C. S. (1992). Measurement of inducible proteins improves the precision of the local pigeon crop-sac bioassay for prolactin. *Life Sci.* **50**, 2019–2024.

Leung, F. C., Taylor, J. E., Steelman, S. L., Bennett, C. D., Rodkey, J. A., Long, R. A., Serio, R., Weppelman, R. M., and Olson, G. (1984). Purification and properties of chicken growth hormone and the development of a homologous radioimmunoassay. *Gen. Comp. Endocrinol.* **56**, 389–400.

Li, C. H., Chung, D., Oelofsen, W., and Naude, R. J. (1978). Adrenocorticotropin 53. The amino acid sequence of the hormone from the ostrich pituitary gland. *Biochem. Biophys. Res. Commun.* **84**, 900–906.

Li, Z., and Johnson, A. L. (1993). Regulation of P_{450} cholesterol side-chain cleavage messenger ribonucleic acid expression and progesterone production in hen granulosa cells. *Biol. Reprod.* **49**, 463–469.

MacKenzie, D. S. (1981). *In vivo* thyroxine in day old cockerels in response to acute stimulation by mammalian and avian pituitary hormones. *Poultry Sci.* **60**, 2136–2143.

Macnamee, M. C., and Sharp, P. J. (1989). The functional activity of hypothalamic dopamine in broody bantam hens. *J. Endocrinol.* **121**, 67–74.

Macnamee, M. C., Sharp, P. J., Lea, W., Sterling, R. J., and Harvey, S. (1986). Evidence that vasoactive intestinal peptide is a physiological prolactin-releasing factor in the bantam hen. *Gen. Comp. Endocrinol.* **62**, 470–478.

Malamed, S., Gibney, J. A., Loesser, K. E, and Scanes, C. G. (1985). Age-related changes of the somatotrophs of the domestic fowl *Gallus domesticus*. *Cell Tissue Res.* **239**, 87–91.

Malamed, S., Gibney, J. A., and Scanes, C. G. (1988). Immunogold identification of the somatotrophs of domestic fowl of different ages. *Cell Tissue Res.* **251**, 581–585.

Malamed, S., Gibney, J. A., Cain, L. D., Perez, F. M., and Scanes, C. G. (1993). Immunocytochemical studies of chicken somatotrophs and somatotroph granules before and after hatching. *Cell Tissue Res.* **272**, 369–374.

Maung, Z. W., and Follett, B. K. (1977). Effects of chicken and ovine luteinizing hormone on androgen release and cyclic AMP production by isolated cells from the quail testis. *Gen. Comp. Endocrinol.* **33**, 242–253.

Maung, S. L., and Follett, B. K. (1978). The endocrine control by luteinizing hormone of testosterone secretion from the testis of the Japanese quail. *Gen. Comp. Endocrinol.* **36**, 79–89.

Mauro, L. J., Elde, R. P., Youngren, O. M., Phillips, R. E., and El Halawani, M. E. (1989). Alterations in hypothalamic vasoactive intestinal peptide-like immunoreactivity are associated with reproduction and prolactin release in the female turkeys. *Endocrinology* **125**, 1795–1804.

Mauro, L. J., Youngren, O. M., Proudman, J. A., Philips, R. E., and El Halawani, M. E. (1992). Effects of reproductive status, ovariectomy, photoperiod on vasoactive intestinal peptide in the female turkey hypothalamus. *Gen. Comp. Endocrinol.* **87**, 481–493.

McCann-Levorse, L. M., Radecki, S. V., Donoghue, D. J., Malamed, S., Foster, D. M., and Scanes, C. G. (1993). Ontogeny of pituitary growth hormone and growth hormone mRNA in the chicken. *Proc. Soc. Exp. Biol. Med.* **202**, 109–113.

McNeilly, A. S., Etches, R. J., and Friesen, H. G. (1978). A heterologous radioimmunoassay for avian prolactin: application to the measurement of prolactin in the turkey. *Acta Endocrinol. (Copenhagen)* **89**, 60–69.

McNichols, M. J., and McNabb, F. M. A. (1988). Development of thyroid function and its pituitary control in embryonic and hatching precocial Japanese quail and altricial ring doves. *Gen. Comp. Endocrinol.* **69**, 109–118.

Mikami, S. (1980). Hypothalamic control of the avian adenohypophysis. In "Biological Rhythms in Birds: Neural and Endocrine Aspects" pp. 17–32. (Y. Tanabe, K. Tanaka and T. Ookawa, Eds.), Springer-Verlag Tokyo.

Millam, J. R., Burke, W. H., and El Halawani, M. E. (1984). Release of gonadotropin-releasing hormone from the Japanese quail hypothalamus *in vitro*. *Gen. Comp. Endocrinol.* **53**, 293–301.

Millam, J. R., Craig-Veit, C. B., and Petitte, J. N. (1993). Brain content of cGnRH I and II during embryonic development in chickens. *Gen. Comp. Endocrinol.* **92**, 311.

Millar, R. P., de L. Milton, R. C., Follett, B. K., and King, J. A. (1986). Receptor binding and gonadotropin-releasing activity of a novel chicken gonadotropin-releasing hormone [His^5, Trp^7, Tyr^8]-GnRH and a D-Arg^6 and analog. *Endocrinology* **119**, 224–231.

Miyamoto, K., Hasegawa, Y., Nomara, M., Igarashi, M., Kangawa, K., and Matsuo, H. (1984). Identification of the second gonadotropin-releasing hormone in chicken hypothalamus: evidence that gonadotropin secretion is probably controlled by two distinct gonadotropin releasing hormones in avian species. *Proc. Natl. Acad. Sci. USA* **81**, 3874–3878.

Murphy, M. J., Brown, P. S., and Brown, S. C. (1986). Osmoregulatory effects of prolactin and growth hormone in embryonic chicks. *Gen. Comp. Endocrinol.* **62**, 485–492.

Naude, R. J., and Oelofsen, W. (1977). The isolation and characterization of corticotropin from the pituitary gland of the ostrich *Struthio camelus*. *Biochem. J.* **165**, 519–523.

Naude, R. J., Chung, D., Li, C. H., and Oelofsen, W. (1981a). β-lipotropin: Primary structure of the hormone from the ostrich pituitary gland. *Int. J. Pept. Protein Res.* **18**, 138–147.

Naude, R. J., Chung, D., Li, C. H., and Oelofsen, W. (1981b). β-endorphin. Primary structure of the hormone from the ostrich pituitary gland. *Biochem. Biophys. Res. Commun.* **98**, 108–114.

Naude, R. J., Oelofsen, W., and Maske, R. (1980). Isolation, characterization and opiate activity of β-endorphin from the pituitary gland of the ostrich, *Struthio camelus*. *Biochem. J.* **187**, 245–248.

Nicoll, S. C. (1967). Bio-assay of prolactin: Analysis of the pigeon crop-sac response to local prolactin injection by an objective and quantitative method. *Endocrinology* **80**, 641–655.

Nicoll, S. C., Hebert, N. J., and Russell, S. M. (1985). Lactogenic hormones stimulate the liver to secrete a factor that acts synergistically with prolactin to promote growth of the pigeon crop-sac mucosal epithelium *in vivo*. *Endocrinology* **116**, 1449–1453.

Nilsson, A. (1975). Structure of the vasoactive intestinal octacosapeptide from chicken intestine: The amino acid sequence. *FEBS Lett.* **60**, 322–326.

Norgren, R. B., and Silver, R. (1990). Distribution of vasoactive intestinal peptide-like and neurophysin-like immunoreactive neurons and acetyl cholinesterase staining in the ring dove hypothalamus with emphasis on the question of an avian suprachiasmatic nucleus. *Cell Tissue Res.* **259**, 331–339.

Nouwen, E. J., Decuypere, E., Kuhn, E. R., Michels, H., Hall, T. R., and Chadwick, A. (1984). Effect of dehydration, hemorrhage and oviposition on serum concentrations of vasotocin, mesotocin, and prolactin in the chicken. *J. Endocrinol.* **102,** 345–351.

Opel, H., and Proudman, J. (1988). Stimulation of prolactin release in turkeys by vasoactive intestinal peptide. *Proc. Soc. Exp. Biol. Med.* **187,** 455–460.

Panksepp, J., and Normansell, L. (1990). Effects of ACTH (1-24) and ACTH/MSH (4-10) on isolation-induced distress vocalization in domestic chicks. *Peptides* **11,** 915–919.

Perez, F. M., Malamed, S., and Scanes, C. G. (1985). Biosynthetic human somatomedin C inhibits hpGRF (1-44) NH_2-induced and TRH-induced GH release in a primary culture of chicken pituitary cells. *IRCS Med. Sci.* **13,** 871–872.

Perez, F. M., Malamed, S., and Scanes, C. G. (1987). Growth hormone secretion from chicken adenohypophyseal cells in primary culture: Effects of human pancreatic growth hormone-releasing factor, thyrotropin-releasing hormone and somatostatin on growth hormone release. *Gen. Comp. Endocrinol.* **65,** 408–414.

Powell, R. C., Jach, H., Millar, R. P., and King, J. A. (1987). Identification of Gln^8-GnRH and His^5, Trp^7, Tyr^8-GnRH in the hypothalamus and extrahypothalamic brain of the ostrich (*Struthio camelus*). *Peptides* **8,** 185–190.

Proudman, J. A., and Opel, H. (1981). Turkey prolactin: Validation of a radioimmunoassay and measurement of changes associated with broodiness. *Biol. Reprod.* **25,** 573–580.

Proudman, J. A., and Opel, H. (1988). Stimulation of prolactin secretion from turkey anterior pituitary cells in culture. *Proc. Soc. Exp. Biol. Med.* **187,** 448–454.

Proudman, J. A., and Opel, H. (1990). Effect of peptide histidine isoleucine on *in vitro* and *in vivo* prolactin secretion in the turkey. *Poult. Sci.* **69,** 1209–1214.

Pukac, L. A., and Horseman, N. D. (1984). Regulation of pigeon crop gene expression by prolactin. *Endocrinology* **114,** 1718–1724.

Pukac, L. A., and Horseman, N. D. (1986). Regulation of cloned prolactin-inducible genes in pigeon crop. *Mol. Endocrinol.* **1,** 188–194.

Radecki, S. V., Deaver, D. R., and Scanes, C. G. (1994). Triiodothyronine reduced growth hormone (GH) secretion and pituitary GH mRNA in the chicken *in vivo* and *in vitro*. *Proc. Soc. Exp. Biol. Med.* **205,** 340–346.

Reichardt, A. K. (1993). Functional differentiation of the pituitary gland during development of the domestic ring dove (*Streptopelia roseogrisea*). *Gen. Comp. Endocrinol.* **92,** 41–53.

Reiner, A., Davis, B. M., Brecha, N. C., and Karten, H. J. (1984). The distribution of enkephalin like immunoreactivity in the telencephalon of the adult and developing domestic chicken. *J. Comp. Neurol.* **228,** 245–262.

Robinson, G. A., Wasnidge, D. C., Floto, F., and Downie, S. E. (1976). Ovarian ^{125}I transference in the laying Japanese quail: Apparent stimulation of FSH and lack of stimulation by TSH. *Poult. Sci.* **55,** 1665–1671.

Robinzon, B., Koike, T. I., and Marks, P. A. (1993). At low dose, arginine vasotocin has vasopressor rather than vasodepressor effect in chickens. *Gen. Comp. Endocrinol.* **91,** 105–112.

Robinzon, B., Koike, T. I., Neldon, H. L., Kinzler, S. L., Hendry, I. R., and El Halawani, M. E. (1988a). Physiological effects of arginine vasotocin and mesotocin in cockerels. *Br. Poult. Sci.* **29,** 639–652.

Robinzon, B., Koike, T. I., Neldon. H. L., and Kinzler, S. L. (1988b). Distribution of immunoreactive mesotocin and vasotocin in the brain and pituitary of chickens. *Peptides* **9,** 829–833.

Rombauts, L., Berghman, L. R., Vanmontfort, D., Decuypere, E., and Verhoeven, G. (1993). Changes in immunoreactive FSH and inhibin in developing chicken embryos and the effects of estradiol and the aromatose inhibitor R76713. *Biol. Reprod.* **49,** 549–554.

Rosebrough, R. W., McMurtry, J. P., and Vasilatos-Younken, J. (1991). Effect of pulsatile or continuous administration of pituitary-derived chicken growth hormone (p-cGH) on lipid metabolism in broiler pullets. *Comp. Biochem. Physiol. A* **99,** 207–214.

Rozenboim, I., Silsby, J. L., Tabibzadeh, C., Pitts, G. R., Youngren, O. M., and El Halawani, M. E. (1993). Hypothalamic and posterior pituitary content of vasoactive intestinal peptide, gonadotropin-releasing hormone I and II in the turkey hen. *Biol. Reprod.* **49,** 622–627.

Rzasa, J. (1978). Effects of arginine vasotocin and prostaglandin E_1 on the hen uterus. *Prostaglandins* **16,** 357–372.

Rzasa, J. (1984). The effect of arginine vasotocin on prostaglandin production of the hen uterus. *Gen. Comp. Endocrinol.* **53,** 260–263.

Saeed, W., and El Halawani, M. E. (1986). Modulation of prolactin response to thyrotropin releasing hormone by ovarian steroids in ovariectomized turkeys (*Meleagris gallopavo*). *Gen. Comp. Endocrinol.* **62,** 129–136.

Saito, N., Kinzler, S., and Koike, T. I. (1990). Arginine vasotocin and mesotocin levels in theca and granulosa levels of the ovary during the oviposition cycle in hens (*Gallus domesticus*). *Gen. Comp. Endocrinol.* **79,** 54–63.

Scanes, C. G. (1974). Some *in vitro* effects of synthetic thyrotrophin releasing factor on the secretion of thyroid stimulating hormone from the anterior pituitary gland of the domestic fowl. *Neuroendocrinology* **15,** 1–9.

Scanes, C. G. (1992). Lipolytic and diabetogenic effects of native and biosynthetic growth hormone in the chicken: A reevaluation. *Comp. Biochem. Physiol. A* **101,** 871–878.

Scanes, C. G., and Harvey, S. (1984). Stimulation of growth hormone secretion by human pancreatic growth-hormone releasing factor and thyrotrophin-releasing hormone in anesthetized chickens. *Gen. Comp. Endocrinol.* **56,** 198–203.

Scanes, C. G., and Harvey, S. (1985). Growth hormone secretion in anesthetized fowl. 2. Influence of heterologous stimuli in birds refractory to human pancreatic growth hormone releasing factor (hpGRF) or thyrotrophin releasing hormone. *Gen. Comp. Endocrinol.* **59,** 10–14.

Scanes, C. G., and Harvey, S. (1989a). Somatostatin inhibition of thyrotropin-releasing hormone and growth hormone-releasing factor-induced growth hormone secretion in young and adult anesthetized chickens. *Gen. Comp. Endocrinol.* **75,** 256–264.

Scanes, C. G., and Harvey, S. (1989b). Triiodothyronine inhibition of thyrotropin-releasing hormone- and growth hormone releasing factor-induced growth hormone secretion in anesthetized chickens. *Gen. Comp. Endocrinol.* **73,** 477–484.

Scanes, C. G., and Johnson, A. L. (1984). Failure of castration to prevent the pre–pubescent decline in the circulating concentration of growth hormone in the domestic fowl. *Gen. Comp. Endocrinol.* **53,** 398–401.

Scanes, C. G., Carsia, R. V., Lauterio, T. J., Huybrechts, L., Rivier, J., and Vale, W. (1984). Synthetic human pancreatic growth hormone releasing factor (GRF) stimulates growth hormone secretion in the domestic fowl (*Gallus domesticus*). *Life Sci.* **34,** 1127–1134.

Scanes, C. G., Chadwick, A., and Bolton, N. J. (1976). Radioimmunoassay of prolactin in the plasma of domestic fowl. *Gen. Comp. Endocrinol.* **30,** 12–20.

Scanes, C. G., Duyka, D. R., Lauterio, T. T., Bowen, S. J., Huybrechts, L. M., Bacon, W. L., and King, D. B. (1986). Effect of chicken growth hormone, triiodothyronine and hypophysectomy in growing domestic fowl. *Growth* **50,** 12–31.

Scanes, C. G., Godden, P. M. M., and Sharp, P. J. (1977). An homologous radioimmunoassay for chicken follicle-stimulating hormone: Observations on the ovulatory cycle. *J. Endocrinol.* **73,** 473–482.

Scanes, C. G., Griminger, P., and Buonomo, F. C. (1981). Effects of dietary protein restriction on circulating concentrations of growth hormone in growing domestic fowl (*Gallus domesticus*). *Proc. Soc. Exp. Biol. Med.* **168**, 334–337.

Shahabi, N. A., Bahr, J. M., and Nalbandov, A. V. (1975). Effect of LH injection on plasma and follicular steroids in the chicken. *Endocrinology* **96**, 969–972.

Sharp, P. J., and Klandorf, H. (1985). Environmental and physiological factors controlling thyroid function in galliformes. In "The Endocrine System and the Environmental" (B. K. Follett, S. Ishii, and A. Chandola, Eds.), pp. 175–181. Springer-Verlag, Berlin.

Sharp, P. J., Chiasson, R. B., El Tounsy, M. M., Klandorf, H., and Radke, W. J. (1979). Localization of cells producing thyroid-stimulating hormone in the pituitary gland of the domestic fowl. *Cell Tissue Res.* **198**, 53–63.

Sharp, P. J., Chiasson, R. B., El Tounsy, M. M., Klandorf, H., and Sharp, P. J., Dunn, I. C., and Talbot. R. L. (1987). Sex differences in the LH response to chicken LHRH-I and -II in the domestic fowl. *J. Endocrinol.* **115**, 323–331.

Sharp, P. J., Macnamee, M. C., Sterling, R. J., Lea, R. W., and Pedersen, H. C. (1988). Relationships between prolactin, LH and broody behavior in bantam hens. *J. Endocrinol.* **118**, 279–286.

Sharp, P. J., Sterling, R. J., Talbot, R. T., and Huskisson, N. S. (1989). The role of hypothalamic vasoactive intestinal polypeptide in the maintenance of prolactin secretion in incubating bantam hens: Observations using passive immunization, radioimmunoassay and immunohistochemistry. *J. Endocrinol.* **122**, 5–13.

Sharp, P. J., Talbot, R. T., Main, G. M., Dunn, I. C., Fraser, H. M., and Huskisson, N. S. (1990). Physiological roles of chicken LHRH-I and -II in the control of gonadotrophin release in the domestic fowl. *J. Endocrinol.* **124**, 291–299.

Shirley, H. V., and Nalbandov, A. V. (1956). Effects of neurohypophysectomy in domestic chickens. *Endocrinology* **58**, 477–483.

Silver, R., Ramos, C. L., and Silverman, A-J. (1992). Sexual behavior triggers the appearance of non-neuronal cells containing gonadotropin-releasing hormone-like immunoreactivity. *J. Neuroendocrinol.* **4**, 207–210.

Simon-Oppermann, C. H., and Simon, E. (1982). Osmotic and volume control of diuresis in conscious ducks (*Anas platyrhynchos*). *J. Comp. Physiol.* **146**, 17–25.

Simon-Oppermann, C., Gray, D., Szczepanska-Sadowski, E., and Simon, E. (1984). Blood volume changes and arginine vasotocin (AVT) blood concentrations in conscious fresh water and salt water adapted ducks. *Pfleugers Arch.* **400**, 151–159.

Simon-Oppermann, C., Simon, E., Deutch, H., Mohring, L., and Schoun, J. (1980). Serum arginine-vasotocin (AVT) and afferent anc central control of osmoregulation in conscious Pekin ducks. *Pfleugers Arch.* **387**, 99–106.

Soares, M. J., and Proudman, J. A. (1991). Turkey and chicken prolactins stimulate the proliferation of rat Nb2 lymphoma cells. *Proc. Soc. Exp. Biol. Med.* **197**, 384–386.

Souza, L. M., Boone, T. C., Murdock, D., Langley, K., Wypych, J., Fenton, D., Johnson, S., Lai, P. H., Everett, R., Hsu, R-H., and Bosselman, R. (1984). Application of recombinant DNA technologies to studies on chicken growth hormone. *J. Exp. Zool.* **232**, 465–474.

Spiess, J., Rivier, J. E., Rodkey, J. A., Bennett, C. D., and Vale, W. (1979). Isolation and characterization of somatostatin from pigeon pancreas. *Proc. Natl. Acad. Sci. USA* **76**, 2974–2978.

Stallone, J. N., and Braun, E. J. (1985). Contributions of glomerular and tubular mechanisms to antidiuresis in conscious domestic fowl. *Am. J. Physiol.* **249**, F842–F850.

Stallone, J. N., and Braun, E. J. (1986). Osmotic and volemic regulation of plasma arginine vasotocin in conscious domestic fowl. *Am. J. Physiol.* **250**, R644–R657.

Sturkie, P. D., and Lin, Y. C. (1966). Release of vasotocin and oviposition in the hen. *J. Endocrinol.* **35**, 325–326.

Talbot, R. T., Hanks, M. C., Sterling, R. J., Sang, H. M., and Sharp, P. J. (1991). Pituitary prolactin messenger ribonucleic acid levels in incubating and laying hens: Effects of manipulating plasma levels of vasoactive intestinal polypeptide. *Endocrinology* **129**, 496–502.

Tanaka, M., Maeda, K., Okubo, T., and Natashima, K. (1992). Double antenna structure of chicken prolactin receptor deduced from the cDNA sequence. *Biochem. Biophys. Res. Commun.* **188**, 490–496.

Tennyson, V. M., Hou-Yu, A., Nilaver, G., and Zimmerman, E. A. (1985). Immunocytochemical studies of vasotocin and mesotocin in the hypothalamo-hypophyseal system of the chicken. *Cell Tissue Res.* **239**, 279–291.

Tennyson, V. M., Nilaver, G., Hou-Yu, A., Valiquette, G., and Zimmerman, E. A. (1986). Immunocytochemical study of the development of vasotocin/mesotocin in the hypothalamo-hypophysial system of the chick embryo. *Cell Tissue Res.* **243**, 15–31.

Thommes, R. C., Caliendo, J., and Woods, J. E. (1985). Hypothalamo-adenohypophyseal-thyroid interrelationships in the developing chick embryo. VII. Immunocytochemical demonstration of thyrotrophin-releasing hormone. *Gen. Comp. Endocrinol.* **57**, 1–9.

Thommes, R. C., Martens, J. B., Hopkins, W. E., Caliendo, J., Sorrentino, M. J., and Woods, J. E. (1983). Hypothalamo-adenohypophyseal-thyroid interrelationships in the chick embryo. IV. Immunocytochemical demonstration of TSH in the hypophyseal pars distalis. *Gen. Comp. Endocrinol.* **51**, 434–443.

Thommes, R. C., Williams, D. J., and Woods, J. E. (1984). Hypothalamo-adenohypophyseal-thyroid interrelationship in the chick embryo. VI. Midgestational adenohypophyseal sensitivity to thyrotrophin-releasing hormone. *Gen. Comp. Endocrinol.* **55**, 275–279.

Tilly, J. L., and Johnson. A. L. (1987). Presence and hormonal control of plasminogen activator in granulosa cells of the domestic hen. *Biol. Reprod.* **37**, 1156–1164.

Tilly, J. L., and Johnson, A. L. (1988a). Effects of a phorbol ester, a calcium ionophore, and 3′,5′-adenosine monophosphate production on hen granulosa cell plasminogen activator activity. *Endocrinology* **123**, 1433–1441.

Tilly, J. L., and Johnson, A. L. (1988b). Attenuation of hen granulosa cell steroidogenesis by a phorbol ester and 1-oleoyl-2-acetylglycenol. *Biol. Reprod.* **39**, 1–8.

Tilly, J. L., and Johnson, A. L. (1989). Regulation of androstenedione production by adenosine 3′,5′-monophosphate and phorbol myristate acetate in ovarian thecal cells of the domestic fowl. *Endocrinology* **125**, 1691–1699.

Tilly, J. L., Kowalski, K. I., and Johnson, A. L. (1991a). Stage of ovarian follicular development associated with the initiation of steroidogenic competence in avian granulosa cells. *Biol. Reprod.* **44**, 305–314.

Tilly, J. L., Kowalski, K. I., and Johnson, A. L. (1991b). Cytochrome P_{450} side-chain cleavage (P_{450scc}) in the hen ovary. II P_{450scc} messenger RNA immunoreactive protein, and enzyme activity in developing granulosa cells. *Biol. Reprod.* **45**, 967–974.

Tonoue, T., and Kitch, J. (1978). Release of cyclic AMP from the chicken thyroid stimulated with TSH *in vitro*. *Endocrinol. Jpn.* **25**, 105–109.

Tsutsui, K., and Ishii, S. (1978). Effects of follicle stimulating hormone and testosterone on receptors of follicle-stimulating hormone in the testis of the immature Japanese quail. *Gen. Comp. Endocrinol.* **36**, 297–305.

Tsuyoshi, H., and Wada, M. (1992). Termination of LH secretion in Japanese quail due to high and low temperatures and short daily photoperiods. *Gen. Comp. Endocrinol.* **85**, 424–429.

Vanderpooten, A., Darras, V. M., Huybrechts, L. M., Rudas, P., Decuypere, E., and Kuhn, E. R . (1991a). Effects of hypophysectomy and acute administration of growth hormone (GH) on GH-receptor binding in chick liver membranes. *J. Endocrinol.* **129,** 275–281.

Vanderpooten, A., Huybrechts, L. M., Decuypere, E., and Kuhn, E. R., (1991b). Differences in hepatic growth hormone receptor binding during development of normal and dwarf chickens. *Reprod. Nutr. Dev.* **31,** 47–55.

Vasilatos-Younken, R., Gravener, T. L., Cogburn, L. A., Mast, M. G., and Wellenreiter, R. H. (1988). Effect of pattern of administration on the response to exogenous pituitary-derived chicken growth hormone by broiler-strain pullets. *Gen. Comp. Endocrinol.* **71,** 268–283.

Vasilatos-Younken, R., Tsao, P. H., Foster, D. N., Smiley, D. L., Bryant, H., and Heiman, M. L. (1992). Restoration of juvenile baseline growth hormone secretion with preservation of the ultradian growth-hormone rhythm by continuous delivery of growth hormone-releasing factor. *J. Endocrinol.* **135,** 371–382.

Watahiki, M., Tanaka, M., Masuda, N., Sugisaki, K., Yamamoto, M. Yamakawa, M., Nagai, J., and Nakashima, K. (1989). Primary structure of chicken pituitary prolactin deduced from the cDNA sequence: Conversed and specific amino acid resides in the domains of the prolactin. *J. Biol. Chem.* **264,** 5535–5539.

Williamson, R. A., and Davison, T. F. (1985). The effects of a single injection of thyrotrophin on serum concentrations of thyroxine, triiodothyronine, and reverse triiodothyronine in the immature chicken (*Gallus domesticus*). *Gen. Comp. Endocrinol.* **58,** 109–113.

Wilson, J. X., and West, N. H. (1986). Cardiovascular responses to neurohormones in conscious chickens and ducks. *Gen. Comp. Endocrinol.* **62,** 268–280.

Wilson, S. C., and Sharp, P. J. (1976a). Induction of luteinizing hormone release by gonadal steriod in the ovariectomized domestic hen. *J. Endocrinol.* **71,** 87–98.

Wilson, S. C., and Sharp, P. J. (1976b). Effects of androgens, oestrogens and deoxycorticosterone acetate on plasma concentrations of luteinizing hormone in laying hens. *J. Endocrinol.* **69,** 93–102.

Wong, E. A., Ferrin, N. H., Silsby, J. L., and El Halawani, M. E. (1991). Cloning of a turkey prolactin mRNA throughout the reproductive cycle of the domestic turkey (*Meleagris gallopavo*). *Gen. Comp. Endocrinol.* **83,** 18–26.

Wong, E. A., Silsby, J. L., and El Halawani, M. E. (1992). Complimentary DNA cloning and expression of Pit-1/GHF-1 from the domestic turkey. *DNA Cell. Biol.* **11,** 651–660.

Yamashiro, D., Ho, C. L., and Li, C. H. (1984). Adrenocorticotropin. 57. Synthesis and biological activity of ostrich and turkey. *Int. J. Peptide Res.* **23,** 42–46.

Yoshimura, Y., Okamoto, T., and Tamura, T. (1993). Electron microscope observations on LH-induced oocyte maturation in Japanese quail (*Coturnix coturnix japonica*). *J. Reprod. Fertil.* **98,** 401–407.

Youngen, O. M., Halawani, M. E., Silsby, J. L., and Phillips, R. E. (1991). Intracranial prolactin perfusion induced incubation behavior in turkey hens. *Biol. Reprod.* **44,** 425–431.

Zhung, X., Silverman, A. J., and Silver, R. (1993). Reproductive behavior, endocrine state, and the distribution of GnRH-like immunoreactive mast cells on dove brain. *Horm. Behav.* **27,** 283–295.

CHAPTER 17

Thyroids

F. M. ANNE MCNABB

Department of Biology
Virginia Polytechnic Institute and State University
Blacksburg, Virginia 24061

I. Anatomy, Embryology, and Histology of Thyroid Glands 461
II. Thyroid Hormones 462
   A. Synthesis and Release of Hormones by the Thyroid Glands 462
   B. Circulating Thyroid Hormones 463
   C. Thyroid Hormone Activation and Degradation 464
III. Hypothalamic–Pituitary–Thyroid Axis 466
IV. Mechanism of Action of Thyroid Hormones 466
V. Effects of Thyroid Hormones 467
   A. Thyroid Hormone Effects on Metabolism 467
   B. Thyroid Hormone Effects on Development 467
   C. Thyroid Hormone Effects on Hatching, Molt, and Reproduction 468
VI. Thyroid Interactions with Other Hormones 468
VII. Environmental Influences on Thyroid Function 469
   References 469

I. ANATOMY, EMBRYOLOGY, AND HISTOLOGY OF THYROID GLANDS

Avian thyroid glands are paired, well-vascularized, oval glands located ventrolaterally to the trachea, just caudal to the junction of the subclavian and common carotid arteries (Figure 1). The histology and ultrastructure of avian thyroids (Fujita and Tanizawa, 1966; French and Hodges, 1977) are like those of other vertebrate classes (Astier, 1980; Dent, 1986). The gland is organized into spherical follicles whose walls are composed of epithelial cells that surround a lumen filled with colloid. The organelles present in the follicle cells correspond to those in the mammalian thyroid. The colloid contains thyroglobulin, the thyroid hormone-containing protein. This extracellular hormone storage in colloid provides for a large stock of reserve thyroid hormone. Such extracellular hormone storage is unique to the thyroid gland and is considered an adaptation related to the scarcity of the trace element iodine which is a component of thyroid hormones (McNabb, 1992). The thyroid receives a rich arterial blood supply via the thyroid arteries which branch off the carotid arteries. The venous drainage from the thyroid enters the jugular veins. The thyroid also contains a lymphatic network and autonomic nerves (Astier, 1980); the latter probably function primarily in vascular control. Calcitonin cells, which are parafollicular in mammals, are lacking in the avian thyroid; instead they are contained in separate ultimobranchial organs (Astier, 1980; see Chapter 18).

The thyroid gland appears very early in incubation (e.g., day 2 of the 21 day incubation period in chicken embryos). As in other vertebrates, the thyroid anlage arises from the ventral pharyngeal wall. This mass of epithelial cells is first attached to the pharynx by a stalk, but then detaches, and the two thyroid glands assume their mature positions (by day 5 in chicken embryos; Romanoff, 1960). Follicle formation is initiated by mid-incubation and proceeds rapidly, along with functional maturation, during the latter half of embryonic life in precocial galliform birds. In contrast, in altricial ring doves, although some follicles appear equally early, little more follicular organization occurs during embryonic life and

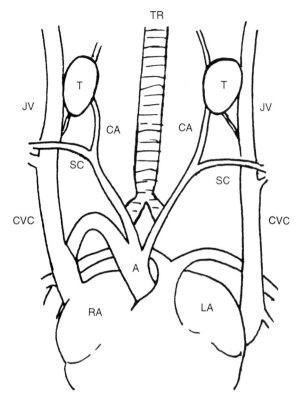

FIGURE 1. Ventral view of the position of avian thyroid glands. A, aorta; CA, common carotid artery; CVC, cranial vena cava; JV, jugular vein; LA, left atrium; RA, right atrium; SC, subclavian artery; T, thyroid gland; TR, trachea.

most further histological (as well as functional) development occurs after hatch (McNabb and McNabb, 1977).

Thyroid growth is in essentially constant proportion to body growth in galliform birds during both embryonic (McNabb et al., 1981) and posthatch growth (review by Wentworth and Ringer, 1986). The thyroid glands comprise about .01–.05% of body weight in chickens and quail. Thyroid gland size varies with the level of pituitary stimulation (Section III).

II. THYROID HORMONES

A. Synthesis and Release of Hormones by the Thyroid Glands

In birds, as in other vertebrates, both thyroxine (tetraiodothyronine or T_4) and triiodothyronine (T_3) are considered thyroid hormones (Figure 2). Iodide is actively transported into the thyroid gland from the blood, resulting in extremely high iodide concentrations in the gland. Uptake of radioiodide by the avian thyroid is rapid (hours) and retention time is prolonged, although both these factors are influenced by the dietary iodide availability. Iodide content of the thyroids is proportional to dietary iodide intake but circulating thyroid hormones and thyroid gland hormone contents are essentially unaffected by a wide range of iodide intakes (Newcomer, 1978; Astier, 1980; McNabb et al., 1985).

Although investigations of avian thyroid function are less comprehensive than those on mammals, both histological and physiological data indicate that the mechanisms of hormone synthesis and release by avian thyroid glands are essentially equivalent to those in mammals. The older literature in support of this generalization was reviewed by Astier in 1980 and there has been relatively little focus on the mechanisms of thyroid gland function in birds since that review. Thyroid hormone synthesis begins with the iodination of tyrosine residues contained in thyroglobulin, the hormone storage protein of the colloid. Thyroglobulin is produced on ribosomes of the endoplasmic reticulum of follicle cells, carbohydrate components are added in the reticular lumen, further modifications and additions occur in the Golgi apparatus, then the thyroglobulin is transported by vesicles to the apical cell border and extruded into the colloid by exocytosis. Iodination of the tyrosines in thyroglobulin requires the action of thyroid peroxidase and an oxidized form of iodide. The exact oxidized form and all the details of the mechanism of iodination are not fully understood. Iodination results in the formation of monoiodotyrosines (MITs) and diiodotyrosines (DITs) within the thyroglobulin. These are coupled by thyroid peroxidase to yield the hormonal thyronines, T_4 (DIT + DIT) and T_3 (DIT + MIT) (Figure 2). Hormone release from the thyroid involves endocytosis of droplets of colloid by the follicle cells, their fusion with lysosomes, digestion of thyroglobulin by lysosomal enzymes, and release of T_4 and T_3 to capillaries at the external surface of the follicle cells (review, McNabb, 1992).

In adult birds with adequate iodide intake, the thyroids contain primarily T_4, with lesser or undetectable amounts of T_3 (Astier, 1980; McNichols and McNabb, 1987). When iodide is limiting, the T_3/T_4 ratio is increased and total hormone stores are decreased (Japanese quail, McNabb et al., 1985; ring doves, McNichols and McNabb, 1987). In mammals, the T_3/T_4 ratio of hormones released is higher than the ratio of stored hormones because of intrathyroidal T_4- to T_3-conversion (Leonard and Visser, 1986) and it seems likely that the same is the case in birds.

Thyroid hormone secretion rates (TSR), measured by a number of methods, are in the range of 1–3 μg T_4/100 g of body weight per day in chickens, quail, and pigeons (review Grosvenor and Turner, 1960; Wentworth and Ringer, 1986). Cold temperatures increase TSR, whereas iodide deficiency and aging tend to decrease TSR (review, Wentworth and Ringer, 1986).

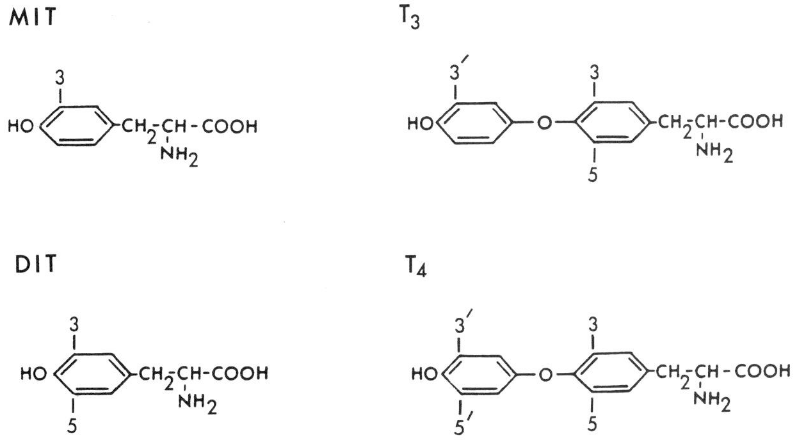

FIGURE 2. Thyroid hormones and their iodinated tyrosine precursors. MIT, monoiodotyrosine; DIT, diiodotyrosine; T_3, triiodothyronine; T_4, thyroxine or tetraiodothyronine. Numbers on the rings indicate the positions of iodine atoms.

In embryos, development of the capability for hormone synthesis precedes organization of the thyroid gland into follicles. In chicken embryos, uptake and concentration of radioiodide by thyroid cells occurs by day 5; colloid droplets and the initiation of follicle formation appear by day 7. Thyroxine is detectable in the plasma by day 6.5 (review, Thommes, 1987), but it is not clear whether this T_4 comes from the embryonic thyroid or reflects uptake of maternal hormones stored in the yolk (Prati *et al.*, 1992; Wilson and McNabb, 1997) or both. Review of the older literature indicates wide variation in the timing of development of events involved in hormone synthesis as reported by different investigators. It should be kept in mind that identification of when an event first occurs is limited by the sensitivity of the technique employed and must be evaluated in that context. Thyroid gland function initially develops independently of hypothalamic-pituitary control. Increases in the rate of radioiodide uptake and increases in thyroid gland hormone content at about 12–13 days of incubation in chicken embryos reflect the establishment of the hypothalamic-pituitary-thyroid (HPT) control axis (Section III; and review, Thommes, 1987).

B. Circulating Thyroid Hormones

Historically, circulating thyroid hormones were first measured as protein-bound iodine, then by competitive binding assays, and now they are determined by radioimmunoassays (RIAs). Of these techniques, RIAs are the most sensitive and the most accurate and have been in common use for analyzing avian thyroid hormones since the late 1970s. Thyroxine concentrations exceed those of T_3 in avian plasma by severalfold although this relationship is much less extreme than in the thyroid gland. Adult birds of many species have plasma or serum T_4 concentrations in the range of 5–15 ng T_4/ml (6–19 pmol/ml) and T_3 concentrations in the range of 0.5–4 ng T_3/ml (0.7–1.5 pmol/ml). In comparison to mammalian plasma, avian plasma contains less T_4 but similar concentrations of T_3. Plasma and serum concentrations of both hormones have been measured in many avian species (Astier, 1980). However, generalizations about species differences seem unwarranted because of the variety of techniques used and because many factors that influence thyroid status typically have not been reported. Factors that influence thyroid function include dietary iodine availability, food availability, food composition, seasonality, age, and time of day of blood collection (reviews, Decuypere *et al.*, 1985; Sharp and Klandorf, 1985). Thus, to have confidence in interspecies comparisons, the different species need to be maintained in equivalent conditions, plasma samples need to be analyzed in the same laboratory (to minimize assay variation), and the RIAs need to be validated for use on plasma or serum from each species.

Diurnal patterns of plasma thyroid hormone concentrations are readily demonstrated in birds (chickens, quail, ducks; reviews, Wentworth and Ringer, 1986; Cogburn and Freeman, 1987). Plasma T_4 concentrations rise and peak during the dark period; T_3 concentrations rise and are highest during the light period. This pattern is consistent with the idea that hormone release from the thyroid is highest during the dark period and extrathyroidal conversion of T_4 to T_3 is highest during the light period. Patterns of food intake are a key factor influencing T_4 to T_3 conversion and its role in diurnal patterns of plasma thyroid hormones (Decuypere and Kühn, 1984). However, several other factors also may be important (Cogburn and Freeman, 1987). For example, cold temperatures increase and warm temperatures depress plasma T_3 concentrations within these diurnal patterns. In general, the influences of temperature change on plasma T_4 are opposite to those on T_3, although the T_4 effects are more complex (Cogburn and Freeman, 1987).

Circulating thyroid hormones are transported by binding proteins. The major transport proteins in birds are albumin and prealbumin/transthyretin (review, Wentworth and Ringer, 1986). Thyroxine-binding globulin (TBG), which is present in the blood of large mammals, is absent in birds, and thus the binding affinity for T_4 is less than that in large mammals (Larsson et al., 1985). Equilibrium dialysis and column chromatography estimates of thyroid hormone binding have suggested higher concentrations of free (i.e., unbound) T_4 in avian than in mammalian plasma. This has led to speculation that the short half lives of thyroid hormones in birds result from higher free thyroid hormone concentrations (i.e., lower hormone binding capacities) in birds than in mammals (Wentworth and Ringer, 1986). However, free hormone RIAs indicate similar free T_4 and T_3 concentrations in avian plasma to those in mammalian plasma (McNabb and Hughes, 1983) and thus argue against this interpretation. It seems likely that the relatively low binding affinity is more important than the binding capacity in the short thyroid hormone half-lives in birds.

Thyroid hormone binding proteins serve several functions in vertebrates: they maintain an extrathyroidal hormone store, they maintain tissue hormone supply if thyroid gland function varies, and they are a factor in the regulation of hormone supply to the tissues (review, Robbins and Bartelena, 1986). That binding proteins play a role in regulating hormone supply has general acceptance, but knowledge of free hormone concentrations alone is not sufficient to fully understand hormone entry into different tissues under different conditions (review, McNabb, 1992). Several studies in birds suggest that binding proteins modulate the picture of thyroid hormone availability; for example, during development in precocial quail and altricial doves (McNabb and Hughes, 1983; McNabb et al., 1984a; Spiers and Ringer, 1984) and in the diurnal cycle in adult ducks (Harvey et al., 1980). In addition to the functions listed above, transthyretin may play a more specific T_4 supply role by transporting T4 through the choroid plexus into the brain (Chanoine et al., 1992) at stages critical to brain development in embryonic chickens (Southwell et al., 1991).

The half-lives of T_3 and T_4 are essentially identical and are short (3–9 hr) in birds (chickens, ducks, Japanese and Bobwhite quail) relative to those in mammals (see, for example, Tata and Shellabarger, 1959; Singh et al., 1967). This subject warrants some reinvestigation, because of problems inherent in a number of the studies: (1) there is a limited time after injection of the labeled hormone when its disappearance from the circulation can be followed (2 hr is required for hormone equilibration, after 24 hr labeled iodide recycling confounds the data; Etta et al., 1972), and (2) most studies have not identified the radioactive compounds in the plasma and have assumed that all label in the plasma is the specific hormone injected, although this assumption is not warranted. Another consideration is that relatively little is known about how hormone half-lives change with alterations in environmental conditions or in the physiological state of birds. Temperature effects on hormone half-lives have been studied; cold temperatures decrease the T_4 half-life both in the laboratory (quail; McFarland et al., 1966) and during winter (Hendrich and Turner, 1967). This decrease in T_4 half-life is consistent with increases in T_4 conversion to T_3 that are a key part of the metabolic response to cold in birds (Rudas and Pethes, 1984).

The ontogenic pattern of plasma thyroid hormones differs in birds with precocial vs altricial development. In precocial species, in which thyroid gland function and its control mature before hatching, plasma T_4 rises several fold during the latter half of embryonic life but plasma T_3 remains very low. Reverse-T_3 (rT_3), an inactive hormone analog, can be detected in the plasma toward the end of embryonic life (Section II,C). During the perihatch period, plasma T_4 and T_3 concentrations rise dramatically to reach some of the highest concentrations measured in avian plasma. This pattern is consistent among precocial galliform birds (e.g., chickens, Thommes and Hylka, 1977; quail, McNabb et al., 1981; turkeys, Christensen et al., 1982) but has not been investigated in precocial species from other avian orders. The perihatch peaks in thyroid hormones are associated with the initiation of thermoregulatory responses to cooling that occur in precocial species during this time (McNabb et al., 1984b). After the perihatch peak, plasma thyroid hormones decrease markedly, then gradually increase during posthatch life to reach adult concentrations.

In altricial species from two avian orders, plasma concentrations of T_3 and T_4 are very low during embryonic life and the perihatch period, then gradually increase during the first 2–3 weeks posthatch to approach adult hormone concentrations (ring doves, McNabb and Cheng, 1985; European starlings, Schew and McNabb, 1996; Vyboh et al., 1996). In these altricial species there is little thyroid development during embryonic life, endothermic responses to cooling first appear several days to a week after hatching, and the young are homeothermic by the time of fledging (McNichols and McNabb, 1988).

C. Thyroid Hormone Activation and Degradation

Fundamental to the interpretation of plasma thyroid hormone concentrations is the need to understand the relative potencies of the hormones. Physiological studies (during the 1950s and 1960s) of the effects of T_3 and T_4 in chickens and quail indicated similar hormone potencies in triggering several responses. These results were in marked contrast to the picture in mammals, in which T_3 is physiologically 3–10× more potent than T_4. Examples of responses that showed equal T_3 and T_4 potency in birds are goiter prevention (Shellabarger, 1955), stimulation of oxygen consumption, heart rate and feather growth in hypothyroid chickens (Newcomer, 1957; Raheja and Snedecor, 1970), and suppression of the effects of the thyroid inhibitor methimazole (Singh *et al.*, 1968). A slightly different picture, namely that T_4 is somewhat more potent than T_3, was suggested by studies of thyroid hormone stimulation of oxygen consumption in the whole animal (Singh *et al.*, 1968) and in cardiac muscle (Newcomer and Barrett, 1960) as well as in the amount of hormone required to prevent goiter after thiouracil treatment (Newcomer, 1957; Mellen and Wentworth, 1959). Yet other studies suggested that T_3 is somewhat more potent than T_4; for example, in stimulating amino acid incorporation into embryonic bone (Adamson and Ingbar, 1967).

The studies above that suggested approximately equal potency of T_3 and T_4 in birds preceded current knowledge indicating avian thyroid receptors are "T_3 receptors" (Section IV) and that apparent T_4 effects may be due to T_3 derived from T_4 by deiodination (see below). More recently, studies of avian thyroid receptors and iodothyronine deiodinases indicate that these aspects of thyroid function are virtually identical to those in mammals. Thus, such studies provide a great deal of circumstantial evidence that T_3 is responsible for most thyroid hormone action in birds, as is the case in mammals (review, Oppenheimer *et al.*, 1987). However, the reason for the *apparent* greater potency of T_4 for triggering some physiological actions in birds has not yet been revealed.

Understanding of the roles of deiodinase enzymes in thyroid hormone activation and degradation in mammals (Leonard and Visser, 1986) has pointed to the value of comparable investigations in birds. In general, birds possess essentially identical deiodinase pathways (Figure 3) to those in mammals and much of the functional significance of these pathways appears to be the same. Three pathways (or types) of deiodinations are present in birds and are important in the relative availability of T_3: (1) type I deiodinase, which can degrade T_4 to either T_3 (by 5'-deiodination (5'D) of the outer ring) or to inactive rT_3 (by 5-deiodination (5D) of the inner ring), is present in liver, kidney, and small intestine of chickens, quail, and doves; (2) type II deiodinase, which converts T_4 to T_3, is present in the central nervous system of chickens; and (3) type III deiodinase, which degrades T_3 by 5D, is present in the liver of young chickens (Borges *et al.*, 1980, 1981; Galton and Hiebert, 1987; Freeman and McNabb, 1991; Darras *et al.*, 1992; Rudas *et al.*, 1993; Suvarna *et al.*, 1993).

Similarities in deiodinase reactions, as well as the predominance of T_4 in the thyroid gland, suggest that most T_3 production is extrathyroidal in birds as it is in humans and some laboratory mammals (Engler and Burger, 1984). Thus, deiodination reactions play an im-

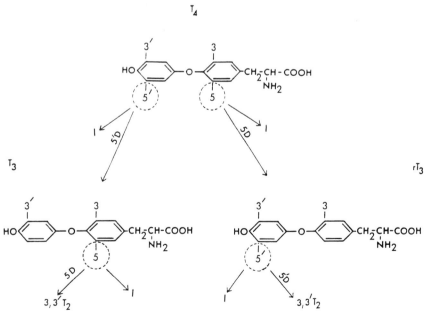

FIGURE 3. Thyroid hormone conversions by deiodinations. Numbers on the rings indicate the positions of iodine atoms. T_4, thyroxine; T_3, triiodothyronine; rT_3, reverse-triiodothyronine.

portant role in the availability of circulating T_3 as well as in local intratissue T_3 concentrations in some cases. Several aspects of the functional significance of deiodinases have been investigated in chickens and quail and appear to play similar roles to those in mammals. During the latter part of embryonic development, T_4 is deiodinated mostly to inactive rT_3 by 5D (Borges et al., 1980) and little 5'D activity is present (Darras et al., 1992). Hepatic type III 5D also is present at this time and presumably degrades the small amount of T_3 that is formed. These patterns explain why plasma T_3 does not increase concurrent with T_3 increases during late embryonic life. Beginning the day before hatching, type I 5'D activity rises, resulting in increased T_3 production (Borges et al., 1980; Galton and Hiebert, 1987; Freeman and McNabb, 1991; Darras et al., 1992) and the type III 5D activity decreases resulting in decreased T_3 degradation (Darras et al., 1992). In combination with the peak of plasma T_4 at this time, these alterations in deiodination reactions result in the perihatch peak in plasma T_3 concentrations (*in vivo* evidence, Decuypere et al., 1982; see citations above for *in vitro* evidence). In brain, which is dependent on adequate thyroid hormone availability for normal development and function, type II 5'D activity plays an important role in protecting T_3 supply to the central nervous system when plasma thyroid hormone concentrations are low. Thus, type II 5'D activity in brain is enhanced in thyroidectomized young chickens and this effect is complemented by enhanced T_3 and T_4 uptake by brain and decreased T_3 and T_4 losses from brain (Rudas and Bartha, 1993; Rudas et al., 1993, 1994).

III. HYPOTHALAMIC–PITUITARY–THYROID AXIS

The avian thyroid gland is primarily under the control of the hypothalamic–pituitary–thyroid (HPT) axis. The avian hypothalamus produces two hormones, thyrotropin releasing hormone (TRH) and somatostatin, that have stimulatory and inhibitory effects, respectively, on the pituitary. Thyroid stimulating hormone (TSH or thyrotropin) produced by thyrotrophs in the anterior pituitary is the major controller of the production and release of thyroid hormones by the thyroid gland. Negative feedback is exerted by thyroid hormones on the pituitary and hypothalamus.

Avian HPT control, based on investigations of chickens (Decuypere and Kühn, 1988), appears to be like that in mammals. The structure of TRH, a tripeptide, is identical in birds and mammals. The structure of TSH, which is a glycoprotein, differs in its beta chain in different vertebrate classes, but heterologous TSHs stimulate thyroid function in birds (McNabb et al., 1986). The lack of antibodies specific to avian TSH, and thus the lack of RIAs for measuring avian TSH, has been a limiting factor in investigations in this area. Antibodies to heterologous TSHs have been used in immunocytochemical studies of HPT axis development (review, Thommes, 1987) and recent reports of antibodies to avian TSH offer promise for future assays (Berghman et al., 1993).

During development, the establishment of HPT axis control over the thyroid gland occurs between embryonic days 10.5 and 11.5 in chickens. The key evidence for this timing is that embryonic decapitation (which includes removal of the pituitary) on day 9.5 does not alter T_4 concentrations in the plasma until day 11.5, but prevents the normal plasma T_4 increase that occurs after day 11.5 in chicken embryos. These findings suggest that HPT control is established on day 11.5 and that prior thyroid gland development must be autonomous. However, the capability for pituitary and thyroid responses is present earlier, because treatment of 6.5 day embryos with exogenous TRH results in increased plasma T_4 concentrations. This implies that by day 6.5 the TRH treatment results in pituitary TSH release, which in turn stimulates increased T_4 release from the thyroid gland (reviews Thommes, 1987; Thommes et al., 1988).

The effects of environmental factors (e.g., temperature) on thyroid function may be mediated partially or completely through the HPT axis. Increased thyroid gland size provides evidence of increased pituitary TSH release during cold exposure or winter conditions. Exposure to high temperatures has the opposite effect (review, Wentworth and Ringer, 1986). The picture of temperature effects on circulating thyroid hormones is more complex, because in addition to thyroid secretion rate, extrathyroidal deiodination (Section II,C) and other aspects of hormone turnover may be affected (review, Wentworth and Ringer, 1986).

IV. MECHANISM OF ACTION OF THYROID HORMONES

In birds, as in other vertebrate classes, thyroid hormone action is mediated through nuclear receptors (review, Bellabarba et al., 1988) that are members of the steroid receptor superfamily (Lazar, 1993). These receptors are referred to as T_3 receptors because they bind T_3 with higher affinity than they bind T_4 or any other functional thyroid hormone analog. The physical and biochemical characteristics of avian T_3 receptors in adult galliform birds are virtually identical to those of other vertebrate classes and recent work suggests the same molecular forms are present in birds and mammals (review, McNabb, 1992). The physiological responses mediated by thyroid hormones are of two types, metabolic and developmental (Sections V,A and V,B).

At present it is reasonable to assume that most thyroid hormone effects are triggered by T_3 in birds as they are in mammals. The types of evidence that support this assumption are: (1) thyroid receptors in birds, as in all other vertebrate classes, have their highest affinity for T_3; (2) the relative proportion of T_3 to T_4 is higher in birds than in mammals, suggesting greater potential for T_3 binding; and (3) deiodinase enzymes, their characteristics, and their functional significance are consistent with T_3 being the primary metabolically active hormone (review, McNabb and King, 1993). Two key types of experiments that have been done in mammals are lacking in birds. Those are: (1) the demonstration that T_3 accounts for >90% of receptor occupancy *in vivo* in some hormone responsive tissues, and (2) the strong correlations between T_3 receptor occupancy and specific thyroid hormone responses (review, Oppenheimer *et al.*, 1987). These types of measurements will be necessary for quantifying the relative importance of T_3 vs T_4 (and other hormone analogs) in triggering physiological responses in birds.

V. EFFECTS OF THYROID HORMONES

A. Thyroid Hormone Effects on Metabolism

Thyroid hormones are considered the key controllers of that part of metabolic heat production that is necessary for the maintenance of high and constant body temperature in homeothermic birds and mammals (Danforth and Burger, 1984). Many specific aspects of thyroid hormone-triggered metabolic effects have been investigated in birds. Administration of exogenous thyroid hormones stimulates and thyroidectomy or goitrogen administration depresses oxygen consumption (review, Wentworth and Ringer, 1986). Likewise, altered thyroid hormone concentrations influence the metabolic energy supply, liver glycogen storage is facilitated by increases in thyroid hormone, and glycogen depletion and plasma glucose decreases are associated with decreases in thyroid hormones. The length of time that oxygen consumption remains elevated after thyroid hormone administration is relatively short (hours) in birds (Singh *et al.*, 1968) and this is consistent with the relatively short half-lives of the hormones in birds (Section II,B).

The timing of thyroid development appears to be critical to the development of thermoregulation in both precocial species with early maturation and in altricial species with much later maturation (precocial quail, Spiers *et al.*, 1974; altricial doves, McNabb and Cheng, 1985; altricial starlings, Schew and McNabb, 1996). During development there are positive correlations between periods of high plasma thyroid hormone concentrations and high metabolic activity (e.g., during the perihatch period in turkeys; Christensen *et al.*, 1982). There also are correlations between oxygen consumption and plasma thyroid hormones in different age groups of chickens (Bobek, 1977).

During late embryonic life, precocial embryos go through several stages in the development of their thermogenic capabilities (Whittow and Tazawa, 1991 and see Chapter 12). Chicken embryos (on day 16.5) increase plasma T_4 in response to cooling (Thommes, 1987) but neither plasma T_3 nor the capability for T_4 to T_3 conversion were measured in this study. Apparent metabolic responses to cooling during late embryonic life in precocial species vary with different studies and different species; there is no evidence of any response to cooling in altricial embryos (review, Whittow and Tazawa, 1991). Thermogenic responses to thyroid hormones are present after hatch in precocial chickens (Freeman, 1970) and posthatching exposure to cold involves both changes in extrathyroidal T_3 production and HPT axis activation (Sections II,C and III).

B. Thyroid Hormone Effects on Development

Thyroid hormones influence both aspects of development (i.e., growth and differentiation/maturation) of birds. Growth (i.e., increase in mass) involves primarily cell proliferation, but also may result from increases in cell size (hypertrophy). In general, thyroid hormones appear to act permissively or indirectly, in concert with other control substances, in their stimulation of growth in birds (review, McNabb and King, 1993). The primary direct hormonal stimulation of body growth results from circulating growth factors (such as insulinlike growth factor-1; IGF1) that are primarily under the control of growth hormone (GH) from the pituitary (see Chapter 23).

A requirement for thyroid hormones in the growth of birds has been demonstrated by the reduced growth that results from thyroidectomy or goitrogen administration. However, within the physiological range of circulating thyroid hormone concentrations, there is little consistent evidence of thyroid hormone stimulation of general body growth (review, McNabb and King, 1993). At hyperthyroid extremes, growth is depressed because of high metabolic rates and a shift toward catabolism (reviews, King and May, 1984; Wentworth and Ringer, 1986). Sex-linked dwarf chickens, which have a deficiency in 5'D activity and consequent very low plasma T_3 concentrations have been an important model for studying the effects of thyroid hormone deficiency on growth (review, Tixier-Boichard et al., 1989).

Thyroid hormones are important in triggering tissue-specific differentiation and maturation processes in many tissues. Typically, this involves the induction of specific structural or enyzmatic proteins that are critical to the development of new tissue functions. Thyroid hormone effects on differentiation/maturation have been extensively studied in mammalian muscle, skeletal, and central nervous tissues (reviews, McNabb, 1992; McNabb and King, 1993). Less extensive information on birds suggests similar thyroid hormone effects on development in avian tissues. For example, in the central nervous system, thyroid hormones are necessary for the development of normal brain architecture and neuronal connections critical to the function of brain regions (e.g., see morphometric studies of Bouvet et al., 1987, in chick brain).

In skeletal muscle and the skeletal system thyroid hormones are required for both growth and differentiation. Posthatching growth in skeletal muscle involves both muscle fiber hypertrophy and satellite cell proliferation and fusion with muscle fibers. Thyroid hormones, perhaps in conjunction with GH, stimulate this cell proliferation posthatch, but do not have this effect in late embryonic life (reviews, King et al., 1987; McNabb and King, 1993). Muscle differentiation, which involves the appearance of specific myosin isoforms critical to functional maturation, also is modulated by T_3 during late embryonic development in chickens (Gardahaut et al., 1992). Hereditary muscular dystrophy of chickens, in which thyroid function is altered, has provided a useful model for studying thyroid hormone effects on muscle development in birds (King and Entrikin, 1991).

In skeletal tissues, in vitro studies of pelvic cartilages from early chicken embryos have been useful in delimiting the roles of thyroid hormones and some growth factors in growth vs differentiation. Thyroid hormones trigger cartilage differentiation/maturation by stimulating matrix production and ossification. In this case T_3 plays a role in growth by stimulating chondrocyte hypertrophy, but does not influence cell proliferation (Burch and Lebovitz, 1982). This effect on growth is by T_3 enhancement of the effects of IGF1 and this growth effect can be separated from the effects of T_3 (even without IGF1) on cartilage maturation (Burch and Van Wyk, 1986). The induction of hepatic malic enzyme in liver by T_3, during early posthatching life, provides an example of how molecular investigations can reveal the exact step-by-step influences of thyroid hormones on the regulation of gene expression during development (review, Goodridge et al., 1989). In some tissues thyroid hormones interact with other hormones in stimulating tissue differentiation and maturation (e.g., in gut and lung development) (review, McNabb and King, 1993).

C. Thyroid Hormone Effects on Hatching, Molt, and Reproduction

Thyroid hormones are important in several organismal level processes such as hatching, molt, and reproduction in birds. The high perihatch thyroid hormone concentrations appear to be stimulating a variety of developmental and metabolic processes necessary for successful hatching (Section V and review; Decuypere et al., 1991).

Thyroid hormones and/or starvation can be used to induce molt in birds and alterations in plasma thyroid hormones occur in conjunction with natural molts. Inhibition of reproductive activity (including cessation of egg laying) and molt occur concurrently in seasonally reproducing wild birds and in commercial poultry practice. Thyroid hormones are necessary for reproductive system development and reproductive function, but high concentrations of thyroid hormones have antigonadal effects. Estrogen decreases appear to be important in the initiation of molt, whereas an increase in the thyroid hormone/estrogen ratio appears to be important in new feather formation (review, Decuypere and Verheyen, 1986).

VI. THYROID INTERACTIONS WITH OTHER HORMONES

Thyroid hormones have important interactions with other hormones (GH, growth factors, glucocorticoids) primarily in growth and development of birds. Thyroid hormones are required for body growth, but it appears that they act in a permissive or indirect way, in conjunction with GH which influences growth factors that directly stimulate cell proliferation (see Chapter 23). Thyroid hormones also modulate GH production and release by the pituitary, by direct inhibition of pituitary somatotropes and by feedback effects on TRH which stimulates somatotropes (Harvey, 1993). This picture differs from that in mammals, in which T_3 consistently stimulates GH production and release and is one of the best understood actions of thyroid hormones. GH also influences thyroid physiology by its effects on extrathyroidal deiodination pathways (Section II,C, and Darras et al., 1993). Increased plasma GH toward the end of incubation is one of the factors that contributes to the increase in plasma T_3 during the perinatal period. Glucocorticoids, which also rise at this time, may increase 5'D activity (Decuypere et al., 1991) and also have important effects on organ differentiation in lung and gut

(Section V,B). These hormonal interactions are conspicuous during the perihatch period (Section V; Decuypere *et al.*, 1991) but continue to be important in a variety of metabolic and developmental events at other ages.

VII. ENVIRONMENTAL INFLUENCES ON THYROID FUNCTION

Changes in temperature and food availability appear to be the environmental factors that have the greatest influence on thyroid function. As mentioned elsewhere in this chapter, environmental cold can act through the HPT axis (Section III) to increase thyroid hormone release (Section II,B), and can alter the peripheral thyroid hormone picture by increasing extrathyroidal T_4 to T_3 conversion (Section II,C) and increasing thyroid hormone turnover (i.e., decrease hormone half-lives; Section II,B). Overall, these changes provide increased thyroid hormone availability and increased T_3 (i.e., hormone activation) in association with the metabolic increases required for homeothermy at cold ambient temperatures. Plasma T_4 concentrations often change relatively little (i.e., less than T_3 changes) with cold exposure, suggesting that the HPT axis response is less important than that of extrathyroidal 5'D (see for example, Rudas and Pethes, 1984). In general, hot temperatures have the opposite effect (i.e., they depress thyroid function) (reviews, Sharp and Klandorf, 1985; Wentworth and Ringer, 1986).

Like temperature change, alterations in food availability and composition alter thyroid status; food restriction decreases circulating T_3, feeding increases T_3 (Sections II,B and II,C). Feeding is one of the key factors that plays a role in the diurnal patterns of plasma thyroid hormone concentrations (Section II,B). Likewise, it may be an important factor in temperature induced changes in thyroid hormone; for example, the depressed plasma T_3 concentrations in chickens exposed to 40°C may result largely from the decreased food consumption at high temperatures (Williamson *et al.*, 1985).

Despite many investigations of photoperiod and thyroid function, no general relationship has emerged (Nicholls *et al.*, 1985; Lal and Thapliyal, 1985). Some of the variation between studies may result from species specific differences. In addition, seasonal influences may be dominated by temperature effects, and overlapping changes in photoperiod, gonad function, and food quality/availability provide a complex set of interactions (review, Wentworth and Ringer, 1986).

References

Adamson, L. F., and Ingbar, S. H. (1967). Some properties of the stimulatory effect of thyroid hormones on amino acid transport by embryonic chick bone. *Endocrinology* **81**, 1372–1378.

Astier, H. (1980). Thyroid gland in birds: Structure and function. *In* "Avian Endocrinology" (A. Epple and M. H. Stetson, eds.), pp. 167–189. Academic Press, New York.

Bellabarba, D., Belisle, S., Gallo-Payet, N. and Lehoux, J-G. (1988). Mechanism of action of thyroid hormones during chick embryogenesis. *Am. Zool.* **28**, 389–399.

Berghman, L. R., Darras, V. M., Chiasson, R. B., Decuypere, E., Kühn, E. R., Buyse, J., and Vandesande, F. (1993). Immunocytochemical demonstration of chicken hypophyseal thyrotropes and development of a radioimmunological indicator for chicken TSH using monclonal and polyclonal homologous antibodies in a substractive strategy. *Gen. Comp. Endocrinol.* **92**, 189–200.

Bobek, S., Jastrzebski, M., and Pietras, M. (1977). Age-related changes in oxygen consumption and plasma thyroid hormone concentration in the young chicken. *Gen. Comp. Endocrinol.* **31**, 169–174.

Borges, M., LaBourene, J., and Ingbar, S. H. (1980). Changes in hepatic iodothyronine metabolism during ontogeny of the chick embryo. *Endocrinology* **107**, 1751–1761.

Borges, M., Eisenstein, Z., Burger, A. G., and Ingbar, S. H. (1981). Immunosequestration: A new technique for studying peripheral iodothyronine metabolism *in vitro*. *Endocrinology* **108**, 1665–1671.

Bouvet, J., Usson, Y., and Legrand, J. (1987). Morphometric analysis of the cerebellar purkinje cell in the developing normal and hypothyroid chick. *Int. J. Dev. Neurosci.* **5**, 345–355.

Burch, W. M., and Lebovitz, H. E. (1982). Triiodothyronine stimulation of *in vitro* growth and maturation of embryonic chick cartilage. *Endocrinology* **111**, 462–468.

Burch, W. M., and Van Wyk, J. J. (1986). Triiodothyronine stimulates cartilage growth and maturation by different mechanisms. *Am. J. Physiol.* **252**, E176–E182.

Chanoine, J-P., Alex, S., Fang, S. L., Stone, S., Leonard, J. L., Korhle, J., and Braverman, L. E. (1992). Role of transthyretin in the transport of thyroxine from the blood to the choroid plexus, the cerebrospinal fluid, and the brain. *Endocrinology* **130**, 933–938.

Christensen, V. L., Biellier, H. V., and Forward, J. F. (1982). Physiology of turkey embryos during pipping and hatching. III. Thyroid function. *Poult. Sci.* **61**, 367–374.

Cogburn, L. A., and Freeman, R. M. (1987). Response surface of daily thyroid hormone rhythms in young chickens exposed to constant ambient temperature. *Gen. Comp. Endocrinol.* **68**, 113–123.

Danforth, E., Jr., and Burger, A. (1984). The role of thyroid hormones in the control of energy expenditure. *Clin. Endocrinol. Metab.* **13**, 581–595.

Darras, V. M., Visser, T. J., Berghman, L. R., and Kühn, E. R. (1992). Ontogeny of Type I and Type II deiodinase activities in embryonic and posthatch chicks: Relationship with changes in plasmaa triiodothyronine and growth hormone levels. *Comp. Biochem. Physiol. A* **103**, 131–136.

Darras, V. M., Rudas, P., Visser, T. J., Hall, T. R., Huybrechts, L. M., Vanderpooten, A., Berghmann, L. R., Decuypere, E., and Kühn, E. R. (1993). Endogenous growth hormone controls high plasma levels of 3,3',5-triiodothyronine (T_3) in growing chickens by decreasing the T_3-degrading type III deiodinase activity. *Dom. Anim. Endocrinol.* **10**, 55–65.

Davison, T. F., Flack, I. H., and Butler, E. J. (1978). The binding of thyroxine and tri-iodothyronine to plasma proteins in the chicken at the physiological pH. *Res. Vet. Sci.* **25**, 280–283.

Decuypere, E., and Kühn, E. R. (1984). Effect of fasting and feeding time on circadian rhythms of serum thyroid hormone concentrations, glucose, liver mondeiodinase activity and rectal temperature in growing chickens. *Dom. Anim. Endocrinol.* **1**, 251–262.

Decuypere, E., and Kühn, E. R. (1988). Thyroid hormone physiology in galliformes: Age and strain related changes in physiological control. *Am. Zool.* **28**, 401–415.

Decuypere, E., and Verheyen, G. (1986). Physiological basis of induced moulting and tissue regeneration in fowls. *World's Poult. Sci. J.* **42**, 56–66.

Decuypere, E., Kühn, E. R., Clijmans, B., Nouwen, E. J. and Michels, H. (1982). Prenatal peripheral monodeiodination in the chick embryo. *Gen. Comp. Endocrinol.* **47**, 15–17.

Decuypere, E., Kühn, E. R., and Chadwick, A. (1985). Rhythms in circulating prolactin and thyroid hormones in the early postnatal life of the domestic fowl: Influence of fasting and feeding on thyroid rhythmicity. *In* "The Endocrine System and the Environment" (B. K. Follett, S. Ishii, and A. Chandola, eds.), pp. 189–200. Japan Scientific Societies Press, Tokyo and Springer-Verlag, Berlin.

Decuypere, E., Dewil, E., and Kühn, E. R. (1991). The hatching process and the role of hormones. *In* "Avian Incubation" (S. G. Tullett, ed.), pp. 239–256. Butterworth-Heinemann, London.

Dent, J. N. (1986). The thyroid gland. *In* "Vertebrate Endocrinology: Fundamentals and Biomedical Implications" (P. K. T. Pang and M. P. Schreibman, eds.), Vol. 1, pp. 175–206. Academic Press, New York.

Engler, D., and Burger, A. (1984). The deiodination of the iodothyronines and their derivatives in man. *Endocr. Rev.* **5**, 151–184.

Etta, K. M., Ringer, R. K., and Reineke, E. P. (1972). Degradation of thyroxine confounded by thyroidal recycling of radioactive iodine. *Proc. Soc. Exp. Biol. Med.* **140**, 462–464.

Freeman, B. M. (1970). Thermoregulatory mechanisms of the neonate fowl. *Comp. Biochem. Physiol.* **33**, 219–230.

Freeman, T. B., and McNabb, F. M. A. (1991). Hepatic 5′-deiodinase activity of Japanese quail using reverse-T_3 as substrate: Assay validation, characterization, and developmental studies. *J. Exp. Zool.* **258**, 212–220.

French, E. I., and Hodges, R. D. (1977). Fine structural studies on the thyroid gland of the normal domestic fowl. *Cell Tissue Res.* **178**, 397–410.

Fujita, H., and Tanizawa, Y. (1966). Electron microscopic studies on the development of the thyroid gland of chick embryo. *Z. Zellforsch. Mikrosk. Anat.* **125**, 132–151.

Galton, V. A., and Hiebert, A. (1987). The ontogeny of the enzyme systems for the 5′- and 5-deiodination of thyroid hormones in chick embryo liver. *Endocrinology* **120**, 2604–2610.

Gardahaut, M. F., Fontaine-Perus, J., Rouaud, T., Bandman, E., and Ferrand, R. (1992). Developmental modulation of myosin expression by thyroid hormone in avian skeletal muscle. *Development* **115**, 1121–1131.

Goodridge, A. G., Crish, J. F., Hillgartner, F. B., and Wilson, S. B. (1989). Nutritional and hormonal regulation of the gene for avian malic enzyme. *J. Nutr.* **119**, 299–308.

Grosvenor, C. E., and Turner, C. W. (1960). Measurement of thyroid secretion rate of individual pigeons. *Am. J. Physiol.* **198**, 1–3.

Harvey, S. (1993). Growth hormone secretion in poikilotherms and homeotherms. *In* "The Endocrinology of Growth, Development, and Metabolism of Vertebrates" (M. P. Schreibman, C. G. Scanes, and P. K. T. Pang, eds.), pp. 151–182. Academic Press, New York.

Harvey, S., Davison, T. F., Klandorf, H., and Phillips, J. G. (1980). Diurnal changes in the plasma concentrations of thyroxine and triiodothyronine and their binding to plasma proteins in the domestic duck (*Anas platyrhynchos*). *Gen. Comp. Endocrinol.* **42**, 500–504.

Hendrich, C. E., and Turner, C. W. (1967). A comparison of the effect of environmental temperature changes and 4.4 C cold on the biological half-life ($t_{1/2}$) of thyroxine-^{131}-I in fowls. *Poult. Sci.* **46**, 3–5.

King, D. B., and Entrikin, R. K. (1991). Thyroidal involvement in the expression of avian muscular dystrophy. *Life Sci.* **48**, 909–916.

King, D. B., and May, J. D. (1984). Thyroidal influence on body growth. *J. Exp. Zool.* **232**, 453–460.

King, D. B., Bair, W. B., and Jacaruso, R. B. (1987). Thyroidal influence on nuclear accumulation and DNA replication in skeletal muscles of young chickens. *J. Exp. Zool. Suppl.* **1**(Suppl.), 291–298.

Larsson, M., Pettersson, T., and Carlstrom. (1985). Thyroid hormone binding in serum of 15 vertebrate species: Isolation of thyroxine-binding globulin and prealbumin analogs. *Gen. Comp. Endocrinol.* **58**, 360–375.

Lazar, M. A. (1993). Thyroid hormone receptors: Multiple forms, multiple possibilities. *Endocr. Rev.* **14**, 184–192.

Lal, P., and Thapliyal, J. P. (1985). Photorefractoriness in migratory red-headed bunting, *Emberiza bruniceps*. *In* "The Endocrine System and the Environment" (B. K. Follett, S. Ishii, and A. Chandola), pp. 127–135. Japan Scientific Societies Press, Tokyo and Springer-Verlag, Berlin.

Leonard, J. L., and Visser, T. J. (1986). Biochemistry of deiodination. *In* "Thyroid Hormone Metabolism" (G. Hennemann, ed.), pp. 189–229. Dekker, New York.

McFarland, L. Z., Yousef, M. R., and Wilson, W. O. (1966). The influence of ambient temperature and hypothalamic lesions on the disappearance rate of thyroxine ^{131}-I in the Japanese quail. *Life Sci.* **5**, 309–315.

McNabb, F. M. A. (1992). "Thyroid Hormones." Prentice Hall, Englewood Cliffs, NJ.

McNabb, F. M. A., and Cheng, M.-F. (1985). Thyroid development in Ring doves, *Streptopelia risoria*. *Gen. Comp. Endocrinol.* **58**, 243–251.

McNabb, F. M. A., and Hughes, T. E. (1983). The role of serum binding proteins in determining free thyroid hormone concentrations during development in quail. *Endocrinology* **113**, 957–963.

McNabb, F. M. A., and King, D. B. (1993). Thyroid hormone effects on growth, development and metabolism. *In* "The Endocrinology of Growth, Development, and Metabolism of Vertebrates" (M. P. Schreibman, C. G. Scanes, and P. K. T. Pang, eds.), pp. 393–417. Academic Press, New York.

McNabb, F. M. A., and McNabb, R. A. (1977). Thyroid development in precocial and altricial avian embryos. *Auk* **94**, 736–742.

McNabb, F. M. A., Weirich, R. T., and McNabb, R. A. (1981). Thyroid development in embryonic and perinatal Japanese quail. *Gen. Comp. Endocrinol.* **43**, 218–226.

McNabb, F. M. A., Lyons, L. J., and Hughes, T. E. (1984a). Free thyroid hormones in altricial (ring doves) versus precocial (Japanese quail) development. *Endocrinology* **115**, 2133–2136.

McNabb, F. M. A., Stanton, F. W., and Dicken, S. G. (1984b). Posthatching thyroid development and body growth in precocial vs. altricial birds. *Comp. Biochem. Physiol. A* **78**, 629–635.

McNabb, F. M. A., Blackman, J. R., and Cherry, J. A. (1985). The effects of different maternal dietary iodine concentrations on Japanese quail. *Dom. Anim. Endocrinol.* **2**, 25–34.

McNabb, F. M. A., McNichols, M. J., and Slack, P. M. (1986). The nature of thyrotropin stimulation of thyroid function in Japanese quail: Prolonged thyrotropin exposure is necessary to increase thyroidal ^{125}-I uptake. *Comp. Biochem. Physiol. A* **85**, 459–463.

McNichols, M. J., and McNabb, F. M. A. (1987). Comparative thyroid function in adult Japanese quail and ring doves: Influence of dietary iodine availability. *J. Exp. Zool.* **244**, 263–268.

McNichols, M. J., and McNabb, F. M. A. (1988). Development of thyroid function and its pituitary control in embryonic and hatchling precocial Japanese quail and altricial Ring doves. *Gen. Comp. Endocrinol.* **69**, 109–118.

Mellon, W. J., and Wentworth, B. C. (1959). Observations on radiothyroidectomized chickens. *Poult. Sci.* **41**, 134–141.

Newcomer, W. S. (1957). Relative potencies of thyroxine and triiodothyronine based on various criteria in thiouracil-treated chickens. *Am. J. Physiol.* **190,** 413–418.

Newcomer, W. S. (1978). Dietary iodine and accumulation of radioiodine in thyroids of chickens. *Am. J. Physiol.* **234,** E168–E176.

Newcomer, W. S., and Barrett, P. A. (1960). Effects of various analogues of thyroxine on oxygen uptake of cardiac muscle from chicks. *Endocrinology* **66,** 409–415.

Nicholls, T. J., Goldsmith, A. R., Dawson, A., Chakraborty, S., and Follett, B. K. (1985). Involvement of the thyroid glands in photorefractoriness in starlings. *In* "The Endocrine System and the Environment" (B. K. Follett, S. Ishii, and A. Chandola), pp. 127–135. Japan Scientific Societies Press, Tokyo and Springer-Verlag, Berlin.

Oppenheimer, J. H., Schwartz, H. L., Mariash, C. N., Kinlaw, W. B., Wong, N. W., and Freake, H. C. (1987). Advances in our understanding of thyroid hormone action at the cellular level. *Endocr. Rev.* **8,** 288–308.

Prati, M., Calvo, R., and Morreale de Escobar, G. (1992). L-thyroxine and 3,5,3′-triiodothyronine concentrations in the chicken egg and in the embryo before and after the onset of thyroid function. *Endocrinology* **130,** 2651–2659.

Raheja, K. L., and Snedecor, J. G. (1970). Comparison of subnormal multiple doses of L-thyroxine and L-triiodothyronine in propylthiouracil-fed and radiothyroidectomized chicks (*Gallus domesticus*). *Comp. Biochem. Physiol.* **37,** 555–563.

Robbins, J., and Bartelena, L. (1986). Plasma transport of thyroid hormones. *In* "Thyroid Hormone Metabolism" (G. Hennemann, ed.), pp. 3–38. Dekker, New York.

Romanoff, A. L. (1960). "The Avian Embryo." Macmillan, New York.

Rudas, P., and Bartha, T. (1993). Thyroxine and triiodothyronine uptake by the brain of chickens. *Acta vet. Hung.* **41,** 395–408.

Rudas, P., and Pethes, G. (1984). The importance of the peripheral thyroid hormone deiodination in adaptation to ambient temperature in the chicken (*Gallus domesticus*). *Comp. Biochem. Physiol.* **77,** 567–571.

Rudas, P., Bartha, T., and Frenyo, V. L. (1993). Thyroid hormone deiodination in the brain of young chickens acutely adapts to changes in thyroid status. *Acta vet. Hung.* **41,** 381–393.

Rudas, P., Bartha, T., and Frenyo, V. L. (1994). Elimination and metabolism of triiodothyronine depend on the thyroid status in the brain of young chickens. *Acta vet. Hung.* **42,** 218–230.

Schew, W. A., McNabb, F. M. A., and Scanes, C. G. (1996). Comparison of the ontogenesis of thyroid hormones, growth hormone, and insulin-like growth factor-1 in *ad libitum* and food-restricted (altricial) European starlings and (precocial) Japanese quail. *Gen. Comp. Endocrinol.* **101,** 304–316.

Sharp, P. J., and Klandorf, H. (1985). Environmental and physiological factors controlling thyroid function in Galliformes. *In* "The Endocrine System and the Environment" (B. K. Follett, S. Ishii, and A. Chandola), pp. 175–188. Japan Scientific Societies Press, Tokyo and Springer-Verlag, Berlin.

Singh, A., Reineke, E. P., and Ringer, R. K. (1967). Thyroxine and triiodothyronine turnover in the chicken and the Bobwhite and Japanese quail. *Gen. Comp. Endocrinol.* **9,** 353–361.

Singh, A., Reineke, E. P., and Ringer, R. K. (1968). Influence of thyroid status of the chick on growth and metabolism with observations on several parameters of thyroid function. *Poult. Sci.* **47,** 212–219.

Shellabarger, C. J. (1955). A comparison of triiodothyronine and thyroxine in the chick goiter-prevention tests. *Poult. Sci.* **34,** 1437–1440.

Southwell, B. R., Duan, W., Tu, G.-F., and Schreiber, G. (1991). Ontogenesis of transthyretin gene expression in chicken choroid plexus and liver. *Comp. Biochem. Physiol. B* **100,** 329–338.

Spiers, D. E., and Ringer, R. K. (1984). Thyroid hormone changes in the Bobwhite (*Colinus virginianus*) after hatching. *Gen. Comp. Endocrinol.* **53,** 302–308.

Spiers, D. E., McNabb, R. A., and McNabb, F. M. A. (1974). The development of thermoregulatory ability, heat-seeking activities, and thyroid function in hatchling Japanese quail (*Coturnix coturnix japonica*). *J. Comp. Physiol.* **89,** 159–174.

Suvarna, S., McNabb, F. M. A., Dunnington, E. A. and Siegel, P. B. (1993). Intestinal 5′deiodinase activity of developing and adult chickens selected for high and low body weight. *Gen. Comp. Endocrinol.* **91,** 259–266.

Tata, J. R., and Shellabarger, C. J. (1959). An explanation for the difference between the responses of mammals and birds to thyroxine and tri-iodothyronine. *Biochem. J.* **72,** 608–613.

Thommes, R. C. (1987). Ontogenesis of thyroid function and regulation in the developing chick embryo. *J. Exp. Zool.* **1**(Suppl.), 273–279.

Thommes, R. C., and Hylka, V. W. (1977). Plasma iodothyronines in the embryonic and immediate post-hatch chick. *Gen. Comp. Endocrinol.* **32,** 417–422.

Thommes, R. C., Hylka, V. W., Tonetta, S. A., Griesbach, D. A., Ropka, S. L., and Woods, J. E. (1988). Hypothalamic regulation of the pituitary-thyroid unit in the developing chick embryo. *Am. Zool.* **28,** 417–426.

Tixier-Boichard, M., Huybrechts, L. M., Kühn, E., Decuypere, E., Charrier, J., and Mongin, P. (1989). Physiological studies on the sex-linked dwarfism of the fowl: A review on the search for the gene's primary effect. *Genet. Sel. Evol.* **21,** 217–234.

Vyboh, P., Zeman, M., Jurani, M., Buyse, J., and Decuypere, E. (1996). Plasma thyroid hormone and growth hormone patterns in precocial Japanese quail and European starlings during postnatal development. *Comp. Biochem. Physiol.* **114,** 23–27.

Wentworth, B. C., and Ringer, R. K. (1986). Thyroids. *In* "Avian Physiology" (P. D. Sturkie, ed.). pp. 452–465. Springer-Verlag, New York.

Whittow, G. C., and Tazawa, H. (1991). The early development of thermoregulation in birds. *Physiol. Zool.* **64,** 1371–1390.

Williamson, R. A., Mission, B. H., and Davison, T. F. (1985). The effect of exposure to 40 C on the heat production and the serum concentrations of triiodothyronine, thyroxine, and corticosterone in immature domestic fowl. *Gen. Comp. Endocrinol.* **60,** 178–186.

Wilson, C. M., and McNabb, F. M. A. (1997). Maternal thyroid hormones in Japanese quail eggs and their influence on embryonic development. *Gen. Comp. Endocrinol.* **107,** 153–165.

CHAPTER 18

The Parathyroids, Calcitonin, and Vitamin D

CHRISTOPHER G. DACKE
Pharmacology Division
School of Pharmacy and Biomedical Science
University of Portsmouth
Portsmouth PO1 2DT, United Kingdom

I. Introduction 473
II. Parathyroid Hormone and Related Peptides 474
    A. The Parathyroid Glands 474
    B. Parathyroidectomy 474
    C. Chemistry of Parathyroid Hormone and Related Peptides 474
    D. Circulating Parathyroid Hormone 475
    E. Actions of Parathyroid Hormone 475
    F. Parathyroid Hormone Related Peptide 477
III. Calcitonin and the Ultimobranchial Glands 479
    A. The Ultimobranchial Glands 479
    B. Calcitonin 479
    C. Circulating Calcitonin Levels 479
    D. Actions of Calcitonin 480
    E. Role of Extracellular Ca 480
    F. Calcitonin Gene-Related Peptide and Amylin 480
IV. The Vitamin D System 481
    A. Renal Vitamin D Metabolism 481
    B. Circulating Levels of Vitamin D Metabolites 482
    C. Actions of Vitamin D 482
V. Prostaglandins and Other Factors 483
VI. Conclusions 484
    References 485

I. INTRODUCTION

Calcium (Ca) is one of the most efficiently regulated plasma constituents in birds. The classical Ca-regulating hormones, parathyroid hormone (PTH), calcitonin (CT), and 1,25-dihydroxy vitamin D_3 (1,25-$(OH)_2$ D_3), are all recognized in this class, although their actions and/or sensitivities can be quite different from those in mammals (Dacke, 1979) and it is clear that avian skeletal metabolism acts at an amplified rate compared with that in mammals (Gay, 1988). We now realize that other putative Ca and bone-regulating factors such as prostaglandins (PGs; Dacke, 1989) and calcitonin gene-related peptide (CGRP; Dacke *et al.*, 1993) also affect avian Ca metabolism in ways which are profoundly different from those in mammals. This may be partly related to enhanced requirements for Ca during eggshell calcification, although the requirements of bone and Ca metabolism in growing birds are essentially similar to those of growing mammals. Recent general reviews of avian Ca and bone metabolism include those by Taylor and Dacke (1984), Soares (1984), Kenny (1986), Hurwitz (1989), and Gay (1996). Avian Ca metabolism shares many features found in other vertebrate classes, but is also typified by several unique characteristics related to the ability of this class to lay large megalecithal eggs with a well-developed calcified eggshell (Romanoff and Romanoff, 1963). Other species, notably extant crocodilia and extinct dinosaurs, the closest relatives of modern birds, also lay or laid eggs with some degree of shell calcification. The amount of Ca in each egg typically represents about 10% of the total body stores of Ca of the bird (Kenny, 1986), an enormous amount by any measure. The Ca metabolism of an egg-laying hen in domesticated species such as chickens and Japanese quail can be compared with that of a human female in 18 months of combined pregnancy and lactation. In order to provide a source of Ca for eggshell calcification to supplement the supply provided directly from the diet, egg-laying hens uniquely possess a highly labile reservoir in the form of medullary bone, which develops within the long bones of hens in response to the activity of gonadal steroids. It is the most overtly estrogen-sensitive form of vertebrate bone (Dacke et al., 1993). A second feature which may have profoundly influenced the evolution of avian Ca metabolism is the ability to fly. This probably led to the development of a light but robust skeleton in which long bones tend to be more hollowed out than in other vertebrates. Similar lightweight structures are found in fossilized bones of the flying pterosaurs (Wellnhoser, 1991). This hollowing implies a high degree of remodeling during the growth phase of the skeleton and has probably influenced the evolution of certain hormonal activities and sensitivities in birds.

Mammals given either an acute intravenous Ca load or ethylene glycol-bis (-amino ethyl ether) N,N,N',N'-tetraacetic acid (EGTA), which lowers plasma Ca levels, respond to this challenge within tens of minutes to a few hours, while 1-week-old chickens will correct such challenges within a few minutes, recovery from the hypocalcaemia being dependent upon the presence of PTH (Koch et al., 1984). An intravenous dose of ^{45}Ca in a week-old chick is rapidly cleared from the plasma pool and by 15 min approximately 40% of the original dose is located within the skeleton (Shaw et al., 1989). By calculating the unidirectional plasma–bone clearance constant ($K_{pb}{}^{45}$Ca) and hence net Ca^{2+} influx, and by estimating total rate of Ca accretion into the skeleton, we find that total skeletal outflux of Ca^{2+} is approximately 80% of influx and that the plasma pool of Ca^{2+} is cleared into the skeleton every few minutes. Typically, net Ca^{2+} accumulation into the chick femur is about 0.28 mole min^{-1}g^{-1} wet weight, although this will vary according to prevailing rates of dietary Ca absorption and other factors. Clearly any factor which modifies either rapid influx or outflux of Ca^{2+} in this system is likely to profoundly affect minute-to-minute plasma Ca modulation in the rapidly growing animal. Bronner and Stein (1992) calculated from our data (Shaw et al., 1989) that the $t_{1/2}$ for ^{45}Ca uptake by the chick femur is less than 10 min, compared with around 30 min in rabbit, dog, and rat.

II. PARATHYROID HORMONE AND RELATED PEPTIDES

A. The Parathyroid Glands

The number of parathyroids varies between two and four in birds; in chickens there are two pairs slightly caudal to the thyroid and often fused together while one pair is found in Japanese quail (Dacke, 1979; Clark and Sassayama, 1981; Kenny, 1986). The parathyroids are derived embryologically from the third and fourth branchial pouches in association with the thymus. They are encapsulated by connective tissue and composed mainly of chief cells, a situation similar to the rat (Roth and Schiller, 1976). Oxyphil cells, which are present in mammalian parathyroid tissues, are absent in birds. While the ultrastructure of the avian chief cell is essentially similar to that in mammals, the low granular content of the avian cell is consistent with a low level of PTH secretion (Kenny, 1986).

B. Parathyroidectomy

Details of the effects of parathyroidectomy (PTX) in birds were described in the previous edition (Kenny, 1986). This procedure has been accomplished in several avian species, including pigeon, duck, starling, chicken, and Japanese quail. Responses include hypocalcemia, tetany and ultimately death, depending on factors such as dietary Ca intake, the presence of accessory parathyroid tissue and reproductive status. Clark and her associates examined the effects of PTX on urinary Ca and phosphate (P_i) excretion in starlings (Clark and Wideman, 1977) and Japanese quail (Clark and Sassayama, 1981) and demonstrated that it leads to increased Ca excretion and decreased P_i excretion. Hypocalcemia also occurred in both species while hyperphosphatemia was observed in starlings.

C. Chemistry of Parathyroid Hormone and Related Peptides

Mammalian PTH is an 84-amino-acid polypeptide in its native form, although there is recent evidence that the chicken hormone consists of 88 amino acids (Khosla et al., 1988; Russell and Sherwood, 1989; Lim et al., 1991). However, only the first 32–34 amino acids from the N terminal are necessary for biological activity and PTH (1–34) has similar activity to PTH (1-84) (Tregear et al., 1973). The nucleotide sequence of chicken pre-pro-PTH mRNA was determined from a 2.3-kb fragment of complementary PTH DNA cloned in Escherichia coli. The mRNA (2.3 kb) for chicken hormone precursor was approximately 3× the size of mRNA for mammalian pre-pro-PTH. The hormone sequence deduced from the DNA showed that chicken pre-pro-PTH mRNA encoded a 119-amino-acid precursor peptide, and an 88-amino-acid hormone which is four residues longer than all known mammalian homologs and includes gene deletions and insertions. There is significant homology of sequence in the biologically active 1-34 region with mammalian hormones, but much less in the middle and carboxyl-terminal regions (Russell and Sherwood, 1989).

D. Circulating Parathyroid Hormone

Few reports exist of the circulating levels of PTH in birds probably reflecting the fact that they are normally extremely low and until recently pure avian PTH was unavailable for purposes of antibody production. Van de Velde *et al.* (1984) measured plasma PTH-like bioactivity during the chickens egg-laying cycle by cytochemical bioassay. This is elevated during the period of eggshell calcification after which it falls to a low level but is slightly raised again 2 hr after ovulation. Singh *et al.* (1986) similarly measured levels of "PTH" during the egg cycle of chickens using an *in vitro* bioassay. They found that PTH bioreactivity was higher in hens fed Ca deficient diets than in those on a high Ca diet. The levels were highest during the phase of shell calcification than shortly after ovulation in both groups of hens, furthermore, they showed an inverse relationship with plasma-ionized Ca levels (Figure 1), suggesting that changes in bioactive PTH play an important role in the Ca metabolism of the chicken during this physiological Ca stress. At least part of the PTH-like activity measured in these studies was probably PTHrP and it will be useful in the future to distinguish between the two peptides using specific antibodies to ascertain their respective roles in the avian egg cycle as well as in their Ca metabolism in general.

E. Actions of Parathyroid Hormone

The importance of PTH in maintaining avian Ca levels has been recognized for many years. PTX causes hypocalcemia, while PTH injections into Japanese quail or chickens increase plasma Ca levels (Kenny and Dacke, 1974). The hypercalcemic effects of PTH are greater in egg laying hens than in cockerels due to either Ca binding by yolk proteins in the plasma or additional PTH receptors, which may be present in medullary bone and oviduct (Dacke, 1979). Immature birds show a very rapid and sensitive reaction to an intravenous dose of PTH, this response being widely used as a bioassay (Dacke and Kenny, 1973, Parsons *et al.*, 1973). Primary targets for PTH in birds as in mammals are bone and kidney. The major physiological stimulus for PTH secretion from the chief cells is a fall in plasma Ca concentration, while a rise in Ca suppresses it (Brown, 1991).

1. Skeletal Actions

PTH has multiple catabolic and anabolic effects on the skeleton. It causes a slow hypercalcemic response in mammals involving recruitment and activation of osteoclasts, mediated via receptors located in osteoblasts (Hurwitz, 1989). It has been recognized for more than 35 years (Polin *et al.*, 1957) that birds are exquisitely sensitive to PTH. Furthermore the hypercalcemic response occurs in egg-laying hens as early as 8 min after administration (Candlish and Taylor, 1970), a time scale

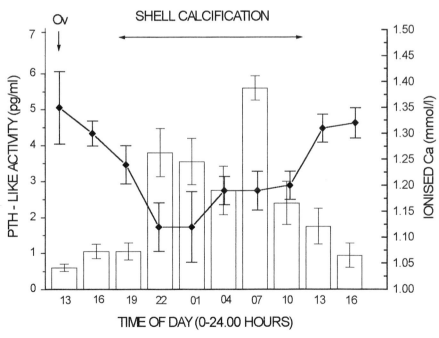

FIGURE 1 Blood levels of PTH-like bioactivity (column graph) and ionised Ca (line graph) during the egglay cycle of the chicken. Ovulation (Ov). (Redrawn from Singh *et al.* (1986), *Gen. Comp. Endocrinol.* with permission.)

too brief for significant osteoclastic resorption (Hurwitz, 1989) and unlikely to be accounted for by the relatively slow changes in intestinal or renal transport mechanisms. Using a technique of acute ^{45}Ca labeling, it was demonstrated that the initial (0–30 min) response to PTH involved inhibition of plasma Ca^{2+} clearance (Kenny and Dacke, 1974). The basis for this response remained obscure for more than a decade, until a method for temporal microwave fixation of injected radioisotopes in skeletal and other tissues was developed (Shaw and Dacke, 1985; Shaw et al., 1989). This enabled the demonstration that decreased plasma ^{45}Ca clearance in chicks is partly accounted for by an inhibition of net Ca uptake into the skeleton. These dose dependent responses were elicited using the active bPTH(1-34) fragment, but could also be seen following intravenous injection of a PGE_2 analog, 16,16-dimethyl PGE_2 (Table 1). They are extremely rapid and most apparent in the femur and, to a lesser extent, calvaria (Dacke and Shaw, 1987). Phosphodiesterase inhibitors (i.e., caffeine or 3-isobutyl-1-methylxanthine) mimic the effects of PTH on skeletal ^{45}Ca uptake in chicks, suggesting a possible role for cAMP in this response (Shaw and Dacke, 1989). Oxidation of bPTH(1-34) with hydrogen peroxide abolishes Ca uptake and reduces the concurrent cAMP activation, but not the hypercalcemic response. The analog [Nle^8, Nle^{18}, Tyr^{34}]-bPTH(1-34) gives a much-reduced hypercalcemic response and slightly reduces effects on plasma ^{45}Ca clearance and bone uptake, which suggests that the initial hypercalcemic response to PTH is not merely a reflection of its acute effects on net skeletal Ca uptake (Dacke and Shaw, 1988).

PTH receptors are located on osteoblast surfaces but have been considered absent in osteoclasts (Hurwitz, 1989), although this is disputed (Pandala and Gay, 1990; Teti et al., 1991; May et al., 1993). Considerable evidence exists to suggest that PTH can induce rapid changes in Ca transfer by osteoblasts and osteocytes. Thus PTH-stimulated increases in Ca uptake by these cells have been observed, while others have reported either no response in intracellular Ca [Ca_i] concentration or a net Ca efflux from bone cells, at least in embryonic chick bone in vitro (Hurwitz, 1989; Malgaroli et al., 1989). At least two types of voltage-controlled ionic channels were described, using patch clamp techniques in cultured embryonic chick osteoblasts by Ypey et al. (1988), who predicted a role for these channels in the response to PTH.

PTH also affects cell spread area in avian osteoclasts as shown by Miller (1977, 1978), who used electron microscopy to determine responses of medullary bone osteoclasts in egg-laying Japanese quail in vivo during the inactive phase of their eggshell calcification cycle. PTH induced the osteoclasts to form ruffled borders bounded by filamentous-rich clear zones within 20 min of hormone treatment, these changes being characteristic of "active" bone resorbing cells found during shell calcification. Similarly, Sugiyama and Kusuhara (1996) demonstrated that PTH induces ruffled border formation in osteoclasts located within hen medullary bone maintained in culture. Zambonin-Zallone et al. (1982) adopted a procedure whereby domestic hens are prefed a Ca-deficient diet in order to increase cell yield and then a prolonged isolation technique is used consisting of unit sedimentation and filtration. More recently, using tibiae from Ca deficient chicks, osteoclasts in situ were shown to increase their cell spread area by 40% within 2–4 min of PTH challenge (Pandala and Gay, 1990). Whether this response represents direct or indirect effects of the hormone on these cells, it is remarkably fast compared with classical responses to PTH (Dacke, 1979; Hurwitz, 1989). This led Bronner (1996) to speculate that an important mechanism underlying the minute-to-minute regulation of blood Ca levels in both birds and mammals is an ability of bone lining cells, osteoclasts as well as osteoblasts, to alter their size and shape and migrate to and from areas of the bone surface where high- or low-affinity binding sites for Ca^{2+} are located (Figure 2). This is an interesting hypothesis

TABLE 1 Acute Effects of Calcitrophic Agents on Chick Ca Metabolism

Agent	Plasma CA	Plasma ^{45}Ca	Bone ^{45}Ca	Osteoclast cell spread area
bPTH(1-34)	↑↑	↑↑	↓↓	↑↑
bPTHrP(1-34)	↑	↑	↓↓↑	—
PGE_2	↑↑↑	↑↑↑	↓↓↓	—
CT	0	—	↓↓	↓↓
CGRP	↑↑	↓↓	↑	—
Amylin	0	0	—	—
Ca^{2+}	↑↑	—	—	↓↓

Note. ↑, increase; ↓, decrease; 0, no change from control value; —, not measured. See text for abbreviations.

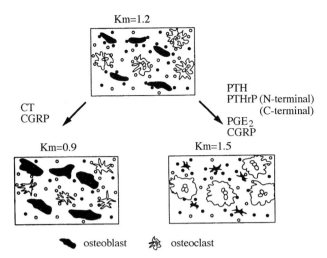

FIGURE 2 Effect of calcitrophic agents on spatial relationships of osteoclasts and osteoblasts on a bone surface. ○ = high, ● = low affinity Ca^{2+} binding sites. K_m=1.2 normocalcemic situation with an equal number of high and low affinity Ca^{2+} binding sites, K_m= 0.9 hypocalcaemic situation where agents cause exposure of high affinity binding sites due to osteoclast shrinkage, and K_m=1.5 hypercalcaemic situation where exposure to agents cause exposure of low affinity binding sites due to shrinkage of osteoblasts. Apparent K_m values are given in terms of total extra cellular Ca concentration (m moles liter^{-1}) for convenience. Modified after Dacke et al. (1993).

which at the time of the writing this chapter remains to be tested experimentally.

The biochemical mechanisms underlying avian osteoclast function are essentially similar to those of mammalian osteoclasts, although in some respects they are functionally distinct from the latter variety (Gay, 1988). Thus the ruffled border contains a proton pump-ATPase and an Na^+,K^+-ATPase. It also contains carbonic anhydrase which is closely associated with the cytoplasmic side of the membrane. Ca^{2+}-ATPase has been found on the plasma membrane of the narrow side of osteoclasts but is absent in the ruffled borders; its role is presumably to direct the outward flow of transmembrane Ca^{2+} flux. Using chick osteoclasts, May et al. (1993) demonstrated a direct effect of PTH resulting in increased acid production by these cells. The mechanism involves activation of adenylate cyclase via a G_s type-protein. Stimulation of acidification by PTH and cAMP is blocked by estradiol. Estradiol was inhibitory to the same extent as CT; these effects were not additive. Estradiol-17 β in micromolar, but not in nanomolar amounts, blocked proton pumping in isolated plasma membrane vesicles (Gay et al., 1993). The actions of PTH and other calciotrophic factors on bone cell function are summarized in Figure 3.

PTH also has potent hypotensive activity in a variety of vertebrates including birds. In the domestic hen a PTH bolus caused a drop in bone blood flow by 3 min, associated with transient hypocalcemia, followed by hyperaemia by 30 min associated with an increased venous/arterial Ca gradient and hypercalcemia (Boelkins et al., 1976).

2. Renal Actions

Mechanisms of renal Ca and P_i regulation were reviewed by Laverty and Clark (1989). In immature males and egg-laying females, intravenous injection of PTH transiently reduces plasma Ca and P_i levels, followed by an increase peaking 20–30 min after the injection. By contrast, adult cockerels seldom show any variation in plasma Ca or P_i levels. In all birds PTH increases glomerular filtration rate, urine flow rate, and P_i and Ca clearance. Infusion of CT decreases plasma Ca only in PTX chickens but in all birds increased GFR, urine volume, and Ca excretion, and in parathyroid-intact birds renal P_i clearance also increased (Figure 4). These data suggest that renal Ca and P_i homeostasis in the chicken, as in the mammal, depends on a balanced release of PTH and CT. However, in the chicken, the response time is faster, thus minimizing the fluctuations in plasma Ca and P_i levels resulting from exogenous hormone administration. This rapid homeostatic response is less effective when Ca demands by the body are high as in growing or laying birds (Sommerville et al., 1987).

F. Parathyroid Hormone Related Peptide

Recent clinical research in bone and Ca metabolism has focused on PTHrP, which is found in the circulation of certain patients with malignancy-associated hypercalcemia and also as the predominant peptide in fetal mammals. PTH and PTHrP genes are located on different chromosomes but share common organizational fea-

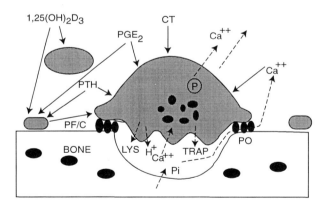

FIGURE 3 Diagrammatic representation of sites of regulation of avian osteoclastic bone resorption. Inorganic phosphate (P_i), preosteoclast (PreOC), osteoclast (OC), osteoblast (OB), tartrate-resistant acid phosphatase (TRAP), pump (P), 1,25-dihydroxy vitamin D_3 (1,25-$(OH)_2D_3$), calcitonin (CT), parathyroid hormone (PTH), prostaglandin E_2 (PGE_2), paracrine and cytokine factors (PF/C), podosomes (PO), lysosymes (LYS), ruffled border (R), solid lines = controlling factors, dotted lines = secretions/ion fluxes.

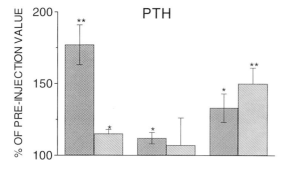

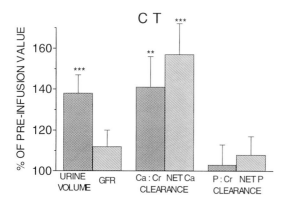

FIGURE 4 Renal responses to injection of 12 IU PTH/kg, or 30 min infusion of 1-12 U CT/Kg. to PTH/CT in adult cockerels. Data are volume of urine excreted, GFR, Ca/creatinine and net Ca and Pi/creatinine and net Pi clearances expressed as % saline injected/ infused control birds. * $P < 0.05$, ** $P < 0.01$, *** $P < 0.001$, v controls. (Redrawn from Sommerville and Fox (1987), *Gen. Comp. Endocrinol.* with permission.)

They localized the cycle-associated fluctuations in PTHrP mRNA and immunoreactive PTHrP levels to the shell gland serosal and smooth muscle layer, which suggests that the peptide may modulate vascular smooth muscle activity. In support of this hypothesis, synthetic chicken PTHrP (1-34)NH$_2$ was found to relax the resting tension of isolated shell gland blood vessels in a dose-dependent manner. Together, these data indicate that expression of the PTHrP gene in the avian oviduct is both temporally and spatially regulated during the egg-laying cycle and that PTHrP may function as an autocrine/paracrine modulator of shell gland smooth muscle activity. The vasorelaxant property of N-terminal fragments of PTHrP supports a role for this molecule in the temporal increase in blood flow to the shell gland during egg calcification.

Both PTH and PTHrP enhance cAMP and inhibit collagen synthesis in avian epiphysial cartilage cells, an effect which is blockadeable by the PTH antagonist PTH (3-34) (Pines, 1990). PTHrP(1-34) was tested in the chick hypercalcemic assay and showed only slight PTH agonist activity with respect to either plasma Ca levels or ^{45}Ca clearance. In femur the peptide caused a substantial decrease in ^{45}Ca uptake while in calvarium the opposite effect apparently occurred (Dacke *et al.*, 1993). Fenton *et al.* (1994) studied carboxyl-terminal peptides from PTHrP for their effect on bone resorption by embryonic chick osteoclasts. Basal bone resorption was directly inhibited by chicken and human PTHrP-(107-139) and by the pentapeptide PTHrP-(107-111). The number of resorption pits and total area resorbed per bone slice were reduced by PTHrP-(107-139) while resorption stimulated by hPTH-(1-34) in cocultured chicken osteoclasts and osteoblasts was also inhibited by cPTHrP-(107-139). These results suggest that C-terminal PTHrP may be a paracrine regulator of bone cell activity.

Schermer *et al.* (1994) studied the functional properties of synthetic chicken PTHrP fragments in avian (chicken renal plasma membranes and 19-day chick embryonic bone cells) and mammalian (canine renal plasma membranes and rat osteosarcoma cells [UMR-106-H5]) systems. The biologic activities of human PTHrP(1-34) and bovine PTH(1-34) are remarkably similar despite marked sequence divergence in their primary binding domain, residues 25–34. In both avian and mammalian systems the binding affinity of [36Tyr]cPTHrP(1-36)NH$_2$ is half that of hPTHrP(1-34)NH$_2$. Potencies of [36Tyr]cPTHrP(1-36)NH$_2$ and hPTHrP(1-34)NH$_2$ for activation of adenylate cyclase were similar in canine renal membranes and chick bone cells. In UMR-106 cells and chicken renal membranes, the potency of [36Tyr[cPTHrP(1-36)NH$_2$ for activation of adenylate cyclase was half that of [36Tyr]hPTHrP(1-36)NH$_2$. Binding of ^{125}I-[36Tyr]cPTHrP(1-36)NH$_2$ to chick bone cells and chicken renal membranes was completely displaced by bPTH(1-34) and hPTHrP(1-

tures, and the peptides show much structural homology, suggesting a common ancestral gene. PTHrP exists in three known forms ranging in size from 139 to 173 amino acids, as the result of alternative gene splicing. It has a wide spectrum of actions in mammals, many in common with PTH; these range from stimulation of osteoclastic bone resorption to enhancement of placental mineral transport (Mallette, 1991). PTHrP is expressed in a variety of tissues in chick embryos, this molecule having a highly conservative structural homology with the human sequence, the first 21 residues being identical (Schermer, 1991). It is also expressed in the isthmus and shell gland of the hens' oviduct where it has a potential role as a local modulator of vascular smooth muscle tension and shell gland motility during the egg-laying cycle (Thiede *et al.*, 1991). They followed the expression of PTHrP in the shell gland at different times in the laying cycle and found levels of PTHrP to transiently increase as the egg moves through the oviduct, gradually returning to basal levels in the 15-hr calcification period (Figure 5).

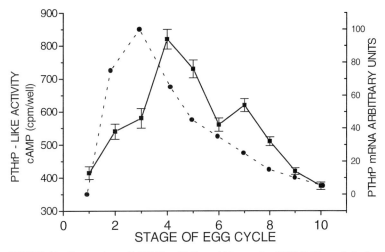

FIGURE 5 Expression of PTHrP mRNA and bioreactive PTHrP-like activity by the oviduct shell gland in chickens during the egglay cycle. PTH mRNA values are redrawn to an arbitrary scale while PTHrP-like activity is represented in terms of a cAMP activation assay. During phases 1-4 the egg was passing through the isthmus and magnum of the oviduct, before entering the shell gland. Phases 5-9 represent the period when the egg is undergoing shell calcification. Phase 10 represents the period following oviposition. [Redrawn from M. A. Thiede, S. C. Harm, R. L. McKee, W. A. Grasser, L. T. Duong, and R. M. Leach; Expression of the parathyroid hormone-related protein gene in the avian oviduct: potential role as a local modulator of vascular smooth muscle tension and shell gland motility during the egg-laying cycle, *Endocrinology*, **129**, 1958–1966 (1991) © The Endocrine Society.]

34)NH$_2$; thus, there is no evidence for a distinct chicken PTHrP N-terminal receptor.

III. CALCITONIN AND THE ULTIMOBRANCHIAL GLANDS

A. The Ultimobranchial Glands

Birds possess anatomically distinct, asymmetrical paired ultimobranchial glands, which in the chicken lie posterior to the parathyroids and caudodorsal to the base of the brachiocephalic artery into the common carotid and subclavian arteries (Kenny, 1986). Embryologically CT secreting C cells derive from the sixth branchial pouch, although their ultimate origin lies in the neuroectoderm of the neural crest. Avian C cells contain poorly developed rough endoplasmic reticulum and specific secretory granules surrounded by a single membrane. Numerous mitochondria, a well defined Golgi apparatus, and abundant free ribosomes are present (Pearse, 1976). The avian ultimobranchial gland is particularly rich in CT, the principal cells having a striking resemblance to mammalian C-cells (i.e., they possess small intracytoplasmic dense-core secretory granules) 150–300 nm in diameter. The gland also contains a second, morphologically distinct endocrine cell type with larger granules, 500–800 nm in diameter. A sensitive immunocytochemical reaction developed using antibodies against salmon CT indicated the presence of CT-immunoreactive molecules in both secretory cell types of the chicken ultimobranchial gland (Treilhou-Lahille *et al.*, 1984)

B. Calcitonin

CT is a polypeptide hormone of 32 amino acids and a seven membered N-terminal ring. The entire chain is necessary for biological activity. There are variations in amino acid sequence of CT from different species which give rise to differences in bioactivity; CT from fish origin appears to be particularly potent (Dacke, 1979). Structures of different CTs including that from chicken are reviewed by Zaidi *et al.* (1990a), who also give detailed consideration to structure activity relationships of the various CTs.

CT secretion is regulated primarily by rising plasma Ca levels leading to increased secretion from the C cells (Dacke, 1979). Eliam *et al.* (1993) studied influences of Ca- and vitamin D-deficient diets on CT gene expression in the ultimobranchial cells of the developing chicken. These chicks exhibited a striking reduction of CT biosynthesis by decreasing the number of secretory cells and not by triggering modifications of the biosynthetic activity of the ultimobranchial endocrine cells.

C. Circulating Calcitonin Levels

High levels of circulating CT are present in submammalian species including birds and are detectable in the plasma of Japanese quail using bioassays (Dacke, 1979).

They are higher in adult males than egg-laying females except for a brief period immediately before the commence of lay. In egg-laying quail hens they are at their highest shortly after ovulation and fall as eggshell calcification proceeds, rising at the end of calcification. In this species at least, gonadal steroids, especially androgens, appear to have a major influence on circulating CT levels (see Taylor and Dacke, 1984; Kenny, 1986). In chickens plasma, CT levels positively correlated with dietary Ca intake and thus to circulating Ca levels (Taylor and Dacke, 1984), while in chick embryos they are extremely low until just prior to hatching (Abbas et al., 1985).

D. Actions of Calcitonin

Despite the discovery of CT more than three decades ago, its role in bone and Ca metabolism remains an enigma. Only in mammalian species has CT been shown to regulate plasma Ca levels, the basis of its hypocalcemic action lying in an ability to inhibit osteoclastic bone resorption (Copp and Kline, 1989). Plasma Ca levels in submammalian vertebrates are refractory to dosage with CT (Dacke, 1979). Whether this is due to the high circulating levels of biologically active hormone causing receptor down-regulation is not clear. Initial studies in avian osteoclasts showed no response to CT in terms of postreceptor events (Miyaura et al., 1981; Ito et al., 1985; Nicholson et al., 1986; Dempster et al., 1987); however, cAMP responses were demonstrated in osteoclasts from chicks maintained on low Ca or rachitogenic diets (Eliam et al., 1988; Rifkin et al., 1988). Moreover, osteoclasts from Ca-deficient chicks responded to CT *in vitro* within 4 min by a 58% reduction in cell spread area (Pandala and Gay, 1990) and also by an inhibition of their bone resorptive activity (de Vernejoul et al., 1988). Calcitonin also causes disappearance of ruffled borders in cultured medullary bone osteoclasts (Sugiyama and Kusuhara, 1996). These findings are consistent with the idea that CT receptors are down regulated under normal physiological conditions in the chick. We recently found that dosing heavily (22-hr) fasted chicks with salmon CT *in vivo* caused a rapid (10 min) but variable inhibition of net ^{45}Ca uptake into the skeleton (Table 1), the femur being most affected (Ancill et al., 1991). This effect is similar to that of PTH and PGE_2, but its physiological significance is obscure.

Farley et al. (1988) studied the acute effects of salmon CT on bone formation indices in chick embryos *in vitro*. It increased calvarial cell proliferation, [^{3}H]thymidine incorporation into DNA, [^{3}H]proline incorporation into bone matrix collagen, and [^{3}H]hydroxyproline in intact calvaria and tibiae. The increased [^{3}H]hydroxyproline incorporation was associated with proportional increases in alkaline phosphatase activity in the bones. [^{3}H]Proline incorporation in embryonic chicken calvaria also increased during 3 days of limited exposure (i.e., 4 hr/day) to CT. The proliferative action/s of CT also occur in cultures of newborn mouse calvaria.

Bouizar et al. (1989) studied the distribution of renal CT binding sites in vertebrates. No renal ^{125}I-sCT binding sites were detected in fish, amphibians, or reptiles. In the rat, chicken, and quail, binding sites were observed in the medulla and in the cortex. The pattern seemed to follow the distribution of the glomeruli and/or the collecting tubules, suggesting that CT renal receptors appeared late in evolution and regulation of renal function by CT is only effective in birds and mammals. As mentioned previously, CT has effects on renal Ca and P_i excretion which are most pronounced in PTX chickens.

E. Role of Extracellular Ca

It was recently reported that freshly isolated quail medullary bone osteoclasts, unlike those from neonatal rats, do not respond to elevated extracellular Ca $[Ca^{2+}]_e$ by a rise in intracellular Ca $[Ca^{2+}]_i$ (Bascal et al., 1992). Unlike rat cells, cultured chicken osteoclasts are sensitive to changes in membrane voltage and to dihydropyridine-type Ca^{2+} channel blockers (Miyauchi et al., 1990). When medullary bone osteoclasts are cultured away from bone substrata for several days they recover an ability to respond to elevated $[Ca^{2+}]_e$ but neither the fresh nor cultured cells exhibit any response to CT in terms of raised $[Ca^{2+}]_i$ (Arkle et al., 1994). Freshly isolated quail medullary osteoclasts are also refractory to $[Ca^{2+}]_e$ in terms of cell spread area. However, unlike neonatal rat cells, they respond to ionomycin (a Ca^{2+} ionophore) with a modest reduction in cell spread area. This suggests that the quail cells retain intracellular mechanisms necessary for elaboration of the aforementioned responses, but lack receptor mechanisms for detecting changes in $[Ca^{2+}]_e$. When medullary bone osteoclasts are cultured for several days, the stimulus causes a reduction in cell spread area similar to that in fresh rat cells (Bascal et al., 1994). These results indicate that the putative Ca^{2+} "receptors" on freshly isolated quail medullary bone osteoclasts are normally down-regulated but that this disappears upon culturing the cells for several days. This suggests that during the eggshell calcification cycle when resorption of medullary bone prevails and raised local Ca^{2+} levels are generated by intense osteoclastic activity, the osteoclasts become insensitive to inhibitory factors such as elevated $[Ca^{2+}]_e$ (Figure 3).

F. Calcitonin Gene-Related Peptide and Amylin

Calcitonin gene-related peptide (CGRP) is a 37-amino acid neuropeptide derived from the same gene as CT, both belonging to the amylin super family (Rosenfield *et al.*, 1983). Unlike CT, which is expressed mainly within the avian ultimobranchial body, it is found within the central and peripheral nervous system, including the chicken retina (Kiyama *et al.*, 1985), carotid body (Kameda, 1989), spinal cord motoneurons of developing embryos and post-hatch chicks (Villar *et al.*, 1988), and bone neurones (Bjurholm *et al.*, 1985), where its distribution corresponds with that of substance P (Mallette, 1991). It is also expressed within the dense distributions of peptidergic neurones found within the chick ultimobranchial gland (Kameda, 1991). CGRP molecules from man, rat, cow, salmon, and chicken have been sequenced; the structure is well conserved with around 90% structural homology between chicken and human CGRPs, compared with only 50% between the respective CT molecules (Zaidi *et al.*, 1990a). CGRP presence in bone neurones coupled with the interaction of this peptide with osteoclastic CT receptors (Goltzman and Mitchell, 1985; Zaidi *et al.*, 1987) suggests a paracrine role, modulating bone turnover and hence Ca homeostasis. A further member of the CT/CGRP family, amylin, a peptide from the pancreatic islet cells, is the most potent non-CT peptide thus far discovered, at least in mammalian assay systems (Zaidi *et al.*, 1990a). CGRP has several putative physiological functions. It is a potent vasodilator and is implicated in central and peripheral neurotransmission and modulation (Zaidi *et al.*, 1990a). Only recently has its role in bone and Ca metabolism been recognized. CGRP shares the acute hypocalcemic effects of CT albeit at around 1000-fold less potency in rodents and in inhibiting bone resorption, stimulating cAMP production in mouse calvaria, and inhibiting neonatal rat osteoclastic spreading (Zaidi *et al.*, 1990a). In the rabbit *in vivo* CGRP causes transient hypocalcemia followed by a more sustained hypercalcemia (Tippins *et al.*, 1984) and in the same paper a preliminary report of the *in vivo* hypercalcemic effect of the peptide in chicks was given. These findings in chicks were repeated and extended (Bevis *et al.*, 1990; Ancill *et al.*, 1990). The former paper gave details of comparative dose–response curves for CGRP and PTH in chicks, the two peptides being approximately equipotent on a molar basis. In the latter report we investigated the effect of the peptide on a simultaneously injected ^{45}Ca-label in the plasma. Intravenous injection of rat CGRP gave a rapid hypercalcemic response which was sustained for at least 1 hr. This was most evident in nonfasted chicks. Fasted chicks by contrast gave a hypophosphatemic response and also showed an increased plasma ^{45}Ca clearance. The effect of CGRP on ^{45}Ca uptake into the chick skeleton was subsequently investigated (Ancill *et al.*, 1991). Both rat and chicken CGRP sequences caused transient (10 min) increases in ^{45}Ca uptake into a variety of bone types (Table 1). These responses were well developed in fasted chicks but absent in fed ones; the greatest were in calvaria and vertebrae. Furthermore, with low doses of CGRP reversal of the response was noticed in calvaria but not other bone types, while in fed animals this was the only response seen, again in calvaria. These findings indicate that CGRP may have a variety of effects on bone and Ca metabolism in the chick, which involve acute effects on net movement of Ca into and out of the skeleton. However, while consistent with changes in plasma ^{45}Ca clearance, they seem too transient to account for them and would not account for the hypercalcemic responses found, although some alternative target site such as the kidney (Zaidi *et al.*, 1990b) may be responsible for these. In a preliminary experiment, the effect of amylin on Ca metabolism was tested in chicks *in vivo* and found to be lacking (Dacke *et al.*, 1993).

Mixed bone cell cultures obtained by sequential collagenase–trypsin digestion of newborn chick and rodent calvaria responded to CGRP with dose-dependent increases in cAMP formation (Michelangeli *et al.*, 1989). This effect was not the result of an action as a weak CT agonist, since in most instances a CT effect was not observed. They concluded that chick, rat, and mouse bones contain cells in osteoblast-rich populations that respond specifically to CGRP by a rise in cAMP.

IV. THE VITAMIN D SYSTEM

Hou (1931) reported that removal of the preen gland from chicks causes rickets even when they are fed a normal diet and exposed to sunlight. He concluded that birds must secrete provitamin D_3 from the preen gland onto their feathers where it is converted into vitamin D_3. It is no surprise that in the intervening period much of our knowledge of vitamin D_3 metabolism in general has been obtained from avian species. The laying hen requires large amounts of Ca for eggshell formation which is met ultimately by the large Ca fluxes which occur from the intestine with subsequent storage and resorption from bone and across the uterus into the eggshell. The active vitamin D_3 metabolite 1,25-$(OH)_2D_3$ is of the utmost importance in controlling these processes.

A. Renal Vitamin D Metabolism

The control of vitamin D_3 metabolism in birds was reviewed by Taylor and Dacke (1984) and more recently by Norman (1987), Hurwitz (1989), and Nys (1993). The renal vitamin D endocrine system of chickens and Japanese quail has been widely studied. It is well established that chickens metabolize vitamin D_3 to 25-(OH)-

D_3 and $1,25\text{-}(OH)_2 D_3$ in their liver and kidneys, respectively (Holick, 1989). However, unlike mammals, birds discriminate between vitamin D_2 and D_3 and chickens are unable to utilize vitamin D_2 as efficiently as D_3 (Taylor and Dacke, 1984). This appears to be due to the fact that in birds plasma vitamin D binding protein has relatively low affinity for vitamin D_2, which is thus more rapidly broken down (Holick, 1989). The avian kidney, like that in mammals and other vertebrates, synthesises and secretes $1,25\text{-}(OH)_2D_3$. Numerous factors stimulate production of $1,25\text{-}(OH)_2D_3$ including PTH, $1,25(OH)_2D_3$ itself, and prolactin (Kenny, 1981; Henry and Norman, 1984). There are contradictory reports as to whether CT influences $1,25\text{-}(OH)_2D_3$ production by avian kidney but Kenny (1986) concludes that CT is not a major regulator of vitamin D_3 in birds.

It has been known for several years that egg yolk is a rich source of vitamin D, of which 90% is in the form of vitamin D_3 and 5% in the form of $25\text{-}(OH)D_3$. More than 90% of this vitamin D is in the yolk and is probably derived from the hen rather than the embryo (Fraser and Emtage, 1976).

B. Circulating Levels of Vitamin D Metabolites

Vitamin D_3 and its metabolites are transported in association with plasma albumins (DBP) and in some avian species with α- and β-globulins as well. The DBP exists as two 4S forms which differ in the number of neuraminic acid residues present. DBP probably has a single binding site for all vitamin D metabolites and its affinities for 25– and $24,25\text{-}(OH)_2D_3$ are similar, whereas $1,25\text{-}(OH)_2D_3$ has a 10–fold lower affinity. The concentration of DBP was more than twice as high in the plasma of laying hens than in immature birds or adult males (Taylor and Dacke, 1984).

Levels of circulating vitamin D_3 metabolites in Japanese quail and egg-laying chickens have been determined. The increase in intestinal Ca absorption occurring after sexual maturity and during eggshell formation is related to renal 25-hydroxycholecalciferol-1-hydroxylase activity (Kenny, 1976), which in turn enhances circulating $1,25\text{-}(OH)_2D_3$ levels (Castillo et al., 1979) and the accumulation of this metabolite in the intestinal mucosa (Bar et al., 1978). Increases in $25\text{-}(OH)D_3$-1-hydroxylase activity are induced by injecting estrogen into immature birds (Baksi and Kenny, 1977; Sedrani et al., 1981). Abe et al. (1979) reported that plasma concentrations of $25\text{-}(OH)D_3$ and $1,25\text{-}(OH)_2D_3$ but not $24,25\text{-}(OH)_2D_3$ in egg-laying hens fluctuate during the eggshell calcification cycle (Table 2).

These results were confirmed by Nys et al. (1986), who also demonstrated that hens laying shell-less eggs do not show the cyclical fluctuation in $1,25\text{-}(OH)_2D_3$ levels. Circulating levels of $1,25\text{-}(OH)_2D_3$ increase in laying hens during the prelaying period and again at the onset of egg production (Nys, 1993). Using Ca-deficient birds, Bar and Hurwitz (1979) demonstrated that the stimulatory effect of estrogen on renal $25\text{-}(OH)D_3$-hydroxylase is eliminated, suggesting that increased $1,25\text{-}(OH)_2D_3$ production results from Ca deficiency induced by estrogens.

C. Actions of Vitamin D

1. Intestine and Oviduct

$1,25\text{-}(OH)_2D_3$ regulates intestinal absorption of Ca by inducing RNA transcription and synthesis of proteins such as calbindin D_{28k}. The physiological function of calbindin D_{28k} is not well established but its concentration in the intestine is reflected by the ability to absorb Ca. Three forms of calbindin D_{28k} with varying size have been identified in the chicken intestine, the smallest being most abundant (Nys, 1993). In vitamin D-deficient chicks, intestinal calbindin mRNA is barely detectable but increases dramatically following $1,25\text{-}(OH)_2D_3$ injection (Mayel-Afshar et al., 1988). The onset of ovulation in hens is associated with increased intestinal Ca absorption and elevated concentrations of calbindin D_{28k} (Nys, 1993) (Table 2), coinciding with increased plasma $1,25\text{-}(OH)_2D_3$ concentrations (Castillo et al., 1979; Nys et al., 1992). A similar or identical protein is present in the avian uterus (Fullmer et al., 1976). The uterus of the laying hen contains receptors for $1,25\text{-}(OH)_2D_3$ although a large part of the calbindin appears to be independent of this metabolite (Cory, 1981). Uterine calbindin and its mRNA concentration increase in immature pullets treated with estrogen and is higher in hens laying shell-less eggs (Navickis et al., 1979). However, the effect of the sex steroid on uterine calbindin appears to be an indirect one, probably related to the general development and maturation of the oviduct (Nys, 1993). Although concentrations of uterine calbindin and its mRNA rise during formation of the first and subsequent eggs in chickens and quail, they are not directly related to the activity of $1,25\text{-}(OH)_2D_3$ (Bar et al., 1992; Nys, 1993). The function of uterine calbindin is not precisely established but its presence is indicated wherever there is a physiological requirement for Ca translocation across the uterine wall (Hurwitz, 1989).

2. Bone Actions

Feeding laying hens vitamin D deficient diets results in resorption of medullary bone while in nonlaying birds osteodystrophy results (Wilson and Duff, 1991). 1,25-

TABLE 2 Time Course for Changes in Circulating Levels of Vitamin D Metabolites and Duodenal Calbindin and Calbindin mRNA during the Eggshell Calcification Cycle of the Egg-Laying Hen

	Hr after ovulation[a]					
	0	4	6	12	15	22
Duodenal calbindin (μ/mg protein)	—	112 ± 7	—	82 ± 4	—	—
Duodenal calbindin mRNA % nonlaying value)	—	838 ± 142	—	1135 ± 181	—	—
Plasma vitamin D metabolites (ng/ml)						
25-(OH)D$_3$	17.0 ± 2.5	12.0 ± 1.5	10.0 ± 1.5	—	16.5 ± 2.5	18.0 ± 3.5
24,25-(OH)$_2$D$_3$	1.5 ± 0.03	1.0 ± 0.02	1.0 ± 0.02	—	1.0 ± 0.02	1.2 ± 0.03
1,25-(OH)$_2$D$_3$	0.13 ± 0.03	0.11 ± 0.03	0.12 ± 0.04	—	0.30 ± 0.04	0.12 ± 0.04

Note. Data are means ± standard errors for 5–6 birds. From Abe *et al.* (1979) and Nys *et al.* (1992).
[a]Shell calcification occurs between 6 and 22 hr.

(OH)$_2$D$_3$ appears to facilitate bone formation by inducing biosynthesis of osteocalcin (bone γ-carboxyglutamic acid protein). The biological function of this non-collagenous vitamin D binding protein in skeletal mineralization is obscure although it appears to be a specific product of osteoblasts during bone formation. This small (MW, 5500) protein has been purified from bone and sequenced for several species including chicken. Its precise function is unknown but it binds Ca and shows affinity for hydroxyapatite, suggesting its involvement in the mineral dynamics of bone (Hauschka *et al.*, 1989). Osteocalcin is released into the circulation where it provides a convenient index of bone turnover, reflecting new osteoblast formation rather than release of matrix protein during bone resorption (Nys, 1993). 1,25-(OH)$_2$D$_3$ stimulates osteocalcin synthesis by binding to promotor elements and enhancing osteocalcin gene transcription. However unlike intestinal calbindin, there is substantial osteocalcin synthesis in vitamin D-deficient chicks (Lian *et al.*, 1982).

Approximately 30–40% of eggshell Ca is derived from medullary bone. Formation of medullary bone matrix is induced by sex steroids regardless of the vitamin D status of the bird, although this bone only becomes fully mineralized when both vitamin D$_3$ and the sex steroids are present (Takahashi *et al.*, 1983). Nys (1993) reports that changes in blood osteocalcin levels parallel those of 1,25-(OH)$_2$D$_3$ in laying hens so that concentrations of blood osteocalcin rise in hens fed a low Ca diet and decrease in hens laying shelless eggs. It is possible that increased osteocalcin levels in response to estrogen are a reflection of increased vitamin D receptor expression by osteoblasts (Liel *et al.*, 1992). However osteoclasts from medullary bone as well as from rat bone appear to be devoid of 1,25-(OH)$_2$D$_3$ receptors and the effects of the metabolite are considered to be mediated via the osteoblasts (Merke *et al.*, 1986). Harrison and Clark (1986) succeeded in culturing medullary bone obtained from egg laying hens. They demonstrated that these cultures could respond to 1,25-(OH)$_2$D$_3$ by a dose-dependent inhibition of [^{3}H]proline uptake.

Clark (1991) measured serum and renal clearance values of P$_i$ and Ca and compared these in vitamin D-deficient and vitamin D-replete chickens She observed that most renal functions studied after Ca loading, PTH administration, or PTX are unaltered by vitamin D deficiency in the chicken. The major significant finding was that vitamin D-deficient chickens do not excrete increased amounts of P$_i$ in response to PTH stimulus.

V. PROSTAGLANDINS AND OTHER FACTORS

The role of prostaglandins (PGs) and other eicosanoids in vertebrate (including avian) Ca and bone metabolism was reviewed by Dacke (1989). PG effects on mammalian bone cells are essentially similar to those of PTH in that they stimulate cAMP production, cause transient increases in Ca^{2+} influx, activate carbonic anhydrase, release lysosomal enzymes, and may inhibit collagen synthesis. They also elicit morphological responses in osteoclasts and osteoblasts similar to those with other osteolytic agents (Dacke, 1989). It was demonstrated that a stable methylated PGE$_2$ analog, 16,16–dimethyl PGE$_2$, is hypercalcemic when injected into chicks, the response being more profound than that in mammals (Kirby and Dacke, 1983). Indomethacin, a drug which interferes with synthesis of PGs, produces hypocalcemia in egg-laying chickens (Hammond and Ringer, 1978) and quail (Dacke and Kenny, 1982). In chickens this is accompanied by a considerable delay in oviposition and thicker eggshells. PGE$_2$ and other eicosanoids rank alongside PTH and 1,25-(OH)$_2$D$_3$ as powerful stimulators of bone resorption (Dacke, 1989).

The effects of CT, PTH, and PGE_2 on cyclic AMP production were studied in osteoclast-rich cultures derived from avian medullary bone and long bones of newborn rats. PGE_2 increased cAMP production in both types of osteoclasts, suggesting essentially similar mechanisms (Nicholson *et al.*, 1986; Arnett and Dempster, 1987). In addition to PTH, CGRP, and possibly CT, PGs can acutely influence Ca exchange between avian blood and bone (Table 1). We previously published a simple model (Shaw *et al.*, 1989) in which the rapid effects of PTH and PGE_2 on skeletal ^{45}Ca uptake could be explained in terms of cAMP mediated inhibitions of outwardly directed Ca^{2+} pumps located in the membranes of bone lining cells. Such a model has its disadvantages; it would be metabolically wasteful since the intracellular free Ca level would have to remain undisturbed in the face of a high rate of bidirectional transcellular Ca^{2+} flux. An alternative model (Figure 2) would involve rapid changes in shape and location of bone-lining cells as hypothesized by Bronner (1996).

VI. CONCLUSIONS

Figure 6 summarizes our present understanding of the regulation of avian Ca and bone metabolism. Within the past decade it has become apparent that a variety of nonclassical factors can influence bone and Ca metabolism in birds as well as in mammals. The avian responses are often more explicit than the equivalent ones in mammals; for example, the hypercalcemic responses to PGs. Other factors such as PTHrP and CGRP are represented in birds and the next few years can be expected to provide a fertile ground for new research on their role in bone and Ca metabolism in general, to which avian models are likely to make an important contribution. The interactions of these novel agents with more classical hormones in avian bone remain to be elucidated. Their interactions with gonadal steroids in forming and maintaining avian medullary bone may prove a particularly rewarding area for future studies. Medullary bone represents the most overtly estrogen sensitive of all vertebrate bone types. It forms in male birds dosed with estrogens within a matter of days and this process can be blocked by the simultaneous administration of antiestrogenic compounds such as tamoxifen (Williams *et al.*, 1991). Upon cessation of estrogen treatment medullary bone resorbs just as rapidly. Medullary bone represents an excellent rapidly responding model for studies of the effects of antiosteoporotic drugs such as bisphosphonates. Preliminary studies indicate that the bisphosphonate alendronate can protect structural bone and inhibit medullary bone formation if given to hens before the commence of egglay. When given during egglay the drug reduces medullary bone volume and, at higher doses, eggshell quality (Thorp *et al.*, 1993). It is anticipated that this model will become exploited to a much greater extent than hitherto.

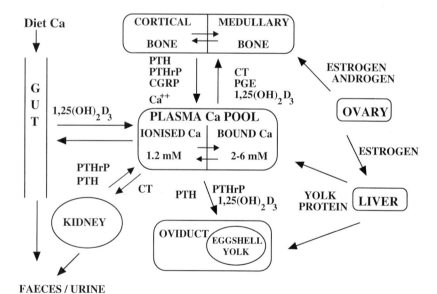

FIGURE 6 Summary of Ca metabolism in egg laying birds. Plasma Ca levels are very high in hens due to binding by yolk protein precursors. Transfer of Ca (and Pi) between dietary, plasma, structural and medullary bone pools and across the oviduct may be influenced by a variety of recognised and putative hormonal factors (see text).

Acknowledgments

The author is grateful to Mrs. Gill Whitaker, who assisted with preparation of the figures, and to his wife Joan for literary advice.

References

Abbas, S. K., Fox, J., and Care, A. D. (1985). Calcium homeostasis in the chick embryo. *Comp. Biochem. Physiol. B* **81,** 975–980.

Ancill, A. K., Bascal. Z. A., Whitaker. G., and Dacke. C. G. (1991). Calcitonin gene-related peptide promotes transient radiocalcium uptake into chick bone *in vivo. Experimental Physiol.* **76,** 143–146.

Arkle, S., Wormstone, I. M., Bascal, Z. A., and Dacke, C. G. (1994). Estimation of intracellular calcium activity in confluent monolayers of primary cultures of quail medullary bone osteoclasts. *Exp. Physiol.* **79,** 975–982.

Arnett, T. R., and Dempster, D. W. (1987). A comparative study of disaggregated chick and rat osteoclasts *in vitro*: Effects of calcitonin and prostaglandins. *Endocrinology* **120,** 602–608.

Baksi, S. N., and Kenny, A. D. (1977). Vitamin D_3 metabolism in immature Japanese quail: Effects of ovarian hormones. *Endocrinology* **101,** 1216–1220.

Bar, A., and Hurwitz, S. (1979). The interaction between dietary calcium and gonadal hormones in their effect on plasma calcium, bone, 25-hydroxycholecalciferol-1-hydroxylase and duodenal calcium-binding protein, measured by a radio immunoassay in chicks. *Endocrinology* **104,** 1455–1460.

Bar, A., Cohen, A., Edelstein, S., Shemesh, M., Montecuccoli, G., and Hurwitz, S. (1978). Involvement of cholecalciferol metabolism in birds in the adaptation of calcium absorption to the needs during reproduction. *Comp. Biochem. Physiol. B* 59, 245–249.

Bar, A., Striem, S., Vax, E., Talpaz, H., and Hurwitz, S. (1992). Regulation of calbindin turnover in intestine and shell gland of the chicken. *Am. J. Physiol.* **262,** R800–R805.

Bascal, Z. A., Moonga, B. S., Zaidi, M., and Dacke, C. G. (1992). Osteoclasts from medullary bone of egg-laying Japanese quail do not possess the putative calcium receptor. *Exp. Physiol.* **77,** 501–504

Bascal, Z. A., Alam, A. S. M. T., Zaidi, M., and Dacke, C. G., (1994). Effect of raised extracellular calcium on cell spread area in quail medullary bone osteoclasts. *Exp. Physiol.* **79,** 15–24.

Bevis, P. J. R., Zaidi, M., and MacIntyre, I. (1990). A dual effect of calcitonin gene-related peptide on plasma calcium levels in the chick. *Biochem. Biophys. Res. Commun.* **169,** 846–850.

Boelkins, J. N., Mazurkiewicz, M., Mazur, P. E. P. E., and Mueller, W. J. (1976). Changes in blood flow to bone during hypocalcemic and hypercalcemic phases of the response to parathyroid hormone. *Endocrinology* **98,** 403–412.

Bouizar, Z., Khattab, M., Taboulet, J., Rostene, W., Milhaud, G., Treilhou-Lahille, F., and Moukhtar, M. S. (1989). Distribution of renal calcitonin binding sites in mammalian and nonmammalian vertebrates. *Gen. Comp. Endocrinol.* **76,** 364–70.

Bronner, F. (1996). Calcium metabolism in birds and mammals. *In* "Comparative Endocrinology of Calcium Regulating Hormones" (C. G. Dacke, J. Danks, G. Flik, and I. Caple, eds.), pp. 131–135. J. Endocrinol. Ltd., Bristol.

Brown, E. M. (1991). Extracellular Ca^{2+} sensing, regulation of parathyroid cell function and role of Ca^{2+} and other ions as extracellular (first) messengers. *Physiol. Revs.* **71,** 371–411.

Bjurholm, A., Kreicbergs, A., Brodin, E., and Schultzberg, M. (1985). Substance P- and CGRP- immunoreactive nerves in bone. *Peptides* 9, 165–171.

Candlish, J. K., and Taylor, T. G. (1970). The response-time to the parathyroid hormone in the laying fowl. *J. Endocrinol.* **48,** 143–144.

Castillo. L., Tanaka. Y., Wineland. M. J., Jowsey, J. O., and Deluca, H. F. (1979). Production of 1,25-dihydroxyvitamin D_3 and formation of medullary bone in the egg laying hen. *Endocrinology* **104,** 1598–1606.

Clark, N. B. (1991). Renal clearance of phosphate and calcium in vitamin D-deficient chicks: Effect of calcium loading, parathyroidectomy, and parathyroid hormone administration. *J. Exp. Zool.* **259,** 188–195.

Clark, N. B., and Wideman, R. F. (1977). Renal excretion of phosphate and calcium in parathyroidectomised starlings. *Am. J. Physiol.* **233,** F138–144.

Clark, N. B., and Sassayama, Y. (1981). The role of parathyroid hormone on renal excretion of calcium and phosphate in the Japanese quail. *Gen. Comp. Endocrinol.* **43,** 234–241.

Copp, D. H., and Kline, L. W. (1989). Calcitonin. *In* "Vertebrate Endocrinology Fundamentals and Biomedical Implications" (P. K. T. Pang and M. P. Schreibman eds.), Vol. 3, pp. 79–103. Academic Press, New York.

Cory, W. A. (1981). A specific, high affinity binding protein for 1,25-dihydroxyvitamin D in the chick oviduct shell gland. *Biochem. Biophys. Res. Commun.* **93,** 285–292.

Dacke, C. G. (1976). Parathyroid hormone and eggshell calcification in Japanese quail. *J. Endocrinol.* **71,** 239–243.

Dacke, C. G. (1979). "Calcium Regulation in Submammalian Vertebrates." Academic Press, London.

Dacke, C. G., (1989). Eicosanoids, steroids and miscellaneous hormones. *In* "Vertebrate Endocrinology: Fundamentals and Biomedical Implications" (P. K. T. Pang and M. P. Schreibman, eds.), Vol. 3, pp. 171–210. Academic Press, New York.

Dacke, C. G., and Kenny, A. D. (1973). Avian bioassay for parathyroid hormone. *Endocrinology* **92,** 463–470.

Dacke, C. G., and Kenny, A. D. (1982). Prostaglandins: Are they involved in avian calcium homeostasis? *In* "Aspects of Avian Endocrinology: Practical and Theoretical Implications" (C. G. Scanes, M. A. Ottinger, A. D. Kenny, J. Balthazart, J. Cronshaw, and I. Chester-Jones, eds.), pp. 255–262. Texas Tech Univ. Press, Lubbock.

Dacke, C. G., and Shaw, A. J. (1988). Effects of synthetic bovine parathyroid hormone (1–34) and its analogues on ^{45}Ca uptake and adenylate cyclase activation in bone and plasma calcium levels in the chick. *Quart. J. Exp. Physiol.* **73,** 573–584.

Dacke, C. G., Ancill, A. K., Whitaker, G., and Bascal, Z. A. (1993a). Calcitrophic peptides and rapid calcium fluxes into chicken bone *in vivo*. *In* "Avian Endocrinology" (P. J. Sharp, ed.), pp. 239–248. J. Endocrinol. Ltd, Bristol.

Dacke, C. G., and Shaw, A. J. (1987). Studies of the rapid effects of parathyroid hormone and prostaglandins on ^{45}Ca uptake into chick and rat bone *in vivo*. *J. Endocrinol.* **115,** 369–377.

Dacke, C. G., Arkle, S. A., Cook, D. J., Wormstone, I. M., Jones, S., Zaidi, M., and Bascal, Z. A. (1993b). Medullary bone and avian calcium metabolism. *J. Exp. Biol.* **184,** 63–88.

Dempster, D. W., Murrils, F. J., Horbert, W. R., and Arnett, T. R. (1987). Biological activity of chicken calcitonin: effects on neonatal rat and embryonic chick osteoclasts. *J. Bone. Miner. Res.* **2,** 443– 448.

Eliam, M. C., Basle', M., Bouziar, B., Bielakoff, J., Moukhtar, M., and de Vernejoul, M. C. (1988). Influence of blood calcium on calcitonin receptors in isolated chick osteoclasts. *J. Endocrinol.* **119,** 243–248.

Eliam-Cisse, M. C., Taboulet, J., Bielakoff, J., Lasmoles, F., de-Vernejoul, M. C., and Treilhou-Lahille, F. (1993). Influence of calcium and vitamin D deficient diet on calcitonin gene expression in the ultimobranchial cells of the developing chicken. *Gen. Comp. Endocrinol.* **89,** 195–205.

Farley, J. R., Tarbaux, N. M., Hall, S. L., Linkhart, T. A., and Baylink, D. J. (1988). The anti-bone-resorptive agent calcitonin also acts in vitro to directly increase bone formation and bone cell proliferation. *Endocrinology.* 123, 159–167.

Fenton, A. J., Martin, T. J., and Nicholson, G. C. (1994). Carboxyl-terminal parathyroid hormone-related protein inhibits bone resorption by isolated chicken osteoclasts. *J. Bone Miner. Res.* **9,** 515–519.

Fraser, D. R., and Emtage, J. S. (1976). Vitamin D in the avian egg: Its molecular identity and mechanism of incorporation into the yolk. *Biochem. J.* **160,** 671–682.

Fullmer, C. S., Bridak, M. E., Bar, A., and Wasserman, R. H. (1976). The purification of calcium binding protein from the uterus of laying hens. *Proc. Roy. Soc. Expl. Biol. Med.* **152,** 237–241.

Gay, C. V. (1988). Avian bone resorption at the cellular level. *CRC Crit. Rev. Poult. Biol.* **1,** 197–210.

Gay, C. V. (1996). Avian bone turnover and the role of bone cells. In "Comparative Endocrinology of Calcium Regulating Hormones" (C. G. Dacke, J. Danks, G. Flik, and I. Caple, eds.), pp. 113–121. J. Endocrinol. Ltd., Bristol.

Gay, C. V., Kief, N. L., and Bekkerm, P. J. (1993). Effect of estrogen on acidification in osteoclasts. *Biochem. Biophys. Res. Commun.* **192,** 1251–1259.

Goltzman, D., and Mitchell, J. (1985). Interaction of calcitonin and calcitonin gene-related peptide at receptor sites in target tissues. *Science* **227,** 1343–1345.

Hammond, R. W., and Ringer, R. K. (1978). Effect of indomethacin on the laying cycle, plasma calcium and shell thickness in the laying hen. *Poult. Sci.* **57,** 1141.

Hauschka, P. V., Lian, J. B., Cole, D. E., and Gundberg, C. M. (1989). Osteocalcin and matrix Gla protein; Vitamin K-dependent protein in bone. *Physiol. Rev.* **69,** 990–1034.

Henry, H. L., and Norman, A. W. (1984). Vitamin D: metabolism and biological actions. *Ann. Rev. Nutr.* **4,** 493–520.

Holick, M. F. (1989). Phylogenetic and evolutionary aspects of vitamin D from phytoplankton to humans. In "Vertebrate Endocrinology" (P. K. T. Pang and M. P. Schreibman, eds.), Vol. 3, pp. 7–43. Acaemic Press, New York.

Hou, H. C. (1931). Relation of preen gland of birds to rickets. III. Site of activation during irradiation. *Chin. J. Physiol.* **5,** 11–18.

Hurwitz, S. (1989). Parathyroid hormone. In "Vertebrate Endocrinology: Fundamentals and Biomedical Implications" (P. K. T. Pang and M. P. Schreibman, eds.), Vol. 3, pp. 45–77. Academic Press, New York.

Hurwitz, S. (1989). Calcium homeostasis in birds. *Vitam. Horm.* **45,** 173–221.

Ito, M. B., Schraer, H., and Gay, C. V. (1985). The effects of calcitonin, parathyroid hormone and prostaglandin E_2 on cyclic AMP levels of isolated osteoclasts. *Comp. Biochem. Physiol. A* **81,** 653–657.

Kameda, Y. (1989). Distribution of CGRP-, somatostatin-, galanin-, VIP-, and substance P-immunoreactive nerve fibers in the chicken carotid body. *Cell. Tissue Res.* **257,** 623–629.

Kameda, Y. (1991). Immunocytochemical localisation and development of multiple kinds of neuropeptides and neuroendocrine proteins in the chick ultimobranchial gland. *J. Comp. Neurol* **304,** 373–386.

Kenny, A. D. (1976). Vitamin D metabolism: Physiological regulation in egg laying Japanese quail. *Am. J. Physiol.* **230,** 1609–1615.

Kenny, A. D. (1981). "Intestinal Absorption of Calcium and Its Regulation." CRS Press, Boca Raton, FL.

Kenny, A. D. (1986). Parathyroid and ultimobranchial glands. In "Avian Physiology," (P. D. Sturkie, ed.), 4th ed., pp. 466–478. Springer-Verlag, New York.

Kenny, A. D., and Dacke, C. G. (1974). The hypercalcaemic response to parathyroid hormone in Japanese quail. *J. Endocrinol.* **62,** 15–23.

Kirby, G. C., and Dacke, C. G. (1983). Hypercalcaemic responses to 16,16-dimethyl prostaglandin E_2, a stable prostaglandin E_2 analogue, in chicks. *J. Endocrinol.* **99,** 115–122.

Kiyama, H., Katayama, Y., Hillyard, C. J., Girgis, S., MacIntyre, I., Emson, P. C., and Tohyama, M. (1985). Occurrence of calcitonin gene-related peptide in the chicken amacrine cells. *Brain Res.* **327,** 367–369.

Koch, J., Wideman, R. F., and Buss, E. G. (1984). Blood ionic calcium response to hypocalcemia in the chicken induced by ethylene glycol-bis-(B-aminoethylether) -N,N'-tetraacetic acid: Role of the parathyroids. *Poult. Sci.* **63,** 167–171.

Khosla, S., Demay, M., Pines, M., Hurwitz, S., Potts, J. T. Jr., and Kronenberg, H. M. (1988). Nucleotide sequence of cloned cDNAs encoding chicken preproparathyroid hormone. *J. Bone Miner. Res.* **3,** 689–698.

Laverty, G., and Clark, N. B. (1989). The kidney. In "Vertebrate Endocrinology: Fundamentals and Biomedical Implications" (P. K. T. Pang & M. P. Schreibman, eds.), Vol. 3, pp. 277–317. Academic Press, New York.

Lian, J. B., Glimcher, M. J., Roufosse, A. H., Hauscha, P. V., Gallop, P. M., Cohen-Solal, L., and Reit, B. (1982). Alterations in the gamma-carboxyglutamic acid and osteocalcin concentrations in vitamin D deficient chick bone. *J. Biol. Chem.* **257,** 4999–5003.

Liel, Y., Kraus, S., Levy, J., and Shany, S. (1992.), Evidence that estrogens modulate activity and increase the number of 1,25-dihydroxyvitamin D receptors in osteoblast-like cells. *Endocrinology* **130,** 2597–2601.

Limm, S. K., Gardella, T., Thompson, A., Rosenberg, J., Keutmann, H., Potts, J., Kronenberg, H., and Nussbaum, S. (1991). Full-length chicken parathyroid hormone: Biosynthesis in *Escherichia coli* and analysis of biologic activity. *J. Biol. Chem.* **266,** 3709–3714.

Malgaroli, A., Meldolesi, J., Zambonin-Zallone, A., and Teti, A. (1989). Control of cytosolic free calcium in rat and chicken osteoclasts: The role of extracellular calcium and calcitonin. *J. Biol. Chem.* **264,** 14342–14347.

Mallette, L. E. (1991). The parathyroid polyhormones: New concepts in the spectrum of peptide hormone action *Endocr. Rev.* **12,** 110–117.

May, L. G., Gilman, V. R., and Gay, C. V. (1993). Parathyroid regulation of avian osteoclasts. In "Avian Endocrinology" (P. J. Sharp, ed.), pp. 227–237. J. Endocrinol. Ltd., Bristol.

Mayel-Afshar, S., Lane, S. M., and Lawson, D. E. M. (1988). Relationship between the levels of calbindin synthesis and calbindin mRNA. *J. Biol. Chem.* **263,** 4355–4361.

Merke, J., Klaus G., Hugel, U., Waldherr, R., and Ritz, E. (1986). No 1,25-dihydroxyvitamin D_3 receptors on osteoclasts of calcium-deficient chickens despite demonstratable receptors on circulating monocytes. *J. Clin. Invest.* **77,** 312–314.

Michelangeli, V. P., Fletcher, A. E., Allan, E. H., Nicholson, G. C., and Martin, T. J. (1989). Effects of calcitonin gene-related peptide on cyclic AMP formation in chicken, rat, and mouse bone cells. *J. Bone Miner. Res.* **4,** 269–272.

Miller, S. C. (1977). Osteoclast cell-surface changes during the egg-laying cycle in Japanese quail. *J Cell Biol.* **75,** 104–118.

Miller, S. C. (1978). Rapid activation of the medullary bone osteoclast cell surface by parathyroid hormone. *J. Cell Biol.* **76,** 615–618.

MiyauchiI, A., Hruska, K. A., Greenfield, E. M., Duncan, R., Alverez, J., Barattolo R, Colucci S, Zambonin-Zallone A., Teitlebaum

S. L., and Teti A. (1990). Osteoclast cytosolic calcium regulated by voltage-gated calcium channels and extracellular calcium, controls podosome assembly and bone resorption. *J. Cell Biol.* **111,** 2543–2552.

Miyaura, C., Nagata, N., and Suda, T. (1981). Failure to demonstrate the stimulatory effect of calcitonin on cyclic AMP accumulation in avian bone *in vitro. Endocrinol. Jpn.* **28,** 403–408.

Navickis, R. J., Katzenellenbogen, B. S., and Nalbandov, A. V. (1979). Effects of the sex steroid and vitamin D_3 on calcium-binding protein in the chick shell gland. *Biol. Reprod.* **21,** 1153–1162.

Nicholson, G. C., Livesey, S. A., Moseley, J. M., and Martin, T. J. (1986). Actions of calcitonin, parathyroid hormone, and prostaglandin E_2 on cyclic AMP formation in chicken and rat osteoclasts. *J. Cell. Biochem.* **31,** 229–241.

Nicholson, G. C., Moseley. J. M., Sexton, P. M., and Martin, T. J. (1987). Chicken osteoclasts do not possess calcitonin receptors. *J. Bone Miner. Res.* **2,** 53–59.

Norman, A. W. (1987). Studies on the vitamin D endocrine system in the avian. *J. Nutr.* **117,** 797–807.

Nys, Y. (1993). Regulation of plasma 1,25-$(OH)_2D_3$, of osteocalcin and of intestinal and uterine calbindin in hens. *In* "Avian Endocrinology" (P. J. Sharp, ed.), pp. 345–357. J. Endocrinol. Ltd., Bristol.

Nys, Y., N'Guyen, T. M., Williams, J., and Etches, R. J. (1986). Blood levels of ionised calcium, inorganic phosphorus, 1,25-dihydroxycholecalciferol and gonadal hormones in hens laying hard shelled or shell-less eggs. *J. Endocrinol.* **111,** 151–157.

Nys, Y., Van Baelen, H., and Bouillon, R. (1992). Plasma 1,25-dihydroxycholecalciferol and its free index are potentiated by the ovulation dependent factors and shell formation induced hypocalcemia in the laying hen. *Dom. Anim. Endocrinol.* **9,** 37–47.

Pandala, S., and Gay, C. V. (1990). Effects of parathyroid hormone, calcitonin, and dibutyryl-cyclic AMP on osteoclast area in cultured chick tibiae. *J. Bone Miner. Res.* **5,** 701–705.

Parsons, J. A., Reit, B., and Robinson, C. J. (1973). Bioassay for parathyroid hormone using chicks. *Endocrinology* **92,** 454–462.

Pearse, A. G. E. (1976). Morphology and cytochemistry of the thyroid and ultimobranchial C cells. *In* "Handbook of Physiology: Parathyroid Gland" (G. D. Aurbach, R. O. Greep, and E. B. Astwood, eds.), Section 7, pp. 411–421. American Physiological Society, Washington, DC.

Pines, M., Granot, I., and Hurwitz, S. (1990). Cyclic AMP-dependent inhibition of collagen synthesis in avian epiphysial cartilage cells: Effect of chicken and human parathyroid hormone and parathyroid hormone-related peptide *Bone Miner.* **9,** 23–34.

Polin, D., Sturkie, P. D., and Hunsaker, W. (1957). The blood calcium response of the chicken to parathyroid extracts. *Endocrinology* **60,** 1–5.

Rifkin, B. R., Auszmann, J. M., Kleckner, A. P., Vernillo, A. T., and Fine, A. S. (1988). Calcitonin stimulates cAMP accumulation in chick osteoclasts. *Life Sci.* **42,** 799–804.

Romanoff, A. L., and Romanoff, A. J. (1963). "The Avian Egg," 2nd ed. Wiley, New York.

Rosenfield, M. G., Mermod, J. J., Amara, S. G., Swanson, L. W., Sawchenko, P. E., Rivier, J., Vale, W. W., and Evans, R. M. (1983). Production of a novel peptide encoded by the calcitonin gene via tissue specific RNA processing. *Nature* **304,** 129–135.

Roth, S. I., and Schiller, A. L. (1976). Comparative anatomy of the parathyroid glands. *In* "Handbook of Physiology: Parathyroid Gland" (G. D. Aurbach, R. O. Greep, and E. B. Astwood, eds.), Section 7, pp. 281–311. American Physiological Society, Washington DC.

Russell, J., and Sherwood, L. M. (1989). Nucleotide sequence of the DNA complementary to avian (chicken) preproparathyroid hormone mRNA and the deduced sequence of the hormone precursor. *Mol. Endocrinol.* **3,** 325–331.

Schermer, D. T., Chan, S. D. H., Bruce, R., Nissenson, R. A., Wood, W. I., and Strewler, G. J. (1991). Chicken parathyroid hormone-related protein and its expression during embryologic development. *J. Bone Miner. Res.* **6,** 149–155.

Schermer, D. T., Bradley, M. S., Bambino, T. H., Nissenson, R. A., and Strewler. G. J. (1994). Functional properties of a synthetic chicken parathyroid hormone-related protein 1–36 fragment. *J. Bone Miner. Res.* **9,** 1041–1046.

Sedrani, S. H., Taylor, T. G., and Akbtar, M. (1981). The regulation of 25-hydroxylase-calciferol metabolism in the kidney of the Japanese quail by sex hormones and by parathyroid extract. *Gen. Comp. Endocrinol.* **44,** 514–523.

Shaw, A. J., and Dacke, C. G. (1985). Evidence for a novel inhibition of calcium uptake into chick bone in response to bovine parathyroid hormone (1–34) or 16,16–dimethyl prostaglandin E_2 *in vivo. J. Endocrinol.* **105,** R5–R8.

Shaw, A. J., Whitaker, G., and Dacke, C. G. (1989). Kinetics of rapid ^{45}Ca uptake into chick skeleton *in vivo:* Effects of microwave fixation *Quart. J. Exp. Physiol.* **74,** 907–915

Shaw, A. J., and Dacke, C. G. (1989). Cyclic nucleotides and the rapid inhibitions of bone ^{45}Ca uptake in response to bovine parathyroid hormone and 16,16–dimethyl prostaglandin E_2 in chicks *Calcif. Tiss. Int.* **44,** 209–213.

Singh, R., Joyner, C. J., Peddie, M. J., and Taylor, T. G. (1986). Changes in the concentrations of parathyroid hormone and ionic calcium in the plasma of laying hens during the egg cycle in relation to dietary deficiencies of calcium and vitamin D. *Gen. Comp. Endocrinol.* **61,** 20–28.

Soares, J. H. (1984). Calcium metabolism and its contro—A review. *Poult. Sci.* **63,** 2075–2083.

Sommerville, B. A., and Fox, J. (1987). Changes in renal function of the chicken associated with calcitonin and parathyroid hormone. *Gen. Comp. Endocrinol.* **66,** 381–386.

Sugiyama, T., and Kusuhara, S. (1996). Morphological changes of osteoclasts on hen medullary bone during the egg-laying cycle and their regulation. *In* "Comparative Endocrinology of Calcium Regulating Hormones" (C. G. Dacke, J. Danks, G. Flik, and I. Caple, eds.) J. Endocrinol. Ltd., Bristol.

Takahashi, N., Shinki, T., Abe, E., Morinchi, N., Yamaguchi, A., Yoshiki, S., and Suda, T. (1983). The role of vitamin D in the medullary bone formation in egg-laying Japanese quails and in immature chicks treated with sex hormone. *Calcif. Tissue Int.* **35,** 465–471.

Taylor, T. G., and Dacke, C. G. (1984). Calcium metabolism and its regulation. *In* "Physiology and Biochemistry of the Domestic Fowl" (B. M. Freeman, ed.), Vol. 5, pp. 125–170. Academic Press, London.

Teti, A., Rizzoi, R., and Zambonin-Zallone, A. (1991). Parathyroid hormone binding to cultured avian osteoclasts *Bioch. Biophys. Res. Commun.* **174,** 1217–1222.

Thiede, M. A., Harm, S. C., McKee, R. L., Grasser, W. A., Duong, L. T., and Leach, R. M. (1991). Expression of the parathyroid hormone-related protein gene in the avian oviduct: potential role as a local modulator of vascular smooth muscle tension and shell gland motility during the egg-laying cycle. *Endocrinology* **129,** 1958–1966.

Thorp, B. H., Wilson, S., Rennie, S., and Solomons, S. (1993). The effect of a bisphosphonate on bone volume and eggshell structure in the hen. *Avian Pathol.* **22,** 671–682.

Tippins, J. R., Morris, H. R., Panico, M., Etienne, T., Bevis, P., Girgis, S., MacIntyre, I., Azria, M., and Attinger, M. (1984). The myotropic and plasma-calcium modulating effects of calcitonin gene-related peptide (CGRP). *Neuropeptides* **4,** 425–434.

Tregear, G. W., van Rietschoten, J., Greene, E., Keutmann, H. T., Niall, H. D., Reit, B., Parsons, J. A., and Potts, J. T. (1973). Bovine parathyroid hormone: Minimum chain length of synthetic peptide required for biological activity. *Endocrinology* **93,** 1349–1353.

Treilhou,-Lahille, F., Lasmoles, F., Taboulet, J., Barlet, J. P., Milhaud, G., and Moukhtar, M. S. (1984). Ultimobranchial gland of the domestic fowl: Two types of secretory cells involved in calcitonin metabolism. *Cell. Tissue Res.* **235,** 439–448.

van de Velde, J. P., Loveridge, N., and Vermeiden, J. P. (1984). Parathyroid hormone responses to calcium stress during eggshell calcification. *Endocrinology* **115,** 1901–1904.

de Vernejoul, M. C., Horowitz, M., Demignon, J., Neff, L., and Baron, R. (1988). Bone resorption by isolated chick osteoclasts in culture is stimulated by murine spleen cell supernatant fluids (osteoclast activating factor) and inhibited by calcitonin and prostaglandin E_2. *J. Bone Min. Res.* **3,** 69–80.

Villar, M. J., Huchet, M., Hokfelt, T., Changeux, J. P., Fahrenkrug, J., and Brown, J. C. (1988). Existence and coexistence of calcitonin gene-related peptide, vasoactive intestinal polypeptide- and somatostatin-like immunoreactivities in spinal cord motoneurons of developing embryos and post-hatch chicks. *Neurosci. Lett.* **86,** 114–118.

Wellnhoser, P. (1991). "The Illustrated Encyclopaedia of Pterosaurs." Salamander Books Ltd., London.

Williams, D. C., Paul, D. C., and Herring, J. R. (1991). Effects of antiestrogenic compounds on avian medullary bone formation. *J. Bone Miner. Res.* **6,** 1249–1256.

Wilson, S., and Duff, S. R. I. (1991). Effects of vitamin or mineral deficiency on the morphology of medullary bone in laying hens. *Res. Vet. Sci.* **50,** 216–221.

Ypes, D. L., Ravesloot, J. H., Buisman, H. P., and Nijweide, P. J. (1988). Voltage activated ionic channels and conductance's in embryonic chick osteoblast cultures. *J. Membr. Biol.* **101,** 141–150.

Zaidi, M., Moonga, B., Bevis, J. R., Bascal, Z. A., and Breimer, M. D. (1990a). The calcitonin gene peptides: Biology and clinical relevance. *Crit. Rev. Clin. Lab. Sci.* **28,** 109–174.

Zaidi, M., Datta, H. K., and Bevis, P. J. R. (1990b). Kidney: A target organ for calcitonin gene-related peptide. *Exp. Physiol.* **75,** 27–32.

Zaidi, M., Chambers, T. J., Gaines-Das, R. E., Morris, H. R., and MacIntyre, I. (1987). A direct action of human calcitonin gene-related peptide on isolated human osteoclasts. *J. Endocrinol.* **155,** 511–518.

Zambonin-Zallone, A., Teti, A., and Primavera, M. V. (1982). Isolated osteoclasts in primary culture: First observations on structure and survival time in culture media. *Anat. Embryol.* **165,** 405–413.

CHAPTER 19

Adrenals

R. V. CARSIA
Department of Cell Biology
University of Medicine and Dentistry of New Jersey
School of Osteopathic Medicine
Stratford, New Jersey 08084

S. HARVEY
Department of Physiology
University of Alberta
Edmonton, Alberta
Canada T69 2H7

I. Anatomy   489
   A. Microanatomy   490
II. Adrenocortical Hormones   492
   A. Corticosteroid Secretory Products   492
   B. Synthesis of Corticosteroids   493
   C. Transport of Corticosteroids   496
   D. Circulating Concentrations of Corticosteroids   496
   E. Secretion, Clearance, and Metabolism of Corticosteroids   497
   F. Other Secretory Products   498
   G. Hypothalamo-Pituitary-Adrenal Axis   498
   H. Stress and Adrenocortical Function   500
   I. General Regulation of Adrenocortical Function   502
   J. Regulation of Aldosterone Secretion   506
   K. Adrenocortical Function in Physiological and Ecological Contexts   509
   L. Changes in Development, Maturation, and Senescence   513
III. Physiology of Adrenocortical Hormones   514
   A. Corticosteroid Receptors   514
   B. Corticosteroid Action in Target Cells   515
   C. Corticosteroids and Intermediary Metabolism   516
   D. Corticosteroids and Immune Function   517
   E. Corticosteroids and Behavior   517
   F. Corticosteroids and Electrolyte Balance   518
IV. Adrenal Chromaffin Hormones   519
   A. Catecholamine Synthesis   519
   B. Control of Catecholamine Synthesis and Release   519
   C. Physiological Actions of Epinephrine and Norepinephrine   520
   D. Changes in Development, Maturation, and Senescence   521
References   522

I. ANATOMY

The paired adrenal glands are located (Figure 1) anterior and medial to the cephalic lobe of the kidney. Although their intraclass shape is highly variable, they are generally flattened and lie close together, even fusing in some species (Hartman and Brownell, 1949). The glands receive direct arterial blood supply from the cranial renal arteries and occasionally from the aorta. Each gland has a single adrenal vein which drains to the posterior (caudal) vena cava (Chester Jones, 1957). In addition, in the chicken, each gland receives one or two lymph vessels (Dransfield, 1945; Payne, 1971). Preganglionic sympathetic fibers from thoracic and synsacral splanchnic nerves (Freedman, 1968) converge on the cranial and caudal ganglia located on the pericapsular sheath or embedded in the gland substance (Figure 3). The preponderance of preganglionic sympathetic fibers (cholinergic) course through the ganglia undisturbed and penetrate the gland to innervate clusters of chromaffin cells. Each cluster consists of 12 or more chromaffin cells (Unsicker, 1973a). Postganglionic sympathetic rami (noradrenergic) of the ganglia innervate adrenal blood vessels and possibly chromaffin cells as well (Unsicker, 1973b), and contribute to renal and gonadal plexuses. There is some suggestion that, as in mammals, the

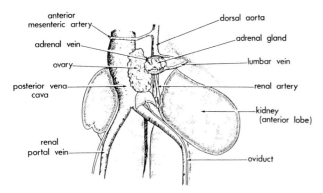

FIGURE 1 The position of the left adrenal gland and its vascular supply in the female gull (*Larus argentatus*). (Preparation and drawing by I. Carthy and J. G. Phillips.)

avian adrenal gland is also innervated by parasympathetic fibers, presumably from the vagus nerve via the celiac plexus (Freedman, 1968; Bennett, 1974). Indeed, blood vessels in the gland receive both cholinergic and adrenergic endings (Unsicker, 1973b). Synaptic endings appear in the vicinity of adrenocortical cells but rarely form direct connections (Unsicker, 1973c).

A. Microanatomy

The avian adrenal parenchyma is composed of an intermingling of adrenocortical and chromaffin tissues (Figures 2A and 3), the latter tending to be more concentrated centrally in the gland where the vasculature is particularly enriched. As in mammals, clusters of 10–12 chromaffin cells are innervated by a single nerve bundle, although in birds, the clusters do not fuse to become a definitive adrenal medulla. One neuronal terminal may synapse with up to three chromaffin cells of the same type, i.e., norepinephrine (NE) or epinephrine (E) chromaffin cells (Unsicker, 1973a).

Estimates of the relative abundance of adrenocortical and chromaffin tissues comprising the avian adrenal gland must be interpreted with caution due to the extrapolation of volume ratios from *camera lucida* area ratios of sections and the shape and dimesions of the gland. Nevertheless, area ratios suggest that adrenocortical and chromaffin tissues roughly comprise 58 and 42%, respectively, of the parenchyma (Holmes and Phillips, 1976). Indeed, in preparations of collagenase-digested chicken adrenal tissue, adrenocortical and chromaffin cells comprise 60 and 40%, respectively, of the total intact adrenal cell isolate (Carsia *et al.*, 1985a).

Based on ultrastructural and cytochemical studies, two chromaffin cell types exist in the avian adrenal gland containing either norepinephrine (NE) or epinephrine (E) (Carmichael, 1987). However, there is some controversy as to the relative proportion of the cell types. Ghosh (1980) reported a greater number of NE chromaffin cells in all birds studied (except passerines), while Unsicker (1973d) considered the E chromaffin cells predominate. Recent work with the chicken adrenal gland

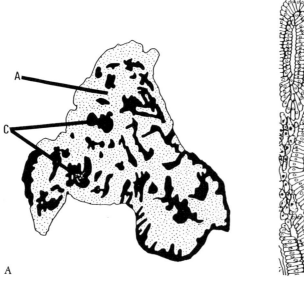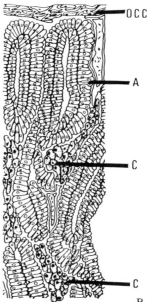

FIGURE 2 Microanatomy of a typical avian adrenal gland (Florida quail, *Colinus virginianus floridanus*). (A) Distribution of chromaffin (C) and adrenocortical (A) tissue. (B) Structure of the looped cords of adrenocortical cells with intermingled chromaffin cell islets. The outer connective tissue (adrenal capsule) is also indicated (OCC). (Taken from Chester Jones and Phillips, 1986.)

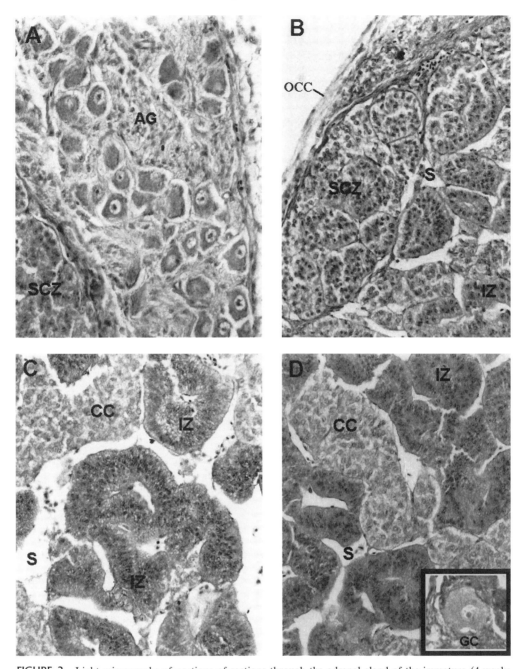

FIGURE 3 Light micrographs of portions of sections through the adrenal gland of the immature (4 weeks old) domestic turkey (*Meleagris gallopavo*). (A) Pericapsular adrenal ganglion (AG) and adjacent subcapsular zonal cell cords (SCZ); (B) outer connective tissue capsule (OCC) with subcapsular zonal cell cords (SCZ), inner zonal cell cords (IZ), and sinusoid (S); (C) inner zonal cell cords (IZ) with sinusoid (S) and chromaffin cell islet (CC); (D) chromaffin cell islet (CC) with ganglion cell (GC, inset), inner zonal cell cords (IZ) and sinusoid (S) (×250). (Micrographs courtesy of R. G. Nagele, Ph.D., Department of Molecular Biology, UMDNJ-School of Osteopathic Medicine.)

indicates that 70% of the chromaffin cells are E cells (Ohmori *et al.*, 1997). There is some consistency in the location of the NE chromaffin cells at the periphery of the chromaffin cell clusters, whereas the E chromaffin cells are more central (Unsicker, 1973d; Holmes *et al.*, 1991).

Adrenocortical cells are arranged in cords (Figures 2B and 3) that radiate from the center of the gland, branching and anastomosing frequently, and at the periphery, looping against the inner surface of the connective tissue capsule. Each cord is composed of a double row of adrenocortical cells orientated with their long axes perpendicular to the cord. Overall, with some intraclass variation, their nuclei are eccentrically located toward the common, absinusoidal basal lamina of the cord axis (Hartman and Brownell, 1949). The microscopic and ultrastructural heterogeneity of these cells along the cords suggests a rudimentary zonation. Cells at the periphery of the gland are said to be large, binucleated and replete with lipid droplets whereas cells toward the center of the gland tend to be smaller, more elongated, and contain less lipid (Hartman and Brownell, 1949; Chester Jones and Phillips, 1986). Additions to this theme are exemplified in the duck (*Anas platyrhynchos*) adrenal gland (Pearce et al., 1978, 1979; Holmes and Cronshaw, 1980, 1984), in which two regions of adrenocortical tissue are apparent: a subcapsular zone (SCZ), 40–60 cells thick, composed of cells having irregularly shaped nuclei, relatively little smooth endoplasmic reticulum, and mitochondria bearing regularly arranged tubular cristae, some of which bridge the inner membranes, reminiscent of mammalian zona glomerulosa cells, and an inner zone (IZ) composed of cells having rounded nuclei, more abundant smooth endoplasmic reticulum and mitochondria bearing tubulovesicular cristae, reminiscent of mammalian zona fasciculata/reticularis cells (Malamed, 1975). To varying extent, these features are also apparent in the brown pelican (*Pelecanus occidentalis*) (Knouff and Hartman, 1951) and chicken (*Gallus gallus domesticus*) (Kondics and Kjaerheim, 1966; Taylor et al., 1970). The IZ cells, but not the SCZ cells, are susceptible to adrenocorticotropic hormone (ACTH) withdrawal (adenohypophysectomy), in which the ultrastructural features convert to the SCZ type. However, IZ ultrastructural features are restored with ACTH replacement (Pearce et al., 1979; Holmes and Cronshaw, 1980).

Functional studies with duck SCZ and IZ tissues corroborate the structural evidence for a zonation and the presence of distinct steroidogenic cell types. These indicate a relatively greater production of aldosterone by SCZ tissue in response to ACTH (Klingbeil et al., 1979) and an exclusive aldosterone response of the SCZ tissue to angiotensin II (Klingbeil, 1985a). By contrast, a greater corticosterone response to ACTH is apparent for the IZ tissue (Klingbeil, 1985a). In addition, recent evidence is provided for functionally distinct subpopulations of adrenocortical cells in the domestic turkey (*Meleagris gallopavo*) (Kocsis et al., 1995a; Carsia and McIlroy, 1998). However, it should be pointed out that this concept of zonation and functionally distinct adrenocortical cells in the avian adrenal gland, in terms of ratios of corticosteroids secreted and responsiveness to tropic hormones, is supported by evidence from studies on a limited number of precocial species (for review, see Holmes and Cronshaw, 1993). Furthermore, the physiological significance of results with tissue fragments and isolated cells should be considered with caution since steroidogenic responses from gland tissue and isolated cells, anatomically removed from neural and polyhormonal influences with disruption of paracrine and juxtacrine interactions, may be considerably different from responses of the intact gland *in vivo* (see Vinson et al., 1985).

II. ADRENOCORTICAL HORMONES

A. Corticosteroid Secretory Products

The glucocorticoid corticosterone is the principal corticosteroid released by the avian adrenal gland (DEROOS, 1961; Assenmacher, 1973). The mineralocorticoid aldosterone is produced in considerably less quantity in mature birds (see Vinson et al., 1979). *In vitro* studies with chicken adrenal tissue (DEROOS, 1961, 1969), duck adrenal tissue (DEROOS, 1961; Klingbeil et al., 1979; Cronshaw et al., 1985; Klingbeil, 1985a), and isolated adrenocortical cells from the duck (Collie et al., 1992), chicken (Rosenberg et al., 1986, 1987, 1988a,b; Carsia et al., 1987a, 1988a; Collie et al., 1992), and turkey (Kocsis and Carsia, 1989; Kocsis et al., 1994a, 1995a) suggest an aldosterone-to-corticosterone secretion ratio of about 1:19. Taking into consideration varying sampling stress and corticosteroid half-life, if is not surprising that the average ratio of basal circulating aldosterone to corticosterone is about 1:100 in mature birds in which both corticosteroids were measured (Table 1).

It should be pointed out that other steroids are synthesized and secreted in significant quantities by the avian adrenal gland, particularly in the embryonic and perinatal periods. These include cortisol and cortisone (Kalliecharan and Hall, 1974, 1976, 1977; Idler et al., 1976; Nakamura et al., 1978; Kalliecharan, 1981; Carsia et al., 1987a). The physiological significance of the appearance and decline of these glucocorticoids is unclear but may reflect their differential efficacy of action or the differential sensitivity of developing and maturing target tissues (Marie, 1981; Kalliecharan and Buffett, 1982). In addition, the adrenal appears to an important source of testosterone (Nakamura et al., 1978; Tanabe et al., 1979, 1983, 1986) and possibly estradiol (Tanabe et al., 1979) during the embryonic and perinatal period. Indeed, the adrenal may be a more important source of testosterone than the gonad in the chick embryo (Tanabe et al., 1979). Furthermore, in the chick embryo, the female adrenal gland synthesizes more testosterone than the male adrenal gland (Tanabe et al., 1986). This, together with the overall greater sex-steroidogenic activity of the ovary (Tanabe et al., 1979, 1983, 1986), may form an important "adrenal-ovarian unit" for the high circulating levels of estradiol needed for feminization (Tanabe *et al.*, 1979, 1986).

B. Synthesis of Corticosteroids

The synthetic pathways of the major steroids secreted by the avian adrenal gland during the life cycle are shown in Figure 4. It is assumed that the pattern of enzymes carrying out these synthetic steps is highly conserved among the vertebrates and homologous to that characterized in the mammal (Miller, 1988; Nebert *et al.*, 1991; Kawamoto *et al.*, 1992; Mukai *et al.*, 1993; Takemori *et al.*, 1995). It is interesting to note that there are substantial differences in the primary structure of steroidogenic enzymes studied thus far between mammalian and avian species. However, indistinguishable conserved domains, thought to be enzymatically active sites, are apparent (Nakabayashi *et al.*, 1995; Nomura *et al.*, 1997). In this steroidogenic scheme, mitochondrial cytochrome $P-450_{SCC}$ (or CYP11A1) mediates 20α-/22-hydroxylation and scission of the C20-22 carbon bond, producing pregnenolone and isocaproaldehyde, a set of reactions traditionally called "20,22-desmolase." It is largely thought that pregnenolone is converted to progesterone by a non-P-450 microsomal dehydrogenase–isomerase complex (HSD3B) which mediates 3β-hydroxysteroid dehydrogenase and isomerase (Δ^5 double bond to Δ^4 double bond) activities. However, recent work indicates that this enzyme also resides in the mitochondria of adrenocortical cells (Cherradi *et al.*, 1994; Sauer *et al.*, 1994). Microsomal $P-450_{C21}$ (CYP21A1) then mediates a 21-hydroxylation step sequence converting progesterone to 11-deoxycorticosterone. As stated previously, 11-deoxycorticosterone is a significant secretory product of the avian adrenal gland. However, corticosterone is the predominant corticosteroid product. Corticosterone is formed from 11-deoxycorticosterone by an 11β-hydroxylation via a mitochondrial enzyme (CYP11B1; $P-450_{11\beta}$). This pathway to corticosterone appears to operate in the avian adrenal gland (Sandor *et al.*, 1963; Kalliecharan and Hall, 1977; Nakamura *et al.*, 1978), albeit avian-specific enzymes await molecular characterization. In this connection, the primary structures of chicken 3β-hydroxysteroid dehydrogenase (Nakabayashi *et al.*, 1995) and chicken adrenal $P-450_{SCC}$ (Nomura *et al.*, 1997) have been determined.

Although most of the synthesized corticosterone is immediately released, some of the corticosterone is the precursor for aldosterone synthesis (Pedernera and Lantos, 1973; Lehoux, 1974; Aupetit *et al.*, 1979). Indeed, with isolated turkey adrenocortical cells, aldosterone synthesis is dependent on the available corticosterone in that specific stimulators of aldosterone synthesis, K^+ and angiotensin II, do not alter cholesterol side-chain cleavage (Kocsis *et al.*, 1995a). The conversion of corticosterone to aldosterone requires two 18-hydroxylations forming an unstable intermediate that decomposes to form a C18-aldehyde (aldosterone) and free water. In mammals, this conversion proceeds via two modes: a zonal-specific alteration in the activity and substrate specificity of $P-450_{11\beta}$ (porcine-bovine mode) or the zonal-specific expression of a distinct mitochondrial enzyme, aldosterone synthase (termed CYP11B2, $P-450_{C18}$, or $P-450_{ALDO}$) (rodent–human mode) (Takemori *et al.*, 1995). It is yet to be determined whether enzymatic counterparts exist in avian species.

Other enzymatic activities are also apparent during the embryonic and neonatal periods. Formation of cortisol is predominantly, if not exclusively, via 17α-hydroxylation of progesterone to form 17α-hydroxyprogesterone (mediated by CYP17A1, a microsomal $P-450_{17\alpha}$), 21α-hydroxylation to form 11-deoxycortisol (CYP21A1; microsomal $P-450_{C21}$), and then an 11β-hydroxylation step (mitochondrial $P-450_{11\beta}$) to form cortisol. In avian species it is not clear whether a CYP11B1 carries out the 11β-hydroxylation for the formation of both cortisol and corticosterone or that corticosterone is preferentially formed from the activity of a CYP11B2, which in man is linked to aldosterone synthesis. There is some evidence for the potential to form 11β-hydroxyprogesterone to serve as a precursor for corticosterone (via a 21-hydroxylation step) and cortisol (via 17α- and 21-hydroxylation steps), albeit these pathways appear to be remote (Sandor *et al.*, 1963; Nakamura *et al.*, 1978; Freeman, 1983). In addition, microsomal $P-450_{17\alpha}$ (CYP17) has 17,20-lyase activity, and thus its further action on 17α-hydroxyprogesterone produces androstenedione, a precursor for testosterone synthesis (via a microsomal, non-P-450 enzyme, 17-ketosteroid reductase) (Nakamura *et al.*, 1978; Tanabe *et al.*, 1979, 1983, 1986). In this regard, a chicken $P-450_{17\alpha}$ has been cloned and is expressed in the neonatal adrenal gland (Ono *et al.*, 1988). Furthermore, there is some evidence that testosterone is aromatized to estradiol (microsomal $P-450_{AROM}$) in the embryonic adrenal gland (Tanabe *et al.*, 1979).

Other pathways collectively regarded as "degradation" pathways deserve attention, because evidence suggests that the resulting steroid metabolites have intra- and extraadrenal function. Two pathways of interest are the conversion of progesterone via 5β- and 5α-reduction to predominantly 5β-pregnanedione and to a lesser extent 5α-pregnanedione (5β- and 5α-pregnan-3,20-dione) during the embryonic period and the conversion of corticosterone to 5α-reduced metabolites (adrenal 5α-reductase), 5α-dihydrocorticosterone

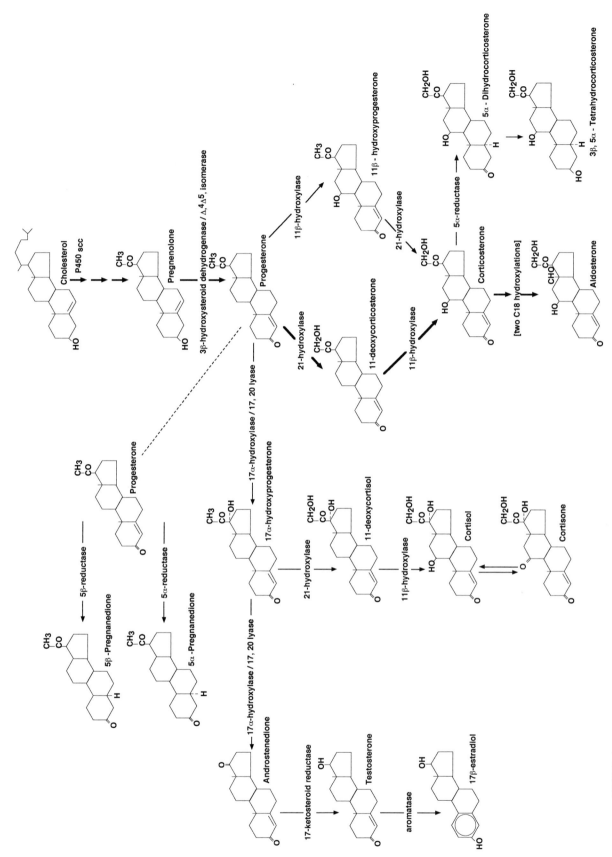

FIGURE 4 Steroidogenesis in the avian adrenal gland. Axis from cholesterol is the predominant pathway posthatching (postnatally). Left pathway branches are significant pathways during the embryonic period. Right pathway branch is a possible degradation pathway. Bolded pathway arrows indicate predominant pathway. See text for details. (Pathway scheme prepared by H. Weber, Department of Cell Biology, UMDNJ-School of Osteopathic Medicine.)

(5α- pregnan-11β, 21-diol-3,20-dione), and 5α-tetrahydrocorticosterone (5α-pregnan-3β,11β,21-triol-20-one). The former pathway may have a role in embryonic synthesis of proteins in developing organs and hemoglobin synthesis (Gonzalez et al., 1983; Aragonés et al., 1991). For example, in the chick embryo, the conversion of progesterone to 5β-pregnanedione is greater than that to corticosterone, but this declines approaching hatch (Pedernera and Lantos, 1973; Gonzalez et al., 1983). In addition, both the former and latter pathways may be inactivating pathways. Such inactivating pathways may serve to coordinate adrenal steroid synthesis with target tissue demands and to prevent hypersecretion of corticosteroids and subsequent deleterious effects at target sites. Equally plausible is that these pathways reduce high corticosteroid concentrations that are thought to inhibit adrenal protein synthesis (Morrow et al., 1967), mitochondrial function (Burrow, 1969), and to damage adrenal steroidogenic enzymes through the induction of free radical formation (Hornsby and Crivello, 1983). For example, in experiments with isolated chicken adrenocortical cells, exogenous corticosterone acutely reduces endogenous corticosterone production (Carsia et al., 1983, 1984, 1987b) through the stimulation of 5α-reductase activity (Carsia et al., 1984). This activity is enhanced by low, subthreshold steroidogenic concentrations of ACTH (Carsia et al., 1987b). It is, however, acutely counteracted by prolactin (Carsia et al., 1984, 1987b). This is in keeping with the observation that various stressful stimuli in domestic fowl (chickens and turkeys) rapidly increase circulating levels of prolactin (Harvey et al., 1979; Harvey et al., 1984b; Opel and Proudman, 1986). Although there appear to be species differences in this prolactin response to various stressors (Opel and Proudman, 1984), the overall pattern of response suggests that in times of stress, this intraadrenal mechanism of corticosterone metabolism can be overridden.

Steroid biosynthesis involves depots of cholesterol. Avian adrenocortical cells are laden with cholesterol-ester-rich lipid droplets (Pearce et al., 1979; Holmes and Cronshaw, 1980; Carsia et al., 1985a; Holmes et al., 1991) and thus, not surprisingly, their intrinsic ability to synthesize cholesterol is normally low (Lehoux et al., 1977). Cholesterol stores for steroidogenesis are apparently initially derived from plasma lipoproteins (Hertelendy et al., 1992; Latour et al., 1995). Although it is generally thought that these cholesterol stores are mobilized directly to the steroidogenic pathway with tropic hormone stimulation, there is evidence suggesting that cholesterol for acute steroidogenesis traffics via the plasma membrane (Freeman, 1989; Nagy and Freeman, 1990). Regardless of trafficking mode, free cholesterol ultimately enters the steroidogenic pathway which is composed of steroidogenic enzymes in membrane domains of the mitochondria and smooth endoplasmic reticulum. Steroidogenic P-450 enzymes require intact membranes for function and studies suggest that they exhibit a directional orientation that orders the flow of steroid products among the membranes of these organelles (see Miller, 1988; Stocco and Clark, 1996). Yet, despite much progress in understanding steroidogenic enzymes at the molecular level, there is little information on the mechanisms of steroid compartmentalization and movement during steroidogenesis. This appears to be accomplished by proteins that bind steroids with relative specificity and that may serve as shuttles and by the close association of the steroidogenic organelle membranes driven by the cytoskeleton (see Stocco and Clark, 1996). For example, there appears to be a family of cholesterol trafficking molecules such as the Niemann-Pick type C disease gene product (see Pennisi, 1997). How this family functions in shuttling cholesterol for steroidogenesis is unclear. Nevertheless, it is generally agreed that the delivery of cholesterol to P-450$_{SCC}$ in the inner mitochondrial membrane is the regulated rate-limiting step of steroidogenesis (Karaboyas and Koritz, 1965; Koritz et al., 1977; Crivello and Jefcoate, 1980; Jefcoate et al., 1987). A promising intracellular transporter of free cholesterol to serve this function is sterol carrier protein 2 (McNamara and Jefcoate, 1989; also see Stocco and Clark, 1996), which appears to have an avidity for mitochondria. Facilitators of movement of cholesterol to the inner mitochondrial membrane include steroidogenesis activator polypeptide, the peripheral benzodiazepine receptor (PBR), and steroidogenic acute regulatory protein (StAR) (Clark and Stocco, 1996; Stocco and Clark, 1996). Although controversy exists concerning the importance of PBR (Amri et al., 1997), StAR has gained preeminence because it is inducible by tropic hormone, has a short half-life, and is sensitive to cycloheximide. Its precursor form has a mitochondrial targeting sequence, and expression of this protein in a competent cell line or in a cell-free system containing mitochondria, in the absence of tropic hormone, is sufficient for steroid synthesis. However, beyond the movement of cholesterol, little is known about the mechanisms regulating the shuttling of steroid intermediates among the membranes of the smooth endoplasmic reticulum and the mitochondria in the synthesis of corticosteroids. It is thought that part of the movement of cholesterol and steroid intermediates is facilitated by the close association of steroidogenic organelles. Indeed, the close association of mitchondria, smooth endoplamic reticulum, and lipid (cholesterol-ester-rich) droplets is conspicuous in mammalian (Malamed, 1975) and avian (Pearce et al., 1979; Holmes and Cronshaw, 1980; Carsia et al., 1984; Holmes et al., 1991) adrenocortical cells. The framework and modulation of this association appears to be directed by the cytoskeleton (reviewed in Feuilloley and Vaudry, 1996) and in avian adrenocortical cells, especially microfilaments (Cronshaw et al., 1984, 1992a) and intermediate filaments (Carsia et al., 1985a, 1987c).

Studies with isolated adrenocortical cells from avian species suggest wide variations in steroidogenic activity, albeit strong conclusions concerning the relevance of these differences in steroidogenic activity to differences in circulating values cannot be drawn, due to the different ages of the birds from which the cells were derived. It might be noted, however, that corticosteroidogenic activity of avian adrenocortical cells is considerably less than that of rat adrenocortical cells. For example, on an equivalent cell concentration basis, maximal ACTH-induced corticosterone production by adult rat adrenocortical cells is 10–40 times that of adult chicken adrenocortical cells (Carsia et al., 1985, 1987b,c).

C. Transport of Corticosteroids

In birds, most circulating corticosterone is transported bound in dynamic equilibrium to plasma proteins. Two components of these transport proteins are apparent: a specific binding protein, transcortin or corticosteroid-binding globulin (CBG) (that has a high affinity and low binding capacity for corticosterone), and a low-affinity, nonspecific binding protein (presumably albumin) with a very high capacity (Wingfield et al., 1984). A CBG, with physicochemical properties similar to that of other vertebrates (for details, see Wingfield et al., 1984), has been identified in the plasma of 23 avian species, in which it has different affinities ($K_d = 10^{-9}$–10^{-7} mol/liter) and capacities (10^{-9}–10^{-8} mol/liter). Avian CBG binds with corticosterone and progesterone and, in some species, there is significant (>20%) cross-reaction with testosterone and 5α-dihydrotesterone. In chickens, it is reported that CBG has a K_d of 5.6×10^{-8} to 8.2×10^{-8} mol/liter and a capacity of 1.4×10^{-8} to 2.2×10^{-7} mol/liter (Gould and Siegel, 1978; Wingfield et al., 1984; Carsia et al., 1988a). The binding of corticosterone to CBG and other plasma proteins influences the availability of the corticosteroid for target cells. Thus, CBG and other plasma proteins may provide stabilization of the concentration of free hormone through inhibition of clearance by the kidney and liver (Siiteri et al., 1982). It has been thought that protein-bound steroids enter cells less rapidly than free steroids and hence appear less active (Slaunwhite et al., 1962; Mendel, 1989). However, recent work suggests that CBG selectively enters and/or is selectively cleaved by glucocorticoid target cells, thus accentuating glucocorticoid bioavailability/action (Hammond, 1995).

A number of factors affect the circulating concentration of CBG, presumably by affecting its synthesis by the liver. For example, it appears to vary with embryonic (Siegel and Gould, 1976; Martin et al., 1977; Gasc and Martin, 1978) and post embryonic (Gould and Siegel, 1974) age, and with sex (Gould and Siegel, 1978). In the sparrow (*Zonotrichia albicollis*) (Meier et al., 1978) and Japanese quail (Kovács and Péczely, 1983), there is a positive correlation between plasma corticosterone and CBG concentrations. However, at night, during the plasma corticosterone nadir, there is a peak in plasma CBG which correlates well with increased locomotory activity in this species (Meier et al., 1978). It is postulated that this CBG peak maintains the delivery of corticosterone to metabolically active tissues, facilitated by the increase in tissue temperature and reduction in tissue pH. In addition, both endocrine and nutritional status affect CBG concentration. For instance, it is decreased by hypophysectomy in embryos (Gasc and Martin, 1978) by the administration of thyroxine or testosterone (Péczely, 1979; Péczely and Daniel, 1979; Kovács and Péczely, 1983) and by dietary protein restriction (Carsia et al., 1988a). However, a largely unexplored area is the extrahepatic synthesis of CBG in avian species during development and posthatch maturation that is becoming increasingly obvious in mammals (Hammond et al., 1995). Extrahepatic synthesis may contribute to the circulating concentration. In addition, differential tissue synthesis may finely regulate the local concentration of glucocorticoids and hence their bioavailability/action. Moreover, an understanding of the developmental/maturational expression of CBG isoforms, having different specificities/affinities for cortisol, corticosterone, and other glucocorticoid metabolites might explain the shifting pattern of glucocorticoid metabolite secretion by the adrenal gland during developmental, neonatal, and maturational periods.

D. Circulating Concentrations of Corticosteroids

Radioimmunoassays have largely replaced complex chromatographic (Phillips and Chester Jones, 1957), double-isotopic (e.g., Stachenko et al., 1964), fluorescent (Nagra et al., 1963a), high-pressure liquid chromatographic (Fowler et al., 1983), and competitive protein-binding (Wingfield and Farner, 1975) methods for the measurement of avian corticosteroids, especially that of aldosterone (see Table 1). In general, the greater precision, sensitivity, and specificity of radioimmunoassay over other techniques has yielded lower corticosterone concentrations. Overall, corticosterone is in the nanograms-per-mililiter range whereas aldosterone is in the picogram-per-mililiter range. However,

TABLE 1 Basal Peripheral Plasma Concentrations of Corticosterone and Aldosterone in Selected Avian Species in which Both Corticosteroids Were Determined in the Same Samples by Radioimmunoassay

Species (age)	Corticosterone (ng/ml)	Aldosterone (pg/ml)	Reference
Chicken			
hatch (0)	11	33	Holmes et al. (1992)
2 weeks	1.5	21	Rosenberg and Hurwitz (1987)
9 weeks	0.6	31.6	Radke et al. (1984)
10 weeks	11.6	51.5	Weber et al. (1990)
adult	2.6	22	Skadhauge et al. (1983)
adult	1.5	7	Arad and Skadhauge (1984)
adult	1.8	33	Radke et al. (1985a)
Turkey			
5 weeks	23.4	100.8	Carsia and McIlroy (1997)
6 weeks	32.2	62.6	Kocsis et al. (1995a)
13 weeks	5.3	21	Rosenberg and Hurwitz (1987)
Japanese quail			
adult	1.5	25	Kobayashi and Takei (1982)
Duck			
hatch (0)	11.7	234	Holmes et al. (1989)
hatch (0)	15–20	35–45	Holmes et al. (1992)
1 day	11.6	460	Holmes et al. (1989)
4 days	10.3	362	Holmes et al. (1989)
1 week	5.8	427	Holmes et al. (1989)
2 weeks	6.8	423	Holmes et al. (1989)
3 weeks	10.3	403	Holmes et al. (1989)
9 weeks	2.6	51.3	Radke et al. (1984)
10–16 weeks	9.6	32	Klingbeil (1985b)
adult	25.8	105.8	Gray et al. (1989)
adult	8.8	250	Holmes et al. (1989)

subnanogram-per-mililiter values of corticosterone have been reported for the chicken (Scott et al., 1983; Radke et al., 1984).

E. Secretion, Clearance, and Metabolism of Corticosteroids

Circulating concentrations of glucocorticoids are maintained by the dynamic balance between metabolic clearance and adrenal secretion. Transmembrane passage of corticosteroids is rapid for all cells ($\sim 10^{-4}$ cm/sec) (Giorgi and Stein, 1981), thus facilitating rapid clearance and metabolism in competent cells as for example the kidney and liver (Hansson et al., 1974; Siiteri et al., 1982). Indeed, the half-life ($t_{1/2}$) of corticosterone and aldosterone is about 15 min in avian species (see Table 2). Not surprisingly, corticosteroid concentrations which are initially high in the adrenal gland (Assenmacher, 1973; Cronshaw et al., 1989) and adrenal venous effluent (Phillips and Chester Jones, 1957; Brown, 1960; Nagra et al., 1960; Urist and Deutsch, 1960; Taylor et al., 1970) (in micrograms per milliliter) and that are secreted at about a microgram per minute per kilogram (Assenmacher, 1973; see Table 2), are then rapidly

TABLE 2 Secretion and Clearance Dynamics of Adrenal Corticosteroids

Species	Half-life (min)	Metabolic clearance rate (ml/min/kg)	Secretion rate (μg/min/kg)	Reference
Corticosterone				
Chicken	22	—	—	Birrenkott and Wiggins (1984)
Chicken	8	10.5	0.4	Carsia et al. (1988a)
Duck	11.2	44.6	1.03	Gorsline and Holmes (1982)
Japanese quail	10	—	—	Kovács and Péczely (1983)
Pigeon	18.4	—	—	Chan et al. (1972)
Aldosterone				
Duck	12.6	87.1	0.2	Thomas and Phillips (1975)

diluted in the circulation and distributed to the extracellular fluid.

The volume in which corticosterone is distributed varies with the physiological status of birds; it is increased in hypophysectomized (Bradley and Holmes, 1971) and saline-loaded (Donaldson and Holmes, 1965) ducks. Similarly, the metabolic clearance rate for corticosterone varies with physiological status; it is decreased with age (Holmes and Kelly, 1976) and by thyroidectomy (Kovács and Péczely, 1983). In addition, it appears to be influenced by nutritional status. For example, in chickens subjected to dietary protein restriction it is increased 85%, which translates to a four fold increase in secretion rate (Carsia et al., 1988a). Clearance from the circulation is facilitated by intracellular binding and sequestration and ultimately metabolism, primarily by the liver. The metabolizing enzymes include 5α-reductase and convert corticosterone to presumably inactive metabolites such as 11-dehydrocorticosterone and 5α-tetrahydrocorticosterone; the activity of this 5α-reductase appears to be augmented by acute stress (Daniel and Assenmacher, 1971).

F. Other Secretory Products

Recent evidence indicates that the embryonic chicken adrenal gland is a major source of inhibin during development (Rombauts et al., 1994; Decuypere et al., 1997). The inhibin content (expressed per milligrams of tissue) of the embryonic chick (day 18) adrenal gland is about four times that of the ovary and nearly half that of the testis. Interestingly, dexamethasone depresses inhibin release *in ovo*, whereas ACTH elicits a dose-dependent release (about five fold over basal) of inhibin that parallels corticosterone release in primary cultures of embryonic chick adrenal cells. It also appears that inhibin may be regulated differently in various body compartments of the chick embryo (Rombauts et al., 1992). Accumulating evidence suggests a role of this dimeric glycoprotein in embryonic and fetal development of birds and mammals (see Rombauts et al., 1994 and Decuypere et al., 1997). Although the ovary is the major source of inhibin in the hen, the adrenal continues to be a significant extragonadal source (Vanmontfort et al., 1997; Decuypere et al., 1997). The role of extragonadal inhibin postnatally remains to be investigated; however, the direct inhibition of inhibin release from chicken adrenal cells by dexamethasone (Vanmontfort et al., 1997) suggests a paracrine role within the adrenal gland (see Figure 5).

G. Hypothalamo-Pituitary-Adrenal Axis

Evidence for the presence of a hypothalamo-pituitary-adrenal (HPA) axis in birds has been reviewed extensively (Holmes, 1978; Baylé, 1980; Harvey et al., 1984a) as well as the mechanisms of its activation in response to stressors (Harvey and Hall, 1990). Activation of adrenocortical function, in terms of corticosterone secretion (Beuving and Vonder, 1978, 1986; Radke et al., 1985a,b) and probably aldosterone secretion as well (Radke et al., 1985a,b), through this axis is mediated primarily by pituitary ACTH. As in mammals, ACTH is a 39-amino-acid peptide released from the avian pituitary in response to stress (Kovács and Péczely, 1991). There is strong neuroanatomical (Jozsa et al., 1984, 1986; Mikami and Yamada, 1984; Péczely and Antoni, 1984; Yamada and Mikami, 1985; Ball et al., 1989) and physiological (Carsia et al., 1986; Romero and Wingfield, 1998; Romero et al., 1998a,b) evidence that corticotropin-releasing factor (CRF) is an avian secretagog for ACTH. It should be pointed out that in addition to CRF, arginine vasotocin (AVT), which has a distribution similar to CRF in the median eminence (Mikami and Yamada, 1984), is a significant ACTH secretagog in some avian species (Castro et al., 1986; Westerhof et al., 1992) and there is evidence that the relative responses of avian pituitary glands to CRF and AVT change with physiological status (e.g., molt) (Romero et al., 1998a). Given what is known in the rat concerning the shifting preeminence of CRF and vasopressin depending on the type of chronic stress (Ma et al., 1997; Chrousos, 1998), the shifting secretagog strengths of CRF and AVT may also operate in avian species under different ecological and physiological conditions. Therefore, it may be necessary to use both exogenous CRF and AVT to test the pituitary component of the avian HPA axis. Mesotocin also appears to be an ACTH secretagog (Castro et al., 1986), but its effect *in vivo* is inconsistent. There is less evidence for inhibitory control of ACTH release (Harvey and Hall, 1990). Recent work with the chicken suggests that ACTH release from the pituitary is negatively modulated by somatostatin (Cheung et al., 1988a) and by opioids (Peebles et al., 1997).

The pituitary content of ACTH in the chicken is estimated to be 1600 ng/gland (Hayashi et al., 1991). The hormone is thought to be exclusively confined to some parenchymal cells of the cephalic lobe of the anterior pituitary (Kalliecharan and Buffet, 1982; Mikami and Yamada, 1984). Apparently, ACTH-like activity is also present in the avian brain (Ng and Ng, 1987; Kovács et al., 1989). In chickens, turkeys, and geese, ostensibly unstressed, plasma ACTH concentrations range from 20 to 150 pg/ml (4.0×10^{-12} M to 3.3×10^{-11} M) and can more than double with various stressors (Carsia et al., 1988; Harvey and Hall, 1990; Kovács and Péczely, 1991; Hendricks et al., 1995a; Kocsis et al., 1995a). Endocrine manipulations appear to influence resting plasma ACTH levels; e.g., it is increased by castration (Kovács and Péczely, 1991) and decreased by thyroidectomy

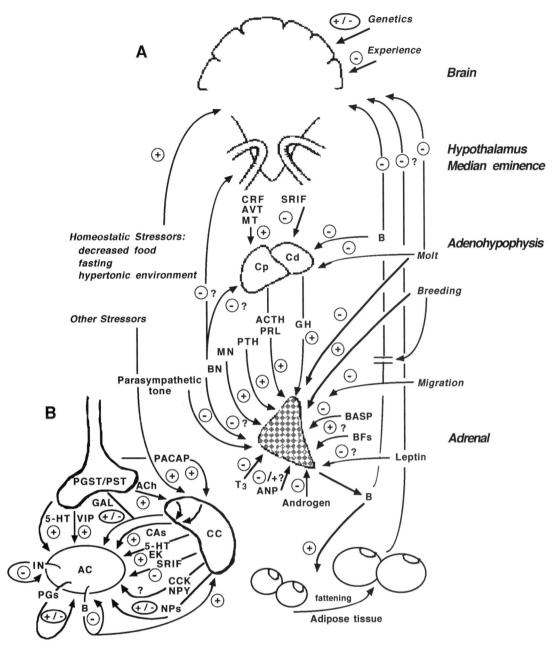

FIGURE 5 (A) Regulation of the hypothalamo–pituitary–adrenal axis in birds. Cp, cephalic lobe of adenohypophysis; Cd, caudal lobe of adenohypophysis; CRF, corticotropin-releasing factor (hormone); AVT, arginine vasotocin; MT, mesotocin; SRIF, somatostatin; GH, growth hormone; ACTH, adrenocorticotropin; PRL, prolactin; PTH, parathyroid hormone; MN, melatonin; BN, bursin; BASP, bursal anti-steroidogenic peptide; BFs, other bursal factors; T_3, 3,5,3'-triiodothyronine; ANP, atrial natriuretic peptide; B, corticosterone. (B) Some intraglandular interactions. PGST, a preganglionic sympathetic terminal or, PST, a postganglionic sympathetic terminal; CC, adrenal chromaffin cell; AC, adrenal steroidogenic (adrenocortical) cell; PACAP, pituitary adenylyl cyclase-activating peptide; ACh, acetylcholine; GAL, galanin; VIP, vasoactive intestinal peptide; 5-HT, serotonin; CAs, catecholamines; EK, enkephalin; SRIF, somatostatin; CCK, cholecystokinin; NPY, neuropeptide-Y; IN, inhibin; NPs, other natriuretic peptides; PGs, prostaglandins; B, corticosterone. See text for details. [From Hormones and stress in birds: Activation of the hypothalamo–pituitary–adrenal axis, S. Harvey and T. R. Hall, *Progress in Comparative Endocrinology* (A. Epple, C. G. Scanes and M. H. Stetson, eds.), Copyright © 1990 by Wiley-Liss, Inc. Reprinted by permission of Wiley-Liss, Inc., a subsidiary of John Wiley & Sons, Inc. Modified by H. Weber, Department of Cell Biology, UMDNJ-School of Osteopathic Medicine.]

(Kovács and Péczely, 1991; Carsia et al., 1997). Peak plasma ACTH concentrations are achieved within 5–10 min after stress (Kovács and Péczely, 1991). In addition, there is an excellent temporal relationship between plasma levels of ACTH and corticosterone in response to stress (Harvey and Hall, 1990; Kovács and Péczely, 1991) and between stress and ACTH infusion-induced plasma corticosterone (Beuving and Vonder, 1978, 1986; Rees et al., 1983; Radke et al., 1985b) and aldosterone (Radke et al., 1985b) responses. Furthermore, in chickens, ACTH infusion stimulates the release of corticosterone and aldosterone rapidly (within 5 min) and in a dose-dependent manner (Radke et al., 1985b; Beuving and Vonder, 1986). By contrast, in turkeys there apears to be a substantial delay (30 min) in plasma corticosterone response to infusion of ACTH (Davis and Siopes, 1987). Interspecies differences in adrenocortical responses to various ACTH peptides may be due to the fact that there may be differences between the primary sequences of ACTH secreted and their respective cognate receptors on adrenocortical cells. For example, the chicken ACTH sequence has greater identity with an amphibian ACTH (*Xenopus laevis*) than with that of another galliform (turkey, *Meleagris gallopavo*) and that of a ratite (ostrich, *Struthio camelus*) (Hayashi et al., 1991). However, it should be pointed out that both chicken and turkey adrenocortical cells exhibit similar differential steroidogenic sensitivities to mammalian and avian ACTH peptides (Carsia et al., 1988a; Kocsis and Carsia, 1989); thus, the reasons for interspecies differences in circulating corticosteroid responses to ACTH infusion require further exploration.

Whereas stimulation of corticosteroidogenesis in the avian adrenal gland by ACTH is well documented, the role of ACTH in adrenal steroidogenic cell growth and proliferation is less clear. For example, various chronic stressors (Holmes et al., 1978; Carsia et al., 1988a; Weber et al., 1990) and exogenous ACTH (Pearce et al., 1979; Davison et al., 1979, 1985) induce adrenal hypertrophy. In addition, studies with the duck suggest that inner zone cells are selectively sensitive to these manipulations and stimuli (Pearce et al., 1978, 1979). Thus, *in vivo*, a pattern for ACTH regulation of avian adrenocortical cell growth is apparent. However, an alternative interpretation of these studies is that exogenous ACTH or stress-induced increases in ACTH augment the trophic action of other factors. One of these "factors" might be the autonomic nervous system; e.g., vagotomy in the pigeon results in hypertrophy of adrenocortical tissue and regression of chromaffin tissue (Verma et al., 1984). Furthermore, while the withdrawal of trophic hormone (ACTH?) support by hypophysectomy induces adrenal atrophy and an associated fall in plasma corticosterone, this is of lesser magnitude than that induced in hypophysectomized rats (Assenmacher, 1973; Carsia et al., 1985c). In contrast, hypophysectomy in the duck results in substantial adrenal atrophy (Pearce et al., 1979) and depression in plasma corticosterone (Holmes et al., 1974; Pearce et al., 1978). Clearly, other hypophyseal-derived factors (prolactin? growth hormone?) and hypophyseal-dependent factors (thyroid hormones?, gonadal hormones?) are also lost with hypophysectomy. In other words, an adequate number of definitive studies with hypophysectomized birds and ACTH replacement is lacking. Thus, the preeminence of ACTH as an avian adrenal maintenance factor is not clear. In keeping with this notion, ACTH does not influence the pattern of cell proliferation in primary cultures of cells derived from embryonic Pekin duck adrenal glands, (Cronshaw et al., 1992b). Moreover, in hypophysectomized chickens, ACTH replacement maintains plasma free fatty acids but does not maintain plasma corticosterone and adrenocortical cell function (R. V. Carsia, C. G. Scanes, and D. B. King, unpublished observations). The roles of growth hormone and the thyroid hormone, 3,5,3'-triiodothyronine (T_3) as possible adrenal trophic hormones is discussed herewith (*vide infra*).

Corticosterone and other glucocorticoids appear to exert a negative feedback influence at all levels of the hypothalamo-pituitary-adrenal axis in birds. For example, exogenous corticosterone suppresses basal and photostimulated corticosterone secretion in ducks (Péczely and Daniel, 1979). Also, dexamethasone blocks basal and exercise stress-induced corticosterone release in chickens (Etches, 1976; Harvey and Hall, 1990). In addition, endogenously high plasma corticosterone concentrations in response to food deprivation stress suppress plasma ACTH levels with chronic dietary protein restriction (Carsia et al., 1988a) and plasma corticosterone response to exercise (Harvey and Hall, 1990). Furthermore, dexamethasone is a potent suppressor of plasma ACTH in chickens (Herold et al., 1992) and of CRH-induced ACTH release by chicken pituitary cells (Carsia et al., 1986a). Evidence for a direct effect of glucocorticoids on adrenocortical function *in vivo* is equivocal. For example, endogenously high plasma corticosterone concentrations and dexamethasone suppress plasma corticosterone response to exogenous ACTH (Harvey and Hall, 1990). However, earlier work suggests that adrenocortical responsiveness to ACTH is unaffected by various stressor paradigms that elevate plasma corticosterone (Rees et al., 1983, 1985).

H. Stress and Adrenocortical Function

Various environmental and metabolic stimuli of sufficient intensity that require major adjustments in or are threats to homeostasis are known as stressors. In addition, psychological stimuli perceived as threats to well-being (i.e., the perception of stress, allesthesia) may be stressors (Figure 5). Birds have a repertoire of central nervous pathways that mediate the integrated stressor information and the activation of the HPA axis (Harvey and Hall, 1990). Homeokinetic mechanisms in response to a particular stressor may be specific for that stressor or may be general; that is, the "nonspecific stress response." Invariably, adrenocortical activation and corticosterone secretion are hallmark characteristics of the nonspecific stress response in birds (Siegel, 1980; and for details see Harvey et al., 1984a). Thus, behavioral and adrenocortical responses tend to be tightly coupled with environmental and metabolic stressors. This appears to be true for psychological stressors, e.g., fear-related behavior and adrenocortical activation in response to immobilization (Jones et al., 1992a,b, 1994; Jones, 1996; Jones and Satterlee, 1996; Satterlee et al., 1993a,b; Satterlee and Jones, 1997), albeit there is evidence that adrenocortical activation and behavioral displays can be experimentally dissociated in response to psychological stressors (Satterlee et al., 1993b, 1994; Jones et al., 1996).

Parallel increases in the plasma concentration of aldosterone occur in response to some stressors; for example, handling (Radke et al., 1985b) and dietary protein restriction (Weber et al., 1990). However, divergent changes in plasma corticosterone and aldosterone concentrations can sometimes occur in response to stressors that affect electrolyte balance and hemodynamics. For example, changes in sodium (Na^+) balance (Kobayashi and Takei, 1982; Skadhauge et al., 1983; Radke et al., 1984; Rosenberg and Hurwitz, 1987), hemorrhage and saline loading (Radke et al., 1985a) and dehydration and hyperthermia (Arad and Skadhauge, 1984) appear to have a differential influence on aldosterone secretion. Presumably, this differential effect on aldosterone secretion is mediated by the renin–angiotensin system (RAS) (Wilson, 1984) (see Regulation of Aldosterone Secretion). In contrast to these divergent mineralocorticoid and glucocorticoid responses, in many anseriform species, changes in RAS status and the HPA axis are integrated to influence an overall stress response (see Holmes et al., 1992). Thus, electrolyte status can influence the pattern of adrenal corticosteroid synthesis and corticosteroid response to stressors (Figure 5). For example, in ducks given hyperosmotic drinking water (simulating adaptation to saltwater or alkali lake environments), there is a hypophysiotropic ACTH response and subsequent increase in circulating corticosterone, which becomes the principal mineral-regulating hormone (Holmes and Phillips, 1985). Thus, activation of the HPA axis to a superimposed novel stressor elicits a larger plasma corticosterone response compared to ducks given fresh water (Redondo et al., 1988).

The adrenocortical stress response is modulated by a number of factors. Apparently, genetic background has a role: in vivo studies with turkeys (Brown and Nestor, 1973, 1974), chickens (Gross and Colmano, 1971; Edens and Siegel, 1975; Dantzer and Mormede, 1985; Carsia and Weber, 1986; van Mourik et al., 1986), and Japanese quail (Satterlee and Johnson, 1988) suggest an enhanced adrenal responsiveness to stressors (Brown and Nestor, 1973; Gross and Colmano, 1971; Dantzer and Mormede, 1985; Carsia and Weber, 1986; Satterlee and Johnson, 1988) and exogenous ACTH (Brown and Nestor, 1973; Edens and Siegel, 1975; van Mourik et al., 1986) with breeding of disparate breeds (van Mourik et al., 1986), selection for high plasma (Brown and Nestor, 1973; Gross and Colmano, 1971; Edens and Siegel, 1975; Dantzer and Mormede, 1985) or serum (Satterlee and Johnson, 1988) corticosterone response to various stressors and selection for dwarfism (Carsia and Weber, 1986). With enhanced or depressed adrenal responsiveness to stress there is respectively an increase or decrease in relative adrenal weight compared to the genetic control (Carsia and Weber, 1986; Carsia et al., 1988b). In addition, there are alterations at the cellular level which may manifest as alterations in corticosteroidogenic capacity (Carsia et al., 1988b) and/or alterations in cellular sensitivity to ACTH (Carsia and Weber, 1986). Further understanding of the co-selection of positive management traits with advantageous adrenocortical responses to stress, such as social stress, may be of value in improving domestic fowl well-being in a commercial setting (Hester et al., 1996).

Elegant studies with the rat indicate that the activity of the hypothalamo-pituitary-adrenal axis is a component of a larger hypothalamic system that regulates caloric flow (Akana et al., 1994; Dallman et al., 1994). For example, whereas activation of both mineralocorticoid and glucocorticoid receptors restores energy acquisition behavior in diabetic, adrenalectomized rats (Santana et al., 1995), it is the activation of the latter that selectively maintains or increases fat stores, probably at the expense of protein stores (Challet et al., 1995; Santana et al., 1995). Therefore, it is not surprising to find that nutritional status modulates adrenocortical activity. With regard to fat stores, the adipose hormone leptin not only inhibits neuropeptide-Y and other factors affecting food intake (Sahu, 1998), but also appears to negatively modulate stress-induced CRF release (Heiman et al., 1997) and adrenocortical cell function directly (Bornstein et al., 1997). Furthermore, circulating levels of leptin are inversely related to levels of ACTH and corticosteroids (Licinio et al., 1997). A similar system may operate in birds (Figure 5), but it is complex and must be viewed in the physiological, behavioral and ecological contexts (*vide infra*). However, some broad generalizations can be made. In diverse avian species, corticosterone enhances feeding behavior (Nagra *et al.*, 1963b; Bartov *et al.*, 1980; Gross *et al.*, 1980; Gray *et al.*, 1990) and fat storage (Nagra *et al.*, 1963b; Pilo *et al.*, 1985; Silverin, 1986; Wingfield and Silverin, 1986; Buyse *et al.*, 1987; Saadoun *et al.*, 1987; Kafri *et al.*, 1988; Siegel *et al.*, 1989). Furthermore, in some species that experience periods of limited food availability, there is an inverse relationship between fat stores and stress response. For example, maximal plasma corticosterone response to capture stress is inversely related to fat depots in snow buntings (*Plectrophenax nivalis*) and Lapland longspurs (*Calcarius lapponicus*) (Wingfield *et al.*, 1994). However, this relationship may not be seen in species with unlimited food supplies [e.g., the Florida scrub-jay, *Aphelocoma coerulescens*) (Schoech *et al.*, 1997)], in species that may prioritize the protection of protein stores [e.g., the American kestrel, *Falco sparverius* (Heath and Duffy, 1988)], and in different life stages of various species (*vide infra*). Given these correlates, it is plausible that a similar relationship between the hormonal regulation of fat stores and the regulation of the HPA axis exists in some birds or is restricted to a particular life stage of some species. Indeed, one of the targets of leptin action, neuropeptide-Y, is emerging as a dominant factor regulating food intake in some avian species (Richardson *et al.*, 1995).

Other nutriture stimuli appear to affect the avian HPA axis as well. In both altricial (ring dove) and precocial species (chicken, quail, duck, king penguin), food restriction, of sufficient duration, almost uniformly increases circulating corticosterone levels (Harvey *et al.*, 1984a; Rees *et al.*, 1985a; Bartov *et al.*, 1988; Cherel *et al.*, 1988a,b; Le Ninan *et al.*, 1988; Lea *et al.*, 1992). Similar responses are seen with isolated protein restriction. In chickens, transient protein restriction not only enhances adrenocortical activity (Carsia *et al.*, 1988a; Weber *et al.*, 1990) but also entrains a long-term enhancement of activity after removal of this nutritional stressor (Weber *et al.*, 1990). Although plasma ACTH is depressed in restricted birds, there is an increase in relative adrenal weight (Carsia *et al.*, 1988a; Weber *et al.*, 1990), enhanced cellular steroidogenic capacity and corticosteroidogenic responsiveness and sensitivity to ACTH (Carsia *et al.*, 1988a; Carsia and Weber, 1988; Weber *et al.*, 1990), and an increase in the affinity and concentration of cellular ACTH receptors (Carsia and Weber, 1988). Furthermore, there are alterations in signal transduction components and a decrease in corticosterone negative feedback in adrenocortical cells derived from protein-restricted chickens (McIlroy *et al.*, 1998). Thus, in the HPA axis of birds it may be that the adrenal gland is the dominant effector that is activated after integration of nutritional information against a background of physiological, behavioral, and ecological factors.

Two circulating hormones, growth hormone and T_3, may be involved in adrenocortical activation and in mediating the adrenocortical response to protein status. These hormones, apparently independently of ACTH, interact to maintain corticosteroidogenic capacity and cellular sensitivity to ACTH (Carsia *et al.*, 1985c). Alterations in adrenocortical function correlate well with anticipated alterations in response to known changes in the circulating concentrations of these hormones with protein restriction (Lauterio and Scanes, 1987a,b). By contrast, protein restriction in the domestic turkey is not associated with changes in circulating T_3. Thus, the influence of this stressor on this galliform species is quite different compared to its influence on chicken adrenocortical function; whereas circulating corticosterone levels are increased, aldosterone levels are depressed. Furthermore, there is remodeling of the adrenocortical cell composition of the turkey adrenal gland and adrenocortical cell hypofunction (Casia and McIlroy, 1998).

I. General Regulation of Adrenocortical Function

1. ACTH

The most efficacious stimulator of avian corticosteroid secretion is ACTH. Studies *in vivo* (Radke *et al.*, 1985b; Beuving and Vonder, 1986) and *in vitro* with isolated adrenocortical cells from several species [e.g., duck (Vinson *et al.*, 1979; Collie *et al.*, 1992; Holmes *et al.*, 1992), chicken (Carsia *et al.*, 1985a, 1987a, 1988a; Collie *et al.*, 1992), turkey (Kocsis and Carsia, 1989; Kocsis *et al.*, 1994a, 1995), and Japanese quail (Carsia *et al.*, 1988b)] demonstrate a dose-dependent and rapid (less than 5 min) corticosteroid response to ACTH. Studies with isolated chicken adrenocortical cells suggested that ACTH operates through specific receptors composed of high- and low-affinity forms and that are present at concentrations of about 2000–3000 sites/cell (Carsia and Weber, 1988). Characterization of a partial complementary DNA encoding an ACTH receptor expressed in the turkey adrenal gland suggests an 89% identity with mammalian forms (R. V. Carsia, K. I. Tilly, and J. L. Tilly, unpublished data). Despite this high identity there are differences in receptor-mediated steroidogenic responses between rat and chicken adrenocortical cells. A notable difference is that, unlike rat adrenocortical cells, chicken adrenocortical cells fail to distinguish between mammalian $\alpha ACTH_{1-24}$ and its 9-

tryptophan-(*O*-nitrophenylsulfenyl) derivative (Carsia *et al.,* 1985a). However, as with rat adrenocortical cells, human $ACTH_{7-38}$ is a complete antagonist of ACTH in chicken adrenocortical cells (Carsia and Weber, 1988). In addition, unlike rat adrenocortical cells, isolated chicken adrenocortical cells undergo sensitization in response to ACTH (McIlroy *et al.,* 1988), which is consistent with the lack of decay in elevated corticosterone values in response to a continuous infusion (1 week) of ACTH (Latour *et al.,* 1996). Longer treatments with exogenous ACTH in chickens have, nevertheless, been shown to cause adrenocortical desensitization to ACTH (Kalliecharan, 1981; Davison *et al.,* 1985). Obviously, cloning and expression of avian ACTH receptor isoforms will aid in elucidating the molecular basis for these differences (Kapas *et al.,* 1996). Dissemination of the ACTH signal in avian adrenocortical cells appears to be via the cyclic AMP (cAMP)-dependent pathway, involving a stimulatory guanine nucleotide binding protein (Rosenberg *et al.,* 1986, 1988a; McIlroy *et al.,* 1998), adenylyl cyclase (McIlroy *et al.,* 1998), cAMP (Rosenberg *et al.,* 1986, 1987, 1988a, 1989a,b; Carsia *et al.,* 1988a; Kocsis and Carsia, 1989; Weber *et al.,* 1990), and cAMP-dependent protein kinase (Rosenberg *et al.,* 1988a). In contrast, the role of the calcium (Ca^{2+})-phosphoinositide/lipid-protein kinase C pathway is controversial. For example, one study suggests that acutely inducible (ACTH) steroidogenesis of chicken adrenocortical cells is refractory to agonists and antagonists of this pathway (Rosenberg *et al.,* 1988b). Conversely, other work with chicken adrenocortical cells suggests that both basal and ACTH-induced corticosterone production, and their augmentation by prolactin, are enhanced by extracellular Ca^{2+} (Carsia *et al.,* 1987b). In addition, physiological concentrations of Ca^{2+} are critical for optimal sensitivity of turkey adrenocortical cells to ACTH (Kocsis *et al.,* 1994a). Furthermore, a background level of calmodulin activity appears to be required for steroidogenesis in turkey adrenocortical cells (Kocsis *et al.,* 1995b).

2. Angiotensins

There is now a strong body of evidence supporting the role of angiotensins in the regulation of avian adrenocortical function (reviewed in Holmes and Cronshaw, 1993). Infusions or bolus injections of angiotensin II (Ang II) increased plasma aldosterone and corticosterone concentrations in the duck (Gray *et al.,* 1989) and Japanese quail (Kobayashi and Takei, 1982). However, there is a paucity of studies with hypophysectomized birds to preclude an indirect effect via the release of ACTH (Chan and Holmes, 1971). For example, infusion of chicken kidney extracts into hypophysectomized chickens failed to change aldosterone concentrations in the adrenal venous effluent (Taylor *et al.,* 1970). Nevertheless, *in vivo* studies and *in vitro* work with adrenal tissue and cells suggest that angiotensins, notably Ang II, play a direct role in adrenocortical regulation. This topic is discussed in detail under the section "Regulation of Aldosterone Secretion".

3. Other Putative Regulators

Since the last edition of this textbook, studies suggest that avian adrenocortical function may be modulated by secretory products of peripheral immune, neural, and endocrine organs and by other circulating factors (Harvey and Hall, 1990). Interpretations of results derived from these studies should be viewed cautiously since these studies are restricted to anseriform and galliform species and in many instances are based on *in vitro* systems that have not been substantiated *in vivo*. Nevertheless, these studies deserve attention. Such peripheral agents may directly stimulate or inhibit adrenocortical function or may act as positive or negative modulators of tropic-hormone induced responses (see Figure 5).

a. Stimulators and Positive Modulators

The pituitary secretory products prolactin and growth hormone appear to have a positive influence on avian adrenocortical function. The effect of prolactin is not surprising, considering its temporal synergism with corticosterone in regulating migratory and reproductive parameters (Meier, 1973). Intravenous injections of recombinant chicken prolactin into chick embryos dramatically raises plasma corticosterone within 2 hr (Kühn *et al.,* 1996). As mentioned above, prolactin counteracts the self-induced degradation of corticosterone and inhibitory actions of corticosterone on corticosteroidogenesis (Carsia *et al.,* 1984, 1987b). In isolated chicken adrenocortical cells this effect of prolactin is acute, Ca^{2+}-dependent, and occurs with a half-effective concentration (EC_{50}) of 55 ng/ml (Carsia *et al.,* 1987b), a concentration that is well within the physiological range (Harvey *et al.,* 1979). Indeed, chicken adrenocortical cells are four times more sensitive to prolactin than are adrenocortical cells from hypophysectomized rats (Carsia *et al.,* 1987b).

The actions of growth hormone are less clear. Intravenous injections of recombinant chicken growth hormone increase circulating concentrations of corticosterone in posthatch chickens (Cheung *et al.,* 1988b) but not in chick embryos (Kühn *et al.,* 1996). In addition, purified native chicken growth hormone fails to acutely induce corticosterone release from isolated adrenocortical cells (Carsia *et al.,* 1985c). Nevertheless, the importance of growth hormone in avian adrenocortical maintenance is indicated by its hyperstimulation of cellular steroidogenic capacity after hypophysectomy in the absence of ACTH replacement (Carsia *et al.*, 1985c). Additionally, both prolactin and growth hormone are increased with various stressors (Harvey *et al.*, 1979; Harvey *et al.*, 1984b) and presumably these hormones may assist in the maintenance of stress-induced corticosterone release.

In mammals, a corticotropic action of parathyroid hormone has long been suspected because of its calciotropic actions on many tissues and because of sequence similarities with mammalian ACTH (cf. Carsia *et al.*, 1981; Rafferty *et al.*, 1983; Rosenberg *et al.*, 1987). Very high concentrations (10 μM) of human parathyroid hormone stimulate cAMP and corticosterone production in rat adrenocortical cells (Rafferty *et al.*, 1983) and more physiological concentrations (0.1–10 nM) stimulate aldosterone secretion and potentiate Ang II-induced aldosterone secretion by bovine adrenal glomerulosa cells, presumably through increases in cytosolic Ca^{2+} (Isales *et al.*, 1991). The action of parathyroid hormone is via receptors distinct from those for ACTH (Isales *et al.*, 1991). Primary sequence differences in avian (chicken) parathyroid hormone preclude any possibility of "spill-over" to the ACTH receptor (Khosla *et al.*, 1988; Russell and Sherwood, 1989) and this has been verified with chicken adrenocortical cells, using chicken parathyroid extracts containing PTH (Rosenberg *et al.*, 1987). In chicken adrenocortical cell preparations, chicken parathyroid hormone is a potent stimulator of cAMP, corticosterone and aldosterone secretion, and although its potency is 0.4 that of ACTH, at physiological concentrations it is as efficacious as ACTH (Rosenberg *et al.*, 1987, 1988a). It is interesting to note that the parathyroid hormone EC_{50} for aldosterone production is one half to one seventh that for corticosterone secretion (Rosenberg *et al.*, 1988a, 1989a). It is important to point out that the latter studies used purified extracts derived from chicken parathyroid glands. A recent study with recombinant chicken PTH did not demonstrate this novel steroidogenic action (Lim *et al.*, 1991). However, overall, these studies suggest that parathyroid hormone is a prominent avian corticotropic agent and that the aldosterone pathway may be the prime target for PTH action. Conceivably, preferential aldosterone release vs corticosterone release may aid to offset any parathyroid hormone-induced natriuresis, which is primary to phosphaturia, during challenges to calcium homeostasis (e.g., egg formation). This and other postulates raised here await *in vivo* investigations.

The varying degree of intermingling of chromaffin and steroidogenic tissues within the adrenal (interrenal) gland of higher vertebrates provides the opportunity for paracrine interactions (see Figure 5) in addition to the "conventional" neural and humoral (ACTH, Ang II, K^+) influences, respectively (for review, see Hinson, 1990; Edwards and Jones, 1993; Ehrhart-Bornstein *et al.*, 1996). Even in the highly organized mammalian adrenal gland, sufficient evidence indicates that catecholamines are important paracrine stimulators of glucocorticoid release from adrenocortical cells (detailed by Haidan *et al.*, 1998). Alternatively, chromaffin cells or adrenal nerve fibers may release other substances that influence corticosteroid production, as for example, vasoactive intestinal peptide (VIP) (Cunningham and Holzwarth, 1988; Leboulenger *et al.*, 1988) and galanin (Gasman *et al.*, 1996). Unfortunately, there is little information concerning avian species. On the one hand, there is some evidence suggesting that catecholamines via β-adrenergic receptors directly increase chicken corticosterone release *in vivo* (Freeman and Manning, 1979; Rees *et al.*, 1985b) and *in vitro* in isolated fowl adrenocortical cells (Mazzocchi *et al.*, 1997a). Along these lines, another amino acid-derived neurotransmitter, serotonin (5-hydroxytryptamine; 5-HT), has been shown to be a potent and efficacious stimulator of corticosteroid secretion by duck adrenocortical cells (Vinson *et al.*, 1979). In addition, VIP stimulates corticosterone production by turkey and chicken adrenocortical cells (three to four times basal) with an EC_{50} of about 30 nM (R. V. Carsia, unpublished observations). In this connection, pituitary adenylyl cyclase-activating peptide (PACAP) may also be a positive modulator (Mazzocchi *et al.*, 1997b). Other recent evidence suggests that opioids facilitate circulating corticosterone response to ACTH in the chicken (Peebles *et al.*, 1997). On the other hand, catecholamines (dopamine, norepinephrine, and epinephrine) and acetylcholine repress basal (but not inducible) corticosterone and aldosterone release from duck adrenal tissue, albeit this effect is not dose-dependent (Holmes *et al.*, 1991). In addition, some evidence suggests that vagal tone directly or indirectly represses corticosterone release (Viswanathan *et al.*, 1987). Another neuropeptide, galanin, has contrasting influences on adrenocortical function in mammalian species (stimulation) and amphibians (inhibition) (see Gasman *et al.*, 1996). Although galanin exists in chromaffin cells, adrenal nerve fibers, or both, depending on avian species (Zentel *et al.*, 1990), its role in the regulation of avian adrenocortical function is unknown.

There is evidence suggesting that the humoral immune system may exert trophic and tropic influences on the avian adrenal gland. Activated immune cells elaborate a potent releaser of ACTH in chickens (Herold *et al.*, 1992), which is presumably interleukin (IL)-1 (Mashaly *et al.*, 1993). Although a large part of the ACTH is from the pituitary, it appears that significant quantities of ACTH are released from activated lymphocytes. For example, an antigen challenge or CRH induces the release of ACTH from chicken lymphocytes that is sufficient to elicit a corticosterone response from co-incubated chicken adrenocortical cells (Siegel *et al.*, 1985; Hendricks *et al.*, 1991; 1995a; Mashaly *et al.*, 1993). This direct effect of CRH on chicken leukocyte ACTH release is inhibited by corticosterone (Hendricks *et al.*, 1995b). It is thought that this humoral immune–endocrine axis facilitates the redistribution of circulating T lymphocytes to secondary lymphoid tissues during the immune response (Mashaly *et al.*, 1993). In addition, the bursa of Fabricius may elaborate substances necessary for adequate corticosteroidogenic capacity during development and maturation (Pedernera *et al.*, 1980) and for normal corticosterone response of adrenocortical cells to ACTH (El-Far *et al.*, 1994). However, a bursal suppressive factor has recently received more attention (*vide infra*).

b. Inhibitors and Negative Modulators

A bidirectional link between the immune system and the adrenal gland, similar to that which exists in mammals (Bateman *et al.*, 1989), is likely to be present in birds. The secretion of ACTH by chicken lymphocytes and the requirement of the bursa of Fabricius for normal adrenal development (*vide supra*) are examples. However, there is evidence for suppressor substances as well. Highly purified extracts of prepubescent chicken bursae suppress inducible steroid synthesis from chicken granulosa and adrenocortical cells and from mammalian adrenocortical cells. The active principle has been termed bursal antisteroidogenic peptide (BASP) (Byrd *et al.*, 1993, 1994). BASP appears to be a late-pathway inhibitor in that it stimulates cyclic AMP production but inhibits cAMP analog stimulated steroidogenesis. This peptide awaits full characterization.

A growing body of work suggests that the thyroid hormone, T_3, is an important modulator of avian adrenocortical function. For example, protein deficiency (Lauterio and Scanes, 1987a,b; Scanes and Griminger, 1990) and genetic selection (Scanes *et al.*, 1983) resulting in altered plasma T_3 concentrations are associated with alterations in relative adrenal weight and adrenocortical function (Carsia and Weber, 1986; Carsia *et al.*, 1988a). In many cases, there is an inverse relationship between circulating T_3 concentrations and adrenocortical function. For example, in hypophysectomized cockerels, T_3 replacement reduces cellular corticosterone production more than that due to hypophysectomy (Carsia *et al.*, 1985c). In addition, thyroidectomy in chickens results in a marked enhancement in adrenocortical function and this is normalized with T_3 replacement (Carsia *et al.*, 1997). At least part of this effect of T_3 appears to be direct at the adrenocortical cell level (Carsia *et al.*, 1997). However, the combined replacement of growth hormone and T_3 in hypophysectomized chickens maintains adrenocortical cell sensitivity to ACTH (as indicated by the ACTH EC_{50}) but results in an exaggerated corticosterone response (Carsia *et al.*, 1985c). Thus, the presence of other pituitary factors may modulate the influence of T_3 on adrenocortical function.

In keeping with this notion that other pituitary–target organ axes influence adrenocortical cell function are studies of the pituitary–gonadal axis. Early work suggests that gonadal steroids tonically suppress adrenocortical function in chickens (Kar, 1947; Nagra *et al.*, 1965). However, other work with orchiectomized chickens has produced equivocal results since the effect of orchiectomy appears to be modified by age, strain, and duration of the orchiectomized state (Chester Jones, 1957; see pp. 203–205). Nevertheless, more recent work with orchiectomized cockerels suggests that although androgen (testosterone and 5α-dihydrotesterone) replacement suppresses adrenocortical cell response to ACTH, it maintains relative adrenal weight (Carsia *et al.*, 1987d). Interestingly, orchiectomy is required to demonstrate the effects of androgens on adrenocortical function in both rats (Kitay *et al.*, 1966) and domestic fowl (Nagra *et al.*, 1965). Thus, it is quite possible that the testis elaborates both inhibitory and trophic substances that influence avian adrenocortical function (Carsia *et al.*, 1987d).

c. Biphasic Modulators

Whereas the use of isolated adrenocortical cells has a caveat when extrapolating findings to *in vivo* physiology, it is amenable to gaining insights into the fine regulation of adrenal steroidogenic function that would be difficult or impossible to obtain from *in vivo* studies. An example is the understanding that adrenal-targeted factors may elicit biphasic influences on steroidogenic function. Thyroid hormones may fall into this category since T_3 suppresses corticosteroidogenic capacity but preserves cellular sensitivity to ACTH in the hypophysectomized chicken (Carsia *et al.*, 1985c). In contrast, T_3 replacement influences cellular steroidogenic capacity but does not alter cellular sensitivity to ACTH in the thyroidectomized chicken (Carsia *et al.*, 1997). Thus, other pituitary factors may be involved in the regulation of cellular sensitivity to ACTH. Along these lines, there is increasing evidence for significant crosstalk between the various endocrine axes in birds, e.g., the recent evidence suggesting that CRF stimulates thyrotropin-releasing activity and increases circulating thyroid hormones (Geris *et al.*, 1996) and prolactin (Hall *et al.*, 1986).

Prostaglandins also seem to fall into this category. Prostaglandins appear to be potent and efficacious stimulators of avian adrenocortical function. In the Japanese quail (*Coturnix coturnix japonica*), prostaglandin E_2 is more than six times greater than immobilization stress in raising circulating corticosterone. In addition, it synergizes with immobilization stress to rapidly increase plasma corticosterone (Satterlee *et al.*, 1989). A large part of the effect of prostaglandins appears to be due to their interaction with prevailing concentrations of ACTH. In both chicken and turkey adrenocortical cells, prostaglandins exert a biphasic influence on corticosterone production, depending on their concentration. At low concentrations (10^{-8} M to 10^{-5} M) they synergize with submaximal concentrations of ACTH, whereas at high concentrations (10^{-5} M to 10^{-3} M) they are inhibitory. The most potent and efficacious prostaglandin for stimulation is prostaglandin E_2, whereas prostaglandin A_2 is the most potent for inhibition. Prostaglandins have a similar but considerably weaker biphasic influence on basal corticosterone secretion (Kocsis *et al.*, 1998).

J. Regulation of Aldosterone Secretion

1. Role of the Renin–Angiotensin System

It should be pointed out that for some time, the role of renin–angiotensin system (RAS) and its active product, Ang II, in the regulation of avian adrenal steroidogenic function was controversial. This was largely due to the fact that essentially all earlier investigations were on the principal avian model, the domestic chicken (*Gallus gallus domesticus*). Experiments *in vivo* and *in vitro* with various adrenal preparations demonstrated that this species exhibits a pronounced aldosteronogenic tachyphylaxis to Ang II (for review, see Holmes and Cronshaw, 1993). This is perplexing since standard ligand binding isotherms with isolated chicken adrenocortical cells (R. V. Carsia, J. F. Kocsis, and P. J. McIlroy, unpublished data) and recent *in situ* hybridization studies with a cloned chicken adrenal Ang II receptor (Kempf *et al.*, 1996) indicate that chicken adrenocortical cells possess high-affinity Ang II receptors. It was presumed that this well-studied species was representative of avian species in general. It is now clear that the misconceived generalizations derived from studies with the chicken concerning the actions of Ang II on avian aldosterone synthesis will require modification.

In the regulation of aldosterone secretion in birds, the RAS does not appear to have the preeminence that it has in mammals (for review, see Holmes and Cronshaw, 1993). One can anticipate a range of its importance when considering differences in avian feeding strategies (Bentley, 1971) and the breadth of environment and behavior that influences heat-promoting evaporation (Willoughby and Peaker, 1979). The impact of this diversity on water and Na^+ balance calls for variations in homeostatic mechanisms, which include the RAS and the control of aldosterone secretion.

As in other vertebrates (see for review Wilson, 1984), the avian kidney contains juxtaglomerular cells capable of producing a reninlike substance and ion-sensitive (Na^+ depletion) and hemodynamic (hemorrhage) responses have been demonstrated (Taylor *et al.*, 1970; Chan and Holmes, 1971; Nishimura, 1980; Nishimura *et al.*, 1981; Nishimura and Bailey, 1982). In chickens, this substance is capable of cleaving fowl angiotensinogen ([Asp^1, Val^5, Ser^9]tetradecapeptide renin substrate) to form the decapeptide angiotensin I (Nakayama *et al.*, 1973). The Ang II octapeptide produced from this decapeptide ([Val^5]Ang II; M_r = 1031.53) is that of most vertebrates including ovine and bovine species, whereas other mammalian species produce [Ile^5]Ang II (Gorbman *et al.*, 1983). In several species studied thus far (Japanese quail, duck, and turkey), plasma Ang II concentrations range from 30 to 70 pg/ml (2.9×10^{-11} M to 6.8×10^{-11} M) but can increase to 100–400 pg/ml (9.7×10^{-11} M to 3.9×10^{-10} M) with perturbations to electrolyte and hemodynamic balance and with adaptation to saltwater (Kobayashi and Takei, 1982; Gray and Simon, 1985a; Gray *et al.*, 1989; Kocsis *et al.*, 1995a).

As mentioned above, the role of the RAS in avian aldosterone secretion appears less precise than that which exists in mammals. This, in part, may be due to the lack of a distinct organization of adrenal steroidogenic tissue and thus a concomitant lesser precision of regulation of mineralocorticoid and glucocorticoid secretion (Vinson *et al.*, 1979). However, ultrastructural and histological studies of the adrenal gland of several avian species suggest a rudimentary organization of steroidogenic tissue, thus providing a basis for a functional segregation of cell types (Holmes and Phillips, 1976). For example, cytologically distinguishing features for subcapsular zonal vs inner zonal steroidogenic cells are evident in the adrenal gland of the brown pelican (*Pelecanus occidentalis*) (Knouff and Hartman, 1951), the domestic chicken (*Gallus gallus domesticus*) (Kondics and Kjaerheim, 1966; Taylor *et al.*, 1970), and the duck (*Anas platyrhynchos*) (Pearce *et al.*, 1978; Klingbeil *et al.*, 1979; Holmes and Cronshaw, 1984). Also, with duck adrenal tissue, there is a relatively greater production of aldosterone compared to corticosterone by the subcapsular tissue and vice versa by inner zone tissue, both basally and in response to ACTH (Klingbeil *et al.*, 1979; Cronshaw *et al.*, 1985; Klingbeil, 1985a).

Clearly, there is some role of the RAS in avian aldosterone secretion. Perturbations in homeostasis that affect the RAS also affect aldosterone secretion. For example, alterations in dietary sodium (Na^+) influences the thickness of the subcapsular zone without affecting the inner zone (Kondics, 1963; Kondics and Kjaerheim, 1966; Taylor *et al.*, 1970), and Na^+ restriction-induced hypertrophy can be correlated with increased juxtaglomerular cell granulation and renal pressor activity (Taylor *et al.*, 1970). In addition, increases in plasma aldosterone (often exclusive of increases in plasma corticosterone) occur in response to stressors against hemodynamics and Na^+ balance, such as dehydration (Kobayashi and Takei, 1982), dehydration plus hyperthermia (Arad and Skadhauge, 1984), hemorrhage (Kobayashi and Takei, 1982; Radke *et al.*, 1985a), dietary Na^+ restriction (Skadhauge *et al.*, 1983; Radke *et al.*, 1984; Klingbeil, 1985b; Rosenberg and Hurwitz, 1987; Kocsis *et al.*, 1995a), and saltwater adaptation (Gray and Simon, 1985a). Some evidence suggests that the effect of thermoregulation perturbation may be mediated by plasma arginine vasotocin, which has a role in thermoregulation and a relationship with plasma aldosterone (Robinzon *et al.*, 1988). Also, in some studies, increases in plasma aldosterone are correlated with increases in plasma Ang II (Kobayashi and Takei, 1982; Gray and Simon, 1985a; Kocsis *et al.*, 1995a). Furthermore, in ducks and chickens, dietary Na^+ restriction or hyponatremia increases plasma aldosterone response (but not plasma corticosterone response) to exogenous ACTH (Radke *et al.*, 1984) and Ang II (Gray *et al.*, 1989).

In addition to this correlative work, there is evidence for the direct regulation of aldosterone production by Ang II. An intraperitoneal injection of Ang II into Japanese quail (Kobayashi and Takei, 1982) and infusions of Ang II into ducks (Gray *et al.*, 1989) increases circulating aldosterone; the latter study showing a dose-dependent aldosterone response exclusive of alterations in circulating corticosterone. In addition, an exclusive, dose-dependent aldosterone response to Ang II has been demonstrated with duck adrenal tissue, specifically from the subcapsular region (Klingbeil, 1985a; Holmes *et al.*, 1991). Interestingly, Ang II stimulates both aldosterone and corticosterone production from isolated adrenal cells prepared from duck adrenal tissue (Vinson *et al.*, 1979; Collie *et al.*, 1992). Thus, as with the rat adrenal gland (Vinson *et al.*, 1985), the level of cytoarchitectural integrity of the avian adrenal gland may influence the specificity of Ang II action. It should be pointed out, however, that isolated turkey adrenocortical cells exhibit an exclusive aldosterone (compared to corticosterone) response to Ang II (Kocsis *et al.*, 1994b, 1995a), largely because Ang II does not stimulate cholesterol side-chain cleavage in these cells (Kocsis *et al.*, 1995a). Presumably, the prevailing basal corticosterone production is sufficient to support aldosterone synthesis. It may be that, as in mammalian adrenocortical cells (Aguilera and Catt, 1979), tropic hormones may stimulate different parts of the steroidogenic pathway in avian adrenocortical cells, resulting in different proportions of corticosteroid end-products.

The manifestation of Ang II action in the avian adrenal gland is most apparent when the predominant target cells are isolated and studied. Thus, it seems clear that the subcapsular zone (SCZ) of the duck contains the zona glomerulosalike cells that give a clear aldosterone response to Ang II (Klingbeil, 1985a; Holmes *et al.*, 1991). In other species, this distinction may be less apparent or absent. For example, in studies with isolated adrenocortical cells from normal turkeys, there are few functional differences between density-separated subpopulations. However, after dietary Na^+ restriction, one subpopulation, which comprises about 5% of the total adrenocortical cell population, exhibits disproportionately exaggerated aldosterone responses to Ang II and potassium (K^+) but not to ACTH (Kocsis *et al.*, 1995a). It may be that in other avian species, perturbations in electrolyte and hemodynamic homeostasis or other stressors are required to induce zona glomerulosalike function in a subpopulation of adrenal steroidogenic cells.

2. Mechanism of Ang II Action

Studies with duck adrenal membrane preparations (Gray *et al.*, 1989) and isolated turkey (Kocsis *et al.*, 1994a,b, 1995a,b,c; Carsia and McIlroy, 1998) and chicken (R. V. Carsia, J. F. Kocsis and P. J. McIlroy, unpublished data) adrenocortical cells indicate that the avian adrenal gland possesses high-affinity Ang II receptors, having a K_d of 0.9–2 nM and a concentration of about 30,000–150,000 sites/cell. Cellular concentrations vary among cell subpopulations comprising the turkey adrenal gland and appear to be differentially modulated by dietary Na^+ restriction (Kocsis *et al.*, 1995a) and protein restriction (Carsia and McIlroy, 1998). Turkey (Carsia *et al.*, 1993; Murphy *et al.*, 1993) and chicken (Kempf *et al.*, 1996) adrenal Ang II receptors have also been cloned. They are both 359-amino-acid proteins (sharing 99.7% identity) and exhibit a 75% sequence identity with the mammalian type 1 receptor, albeit there are frank differences in pharmacological and physicochemical properties between mammalian and avian forms. Both avian forms have been shown to couple to the phospholipase C signaling pathway (Murphy *et al.*, 1993; Kempf *et al.*, 1996). However, several differences in amino acid sequence and tissue expression exist between turkeys and chickens. For example, the turkey receptor appears to have two unique cysteine residues at positions 92 and 99 of the first extracellular loop and is exclusively expressed in the adrenal gland (Carsia *et al.*, 1993), whereas the chicken receptor has a broader tissue expression (Kempf *et al.*, 1996). In support of ligand binding isotherms with turkey adrenal cells (Kocsis *et al.*, 1995b), the galliform adrenal Ang II receptor is not expressed in adrenal chromaffin cells but is expressed in the postganglionic catecholaminergic cells of the ganglia adjacent to the gland (Kempf *et al.*, 1996). Furthermore, it is expressed in chicken vascular endothelial cells but not in vascular smooth muscle cells (Kempf *et al.*, 1996). These observations support the model proposed by Nishimura *et al.* (1994) to explain the biphasic action of Ang II on chicken vascular tonus: the rapid vasorelaxation after infusion is due to the response of vascular smooth muscle cells to cGMP subsequent to nitric oxide generation by Ang II-stimulated endothelial cells; the later, α-adrenergic-dependent vasoconstriction is due to the more delayed catecholamine release from Ang II-stimulated postganglionic neuronal endings.

In turkey, adrenocortical cells Ang II induces a rapid and sustained increase in cytosolic Ca^{2+} (Kocsis *et al.*, 1994b; Kocsis *et al.*, 1995b). However, demonstrating a link between increases in cytosolic Ca^{2+} and aldosterone production has been difficult. As stated previously, avian corticosteroidogenesis is fairly refractory to agonist and antagonists of the calcium (Ca^{2+})-phosphoinositide/lipid-protein kinase C pathway (Rosenberg *et al.*, 1988b; Kocsis *et al.*, 1995b). However, extracellular Ca^{2+} is required for optimal Ang II-induced aldosterone secretion (Kocsis *et al.*, 1994a) and calmodulin activity appears to be essential for aldosterone synthesis in turkey adrenocortical cells (Kocsis *et al.*, 1995b). Even more puzzling is the action of K^+. Extracellular K^+ does nothing to aldosterone secretion in chicken adrenocortical cells (Rosenberg *et al.*, 1988b). By contrast, it is an absolute requirement for Ang II-induced aldosterone production by turkey adrenocortical cells (Kocsis *et al.*, 1994a; Kocsis *et al.*, 1995b), having a maximal effect at a physiological concentration (about 5 mM; Rosenberg and Hurwitz, 1987). However, this effect of K^+ in turkey adrenocortical cells is not linked to changes in cytosolic Ca^{2+} (Kocsis *et al.*, 1995b). In addition, in turkey adrenocortical cells, Ang II-induced increases in cytosolic Ca^{2+} and aldosterone production appear to be dissociated (Kocsis *et al.*, 1995b). Thus, the scheme of Ang II action in mammalian zona glomerulosa cells (for review, see Holmes and Cronshaw, 1993) may have to be significantly modified when extrapolated to avian adrenal steroidogenic cells.

3. Role of ACTH and Other Positive Modulators

Studies comparing the efficacy of ACTH and Ang II in increasing circulating aldosterone in avian species are lacking. Nevertheless, some work suggests that aldosterone secretion is influenced by the hypothalamo-pituitary unit. Immobilization (Klingbeil, 1985b), handling, and bleeding (Radke *et al.*, 1985b) stressors increase plasma aldosterone in ducks (Klingbeil, 1985b) and chickens (Radke *et al.*, 1985b). In addition, with the exception of an earlier study with chickens (Taylor *et al.*, 1970), exogenous ACTH in a dose-department manner increases circulating concentrations of aldosterone in ducks (Radke *et al.*, 1984) and chickens (Radke *et al.*, 1984, 1985b). Furthermore, an augmentation in the aldosterone response but not the corticosterone response to stress or exogenous ACTH occurs with perturbations in electrolyte balance (Radke *et al.*, 1984; Klingbeil, 1985b). Also, in chickens, pretreatment with dexamethasone blunts basal plasma aldosterone and hemorrhage-induced increases in plasma aldosterone (Radke *et al.*, 1985a).

This apparent role of ACTH in aldosterone secretion is supported by *in vitro* studies with duck adrenal tissue (Klingbeil, 1985a; Holmes *et al.*, 1991) and adrenocortical cells from duck (Collie *et al.*, 1992; Holmes *et al.*, 1992) and turkey (Kocsis *et al.*, 1994a,b, 1995a; Carsia and McIlroy, 1998) adrenal glands. Overall, there is a 2- to 10-fold greater aldosterone response to ACTH than to Ang II. However, the potency of Ang II is about 3 times that of full-length ACTH (Kocsis *et al.*, 1994a,b, 1995a; Carsia and McIlroy, 1998). In addition, with chicken adrenocortical cells (Rosenberg *et al.*, 1988a; Carsia *et al.*, 1988a, 1997) and possibly duck adrenocortical cells (Collie *et al.*, 1992), the potency of ACTH for stimulating aldosterone production is about 2.5–3 times that for stimulating corticosterone production. These *in vivo* and *in vitro* studies showing the actions of ACTH on avian aldosterone secretion raise further questions concerning the importance of Ang II. However, the specificity and importance of Ang II reside in its ability to link alterations in electrolyte and hemodynamic balance to aldosterone secretion, a role that is not fulfilled by ACTH. Furthermore, with turkey adrenocortical cells, threshold and suboptimal concentrations of ACTH greatly augment (nearly 7 times) the efficacy of Ang II in stimulating aldosterone secretion (Kocsis *et al.*, 1994a). Thus, *in vivo* (against an ACTH background), Ang II probably stimulates aldosterone secretion to a magnitude that exceeds what is revealed *in vitro*.

4. Inhibitors and Negative Modulators of Aldosterone Secretion

In general, there are few studies on the regulation of aldosterone secretion in avian species, but even less is known about interacting factors in its regulation. The *in vitro* effects of serotonin (Vinson *et al.*, 1979) and catecholamines (Holmes *et al.*, 1991) on corticosteroid release, mentioned previously, also include aldosterone release. In addition, somatostatin has been shown to inhibit Ang II-stimulated, but not ACTH-stimulated, aldosterone production by turkey adrenocortical cells (Mazzocchi *et al.*, 1997c). However, more attention has been paid to the role of atrial natriuretic peptide (ANP) in avian aldosterone secretion.

Avian ANP was originally demonstrated as a diuretic agent in extracts from chicken hearts (Gregg and Wideman, 1986). It is a 29-amino-acid peptide ($M_r = 3158.45$) that has significant structural homology to mammalian ANPs (Miyata *et al.*, 1988). Avian ANP is stored and released from both atrial and ventricular cardiocytes (Toshimori *et al.*, 1990). In ducks and chickens, the plasma concentration of ANP is about 70–80 pg/ml ($\sim 2.5 \times 10^{-11} M$) and is inversely related to changes in blood volume in response to hemodynamic and electrolyte perturbations (Gray *et al.*, 1991a; Gray, 1993). ANP receptors are found in both renal and adrenal tissue of the duck (Gray *et al.*, 1991b). In turkey adrenocortical cells, it is present at a concentration of $\sim$90,000 sites/cell (Kocsis *et al.*, 1995c). The receptors have an apparent K_d of 1–3 nM (Gray *et al.*, 1991b; Kocsis *et al.*, 1995c). In the duck (*Anas platyrhynchos*), infusions of avian ANP inhibit plasma aldosterone responses to Ang II without affecting circulating corticosterone (Gray *et al.*, 1991b). In addition, in chicken adrenocortical cells, ANP is a potent (EC$_{50}$ $\sim$1 nM) and efficacious inhibitor (>90%) of maximal ACTH and parathyroid hormone-induced aldosterone secretion without affecting corticosterone secretion (Rosenberg *et al.*, 1988b, 1989a). ANP appears to act at several intracellular loci (Rosenberg *et al.*, 1989a) and cyclic GMP appears to be its second messenger (Rosenberg *et al.*, 1988b, 1989a). Interestingly, in turkey adrenocortical cells, ANPs are as efficacious as Ang II in stimulating aldosterone production and actually augment maximal aldosterone production induced by Ang II, K$^+$, and ACTH (Kocsis *et al.*, 1995c). Obviously, further work with diverse avian species are required to resolve this inconsistency and to understand its physiological relevance.

K. Adrenocortical Function in Physiological and Ecological Contexts

Activation of the HPA axis in birds is not an "all-or-none" event. The HPA axis, through its interaction with other neuroendocrine axes, plays an essential role in supporting mechanisms by which birds adjust their morphology, physiology, and behavior in response to overarching environmental cues during their life history stages. Whereas it must retain some plasticity to respond to acute threats to homeostasis, the response parameters must not compromise the unique functions of the stages within the life cycle of a particular species (Wingfield, 1984; Wingfield *et al.*, 1995, 1997a,b). In some cases, the response parameters are unchanged, but at the CNS level, there are changes (reduction) in the perception of stress (allesthesia). In other cases, the setpoints of the HPA axial components that integrate and respond to normal and stressful stimuli are changed. In more extreme cases, there is an uncoupling of the HPA axis and temporary failure (Wingfield *et al.*, 1992) or an augmentation (Astheimer *et al.*, 1994; Wingfield *et al.*, 1995a,b) of stress response; e.g., during the expression of physiological and behavioral traits associated with reproduction in some avian species. This is necessary because each HPA axial component "crosstalks" with other neuroendocrine axes, the latter having their own parameters and limits of stimulation. Work with mammalian species suggests that supervention of the balance of component setpoints between the HPA axis and the other neuroendocrine axes may result in disruption in reproduction or energy flow and partitioning, early component demise, or acute component failure during the life cycle (see Sapolsky, 1996, 1997; and De Kloet *et al.*, 1997). Still another function of setpoint alteration of the HPA axis components during the avian life cycle may be to permit responses to novel stressors (Gross and Siegel, 1985; Rees *et al.*, 1985a). Given the vast diversity of avian species, the integrated setpoint threshold at which the HPA axis responds to novel stressors, triggering a corticosteroid stress response, will vary greatly (Wingfield *et al.*, 1997). However, it follows that a given novel stressor, of sufficient intensity and/or duration, will reach the threshold and elicit a corticosteroid stress response. This section addresses some of these issues regarding the modulation of the avian HPA axis.

1. Diurnal Rhythm of Plasma Corticosterone Concentrations

Circulating corticosterone concentrations show a distinct daily pattern of variation in birds [e.g., Japanese quail (Boissin and Assenmacher, 1968; Assenmacher and Boissin, 1972; Kovács and Péczely, 1983), pigeons (Joseph and Meier, 1973; Sato and George, 1973; Westerhof *et al.*, 1994), ducks (Wilson *et al.*, 1982), chickens (Siegel *et al.*, 1976; Etches, 1979; Koudela *et al.*, 1980; Tanaka and Kamiyoshi, 1980; Johnson and van Tienhoven, 1981; Wilson and Cunningham, 1981; Wilson *et al.*, 1984; Webb and Mashaly, 1985; Smoak and Birrenkott, 1986; Lauber *et al.*, 1987), and turkeys (El Halawani *et al.*, 1973; Davis and Siopes, 1989)]. Diurnal maximal and minimal concentrations approximate a sine function or otherwise, a unimodal increase over baseline is observed without a detectable trough. In general, maxima are found at the dark–light (scotophase–photophase) interface or at some time during the scotophase. This acrophase (increase over baseline to maximum) in circulating corticosterone during scotophase is thought to serve an energy flow function in homeotherms during nocturnal periods of locomotory inactivity and nonfeeding (Wilkinson et al., 1979; Miyabo et al., 1980). However, it should be pointed out that there are discrepancies in reports of the time of the acrophase in domestic fowl. On the one hand, acrophases during photophase have been reported for chickens (Siegel et al., 1976; Johnson and van Tienhoven, 1981; Smoak and Birrenkott, 1986; Lauber et al., 1987) and turkeys (El Halawani et al., 1973), which correlate well with a circulating ACTH acrophase during photophase in chickens (Herold et al., 1992). On the other hand, acrophases during scotophase have also been reported for chickens (Etches, 1977; Koudela et al., 1980; Tanaka and Kamiyoshi, 1980; Wilson and Cunningham, 1981; Wilson et al., 1984) and turkeys (Davis and Siopes, 1989). These discrepancies might be explained by the posthatch ontogeny of the diurnal rhythm. For example, in immature chickens (<11 weeks old), there are reports of no or weak diurnal variations in circulating corticosterone (Freeman and Flack, 1980; Freeman et al., 1981; Webb and Mashaly, 1985; Renden et al., 1994). However, contrasting reports suggest diurnal variations do exist in immature chickens [16 days (Koudela et al., 1980); 6 weeks (Siegel et al., 1976)] and as early as 10 days in turkeys (Davis and Siopes, 1989). In addition, one report suggests that the occurence of the acrophase shifts from the photophase to the scotophase in chickens approaching sexual maturity (Webb and Mashaly, 1985). Another variable adding to the complexity in understanding the circadian rhythm of circulating corticosterone is the evidence that the sensitivity of the chicken adrenal to ACTH (Beuving and Vonder, 1986; Smoak and Birrenkott, 1986) and to stress (Smoak and Birrenkott, 1986) varies diurnally, being nearly two times greater in the morning than in the evening.

The daily variation in circulating concentrations of corticosterone is not only dependent on corticosterone secretion but also on the peripheral disappearance. Variation in corticosterone secretion is presumed to reflect changes in ACTH (Herold et al., 1992), CRF (Sato and George, 1973), and the activity of hypothalamic aminergic neurons (Martin et al., 1982). Indeed, inhibition of the adrenergic and serotoninergic systems disrupts the corticosterone diurnal rhythm (Boissin and Assenmacher, 1971; Martin et al., 1982). This central neuronal regulation is in turn driven by a highly conserved, endogenous circadian clock (Reppert and Weaver, 1997). In Aves, the light–dark cycle may be the predominant factor regulating the circadian rhythm of circulating corticosterone (Assenmacher and Boissin, 1972; Assenmacher and Jallageas, 1980; Koudela et al., 1980; Davis and Siopes, 1989). The rhythm persists for about 24 hr in "free-running" situations (i.e., in constant light or dark) (see Assenmacher and Jallageas, 1980). In addition to the light–dark cycle, duration of food access may also be an important zeitgeber as it appears to be for behavioral rhythms (Hau and Gwinner, 1996, 1997). For example, in chickens, fasting-induced increases in circulating corticosterone are rapidly restored to normal values with presentation of food, irrespective of nutritional value (S. Harvey, unpublished observations). In addition, the diurnal rhythm of corticosterone is nearly eliminated with severe food restriction (Savory and Mann, 1997) or fasting (S. Harvey, unpublished observations). Under constant light, meal-fed birds, but not birds having free access to food, regain a robust corticosterone diurnal rhythm (S. Harvey, unpublished observations). However, at least for behavioral rhythms, there does not appear to be a distinct food-entrainable circadian oscillator in birds that exists in mammals (Hau and Gwinner, 1997); thus, both light and food are integrated by an overarching circadian system. These findings suggest that the circadian system of birds can be finely adjusted by simultaneous periodic environmental cues. Thus, the disparate ecological requirements of various avian species during their life stages will establish the important zeitgebers that impinge on the integrating circadian system. In this connection, in some nocturnal species (e.g, the western screech-owl, *Otus kennicottii*), the daily variation in circulating corticosterone is correlated with activity period rather than with the light–dark cycle (Dufty and Belthoff, 1997).

The dominant avian hypothalamic pacemaker is thought to be the visual suprachiasmatic nucleus which is highly retinorecipient and demonstrates circadian rhythmic activities (see King and Follett, 1997). Indeed, in mammals, the circadian drive to CRF neurons is supplied by the suprachiasmatic nucleus and ablation of this region destroys the circulating corticosterone acrophase (see Dallman et al., 1987a,b). The visual suprachiasmatic nucleus, the retina, and the pineal gland are thought to comprise the overarching avian circadian system (see King and Follett, 1997) and evidence suggest that melatonin (N-acetyl-5-methoxytryptamine) is the effector molecule coordinating this system (Cassone and Menaker, 1984; Chabot and Menaker, 1992; Heigl and Gwinner, 1995; Kumar, 1996; Kumar et al., 1996; Tosini and Menaker, 1996). Circulating concentrations of melatonin, depending on species, range from 150 to 300 pM in the midphotophase to 550 to 2400 pM in the midscotophase [chicken (Doi et al., 1995); duck (Song et al., 1993); pigeon (Poon et al., 1993)]. Circumstantial evidence suggests that this system regulates the circadian rhythm of circulating corticosterone in birds (also see Chapter 21). First, the visual suprachiasmatic nucleus exhibits circadian rhythmicity of melatonin binding (Lu and Cassone, 1993). Second, the night rise in melatonin is associated with a coordinated increase in melatonin uptake in the hypothalamus and concomitant increase in hypothalamic serotonin synthesis (Cassone *et al.*, 1983; 1986). This in turn may drive the HPA axis. In support of this postulate, cranial sympathectomy and superior cervical ganglionectomy suppress the circadian ACTH rhythm in rats (Siaud *et al.*, 1994) and circulating ACTH in pigs (Minton and Cash, 1990). In addition, the avian adrenal gland possesses melatonin receptors (Brown *et al.*, 1994; Pang *et al.*, 1994) that may positively modulate the function of the gland (John *et al.*, 1986). Furthermore, exogenous melatonin results in acute and long-lasting increases in adrenal corticosterone content in passeriform and columbiform species (Mahata and De, 1991). However, this postulated role of melatonin in the modulation of the avian HPA axis requires further investigation.

It is thought that the diurnal rhythmicity in circulating corticosterone is also modulated, in part, by corticosterone feedback. Glucocorticoid binding sites are present in several brain regions of the Japanese quail and the daily fluctuation in available sites is antiphase with and synchronized by the plasma corticosterone concentrations (Kovács and Péczely, 1986; Kovács *et al.*, 1989). Furthermore, in the chicken, dexamethasone treatment destroys the plasma ACTH rhythm (Herold *et al.*, 1992). In addition, the peripheral disappearance in corticosterone, which also exhibits a rhythm, cannot be overlooked and has been poorly studied. Clearly, this can play a role in the development of the corticosterone daily rhythm (Kovács *et al.*, 1983). Indeed, phase shifts in the circadian rhythmicity of circulating corticosterone induced by perturbations in physiological status [e.g., thyroid status (Kovács and Péczely, 1983)] may be largely due to changes in metabolic clearance of corticosterone.

2. Reproductive Parameters and Plasma Corticosterone Concentrations

The notion of corticosteroids being inhibitors of reproductive parameters is a simplistic one (see Wingfield *et al.*, 1992; and Astheimer *et al.*, 1995). Whether high corticosteroids are detrimental has to be viewed in the ecological and physiological contexts. For example, reproductive strategies in wild avian species sometimes require adjustments in the interaction with the environment. Pressures (stressors?) will vary depending on species-related mating behavior, territoriality, nest building, postnuptial molting, and neonatal rearing (altricial vs precocial young). The possible increases in corticosterone with these pressures may require either a facilitating role of corticosterone in reproduction or an uncoupling of changes in circulating corticosterone from various reproductive parameters in order for the reproductive phase to survive. There is evidence supporting each of these strategies.

Corticosterone clearance and glucocorticoid receptor function and the interaction of various steroid receptors at the target organs will contribute to the diverse effects of any one steroid and whether or not physiological "costs" are incurred (see Wingfield *et al.*, 1997a,b). Given the widely broadcasted effects of glucocorticoids and their central role in energy flow, it is not surprising to find a role for corticosterone in reproduction, sometimes functioning as an auxiliary reproductive hormone. For example, in mammals, elevated corticosteroids may be necessary to enhance or maintain excitatory amino acid receptors in the brain that drive the hypothalamo-pituitary-gonadal axis in the face of negative feedback from elevating circulating reproductive steroids (see Brann and Mahesh, 1997). In this connection, earlier work assigned the importance of changes in the circadian rhythm of plasma corticosterone in modulating seasonal reproductive and migratory parameters (Meier and Fivizzani, 1975). In addition, there is evidence for a parallel relationship between plasma corticosterone and plasma luteinizing hormone (LH) (Lea *et al.*, 1986; Meijer and Schwabl, 1989) and gonadal function (Ghosh *et al.*, 1996) and augmentation of corticosterone levels with breeding (Hegner and Wingfield, 1986a,b; Astheimer *et al.*, 1994; Logan and Wingfield, 1995; Romero *et al.*, 1997; Romero and Wingfield, 1998). This breeding-associated augmentation in circulating corticosterone is, in part, due to reduced glucocorticoid negative feedback on the HPA axis (Astheimer *et al.*, 1994) and enhanced adrenocortical response to ACTH (Romero and Wingfield, 1998). Furthermore, there is evidence that high plasma concentrations of corticosterone can differentially inhibit behavioral components of reproduction (e.g., territoriality) without affecting the adenohypophyseal-gonadal axis (e.g., LH and testosterone) (Wingfield and Silverian, 1986). Even stress levels that decrease body weight may leave LH and reproductive hormones unaffected (Wingfield, 1984). By contrast, during the nonbreeding migration period, adrenocortical response is limited (Romero and Wingfield, 1998). In addition, during normal molt (which involves the energetically costly replacement of feathers), the regression of the reproductive system is associated with a decreased adrenocortical function via a variety of mechanisms: a decreased hypothalamic capacity or central nervous upstream signals to secrete CRF and AVT (Romero et al., 1998a,b), a decreased pituitary ACTH response to hypothalamic signals (Romero et al., 1998c), and a decreased adrenal capacity to secrete corticosterone in response to ACTH (Romero et al., 1998a,b,c). Furthermore, in the king penguin (*Aptenodytes patagonica*) in which molting is associated with long-term fasting, adrenocortical function is suppressed presumably to spare protein (Cherel et al., 1988b). However, there is some suggestion that high-enough circulating corticosterone can inhibit the adenohypophyseal–gonadal axis (Wingfield and Silverin, 1986). For example, with food/water deprivation-induced molting in the domestic hen, regression of the reproductive system (Heryanto et al., 1997; Yoshimura et al., 1997) is associated with an increase in circulating corticosterone (Brake et al., 1979; Hoshino et al., 1988). In addition, under severe environmental stresses that exceed the limits of adaptation of some wild avian species, there is disruption of the endocrine and behavioral components of reproduction (Astheimer et al., 1995). Furthermore, physiological concentrations of corticosterone suppress chicken luteinizing hormone-releasing hormone-induced LH release from quail pituitary cells (Connolly and Callard, 1987). Not surprisingly, some species that experience the synergistic insult of more than one stressor (e.g., excessive heat and restricted access to water) possess an additional adaptation of uncoupling the stress–response components of the HPA axis and reducing the corticosterone response to stress (Wingfield et al., 1992). However, it should be pointed out that the postulate that high circulating corticosterone, in the appropriate physiological context, augments the hypothalamo-pituitary-gonadal axis of some avian species, has not been directly tested.

This relationship between the HPA axis and the HP–gonadal axis is also evident in the domestic fowl, although the nature of the interaction is unclear. Most notably, in the hen, sexual maturity is associated with a near abrogation of the corticosterone diurnal rhythm, especially during the time of maximal plasma LH (Wilson et al., 1984). Indeed, there appears to be an inverse relationship between the plasma concentrations of LH and corticosterone (Wilson and Cunningham, 1980). This relationship is thought to be mediated by progesterone, since exogenous progesterone (but not estradiol or testosterone preparations) rapidly reduces plasma corticosterone (Wilson and Cunningham, 1980). However, pharmacological doses of corticosterone (van Tienhoven, 1961; Etches and Cunningham, 1976; Sharp and Beuving, 1978) and ACTH (van Tienhoven, 1961) can induce a preovulatory LH surge (Sharp and Beuving, 1978) and ovulation (Etches and Cunningham, 1976; van Tienhoven, 1961). In addition, the potent synthetic glucocorticoid dexamethasone blocks ovulation when given at least 14 hr prior to ovulation (Soliman and Huston, 1974). Even though these studies are pharmacological, they raise the possibility that a physiological corticosterone surge precedes an LH surge, ovulation, or both. This postulate has evaded resolution. There are contrasting reports even from the same investigators. On the one hand, there appears to be no physiological role of corticosterone for these parameters of the ovulatory cycle (Sharp and Beuving, 1978; Etches, 1979; Tanaka and Kamiyoshi, 1980). On the other hand, other studies suggest the presence of ovulatory cycle-related surges or more gradual increases in circulating corticosterone (Beuving and Vonder, 1977, 1981; Johnson and van Tienhoven, 1981; Wilson and Cunningham, 1981). In addition, it should be pointed out that a more robust surge in corticosterone occurs with the act of oviposition (i.e., irrespective of the presence of an egg) (Beuving and Vonder, 1981).

3. Seasonal Parameters and Plasma Corticosterone Concentrations

Niche diversity requires that the physiological parameters of a bird's life-history stages adapt to accommodate the seasonal parameters of that ecological niche. An important component of season that entrains circulating corticosterone parameters appears to be photoperiod (Lea et al., 1996), albeit food availability (Hau and Gwinner, 1996, 1997) may also be important. In addition, cold ambient temperature may potentiate the photoperiod-entrained, season-related pattern of plasma corticosterone (Rintamaki et al., 1986). However, once established, season-related patterns of corticosterone secretion are driven by seasonal physiological changes and not by changing local weather conditions (Romero et al., 1997).

Not surprisingly, season-related changes in corticosterone reflect and help support reproductive status, diurnal rhythm shifts, feed availability/foraging, migration, molting, and territoriality (reviewed by Meier, 1975; Assenmacher and Jallageas, 1980; Deviche, 1983; Wingfield, 1980, 1984; Wingfield et al., 1995a, 1997a,b). For example, corticosterone may play an essential role in the season-related changes in physiology and behavior (Meier and Fivizzani, 1975; Wingfield et al., 1997b). In addition, the seasonal interaction of corticosterone and prolactin acrophases in controlling gonadal function, fat stores, and migratory parameters in the white-throated sparrow (*Zonotrichia albicollis*) is an example (Meier, 1973). However, during molt in such diverse species as the king penguin (*Aptenodytes patagonica*) (Cherel et al., 1988b), an arctic-breeding song bird (*Carduelis flammea*) (Romero et al., 1998a), Lapland longspurs (*Calcarius lapponicus*) (Romero et al., 1998b), and the snow bunting (*Plectrophenax nivalis*) (Romero *et al.*, 1998c), in which there are extreme adjustments in energy partitioning for feathering, there is dampening of the HPA axis a different points, depending on the species, presumbably to spare protein. Some juvenile stages also have a dampened HPA axis (Le Ninan *et al.*, 1988a,b; Romero *et al.*, 1988a), even with associated extreme fasting (Le Ninan *et al.*, 1988a,b). The previous sections of this chapter have addressed other important functions of corticosteroids in life-history stages of some avian species.

Against this background of the supporting role of corticosteroids, an avian species, irrespective of life-history stage, must respond to acute seasonal stressors that are of sufficient intensity and/or duration to disrupt the life-history stage. Apparently, different strategies of HPA axis activation and corticosterone response maximize the chances of survival and preserve the mechanisms supporting that life-history stage (Wingfield *et al.*, 1997b). Recent evidence suggests that there is restricted plasticity in the genetic changes supporting this adaptation. It appears that genetic changes that supported adaptation and speciation early in evolution, under ecological pressures that were obviously different from those of the present, are retained if they fit well with recent ecological pressures (Klicka and Zink, 1997).

L. Changes in Development, Maturation, and Senescence

1. Changes in Development

In vertebrates, the development of the hypothalamo-pituitary-gonadal and -adrenal axes is initiated by a highly conserved, orphan nuclear receptor, steroidogenic factor-1 (SF-1) (Parker and Schimmer, 1997), whereas the final pattern of adrenocortical development is dependent on additional ACTH-regulated transcription factors (Li and Lau, 1997). It is reasonable to expect that these conserved factors operate in the development of the avian adrenal gland. However, much of what is known about the development of the avian adrenal gland, and maturation and senescence as well, is restricted to precocial species.

In the chick embryo, presumptive adrenocortical cells arise from the dorsal coelomic epithelium by day 4 of incubation and form paired solid masses on each side of the aorta by day 6 (Bohus *et al.*, 1965; Adjovi, 1970; Domm and Erickson, 1970). Earlier work with the chick embryo suggests that presumptive adrenocortical tissue has the steroidogenic machinery to secrete corticosterone as early as day 4 in the chick (Pedernera, 1971, 1972; Siegel and Gould, 1976) and by day 9 in the duck (*Anas platyrhynchos*) (Cronshaw *et al.*, 1989), but sustained circulating levels are not apparent until later. It is interesting to note that the ability to synthesize corticosteroids antedates histological identification of the adrenal anlage (Cronshaw *et al.*, 1989). At around day 9 in the chick embryo, the presumptive adrenocortical tissue undergoes disproportionately rapid growth. The cells show ultrastructural features indicative of steroidogenic potential such as increases in smooth endoplasmic reticulum and lipid droplets (Hall and Hughes, 1970), the latter containing cholesterol stores (Sivaram, 1969). Similar ultrastructural features are apparent in the duck adrenal anlage at about days 12 or 13; however, definitive cords are not apparent until day 15 (Cronshaw *et al.*, 1989; Holmes *et al.*, 1992). Up to about days 12–16 in the chick embryo, these changes in adrenocortical tissue, and a sustained concomitant low level of steroidogenesis which begins by day 8 (Pedernera, 1972), occur in both intact and decapitated embryos and thus are thought to be independent of ACTH (Woods *et al.*, 1971; Domm and Erickson, 1972; Girouard and Hall, 1973; Wise and Frye, 1973; Kalliecharan and Hall, 1974, 1976; Avrutina *et al.*, 1985), even though pituitary ACTH is present from day 8 (Pedernera, 1972). This also seems to be true for the turkey embryo (Wentworth and Hussein, 1985) and the duck embryo (Cronshaw *et al.*, 1989) up to about day 15. Not surprisingly, given the early appearance of ACTH, the precise time of transition to pituitary control of the adrenocortical tissue is equivocal, but it is postulated to occur between 12 and 16 days in these precocial species and is associated with an increase in adrenocortical mitotic activity (Hall, 1970; Girouard and Hall, 1973; Cronshaw *et al.*, 1989). A number of studies of adrenal corticosteroid content *in ovo* (Kalliecharan and Hall, 1976; Marie, 1981; Tanabe, 1982; Tanabe *et al.*, 1986; Cronshaw *et al.*, 1989; Holmes *et al.*, 1992), corticosteroid concentration in allantoic fluid (Woods *et al.*, 1971), and the embryonic circulation (Woods *et al.*, 1971; Kalliecharan and Hall, 1974; Nakamura *et al.*, 1978; Marie, 1981; Tanabe, 1982; Wentworth and Hussein, 1985; Tanabe *et al.*, 1986; Holmes *et al.*, 1992) and *in vitro* with embryonic adrenal tissue (Gonzalez *et al.*, 1983; Tanabe 1982; Tanabe *et al.*, 1983, 1986; Cronshaw *et al.*, 1989; Holmes *et al.*, 1992) and with isolated adrenocortical cells (Carsia *et al.*, 1987a) indicate a robust increase in adrenocortical activity approaching hatch, with a disproportionate surge at day 15 in the chick embryo (Woods *et al.*, 1971; Marie, 1981), days 17–19 in the turkey embryo (Wentworth and Hussein, 1985) and day 21 in the duck embryo (Cronshaw *et al.*, 1989). This enhanced activity is accompanied by a decrease in corticosterone binding capacity (Siegel and Gould, 1976; Martin *et al.*, 1977; Gasc and Martin, 1978), which is thought to permit more free corticosterone necessary for organ differentiation and maturation, as, for example, the intestinal epithelium (Siegel and Gould, 1976), in preparation for hatching and neonatal life (Wentworth and Hussein, 1985). The chicken and duck embryos have similar patterns of adrenal growth (about 0.2 mg/day) and both show a disproportionate adrenal growth spurt in the last 24 hr prior to hatch (Holmes et al., 1992). The embryonic period of adrenocortical activity enhancement appears to be under the control of ACTH (Woods et al., 1971) and the embryonic adrenal is responsive to stressful stimuli (Wise and Frye, 1973; Avrutina et al., 1985), indicating a period of integration of the adrenal gland into HPA axis. During the embryonic period, 21-hydroxylase and attendant corticosterone secretion predominate with the appearance of aldosterone secretion somewhat delayed (see Carsia et al., 1987a; Cronshaw et al., 1989; Holmes et al., 1992). In addition 17α-hydroxylase activity and attendant cortisol secretion are significant (see Carsia et al., 1987a and references therein). Work with cellular or tissue preparations of chicken and duck adrenal glands indicate an order of secretion: corticosterone > aldosterone > cortisol. The significance of other steroid metabolites, such as the products of 5β-reductase activity and the role of the embryonic adrenal as a significant extragonadal source of testosterone, have been mentioned previously (see previous sections under "Adrenocortical Hormones").

2. Changes in Maturation

Here again, most of what is known about the posthatch changes in adrenocortical function is restricted to the precocial neonates; that is, the duck and chicken. Even between these two precocial species there are differences in the first few days after hatch. In the chicken (21-day embryonic period), there appears to be a stress "nonresponsive" period for about 48 hr after hatch that is thought to be largely due to transient hypothalamo–hypophyseal deficiency (Wise and Frye, 1973; Freeman and Manning, 1977, 1984; Freeman and Flack, 1980, 1981; Freeman, 1982). Careful reexamination indicates that the stress response is actually severely blunted, thus, hyporesponsive (Holmes et al., 1990). In support of hypothalamo–hypophyseal deficiency is the fact that adrenocortical cells from 1-day-old chicks exhibit the greatest sensitivity and corticosteroid responsiveness to ACTH compared to other posthatch (postnatal) ages (Carsia et al., 1987a). In contrast, a hyporesponsive period does not occur in the duck (27-day embryonic period) (Holmes et al., 1990). Interestingly, the plasma corticosterone response to stress in a 1-week-old chicken is similar to that of a neonatal duck (Holmes et al., 1990). These differences may reflect the respective developmental ages of the duck and chicken at the end of the embryonic, in ovo period (Holmes et al., 1992).

Overall, a limited number of studies indicate a gradual decline in adrenocortical function with posthatch age (Avrutina et al., 1985). For about 1–3 weeks after hatching, adrenocortical function remains fairly stable or shows a slight increase in function in the duck (Holmes and Kelly, 1976). Indeed, adrenocortical cells derived from chickens over this period have similar patterns of corticosteroid response to ACTH (Carsia et al., 1985b). However, from about 3 weeks of age to sexual maturity, there is a decrease in relative adrenal weight (about 80%) and adrenal weight-specific secretion rate (about 45%). This is compensated for by a decrease in metabolic clearance rate (about 80%) (Holmes and Kelly, 1976). Nevertheless, resting (Schmeling and Nockels, 1978; Wentworth and Hussein, 1985; Carsia and Weber, 1986; Davis and Siopes, 1987; Holmes et al., 1989), stress-induced (Davis and Siopes, 1987; Gorsline, and Holmes, 1982; Holmes et al., 1989) and ACTH-induced (Schmeling and Nockels, 1978; Davis and Siopes, 1987) plasma corticosterone concentrations, and the rate of plasma corticosterone response to ACTH (Beuving and Vonder, 1978) tend to decline with posthatching age. In chickens, this decline, in part, may be due to an age-dependent decrease in adrenocortical cell sensitivity to ACTH (Carsia et al., 1985b). Despite this apparent decline in function, there is an apparent potential temporal window for imprinting of the adrenocortical tissue to an enhanced setpoint of function (Avrutina et al., 1985). In the chicken, this appears to be at about 2–4 weeks posthatching. For example, subjecting cockerels to transient dietary protein restriction over this time period induces long-term potentiation of adrenocortical function (Weber et al., 1990).

3. Changes in Senescence

After sexual maturity, the status of adrenocortical function is unclear. It is thought that the decline in function, if any, is modest. Furthermore, in the chicken, there appear to be no senescent changes in adrenocortical function. Indeed, in a comparison of chickens, 36 weeks old and 126 weeks old, there was no difference in the maximal plasma corticosterone response to infusions of ACTH (Beuving and Vonder, 1978).

III. PHYSIOLOGY OF ADRENOCORTICAL HORMONES

A. Corticosteroid Receptors

Corticosteroids pass freely through the plasma membrane of cells. Although there is evidence for plasma membrane receptors for corticosteroids (Trueba et al.,

1987; Wehling, 1997), it is largely assumed that the major effects of mineralocorticoids (e.g., aldosterone) and glucocorticoids (e.g., corticosterone and cortisol) on target cells are mediated by cognate nuclear receptors: the mineralocorticoid receptors (MRs) and the glucocorticoid receptors (GRs). Unliganded GRs and MRs are primarily in the cytoplasm and exist as monomers (~94 kDa) in a multiprotein complex with several chaperones, such as heat-shock proteins (hsp) (e.g., hsp90 and hsp70) and immunophilins (e.g., hsp56) (Bamberger et al., 1996). Thus, the cytosolic multiprotein complex could be quite large (6S-9S form; ~300 kDa). The chaperone proteins have a role in stabilizing multiprotein-complexing with the receptor, assisting in high-affinity binding and appear to assist in receptor turnover and trafficking (Pratt and Toft, 1997). However, most of these proteins dissociate from the receptor with corticosteroid binding (3.5S-4.6S; ~94 kDa), which is a requirement for further downstream events including binding to corticosteroid response elements in DNA (Pratt and Toft, 1997). The hormone–receptor complex translocates to the nucleus, and after forming homo- or heterodimers or associating with other proteins (Bamberger et al., 1996), initiates either activation or suppression of specific genes (Boumpas et al., 1993).

Many of these features of receptor–protein interactions also appear to be true for the highly conserved corticosteroid receptors found in avian species (Rafestin-Oblin et al., 1989; Cadepond et al., 1994). Specific receptors have been identified in preparations from a number of avian tissues including the brain (Koehler and Moscona, 1975; Saad and Moscona, 1985), liver (Tu and Moudrianakis, 1973), kidney (Charest-Boule et al., 1978, 1980; Beaudry et al., 1983; Bellabarba et al., 1983; Gendreau et al., 1987; Lehoux et al., 1984, 1988), lung (Hylka and Doneen, 1983), intestine (DiBattista et al., 1985, 1989; Rafestin-Oblin et al., 1989), muscle (Sabeur et al., 1993), thymus (Gould and Siegel, 1984; Krajcsi and Aranyi, 1986), cloacal bursa of Fabricius (Sullivan and Wira, 1979; Fassler et al., 1986), and salt-activated nasal gland (Sandor et al., 1977, 1983). In the bursa, the resident lymphocytes express the glucocorticoid receptors whereas the epithelial cells express gonadal steroid receptors (Sullivan and Wira, 1979). As in mammals, avian corticosteroid receptors appear to group into two types: the type I, MR and the type II, GR. The high amino acid identity of the ligand binding domains permits cross-binding (White et al., 1997). Thus, it is not surprising to find both high- and low-affinity sites, albeit under more stringent conditions, with purer receptor preparations and with appropriate antagonists [e.g., blocking the GR in order to isolate the MR (Lehoux et al., 1988; Rafestin-Oblin et al., 1989)], a single class of sites is observed with an apparent K_d of $10^{-11}-10^{-9}$ M. In this connection, the mammalian GR and progesterone receptor antagonist, RU486, exclusively renders the avian GR inactive due to a mutation in the avian progesterone receptor (Groyer et al., 1985; Benhamou et al., 1992).

B. Corticosteroid Action in Target Cells

Target cells of corticosteroid action exhibit disparate biological response. The mechanisms underlying this diversity of response is complex. What is clear is that they are downstream from nuclear translocation. Avian and mammalian components of the cytosolic, multiprotein–receptor complex can be mixed to confer full binding function (Cadepond et al., 1994). In addition, nuclear translocation of avian GRs is neither tissue specific or age dependent (Charest-Boule et al., 1978; Saad and Moscona, 1985). Also, during conditions and treatments that facilitate receptor downregulation (Gould and Siegel, 1984; DiBattista et al., 1985) during development (Bellabarba et al., 1983; Lehoux et al., 1984; Fassler et al., 1986; Gendreau et al., 1987) in response to electrolyte stressor (Charest-Boule et al., 1978; DiBattista et al., 1985) and in avian muscular dystrophy (Sabeur et al., 1993), changes in the magnitude of corticosteroid responses are due to changes in receptor concentration and not to changes in the affinity and specificity of receptors. Furthermore, although corticosteroid receptors from diverse avian tissues exhibit very similar binding characteristics, corticosteroids elicit disparate responses in avian organs and target cells. For example, the potent glucocorticoid cortisol cannot substitute for corticosterone in the induction of lung surfactant formation for neonatal life (Hylka and Doneen, 1983) and is 10 times less potent in stimulating transcecal Na^+ transport (Grubb and Bentley, 1992). By contrast, cortisol may be more important than corticosterone in inducing the differentiation of the gastrointestinal tract (Tur et al., 1989) and the retina (Saad and Moscona, 1985; Vardimon et al., 1988). Progesterone is an inhibitor of glucocorticoid-induced chick osteogenesis (Tenenbaum et al., 1995), but converges with corticosterone on the GR to induce somatotroph differentiation (Morpurgo et al., 1997). Other examples are the differential effects of two glucocorticoids, corticosterone and dexamethasone, on organ growth (Joseph and Ramachandran, 1993) and carbohydrate metabolism (Joseph and Ramachandran, 1992) in the posthatch chick. It appears that target cells for corticosteroid action have different interactions of nuclear transactivating factors and transcription initiation complexes in response to subtle changes in receptor conformation upon ligand binding and that recognize and activate specific deoxyribonucleic acid (DNA) hormone response elements for the initiation of gene transcription (Bamberger et al., 1996; Beato and Sánchez-Pacheco, 1996). Furthermore, each target tissue may have unique cellular factors that confer specificity of glucocorticoids vs mineralocorticoids in various target tissues (Lim-Tio et al., 1997). Finally, the interaction of thyroid, corticosteroid, and sex steroid receptors in certain target tissues, e.g., the uropygial (preen) gland, can be quite complicated in the diverse life stages of avian species (Daniel et al., 1977; Bandyopadhyay et al., 1990; Asnani and Ramachandran, 1993).

The interaction of GRs and MRs in target tissues deserves further attention. The "spillover" of corticosterone to the MR has survival value, yet the unfettered, specific responses of tissues to aldosterone is important for optimal water and electrolyte balance. The interplay between corticosterone and aldosterone in the regulation of kidney, intestine, and possibly nasal gland function is exemplified in anseriforms that encounter hypoosmotic and hyperosmotic water environments (for discussion, see Holmes et al., 1992). However, in the face of much greater circulating concentrations of corticosterone compared to aldosterone (*vide supra*), what confers mineralocorticoid specificity in avian target tissues? First, there is some suggestion that the avian MR has a greater affinity than the GR for aldosterone (Lehoux et al., 1984, 1988; Rafestin-Oblin et al., 1989). Second, the avian MR appears to have about a half-log order greater affinity for corticosteroids compared to the GR (DiBattista et al., 1985, 1989). Thus, if there were mechanisms to isolate the MR from glucocorticoids, e.g., by degrading the local concentration of glucocorticoid, the MR would be able to specifically respond to low circulating concentrations of aldosterone. In the rat, which also secretes corticosterone as the predominant glucocorticoid, and in other mammals, an 11β-hydroxysteroid dehydrogenase enzyme fulfills this role in the kidney and is colocalized with the MR (Funder, 1995; White et al., 1997). A similar enzymatic activity exists in avian tissues, e.g., the duck nasal gland (Sandor et al., 1983) and chicken cecum (Grubb and Bentley, 1992). In the nasal gland, this activity may serve to terminate corticosterone action since aldosterone is without effect (Holmes and Phillips, 1985). In the cecum, it is unclear whether this activity assists MR action. However, in the duck intestine, a cytoplasmic C-20 oxidoreductase enzyme also appears to be a good candidate for this role. The resultant corticosterone degradation product, 20β-dihydrocorticosterone, binds to the GR but not to the MR and is not competed off the GR by aldosterone. Furthermore, the 20β-dihydrocorticosterone-GR complex is translocated to the nucleus. Thus, it is thought that this mechanism effectively removes corticosterone, "silences" the GR, and permits aldosterone binding to the MR (DiBattista et al., 1989). It should be pointed out that aldosterone also undergoes extensive, cytosolic 5β-reduction in the avian kidney (Charest-Boule et al., 1978), albeit the role of this activity in modulating GR and MR action is unclear.

In summary, there is little known about the modulation of steroids and their receptors in avian target tissues. As pointed out by Wingfield and co-workers, information on this modulation is important for understanding the diverse patterns of HPA axis and HP–gonadal axis regulation for appropriate behavioral and physiological responses expressed in the life stages of disparate avian species (Wingfield et al., 1997a,b).

C. Corticosteroids and Intermediary Metabolism

Some studies on the role of glucocorticoids in intermediary metabolism should be viewed with caution because of the differential effects of the molecular species of glucocorticoid employed (Joseph and Ramachandran, 1992, 1993) and the route of administration. As mentioned previously, exogenous corticosterone and other glucocorticoids increase food intake. In addition, they increase gastrointestinal transit time (Tur et al., 1989). Despite these increases in alimentation parameters, there is a dramatic decrease in growth and body weight (Davison et al., 1983a, 1985; Saadoun et al., 1987; Kafri et al., 1988; Hayashi et al., 1994) that is largely due to an increase in net muscle protein catabolism (De La Cruz et al., 1981; Simon, 1984; Saadoun et al., 1987; Takahashi et al., 1993; Hayashi et al., 1994). Similar effects are seen in immature chickens undergoing long-term (weeks) administration of ACTH (Davison et al., 1979; Davison et al., 1983a). Interestingly, the suppressive effect of exogenous glucocorticoids on the rate of body growth can be passed on to the offspring of treated breeders (De La Cruz et al., 1987).

Accompanying this decrease in body weight is an increase in fat deposition (*vide supra*), largely in the abdomen (Davison et al., 1983a, 1985; Pilo et al., 1985; Saadoun et al., 1987; Kafri et al., 1988; Takahashi et al., 1993; Hayashi et al., 1994; Buyse et al., 1997). It appears that peripheral, extraabdominal fat and abdominal fat are regulated differently, the latter being more sensitive to β-adrenergic agonists (Takahashi et al., 1993). This increase in abdominal fat is supported by slight peripheral lipolysis (Harvey et al., 1977) and enhanced lipogenesis (Pilo et al., 1985; Saadoun et al., 1987; Kafri et al., 1988) by an enlarged liver (Davison et al., 1983a, 1985; Buyse et al., 1987; Kafri et al., 1988).

Exogenous corticosterone in the chicken increases plasma glucose concentrations (Davison et al., 1983a, 1985; Simon, 1984; Saadoun et al., 1987; Joseph and

Ramachandran, 1992; Thurston et al., 1993). However, this does not appear to be true for the domestic turkey in which an adrenergic control may predominate (Thurston et al., 1993). Corticosterone-induced hyperglycemia in chickens is, in part, supported by enhanced tissue glycogenolysis and hepatic glucose-6-phosphatase (Joseph and Ramachandran, 1992). Corticosterone also increases circulating insulin (Simon, 1984; Bisbis et al., 1994), but this is counteracted by downregulation of insulin receptors in the liver (Taouis et al., 1993) and suppression of postreceptor (after kinase activity) events in muscles and kidney (Taouis et al., 1993; Bisbis et al., 1994).

These effects of exogenous corticosterone on intermediary metabolism may be mediated directly by specific receptors in the relevant target tissues. In addition, the effects may be indirect in that exogenous corticosterone produces changes in other hormones of intermediary metabolism such as thyroid hormones, growth hormone, prolactin, somatomedin C, and norepinephrine (Buyse et al., 1987; John et al., 1987; Saadoun et al., 1987).

D. Corticosteroids and Immune Function

Some aspects of the interaction between the humoral immune system and adrenocortical function have been addressed elsewhere (see Chapter 26). Other neuroendocrine–immune interactions have been recently reviewed (Marsh and Scanes, 1994). The present section addresses the influence of glucocorticoids on the avian immune system. Since the components of the system possess glucocorticoid receptors (*vide supra*), it is not surprising that they are influenced by circulating glucocorticoids. Exogenous glucocorticoids or increases in circulating concentrations of corticosterone in response to ACTH uniformly cause involution of the thymus, spleen, and cloacal bursa (of Fabricius) (Davison et al., 1979, 1983a, 1985; Gross et al., 1980; Edens, 1983; Donker and Beuving, 1989; Joseph and Ramachandran, 1993). Much of the involution occurs by programmed cell death (apoptosis) (Compton et al., 1990a,b, 1991). Noteworthy is the effect of glucocorticoids on the bursa in that the bursal hormone, bursin, or bursopoetin, an amidated tripeptide (Lys-His-Gly-NH$_2$), influences the development of the pituitary–adrenal axis. Embryonic bursectomy results in precocious stress response in chick embryos and loss of the stress hyporesponsive period in neonatal chicks; however, in early adult life, chickens are stress hyporesponsive. Normal development of HPA axis function can be restored with exogenous bursin (Guellati et al., 1991).

Exogenous glucocorticoids influence the composition of circulating leukocytes. In general, the numbers of all circulating leukocytes are reduced except heterophils (Gross et al., 1980; Davison and Flack, 1981; Trout et al., 1988). Thus, heterophil/lymphocyte ratios increase with glucocorticoid treatment (Gross et al., 1980; Davison et al., 1983b; Donker and Beuving, 1989; Gross, 1992) and vary directly with the level of glucocorticoid treatment (Gross, 1992). Moderate levels of circulating glucocorticoids favor a decrease in inducible cellular cytotoxicity (Thompson et al., 1980), a decrease in lymphoproliferation, and a decrease in IL-2 and γ-interferon production (Isobe and Lillehoj, 1992, 1993). The overall reduction in circulating leukocytes is, in part, explained by the redistribution or trapping of cells in secondary lymphoid organs. For example, in the chicken spleen, although the total number of lymphocytes is reduced, the percentage of cytotoxic-suppressor T cells is reduced, whereas the percentage of helper T cells is increased (Isobe and Lillehoj, 1992, 1993; Mashaly et al., 1993), favoring antibody production. However, since antibody production follows lymphocyte numbers, there is an overall suppression in antibody response (Gross, 1992). This, together with a decrease in cell-mediated immunity, may increase susceptibility to viral infections (Gross et al., 1980; Thompson et al., 1980) and coccidiosis (Isobe and Lillehoj, 1993). By contrast, the relative increase in heterophils may confer increased resistance to bacterial infections (Gross, 1992). In the respiratory system, resistance to such infections may be assisted by the glucocorticoid-induced enhancement of tracheal mucociliary clearance (Kai et al., 1990).

E. Corticosteroids and Behavior

It is clear that corticosteroids influence behavior. The mechanism of this influence is less clear. Exogenous corticosterone can disrupt behavior associated with breeding, rearing of young, and territoriality in several species and this effect is independent of the HP–gonadal axis (Silverin, 1986; Wingfield and Silverin, 1986; also, see Wingfield et al., 1997a,b). However, as mentioned previously, several aspects of reproduction and associated behavior are unaffected by high circulating concentrations of corticosterone. Possibly, treatment with exogenous corticosterone at times when the CBG is falling (Silverin, 1986) results in supraphysiological concentrations of free corticosterone that can impinge on the neural components of behavior. Indeed, exogenous testosterone (Klukowski et al., 1997), and possibly other sex steroids, increase CBG and thus total plasma corticosterone. Presumably, this bound corticosterone can be made available in metabolically active target tissues but is less available for neural centers controlling behavior.

Corticosteroids also have an influence on learning-induced neural plasticity. They have a positive influence on long-term avoidance behavior in chicks (Sandi and Rose, 1994a,b). Furthermore, MRs and GRs appear to mediate different aspects of information processing and storage (Sandi and Rose, 1994b). The learning of avoidance strategies in response to stressors (i.e., experience), should, in turn, eliminate or reduce corticosterone stress responses to those stressors.

Ample evidence indicates a relationship between plasma corticosterone and fearfulness (see Jones, 1996). For example, in the Japanese quail (*Coturnix coturnix japonica*), selection for divergent plasma corticosterone responses to immobilization stress is directly related to the magnitude of expression of fear-related behavior (e.g., freezing, crouching, and tonic immobility) (Jones *et al.*, 1992a,b, 1994; Jones and Satterlee, 1996; Satterlee *et al.*, 1993a,b; Satterlee and Jones, 1997). Interestingly, dietary supplementation with vitamin C can attenuate fear-related behavior without affecting adrenocortical response to stress (Satterlee *et al.*, 1993b, 1994; Jones *et al.*, 1996). Furthermore, exogenous corticosterone appears to enhance fearfulness (Jones *et al.*, 1988).

Exogenous corticosterone also enhances oral stereotypic behavior, e.g., drinking and pecking (Savory and Mann, 1997). In addition, increased feather pecking is correlated with stress-induced increases in circulating corticosterone (Vestergaard *et al.*, 1997). It is unclear whether this pecking behavior has survival value or is displacement behavior used in an attempt to reduce plasma corticosterone and attendant corticosterone-induced fearfulness. In this connection, chickens, selected for low tendency for feather pecking, have significantly greater resting and stress-induced plasma corticosterone concentrations compared to a high feather-pecking line (Korte *et al.*, 1997). However, exogenous corticosterone, which induces protein catabolism, modifies the food pattern of chickens toward a high-protein diet (Covasa and Forbes, 1995). Thus, pecking, associated with high circulating corticosterone levels, may facilitate appropriate nutriture to redress impeding protein loss. A similar behavior pattern is seen with aldosterone, that in synergy with Ang II, evokes salt appetite in pigeons (Massi and Epstein, 1990).

F. Corticosteroids and Electrolyte Balance

Organs that regulate electrolyte balance (see Chapter 11) include the kidney, large intestine, and cloaca (Holmes and Phillips, 1976; Skadhauge, 1980; Holmes *et al.*, 1983). In addition, numerous anseriform species have a salt-activated nasal gland that serves to excrete electrolyte in response to hyperosmotic water environments (Harvey and Phillips, 1982; Butler, 1984; Holmes and Phillips, 1985). The adrenocortical hormones play an important role in supporting these sites of osmoregulation. Their importance for electrolyte balance is evinced by the profound negative Na^+ balance (and eventually death) in adrenalectomized birds and its restoration by their replacement (Thomas and Phillips, 1975; Butler and Wilson, 1985). Whereas corticosteroids increase Na^+ reabsorption in the renal tubule, aldosterone favors potassium excretion and this is reduced by corticosterone.

The distal intestine and cloaca are important extrarenal sites for osmoregulation in birds. Corticosteroids increase the level of solute-linked water uptake across the intestinal and hindgut mucosa (Crocker and Holmes, 1971, 1976; Skadhauge *et al.*, 1983; Thomas *et al.*, 1980; Thomas and Skadhauge, 1982, 1989). With the exception of a recent study (Elbrønd and Skadhauge, 1992), the balance of studies suggests that aldosterone is the predominant corticosteroid inducer of Na^+ transport across the intestine and hindgut (Thomas *et al.*, 1980; Clauss *et al.*, 1987; Grubb and Bentley, 1992), albeit different regions of the gut exhibit different sensitivities to aldosterone (Skadhauge *et al.*, 1983; Grubb and Bentley, 1987). As in the kidney, aldosterone appears to favor potassium secretion in the lower intestine (Thomas and Skadhauge, 1989).

In birds that have hyperosmotic water environments as one of their habitats, the salt-activated nasal gland plays an important role in osmoregulation. In these environments, birds must maintain water uptake that is linked to electrolyte uptake. Interestingly, in the duck, there is a positive correlation between plasma osmolality and plasma Ang II concentrations (Gray and Simon, 1985a). This is due to the fact that maintenance of the extracellular fluid compartment, which is decreased with increased plasma osmolality in saltwater-adapted ducks (Gray and Simon, 1985b), prevails over maintenance of plasma osmolality. However, the ratio of aldosterone to corticosterone secretion in this state is low (Klingbeil, 1985b; Redondo *et al.*, 1988). In addition, the increase in plasma osmolality is a homeostatic stressor that leads to an increase in plasma ACTH and hence corticosterone. Thus, there is a shift in corticosteroid osmoregulatory predominance to corticosterone in the hyperosmotic water environment (see Holmes *et al.*, 1992). By secreting Na^+ solution through the salt-activated nasal gland, which is supported predominantly by corticosterone (Holmes and Phillips, 1985), the birds maintain electrolyte and hemodynamic balance and retain free water for evaporation and for excretion of nitrogenous wastes (see Gray snd Simon, 1985a). Although the nasal gland possesses corticosteroid receptors (Sandor *et al.*, 1977, 1983), the action of corticosterone appears to be indirect. In salt-adapted, adrenalectomized ducks, the gland undergoes hypertrophy and maintains a fairly normal pattern of sodium chloride secretion (Butler, 1984, 1987). However, with cardiovascular deterioration there is secretory failure. Thus, corticosterone is required to maintain adequate perfusion pressure of the nasal gland for normal function.

IV. ADRENAL CHROMAFFIN HORMONES

A. Catecholamine Synthesis

NE is widely synthesized throughout the postganglionic sympathetic nervous system which includes the adrenal chromaffin tissue. Studies with adrenalectomized ducks suggest that about half of the circulating NE and essentially all of the circulating E is derived from the adrenal gland (Wilson and Butler, 1983a,b; Butler and Wilson, 1985). The biosynthetic pathway for the adrenal catecholamines is outlined in Figure 6.

The three methods employed [fluorometric method, radioenzymatic assay, and high performance liquid chromatography followed by electrochemical detection (HPLC-ECD)] to estimate the circulating concentrations of catecholamines in avian species have yielded disparate and conflicting values. However, the consensus of opinion is that the latter two methods are more specific and verifiable. Table 3 lists the basal circulating concentrations of E and NE in some avian species as determined by radioenzymatic assay or HPLC-ECD. It should be pointed out that dopamine also exists at variable but relatively significant concentrations (Wilson and Butler, 1983a,c; Rees *et al.*, 1984; Butler and Wilson, 1985).

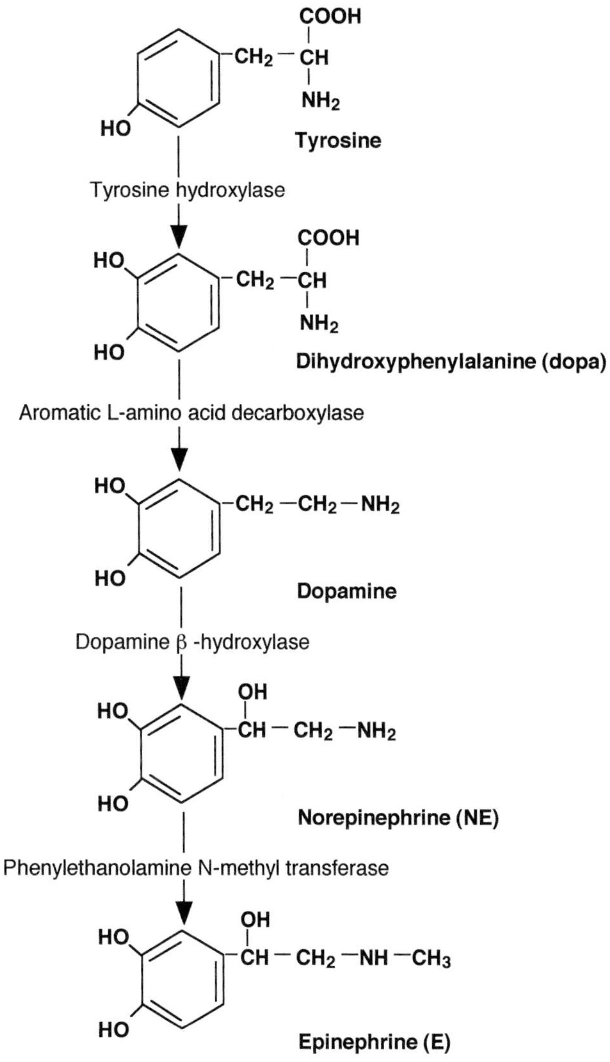

B. Control of Catecholamine Synthesis and Release

There is no preferential release of E or NE from isolated chicken chromaffin cells in response to muscarinic activation by acetylcholine analogs (Knight and Baker, 1986). *In vivo*, however, the secretion of E and NE from the avian adrenal gland is finely regulated by a number of neural-derived and blood-borne factors and hormones. Although well established in mammals (Axelrod and Reisine, 1984), a predominant role of preganglionic (splanchnic) neuronal activity in regulating tyrosine hydroxylase (TH), the rate-limiting enzyme in catecholamine synthesis in the avian adrenal is less clear. For example, there is evidence that ACTH can directly influence TH of the avian adrenal chromaffin cell (Manelli *et al.*, 1973; Accordi *et al.*, 1975). In addition, adrenal catecholamine content is not appreciably altered with splanchnic denervation (Mahata and

FIGURE 6 Biosynthesis of norepinephrine and epinephrine.

Ghosh, 1985). However, splanchnic nerve integrity is essential for NE synthesis and NE and E resynthesis after secretion (Mahata *et al.*, 1988; Mahata and Ghosh, 1990, 1991a,b,c). Furthermore, depending on the type of secretory stimulus, splanchnic nerve activity differentially directs NE and E content status (Mahata *et al.*, 1988; Mahata and Ghosh, 1985, 1989, 1990, 1991a,b,c).

In addition to its possible direct effect on TH activity, ACTH, largely via intraglandular glucocorticoid production, increases phenylethanolamine *N*-methyltransferase (PNMT) activity and thus the conversion of NE to E (Wassermann and Bernard, 1971; Zachariasen and Newcomer, 1974). Here again, splanchnic nerve activity facilitates glucocorticoid-induced E synthesis (Mahata and Ghosh, 1991a), albeit vagal tone may also play an important role in regulating PNMT activity (Viswanathan *et al.*, 1987). The picture is further complicated in that a bursal factor may also play a role in

TABLE 3 Basal Peripheral Plasma Concentrations of Epinephrine (E) and Norepinephrine (NE) in Selected Avian Species

Species	Sex	Plasma concentration (ng/ml)		Method	Reference
		E	NE		
Chicken	F	0.19	0.44	Radioenzymatic	Nishimura et al. (1981)
	M	0.47	0.71	Radioenzymatic	Rees et al. (1984)
	F	0.53	1.2	HPLC–ECD	John et al. (1987)
	F	0.35	1.7	HPLC–ECD	Fujita et al. (1992)
	F	0.09	0.23	HPLC–ECD	Korte et al. (1997)
Duck					
		0.68	1.4	Radioenzymatic	Hudson and Jones (1982)
	M	ND[a]–0.03	0.25	Radioenzymatic	Wilson and Butler (1983a)
	M	0.08	0.81	Radioenzymatic	Wilson and Butler (1983c)
	M	0.12	0.66	Radioenzymatic	Butler and Wilson (1985)
	M	0.27	0.76	HPLC–ECD	Lacombe and Jones (1990)
Pigeon					
		0.60	0.73	Radioenzymatic	Hissa et al. (1982)
		5.1	1.2	HPLC–ECD	Viswanathan et al. (1987)
		6.6	9.2	HPLC–ECD	John et al. (1990)

[a]ND, not detectable.

facilitating early, stress-induced catecholamine release and subsequent catecholamine resynthesis, especially E synthesis (Mahata et al., 1990).

Not surprisingly, stressors in the environment will elicit differential activation of the HPA and sympathochromaffin axes and probably the RAS as well. Thus, the short-term stress response involves increases in both corticosteroids and catecholamines (Brown and Nestor, 1973; El-Halawani et al., 1974; Buckland et al., 1974; Howard, 1974; Davison, 1975; Edens and Siegel, 1975; Zachariasen and Newcomer, 1975; Jurani et al., 1978; Harvey et al., 1984a; Rees et al., 1984). Given extraadrenal sources of catecholamines, what proportion of circulating catecholamines is contributed by the adrenal gland in the short-term stress response? First, as mentioned previously, studies with adrenalectomized ducks suggest that about half of the circulating NE and essentially all of the circulating E is derived from the adrenal gland (Wilson and Butler, 1983a,b; Butler and Wilson, 1985). Second, ACTH stimulates E and NE release (Zachariasen and Newcomer, 1974). Third, with some stressors (e.g., immobilization stress), plasma E response is attenuated with blockade of ACTH secretion (Zachariasen and Newcomer, 1975). Fourth, there is a tight temporal relationship between the rise in circulating catecholamines and their depletion from the adrenal gland with exercise stress (Rees et al., 1984). Finally, studies with intact and adrenalectomized ducks, subjected to brief forced submergence, indicate that essentially all of the circulating E and 70–80% of NE, in response to brief forced submergence, come from the adrenal gland (Lacombe and Jones, 1990). Yet, in mammals, avians, and other vertebrates as well it is not known how the differential adrenal secretion of E and NE occurs in response to disparate stressors. For example, avian genetic lines, in which there is more active coping strategy upon exposure to alarming stimulation, have greater circulating concentrations of NE compared to E (Satterlee and Edens, 1987; Jones and Satterlee, 1996; Korte et al., 1997). In addition, other contrasting responses have been noted with anesthesia in ducks (drop in plasma NE; Wilson and Butler, 1983c) and hypoglycemic stress after insulin administration (Pittman and Hazelwood, 1973).

C. Physiological Actions of Epinephrine and Norepinephrine

A discussion of the diverse effects of catecholamines on avian tissues is beyond the scope of this chapter (see Chapter 7). However, their role in coordinated stress response and in an effector limb of the RAS deserve some attention. It should be pointed out, however, that studies with exogenous catecholamines should be viewed with caution in that the circulating levels achieved may exceed physiologic limits and thus generate effects that are not representative of true endocrine events.

One effect of catecholamines is the augmentation of the HPA axis. Catecholamines, through the activation of β- and α-adrenergic receptors, can, respectively, augment or suppress plasma corticosterone responses to both endogenous and exogenous ACTH (Freeman and Manning, 1978, 1979; Rees *et al.*, 1985b). There is some suggestion that this effect is direct at the level of the adrenocortical tissue (Rees *et al.*, 1985b). As mentioned previously, *in vitro* studies, with isolated chicken adrenocortical cells, support a direct stimulation of corticosteroid secretion (Mazzocchi *et al.*, 1997a,b), albeit in duck adrenocortical tissue there is a suggestion that catecholamines exert a suppressive effect (Holmes *et al.*, 1991).

Although postganglionic sympathetic NE release plays a role in mediating the pressor response to Ang II (Nishimura *et al.*, 1994), blood-borne adrenal catecholamines also play a significant role (Nishimura *et al.*, 1981; Wilson and Butler, 1983a,c). Adrenalectomy abolishes plasma E (Wilson and Butler, 1983a; Butler and Wilson, 1985) and nearly abolishes the rise in plasma NE in response to exogenous Ang II (Wilson and Butler, 1983a). However, the α-adrenergic-mediated increase in blood pressure (Bulter *et al.*, 1986) is only attenuated (Wilson and Butler, 1983a) because of the Ang II-induced NE release from adrenergic nerve endings. A similar attenuation of Ang II-induced pressor response is seen after adrenal catecholamine depletion with reserpine (Wilson and Butler, 1983c). Complete blockade of the pressor effect elicited by exogenous Ang II is accomplished with adrenalectomy and treatment with the adrenergic-ending, NE-depleting agent 6-hydroxydopamine (Wilson and Butler, 1983b).

Catecholamines have diverse effects on immune cells. For example, continuous infusion of physiological concentrations of NE and E enhance chicken leukocyte migration and differentially affect the phytohemagglutinin wattle response (McCorkle and Taylor, 1993). In addition, NE and E, *in vivo* and *in vitro*, have contrasting effects on immunoglobulin plaque-forming cells from splenic lymphocytes, and these contrasting effects are probably mediated by α- and β-adrenergic receptors (Denno *et al.*, 1994). Catecholamines also differentially affect macrophage effector functions *in vitro* (Ali *et al.*, 1994).

Catecholamines also influence carbohydrate and lipid metabolism (see Chapter 13). For example, exogenous insulin dramatically increases plasma E in the chicken (Pittman and Hazelwood, 1973) and stimulates adrenal catecholamine release and synthesis in the pigeon (Mahata and Ghosh, 1990). E stimulates glycogenolysis in chicken hepatocytes (Cramb *et al.*, 1982; Picardo and Dickson, 1982) via β-adrenergic receptor activation and cyclic AMP production, which, in turn, activates glycogen phosphorylase. Catecholamines (NE and presumably E) also stimulate gluconeogenesis (Cramb *et al.*, 1982; Sugano *et al.*, 1982) via α-adrenergic receptor activation of intracellular calcium mobilization. It is probable that the effects of E on glycogenolysis, gluconeogenesis, and also lipogenesis (*vide infra*) are physiological roles for adrenal E, as these metabolic parameters are influenced by concentrations of E that are found in the circulation of a stressed bird. However, the effect of catecholamines in ultimately increasing blood glucose levels may be species specific, in that turkey appears to have a greater glycemic response compared to the chicken (Thurston *et al.*, 1993). Furthermore, in some species, melatonin plays a vital role in glucose homeostasis in that it influences adrenal catecholamine status and, depending on species and age, has differential effects on circulating glucose (Mahata *et al.*, 1988).

In chickens, during periods of behavioral inactivity, there is a positive relationship between circulating catecholamines and free fatty acids. Thus, as mentioned above, catecholamines, at least in the liver, may be physiological regulators of lipid metabolism. The synthesis of fatty acid (lipogenesis) by liver tissue and cells is inhibited by E and to a lesser extent by NE (Prigge and Grande, 1971; Capuzzi *et al.*, 1975; Cramb *et al.*, 1982; Campbell and Scanes, 1985). This effect is mediated via both α- and β-adrenergic receptors and, at least partially, by cyclic AMP acting as the intracellular messenger (Campbell and Scanes, 1985). Lipolysis in fat cells is stimulated by E in several species of birds including chickens (Langslow and Hales, 1969), pigeons (Goodridge and Ball, 1975), geese, and owls, but not in ducks (Prigge and Grande, 1971). The effect of E appears to be mediated via β-adrenergic receptors and cyclic AMP (Langslow and Hales, 1969; Campbell and Scanes, 1985). There is some evidence suggesting that different regional fat depots (e.g., abdominal vs. peripheral fat) respond differently to β-adrenergic stimulation. In view of the very high concentrations of E required ($10\mu M$ or ~1,750 ng/ml) to stimulate lipolysis *in vitro*, it is unlikely that E and NE of adrenal origin are involved in the physiological control of lipolysis. However, it is quite possible, that adrenergic innervation of fat (Freeman, 1971; Langslow, 1972) could produce regionally high local concentrations of NE.

D. Changes in Development, Maturation, and Senescence

1. Development

Experiments with precocial species indicate that the adrenal chromaffin cells are derived from caudal thoracic (region of somites 18–24) populations of neural crest cells (sympathoadrenal progenitor cells) (Coupland, 1965; LeDouarin and Teillet, 1971; Teillet and

LeDouarin, 1974). The primitive sympathetic cells (lacking chromaffin granules) and their progeny are directed to the cortical primordia through interactions with the somitic and ventral neural tubular cells (Norr, 1973; Teillet and LeDouarin, 1974; Black, 1982). In chick (Hall and Hughes, 1970) and duck (Cronshaw et al., 1989) embryos, the primitive sympathetic cells become associated with the cortical primordia at about day 10 and day 14, respectively. With migration, cell interaction, and intraadrenal clumping, the primitive sympathetic cells acquire increasing transmitter specificity and lose much of their mutability (Black, 1982). About 2 days later in each species, the early chromaffin cells form synapses with preganglionic neuronal terminals. There is some suggestion that the terminals may be different for NE and E cells (Unsicker, 1973a). Initially, as in other vertebrate species, it appears that the pheochromoblasts (initiation of chromaffin granule formation) and early pheochromocytes (definitive chromaffin cells) are of the norepinephrine (NE) type (Hall and Hughes, 1970; Black, 1982; Cronshaw et al., 1989). In chick (Hall and Hughes, 1970) and duck (Cronshaw et al., 1989) embryos, ultrastructurally definitive NE and epinephrine (E) cells are not apparent until day 12 and day 18, respectively. In general, in NE cells, the granules are compact, electron opaque, and eccentrically placed, not completely filling the vesicles, whereas in E cells, the granules are more rarefied, less electron opaque, and fill the vesicles (Lewis and Shute, 1968; Kuramoto, 1987; Holmes et al., 1991). Many of these developmental events are driven by fibroblast growth factor-2 (FGF-2), which is elaborated by the chromaffin cells. FGF-2 serves as a differentiation factor for the primitive sympathetic cells, as a target-derived neurotrophic factor for the preganglionic sympathetic terminals and as an autocrine/paracrine factor (Grothe and Meisinger, 1997). The intraglandular environment directs a pattern of neurotransmitter/neuropeptide expression in the sympathoadrenal progenitor cells that is distinct from that of cells destined to form the paravertebral sympathetic ganglia (García-Arrarás and Martinez, 1990; García-Arrarás et al., 1992; Barreto-Estrada et al., 1997). In the chick, in addition to catecholamines, a variety of peptides are elaborated by the adrenal chromaffin cells postnatally: serotonin (5-HT), galanin (GAL), cholecystokinin (CCK), methionine-enkephalin (EK) (Ohmori et al., 1997), somatostatin (SRIF) (García-Arrarás and Martinez, 1990), and neuropeptide-Y (NPY) (García-Arrarás et al., 1992) (see Figure 5).

The temporal ultrastructural identification of distinct cell types does not preclude earlier synthesis and release of E in that, in the chick embryo, E is detected in the adrenal gland as early as day 9, and in the blood, at day 10 at the latest (Epple et al., 1992). In contrast, another report indicates that E in the chick embryonic circulation is not detectable before day 13 (Dragon et al., 1996). Nevertheless, evidence suggests that PNMT activity and hence E synthesis increases during development in response to the maturation of the HPA axis (Wassermann and Bernard, 1971). The importance of the adrenal gland in providing catecholamines for embryonic homeostasis is unclear since catecholamines can come from both adrenal and neuronal sources. However, a consistent observation is that NE predominates over dopamine and E in the plasma of chicken and turkey embryos (Christensen and Edens, 1989; Epple et al., 1992; Dragon et al., 1996) and dramatically increases approaching hatch (Dragon et al., 1996). This rise is essential for mediating the hypoxic stimulation of red blood cell enzymes (Dragon et al., 1996). In addition, NE appears to be an important catecholamine that assists the embryo during the hypoxic transition from diffusive to convective respiration and that activates intestinal glycolytic enzymes for the change from lipid-based to carbohydrate-based metabolism during the perihatch period (see Christensen and Edens, 1989). The importance of NE is also suggested by its differential recruitment from the allantois in response to various stressors (Epple et al., 1992; Gill et al., 1994; Tenbusch et al., 1997).

2. Maturation and Senescence

There appears to be an extreme range of variation in the adrenal NE and E content of birds of different phylogenetic groups (Ghosh, 1977, 1980). In addition, the content varies in relation to age (Mahata and Ghosh, 1986). Furthermore, the catecholamine content response of the chromaffin tissue to endocrine alterations, as for example, in response to melatonin (Mahata et al., 1988), may vary among different species with age. However, a consistent finding that is shared with the limited number of vertebrate species studied thus far (see Carsia and Malamed, 1989) is that sympatho-adrenomedullary (-chromaffin) activity increases with age in disparate avian species (Mahata and Ghosh, 1986).

References

Accordi, F., Rossi, A., Manelli, H., and Toschi, G. (1975). PNMT activity of chick-embryo adrenals, cultivated in vitro and the action of corticoids and ACTH. Acta Embryol. Exp. **3**, 243–248.

Adjovi, Y. (1970). Morphogenèse et activité de la glands corticosurrénale de l'embryon de poulet normal et décapité. Arch. Anat. Microsc. Morphol. Exp. **59**, 185–200.

Aguilera, G., and Catt, K. J. (1979). Loci of action of regulators of aldosterone biosynthesis in isolated glomerulosa cells. Endocrinology **104**, 1046–1052.

Akana, S. F., Strack, A. M., Hanson, E. S., and Dallman, M. F. (1994). Regulation of activity in the hypothalamo–pituitary–adrenal axis is integral to a larger hypothalamic system that determines caloric flow. *Endocrinology* **135**, 1125–1134.

Ali, R. A., Qureshi, M. A., and McCorkle, F. M. (1994). Profile of chicken macrophage functions after exposure to catecholamines *in vitro*. *Immunopharmacol. Immunotoxicol.* **16**, 611–625.

Amri, H., Drieu, K., and Papadopoulos, V. (1997). *Ex vivo* regulation of adrenal cortical cell steroid and protein synthesis, in response to adrenocorticotropic homone stimulation, by the *Ginkgo biloba* extract EGb 761 and isolated ginkgolide B. *Endocrinology* **138**, 5415–5426.

Arad, Z., and Skadhauge, E. (1984). Plasma hormones (arginine vasotocin, prolactin, aldosterone, and corticosterone) in relation to hydration state, NaCl intake, and egg laying in fowls. *J. Exp. Zool.* **232**, 707–714.

Aragonés, A., Gonzalez, C. B., Spinedi, N. C., and Lantos, C. P. (1991). Regulatory effects of 5β-reduced steroids. *J. Steroid Biochem. Mol. Biol.* **39**, 253–263.

Asnani, M. V., and Ramachandran, A. V. (1993). Roles of adrenal and gonadal steroids and season in uropygial gland function in male pigeons, *Columba livia. Gen. Comp. Endocrinol.* **92**, 213–224.

Assenmacher, I. (1973). The peripheral endocrine glands. *In* "Avian Biology" (D. S. Farner, J. R. King and K. C. Parkes, eds.), Vol. III, pp. 183–286. Academic Press, New York.

Assenmacher, I., and Boissin, J. (1972). Circadian endocrine rhythms in birds. *Gen. Comp. Endocrinol.* **3**, 489–498.

Assenmacher, I., and Jallageas, M. (1980). Circadian and circannual hormonal rhythms. *In* "Avian Endocrinology" (A. Epple and M. H. Stetson, eds.), pp. 391–411. Academic Press, New York.

Astheimer, L. B., Buttemer, W. A., and Wingfield, J. C. (1994). Gender and seasonal differences in the adrencortical response to ACTH challenge in an arctic passerine, *Zonotrichia leucophrys gambelii. Gen. Comp. Endocrinol.* **94**, 33–43.

Astheimer, L. B., Buttemer, W. A., and Wingfield, J. C. (1995). Seasonal and acute changes in adrenocortical responsiveness in an arctic-breeding bird. *Horm. Behav.* **29**, 442–457.

Aupetit, B., Bastien, C., and Legrand, J. C. (1979). Cytochrome P_{450} et transformation de la 18 hydrocorticosterone en aldosterone. *Biochemie* **61**, 1085–1089.

Avrutina, A. J., Galpern, I. L., and Kisljuk, S. M. (1985). Stimulation of adrenals during the critical periods of development and production in fowls. *World's Poult. Sci. J.* **41**, 198–114.

Axelrod, J., and Reisine, T. D. (1984). Stress hormones: Their interaction and regulation. *Science* **224**, 452–459.

Ball, G. F., Faris, P. L., and Wingfield, J. C. (1989). Immunohistochemical localization of corticotropin-releasing factor in selected brain areas of the European starling (*Sturnus vulgaris*) and the song sparrow (*Melospiza melodia*). *Cell Tissue Res.* **257**, 155–161.

Bamberger, C. M., Schulte, H. M., and Chrousos, G. P. (1996). Molecular determinants of glucocorticoid receptor function and tissue sensitivity to glucocorticoids. *Endocr. Rev.* **17**, 245–261.

Bandyopadhyay, A., Deadhikari, H., Ranjit, M., and Bhattacharyya, S. P. (1990). Adrenocortical influence on histokinetics and lipid components of uropygial gland of immature chick. *Ind. J. Exp. Biol.* **28**, 915–919.

Barreto-Estrada, J. L., Medinavera, L., De Jesús-Escobar, J. M., and García-Arrarás, J. E. (1997). Development of galanin- and enkephalin-like immunoreactivities in the sympathoadrenal lineage of the avian embryo—*in vivo* and *in vitro* studies. *Dev. Neurosci.* **19**, 328–336.

Bartov, I., Jensen, L. S., and Veltmann, J. R. (1980). Effect of dietary protein and fat levels on fattening of corticosterone-injected broiler chicks. *Poult. Sci.* **59**, 1328–1338.

Bartov, I., Bornstein, S., Lev, Y., Pines, M., and Rosenberg, J. (1988). Feed restriction in broiler breeder pullets: Skip-a-day versus skip-two-days. *Poult. Sci.* **67**, 809–813.

Bateman, A., Singh, A. Kral, T., and Solomon, S. (1989). The immune–hypothalamic–pituitary–adrenal axis. *Endocr. Rev.* **10**, 92–112.

Baylé, J.-D. (1980). The adenohypophysiotropic mechanisms. *In* "Avian Endocrinology" (A. Epple and M. H. Stetson, eds.)., pp. 117–145. Academic Press, New York.

Beato, M., and Sánchez-Pacheco, A. (1996). Interaction of steroid hormone receptors with transcription initiation complex. *Endocr. Rev.* **17**, 587–609.

Beaudry, C., Bellabarba, D., and Lehoux, J.-G. (1983). Corticosterone receptors in the kidney of chick embryo. I. Nature and properties of corticosterone receptor. *Gen. Comp. Endocrinol.* **50**, 292–304.

Bellabarba, D., Beaudry, C., and Lehoux, J.-G. (1983). Corticosterone receptors in kidney of chick embryo. II. Ontogeny of corticosterone receptor and cellular development. *Gen. Comp. Endocrinol.* **50**, 305–312.

Benhamou, B., Garcia, T., Lerouge, T., Vergezac, A., Gofflo, D., Bigogne, C., Chambon, P., and Gronemyer, H. (1992). A single amino acid that determines the sensitivity of progesterone receptor to RU486. *Science* **25**, 206–209.

Bennett, T. (1974). Peripheral and autonomic nervous system. *In* "Avian Biology" (D. S. Farner and J. R. King, eds.), Vol. 4, pp. 1–77, Academic Press, New York.

Bentley, P. J. (1971). The birds. *In* "Endocrines and Osmoregulation. A Comparative Account of the Regulation of Water and Salt in Vertebrates," pp. 111–134. Springer-Verlag, New York/Berlin/Heidelberg.

Beuving, G., and Vonder, G. M. A. (1977). Daily rhythm of corticosterone in laying hens and the influence of egg laying. *J. Reprod. Fertil.* **51**, 169–173.

Beuving, G., and Vonder, G. M. A. (1978). Effect of stressing factors on corticosterone levels in the plasma of laying hens. *Gen. Comp. Endocrinol.* **35**, 153–159.

Beuving, G., and Vonder, G. M. A. (1981). The influence of ovulation and oviposition on corticosterone levels in the plasma of laying hens. *Gen Comp. Endocrinol.* **44**, 382–388.

Beuving, G., and Vonder, G. M. A. (1986). Comparison of the adrenal sensitivity to ACTH of laying hens with immobilization and plasma baseline levels of corticosterone. *Gen. Comp. Endocrinol.* **62**, 353–358.

Birrenkott, G. P., and Wiggins, M. E. (1984). Determination of dexamethasone and corticosterone half-lives in male broilers. *Poult. Sci.* **63**, 1064–1068.

Bisbis, S., Taouis, M., Derouet, M., Chevalier, B., and Simon, J. (1994). Corticosterone-induced insulin resistance is not associated with alterations of insulin receptor number and kinase activity in chicken kidney. *Gen. Comp. Endocrinol.* **96**, 370–377.

Black, I. B. (1982). Stages of neurotransmitter development in autonomic neurons. *Science* **215**, 1198–1204.

Bohus, B., Straznicky, K., and Hajós, F. (1965). The development of 3β-hydroxysteroid-dehydrogenase activity in embryonic adrenal gland of chickens. *Gen. Comp. Endocrinol.* **5**, 665. [Abstract]

Boissin, J., and Assenmacher, I. (1968). Rythmes circadiens dex taux sanguin et surrenalien de la corticosterone chez la caille. *C. R. Hebd. Seances Acad. Sci.* **267**, 2193–2196.

Boissin, J., and Assenmacher, I. (1971). Implication des mécanismes aminergiques centraux dans le determinisme du rythme circadien de la corticosteronemie chez la caille. *Cr. R. Hebd. Seances Acad. Sci.* **273**, 1744–1747.

Bornstein, S. R., Uhlmann, K., Haidan, A., Ehrhart-Bornstein, M., and Scherbaum, W. A. (1997). Evidence for a novel peripheral action of leptin as a metabolic signal to the adrenal gland. *Diabetes* **46**, 1235–1238.

Boumpas, D. T., Chrousos, G. P., Wilder, R. L., Cupps, T. R., and Balow, J. E. (1993). Glucocorticoid therapy for immune-mediated diseases: Basic and clinical correlations. *Ann. Inter. Med.* **119**, 1198–1208.

Bradley, E. L., and Holmes, W. N. (1971). The effects of hypophysectomy on adrenocortical function in the duck (*Anas platyrhynchos*). *J. Endocrinol.* **49**, 437–457.

Brake, J., Thaxton, P., and Benton, E. H. (1979). Physiological changes in caged layers during a forced molt. 3. Plasma thyroxin, plasma triiodothyronine, adrenal cholesterol, and total adrenal steroids. *Poult. Sci.* **58**, 1345–1350.

Brann, D. W., and Mahesh, V. B. (1997). Excitatory amino acids: Evidence for a role in the control of reproduction and anterior pituitary hormone secretion. *Endocr. Rev.* **18**, 678–700.

Brown, K. I. (1960). Response of turkey adrenals to ACTH and stress measured by plasma corticosterone. *Anat. Rec.* **137**, 344.

Brown, K. I., and Nestor, K. E. (1973). Some physiological responses of turkeys selected for high and low adrenal response to cold stress. *Poult. Sci.* **52**, 1948–1954.

Brown, K. I., and Nestor, K. E. (1974). Implications of selection for high and low adrenal response to stress. *Poult. Sci.* **53**, 1297–1306.

Brown, G. M., Pang, C. S., and Pang, S. F. (1994). Binding sites for 2-[^{125}I]iodomelatonin in the adrenal gland. *Biol. Signals* **3**, 91–98.

Buckland, R. B., Blagrave, K., and Lague, P. C. (1974). Competitive protein binding assay for corticoids in peripheral plasma of the immature chicken. *Poult. Sci.* **53**, 241–245.

Burrow, G. N. (1969). A steroid inhibitory effect on adrenal mitochondria. *Endocrinology* **84**, 979–985.

Butler, D. G. (1984). Endocrine control of the nasal salt glands in birds. *J. Exp. Zool.* **232**, 725–736.

Butler, D. G. (1987). Adrenalectomy fails to block salt gland secretion in Pekin ducks (*Anas platyrhynchos*) adapted to 0.9% saline drinking water. *Gen. Comp. Endocrinol.* **66**, 171–181.

Butler, D. G., and Wilson, J. X. (1985). Cardiovascular function in adrenalectomized Pekin ducks (*Anas platyrhynchos*). *Comp. Biochem. Physiol. A* **81**, 353–358.

Butler, D. G., Wilson, J. X., and Graves, L. E. (1986). α- and β-adrenergic mechanisms mediate blood pressure control by norepinephrine and angiotensin in ducks. *Gen. Comp. Endocrinol.* **61**, 323–329.

Buyse, J., Decuypere, E., Sharp, P. J., Huybrechts, L. M., Kühn, E. R., and Whitehead, C. (1987). Effect of corticosterone on circulating concentrations of corticosterone, prolactin, thyroid hormones and somatomedin C and fattening in broilers selected for high or low fat content. *J. Endocrinol.* **112**, 229–237.

Byrd, J. A., Hayes, T. K., Wright, M. S., Dean, C. E., and Hargis, B. M. (1993). Detection and partial characterization of an anti-steroidogenic peptide from the humoral immune system of the chicken. *Life Sci.* **52**, 1195–1207.

Byrd, J. A., Dean, C. E., and Hargis, B. M. (1994). The effect of the humoral immune system-derived bursal anti-steroidogenic peptide (BASP) on corticosteroid biosynthesis in avian, porcine, and canine adrenal cortical cells. *Comp. Biochem. Physiol. C* **108**, 221–227.

Cadepond, F., Jibard, N., Binart, N., Schweizer-Groyer, G., Segard-Maurel, I., and Baulieu, E. E. (1994). Selective deletions in the 90 kDa heat shock protein (hsp90) impede heterooligomeric complex formation with glucocorticosteroid receptor (GR) or hormone binding by GR. *J. Steroid Biochem. Mol. Biol.* **48**, 361–367.

Campbell, R. M., and Scanes, C. G. (1985). Adrenergic control of lipogenesis and lipolysis in the chicken *in vitro*. *Comp. Biochem. Physiol. C* **82**, 137–142.

Carmichael, S. W. (1987). Morphology and innervation of the adrenal medulla. *In* "Stimulus–Secretion Coupling in Chromaffin Cells" (K. Rosenheck and P. I. Lelkes, eds.), Vol. I, pp. 1–29. CRC Press, Boca Raton, FL.

Capuzzi, D. M., Lackman, R. D., and Reed, M. A. (1975). Species differences in the hormonal control of lipogenesis in rat and chicken hepatocytes. *Comp. Biochem. Physiol. B* **50**, 169–175.

Carsia, R. V., and Malamed, S. (1989). The adrenals. *In* "Development, Maturation, and Senescence of Neuroendocrine Systems: A Comparative Approach" (M. P. Schreibman and C. G. Scanes, eds.), pp. 353–380. Academic Press, San Diego.

Carsia, R. V., and McIlroy, P. J. (1998). Dietary protein restriction stress in the domestic turkey (*Meleagris gallopavo*) induces hypofunction and remodeling of adrenal steroidogenic tissue. *Gen. Comp. Endocrinol.* **109**, 140–153.

Carsia, R. V., and Weber, H. (1986). Genetic-dependent alterations in adrenal stress response and adrenocortical cell function of the domestic fowl (*Gallus domesticus*). *Proc. Soc. Exp. Biol. Med.* **183**, 99–105.

Carsia, R. V., and Weber, H. (1988). Protein malnutrition in the domestic fowl induces alterations in adrenocortical cell adrenocorticotropin receptors. *Endocrinology* **122**, 681–688.

Carsia, R. V., Segre, G. V., Clark, I., and Malamed, S. (1981). Corticosteroidogenesis *in vitro*: Effects of parathyroid hormone, ACTH, and calcium. *Proc. Soc. Exp. Biol. Med.* **167**, 402–406.

Carsia, R. V., Macdonald, G. J., and Malamed, S. (1983). Steroid control of steroidogenesis in isolated adrenocortical cells: Molecular and species specificity. *Steroids* **41**, 741–755.

Carsia, R. V., Scanes, C. G., and Malamed, S. (1984). Self-suppression of corticosteroidogenesis: Evidence for a role of adrenal 5α-reductase. *Endocrinology* **115**, 2464–2472.

Carsia, R. V., Scanes, C. G., and Malamed, S. (1985a). Isolated adrenocortical cells of the domestic fowl (*Gallus domesticus*): Steroidogenic and ultrastructural properties. *J. Steroid Biochem.* **22**, 273–279.

Carsia, R. V., Scanes, C. G., and Malamed, S. (1985b). Loss of sensitivity to ACTH of adrenocortical cells isolated from maturing domestic fowl. *Proc. Soc. Exp. Biol. Med.* **179**, 279–282.

Carsia, R. V., Weber, H., King, D. B., and Scanes, C. G. (1985c). Adrenocortical cell function in the hypophysectomized domestic fowl: Effects of growth hormone and 3,5,3′-triiodothyronine replacement. *Endocrinology* **117**, 928–933.

Carsia, R. V., Weber, H., and Perez, Jr., F. M. (1986). Corticotropin-releasing factor stimulates the release of adrenocorticotropin from domestic fowl pituitary cells. *Endocrinology* **118**, 143–148.

Carsia, R. V., Morin, M. E., Rosen, H. D., and Weber, H. (1987a). Ontogenic corticosteroidogenesis of the domestic fowl: Response of isolated adrenocortical cells. *Proc. Soc. Exp. Biol. Med.* **184**, 436–445.

Carsia, R. V., Scanes, C. G., and Malamed, S. (1987b). Polyhormonal regulation of avian and mammalian corticosteroidogenesis *in vitro*. *Comp. Biochem. Physiol. A* **88**, 131–140.

Carsia, R. V., Schwarz, L. A., and Weber, H. (1987c). Effect of 3, 3′-iminodiproprionitrile (IDPN) on corticosteroidogenesis of isolated adrenocortical cells. *Proc. Soc. Exp. Biol. Med.* **184**, 461–467.

Carsia, R. V., Reisch, N. M., Fennell, M. J., and Weber, H. (1987d). Adrenocortical function of the domestic fowl: Effects of orchiectomy and androgen replacement. *Proc. Soc. Exp. Biol. Med.* **185**, 223–232.

Carsia, R. V., Weber, H., and Lauterio, T. J. (1988a). Protein malnutrition in the domestic fowl induces alterations in adrenocortical function. *Endocrinology* **122**, 673–680.

Carsia, R. V., Weber, H., and Satterlee, D. G. (1988b). Steroidogenic properties of isolated adrenocortical cells from Japanese quail selected for high serum corticosterone response to immobilization. *Dom. Anim. Endocrinol.* **5**, 231–240.

Carsia, R. V., McIlroy, P. J., Kowalski, K. I., and Tilly, J. L. (1993). Isolation of turkey adrenocortical cell angiotensin II (AII) receptor partial cDNA: Evidence for a single-copy gene expressed predominantly in the adrenal gland. *Biochem. Biophys. Res. Commun.* **191,** 1073–1080.

Carsia, R. V., Lamm, E. T., Marsh, J. A., Scanes, C. G., and King, D. B. (1997). The thyroid hormone, 3,5,3′-triiodothyronine, is a negative modulator of domestic fowl (*Gallus gallus domesticus*) adrenal steroidogenic function. *Gen. Comp. Endocrinol.* **170,** 251–261.

Cassone, V. M., and Menaker, M. (1984). Is the avian circadian system a neuroendocrine loop? *J. Exp. Zool.* **232,** 539–549.

Cassone, V. M., Lane, R. F., and Menaker, M. (1983). Daily rhythms of serotonin metabolism in the medial hypothalamus of the chicken: Effects of pinealectomy and exogenous melatonin. *Brain Res.* **289,** 129–134.

Cassone, V. M., Lane, R. F., and Menaker, M. (1986). Melatonin-induced increases in serotonin concentrations in specific regions of the chicken brain. *Neuroendocrinology* **42,** 38–43.

Castro, M. G., Estivariz, F. E., and Iturriza, M. J. (1986). The regulation of the corticomelanotropic cell activity in Aves-II. effect of various peptides on the release of ACTH from dispersed, perfused duck pituitary cells. *Comp. Biochem. Physiol. A* **83,** 71–75.

Chabot, C. C., and Menaker, M. (1992). Effect of physiological cycles of infused melatonin on circadian rhythmicity in pigeons. *J. Comp. Physiol. A* **170,** 615–622.

Challet, E., Le Maho, Y., Robin, J. P., Malan, A., and Cherel, Y. (1995). Involvement of corticosterone in the fasting-induced rise in protein utilization and locomotor activity. *Pharm. Biochem. Behav.* **50,** 405–412.

Chan, M. Y., and Holmes, W. N. (1971). Studies on a 'renin-angiotensin' system in the normal and hypophysectomized pigeon (*Columba livia*). *Gen. Comp. Endocrinol.* **16,** 304–311.

Chan, M. Y., Bradley, E. L., and Holmes, W. N. (1972). The effects of hypophysectomy on the metabolism of adrenal steroids in the pigeon. (*Columba livia*). *J. Endocrinol.* **52,** 435–450.

Charest-Boule, L., Mehdi, A. Z., Sandor, T. (1978). Corticosteroid receptors in the avian kidney. *Steroids* **32,** 109–126.

Charest-Boule, L., Mehdi, A. Z., Sandor, T. (1980). Corticosterone receptors in the avian kidney. *J. Steroid Biochem.* **13,** 897–905.

Cherel, Y., Robin, J. P., Walch, O., Karmann, H., Netchitailo, P., and Le Maho, Y. (1988a). Fasting in king penguin. I. Hormonal and metabolic changes during breeding. *Am. J. Physiol.* **254,** R170–R177.

Cherel, Y., Leloup, J., and Le Maho, Y. (1988b). Fasting in king penguin. II. Hormonal and metabolic changes during molt. *Am. J. Physiol.* **254,** R178–R184.

Cherradi, N., Defaye, G., and Chambaz, E. M. (1994). Characterization of the 3β-hydroxysteroid dehydrogenase activity associated with bovine adrenocortical mitochondria. *Endocrinology* **134,** 1358–1364.

Chester Jones, I. (1957). "The Adrenal Cortex." Cambridge Univ. Press, London/New York.

Chester Jones, I., and Phillips, J. G. (1986). The adrenal and interrenal glands. *In* "Vertebrate Endocrinology: Fundamental and Biomedical Implications" (P. K. T. Pang and M. P. Schreibman, eds.), Vol. 1, pp. 319–350. Academic Press, New York.

Cheung, A., Harvey, S., Hall, T. R., Lam, S. K., and Spencer, G. S. G. (1988a). Effect of passive immunization with anti-somatostatin serum on plasma corticosterone concentrations in young domestic cockerels. *J. Endocrinol.* **116,** 179–183.

Cheung, A., Hall, T. R., and Harvey, S. (1988b). Stimulation of corticosterone release in the fowl by recombinant DNA-derived chicken growth hormone. *Gen. Comp. Endocrinol.* **69,** 128–132.

Christensen, V. L., and Edens, F. W. (1989). Blood plasma catecholamine concentration of poult embryos during the transition from diffusive to convective respiration. *Comp. Biochem. Physiol. B* **92,** 549–553.

Chrousos, G. P. (1998). Ultradian, circadian, and stress-related hypothalamic–pituitary–adrenal axis activity—a dynamic digital-to-analogue modulation. *Endocrinology* **139,** 437–440.

Clark, B. J., and Stocco, D. M. (1996). StAR—A tissue specific acute mediator of steroidogenesis. *Trends Endocrinol. Metab.* **7,** 227–233.

Clauss, W., Durr, J. E., Guth, D., and Skadhauge, E. (1987). Effects of adrenal steroids on Na transport in the lower intestine (coprodeum) of the hen. *J. Membrane Biol.* **96,** 141–152.

Collie, M. A., Holmes, W. N., and Cronshaw, J. (1992). A comparison of the responses of dispersed steroidogenic cells derived from embryonic adrenal tissue from the domestic chicken (*Gallus domesticus*), the domestic Pekin duck and the wild mallard duck (*Anas platyrhynchos*), and the domestic muscovy duck (*Cairina moschata*). *Gen. Comp. Endocrinol.* **88,** 375–387.

Compton, M. M., Gibbs, P. S., and Swicegood, L. R. (1990). Glucocorticoid-mediated activation of DNA degradation in avian lymphocytes. *Gen. Comp. Endocrinol.* **80,** 68–70.

Compton, M. M., Gibbs, P. S., and Johnson, L. R. (1990). Glucocorticoid activation of deoxyribonucleic acid degradation in bursal lymphocytes. *Poult. Sci.* **69,** 1292–1298.

Compton, M. M., Johnson, L. R., and Gibbs, P. S. (1991). Activation of thymocyte deoxyribonucleic acid degradation by endogenous glucocorticoids. *Poult. Sci.* **70,** 521–529.

Connolly, P. B., and Callard, I. P. (1987). Steroids modulate the release of luteinizing hormone from quail pituitary cells. *Gen. Comp. Endocrinol.* **68,** 466–472.

Covasa, M., and Forbes, J. M. (1995). Selection of foods by broiler chickens following corticosterone administration. *Br. Poult. Sci.* **36,** 489–501.

Coupland, R. E. (1965). "The Natural History of the Chromaffin Cell." Longmans, London.

Cramb, G., Langslow, D. R., and Phillips, J. H. (1982). Hormonal effects on cyclic nucleotides and carbohydrate and lipid metabolism in isolated chicken hepatocytes. *Gen. Comp. Endocrinol.* **46,** 310–321.

Crivello, J. F., and Jefcoate, C. R. (1980). Intracellular movement of cholesterol in rat adrenal cells. *J. Biol. Chem.* **255,** 8144–8151.

Crocker, A. D., and Holmes, W. N. (1971). Intestinal absorption in duckling (*Anas platyrhynchos*) maintained on freshwater and hypertonic saline. *Comp. Biochem. Physiol. A* **40,** 203–211.

Crocker, A. D., and Holmes, W. N. (1976). Factors affecting intestinal absorption in duckling (*Anas platyrhynchos*). *J. Endocrinol.* **71,** 88P–89P.

Cronshaw, J., Holmes, W. N., and West, R. D. (1984). The effects of colchicine, vinblastine and cytochalasins on the corticotropic responsiveness and ultrastructure of inner zone adrenocortical tissue in the Pekin duck. *Cell Tissue Res.* **236,** 333–338.

Cronshaw, J., Ely, J. A., and Holmes, W. N. (1985). Functional differences between two structurally distinct types of steroidogenic cell in the avian adrenal gland. *Cell Tissue Res.* **240,** 561–567.

Cronshaw, J., Holmes, W. N., Ely, J. A., and Redondo, J. L. (1989). Pre-natal development of the adrenal gland in the mallard duck (*Anas platyrhynchos*). *Cell Tissue Res.* **258,** 593–601.

Cronshaw, J., Reese, B. K., Collie, M. A., and Holmes, W. N. (1992a). Cytoskeletal changes accompanying ACTH-induced steroidogenesis in cultured embryonic adrenal gland cells from the Pekin duck. *Cell Tissue Res.* **268,** 157–165.

Cronshaw, J., Collie, M. A., and Holmes, W. N. (1992b). Functional and morphological changes associated with the ageing of primary cultures of embryonic adrenal gland cells derived from the Pekin duck. *Cell Tissue Res.* **269,** 535–545.

Cunningham, L. A., and Holzwarth, M. A. (1988). Vasoactive intestinal peptide stimulates adrenal aldosterone and corticosterone secretion. *Endocrinology* **122,** 2090–2097.

Dallman, M. F., Akana, S. F., Cascio, C. S., Darlington, D. N., Jacobson, L., and Levin, N. (1987a). Regulation of ACTH secretion: Variations on a theme of B. *Rec. Prog. Horm. Res.* **43,** 113–167.

Dallman, M. F., Akana, S. F., Jacobson, L., Levin, N., Cascio, C. S., and Shinsako, J. (1987b). Characterization of corticosterone feedback regulation of ACTH secretion. *Ann. N. Y. Acad. Sci.* **512,** 402–414.

Dallman, M. F., Akana, S. F., Bradbury, M. J., Strack, A. M., Hanson, E. S., and Scribner, K. A. (1994). Regulation of the hypothalamo–pituitary–adrenal axis during stress: Feedback, facilitation and feeding. *Semin. Neurosci.* **6,** 205–213.

Daniel, J. Y., and Assenmacher, I. (1971). Early appearance of metabolites after single i. v. injection of 3H-corticosterone in rabbit and duck. *Steroids* **18,** 325–340.

Daniel, J. Y., Assenmacher, I., and Rochefort, H. (1977). Evidence of androgen and estrogen receptors in the preen gland of male ducks. *Steroids* **30,** 703–709.

Dantzer, R., and Mormede, P. (1985). Stress in domestic animals: A psychoneuroendocrine approach. In "Animal Stress" (G. P. Moberg, ed.), pp. 81–95. American Physiological Society, Bethesda.

Davis, G. S., and Siopes, T. D. (1987). Plasma corticosterone response of turkeys to adrenocorticotropic hormone: Age, dose and route of administration effects. *Poult. Sci.* **66,** 1727–1732.

Davison, T. F. (1975). The effects of multiple samplings by cardiac puncture and diurnal rhythm on plasma glucose and hepatic glycogen of the immature chicken. *Comp. Biochem. Physiol. A* **50,** 569–573.

Davison, T. F., and Flack, I. H. (1981). Changes in the peripheral blood leucocyte populations following an injection of corticotrophin in the mature chicken. *Res. Vet. Sci.* **30,** 79–82.

Davison, T. F., Scanes, C. G., Flack, I. H., and Harvey, S. (1979). Effect of daily injections of ACTH on growth and on the adrenal and lymphoid tissues of two strains of immature fowls. *Br. Poult. Sci.* **20,** 575–585.

Davison, T. F., Rea, J., and Rowell, J. G. (1983a). Effects of dietary corticosterone on the growth and metabolism of immature *Gallus domesticus*. *Gen. Comp. Endocrinol.* **50,** 463–468.

Davison, T. F., Rowell, J. G., and Rea, J. (1983b). Effects of dietary corticosterone on peripheral blood lymphocytes and granulocyte populations in immature domestic fowl. *Res. Vet. Sci.* **34,** 236–239.

Davison, T. F., Freeman, B. M., and Rea, J. (1985). Effects of continuous treatment with synthetic ACTH$^{1-24}$ or corticosterone on immature *Gallus domesticus*. *Gen. Comp. Endocrinol.* **59,** 416–423.

Decuypere, E., Rombauts, L., Vanmontfort, D., and Verhoeven, G. (1997). Inhibin from embryo to adult hen. In "Perspectives in Avian Endocrinology" (S. Harvey and R. J. Etches, eds.), pp. 71–80. Journal of Endocrinology Ltd., Bristol.

De Kloet, E. R., Vreugdenhil, E., Oitzl, M. S., and Joëls, M. (1997). Glucocorticoid feedback resistance. *Trends Endocrinol. Metab.* **8,** 26–33.

De La Cruz, L. F., Mataix, F. J., and Illera, M. (1981). Effects of glucocorticoids on protein metabolism in laying quails (*Coturnix coturnix japonica*). *Comp. Biochem. Physiol. A* **70,** 649–652.

De La Cruz, L. F., Illera, M., and Mataix, F. J. (1987). Developmental changes induced by glucocorticoids treatment in breeder quail (*Coturnix coturnix japonica*). *Horm. Metab. Res.* **19,** 101–104.

Denno, K. M., McCorkle, F. M., and Taylor, R. L., Jr. (1994). Catecholamines modulate immunoglobulin M and immunoglobulin G plaque-forming cells. *Poult. Sci.* **73,** 1858–1866.

DeRoos, R. (1961). The corticoids of the avian adrenal gland. *Gen. Comp. Endocrinol.* **1,** 494–512.

DeRoos, R. (1969). Effect of mammalian corticotropin and progesterone on corticoid production by chicken adrenal tissue *in vitro*. *Gen. Comp. Endocrinol.* **13,** 455–459.

DiBattista, J. A., Mehdi, A. Z., and Sandor, T. (1985). A profile of the intestinal mucosal corticosteroid receptors in the domestic duck. *Gen. Comp. Endocrinol.* **59,** 31–49.

DiBattista, J. A., Mehdi, A. Z., and Sandor, T. (1989). Steroid C-20 oxidoreductase activity of duck intestinal mucosa: The interrelations of the enzymatic activity with steroid binding. *Gen. Comp. Endocrinol.* **74,** 136–147.

Doi, O., Iwasawa, A., Nakamura, T., and Tanabe, Y. (1995). Effects of different photoperiods on plasma melatonin rhythm of the chicken. *Anim. Sci. Technol. Jpn.* **66,** 16–26.

Domm, L. V., and Erickson, G. C. (1972). 3β-hydroxysteroid dehydrogenase activity in the adrenals of normal and hypophysectomized chick embryos. *Proc. Soc. Exp. Biol. Med.* **140,** 1215–1220.

Donaldson, E. M., and Holmes, W. N. (1965). Corticosteroidogenesis in the fresh-water and saline-maintained duck (*Anas platyrhynchos*). *J. Endocrinol.* **32,** 329–336.

Donker, R. A., and Beuving, G. (1989). Effect of corticosterone infusion on plasma corticosterone concentration, antibody production, circulating leukocytes and growth in chicken lines selected for humoral immune responsiveness. *Br. Poult. Sci.* **30,** 361–369.

Dragon, S., Glombitza, S., Gotz, R., and Baumann, R. (1996). Norepinephrine-mediated hypoxic stimulation of embryonic red cell carbonic anhydrase and 2,3-DPG synthesis. *Am. J. Physiol.* **40,** R982–R989.

Dransfield, J. W. (1945). The lymphatic system of the domestic fowl. *Vet. J.* **101,** 171–179.

Dufty, A. M., and Belthoff, J. R. (1997). Corticosterone and the stress response in young western screech-owls—Effects of captivity, gender, and activity period. *Physiol. Zool.* **70,** 143–149.

Edens, F. W. (1983). Effect of environmental stressors on male reproduction. *Poult. Sci.* **62,** 1676–1689.

Edens, F. W., and Siegel, H. S. (1975). Adrenal responses in high and low ACTH response lines of chickens during acute heat stress. *Gen. Comp. Endocrinol.* **25,** 64–73.

Edwards, A. V., and Jones, C. T. (1993). Autonomic control of adrenal function. *J. Anat.* **183** (2), 291–307.

Ehrhart-Bornstein, M., Bornstein, S. R., and Scherbaum, W. A. (1996). Sympathoadrenal system and immune system in the regulation of adrenocortical function. *Eur. J. Endocrinol.* **135,** 19–26.

Elbrønd, V. S., and Skadhauge, E. (1992). Na-transport during long-term incubation of the hen lower intestine: No aldosterone effect. *Comp. Biochem. Physiol. A* **101,** 203–208.

El-Far, A. A., Mashaly, M. M., and Kamar, G. A. (1994). Bursectomy and *in vitro* response of adrenal gland to adrenocorticotropic hormone and testis to human chorionic gonadotropin in immature male chickens. *Poult. Sci.* **73,** 113–117.

El-Halawani, M. E., Waibel, P. E., Appel, J. R., and Good, A. L. (1974). Effects of temperature stress on catecholamines and corticosterone of male turkeys. *Am. J. Physiol.* **224,** 384–388.

Epple, A., Gill, T. S., and Nibbio, B. (1992). The avian allantois: A depot for stress-released catecholamines. *Gen. Comp. Endocrinol.* **85,** 462–476.

Etches, R. J. (1979). Plasma concentrations of progesterone and corticosterone during the ovulation cycle of the hen (*Gallus domesticus*). *Poult. Sci.* **58,** 211–216.

Etches, R. J., and Cunningham, F. J. (1976). The effect of pregnenolone, progesterone, deoxycorticosterone or corticosterone on the time of ovulation and oviposition in the hen. *Br. Poult. Sci.* **17,** 637–642.

Fassler, R., Schwarz, S., Dietrich, H., and Wick, G. (1986). Sex steroid and glucocorticoid receptors in the bursa of Fabricius of obese strain chickens spontaneously developing autoimmune thyroiditis. *J. Steroid Biochem.* **24,** 405–411.

Feuilloley, M., and Vaudry, H. (1996). Role of the cytoskeleton in adrenocortical cells. *Endocr. Rev.* **17,** 269–288.

Fowler, K. C., Pesti, G. M., and Howarth, B. (1983). The determination of plasma corticosterone of chickens by high pressure liquid chromatography. *Poult. Sci.* **62,** 1075–1079.

Freedman, S. L. (1968). The innervation of the suprarenal gland of the fowl (*Gallus domesticus*). *Acta Anat.* **69,** 18–25.

Freeman, B. M. (1971). Non-shivering thermogenesis in birds. *In* "Non-shivering Thermogenesis" (L. Jansky, ed.), pp. 83–96. Academia, Prague.

Freeman, B. M. (1982). Stress non-responsiveness in the newly-hatched fowl. *Comp. Biochem. Physiol. A* **72,** 251–253.

Freeman, B. M. (1983). Adrenal glands. *In* "Physiology and Biochemistry of the Domestic Fowl" (B. M. Freeman, ed.), Vol. 4, Chapt. 11, pp. 191–209. Academic Press, London.

Freeman, B. M., and Flack, I. H. (1980). Effects of handling on plasma corticosterone concentrations in the immature domestic fowl. *Comp. Biochem. Physiol. A* **66,** 77–81.

Freeman, B. M., and Flack, I. H. (1981). The sensitivity of the newly hatched fowl to corticotrophin. *Comp. Biochem. Physiol. A* **70,** 257–259.

Freeman, B. M., and Manning, A. C. C. (1977). Responses of the immature fowl to a single injection of adrenocorticotrophic hormone. *Br. Poult. Sci.* **18,** 561–567.

Freeman, B. M., and Manning, A. C. C. (1978). Short-term stressor effects of reserpine. *Br. Poult. Sci.* **19,** 623–630.

Freeman, B. M., and Manning, A. C. C. (1979). The effects of repeated injections of adrenaline on the response of the fowl to further alarm stimulation. *Res. Vet. Sci.* **27,** 76–81.

Freeman, B. M., and Manning, A. C. C. (1984). Re-establishment of the stress response in *Gallus domesticus* after hatching. *Comp. Biochem. Physiol. A* **78,** 267–270.

Freeman, D. A. (1989). Plasma membrane cholesterol: Removal and insertion into the membrane and utilization as substrate for steroidogenesis. *Endocrinology* **124,** 2527–2534.

Fujita, M., Nishibori, M., and Yamamoto, S. (1992). Changes in plasma catecholamine, free fatty acid, and glucose concentrations, and plasma monoamine oxidase activity before and after feeding in laying hens. *Poult. Sci.* **71,** 1067–1072.

Funder, J. W. (1995). Apparent mineralocorticoid excess, 11β hydroxysteorid dehydrogenase and aldosterone action. *Trends Endocrinol. Metab.* **6,** 248–251.

García-Arrarás, J. E., and Martinez, R. (1990). Developmental expression of serotonin-like immunoreactivity in the sympathoadrenal system of the chicken. *Cell Tiss. Res.* **262.** 363–372.

García-Arrarás, J. E., Lugo-Chinchilla, A. M., and Chévere-Colón, I. (1992). The expression of neuropeptide Y immunoreactivity in the avian sympathoadrenal system conforms with two models of co-expression development for neurons and chromaffin cells. *Development* **115,** 617–627.

Gasc, J. M., and Martin, B. (1978). Plasma corticosterone binding capacity in the partially decapitated chick embryo. *Gen. Comp. Endocrinol.* **35,** 274–279.

Gasman, S., Vaudry, H., Cartier, F., Tramu, G., Fournier, A., Conlon, J. M., and Delarue, C. (1996). Localization, identification, and action of galanin in the frog adrenal gland. *Endocrinology* **137,** 5311–5318.

Gendreau, P., Lehoux, J.-G., Belisle, S., and Bellabarba, D. (1987). Glucocorticoid receptors in chick embryos: Properties and ontogeny of the nuclear renal receptor. *Gen. Comp. Endocrinol.* **67,** 58–66.

Geris, K. L., Kotanen, S. P., Berghman, L. R., Kühn, E. R., and Darras, V. M. (1996). Evidence of a thyrotropin-releasing activity of ovine corticotropin-releasing factor in the domestic fowl (*Gallus domesticus*). *Gen. Comp. Endocrinol.* **104,** 139–146.

Ghosh, A. (1977). Cytophysiology of the avian medulla. *Int. Rev. Cytol.* **49,** 253–284.

Ghosh, A. (1980). Avian adrenal medulla: Structure and function. *In* "Avian Endocrinology" (A. Epple and M. H. Stetson, eds.), pp. 301–318. Academic Press, New York.

Ghosh, S., Sengupta, S., Dasadhikari, S., and Ghosh, A. (1996). Relation between the adrenals and thyroid and their hormones and the testicular cycle of a subtropical avian species. *Biol. Rhythm Res.* **27,** 216–226.

Gill, T. S., Porta, S., Nibbio, B., and Epple, A. (1994). Sulfate conjugates of catecholamines in the allantoic fluid of the chicken embryo. *Gen. Comp. Endocrinol.* **96,** 255–258.

Giorgi, E. P., and Stein, W. D. (1981). The transport of steroids into animal cells in culture. *Endocrinology* **108,** 688–697.

Girouard, R. J., and Hall, B. K. (1973). Pituitary-adrenal interaction and growth of the embryonic adrenal gland. *J. Exp. Zool.* **183,** 323–332.

Gonzalez, C. B., Cozza, E. N., De Bedners, M. E. O., Lantos, C. P., and Aragonés, A. (1983). Progesterone and its reductive metabolism in steroidogenic tissues of the developing hen embryo. *Gen. Comp. Endocrinol.* **51,** 384–393.

Goodridge, A. G., and Ball, E. G. (1975). Studies on the metabolism of adipose tissue. XVIII. *In vitro* effects of insulin, epinephrine and glucagon on lipolysis and glycolysis in pigeon adipose tissue. *Comp. Biochem. Physiol. B* **46,** 367–373.

Gorbman, A., Dickhoff, W. W., Vigna, W. R., Clark, N. B., and Ralph, C. L. (1983). "Comparative Endocrinology," pp. 415–450. Wiley, New York.

Gorsline, J., and Holmes, W. N. (1982). Variations with age in the adrenocortical responses of mallard ducks (*Anas platyrhynchos*) consuming petroleum-contaminated food. *Bull. Environ. Contam. Toxicol.* **29,** 146–152.

Gould, N. R., and Siegel, H. S. (1974). Age variation in corticosteroid binding by serum proteins of growing chickens. *Gen. Comp. Endocrinol.* **24,** 177–182.

Gould, N. R., and Siegel, H. S. (1978). Effect of age and sex on the association constant and binding capacity of chicken serum for corticosteroid. *Poult. Sci.* **57,** 778–784.

Gould, N. R., and Siegel, H. S. (1984). Effect of adrenocorticotropin hormone injections on glucocorticoid receptors in chicken thymocytes. *Poult. Sci.* **63,** 373–377.

Gray, D. A. (1993). Plasma atrial natriuretic factor concentrations and renal actions in the domestic fowl. *J. Comp. Physiol. B* **163,** 519–523.

Gray, D. A., and Simon, E. (1985a). Control of plasma angiotensin II in a bird with salt glands (*Anas platyrhynchos*). *Gen. Comp. Endocrinol.* **60,** 1–13.

Gray, D. A., and Simon, E. (1985b). Extracellular fluid volume and sodium handling in salt water adapted Pekin ducks. *Pflügers Arch.* **403** (Suppl.), R19.

Gray, D. A., Gerstberger, R., and Simon, E. (1989). Role of angiotensin II in aldosterone regulation in the Pekin duck. *J. Endocrinol.* **123,** 445–452.

Gray, D. A., Schütz, H., and Gerstberger, R. (1991a). Plasma atrial natriuretic factor responses to blood volume changes in the Pekin duck. *Endocrinology* **128,** 1655–1660.

Gray, D. A., Schütz, H., and Gerstberger, R. (1991b). Interaction of atrial natriuretic factor and osmoregulatory hormones in the Pekin duck. *Gen. Comp. Endocrinol.* **81,** 246–255.

Gray, J. M., Yarian, D., and Ramenofsky, M. (1990). Corticosterone, foraging behavior, and metabolism in dark-eyed juncos, *Junco hyemalis*. *Gen. Comp. Endocrinol.* **79**, 375–384.

Gregg, C. M., and Wideman, Jr., R. F. (1986). Effects of atriopeptin and chicken heart extract in *Gallus domesticus*. *Am. J. Physiol.* **251**, R543–R551.

Gross, W. B. (1992). Effect of short-term exposure of chickens to corticosterone on resistance to challenge exposure with *Escherichia coli* and antibody response to sheep erythrocytes. *Am. J. Vet. Res.* **53**, 291–293.

Gross, W. B., and Colmano, G. (1971). Effect of infectious agents on chickens selected for plasma corticosteroid response to social stress. *Poult. Sci.* **50**, 1213–1217.

Gross, W. B., and Siegel, P. B. (1985). Selective breeding of chickens for corticosterone response to social stress. *Poult. Sci.* **64**, 2230–2233.

Gross, W. B., Siegel, P. B., and DuBose, R. T. (1980). Some effects of feeding corticosterone to chickens. *Poult. Sci.* **59**, 516–522.

Grothe, C., and Meisinger, C. (1997). The multifunctionality of FGF-2 in the adrenal medulla. *Anat. Embryol.* **195**, 103–111.

Groyer, A., Le Bouc, Y., Joab, I., Radanyi, C., Renoir, J. M., Robel, P., and Baulieu, E. E. (1985). Chick oviduct glucocorticosteroid receptor. *Eur. J. Biochem.* **149**, 445–451.

Grubb, B. R., and Bentley, P. J. (1987). Aldosterone-induced, amiloride-inhibitable short-circuit current in the avian ileum. *Am. J. Physiol.* **253**, G211–G216.

Grubb, B. R., and Bentley, P. J. (1992). Effects of corticosteroids on short-circuit current across the cecum of the domestic fowl, *Gallus domesticus*. *J. Comp. Physiol. B* **162**, 690–695.

Guellati, M., Ramade, F., Le Nguyen, D., Ibos, F., and Baylé, J.-D. (1991). Effects of early embryonic bursectomy and opotherapic substitution on the functional development of the adrenocorticotropic axis. *J. Dev. Physiol.* **15**, 357–363.

Haidan, A., Bornstein, S. R., Glasow, A., Uhlmann, K., Lübke, C., and Ehrhart-Bornstein, M. (1998). Basal steroidogenic activity of adrenocortical cells is increased 10-fold by coculture with chromaffin cells. *Endocrinology* **139**, 772–780.

Hall, B. K. (1970). Response of the host embryonic chicks to grafts of additional adrenal glands. I. Response of host adrenal glands, gonads and kidneys. *Can. J. Zool.* **49**, 381–384.

Hall, B. K., and Hughes, H. P. (1970). Response of host embryonic chicks to grafts of additional adrenal glands. II. Ultrastructure of normal and host adrenal glands. *Cell Tissue Res.* **108**, 1–16.

Hall, T. R., Harvey, S., and Chadwick, A. (1986). Control of prolactin secretion in birds: A review. *Gen. Comp. Endocrinol.* **62**, 171–184.

Hammond, G. L. (1995). Potential functions of plasma steroid-binding proteins. *Trends Endocrinol. Metab.* **6**, 298–304.

Hansson, V., Ritzen, E. M., Weddington, S. C., McLean, W. S., Tindall, D. J., Nayfeh, S. N., and French, F. S. (1974). Preliminary characterization of a binding protein for androgen in rabbit serum: Comparison with the testosterone-binding globulin (TBG) in human serum. *Endocrinology* **95**, 690–700.

Hartman, F. A., and Brownell, K. A. (1949). "The Adrenal Gland." Lea and Febiger, Philadelphia.

Harvey, S., and Hall, T. R. (1990). Hormones and stress in birds: Activation of the hypothalamo–pituitary–adrenal axis. *In* "Progress in Comparative Endocrinology" (A. Epple, C. G. Scanes, and M. H. Stetson, eds.), pp. 453–460. Wiley-Liss, New York.

Harvey, S., and Phillips, J. G. (1982). Endocrinology of salt gland function. *Comp. Biochem. Physiol. A* **71**, 537–546.

Harvey, S., Scanes, C. G., and Howe, T. (1977). Growth hormone effects on *in vitro* metabolism of avian adipose and liver tissue. *Gen. Comp. Endocrinol.* **33**, 322–328.

Harvey, S., Scanes, C. G., Chadwick, A., and Bolton, N. J. (1979). Growth hormone and prolactin secretion in growing domestic fowl: Influence of sex and breed. *Br. Poult. Sci.* **20**, 9–17.

Harvey, S., Phillips, J. G., Rees, A., and Hall, T. R. (1984a). Stress and adrenal function. *J. Exp. Zool.* **232**, 633–645.

Harvey, S., Hall, T. R., and Chadwick, A. (1984b). Growth hormone and prolactin secretion in water-deprived chickens. *Gen. Comp. Endocrinol.* **54**, 46–50.

Hau, M., and Gwinner, E. (1996). Food as a circadian zeitgeber for house sparrows: The effect of different food access durations. *J. Biol. Rhythms* **11**, 196–207.

Hau, M., and Gwinner, E. (1997). Adjustment of house sparrow circadian rhythms to a simultaneously applied light and food zeitgeber. *Physiol. Behav.* **62**, 973–981.

Hayashi, H., Imai, K., and Imai, K. (1991). Characterization of chicken ACTH and α-MSH: The primary sequence of chicken ACTH is more similar to *Xenopus* ACTH than to other avian ACTH. *Gen. Comp. Endocrinol.* **82**, 434–443.

Hayashi, K., Nagai, Y., Ohtsuka, A., and Tomita, Y. (1994). Effects of dietary corticosterone and trilostane on growth and skeletal muscle protein turnover in broiler cockerels. *Br. Poult. Sci.* **35**, 789–798.

Heath, J. A., and Dufty, A. M. (1998). Body condition and the adrenal stress response in captive American kestrel juveniles. *Physiol. Zool.* **71**, 67–73.

Hegner, R. E., and Wingfield, J. C. (1986a). Behavioral and endocrine correlates of multiple brooding in the semicolonial house sparrow *Passer domesticus*. I. Males. *Horm. Behav.* **20**, 294–312.

Hegner, R. E., and Wingfield, J. C. (1986b). Behavioral and endocrine correlates of multiple brooding in the semicolonial house sparrow *Passer domesticus*. II. Females. *Horm. Behav.* **20**, 313–326.

Heigl, S., and Gwinner, E. (1995). Synchronization of circadian rhythms of house sparrows by oral melatonin: Effects of changing period. *J. Biol. Rhythms* **10**, 225–233.

Heiman, M. L., Ahima, R. S., Craft, L. S., Schoner, B., Stephens, T. W., and Flier, J. S. (1997). Leptin inhibition of the hypothalamic-pituitary-adrenal axis in response to stress. *Endocrinology* **138**, 3859–3863.

Hendricks, III, G. L., Siegel, H. S., and Mashaly, M. M. (1991). Ovine corticotropin-releasing factor increases endocrine and immunological activity of avian leukocytes *in vitro*. *Proc. Soc. Exp. Biol. Med.* **196**, 390–395.

Hendricks, III, G. L., Mashaly, M. M., and Siegel, H. S. (1995a). Validation of an assay to measure adrenocorticotropin in plasma and from chicken leukocytes. *Poult. Sci.* **74**, 337–342.

Hendricks, III, G. L., Mashaly, M. M., and Siegel, H. S. (1995b). Effect of corticosterone *in vivo* and *in vitro* on adrenocorticotropic hormone production by corticotropin releasing factor-stimulated leukocytes. *Proc. Soc. Exp. Biol. Med.* **209**, 382–386.

Herold, M., Brezinschek, H. P., Gruschwitz, M., Dietrich, H., and Wick, G. (1992). Investigation of ACTH responses of chickens with autoimmune disease. *Gen. Comp. Endocrinol.* **88**, 188–198.

Hertelendy, F., Todd, H., and Molnar, M. (1992). Influence of chicken and human lipoproteins on steroidogenesis in granulosa cells of the domestic fowl (*Gallus domesticus*). *Gen. Comp. Endocrinol.* **85**, 335–340.

Heryanto, B., Yoshimura, Y., Tamura, T., and Okamato, T. (1997). Involvement of apoptosis and lysosomal hydrolase activity in the oviductal regression during induced molting in chickens: A cytochemical study for end labeling of fragmented DNA and acid phosphatase. *Poult. Sci.* **76**, 67–72.

Hester, P. Y., Muir, W. M., Craig, J. V., and Albright, J. L. (1996). Group selection for adaptation to multiple-hen cages: Hematology and adrenal function. *Poult. Sci.* **75**, 1295–1307.

Hinson, J. P. (1990). Paracrine control of adrenocortical function: A new role for the medulla? *J. Endocrinol.* **124,** 7–9.

Hissa, R., George, J. C., John, T. M., and Etches, R. J. (1982). Propranolol induced changes in plasma catecholamine, corticosterone, T_4, T_3, and prolactin levels in the pigeon. *Horm. Metab. Res.* **14,** 606–610.

Holmes, W. N. (1978). Control of adrenocortical function in birds. *Pavo* **16,** 105–121.

Holmes, W. N., and Cronshaw, J. (1980). Adrenal cortex: Structure and function. *In* "Avian Endocrinology" (A. Epple and M. Stetson, eds.), pp. 271–299. Academic Press, New York.

Holmes, W. N., and Cronshaw, J. (1984). The adrenal gland: Some evidence for the structural and functional zonation of the steroidogenic tissues. *J. Exp. Zool.* **232,** 627–631.

Holmes, W. N., and Cronshaw, J. (1993). Some actions of angiotensin II in gallinaceous and anseriform birds. *In* "Avian Endocrinology" (P. J. Sharp, ed.), pp. 201–216. Journal of Endocrinology Ltd., Bristol.

Holmes, W. N., and Kelly, M. E. (1976). The turnover and distribution of labelled corticosterone during post-natal development of the duckling (*Anas platyrhynchos*). *Pflügers Arch.* **365,** 145–150.

Holmes, W. N., and Phillips, J. G. (1976). The adrenal cortex of birds. *In* "General, Comparative and Clinical Endocrinology of the Adrenal Cortex" (I. Chester Jones and I. W. Henderson, eds.), Vol. 1, pp. 293–413. Academic Press, New York.

Holmes, W. N., and Phillips, J. G. (1985). The avian salt gland. *Biol. Rev.* **60,** 213–256.

Holmes W. N., Brock R. L., and Devlin, J. (1974). Tritiated corticosterone metabolism in intact and adenohypophysectomized ducks (*Anas platyrhynchos*). *Gen. Comp. Endocrinol.* **22,** 417–427.

Holmes, W. N., Cronshaw, J., and Gorsline, J. (1978). Some effects of ingested petroleum on seawater-adapted ducks (*Anas platyrhynchos*). *Environ. Sci.* **17,** 177–190.

Holmes, W. N., Wright, A., and Gorsline, J. (1983). Effect of aldosterone and corticosterone on cloacal water and electrolyte excretion of constantly-loaded intact and colostomized ducks (*Anas platyrhynchos*). *Comp. Biochem. Physiol. A* **74,** 795–805.

Holmes, W. N., Redondo, J. L., and Cronshaw, J. (1989). Changes in the adrenal steroidogenic responsiveness of the Mallard duck (*Anas platyrhynchos*) during early post-natal development. *Comp. Biochem. Physiol. A* **92,** 403–408.

Holmes, W. N., Cronshaw, J., and Redondo, R. L. (1990). Stress-induced adrenal steroidogenesis in neonatal mallard ducklings and domestic chickens. *Zool. Sci.* **7,** 723–730.

Holmes, W. N., Al-Ghawas, S. C., Cronshaw, J., and Rohde, K. E. (1991). The structural organization and the steroidogenic responsiveness *in vitro* of adrenal gland tissue from the neonatal mallard duck (*Anas platyrhynchos*). *Cell Tissue Res.* **263,** 557–566.

Holmes, W. N., Cronshaw, J., Collie, M. A., and Rohde, K. E. (1992). Cellular aspects of the stress response in precocial neonates. *Ornis Scand.* **23,** 388–397.

Hornsby, P. J., and Crivello, J. F. (1983). The role of lipid peroxidation and biological antioxidants in the function of the adrenal cortex. Part 2. *Mol. Cell. Endocrinol.* **30,** 123–147.

Hoshino, S., Suzuki, M., Kakegawa, T., Imai, K., Wakita, M., Kobayashi, Y., and Yamada, Y. (1988). Changes in plasma thyroid hormones, luteinizing hormone (LH), estradiol, progesterone, and corticosterone of laying hens during a forced molt. *Comp. Biochem. Physiol. A* **90,** 355–359.

Howard, B. R. (1974). The assessment of stress in poultry. *Br. Vet. J.* **130,** 88.

Hudson, D. M., and Jones, D. R. (1982). Remarkable blood catecholamine levels in forced dived ducks. *J. Exp. Zool.* **224,** 451–456.

Hylka, V. W., and Doneen, B. A. (1983). Ontogeny of embryonic chicken lung: Effects of pituitary gland, corticosterone, and other hormones upon pulmonary growth and synthesis of surfactant phospholipids. *Gen. Comp. Endocrinol.* **52,** 108–120.

Idler, D. R., Walsh, J. M., Kalliecharan, R., and Hall, B. K. (1976). Identification of cortisol in plasma of the embryonic chick. *Gen. Comp. Endocrinol.* **30,** 539–540.

Isales, C. M., Barrett, P. Q., Brines, M., Bollag, W., and Rasmussen, H. (1991). Parathyroid hormone modulates angiotensin II-induced aldosterone secretion from the adrenal glomerulosa cell. *Endocrinology* **129,** 489–495.

Isobe, T., and Lillehoj, H. S. (1992). Effects of corticosteroids on lymphocyte subpopulations and lymphokine secretion in chickens. *Avian Dis.* **36,** 590–596.

Isobe, T., and Lillehoj, H. S. (1993). Dexamethasone suppresses T cell-mediated immunity and enhances disease susceptibility to *Eimeria mivati* infection. *Vet. Immunol. Immunopathol.* **39,** 431–436.

Jefcoate, C. R., DiBartolomeos, M. J., Williams, C. A., and McNamara, B. C. (1987). ACTH regulation of cholesterol movement in isolated adrenal cells. *J. Steroid Biochem.* **27,** 721–729.

John, T. M., George, J. C., and Etches, R. J. (1986). Influence of subcutaneous melatonin implantation on gonadal development and on plasma levels of luteinizing hormone, testosterone, estradiol, and corticosterone in the pigeon. *J. Pineal Res.* **3,** 169–179.

John, T. M., Viswanathan, M., Etches, R. J., Pilo, B., and George, J. C. (1987). Influence of corticosterone infusion on plasma levels of catecholamines, thyroid hormones, and certain metabolites in laying hens. *Poult. Sci.* **66,** 1059–1063.

John, T. M., Viswanathan, M., George, J. C., and Scanes, C. G. (1990). Influence of chronic melatonin implantation on circulating levels of catecholamines, growth hormone, thyroid hormones, glucose, and free fatty acids in the pigeon. *Gen. Comp. Endocrinol.* **79,** 226–232.

Johnson, A. L., and van Tienhoven, A. (1981). Plasma concentrations of corticosterone relative to photoperiod, oviposition, and ovulation in the domestic hen. *Gen. Comp. Endocrinol.* **43,** 10–16.

Jones, R. B. (1996). Fear and adaptability in poultry: Insights, implications and imperatives. *World's Poult. Sci. J.* **52,** 131–174.

Jones, R. B., and Satterlee, D. G. (1996). Threat-induced behavioural inhibition in Japanese quail genetically selected for contrasting adrenocortical response to mechanical restraint. *Br. Poult. Sci.* **37,** 465–470.

Jones, R. B., Beuving, G., and Blokhuis, H. J. (1988). Tonic immobility and heterophil/lymphocyte responses of domestic fowl to corticosterone infusion. *Physiol. Behav.* **42,** 249–253.

Jones, R. B., Satterlee, D. G., and Ryder, F. H. (1992a). Research note: Open-field behavior of Japanese quail chicks genetically selected for low or high plasma corticosterone response to immobilization stress. *Poult. Sci.* **71,** 1403–1407.

Jones, R. B., Satterlee, D. G., and Ryder, F. H. (1992b). Fear and distress in Japanese quail chicks of two lines genetically selected for low or high adrenocortical response to immobilization stress. *Horm. Behav.* **26,** 385–393.

Jones, R. B., Satterlee, D. G., and Ryder, F. H. (1994). Fear of humans in Japanese quail selected for low or high adrenocortical response. *Physiol. Behav.* **56,** 379–383.

Jones, R. B., Satterlee, D. G., Moreau, J., and Waddington, D. (1996). Vitamin C supplementation and fear-reduction in Japanese quail: Short-term cumulative effects. *Br. Poult. Sci.* **37,** 33–42.

Joseph, M. M., and Meier, A. H. (1973). Daily rhythms of plasma corticosterone in the common pigeon *Columba livia*. *Gen. Comp. Endocrinol.* **20,** 326–330.

Joseph, J., and Ramachandran, A. V. (1993). Effect of exogenous dexamethasone and corticosterone on weight gain and organ growth in post-hatched White Leghorn chicks. *Ind. J. Exp. Biol.* **31,** 858–860.

Joseph, J., and Ramachandran, A. V. (1992). Alterations in carbohydrate metabolism by exogenous dexamethasone and corticosterone in post-hatched White Leghorn chicks. *Br. Poult. Sci.* **33,** 1085–1093.

Jozsa, R., Vigh, S., Schally, A. V., and Mess, B. (1984). Localization of corticotropin-releasing factor-containing neurons in the brain of the domestic fowl: An immunohistochemical study. *Cell Tissue Res.* **236,** 245–248.

Jozsa, R., Vigh, S., Mess, B., and Schally, A. V. (1986). Ontogenic development of corticotropin-releasing factor (CRF)-containing neural elements in the brain of the chicken during incubation and after hatching. *Cell Tissue Res.* **244,** 681–685.

Jurani, M., Vyboh, P., Lamosova, D., and Nvota, J. (1978). Effect of restraint upon hypothalamic and adrenal catecholamines in Japanese quail. *Br. Poult. Sci.* **19,** 321–325.

Kafri, I., Rosebrough, R. W., McMurtry, J. P., and Steele, N. C. (1988). Corticosterone implants and supplemental dietary ascorbic acid effects on lipid metabolism in broiler chicks. *Poult. Sci.* **67,** 1356–1359.

Kai, H., Yamamoto, S., Takahama, K., and Miyata, T. (1990). Influence of corticosterone on tracheal mucociliary transport in pigeons. *Jpn. J. Pharmacol.* **52,** 496–499.

Kalliecharan, R. (1981). The influence of exogenous ACTH on the levels of corticosterone and cortisol in the plasma of young chicks (*Gallus domesticus*). *Gen. Comp. Endocrinol.* **44,** 249–251.

Kalliecharan, R., and Buffet, B. R. (1982). The influence of cortisol on the ACTH-producing cells in the pituitary gland of the chick embryo: An immunocytochemical study. *Gen. Comp. Endocrinol.* **46,** 435–443.

Kalliecharan, R., and Hall, B. K. (1974). A developmental study of the levels of progesterone, corticosterone, cortisol and cortisone circulating in plasma of chick embryos. *Gen. Comp. Endocrinol.* **24,** 364–372.

Kalliecharan, R., and Hall, B. K. (1976). A developmental study of progesterone, corticosterone, cortisol and cortisone in the adrenal gland of the chick embryo. *Gen. Comp. Endocrinol.* **30,** 404–409.

Kalliecharan, R., and Hall, B. K. (1977). The in vitro biosynthesis of steroids from pregnenolone and cholesterol and the effects of bovine ACTH on corticoid production by adrenal glands of embryonic chicks. *Gen. Comp. Endocrinol.* **33,** 147–159.

Kapas, S., Cammas, F. M., Hinson, J. P., and Clark, A. J. L. (1996). Agonist and receptor binding properties of adrenocorticotropin peptides using the cloned mouse adrenocorticotropin receptor expressed in a stably transfected HeLa cell line. *Endocrinology* **137,** 3291–3294.

Kar, A. B. (1947). The adrenal cortex testicular relations in the fowl: The effect of castration and replacement therapy on the adrenal cortex. *Anat. Rec.* **99,** 177–197.

Karaboyas, G. C., and Koritz, S. B. (1965). Identity of the site of action of cAMP and ACTH in corticosteroidogenesis in rat adrenal and beef adrenal cortex slices. *Biochemistry* **4,** 462–468.

Kawamoto, T., Mitsuuchi, Y., Toda, K., Yokoyama, Y., Miyahara, K., Miura, S., Ohnishi, T., Ichikawa, Y., Nakao, K., Imura, H., Ulick S., and Shizuta, Y. (1992). Role of steroid 11β-hydroxylase and steroid 18-hydroxylase in the biosynthesis of glucocorticoids and mineralocorticoids in humans. *Proc. Natl. Acad. Sci. USA* **89,** 1458–1462.

Kempf, H., le Moullec, J.-M., Corvol, P., and Gasc, J.-M. (1996). Molecular cloning, expression and tissue distribution of a chicken angiotensin II receptor. *FEBS Lett.* **399,** 198–202.

Khosla, S., Demay, M., Pines, M., Hurwitz, S., Potts, Jr., J. T., and Kronenberg, H. M. (1988). Nucleotide sequences of cloned cDNAs encoding chicken preproparathyroid hormone. *J. Bone Mineral Res.* **3,** 689–698.

King, V. M., and Follett, B. K. (1997). c-fos expression in the putative avain suprachiasmatic nucleus. *J. Comp. Physiol. A* **180,** 541–551.

Kitay, J. I., Coyne, M. D., Nelson, R., and Newsom, W. (1966). Relation of the testis to adrenal enzyme activity and adrenal corticosterone production in the rat. *Endocrinology* **78,** 1061–1066.

Klicka, J., and Zink, R. M. (1997). The importance of recent ice ages in speciation—a failed paradigm. *Science* **277,** 1666–1669.

Klingbeil, C. (1985a). Corticosterone and aldosterone dose-dependent responses to ACTH and angiotensin II in the duck (*Anas platyrhynchos*). *Gen. Comp. Endocrinol.* **59,** 382–390.

Klingbeil, C. (1985b). Effects of chronic changes in dietary electrolytes and acute stress on plasma levels of corticosterone and aldosterone in the duck (*Anas platyrhynchos*). *Gen. Comp. Endocrinol.* **58,** 10–19.

Klingbeil, C. K., Holmes, W. N., Pearce, R. B., and Cronshaw, J. (1979). Functional significance of interrenal cell zonation in the adrenal gland of the duck. (*Anas platyrhynchos*). *Cell Tissue Res.* **201,** 23–36.

Klukowski, L. A., Cawthorn, J. M., Ketterson, E. D., and Nolan, V. (1997). Effects of experimentally elevated testosterone on plasma corticosterone and corticosterone-binding globulin in dark-eyed juncos (*Junco hyemalis*). *Gen. Comp. Endocrinol.* **108,** 141–151.

Knight, D. E., and Baker, P. F. (1986). Observations on the muscarinic activation of catecholamine secretion in the chicken adrenal. *Neuroscience* **19,** 357–366.

Knouff, R. A., and Hartman, F. A. (1951). A microscopic study of the adrenal of the brown pelican. *Anat. Rec.* **109,** 161–187.

Kobayashi, H., and Takei, Y. (1982). Mechanisms for induction of drinking with special reference to angiotensin II. *Comp. Biochem. Physiol. A* **71,** 485–494.

Kocsis, J. F., and Carsia, R. V. (1989). Steroidogenic properties of isolated turkey adrenocortical cells. *Dom. Anim. Endocrinol.* **6,** 121–131.

Kocsis, J. F., Boyette, M. H., McIlroy, P. J., and Carsia, R. V. (1994a). Regulation of aldosteronogenesis in domestic turkey. (*Meleagris gallopavo*) adrenal steroidogenic cells. *Gen. Comp. Endocrinol.* **96,** 108–121.

Kocsis, J. F., McIlroy, P. J., Chiu, A. T., Schimmel, R. J., and Carsia, R. V. (1994b). Properties of angiotensin II receptors of domestic turkey (*Meleagris gallopavo*) adrenal steroidogenic cells. *Gen. Comp. Endocrinol.* **96,** 92–107.

Kocsis, J. F., Lamm, E. T., McIlroy, P. J., Scanes, C. G., and Carsia, R. V. (1995a). Evidence for functionally distinct subpopulations of steroidogenic cells in the domestic turkey (*Meleagris gallopavo*) adrenal gland. *Gen. Comp. Endocrinol.* **98,** 57–72.

Kocsis, J. F., Schimmel, R. J., McIlroy, P. J., and Carsia, R. V. (1995b). Dissociation of increases in intracellular calcium and aldosterone production induced by angiotensin II (AII): Evidence for regulation by distinct AII receptor subtypes or isomorphs. *Endocrinology* **136,** 1626–1634.

Kocsis, J. F., McIlroy, P. J., and Carsia, R. V. (1995c). Atrial natriuretic peptide stimulates aldosterone production by turkey (*Meleagris gallopavo*) adrenal steroidogenic cells. *Gen. Comp. Endocrinol.* **99,** 364–372.

Kocsis, J. F., Rinkardt, N. E., Satterlee, D. G., and Carsia, R. V. (1998). Concentration-dependent, biphasic effect of prostaglandins on avian corticosteroidogenesis *in vitro. Gen. Comp. Endocrinol.,* in press.

Koehler, D. E., and Moscona, A. A. (1975). Corticosteroid receptors in the neural retina and other tissues of the chick embryo. *Arch. Biochem. Biophys.* **170,** 102–113.

Kondics, L. (1963). Adrenal zone controlling mineral metabolism in the pigeon. *Acta Biol. Acad. Sci. Hung.* **13**, 233–240.

Kondics, L., and Kjaerheim, A. (1966). The zonation of interrenal cells in fowls (an electron microscopical study). *Z. Zellforsch.* **70**, 81–90.

Koritz, S. B., Bhargava, G., and Schwartz, E. (1977). ACTH action on adrenal steroidogenesis. *Ann. N.Y. Acad. Sci.* **297**, 329–335.

Korte, S. M., Beuving, G., Ruesink, W., and Blokhuis, H. J. (1997). Plasma catecholamine and corticosterone levels during manual restraint in chicks from a high and low feather pecking line of laying hens. *Physiol. Behav.* **62**, 437–441.

Kovács, K. J., and Péczely, P. (1983). Phase shifts in circadian rhythmicity of total, free corticosterone and transcortine plasma levels in hypothyroid male Japanese quails. *Gen. Comp. Endocrinol.* **50**, 483–489.

Kovács, K. J., and Péczely, P. (1986). Distribution and daily fluctuation of the available glucocorticoid binding sites in different areas of the quail brain. *Acta Physiol. Hung.* **67**, 423–428.

Kovács, K. J., Westphal, H. M., and Péczely, P. (1989). Distribution of glucocorticoid receptor-like immunoreactivity in the brain, and its relation to CRF and ACTH immunoreactivity in the hypothalamus of the Japanese quail. *Coturnix coturnix japonica*. *Brain Res.* **505**, 239–245.

Kovács, K. J., and Péczely, P. (1991). Plasma adrenocorticotropin in domestic geese: Effects of ether stress and endocrine manipulations. *Gen. Comp. Endocrinol.* **84**, 192–198.

Krajcsi, P., and Aranyi, P. (1986). Characterization of the partially purified, ligand-free glucocorticoid receptor. *Biochim. Biophys. Acta* **883**, 215–224.

Kühn, E. R., Shimada, K., Ohkubo, T., Vleurick, L. M., Berghman, L. R., and Darras, V. M. (1996). Influence of recombinant chicken prolactin on thyroid hormone metabolism in the chick embryo. *Gen. Comp. Endocrinol.* **103**, 349–358.

Kumar, V. (1996). Effects of melatonin in blocking the response to a skeleton photoperiod in the blackheaded bunting. *Physiol. Behav.* **59**, 617–620.

Kumar, V., Jain, N., and Follett, B. K. (1996). The photoperiodic clock in blackheaded buntings (*Emberiza melanocephala*) is mediated by a self-sustaining circadian system. *J. Comp. Physiol. A* **179**, 59–64.

Kuramoto, H. (1987). An immunohistochemical study of chromaffin cells and nerve fibers in the adrenal gland of the bullfrog, *Rana catesbeiana*. *Arch. Histol. Jpn.* **50**, 15–38.

Lacombe, A. M. A., and Jones, D. R. (1990). The source of circulating catecholamines in force dived ducks. *Gen. Comp. Endocrinol.* **80**, 41–47.

Langslow, D. R. (1972). The development of lipolytic sensitivity in the isolated fat cells of *Gallus domesticus* during the foetal and neonatal period. *Comp. Biochem. Physiol. B* **43**, 689–701.

Langslow, D. R., and Hales, C. N. (1969). Lipolysis in chicken adipose tissue *in vitro*. *J. Endocrinol.* **43**, 285–294.

Latour, M. A., Peebles, E. D., Boyle, C. R., Brake, J. D., and Kellogg, T. F. (1995). Changes in serum lipid, lipoprotein and corticosterone concentrations during neonatal chick development. *Biol. Neonate* **67**, 381–386.

Lauber, J. K., Vriend, J., and Oishi, T. (1987). Plasma corticosterone in chicks raised under several lighting schedules. *Comp. Biochem. Physiol. A* **86**, 73–78.

Lauterio, T. J., and Scanes, C. G. (1987a). Hormonal responses to protein restriction in two strains of chickens with different growth characteristics. *J. Nutr.* **117**, 758–763.

Lauterio, T. J., and Scanes, C. G. (1987b). Time courses of changes in plasma concentrations of the growth related hormones during protein restriction in the domestic fowl (*Gallus domesticus*). *Proc. Soc. Exp. Biol. Med.* **185**, 420–426.

Lea, R. W., Sharp, P. J., Klandorf, H., Harvey, S., Dunn, I. C., and Vowles, D. M. (1986). Seasonal changes in concentrations of plasma hormones in the male ring dove (*Streptopelia risoria*). *J. Endocrinol.* **108**, 385–391.

Lea, R. W., Klandorf, H., Harvey, S., and Hall, T. R. (1992). Thyroid and adrenal function in the ring dove (*Streptopelia risoria*) during food deprivation and a breeding cycle. *Gen. Comp. Endocrinol.* **86**, 138–146.

Leboulenger, F., Benyamina, M., Delarue, C., Netchitaïlo, P., Saint-Pierre, S., and Vaudry, H. (1988). Neuronal and paracrine regulation of adrenal steroidogenesis: Interactions between acetylcholine, serotonin and vasoactive intestinal peptide (VIP) on corticosteroid production by frog interrenal tissue. *Brain Res.* **453**, 103–109.

LeDouarin, N. M., and Teillet, M.-A. (1971). Localisation, par la méthode des greffes interspécifiques, du territoire neural dont dérivent les cellules adrenals surréniennes chez l'embryon d'Oiseau. *C. R. Acad. Sci.* **272**, 481–484.

Lehoux, J.-G. (1974). Aldosterone biosynthesis and presence of cytochrome P-450 in the adrenocortical tissue of the chick embryo. *Mol. Cell. Endocrinol.* **2**, 43–58.

Lehoux, J.-G., Tan, L., and Preiss, B. (1977). A comparative study of 3-hydroxy-3-methylglutaryl coenzyme A reductase activity in vertebrate adrenocortical cells. *Gen. Comp. Endocrinol.* **33**, 133–138.

Lehoux, J.-G., Bellabarba, D., and Beaudry, C. (1984). Corticosteroid receptors in the kidney of chick embryo. III. Nature, properties, and ontogeny of aldosterone receptor. *Gen. Comp. Endocrinol.* **53**, 116–125.

Lehoux, J.-G., Allard, C., Bouthillier, F., Belisle, S., and Bellabarba, D. (1988). Aldosterone nuclear receptors in kidneys of chick embryo. *J. Steroid Biochem.* **30**, 295–300.

Le Ninan, F., Cherel, Y., Sardet, C., and Le Maho, Y. (1988). Plasma hormone levels in relation to lipid and protein metabolism during prolonged fasting in king penguin chicks. *Gen. Comp. Endocrinol.* **71**, 331–337.

Lewis, P. R., and Shute, C. C. D. (1968). An electron-microscopic study of cholinesterase distribution in the rat adrenal medulla. *J. Microscopy* **89**, 181–193.

Li, Y., and Lau, L. F. (1997). Adrenocorticotropic hormone regulates the activities of the orphan nuclear receptor Nur77 through modulation of phosphorylation. *Endocrinology* **138**, 4138–4146.

Licinio, J., Mantzoros, C., Negrao, A. B., Cizza, G., Wong, M.-L., Bongiorno, P. B., Chrousos, G. P., Karp, B., Allen, C., Flier, J. S., and Gold, P. W. (1997). Human leptin levels are pulsatile and inversely related to pituitary-adrenal function. *Nature Med.* **3**, 575–579.

Lim, S. K., Gardella, T., Thompson, A., Rosenberg, J., Keutmann, H., Potts, J. Jr., Kronenberg, H., and Nussbaum, S. (1991). Full-length chicken parathyroid hormone: Biosynthesis in *Escherichia coli* and analysis of biological activity. *J. Biol. Chem.* **266**, 3709–3714.

Lim-Tio, S. S., Keightley, M.-C., and Fuller, P. J. (1997). Determinants of specificity of transactivation by the mineralocorticoid or glucocorticoid receptor. *Endocrinology* **138**, 2537–2543.

Logan, C. A., and Wingfield, J. C. (1995). Hormonal correlates of breeding status, nest construction, and parental care in multiple-brooded northern mockingbirds, *Mimus polyglottos*. *Horm. Behav.* **29**, 12–30.

Lu, J., and Cassone, V. M. (1993). Pineal regulation of circadian rhythms of 2-deoxy[^{14}C]glucose uptake and 2[^{125}I]iodomelatonin binding in the visual system of the house sparrow, *Passer domesticus*. *J. Comp. Physiol. A* **173**, 765–774.

Ma, X.-M., Levy, A., and Lightman, S. L. (1997). Emergence of an isolated arginine vasopression (AVP) response to stress after repeated restraint: A study of both AVP and corticotropin-releasing hormone messenger ribonucleic acid (RNA) and heteronuclear RNA. *Endocrinology* **138,** 4351–4357.

Mahata, S. K., and De, K. (1991). Effect of melatonin on norepinephrine, epinephrine, and corticosterone contents in the adrenal gland of three avian species. *J. Comp. Physiol. B* **161,** 81–84.

Mahata, S. K., and Ghosh, A. (1985). Effect of denervation and/or reserpine-induced changes on adrenomedullary catecholamines in pigeon: A fluorescence histochemical study. *Basic Appl. Histochem.* **29,** 331–336.

Mahata, S. K., and Ghosh, A. (1986). Influence of age on adrenomedullary catechol hormones content in twenty-five avian species. *Boll. Zool.* **53,** 63–67.

Mahata, S. K., and Ghosh, A. (1989). Influence of splanchnic nerve on reserpine action in avian adrenal medulla. *Gen. Comp. Endocrinol.* **73,** 165–172.

Mahata, S. K., and Ghosh, A. (1990). Neural influence on the action of insulin in the adrenomedullary catecholamine content in the pigeon. *Neurosci. Lett.* **116,** 336–340.

Mahata, S. K., and Ghosh, A. (1991a). Role of splanchnic nerve on steroid-hormone-induced alteration of adrenomedullary catecholamines in untreated and reserpenized pigeon. *J. Comp. Physiol. B* **161,** 598–601.

Mahata, S. K., and Ghosh, A. (1991b). Neural modulation of lysine vasopressin-induced changes of catecholamines in the adrenal medulla of the pigeon. *Neuropeptides* **18,** 29–33.

Mahata, S. K., and Ghosh, A. (1991c). Neural influence on oxytocin-induced changes of adrenomedullary catecholamines in the pigeon. *Reg. Peptides* **33,** 183–190.

Mahata, S. K., Mandal, A., and Ghosh, A. (1988). Influence of age and splanchnic nerve on the action of melatonin in the adrenomedullary catecholamine content and blood glucose level in the avian group. *J. Comp. Physiol. B* **158,** 601–607.

Mahata, S. K., De, M., Pal, D., and Ghosh, A. (1990). Effect of stress on the catecholamine content of the adrenal gland of intact and bursectomized chicks. *Clin. Exp. Pharmacol. Physiol.* **17,** 805–808.

Malamed, S. (1975). Ultrastructure of the mammalian adrenal cortex in relation to secretory function. In "Handbook of Physiology; Adrenal Cortex" (G. Sayers, ed.), Section 7: Endocrinology, Vol. 6, pp. 25–40. American Physiological Society, Bethesda.

Manelli, H., Accordi, F., Grassi Milano, E., and Mastrolia, L. (1973). Action of corticosteroids and ACTH on the chromaffin cells of chick embryo's adrenal glands in organ culture. *Acta Embryol. Exp.* **3,** 259–265.

Marie, C. (1981). Ontogenesis of the adrenal glucocorticoids and of the target function of the enzymatic tyrosine transaminase activity in the chick embryo. *J. Endocrinol.* **90,** 193–200.

Marsh, J. A., and Scanes, C. G. (1994). Neuroendocrine-immune interactions. *Poult. Sci.* **73,** 1049–1061.

Martin, B., Gasc, J. M., and Thibier, M. (1977). C_{12}-steroid binding protein and progesterone levels in chicken plasma during ontogenesis. *J. Steroid Biochem.* **8,** 161–166.

Martin, J. T., El-Halawani, M. E., and Phillips, R. E. (1982). Diurnal variation in hypothalamic monoamines and plasma corticosterone in the turkey after inhibition of tyrosine hydroxylase or tryptophan hydroxylase. *Neuroendocrinology* **34,** 191–196.

Mashaly, M. M., Trout, J. M., and Hendricks, III, G. M. (1993). The endocrine function of the immune cells in the initiation of humoral immunity. *Poult. Sci.* **72,** 1289–1293.

Massi, M., and Epstein, A. N. (1990). Angiotensin/aldosterone synergy governs the salt appetite of the pigeon. *Appetite* **14,** 181–192.

Mazzocchi, G., Gottardo, G., and Nussdorfer, G. G. (1997a). Catecholamines stimulate steroid secretion of dispersed fowl adrenocortical cells, acting through the β-receptor subtype. *Horm. Metab. Res.* **29,** 190–192.

Mazzocchi, G., Gottardo, G., and Nussdorfer, G. G. (1997b). Pituitary adenylate cyclase-activating peptide enhances steroid production by interrenal glands in fowls; evidence for an indirect mechanism of action. *Horm. Metab. Res.* **29,** 86–87.

Mazzocchi, G., Gottardo, G., and Nussdorfer, G. G. (1997c). Effects of somatostatin on steroid production by adrenocortical cells of the domestic turkey and fowl. *Zool. Sci.* **14,** 359–361.

McIlroy, P. J., Kocsis, J. F., Weber, H., and Carsia, R. V. (1998). Dietary protein restriction stress in the domestic fowl (*Gallus gallus domesticus*) alters adrenocorticotropin (ACTH)-transmembranous signaling and corticosterone negative feedback in adrenal steroidogenic cells. *Gen. Comp. Endocrinol.* in press.

McCorkle, F. M., and Taylor, R. L., Jr. (1993). Biogenic amines regulate avian immunity. *Poult. Sci.* **72,** 1285–1288.

McNamara, B. C., and Jefcoate, C. R. (1989). The role of sterol carrier protein 2 in stimulation of steroidogenesis in rat adrenal mitochondria by adrenal cytosol. *Arch. Biochem. Biophys.* **275,** 53–62.

Meier, A. H. (1973). Daily hormone rhythms in the White-throated sparrow. *Am. Scientist* **61,** 184–187.

Meier, A. H., and Fivizzani, A. J. (1975). Changes in the daily rhythm of plasma corticosterone concentration related to seasonal conditions in the white-throated sparrow, *Zonotrichia albicollis*. *Proc. Soc. Exp. Biol. Med.* **150,** 356–362.

Meier, A. H., Fivizzani, A. J., and Ottenweller, J. E. (1978). Daily rhythm of plasma corticosterone binding activity in the white-throated sparrow, *Zonotrichia albicollis*. *Life Sci.* **22,** 401–406.

Meijer, T., and Schwabl, H. (1989). Hormonal patterns in breeding and nonbreeding kestrels, *Falco tinnunculus:* Field and laboratory studies. *Gen. Comp. Endocrinol.* **74,** 148–160.

Mendel, C. M. (1989). The free hormone hypothesis: A physiological based mathematical model. *Endocr. Rev.* **10,** 232–274.

Mikami, S., and Yamada, S. (1984). Immunohistochemistry of the hypothalamic neuropeptides and anterior pituitary cells in the Japanese quail. *J. Exp. Zool.* **232,** 405–417.

Miller, W. L. (1988). Molecular biology of steroid hormone synthesis. *Endocr. Rev.* **9,** 295–318.

Minton, J. E., and Cash, W. C. (1990). Effect of cranial sympathectomy on circadian rhythms of cortisol, adrenocorticotropic hormone and melatonin in boars. *J. Anim. Sci.* **68,** 4277–4282.

Miyata, A., Minamino, N., Kangawa, K., and Matsuo, H. (1988). Identification of a 29-amino acid natriuretic peptide in chicken heart. *Biochem. Biophys. Res. Commun.* **155,** 1330–1337.

Morpurgo, B., Dean, C. E., and Porter, T. E. (1997). Identification of the blood-borne somatotroph-differentiating factor during chicken embryonic development. *Endocrinology* **138,** 4530–4535.

Morrow, L. B., Burrow, G. N., and Mulrow, P. J. (1967). Inhibition of adrenal protein synthesis by steroids *in vitro*. *Endocrinology* **80,** 833–888.

Mukai, K., Imai, M., Shimada, H., and Ishimura, Y. (1993). Isolation and characterization of rat CYP11B genes involved in late steps of mineralo- and glucocorticoid syntheses. *J. Biol. Chem.* **268,** 9130–9137.

Murphy, T. J., Nakamura, Y., Takeuchi, K., and Alexander, R. W. (1993). A cloned angiotensin receptor isoform from the turkey adrenal gland is pharmacologically distinct from mammalian angiotensin receptors. *Mol. Pharmacol.* **44,** 1–7.

Nagra, C. L., Baum, G. J., and Meyer, R. K. (1960). Corticosterone levels in adrenal effluent blood of some gallinaceous birds. *Proc. Soc. Exp. Biol. Med.* **105,** 68–70.

Nagra, C. L., Birnie, J. G., and Meyer, R. K. (1963a). Supression of the output of corticosterone in the pheasant by methopyrapone (metopirone). *Endocrinology* **73**, 835–837.

Nagra, C. L., Breitenbrach, R. P., and Meyer, R. V. (1963b). Influence of hormones on food intake and lipid deposition in castrated pheasants. *Poult. Sci.* **42**, 770–775.

Nagra, C. L., Sauers, A. K., and Wittmaier, H. N. (1965). Effect of testosterone, progestagens, and metopirone on adrenal activity in cockerels. *Gen. Comp. Endocrinol.* **5**, 69–73.

Nagy, L., and Freeman, D. A. (1990). Effect of cholesterol transport inhibitors on steroidogenesis and plasma membrane cholesterol transport in cultured MA-10 Leydig tumor cells. *Endocrinology* **126**, 2267–2276.

Nakabayashi, O., Nomura, O., Nishimori, K., and Mizuno, S. (1995). The cDNA cloning and transient expression of a chicken gene encoding a 3β-hydroxysteroid dehydrogenase/Δ^{5-4}isomerase unique to major steroidogenic tissues. *Gene* **162**, 261–265.

Nakamura, T., Tanabe, Y., and Hirano, H. (1978). Evidence for the *in vitro* formation of cortisol by adrenal glands of embryonic and young chicken. *Gen. Comp. Endocrinol.* **35**, 302–308.

Nakayama, T., Nakajima, T., and Sokabe, H. (1973). Comparative studies on angiotensins. III. Structure of fowl angiotensin and its identification by DNS-method. *Chem. Pharmacol. Bull.* **21**, 2085–2087.

Nebert, D. W., Nelson, D. R., Coon, M. J., Estabrook, R. W., Feyereisen, R., Fujii-Kuriyama, Y., Gonzalez, F. J., Guengerich, F. P., Gunsalus, I. C., Johnson, E. F., Loper, J. C., Sato, R., Waterman, M. R., and Waxman, D. J. (1991). The P450 superfamily: Update on new sequences, gene mapping, and recommended nomenclature. *DNA Cell Biol.* **10**, 1–14.

Ng, S. L., and Ng, T. B. (1987). Corticotropin-like material in pigeon brains. *Comp. Biochem. Physiol. C* **87**, 453–455.

Nishimura, H. (1980). Comparative endocrinology of renin and angiotensin. *In* "The Renin–Angiotensin System" (J. A. Johnson and R. R. Anderson, eds.), pp. 29–77. Plenum, New York.

Nishimura, H., and Bailey, J. R. (1982). Intrarenal renin–angiotensin system in primitive vertebrates. *Kid. Int. Suppl.* **12**, S185–S192.

Nishimura, H., Nakamura, Y., Taylor, A. A., and Madey, M. A. (1981). Renin-angiotensin and adrenergic mechanisms in control of blood pressure in fowl. *Hypertension* **3**, 141–149.

Nishimura, H., Walker, O. E., Patton, C. M., Madison, A. B., Chiu, A. T., and Keiser, J. (1994). Novel angiotensin receptor subtypes in fowl. *Am. J. Physiol.* **267**, R1174–R1181.

Nomura, O., Nakabayashi, O., Nishimori, K., and Mizuno, S. (1997). The cDNA cloning an transient expression of a chicken gene encoding cytochrome P-450scc. *Gene* **185**, 217–222.

Norr, S. C. (1973). *In vitro* analysis of sympathetic neuron differentiation from chick neural crest cells. *Dev. Biol.* **34**, 16–38.

Ohmori, Y., Okada, Y., and Watanabe, T. (1997). Immunohistochemical localization of serotonin, galanin, cholecystokinin and methionine-enkephalin in adrenal medullary cells of the chicken. *Tissue Cell* **29**, 199–205.

Ono, H., Iwasaki, M., Sakamoto, N., and Mizuno, S. (1988). cDNA cloning and sequence analysis of a chicken gene expressed during the gonadal development and homologous to mammalian cytochrome P-450c17. *Gene* **66**, 77–85.

Opel, H., and Proudman, J. A. (1984). Two methods for serial blood sampling from unrestrained, undisturbed turkeys with notes on the effects of acute stressors on plasma levels of prolactin. *Poult. Sci.* **63**, 1644–1652.

Opel, H., and Proudman, J. A. (1986). Plasma prolactin responses to serial bleeding in turkeys. *Dom. Anim. Endocrinol.* **3**, 199–207.

Pang, C. S., Tsang, K. F., Brown, G. M., and Pang, S. F. (1994). Specific 2-[^{125}I]iodomelatonin binding sites in the duck adrenal gland. *Neurosci. Lett.* **165**, 55–58.

Parker, K. L., and Schimmer, B. P. (1997). Steroidogenic factor 1: A key determinant of endocrine development and function. *Endocr. Rev.* **18**, 361–377.

Payne, L. N. (1971). The lymphoid system. *In* "The Physiology and Biochemistry of the Domestic Fowl" (D. J. Bell and B. M. Freeman, eds.), pp. 985–1037. Academic Press, New York.

Pearce, R. B., Cronshaw, J., and Holmes, W. N. (1978). Evidence for the zonation of interrenal tissue in the adrenal gland of the duck (*Anas platyrhynchos*). *Cell Tissue Res.* **192**, 363–379.

Pearce, R. B., Cronshaw, J., and Holmes, W. N. (1979). Structural changes occurring in interrenal tissue of the duck (*Anas platyrhynchos*) following adenohypophysectomy and treatment *in vivo* and *in vitro* with corticotropin. *Cell Tissue Res.* **196**, 429–447.

Péczely, P. (1979). Effect of testosterone and thyroxine on corticosterone and transcortin plasma levels in different bird species. *Acta Physiol. Sci. Hung.* **53**, 9–15.

Péczely, P., and Antoni, F. A. (1984). Comparative localization of neurons containing ovine corticotropin releasing factor (CRF)-like and neurophysin-like immunoreactivity in the diencephalon of the pigeon (*Columba livia domestica*). *J. Comp. Neurol.* **228**, 69–80.

Péczely, P., and Daniel, J. Y. (1979). Interactions reciproques testothyroido-surrenaliennes chez la Caille male. *Gen. Comp. Endocrinol.* **39**, 164–173.

Pedernera, E. A. (1971). Development of the secretory capacity of the chick embryo adrenal glands. *J. Embryol. Exp. Morphol.* **25**, 213–222.

Pedernera, E. A. (1972). Adrenocorticotropic activity *in vitro* of the chick embryo pituitary gland. *Gen. Comp. Endocrinol.* **19**, 589–590.

Pedernera, E. A., and Lantos, C. P. (1973). The biosynthesis of adrenal steroids by the 15-day-old chick embryo. *Gen. Comp. Endocrinol.* **20**, 331–341.

Pedernera, E. A., Romano, M., Besedovsky, H. O., and Aguilar, M. D. C. (1980). The bursa of Fabricius is required for normal endocrine development in chicken. *Gen. Comp. Endocrinol.* **42**, 413–419.

Peebles, E. D., Pond, A. L., Thompson, J. R., McDaniel, C. D., Cox, N. M., and Latour, M. A. (1997). Naloxone attenuates serum corticosterone and augments serum glucose concentrations in broilers stimulated with adrenocorticotropin. *Poult. Sci.* **76**, 511–515.

Pennisi, E. (1997). Newfound gene holds key to cell's cholesterol traffic. *Science* **277**, 180–181.

Phillips, J. G., and Chester Jones, I. (1957). The indentity of adrenocortical secretion in lower vertebrates. *J. Endocrinol.* **16**, iii.

Picardo, M., and Dickson, A. J. (1982). Hormonal regulation of glycogen metabolism in hepatocyte suspensions isolated from chicken embryos. *Comp. Biochem. Physiol. B* **71**, 689–693.

Pilo, B., Etches, R. J., and George, J. C. (1985). Effects of corticosterone infusion on the lipogenic activity and ultrastructure of the liver of laying hens. *Cytobios* **44**, 273–285.

Pittman, R. P., and Hazelwood, R. L. (1973). Catecholamine response of chickens to exogenous insulin and tolbutamide. *Comp. Biochem. Physiol. A* **45**, 141–147.

Poon, A. M., Wang, X. L., and Pang, S. F. (1993). Characteristics of 2-[^{125}I]iodomelatonin binding sites in the pigeon spleen and modulation of binding by guanine nucleotides. *J. Pineal Res.* **14**, 169–177.

Pratt, W. B., and Toft, D. O. (1997). Steroid receptor interactions with heat shock protein and immunophilin chaperones. *Endocr. Rev.* **18**, 306–360.

Prigge, W. F., and Grande, F. (1971). Effects of glucagon, epinephrine and insulin on *in vitro* lipolysis of adipose tissue from mammals and birds. *Comp. Biochem. Physiol. B* **39**, 69–82.

Radke, W. J., Albasi, C. M., and Harvey, S. (1984). Dietary sodium and adrenocortical activity in ducks (*Anas platyrhynchos*) and chickens (*Gallus domesticus*). *Gen. Comp. Endocrinol.* **56**, 121–129.

Radke, W. J., Albasi, C. M., and Harvey, S. (1985a). Haemorrhage and adrenocortical activity in the fowl (*Gallus domesticus*). *Gen. Comp. Endocrinol.* **60**, 204–209.

Radke, W. J., Albasi, C. M., and Harvey, S. (1985b). Stress and ACTH stimulated aldosterone secretion in the fowl (*Gallus domesticus*). *Comp. Biochem. Physiol. A* **82**, 285–288.

Rafestin-Oblin, M. E., Couette, B., Radanyi, C., Lombes, M., and Baulieu, E. E. (1989). Mineralocorticoid receptor of the chick intestine: Oligomeric structure and transformation. *J. Biol. Chem.* **264**, 9304–9309.

Rafferty, B., Zanelli, J., Rosenblatt, M., Schulster, D. (1983). Corticosteroidogenesis and adenosine 3′,5′-monophosphate production by the animo-terminal (1–34) fragment of human parathyroid hormone in rat adrenocortical cells. *Endocrinology* **113**, 1036–1042.

Redondo, J. L., Holmes, W. N., and Cronshaw, J. (1988). Effects of short-term changes in electrolyte intake on the adrenal steroidogenic responses of juvenile mallard ducks (*Anas platyrhynchos*). *J. Comp. Biochem. Physiol. A* **91**, 513–518.

Redondo, J. L., Cronshaw, J., Holmes, W. N., and Lin, F.-L. (1989). Structural and functional characteristics of primary cell cultures derived from the adrenal gland tissue of neonatal mallard ducklings (*Anas platyrhynchos*). *Cell Tissue Res.* **257**, 389–397.

Rees, A., Harvey, S., and Phillips, J. G. (1983). Habituation of the corticosterone response of ducks (*Anas platyrhynchos*) to daily treadmill exercise. *Gen. Comp. Endocrinol.* **49**, 485–489.

Rees, A., Hall, T. R., and Harvey, S. (1984). Adrenocortical and adrenomedullary responses of fowl to treadmill exercise. *Gen. Comp. Endocrinol.* **55**, 488–492.

Rees, A., Harvey, S., and Phillips, J. G. (1985a). Adrenocortical responses to novel stressors in acutely or repeatedly starved chickens. *Gen. Comp. Endocrinol.* **59**, 105–109.

Rees, A., Harvey, S., and Phillips, J. G. (1985b). Adrenergic stimulation of adrenocortical secretion in immature fowl. *Comp. Biochem. Physiol. C* **81**, 387–389.

Reppert, S. M., and Weaver, D. R. (1997). Forward genetic approach strikes gold: Cloning of a mammalian *Clock* gene. *Cell* **89**, 487–490.

Richardson, R. D., Boswell, T., Raffety, B. D., Seeley, R. J., Wingfield, J. C., and Woods, S. C. (1995). NPY increases food intake in white-crowned sparrows: Effect in short and long photoperiods. *Am. J. Physiol.* **268**, R1418–R1422.

Robinzon, B., Koike, T. I., Neldon, H. L., Kinzler, S. L., Hendry, I. R., and El-Halawani, M. E. (1988). Physiological effects of arginine vasotocin and mesotocin in cockerels. *Br. Poult. Sci.* **29**, 639–652.

Rombauts, L., Vanmontfort, D., Verhoeven, G., and Decuypere, E. (1992). Immunoreactive inhibin in plasma, amniotic fluid, and gonadal tissue of male and female chick embryos. *Biol. Repro.* **46**, 1211–1216.

Rombauts, L., Vanmontfort, D., Berghman, L. R., Decuypere, E., and G. Verhoeven (1994). Contribution of the fetal adrenal to circulating immunoactive inhibin in the chicken embryo. *Biol. Repro.* **51**, 926–933.

Romero, L. M., and Wingfield, J. C. (1998). Seasonal changes in adrenal sensitivity alter corticosterone levels in Gambel's white-crowned sparrows (*Zonotrichia leucophrys gambelli*). *Comp. Biochem Physiol. C* **119**, 31–36.

Romero, L. M., Ramenofsky, M., and Wingfield, J. C. (1997). Season and migration alters the corticosterone response to capture and handling in an Arctic migrant, the white-crowned sparrow (*Zonotrichia leucophrys gambelii*). *Comp. Biochem. Physiol. C* **116**, 171–177.

Romero, L. M., Soma, K. K., and Wingfield, J. C. (1998a). The hypothalamus and adrenal regulate modulation of corticosterone release in redpolls (*Carduelis flammea*—an arctic-breeding song bird). *Gen. Comp. Endocrinol.* **109**, 347–355.

Romero, L. M., Soma, K. K., and Wingfield, J. C. (1998b). Hypothalamic-pituitary-adrenal axis changes allow seasonal modulation of corticosterone release in a bird. *Am. J. Physiol.*, **43**, R1338–R1344.

Romero, L. M., Soma, K. K., and Wingfield, J. C. (1998c). Changes in pituitary and adrenal sensitivities allow the snow bunting (*Plectrophenax nivalis*), an arctic-breeding song bird, to modulate corticosterone release seasonally. *J. Comp. Physiol. B*, **168**, 353–358.

Rosenberg, J., and Hurwitz, S. (1987). Concentration of adrenocortical hormones in relation to cation homeostasis in birds. *Am. J. Physiol.* **253**, R20–R24.

Rosenberg, J., Pines, M., and Hurwitz, S. (1986). Relationship between endogenous cyclic AMP production and steroid hormone secretion in chick adrenal cells. *Comp. Biochem. Physiol. B* **84**, 71–75.

Rosenberg, J., Pines, M., and Hurwitz, S. (1987). Response of adrenal cells to parathyroid hormone stimulation. *J. Endocrinol.* **112**, 431–437.

Rosenberg, J., Pines, M., and Hurwitz, S. (1988a). Stimulation of chick adrenal steroidogenesis by avian parathyroid hormone. *J. Endocrinol.* **116**, 91–95.

Rosenberg, J., Pines, M., and Hurwitz, S. (1988b). Regulation of aldosterone secretion by avian adrenocortical cells. *J. Endocrinol.* **118**, 447–453.

Rosenberg, J., Pines, M., and Hurwitz, S. (1989a). Inhibition of aldosterone secretion by atrial natriuretic peptide in chicken adrenocortical cells. *Biochim. Biophys. Acta* **1014**, 189–194.

Rosenberg, J., Pines, M., Levy, J. J., Nutt, R. F., Caulfield, M. P., Russell, J., Sherwood, L. M., and Hurwitz, S. (1989b). Renal and adrenal adenosine 3′,5′-monophosphate production and corticosteroid secretion in response to synthetic chicken parathyroid hormone-(1-34). *Endocrinology* **125**, 1082–1089.

Russell, J., and Sherwood, L. M. (1989). Nucleotide sequence of DNA complementary to avian (chicken) preproparathyroid hormone mRNA and the deduced sequence of the hormone precursor. *Mol. Endocrinol.* **3**, 325–331.

Saad, A. D., and Moscona, A. A. (1985). Cortisol receptors and inducibility of glutamine synthetase in embryonic retina. *Cell Differ.* **16**, 241–250.

Saadoun, A., Simon, J., and Leclercq, B. (1987). Effect of exogenous corticosterone in genetically fat and lean chickens. *Br. Poult. Sci.* **28**, 519–528.

Sabeur, K., King, D. B., and Entrikin, R. K. (1993). Differential effects of methimazole and dexamethasone in avian muscular dystrophy. *Life Sci.* **52**, 1149–1159.

Shau, A. (1988). Evidence suggesting that galanin (GAL), melanin-concentrating hormone (MCH), neurotensin (NT), proopiomelanocorticn (POMC) and neuropeptide Y (NPY) are targets of leptin signaling in the hypothalamus. *Endocrinology* **139**, 795–798.

Sandi, C., and Rose, S. P. (1994a). Corticosterone enhances long-term retention in one-day-old chicks trained in a weak passive avoidance learning paradigm. *Brain Res.* **647**, 106–112.

Sandi, C., and Rose, S. P. (1994b). Corticosteroid receptor antagonists are amnestic for passive avoidance learning in day-old chicks. *Eur. J. Neurosci.* **6**, 1292–1297.

Sandor, T., Lamoureux, J., and Lanthier, A. (1963). Adrenocortical function in birds: *In vitro* biosynthesis of radioactive corticosteroids from pregnenolone-7-H^3 and progesterone-4-C^{14} by adrenal glands of the domestic duck (*Anas platyrhynchos*) and the chicken (*Gallus domesticus*). *Endocrinology* **73,** 629–636.

Sandor, T., Mehdi, A. Z., and Fazekas, A. G. (1977). Corticosteroid-binding macromolecules in the salt-activated nasal gland of the domestic duck (*Anas platyrhynchos*). *Gen. Comp. Endocrinol.* **32,** 348–359.

Sandor, T., Mehdi, A. Z., and DiBattista, J. A. (1983). Further studies on the corticosteroid–receptor system of the nasal gland of the domestic duck (*Anas platyrhynchos*). *Can. J. Biochem. Cell Biol.* **61,** 731–743.

Santana, P., Akana, S. F., Hanson, E. S., Strack, A. M., Sebastian, R. J., and Dallman, M. F. (1995). Aldosterone and dexamethasone both stimulate energy acquisition whereas only the glucocorticoid alters energy storage. *Endocrinology* **136,** 2214–2222.

Sapolsky, R. M. (1996). Stress, glucocorticoids and damage to the nervous system: The current state of confusion. *Stress* **1,** 1–19.

Sapolsky, R. M. (1997). The importance of a well-groomed child. *Science* **227,** 1620–1621.

Sato, T., and George, J. C. (1973). Diurnal rhythm of corticotropin-releasing factor activity in the pigeon hypothalamus. *Can. J. Physiol. Pharmacol.* **51,** 743–747.

Satterlee, D. G., and Edens, F. W. (1987). Plasma catecholamine releases post-immobilization in Japanese quail high and low blood corticosterone stress response lines. *Poult. Sci.* **66** (Suppl. 1), 38–39.

Satterlee, D. G., and Johnson, W. A. (1988). Selection of Japanese quail for contrasting blood corticosterone response to immobilization. *Poult. Sci.* **67,** 25–32.

Satterlee, D. G., and Jones, R. B. (1997). Ease of capture in Japanese quail of two lines divergently selected for adrenocortical response to immobilization. *Poult. Sci.* **76,** 469–471.

Satterlee, D. G., Comeaux, A. P., Johnson, W. A., and Munn, B. J. (1989). Influence of exogenous prostaglandin E_2 and stress on circulating corticosterone concentrations in Coturnix. *Poult. Sci.* **68,** 1289–1293.

Satterlee, D. G., Jones, R. B., and Ryder, F. H. (1993a). Short-latency stressor effects on tonic immobility fear reactions of Japanese quail divergently selected for adrenocortical responsiveness to immobilization. *Poult. Sci.* **72,** 1132–1136.

Satterlee, D. G., Jones, R. B., and Ryder, F. H. (1993b). Effect of vitamin C supplementation on the adrenocortical and tonic immobility fear reactions of Japanese quail genetically selected for high corticosterone response to stress. *Appl. Anim. Behav. Sci.* **35,** 347–357.

Satterlee, D. G., Jones, R. B., and Ryder, F. H. (1994). Effects of ascorbyl-2-polyphosphate on adrenocortical activation and fear-related behavior in broiler chickens. *Poult. Sci.* **73,** 194–201.

Sauer, L. A., Chapman, J. C., and Dauchy, R. T. (1994). Topology of 3β-hydroxy-5-ene-steroid dehydrogenase/Δ5-Δ4-isomerase in adrenal cortex mitochondria and microsomes. *Endocrinology* **134,** 751–759.

Savory, C. J., and Mann, J. S. (1997). Is there a role for corticosterone in expression of abnormal behaviour in restricted-fed fowls? *Physiol. Behav.* **62,** 7–13.

Scanes, C. G., and Griminger, P. (1990). Endocrine-nutrition interactions in birds. *J. Exp. Zool. Suppl.* **4,** 98–105.

Scanes, C. G., Marsh, J., Decuypere, E., Rudas, P. (1983). Abnormalities in the plasma concentrations of thyroxine, triiodothyronine and growth hormone in sex-linked and autosomal dwarf White Leghorn domestic fowl (*Gallus domesticus*). *J. Endocrinol.* **97,** 127–135.

Scott, T. R., Satterlee, D. G., and Jacobs-Perry, L. A. (1983). Circulating corticosterone responses of feed and water deprived broilers and Japanese quail. *Poult. Sci.* **62,** 290–297.

Schoech, S. J., Mumme, R. L., and Wingfield, J. C. (1997). Corticosterone, reproductive status, and body mass in a cooperative breeder, the Florida scrub-jay (*Aphelocoma coerulescens*). *Physiol. Zoo.* **70,** 68–73.

Schmeling, S. K., Nockels, C. F. (1978). Effects of age, sex, and ascorbic acid ingestion on chicken plasma corticosterone levels. *Poult. Sci.* **57,** 527–533.

Sharp, P. J., and Beuving, G. (1978). The role of corticosterone in the ovulatory cycle of the hen. *J. Endocrinol.* **78,** 195–200.

Siaud, P., Mekaouche, M., Maurel, D., Givalois, L., and Ixart, G. (1994). Superior cervical ganglionectomy suppresses circadian corticotropic rhythms in male rats in the short term (5 days) and long term (10 days). *Brain Res.* **652,** 273–278.

Siegel, H. S. (1980). Physiological stress in birds. *Bioscience* **30,** 529–534.

Siegel, H. S., and Gould, N. R. (1976). Chick embryonic plasma proteins and binding capacity for corticosterone. *Dev. Biol.* **50,** 510–516.

Siegel, H. S., Gould, N. R., and Latimer, J. W. (1985). Splenic leukocytes from chickens injected with *Salmonella pullorum* antigen stimulate production of corticosteroids by isolated adrenal cells. *Proc. Soc. Exp. Biol. Med.* **178,** 523–530.

Siegel, P. B., Gross, W. B., and Dunnington, E. A. (1989). Effects of dietary corticosterone in young Leghorn and meat-type cockerels. *Br. Poult. Sci.* **30,** 185–192.

Siiteri, P. K., Murai, J. T., Hammond, G. L., Nisker, J. A., Raymoure, W. J., and Kuhn, R. W. (1982). The serum transport of steroid hormones. *Rec. Prog. Horm. Res.* **38,** 457–510.

Silverin, B. (1986). Corticosterone-binding proteins and behavior effects of high plasma levels of corticosterone during the breeding period in the pied flycatcher. *Gen. Comp. Endocrinol.* **64,** 67–74.

Simon, J. (1984). Effects of daily corticosterone injections upon plasma glucose, insulin, uric acid and electrolytes and food intake pattern in the chicken. *Diab. Metab.* **10,** 211–217.

Sivaram, S. (1969). Nature and distribution of lipids in the developing adrenal cortex of *Gallus domesticus*. *Acta Histochem.* (*Jena*) **32,** 253–261.

Skadhauge, E. (1980). Intestinal osmoregulation. In "Avian Endocrinology" (A. Epple and M. H. Stetson, eds.), pp. 481–498. Academic Press, New York.

Skadhauge, E., Thomas, D. H., Chadwick, A., and Jallageas, M. (1983). Time course of adaptation to low and high NaCl diets in the domestic fowl. *Pflügers Arch.* **396,** 301–307.

Slaunwhite, W. R., Lockie, G. N., Back, N., and Sandberg, A. A. (1962). Inactivity *in vivo* of transcortin-bound cortisol. *Science* **135,** 1062–1063.

Soliman, K. F. A., and Huston, T. M. (1974). Involvement of the adrenal gland in ovulation of the fowl. *Poult. Sci.* **53,** 1664–1667.

Song, Y., Ayre, E. A., and Pang, S. F. (1993). [^{125}I]iodomelatonin binding sites in mammalian and avian kidneys. *Biol. Signal* **2,** 207–220.

Smoak, K. D., and Birrenkott, G. P. (1986). *In vivo* responsiveness of the broilers' adrenocortical system. *Poult. Sci.* **65,** 194–196.

Stachenko, J., Laplante, C., and Giroud, C. J. P. (1964). Double isotope derivative assay of aldosterone, corticosterone, and cortisol. *Can. J. Biochem.* **42,** 1275–1291.

Stocco, D. M., and Clark, B. J. (1996). Regulation of acute production of steroids in steroidogenic cells. *Endocr. Rev.* **17,** 221–244.

Sugano, T., Shiota, M., Khono, H., and Shimada, M. (1982). Stimulation of gluconeogenesis by glucagon and norepinephrine in the perfused chicken liver. *J. Biochem.* **92,** 111–120.

Sullivan, D. A., and Wira, C. R. (1979). Sex hormone and glucocorticoid receptors in the bursa of Fabricius of immature chicks. *J. Immunol.* **122,** 2617–2623.

Takahashi, K., Akiba, Y., and Horiguchi, M. (1993). Effects of a β-adrenergic agonist (clenbuterol) on performance, caracass composition, hepatic microsomal mixed function oxidase and antibody production in female broilers treated with or without corticosterone. *Br. Poult. Sci.* **34**, 167–175.

Takemori, S., Kominami, S., Yamazaki, T., and Ikushiro, S. (1995). Molecular mechanism of cytochrome P-450-dependent aldosterone biosynthesis in the adrenal cortex. *Trends Endocrinol. Metab.* **6**, 267–273.

Tanabe, Y. (1982). Ontogenetic aspect of steroidogenesis with special reference to corticosteroidogenesis of the chicken (*Gallus domesticus*). In "Aspects of Avian Endocrinology: Practical and Theoretical Implications" (C. G. Scanes, M. A. Ottinger, A. D. Kenny, J. Balthazart, J. Cronshaw, and I. Chester Jones, eds.), pp. 337–348. Grad. Studies, Texas Tech. Univ. **26**, 1–411, Texas Tech. Press, Lubbock.

Tanabe, Y., Nakamura, T., Fujioka, K., and Doi, O. (1979). Production and secretion of sex steroid hormones by the testes, the ovary, and the adrenal glands of embryonic and young chickens (*Gallus domesticus*). *Gen Comp. Endocrinol.* **39**, 26–33.

Tanabe, Y., Yano, T., and Nakamura, T. (1983). Steroid hormone synthesis and secretion by testes, ovary, and adrenals of embryonic and postembryonic ducks. *Gen. Comp. Endocrinol.* **49**, 144–153.

Tanabe, Y., Saito, N., and Nakamura, T. (1986). Ontogenetic steroidogenesis by testes, ovary, and adrenals of embryonic and postembryonic chickens (*Gallus domesticus*). *Gen. Comp. Endocrinol.* **63**, 456–463.

Tanaka, K., and Kamiyoshi, M. (1980). Rhythms of steroid hormone secretion in the domestic fowl. In "Biological Rhythms in Birds: Neural and Endocrine Aspects" (Y. Tanabe, K. Tanaka, and T. Ookawa, eds.), pp. 169–177. Japan Sci. Soc. Press/Springer-Verlag, Berlin.

Taouis, M., Derouet, M., Chevalier, B., and Simon, J. (1993). Corticosterone effects on insulin receptor number and kinase activity in chicken muscle and liver. *Gen. Comp. Endocrinol.* **89**, 167–175.

Taylor, A. A., Davis, J. O., Breitenbach, R. P., and Hartroft, P. M. (1970). Adrenal steroid secretion and a renal pressor system in the chicken. *Gen. Comp. Endocrinol.* **14**, 321–333.

Teillet, M.-A., and LeDouarin, N. M. (1974). Détermination par la méthode des greffes hétérospécifiques d'ébauches neurales de Caille sur l'embryon de poulet, du niveau du névraxe dont dérivent les cellules médullosurrénaliennes. *Arch. Anat. Microsc. Morphol. Exp.* **63**, 51–62.

Tenbusch, M., Milakofsky, L., Hare, T., Nibbio, B., and Epple, A. (1997). Impact of ethanol stress on components of the allantoic fluid of the chicken embryo. *Comp. Biochem. Physiol. A* **116**, 125–129.

Tenenbaum, H. C., Kamalia, N., Sukhu, B., Limeback, H., McCulloch, C. A. G. (1995). Probing glucocorticoid-dependent osteogenesis in rat and chick cells *in vitro* by specific blockade of osteoblastic differentiation with progesterone and RU38486. *Anat. Rec.* **242**, 200–210.

Thomas, D. H., and Phillips, J. G. (1975). Studies in avian adrenal steroid function. IV: Adrenalectomy and the response of domestic ducks (*Anas platyrhynchos* L.) to hypertonic NaCl loading. *Gen. Comp. Endocrinol.* **26**, 427–439.

Thomas, D. H., and Skadhauge, E. (1982). Regulation of electrolyte transport in the lower intestine of birds. In "Electrolyte and Water Transport Across Gastrointestinal Epithelia" (M. Case, A. Garner, L. Turnberg, and J. A. Young, eds.), pp. 295–303. Raven, New York.

Thomas, D. H., and Skadhauge, E. (1989). Function and regulation of the avian caecal bulb: Influence of dietary NaCl and aldosterone on water and electrolyte fluxes in the hen (*Gallus domesticus*) perfused *in vivo*. *J. Comp. Physiol. B* **159**, 51–60.

Thomas, D. H., Jallageas, M., Munck, B. G., and Skadhauge, E. (1980). Aldosterone effects on electrolyte transport of the lower intestine (coprodeum and colon) of the fowl (*Gallus domesticus*) in vitro. *Gen. Comp. Endocrinol.* **40**, 44–51.

Thompson, D. L., Elgert, K. D., Gross, W. B., and Siegel, P. B. (1980). Cell-mediated immunity in Marek's disease virus-infected chickens genetically selected for high and low concentrations of plasma corticosterone. *Am. J. Vet. Res.* **41**, 91–96.

Thurston, R. J., Bryant, C. C., and Korn, N. (1993). The effects of corticosterone and catecholamine infusion on plasma glucose levels in chicken (*Gallus domesticus*) and turkey (*Meleagris gallopavo*). *Comp. Biochem. Physiol. C* **106**, 59–62.

Toshimori, H., Toshimori, K., Minamino, N., Kangawa, K., Oura, C., Matsukura, S., and Matsuo, H. (1990). Chicken atrial natriuretic peptide (chANP) and its secretion. *Cell Tissue Res.* **259**, 293–298.

Tosini, G., and Menaker, M. (1996). Circadian rhythms in cultured mammalian retina. *Science* **272**, 419–421.

Trout, J. M., Mashaly, M. M., and Siegel, H. S. (1988). Changes in the profiles of circulating white blood cells, corticosterone, T_3 and T_4 during the initiation of humoral immunity in immature male chickens. *Dev. Comp. Immunol.* **12**, 331–346.

Trueba, M., Guantes, J. M., Vallejo, A. I., Sancho, M. J., Marino, A., and Macarulla, J. M. (1987). Characterization of cortisol binding sites in chicken liver plasma membrane. *Int. J. Biochem.* **19**, 957–962.

Tu, A. S., and Moudrianakis, E. N. (1973). Purification and characterization of a steroid receptor from chick embryo liver. *Biochemistry* **12**, 3692–3700.

Tur, J., Esteban, S., Rayo, J. M., Moreno, M., Miralles, A., and Tur, J. A. (1989). Effect of glucocorticoids on gastrointestinal emptying in young broilers. *Br. Poult. Sci.* **30**, 693–698.

Unsicker, A. (1973a). Fine structure and innervation of the avian adrenal gland. II. Cholinergic innervation of adrenal chromaffin cells. *Z. Zellforsch.* **145**, 417–442.

Unsicker, A. (1973b). Fine structure and innervation of the avian adrenal gland. III. Noncholinergic nerve fibers. *Z. Zellforsch.* **145**, 557–575.

Unsicker, K. (1973c). Fine structure and innervation of the avian adrenal gland. V. Innervation of interrenal cells. *Z. Zellforsch.* **146**, 403–416.

Unsicker, A. (1973d). Fine structure and innervation of the avian adrenal gland. I. Fine structure of adrenal chromaffin cells and ganglion cells. *Z. Zellforsch.* **145**, 389–416.

Urist, M. R., and Deutsch, N. M. (1960). Influence of ACTH upon avian species and osteoporosis. *Proc. Soc. Exp. Biol. Med.* **104**, 35–39.

Vamontfort, D., Room, G., Bruggeman, V., Rombauts, L., Berghman, L. R., Verhoeven, G., and Decuypere, E. (1997). Ovarian and extraovarian sources of immunoreactive inhibin in the chicken—effects of dexamethasone. *Gen. Comp. Endocrinol.* **105**, 333–343.

van Tiehoven, A. (1961). The effect of massive doses of corticotrophin and of corticosterone on ovulation of the chicken (*Gallus domesticus*). *Acta Endocrinol.* **38**, 407–412.

van Mourik, S., Outch, K. H., Cumming, R. B., and Stelmasiak, T. (1986). Incidence of a heterosis effect in adrenal responsiveness to ACTH stimulation in the chicken. *Comp. Biochem. Physiol. A* **84**, 397–399.

Vardimon, L., Fox, L. L., Degenstein, L., and Moscona, A. A. (1988). Cell contacts are required for induction by cortisol of glutamine synthetase gene transcription in the retina. *Proc. Natl. Acad. Sci. USA* **85**, 5981–5985.

Verma, R. J., Patel, P. V., and Pilo, B. (1984). Alterations in structure and function of adrenal gland due to bilateral vagotomy in domestic pigeons. *Curr. Sci.* **53**, 409–412.

Vestergaard, K. S., Skadhauge, E., and Lawson, L. G. (1997). The stress of not being able to perform dustbathing in laying hens. *Physiol. Behav.* **62**, 413–419.

Vinson, G. P., Whitehouse, B. J., Goddard, C., and Sibley, C. P. (1979). Comparative and evolutionary aspects of aldosterone secretion and zona glomerulosa function. *J. Endocrinol.* **81**, 5P–24P.

Vinson, G. P., Hinson, J. P., and Raven, P. W. (1985). The relationship between tissue preparation and function; methods for the study of control of aldosterone secretion: A review. *Cell Biochem. Funct.* **3**, 235–253.

Viswanathan, M., Pilo, B., George, J. C., and Etches, R. J. (1987). Effect of vagotomy on circulating levels of catecholamines and corticosterone in the pigeon. *Comp. Biochem. Physiol. C* **86**, 7–9.

Wassermann, G. F., and Bernard, E. A. (1971). The influence of corticoids on the phenylethanolamine-N-methyltransferase activity in the adrenal glands of *Gallus domesticus. Gen. Comp. Endocrinol.* **17**, 83–93.

Weber, H., Kocsis, J. F., Lauterio, T. J., and Carsia, R. V. (1990). Dietary protein restriction stress and adrenocortical function: Evidence for transient and long-term induction of enhanced cellular function. *Endocrinology* **127**, 3138–3150.

Wehling, M. (1997). Specific, nongenomic actions of steroid hormones. *Ann. Rev. Physiol.* **59**, 365–393.

Wentworth, B. C., and Hussein, M. O. (1985). Serum corticosterone levels in embryos, newly hatched, and young turkey poults. *Poult. Sci.* **64**, 2195–2201.

Westerhof, I., Lumeij, J. T., Mol, J. A., van den Brom, W. E., and Rijnberk, A. (1992). *In vivo* studies on the effects of ovine corticotrophin-releasing hormone, arginine vasotocin, arginine vasopressin, and haloperidol on adrenocortical function in the racing pigeon (*Columba livia domestica*). *Gen. Comp. Endocrinol.* **88**, 76–82.

Westerhof, I., Mol, J. A., van den Brom, W. E., Lumeij, J. T., and Rijnberk, A. (1994). Diurnal rhythms of plasma corticosterone concentrations in racing pigeons (*Columba livia domestica*) exposed to different light regimens, and the influence of frequent blood sampling. *Avian Dis.* **38**, 428–434.

White, P. C., Mune, T., and Agarwal, A. K. (1997). 11β-hydroxysteroid dehydrogenase and the syndrome of apparent mineralocorticoid excess. *Endocr. Rev.* **18**, 135–156.

Willoughby, E. J., and Peaker, M. (1979). Birds. In "Comparative Physiology of Osmoregulation in Animals" (G. M. O. Maloiy, ed.), Vol. 2, pp. 1–55. Academic Press, London.

Wilson, J. X. (1984). The renin-angiotensin system in nonmammalian vertebrates. *Endocr. Rev.* **5**, 45–61.

Wilson, J. X., and Butler, D. G. (1983a). Adrenalectomy inhibits noradrenergic, adrenergic, and vasopressor responses to angiotensin II in the Pekin duck (*Anas platyrhynchos*). *Comp. Biochem. Physiol. A* **81**, 353–358.

Wilson, J. X., and Butler, D. G. (1983b). 6-Hydroxydopamine treatment diminishes noradrenergic and pressor responses to angiotensin II in adrenalectomized ducks. *Endocrinology* **112**, 653–658.

Wilson, J. X., and Butler, D. G. (1983c). Catecholamine-mediated pressor responses to angiotensin II in the Pekin duck, *Anas platyrhynchos. Gen. Comp. Endocrinol.* **51**, 477–489.

Wilson, S. C., and Cunningham, F. J. (1980). Concentrations of corticosterone and luteinizing hormone in plasma during the ovulatory cycle of the domestic hen and after the administration of gonadal steroids. *J. Endocrinol.* **85**, 209–218.

Wilson, S. C., and Cunningham, F. J. (1981). Effect of photoperiod on the concentrations of corticosterone and luteinizing hormone in the plasma of the domestic hen. *J. Endocrinol.* **91**, 135–143.

Wilson, S. C., Cunningham, F. J., and Morris, T. R. (1982). Diurnal changes in the plasma concentrations of corticosterone, luteinizing hormone and progesterone during sexual development and the ovulatory cycle of Khaki Campbell Ducks. *J. Endocrinol.* **93**, 267–277.

Wilson, S. C., Jennings, R. C., and Cunningham, F. J. (1984). Developmental changes in the diurnal rhythm of secretion of corticosterone and LH in the domestic hen. *J. Endocrinol.* **101**, 299–304.

Wingfield, J. C. (1984). Influence of weather on reproduction. *J. Exp. Zool.* **232**, 589–594.

Wingfield, J. C., and Silverin, B. (1986). Effects of corticosterone on territorial behavior of free-living male song sparrows *Melospiza melodia. Horm. Behav.* **20**, 405–417.

Wingfield, J. C., Matt, K. S., and Farner, D. S. (1984). Physiologic properties of steroid hormone-binding proteins in avian blood. *Gen. Comp. Endocrinol.* **53**, 281–292.

Wingfield, J. C., Vleck, C. M., and Moore, M. C. (1992). Seasonal changes of the adrenocortical response to stress in birds of the Sonoran Desert. *J. Exp. Zool.* **264**, 419–428.

Wingfield, J. C., Suydam, R., and Hunt, K. (1994). The adrenocortical response to stress in snow buntings (*Plectrophenax nivalis*) and Lapland longspurs (*Calcarius lapponicus*) at Barrow, Alaska. *Comp. Biochem. Physiol. C* **108**, 299–306.

Wingfield, J. C., O'Reilly, K. M., and Astheimer, L. B. (1995a). Ecological bases of the modulation of adrenocortical responses to stress in Arctic birds. *Am. Zool.* **35**, 285–294.

Wingfield, J. C., Kubokawa, K., Ishida, S., and Wada, M. (1995b). The adrenocortical response to stress in male bush warblers, *Cettia diphone*: A comparison of breeding populations in Honshu and Hokkaido, Japan. *Zool. Sci.* **12**, 615–621.

Wingfield, J. C., Jacobs, J., and Hillgarth, N. (1997a). Ecological constraints and the evolution of hormone-behavior interrelationships. *Ann. N. Y. Acad. Sci.* **807**, 22–41.

Wingfield, J. C., Breuner, C., and Jacobs, J. (1997b). Corticosterone and behavioral responses to unpredictable events. In "Perspectives in Avian Endocrinology" (S. Harvey and R. J. Etches, eds.), pp. 267–278. Journal of Endocrinology Ltd., Bristol.

Wise, P. M., and Frye, B. E. (1973). Functional development of the hypothalamo–hypophyseal–adrenal cortex axis in chick embryo. *J. Exp. Zool.* **185**, 277–292.

Woods, J. E., De Vries, G. W., and Thommes, R. C. (1971). Ontogenesis of the pituitary–adrenal axis in the chick embryo. *Gen. Comp. Endocrinol.* **17**, 407–415.

Yamada, S., and Mikami, S. (1985). Immunohistochemical localization of corticotropin-releasing factor (CRF)-containing neurons in the hypothalamus of the Japanese quail, *Coturnix coturnix. Cell Tissue Res.* **239**, 299–304.

Yoshimura, Y., Heryanto, B., and Tamura, T. (1997). Changes in the population of proliferating cells in chicken anterior pituitary during induced molting: An immunocytochemical analysis for proliferating cell nuclear antigen. *Poult. Sci.* **76**, 1569–1573.

Zachariasen, R. D., and Newcomer, W. S. (1974). Phenylethanolamine-N-methyl transferase activity in the avian adrenal following immobilization or adrenocorticotropin. *Gen. Comp. Endocrinol.* **23**, 193–198.

Zachariasen, R. D., and Newcomer, W. S. (1975). Influence of corticosterone on the stress-induced elevation of phenylethanolamine-N-methyl transferase in the avian adrenal. *Gen. Comp. Endocrinol.* **25**, 332–338.

Zentel, H. J., Nohr, D., Muller, S., Yanaihara, N., and Weihe, E. (1990). Differential occurrence and distribution of galanin in adrenal nerve fibres and medullary cells in rodent and avian species. *Neurosci. Lett.* **120**, 167–170.

CHAPTER 20

Pancreas

ROBERT L. HAZELWOOD

Department of Biology
University of Houston
Houston, Texas 77204

I. Glucoregulation: Why Glucoregulate? 539
II. Organs of Importance in Glucoregulation 540
    A. Liver: Anatomical Considerations and Function 540
    B. Pancreas: Anatomical Considerations and Function 540
III. Avian Carbohydrate Metabolism: Different from Mammals? 541
IV. Central Role of the Pancreatic Organ 542
    A. Embryogenesis 542
    B. Cytodifferentiation and Cell Distribution 542
    C. Hormonogenesis and Secretogenesis 543
    D. The Insulin/Glucagon [I/G] Molar Ratio 544
V. Mechanisms of Pancreatic Hormone Action: Molecular Events 544
    A. Insulin 544
    B. Glucagon 544
    C. Other Hormones 546
    D. Receptor Considerations 546
VI. Pancreatic–Enteric Regulation of Carbohydrate Metabolism 548
    A. General Considerations 548
    B. Tissue Effects of Pancreatic Hormones in Birds 549
    C. Role of the Insulin/Glucagon [I/G] Molar Ratio 550
VII. Carbohydrate Metabolism in Other Tissues 551
    A. Heart 551
    B. Erythrocytes (RBC) 551
    C. The Glycogen Body 552
VIII. Altered Avian Carbohydrate Metabolism 552
    A. Spontaneous Diabetes Mellitus 552
    B. Pancreatic Chemocytotoxins 552
    C. Surgical Extirpation of the Pancreas 553
    D. Pancreatropic Agents 553
    E. Role of Other Hormones 554
References 554

I. GLUCOREGULATION: WHY GLUCOREGULATE?

Why not regulate protein metabolism or lipid metabolism? Simple questions with fairly simple answers: glucose is the fuel substance used by most tissues under normal circumstances. Glucose is the "easiest" substrate for cells to utilize to release needed energy. In times of need, glucose is readily synthesized from noncarbohydrate sources. Glucose is the only substrate employed by certain tissues for energy purposes, although these tissues under duress frequently switch to other substrate forms (e.g., brain tissue). Finally, effective homeostatic regulation of glucose metabolism invariably adjusts both protein and lipid metabolism to normalcy also. The latter two "depend" on normal metabolic disposition and/or production of glucose, glucose provided by various extracellular fluid (ECF) compartments for transfer into the cytosolic compartment of all cells.

As stated above, certain tissues (neural, retina, and adrenal medulla, for example) require glucose as their only substrate for maintaining normal function. This demand is constant, uninterrupted, 24 hr a day, 7 days a week. Over this span of time in normal humans, the adult brain, for example, requires approximately 5.6 to 6.4 gm of glucose per hour. Delivery of this amount to brain cells cannot be compromised in any way regardless of the activity of the human organism. If one is fasting overnight in the postabsorptive state, playing volleyball, engaged in a chess match, doing arithmetic computations, or sleeping, the healthy brain will require about 6 g of glucose per hour. Because the demand for fuel is constant, and our ability to deliver it is intermittent (due largely to our meal-eating habits), we must rely on a highly integrated series of regulatory mechanisms to sustain normal tissue-cell activity. The mammalian and avian organism must store fuel in various body compartments for future, though constant, use. Thus, potential fuel is stored for subsequent utilization when dietary intake does not meet the demands of the tissues and when prolonged absence of fuel obtains (as during an overnight fast). Such stored fuel takes the form of fat (by far the greatest source of energy), protein (at one-sixth the amount of potential fuel as that of fat), and glycogen, both hepatic and muscle depots (collectively, the total being less than 0.5% the energy of fat depots). Clearly in humans, glycogen stores are adequate at best to support (the otherwise falling) blood glucose levels during the night after the evening meal but before breakfast. Such glucose production is the responsibility of the liver (total glycogen available about 300 kcal or 1255 kJ energy), not the skeletal muscles (about 500 kcal or 2072 kJ total).

Glucoregulation, therefore, is essential for the foregoing reasons in all higher mammals, and the overwhelming evidence at hand indicates that this dictum is true for birds also.

II. ORGANS OF IMPORTANCE IN GLUCOREGULATION

While all tissues rely on glucose metabolism to a large extent for proper function, the liver and pancreas play a major role in the ultimate distribution and utilization of nutrients, including glucose. The liver is an important site of glycogen synthesis and storage; it also is responsible for glucose production and accelerated gluconeogenesis during times of nutrient deprivation. In humans, the liver appears to make available approximately 70–75% of all glucose released to the plasma. This figure is slightly higher in Aves. Contrary to the liver, the pancreas does not serve as a site for glycogenesis. Rather, its task appears to be that of releasing the proper mixtures of insulin and glucagon to the portal vein added to the most recently absorbed nutrients (potential fuel) from the gut. Thus, the liver gets "first chance" at distributing the nutrients while the coexistent pancreatic hormones act at the liver to direct the end-products of digestion (amino acids, hexoses, monoglycerides and glycerol) into metabolic (anabolic and catabolic) pathways important to the organism at the moment.

A. Liver: Anatomical Considerations and Function

As in mammals, the avian liver plays a near central role to the regulation of carbohydrate metabolism in birds. It's strategic location, between absorptive gut and the ascending vena cava, allows the hepatocyte to absorb the newly digested food products from an extension of the portal vein before this nutrient-rich perfusate reaches any other organ or tissue complex in the body. Similarly, after hepatic modulation the nutrients or their metabolites are sent quickly to the right heart via the hepatic veins and vena cava for general, systemic distribution. As seen in Chapter 12, the avian liver differs little from that of mammals, although it does, indeed, receive an extra allotment of blood from vessels that leave the renal portal system, thereby dumping blood directly into the portal vein. This bypass of the lower vena cava, therefore, dilutes the otherwise nutrient-rich perfusate from the gut prior to entering the liver. A complete battery of hepatic enzymes exists to carry out the mechanisms of glycogenesis, protein synthesis, and lipogenesis in the well-fed bird. In contrast to the situation in mammals, most triglycerides (TG) are synthesized in the avian liver before being transported to various peripheral deposit sites. The high plasma glucose levels resulting from feeding trigger release of insulin from the pancreatic B-cell, which in turn aids the enzymatic machinery of the liver to carry out the anabolic processes just mentioned, as well as similar ones in both muscle and fat tissue (Hazelwood, 1986a).

During periods of fasting, nutrients must be retrieved from deposits built up around the body by former feasting periods. During such fasting periods in both mammals and birds, insulin levels are low and glucagon levels are high. Thus, a catabolic mode exists. The enzymatic machinery activated by glucagon at the liver under these conditions is directed toward producing glucose, glucose that will be released to the circulation and ultimately serve the various peripheral tissues. Such metabolic pathways include glycogenolysis, gluconeogenesis, glucose-6-phosphatase systems, and lipolytic pathways. All tend to buffer a decreasing blood glucose by releasing the hexose to the circulation. Obviously, the liver plays a central role in glucose production in both birds and mammals. Degradation of lipid (TG) in adipose tissue and muscle protein and glycogen both contribute to the substrate that the liver employs to form *de novo* glucose.

B. Pancreas: Anatomical Considerations and Function

Strategically, it seems, the avian pancreas lies suspended by vascular elements between the limbs of the descending and ascending duodenum, poised, as it were, to contribute a hormonal mix to the portal vein that has just been loaded with recently absorbed nutrients (Figure 1). The discrete, lobular avian pancreas (Hazelwood, 1986b) supplies ample digestive juice through ducts to aid small intestinal digestion, a process that follows initial chemical breakdown that started in the proventriculus (see Chapter 12). The endocrine portion of the pancreas of birds occupies considerably more tissue mass than it does in mammals, and the distribution of cell types differs considerably also. A-, B-, D-, and F- (i.e., PP) cells comprise the avian endocrine islet, but the distribution within the islet as well as within different pancreatic lobes appears more random than the almost logical distribution found in discrete mammalian pancreata. A-cells synthesize and release glucagon; B-cells synthesize and release insulin; D-cells somatostatin; and F-cells (sometimes called PP-cells) pancreatic polypeptide (PP). Stimuli to the A- and B-cells have been hinted at above, namely increasing blood glucose levels stimulate B-cell activity; decreasing blood glucose levels increase the release of the "retrieval" hormone glucagon. Absorbed nutrients stimulate the D-cell as does A-cell activity. And gut peptides such as cholecystokinin (CCK), secretin, gastrin, as well as absorbed amino acids are excellent stimuli to the F-cell. Further, by sharing the same ECF within the islet itself, paracrine modulation of endocrine secretion also occurs in the avian pancreas. Thus, the A-cell stimulates both B-cell and D-cell activities and thereby closes a short-loop negative feedback, and D-cells appear to inhibit all other islet cells. In fact, the D-cell most probably regulates the proportion of insulin (I) secreted to that of glucagon (G) secreted simultaneously; that is, it adjusts the I/G molar ratio to meet the needs of the organism at any given moment (Hazelwood, 1989). Avian pancreatic insulin is a very powerful anabolic hormone; in fact, on a per unit weight basis it is many times more potent than equal amounts of mammalian insulins in producing glycogenesis, hypoglycemia, and lipid formation.

Glucagon is a linear 29-amino-acid polypeptide, differing by one or two (duck) residues from that of all mammalian glucagons and is a powerful catabolic hormone in birds as well as in experimental systems. The hormone circulates at levels 6–8 times higher than those observed in humans. Somatostatin concentrations in avian pancreas range from 2 to 150 times greater than that found in mammalian pancreata (Gerich, 1983), but its function remains identical in both classes of animals, namely the suppression of islet hormone release, adjusting the I/G molar ratio, and slowing nutrient absorption (especially that of glucose and lipid) in the small intestine.

Pancreatic polypeptide, the product of the avian F(PP)-cell, circulates at levels 20–40 times greater than that of human PP (HPP). Its action is primarily to inhibit GI motility and secretion, and gallbladder and exocrine pancreatic secretion, as well as to induce a sense of satiety (see below) by an action on the central nervous system (CNS).

III. AVIAN CARBOHYDRATE METABOLISM: DIFFERENT FROM MAMMALS?

This question has been raised by numerous physiologists over the past 100 years, largely due to two observations: early—and most—attempts to depancreatize birds did not result in a diabetic condition, and second, the fasting blood glucose levels of most Aves is 150–300% above that of fasting mammals. Attempts to explain these differences, as well as to explain how normal birds can survive "chronic hyperglycemia" equal to that of uncontrolled diabetes in mammals, have not been plentiful. Neither have they been convincing. Yet differences do exist, and one should be soundly aware of them. Embryonic Aves present quite a different metabolic ambiance; protected in the shell with a large yolk sac, as opposed to the human embryo/fetus, which is attached to its mother via the placenta and thereby shares its mother's circulating nutrients. Additionally, in adult forms, insulin has been established to be the dominate pancreatic hormone in mammals while current evidence would indicate that glucagon is the dominate hormone in Aves. This is not to say that glucagon is impotent in mammals; to wit, notice should be made of the hyperglucagonemia that persists after B-cell reduction in most mammals. Similarly, insulin is not to be ignored in normal avian metabolism as indicated by the observation that anti-insulin antibodies pro-

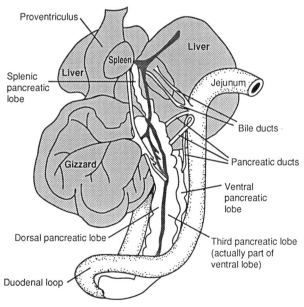

FIGURE 1 Anatomy of the avian pancreas.

duce long-lasting hyperglycemia in ducks. Fasting glucose metabolism in birds (10–13 mg/min/kg body wt) is approximately twice that of fasting mammals (Hazelwood, 1986a).

In addition to use of the uridine diphosphate (UDP) pathway in glycogenesis, birds use this path to synthesize a related sugar acid, namely ascorbic acid (vitamin C). Mammals are incapable of such synthesis and most rely on dietary intake of the vitamin to prevent connective tissue abnormalities and the onset of increased capillary fragility.

Gluconeogenic pathways may appear less active in birds than in mammals, but, still, they are extremely important as is evidenced by the fact that a 24- to 48-hr fast lowers plasma glucose only 10–15% from nonfasted levels of 210–250 mg/dl. Actually, even though there is an absolute decrease in plasma glucose, when the latter is expressed in terms of unit body weight, glucose levels remain almost constant. Thus, considerable recycling of glucose and glucose substrate must occur (Simon, 1989). Kidney may provide up to 30% of the glucose by gluconeogenesis, employing pyruvate, lactate, amino acids, and glycerol as substrates. Cytosolic phosphoenolpyruvate carboxykinase (PEPK), the most important gluconeogenic enzyme in kidney, is an adaptive form that adjusts (is induced by) to the nutritional state of the bird. Also, there may be a low utilization of citric acid cycle intermediates for *de novo* glucose formation, and that from pyruvate is modest at best. Therefore, inavailability of cytosolic-reducing equivalents may have a significant impact. In contrast to kidney, liver cells use mainly dihydroxyacetone and lactate in gluconeogenic production of glucose. Liver PEPK enzyme is restricted to the mitochondria and thereby is not adaptive.

Probably of all metabolic pathways studied in birds in comparison with mammals that of the hexomonophosphate [pentose] shunt is best understood. This "direct oxidative" pathway makes available pentoses, three-carbon phosphates, NADPH, and CO_2 for synthesis of nucleotides and nucleic acids, recombination into glucose-6-phosphate, and both steroidogenesis and lipogenesis, respectively. While this shunt activity varies from embryonic tissue to embryonic tissue, it peaks during the period of days 8–15 of embryonic life, only to decline to barely detectable levels by day 21. After hatching, it contributes very little to the metabolic economy of the growing bird. Neural, cardiac, and intestinal embryonic tissue appear to be most dependent on the activity of the pentose shunt in birds. In contrast, the mammalian liver relies heavily on pentose shunt activity, as does the mature RBC (Hazelwood, 1986a). Even though over 95% of the bird's lipid synthesis occurs in the liver, it does so in the relative absence of the pentose pathway. The importance of malic enzyme thus is obvious. Other metabolic differences, of a less significant nature, occur between chick and mammals and the reader is referred to an earlier edition of this text for a review (Hazelwood, 1986a).

IV. CENTRAL ROLE OF THE PANCREATIC ORGAN

A. Embryogenesis

As in mammals, the embryonic avian pancreas arises from two outpocketings of the early gut. The dorsal evagination appears first, usually on day 3, arising from the embryonic duodenum just anterior to the area that will give rise to the liver. The ventral pancreatic evagination also arises from the duodenum on day 4 and is more ventral and posterior to that of the liver. A single and discrete lobular pancreatic organ is formed by the fusion of the two embryonic analgen. Pancreatic ducts that carry digestive juices to the duodenum are larger and better developed in the ventral lobes of the fused pancreas. These ducts are lined with columnar-type epithelium, interspersed with mucus-secreting type cells. However, unlike mammals, there is no evidence that ductile cells ultimately give rise to endocrine-type cells. Whether future generations of islet cells arise from ductile elements in birds is yet to be established.

B. Cytodifferentiation and Cell Distribution

The larger of two cell masses occupying the epithelial bud appears uncommitted either to endocrine or to acinar cell types. Presumably, these are stem cells that may develop into either of the two basic pancreatic types of cells. The smaller, distinct embryologic cell type consists of obvious endocrine (islet)-type cells and other cells associated with them. As far as acinar tissue morphogenesis is concerned, evidence exists indicating digestive juice secretion as early as days 3–4 in duck embryos. Single-membrane bound zymogen granules are present at the same time. Of particular interest is the appearance of clear "glycogen cells," which are very numerous during days 3–7. Their function has never been established. Gradually, during embryonation, they disappear as an entity. In fact, in chicks and ducks, they vanish almost completely by days 14–16.

The islet endocrine-type cells appear before the acinar cells in avian pancreata, the first of which being the A-cells that release glucagon. Using the chick as a reference, the A-cell appears early on the third day of incubation, while the B-cells appear in the dorsal bud late on the same day. Late on day 4, more commonly on day 5, both D-cells and PP(F)-cells appear. By day 7, the two analgen have fused and the splenic lobe of the pancreas exhibits true endocrine islet formation. Islet formation in the other three lobes takes place during the period of days 8 to 13 (see Table 23-1 in Hazelwood, 1986b).

Islet tissue appears to be more sparsely distributed among the sea of acinar cells in avian pancreata than is found in mammalia. Partly, this is due to the fact that of all vertebrates, Aves have much more PP-cell tissue arranged as aggregates of F-cells located outside of the islet, buried in the exocrine tissue. Another consideration is the splenic lobe, which appears to be quite different in many ways from other regions of the avian pancreas. Here, islet cell population may be very great (40% of total mass) and cell size, especially of the A-cell, is exceptionally large. Species variations exist.

Classically, two types of islets have been described for the avian pancreas: large, dark islets containing many A-cells, and smaller, lighter (staining) islets that contain predominately B-cells. The "dark" islets appear largely restricted to one longitudinal half of the organ. A paucity of PP(F)-cells is found wherever the A-cell mass is large. Frequently, D-cells occupy a central position in the dark islet. Unlike the dark islets, the "light" islets are scattered much more uniformly throughout the four lobes of the pancreas. Again, splenic lobe B-cells are very large, resulting frequently in islet dimensions 5–6 times that found in other regions.

PP-cells appear to be fairly uniformly distributed throughout the pancreas as clusters, single cells, or incorporated within the islet itself. Extraislet-located PP-cells are found frequently throughout the acinar mass. One striking similarity of Aves with mammals is the fact that wherever there is a heavy preponderance of A-cells, PP-cells are few in number (Table 23-2 in Hazelwood, 1986b).

C. Hormonogenesis and Secretogenesis

Plasma levels of glucagon are approximately the same as that of insulin from day 5 through day 12–15 in the embryonic chick, even though whole pancreas values may increase 10 fold during this time, and pancreatic glucagon content may increase 25 times (insulin: 9 ng to 100 ng; glucagon: 11 ng to 250 ng). Secretory granules in the islet cells appear as early as the third day of incubation (A-cells followed by B-cells), and as late as the seventh day (PP-cells). The D-cell falls in step on day 4. By the time of hatching (+1 day), the pancreatic content of the four polypeptide hormones (chick data) has been estimated as: glucagon, 750–800 ng; insulin, 250–300 ng; APP, 225–260 ng; and somatostatin, 210–240 ng.

All four hormones are synthesized as "giant" molecules that lack biological activity. From the rough endoplasmic reticulum, each is conveyed as a prohormone in membrane-lined vesicles to the Golgi area, where cleaving, insertions, folding, and final conformational changes to the molecule take place. Most of the prohormone at this stage has been modified to a smaller, biologically active hormone and incorporated into secretory granules that are membrane-lined and as such move to the plasma membrane. The vesicular membranes fuse with the cell membrane and become incorporated within them. In doing so, vesicular contents are extruded via exocytosis into the ECF and subsequently become intravascular. Glucagon levels in avian plasma are reported to be 10–80 times that of mammals (0.05–0.10 ng/ml vs 1–4 ng/ml plasma). Structurally, and from an evolutionary point of view, the glucagon molecule has been remarkably conserved over millions of years.

Insulin has been identified in homogenized 1- and 2-day-old chick embryos. This observation suggests that maternal contributions may be important in the early developing chick embryo. In particular, the process of myogenesis is known to require insulin; this process is initiated on days 1–2. Once the preprohormone is formed, it is cleaved to a prohormone (84 amino acids) and thence at the Golgi area to a biologically active, two-chain hormone of 51 amino acids. When exocytosed, insulin molecules are accompanied by other vesicular components such as equimolar concentrations of connecting peptide, small amounts of uncleaved proinsulin, ATP, Zn^{2+}, and so on. Although plasma levels of insulin vary somewhat from avian species to species, they are consistently much higher than levels found in normal mammals.

PP in birds emanates only from the pancreatic PP(F)-cell. (In mammals, there exist small, auxiliary sources of PP.) Avian PP (APP) is synthesized as the other peptide hormones, namely as a preprohormone, then modified to a prohormone, and finally as a biologically active, 36-amino-acid hormone released to the plasma. The hormone circulates at levels some 40–80 times higher than those found in mammals, including man. Fed chickens have plasma levels of APP approximating 6–8 ng/ml.

Somatostatin (SRIF), like insulin, remains at low levels in the embryonic pancreas from day 7 to day 13, after which both hormones increase in concentration greatly. Pancreatic tissue levels of SRIF vary from lobe to lobe, for example from 3 ng/mg tissue in the ventral lobe to as high as 22 ng/mg tissue in the splenic lobe (Hazelwood, 1986b). D-cells make up as much as 30% of the dark islets and are amply supplied with cholinergic neural elements. Pigeon pancreatic SRIF levels have been reported to be 100–200 times greater than human levels, and the very few reports on various avian plasma levels of this tetradecapeptide indicate levels (about

1.3 ng/ml) that also are much higher than those reported for mammals.

D. The Insulin/Glucagon [I/G] Molar Ratio

Although discussed more thoroughly below, the I/G molar ratio is more than a concept and should be introduced at this point of the chapter. The fact that all islet cells may be contiguous with each other and, additionally, that each shares *intraislet* ECF with the others makes probable the direct influence that one cell type could exert on the others. Gap junctions as well as paracrine mechanisms, therefore, allow interactions between and among islet A-, B-, D-, and F-cells. SRIF from the D-cell inhibits secretion from all other islet cells, with the A-cell being most sensitive and the F-cell the least sensitive to D-cell action. As a result, the D-cell is thought to play a critical role in controlling the *relative* secretion of other islet hormones. Of most importance in glucoregulation are the hormones insulin and glucagon, and thus the D-cell through SRIF regulates the I/G molar ratio. Adjusting this ratio to meet the momentary needs of the avian organism is critical to glucohomeostasis.

V. MECHANISMS OF PANCREATIC HORMONE ACTION: MOLECULAR EVENTS

A. Insulin

The biologically active insulin molecule released from the avian B-cell is a two-chained, 51-amino-acid polypeptide wherein the A-chain (20 aa) is connected to the B-chain (31 aa) by two disulfide bonds that must remain intact for full biological action. Additionally, the A-chain has an intrachain disulfide bond connecting a CYS at position 6 with another CYS at position 11 (Figure 2).

Overall, B-cell release of insulin in birds is similar to that reported in mammals, though a few exceptions exist. Glucose is not a major trigger of insulin release in birds, the B-cell being far more sensitive to cholecystokinin, glucagon, and a mixture of absorbed amino acids. Actually, a certain synergism appears to exist between glucose and amino acids in regulating insulin secretion in birds. Epinephrine, fatty acids, and the hormone secretin appear to be without significant effect on avian insulin release. In summary, the release of insulin from the avian islet is the result of a complex of inputs, namely endocrine, exocrine, paracrine, neural, and humoral. Circulating insulin levels of insulin in ducks and chickens approach 3–4 ng/ml plasma and are little affected by short-term fasting.

At its receptor, avian insulin interacts with the external α-subunit of the tetrameric receptor protein, similar to events that occur in mammals. The two α-subunits are connected to two β-subunits that traverse the membrane and which are connected by disulfide bonds to the α-units. That portion of the β-subunit that is cytosolic contains a "kinase domain" that undergoes autophosphorylation as a result of conformational changes induced by insulin attachment to the external α-subunits (Simon and Taouis, 1993). In general, properties of the insulin receptor in avian tissues are similar to those described in mammals (see Section V,D below).

Autophosphorylation of tyrosine [and possibly of serine] residues on the avian insulin receptor results in a widely diverse response, depending on the tissue involved. Increased transfer of glucose from the ECF to the cytosol (muscle, adipose tissue), increased protein and glycogen synthesis (muscle, liver), increased membrane potentials (most tissues), and increased facilitative diffusion of glucose (muscle; liver indirectly), occurs. Thus, insulin is an anabolic hormone, a hormone of feasting. At the muscle and adipocyte level, insulin exerts its effects by increasing the availability of (protein) glucose transport carriers, called GLUT-4. These carriers, of a proteinacious nature, are normally sequestered to a large degree away from the plasma membrane and deep within the cytosol and are recruited to the plasma membrane as a result of insulin action where they increase the transport of free glucose (from the ECF) to the inside of the cell, where the sugar is released. In liver, avian insulin increases the activity of glucokinase, thus phosphorylating glucose at the 6-position and thereby reducing the amount of free glucose that had freely entered the hepatocyte. In doing so, the hormone increases the downhill gradient, from ECF to ICF, of free glucose, making it easier for ECF-located glucose to enter the cell. The ultimate result is a reduction of ECF glucose levels, that is, hypoglycemia. Although insulin is also antigluconeogenic in birds, it is not antilipolytic.

B. Glucagon

The hormone of fasting, glucagon, is a linear polypeptide containing 29 amino acids and is the final product of the A-cell's synthesis, originally as preproglucagon (179 aa), then cleaved to proglucagon (96-100 aa), and finally to the biologically active 29-residue structure. The structure is invariant in mammals; no substitutions can be made without considerable loss of biological activity. The structure of chicken glucagon is presented in Figure 3. All mammals have an ASN at position 28, and the duck differs from most Aves by also having a THR replace the SER at position 16.

As released from the avian pancreas, glucagon circulates in the plasma of most birds at levels of 2–4 ng/ml

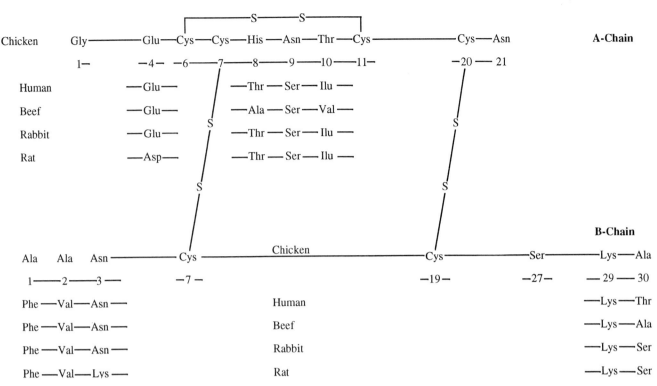

FIGURE 2 Structure of chicken insulin. Several mammalian insulin structures also are shown for comparison. Turkey insulin is identical with that of chicken, and duck insulin differs only at positions 8 and 10 where GLU and PRO, respectively, are substituted, as well as at position 30 on the B-chain where THR is inserted.

plasma (chicken, duck, pigeon, and goose). These levels increase 100–200% during a 48- to 72-hr fast. Triggers of glucagon release from the avian pancreas include free fatty acids and cholecystokinin (CCK). SRIF has been reported to *increase* duck glucagon as has insulin. Both of these latter observations are in stark contrast to findings in other avian species and, especially, in mammals. Glucose has a profound inhibitory effect on glucagon release from the avian A-cell. A possible feedback regulation involving plasma free fatty acids and glucagon release has been suggested (Hazelwood, 1986a).

Glucagon exerts its major action via a classic second messenger system, as described in mammals. After binding to its fixed membrane receptor, the hormone–receptor complex interacts with a regulator protein of the G-series which increases activity of adenylate cyclase, an enzyme system that appears to be bound to the interior margin of the plasma membrane. This enzyme cyclizes ATP to cAMP which, in turn, activates a specific protein kinase, usually protein kinase A. A cascade of events follows, all based on phosphorylation–dephosphorylation reactions that ultimately lead to specific biological effects. Thus, at the liver, glucagon activates glycogen phosphorylase while inhibiting glycogen synthase. Glycogenolysis and hyperglycemia follow. In liver and muscle, glucagon activates gluconeogenic enzymes, resulting in conversion of amino acid substrates to glucose and hyperglycemia. And in adipose

```
              1               5                  10
             His—Ser—Gln—Gly—Thr—Phe—Thr—Ser—Asp—Tyr—Ser—Lys—Tyr—Leu
                                                                    |
                                                                    |
             29              25                 20                 15
             Thr—Ser—Met—Leu—Trp—Gln—Val—Phe—Asp—Gln—Ala—Arg—Arg—Ser—Asp
                 (Asn:                                      (Thr:
                  All mammals)                               Duck)
```

FIGURE 3 Structure of avian glucagon. Substitutions in mammalian glucagon are shown.

tissue, the hormone inhibits LPLipase while stimulating the hormone-sensitive lipase, thereby adding more fatty acids to the plasma. Interestingly, glucagon receptors appear in far greater numbers than do insulin receptors on the same tissue, an observation that possibly supports the suggestion that glucagon is the dominate hormone of avian metabolism (also see below).

C. Other Hormones

Somatostatin (SRIF) from the avian D-cell appears as a tetradecapeptide, with an internal disulfide bond connecting positions 3 and 14 (Figure 4). The basic 14-residue structure appears to be invariant among mammals, although a larger, 28-residue molecular species is known to have slight modifications throughout the animal kingdom.

Avian pancreata contain from 20 (chicken) to 300 (pigeon) times the amount of SRIF found in mammalian tissue, and plasma levels in birds are generally 200–400% greater than those in mammals. The role of SRIF in avian metabolism lies outside that of a direct action on carbohydrate metabolism. Most likely, the hormone effectively suppresses the activity of neighboring pancreatic islet cells. The task of adjusting the I/G molar ratio falls to SRIF, thereby preparing the avian organism for a catabolic or an anabolic mode as the metabolic situation of the moment dictates. The precise mechanism by which SRIF exerts its suppressive action at the cellular level is yet to be established. At the gut, where this hormone delays the absorption of glucose and lipids, cAMP does not appear to be the means by which SRIF exerts its effect(s). However, the hormone may lower cAMP levels in adipocytes to decrease lipolysis.

Avian pancreatic polypeptide was the first of the pancreatic polypeptide family to be isolated, characterized, and studied metabolically by Kimmel *et al.* (1968). Synthesized originally as a preprohormone, the biologically active 36-amino-acid structure is encased in secretory vesicles at the Golgi apparatus of the F-cell and subsequently released via exocytosis to the ECF. The primary structure of APP is totally distinct from all other gut and pancreatic hormones (Figure 5) and, as such, precludes it from being included in the gastrin, secretin, or insulin family of polypeptides. The other two members of the PP-family, PPY and NPY, emanate from the L-cell of the lower colon and various hypothalamic nuclei and some sympathetic neural elements, respectively.

All three of these PP-type hormones have receptors on the exocrine pancreas, intestinal crypts, gall bladder, and various brain structures. Only APP has been demonstrated to have a metabolic effect, exercised via receptors and cAMP on the avian adipocyte, where the polypeptide acts as an antilipolytic agent. This may be very important since insulin is not antilipolytic in birds. The precise mechanism of action of APP relative to its enteric, vascular, and hypothalamic effects awaits further investigation (also see below).

D. Receptor Considerations

Unlike investigations of a similar nature in mammals, intensive studies of avian receptors to the four pancreatic polypeptides have not been carried out or reported. An exception is that of the avian insulin receptor in a limited number of avian tissues and where several good review papers exist (Simon and LeRoith, 1986; Adamo *et al.*, 1987, 1988; Simon and Taouis, 1993). Most such work has been done on chickens only and that of intensity restricted to liver, muscle, and brain tissue.

Insulin receptors have been described for avian liver, brain, cardiac and skeletal muscle tissue, myoblasts, chondrocytes, fibroblasts, and erythrocytes. Their ontogeny during embryogenesis differs with each tissue but usually they can be detected as early as days 2–3 in

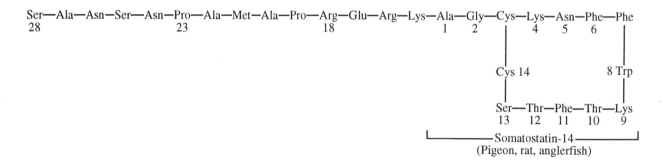

FIGURE 4 Structure of pigeon somatostatin-14 (SRIF). An additional 14-amino-acid extension to the N-terminus of SRIF-14 is shown also.

	1			5					10					15			18
APP:	Gly—Pro—Ser—Gln—Pro—Thr—Tyr—Pro—Gly—Asp—Asp—Ala—Pro—Val—Glu—Asp—Leu—Ile																	
BPP:	Ala		Leu—Glu		Glu				Asn			Thr—Pro			Gln—Met—Ala		
HPP:	Ala		Leu—Glu		Val				Asn			Thr—Pro			Gln—Met—Ala		
RPP:	Ala		Leu			Met				Tyr			Thr—His			Gln—Arg—Ala		

	NH$_2$ 35				30					25					20			
APP:	Tyr—Arg—His—Arg—Thr—Val—Val—Asn—Leu—Tyr—Gln—Gln—Leu—Asn—Asp—Tyr—Phe—Arg																	
BPP:		Pro			Leu—Met			Ile			Arg—Arg			Glu—Ala—Ala—Tyr—Gln			
HPP:		Pro			Leu—Met			Ile			Arg—Arg			Asp—Ala—Ala—Tyr—Gln			
RPP:		Pro			Leu—Thr			Ile			Arg—Arg			Gln—Thr—Glu—Tyr—Gln			

FIGURE 5 Structure of avian pancreatic polypeptide (APP). A comparison with other species is shown. BPP, bovine PP; HPP, human PP; RPP, rat PP.

the embryo. Insulin binding is highest in neural-related structures by day 8, but its distribution throughout is not in accord with that of IGF-1, indicating possibly a different role for insulin during the development of the nervous system (Simon and Taouis, 1993; Goddard et al., 1993). Injection of insulin antibodies or receptor antibodies as early as days 2–3 impairs development of those structures normally aided by insulin presence. The relatively low number of insulin receptors in liver and other tissues in birds probably contribute to the documented insulin insensitivity that exists as compared to mammals. Fasting increases insulin receptor number (in liver but not brain), but not affinity; feeding lowers the receptor number to normal (Simon and Taouis, 1993). Inhibitors of lysosomal proteases do not alter the half-life of chicken insulin receptors (about 10 hr), indicating that after internalization the receptor remains intact and is recycled back to the plasma membrane. This is in contrast to internalized insulin, which is degraded readily, indicating separate pathways for handling the hormone and the receptor in the cytosol (Simon and Taouis, 1993).

The structure of the avian insulin receptor has been studied thoroughly (Simon and LeRoith, 1986; Adamo et al., 1987) and has been found to be essentially the same as that described in mammals; that is, a heterotetrameric form consisting of two α- and two β-subunits. From a biosynthesis point of view, the proreceptor precursor is first glycosylated in the endoplasmic reticulum, only to be glycosylated further in the Golgi region (following cleaving to form the final α- and β-subunits). Different sizes of the α-subunit exist, and that for the muscle receptor is intermediate in size between those of brain and liver.

Insulin attaches to the externally exposed α-subunit, initiating autophosphorylation processes in the β-subunits. Activation of constitutive tyrosine (TYR) kinase follows, thus giving rise to the "insulin signal" (Simon and Taouis, 1993). Receptor kinase has been reported to be active in the chick embryo as early as day 2. The autophosphorylation process occurs within seconds (even at room temperature) and probably involves TYR 1146, 1150, and 1151 in the kinase domain (Adamo et al., 1987). Interestingly, fasting—which is likely to induce increased resistance to insulin—reduces insulin-stimulated phosphorylation by the liver receptor, but is without effect on the muscle or brain insulin receptor.

Avian insulin has been reported by many laboratories to be 2–4 times more potent than equivalent amounts of other (species) insulins employed under identical conditions. Also, the bird has been reported to be very resistant to exogenous insulin, both of avian and mammalian sources, though still the avian hormone is more effective. Growing chicks have decreased insulin receptors on their hepatocytes and thymocytes (Simon, 1989). However, there exists a very high affinity for insulin binding concomitant with a very prolonged delay in dissociation of the hormone-receptor complex, features that may well explain a good part of the so-called "insulin resistance" to exogenous hormone. The latter phenomenon appears to be due to the HIS occupying position 8 on the A-chain of the avian insulin molecule (Simon, 1989).

Genetic obese chickens are more sensitive to insulin than are lean chickens, a situation quite opposite to that found in most mammals. Further studies in obese chickens (Simon and Taouis, 1993) have indicated that functional coupling between insulin binding to the α-receptor subunit and the activation of the β-subunit kinase is not always the same, that is, it may vary among different tissues. Overall, it would appear that insulin resistance, as observed and expressed in Aves, involves both altered receptor/receptor–kinase as well as postreceptor events intimate to the signal transduction process. Despite the differences cited above, the avian insulin receptor appears to exist and function in a manner much similar to that described for mammals (Adamo *et al.*, 1987, 1988; Simon and Taouis, 1993).

Insulinlike growth factors (IGF-1, IGF-2) have been identified and characterized in birds, as have their receptors. The cloning, isolation, and characterization of chicken IGF-1 cDNAs have been described, and the nucleotide sequence encoding the mature peptide has been established (Goddard *et al.*, 1993). IGF-1 is prominent during embryogenesis and during the rapid-growth post-hatch phase, but its expression differs according to tissue. Thus, after day 10 of embryogenesis, IGF-1 mRNA is readily detectable in both pancreas and brain, but not liver. However, by 50 days post-hatch, liver IGF-1 mRNA is spectacularly increased (Goddard *et al.*, 1993). A certain amount of IGF-1 expression, but not all, is growth hormone dependent. In mammals, two types of receptors for the IGFs have been described, with the type 1 receptor being structurally related to the insulin receptor. Type 1 receptor definitely has been identified in chickens, even as early as day 2 of development. The biological action of IGF peptides in the embryonic chicken is mediated by the type 1 receptor and probably in the early post-hatch period as well. No evidence of the existence of a type 2 receptor in chickens has been offered, although a chicken liver cation-independent mannose-6-phosphate receptor has been isolated and purified. However, it does not bind IGF-2. Apparently, the biological actions of IGF-2 in Aves are initiated by the type 1 or related receptor (Goddard *et al.*, 1993).

Finally, the only other receptor data available germane to binding pancreatic islet hormones appears to be the chicken APP receptor, particularly the brain plasma membrane receptor (Adamo and Hazelwood, 1990). (Considerably more work in this arena has come from laboratories studying receptor requirements and characteristics of other members of the PP-family, namely that of PYY and NPY. Again, of all tissues, the brain receptor appears most studied (Hazelwood, 1993a,b). Characteristics of the APP receptor in plasma membranes prepared from chicken cerebellum include possession of disulfide bonds that are important in ligand binding. Further, thiol groups are essential for specific binding to the membranes. Considerably more work is needed in this area to complete our knowledge of islet cell receptor-signaling phenomena in Aves.

VI. PANCREATIC–ENTERIC REGULATION OF CARBOHYDRATE METABOLISM

The fact that embryogenesis of the pancreas commences with two evaginations from the foregut (near the hepatic anlagen) suggests strongly that an anatomical relationship between gut (enteric) and pancreatic tissue mass exists. The fact that exocrine pancreatic secretions are released (via ducts) to the small intestine to aid the process of digestion suggests a functional relationship between the two organs, also. That a functional interrelationship exists between gut and pancreas is even more obvious when one considers the humoral secretogogs from the upper gut that release a battery of pancreatic digestive enzymes. And tying all of this together would be absorbed nutrient stimulation of the islet cells as well as gut peptides (especially gastric inhibitory peptide (GIP)) releasing islet hormones even before (that is, in anticipation of) nutrients have been absorbed into the blood stream. Indeed, a true pancreatic-enteric functional axis exists and, to a great extent, it is oriented toward the organism's glucoregulation.

A. General Considerations

Glucose homeostasis requires organismic cooperation in all vertebrates. In particular, the gut and the pancreas must work together in evaluating the needs of the organism at any given time, and then respond with the appropriate adjustments to effect glucoregulation. Such maneuvers range from modulating the absorptive process to arranging the proper mix of islet hormones to be added to the freshly absorbed nutrients in the portal vein. Passing through the liver, then, this admixture of islet hormones, amino acids, free fatty acids and glycerol, and glucose are presented to the hepatocyte for subsequent handling. In both mammals and birds, insulin is a powerful anabolic hormone, a "hormone of feasting." Yet only in mammals is this hormone the dominant hormone of metabolism, of glucoregulation. Glucagon is a powerful catabolic hormone in both mammals and birds. It is a "hormone of fasting." While it plays a significant role in mammals in buffering decreasing plasma glucose levels, glucagon's role appears to be secondary to that of insulin in overall glucoregulation. Contrarily, in Aves, glucagon plays *the* major role in directing the distribution and destination of nutrient substrates, and insulin is secondary. Possibly, this is due to the fact that the rate of avian metabolism (particularly in passerines) demands a constant (high) supply of carbohydrate, regardless of physical activity, and glucagon plays a major role in retrieving/converting potential fuel. Totally depancreatized birds usually die in *hypoglyce-*

mic crises, advising us once again that secretions of the A-cell are life preserving in Aves.

Interestingly, the peripheral administration to birds of substances that inhibit glucose metabolism, or centrally of heavy metal-toxic substances that lesion specific hypothalamic nuclei associated with hunger-satiety, do not cause hyperphagia and/or obesity as they do in mammals (Simon, 1989). Quite possibly, hypothalamic, glucostatic control over food intake may not be important in birds. Frequently, exogenous insulin has little or no effect on food intake in birds. Recent studies indicate that portal vein infusion of glucose or glucagon clearly depresses food intake, an observation that may suggest the liver to be a prominent organ which signals the CNS of the nutritional state of the organism. This may involve the vagus nerves, as it has been known for over 50 years that hepatic glycogen levels and glucose production in birds are seriously altered by vagal activity. The avian hypothalamus may, indeed, exert direct control over the pancreatic islets. Lesions of the ventral medial hypothalamic (VMH) nuclei lead to a hyperinsulinemia in geese and chickens, as well as hyperphagia and occasional obesity (Simon, 1989).

Somatostatin's regulatory role is not a direct role as far as carbohydrate is concerned, but it may play a subsidiary role at the absorbing surface of the avian gut where it tends to smooth out over a protracted period of time the absorption of both glucose and lipid fragments. Pancreatic polypeptide (APP) remains somewhat of an enigma to students of avian carbohydrate metabolism, even though the first reports on it indicated a prominent glycogenolytic and hypocholesterolemic action. Its precise role in avian carbohydrate metabolism is yet to be established.

B. Tissue Effects of Pancreatic Hormones in Birds

As in mammals, most of the pancreatic islet hormones' effects are expressed through actions at the level of muscle, liver, adipose tissue, and to a much lesser extent, the erythrocyte, chondrocyte, thymocyte, and other tissues.

1. Plasma Effects

Plasma glucose is the ultimate source of intracellular glucose in the aforementioned tissues. Insulin is a powerful hypoglycemic agent in mammals but birds usually require administration of considerably larger amounts of the hormone to achieve the same (quantitative percentage) degree of hypoglycemia. In particular, convulsive doses of insulin in mammals are usually found to be moderately hypoglycemic in Aves, although there exists species differences in susceptibility. The chicken, for example, is extremely resistant to insulin-induced convulsions, the pigeon much less so. Fasting increases insulin sensitivity in birds. In addition to the insulin receptor characteristics discussed above, a plasma proteinaceous component in all probability contributes to avian insulin resistance. Nonetheless, the resulting hypoglycemia is the result of increased uptake and utilization of free glucose at the level of liver, muscle, and adipose tissue. The mechanisms by which insulin induces contramembrane glucose transport are discussed above (Section V,A).

Similar to its effects on plasma glucose, insulin aids the transport of both metabolizable and nonmetabolizable amino acids into cells of a wide variety of tissues ranging from myocardial cells to osteoblasts. Its effect appears to be mainly an action directed at the influx processes, not the efflux counterparts. Although the hormone also increases plasma uric acid levels, it still reduces circulating amino acid levels by about 30–40% and thereby induces a positive nitrogen balance. Apparently, this anabolic action requires *de novo* protein synthesis (Simon, 1989). Protein anabolic effects have been demonstrated in depancreatized ducks and geese, also. Additionally, although delayed in response, insulin induces an increase in plasma free fatty acids, a response that is largely attributable to glucagon release. It is debatable whether the release of glucagon is a direct result of insulin presence or whether it is a homeostatic response to the induced hypoglycemia. Increased plasma free fatty acids can easily be inhibited by glucose loading, indicating very likely a feedback mechanism involving glucose–glucagon–fatty acids. In summary, therefore, insulin has direct effects in removing both glucose and amino acids from the circulation and by an indirect action adds lipid substrates to the plasma.

2. Liver

Liver cells are not dependent on insulin presence for ECF free (nonphosphorylated) glucose to cross the plasma membrane; that is, glucose transport is not a rate-limiting feature of the hepatocyte. Evidence at hand strongly indicates that in chickens, as in mammals, insulin exerts its major effect on liver by increasing the activity of the enzyme glucokinase, not hexokinase. In doing so, the phosphorylated form of glucose, namely glucose-6-phosphate, reduces the level of free glucose in the cytosol, which in turn steepens the downhill concentration gradient for free glucose from outside to inside the cell. Simultaneously, the phosphorylated glucose is at the metabolic crossroads, ready to move in any one of three directions, two of which are most important to our discussion. In embryos as well as in growing birds, physiological concentrations of insulin favor glycogen synthesis and storage by an action on the rate-limiting enzyme, glycogen synthase. At the same time, the hormone inhibits phosphorylase activation (to break down glycogen, or glycogenolysis) and glucose-6-phosphatase, thereby keeping glucose-6-phosphate within the metabolic arena. IGF-1 has similar properties (as above) to that of insulin. And glucagon has strong counter action in that it inhibits glucokinase and glycogen synthase activities while augmenting phosphorylase and glucose-6-phosphatase activities. Thus, glucagon promotes hyperglycemia by its action on the liver. Insulin's contribution to the usual induced hypoglycemia, therefore, is not due to liver membrane transport phenomena, but rather due to shutting down hepatic glucose production.

Insulin's effect on avian liver to reduce gluconeogenesis, as is well established in mammals, has not been established with certainty. However, by increasing the activity of liver phosphofructokinase, insulin maintains the flow of glucose through the glycolytic pathway, an antigluconeogenic action. On the other hand, glucagon decreases the flux through this pathway, thereby favoring the formation of "new" glucose and releasing it to the plasma.

Hepatic lipid metabolism during embryogenesis is relatively low, the lipids being provided by oocytes, yolk sac, and related tissues. With the change to dry food after hatching, hepatic lipogenic enzymes increase in activity and are responsible for producing at least 95% of all lipids in birds, preparatory to their transfer to peripheral deposit sites. Adipocytes, therefore, and unlike the situation in mammals, synthesize less than 5% of the usable lipid in birds. Again, the islet hormones of importance in the lipogenic–lipolytic sequence are insulin and glucagon. Insulin stimulates synthesis and activation of the lipogenic enzymes malic enzyme, acetyl CoA carboxylase, and fatty acid synthase leading to lipid formation in the avian hepatocyte. Thyroid hormones potentiate these effects, particularly in the embryo. After hatching, insulin's effects on lipogenesis are more difficult to observe. Still, it would appear that in the liver, insulin is essential for the induction of lipogenic enzymes, maintenance of active lipogenesis, and the latter's restimulation following inhibition induced by a fasting period (Hazelwood, 1986a; Simon, 1989). On the other hand, avian lipolysis is under the control of glucagon, especially during fasting, where this hormone blocks synthesis of malic enzyme mRNA, activates hormone-sensitive lipase, and blocks LPL activities.

3. Adipose Tissue

Adipocytes in birds are under an enormous "metabolic pressure" to assist the liver to produce/make available adequate lipid-derived substrate to meet the needs of the bird. Egg laying, simple fasting, and migratory flight of hundreds of miles all place special demands on the organism to provide fuel to satisfy the metabolic needs of the moment. Insulin stimulation of lipogenesis in avian adipocytes appears quite modest, indeed, when compared to that seen in mammals. Further, epinephrine, a powerful lipolytic agent in mammals, is weakly lipolytic in Aves. Again, the major contribution to fat metabolism in birds appears to be that of glucagon, especially during times of metabolic stress and irrespective of the presence of insulin. Recall that insulin is not antilipolytic in birds; however, APP is antilipolytic *in vitro*. Prolonged flight, even in trained pigeons, invariably releases free fatty acids, epinephrine, norepinephrine, glucagon, and growth hormone (George *et al.*, 1989; John *et al.*, 1989). Mobilized fatty acids form the chief fuel during sustained flight and because the catecholamines are not lipolytic in birds to any great extent, their role may be to stimulate glucagon release. As a stress hormone, growth hormone elevation would support the lipolytic action of glucagon.

4. Muscle Tissue

Muscle metabolism in birds has not received the attention that other tissues have and in part this is due to the difficulty in preparing and working with this system experimentally. The role of insulin in transportation of amino acids is mentioned above. Additionally, it should be stated that the *de novo* synthesis of protein is enhanced by insulin presence, as is the inhibition of protein degradation. At low concentrations, insulin has been shown to increase hepatic production of albumin, alpha-1 globulin, fibrinogen, lipoproteins, and so on, while at the muscle level the hormone is responsible for rapid myogenesis in the embryo and muscle structural proteins in the adult. In final analysis, however, the anabolic effects of insulin via protein metabolism appear greatly overshadowed by the anabolic effect of the hormone on lipid metabolism, wherein lipogenesis and fat deposition are highly stimulatable (Simon, 1989).

C. Role of the Insulin/Glucagon [I/G] Molar Ratio

Considering the vagaries of the metabolic demands confronting birds such as being fully fed, or confronted by scarce availability of food during various annual seasons or the breeding season, clutch egg-laying periods, and prolonged migratory flight patterns, to name but a few common examples, it is reasonable to believe that homeostatic regulators must exist to ensure the organism of adequate energy (fuel) to carry out its activity. To the islet D-cell falls the onus of regulating the proper mix of hormones entering the portal vein that will ensure the proper utilization, distribution and/or channeling of nutrients to meet the required metabolic needs (Hazelwood, 1989). It is obvious that the requirements for metabolism (especially of carbohydrate) differ during a feeding period from that of a moderate fasting period. In the first case, the bird is metabolically "concerned" with the storage of excess calories for future use. The organism is in an anabolic mode. In the case of fasting, however, the concern is one of providing adequate fuel to sustain the bird even though little or no nutrients are entering the organism. In a sense, the bird must go to its (endogenous) nutrient bank and withdraw deposits made at an earlier time. It is in a catabolic mode. As stated earlier, insulin is a hormone of feasting; glucagon is a hormone of fasting. And the molar ratio of these two hormones largely dictates the proportional mixture added to the portal vein, blood that first perfuses the liver before being distributed elsewhere in the body.

Release of somatostatin (SRIF) from the avian pancreatic D-cell is under the regulation of neural, hormonal, humoral, and paracrine control. SRIF suppresses all pancreatic hormones (as well as delaying intestinal absorption), but to varying degrees. Glucagon appears to be most sensitive to the inhibitory action of SRIF, and APP appears to be the least sensitive. The adjustment of the appropriate insulin to glucagon (i.e., I/G) molar ratio by SRIF, therefore, assures the bird of hormones necessary to meet the needs of an anabolic or a catabolic mode at that time. Normal I/G ratios in the postabsorptive state in primates approximate 2.5 to 3.5. Normal avian (duck, goose, chicken, pigeon) I/G molar ratios are half or less than half of those of mammals. Any catabolic state in mammals (such as fasting, exercise, injury, diabetes mellitus) lowers the I/G ratio as insulin decreases in face of increased glucagon secretion. (Most catabolic states are stressful; glucagon is one of several stress hormones.) Such a change favors the retrieval of nutrients previously stored but needed at the current moment. By comparison, then, *normal* birds appear to be in a continuous catabolic mode, that is, they appear to be very similar to diabetic mammals. Studies in 99% depancreatized ducks, geese, and chickens appear to verify such a conclusion (Laurant and Mialhe, 1978; Hazelwood, 1986c). Such animals have an immediate decrease in I/G ratio for 4–5 days following surgery, followed by reattainment of normalcy thereafter. The role of SRIF, therefore, is not a direct one on carbohydrate metabolism, but rather is reflected by its ability to modulate other islet hormone secretions in order to achieve the desired and most efficacious mixture of insulin and glucagon.

VII. CARBOHYDRATE METABOLISM IN OTHER TISSUES

The effects of various avian pancreatic hormones, especially insulin, have been discussed in relation to effects on liver, muscle, and adipose tissue metabolism. Casual reference also has been made to effects of these hormones and the IGF's on fibroblasts, myoblasts, lens cells, chondrocytes, neural cells, and thymocytes as well. Three other structures of interest should now be mentioned, namely, heart, erythrocytes, and the unique glycogen body.

A. Heart

Embryonic chick hearts have exceedingly high glycogen levels as compared with other tissues at the same stage of development. At day 3, the myocardiocyte may be 15–20 times richer in glycogen than either the myocyte or the hepatocyte. Up to day 7 of embryogenesis, cardiac muscle is insensitive to the presence of insulin, glucose uptake being a function of its availability. After day 8, however, chick hearts are remarkably sensitive to insulin, as the hormone greatly increases glucose uptake and glycogen synthesis. Thus, between days 7 and 8 of chick development, receptor binding of insulin by the cardiocyte is functionally active. At later stages of embryonic development, exogenous insulin has a negative chronotropic effect on the heart, probably due to the hormone's ability to hyperpolarize cardiac muscle cells. After hatching, the hormone has little effect on cardiac glycogen levels.

Levels of cardiac glycogen continue to increase until about day 13 in the chick embryo, after which they gradually decrease to "adult" levels of approximately 180–200 mg/100 g wet tissue weight at hatching. Interestingly, these glycogen levels double or even triple during a 48- to 96-hr fast. As seen also in fasted mammals, virtually all other tissues' glycogen deposits are depleted while those of the heart increase. The pituitary gland must be functionally intact for the increase in cardiac glycogen to occur, growth hormone being central to this effect. Metabolism of avian cardiac tissue is similar to that of mammalian cardiac cells; that is, birds draw a disproportionate amount of their metabolic fuel from plasma free fatty acids and lactate. Fasting increases this "sparing" of carbohydrate reserves, largely by the lipolytic action of both growth hormone and glucagon, the secretions of which are increased in the fasting state. Insulin has little or no effect on adult avian heart glycogen, and neither SRIF nor APP have been shown to affect the heart in any way.

B. Erythrocytes (RBC)

Turkey erythrocytes have been candidates for considerable study mainly in characterizing the insulin receptor. (Turkey insulin and chicken insulin are identical in structure, function, and potency.) *In vitro,* insulin hyperpolarizes the erythrocytic membrane; it also encourages the uptake of potassium but has no effect on glucose transport. Insulin decreases plasma levels of phosphorous, magnesium, and calcium in turkeys, but in chickens the effect on Ca^{2+} is one of hypercalcemia due (probably) to increased bone resorption (Simon, 1989). Species differences and physiological status make direct comparisons difficult. Studies on the effects of glucagon, SRIF, and APP on various avian erythrocytes are yet to be reported.

C. The Glycogen Body

Originally described by Italian anatomists over 120 years ago, the avian glycogen body was initially called "the corpora sciatica" because, anatomically, it is positioned dorsal to the spinal cord at the level of the emergence of the sciatic plexus in all birds examined (see Chap. 1 in Hazelwood, 1986a, for details). The gland is somewhat pea shaped, gelatinously opaque in texture and color, and is 80% glycogen on a lipid-free basis. Cells are so loaded with glycogen that the nucleus is pushed entirely to the side in juxtaposition with the plasma membrane. Glycolytic enzymes common to other avian tissues are located within the glycogen body but attempts to alter the glycogen content by nutritional means, hormones, temperature, or other techniques have failed. In contrast with other tissues in the body, the hexosemonophosphate shunt appears to be extremely active. Could this mean that the NADPH produced by this oxidative pathway is of significant functional value to the nearby CNS? Surgical attempts to remove the glycogen body have been made, but the surviving birds do not appear to suffer any defects. Despite its name, the avian glycogen body remains an enigma to students of avian carbohydrate metabolism. Possibly, studies of the anatomical counterpart area in related but lower forms, such as fish, amphibia, and reptiles, may divulge the secret that the avian glycogen body holds.

VIII. ALTERED AVIAN CARBOHYDRATE METABOLISM

A. Spontaneous Diabetes Mellitus

This condition, well known in humans, appears spontaneously in a wide variety of animals, mostly mammals. Classical signs of Type 1 diabetes (insulin dependent) have been reported in monkeys, dogs, cats, sheep, pigs, beef, and horses to name but a few examples. In birds, however, the disease appears to be a clinical rarity. What case reports do exist, usually come to the attention of a veterinarian who is treating a pet for polyuria, polydipsia, and weight loss. Invariably, when found, this condition occurs in budgies (*Melopsittacus undulatus,* the so-called "love bird"). The afflicted birds have an exceedingly high blood glucose level, often well above 800 mg/dl plasma. Concomitant with this severe hyperglycemia is a marked ketonemia, one that is readily lethal if the victim is not treated daily with insulin. Data are severely missing on the nature and the progress of this disease in birds.

B. Pancreatic Chemocytotoxins

Early reports that cobaltus chloride and related substances selectively destroyed the avian A-cell have come under close scrutiny and re-evaluation. These metals cause a general toxic effect, affecting liver and adrenal medulla as well as causing spurious pancreatic damage, and thereby induce an hyperglycemic state. Use of Synthalin-A, a biguanide, to achieve a similar effect also has come under severe review, largely due to its toxic effects on the glycolytic pathway. Surgical extirpation of the third pancreatic lobe *and* the splenic lobe appears to be the best way to achieve selective A-cell deficiency, one that leads to severe hypoglycemia and early death in birds.

Chemical destruction of the insulin-secreting cells of the pancreas is a favorite experimental tool used in mammals to examine various aspects of Type 1 diabetes mellitus. However, birds appear to be impervious to the beta-cytotoxic action of these agents. Insulin release and early B-cell depletion does occur in young chickens treated with ascorbic acid. Glucose intolerance results and attempts to release additional insulin from the pancreas by use of pancreatropic drugs are without success. Thus, the "diabetic state," transitory as it is, appears to be due to the parent compound, not to its oxidation product, dehydroascorbic acid. The latter is known to be diabetogenic in rats, guinea pigs, and humans.

Successful experimental diabetes has been established in mammals by use of alloxan, a derivative of pyrimidine, since the mid-1940's, and streptozotocin, an antibiotic, since the mid-1960's. To a large extent, use of the latter in mammals has replaced use of the former due to its higher degree of selectivity in destroying the pancreatic B-cell. Alloxan actually damages several endocrine tissues, as well as the liver, but with time, only that of the pancreas appears permanent. Alloxan is without diabetogenic effect in birds such as chickens, pigeons, geese, crows, kingfisher, black munia, parakeets, etc. (Hazelwood, 1986a; Ghosh, 1991). Regardless of dose, alloxan is ineffective in altering basal insulin levels, glucose-stimulated insulin release patterns, or glucose tolerance (Hazelwood, 1986a). Similarly, streptozotocin equals alloxan's ineffectiveness in birds, although it does hinder the insulin response to a glucose load. Obviously, different cellular mechanisms underlie the avian response to alloxan and streptozotocin, yet neither of these potentially diabetogenic agents leads to diabetes in Aves. Infusion of glucose along with, or shortly before, the injection of alloxan to rats prevents B-cell damage. Could the normally high (by mammalian standards) avian blood glucose (200-250 mg/dl, i.e., double that of normal mammals) act to buffer the otherwise damaging effects of alloxan? Or could the excessively high levels of glutathione in the avian pancreas (about 10 times that of mammals) offer protection against the beta-cytotoxic agents? The only mammal reported to be refractive to alloxan and streptozotocin is the guinea pig and, interestingly, this specie has a very high pancreatic content of glutathione (Ghosh, 1991). There are no reports on the attempts (or effects of) at destroying either the pancreatic D- or F-cells.

C. Surgical Extirpation of the Pancreas

Experimental removal of the avian pancreas dates back at least to the classic studies of vonMering and Minkowski in the 1890's when they included several species of birds in their survey on the effects of pancreatectomy in a wide variety of vertebrates. In fact, these early studies identified the pancreas as the major organ controlling normal carbohydrate metabolism in mammals. vonMering and Minkowski's efforts led to the belief that birds could not be made diabetic permanently by surgical removal of the pancreas. Either their birds did not become diabetic at all after surgery, or they demonstrated a temporary diabetic state and died (after a severe bout of anorexia) within 10 days of the operation. The suggestion was made, however, that carnivorous birds were more likely to show "diabetic symptoms" than other avian forms. This perception persists even today, although there is little concrete evidence to support such a suggestion. The functional role of the pancreatic splenic lobe (see Fig. 1) appears to be the basis of most discrepant reports over the last 70-90 years relative to inducing diabetes in birds by removal of the organ (Hazelwood, 1986b; Simon, 1989). In some species (duck), removal of this lobe of the pancreas is relatively simple and the resulting pancreatectomy is complete. In other avian forms (goose), the splenic lobe is moderately accessible and pancreatectomy is usually 95-100% complete. In chickens, however, this lobe is very inaccessible and complete pancreatectomy is rarely achieved. Thus, discordant results over the years emanating from attempts at pancreatectomy are most likely due to the degree of completeness of surgery. Remnants of pancreatic tissue left behind after surgery, rapidly hypertrophy and experience altered sensitivity to islet challenges. Summarizing, in those preparations with assured 100% pancreas removal, the bird suffers a severe hypoglycemic crises, requiring repeated glucagon and/or glucose infusions to prevent moribund convulsions. Glucagon is essential to life, especially during a fasting period. Still, anti-insulin antibodies cause hyperglycemia in depancreatized ducks, an observation indicating that insulin, also, is essential for glucoregulation (Hazelwood, 1986a, 1986b; Simon, 1989). This is emphasized further when periods of feeding are studied in such operated ducks. Also, these observations suggest that a secondary source of insulin, other than the pancreas, may exist in birds, especially since there is a persistence of circulating insulin for as long as 10 days following surgery (Colca and Hazelwood, 1982; Hazelwood, 1986a). This suggestion has been made many times by many workers but it is a statement still awaiting final verification (Colca and Hazelwood, 1982; Simon, 1989).

In birds that undergo 99% surgical pancreatectomy, metabolic alterations vary with the specie involved. Subtotal pancreatectomy apparently results in a transient diabetic state in ducks, the birds becoming euglycemic two weeks later. In geese, however, permanent diabetes results and is attended by low plasma insulin levels and progressively rising glucagon levels (Laurant and Mialhe, 1978; Karmann and Mialhe, 1982; Hazelwood, 1989). The diabetic state is characterized by hyperglycemia, low plasma insulin and glucagon levels, and high amino acid levels. The maintenance of normal responsiveness to glucose by the avian islet A-cell requires the presence of insulin (Karmann and Mialhe, 1982; Hazelwood, 1989; Simon, 1989).

Of all experimental results obtained from surgical extirpation of the pancreas in birds, the one basic, underlying feature all data emphasize, appears to be that glucagon is very important to Aves for proper glucoregulation, although insulin is necessary for complete normalcy. Further, glucagon is more important in the fasting bird, while insulin is more important in the fed state.

D. Pancreatropic Agents

Oral anti-diabetic agents, used clinically in diabetic humans, are effective hypoglycemic and hypofattyacidemic agents in birds, also. Actually, the plasma free fatty acid levels decrease prior to glycemic levels when tolbutamide or chlorpromamide is injected into domestic fowl. Interestingly, a hypoglycemic response to tolbutamide administration occurs also in pancreatectomized–hepatectomized chickens, once again raising the question of the possible existence of a secondary source of insulin in Aves. Apparently, there is no peripheral inhibition of lipid synthesis in depot fat sites when birds are given tolbutamide.

E. Role of Other Hormones

Of all other avian hormones, growth hormone (GH) and prolactin (PRL) appear to have the most impact on normal avian carbohydrate metabolism. The injection of GH is without effect on glycemic levels in birds, even though it does suppress the islet A-cell (ducks) while stimulating the B-cell (ducks and geese). Hypophysectomy in birds leads to increased hepatic lipogenesis and generalized, systemic obesity. Avian GH is a very potent agent—much more so than equivalent amounts of mammalian GH—in promoting lipolysis and glycerol release in turkey, chicken, and pigeon adipocytes, even in the absence of supportive glucocorticoids. Fundamentally, GH stimulates basal lipolysis, inhibits hormone-stimulated lipolysis [similar to that of insulin in mammals], and stimulates glucose uptake in birds (Rudas and Scanes, 1983; Scanes, 1992; Scanes et al., 1993). Thus, the absence of insulin's anti-lipolytic effect is alleviated by the action of both GH and APP in birds. In all probability, the different effects of GH in Aves are modulated by different signal transduction systems (Scanes et al., 1993). Furthermore, as has been found with different species of mammalian GH, different regions (specific domains) of the avian molecule are responsible for the many different effects that are observed when the entire structure is administered to birds (Scanes et al., 1993).

Prolactin (PRL) injection into birds results in diverse biological effects that probably are mediated by PRL variants. Some birds have more than 10 forms of plasma PRL, ranging in size from 23 to 75 kDa (Nicoll et al., 1986). Glucagon injections increase PRL release from the pituitary gland, simultaneously with a decrease in GH secretion. These opposite effects are attributed to the hyperglycemia induced by glucagon. Fasting has the opposite effect on these two pituitary hormones. PRL is lipogenic in birds, especially in liver tissue. PRL, too, is anti-insulin in most of its metabolic actions. Its fluctuating plasma levels during stressful times in a bird's (normal) life suggest that PRL is important in shifting "metabolic gears" to provide fuel at times of crises (Hazelwood, 1986a). Thus, PRL is to be added to the fraternity of stress hormones in birds.

In addition to GH and PRL, other hormones have counter-regulatory actions (to those of insulin) in birds, and therefore fall into that group of hormones called "stress hormones." A brief treatment of these (glucocorticoids, catecholamines, thyroxine) is presented in Hazelwood (1986a).

In summary, avian carbohydrate metabolism is regulated by a complex integration of neural, hormonal, humoral, and nutrient components. In this context, mammals and birds are similar. Pancreatic hormones play a dominant role in this regulation, the I/G molar ratio being central to maintaining metabolic homeostasis. This ratio of insulin to glucagon appears set in birds to "favor" catabolic, or retrieval, reactions, thereby assuring the animal of plentiful fuel supplies to sustain a high rate of metabolism. This set point also is such that adequate energy substrate is available to meet such stressful and diverse metabolic demands as presented by severe fasting, egg laying, and/or sustained migratory flight.

References

Adamo, M. L., and Hazelwood, R. L. [1990]. Characterization of liver and cerebellar binding sites for avian pancreatic polypeptide. *Endocrinology* **126**, 434–440.

Adamo, M., Simon, J., Rosebrough, R. W.., McMurtry, J. P., Steele, N. C., and LeRoith, D. (1987). Characterization of the chicken muscle insulin receptor. *Gen. Compar. Endocrinol.* **68**, 456–465.

Adamo, M. L., LeRoith, D., Simon, J., and Roth, J. (1988). Effect of altered nutritional states on insulin receptors. *Ann. Rev. Nutr.* **8**, 149–166.

Colca, J. R., and Hazelwood, R. L. (1982). Persistence of immunoreactive insulin, glucagon, and pancreatic polypeptide in the plasma of depancreatized chickens. *J. Endocrinol.* **92**, 317–326.

George, J. C., John, T. M., and Mitchell, M. A. (1989). Flight effects on plasma levels lipid, glucagon, and thyroid hormones in homing pigeons. *Horm. Metab. Res.* **21**, 542–545.

Gerich, J. (1983). Somatostatin and analogues. In "Diabetes Mellitus" (M. Ellenberg and H. Rifkin, eds.], 3rd ed., pp. 225–254. Med. Exam. Pub., New York.

Ghosh, A. (1991).New therapeutic modalities of diabetes: Avian endocrinological approach. *Proc. Zool. Soc.* (Calcutta) **44**, 65–68.

Goddard, C., Butterwith, S. C., Roberts, R. D., and Duclos, M. J. (1993). Insulin-like growth factors and IGF binding proteins. In "Avian Endocrinology" (P. J. Sharp, ed.), pp. 275–284. J. Endocrinology, Bristol.

Hazelwood,R. L. (1986a). Carbohydrate metabolism. In"Avian Physiology" (P. D. Sturkie, ed.), 4th ed., pp. 303–325. Springer-Verlag, New York.

Hazelwood, R. L. (1986b). Pancreas. In "Avian Physiology" (P. D. Sturkie, ed.), 4th ed., pp. 494–500. Springer-Verlag, New York.

Hazelwood, R. L. (1986c). Are birds normally diabetic? In"ACTA XIX Congress Intl. Ornithologica" (H. Ouellet, ed.), Vol. 2, pp. 2223–2233, Univ. Ottawa Press, Ottawa.

Hazelwood, R. L. (1989). Somatostatin: Regulator of I/G molar ratios. In "The Endocrine Pancreas," pp. 144–146. Prentice-Hall, New Jersey.

Hazelwood, R. L. (1993a). The pancreatic polypeptide [PP-fold] family: vascular, gastrointestinal, and feeding behavioral implications. *Proc. Soc. Exp. Biol. Med.* **202**, 44–63.

Hazelwood, R. L. (1993b). From avian pancreatic polypeptide to mammalian neuropeptides: Carbohydrate implications. In "Avian Endocrinology" (P. J. Sharp, ed.), pp. 189–201. J. Endocrinology Ltd., Bristol.

John, T. M., Viswanathan, M., George, J. C., and Scanes, C. G. (1988). Flight effects on plasma levels of free fatty acids, growth hormone, and thyroid hormones in homing pigeons. *Horm. Metabol. Res.* **20,** 271–273.

Karmann, H., and Miahle, P. [1982]. Progressive loss of sensitivity of the A-cell to insulin in geese made diabetic by subtotal pancreatectomy. *Horm. Metab. Res.* **14,** 452–458.

Kimmel, J. R., Pollock, H. G., and Hazelwood, R. L. (1968). Isolation and characterization of chicken insulin. *Endocrinology* **83,** 1323–1330.

Laurant, F., and Mialhe, P. (1978). Effect of free fatty acids and amino acids on glucagon and insulin secretions in normal and diabetic ducks. *Diabetologia* **15,** 313–321.

Nicoll, C. S., Mayer, G. L., and Russell, S. M. (1986). Structural features of prolactins and growth hormones that can be related to their biological properties. *Endocr. Rev.* **7,** 169–203.

Rudas, P., and Scanes, C. G. (1983). Influences of growth hormone on glucose uptake by avian adipose tissue. *Poult. Sci.* **62,** 1838–1845.

Scanes, C. G. (1992). Lipolytic and diabetogenic effects of native and biosynthetic growth hormone in the chicken: A re-evaluation. *Comp. Biochem. Physiol. A* **101,** 871–878.

Scanes, C. G., Aramburo, C., Campbell, R. M., Kopchick, J. J., and Radecki, S. V. (1993). Chemistry and physiology of poultry growth hormone. In "Avian Endocrinology" (P. J. Sharp, ed.), pp. 261–274. J. Endocrinology Ltd., Bristol.

Simon, J. (1989). Chicken as a useful species for the comprehension of insulin action. In "CRC Critical Reviews in Poultry Biology," Vol. 2, pp. 121–148. CRC Publ., Boca Raton, FL.

Simon, J., and LeRoith, D. (1986). Insulin receptors of chicken liver and brain. Characterization of alpha and beta subunit properties. *Eur. J. Biochem.* **158,** 125–132.

Simon, J., and Taouis, M. (1993). The insulin receptor in chicken tissues. In "Avian Endocrinology" (P. J. Sharp, ed.), pp. 177–188. J. Endocrinology Ltd., Bristol.

CHAPTER 21

The Pineal Gland, Circadian Rhythms, and Photoperiodism

EBERHARD GWINNER AND MICHAELA HAU
Max-Planck-Institut für Verhaltensphysiologie
D-82346 Andechs
Germany

I. Anatomy of the Pineal Gland   557
II. Pineal Hormones   558
   A. Melatonin   558
   B. Extrapineal Melatonin Production   559
III. Physiological Effects of the Pineal Gland   560
   A. Circadian System   560
   B. Photoperiodism and Endogenous Circannual Cycles   564
   C. Thermoregulation   565
   D. Sleep   565
   References   565

I. ANATOMY OF THE PINEAL GLAND

The pineal gland (epiphysis cerebri) of birds is located at the dorsal surface of the brain, where it is embedded in a triangular space between the two hemispheres of the telencephalon and the cerebellum (Figure 1, top). During embryonic development, the pineal gland is formed as an evagination of the neuroepithelium of the posterior diencephalic roof between the habenula and the posterior commissure. In the course of phylogeny, it has changed from an organ that was both secretory and photosensitive (e.g., in fish and amphibians) to a purely neuroendocrine gland (in mammals) to which light information is transmitted only indirectly through neuronal pathways. These phylogenetic changes of the pineal were accompanied by alterations in the structure and neuronal organization of the gland. In birds, a striking interspecific variability in pineal anatomy and function is apparent (Quay, 1965; Menaker and Oksche, 1974; Binkley, 1988).

The majority of diurnally active birds examined have a well-developed pineal organ. However, rudimentary pineals have been described for some nocturnal and crepuscular species; for example, owls (*Strigiformes*) and night-active shearwaters and petrels (*Procellariiformes*; Quay, 1972).

In general, the avian pineal gland consists of a distal enlargement, the pineal vesicle, which is exposed to the skull and adheres to the dura mater, and a proximal slender pineal stalk, which is connected to the dorsal wall of the third ventricle (Figure 1, bottom). In the domestic chicken (*Gallus domesticus*), the pineal is approximately 1.8 mm wide and 2.8 mm long (Sato and Wake, 1983). According to morphological criteria, avian pineals have been classified as (1) saccular (passerine birds), (2) tubulofollicular (pigeon, *Columba livia* and duck, *Anas platyrynchus*), and (3) lobular glands (adult chicken and Japanese quail, *Coturnix coturnix japonica*; Figure 1, bottom).

The pineal gland of birds contains photoreceptorlike and ependymal (interstitial) cells as well as neurons (Menaker and Oksche, 1974; Collin and Oksche, 1981; Ariens Kappers, 1983; Binkley, 1988). The receptorlike pinealocytes represent modified photoreceptor cells with reduced outer segments, which in many avian species lack any synaptic connections to intrinsic nerve cells. However, the immunohistochemical detection of photosensitive pigments in these outer segments (e.g., pinopsin, Okano *et al.*, 1994) structurally related to rhodopsin

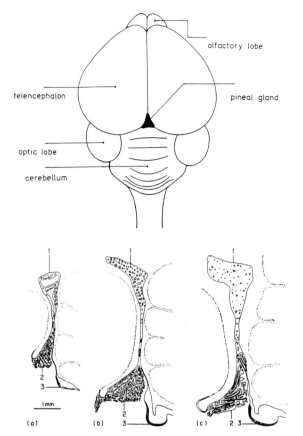

FIGURE 1 (Top) The pineal organ of the pigeon, embedded between the two hemispheres of the telencephalon and the cerebellum. (Bottom) Various types of avian pineal organs: (a) Saccular (e.g., house sparrow); (b) tubulofollicular (e.g., pigeon); (c) lobular (e.g., chicken). (1) Pineal organ; (2) choroid plexus of the third ventricle; (3) subcommissural organ. (After Matsumoto and Ishii (1992), and Menaker and Oksche (1974), with permission.)

rons which exit the pineal via the stalk and project into the habenular nuclei and the periventricular layer of the hypothalamus. In contrast to passerines, only a few pineal projections to the brain have been found in galliforms.

II. PINEAL HORMONES

A. Melatonin

Most pinealocytes produce various indoleamines among which melatonin (5-methoxy-*N*-acetyltryptamine) is of major significance. Melatonin is synthesized from tryptophan in a multistep process (Figure 2). First, tryptophan is actively taken up from the blood and converted into serotonin. Next, *N*-acetylserotonin is formed by the enzyme *N*-acetyltransferase (NAT), the rate-limiting step in melatonin synthesis. From there, hydroxyindole-*O*-methyltransferase (HIOMT) finally synthesizes melatonin. Melatonin is a small and highly lipophilic molecule,

suggests that avian pinealocytes are photoreceptive. Indeed, melatonin production in isolated pineal cells is strongly light dependent (Robertson and Takahashi 1988b; Takahashi et al., 1989). Consistent with the neuroendocrine function of the gland, pinealocytes contain secretory granules (dense core vesicles) in the golgi regions and within the basal processes. The nerve cells are mainly of the bipolar and in some species also of the pseudounipolar type.

The pineal gland is connected with the brain via both efferent and afferent pathways (see e.g., Korf et al., 1982; Ariens Kappers, 1983; Sato and Wake, 1983). A rich innervation is typical of many avian pineal glands and is mainly accomplished by sympathetic (noradrenergic) postganglionic fibers originating in the superior cervical ganglia (SCG). Additional nervous inputs from the habenular complex and the area of the paraventricular nucleus have been shown in house sparrows (*Passer domesticus*). The most conspicuous afferent pathway in passerines consists of acetylcholinesterase-positive neu-

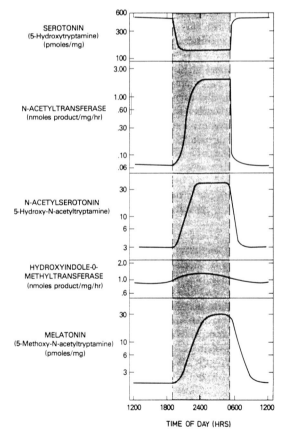

FIGURE 2 Rhythms in indole metabolism in the pineal gland (data from the rat). (Left) Metabolic pathway from serotonin to melatonin. (Right) Daily variations in the concentrations of metabolites and enzymes. Shaded portion indicates the dark fraction of the daily light–dark cycle. (After Klein, 1985.)

which rapidly diffuses through membranes and into the blood stream to exert its actions throughout the body.

Once in the circulation, melatonin is rapidly catabolized and inactivated in the liver through hydroxylation (Kopin et al., 1961; Cardinali et al., 1983; Binkley, 1988). Another enzymatic catabolic mechanism which often occurs in retina and skin of vertebrates involves deacetylation and deamination (Grace et al., 1991). Finally, nonenzymatic degradation takes place when melatonin scavenges free radicals (e.g., Reiter et al., 1995; Filadelfi and Castrucci, 1996; see below).

A characteristic feature of pineal melatonin production and secretion is a 24-hr rhythm, where levels are high at night and low during day. Although the amplitude of the nocturnal rise in melatonin may differ conspicuously among species, this general pattern has been found in literally every vertebrate studied so far, irrespective of whether it is day- or night-active. In most species investigated, the rhythm in melatonin synthesis persists under constant laboratory conditions both *in vivo* and *in vitro*, indicating that it is controlled by an endogenous ciradian oscillator (Figure 3). Among the enzymes involved in melatonin synthesis, only the rate-limiting NAT exhibits a circadian rhythm in enzymatic activity (e.g., Binkley, 1988).

Melatonin exerts many of its actions by binding to specific high-affinity membrane-bound receptors coupled to a G-protein (Morgan et al., 1994; Reppert et al., 1995). Melatonin binding, as revealed by 2-[^{125}I]iodomelatonin (^{125}I-Mel), a potent melatonin agonist (Dubocovich and Takahashi, 1987), occurs throughout the avian brain, but is highest in areas associated with vision. In particular, ^{125}I-Mel binds to retinorecipient structures and integrative nuclei of the visual system, including the retina and circadian, tectofugal, thalamofugal, and accessory optic visual pathways (e.g., Cassone et al., 1995). ^{125}I-Mel binding was also found in the song nuclei of the zebra finch (*Poephila guttata*) and house sparrow (Gahr and Kosar, 1996; Whitfield-Rucker and Cassone, 1996), as well as in auditory relay nuclei and structures of the limbic system associated with arousal and vocalizations (Rivkees et al., 1989; Cozzi et al., 1993). In the quail, ^{125}I-Mel binding was also observed in the nuclei of the cranial nerves that coordinate eye movements (Cozzi et al., 1993). In contrast to mammals, the avian pineal does not bind ^{125}I-Mel (e.g., Cassone et al., 1995). In many brain areas the density of melatonin binding sites undergoes a circadian rhythm in a phase opposite to that of melatonin, with highest densities during the late subjective day (Brooks and Cassone, 1992; Lu and Cassone, 1993a,b).

So far, two melatonin receptor subtypes have been identified in birds: CKA, which is similar to the mammalian Mel$_{1a}$ receptor, and CKB, which is similar to the *Xenopus* Mel$_{1c}$ receptor. The relative contribution of the two receptor subtypes to melatonin binding sites in the brain differs (Reppert et al., 1995): CKA mRNA is expressed in high amounts in the optic tectum and the retina (to a lesser extent also in the neostriatum, the hypothalamus, and the thalamus) and CKB mRNA is expressed to a moderate extent in the optic tectum, the neostriatum, the hypothalamus, the thalamus, and the pineal (in smaller amounts also in the cerebellum and the retina).

Recent evidence from mammals suggests that melatonin may also exert some actions independent of receptors—a result of its ability to rapidly penetrate cell membranes. There, it may act as a scavenger of free radicals, preventing oxidative damage within cells (Reiter et al., 1995, 1996). Futhermore, in mammals a nuclear melatonin receptor has been found in the central nervous system as well as in peripheral organs. ^{125}I-Mel binding outside the central nervous system includes sites in the retinae, the gastrointestinal system, the Harderian gland, the gonads, the spleen, the kidney, and the spinal cord (Ayre et al., 1992, 1994; Song et al., 1992; Huether, 1993; Lee and Pang, 1993; Pontoire et al., 1993; Poon et al., 1993; Wan and Pang, 1993).

B. Extrapineal Melatonin Production

Originally identified in the pineal organ, melatonin was subsequently found to be also synthesized in at least three other structures: the retina, the Harderian gland,

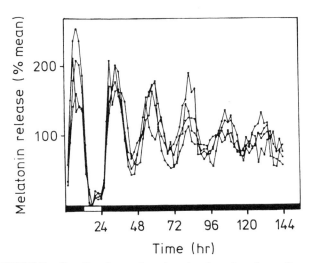

FIGURE 3 Circadian changes in melatonin release from four cultures of dispersed chick pineal cells. Cultures were kept at 37°C and initially exposed to a 12:12-hr light–dark cycle before being transferred to constant darkness. Data were normalized relative to the individual means of the four cultures. A circadian rhythm with a period slightly shorter than 24 hr persisted in all four cultures. (Reprinted from *J. Neurosci.* **8**, L. M. Robertson and J. S. Takahashi, Circadian clock in cell culture. I. Oscillation of melatonin release from dissociated chick pineal cells in flow-through microcarrier culture, pp. 12–21, Copyright (1988a) with permission from Elsevier Science.)

and the gastrointestinal tract (reviewed in Huether, 1993). While pineal melatonin is released into the circulation to exert most of its known functions in various organs of the body, retinal, Harderian gland, and gastrointestinal melatonin is usually not or only in some species and/or under certain circumstances released into the vascular system. Extrapineal melatonin presumably acts mainly locally near the site of production.

Retinal melatonin production is widespread in birds and other vertebrates. It resembles pineal melatonin production to the extent that it exhibits a circadian rhythm which persists not only *in vivo* but also *in vitro* in cultured eyecups (Pierce *et al.*, 1993). Depending on the species, retinal melatonin may or may not be released into the circulation. In house sparrows and European starlings (*Sturnus vulgaris*), little or no retinal melatonin is released into the blood (Janik *et al.*, 1992), whereas in pigeons (Foa and Menaker, 1988) and quails Underwood and Siopes, 1984) about one-third of the plasma melatonin in the night stems from the retina.

High-affinity melatonin binding sites have been found in the amacrine cells of the inner plexiform layer of the retina (Dubocovich, 1988). Possibly retinal melatonin acts as a modulator of neural transmission and neuronal excitability in the retina, counteracting dopamine (Huether, 1993; Dubocovich, 1988). Retinal melatonin is also involved in, for example, photoreceptor outer segment disk shedding, retinomotor movements of cones and rods, and retinal pigment aggregation (e.g., Cahill and Besharse, 1995).

The Harderian gland of the chicken and pigeon (Pang *et al.*, 1977; Vakkuri *et al.*, 1985a) contains high levels of melatonin during darkness. Since HIOMT activity was also detected in the Harderian gland, it is likely that melatonin is indeed produced in this organ. The function of Harderian gland melatonin, which is presumably not released into the circulation, is unknown.

In the gastrointestinal tract of birds and mammals high concentrations of melatonin (about 400 times as much as in the pineal; Huether, 1993) have been measured in the enterochromaffin cells of the mucosal epithelium of the gastrointestinal tract. In the pigeon (Vakkuri *et al.*, 1985a) gut melatonin content varies in a pronounced 24-hr rhythm; levels increase at night and decline during day. In pigeons, this rhythm persists with an unaltered amplitude after pinealectomy (Vakkuri *et al.*, 1985b), suggesting that melatonin is produced rhythmically in the gut. This contention is supported by the fact that, in quail, NAT, the rate-limiting enzyme of melatonin synthesis, is present in the duodenum and that its activity shows a 24-hr rhythm similar to that found in the pineal and the retina (Lee *et al.*, 1991). To what extent this rhythm is autonomous (based on an endogenous circadian mechanism) is not clear.

High-affinity melatonin bindig sites have been found throughout the gastrointestinal tract of ducks (Lee *et al.*, 1991; Lee and Pang, 1993; Lee *et al.*, 1995) and in the duodenum of chicken (Pontoire *et al.*, 1993), suggesting that melatonin exerts paracrine or autocrine functions within the gastrointestinal tract itself. In mammals, melatonin may reduce gastrointestinal motility and thus food transit time, thereby increasing food utilization (Harlow and Weekly, 1986). Gut melatonin has also been implicated as a modulator of intestinal crypt cell proliferation. Finally, intestinal melatonin may play a role as a free-radical scavenger and antioxidant (Pentney and Bubenik, 1995; Reiter *et al.*, 1995). Whether melatonin, apart from acting locally in the gut, is also released into the portal blood and thus contributes to circulating melatonin levels is not clear yet. However, recent studies in mammals have found transient daytime increases in peripheral melatonin levels after food ingestion suggesting that gastrointestinally produced melatonin may affect systemic functions (summarized e.g., in Bubenik *et al.*, 1996).

III. PHYSIOLOGICAL EFFECTS OF THE PINEAL GLAND

First isolated by Lerner *et al.* in 1958, melatonin was named after its lightening action of the skin of amphibians and fish, resulting from pigment (melanin) aggregation. Today many other endocrine effects of melatonin are known; in vertebrates they include involvement in themoregulation, skin color changes, reproduction, immune response, puberty, carcinostasis, aging, and circadian organization (Touitou *et al.*, 1993; Yu and Reiter, 1993; Arendt, 1995). However, many of these functions have not yet been established in birds. In the following, we will concentrate on the four major areas in which research on birds has contributed substantially to our understanding of pineal function.

A. Circadian System

1. The Pineal Gland as a Component of the Circadian Pacemaker

Numerous processes in organisms show regular 24-hr variations, reflecting the pronounced daily pattern of their environment. In many organisms, these 24-hr rhythms in biological function are partly based on an endogenous circadian rhythmicity that persists with a period slightly deviating from 24 hr even under constant environmental conditions. In 1968, Gaston and Menaker discovered that this "free-running" circadian locomotor activity rhythm of house sparrows held in con-

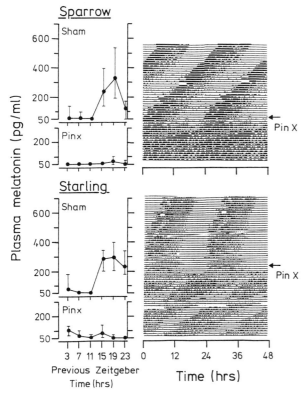

essential component of the avian circadian pacemaking system (reviews in Menaker and Zimmerman, 1976; Rusak, 1982; Underwood and Goldman, 1987; Binkley, 1988; Gwinner, 1989; Cassone, 1990a,b; Gwinner et al., 1994, 1997). This hypothesis was strongly supported by an experiment in which pineal organs of house sparrows were transplanted into the anterior chamber of the eye of pinealectomized arrhythmic sparrows kept in DD (Figure 5): the pineal transplant restored rhythmicity in the host bird with a phase characteristic of the rhythm of the donor (Zimmerman and Menaker, 1979). Subsequent studies lent further support to the idea that the pineal exerts its rhythm-sustaining function through the periodic production and secretion of melatonin. In particular, it was shown that periodic (approximately 24 hr) injections or infusions of melatonin or its periodic application via the drinking water were capable of inducing and synchronizing circadian functions like locomotor activity or body temperature (summaries in Gwinner et al., 1994, 1997; Figure 6).

FIGURE 4 (Left) Plasma melatonin levels (medians with quartiles) in sham pinealectomized (Sham) and pinealectomized (Pinx) house sparrows (upper) and European starlings (lower) measured 2 days after transfer from a 24-hr light–dark cycle to constant darkness (DD) and plotted relative to the previous light–dark cycle ("zeitgeber time"; lights on from 0 to 12 hr). (Right) Changes in the pattern of feeding of an individual house sparrow and starling held in DD (sparrow) or dim light (LL; starling). Each horizontal line from hr 0 to 24 represents the activity recording of 1 day. Recording of successive days are displayed underneath each other and the original recordings (hr 0 to 24) have been double-plotted (hr 24 to 48) for more convenient inspection of the data. Initially both birds showed a free-running circadian rhythm of feeding activity with a period considerably (sparrow) or slightly (starling) shorter than 24 hr, as indicated by the progressively earlier beginning of activity on successive days. Following pinealectomy (and the resulting abolition of the plasma melatonin rhythm) feeding activity became arrhythmic in the sparrow and more irregular in the starling. (Reprinted from *Brain Res. Bull.* **44,** E. Gwinner, M. Hau, and S. Heigl, Melatonin: Generation and modulation of avian circadian rhythms, pp. 439–444, Copyright (1994), with permission from Elsevier Science.)

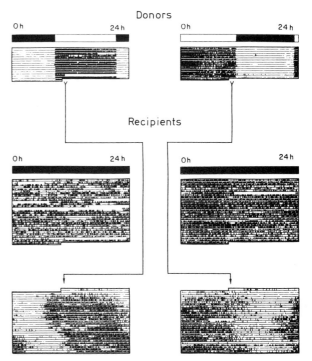

FIGURE 5 Transplantation of pineal organs from two house sparrow donors (upper) into two pinealectomized recipients (middle). The donors were on the two different light–dark cycles indicated at the top of their respective activity records; their perch-hopping rhythms were synchronized accordingly. The pinealectomized recipients lived in constant darkness and showed arrhythmic activity patterns. Following transplantation of donor pineals, rhythmicity was resumed by the previously arrhythmic recipients (lower). The emerging rhythm had the phase of the rhythm of the respective donor. For details on actograms see the legend to Figure 4. (After Zimmerman and Menaker, 1979.)

stant darkness (DD) was abolished following surgical removal of the pineal organ (Figure 4, upper right). Sympathetic denervation of the pineal did not result in a corresponding impairment of circadian rhythmicity; hence, the pineal appeared to exert its rhythm-sustaining function humorally rather than neuronally (Menaker and Zimmerman, 1976). Since the avian pineal is capable of periodically synthesizing melatonin even *in vitro* and under constant environmental conditions (Figure 3), it was proposed that the pineal is an

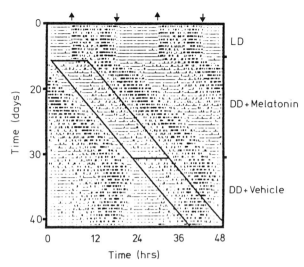

FIGURE 6 Feeding activity of a pinealectomized pigeon that was first synchronized with a light–dark (LD) cycle with light from 6:00 to 18:00 (arrows) and subsequently released into constant darkness (DD). On the first day in DD and each day thereafter the bird received a 10-hr continuous infusion of a melatonin/vehicle solution (DD + melatonin) or a vehicle solution only (DD + vehicle). The boxed area represents infusion time. Onsets of infusions occurred 0.77 hr later each day than the previous day. The bird's activity synchronized with the melatonin cycle such that high feeding activity coincided with the time of day at which plasma melatonin levels were low; during vehicle infusion the rhythm free-ran with a period shorter than 24 hr. Note that in pigeons plasma melatonin rhythmicity is not completely eliminated by pinealectomy because in this species retinal melatonin is secreted into the bloodstream with a circadian pattern. (After *J. Comp. Physiol. A,* Effects of physiological cycles of infused melatonin on circadian rhythmicity in pigeons, C. C. Cabot and M. Menaker, **170,** 615–622, (1992), © Springer-Verlag.)

Within the circadian system, the pineal melatonin rhythm interacts with at least one other circadian oscillator. This is indicated, for instance, by the observation that following pinealectomy or the discontinuation of periodic melatonin substitution in pinealectomized birds, it usually takes the behavioral rhythms of a bird several cycles to damp out (Figure 4, upper right; Chabot and Menaker, 1992; Heigl and Gwinner, 1994). The same holds true for the rhythms of pinealectomized birds transferred from light–dark cycles (which are capable of inducing and synchronizing rhythmicity in birds without a pineal; Gaston and Menaker, 1968) to constant light or darkness. Moreover, when melatonin rhythms with different periods are exogenously applied, the range of periods to which the behavioral rhythms of pinealectomized house sparrows can be synchronized is limited (Heigl and Gwinner, 1995). This is indicative of an oscillator operating under the influence of a synchronizing agent. Melatonin acts on an oscillatory system that is at least partly located in the avian equivalent of the mammalian suprachiasmatic nucleus (SCN). Recent studies have identified two paired brain regions in the hypothalamus that might contain the avian SCN. First, an area in the medial hypothalamus adjacent to the third ventricle, the periventricular preoptic nucleus or medial SCN (mSCN), seems to represent the anatomical equivalent of the mammalian SCN (e.g., Hartwig, 1974). Large lesions of the mSCN region usually abolish circadian rhythms (reviewed in Cassone and Menaker, 1984; Gwinner, 1989; Cassone, 1990a,b). Second, recent studies have pointed to a region caudal to the mSCN, the lateral hypothalamic retinorecipient nucleus or visual SCN (vSCN) which is situated between the supraoptic decussations and the ventral lateral geniculate body (Cassone and Moore, 1987; Norgren and Silver, 1989a,b). The vSCN are densely innervated by retinal cells via the retinohypothalamic tract (RHT; Cassone and Moore, 1987; Norgren and Silver, 1989a,b). Furthermore, the vSCN contain melatonin binding sites (e.g., Rivkees *et al.,* 1989; Cassone *et al.,* 1995) and the rhythmic metabolic activity of the nuclei (as measured by 2-deoxyglucose uptake) depends on the presence of an intact pineal or periodic exogenous melatonin administration (Lu and Cassone, 1993a,b). Also, the circadian rhythm in noradrenergic turnover in the avian pineal gland is disrupted by lesion of the vSCN (Cassone *et al.,* 1990). The vSCN are thought to provide feedback to the pineal organ via a polysynaptic pathway, whose interruption (e.g., by surgical removal of the superior cervical ganglion) damps circadian rhythms in plasma melatonin of chicken in DD (Cassone and Menaker, 1983).

In addition to the pineal and the SCN, the eyes are also components of the avian circadian pacemaking system—at least in some galliform and columbiform species. In the pigeon, the eyes contribute to the circadian system via periodic melatonin secretion by the retina. Pinealectomy merely leads to an incomplete abolition of the plasma melatonin rhythm; complete behavioral arrhythmia can only be achieved by simultaneous pinealectomy and optical enucleation (Ebihara *et al.,* 1984; Foa and Menaker, 1988; Oshima *et al.,* 1989). Likewise, in quail, part of the nocturnal peak in plasma melatonin stems from melatonin secretion from the retinae. In this species, however, the main contribution of the eyes to circadian performance is neuronal: circadian rhythms become abolished after optic-tract impairment, a treatment that leaves the plasma melatonin rhythm intact (Underwood *et al.,* 1990).

On the basis of these and other findings, Cassone and Menaker (1984) have suggested that the pineal, the SCN, and (at least in some species) the eyes are connected with each other in a neuroendocrine feedback loop and that the gross output of this composite pacemaking system depends on inhibitory interactions among these three oscillatory components. The interactions between these oscillators can also be described in more general terms by a model of "internal resonance" (Gwinner, 1989) in which the oscillatory components of the circadian pacemaking system synchronize and amplify each other through resonance rather than interacting with each other by mutual inhibition. The known or suspected interactions between the components described above are shown in Figure 7.

2. Inter- and Intraspecific Diversity of Avian Circadian Systems

Some of the examples presented above suggest that species differ with regard to the extent to which the pineal and/or melatonin control avian circadian rhythms: in the house sparrow the pineal is an obligatory component of the sytem while in the quail (Simpson and Follett, 1981; Underwood and Siopes, 1984) and the chicken (MacBride, 1973) the pineal and its melatonin rhythm seem to be almost redundant for circadian function. The European starling assumes an intermediary position insofar as pinealectomy may or may not abolish the rhythm of perch-hopping (Figure 4, lower right, see also Gwinner, 1978). Species differences of this type can be accommodated by the above model if one assumes that the relative strengths of the oscillators, the pineal, SCN, and (only in some species) eyes vary among species. There are also differences between species with regard to the significance of the melatonin rhythm in the control of different circadian functions. In European starlings, for instance, although pinealectomy disturbs and in some cases disrupts circadian rhythms of locomotor activity, the same operation has no effect on the rhythm of feeding (Gwinner et al., 1987). Even more extreme examples of this kind are known from reptiles (Menaker and Tosini, 1996).

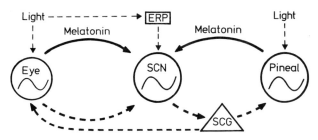

FIGURE 7 Interactions between oscillatory components in the pineal, the SCN, and the eyes in the circadian pacemaking system of birds. (→) Endocrine connections (melatonin); (••▶) neuronal connections; (--▶) action of light on the eye, the pineal, and extraretinal/extrapineal photoreceptors (ERP).

3. The Pineal and Melatonin as Modulators of Synchronized Circadian Rhythms

The effects of the pineal and melatonin described above concern circadian rhythms of birds kept under constant conditions, where unimpaired circadian rhythms free-run with periods slightly deviating from 24 hr. Under natural conditions, however, the endogenous rhythms are synchronized with the 24-hr periodicity of the environment. Periodically changing environmental factors like food availability, temperature, or social stimuli can synchronize avian circadian rhythms, but the most important synchronizer ("zeitgeber") of circadian rhythms is the periodic alteration of light intensity (Aschoff, 1981; Pittendrigh, 1981). In birds, light for circadian synchronization is perceived in part by the eyes from which the photic information is transmitted to the SCN via the RHT. In addition, light perceived by the pineal and by deep encephalic photoreceptors may also play a role in synchronization (Cassone and Menaker, 1984; Foster et al., 1994; Saldanha et al., 1994).

When activities are synchronized with light, pinealectomy does not abolish their rhythmicity, even in species in which the same operation causes arrhythmicity under constant conditions (Gaston and Menaker, 1968). Obviously light perceived by the eyes and deep encephalic photoreceptors still reaches whatever is left of the circadian pacemaker. Slight modifications of the pattern of activity do, however, occur as a result of pinealectomy (e.g., Gaston, 1971). Likewise, exogenous melatonin treatment changes the properties of the synchronized circadian system as has been documented for house sparrows: when the birds are carrying subcutaneous silastic implants filled with melatonin (resulting in constantly elevated levels of plasma melatonin) their circadian rhythms of locomotor activity adjusted more quickly to a phase-shifted light–dark cycle as compared to birds carrying empty implants (Hau and Gwinner, 1995). Also, the circadian rhythm had a larger "range of entrainment" (i.e., it could be synchronized with a wider range of zeitgeber periods) in birds carrying a melatonin implant than in birds carrying an empty one (Hau and Gwinner, 1994). When house sparrows were kept in constant conditions, melatonin implantation resulted in the destabilization or even elimination of free-running circadian rhythms (Turek et al., 1976; Beldhuis et al., 1988), suggesting that melatonin treatment—by eliminating or greatly reducing the amplitude of the endogenous melatonin rhythm—renders the circadian system more passive (less self-sustained). In the synchronized state, less self-sustained rhythms are known to have shorter resynchronization times and larger ranges of entrainment than more self-sustained oscillators (Wever, 1965). In light of these results, the seasonal changes in melatonin amplitude which have been described for several species of birds may have a biological function: by modulating the melatonin amplitude, the avian circadian system may become more or less responsive to zeitgeber stimuli and thereby more or less adjustable to environmental synchronizing cues. The reduced melatonin amplitude observed in high arctic birds during midsummer and in migratory birds during the migratory seasons may serve the function of rendering birds more susceptible to zeitgeber stimuli during times of the year at which reliable synchronization to weak or changing zeitgeber conditions is particularly important (Gwinner et al., 1997).

B. Photoperiodism and Endogenous Circannual Cycles

The life of most organisms is organized not only on a daily but also on a yearly basis. Reproduction, in particular, is usually restricted to favorable times of the year, thus guaranteeing optimal conditions for offspring development and parent survival (Immelmann, 1973; Murton and Westwood, 1977; Farner and Follet, 1979; Farner and Gwinner, 1980). In most animals, annual cycles in reproduction and other processes are strongly controlled by the seasonal changes in day length (photoperiod). In some species, at least, photoperiod acts on an endogenous circannual rhythm, which is capable of oscillating even in the absence of photoperiodic variations, but which is normally synchronized by photoperiod (Gwinner, 1986). Hence, in these cases, the yearly photoperiodic cycle and the internal circannual rhythm interact with each other in much the same way as the daily cycle of light intensity with the endogenous circadian clock. As with circadian rhythms, we may thus ask whether the pineal and its hormone melatonin affect the basic oscillatory system, its synchronization mechanism, or both.

While periodic melatonin secretion by the pineal and other structures represents a basic component of avian *circadian* pacemaking systems (Figure 7), there is little evidence—either from birds or from mammals—in favor of a significant involvement of the pineal in generating *circannual* cycles. A study in European starlings has indicated that the circannual rhythms of reproduction and molt exhibited in LD 12:12 were abolished by pinealectomy in most individuals, but rhythmicity persisted in others (Gwinner and Dittami, 1980). For the interpretation of these results it must be considered that starling circannual rhythms are labile and expressed only under constant photoperiods very close to 12 hr, hence, the effects of pinealectomy may have been unspecific. The circannual rhythms in food intake, body mass, and testicular size of spotted munias (*Lonchura punctulata*) held in constant light were not affected by pinealectomy (Chandola-Saklani et al., 1991).

In contrast to its apparent redundancy for circannual rhythm generation in both birds and mammals, the mammalian pineal is involved in the transduction of photoperiodic information required for synchronization of annual reproductive cycles. In hamsters and sheep, for instance, the duration of the daily dark phase is reflected in the pattern of melatonin secretion: the longer the night length, the longer is the duration of elevated melatonin levels. This duration of the phase of elevated melatonin "informs" the brain about whether a day is long or short. This is shown, for instance, in pinealectomized ferrets or sheep, where the hypothalamo–pituitary–gonadal axis can no longer respond to photoperiodic changes. Timed daily infusions of melatonin resulted in predicted photoperiodic responses: infusions mimicking long photoperiods (i.e., short nights) induced long-day responses while infusions mimicking short photoperiods induced short-day responses (e.g., Bartness and Goldman, 1989; Bartness et al., 1993; Bittman, 1993; Woodfill et al., 1994).

In birds, photoperiod is likewise reflected in the duration of the daily melatonin peak (e.g., Kumar and Follett, 1993). However, in contrast to the situation in mammals, this signal seems to be redundant for photoperiodic time measurement in several species of birds, including the quail, in which pinealectomy leaves photoperiodic responses of the hypothalamo–pituitary–gonadal axis intact (summaries in Gwinner et al., 1981; Kumar et al., 1993). Moreover, in quail (Juss et al., 1993) and the black-headed bunting (*Emberiza melanocephala*) (Kumar, 1966), the extension of the duration of elevated nocturnal melatonin by exogenous melatonin did not result in a corresponding change of the birds' interpretation of night length. Only for a few bird species (baya weaver, *Ploceus phillipinus;* spotted munia) have pineal effects on the reproductive system been reported (summary in Chandola-Saklani et al., 1991).

While the melatonin signal does not usually seem to be involved in the regulation of the hypothalamo–pituitary–gonadal axis, it may play a role in controlling seasonal phenomena such as birdsong. In passerine birds, song is known to depend on certain song control nuclei in the brain, including the high vocal center (HVC). The size of some of these nuclei changes seasonally, concomitantly with the waxing and waning of song activity (summaries: Brenowitz et al., 1991; Nottebohm, 1981). Androgens regulate these anatomical changes but a certain long-day-induced increase in the HVC has been observed even in castrated American tree sparrows (*Spizella arborea;* Bernard *et al.,* 1997). The presence of high-affinity ^{125}I-Mel binding sites in the HVC of male zebrafinches (Gahr and Kosar, 1996) and house sparrows (Whitfield-Rucker and Cassone, 1996) suggests a possible role for melatonin in regulating song nucleus size in these birds.

C. Thermoregulation

Removal of the pineal gland abolishes the circadian rhythm in body temperature of house sparrows kept in constant light (Binkley *et al.,* 1971). If kept in a LD cycle, pinealectomized house sparrows, pigeons, quails, and chickens continue to exhibit a circadian rhythm in body temperature but nocturnal minimal body temperatures are higher than normal (summarized in e.g., Heldmaier, 1991). The resulting decrease of the amplitude of the circadian temperature rhythm can be reversed by melatonin treatment (John *et al.,* 1978; Saarela and Heldmaier, 1987). Furthermore, the pineal, presumably via its hormone melatonin, can cause direct hypothermic effects. Melatonin injections result in a considerable decrease of house sparrow body temperature (by 4.7°C) within 30 min (Binkley 1986). On the other hand, chicken body temperature remains unaltered by pineal removal, but the birds exhibit a decreased heat tolerance during scotophase (Cogburn and Harrison, 1980).

D. Sleep

Melatonin has been proposed to be involved in sleep regulation of birds (pigeons), as daily infusions of the hormone restored REM and NREM sleep disturbed by exposure to continuous bright light (Phillips and Berger, 1992). The sleep impairment is mediated by the light-induced suppression of endogenous melatonin production. Because melatonin decreases body temperature and metabolic rate as well as facilitating sleep at night, it has been suggested that it acts by synchronizing the physiology of birds to the external light–dark cycle, reducing energy expenditure during periods of inactivity (Berger and Phillips, 1994).

References

Arendt, J. (1995). "Melatonin and the Mammalian Pineal Gland." Chapman and Hall, London.
Ariens Kappers, J. (1993). Innervation of the vertebrate pineal gland. *In* "The Pineal Gland and Its Endocrine Role" (J. Axelrod, F. Fraschini, G. P. Velo, eds.), pp. 87–107. Plenum, New York/London.
Ashoff, J. (1981). Freerunning and entrained circadian rhythms. *In* "Handbook of Behavioural Neurobiology" (J. Aschoff, ed.), Vol. 4, pp. 81–93. Plenum, New York/London.
Ayre, E. A., Yuan, H., and Pang, S. F. (1992). The identification of ^{125}I-labelled iodomelatonin-binding sites in the testes and ovaries of the chicken (*Gallus domesticus*). *J. Endocrinol.* **133,** 5–11.
Ayre, E. A., Wang, Z. P., Brown, G. M., and Pang, S. F. (1993). [^{125}I] Iodomelatonin binding-sites in avian gonads. *FASEB J.* **7,** 181.
Ayre, E. A., Wang, Z. P., Brown, G. M., and Pang, S. F. (1994). Localization and characterization of [^{125}I] iodomelatonin binding-sites in duck gonads. *J. Pineal Res.* **17,** 39–47.
Bartness, T. J., Bradley Powers, J., Hastings, M. H., Bittman, E. L., and Goldman, B. D. (1993). The timed infusion paradigm for melatonin delivery: What has it taught us about the melatonin signal, its reception, and the photoperiodic control of seasonal responses? *J. Pineal Res.* **15,** 161–190.
Bartness, T. J., and Goldman, B. D. (1989). Mammalian pineal melatonin: A clock for all seasons. *Experientia* **45,** 939–945.
Beldhuis, H., Dittami, J., and Gwinner, E. (1988). Melatonin and the circadian rhythms of feeding and perch-hopping in the European starling, *Sturnus vulgaris*. *J. Comp. Physiol. A* **164,** 7–14.
Berger, R. J., and Phillips, N. H. (1994). Constant light suppresses sleep and circadian rhythms in pigeons without consequent sleep rebound in darkness. *Am. J. Physiol.* **36,** R945–R952.
Bernard, D. J., Wilson, F. E., and Ball, G. F. (1997). Testis-dependent and -independent effects of photoperiod on volumes of song control nuclei in American tree sparrows (*Spizella arborea*). *Brain Res.* **760,** 163–169.
Binkley, S. (1988). "The Pineal: Endocrine and Neuroendocrine Function." Prentice-Hall, Englewood Cliffs, NJ.
Binkley, S., Kluth, E., and Menaker, M. (1971). Pineal function in sparrows: Circadian rhythms and body temperature. *Science* **174,** 311–314.
Bittman, E. L. (1993). The sites and consequences of melatonin binding in mammals. *Am. Zool.* **33,** 200–211.
Brenowitz, E. A., Nalls, B., Wingfield, J. C., and Kroodsma, D. E. (1991). Seasonal changes in avian song nuclei without seasonal changes in song repertoire. *J. Neurosci.* **11,** 1367–1374.
Brooks, D. S., and Cassone, V. M. (1992). Daily and circadian regulation of 2-[^{125}I]iodomelatonin binding in the chick brain. *Endocrinology* **131,** 1297–1304.
Bubenik, G. A., Pang, S. F., Hacker, R. R., and Smith, P. S. (1996). Melatonin concentrations in serum and tissues of porcine gastrointestinal tract and their relationship to the intake and passage of food. *J. Pineal Res.* **21,** 251–256.
Cahill, G. M., and Besharse, J. C. (1995). Circadian rhythmicity in vertebrate retinas—Regulation by a photoreceptor oscillator. *Prog. Ret. Eye Res.* **14,** 267–291.
Cardinali, D. O., Vacas, M. I., Keller Sarmiento, M. I., and Morguenstern, E. (1983). Melatonin action: Sites and possible mechanisms in brain. *In* "The Pineal Gland and Its Endocrine Role" (J. Axelrod, F. Fraschini, and G. P. Velo, eds.), pp. 277–301. Plenum, New York/London.
Cassone, V. M. (1990a). Effects of melatonin on vertebrate circadian systems. *Trends Neurosci.* **13,** 457–464.
Cassone, V. M. (1990b). Melatonin: Time in a bottle. *In* "Oxford Reviews of Reproductive Biology" (S. R. Milligan, ed.), Vol. 12, pp. 319–367. Oxford Univ. Press, Oxford.
Cassone, V. M., and Menaker, M. (1983). Sympathetic regulation of chicken pineal rhythms. *Brain Res.* **272,** 311–318.
Cassone, V. M., and Menaker, M. (1984). Is the avian circadian system a neuroendocrine loop? *J. Exp. Zool.* **232,** 539–549.
Cassone, V. M., and Moore, R. Y. (1987). Retinohypothalamic projection and suprachiasmatic nucleus of the house sparrow, *Passer domesticus*. *J. Comp. Neurol.* **266,** 171–183.
Cassone, V. M., Forsyth, A. M., and Woodles, G. L. (1990). Hypothalamic regulation of circadian noradrenergic input to the chick pineal gland. *J. Comp. Physiol. A* **167,** 187–192.

Cassone, V. M., Brooks, D. S., and Kelm, T. A. (1995). Comparative distribution of 2[^{125}I]iodomelatonin binding in the brains of diurnal birds—Outgroup analysis with turtles. *Brain Behav. Evol.* **45,** 241–256.

Chabot, C. C. (1990). "Melatonin regulation of feeding and locomotor rhythms in the pigeon, *Columba livia*." Unpublished Dissertation thesis.

Chabot, C. C., and Menaker, M. (1992). Effects of physiological cycles of infused melatonin on circadian rhythmicity in pigeons. *J. Comp. Physiol. A* **170,** 615–622.

Chandola-Saklani, A., Pant, K., and Lal, P. (1991). Involvement of the pineal gland in avian reproductive and other seasonal rhythms. *Acta XX Congr. Int. Ornithol.* **4,** 2030–2041.

Cogburn, L. A., and Harrison, P. C. (1980). Adrenal, thyroid, and rectal temperature responses of pinealectomized cockerels to different ambient temperatures. *Poult. Sci.* **59,** 1132–1141.

Collin, J.-P., and Oksche, A. (1981). Structural and functional relationships in the nonmammalian pineal gland. In "The Pineal Gland" (R. J. Reiter, ed.), Vol. I, pp. 27–67. CRC Press, Boca Raton, FL.

Cozzi, B., Stankov, B., Viglietti-Panzica, C., Capsoni, S., Aste, N., Lucini, V., Fraschini, F., and Panzica, G. (1993). Distribution and characterization of melatonin receptors in the brain of the Japanese quail, *Coturnix japonica*. *Neurosci. Lett.* **150,** 149–152.

Dubocovich, M. L. (1988). Pharmacology and function of melatonin receptors. *FASEB J.* **2,** 2765–2773.

Dubocovich, M. L., and Takahashi, J. S. (1987). Use of 2-[^{125}I]iodomelatonin to characterize melatonin binding sites in chicken retina. *Proc. Natl. Acad. Sci. USA* **84,** 3916–3920.

Ebihara, S., Uchiyama, K., and Oshima, I. (1984). Circadian organization in the pigeon *Columbia livia*: The role of the pineal organ and the eye. *J. Comp. Physiol. A* **154,** 59–69.

Farner, D. S., and Follett, B. K. (1979). Reproductive periodicity in birds. In "Hormones and Evolution" (E. J. W. Barrington, ed.), pp. 129–148. Academic Press, London.

Farner, D. S., and Gwinner, E. (1980). Photoperiodicity, circannual and reproductive cycles. In "Avian Endocrinology" (A. Epple and M. H. Stetson, eds.), pp. 331–366. Academic Press, New York.

Filadelfi, A. M. C., and Castrucci, A. M. D. L. (1996). Comparative aspects of the pineal/melatonin system of poikilothermic vertebrates. *J. Pineal Res.* **20,** 175–186.

Foà, A., and Menaker, M. (1988). Blood melatonin rhythms in the pigeon. *J. Comp. Physiol.* **164,** 25–30.

Foster, R. G., Grace, M. S., Provencio, I., Degrip, W. J., and Garcia-Fernandez, J. M. (1994). Identification of vertebrate deep brain photoreceptors. *Neurosci. Biobehav. Rev.* **18,** 541–546.

Gahr, M., and Kosar, E. (1996). Identification, distribution, and developmental changes of a melatonin binding site in the song control system of the zebra finch. *J. Comp. Neurol.* **367,** 308–318.

Gaston, S. (1971). The influence of the pineal organ on the circadian activity rhythm in birds. In "Biochronometry" (M. Menaker, ed.), pp. 541–549. National Acad. Sciences, Washington, DC.

Gaston, S., and Menaker, M. (1968). Pineal function: The biological clock in the sparrow? *Science* **160,** 1125–1127.

Grace, M. S., Cahill, G. M., and Besharse, J. C. (1991). Melatonin deacetylation: Retinal vertebrate class distribution and *Xenopus laevis* tissue distribution. *Brain Res.* **559,** 56–63.

Gwinner, E. (1978). Effects of pinealectomy on circadian locomotor activity rhythms in European starlings, *Sturnus vulgaris*. *J. Comp. Physiol.* **126,** 123–129.

Gwinner, E. (1986). "Circannual Clocks." Springer-Verlag, Berlin/Heidelberg.

Gwinner, E. (1989). Melatonin and the circadian system of birds: Model of internal resonance. In "Circadian Clocks and Ecology" (T. Hiroshige and K. Honma, eds.), pp. 127–145. Hokkaido Univ. Press, Sapporo.

Gwinner, E., and Dittami, J. (1980). Pinealectomy affects the circannual testicular rhythm in European starlings (*Sturnus vulgaris*). *J. Comp. Physiol.* **136,** 345–348.

Gwinner, E., Wozniak, J., and Dittami, J. (1981). The role of the pineal organ in the control of annual rhythms in birds. In "The Pineal Organ: Photobiology, Biochronometry, Endocrinology" (Proceedings of the 2nd Coll. of the European Pineal Study Group) (A. Oksche and H. Wallraff, eds.), pp. 99–121. Springer-Verlag, Heidelberg.

Gwinner, E., Subbaraj, R., Bluhm, C. K., and Gerkema, M. (1987). Differential effects of pinealectomy on circadian rhythms of feeding and perch-hopping in the European starling. *J. Biol. Rhythms* **2,** 109–120.

Gwinner, E., Hau, M., and Heigl, S. (1994). Phasic and tonic effects of melatonin on avian circadian systems. In "Circadian Clocks and Evolution" (T. Hiroshige and K. Honma, eds.), pp. 127–137. Hokkaido Univ. Press, Sapporo.

Gwinner, E., Hau, M., and Heigl, S. (1997). Melatonin: Generation and modulation of avian circadian rhythms. *Brain Res. Bull.* **44,** 439–444.

Hadley, M. E. (1992). "Endocrinology." Prentice-Hall, Englewood Cliffs, NJ.

Harlow, H. J., and Weekly, B. I. (1986). Effect of melatonin on the force of spontaneous contractions of *in vitro* rat small and large intestine. *J. Pineal Res.* **3,** 277–284.

Hartwig, H. G. (1974). Electron microscopic evidence for a retinohypothalamic projection to the suprachiasmatic nucleus of *Passer domesticus*. *Cell Tissue Res.* **153,** 89–99.

Hau, M., and Gwinner, E. (1994). Melatonin facilitates synchronization of sparrow circadian rhythms to light. *J. Comp. Physiol. A* **175,** 343–347.

Hau, M., and Gwinner, E. (1995). Continuous melatonin administration accelerates resynchronization following phase shifts of a light-dark cycle. *Physiol. Behav.* **58,** 89–95.

Heigl, S., and Gwinner, E. (1994). Periodic melatonin in the drinking water synchronizes circadian rhythms in sparrows. *Naturwissenschaften* **81,** 83–85.

Heigl, S., and Gwinner, E. (1995). Synchronization of circadian rhythms of house sparrows by oral melatonin: Effects of changing period. *J. Biol. Rhythms* **10,** 225–233.

Heldmaier, G. (1991). Pineal involvement in avian thermoregulation, *Acta XX Congr. Int. Ornithol, Christchurch, New Zealand*, pp. 2042–2051.

Huether, G. (1993). The contribution of extrapineal sites of melatonin synthesis to circulating melatonin levels in higher vertebrates. *Experientia* **49,** 665–670.

Immelmann, K. (1973). Role of the environment in reproduction as source of "predictive" information. In "Breeding Biololgy of Birds" (D. S. Farner, ed.), pp. 121–147. Natl. Acad. Sci. USA, Washington, DC.

Janik, D., Dittami, J., and Gwinner, E. (1992). The effect of pinealectomy on circadian plasma melatonin levels in house sparrows and European starlings. *J. Biol. Rhythms* **7,** 277–286.

John, T. M., Itoh, S., and George, J. C. (1978). On the role of the pineal in thermoregulation in the pigeon. *Horm. Res.* **9,** 41–56.

Juss, T. S., Meddle, S. L., Servant, R. S., and King, V. M. (1993). Melatonin and photoperiodic time measurement in Japanese quail (*Coturnix coturnix japonica*). *Proc. R. Soc. Lond. B* **254,** 21–28.

Klein, D. C. (1985). Photoneural regulation of the mammalian pineal gland. In "Photoperiodism, Melatonin and the Pineal," Ciba Foundation Symposium 117, pp. 38–51.

Kopin, I. J., Pare, C. M. B., Axelrod, J., and Weissbach, H. (1961). The fate of melatonin in animals. *J. Biol. Chem.* **236**, 3072–3075.

Korf, H-W., Zimmerman, N. H., and Oksche, A. (1982). Intrinsic neurons and neural connections of the pineal organ of the house sparrow, *Passer domesticus*, as revealed by anterograde and retrograde transport of horseradish peroxidase. *Cell Tissue Res.* **222**, 243–260.

Kumar, V. (1996). Effects of melatonin in blocking the response to a skeleton photoperiod in the blackheaded bunting. *Physiol. Behav.* **59**, 617–620.

Kumar, V., and Follett, B. K. (1993). The circadian nature of melatonin secretion in Japanese quail (*Coturnix coturnix japonica*). *J. Pineal Res.* **14**, 192–200.

Kumar, V., Juss, T. S., and Follett, B. K. (1993). Melatonin secretion in quail provides a seasonal calendar but not one used for photoperiodic time-measurement. *In* "Melatonin and the Pineal Gland" (Y. Touitou, J. Arendt, and B. K. Follett, eds.), pp. 163–168. Elsevier, Amsterdam.

Lee, P. P. N., and Pang, S. F. (1993). Melatonin and its receptors in the gastrointestinal tract. *Biol. Signals* **2**, 181–193.

Lee, P. P. N., Hong, G. X., and Pang, S. F. (1991). Melatonin in the gastrointestinal tract. *In* "Role of Melatonin and Pineal Peptides in Neuroimmunomodulation" (F. Fraschini and R. J. Reiter, eds.), pp. 127–138. Plenum, New York.

Lee, P. P. N., Shiu, S. Y. W., and Pang, S. F. (1995). Regional and diurnal studies of melatonin and melatonin binding sites in the duck gastrointestinal tract. *Biol. Signals* **4**, 212.

Lerner, A. B., Case, J. D., Takahashi, Y., Lee, T. H., and Mori, W. (1958). Isolation of melatonin, the pineal gland factor that lightens melanocytes. *Am. J. Chem. Soc.* **80**, 2587.

Lu, J., and Cassone, V. M. (1993a). Daily melatonin administration synchronizes circadian patterns of brain metabolism and behavior in pinealectomized house sparrows, *Passer domesticus*. *J. Comp. Physiol. A* **173**, 775–782.

Lu, J., and Cassone, V. M. (1993b). Pineal regulation of circadian rhythms of 2-deoxy[^{14}C]glucose uptake and 2[^{125}I]iodomelatonin binding in the visual system of the house sparrow, *Passer domesticus*. *J. Comp. Physiol. A* **173**, 765–774.

MacBride, S. E. (1973). "Pineal biochemical rhythms of the chicken (*Gallus domesticus*): Light cycle and locomotor activity correlates" Unpublished Ph. D. thesis.

Matsumoto, A., and Ishii, S. (1987). "Atlas of Endocrine Organs." Springer-Verlag, Berlin.

Menaker, M., and Oksche, A. (1974). The avian pineal organ. *In* "Avian Biology" (D. S. Farner and J. R. King, eds.), Vol. IV, pp. 79–118. Academic Press, New York/San Francisco/London.

Menaker, M., and Zimmerman, N. (1976). Role of the pineal in the circadian system of birds. *Am. Zool.* **16**, 45–55.

Menaker, M., and Tosini, G. (1996). The evolution of vertebrate circadian systems. *In* "Circadian Organization and Oscillatory Coupling" (K. I. Honma and S. Honma, eds.), pp. 39–52. Hokkaido Univ. Press, Sapporo.

Morgan, P. J., Barrett, P., Howell, H. E., and Helliwell, R. (1994). Melatonin receptors: Localization, molecular pharmacology and physiological significance. *Neurochem. Int.* **24**, 101–146.

Murton, R. K., and Westwood, N. J. (1977). "Avian Breeding Cycles." Clarendon Press, Oxford.

Norgren, R. B., Jr., and Silver, R. (1989a). Retinal projections in quail. *Vis. Neurosci.* **3**, 377–387.

Norgren, R. B., Jr., and Silver, R. (1989b). Retinohypothalamic projections and the suprachiasmatic nucleus in birds. *Brain Behav. Evol.* **34**, 73–83.

Nottebohm, f. (1981). A brain for all seasons: Cyclical anatomical changes in song-control nuclei of the canary brain. *Science* **214**, 1368–1370.

Okano, T., Yoshizawa, T., and Fukada, Y. (1994). Pinopsin is a chicken pineal photoreceptive molecule. *Nature* **372**, 94–97.

Oshima, I., Yamada, H., Sato, K., and Ebihara, S. (1989). The role of melatonin in the circadian system of the pigeon, *Columba livia*. *In* "Circadian Clocks and Ecology" (T. Hiroshige and K. Honma, eds.), pp. 118–126. Hokkaido Univ. Press, Sapporo.

Pang, S. F., Brown, G. M., Grota, L., Chambers, J. W., and Rodman, R. L. (1977). Determination of *N*-acetylserotonin and melatonin activities in the pineal gland, retina, Harderian gland, brain, and serum of rats and chickens. *Neuroendocrinology* **12**, 1–13.

Partridge, W. M., and Mietus, L. J. (1980). Transport of albumin-bound melatonin through blood–brain barrier. *J. Neurochem.* **34**, 1761–1763.

Pentney, P. T., and Bubenik, G. A. (1995). Melatonin reduces the severity of dextran-induced colitis in mice. *J. Pineal Res.* **19**, 31–39.

Phillips, N. H., and Berger, R. J. (1992). Melatonin infusions restore sleep suppressed by continuous bright light in pigeons. *Neurosci. Lett.* **145**, 217–220.

Pierce, M. E., Sheshberadaran, H., Zhang, Z., Fox, L. E., Applebury, M. L., and Takahashi, J. S. (1993). Circadian regulation of iodopsin gene expression in embryonic photoreceptors in retinal cell culture. *Neuron* **10**, 579–584.

Pittendrigh, C. S. (1981). Circadian systems: General perspective. *In* "Handbook of Behavioural Neurobiology" (J.. Aschoff, ed.), Vol. 4, pp. 57–80. Plenum, New York/London.

Pontoire, C., Bernard, M., Silvain, C., Collin, J.-P., and Voisin, P. (1993). Characterization of melatonin binding sites in chicken and human intestines. *Eur. J. Pharmacol.* **247**, 111–118.

Poon, A. M. S., Wang, X. L., and Pang, S. F. (1993). Characteristics of 2-[^{125}I]iodomelatonin binding-sites in the pigeon spleen and modulation of binding by guanine-nucleotides. *J. Pineal Res.* **14**, 169–177.

Quay, W. B. (1965). Histological structure and cytology of the pineal organ in birds and mammals. *Progr. Brain Res.* **10**, 49–86.

Quay, W. B. (1972). Infrequency of pineal atrophy among birds and its relation to nocturnality. *Condor* **74**, 33–45.

Reiter, R. J., Melchiorri, D., Sewerynek, E., Poeggeler, B., Barlow-Walden, L., Chuang, J. I., Ortiz, G. G., and Acuña-Castroviejo, D. (1995). A review of the evidence supporting melatonin's role as an antioxidant. *J. Pineal Res.* **18**, 1–11.

Reiter, R. J., Oh, C. S., and Fujimori, O. (1996). Melatonin: Its intracellular and genomic actions. *Trends Endocrinol. Metab.* **7**, 22–27.

Reppert, S. M., and Weaver, D. R., Cassone, V. M., Godson, C., and Kolakowski, L. F. Jr. (1995). Melatonin receptors are for the birds: Molecular analysis of two receptor subtypes differentially expressed in chick brain. *Neuron* **15**, 1003–1015.

Rivkees, S. A., Cassone, V. M., Weaver, D. R., and Reppert, S. M. (1989). Melatonin receptors in chick brian: Characterization and localization. *Endocrinology* **125**, 363–368.

Robertson, L. M., and Takahashi, J. S. (1988a). Circadian clock in cell culture. I. Oscillation of melatonin release from dissociated chick pineal cells in flow-through microcarrier culture. *J. Neurosci.* **8**, 12–21.

Robertson, L. M., and Takahashi, J. S. (1988b). Circadian clock in cell culture. II. *In vitro* photic entrainment of melatonin oscillation from dissociated chick pineal cells. *J. Neurosci.* **8**, 22–30.

Rusak, B. (1982). Circadian organization in mammals and birds: Role of the pineal gland. *In* "The Pineal Gland: Extrareproductive Effects" (R. J. Reiter, ed.), Vol. 3, pp. 29–51. CRC Press, Boca Raton, FL.

Saarela, S., and Heldmaier, G. (1987). Effect of photoperiod and melatonin on cold resistance, thermoregulation and shivering/

nonshivering thermogenesis in Japanese quail. *J. Comp. Physiol. A* **157,** 625–633.

Saldanha, C. J., Leak, R. K., and Silver, R. (1994). Detection and transduction of daylength in birds. *Psychoneuroendocrinology* **19,** 641–656.

Sato, T., and Wake, K. (1983). Innervation of the avian pineal organ. *Cell Tissue Res.* **233,** 237–264.

Simpson, S. M., and Follett, B. K. (1981). Pineal and hypothalamic pacemakers: Their role in regulating circadian rhythmicity in Japanese quail. *J. Comp. Physiol.* **114,** 381–389.

Song, Y., Ayre, E. A., and Pang, S. F. (1992). The identification and characterization of ^{125}I labeled iodomelatonin-binding sites in the duck kidney. *J. Endocrinology* **135,** 353–359.

Takahashi, J. S., Murakami, M., Nikaido, S. S., Pratt, B. L., and Robertson, L. M. (1989). The avian pineal, a vertebrate model system of the circadian oscillator: Cellular regulation of circadian rhythms by light, second messengers, and macromolecular synthesis. *Rec. Progr. Horm. Res.* **45,** 279–352.

Touitou, Y., Arendt, J., and Follett, B. K. (1993). "Melatonin and the Pineal Gland." Elsevier, Amsterdam.

Turek, F. W., McMillan, J. P., and Menaker, M. (1976). Melatonin: Effects on the circadian locomotor rhythm of sparrows. *Science* **194,** 1441–1443.

Underwood, H., Barrett, R. K., and Siopes, T. (1990). Melatonin does not link the eyes to the rest of the circadian system in quail: A neural pathway is involved. *J. Biol. Rhythms* **5,** 349–361.

Underwood, H., and Goldman, B. D. (1987). Vertebrate circadian and photoperiodic systems: Role of the pineal gland and melatonin. *J. Biol. Rhythms* **2,** 279–315.

Underwood, H., and Siopes, T. (1984). Circadian organization in Japanese quail. *J. Exp. Zool.* **232,** 557–566.

Vakkuri, O., Rintamäki, H., and Leppäluoto, J. (1985a). Presence of immunoreactive melatonin in different tissues of the pigeon (*Columba livia*). *Gen. Comp. Endocrinol.* **58,** 69–75.

Vakkuri, O., Rintamäki, H., and Leppäluoto, J. (1985b). Plasma and tissue concentrations of melatonin after midnight light exposure and pinealectomy in the pigeon. *J. Endocrinol.* **105,** 263–268.

Wan, Q., and Pang, S. F. (1993). (^{125}I)iodomelatonin binding sites in the chicken spinal cord: Binding characteristics and diurnal variation. *Neurosci. Lett.* **163,** 101–104.

Wever, R. (1965). Pendulum versus relaxation oscillation. *In* "Circadian Clocks" (J. Aschoff, ed.), pp. 74–83. North-Holland, Amsterdam.

Whitfield-Rucker, M. G., and Cassone, V. M. (1996). Melatonin binding sites in the house sparrow song control system: Sexual dimorphism and the effect of photoperiod. *Horm. Behav.* **30,** 528–537.

Woodfill, C. J. I., Wayne, N. L., Moenter, S. M., and Karsch, F. J. (1994). Photoperiodic synchronization of a circannual reproductive rhythm in sheep: Identification of season-specific time cues. *Biol. Reprod.* **50,** 965–976.

Yu, H. S., and Reiter, R. J. (1993). "Melatonin: Biosynthesis, Physiological Effects, and Clinical Applicaton." CRC Press, Boca Raton, FL.

Zimmerman, N. H., and Menaker, M. (1979). The pineal gland: A pacemaker within the circadian system of the house sparrow. *Proc. Natl. Acad. Sci. USA* **76,** 999–1003.

CHAPTER 22

Reproduction in the Female

A. L. JOHNSON
Department of Biological Sciences
University of Notre Dame
Notre Dame, Indiana 46556

I. Anatomy of the Female Reproductive System 569
    A. Ovary 569
    B. Oviduct and Sperm Storage Glands 574
II. Breeding and Ovulation–Oviposition Cycles 575
    A. Ovulation–Oviposition Cycle and Rate of Lay 575
    B. Parthenogenesis 577
III. Ovarian Hormones 577
    A. Steroid Production and Secretion 578
    B. Nonsteroidal Hormones and Growth Factors 580
IV. Hormonal and Physiologic Factors Affecting Ovulation 580
    A. Hormones and Ovulation 580
    B. Hormones during Sexual Maturation and Molt 583
    C. Effect of Light on the Ovary and Ovulation 583
    D. Photorefractoriness 584
    E. Molt 584
V. Oviposition 584
    A. Neurohypophyseal Hormones 584
    B. Prostaglandins 584
    C. Postovulatory Follicle 585
    D. Preovulatory Follicle 585
    E. Other Factors 585
    F. Broodiness and Nesting Behavior 585
VI. Composition and Formation of Yolk, Albumen, Organic Matrix, and Shell 586
    A. Yolk 586
    B. Albumen 586
    C. Organic Matrix 587
    D. Layers of Crystallization 588
    E. Calcium Metabolism 589
References 591

I. ANATOMY OF THE FEMALE REPRODUCTIVE SYSTEM

The right ovary and oviduct are present in embryonic stages of all birds, but the distribution of primordial germ cells to the ovaries of the chicken becomes asymmetrical by day 4 of incubation; by day 10 regression of the right oviduct is initiated under the influence of Mullerian inhibiting substance (Hutson *et al.,* 1985). The reproductive system of galliformes consists of a single left ovary and its oviduct, although on occasion a functional right ovary and oviduct may be present. Among the falconiformes and in the brown kiwi, both left and right gonads and associated oviducts are commonly functional, although the ovaries may be asymmetrical in size; in sparrows and pigeons, about 5% of specimens have two developed ovaries (see Romanoff and Romanoff, 1949; Kinsky, 1971).

A. Ovary

The left ovary is attached by the mesovarian ligament at the cephalic end of the left kidney. The number of oocytes of the chick embryo increases from approximately 28,000 on the 9th day of development to 680,000 on the 17th day and subsequently decreases to 480,000 by the time of hatching, when oogenesis is terminated. The ovary of the immature bird consists of a mass of small ova, of which at least 2,000 are visible to the naked eye. Only a relatively few of these (250–500) reach maturity and are ovulated within the life span of most domesticated species, and considerably fewer mature in wild species. The functionally mature ovary of the hen is arranged with an obvious hierarchy of follicles and in its entirety weighs 20–30 g. Commonly, there are four to six large yolk-filled follicles 2–4 cm in diameter accompanied by a greater number of 6- to 12-mm follicles in which yellow yolk deposition has been initiated and numerous small white follicles <6 mm (Figure 1).

The ovary receives its blood supply from the ovarian artery, which usually arises from the left renolumbar artery but may branch directly from the dorsal aorta (Hodges, 1965). Within the ovary, bloodflow is greatest to the five largest preovulatory follicles (Scanes et al., 1982; see Chapter 7 for further details). The ovarian artery divides into many branches and leads to a single follicular stalk, usually from two to four separate arterial branches. All veins from the ovary unite into the two main anterior and posterior veins, which subsequently drain into the posterior vena cava.

Several workers have shown that the ovary is well innervated by both adrenergic and cholinergic fibers (Gilbert, 1969; Dahl, 1970b; Unsicker et al., 1983) and that a greater number of neurons are present within the theca layer as a follicle progressively matures. It is now clear that such innervation provides a variety of neurochemical (e.g., catecholamines) and neurohumeral factors (e.g., neurotropins, vasoactive intestinal peptide, substance P, calcitonin gene-related peptide) to the ovary and its follicles; such factors have been proposed to function in diverse roles such as in the organization of primordial follicles in the embryonic and early posthatch ovary and the differentiation of preovulatory follicles in the adult ovary.

1. Ovarian Follicle

The follicle consists of concentric layers of tissue that surround the oocyte and yolk, including: (1) the oocyte plasma membrane, (2) the perivitelline layer, (3) granulosa cells, (4) basal lamina (basement membrane), and (5) the theca (interna and externa) (Figure 2). The follicle is highly vascularized except for the stigma (the point of rupture during ovulation), which contains a lesser number of underlying small veins and arteries (Nalbandov and James, 1949). The main arteries from the stalk are directed toward the fastest growing follicles, branch into arterioles, and pass through the theca to the basal lamina to form arterial capillaries (Dahl, 1970a).

The germinal vesicle (blastoderm) is localized within the germinal disc region of the granulosa layer. The avian oocyte remains arrested at the first meiotic prophase stage during follicle development and resumes meiosis (undergoes germinal vesicle breakdown) 4–6 hr prior to ovulation. Granulosa cells of prehierarchal follicles are closely packed and cuboidal, and the granulosa cell layer may be several cells deep. As the follicle grows exponentially in diameter during the rapid growth phase, the granulosa cells become squamous in shape and form a single cell layer. Spaces and cell junctions form between granulosa cells to facilitate transport of yolk. Despite the dramatic increase in the circumference of the granulosa layer during this time relatively little cell proliferation occurs, except within the germinal disk region. Granulosa cell DNA synthesis and proliferation are 5- to 10-fold higher in prehierarchal follicles when compared to granulosa cells from hierarchal follicles; moreover, DNA synthesis is 2-fold higher in granulosa cells within the germinal disk region when compared to cells from the outer layer region (Tilly et al., 1992).

The basal lamina of the hen follicle is approximately 1 mm thick and consists of collagen and the glycoprotein, fibronectin, among other components; these are deposited by the adjacent granulosa and theca layers. Fibronectin production and secretion is regulated by a variety of gonadal hormones and gonadotropins (Conkright and Asem, 1995).

2. Postovulatory and Atretic Follicles

The postovulatory follicle (POF) contains the granulosa and thecal layers which remain subsequent to ovulation (Figure 1). The POF of the hen is metabolically active at a progressively lower rate for several days after ovulation, as indicated by the decreasing presence of enzymatic activity (Chalana and Guraya, 1978). Struc-

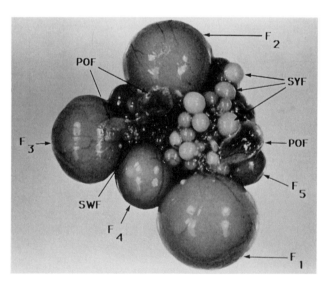

FIGURE 1 Prehierarchal and hierarchal follicles of the laying hen ovary. F_1, F_2, F_3, F_4, F_5, five largest hierarchal follicles; POF, postovulatory follicle; SWF, small white follicles, SYF, small yellow follicles. (Reprinted with permission from A. L. Johnson (1990), *CRC Crit. Rev. Poult. Biol.* **2**, 319–346. Copyright CRC Press, Boca Raton, Florida.)

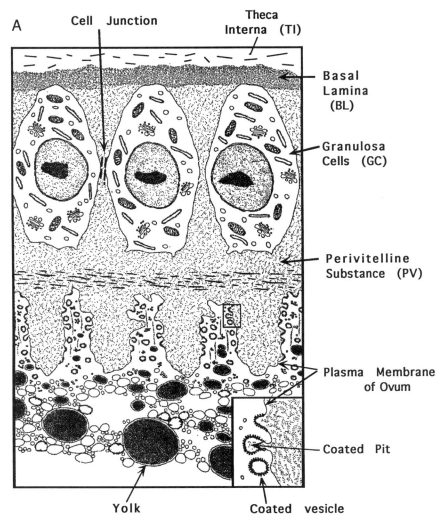

FIGURE 2A Diagrammatic relationship of follicle tissue layers immediately surrounding the oocyte with special emphasis on the coated vesicles of the perivitelline layer. (Redrawn from Wyburn *et al.*, 1966, with permission of Cambridge University Press.)

tural resorption of the POF occurs, in part, via the process of apoptosis (Tilly *et al.*, 1991b) (Figure 3) over a period of 6–10 days in the chicken and several months in the mallard. The POF may influence nesting behavior and has been demonstrated to be functional in timing oviposition (Gilbert *et al.*, 1978). There is no structure in birds functionally analogous to the corpus luteum of mammals.

Follicles that initially begin to grow but fail to reach the fully differentiated stage at which they are ovulated become atretic. The death and resorption of ovarian follicles also occurs via apoptosis (Tilly *et al.*, 1991b); atresia is characterized by pronounced oligonucleosome formation within the granulosa cell layer, with considerably less apoptotic cell death within the theca layer (Figure 3). The subsequent reabsorption of the oocyte and yolk occurs either via the vascular system (involu-

tion atresia) or by the rupture of the thecal layers and slow loss of the yolk into the peritoneal cavity (bursting atresia). Atresia occurs with high incidence among prehierarchal follicles (<9 mm in diameter) and under normal physiological conditions is absent among preovulatory hierarchal follicles (Gilbert *et al.*, 1983). Natural instances of atresia among hierarchal follicles occur, for example, subsequent to a change from the egg-laying state to incubation and broody behavior and with the onset of molt. Atresia of hierarchal follicles can also be induced by prolonged daily administration of equine chorionic gonadotropin (eCG) (Johnson and Leone, 1985), by hypophysectomy (Yoshimura *et al.*, 1993a), or by destruction of the germinal disk region (Yoshimura *et al.*, 1994).

The high rate of atresia in prehierarchal follicles correlates with the susceptibility of the granulosa cell layer

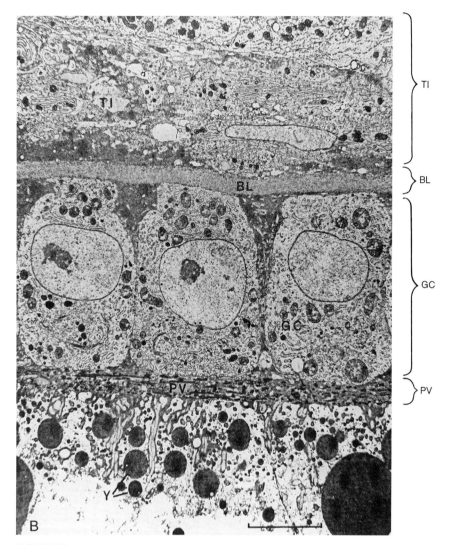

FIGURE 2B Electron micrograph of tissue layers depicted in Fig. 2A. Abbreviations are as labeled in Fig. 2A. (From Perry and Gilbert, 1979, and from Agricultural Research Council Poultry Research Centre Report, Roslin, Midlothian, Scotland, 1980.)

to undergo apoptosis (Figure 3); by contrast, granulosa cells from hierarchal follicles are inherently resistant to undergoing cell death. A current area of intense research centers on investigations of cellular mechanisms mediating follicle atresia via apoptosis in vertebrate species. Apoptosis (sometimes inappropriately referred to as programmed cell death; see Wyllie, 1992) is an evolutionarily conserved process of cell death which serves to selectively eliminate cells/tissues during tissue remodeling or in the course of normal tissue turnover. Among the factors proposed to mediate cell viability, or alternatively, cell death are the Bcl-2-related family of proteins, the interleukin converting enzyme (ICE)-related family of enzymes, as well as a select group of protooncogenes (e.g., c-*myc*) and tumor suppressor genes (e.g., p53) (Williams and Smith, 1993; Wyllie, 1994; Tilly *et al.*, 1995a,b). It is significant to note that: (1) genes from each of these groups are expressed in developing chicken ovarian follicles, (2) the level of constitutive expression for each gene (death inducing/death suppressing) is correlated to inherent granulosa cell susceptibility or resistance to cell death, and (3) the level of mRNA expression for such genes is generally regulated by the same physiological factors (e.g., growth factors and gonadotropins) that promote hen follicle viability (Johnson *et al.*, 1993, 1997a,b).

3. Growth of the Follicle and Deposition of Lipid

Growth of the follicle can be divided into three phases: (1) slow growth of follicles 60–100 μm in diameter, lasting months to years; (2) a several-month period

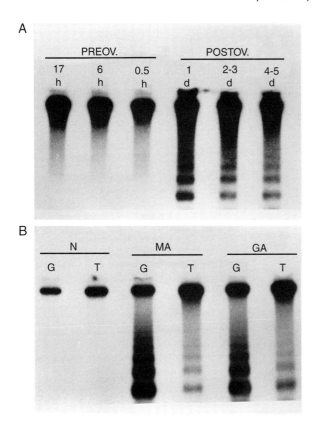

FIGURE 3 (A) Oligonucleosome formation, the hallmark of apoptosis, in preovulatory (Preov) and postovulatory (Postov) hen follicles. Abbreviations, h, hours before ovulation; d, days following ovulation. (B) Presence of oligonucleosomes in granulosa (G) and theca (T) tissues from morphologically normal (N), mildly atretic (MA), and grossly atretic (GA) follicles. (Adapted from Tilly *et al.*, 1991b.)

of increasingly rapid growth that consists mainly of deposition of yolk protein; and (3) a rapid growth phase during the final 6–11 days prior to ovulation in domestic fowl, ducks, and pigeons (up to 16 days in some penguins; Grau, 1982) when the majority of yolk protein and lipids are deposited. In this final phase of development, the follicle of the domestic hen will, on average, transport up to 2 g of yolk protein per day, grow in diameter from 8 to 37 mm, and increase in volume by a factor of 3500 to 8000.

Yolk protein formation occurs in the liver and is regulated primarily by gonadotropin and steroid hormones. Transport from the liver to the ovary occurs by the blood. Following release from capillaries within the theca layer, the plasma-borne precursors vitellogenin (VTG, a phosphoglycolipoprotein) and very low density lipoprotein (VLDL, functions mainly to transport triglycerides, phospholipids, and cholesterol) pass across the basement membrane and through gaps between granulosa cells to the plasma membrane of the oocyte (Figure 2). Translocation of VTG and VLDL across the oocyte plasma membrane is a receptor-mediated event (Shen *et al.*, 1993). The receptor for VTG/VLDL is localized within coated vessicles of the oocyte plasma membrane but is not expressed in extraovarian tissues. Moreover, this receptor is absent in a genetically altered, nonlaying strain of chickens (the restricted ovulator, R/O; Barber *et al.*, 1991). The lack of VTG/VLDL receptor expression in this strain results in large accumulations of VTG and VLDL in the blood and apparently prevents ovarian follicles from entering the rapid growth phase of development.

Following transport across the plasma membrane, VTG and VLDL become localized to yolk spheres where proteolytic processing to phosvitin, lipovitellin, triglycerides, cholesterol, and phospholipids by cathepsin D occurs (Retzek *et al.*, 1992). Lipids and protein are deposited into the growing follicle at about the same ratio for most of the growth phase, but during the final rapid growth phase, relatively more lipid is incorporated. It has been suggested that deposition of yolk into the maturing follicle is terminated by 24 hr before ovulation. The ultrastructure of developing follicles has been well described (Wyburn *et al.* (1965), Rothwell and Solomon (1977), Perry *et al.* (1978a,b), and Gilbert *et al.* (1980)).

It has been proposed that in chickens the decrease in egg production with age is in part caused by both an increase in the incidence of atresia and a reduction in the number of follicles reaching the final phase of rapid growth; that is, fewer follicles receive a proportionately greater quantity of yolk, resulting in larger sized eggs (Williams and Sharp, 1978; Palmer and Bahr, 1992).

Mechanisms controlling development and maintenance of the follicular hierarchy are poorly understood. Pituitary gonadotropin involvement is indicated by the finding that hypophysectomy rapidly leads to follicular atresia, whereas ablation and treatment with a chicken pituitary extract results in the maintenance of the normal follicular hierarchy. Moreover, daily injections of follicle-stimulating hormone (FSH; 200 μg porcine FSH) or eCG (25 IU) were found to decrease the rate of atresia and increase the number of growing follicles in aging hens (Palmer and Bahr, 1992). On the other hand, daily injections of eCG at higher doses (75 IU) to intact laying hens stimulates numerous small follicles to enter the rapid growth phase and results in the elimination of the follicular hierarchy and the cessation of ovulation (Johnson and Leone, 1985).

Thus, levels of circulating FSH would appear to play a critical role in establishing and maintaining the preovulatory hierarchy as well as in regulating the rate of follicle atresia. Inhibin is one hormone that has recently been shown to directly influence circulating levels of FSH. This biologically active glycoprotein is a member of the transforming growth factor beta (TGFβ) superfamily of peptides and is composed of an α- and β-

subunit. In the hen ovary, inhibin is produced at highest levels by the four largest preovulatory follicles; surgical removal of preovulatory follicles results in a 80% decline in circulating inhibin levels and in a significant elevation in plasma FSH, but not luteinizing hormone (LH) (Johnson, P.A. et al., 1993).

B. Oviduct and Sperm Storage Glands

The oviduct of the hen is derived from the left Mullerian duct. In the chicken, the left oviduct develops rapidly after 16 weeks of age and becomes fully functional just prior to the onset of egg production. The oviduct consists of five distinguishable regions: infundibulum, magnum, isthmus, shell gland, and vagina. (For details on general oviduct histology, see Aitken (1971) and King (1975); on the infundibulum and magnum, see Wyburn et al. (1970); for the isthmus, see Hoffer (1971) and Solomon (1975); and for the shell gland, see Breen and DeBruyn, 1969, Nevalainen (1969), and Wyburn et al. (1973).) A dorsal and a ventral ligament suspend the oviduct within the peritoneal cavity. Over recent years, the chicken oviduct has provided the molecular biologist with a superb model for the study of steroid hormone regulation of gene expression in eukaryotes, and in particular, the regulation of ovalbumin synthesis by steroids (see, for example, Schrader et al., 1981).

1. Infundibulum

Subsequent to ovulation, the ovum is engulfed by the infundibulum (a structure not directly connected to the ovary) where it resides for approximately 18 min (range, 15–30 min). Occasionally, the ovum fails to be picked up by the infundibulum (an "internal ovulation") and is reabsorbed in 24 hr or less. Infundibular activity appears not to be controlled by ovulation per se, as foreign objects placed into the abdominal cavity prior to ovulation will also be taken up, and there are reports that entire unovulated follicles may be engulfed and later laid as fully developed eggs. Fertilization of the ovum occurs in the infundibulum, and it is here that the first layer of albumen is produced in the chicken.

2. Magnum

The ovum next passes to the largest portion of the oviduct, the magnum (a length of 33 cm in the chicken), where in the mature female the majority of albumen is formed. Estrogen stimulates epithelial stem cells to develop into three morphologicaly different cell types: tubular gland cells, ciliated cells, and goblet cells. Tubular glands are responsible for the production of ovalbumin, lysozyme, and conalbumin under the stimulatory control of estrogen, while goblet cells synthesize avidin following exposure to progesterone and estrogen (Tuohimaa et al., 1989). A measurable amount of calcium secretion occurs within the magnum, but at no time during the laying cycle is calcium secretion greater than the basal rate of secretion found in the shell gland (Eastin and Spaziani, 1978a). The ovum remains in the magnum approximately 2–3 hr.

3. Isthmus

The isthmus is clearly distinguishable from the magnum. It has a thick circular muscle layer muscle, and glandular tissue is less developed compared to the magnum. This tissue is characterized by a layer of epithelial cells with underlying tubular gland cells. Both inner and outer shell membranes are formed during the 1- to 2-hr (mean time, 1 hr and 14 min) passage through the isthmus. There is some evidence to suggest that shell formation, particularly the mammillary cores, is initiated in the distal portion of the isthmus.

4. Shell Gland

The shell gland (uterus) is characterized by a prominent longitudinal muscle layer lined medially with both tubular gland and unicellular goblet cells. Prior to calcification, the egg takes up salts and approximately 15 g fluid into the albumin from the tubular glands, a process termed "plumping"; this fluid contains carbonic anhydrase, acid phosphatase, and esterase activty plus bicarbonate and a variety of additional ions (Salevsky and Leach, 1980). The ovum remains in the shell gland for 18–26 hr depending on cycle length.

Calcification within the shell gland is associated with stimuli initiated by ovulation or by neuroendocrine factors that control and coordinate both ovulation and calcium secretion. Additional evidence suggests that distension of the shell gland by the egg is not a stimulus for initiating a high rate of calcium secretion, which is characteristic of calcification, nor is autonomic innervation involved (Eastin and Spaziani, 1978a). Calcification of the egg at first occurs slowly, increases to a rate of up to 300 mg/hr over a duration of 15 hr, then again slows during the last 2 hr before oviposition; shell pigments (primarily protoporphyrin and biliverdin) are deposited via ciliated cells of the shell gland epithelium during the period of 3 hr through 0.5 hr prior to oviposition.

5. Vagina

The vagina is separated from the shell gland by the uterovaginal sphincter muscle and terminates at the cloaca. There are numerous folds of mucosa, which are lined by ciliated and nonciliated cells, and a few tubular mucosal glands which may have a secretory function. The vagina has no role in the formation of the egg, but, in coordination with the shell gland, participates in the expulsion of the egg.

In domestic fowl, spermatozoa are stored in specialized sperm-storage tubules located in the uterovaginal region and remain viable for a period of 7–14 days in the chicken and for greater than 21 days in the turkey hen. Uterovaginal junction glands are apparently devoid of innervation and contractile tissue, but possess a well-developed vascular system (Gilbert et al., 1968; Burke et al., 1972; Tingari and Lake, 1973). Evidence suggests that spermatozoa fill the uterovaginal glands in a sequential fashion without mixing so that with successive inseminations, sperm from the latest insemination is most likely to fertilize an ovum. Following oviposition of each egg, spermatozoa are released from these tubules by an unknown mechanism and migrate to the infundibulum for fertilization (Zavaleta and Ogasawara, 1987).

6. Blood Flow and Innervation to the Oviduct

The blood supply to the oviduct and shell gland of the domestic hen has been described by Freedman and Sturkie (1963) and Hodges (1965). For a review of the vasculature to the oviduct of additional avian species, see Gilbert (1979). Blood flow to the shell gland of the hen is increased during the presence of a calcifying egg compared to times when no egg is present (Scanes et al., 1982). For additional details on blood flow, see Chapter 7.

The oviduct is innervated by both sympathetic and parasympathetic nerves. Innervation of the infundibulum is via the aortic plexus and the magnum by the aortic and renal plexuses (Hodges, 1974). Sympathetic innervation of the shell gland is via the hypogastric nerve, which is the direct continuation of the aortic plexus. Parasympathetic pelvic nerves, which constitute the left pelvic plexus, arise from the pelvic visceral rami of spinal nerves 30–33.

7. Oviduct Motility

Cilia are found along the entire length of the oviduct, and a likely function of these cilia is that of sperm transport. Egg transport is primarily accomplished by contractions of the oviduct; oviduct musculature functions as a stretch receptor and the mechanical stimulus is produced by the ovum itself (Ariamaa and Talo, 1983). Changes in electrical activity and oviduct motility have been recorded in the magnum, isthmus, and shell gland during the ovulatory cycle, with the greatest frequency of electrical activity and contractions occurring in the shell gland at the time of oviposition (Shimada, 1978; Shimada and Asai, 1978). Both α- and β-adrenergic receptors are present throughout the length of the oviduct and have been shown to affect oviduct motility (Crossley et al., 1980).

II. BREEDING AND OVULATION–OVIPOSITION CYCLES

Breeding cycles among species of birds can be classified according to the length of the cycle and the time of the year at which each species becomes reproductively active. Continuous breeders, such as the domestic hen or Khaki Campbell duck, are, under optimal conditions, reproductively active throughout the year. Most wild species that breed in temperate, subarctic, and arctic zones display yearly cycles, while birds adapted to tropical or desert climates may breed with cycles less than a year, at 6-month intervals, or when favorable conditions exist (opportunistic breeders) (for further details, see Lofts and Murton, 1973; Wingfield and Farner, 1993).

Wild birds usually lay one or more eggs in a clutch and then terminate laying to incubate the eggs. The number of eggs per clutch and total number of clutches vary with the species and season. For example, some birds, such as the auk or sooty tern (*Sterna fuscata*), lay a single egg to incubate. The king penguin (*Aptenodytes patogonica*) will lay but one egg and breeding will not necessarily occur on a yearly basis. In contrast, the European partridge will lay a single clutch per year consisting of as many as 12–20 eggs. The pigeon usually lays two eggs per clutch and averages eight clutches per year. The interval between clutches is approximately 45 days in the fall and winter and from 30 to 32 days in the spring and early summer. Finally, clutch size of the bobwhite quail declines with advancing season, from a mean high of 19.2 eggs in early May to 11.3 eggs in late July.

In some species, when eggs are destroyed or removed from the nest early enough in the breeding season birds may produce an additional clutch (indeterminate layers; an example is the duck *Anas platyrhynchos*) (Donham et al., 1976). Determinate layers fail to lay additional eggs on the removal or destruction of eggs in the nest. In seasonal (or noncontinuous) breeders, the ovary undergoes periods of growth and regression. The weight of the European starling ovary may vary from 8 mg during the regression phase to 1400 mg at the height of the breeding season.

A. Ovulation–Oviposition Cycle and Rate of Lay

The ovulation–oviposition cycle (the time from ovulation of an ovum to the oviposition of the egg) of the domestic hen generally ranges from somewhat longer than 24 hr (24+) to 28 hr in length, and ovulations proceed uninterrupted for several days or as long as 1 year or more in extreme instances. The number of eggs laid on successive days is called a sequence, and each sequence is separated by 1 or more pause days on which no egg is laid. The term clutch is sometimes used synonomously with sequence, although the former term is generally considered more appropriate to describe the group of eggs laid by nondomesticated species prior to incubation behavior. The 24+-hr ovulation–oviposition cycle is characteristic of the chicken, turkey, bobwhite quail, and several other species including the Japanese quail. By contrast, the interval between successively laid eggs in the pigeon has been reported to be 40–44 hr and in the Khaki Campbell duck is 23.5–24.5 hr (see Johnson, 1986).

In the domestic hen, the longer the sequence, the shorter the duration of the ovulation–oviposition cycle. The delay, or "lag," in hours between the oviposition of successive eggs in a sequence is not constant (Table 1). The differences in lag time within a sequence represent mainly variations in the amount of time between oviposition and the subsequent ovulation; this interval ranges from 15 to 75 min. It is clear that even when the lag between the oviposition of successive eggs approaches 24 hr there remains a progressive shift toward laying the egg later and later in the day. As the sequence becomes shorter, the lag becomes greater and the length of the ovulation–oviposition cycle deviates more from 24 hr. The total lag time between the first ovulation of a sequence (C_1 ovulation) and last ovulation of a sequence (C_t ovulation) is greater in chickens (4–8 hr depending on sequence length) than in Japanese quail (1.5–5 hr). The normal release of LH in the hen is restricted to a 4- to 11-hr period (the "open period") beginning at the onset of the scotophase (dark phase). The timing and regulation of the "open period" is as yet incompletely understood. For further details concerning the ovulation–oviposition cycle and the sequence, see Fraps (1955), van Tienhoven (1981), and Etches (1990).

A chicken typically ovulates the first egg of a sequence early in the photophase, and the timing of ovulation is synchronized by the onset of the scotophase. In comparison, Japanese quail ovulate the C_1 egg 8–9 hr after the onset of the photophase, and this appears to be synchronized by the timing of the light phase.

The rate of lay refers to the number of eggs laid in a given period of time, irrespective of the regularity or pattern of laying. For example, a chicken laying at the rate of 50% for 60 days produces 30 eggs, the same number of eggs as another that lays at a rate of 75% for 40 days. On reaching sexual maturity (at ~20–22 weeks of age), the domestic hen lays with an erratic pattern (sequences with 2 or more pause days or occasionally more than 1 egg per day) and with a high incidence of abnormal (soft-shelled and double-yolked) eggs for the first 2 weeks. Six to 10 weeks after the onset of egg laying, production peaks (frequently at a rate of 90% or more) and gradually decreases over a period of 40–50 weeks, depending on whether the chicken is an egg- or meat-type breed. Subsequently, egg production progressively declines and the hen begins a period of

TABLE 1 Egg Production in the Domestic Hen Housed under a Photoperiod of 14 hr L:10 hr D[a]

	Rate of egg production (%)							
	66	75	80	83	86	88	89	90
	Number of eggs in continuous sequence							
	2	3	4	5	6	7	8	9
Position in sequence								
1	09:05	08:41	08:01	07:24	07:33	07:45	07:36	07:18
2	13:30	11:26	10:09	09:22	09:20	09:12	08:59	08:39
3		15:15	12:03	10:26	10:34	10:10	09:57	09:25
4			15:23	11:46	11:30	10:51	10:26	09:44
5				15:05	12:43	11:39	10:55	09:53
6					15:40	12:38	11:43	10:26
7						15:28	12:49	10:59
8							15:44	12:02
9								15:26
Average interval between consecutive eggs:								
	28:25	27:17	26:27	25:55	25:37	25:17	25:09	25:0

[a] Lights on 06:00 hr. From Etches and Schoch (1984).

molt. A typical laying strain of chicken will produce 280 or more eggs in a 50-week production period.

B. Parthenogenesis

Parthenogenesis (development from an unfertilized oocyte) has been documented in turkeys; 32–49 % of infertile eggs may initiate development, but the embryos usually die at an early stage (Olsen, 1960). Genetic selection can increase the incidence of parthenogenesis in turkeys, and viable poults are homozygous diploid males that are often sexually competent. Cytologic studies indicate that parthenogenesis in turkeys is initiated from a haploid oocyte and proceeds to the diploid state after a meiosis that is unaccompanied by cytokinesis (mitosis) (see van Tienhoven, 1983). Parthenogenesis occurs much less frequently in chickens, while its occurrence in nondomesticated species has apparently not been documented.

III. OVARIAN HORMONES

While much work has been published concerning the expression of steroidogenic enzymes and the production and metabolism of ovarian steroids in preovulatory, hierarchal follicles of the domestic fowl, only recently has the steroidogenic profile of prehierarchal (<9 mm) follicles been described (Figures 4 and 5; see, for instance, Tilly et al., 1991a; Kato et al., 1995; Tabibzadeh et

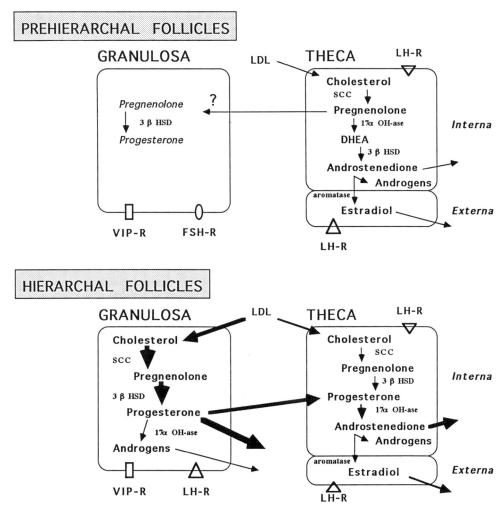

FIGURE 4 Steroid synthesis by granulosa and theca tissues from prehierarchal (<9-mm diameter) and hierarchal follicles of the laying hen. Note that while a three-cell model of steroidogenesis is proposed for hierarchal follicles, the granulosa layer of prehierarchal follicles produces virtually no steroids. 17α-OH-ase, cytochrome P450 17α-hydroxylase; 3β-HSD, 3β-hydroxysteroid dehydrogenase; FSH-R, follicle-stimulating hormone receptor; LDL, low density lipoprotein; LH-R, luteinizing hormone receptor; SCC, cytochrome P450 cholesterol side-chain cleavage; VIP-R, vasoactive intestinal peptide receptor.

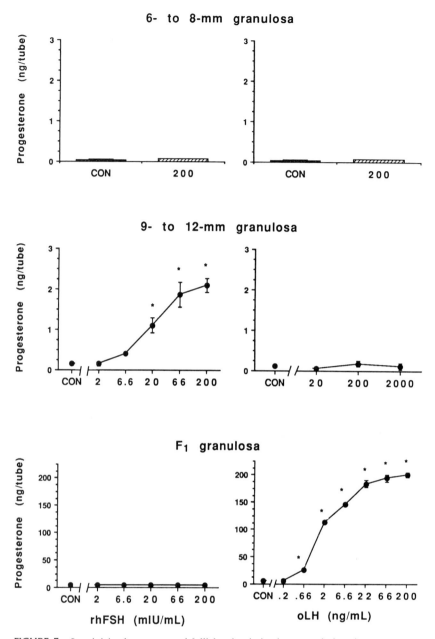

FIGURE 5 Luteinizing hormone- and follicle-stimulating hormone-induced progesterone production in granulosa cells from follicles at different stages of development. Abbreviations: rhFSH, recombinant human FSH; oLH, ovine LH. (From Johnson, 1993.)

al., 1995). Moreover, while numerous additional factors (i.e., prostaglandins, neurohormones, growth factors) and/or their receptors are known to be expressed in ovarian tissues, the function(s) for many of these has yet to be fully established. The present discussion emphasizes recent research (derived almost entirely from domesticated fowl) related to steroid synthesis and the process of ovarian follicle differentiation. For more detailed information, refer to Johnson (1986,1990,1994). There is comparatively less information about the endocrinology of reproduction in wild species; however, for a comprehensive discussion of annual reproductive cycles see Wingfield and Farner (1993).

A. Steroid Production and Secretion

1. Hierarchal Follicles

Fully potentiated steroid production by preovulatory follicles is best described by a three-cell model (Porter et al., 1989; Nitta et al., 1991) and occurs predominantly via the Δ^4 steroidogenic pathway. The granulosa layer produces predominantly progesterone which serves as a precursor for androstenedione and testosterone synthesis by the theca layer and to a lesser extent by granulosa cells (Figure 4). Progesterone is also the primary steroid involved in potentiating the preovulatory surge of LH that precedes ovulation by 4–6 hr. The theca layer also expresses cytochrome P_{450} cholesterol side-chain cleavage (P_{450}scc) activity; however, the predominant steroid product of the theca interna layer is androstenedione while cells in the theca externa synthesize estrogen.

Steroid production by granulosa and theca cells is predominantly under the regulatory control of LH acting via the adenylyl cyclase/cAMP second messenger signaling pathway (Bahr and Calvo, 1984; Tilly and Johnson, 1989). LH receptor mRNA is expressed in granulosa and theca tissue of all hierarchal follicles (Johnson et al., 1996), and LH-induced adenylyl cyclase responsiveness increases within the granulosa layer as a follicle approaches the time of ovulation. By contrast, FSH binding to membranes and FSH-induced cAMP formation and steroidogenesis is low to nonexistent in hierarchal follicles (Ritzhaupt and Bahr, 1987; Johnson, 1993). The theca cell adenylyl cyclase system is highly sensitive to LH, but is essentially unresponsive to FSH. Finally, there is some evidence for LH-stimulated steroid production occurring within both follicle layers via inositol 1,4,5-trisphosphate (IP_3) formation and calcium mobilization (Hertelendy, 1989; Levorse et al., 1991).

In contrast, activation of the diacylglycerol/protein kinase C second messenger system attenuates LH-induced steroid production by granulosa and theca cells (Tilly and Johnson, 1991). Among the potential physiological factors that may act via protein kinase C are included growth factors [e.g., transforming growth factor α (TGFα) and epidermal growth factor (EGF)] and prostaglandins. Nevertheless, it is significant to note that while TGFα attenuates LH-induced steroid production in the F1 granulosa cells, the extent of this inhibition is limited to approximately 50% of maximal stimulation. It is proposed that this inhibitory action is related to the termination of the preovulatory progesterone surge. Of additional interest is the observation that LH-induced cAMP accumulation and androstenedione production from the theca layer dramatically decline within the F1 follicle (Marrone and Hertelendy, 1985); this "desensitization" may be mediated via protein kinase C activation. Presumably, the decrease in the rate of progesterone metabolism to androstenedione ensures maximal progesterone secretion at the time of the preovulatory LH surge.

Arachidonic acid may function as an additional hormone-induced second messenger and has been determined to inhibit LH-induced progesterone production from F1 granulosa cells. While this inhibitory action occurs independently from protein kinase C activation, the specific endocrine/paracrine factor(s) that activate this system has yet to be identified (Johnson and Tilly, 1990a,b). Finally, the requirement and involvement of a number of ions (e.g., Ca^{2+}, Na^+, K^+, Cl^-) and ion channels in granulosa cell steroid production has been investigated (e.g., Morley et al., 1991; Asem and Tsang, 1989).

2. Prehierarchal Follicles and Stromal Tissue

The theca layer of prehierarchal (<8-mm diameter) follicles, as well as stromal tissue from prepubertal pullets and laying hens, is steroidogenically active, both occurring predominantly via the Δ^5 steroidogenic pathway (Figure 4) (Kowalski et al., 1991; Levorse and Johnson, 1994). By contrast, granulosa cells from prehierarchal follicles are incapable of synthesizing progesterone due to the lack of P_{450}scc activity, although they do possess the ability to produce cAMP in response to FSH treatment (Tilly et al., 1991a). While granulosa cells from prehierarchal follicles express 3β-hydroxysteroid dehydrogenase (3β-HSD) activity, it is unclear whether any significant amount of progesterone is produced by this cell layer in vivo.

As selection of a follicle into the hierarchy is proposed to occur (presumably at a rate of one per day) from a pool of approximately 8 to 14 6- to 8-mm diameter follicles, it has been proposed that the process of follicle recruitment is coupled to the initial expression of LH receptor mRNA and the acquisition of functional P_{450}scc enzyme activity within the granulosa layer. Expression of P_{450}scc mRNA and the initiation of enzyme activity in vitro is induced by culture with FSH, and this induction is prevented by coculture with TGFα or EGF (Li and Johnson, 1993). This latter finding is of significance as such results may help to explain how prehierarchal follicles are maintained in a relatively undifferentiated state in vivo despite their continued, daily exposure to circulating concentrations of FSH. Once the selection of a follicle occurs, granulosa cells begin the transition from being predominantly FSH and VIP dependent (granulosa cells from follicles up to approximately 12-mm in diameter) to becoming primarily LH dependent (hierarchal follicles) (Figure 5).

3. Postovulatory and Atretic Follicles

Progesterone content of the most recent postovulatory follicle declines during the first 15 hr after ovulation (Dick et al., 1978), although LH-stimulable adenylyl cyclase activity is present for at least 24 hr after ovulation

(Calvo et al., 1981). Both the enzymes 3β-HSD and 17β-hydroxysteroid dehydrogenase have been identified in follicles during the early stages of atresia; however, this activity is rapidly lost as atresia progresses.

B. Nonsteroidal Hormones and Growth Factors

Prostaglandins E and F have been identified in preovulatory and postovulatory follicles (Hammond et al., 1980; Shimada et al., 1984); however, prostaglandin production is not stimulated by gonadotropins in vitro (Hertelendy and Hammond, 1980). Norepinephrine (NE) and to a lesser extent epinephrine (EPI) and dopamine are found predominantly within the theca layer of preovulatory follicles, and tissue levels of EPI and NE peak approximately 6–9 hr prior to ovulation (Bahr et al., 1986).

Vasoactive intestinal peptide (VIP) is localized within nerve terminals of the theca layer and is proposed to act on the adjacent granulosa layer via the adenylyl cyclase/cAMP second messenger system. For instance, VIP has been found to stimulate steroidogenesis and plasminogen activator activity in preovulatory (F1) follicle granulosa and theca cells (Johnson and Tilly, 1988; Tilly and Johnson, 1989) and to promote differentiation and suppress apoptosis in 6- to 8-mm follicle granulosa cells (Johnson et al., 1994; Flaws et al., 1995). The neurohypophyseal hormones arginine vasotocin and mesotocin are produced in granulosa and theca tissue of hierarchal follicles, and levels of each vary with a differential pattern during the ovulatory cycle (Saito et al., 1990).

Several locally produced growth factors and their receptors have recently been implicated in regulating ovarian function via paracrine effects, including TGFα and its receptor, the EGF receptor (Johnson, 1994; Onagbesan et al., 1994), TGFβ (Law et al., 1995), nerve growth factor (NGF) and the p75 (low affinity) NGF receptor (Johnson, unpublished data), stem cell factor and its receptor, c-kit (Johnson, 1993; Jensen and Johnson, 1995), and insulinlike growth factor I (IGF-I) (Roberts et al., 1994). It is inevitable that many additional physiologically relevant growth factors and receptors have yet to be identified within avian ovarian tissue.

IV. HORMONAL AND PHYSIOLOGIC FACTORS AFFECTING OVULATION

Ovulation in the domestic hen generally follows an oviposition within 15–75 min except for the first ovulation of a sequence, which is not associated with oviposition. Neither the premature expulsion of the egg (which can be induced with prostaglandins or other agents) nor a delayed oviposition (affected by epinephrine or progesterone) influences the time of ovulation. Ovulation occurs via a rupture along the follicle stigma, and this process likely involves multiple, often redundant, factors which include proteolytic enzymes (e.g., collagenase, plasminogen activator), vasoactive substances and the process of apoptosis. For additional information, see Johnson (1986).

A. Hormones and Ovulation

1. Luteinizing Hormone Releasing Hormone

Avian species express two different forms of luteinizing hormone-releasing hormone (LHRH-I and LHRH-II; see Chapter 14); however, only LHRH-I appears to be directly involved in regulating LH secretion (Sharp et al., 1990). Injection of LHRH-I induces ovarian steroid production and premature ovulation, whereas in vivo administration of LHRH antiserum has been shown to block ovulation. These effects, however, are mediated via the pituitary and LH secretion, and there is no evidence of a direct ovulation-inducing effect of either LHRH moiety within the ovary.

2. Luteinizing Hormone

Plasma concentrations of LH in the domestic hen (and most other birds studied to date) peak 4–6 hr prior to ovulation (Figure 6) (Johnson and van Tienhoven, 1980). It is likely that this preovulatory surge provides a direct stimulus for germinal vesicle breakdown and for subsequent ovulation. Some workers have reported an additional peak of LH at 14–11 hr prior to ovulation

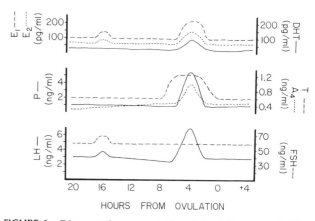

FIGURE 6 Diagramatic representation of circulating concentrations of ovarian steroids and pituitary gonadotropins during the ovulatory cycle of the hen. A_4, androstenedione; DHT, 5α-dihydrotestosterone; E_1, estrone; E_2, estradiol-17β; FSH, follicle-stimulating hormone; LH, luteinizing hormone; P, progesterone; T, testosterone. (Reprinted with permission from A. L. Johnson (1990), CRC Crit. Rev. Poult. Biol. **2**, 319–346.)

(Etches and Cheng, 1981); however, the significance of this second peak has yet to be determined. In addition to these ovulatory–oviposition cycle-related peaks there occurs a crepuscular (occurring at the onset of darkness) peak of LH, which has a periodicity of 24 hr and has been proposed to act as a timing cue for the subsequent preovulatory LH surge (Johnson and van Tienhoven, 1980; Wilson et al., 1983).

Injection of LH into laying hens almost always increases plasma concentrations of progesterone, estrogens and androgens; however, the ovulatory response is dependent on the stage within the sequence. For instance, LH treatment 14–11 hr prior to the first ovulation of a sequence results in premature ovulation, whereas the same treatment before a midsequence ovulation commonly results in follicle atresia and blocked ovulation (Gilbert et al., 1981). Passive immunization of laying hens with antiserum generated against partially purified chicken LH results in the cessation of ovulation for approximately 5 days and extensive atresia of existing follicles (Sharp et al., 1978).

3. Follicle-Stimulating Hormone

A rise in plasma follicle-stimulating hormone (FSH) has been observed 15 hr prior to ovulation in the domestic hen (Scanes et al., 1977a) (Figure 6), and this occurs coincident with an increase in FSH binding to ovarian tissue (Etches and Cheng, 1981). Other researchers find less pronounced cycle-related fluctuations in circulating FSH (Krishnan et al., 1993). Endocrine mechanisms regulating pituitary release of avian FSH are not clear. For instance, Hattori et al. (1986) have reported that LHRH-I induces FSH secretion *in vivo* and *in vitro* in the quail, while Krishnan et al. (1993) failed to detect any stimulatory effect of LHRH-I. The reasons for this discrepancy are not apparent.

A primary role for FSH is related to granulosa cell differentiation and the induction of steroidogenesis in prehierarchal follicle granulosa cells (see above). Although recombinant FSH is capable of inducing ovulation in mammals, this apparently has not been tested in avian species.

4. Progesterone

Highest concentrations of plasma progesterone are found 6–4 hr before ovulation and coincide with the preovulatory LH peak (Figure 6). This increase is predominantly the result of progesterone secretion by the largest preovulatory follicle (Etches, 1990).

Systemic and intraventricular injection of progesterone can induce both a preovulatory surge of LH and premature ovulation, while administration of progesterone antiserum prior to the preovulatory surge of progesterone can block ovulation. Evidence suggests that the preovulatory LH surge is preceded by an increase in progesterone in view of the findings that: (1) in the absence of the normal preovulatory rise in progesterone (blocked by the steroidogenesis inhibitor, aminoglutethimide), there is no preovulatory increase in plasma LH; and (2) intramuscular injection of progesterone in aminoglutethimide-treated hens induces a normal preovulatory LH surge in the absence of any increase in plasma testosterone or estrogen (Johnson and van Tienhoven, 1984). Moreover, results from a study of hypophysectomized hens suggest that progesterone, in the absence of preovulatory gonadotropin, can induce ovulation (Nakada et al., 1994).

Progesterone circulates in the blood bound to a high-affinity corticoid-binding globulin or to albumin and other γ-globulins. Progesterone-specific receptors have been demonstrated in the hypothalamus and pituitary of the hen and the number of receptors is influenced by the reproductive state. In the oviduct, progesterone receptors are present in surface epithelial cells, tubular gland cells, stromal fibroblasts, and smooth muscle cells in the arterial walls and myometrium (Yoshimura and Bahr, 1991a). These findings provide evidence for a role of progesterone in the production of avidin, contraction of the myometrium, and shell formation. Moreover, progesterone receptors are expressed within granulosa, theca, and germinal epithelial cells of preovulatory follicles (Isola et al., 1987; Yoshimura and Bahr, 1991b). The localization of such receptors associated with the stigma region provides a physiological mechanism by which progesterone may mediate a direct effect on ovulation.

5. Androgens

Peak preovulatory concentrations of testosterone occur 10–6 hr prior to ovulation, whereas highest levels of 5α-dihydrotestosterone (DHT) occur approximately 6 hr before ovulation (Figure 6). The increase in testosterone at this time is the result of secretion from at least the four largest follicles.

The role of androgens in ovulation is, as yet, unclear. Intraventricular injections of testosterone fail to release LH or induce premature ovulation, and peripheral injections of testosterone that generally result in unphysiologically high circulating concentrations of testosterone are required to stimulate LH secretion and induce ovulation. Moreover, ovulation can occur in the absence of any preovulatory increase in plasma testosterone (Johnson and van Tienhoven, 1984).

On the other hand, *in vivo* administration of antitestosterone serum effectively blocks ovulation, while testosterone treatment induces ovulation *in vitro* (Tanaka and Inoue, 1990). Both granulosa and theca cells of prehierarchal and hierarchal follicles express an androgen receptor, and androgens regulate steroidogenesis and plasminogen activator activity via a paracrine/autocrine action (Tilly and Johnson, 1987; Lee and Bahr, 1990; Yoshimura *et al.*, 1993b).

As in the male, androgens are responsible for the growth and coloring of the comb and wattles at the time of sexual maturation and synergize with estrogen to induce medullary ossification. In addition, androgens induce protein synthesis in the oviduct of estrogen-primed birds; however, the mechanism has been less intensively studied than that for progesterone.

6. Estrogens

Both estradiol-17β and estrone show highest plasma concentrations 6–4 hr before ovulation with a smaller, more inconsistent rise in estrogens 23–18 hr before ovulation (Figure 6). Preovulatory follicle secretion of estradiol increases in each of the four largest follicles 6–3 hr prior to ovulation and is greatest in the third- and fourth-largest follicles. Overall, however, the majority of estrogen produced by the ovary originates from prehierarchal follicles.

Like testosterone, estrogens are unlikely to be involved in the direct induction of LH secretion or ovulation, as injection of estradiol in laying hens either inhibits or has no effect on ovulation or LH secretion. Moreover, ovulation can occur in the absence of a preovulatory increase in plasma estrogens. Estrogens, together with progesterone, are required for priming of the hypothalamus and pituitary in order that progesterone can induce LH release (Wilson and Sharp, 1976).

Ovarian estrogens have a variety of additional functions related to reproduction, including the regulation of calcium metabolism for shell formation (Etches, 1987), induction of its own receptor in the oviduct (Moen and Palmiter, 1980), and induction of progesterone receptor expression in the ovary and reproductive tract (Pageaux *et al.*, 1983; Isola *et al.*, 1987). Treatment of immature Japanese quail and young female chickens with estradiol enhances growth of the oviduct and promotes the formation of tubular secretory glands and epithelial differentiation. Moreover, estradiol induces the synthesis of ovalbumin, conalbumin, ovomucoid, and lysozyme in the oviduct (Palmiter, 1972) and vitellogenin in the liver (Deely *et al.*, 1975).

Secondary sex characteristics, such as color and shape of plumage and sexual behavior, are also under the control of estrogens. Sexual differentiation of the female brain is under the influence of estrogen, as indicated by the finding that treatment of female quail embryo with an antiestrogen results in behavioral masculinization.

7. Corticosterone

Plasma concentrations of corticosterone display both a daily (photoperiod-related) rhythm and a peak coincident with oviposition. The role of corticoids on the ovary and in the ovulatory process is unclear, as both facilitory and inhibitory actions have been described. While there is some evidence to suggest that the adrenal gland, via corticosterone secretion, regulates the timing of the preovulatory LH surge (Wilson and Cunningham, 1980), there is no ovulation-related increase in circulating corticosterone. Injection of corticosterone, deoxycorticosterone, or ACTH induces premature ovulation; however, it is unlikely to act at the level of the hypothalamus and/or pituitary to directly induce LH release (Etches *et al.*, 1984a,b). On the other hand, infusion of corticosterone blocks the photoperiod-induced rise in LH as well as the gonadotropic effects of eCG on follicle growth (Petitte and Etches, 1988, 1989).

In several wild species seasonal increases of plasma corticosterone are associated with egg laying (e.g., White-crowned sparrow, European starling, Western meadowlark) or with brooding behavior (e.g., Pied flycatcher), whereas in others (e.g., Canada goose) there is little change or a decrease in corticosterone during the breeding season. Absolute plasma levels of corticosterone can be misleading, however, as the capacity of the corticosteroid binding globulin may also fluctuate with season (Wingfield and Farner, 1993).

8. Prolactin

It is well accepted that prolactin is associated with parental behavior and may also play a role in osmoregulation, especially in marine birds. Prolactin secretion from the anterior pituitary is primarily under the stimulatory control of vasoactive intestinal peptide (see Chapter 14). Circulating levels of prolactin increase at the onset of egg laying in many species of birds, including the ruffed grouse (Etches *et al.*, 1979), bantam fowl (Sharp *et al.*, 1979), turkey hen (Proudman and Opel, 1981), Japanese quail (Goldsmith and Hall, 1980), pied flycatcher (Silverin and Goldsmith, 1990), and song sparrow (Wingfield and Goldsmith, 1990). During the ovulation–oviposition cycle of the domestic hen, prolactin is highest approximately 10 hr and lowest 6 hr prior to ovulation, though it is unlikely that prolactin plays a direct role in the ovulatory process. Prolactin levels are also elevated during the summer months in response to increasing photoperiod.

Expression of prolactin receptor mRNA and prolactin binding has been detected in crop-sac mucosa, liver, brain, ovary, and oviduct (Tanaka et al., 1992). Recent data indicate that prolonged elevated prolactin levels (as occurs during incubation; see below) have an antisteroidogenic effect on the ovary, in part via inhibition of steroidogenic enzyme gene expression (Tabibzadeh et al., 1995), and act on the neuroendocrine system to reduce hypothalamic GnRH levels and inhibit LH secretion (Rozenboim et al., 1993).

9. Other Factors

Preovulatory follicles produce prostaglandins, and the secretion of prostaglandin (PG) F2α by the F1 follicle is greatest around the time of ovulation. Although prostaglandins do not affect granulosa cell steroid production, both PGE_1 and PGE_2 enhance follicle plasminogen activator activity (Tilly and Johnson, 1987). This latter action might suggest a role for PGs in mediating the enzymatic rupture of the stigma. In addition, $PGF_{2\alpha}$, PGE_2, acetylcholine, and oxytocin increase the contraction of smooth muscles in the connective tissue wall of the follicle *in vitro* and may, in combination with proteolytic enzymes, play a role in the rupture of the stigma at ovulation (Yoshimura et al., 1983). Nevertheless, it is clear that PGs are not obligatory for ovulation to occur as neither the administration of prostaglandins nor the injection of indomethacin into the follicle affects the timing of ovulation.

Administration of serotonin (5-hydroxytryptamine) has been reported to block ovulation, but this result is probably mediated via its inhibitory effect on LH release. The catecholamine-synthesis inhibitor, α-methylmetatyrosine and the α-adrenergic blocking agents phentolamine and dibenzyline block ovulation when injected intrafollicularly, whereas anti-β-adrenergic agents are ineffective. Subsequent administration of EPI or NE overcomes the blocking effect of phentolamine, while only EPI overcomes the effect of α-methylmetatyrosine, and neither catecholamine could reverse the blocking effect of dibenzyline. While NE has no effect on basal or LH-stimulated progesterone production in granulosa cells, preculture with NE for 2 days enhances LH-responsiveness via α-adrenergic receptors (Chotesangasa et al., 1988). Nevertheless, it remains to be demonstrated that catecholamines play a physiologic role in the process of ovulation.

B. Hormones during Sexual Maturation and Molt

Investigation of the changes in gonadotropin concentrations between the time of hatching and the onset of lay in the domestic hen shows that there is an early LH peak at about 1 week of age and a prepubertal LH rise beginning approximately 15 weeks posthatch that is highest about 3 weeks prior to sexual maturity. Ovarian steroidogenesis can be demonstrated as early as day 3.5 of incubation. Plasma estradiol increases from less than 100 pg/ml 6 weeks before lay to a peak of about 350 pg/ml 2–3 weeks before lay; it then decreases to basal concentrations (100–150 pg/ml) at the time of first lay. Plasma progesterone concentrations are low (0.1–0.5 ng/ml) until about 1 week before laying, at which time baseline levels increase to 0.4-0.6 ng/ml (Williams and Sharp, 1977). The hen's ovary is unresponsive to stimulation by either mammalian gonadotropins or avian pituitary extracts until 16–18 weeks of age. In contrast, the pituitary becomes less responsive to exogenous LHRH as the onset of puberty approaches (Wilson and Sharp, 1975); this change is probably the result of negative feedback by increasing steroid secretion by the ovary.

C. Effect of Light on the Ovary and Ovulation

Photoperiod influences the reproductive activity of birds relative to both seasonal changes (signaling the onset and termination of breeding seasons in temperate latitudes) and daily changes (entraining the time of ovulation and/or oviposition). Ovarian development and egg laying appear to be most stimulated by increasing photoperiod, as normally occurs during the spring in the northern hemisphere. A notable exception is the Emperor penguin (*Aptenodytes forsteri*), which breeds during the short photoperiod of antarctic winters (Groscolas et al., 1986). Migratory species such as the junco or white-crowned sparrow, as well as the domestic turkey, fail to show sexual development when held under constant short-day conditions. In contrast, domestic fowl raised under a continuous photoperiod of either 6 hr of light or 22 hr of light will reach sexual maturity at about 21 weeks of age. Even in this species, however, a progressively increasing photoperiod from hatch to 18 weeks of age advances the onset of egg laying by 2–3 weeks when compared to birds raised under a constant photoperiod (Morris, 1978).

Maximal photostimulation is generally considered to occur with a photoperiod of 12–14 hr of light; however, normal egg production can be obtained with a photoperiod between 12 and 18 hr, and the domestic hen will continue to lay (although more sporadically) even if kept in continuous darkness. Moreover, an intermittent photoperiod consisting of a total of 10 hr of light in the pattern 8 hr light (L):10 hr dark (D):2 hr L:4 hr D has proven as effective in maintaining egg production as a 14L:10D photoperiod.

D. Photorefractoriness

The seasonal termination of reproduction in birds is an adaptive mechanism to insure that the newly hatched young are raised under optimal conditions. For instance, many species that breed in northern latitudes end the nesting phase (and undergo spontaneous gonadal regression) early enough in the summer to insure sufficient time for development of fledglings prior to fall migration. A common (but not universal) mechanism responsible for ending the breeding season is photorefractoriness; this state is characterized by the inability of the reproductive system to respond to the otherwise stimulatory effects of increasing daylength. Photorefractoriness is often initiated by long days, and the number of long days required for induction is dependent upon day length. In the White-crowned sparrow (*Zonotrichia leucophrys gambelii*), gonads will remain regressed for several years when continuously maintained on long days, and recovery of photosensitivity occurs only after birds are exposed to short photoperiods for 40–60 days.

Considerable attention has been focused on the hypothalamus as the anatomical location of photorefractoriness. For example, there is evidence for increased hypothalamic sensitivity to the negative feedback effects of sex steroids as the breeding season progresses and for a decline in hypothalamic concentrations of LHRH. By comparison, neither the gonads nor the anterior pituitary appear to become refractory. For additional information and references, see Wingfield and Farner (1993).

E. Molt

Molt refers to the orderly replacement of feathers and is accompanied by the total regression of the reproductive organs and the cessation of lay. In wild, temperate species a prenuptial molt may occur prior to the breeding season, while a postnuptial molt occurs between the end of the reproductive season and the onset of autumn or the autumn migration. In the domestic fowl, molt occurs after approximately 1 year of egg production.

The physiologic mechanisms that cause molt remain obscure; however, the pituitary, adrenal, and particularly the thyroid glands have been suggested as mediators (Jallageas and Assenmacher, 1985). Forced molt in domestic fowl can be induced by restriction of feed and water, decreased day length, or by administration of prolactin or large doses of progesterone. In addition, injection of thyroxine induces molt in the domestic hen and promotes feather growth in redheaded buntings (Pati and Pathak, 1986), while thyroidectomy retards molt in the spotted munia and redheaded bunting. This relationship is not consistent among birds, however, as thyroidectomy of juvenile mallard ducks fails to alter timing of the molt to breeding plumage.

Circulating concentrations of plasma prolactin, growth hormone, LH, and ovarian steroids are lower in molting females than in laying hens (Scanes *et al.*, 1979b; Hoshino *et al.*, 1988). In turkey hens, plasma prolactin levels during molt are less than half those found in laying hens (Proudman and Opel, 1981). On the other hand, plasma levels of corticosterone, testosterone, and triiodothyronine (T3), but not thyroxine (T4) increase during the molting period. The elevation in plasma thyroxine may not only play a role in the induction of molt, but may also be required for thermoregulation during the period of heavy feather loss.

V. OVIPOSITION

Expulsion of the egg (oviposition) involves the relaxation of abdominal muscles and sphincter between the shell gland and vagina and the muscular contraction of the shell gland. The majority of research to elucidate the mechanisms controlling oviposition has been performed in the quail and domestic hen and has implicated neurohypophyseal hormones, prostaglandins, and hormones of the pre- and postovulatory follicles.

A. Neurohypophyseal Hormones

Both oxytocin and arginine vasotocin induce premature oviposition in hens. While arginine vasotocin activity in the blood of the laying hen was reported to be highest during oviposition, removal of the neurohypophysis failed to affect the pattern of timing of oviposition. On the other hand, it is possible that an additional source of arginine vasotocin is the ovary itself (Saito *et al.*, 1990). Finally, there is evidence that the oviposition-inducing actions of oxytocin may be mediated via prostaglandins, as administration of indomethacin, a prostaglandin synthesis inhibitor, blocks oxytocin-induced premature oviposition.

B. Prostaglandins

There is much experimental data suggesting that prostaglandins are directly involved in oviposition. Exogenous administration of $PGF_{2\alpha}$ stimulates shell gland contractility, relaxes the vagina, and induces premature oviposition (Shimada and Asai, 1979), while treatment with indomethacin or aspirin depresses the peak of prostaglandins in the plasma and in pre- and postovulatory follicles, suppresses uterine contractility, and delays oviposition (Hertelendy and Biellier, 1978). In addition, injection of PGE_1 induces premature oviposition in chickens and Japanese quail, while passive immunization with PGE_1 antiserum delays oviposition in the hen. Further evidence to suggest that prostaglandins mediate the oviposition-inducing activity of arginine vasotocin is provided by the finding that plasma prostaglandins are significantly elevated at the time of arginine vasotocin-induced premature oviposition and that arginine vasotocin stimulates the biosynthesis and release of prostaglandin from the shell gland (Rzasa, 1984). Arginine vasotocin administered to hen shell gland muscle *in vitro* increases tissue content of prostaglandins and stimulates shell gland muscle contractility, while concomitant treatment with indomethacin blocks the production of prostaglandins and decreases the rate of muscle contraction.

Endogenous concentrations of prostaglandins E and F increase in the largest preovulatory follicle beginning 6–4 hr prior to the expected time of ovulation (or oviposition). In contrast, postovulatory levels of prostaglandin F increase about 100-fold approximately 24 hr after that ovulation or 2 hr before the next expected oviposition (Day and Nalbandov, 1977). Increased prostaglandin production and secretion from one or more of these sources is reflected by significantly elevated plasma concentrations of the prostaglandin metabolite 13,14-dihydro-15-keto prostaglandin $F_{2\alpha}$, coincident with the time of oviposition. $TGF\alpha$ and $TGF\beta$ are among factors that have been shown to regulate granulosa cell PG production (Li *et al.*, 1994).

Administration of the prostaglandin precursor, arachidonic acid, or prostaglandin F2α ($PGF_{2\alpha}$) increases intraluminal pressure of the infundibulum, magnum, isthmus, and shell gland. Substance P, $PGF_{2\alpha}$, and arginine vasotocin stimulate shell gland contractility *in vitro* (Olson *et al.*, 1978; De Saedeleer *et al.*, 1989), while $PGF_{2\alpha}$ has been demonstrated to be effective *in vivo* (Shimada and Asai, 1979). Specific binding sites for $PGF_{2\alpha}$ and arginine vasotocin have been identified and characterized in the membranes of shell gland muscle (Toth *et al.*, 1979; Takahashi *et al.*, 1992). In contrast, prostaglandins of the E series (PGE_1, PGE_2) suppress spontaneous motility and decrease intraluminal pressure in the vagina (Wechsung and Houvenaghel, 1978). These observations suggested that ovum transport through the oviduct and expulsion of the egg from the shell gland may involve prostaglandin-stimulated contractility of oviduct smooth muscle.

C. Postovulatory Follicle

Removal of the most recent postovulatory follicle results in a 1- to 7-day delay in oviposition; this effect has been attributed to the removal of factors produced by the granulosa cell layer (Gilbert *et al.*, 1978); the factor(s) may be a peptide, possibly analagous or identical to the neurohypophyseal hormones or prostaglandins.

D. Preovulatory Follicle

A role for the preovulatory follicle in the process of oviposition was first suggested by Fraps (1942), who found that a premature midsequence ovulation induced by the injection of LH was accompanied by premature expulsion of the egg within the oviduct. Soon after, Rothchild and Fraps (1944) demonstrated that removal of the most mature preovulatory follicle resulted in the delayed oviposition of the egg in the shell gland, although the delay was not as long as that found after removal of the postovulatory follicle.

E. Other Factors

As mentioned above, plasma concentrations of corticosterone increase dramatically at the time of oviposition. Premature oviposition (induced with vasopressin) or delayed oviposition (resulting from the removal of the postovulatory follicle) failed to alter the timing of the corticosterone peak from when the egg would normally have been laid; an additional increase in corticosterone occurs at the time of vasopressin-induced oviposition (Beuving and Vonder, 1981). Nevertheless, these results do not necessarily indicate a causative relationship between corticosterone secretion and the induction of oviposition.

Despite the fact that the shell gland contractility is influenced by epinephrine and norepinephrine, there is no evidence that these catecholamines influence the timing of oviposition. Many other factors, including acetylcholine, histamine, pentobarbital, and lithium, induce premature oviposition, but the mechanisms and physiological relevance of their actions are unclear. For additional information, see Shimada and Saito (1989).

F. Broodiness and Nesting Behavior

Broody behavior consists of termination of egg production, the incubation of eggs, and care of the young. The onset of incubation occurs coincident with complete regression of the ovary and accessory reproductive tissues such as the oviduct and comb. In present-day commercial egg-producing hens, broodiness is virtually nonexistent, but is still present in the Bantam hen and broiler-breeder and is common in turkeys. Broody behavior is often accompanied by the development of an incubation (or brood) patch, and the increased vascularity, edema, and thickening of the epidermis occurs under the regulation of estrogen and prolactin. Induction of incubation behavior in turkey hens can be initiated by administration of prolactin, and active immunization against prolactin reduced the incidence or delayed the onset of broodiness in Bantam hens (Youngren et al., 1991; March et al., 1994). Incubation activity stimulates the development of the crop sac in the pigeon; proliferation of this gland is under the direct control of prolactin.

In the ring dove, endogenous estrogen production, stimulated by FSH secretion, is thought to induce the female to build nests. In turn, the initiation of nestbuilding activity in the male is dependent on the hormonal environment and behavior of the female.

In general, behavioral broodiness in the female is preceded by decreases in circulating concentrations of LH, progesterone, testosterone, and estradiol, while growth hormone tends to be lower only while hens are caring for the young (Harvey et al., 1981). As discussed above, concentrations of prolactin are frequently increased during egg laying and around the onset of incubation and remain elevated throughout incubation. Among species in which the male assumes the predominant role of incubation, levels of prolactin are higher in the male than in the female (Oring et al., 1988). For further information, see El Halawani et al. (1988).

VI. COMPOSITION AND FORMATION OF YOLK, ALBUMEN, ORGANIC MATRIX, AND SHELL

Components of the egg include the yolk, albumen, the organic matrix, and the crystalline shell. These components have been reviewed by a number of workers, and the reader is referred to the following literature for more details: Gilbert (1971a,b), Simkiss and Taylor (1971), Parsons (1982), and Burley and Vadehra (1989). There appears to be a remarkable uniformity among many species of birds as to the density of yolk and albumen and the proportion of egg yolk relative to egg weight. However, variations do occur in the lipid, amino acid and carbohydrate content of albumen and yolk, with precocial species (e.g., from the orders Anseriformes and Galliformes) having a higher energy density than altricial species (e.g., the pigeon *Columba livia* and pelican *Pelecanus occidentalis*) (Roca et al., 1984; Bucher, 1987).

A. Yolk

Yolk serves to provide lipids and many of the proteins required for embryonic growth. The final composition of yolk in the hen's egg consists of approximately 33% lipid (by wet weight) compared to 17% protein. The greatest proportion of yolk is water (48%) with lesser amounts of free carbohydrates (0.2%) and inorganic elements (1%). Of the major classes of lipids, approximately 70% are triacylglycerols, 25% are phospholipids, and 5% cholesterol and cholesterol esters. Yolk is usually deposited in concentric bands which result from a nonuniform pattern of daily feed intake.

B. Albumen

There are four distinct layers of albumen in the fully formed egg: (1) the chalaziferous (inner thick) layer, attached to the yolk; (2) the inner thin (liquid) layer; (3) the outer thick layer; and (4) the outer thin (fluid) layer. Approximately one-fourth of the total albumen by weight is found in the outer thin layer and slightly greater than one-half in the outer thick layer. The inner layer represents 16.8% of the total albumen, while the chalaziferous layer plus chalazae 2.7% of the total (Table 2). The function of the egg albumen layer is likely to prevent invasion of microorganisms into the yolk (see below), as well as to serve as a source of water, protein, and minerals to the embryo during development. The

TABLE 2 Composition of the Hen's Egg[a]

	Yolk	Albumen			Chalaziferous	Shell
		Outer	Middle	Inner		
Weight (g)	18.7	7.6	18.9	5.5	0.9	6.2
Water (%)	48.7	88.8	87.6	86.4	84.3	1.6
Solids (%)	51.3	11.2	12.4	13.6	15.7	98.4

	All layers		
	Yolk	Albumen	Shell
Proteins (%)	16.0	10.6	3.3
Carbohydrates (%)	1.0	0.9	—
Fats (%)	32.6	Trace	0.03
Minerals (%)	1.1	0.6	95.10

[a] From Romanoff and Romanoff (1949).

chalazae are two fiberlike structures at each end of the egg permit limited rotation but little lateral displacement of the yolk.

In the chicken, the initial albumen layer is deposited by the caudal region of the infundibulum. However, the majority of albumen proteins are deposited by tubular gland cells of the magnum, while avidin is synthesized by the goblet cells. The discrete layers of albumen are the result of either its consecutive deposition by different regions of the magnum or changes that occur during plumping within the shell gland and with movement of the egg through the oviduct.

The major proteins found in albumen include: (1) ovalbumen, 54%; (2) ovotransferrin (conalbumin), 13%; (3) ovomucoid, 11%; (4) ovoglobulins (G2 and G3), 8%; (5) lysozyme, 3.5%; and (6) α- and β-ovomucin, 1.5–3.0%. There are also several characteristic proteins present in lesser concentrations, including avidin, flavoprotein- and thiamine-binding proteins; ovomacroglobulin; and cystatin.

Ovalbumin serves as a source of amino acids for the embryo during development and, based on its homology to the serpin family of protease inhibitors, may also suppress enzyme activity in the egg. Ovotransferrin acts primarily as an iron-chelator, thereby preventing bacterial growth within the egg. Ovomucoid is a serine protease inhibitor and represents the major inhibitor of protease (primarily trypsin) activity in the egg; target enzymes may be produced by invading microorganisms or, alternatively, ovomucoid may serve to modulate the enzymatic degradation of albumin during embryo development.

Alpha-ovomucin (a glycoprotein) and β-ovomucin (60% carbohydrate by weight) are insoluble, fibrous proteins that are responsible for the gel-like qualities of egg white, particularly in the thick layers. Ovomucins block invasion of microorganisms and may express antiviral properties. The main biologic function egg albumen lysozyme is lytic activity against Gram-negative bacterial cell walls.

Several egg proteins are known for their ability to bind specific substrates (e.g., ovotransferrin binds iron, copper and zinc; flavoprotein, riboflavin; and avidin, biotin), and others act as protease inhibitors (ovomucoid, ovoinhibitor, cystatin, ovomacroglobulin). Avidin is a progesterone-dependent secretory protein; it's biologic role in the hen egg is likely related to its ability to decrease biotin levels, thus inhibiting bacterial growth. For more detailed information on albumen synthesis and properties, see Burley and Vadehra (1989).

C. Organic Matrix

The organic fraction of the eggshell consists of shell membranes, the mammillary cores, the shell matrix, and the cuticle. Although these components constitute only a small fraction of the entire eggshell, their integrity is critical to its formation and strength.

1. Shell Membranes

Shell membranes, organized into an inner ($\sim$50 to 70-μm thick) and outer ($\sim$15 to 25-μm thick) membrane, are produced by the isthmus region of the oviduct. The membranes lie adjacent to one another except at one end where they part to form the air cell. There is some question as to whether epithelial cells or tubular gland cells of the isthmus are responsible for the secretion of the membranes. The membranes consist of a meshwork of protein fibers, cross-linked by disulfide and lysine-derived bonds, with small fibrous protuberances of uncertain function.

The membranes are composed in part of collagen (10%), inasmuch as both hydroxyproline and hydroxylysine can be identified. The remainder of the fibrous component is composed of protein (70–75%) and glycoprotein (Candlish, 1972). The membranes are semipermeable and permit the passage of gases, water, and crystalloids but not albumen; there is no relationship between thickness of the membranes and thickness of the shell, but membrane thickness does decrease with the age of the hen.

2. Mammillary Cores

The mammillary cores, which are projections from the outer membrane surface (Figure 7), are proposed as the initial sites of calcification. They are composed largely of protein, but also contain carbohydrate and mucopolysaccharides and are thought to be formed by the epithelial cells of the isthmus. The mammillary cores represent the greatest proportion of organic material in the eggshell.

3. Shell Matrix

The organic shell matrix is a series of layers of protein plus acid mucopolysaccharide on which calcification takes place. It represents approximately 2% of the total organic composition of the eggshell. The matrix plus calcified crystals make up the palisade layer of the shell. The innermost region of the shell has a greater density of matrix compared to the outer regions. Both calcium-binding protein and carbonic anhydrase have been identified within the matrix. Deposition of the matrix occurs soon after the egg reaches the shell gland.

4. Cuticle

The outermost surface of the egg is often (but not always) covered by a thin waxy cuticle composed of protein, polysaccharide, and lipid. The cuticle may be

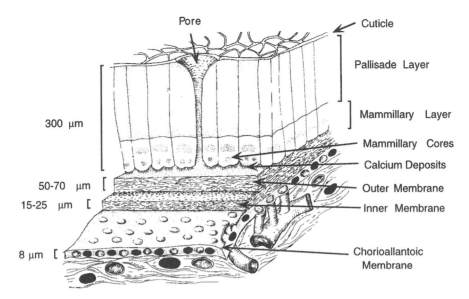

FIGURE 7 Diagram of section through the shell and membranes of a hen's egg, showing crystalline structure and organic material that remains after calcium carbonate has been dissolved. (Adapted from *How eggs breathe*, H. Rahn, A. Ar, and C. Paganelli. Copyright © 1979 by Scientific American, Inc. All rights reserved.)

unevenly distributed over the entire surface, ranging in depth from 0.5 to 12.8 mm; the average dry weight of the cuticle in a 60 g chicken egg is 12 mg (Parsons, 1982). Its function is likely to protect the egg from water evaporation and microbial invasion, although it is unlikely to add to the structural integrity of the shell. The cuticle, when present, is formed during the last 30 min prior to oviposition.

D. Layers of Crystallization

The calcified portion of the shell can be arbitrarily divided into the mammillary knob layer, the palisade layer, and the outer surface crystal layer (see Figure 7). Collectively, these layers represent the major portion of the avian egg shell, are largely responsible for its mechanical strength, and consist of approximately 97% inorganic material. While calcium is the predominant cation, the shell also sequesters a significant amount of magnesium in the form of magnesium carbonate; manganese is likely required for the development of the mammillae because of its role in the synthesis of mucopolysaccharides (Leach and Gross, 1983).

1. Mammillary Knob Layer

Outward crystallization of the mammillary cores results in the formation of the mammillary knob layer. In addition, calcium deposits radiate from the bottom of the mammillary cores to penetrate the outer shell membrane. Formation of the mammillary knobs occurs in the shell gland during the first 5 hr of calcification. There is some disagreement as to whether or not the density of mammillary knobs is related to the structural integrity of the eggshell. The passage of water during plumping stretches the shell membranes and increases the distance between the tips of the mammillae. Crystals that form laterally grow and eventually abut with crystals from other mammillae, whereas those that grow outward may extend to the shell surface. At some points the crystals do not grow completely together, leaving pores with a diameter of 0.3–0.9 μm (Figure 7).

2. Palisade Layer

The crystallized palisade (or spongy) layer (approximately 200 μm thick) is composed mainly of crystalline calcium carbonate in the form of calcite and represents the greatest portion of the shell. Calcification of this layer is initiated 5–6 hr after the entrance of the egg into the shell gland, during the process of plumping. The palisade layer is arranged as columns situated directly over mammillary knobs that are perpendicular to the surface (Silyn-Roberts and Sharp, 1986).

3. Surface Crystal Layer

The outermost layer of calcification is designated the surface crystal layer. The crystalline structure is more dense than that of the palisade region and lies perpendicular to the shell surface. The overall thickness of this layer ranges from 3 to 8 μm.

4. Respiration via the Eggshell

Shell pores of the hen egg are simple funnel-shaped openings that arise at the shell surface and protrude through to the mammillary knob layer. In most species, however, pores traverse the shell radially and branch longitudinally along the axis of the egg. The pores are the result of areas of incomplete crystallization. The number of pores is generally related to metabolic demand prior to the initiation of lung function, the number of pores per unit area decreasing with increasing egg weight (Tullett, 1978). The function of eggshell pores is to serve as a mechanism of chemical communication between the air cell of the egg and the external environment during embryo development (Tullett, 1984; Rahn and Paganelli, 1990). The exchange of oxygen, carbon dioxide, and water occurs predominantly via passive diffusion, and a 79 gm turkey egg is estimated to exchange the equivalent of 27 l of gas (oxygen, carbon dioxide and water vapor) during the 28-day incubation period (Paganelli *et al.*, 1987). The average diffusive water loss during incubation for 117 species of birds is 15%; the loss is determined by the difference in water vapor pressure between the egg and nest environment (see Chapter 21). Inner and outer shell membranes play a minor role in limiting the flow of oxygen, carbon dioxide, and water vapor, while the major resistance to passage of these substances comes from the inner end of the pore. Therefore, air-cell gas tensions are determined predominantly by the number and size of the pores and the thickness of the shell, relative to the metabolic rate of the embryo. The total area of pores on a turkey egg with a surface area of 90 cm^2 is calculated to be 2.2 mm^2 (Paganelli *et al.*, 1987). Furthermore, experimental data suggest that a change in environmental conditions may be reflected by an alteration in shell conductance (see Tullett, 1978).

E. Calcium Metabolism

The shell gland transports 2–2.5 g of calcium within a period of 15 hr for the calcification of a single egg. A hen that lays 280 eggs in a production year will use a quantity of calcium for the purpose of shell formation corresponding to 30 times the calcium content of the entire body. A detailed description concerning mechanisms of absorption and changes in uptake by the intestine relative to age and egg production are beyond the scope of the present discussion, and the reader is referred to Bar *et al.* (1977), Gilbert (1983), and Etches (1987).

1. Sources of Eggshell Calcium

Calcium for eggshell formation is provided via the blood following absorption from the intestine (duodenum and upper jejunum) or resorption from bone (principally medullary bone, but also cortical bone under conditions of calcium deficiency). Resorption of calcium from bone is regulated by both parathyroid hormone and 1,25-dihydroxyvitamin D_3 (1,25-$(OH)_2D_3$), while absorption through the gut is facilitated by 1,25 $(OH)_2D_3$. The relative importance of the two organs as sources of calcium depends on the concentration of dietary calcium. Hens consume approximately 25% more feed on days when eggshell formation occurs than on days when it does not. If concentrations of calcium in the food are 3.6% or higher, most of the eggshell calcium is derived directly from the intestine. If the concentration is 2.0%, bone supplies 30–40% of the shell calcium, and on calcium-free diets, the skeleton is the principal source. However, it is likely that these relationships vary depending on the time of day. When hens are provided constant free access to feed, much of the daily intake occurs early in the photoperiod; the remainder is consumed at the end of the photoperiod. However, much of the shell is formed during the night, when generally no calcium is consumed and when the calcium content of the digestive tract is gradually decreasing. Therefore, medullary bone is the primary source of shell calcium during the latter hours of darkness.

Blood calcium circulates in two forms, as nondiffusible protein-bound calcium and as diffusible ionized calcium. Nondiffusible calcium is bound by plasma calcium-binding proteins vitellogenin and albumin. Estrogen treatment increases total plasma calcium, in part by stimulating the production of blood calcium-binding proteins. Similarly, total plasma calcium increases several weeks before laying, and the increase is attributable to an increase in the protein-bound, but not in the diffusible fraction. During the ovulation–oviposition cycle, concentrations of ionized calcium peak (0.057 mg/ml) 4 hr after oviposition, then decrease significantly during the period of shell calcification (minimum of 0.049 mg/ml). On the other hand, plasma concentrations of total calcium (ionized plus inionized) fluctuate only minimally (between 0.2 and 0.26 mg/ml) during the ovulation–oviposition cycle.

Feeding a calcium-deficient diet to laying hens causes a significant decrease in plasma ionized calcium concentrations, a significant decrease or complete cessation of lay, and regression of the ovary within 6 to 9 days. However, birds on the same diet injected with crude chicken pituitary extracts continued to lay for a more prolonged period, despite a comparable decrease in ionized calcium (Luck and Scanes, 1979a). Basal concentrations of LH in birds fed a calcium-deficient diet (0.2% Ca) are significantly lower compared to birds on a calcium-rich diet (3.0% Ca), but between the two groups the LH-releasing activity of exogenously administered LHRH is similar (Luck and Scanes, 1979b). These results suggest that pituitary LH synthesis and release and the ability of the ovary to respond to gonadotropins are resistant to a calcium deficiency.

Mongin and Sauveur (1974) determined that voluntary calcium consumption by laying hens increases at the time the egg enters the shell gland and given a choice between a calcium-deficient diet (1.0% Ca) and a diet supplemented with calcium-rich oyster shell, hens will preferentially consume the calcium-rich diet during the period of eggshell calcification. For further information, see Etches (1987).

2. Vitamin D Metabolism

Vitamin D plays an important role in the regulation of calcium metabolism via the metabolically active metabolite $1,25\text{-}(OH)_2D_3$. Conversion to this form occurs by the 1-hydroxylation of 25-hydroxyvitamin D_3 (25-OH-D_3) in the kidney under the hormonal control of estradiol and PTH (Soares, 1984). For instance, renal 1-hydroxylase activity increases just prior to the initiation of egg laying, at a time corresponding with an increase in circulating estrogen and total plasma calcium. In addition, 1-hydroxylase activity during the ovulation–oviposition cycle increases at the time of ovulation. The increase in activity is followed by an increase in circulating concentrations of $1,25\text{-}(OH)_2D_3$ 4 hr after ovulation, and elevated levels persist until 10 hr after ovulation or after the initiation of eggshell formation (Castillo et al., 1979). For additional information, see Norman (1987).

3. Calcium Mobilization from Medullary Bone

In the long bones (e.g., femur and tibia), medullary bone lines the endosteal surface and grows with a system of interconnecting spicules that may completely fill the narrow spaces. Medullary bone forms in female birds during the final 10 days before egg laying under the influence of estrogen and testosterone; it can be readily induced in intact male birds by administration of estrogen or in castrated males by estrogen and testosterone. Gonadal steroids apparently act directly on the cells in the medullary cavity and independently of calcium intake. In addition, it is unlikely that $1,25\text{-}(OH)_2D_3$ mediates the transfer of calcium to medullary bone at this time, as the appearance of this bone occurs 1–2 weeks prior to the increase in renal 1-hydroxylase activity and the elevation of plasma total calcium.

During the ovulation–oviposition cycle, periods of intense medullary bone formation alternate with periods of severe bone depletion. Hens fed a high-calcium diet are generally able to replenish the calcium lost from medullary bone during shell calcification when shell formation is not taking place, but on a low-calcium diet the cortical bone of the femur is eroded, while medullary bone is maintained in a fairly constant amount. Under these conditions the new medullary bone that forms is only partially calcified, and an increase in the number of osteoblasts is indicative of a more rapid turnover rate.

4. Calcium Absorption and Secretion by the Shell Gland

The mechanism of calcium transfer across the intestine and shell gland, as well as the hormonal regulation of the process, is poorly understood. As discussed above, the initiation of egg production is associated with increased $1,25\text{-}(OH)_2D_3$ synthesis and accumulation in the intestine and shell gland, and calcium transport as well as concentrations of the calcium binding protein, calbindin-D_{28K}, increase in both organ systems during egg production. Calbindin is expressed in tubular gland cells of the shell gland and in the distal portion of the isthmus, but not in the proximal isthmus or magnum (Wasserman et al., 1991). Calbindin synthesis in the intestine is regulated primarily by $1,25\text{-}(OH)_2D_3$, yet its production in the shell gland apparently requires ovarian steroids and other, as-yet-unidentified factors in addition to $1,25\text{-}(OH)_2D_3$ (Bar et al., 1990). While levels of calbindin mRNA increase during the ovulatory cycle at a time coincident with shell calcification there is little to no change in tissue levels of calbindin protein; such results indicate posttranslational control of calbindin levels (Nys et al., 1989).

Calcium secretion from the shell gland increases approximately 7 hr after ovulation, reaches a maximum as the shell is being formed, and decreases to the basal secretion rate after shell formation is complete but before the expulsion of the egg (Eastin and Spaziani, 1978a). The presence of an egg within the shell gland is not likely to be the major stimulus for the initiation of calcium secretion but appears to be more closely related to ovulation. The hormonal signal(s) that affects changes in the rate of calcium secretion is unknown, although estrogen involvement has been suggested (Eastin and Spaziani, 1978a). Similarly, termination of a high rate of calcium secretion is not related to egg removal, as calcium deposition decreases fully 2 hr before eggs are laid. Movement of calcium across the shell gland, which may occur by both diffusion and active transport (Eastin and Spaziani, 1978b), involves expenditure of metabolic energy. In the aged hen, calcium deposition and egg shell weight and density are related to shell gland levels of calbindin (Bar et al., 1992).

5. Carbonate Formation and Deposition

The mineral content of the eggshell consists of about 97–98% calcium carbonate while the remainder consists of magnesium carbonate and tricalcium phosphate. Carbonate in the shell is derived from blood bicarbonate or is synthesized from CO_2 by carbonic anhydrase located in oviduct tissue and the shell. In addition, luminal carbonate ion content is supplemented by bicarbonate in fluid flowing from the magnum to the shell gland. The secretion from the shell gland of HCO_3^-, like that of calcium, is accomplished by active transport (Eastin and Spaziani, 1978b) and may be influenced by luminal HCO_3^- concentrations. The net amount of calcium secretion is functionally linked to HCO_3^- production and to luminal HCO_3^- concentrations. For excellent reviews of carbonate/bicarbonate production and calcium–bicarbonate interactions and secretion, see Simkiss (1967) and Eastin and Spaziani (1978b).

References

Adkins, K. K. (1978). Sex steroids and the differentiation of behavior. *Am. Zool.* **18,** 501–509.

Aitken, R. N. C. (1971). The oviduct. In "Physiology and Biochemistry of the Domestic Fowl" (J. Bell and B. M. Freeman, eds.), Vol. 3. Academic Press, London/New York.

Ariamaa, O., and Talo, A. (1983). The membrane potential of the Japanese quail's oviductal smooth muscle during ovum transport. *Acta Physiol. Scand.* **118,** 349–353.

Asem, E. K., Molnar, M., and Hertelendy, F. (1987). Luteinizing hormone-induced intracellular calcium mobilization in granulosa cells: Comparison with forskolin and 8-bromo-adenosine 3',5'-monophosphate. *Endocrinology* **120,** 853–859.

Asem, E. K., and Tsang, B. K. (1989). Na^+/H^+ exchange in granulosa cells of the three largest preovulatory follicles of the domestic hen. *Gen. Comp. Endocrinol.* **75,** 129–134.

Bahr, J. M., and Calvo, F. O. (1984). A correlation between adenylyl cyclase activity and responsiveness to gonadotrophins during follicular maturation in the domestic hen. In "Reproductive Biology of Poultry" (F. J. Cunningham, P. E. Lake, and D. Hewitt, eds.), pp. 75–87. British Poultry Science Ltd., Harlow.

Bahr, J. M., Ritzhaupt, L. K., McCullough, S., Arbogast, L. A., and Ben-Jonathan, N. (1986). Catecholamine content of the preovulatory follicles of the domestic hen. *Biol. Reprod.* **34,** 502–506.

Bar, A., Cohen, A., Montecuccoli, G., Edelstein, S., and Hurwitz, S. (1977). Relationship of intestinal calcium and phosphorus absorption to vitamin D metabolism during reproductive activity of birds. In "Vitamin D Biochemical, Chemical and Clinical Aspects Related to Calcium Metabolism" (A. W. Norman et al., eds.), p. 93. de Gruyter, Berlin.

Bar, A., Striem, S., Mayel-Afshar, S., and Lawson, D.E.M. (1990). Differential regulation of calbindin-D_{28K} mRNA in the intestine and eggshell gland of the laying hen. *J. Mol. Endocrinol.* **4,** 93–99.

Bar, A., Vax, E., and Striem, S. (1992). Relationships between calbindin (M_r 28,000) and calcium transport by the eggshell gland. *Comp. Biochem. Physiol. A* **101,** 845–848.

Barber D. L., Sanders, E. J., Aebersold, R., and Schneider, W. J. (1991). The receptor for yolk lipoprotein deposition in the chicken oocyte. *J. Biol. Chem.* **266,** 18761–18770.

Beuving, G., and Vonder, G. M. A. (1981). The influence of ovulation and oviposition on corticosterone levels in the plasma of laying hens. *Gen. Comp. Endocrinol.* **44,** 382–388.

Breen, P. C., and DeBruyn, P. P. H. (1969). The fine structure of the secretory cells of the uterus (shell gland) of the chicken. *J. Morphol.* **128,** 35–66.

Bucher T. L. (1987). Patterns in the mass-independent energetics of avian development. *J. Exp. Zool.* **1,** (Suppl.), 139–150.

Burke, W. H., Ogasawara, F. X., and Fuqua, C. L. (1972). A study of the ultrastructure of the uterovaginal sperm-storage glands of the hen, *Gallus domesticus,* in relation to a mechanism for the release of spermatozoa. *J. Reprod. Fertil.* **29,** 29–36.

Burley, R. W., and Vadehra, D. V. (1989). "The Avian Egg." Wiley, New York.

Calvo, F. O., Wang, S.-C., and Bahr, J. M. (1981). LH–stimulable adenylyl cyclase activity during the ovulatory cycle in granulosa cells of the three largest follicles and the postovulatory follicle of the domestic hen *(Gallus domesticus).* *Biol. Reprod.* **25,** 805–812.

Candlish, J. K. (1972). The role of the shell membranes in the functional integrity of the egg. In "Egg Formation and Production" (B. M. Freeman and P. E. Lake, eds.). British Poultry Science Ltd., Edinburgh.

Castillo, L., Tanaka, Y., Wineland, M. J., Jowsey, J. O., and DeLuca, H. F. (1979). Production of 1,25-dihydroxyvitamin D_3 and formation of medullary bone in the egg–laying hen. *Endocrinology* **104,** 1598–1601.

Chalana, R. K., and Guraya, S. S. (1978). Histophysiological studies on the postovulatory follicles of the fowl ovary. *Poult. Sci.* **57,** 814–817.

Chotesangasa, R., Kamiyoshi, M., Tanaka, K., Tomogane, H., and Yokoyama, A. (1988). Enhancement of the response of hen granulosa cells to LH with norepinephrine *in vitro.* *Comp. Biochem. Physiol. C* **90,** 225–229.

Conkright, M. D., and Asem, E. K. (1995). Intracrine role of progesterone in FSH– and cAMP–induced fibronectin production and deposition by chicken granulosa cells: Influence of follicular development. *Endocrinology* **136,** 2641–2651.

Crossley, J., Ferrando, G., and Eiler, H. (1980). Distribution of adrenergic receptors in the domestic fowl oviduct. *Poult. Sci.* **59,** 2331–2335.

Dahl, E. (1970a). Studies on the fine structure of ovarian interstitial tissue. 2. The ultrastructure of the thecal gland of the domestic fowl. *Z. Zellforsch. Mikrosk. Anat.* **109,** 195–211.

Dahl, E. (1970b). Studies of the fine structure of ovarian interstitial tissue. 3. The innervation of the thecal gland of the domestic fowl. *Z. Zellforsch. Mikrosk. Anat.* **109,** 212–226.

Day, S. L., and Nalbandov, A. V. (1977). Presence of prostaglandin F (PGF) in hen follicles and its physiological role in ovulation and oviposition. *Biol. Reprod.* **16,** 486–494.

De Saedeleer, V., Wechsung, E., and Houvenaghel, A. (1989). Influence of substance P on *in vitro* motility of the oviduct and intestine in the domestic hen. *Anim. Reprod. Sci.* **21,** 293–300.

Deely, R. G., Mullinix, K. P., Wetekam, W., Kronenberg, H. M., Meyers, M., Eldridge, J. D., and Goldberger. R. F. (1975). Vitellogenin synthesis in the avian liver. Vitellogenin is the precursor of the egg yolk phosphoproteins. *J. Biol. Chem.* **250,** 9060–9066.

Dick, H. R., Culbert, J., Wells, J. W., Gilbert, A. B., and Davidson, M. F. (1978). Steroid hormones in the postovulatory follicle of the domestic fowl *(Gallus domesticus).* *J. Reprod. Fertil.* **53,** 103–107.

Donham, R. S., Dane, C. W., and Farner, D. S. (1976). Plasma luteinizing hormone and the development of ovarian follicles after loss of clutch in female mallards (*Anas platyrhynchos*). *Gen. Comp. Endocrinol.* **29,** 152–155.

Eastin, W. C., and Spaziani, E. (1978a). On the control of calcium secretion in the avian shell gland (uterus). *Biol. Reprod.* **19,** 493–504.

Eastin, W. C., and Spaziani, E. (1978b). On the mechanism of calcium secretion in the avian shell gland (uterus). *Biol. Reprod.* **19,** 505–518.

El Halawani, M. E., Fehrer, S., Hargis, B. M., and Porter, T. E. (1988). Incubation behavior in the domestic turkey. *CRC Crit. Rev. Poult. Biol.* **1,** 285–314.

El Halawani, M. E., Silsby, J. L., Youngren, O. M., and Phillips, R. E. (1991). Exogenous prolactin delays photoinduced sexual maturity and suppresses ovriectomy induced luteinizing hormone secretion in the turkey (*Meleagris gallopavo*). *Biol. Reprod.* **44,** 420–424.

Etches, R. J. (1987). Calcium logistics in the laying hen. *J. Nutr.* **117,** 619–628.

Etches, R. J. (1990). The ovulatory cycle of the hen. *CRC Crit. Rev. Poult. Biol.* **2,** 293–318.

Etches, R. J., and Cheng, K. W. (1981). Changes in the plasma concentrations of luteinizing hormone, progesterone, oestradiol and testosterone and in the binding of follicle-stimulating hormone to the theca of follicles during the ovulation cycle of the hen (*Gallus domesticus*). *J. Endocrinol.* **91,** 11–22.

Etches, R. J., and Schoch, J. P. (1984). A mathematical representation of the ovulatory cycle of the domestic hen. *Br. Poult. Sci.* **25,** 65–76.

Etches, R. J., Garbutt, A., and Middleton, A. L. (1979). Plasma concentrations of prolactin during egg laying and incubation in the ruffed grouse (*Bonasa umbellus*). *Can. J. Zool.* **57,** 1624–1627.

Etches, R. J., Petitte, J. N., and Anderson-Langmuir, C. E. (1984a). Interrelationships between the hypothalamus, pituitary gland, ovary, adrenal gland, and the open period for LH release in the hen (*Gallus domesticus*). *J. Exp. Zool.* **232,** 501–511.

Etches, R. J., Williams, J. B., and Rzasa, J. (1984b). Effect of corticosterone and dietary changes in the hen on ovarian function, plasma LH and steroids and the response to exogenous LHRH. *J. Reprod. Fertil.* **70,** 121–130.

Flaws, J. A., DeSanti, A., Tilly, K. I., Javid, R. O., Kugu, K., Johnson, A. L., Hirshfield, A. N., and Tilly, J. L. (1995). Vasoactive intestinal peptide–mediated suppression of apoptosis in the ovary: Mechanisms of action and evidence of a conserved anti-atretogenic role through evolution. *Endocrinology* **136,** 4351–4359.

Fraps, R. M. (1942). Synchronized induction of ovulation and premature oviposition in the domestic fowl. *Anat. Rec.* **84,** 521.

Fraps, R. M. (1955). Egg production and fertility in poultry. *In* "Progress in the Physiology of Farm Animals" (J. Hammond, Ed.), Vol. II. Butterworth, London.

Freedman, S. L., and Sturkie, P. D. (1963). Blood vessels of the chicken's uterus (shell gland). *Am. J. Anat.* **113,** 1–7.

Gilbert, A. B. (1969). Innervation of the ovary of the domestic hen. *Quart. J. Exp. Physiol.* **54,** 404–411.

Gilbert, A. B. (1971a). Egg albumin and its formation. *In* "Physiology and Biochemistry of the Domestic Fowl" (D. J. Bell and B. M. Freeman, eds.), Vol. 3, pp. 1291–1329. Academic Press, London/New York.

Gilbert, A. B. (1971b). The egg: Its physical and chemical aspects. *In* "Physiology and Biochemistry of the Domestic Fowl" (D. J. Bell and B. M. Freeman, eds.), Vol. 3, pp. 1379–1399. Academic Press, London/New York.

Gilbert, A. B. (1979). Female genital organs. *In* "Form and Function in Birds" (A. S. King and J. McLelland, eds.), Vol. I, pp. 237–360. Academic Press, London/New York.

Gilbert, A. B. (1983). Calcium and reproductive function in the hen. *Proc. Nutr. Soc.* **42,** 195–212.

Gilbert, A. B., Reynolds, M. E., and Lorenz, F. W. (1968). Distribution of spermatozoa in the oviduct and fertility in domestic birds. VII. Innervation and vascular supply of the uterovaginal sperm–host glands of the domestic hen. *J. Reprod. Fertil.* **17,** 305–310.

Gilbert, A. B., Davidson, M. F., and Wells, J. W. (1978). Role of the granulosa cells of the postovulatory follicle of the domestic fowl in oviposition. *J. Reprod. Fertil.* **52,** 227–229.

Gilbert, A. B., Hardie, M. A., Perry, M. M., Dick, H. R., and Wells, J. W. (1980). Cellular changes in the granulosa layer of the maturing ovarian follicle of the domestic fowl. *Br. Poult. Sci.* **21,** 257–263.

Gilbert, A. B., Davidson, M. F., Hardie, M. A., and Wells, J. W. (1981). The induction of atresia in the domestic fowl (*Gallus domesticus*) by ovine LH. *Gen. Comp. Endocrinol.* **44,** 344–349.

Gilbert, A. B., Perry, M. M., Waddington, D., and Hardie, M. A. (1983). Role of atresia in establishing the follicular hierarchy in the ovary of the domestic hen (*Gallus domesticus*). *J. Reprod. Fertil.* **69,** 221–227.

Goldsmith, A. R., and Hall, M. (1980). Prolactin concentrations in the pituitary gland and plasma of Japanese quail in relation to photoperiodically induced sexual maturation and egg laying. *Gen. Comp. Endocrinol.* **42,** 449–454.

Grau, C. R. (1982). Egg formation in Fiordland crested penguins (*Eudyptes pachyrhynchus*). *Condor* **84,** 172–177.

Groscolas, R., Jallageas, M., Goldsmith, A., and Assenmacher, I. (1986). The endocrine control of reproduction and molt in male and female emperor (*Aptenodytes forsteri*) and Adelie (*Pygoscelis adeliae*) penguins. I. Annual changes in plasma levels of gonadal steroids and LH. *Gen. Comp. Endocrinol.* **62,** 43–53.

Hammond, R. W., Olson, D. M., Frenkel, R. B., Biellier, H. V., and Hertelendy, F. (1980). Prostaglandins and steroid hormones in plasma and ovarian follicles during the ovulation cycle of the domestic hen (*Gallus domesticus*). *Gen. Comp. Endocrinol.* **42,** 195–202.

Harvey, S., Bedrak, E., and Chadwick, A. (1981). Serum concentrations of prolactin, luteinizing hormone, growth hormone, corticosterone, progesterone, testosterone and oestradiol in relation to broodiness in domestic turkeys (*Meleagris gallopavo*). *J. Endocrinol.* **89,** 187–195.

Hattori, A., Ishii, S., and Wada, M. (1986). Effects of two kinds of chicken luteinizing hormone-releasing hormone (LH–RH) and its analogs on the release of LH and FSH in Japanese quail and chicken. *Gen. Comp. Endocrinol.* **64,** 446–455.

Hertelendy, F. (1973). Block of oxytocin-induced parturition and oviposition by prostaglandin inhibitors. *Life Sci.* **13,** 1581–1589.

Hertelendy, F., and Biellier, H. V. (1978). Evidence for a physiological role of prostaglandins in oviposition by the hen. *J. Reprod. Fertil.* **53,** 71–74.

Hertelendy, F., and Hammond, R. W. (1980). Prostaglandins do not affect steroidogenesis and are not being produced in response to oLH in chicken granulosa cells. *Biol. Reprod.* **23,** 918–923.

Hertelendy, F., Nemecz, G., and Molnar, M. (1989). Influence of follicular maturation on luteinizing hormone and guanosine 5'-O-thiotriphosphate-promoted breakdown of phosphoinositides and calcium mobilization in chicken granulosa cells. *Biol. Reprod.* **40,** 1144–1151.

Hodges, R. D. (1965). The blood supply to the avian oviduct, with special reference to the shell gland. *J. Anat.* **99,** 485–506.

Hodges, R. D. (1974). *In* "The Histology of the Fowl." Academic Press, New York/London.

Hoffer, A. P. (1971). The ultrastructure and cytochemistry of the shell membrane-secreting region of the Japanese quail oviduct. *Am. J. Anat.* **131,** 253–288.

Hoshino, S., Suzuki, M., Kakegawa, T., Imai, K., Kobayashi, Y., and Yamada, Y. (1988). Changes in plasma thyroid hormones, luteinizing hormone (LH), estradiol, progesterone and corticosterone of laying hens during a forced molt. *Comp. Biochem. Physiol. A* **90**, 355–359.

Hutson, J. M., Donahoe, P. K., and MacLaughlin, D. T. (1985). Steroid modulation of Mullerian duct regression in the chick embryo. *Gen. Comp. Endocrinol.* **57**, 88–102.

Isola, J., Korte, J.-M., and Tuohimaa, P. (1987). Immunocytochemical localization of progesterone receptor in the chick ovary. *Endocrinology* **121**, 1034–1040.

Jallageas, M., and Assenmacher, I. (1985). Endocrine correlates of molt and reproduction functions in birds. In "Acta XVIII Congress of International Ornithology" (V. D. Ilyichev and V. M. Gavrilov, eds.), pp. 935–945. Nauka, Moskow.

Jensen, T., and Johnson, A. L. (1995). Expression of c-*kit* mRNA in the embryonic and adult chicken ovary. *Biol. Reprod.* **52** (Suppl. 1), 194.

Johnson, A. L. (1984). Interactions of progesterone and LH leading to ovulation in the domestic hen. In "Reproductive Biology of Poultry" (F. J. Cunningham, P. E. Lake, and D. Hewitt, eds.), pp. 133–143. British Poultry Science Ltd, Harlow.

Johnson, A. L. (1986). Reproduction in the Female. In "Avian Physiology" (P. D. Sturkie, ed.), 4th Ed., pp. 403–431. Springer-Verlag, New York.

Johnson, A. L. (1990). Steroidogenesis and actions of steroids in the hen ovary. *CRC Crit. Rev. Poult. Biol.* **2**, 319–346.

Johnson, A. L. (1993). Regulation of follicle differentiation by gonadotropins and growth factors. *Poult. Sci.* **72**, 867–873.

Johnson, A. L. (1994). Gonadotropin and growth factor regulation of avian granulosa cell differentiation. In "Perspectives in Comparative Endocrinology" (K. G. Davey, R. E. Peter, and S. S. Tobe, Eds.), pp. 613–618. National Research Council Canada, Ottawa.

Johnson, A. L., and Leone, E. W. (1985). Ovine luteinizing hormone–induced steroid and luteinizing hormone secretion, and ovulation in intact and pregnant mare serum gonadotropin-primed hens. *Poult. Sci.* **64**, 2171–2179.

Johnson, A. L., and Tilly, J. L. (1988). Effects of vasoactive intestinal peptide on steroid secretion and plasminogen activator activity in granulosa cells of the hen. *Biol. Reprod.* **38**, 296–303.

Johnson, A. L., and Tilly, J. L. (1990a). Arachidonic acid inhibits luteinizing hormone-stimulated progesterone production in hen granulosa cells. *Biol. Reprod.* **42**, 458–464.

Johnson, A. L., and Tilly, J. L. (1990b). Evidence that arachidonic acid influences granulosa cell steroidogenesis and plasminogen activator activity by a protein kinase C-independent mechanism. *Biol. Reprod.* **43**, 922–928.

Johnson, A. L., Bridgham, J. T., Witty, J., and Tilly, J. L. (1996). Susceptibility of avian granulosa cells to apoptosis is dependent upon stage of follicle development and is related to endogenous levels of *bcl*-xlong gene expresssion. *Endocrinology* **137**, 2059–2066.

Johnson, A. L., and van Tienhoven, A. (1980). Plasma concentrations of six steroids and LH during the ovulatory cycle of the hen, *Gallus domesticus*. *Biol. Reprod.* **23**, 386–393.

Johnson, A. L., and van Tienhoven, A. (1984). Effects of aminoglutethimide on luteinizing hormone and steroid secretion, and ovulation in the hen, *Gallus domesticus*. *Endocrinology* **114**, 2276–2283.

Johnson, A. L., Bridgham, J. T., Witty, J. P., and Tilly, J. L. (1997a). Expression of *bcl*-2 and *nr*-13 in hen ovarian follicles during development. *Biol. Reprod.* **57**, 1096–1103.

Johnson, A. L., Li, Z., Gibney, J. A., and Malamed, S. (1994). Vasoactive intestinal peptide-induced expression of cytochrome P450 cholesterol side-chain cleavage and 17α-hydroxylase enzyme activity in hen granulosa cells. *Biol. Reprod.* **51**, 327–333.

Johnson, A. L., Bridgham, J. T., Bergeron, L., and Yuan, J. (1997b). Characterization of the avian *ich*-1 cDNA and expression of *ich*-1 long mRNA in the hen ovary. *Gene* **192**(2), 227–233.

Johnson, A. L., Bridgham, J. T., and Wagner, B. (1996). Characterization of a chicken luteinizing hormone receptor (cLH-R) cDNA, and expression of cLH-R mRNA in the ovary. *Biol. Reprod.* **55**, 304–309.

Johnson, P. A., Brooks, C., Wang, S.-Y., and Chen, C.-C. (1993). Plasma concentrations of immunoreactive inhibin and gonadotropins following removal of ovarian follicles in the domestic hen. *Biol. Reprod.* **49**, 1026–1031.

Kato, M., Shimada, K., Saito, N., Noda, K., and Ohta, M. (1995). Expression of P450 17α-hydroxylase and P450 aromatase genes in isolated granulosa, theca interna, and theca externa layers of chicken ovarian follicles during follicular growth. *Biol. Reprod.* **52**, 405–410.

King, A. S. (1975). Aves urogenital system. In "Sisson and Grossman's The Anatomy of Domestic Animals" (R. Getty, ed.), 5th ed., Vol. 2. Saunders, Philadelphia.

Kinsky, F. C. (1971). The consistent presence of paired ovaries in the Kiwi (*Aptryx*) with some discussion of this condition in other birds. *J. Ornithol.* **112**, 334–357.

Kowalski, K. I., Tilly, J. L., and Johnson, A.L. (1991). Cytochrome P450 side–chain cleavage (P450scc) in the hen ovary. I. Regulation of P450scc messenger RNA levels and steroidogenesis in theca cells of developing follicles. *Biol. Reprod.* **45**, 955–966.

Krishnan, K. A., Proudman, J. A., Bolt, D. J., and Bahr, J. M. (1993). Development of an homologous radioimmunoassay for chicken follicle-stimulating hormone and measurement of plasma FSH during the ovulatory cycle. *Comp. Biochem. Physiol. A* **105**, 729–734.

Law, A. S., Burt, D. W., and Armstrong, D. G. (1995). Expression of transforming growth factor–β mRNA in chicken ovarian follicular tissue. *Gen. Comp. Endocrinol.* **98**, 227–233.

Levorse, J. M., Tilly, J. L., and Johnson, A. L. (1991). Role of calcium in the regulation of theca cell androstenedione production in the domestic hen. *J. Reprod. Fertil.* **92**, 159–167.

Levorse, J. M., and Johnson, A. L. (1994). Regulation of steroid production in ovarian stromal tissue from 5- to 8-week-old pullets and laying hens. *J. Reprod. Fertil.* **100**, 195–202.

Leach, R. M. (1982). Biochemistry of the organic matrix of the eggshell. *Poult. Sci.* **61**, 2040–2047.

Leach, R. M., and Gross, J. R. (1983). The effect of manganese deficiency upon the ultrastructure of the eggshell. *Poult. Sci.* **62**, 499–504.

Lee, H. T., and Bahr, J. M. (1990). Inhibition of the activities of P450 cholesterol side–chain cleavage and 3β-hydroxysteroid dehydrogenase and the amount of P450 cholesterol side–chain cleavage by testosterone and estradiol–17β in hen granulosa cells. *Endocrinology* **126**, 779–786.

Li, Z., and Johnson, A. L. (1993). Regulation of P450 cholesterol side chain cleavage messenger ribonucleic acid expression and progesterone production in hen granulosa cells. *Biol. Reprod.* **49**, 463–469.

Li, J., Li, M., Lafrance, M., and Tsang, B. K. (1994). Avian granulosa cell prostaglandin secretion is regulated by transforming growth factor α and β and does not control plasminogen activator activity during follicular development. *Biol. Reprod.* **51**, 787–794.

Lofts, B., and Murton, R. K.(1973). Reproduction in Birds. In "Avian Biology" (D. S. Farner and J. R. King, eds.), Vol. 3, pp. 1–107. Academic Press, London/New York.

Luck, M. R., and Scanes, C. G. (1979a). Plasma levels of ionized calcium in the laying hen (*Gallus domestics*). *Comp. Biochem. Physiol. A* **63**, 177–181.

Luck, M. R., and Scanes, C. G. (1979b). The relationship between reproductive activity and blood calcium in the calcium–deficient hen. *Br. Poult. Sci.* **20,** 559–564.

March, J. B., Sharp, P. J., Wilson, P. W., and Sang, H. M. (1994). Effect of active immunization against recombinant-derived chicken prolactin fusion protein on the onset of broodiness and photoinduced egg laying in bantam hens. *J. Reprod. Fertil.* **101,** 227–233.

Marrone, B. L., and Hertelendy, F. (1985). Decreased androstenedione production with increased follicular maturation in theca cells from the domestic hen (*Gallus domesticus*). *J. Reprod. Fertil.* **74,** 543–550.

Moen, R. C., and Palmiter, R. D. (1980). Changes in hormone responsiveness of chick oviduct during primary stimulation with estrogen. *Dev. Biol.* **78,** 450–463.

Mongin, P., and Sauveur, B. (1974). Voluntary food and calcium intake by the laying hen. *Br. Poult. Sci.* **15,** 349–359.

Morley, P., Schwartz, J. L., Whitfield, J. F., and Tsang, B. K. (1991). Role of chloride ions in progesterone production by chicken granulosa cells. *Mol. Cell. Endocrinol.* **82,** 107–115.

Morris, T. R. (1978). The influence of light on ovulation in domestic birds. In "Animal Reproduction" (H. Hawk, ed.), BARC Symposium Number 3, Chapter 19, p. 307. Allenheld, Montclair.

Nakada, T., Koja, Z., and Tanaka, K. (1994). Effect of progesterone on ovulation in the hypophysectomized hen. *Br. Poult. Sci.* **35,** 153–156.

Nalbandov, A. V., and James, M. F. (1949). The blood–vascular system of the chicken ovary. *Am. J. Anat.* **85,** 347–377.

Nevalainen, T. J. (1969). Electron microscope observations on the shell gland mucosa of calcium–deficient hens (*Gallus domesticus*). *Anat. Rec.* **164,** 127–140.

Nitta, H., Osawa, Y., and Bahr, J. M. (1991). Multiple steroidogenic cell populations in the theca layer of preovulatory follicles of the chicken ovary. *Endocrinology* **129,** 2033–2040.

Norman, A. W. (1987). Studies on the vitamin D endocrine system in the avian. *J. Nutr.* **117,** 797–807.

Nys, Y., Mayel-Afshar, S., Bouillon, R., Van Baelen, H., and Lawson, D. E. M. (1989). Increases in calbindin D 28K mRNA in the uterus of the domestic fowl induced by sexual maturity and shell formation. *Gen. Comp. Endocrinol.* **76,** 322–329.

Olsen, M. W. (1960). Performance record of a parthenogenetic turkey male. *Science* **132,** 1661–1662.

Olson, D. M., Biellier, H. V., and Hertelendy, F. (1978). Shell gland responsiveness to prostaglandins F2α and E, and to arginine vasotocin during the laying cycle of the domestic hen (*Gallus domesticus*). *Gen. Comp. Endocrinol.* **36,** 559–565.

Onagbesan, O. M., Gullick, W., Woolveridge, I., and Peddie, M. J. (1994). Immunohistochemical localization of epidermal growth factor receptors, epidermal-growth-factor-like and transforming-growth-factor-α-like peptides in chicken ovarian follicles. *J. Reprod. Fertil.* **102,** 147–153.

Oring, L. W., Fivizzani, A. J., Colwell, M. A., and El Halawani, M. E. (1988). Hormonal changes associated with natural and manipulated incubation in the sex-role reversed Wilson's phalarope. *Gen. Comp. Endocrinol.* **72,** 247–256.

Ottinger, M. A., Adkins-Regan, E., Buntin, J., Cheng, M. F., DeVoogt, T., Harding, C., and Opel. H. (1984). Hormonal mediation of reproductive behavior. *J. Exp. Zool.* **232,** 605–616.

Paganelli, C. V., Ar, A., and Rahn, H. (1987). Diffusion-induced convective gas flow through the pores of the eggshell. *J. Exp. Zool.* **1,**(Suppl.), 173–180.

Pageaux, J. F., Laugier, C., Pal, D., and Pacheco, H. (1983). Analysis of progesterone receptor in the quail oviduct. Correlation between plasmatic estradiol and cytoplasmic progesterone receptor concentrations. *J. Steroid Biochem.* **18,** 209–214.

Palmer, S. S., and Bahr, J. M. (1992). Follicle stimulating hormone increases serum oestradiol-17β concentrations, number of growing follicles and yolk deposition in aging hens (*Gallus gallus domesticus*) with decreased egg production. *Brit. Poult. Sci.* **33,** 403–414.

Palmiter, R. D. (1972). Regulation of protein synthesis in chick oviduct. I. Independent regulation of ovalbumin, conalbumin, ovomucoid and lysozyme induction. *J. Biol. Chem.* **247,** 6450–6461.

Parsons, A. H. (1982). Structure of the eggshell. *Poult. Sci.* **61,** 2013–2021.

Parsons, A. H., and Combs, G. F. (1981). Blood ionized calcium cycles in the chicken. *Poult. Sci.* **60,** 1520–1524.

Pati, A. K., and Pathak, V. K. (1986). Thyroid and gonadal hormones in feather regeneration of the redheaded bunting, *Emberiza bruniceps*. *J. Exp. Zool.* **238,** 175–181.

Peccarelli, L. L., Johnson, A. L., Gibney, J. A., and Malamed, S. (1994). Catecholaminergic fibers, nerve growth factor and its receptor are found in the pre-follicular chick ovary. Endocr. Soc. 617. (Abstract)

Perry, M. M., Gilbert, A. B., and Evans, A. J. (1978a). Electron microscope observations on the ovarian follicle of the domestic fowl during the rapid growth phase. *J. Anat.* **125,** 481–497.

Perry, M. M., Gilbert, A. B., and Evans, A. J. (1978b). The structure of the germinal disc region of the hen's ovarian follicle during the rapid growth phase. *J. Anat.* **127,** 379–392.

Perry, M. M., and Gilbert, A. B. (1979). Yolk transport in the ovarian follicle of the hen (*Gallus domesticus*): Lipoprotein-like particles at the periphery of the oocyte in the rapid growth phase. *J. Cell Sci.* **39,** 257–272.

Petitte, J. N., and Etches, R. J. (1988). The effect of corticosterone on the photoperiodic response of immature hens. *Gen. Comp. Endocrinol.* **69,** 424–430.

Petitte, J. N., and Etches, R. J. (1989). The effect of corticosterone on the response of the ovary to pregnant mare's serum gonadotropin in sexually mature hens. *Gen. Comp. Endocrinol.* **74,** 377–384.

Porter, T. E., Hargis, B. M., Silsby, J. L., and El Halawani, M. E. (1989). Differential steroid production between theca interna and theca externa cells: a three cell model for follicular steroidogenesis in avian species. *Endocrinology* **125,** 109–116.

Proudman, J. A., and Opel, H. (1981). Turkey prolactin: Validation of a radioimmunoassay and measurement of changes associated with broodiness. *Biol. Reprod.* **25,** 573–580.

Rahn, H., Ar, A., and Paganelli, C. V. (1979). How eggs breathe. *Sci. Am.* **240,** 46–55.

Rahn, H., and Paganelli, C. V. (1990). Gas fluxes in avian eggs: Driving forces and the pathway for exchange. *Comp. Biochem Physiol. A* **95,** 1–15.

Retzek, H., Steyrer, E., Sanders, E. J., Nimpf, J., and Schneider, W. J. (1992). Cathepsin D: A key enzyme for yolk formation in the chicken oocyte. *DNA Cell Biol.* **11,** 661–672.

Ritzhaupt, L. K., and Bahr, J. M. (1987). A decrease in FSH receptors of granulosa cells during follicular maturation in the domestic fowl (*Gallus domesticus*). *J. Endocrinol.* **115,** 303–310.

Roberts, R. D., Sharp, P. J., Burt, D. W., and Goddard, C. (1994). Insulin-like growth factor-I in the ovary of the laying hen: Gene expression and biological actions on granulosa and theca cells. *Gen. Comp. Endocrinol.* **93,** 327–336.

Roca, P., Sainz, F., Gonzalez, M., and Alemany, M. (1984). Structure and composition of the eggs from several avian species. *Comp. Biochem Physiol. A* **77,** 307–310.

Romanoff, A. L., and Romanoff, A. J. (1949). *In* "The Avian Egg." Wiley, New York.

Rothchild, I., and Fraps, R. M. (1944). On the function of the ruptured ovarian follicle of the domestic fowl. *Proc. Soc. Exp. Biol. Med.* **56,** 79–82.

Rothwell, B., and Solomon, S. E. (1977). The ultrastructure of the follicle wall of the domestic fowl during the phase of rapid growth. *Br. Poult. Sci.* **18**, 605–610.

Rozenboim, I., Tabibzadeh, C., Silsby, J. L., and El Halawani, M. E. (1993). Effect of ovine prolactin administration on hypothalamic vasoactive intestinal peptide (VIP), gonadotropin releasing hormone I and II content, and anterior pituitary VIP receptors in laying turkey hens. *Biol. Reprod.* **48**, 1246–1250.

Rzasa, J. (1984). The effect of arginine vasotocin on prostaglandin production of the hen uterus. *Gen. Comp. Endocrinol.* **53**, 260–263.

Saito, N., Kinzler, S., and Koike, T. I. (1990). Arginine vasotocin and mesotocin levels in theca and granulosa layers of the ovary during the oviposition cycle in hens (*Gallus domesticus*). *Gen. Comp. Endocrinol.* **79**, 54–63.

Salevsky, E., and Leach, R. M. (1980). Studies on the organic components of shell gland fluid and the hen's egg shell. *Poult. Sci.* **59**, 438–443.

Scanes, C. G., Godden, P. M. M., and Sharp, P. J. (1977a). An homologous radioimmunoassay for chicken follicle-stimulating hormone: Observations on the ovulatory cycle. *J. Endocrinol.* **73**, 473–481.

Scanes, C. G., Sharp, P. J., and Chadwick, A. (1977b). Changes in plasma prolactin concentration during the ovulatory cycle of the chicken. *J. Endocrinol.* **72**, 401–402.

Scanes, C. G., Mozelic, H., Kavanagh, E., Merrill, G., and Rabii, J. (1982). Distribution of blood flow in the ovary of domestic fowl (*Gallus domesticus*) and changes after prostaglandin F-2α treatment. *J. Reprod. Fertil.* **64**, 227–231.

Schrader, W. T., Birnbaumer, M., Hughes, M. R., Weigel, N. L., Grody, W. W., and O'Malley, B. W. (1981). Studies on the structure and function of the chicken progesterone receptor. *Rec. Prog. Horm. Res.* **37**, 583–633.

Sharp, P. J., Scanes, C. G., and Gilbert, A. B. (1978). *In vivo* effects of an antiserum to partially purified chicken luteinizing hormone (CM2) in laying hens. *Gen. Comp. Endocrinol.* **34**, 296–299.

Sharp, P. J., Scanes, C. G., Williams, J. B., Harvey, S., and Chadwick, A. (1979). Variations in concentrations of prolactin luteinizing hormone, growth hormone and progesterone in the plasma of broody bantams (*Gallus domestics*). *J. Endocrinol.* **80**, 51–57.

Sharp, P. J., Talbot, R. T., Main, G. M., Dunn, I. C., Fraser, H. M., and Huskisson, N. S. (1990). Physiological roles of chicken LHRH-I and -II in the control of gonadotropin release in the domestic chicken. *J. Endocrinol.* **124**, 291–299.

Shen, X., Steyrer, E., Retzek, H., Sanders, E. J., and Schneider, W. J. (1993). Chicken oocyte growth: Receptor-mediated yolk deposition. *Cell Tissue Res.* **272**, 459–471.

Shimada, K. (1978). Electrical activity of the oviduct of the laying hen during egg transport. *J. Reprod. Fertil.* **53**, 223–230.

Shimada, K., and Asai, I. (1978). Uterine contraction during the ovulatory cycle of the hen. *Biol. Reprod.* **19**, 1057–1062.

Shimada, K., and Asai, I. (1979). Effects of prostaglandin F2α and indomethacin on uterine contraction in hens. *Biol. Reprod.* **21**, 523–527.

Shimada, K., and Saito, N. (1989). Control of oviposition in poultry. *CRC Crit. Rev. Poult. Biol.* **2**, 235–253.

Shimada, K., Olson, D. M., and Etches, R. J. (1984). Follicular and uterine prostaglandin levels in relation to uterine contraction and the first ovulation of a sequence in the hen. *Biol. Reprod.* **31**, 76–82.

Silverin, B., and Goldsmith, A. R. (1990). Plasma prolactin concentrations in breeding pied flycatchers (*Ficedula hypoleuca*) with an experimentally prolonged brooding period. *Horm. Behav.* **24**, 104–113.

Silyn-Roberts, H., and Sharp, R. M. (1986). Crystal growth and the role of the organic network in eggshell biomineralization. *Proc. R. Soc. London B* **227**, 303–324.

Simkiss, K. (1967). "Calcium in Reproductive Physiology." Reinhold, New York.

Simkiss, K., and Taylor, T. G. (1971). Shell formation. *In* "Physiology and Biochemistry of the Domestic Fowl" (D. J. Bell and B. M. Freeman, eds.), Vol. 3, pp. 1331–1343. Academic Press, New York.

Soares, J. H. (1984). Calcium metabolism and its control—Review. *Poult. Sci.* **63**, 2075–2083.

Solomon, S. E. (1975). Studies on the isthmus region of the domestic fowl. *Br. Poult. Sci.* **16**, 255–258.

Sturkie, P. D., and Mueller, W. J. (1976). Reproduction in the female and egg production. *In* "Avian Physiology," 3rd ed., pp. 302–330. Springer-Verlag, New York.

Tabibzadeh, C., Rozenboim, I., Silsby, J. L., Pitts, G. R., Foster, D. N., and El Halawani, M. E. (1995). Modulation of ovarian cytochrome P450 17α-hydroxylase and cytochrome aromatase messenger ribonucleic acid by prolactin in the domestic turkey. *Biol. Reprod.* **52**, 600–608.

Takahashi, T., Kawashima, M., Kamiyoshi, M., and Tanaka, K. (1992). Arginine vasotocin binding component in the uterus (shell gland) of the chicken. *Acta Endocrinol.* **127**, 179–184.

Tanaka, K., and Inoue, T. (1990). Role of steroid hormones and catecholamines in the process of ovulation in the domestic fowl. *In* "Endocrinology of Birds: Molecular to Behavioral" (M. Wada, S. Ishii, and C. G. Scanes, eds.), pp. 59–68. Springer-Verlag, Berlin.

Tanaka, M., Maeda, K., Okubo, T., and Nakashima, K. (1992). Double antenna structure of chicken prolactin receptor deduced from the cDNA sequence. *Biochem. Biophys. Res. Commun.* **188**, 490–496.

Tilly, J. L., and Johnson, A. L. (1987). Presence and hormonal control of plasminogen activator in granulosa cells of the domestic hen. *Biol. Reprod.* **37**, 1156–1164.

Tilly, J. L., and Johnson, A. L. (1989). Regulation of androstenedione production by adenosine 3', 5'-monophosphate and phorbol myristate acetate in ovarian thecal cells of the domestic hen. *Endocrinology* **125**, 1691–1699.

Tilly, J. L., and Johnson, A. L. (1991). Protein kinase C in preovulatory follicles from the hen ovary. *Dom. Anim. Endocrinol.* **8**, 1–13.

Tilly, J. L., Kowalski, K. I., and Johnson, A. L. (1991a). Stage of follicular development associated with the initiation of steroidogenic competence in avian granulosa cells. *Biol. Reprod.* **44**, 305–314.

Tilly, J. L., Kowalski, K. I., Johnson, A.L., and Hsueh, A. J. W. (1991b). Involvement of apoptosis in ovarian follicular atresia and postovulatory regression. *Endocrinology* **129**, 2799–2801.

Tilly, J. L., Kowalski, K. I., and Johnson, A. L. (1991c). Cytochrome P450 side-chain cleavage (P450scc) in the hen ovary. II. P450scc messenger RNA, immunoreactive protein, and enzyme activity in developing granulosa cells. *Biol. Reprod.* **45**, 967–974.

Tilly, J. L., Kowalski, K. I., Li, Z., Levorse, J. M., and Johnson, A. L. (1992). Plasminogen activator activity and thymidine incorporation in avian granulosa cells during follicular development and the periovulatory period. *Biol. Reprod.* **46**, 195–200.

Tilly, J. L., Kowalski, K. I., Kenton, M. L., and Johnson, A. L. (1995a). Expression of members of the Bcl-2 gene family in the immature rat ovary: Equine chorionic gonadotropin-mediated inhibition of granulosa cell apoptosis is associated with decreased Bax and constitutive Bcl-2 and Bcl-x_{long} messenger ribonucleic acid levels. *Endocrinology* **136**, 232–241.

Tilly K. I., Banerjee, S., Banerjee, P. P., and Tilly, J. L. (1995b). Expression of the p53 and Wilms' tumor suppressor genes in the rat ovary: Gonadotropin repression *in vivo* and immunohistochemical localization of nuclear p53 protein to apoptotic granulosa cells of atretic follicles. *Endocrinology* **136**, 1394–1402.

Tingari, M. D., and Lake, P. E. (1973). Ultrastructural studies on the uterovaginal sperm-host glands of the domestic hen, *Gallus domesticus*. *J. Reprod. Fertil.* **34**, 423–431.

Toth, M., Olson, D. M., and Hertelendy, F. (1979). Binding of prostaglandin F2α to membranes of shell gland muscle of laying hens: Correlations with contractile activity. *Biol. Reprod.* **20,** 390–398.

Tullett, S. G. (1978). Pore size versus pore number in avian eggshells. In "Respiratory Function in Birds, Adult and Embryonic" (J. Piper, ed.), pp. 219–234. Springer-Verlag, New York.

Tullett, S. G. (1984). The porosity of avian eggshells. *Comp. Biochem. Physiol. A* **78,** 5–13.

Tuohimaa, P., Joensuu, T., Isola, J., Keinanen, R., Kunnas, T., Niemala, A., Pekki, A., Wallen, M., Ylikomi, T., and Kulomaa, M. (1989). Development of progestin-specific response in the chicken oviduct. *Int. J. Dev. Biol.* **33,** 125–134.

Unsicker, K., Seidel, F., Hofmann, H.-D., Muller, T. H., Schmidt, R., and Wilson, A. (1983). Catecholaminergic innervation of the chicken ovary. *Cell Tissue Res.* **230,** 431–450.

van Tienhoven, A. (1981). Neuroendocrinology of avian reproduction, with special emphasis on the reproductive cycle of the fowl (*Gallus domestics*). *World's Poult. Sci. J.* **37,** 156–176.

van Tienhoven, A. (1983). "Reproductive Physiology of Vertebrates," 2d ed. Cornell Univ. Press, Ithaca/London.

Verma, O. P., Prasad, B. K., and Slaughter, J. (1976). Avian oviduct motility induced by prostaglandin E_1. *Prostaglandins* **12,** 217–227.

Wasserman, R. H., Smith, C. A., Smith, C. M., Brindak, M. E., Fullmer, C. S., Krook, L., Penniston, J. T., and Kumar, R. (1991). Immunohistochemical localization of a calcium pump and calbindin-D_{28K} in the oviduct of the laying hen. *Histochemistry* **96,** 413–418.

Wechsung, E., and Houvenaghel, A. (1978). Effect of prostaglandins on oviduct tone in the domestic hen *in vivo*. *Prostaglandins* **15,** 491–505.

Williams, J. B., and Sharp, P. J. (1977). A comparison of plasma progesterone and luteinizing hormone in growing hens from eight weeks of age to sexual maturity. *J. Endocrinol.* **75,** 447–448.

Williams, J. B., and Sharp, P. J. (1978). Ovarian morphology and rates of ovarian follicular development in laying broiler breeders and commercial egg–producing hens. *Br. Poult. Sci.* **19,** 387–396.

Williams, G. T., and Smith, C. A. (1993). Molecular regulation of apoptosis: genetic controls on cell death. *Cell* **74,** 777–779.

Wilson, S. C., and Cunningham, F. J. (1980). Modification by metyrapone of the "open period" for pre-ovulatory LH release in the hen. *Br. Poult. Sci.* **21,** 351–361.

Wilson, S. C., and Sharp, P. J. (1975). Effects of progesterone and synthetic luteinizing hormone-releasing hormone on the release of luteinizing hormone during sexual maturation in the hen (*Gallus domesticus*). *J. Endocrinol.* **67,** 359–369.

Wilson, S. C., and Sharp, P. J. (1976). Induction of luteinizing hormone release by gonadal steroids in the ovariectomised domestic hen. *J. Endocrinol.* **71,** 87–98.

Wilson, S. C., Jennings, R. C., and Cunningham, F. J. (1983). An investigation of diurnal and cyclic changes in the secretion of luteinizing hormone in the domestic hen. *J. Endocrinol.* **98,** 137–145.

Wingfield, J. C., and Goldsmith, A. R. (1990). Plasma levels of prolactin and gonadal steroids in relation to multiple-brooding and renesting in free-living populations of the song sparrow, *Melospiza melodia*. *Horm. Behav.* **24,** 89–103.

Wingfield, J. C., and Farner, D. S. (1993). Endocrinology of reproduction in wild species. In "Avian Biology" (D. S. Farner, J. R. King, and K. C. Parkes, eds.), Vol IX, pp. 163–327. Academic Press, Orlando.

Wyburn, G. M., Aitken, R. N. C., and Johnsron, H. S.(1965). The ultrastructure of the zona radiata of the ovarian follicle of the domestic hen. *J. Anat.* **99,** 469–484.

Wyburn, G. M., Johnston, H. S., and Aitken, R. N. C. (1966). Fate of the granulosa cells in the hen's follicle. *Zeitsch. Zellfors.* **72,** 53–65.

Wyburn, G. M., Johnston, H. S., Draper, M. H., and Davidson, M. F. (1970). The fine structure of the infundibulum and magnum of the oviduct of *Gallus domesticus*. *Quart. J. Exp. Physiol.* **55,** 213–232.

Wyburn, G. M., Johnston, H. S., Draper, M. H., and Davidson, M. F. (1973). The ultrastructure of the shell-forming region of the oviduct and the development of the shell of *Gallus domesticus*. *Quart. J. Exp. Physiol.* **58,** 143–151.

Wyllie, A. H. (1992). Apoptosis and the regulation of cell numbers in normal and neoplastic tissues: An overview. *Cancer Metastat. Rev.* **11,** 95–103.

Wyllie, A. H. (1994). Death gets a brake. *Nature* **369,** 272–273.

Yoshimura, Y., Tanaka, K., and Koga, O. (1983). Studies on the contractility of follicular wall with special reference to the mechanism of ovulation in hens. *Br. Poult. Sci.* **24,** 213–218.

Yoshimura, Y., and Bahr, J. M. (1991a). Localization of progesterone receptors in the shell gland of laying and nonlaying chickens. *Poult. Sci.* **70,** 1246–1251.

Yoshimura, Y., and Bahr, J. M. (1991b). Localization of progesterone receptors in pre- and postovulatory follicles of the domestic hen. *Endocrinology* **128,** 323–330.

Yoshimura, Y., Bahr, J. M., Okamoto, T., and Tamura, T. (1993a). Effects of progesterone on the ultrastructure of preovulatory follicles of hypophysectomized chickens: Possible roles of progesterone in the regulation of follicular function. *Jpn. Poult. Sci.* **30,** 270–281.

Yoshimura, Y., Chang, C., Okamoto, T., and Tamura, T. (1993b). Immunolocalization of androgen receptor in the small, preovulatory, and postovulatory follicles of laying hens. *Gen. Comp. Endocrinol.* **91,** 81–89.

Yoshimura, Y., Tischkau, S. A., and Bahr, J. M. (1994). Destruction of the germinal disc region of an immature preovulatory follicle suppresses follicular maturation and ovulation. *Biol. Reprod.* **51,** 229–233.

Youngren, O. M., El Halawani, M. E., Silsby, J. L., and Phillips, R. E. (1991). Intracranial prolactin perfusion induces incubation behavior in turkey hens. *Biol. Reprod.* **44,** 425–431.

Zavaleta, D., and Ogasawara, F. (1987). A review of the mechanism of the release of spermatozoa from storage tubules in the fowl and turkey oviduct. *World's Poult. Sci. J.* **43,** 132–139.

CHAPTER 23

Reproduction in Male Birds

JOHN D. KIRBY
Department of Poultry Science
University of Arkansas
Fayetteville, Arkansas 72701

DAVID P. FROMAN
Department of Animal Sciences
Oregon State University
Corvallis, Oregon 97331

I. Introduction 597
II. Reproductive Tract Anatomy 597
    A. Testis 597
    B. Excurrent Ducts 599
    C. Accessory Organs 600
III. Ontogeny of the Reproductive Tract 600
    A. Overview 600
    B. Formation of the Undifferentiated Gonad 601
    C. Gonadal Differentiation and Müllerian Duct Regression 602
    D. Formation of the Excurrent Ducts 602
IV. Development and Growth of the Testis 603
    A. Proliferation of Somatic and Stem Cells in the Testis 603
    B. Differentiation of Somatic Cells within the Testis 603
    C. Initiation of Meiosis 605
    D. Altering the Pattern of Testis Growth and Maturation 605
V. Hormonal Control of Testicular Function 605
    A. Central Control of Testicular Function 605
    B. Control of Adenohypophyseal Function in Males 606
    C. Effects of Gonadotropins on Testicular Function 607
VI. Spermatogenesis and Extragonadal Sperm Maturation 607
    A. Spermatogenesis 607
    B. Extragonadal Sperm Transport and Maturation 609
    References 612

I. INTRODUCTION

Reproduction is a process which can be organized into distinct developmental and functional phases. In the case of the male, these include fertilization, formation of a patent reproductive tract, production of sperm, manifestation of male-specific behavioral patterns, and expulsion of sperm from the body. This perspective can provide insights that may be missed if reproduction is viewed primarily as an isolated act. For example, while the reproductive tract is fully functional only in the adult, it is formed, for the most part, prior to hatching. Furthermore, while spermatogenesis is associated with puberty, spermatogenesis is not constrained by chronological age, but rather by by the extent to which testicular cells proliferate and differentiate, which in turn, is coupled to the developmental limitations of gonadotropin secretion. Finally, androgens essential to the function of the reproductive tract, the appearance of secondary sexual attributes, and male behavior may have detrimental effects upon the development of immune and connective tissue if the hormonal signal appears during the period of rapid prepubertal growth and differentiation. Thus, this chapter will discuss the process of reproductive system development and function in the male galliform bird.

II. REPRODUCTIVE TRACT ANATOMY

A. Testis

The gross morphology and relative position of the male reproductive organs are shown in Figure 1 (for detailed reviews see Nickel *et al.*, 1977; King, 1979; Lake,

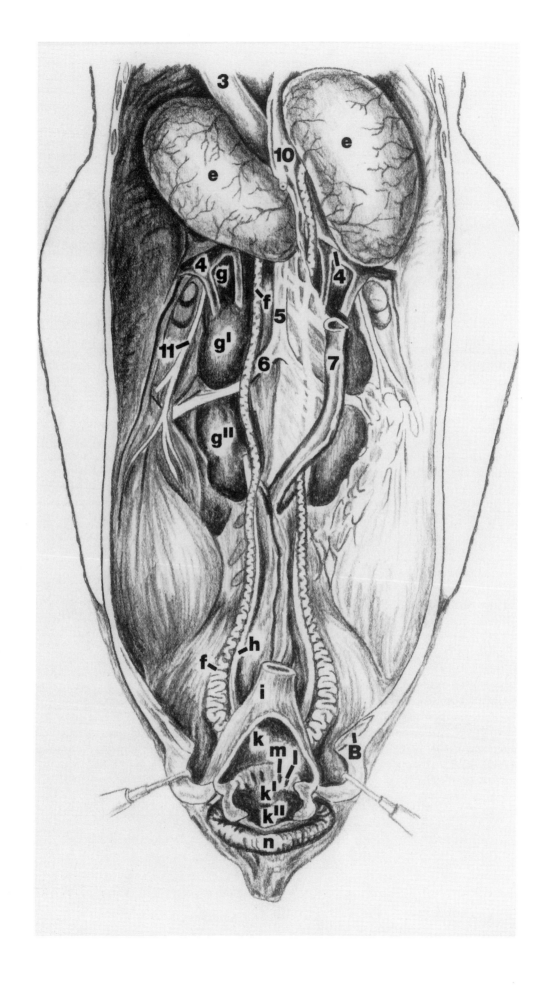

1981). Paired reproductive tracts lie along the dorsal body wall. Each tract consists of a testis, an epididymis, and a highly convoluted deferent duct running alongside the ureter. The testes are connected to the body wall by a mesorchium. This peritoneal fold not only serves as an attachment for the testis but also as a conduit for nerves and blood vessels as well. Each testis is an aggregate of anastomosing seminiferous tubules with associated interstitial tissue enveloped by a connective tissue capsule. Thus, the testis contains two types of parenchymal tissue: the interstitial tissue and the seminiferous epithelium. The interstitial tissue contains blood and lymphatic vessels, nerves, peritubular epithelial cells, and Leydig cells. Thin concentric layers of myoepithelial cells, fibroblasts, and connective tissue fibrils overlie the basal lamina of the seminiferous tubule (Rothwell and Tingari, 1973). The seminiferous epithelium within the seminiferous tubules of sexually mature males is compartmentalized into basal and adluminal regions via tight junctions between adjacent Sertoli cells (Osman et al., 1980; Bergmann and Schindelmeiser, 1987) and it contains developing germ cells in distinct associations referred to as *stages*. The stages are arranged sequentially in a helix that extends along the length of the seminiferous tubule (Lin and Jones, 1990).

B. Excurrent Ducts

The epididymis is located on the dorsomedial aspect of the testis, which is referred to as the hilus (Figure 2). The epididymis is actually a series of ducts that ultimately empty into the deferent duct. The ducts within the epididymis include the rete testis, efferent ducts, connecting ducts, and the epididymal duct. The epididymal ducts and the deferent duct are referred to collectively as the excurrent ducts of the testis. Seminiferous tubules connect with the rete testis at discrete sites along the testis–epididymal interface (Tingari, 1971). Osman (1980) identified three distinct types of junctures between the seminiferous tubules and rete testis of the rooster: (1) a germ cell free epithelium of modified Sertoli cells that abruptly gives way to a rete-like epithelium, referred to as a terminal segment and tubulus rectus, respectively; (2) a terminal segment that connects directly to the rete testis; and (3) anastomosis of a seminiferous tubule with the rete testis. Tubuli recti have also been observed in quail testes (Aire, 1979b).

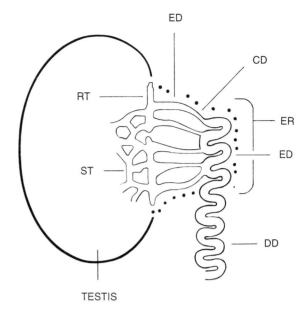

FIGURE 2 Schematic of the excurrent ducts of the testis. ST, seminiferous tubules; RT, rete testis; ED, efferent duct; CD, connecting duct; ER, epididymal region; ED, epididymal duct; DD, deferent duct. Reproduced with permission from Academic Press, London. Redrawn from Lake (1981).

The rete testis has intratesticular, intracapsular, and extratesticular regions (Aire, 1982). As depicted in Figure 2, the rete testis exist as lacunae. The rete testis is lined with simple cuboidal and simple squamous epithelial cells (Aire, 1982). Unlike the rete testis, the efferent ducts are wide at their proximal ends and narrow at their distal ends. On a volumetric basis (Table 1), the efferent ducts are the principal excurrent duct within the epididymis (Aire, 1979a). The efferent duct mucosa is highly folded, especially in the proximal portion (Tingari, 1971; Aire, 1979b; Bakst, 1980). It is characterized by a pseudostratified columnar epithelium that contains ciliated and nonciliated epithelial cells (Tingari, 1972; Budras and Sauer, 1975; Hess and Thurston, 1977; Aire, 1980) as well as sparsely distributed intraepithelial lymphocytes (Aire and Malmqvist, 1979a). The epithelium of the connecting ducts is also composed of pseudostratified columnar cells (Tingari, 1971; Tingari, 1972; Hess and Thurston, 1977; Aire, 1979b; Bakst, 1980). The distinguishing attributes of the connecting tubules are a narrow lumenal diameter relative to that of adjacent

FIGURE 1 Topography of the dorsal body wall of the rooster. Paired testes (*e*) are located anterior to the cranial lobe of the kidneys (*g*). The deferent ducts (*f*) run alongside the ureters (*h*) towards the cloaca (*k-k″*), ultimately opening (*l*) into the urodeum (*k′*). Abbreviations: b, pubis; e, testis; f, ductus deferens; g, cranial; g′, middle, and g″, caudal lobes of the kidney; h, ureter; i, colon; k, coprodeum; k′, urodeum; k″, proctodeum; l, opening of left ductucsdeferens; m, opening of left ureter; n, anus. (Modified from *Anatomy of the Domestic Birds*, Urogenital system, R. Nikel, A. Schummer, E. Seiferle, W. G. Siller, and P. A. L. Wight, pp. 70–84, 1977.)

TABLE 1 Species Variation in Volumetric Proportions (%) of Epididymal Structures[a]

Structures	Chicken	Japanese quail	Guinea fowl
Rete testis	13.3	9.9	10.7
Proximal efferent ductules	27.6	40.8	45.7
Distal efferent ductules	7.7	15.2	16.2
Connecting ducts	2.3	1.7	0.7
Epididymal duct	7.6	2.4	1.8
Connective tissue	38.7	27.3	22.6
Blood vessels	2.5	2.7	2.3
Aberrant ducts	0.3	—	—

[a]Adapted from Aire (1979a).

excurrent ducts and the smooth contour of the mucosal surface (Tingari, 1971; Aire, 1979a; Bakst, 1980).

The tortuous epididymal and deferent ducts are characterized by low mucosal folds covered with nonciliated pseudostratified columnar epithelial cells (Tingari, 1971; Aire, 1979a; Bakst, 1980). Based upon histological and morphological evidence, the deferent duct is a continuation of the epididymal duct. Lumenal diameter gradually increases by three-fold between the cranial epididymal duct and the distal deferent duct (Tingari, 1971). One notable difference between the epididymal duct and the deferent duct are the layers of dense connective tissue and smooth muscle surrounding the mucosa of the latter (Tingari, 1971). A dense capillary network envelopes the excurrent ducts from the level of the proximal efferent ducts to the deferent duct (Nakai et al., 1988). The distal deferent duct straightens and then abruptly widens at its juncture with the cloaca. This structure, known as the receptacle of the deferent duct, has a bean-shaped appearance when it is engorged with semen. Each deferent duct terminates in the cloacal urodeum as a papilla immediately below the ostium of a ureter (Figure 3).

C. Accessory Organs

Accessory reproductive organs include the paracloacal vascular bodies, dorsal proctodeal gland, and lymphatic folds (Fujihara, 1992). The accessory reproductive organs are either in proximity to or are an integral part of the cloaca. As shown in Figure 3, the paracloacal vascular body is found alongside the receptacle of the deferent duct, and lymphatic folds exist within the wall of the proctodeum. During copulation or in response to massage, a nonintromittent phallus forms as tumescent lymphatic tissue is everted through the vent immediately before ejaculation. As reviewed by Fujihara (1992), the paracloacal vascular bodies are essential for lymphatic

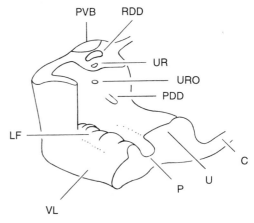

FIGURE 3 Schematic of the lower left quadrant of the cloaca. Before entering the cloacal wall, the deferent duct straightens then widens to form the receptacle of the deferent duct. The paracloacal body, also seen sectioned but shaded, is attached to the lateral aspect of the receptacle. The papilla of the deferent duct is within the urodeum of the cloaca. *PVB, paracloacal vascular body; RDD, receptacle of the deferent duct; UR, ureter; URO, ureter orifice; PDD, papilla of deferent duct; C, ventral wall of coprodeum; U, ventral wall of urodeum; P, ventral wall of proctodeum; VL, ventral lip of vent; LF, lymphatic folds.* (Reproduced from King (1981), with permission from Academic Press.)

tissue tumescence, as they are the sites where lymph is formed by ultrafiltration of blood.

III. ONTOGENY OF THE REPRODUCTIVE TRACT

A. Overview

The adult male reproductive tract is derived from two embryonic organs: a gonad and its associated mesonephros. However, the formation of the mesonephros precedes that of the gonad (Figure 4). In fact, the gonad arises from the ventromesial surface of the mesonephros (Figure 5). As reviewed by Romanoff (1960), three distinct pairs of excretory organs appear sequentially during embryonic development: the pronephros, mesonephros, and metanephros. The most anterior organ, the pronephros, disappears by day 4 of incubation in the chicken. Nonetheless, the pronephric duct, also known as the Wolffian duct, persists through time to: (1) induce the formation of the mesonephric tubules, (2) induce the formation of the Müllerian duct, (3) temporarily link the mesonephros with the cloaca, and (4) ultimately serve as the deferent duct in males. The critical point in the ontogeny of the reproductive tract is gonadal differentiation; for prior to this point in time, the embryo has bipotential gonads and the rudiments

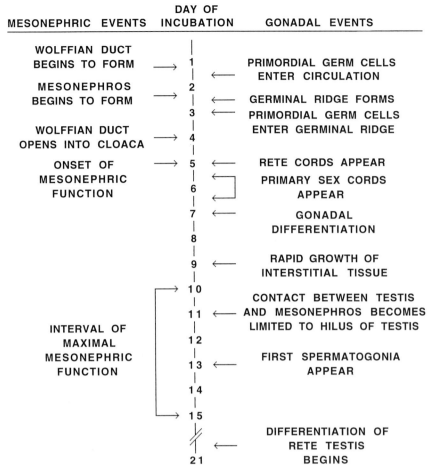

FIGURE 4 Chronology of developmental events *in ovo* associated with the formation of the male reproductive tract in *Gallus domesticus*. See text for details.

of oviducts and deferent ducts in the form of the Müllerian and Wolffian ducts, respectively.

Halverson and Dvorak (1993) have proposed that sex determination in nonratite birds is dependent upon the ratio of Z chromosomes to autosomes. Apparently, this ratio determines whether aromatase, the enzyme responsible for converting testosterone to estradiol, will appear within the embryo at the time of gonadal differentiation. Rashedi *et al.* (1983) demonstrated that genotypic females can be converted into phenotypic males via a testis graft prior to gonadal differentiation. The effect was attributed to Müllerian inhibiting substance (MIS), which inhibits the synthesis of aromatase (Vigier *et al.*, 1989). Elbrecht and Smith (1992) have demonstrated that aromatase activity is essential for development of the female phenotype; for if an aromatase inhibitor is administered to genotypic female (ZW) chick embryos prior to day 7 of incubation, chicks will be phenotypic males at hatch. Mizuno *et al.* (1993) have reported that mRNA for aromatase appears in female chick embryos between 5 and 7 days of incubation. Likewise, H-Y antigen in chickens is sex-specific for females, does not appear until the time of gonadal differentiation, and can be induced by estrogen in cultures of embryonic male gonadal tissue (Ebensperger *et al.*, 1988a, 1988b). Therefore, the ontogeny of the male reproductive tract in galliform birds appears to be dependent upon the limited ability of genotypic males (ZZ) to synthesize estrogens at the time of gonadal differentiation.

B. Formation of the Undifferentiated Gonad

Even though the gonad forms upon the ventromesial surface of the mesonephros, a fully differentiated mesonephros is not required for gonadal development (Merchant-Larios *et al.*, 1984). As reviewed by Romanoff (1960), the gonad is formed by the invasion of primordial germ cells into coelomic epithelium overlying a portion of the mesonephros, known as the germinal ridge (Figure 4). Primordial germ cells arise from endoderm along the anterior interface of the blastoderm's

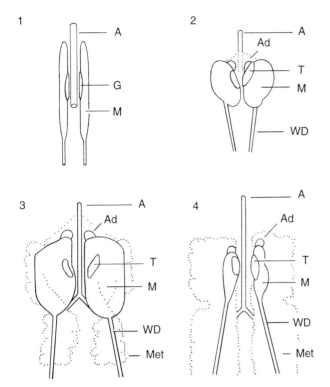

FIGURE 5 Stages in the development of the male urogenital system. Numbers denote the first, second, third, or fourth quarter of incubation. The mesonephros forms before the differentiation of the gonad. After the metanephros becomes functional midway through incubation, mesonephric function declines and the mesonephros recedes apart from where it contacts the testis. A, aorta; G, undifferentiated gonad; M, mesonephros; Ad, adrenal gland; T, testis; WD, Wolffian duct; Met, metanephros. (Reproduced from Romanoff (1960), with permission from the MacMillan Co., New York.)

area opaca and area pellucida. After entering the embryonic circulation, they are distributed randomly throughout the embryo's vasculature (Meyer, 1964) prior to colonizing the germinal ridge (Fujimoto et al., 1976). The undifferentiated gonad contains rete cords and primary sex cords in its anterior and middle regions, respectively (Romanoff, 1960).

C. Gonadal Differentiation and Müllerian Duct Regression

Gonadal differentiation in chick embryos occurs at approximately 6.5–7 days of incubation (Figure 4). In comparison to the ovary, the embryonic testis is characterized by a germinal epithelium that recedes with time, a thicker capsule, the absence of secondary or cortical sex cords, as well as the presence of primary sex cords surrounded by stroma (Romanoff, 1960). The primary sex cords contain numerous primordial germ cells and are the anlage of the seminiferous tubules. Biochemically, gonadal differentiation is evident in terms of increased cyclic nucleotide concentrations (Teng, 1982), increased protein synthesis (Samsel et al., 1986), and the pattern of sex steroid synthesis (Guichard et al., 1977; Imataka et al., 1988a,b; Mizuno et al., 1993). It is noteworthy that these traits become evident prior to the development of the pituitary–gonadal axis, which appears to become functional at 13.5 days of incubation in the chick (Woods and Weeks, 1969). It must also be noted that the interval of sexual bipotentiality differs among species.

Once the testes have formed, the Müllerian ducts cease development and undergo regression in a caudocranial direction (Romanoff, 1960; Hutson et al., 1985). Müllerian duct regression is attributable to MIS. This hormone is a 7.4 kDa gonad-specific glycoprotein (Teng et al., 1987), which, in the male, is derived from the progenitors of Sertoli cells found within the primary sex cords (Stoll et al., 1973). Regression of the Müllerian ducts in males occurs because they contain insufficient circulating estrogen levels to counteract the effect of MIS (MacLaughlin et al., 1983; Hutson et al., 1985).

D. Formation of the Excurrent Ducts

The excurrent ducts are derived from the mesonephros. As shown in Figures 4 and 5, the mesonephros is active throughout the second and third quarters of incubation. The functional unit of the mesonephros includes a malphigian corpuscle, proximal, intermediate, and distal tubules, as well as a connecting tubule. The latter connects the series of mesonephric tubules with the Wolffian duct. The malphigian corpuscle is a capillary tuft or glomerulus within a double-walled epithelial enclosure known as Bowman's capsule. The visceral epithelial layer adheres to the glomerulus. At the pole of the corpuscle where the afferent and efferent arterioles are attached to the glomerulus within, the visceral epithelium folds back upon itself to form the parietal epithelium. As the parietal epithelium extends away from the vascular pole, it is separated from the visceral epithelium by a lumen. While the mesonephros is functional, glomerular filtrate enters the lumen of Bowman's capsule prior to entering the lumen of the proximal tubule. After the metanephros becomes functional during midincubation (Romanoff, 1960), the mesonephros begins to degenerate except where there is contact with the embryonic testis. Consequently, the mesonephros undergoes a profound change in size during embryonic development (Figure 5).

During the last third of incubation, a subset of mesonephric tubules are converted into the excurrent ducts of the testis (Budras and Sauer, 1975) and the malphighian corpuscles in proximity to the rete cords fuse. The glomeruli undergo progressive degeneration while the simple squamous epithelial cells of the parietal epithelium differentiate into simple columnar cells. This tissue becomes the epithelium of the proximal efferent ducts. Proximal, intermediate, and distal mesonephric tubules are transformed into distal efferent ducts, connecting ductules become the connecting ducts, the Wolffian duct associated with the mesonephros becomes the epididymal duct, and the distal Wolffian duct becomes the deferent duct.

The transformation of the mesonephric ductules into the excurrent ducts of the testis is androgen dependent (Maraud and Stoll, 1955; Stoll and Maraud, 1974); initiated shortly after the concentration of blood plasma testosterone has peaked *in ovo* (Woods et al., 1975). However, the entire process of ductule conversion in the chicken requires an interval of 8–10 weeks (Marvan, 1969; Budras and Sauer, 1975). During this time, mean plasma testosterone levels stay constant at about 12–13% of those observed in sexually mature males (Driot et al., 1979; Tanabe et al., 1979). In contrast to the mesonephric tubules, differentiation of the rete cords begins at about the time of hatching and is complete by 5 weeks of age (Budras and Sauer, 1975). The convolutions of the newly formed epididymal and deferent ducts increase until sexual maturity (Budras and Sauer, 1975). The final length of the deferent duct has been estimated to be 4× greater than the linear distance between the epididymis and cloaca due to these convolutions (Marvan, 1969).

IV. DEVELOPMENT AND GROWTH OF THE TESTIS

A. Proliferation of Somatic and Stem Cells in the Testis

The testis of the mature bird is organized into discrete, easily discernible cellular associations and functional compartments (Figure 6D). However, during embryonic and early posthatch development this organization is less apparent. The posthatch development of the fowl's testis can be divided in to three distinct phases: (1) proliferation of spermatogonia and the somatic cells that support spermatogenesis (Sertoli, peritubular myoid, and interstitial cells); (2) differentiation and the acquisition of functional competence by somatic support cells; and (3) Spermatogonial differentiation resulting in the initiation of meiosis. While the boundaries of these phases are not clearly defined, this three-step process results in functional seminiferous tubules that can sustain spermatogenesis when the appropriate hormonal cues are present. It should be noted that the timing of sexual maturation can vary widely among strains of domestic fowl.

The testis of the young cockerel (approximately 0–6 weeks of age) is characterized by an abundance of interstitial tissue and seminiferous tubules with only a single layer of cells within the basal lamina (Figure 6a) (Kumaran and Turner, 1949). The majority of cells located within the seminiferous tubules at this time are Sertoli cells and spermatogonia, with macrophages and mast cells observed as well (de Reviers, 1971b). The seminiferous tubules at this time are narrow (approximately 40 μm in diameter) with either a poorly formed lumen or no apparent lumen at all. This is a period of rapid cellular proliferation, as while the tubules are only slowly growing in diameter, they are growing rapidly in length (de Reviers, 1971a). Even though the somatic and stem cells are proliferating, the absolute weight and volume of the testis is increasing slowly as the seminiferous tubules displace interstitial cells (Marvan, 1969; de Reviers, 1971b).

B. Differentiation of Somatic Cells within the Testis

Following a posthatching period of Sertoli cell proliferation these cells become mitotically quiescent and differentiate (for a review see Russell and Griswold, 1993). While neither the exact period nor the kinetics of Sertoli cell proliferation are known for the fowl, the work of de Reviers and Williams (1984) provides some insight. In that study, hemicastration was shown to result in compensatory hypertrophy of the remaining testis of males up to 8 weeks of age. As hemicastration typically results in significant hypertrophy only during the period of Sertoli cell proliferation in mammals (for complete details see, Russell and Griswold, 1993), it is possible to postulate that Sertoli cells are proliferating in the fowl to about 8 weeks of age.

Sertoli cells differentiate into highly polarized cells that extend from the basal lamina to the lumen of the seminiferous tubule. The mature Sertoli cell is a complex, columnar cell that contains numerous crypts into which developing sperm cells are embedded (Sertoli, 1865; 1878; Nagano, 1962). Sertoli cell functions are thought to be regulated by a myriad of endocrine and paracrine factors, although FSH and testosterone have been most thoroughly studied (Brown et al., 1975; Brown and Follett, 1977). A primary function of Sertoli cells is to provide a carefully regulated environment, including the sequestration of postmeiotic germ cells into the adluminal compartment of the seminiferous tubule, an immunologically privileged location. The formation of the adluminal compartment is accomplished by the formation of tight junctions between adjacent

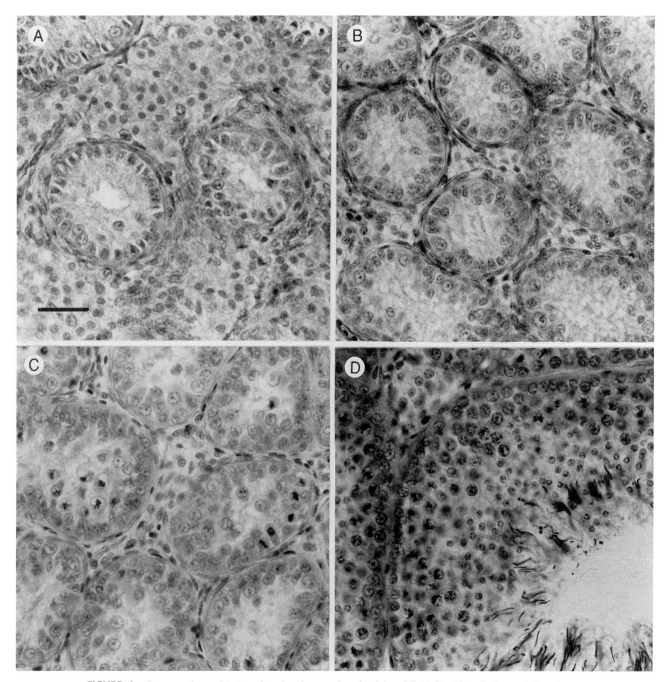

FIGURE 6 Cross sections of testes of cockerels reared under 8 hr of light (A–C) and of an adult male on 14 hr of light (D). (A) Cross section of a testis from a 14-day-old male, notice the simple single layer of spermatogonia and Sertoli cells in the tubules and the relative abundance of interstitial space; (B) cross section of a testis from a 56-day-old cockerel, the single layer of cells within the seminiferous tubule has become taller and more complex; (C) cross section of a testis from a 140-day old cockerel, notice the appearance of spermatocytes (cells with darkly stained chromatin) which have moved away from the basement membrane; and (D) cross section of a testis from an adult male which clearly demonstrates active spermatogenesis and the dramatic reduction in the relative area of the interstitium. The bar in A equals 20 μm; all of the micrographs are of equivalent magnification. Photographs by the author.

Sertoli cells at a point basal to the maturing, meiotic, germ cell (Bergmann and Schidelmeiser, 1987; Pelletier, 1990). While tight junction formation appears to be androgen dependent in the fowl, tight junctions are maintained during periods of spermatogenic quiescence in the mallard (Pelletier, 1990). Another indicator of mature Sertoli cell function is the complex complement of proteins synthesized and secreted in response to androgen and/or FSH stimulation. While studied only superficially in male birds, spermatogenic stage-specific protein synthesis and secretion is well documented in other groups (see Russell and Griswold, 1993). As observed in other species, the close association of Sertoli and developing germ cells in the fowl suggests that they are intimately involved in the regulation of germ cell development (Cooksey and Rothwell, 1973).

The growth of the seminiferous tubules is associated with a reorganization and reduction in the relative area of the testicular interstitium (Figure 6). While the seminiferous tubule is the compartment where spermatogenesis will occur in the adult, Leydig cells of the interstitium are responsible for testicular androgen production. In the interstitium, mesenchymal cell-like Leydig cell precursors (Connell, 1972), which have vesicles of smooth endoplasmic reticulum and lipid droplets, differentiate into their mature form under the influence of LH (Narbaitz and Adler, 1966; Brown *et al.*, 1975; Driot *et al.*, 1979). Mature Leydig cells are formed following a continuum of cellular differentiation occurs during the period testicular growth leading to the prepubertal increases in circulating LH levels (Rothwell, 1973). While limited testosterone production is observed during embryogenesis and early development, the capacity to produce testosterone at sexually mature levels requires the presence of a fully differentiated Leydig cell population (for a review of testicular steroidogenesis see Johnson, 1986). The mature Leydig cell of the rooster has extensive smooth endoplasmic reticulum, a prominent Golgi complex, mitochondria with tubular cristae, abundant lipid droplets, and lysosomal elements (Rothwell, 1973).

C. Initiation of Meiosis

During the period of Sertoli and Leydig cell differentiation, the population of spermatogonia increases in size and complexity. As described later in the chapter, spermatogonia undergo a series of transformations prior to committing to meiosis. However, once a spermatogonium is transformed into a preleptotene spermatocyte, it separates from the basement membrane and shifts toward the adluminal compartment of the seminiferous tubule. The onset of active meiosis is marked by the presence of pachytene and zygotene spermatocytes, which are easily discernible due to their highly condensed and darkly stained chromatin (Figure 6c). The initiation of meiosis appears to occur only after the completion of Sertoli cell proliferation and an increase in circulating androgens is encountered (de Reviers, 1971a,b).

D. Altering the Pattern of Testis Growth and Maturation

The temporal component of testicular maturation can be altered by manipulation of the endocrine milieu. Specifically, onset of meiosis and sustained spermatogenesis can be altered dramatically in the fowl by manipulating the photoperiod (Ingkasuwan and Ogasawara, 1966). Male fowl reared on a long photoperiod (14 hr light or more per day) will typically have reached sexual maturity by 16–20 weeks of age (Ingkasuwan and Ogasawara, 1966; Sharp and Gow, 1983). However, as shown in Figure 6c, when males are reared on short days (in this case 8 hr of light per day) the onset of spermatogenesis is delayed. Eventually, male fowl become refractory to short days, FSH and LH levels increase, and spermatogenesis commences. Conversely, precocious puberty and the onset of spermatogenesis can be accelerated in the fowl by chronic treatment with tamoxifen, an estrogen receptor antagonist (Rozenboim *et al.*, 1986). Males treated with tamoxifen exhibited adult behaviors and produced viable spermatozoa by 9 weeks of age, months ahead of their nontreated siblings (Rozenboim *et al.*, 1986).

V. HORMONAL CONTROL OF TESTICULAR FUNCTION

A. Central Control of Testicular Function

Spermatogenesis is the process in which the division of spermatogonial stem cells ultimately yields sperm cells while a population of stem cells is maintained. This complex phenomenon occurs within the seminiferous epithelium and depends upon the availability of testosterone and FSH, the activity of Sertoli cells, as well as interactions between germ cells and Sertoli cells (Sharpe, 1994). While spermatogenesis is controlled, in part, by cells within the testis, the process is ultimately controlled by neurons within the CNS (see review by Sharp and Gow, 1983). Neurosecretory activity, in turn, is affected by somatic and environmental stimuli. Wingfield *et al.* (1992) have proposed a model accounting for the central integration of environmental information that affects gonadotropin secretion on a seasonal basis. The model is based upon the interplay of endogenous "predictive" rhythms and supplementary environmental information. The duration of the period of light in a subjective day (24-hr period) is the principal environmental factor that stimulates spermatogenesis in galliform birds. As reviewed by Kuenzel (1993), photoreceptors that affect spermatogenesis are neither found within the retina nor the pineal gland. Rather, they are thought to be found within the brain itself. The two most plausible sites include the medial basal hypothalamus and the lobus parolfactorius within the ventral forebrain. While the nature of the deep encephalic photoreceptors remains unknown, they are known to respond to light having wavelengths in the green and orange range of the visible spectrum (Kuenzel, 1993). Ultimately, the deep encephalic photoreceptors appear to modulate the activity of gonadotropin-releasing hormone secreting neurons within the hypothalamus.

Gonadotropin-releasing hormone (GnRH) is the CNS secretory product that stimulates the release of LH and FSH which promote gonadal maturation and function. In the chicken, GnRHergic neurons arise within the olfactory epithelium on day 4.5 of embryonic development, reach the CNS by migrating along the olfactory nerve, and stop migrating by day 12 of embryonic development (Sullivan and Silverman, 1993). GnRHergic neurons are found within extrahypothalamic and hypothalamic sites (Kuenzel and Blähser, 1991). However, only those neurons whose axons terminate within the median eminence are believed to induce gonadotropin secretion *in vivo* (Mikami *et al.*, 1988). Based upon immunocytochemical analysis, GnRH-positive axons are found within the median eminence by day 14 of embryonic development (Sullivan and Silverman, 1993). The significance of GnRHergic neurons was demonstrated well in advance of the purification of chicken GnRH with experiments that induced functional castration by either electrolytic lesions or deafferentation within the hypothalamus (Ravona *et al.*, 1973; Davies and Follett, 1980).

The structure of chicken GnRH was deduced by Miyamoto *et al.* (1982). The hormone is a decapeptide with the following sequence: pGlu-His-Trp-Ser-Tyr-Gly-Leu-Gln-Pro-Gly-NH$_2$. While a second decapeptide, [His^5, Trp^7, Tyr^8]-GnRH (GnRH-II), has been isolated from chicken hypothalami, only [Gln^8]-GnRH (GnRH-I) is found within and secreted from the median eminence (Katz *et al.*, 1990). Consequently, even though the biological activities of GnRH-I and GnRH-II are comparable (Hattori *et al.*, 1986), GnRH-I is believed to be the hormone that directly affects gonadotropin synthesis and secretion.

Onset of sexual maturation is dependent upon an increased rate of GnRH neurosecretion, that is either the frequency and/or the amplitude of GnRH pulses increases (Knight, 1983; Sharp *et al.*, 1990). GnRH induces the secretion of both LH and FSH (Sterling and Sharp, 1984; Hattori *et al.*, 1986). The response to a GnRH pulse is both rapid and short-lived in the male (Sterling and Sharp, 1984). The secretory activity of GnRHergic neurons is dampened by testosterone, which exerts its effect following central aromatization to estrogen (Davies *et al.*, 1980; Wilson *et al.*, 1983). Negative feedback may also be mediated by dopaminergics (El Halawani *et al.*, 1980; Knight *et al.*, 1981) and endogenous opioids (Stansfield and Cunningham, 1987). In contrast, noradrenergic neurons promote GnRH secretion (El Halawani *et al.*, 1980). Finally, alterations in GnRH secretion have been implicated in reproductive senescence (Ottinger, 1992).

B. Control of Adenohypophyseal Function in Males

Axons of GnRHergic neurons terminate in proximity to capillaries within the median eminence of the hypothalamus. GnRH secreted from these axons reaches target cells within the adenohypophysis via the hypothalamo-hypophyseal portal vessels (Gilbert, 1979). GnRH binds to a single receptor type (King *et al.*, 1988) on gonadotropes scattered throughout the adenohypophysis (Mikami and Yamada, 1984). Secretory granules found within gonadotropes contain FSH and LH (Mikami and Yamada, 1984), and both gonadotropins are secreted in response to GnRH (Hattori *et al.*, 1986). FSH and LH are glycoprotein hormones (Ishii, 1993) with molecular weights of 37–38 kDa and 23.5–25 kDa, respectively (Hattori and Wakabayashi, 1979; Sakai and Ishii, 1980; Krishnan *et al.*, 1992). Gonadotropin secretion in sexually mature males is episodic based upon studies of LH secretion in chickens, turkeys (Figure 11), and quail (Wilson and Sharp, 1975; Gledhill and Follett, 1976; Bacon *et al.*, 1991). Gonadotropes are essential for reproduction, as adenohypophysectomy results in a depletion of gonadotropins within the bloodstream, a collapse of the seminiferous epithelium, testicular regression, and atrophy of the excurrent ducts (Hill and Parkes, 1935; Brown and Follett, 1977; Tanaka and Yasuda, 1980).

In addition to GnRH, gonadotropin secretion is also directly regulated by steroid hormones. To date, studies with sexually mature male Japanese quail provide the most comprehensive evaluation of this phenomenon. Even though the adenohypophysis, like the hypothalamus, contains aromatase activity, preincubation of dispersed adenohypophyseal cells with a physiological concentration of 17β-estradiol was without effect on the GnRH-induced secretion of LH from cultured quail gonadotropes (Connolly and Callard, 1987). In contrast, both testosterone and 5α-dihydrotestosterone inhibited GnRH-induced LH secretion. 5α-Dihydrotestosterone was more effective than testosterone and equipotent with corticosterone (Connolly and Callard, 1987). These results are consonant with androgen metabolism in Japanese quail and the effect of glucocorticoids on reproduction (Martin *et al.*, 1984). Gonadotropes *in vivo* might be affected by 5α-dihydrotestosterone synthesized within the adenohypophysis (Davies *et al.*, 1980), by 5α-dihydrotestosterone from the systemic circulation or corticosterone secreted in response to ACTH. King *et al.* (1988) demonstrated that testosterone inhibited GnRH-induced secretion of LH from cultured rooster gonadotropes. Thus, sex steroids in male galliform birds appear to decrease the rate of gonadotropin secretion following GnRH stimulation.

C. Effects of Gonadotropins on Testicular Function

Gonadotropins exert their effects on the testis by binding to specific cell-surface receptors on two distinct types of testicular parenchymal cells: Leydig and Sertoli cells. These cells have been described in detail by Rothwell (1973) and by Cooksey and Rothwell (1973), respectively. The principal role played by each type of cell has been determined by experiments in which one type of gonadotropin as been administered to hypophysectomized males, sexually immature males, or, in the case of Leydig cells, cells in culture (Brown *et al.*, 1975; Ishii and Furuya, 1975; Ishii and Yamamoto, 1976). Such experiments have demonstrated that the principal effects of LH and FSH are exerted upon Leydig cells and Sertoli cells, respectively.

Leydig cells contain the steroidogenic enzymes necessary for the production of androgens (reviewed in Johnson, 1986) and respond rapidly to LH through rapid increases in the second messenger cAMP (Maung and Follett, 1977). The principal steroids secreted by Leydig cells include testosterone and androstenedione, a precursor of testosterone (Nakamura and Tanabe, 1972; Galli *et al.*, 1973; Sharp *et al.*, 1977). The concentration of testosterone in the general circulation is in the range of 5–15 nM, secreted as discrete pulses (Figure 7), which closely follow LH pulses, (Driot *et al.*, 1979; Bacon *et al.*, 1991), and are several times less than that in the testicular vein (Ottinger and Brinkley, 1979). Testosterone in the adult male is essential for spermatogenesis, maintenance of the excurrent ducts and secondary sexual attributes, the expression of specific behaviors, and, as mentioned above, altering the pattern of GnRH secretion. These actions, as well as the inactivation of the hormone, depend upon transformation of testosterone by enzymes such as aromatase, 5α-reductase, and 5β-reductase (for reviews, see Ottinger, 1983; Balthazart, 1989).

While FSH is known to affect Sertoli cells, the means by which FSH works is only poorly understood. However, the effect of FSH, unlike that of LH, is potentiated by testosterone (Tsutsui and Ishii, 1978). As evidenced by the inability of exogenous testosterone to maintain spermatogenesis in hypophysectomized quail (Brown

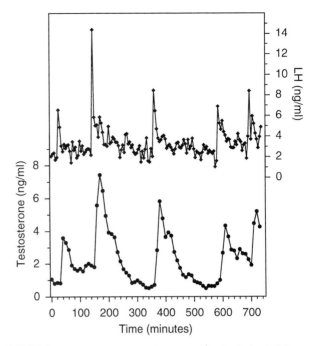

FIGURE 7 The pattern of pulsatile changes in circulating LH (upper panel) and testosterone (lower panel) in an adult male domestic turkey (*Meleagris gallopavo*). The peaks of the discrete LH pulses occur prior to those of testosterone, demonstrating the close relationship between LH secretion and that of testosterone. Figure and data kindly provided by Dr. Wayne Bacon, Ohio State University.

and Follett, 1977), LH and FSH appear to be essential for spermatogenesis in galliform birds.

VI. SPERMATOGENESIS AND EXTRAGONADAL SPERM MATURATION

A. Spermatogenesis

To date, spermatogenesis in galliform birds has been most fully characterized in the Japanese quail (*Coturnix coturnix japonica*). Four types of spermatogonial cells have been described in *Coturnix* (Lin *et al.*, 1990; Lin and Jones, 1992). These have been designated as spermatogonia A_d, A_{p1}, A_{p2}, and B. The letter subscripts indicate intensity of staining and denote "dark" and "pale," respectively. The numerical subscripts denote ultrastructural differences (Lin and Jones, 1992). Spermatogonia A_d are stem cells. Based upon the model of stem cell renewal and spermatogonial proliferation proposed by Lin and Jones (1992), each division of an A_d spermatogonium yields A_d and A_{p1} daughter cells. Thus, spermatogenesis in *Coturnix* commences with the mitosis of an A_d spermatogonium. Cell types derived from each A_{p1} spermatogonium include A_{p2} spermatogonia ($n = 2$), B spermatogonia ($n = 4$), primary spermatocytes ($n = 8$), secondary spermatocytes ($n = 16$), and spermatids ($n = 32$). Secondary spermatocytes and spermatids are formed by the first and second meiotic divisions, respectively.

Spermiogenesis is the transformation of spermatids into sperm cells without further cell division. Spermiogenesis in *Coturnix* entails 12 distinct morphological *steps* (Lin *et al.*, 1990; Lin and Jones, 1993). In comparison, 8 to 10 steps have been reported for *Gallus* (de Reviers, 1971; Gunawardana, 1977; Tiba *et al.*, 1993) and 10 steps for the guinea fowl (Aire *et al.*, 1980). Spermiogenesis entails formation of an acrosome and an axoneme, loss of cytoplasm, and replacement of nucleohistones with nucleoprotamine, which accompanies nuclear condensation (Nagano, 1962; Tingari, 1973; Okamura and Nishiyama, 1976; Gunawardana, 1977; Gunawardana and Scott, 1977; Oliva and Mezquita, 1986; Sprando and Russell, 1988). The reductions in cytoplasmic and nuclear volumes are striking; mature rooster sperm embody only 3% of the initial spermatid cell volume (Sprando and Russell, 1988). In summary, the seminiferous tubules contain Sertoli cells and a broad array of differentiating germinal cells including the various types of spermatogonia, spermatocytes, and spermatids.

However, germinal cells are not found in a physical continuum of differentiation within the seminiferous epithelium. Rather, they are found in distinct cellular associations. The cellular associations in *Coturnix* occupy an average of 17, 902 μm^2 and contain an average of 13.5 Sertoli cells per association (Lin and Jones, 1990). There are 10 cellular associations in *Coturnix*, and each is referred to as a *stage* of the seminiferous epithelium (Figure 8). The seminiferous epithelium passes through each successive stage as a function of time at any fixed point. Once a complete series has been finished, a new series is initiated. Consequently, the series of stages is referred to as the cycle of the seminiferous epithelium. The duration of the cycle in *Coturnix* is 2.69 days (Lin *et al.*, 1990). In comparison, the duration of the cycle of the seminiferous epithelium in *Gallus* has been estimated to be 3–4 days (de Reviers, 1968; Tiba *et al.*, 1993). The time between the division of an A_d spermatogonium and spermiation, known as the duration of spermatogenesis is 12.8 days in *Coturnix* (Lin and Jones, 1992). Thus, 4.75 cycles of the seminiferous epithelium are required to produce 32 sperm cells from a single A_d spermatogonium at any fixed point within the seminiferous epithelium.

A complete series of stages in the context of a seminiferous tubule is referred to as a *spermatogenic wave*. As shown in Figure 9, these waves in *Coturnix* are arranged helically along the length of seminiferous tubules (Lin and Jones, 1990). It must be noted that while sequential stages may be contiguous in space (Figure 9),

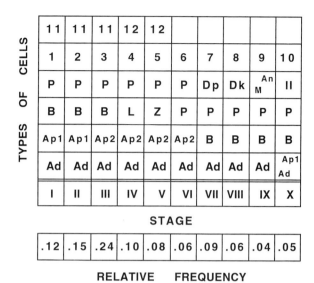

FIGURE 8 The cycle of the seminiferous epithelium of *Coturnix coturnix japonica*. The relative frequency of each cellular association, or stage, is shown below the stage's number. Ad, dark type A spermatogonia (stem cell); Ap1 and 2, pale type A spermatogonia; B, type B spermatogonia; L, leptotene primary spermatocytes; Z, zygotene primary spermatocytes; P, pachytene primary spermatocytes; Dp, diplotene primary spermatocytes; Dk, diakinesis of primary spermatocytes; M, metaphase primary spermatocytes; An, anaphase primary spermatocytes; II, secondary spermatocytes; 1–12, step 1 through step 12 spermatids. (Adapted from Lin *et al.* (1990), *J. Reprod. Fertil.* with permission.)

they are not observed at a common frequency (Figure 8) because duration of each of the stages ranges from 2.5 to 15.5 hr (Lin *et al.*, 1990). Consequently, the prevalence of any given stage is directly proportional to the stage's duration (Lin *et al.*, 1990).

Daily sperm production may be defined as the number of sperm produced per gram of testis per day. Daily sperm production in *Coturnix* has been estimated to be 92.5×10^6 sperm per gram of testis per day (Clulow and Jones, 1982). This estimate is equivalent to the DSP of $80–120 \times 10^6$ sperm per gram of testis reported for *Gallus* (de Reviers and Williams, 1984). These values denote the number of fully formed sperm released per day from the seminiferous epithelium into the lumen of the seminiferous tubules. This phenomenon is defined as spermiation, which in *Coturnix* would be limited to the seminiferous epithelium in stage V (Figure 8). At the time of spermiation, superfluous cytoplasm found alongside the sperm cell's head is jettisoned as a residual body (Sprando and Russell, 1988; Lin and Jones, 1993). Sperm cells released from the seminiferous epithelium are immotile (Ashizawa and Sano, 1990). Galliform sperm are vermiform cells (Figure 10) with a maximum width of 0.5–0.7 μm and a length of 75–90 μm (Thurston and Hess, 1987). They contain a conical acrosome, a

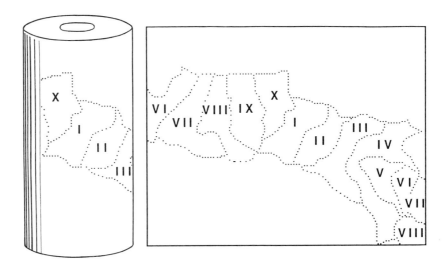

FIGURE 9 Spatial arrangement of the wave of spermatogenesis within a seminiferous tubule of *Coturnix coturnix japonica*. The cylinder to the right represents a section of a seminiferous tubule. The rectangle to its left denotes the two-dimensional representation of two contiguous cycles of the seminiferous epithelium. The cycle is arranged helically along the length of the tubule. Roman numerals denote stages. (Adapted from Lin and Jones (1990), *J. Reprod. Fertil.* with permission.)

slightly bent cylindrical nucleus, and a helix of 25–30 mitochondria surrounding the proximal portion of a long flagellum, which accounts for approximately 84% of the cell's length (for review, see Thurston and Hess, 1987).

B. Extragonadal Sperm Transport and Maturation

Following spermiation, sperm cells are suspended within fluid secreted by Sertoli cells. Sperm passage through the labyrinth of seminiferous tubules most likely depends upon the hydrostatic pressure of seminiferous tubule fluid and the contractility of the myoepithelial cells overlying the outer surface of the seminiferous tubule (Rothwell and Tingari, 1973). Sperm passage through the distal excurrent ducts, in particular the deferent duct, presumably depends upon peristalsis (de Reviers, 1975). Like spermatogenesis, sperm transport through the excurrent ducts has been characterized most fully in the Japanese quail. Thus, an analysis of the phenomena and time-course of sperm transport through the excurrent ducts of *Coturnix* provides a useful context for interpreting data from other galliform species. As shown in Table 2, sperm pass through the excurrent ducts of *Coturnix* in approximately 24 hr (Clulow and Jones, 1988). In comparison, sperm transport through the excurrent ducts of *Gallus* has been estimated to take several days (Munro, 1938; de Reviers, 1975). Perhaps the most evident phenomenon that occurs during sperm transport through the excurrent ducts, as evidenced by

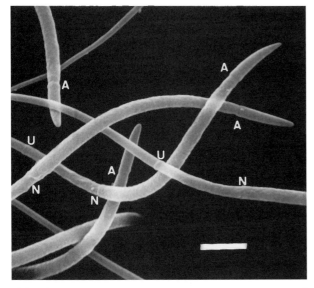

FIGURE 10 Scanning electron micrograph of chicken sperm cells. The constriction at the anterior end of the sperm cell (A) marks posterior boundary of the acrosome. The nucleus extends posteriorly from (A) to the neck region (N), which marks the anterior end of the midpiece. The midpiece, site of the anterior portion of the axoneme and the highly modified mitochondria, extends back to the raised annulus (U). The tail of the sperm extends from the annulus to the cell's termination. At the nucleus, chicken sperm are about 0.5 μm in diameter, with the overall length of the cell approximately 90 μm. Bar = 2 μm. (Adapted from J. M. Bahr and M. R. Bakst, *Reproduction in Farm Animals*, 5th ed., © Lea & Febiger, 1987.)

TABLE 2 Estimates of Plasma Flux (Reabsorption) across the Epithelium Lining the Excurrent Ducts of Japanese Quail[a]

Region	Testicular plasma Output (%)	$\mu l/cm^2/hr$
Rete testis to proximal efferent ducts	6.3	8.0
Proximal to distal efferent ducts	85.8	100.4
Distal efferent ducts to connecting ducts	6.5	21.6
Connecting ducts to epididymal duct	0.4	2.1
Epididymal duct to proximal deferent duct	0.2	0.1
Proximal to distal deferent duct	0.2	0.2

[a]Adapted from Clulow and Jones (1988).

a change in sperm concentration, is the absorption of seminiferous tubule fluid (Table 3). The concentration of sperm in the seminiferous tubule fluid of *Coturnix* is 3.8×10^4 per μl whereas the concentration of sperm within the distal deferent duct is 2.3×10^6 per μl (Clulow and Jones, 1988). This 60-fold increase in sperm concentration is largely due to seminiferous tubule fluid absorption at the level of the efferent ducts (Table 3). Nakai *et al.* (1989) demonstrated that the nonciliated epithelial cells in the proximal efferent ducts of *Gallus* incorporate fluid by pinocytosis.

The efferent ducts, which represent the principal excurrent duct within the epididymis (Table 1), may be a critical site for sperm maturation. Due to extensive mucosal folding, apocrine secretion, the presence of ciliated cells, and epithelial cells with abundant microvilli, the efferent ducts appear to be a site where sperm are mixed with secretions as they are concentrated. Additional evidence of the importance of the efferent duct comes from studies of reproductive anomalies in turkeys and chickens. For example, turkeys may ejaculate yellow rather than white semen (Thurston *et al.*, 1982a).

TABLE 3 Estimates of Duration of Spermatozoal Transit and Velocity through the Excurrent Ducts of Japanese Quail[a]

Duct	Duration	Velocity (mm/min)
Rete testis	25 sec	—
Efferent ducts		
Proximal	3 min	—
Distal	5 min	4.00
Connecting ducts	22 min	0.16
Epididymal duct	80 min	1.04
Deferent duct	22.2 hr	0.37

[a]Adapted from Clulow and Jones (1988).

Epithelial cells within the efferent ducts of such turkeys have been shown to be hypertrophied, engorged with lipid droplets, and characterized by heightened phagocytosis of sperm (Hess *et al.*, 1982). The latter phenomenon appears to be mediated by macrophages within the rete testis under normal conditions (Aire and Malmqvist, 1979b; Nakai *et al.*, 1989), but can be mediated by epithelial cells in the excurrent ducts when the deferent duct is not patent (Tingari and Lake, 1972). Subfertile roosters with malformed proximal efferent ducts (Figure 11) provide a second example of the relationship between efferent duct function and reproductive performance (Kirby *et al.*, 1990). The biochemical imbalances observed in semen from such roosters have been attributed to excurrent duct dysfunction (Al-Aghbari *et al.*, 1992).

In review, sperm are suspended in seminiferous tubule fluid prior to their passage through the excurrent ducts. The volume of this suspensory fluid (Table 3) and its composition (Esponda and Bedford, 1985) are altered prior to sperm entry into the epididymal duct. The resultant medium is seminal plasma. The chemical composition of seminal plasma is distinct from blood plasma as evidenced by differences in electrolyte, free amino acid, and protein composition (Lake, 1966; Lake and Hatton, 1968; Stratil, 1970; Thurston *et al.*, 1982b; Freeman, 1984). The maintenance of these differences along the length of the excurrent ducts has been attributed to tight junctions between adjacent epithelial cells lining the ducts (Nakai and Nasu, 1991).

Apart from proteins specific to the reproductive tract (Stratil, 1970; Esponda and Bedford, 1985), the principal differences between seminal and blood plasma involve glucose, glutamate, K^+, Cl^-, and Ca^{2+} (Table 4). At present, it is not understood how the composition of excurrent duct fluid affects sperm motility. While Esponda and Bedford (1985) have demonstrated that sperm maturation proteins do exist in *Gallus* and while Ashizawa *et al.* (1988) have reported that the Na^+-to-K^+ ratio associated with the midpiece of rooster sperm increases as a function of sperm passage through the deferent duct, such changes do not induce motility; for sperm within the deferent duct are immotile (Ashizawa and Sano, 1990). Nonetheless, sperm acquire the potential for motility as they pass through the excurrent ducts (Munro, 1938; Clulow and Jones, 1982; Howarth, 1983; Ashizawa and Sano, 1990). This potential is distinct from fertilizing ability in that testicular sperm have the capacity to fertilize oocytes, providing they are placed in the oviduct above the vaginal sphincter (Howarth, 1983). As shown by Allen and Grigg (1957), sperm need not be motile to ascend the oviduct above the vaginal sphincter. Thus, the fertilizing ability of testicular sperm may not depend upon motility as much as the ability of

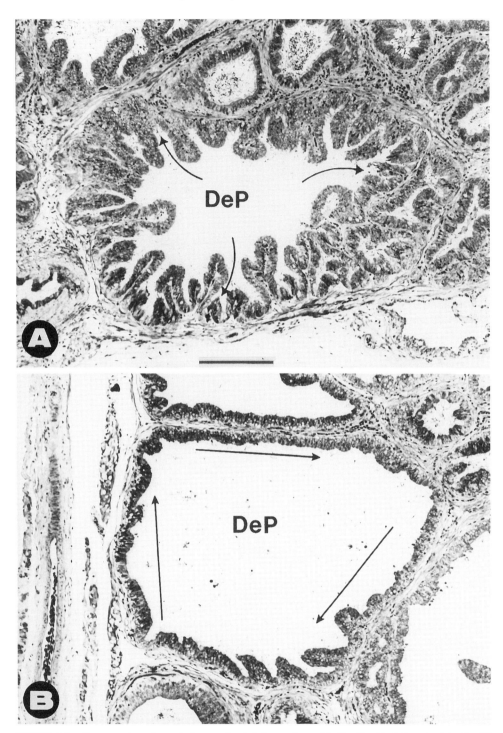

FIGURE 11 Cross sections through the proximal deferent ducts (DeP) of fertile (A) and subfertile (B) roosters. The presence of the *sd* allele in crossbred roosters (B) resulted in a twofold increase in DeP cross-sectional area and a 75% decrease in luminal surface area as compared to the wild-type (A) male. Males exhibiting this genetic defect of the DeP are characterized by poor semen quality, frequently exhibiting 30–90% dead and degenerate sperm in an ejaculate. Reproduced with permission from the Society for the Study of Reproduction. From Kirby *et al.* (1990).

TABLE 4 Comparison of Selected Substances in Rooster Blood Plasma and Deferent Duct Fluid[a]

Substance	Blood plasma	Deferent duct fluid
Glucose	15	—
Glutamate	>1	100
Sodium	160	142
Potassium	5	34
Chloride	115	40
Calcium	5	1

[a]Adapted from appendices in "Physiology and Biochemistry of the Domestic Fowl", Vol. 5, by B. M. Freeman.

sperm cells to undergo an acrosome reaction in response to contact with the oocyte's inner perivitelline layer (Okamura and Nishiyama, 1978).

The deferent duct, particularly the distal portion, contains the bulk of the extragonadal sperm reserve (ESR). In *Coturnix,* the number of sperm in the ESR is equivalent to the number produced daily by the testes and 92% of these are found within the deferent duct (Clulow and Jones, 1982). In contrast, the ESR in *Gallus* is equivalent to ≤3.5 times the daily sperm production, but 95% of the ESR is found in the deferent duct (de Reviers, 1975). In either case, the duration of sperm storage within the deferent duct is relatively brief. As stated above, these sperm are immotile prior to ejaculation. While Ashizawa and Sano (1990) have proposed that a temperature change accompanying ejaculation may initiate sperm motility in *Gallus,* it is also likely that accessory reproductive fluids play a role in the initiation of sperm motility (for review, see Fujihara, 1992). In this regard, Ca^{2+} and HCO_3^- have been shown to be motility agonists *in vitro* (Ashizawa and Wishart, 1987; Ashizawa and Sano, 1990).

References

Aire, T. A. (1979a). Micro-stereological study of the avian epididymal region. *J. Anat.* **129,** 703–706.

Aire, T. A. (1979b). The epididymal region of the Japanese quail (*Coturnix coturnix japonica*). *Acta Anat.* **103,** 305–312.

Aire, T. A. (1980). The ductuli efferentes of the epididymal region of birds. *J. Anat.* **130,** 707–723.

Aire, T. A. (1982). The rete testis of birds. *J. Anat.* **135,** 97–110.

Aire, T. A., and Malmqvist, M. (1979a). Intraepithelial lymphocytes in the excurrent ducts of the testis of the domestic fowl (*Gallus domesticus*). *Acta Anat.* **103,** 142–149.

Aire, T. A., and Malmqvist, M. (1979b). Macrophages in the excurrent ducts of the testes of normal domestic fowl (*Gallus domesticus*). *Zbl. Vet. Med. C. Anat. Histol. Embryol.* **8,** 172–176.

Aire, T. A., Olowo-okorun, M. O., and Ayeni, J. S. (1980). The seminiferous epithelium in the guinea fowl (*Numidia meleagris*). *Cell Tissue Res.* **205,** 319–325.

Al-Aghbari, A., Engel, H. N., Jr., and Froman, D. P. (1992). Analysis of seminal plasma from roosters carrying the *Sd* (sperm degeneration) allele. *Biol. Reprod.* **47,** 1059–1063.

Allen, T. E., and Grigg, G. W. (1957). Sperm transport in the fowl. *Austral. J. Ag. Res.* **8,** 788–799.

Ashizawa, K., and Sano, R. (1990). Effects of temperature on the immobilization and the initiation of motility of spermatozoa in the male reproductive tract of the domestic fowl, *Gallus domesticus. Comp. Biochem. Physiol. A* **96,** 297–301.

Ashizawa, K., and Wishart, G. J. (1987). Resolution of the sperm motility-stimulating principle of fowl seminal plasma into Ca^{+2} and an unidentified low molecular weight factor. *J. Reprod. Fert.* **81,** 495–499.

Ashizawa, K., Ozawa, Y., and Okauchi, K. (1988). Changes of elemental concentrations around and on the surface of fowl sperm membrane during maturation in the male reproductive tract and after in vitro storage. *Gamete Res.* **21,** 23–28.

Bacon, W. L., Proudman, J. A., Foster, D. N., and Renner, P. A. (1991). Pattern of secretion of luteinizing hormone and testosterone in the sexually mature male turkey. *Gen. Comp. Endocrinol.* **84,** 447–460.

Bahr, J. M., and Bakst, M. R. (1987) Poultry. In "Reproduction in Farm Animals" (E. S. E. Hafez, ed.), 5th ed., pp. 379–398. Lea & Febiger, Philadelphia.

Bakst, M. R. (1980). Luminal topography of the male chicken and turkey excurrent duct system. *Scanning Electron Microsc.* **III,** 419–426.

Balthazart, J. (1989). Steroid metabolism and the activation of social behavior. In "Advances in Comparative and Environmental Physiology" (J. Balthazart, ed.), Vol. 3, pp. 103–159. Springer-Verlag, Berlin.

Bergmann, M., and Schindelmeiser, J. (1987). Development of the blood-testis barrier in domestic fowl (*Gallus domesticus*). *Int. J. Androl.* **10,** 481–488.

Brown, N. L., Baylé, J-D., Scanes, C. G., and Follett, B. K. (1975). Chicken gonadotrophins: Their effects on the testes of immature and hypophysectomized Japanese quail. *Cell Tissue Res.* **156,** 499–520.

Brown, N. L., and Follett, B. K. (1977). Effects of androgens on the testes of intact and hypophysectomized Japanese quail. *Gen. Comp. Endocrinol.* **33,** 267–277.

Budras, K.-D., and Sauer, T. (1975). Morphology of the epidydimis of the cock (*Gallus domesticus*) and its effect upon the steroid sex hormone synthesis. I.Ontogenesis, morphology, and distribution of the epididymis. *Anat. Embryol.* **148,** 175–196.

Clulow, J., and Jones, R. C. (1982). Production, transport, maturation, storage and survival of spermatozoa in the male Japanese quail, *Coturnix coturnix. J. Reprod. Fert.* **64,** 259–266.

Clulow, J., and Jones, R. C. (1988). Studies of fluid and spermatozoal transport in the extratesticular genital ducts of the Japanese quail. *J. Anat.* **157,** 1–11.

Connell, C. J. (1972). The effect of luteinizing hormone on the ultrastructure of the Leydig cell of the chick. *Zeitsch. Zellfors. Mikrosk. Anat.* **128,** 139–151.

Connolly, P. B., and Callard, I. P. (1987). Steroids modulate the release of luteinizing hormone from quail pituitary cells. *Gen. Comp. Endocrionol.* **68,** 466–472.

Cooksey, E. J., and Rothwell, B. (1973). The ultrastructure of the Sertoli cell and its differentiation in the domestic fowl (*Gallus domesticus*). *J. Anat.* **114,** 329–345.

Davies, D. T., and Follett, B. K. (1980). Neuroendocrine regulation of gonadotropin releasing hormone secretion in the Japanese quail. *Gen. Comp. Endocrinol.* **40,** 220–225.

Davies, D. T., Massa, R., and James, R. (1980). Role of testosterone and its metabolites in regulating gonadotropin secretion in the Japanese quail. *J. Endocr.* **84,** 211–222.

Driot, F. J. M., de Reviers, M., and Williams, J. B. (1979). Plasma testosterone levels in intact and hemicastrated growing cockerels. *J. Endocr.* **81,** 169–174.

Ebensperger, C., Drews, U., Mayerova, A., and Wolf, U. (1988a). Sereological H-Y antigen in the female chickens occurs during gonadal differentiation. *Differentiation* **37,** 186–191.

Ebensperger, C., Drews, U., and Wolf, U. (1988b). An in vitro model of gonad differentiation in the chicken. *Differentiation* **37,** 192–197.

Elbrecht, A., and Smith, R. G. (1992). Aromatase enzyme activity and sex determination in chickens. *Science* **255,** 467–470.

El Halawani, M. E., Burke, W. H., and Ogren, L. A. (1980). Involvement of catecholaminergic mechanisms in the photoperiodically induced rise in serum luteinizing hormone of Japanese quail (*Coturnix coturnix japonica*). *Gen. Comp. Endocrinol.* **41,** 14–21.

Esponda, P., and Bedford, J. M. (1985) Surface of the rooster spermatozoan changes in passing through the Wolffian duct. *J. Exp. Zool.* **234,** 441–449.

Freeman, B. M. (1984). Appendix X. Reproduction: Semen. In "Physiology and Biochemistry of the Domestic Fowl" (B. M. Freeman, ed.), Vol. 5., pp. 422–423. Academic Press, London.

Fujihara, N. (1992). Accessory reproductive fluids and organs in male domestic birds. *World's Poult. Sci. J.* **48,** 39–56.

Fujimoto, T., Ukeshima, A, and Kiyofuji, R. (1976). The origin, migration, and morphology of the primordial germ cells in the chick embryo. *Anat. Rec.* **185,** 139–154.

Galli, F. E., Irusta, O., and Wasserman, G. F. (1973). Androgen production by testes of *Gallus domesticus* during post-embryonic development. *Gen. Comp. Endocrinol.* **21,** 262–266.

Gilbert, A. B. (1979). *Glandulae endocrinae. In* "Nomina Anatomica Avium" (J. J. Baumel, ed.), pp. 337–342. Academic Press, London.

Gledhill, B., and Follett, B. K. (1976). Diurnal variation and the episodic release of plasma gonadotropins in Japanese quail during a photoperiodically induced gonadal cycle. *J. Endocr.* **71,** 245–247.

Guichard, A., Cedard, L., Mignot, T. M., Scheib, D., and Haffen, K. (1977). Radioim munoassay of steroids produced by cultured chick embryonic gonads: differences according to age, sex, and side. *Gen. Comp. Endocrinol.* **32,** 255–265.

Gunawardana, V. K. (1977). Stages of spermatids in the domestic fowl: A light microscope study using araldite sections. *J. Anat.* **123,** 351–360.

Gunawardana, V. K., and Scott, M. G. A. D. (1977). Ultrastructural studies on the differentiation of spermatids in the domestic fowl. *J. Anat.* **124,** 741–755.

Halverson J. L., and Dvorak, J. (1993). Genetic control of sex determina- tion in birds and the potential for its manipulation. *Poult. Sci.* **72,** 890–896.

Hattori, M., and Wakabayashi, K. (1979). Isoelectric focusing and gel filtration studies on the heterogeneity of avian luteinizing hormone. *Gen. Comp. Endocrinol.* **39,** 215–221.

Hattori, A., Ishii, S., and Wada, M. (1986). Effects of two kinds of chicken luteinizing hormone-releasing hormone (LH-RH), mammalian LH-RH and its analogs on the release of LH and FSH in Japanese quail and chicken. *Gen. Comp. Endocrinol.* **64,** 446–455.

Hess, R. A., and Thurston, R. J. (1977). Ultrastructure of the epithelial cells in the epididymal region of the turkey (*Meleagris gallopavo*). *J. Anat.* **124,** 765–778.

Hess, R. A., Thurston, R. J., and Biellier, H. V. (1982). Morphology of the epididymal region of turkeys producing abnormal yellow semen. *Poult. Sci.* **61,** 531–539.

Hill, R. T., and Parkes, A. S. (1935). Hypophysectomy of birds III. Effect on gonads, accessory organs and head furnishings. *Proc. Roy. Soc. B.* **116,** 221–236.

Howarth, B., Jr. (1983). Fertilizing ability of cock spermatozoa from the testis, epididymis, and vas deferens following intramagnal insemination. *Biol. Reprod.* **28,** 586–590.

Hutson, J. M., Donahoe, P. K., and MacLaughlin, D. T. (1985). Steroid modulation of Müllerian duct regression in the chick embryo. *Gen. Comp. Endocrinol.* **57,** 88–102.

Imataka, H., Suzuki, K., Inano, H., Kohmoto, K., and Tamaoki, B. I. (1988). Sexual differences of steroidogenic enzymes in embryonic gonads of the chicken (*Gallus domesticus*). *Gen. Comp. Endocrinol.* **69,** 153–162.

Ingkasuwan, P., and Ogasawara, F. X. (1966) The effect of light and temperature and their interaction on the semen production of White Leghorn males. *Poult. Sci.* **45,** 1199–1204.

Ishii, S. (1993). The molecular biology of avian gonadotropin. *Poult. Sci.* **72,** 856–866.

Ishii, S., and Furuya, T. (1975). Effects of purified chicken gonadotropins on the chick testis. *Gen. Comp. Endocrinol.* **25,** 1–8.

Ishii, S., and Yamamoto, K. (1976). Demonstration of follicle stimulating hormone (FSH) activity in hypophyseal extracts of various vertebrates by the response of the Sertoli cells of the chick. *Gen. Comp. Endocrinol.* **29,** 506–510.

Johnson, A. L. (1986). Reproduction in the male. *In* "Avian Physiology" (P. D. Sturkie, ed.), pp. 432–451. Springer-Verlag, New York.

Katz, I. A., Millar, R. P., and King, J. A. (1990). Differential regional distribution and release of two forms of gonadotropin-releasing hormone in the chicken brain. *Peptides* **11,** 443–450.

King, A. S. (1979). *Systema urogenitale. In* "Nomina Anatomica Avium" (J. J. Baumel, ed.), pp. 289–335. Academic Press, London.

King, A. S. (1981). Phallus. *In* "Form and Function in Birds" (A. S. King and J. McLelland, eds.), Vol. 2, pp. 107–147. Academic Press, London.

King, J. A., Davidson, J. S., and Millar, R. P. (1988). Interaction of endogenous chicken gonadotrophin-releasing hormone-I and -II on chicken pituitary cells. *J. Endocr.* **117,** 43–49.

Kirby, J. D., Froman D. P., Engel, H. N., Jr., Bernier, P. E., and Hess, R. A. (1990). Decreased spermatozoal survivability associated with aberrant morphology of the ductuli efferentes proximales of the chicken (*Gallus domesticus*). *Biol. Reprod.* **42,** 383–389.

Knight, P. G. (1983). Variations in hypothalamic luteinizing hormone content and release *in vitro* and plasma concentrations of luteinizing hormone and testosterone in developing cockerels. *J. Endocr.* **99,** 311–319.

Knight, P. G., Gladwell, R. T., and Cunningham, F. J. (1981). Effect of gonadectomy on the concentrations of catecholamines in discrete areas of the diencephalon of the domestic fowl. *J. Endocr.* **89,** 389–397.

Krishnan, K. A., Proudman, J. A., and Bahr, J. M. (1992). Purification and characterization of chicken follicle-stimulating hormone. *Comp. Biochem. Physiol. B* **102,** 67–75.

Kuenzel, W. J. (1993). The search for deep encephalic photoreceptors within the avian brain, using gonadal development as a primary indicator. *Poult. Sci.* **72,** 959–967.

Kuenzel, W. J., and Blähser, S. (1991). The distribution of gonadotropin- releasing hormone (GnRH) neurons and fibers throughout the chick brain (*Gallus domesticus*). *Cell Tissue Res.* **264,** 481–495.

Kumaran, J. D. S., and Turner, C. W. (1949). The normal development of testes in the White Plymouth Rock. *Poult. Sci.* **28,** 511–520.

Lake, P. E. (1966). Physiology and biochemistry of poultry semen. *In* "Advances in Reproductive Physiology" (A. McLaren, ed.), Vol. 1, pp. 93–123. Academic Press, London.

Lake, P. E. (1981). Male genital organs. *In* "Form and Function in Birds" (A. S. King and J. McLelland, eds.), Vol. 2, pp. 2–61. Academic Press, London.

Lake, P. E., and Hatton, M. (1968). Free amino acids in the vas deferens, semen, transparent fluid and blood plasma of the domestic rooster, *Gallus domesticus. J. Reprod. Fertil.* **15,** 139–143.

Lin, M., and Jones, R. C. (1990). Spatial arrangement of the stages of the cycle of the seminiferous epithelium in the Japanese quail, *Coturnix coturnix japonica*. *J. Reprod. Fertil.* **90**, 361–367.

Lin, M., and Jones, R. C. (1992). Renewal and proliferation of spermatogonia during spermiogenesis in the Japanese quail, *Coturnix coturnix japonica*. *Cell Tissue Res.* **267**, 591–601.

Lin, M. and Jones, R. C. (1993). Spermiogenesis and spermiation in the Japanese quail (*Coturnix coturnix japonica*). *J. Anat.* **183**, 525–535.

Lin, M., Jones, R. C., and Blackshaw, A. W. (1990). The cycle of the seminiferous epithelium in the Japanese quail (*Coturnix coturnix japonica*) and estimation of its duration. *J. Reprod. Fertil.* **88**, 481–490.

MacLaughlin, D. T., Hutson, J. M., and Donahoe, P. K. (1983). Specific estradiol binding in embryonic Müllerian ducts: A potential modulator of regression in the male and female chick. *Endocrinology* **113**, 141–145.

Maraud, R., and Stoll, R. (1955). Action de la testostérone sur la constitution de l'épididme du Coq. *C. R. Soc. Biol.* **149**, 704–707.

Martin, J. T., Balthazart, J., and Burke, W. J. (1984). Adrenocortical hormones and reproductive function in Japanese quail. *J. Steroid Biochem.* **20**, 1561.

Marvan, F. (1969). Postnatal development of the male genital tract of *Gallus domesticus*. *Anat. Anz. Bd.* **124**, 443–462.

Maung, Z. W., and Follett, B. K. (1977). Effects of chicken and ovine luteinizing hormone on androgen release and cyclic AMP production by isolated cells from the quail testis. *Gen. Comp. Endocrinol.* **33**, 242–253.

Merchant-Larios, H., Popova, L., and Reyss-Brion, M. (1984). Early morphogenesis of chick gonad in the absence of mesonephros. *Dev. Growth Differ.* **26**, 403–417.

Meyer, D. (1964). The migration of primordial germ cells in the chick embryo. *Dev. Biol.* **10**, 154–190.

Mikami, S.-I., and Yamada, S. (1984). Immunohistochemistry of the hypothalamic neuropeptides and anterior pituitary cells in the Japanese quail. *J. Exp. Zool.* **232**, 405–417.

Mikami, S.-I., Yamada, S., Hasegawa, Y., and Miyamoto, K. (1988). Localization of avian LHRH-immunoreactive neurons in the hypothalamus of the domestic fowl, *Gallus domesticus*, and the Japanese quail, *Coturnix coturnix*. *Cell Tissue Res.* **251**, 51–58.

Miyamoto, K., Hasegawa, Y., Minegishi, T., Nomura, M., Takahashi, Y., Igarashi, M., Kangawa, K., and Matsuo, H. (1982). Isolation and characterization of chicken hypothalamic luteinizing hormone releasing hormone. *Biochem. Biophys. Res. Commun.* **107**, 820–827.

Mizuno, S., Saitoh, Y., Nomura, O., Kunita, R., Ohtomo, K., Nishimori, K., Ono, H., and Saitoh, H. (1993). Sex-specific DNA sequences in galliformes and their application to the study of sex differentiation. *In* "Manipulation of the Avian Genome" (R. J. Etches and A. M. V. Gibbins, eds.), pp. 257–274. CRC Press, Boca Raton, FL.

Munro, S. S. (1938). Functional changes in fowl sperm during their passage through the excurrent ducts of the male. *J. Exp. Zool.* **79**, 71–92.

Nagano, T. (1962). Observations on the fine structure of the developing spermatid in the domestic chicken. *J. Cell Biol.* **14**, 193–205.

Nakai, M., and Nasu, T. (1991). Ultrastructural study on junctional complexes of the excurrent duct epithelia in the epididymal region in the fowl. *J. Vet. Med. Sci.* **53**, 677–681.

Nakai, M., Hashimoto, Y., Kitagawa, H., Kon, Y., and Kudo, N. (1988). Microvasculature of the epididymis and ductus deferens of domestic fowls. *Jpn. J. Vet. Sci.* **50**, 371–381.

Nakai, M., Hashimoto, Y., Kitagawa, H., Kon, Y., and Kudo, N. (1989). Histological study on seminal plasma absorption and spermiophagy in the epididymal region of domestic fowl. *Poult. Sci.* **68**, 582–589.

Nakamura, T., and Tanabe, Y. (1972). *In vitro* steroidogenesis by testes of the chicken (*Gallus domesticus*). *Gen. Comp. Endocrinol.* **19**, 432–440.

Narbaitz, R., and Adler, R. (1966) Submicroscopic observations on the differentiation of the chick gonads. *J. Embryol. Exp. Morphol.* **15**, 41–47.

Nickel, R., Schummer, A., Seiferle, E., Siller, W. G., and Wight, P. A. L. (1977). Urogenital system. *In* "Anatomy of the Domestic Birds," pp. 70–84. Springer-Verlag, Berlin.

Okamura, F., and Nishiyama, H. (1976). The early development of the tail and the transformation of the shape of the nucleus of the spermatid of the domestic fowl, *Gallus gallus*. *Cell Tissue Res.* **169**, 345–359.

Okamura, F., and Nishiyama, H. (1978). The passage of spermatozoa through the vitelline membrane in the domestic fowl, *Gallus gallus*. *Cell Tissue Res.* **188**, 497–508.

Oliva, R., and Mezquita, C. (1986). Marked differences in the ability of distinct protamines to disassociate nucleosomal core particles in vitro. *Biochemistry* **25**, 6508–6511.

Osman, D. I. (1980) The connection between the seminiferous tubules and the rete testis in the domestic fowl (*Gallus domesticus*). *Int. J. Androl.* **3**, 177–187.

Osman, D. I., Ekwall, H., and Ploen, L. (1980) Specialized cell contacts and the blood-testis barrier in the seminiferous tubules of the domestic fowl (*Gallus domesticus*). *Int. J. Androl.* **3**, 553–562.

Ottinger, M. A. (1983). Hormonal control of reproductive behavior in the avian male. *Poult. Sci.* **62**, 1690–1699.

Ottinger, M. A. (1992). Altered neuroendocrine mechanisms during reproductive aging. *Poult. Sci. Rev.* **4**, 235–248.

Ottinger, M. A., and Brinkley, H. J. (1979). Testosterone and sex-related physical characteristics during maturation of the male Japanese quail (*Coturnix coturnix japonica*). *Biol. Reprod.* **20**, 905–909.

Pelletier, R. M. (1990) A novel perspective: the occluding zonule encircles the apex of the Sertoli cell as observed in birds. *Am. J. Anat.* **188**, 87–108.

Rashedi, M., Maraud, R., and Stoll, R. (1983). Development of the testis in female domestic fowls submitted to an experimental sex reversal during embryonic life. *Biol. Reprod.* **29**, 1221–1227.

Ravona, H., Snapir, N., and Perek, M. (1973). The effect on the gonadal axis in cockerels of electrolytic lesions in various regions of the basal hypothalamus. *Gen. Comp. Endocrinol.* **20**, 112–124.

de Reviers, M. (1968). Determination de la durée des processus spermatogenetiques chez le coq a l'aide de thymidine tritice. *6th Int. Congr. Anim. Reprod. Paris* **1**, 183–185.

de Reviers, M. (1971a). Le développement testiculaire chez le coq. I. Criissance Ponderale des testicules et développement des tubes seminiferes. *Ann. Biol. Anim. Biochim. Biophys.* **11**, 519–530.

de Reviers, M. (1971b). Le développement testiculaire chez le coq. II. Morphologie de l'épithélium séminifère et établissement de la spermatogenèse. *Ann. Biol. Anim. Biochim. Biophys.* **11**, 531–546.

de Reviers, M. (1975). Sperm transport and survival in male birds. *In* "The Biology of Spermatozoa" (E. S. E. Hafez and C. G. Thibault, eds.), pp. 10–16. Karger, Basel.

de Reviers, M., and Williams, J. B. (1984). Testis development and production of spermatozoa in the cockerel (*Gallus domesticus*). *In* "Reproductive Biology of Poultry" (F. J. Cunningham, P. E. Lake, and D. Hewitt, eds.), pp. 183–202. British Poultry Science, Ltd., Harlow, UK.

Romanoff, A. L. (1960). The urogenital system. *In* "The Avian Embryo," pp. 783–782. MacMillan, New York.

Rothwell, B. (1973). The ultrastructure of Leydig cells in the testis of the domestic fowl. *J. Anat.* **116**, 245–253.

Rothwell, B., and Tingari, M. D. (1973). The ultrastructure of the boundary tissue of the seminiferous tubule in the testis of the domestic fowl (*Gallus domesticus*). *J. Anat.* **144**, 321–328.

Rozenboim, I., Gvaryahu, G., Robinzon, B., Sayag, N., and Snapir, N. (1986) Induction of precocious development of reproductive function in cockerels by Tamoxifen administration. *Poult. Sci.* **65**, 1980–1983.

Russell, L. D., and Griswold, M. D. (Eds.) (1993) "The Sertoli Cell." Cache River Press, Clearwater, FL.

Sakai, H., and Ishii, S. (1980). Isolation and characterization of chicken follicle-stimulating hormone. *Gen. Comp. Endocrinol.* **42**, 1–8.

Samsel, J., Lorber, B., Petit, A, and Weniger, J. P. (1986). Analysis of the cytosolic proteins of chick embryo gonads by two-dimensional gel electrophoresis. *J. Embryol. Exp. Morphol.* **94**, 221–230.

Sertoli, E. (1865). De l'esistenza di particulari cellule ramificate nei canalicoli seminiferi dell'testicolo umano. *Morgagni* **7**, 31–40.

Sertoli, E. (1878). Sulla sturttura dei canalicoli seminiferi dei testicolo. *Arch. Sci. Med.* **2**, 107–146, 267–295.

Sharp, P. J., and Gow, C. B. (1983) Neuroendocrine control of reproduction in the cockerel. *Poult. Sci.* **62**, 1671–1677.

Sharp, P. J., Culbert, J., and Wells, J. W. (1977). Variations in stored and plasma concentrations of androgen and luteinizing hormone during sexual development in the cockerel. *J. Endocr.* **74**, 467–476.

Sharp, P. J., Talbot, R. T., Main, G. M., Dunn, I. C., Fraser, H. M., and Huskisson, N. S. (1990). Physiological roles of chicken LHRH-I and -II in the control of gonadotrophin release in the domestic chicken. *J. Endocr.* **124**, 291–299.

Sharpe, R. M. (1994). Regulation of spermatogenesis. *In* "The Physiology of Reproduction" (E. Knobil and J. D. Neill, eds.), 2nd edition, Vol. 1, pp. 1363–1434. Raven Press, NY.

Sprando, I.L., and Russell, L.D. (1988). Spermiogenesis in the red-eared turtle (*Pseudomys scripta*) and the domestic fowl (*Gallus domesticus*): A study of cytoplasmic events including cell volume changes and cytoplasmic elimination. *J. Morphol.* **108**, 95–118.

Stansfield, S. C., and Cunningham, F. J. (1987). Modulation by endogenous opioid peptides of the secretion of LHRH from cockerel (*Gallus domesticus*) mediobasal hypothalamic tissue. *J. Endocr.* **114**, 103–110.

Sterling, R. J., and Sharp, P. J. (1984). A comparison of the luteinizing hormone- releasing activities of synthetic chicken luteinizing hormone-releasing hormone (LH-RH), synthetic porcine LH-RH, and buserelin, an LH-RH analogue, in the domestic fowl. *Gen. Comp. Endocrinol.* **55**, 463–471.

Stoll, R., Lafitan, L., and Maraud, R. (1973). Sur l' origine de l'hormone testiculaire responsable de la régression des canaux de Müller de l' embryon de Poulet. *C. R. Soc. Biol.* **167**, 1092–1096.

Stoll, R., and Maraud, R. (1974). Le rôle du testicule dans la différenciation sexuell des gonoductes chez l'embryon des vertébrés amniotes. *Bull. Assoc. Anat.* **58**, 699–674.

Stratil, A. (1970). Studies on proteins of seminal fluid from the vasa deferentia of the cock, *Gallus gallus* L. *Int. J. Biochem.* **1**, 728–734.

Sullivan, K. A., and Silverman, A.-J. (1993). The ontogeny of gonadotropin-releasing hormone neurons in the chick. *Neuroendocrinology* **58**, 597–608.

Tanabe, Y., Nakamura, T., Fujioka, K., and Doi, O. (1979). Production and secretion of sex steroid hormones by the testes, the ovary, and the adrenal glands of embryonic and young chickens (*Gallus domesticus*). *Gen. Comp. Endocrinol.* **39**, 26–33.

Tanaka, S., and Yasuda, M. (1980). Histological changes in the testis of the domestic fowl after adenohypophysectomy. *Poult. Sci.* **59**, 1538–1545.

Teng, C. (1982). Ontogeny of cyclic nucleotides in embryonic chick gonads. *Biol. Neonate* **41**, 123–131.

Teng, C. S., Wang, J. J., and Teng, J. I. N. (1987). Purification of chicken testicular Müllerian inhibiting substance by ion exchange and high-performance liquid chromatography. *Dev. Biol.* **123**, 245–254.

Thurston, R. J., and Hess, R. A. (1987). Ultrastructure of spermatozoa from domesticated birds: Comparative study of turkey, chicken, and guinea fowl. *Scanning Electron Microsc.* **1**, 1829–1838.

Thurston, R. J., Hess, R. A., Froman, D. P., and Biellier, H. V. (1982a). Elevated seminal plasma protein: A characteristic of yellow turkey semen. *Poult. Sci.* **61**, 1905–1911.

Thurston, R. J., Hess, R. A., Hughes, B. L., and Froman, D. P. (1982b). Seminal plasma free amino acids and seminal plasma and blood plasma proteins of the guinea fowl (*Numidia meleagris*). *Poult. Sci.* **61**, 1744–1747.

Tiba, T., Yoshida, K., Miyake, M., Tsuchiya, K., Kita, I., and Tsubota, T. (1993). Regularities and irregularities in the structure of the seminiferous epithelium in the domestic fowl (*Gallus domesticus*) I. Suggestion of the presence of the seminiferous epithelial cycle. *Anat. Histol. Embryol.* **21**, 241–253.

Tingari, M. D. (1971). On the structure of the epididymal region and ductus deferens of the domestic fowl (*Gallus domesticus*). *J. Anat.* **109**, 423–435.

Tingari, M. D. (1972). The fine structure of the epithelial lining of the excurrent duct system of the testis of the domestic fowl (*Gallus domesticus*). *Quart. J. Exp. Physiol.* **57**, 271–295.

Tingari, M. D. (1973). Observations on the fine structure of spermatozoa in the testis and excurrent ducts of the male fowl, *Gallus domesticus*. *J. Reprod. Fertil.* **34**, 255–265.

Tsutsui, K., and Ishii, S. (1978). Effects of follicle-stimulating hormone and testosterone on receptors of follicle-stimulating hormone in the testis of the immature Japanese quail. *Gen. Comp. Endocrinol.* **36**, 297–305.

Vigier, B., Forest, M. G., Eychenne, B., Bézard, J., Garrigou, O., Roel, P., and Josso, N. (1989). Anti-Müllerian hormone produces endocrine sex reversal of fetal ovaries. *Proc. Natl. Acad. Sci. USA* **86**, 3684–3688.

Wilson, S. C. (1978). LH secretion in the cockerel and the effects of castration and testosterone injections. *Gen. Comp. Endocrinol.* **35**, 481–490.

Wilson, S. C., and Sharp, P. J. (1975). Episodic release of luteinzing hormone in the domestic fowl. *J. Endocr.* **64**, 77–86.

Wilson, S. C., Knight, P. G., and Cunningham, F. J. (1983). Evidence for the involvement of central conversion of testosterone to oestradiol-17β in the regulation of luteinizing hormone secretion in the cockerel. *J. Endocr.* **99**, 301–310.

Wingfield, J. C., Hahn, T. P., Levin R., and Honey, P. (1992). Environmental predictability and control of gonadal cycles in birds. *J. Exp. Zool.* **261**, 214–231.

Woods, J. E., and Weeks, R. L. (1969). Ontogenesis of the pituitary-gonadal axis in the chick embryo. *Gen. Comp. Endocrinol.* **13**, 242–254.

Woods, J. E., Simpson, R. M., and Moore, P. L. (1975). Plasma testosterone levels in the chick embryo. *Gen. Comp. Endocrinol.* **27**, 543–547.

CHAPTER 24

Incubation Physiology

HIROSHI TAZAWA
Department of Electrical and Electronic Engineering
Muroran Institute of Technology
Muroran 050, Japan

G. CAUSEY WHITTOW
Department of Physiology
John A. Burns School of Medicine
University of Hawaii
Honolulu, Hawaii 96822

I. Introduction   617
II. Composition of the Freshly Laid Egg   617
III. Changes in the Composition of the Egg during Incubation   618
   A. Water Content   618
   B. Energy Content   619
IV. Heat Transfer to the Egg   620
V. Development of Physiological Functions   621
   A. Respiration   621
   B. Acid–Base Regulation   624
   C. Cardiovascular Function   625
   D. Thermoregulation   627
VI. Requirements and Procedures for Incubation   629
   A. Incubation Period   629
   B. Preincubation Egg Storage   629
   C. Egg Turning   630
   D. Ambient Temperatures and Embryonic Tolerance   631
   References   632

I. INTRODUCTION

The freshly laid avian egg contains everything that the embryo needs for its growth and development—except for oxygen and heat. The O_2 diffuses into the egg from the surrounding air through microscopic pores in the eggshell. The pores allow the CO_2 produced by the embryo to diffuse out of the egg but they also permit the loss of water vapor from the egg. Thus the regulation of gas exchange between the egg and its environment is closely related to its water balance; both are governed by the diffusive conductance of the eggshell until the embryo penetrates (pips) the chorioallantoic membrane or eggshell with the aid of its egg tooth. The shell diffusive conductance is a measure of the diffusibility of gas/water molecule through the pores; it depends on the shell geometry (porosity and thickness) and diffusion coefficient of the diffusing molecules. The adult bird has a key role in incubation providing not only the heat necessary for embryonic development but also the microclimate of the egg. In the poultry industry and for research purposes, the adult bird is conveniently replaced by an incubator. The physiological functions of developing embryos were elucidated using artificially incubated chicken eggs.

II. COMPOSITION OF THE FRESHLY LAID EGG

The composition of the freshly laid egg is related to the maturity of the hatchling. Hatchling maturity differs considerably in different species (Nice, 1962). Hatchlings fall into four major categories based on criteria such as mobility, amount of down, and ability to feed themselves: precocial (most mature) which can walk, swim, or dive soon after hatching; semiprecocial; semialtricial; and altricial (least mature), the latter being naked, their eyes closed, and incapable of locomotion. The amount of yolk in the freshly laid egg is much greater in precocial species than it is in altricial species (Sotherland and Rahn, 1987). For example, yolk makes up 69% by mass of the egg of the highly precocial Kiwi (*Apteryx australis*), but only 16% of that of the altricial Red-footed Booby (*Sula sula*). As most of the energy in the egg contents is in the lipid fraction and as most of the lipids are in the yolk, it follows that the energy content of precocial eggs is higher than that of altricial eggs. There is a further correlation: most of the water in the egg is in the albumen and as the yolk/albumen ratio is lower in altricial than in precocial eggs, it follows that precocial eggs have relatively little water. Conversely, in altricial eggs, the water content is relatively high (Sotherland and Rahn, 1987). The relationship between the yolk, water, and energy content of the egg and hatchling maturity is illustrated in Figure 1.

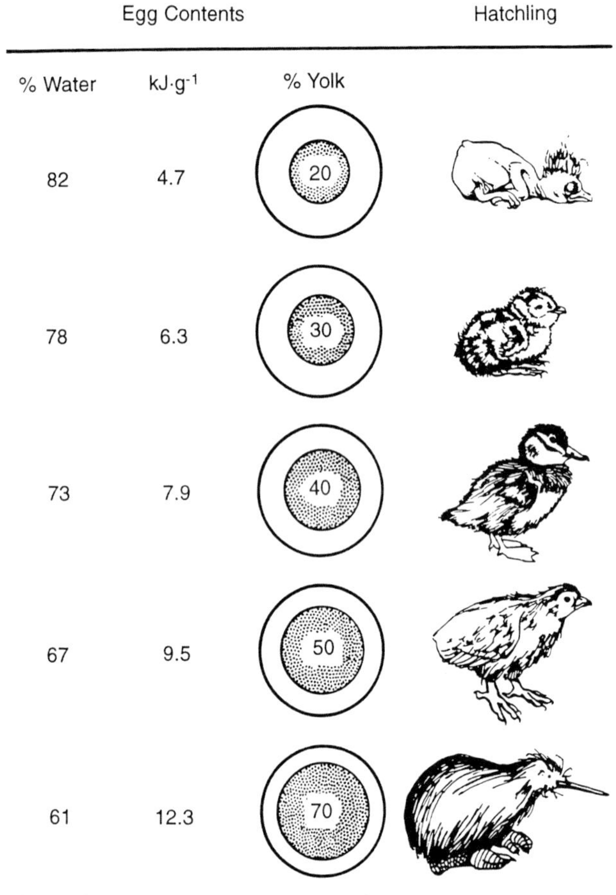

Egg Contents			Hatchling
% Water	kJ·g⁻¹	% Yolk	
82	4.7	20	
78	6.3	30	
73	7.9	40	
67	9.5	50	
61	12.3	70	

FIGURE 1 Water (%), energy (kJ · g⁻¹), and yolk content (%) of the freshly laid eggs of birds with varying degrees of hatchling maturity. Hatchling maturity increases from above downward. The species depicted are, from top: altricial Brown Creeper (*Certhia familiaris*), semiprecocial Least Tern (*Sterna albifrons*), precocial Ruddy Duck (*Oxyura jamaicensis*), precocial Mallee Fowl (*Leipoa ocellata*), precocial Kiwi (*Apteryx australis*). (From *The Condor* **89**, 48–65, 1987, with permission and courtesy of Dr. P. R. Sotherland and Dr. C. V. Paganelli.)

III. CHANGES IN THE COMPOSITION OF THE EGG DURING INCUBATION

A. Water Content

As the embryo grows within the egg the composition of the egg changes. Both yolk and albumen diminish, the yolk providing energy for the growth and maintenance of the embryo, the albumen providing protein for the embryo but also giving up water, which is lost by diffusion through the pores in the shell, to the microclimate of the egg.

Water loss from the egg is inevitable, given the presence of pores in the eggshell. The rate of water loss ($\dot{M}_{H_2O}$, mg · day⁻¹) is determined by two factors: (1) the water vapor conductance of the shell and shell membranes (G_{H_2O}, mg · day⁻¹ · torr⁻¹), and (2) the difference in water vapor pressure between the contents of the egg and the egg's microenvironment (ΔP_{H_2O}, torr; Rahn and Ar, 1974):

$$\dot{M}_{H_2O} = G_{H_2O} \cdot \Delta P_{H_2O}$$

The G_{H_2O} is largely a function of the number of pores in the shell and the shell thickness. Both the number of pores and the shell thickness increase in larger eggs (Ar *et al.*, 1974; Tullett and Board, 1977). While an increase in the number of pores increases the G_{H_2O}, an increase in shell thickness has the opposite effect because it lengthens the diffusion pathway for water vapor. As G_{H_2O} is greater in larger eggs, the increase in the number of pores with increasing egg size is the predominant factor.

The water vapor pressure difference (ΔP_{H_2O}) is the difference between the water vapor pressure of the egg contents ($P_{H_2O,egg}$) and the water vapor pressure in the microclimate of the egg ($P_{H_2O,nest}$). The $P_{H_2O,egg}$ is largely a function of egg temperature, while $P_{H_2O,nest}$ is affected by (1) the nature of the nest or the substrate on which the egg is laid, and (2) the "tightness" with which the incubation patch of the incubating bird is applied to the egg (see Section IV below) and the frequency with which the adult rises from and resettles on the egg ("nest ventilation," Rahn *et al.*, 1976; Ar and Rahn, 1978; Rahn, 1984). When the embryo fractures the shell (external pipping), the rate of water loss from the egg increases, as water vapor can then diffuse through the cracks in the shell (Whittow, 1984). Formation of a piphole results in a further acceleration of water loss from the egg. Not all of the water in the newly laid egg is lost in this way; some is retained in the embryo and in the yolk sac and part in the residual tissues (mainly membranes) left in the egg (Figure 2). In addition, water

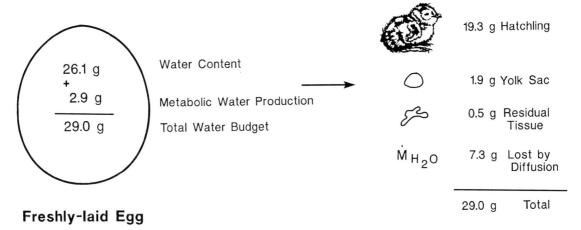

FIGURE 2 Disposition of the water contained in the freshly laid egg of the Bonin Petrel (*Pterodroma hypoleuca*) and of the water produced by metabolism. Redrawn from Pettit et al. (1984).

is actually produced when the fat in the yolk is oxidized ("metabolic water," Figure 2). The water lost from the egg is replaced by air, creating the aircell at the blunt pole of the egg. During "internal pipping," the embryo punctures the chorioallantoic and internal shell membranes and it is then able to begin pulmonary ventilation, rebreathing the air cell gas. This allows it to begin the transition from diffusive respiration through the chorioallantois to convective breathing through the lungs.

The water loss from the egg over the entire incubation period amounts to 18% of the mass of the freshly laid egg (Rahn, 1984).

B. Energy Content

Most of the energy contained in the freshly laid egg is incorporated into the tissues of the hatchling. Of the remaining energy, the greater part is utilized by the embryo to synthesize its new tissues during growth and to meet the physiological demands for energy (maintenance) of the living embryo during incubation (Figure 3). This energy usage increases as the embryo grows and it is reflected in an increase in the oxygen consumption ($\dot{M}_{O_2}$) of the embryo (Figure 4). The $\dot{M}_{O_2}$ increases greatly after the egg has pipped. There is a difference in the pattern of embryonic $\dot{M}_{O_2}$ in species with a preco-

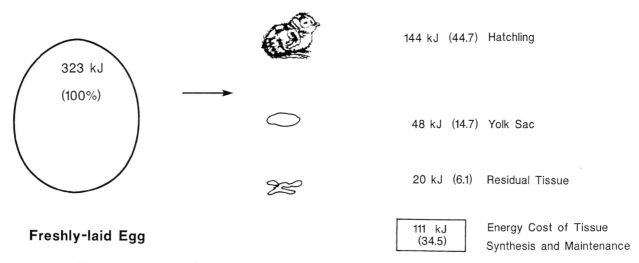

FIGURE 3 Disposition of the energy contained in the freshly laid egg of the Bonin Petrel. Redrawn from Pettit et al. (1984).

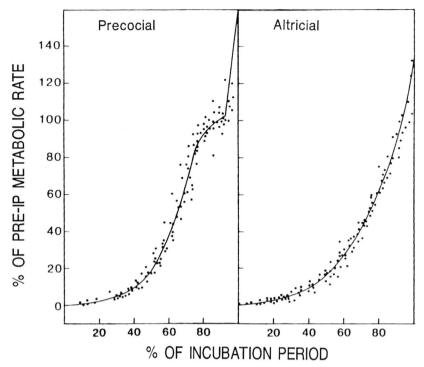

FIGURE 4 Metabolic rate as a function of duration of incubation in eggs of precocial and altricial birds. Metabolic rate is expressed as a percentage of the rate (Pre-IP) just prior to internal pipping (penetration of the egg's aircell by the embryo's beak, permitting the lungs to fill with air). Incubation duration is expressed as a percentage of the total incubation period. (From Vleck and Vleck (1987), *J. Exp. Zool., Suppl.* **1,** 111–125, with permission.)

cial mode of development and in those that hatch in a very immature state (altricial) (Figure 4). In the latter, $\dot{M}_{O_2}$ increases at an increasing rate throughout incubation; in precocial species, the rate of increase in $\dot{M}_{O_2}$ diminishes before the egg is pipped and then increases again after pipping (Figure 4). What is left of the energy is retained in the yolk sac and residual tissues (Figure 3). Much of the energy in the yolk sac is drawn on by the nestling after hatching.

IV. HEAT TRANSFER TO THE EGG

The transfer of heat from its body to the egg is the most important contribution of the incubating adult to incubation. Surprisingly little is known about it, partly because it is difficult to measure (Turner, 1991). Most birds develop a seasonal bare patch of skin, the brood patch, on part of the thorax and abdomen. This is in direct contact with the egg(s) permitting a greater rate of heat transfer than if the patch were covered with plumage. Accompanying the loss of feathers is an increase in the size and number of blood vessels in the bare skin. The adult can adjust the rate of heat transfer by standing or leaving the egg, but also by the closeness with which the bird applies its patch to the egg. In addition, the bird responds physiologically to variations in the temperature of the egg, increasing its metabolic heat production in response to cooling of the egg (Tøien et al., 1986; Rahn, 1991). The heat production of the bantam hen increases 4.8 times after sitting on a clutch of eight eggs at a temperature of 20°C. The amount of heat transferred to the eggs is directly proportional to the increase in heat production (Tøien et al., 1986). The efficiency of heat transfer to the eggs diminishes with decreasing ambient temperature and clutch size. Biebach (1986) pointed out that heat stored in the incubating bird while flying and foraging (i.e., while not incubating the eggs) is transferred to the eggs on return to the nest. If the egg is very cold, "cold vasodilatation" occurs in the brood patch, increasing the patch blood flow and temperature and, as a result, the heat transfer to the egg (Midtgard et al., 1985). The brood patch temperature varies from 34.9°C in the Bonin Petrel (*Pterodroma hypoleuca*) to 42.4°C in the Dusky Flycatcher (*Empidonax oberholseri*). The brood patch temperature is 1.1–5.5°C higher than the egg temperature in different species (Rahn, 1991).

At the beginning of incubation, heat flows *through* the egg by conduction and the surface of the egg diametrically opposed to the incubation patch may be 4°C or more below the incubation patch temperature (Rahn, 1991). As incubation proceeds this temperature difference diminishes because the embryo's developing circulation assists in the distribution of heat and its increasing metabolism is an additional source of heat (Turner, 1987; Rahn, 1991). The result is that the temperature of the egg remote from the brood patch approaches that of the brood patch, which remains relatively constant throughout incubation. The effect of blood flow on heat flow is more important in large eggs than in small ones (Tazawa et al., 1988a).

The main barrier to heat loss *from* the egg is a thin layer of air immediately adjacent to the shell—the boundary layer (Sotherland et al., 1987). If the egg is in a nest, the nest itself imposes an additional resistance to heat loss.

There is a striking variety in the shape, size, and orientation of the nest (Skowron and Kern, 1980). In the Hawaiian honeycreeper Amakihi (*Loxops v. virens*), Whittow and Berger (1977) presented evidence that the thermal conductance of the nest is similar to that of the tissues and plumage of the bird. Kern and Van Riper (1984) produced the additional information that the thermal conductance of the wall of the Amakihi's nest is slightly higher than that of the floor and that it is correlated with the altitude at which the birds nest. Nests not only protect against excessive cooling of the eggs or chicks, buy may also mitigate the effects of solar radiation. An egg laid directly on the ground loses heat to the substrate and also to the surrounding air.

V. DEVELOPMENT OF PHYSIOLOGICAL FUNCTIONS

A. Respiration

The egg with its hard shell lacks ventilatory movements and thus there is no convective gas exchange with atmospheric air until the lungs begin to function. There are three different gas exchangers in the egg during embryonic development; the area vasculosa, the chorioallantois and the lungs. Figure 5 shows the growth rate of the functional surface area of the area vasculosa and the chorioallantois (Ackerman and Rahn, 1981) and the oxygen uptake of the egg in the domestic fowl. The area vasculosa is a well-vascularized region of the yolk sac which fans out from the embryo and surrounds the yolk by rapid growth during days 3 to 5 of incubation. The blood vessels of the yolk sac connect with the dorsal aorta of the embryo by day 2 of incubation and blood begins to circulate through the embryo and the area vasculosa. The fine reticulation of the vitelline circulatory system plays the role of the main gas exchanger until the chorioallantois makes contact with the inner shell membrane around day 6 (Ackerman and Rahn, 1981). After that, there is a transition of respiratory function from the area vasculosa to the chorioallantois. The area vasculosa presumably ceases to function as the gas exchanger prior to day 8 of incubation.

Beginning on the fifth day of incubation, the mesenchyme covering the fundus of the allantoic sac comes into contact with the mesenchyme lining the chorion. The two membranes begin to fuse and the growing allantoic sac flattens out beneath the chorion, which lies close to the eggshell. The outer limb of the flattened allantois, composed of the cohesive chorion and allantois, is the chorioallantois. The growth rate of the chorioallantois is fast (Figure 5). It grows to almost the same size as the embryo by the time it makes contact with the shell membranes and begins to replace the respiratory function of the area vasculosa on day 6. By day 12, the chorioallantois extends to envelop the contents of the whole egg, lining the entire surface of the inner shell membrane.

The outer surface of the chorioallantois is well vascularized (Wangensteen et al., 1970/71; Tazawa and Ono, 1974; Wangensteen and Weibel, 1982). Early in incubation the capillaries lie on the mesenchymal surface of the chorioallantoic membrane. They begin to migrate through the ectoderm on day 10 and lie on its thin layer late in incubation. In addition, relocation of the nuclei of the chorioallantoic capillaries occurs (Mayer et al., 1995). The endothelial nuclei randomly distribute around the capillary lumen early in incubation and they are located progressively on the portion of the capillaries away from the shell membrane after the chorioallantois envelops the whole contents of the egg. Together with capillary migration the relocation of endothelial nuclei results in progressive thinning of the diffusion pathway for gases between the interstices of the inner shell membrane (air space) and the capillary blood.

The function of the chorioallantois as a gas exchanger lasts until almost the final stages of embryonic life. The chorioallantois degenerates when embryos pip the chorioallantoic membrane and the inner shell membrane with the beak (internal pipping). Then, the chick breaks the shell and breathes atmospheric air by lung ventilation (external pipping). As the lungs are aerated, they take over the respiratory function of the chorioallantois (Figure 5). The developmental stage of the embryo is divided into two; prenatal stage until internal pipping occurs and paranatal (perinatal) stage from internal pipping to hatching.

The oxygen uptake of the egg ($\dot{M}_{O_2}$) increases geometrically as the chick embryo grows rapidly during the first 2 weeks of incubation. Then, the increase in $\dot{M}_{O_2}$ becomes asymptotic in the chicken, reaching a plateau

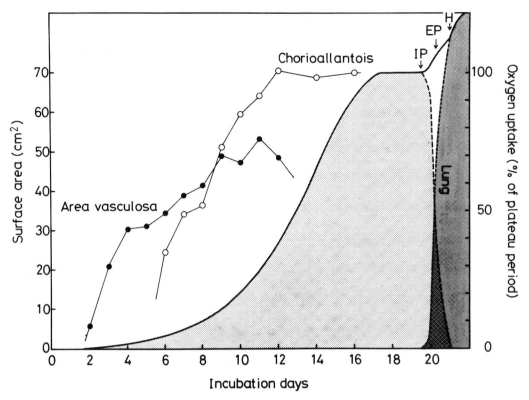

FIGURE 5 Daily changes in functional surface area of the area vasculosa and the chorioallantoic membrane (left ordinate), and developmental pattern of O_2 uptake ($\dot{M}_{O_2}$) during prenatal period (until internal pipping, IP) and paranatal period (from IP to hatching, H) (right ordinate). External pipping (EP) ocurrs during the paranatal stage. The $\dot{M}_{O_2}$ is drawn diagramatically and the plateau value represented as 100%. The lightly shaded area indicates the $\dot{M}_{O_2}$ by diffusion through the area vasculosa/chorioallantois and the heavily shaded area, that by the lungs. (Reprinted from *Respir. Physiol.* **45**, R. A. Ackerman and H. Rahn, In vivo O_2 and water vapor permeability of the hen's eggshell during early development, pp. 1–8, Copyright (1981), with permission from Elsevier Science.)

prior to pulmonary respiration and increasing again as a result of lung ventilation (Figure 5). The developmental pattern of $\dot{M}_{O_2}$ during the late incubation and hatching periods is different in precocial and altricial birds and even among altricial or precocial birds (see Section III,B and Hoyt and Rahn, 1980; Prinzinger and Dietz, 1995). However, in all birds, the gas exchange of embryos before pipping the chorioallantois or shell (i.e., prenatal gas exchange) takes place by diffusive transport between the external environment and capillary blood, across a porous shell, two shell membranes and a membranegas exchanger (Wangensteen and Rahn, 1970/71; Wangensteen et al., 1970/71).

When the chorioallantois envelops all contents of the egg, the gas exchange by diffusive transport is expressed as follows,

$$\dot{M}_{O_2} = G_{O_2} \cdot (P_{I_{O_2}} - P_{A_{O_2}})$$
$$= D_{O_2} \cdot (P_{A_{O_2}} - P_{\bar{C}_{O_2}})$$

where $\dot{M}_{O_2}$ oxygen flux (ml · day^{-1}); G_{O_2} = oxygen conductance of the shell and shell membranes (ml O_2 day^{-1} · torr^{-1}); D_{O_2} = diffusing capacity of the chorioallantoic membrane and capillary blood (ml O_2 day^{-1} · torr^{-1}); $P_{I_{O_2}}$ = effective ambient oxygen tension (torr); $P_{A_{O_2}}$ = air space oxygen tension (torr); and $P_{\bar{C}_{O_2}}$ = mean capillary oxygen tension (torr).

The G_{O_2} is a function of the shell geometry (effective pore area, A_p, and thickness of the shell and shell membranes, L), oxygen diffusion coefficient (d_{O_2}) and the inverse product of the gas constant (R) and absolute temperature (T),

$$G_{O_2} = [(A_p/L) \cdot d_{O_2}]/RT.$$

The G_{O_2} is related to G_{H_2O} by the diffusion coefficient ratio ($d_{O_2}/d_{H_2O} = 0.23/0.27$) and G_{O_2} in ml · day^{-1} · torr^{-1} is derived from G_{H_2O} in mg · day^{-1} · torr^{-1} by multiplying by a factor of 1.06. Because the G_{O_2} in air is constant during the last half of prenatal incubation, the $P_{A_{O_2}}$ decreases as the embryo grows and consumes more O_2. The increased difference of O_2 tension between the ambient air and air space is the driving force to meet the increased O_2 demand by the developing embryo.

In naturally laid chicken eggs, the variability of shell conductance is large, higher than that of egg mass. The mass-specific $\dot{M}_{O_2}$, measured on days 16–19 of incubation, is at a maximum at medium conductance, decreasing at both lower and higher conductance (Visschedijk et al., 1985). The maximum $\dot{M}_{O_2}$ at medium conductance values is considered to be optimal for chick development and the decrease in $\dot{M}_{O_2}$ at both higher and lower conductance as a sign of compromised development. In fact, when the G_{O_2} is altered widely by partially covering the shell with impermeable material and by partially removing the shell over the air cell at the beginning of incubation, the $\dot{M}_{O_2}$ of 16-day embryos increases hyperbolically with increasing G_{O_2}, reaching a maximum at the control G_{O_2} of intact eggs and decreasing with further increase in G_{O_2} (Okuda and Tazawa, 1988). The wet mass of embryos changes similarly with G_{O_2}; it is lowered at both low and high G_{O_2} and maximum at natural G_{O_2}. It is suggested that embryonic development is retarded by lowered P_{O_2} of arterialized blood at decreased G_{O_2}. At increased G_{O_2}, on the other hand, the excess water loss accounts for the reduced mass of embryos. However, when G_{O_2} is increased by removing the shell, for a short period (e.g., 5 hr) in order to obviate the effect of excess water loss, the $\dot{M}_{O_2}$ does not increase significantly, contrary to expectation (Tazawa et al., 1988b). This is due to the decreased D_{O_2} of the inner diffusion barrier brought about by removing the eggshell.

Because the d_{O_2} affects the shell conductance, the G_{O_2} can be changed by replacing N_2 in air with an inert background gas whose density is different from that of N_2 (e.g., He and SF_6) (Erasmus and Rahn, 1976; Ar et al., 1980). Thus, the gas exchange of the egg can be manipulated by changing G_{O_2} with He or SF_6. The d_{O_2} is also inversely related to atmospheric pressure (P_B); thus the gas exchange of the egg is increased at altitude because the shell conductance increases in inverse proportion to the change in P_B. The reduction of G_{O_2} in eggs laid by birds incubating at altitude occurs as a natural adaptation to altitude (Rahn et al., 1977).

As the $P_{A_{O_2}}$ decreases with embryonic development (Ackerman and Rahn, 1981), the $P_{\bar{c}_{O_2}}$ also decreases almost in parallel with it (Figure 6). The O_2 flux from the air space to the hemoglobin of capillary blood (inner diffusion barrier) is facilitated mainly by an increase in the diffusing capacity (D_{O_2}) (Figure 6). It should be noted that the developmental patterns of $\dot{M}_{O_2}$ and D_{O_2} are similar. The D_{O_2} of the inner diffusion barrier is expressed by

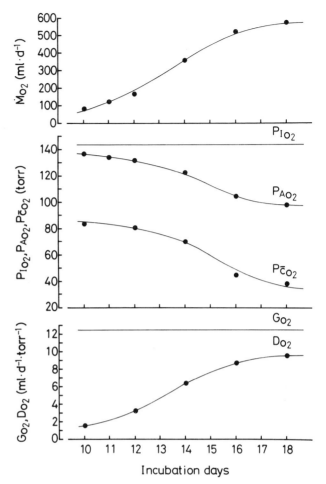

FIGURE 6 Developmental patterns of oxygen uptake ($\dot{M}_{O_2}$), O_2 partial pressure in air ($P_{I_{O_2}}$), air cell ($P_{A_{O_2}}$) and capillary ($P_{\bar{c}_{O_2}}$, mean capillary O_2 pressure), and shell conductance (G_{O_2}) and diffusing capacity of inner barrier (D_{O_2}) during the last half of prenatal incubation. The data of $\dot{M}_{O_2}$, $P_{I_{O_2}}$, $P_{A_{O_2}}$, and G_{O_2} are based on the paper by Ackerman and Rahn (1981). The D_{O_2} is from the paper by Tazawa and Mochizuki (1976). $P_{\bar{c}_{O_2}}$ was calculated from $\dot{M}_{O_2}$, $P_{A_{O_2}}$, and G_{O_2}.

$$D_{O_2} = V_c \cdot F_{\bar{c}_{ox}} \cdot \text{Hct}$$
$$= \dot{Q}_a \cdot t_c \cdot F_{\bar{c}_{ox}} \cdot \text{Hct},$$

where V_c is the capillary volume of the chorioallantoic gas exchanger (in μl), $\dot{Q}_a$ is the blood flow through it (in ml · min^{-1}), t_c is the contact time of erythrocytes with O_2 when they pass through the chorioallantoic capillaries (in sec), $F_{\bar{c}_{ox}}$ is the mean corpuscular oxygenation velocity during the contact time (in sec^{-1} · torr^{-1}), and Hct is the hematocrit. Thus, the D_{O_2} is increased by an increase in V_c, which depends on the $\dot{Q}_a$ and t_c (Figure 7). While the $\dot{Q}_a$ increases about sixfold during the period from day 10 to day 18, the t_c halves, probably because of shortening of the blood circulation time, with cardiac output increasing more than the total blood volume. Consequently, the V_c increases even after the

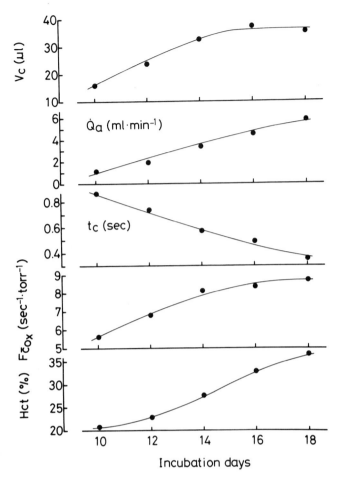

FIGURE 7 Developmental patterns of variables governing the diffusing capacity of the inner diffusion barrier (D_{O_2} shown in Figure 6); chorioallantoic capillary blood volume (V_c), allantoic blood flow ($\dot{Q}_a$), contact time (t_c), mean corpuscular oxygenation velocity ($F_{\bar{c}_{ox}}$), and hematocrit value (Hct). Data from paper by Tazawa and Mochizuki (1976).

chorioallantois spreads over the whole surface of the inner shell membrane on day 12 and reaches a maximum volume on days 14 and 15. Nevertheless, the D_{O_2} increases further after day 14. This is due to the increase in the velocity of blood oxygenation in the chorioallantoic capillary bed. The oxygenation velocity of blood passing through the chorioallantoic capillary depends on the $F_{\bar{c}_{ox}}$ and Hct. Both variables, particularly Hct, increase after the chorioallantoic membrane spreads over the inner shell membrane (Figure 7). Toward the end of prenatal development, the increase in D_{O_2} slows down and the $\dot{M}_{O_2}$ reaches a plateau (Figure 6).

B. Acid–Base Regulation

While embryos consume O_2, they produce CO_2 which is partially dissolved in the blood and body fluids but most of which must be eliminated through the eggshell to atmosphere. As with $\dot{M}_{O_2}$, CO_2 elimination ($\dot{M}_{CO_2}$) depends upon the eggshell conductance (G_{CO_2}) and CO_2 pressure difference between air cell ($P_{A_{CO_2}}$) and atmosphere ($P_{I_{CO_2}}$).

$$\dot{M}_{CO_2} = G_{CO_2} \cdot (P_{A_{CO_2}} - P_{I_{CO_2}}).$$

The G_{CO_2} is a function of eggshell geometry (Ap and L), CO_2 diffusion coefficient and the inverse product of R and T. The $P_{I_{CO_2}}$ is approximately zero if the egg is in air.

As embryos develop and produce more CO_2, CO_2 accumulates in the egg, $P_{A_{CO_2}}$ is increased (and thus arterialized blood P_{CO_2}, $P_{a_{CO_2}}$), and blood pH is lowered.

However, the decrease in pH during embryonic development is slowed late in incubation. Although the plasma bicarbonate is increased with development, the amount of increase is more than would be expected from changes in pH and the buffer value. The pH change due to an accumulation of CO_2 is mitigated by an increase in nonrespiratory bicarbonate (Tazawa, 1986, 1987).

In nature, the natural variations in eggshell conductance cause large differences in $P_{a_{CO_2}}$ among eggs, but the blood seems able to keep pH variations to a minimum (Tazawa et al., 1983). In eggs whose shell conductance is low, the Hct (and hemoglobin) increases in response to hypoxia. In part this must be responsible for the minimum change in pH. When the shell conductance was lowered by covering partially the eggshell with impermeable material, $P_{a_{CO_2}}$ increased and it had already reached a plateau at the time the blood was sampled, 10 min after reducing conductance (Tazawa, 1981a). Concurrently, blood pH decreased, and subsequently it changed toward the control level, though incompletely, while $P_{a_{CO_2}}$ was maintained at a constant value. The respiratory acidosis produced by impeding elimination of CO_2 induced a nonrespiratory compensation during the first 3 to 6 hr (Tazawa, 1981a). Exposure of the egg to a SF_6/O_2 gas mixture has the same effect on the shell conductance. The respiratory acidosis occurs soon after exposure to SF_6/O_2 gas mixture (Tazawa et al., 1981).

The time course of the acid–base changes after respiratory disturbances examined in four chick embryos is presented on the pH–[HCO_3^-] diagram (Figure 8). The embryo shown by, for instance, left-half closed circles undergoes noncompensated respiratory acidosis 30 min after adding CO_2 to the environment (marked by 30 min), which is partially compensated after 3 hr (marked by 3 hr). The increase in [HCO_3^-] is not large enough to resume the control acid–base status. Then air exposure resumes (marked by air), and 30 min after

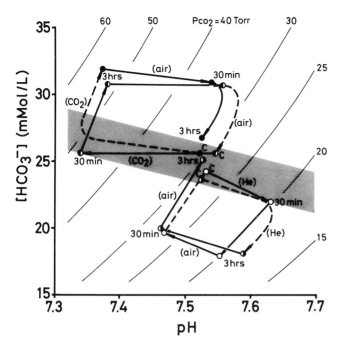

FIGURE 8 Time course of changes in respiratory acid–base disturbances produced by 4% CO_2/21% O_2 in N_2 (referred to as CO_2) or 21% O_2 in He atmosphere (He) and recovery from the disturbances by switching these foreign gas mixtures to air (air). Four embryos were tested in this experiment, shown by different four symbols and connected with solid and dotted lines. c indicates the control acid–base status of the individual embryo in air. (From Tazawa, H., 1982, J. Appl. Physiol. **53**, 1449–1454. With permission.)

air exposure (marked by 30 min) blood P_{CO_2} almost recovers air control value. It is shown by other embryos marked by closed circles that the acid–base equilibrium returns to the control level 3 hr after exposure to air. Nonrespiratory compensation for the respiratory alkalosis produced by exposing the egg to a He/O_2 atmosphere (marked by He) is also shown.

The chick embryo also reacts promptly to metabolic disturbances in the acid–base balance; this is shown by the time course of changes in metabolic acid–base alterations made by infusion of electrolyte solution ($NaHCO_3$ and NH_4Cl) (Figure 5 in Tazawa, 1982). For instance, the infusion of 15 μl 1M $NaHCO_3$ increases plasma [HCO_3^-] and blood P_{CO_2}. Because the embryo has no convective ventilation, the metabolic disturbance is not subjected to respiratory compensation. The increase in P_{CO_2} 1 hr after infusion indicates that the infused HCO_3^- is partly eliminated as dissolved CO_2. The acid–base status returns to control values 6 hr after infusion. Besides elimination of CO_2 from the chorioallantois, the increased fluid volume and hypertonicity of the infused $NaHCO_3$ solution are partially responsible for the decrease in [HCO_3^-]. Additionally, penetration of HCO_3^- into the intracellular space and urinary excretion of bicarbonate contribute to the regulation (Tazawa, 1982; 1986).

C. Cardiovascular Function

The primordial chick's heart is a paired tubular structure that soon becomes a single tube. It begins to elongate more rapidly than the pericardial cavity containing it. The limitation imposed upon the growing heart by lack of space forces the tubular heart to bend. It represents only the ventricle and the bulbus on days 1.5–2 of incubation. The impact of blood streams upon the inner surface of the contorted tube is another force forming the external configuration and internal structure of the heart. The structural alterations separate atrium from ventricle, ventricle from aorta, and the left from the right chambers; this takes place during the period from day 3 to day 8 of incubation. The chick's heart begins to beat at about 30 hr of incubation; sometimes the primordial double heart beats asynchronously at about 20 hr. Blood begins to circulate after about 40 hr of incubation when the connections between the dorsal aorta and the vessels of the yolk sac complete the circuit.

As chick embryos grow, their mass increases in a geometrical fashion until growth rate slows down during the last stages of prenatal development (Romanoff, 1967; Van Mierop and Bertuch, 1967; Tazawa et al., 1971; Lemez, 1972; Clark et al., 1986; Haque et al., 1996; Figure 9a). The geometrical increase in embryo wet mass until day 18 of incubation is well expressed by a power function of incubation time. For the mean values shown in Figure 9a, the following equation was derived,

$$\text{body} = 0.24 \cdot I^4,$$

where body indicates embryo body mass (in mg) and I, incubation time (in days). The mass of the heart also increases geometrically with incubation time (Romanoff, 1967; Clark et al., 1986). The growth of the heart relative to that of whole body is greater during the early than during the later period of embryonic development. The ratio of the heart mass to the whole body mass falls from 1.8% on day 4 to 0.7 % on day 18 of incubation.

The blood volume also increases as a power function of incubation time (Yosphe-Purer et al., 1953; Barnes and Jensen, 1959; Lemez, 1972; Kind, 1975; Figure 9b), but the rate of the increase slows compared with that for embryo growth. The mass (embryo)-specific blood volume decreases rapidly during the early period of incubation and then decreases slowly during the last half of embryo growth.

As both the heart mass and the blood volume increase, the stroke volume of the heart increases (Figure 9c). The stroke volume was determined by taking moving pictures of the exposed heart with a cine camera (Hughes, 1949; Faber et al., 1974). During the early period of development (3 to 5 days), the stroke volume was found to depend on blood volume and increases in

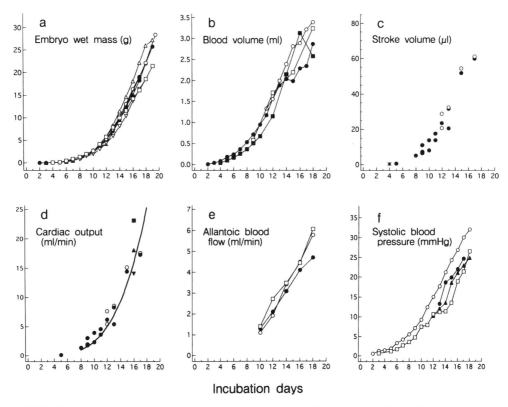

FIGURE 9 Developmental patterns of embryo wet mass (a), total blood volume (b), cardiac stroke volume (c), cardiac output (d), allantoic blood flow (e), and arterial systolic blood pressure (f) during the prenatal stage. Symbols connected by solid lines in a, b, e, and f indicate the developmental patterns cited from the papers shown in the text.

parallel with embryonic growth (Faber et al., 1974). Even after the 2nd week of development, the stroke volume seems to increase as a power function of incubation time (Hughes, 1949).

For early embryos, the dorsal aortic blood flow was determined with a doppler velocity meter (Hu and Clark, 1989). It increases in a power function of incubation time during the period from day 2 and day 6. The mass (embryo)-specific value results in 0.5-1 ml · min^{-1} · g^{-1}. The mass-specific cardiac output of 3–5 days embryos, determined from the stroke volume, is similar: 1 ml · min^{-1} · g^{-1} (Faber et al., 1974). The cardiac output, calculated from the stroke volume of young embryos shown in Figure 9c (Hughes, 1949), is presented in Figure 9d. The cardiac output of 16-day chick embryos, estimated by model analysis and blood O_2 measurement (White, 1974; Rahn et al., 1985; Tazawa and Takenaka, 1985), is also shown in Figure 9d. The mass-specific cardiac output of 16-day embryos was 0.9 to 1.5 ml · min^{-1} · g^{-1}. The cardiac output of young/late embryos seems to increase as a power function of incubation day, while the mass-specific cardiac output results in a narrow range from 0.5 to 1.5 ml · min^{-1} · g^{-1} throughout prenatal development. Eventually, the cardiac output may increase almost in parallel with embryonic growth. If it is assumed that the mass-specific cardiac output during the last 2 weeks of prenatal development is 1 ml · min^{-1} · g^{-1}, the cardiac output of young embryos is related to incubation time as follows:

$$CO = 0.24 \cdot I^4,$$

where CO is cardiac output (in μl · min^{-1}). This power function equation is presented by the solid line in Figure 9d.

In contrast to the difficulty of determination of cardiac output, the blood flow through the chorioallantoic gas exchanger can be determined from the $\dot{M}_{O_2}$ and blood gas analysis (Tazawa and Mochizuki, 1976, 1977; Bissonnette and Metcalfe, 1978) (Figure 9e). It increases with embryonic growth, but the mass-specific value decreases from about 0.5 ml · min^{-1} · g^{-1} on day 10 to 0.25 ml · min^{-1} · g^{-1} on day 18. Although the cardiac output tends to increase in parallel with embryonic growth, its partition to the chorioallantoic gas exchanger decreases as embryos grow. Probably, on day 10, about half of the cardiac output goes to the chorioallantoic

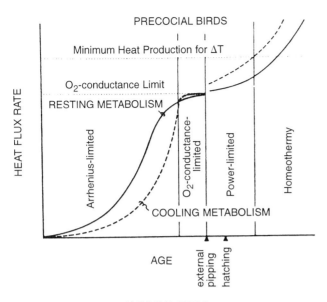

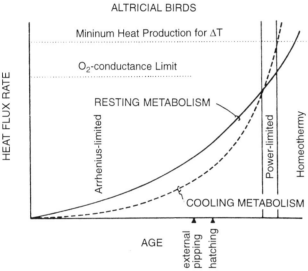

FIGURE 12 Models of the development of homeothermy in precocial and altricial birds. "Minimum Heat Production for ΔT" equals the amount of heat to keep the egg temperature warmer than ambient temperature by ΔT, and, by definition, homeothermy occurs when the embryo heat production reaches this level. (Reprinted by permission from Tazawa *et al.*, 1988c, *Comp. Biochem. Physiol.* **89A**, 125–129. Copyright 1988 by Elsevier Science Inc.)

tion mechanisms occur during paranatal period for thermoregulation of hatchling (Nichelmann *et al.*, 1994).

VI. REQUIREMENTS AND PROCEDURES FOR INCUBATION

A. Incubation Period

The duration of incubation (I, in days) is greater the larger the egg. The relationship can be represented by the following equation (Rahn and Ar, 1974):

$$I = 12 \, egg^{0.22}$$

where *egg* is mass of the freshly-laid egg in g. In addition, as a first approximation, the product of incubation time and $\dot{M}_{O_2}$ at the plateau stage is proportional to egg mass (Rahn *et al.*, 1974):

$$I \cdot \dot{M}_{O_2} = c \cdot egg,$$

where c is a constant. This indicates that for a given egg mass an egg which consumes less O_2 at the plateau stage needs a longer incubation period. Furthermore, because the $\dot{M}_{O_2}$ at the plateau stage is matched to the shell conductance, then for a given egg mass an egg with lower shell condutance needs a longer incubation time.

Apart from these general trends, eggs that are abandoned by the incubating parent bird for a period of time cool, and this results in slower embryonic growth and a longer incubation period. Some birds, notably tropical seabirds and those of the Order Procellariiformes, have much longer incubation periods than that suggested by the general relationship (Whittow, 1980), even though they are incubated continously.

B. Preincubation Egg Storage

Avian incubation is unique in terms of preincubation egg storage. Many precocial birds which lay multiple eggs in a clutch start their incubation with the penultimate or ultimate eggs. Consequently, first laid eggs are stored in the nest until incubation starts.

Egg storage is commonly practiced in artificial incubation of domesticated birds. If the storage temperature for freshly laid chicken eggs is kept below the physiological zero (25–27°C), dormancy of the embryo can be maintained and fertile eggs can be stored for several days without a major loss of hatchability (Butler, 1991; Wilson, 1991). The optimal temperature for 3–7 days storage of chicken eggs is 16–17°C and it drops to 10–12°C if eggs are stored for more than 7 days (Butler, 1991; Wilson, 1991). However, prolonged preincubation egg storage results in malformations and retarded growth of the early embryo, decreased hatchability, increased incubation period, and it even influences growth of the hatchlings. These deleterious effects of preincubation storage are related to not only the length of storage, but also the environmental and physical conditions, such as temperature, relative humidity, atmospheric gas composition, orientation, and positional changes during storage. In addition, parental age changes these effects and varied development of embryos at oviposition results in differences in the ability of the blastoderm to withstand storage.

Prolonged preincubation egg storage also affects the physiological function of developing chick embryos

(Haque et al., 1996). Figure 13 shows comparisons of the developmental patterns of $\dot{M}_{O_2}$ between stored and unstored (control) eggs. The eggs were stored at 10–11°C for 20 days (broken lines) and 30 days (dotted lines) before incubation at 38°C. The $\dot{M}_{O_2}$ was measured daily from day 12 of incubation in six eggs for control (solid lines) and stored eggs. While the developmental patterns of $\dot{M}_{O_2}$ are consistent between the six eggs in the control group, those in the 20- and 30-day storage groups vary between eggs. The $\dot{M}_{O_2}$ of several eggs in the storage groups decreases during the last days of incubation, resulting in the death of the embryos. The levels of $\dot{M}_{O_2}$ are markedly lower in stored eggs than in the control eggs, but they reach a plateau before day 20 of incubation. Thus, the developmental pattern of $\dot{M}_{O_2}$ is not only shifted to the right, but also depressed during the last days of incubation; depression of the $\dot{M}_{O_2}$ curve becomes severe as the storage duration increases. It is suggested that the prolonged preincubation storage decreases the $\dot{M}_{O_2}$ of embryos as it retards embryonic growth and, in addition, the prolonged storage further depresses the $\dot{M}_{O_2}$ during the late incubation periods by other unknown factors. This is also shown by additional experiment on embryo mass and $\dot{M}_{O_2}$ of stored eggs (Haque et al., 1996). The $\dot{M}_{O_2}$ of embryos in control and 20-day storage groups increases in a curvilinear relationship with increasing embryo dry mass. The rate of increase in $\dot{M}_{O_2}$ in both groups decreases at embryo dry mass greater than 0.5 g and becomes asymptotic at 3–4 g dry mass. The asymptotic value is significantly lower in the 20-day storage group than in the control group (Haque et al., 1996).

Preincubation storage causes blastoderm shrinkage and a decrease in the rate of embryonic development during the early incubation period (Mather and Laughlin, 1979). The changes in embryo morphology result in preincubation or early incubation mortality (Arora and Kosin, 1966). Late incubation failure observed in some prolonged storage eggs may be partly due to malfunctions or malformations in embryonic development as a result of irreversible changes in the blastoderm. The prolonged storage may result in defects at the cellular or tissue level, causing embryos not to sustain or increase their metabolism to complete hatching during the last days of incubation.

The developmental patterns of heart rate in chicken embryos are also changed by preincubation egg storage (Haque et al., 1996). They are flattened compared with those of the control eggs. However, the average HR during the second half of incubation is the same as in the control eggs. As a result, the O_2 pulse (O_2 uptake every heartbeat) is markedly lowered by preincubation storage throughout the last half of incubation and especially during the last days of incubation. The O_2 transport by the blood may also be affected by the preincubation storage.

C. Egg Turning

In natural incubation, parent birds actively move their eggs in the nest (egg-turning). The importance of egg-turning has been examined during artificial incubation of domesticated birds. The critical period for lack of egg-turning ranges from day 3 to day 7 of incubation in the domestic fowl (New, 1957; Deeming, 1984a). A minimum number would be about 3 times a day and more than 24 times a day is unnecessary for producing maximum hatchability. Lack of egg-turning produces detrimental effects not only on hatchability, but also incubation period, subembryonic fluid formation, development of the chorioallantois and growth of embryos (New, 1957; Tazawa, 1980; Deeming et al., 1987; Tullett and Deeming, 1987; Deeming, 1989a,b,c).

Failure to turn eggs during incubation also produces adverse effects on gas exchange through the chorioallantoic membrane (Tazawa, 1980). It retards the movement of albumen into the amnion and possibly absorption from the amniotic fluid. The unabsorbed albumen which becomes more viscid and heavy by losing water early in incubation sinks towards the lower end of the egg and is left there. The chorioallantois fails to fold around the unabsorbed albumen. The albumen is interposed between the chorioallantoic membrane and inner shell

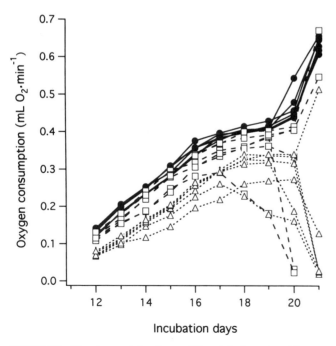

FIGURE 13 Developmental patterns of O_2 uptake in unstored (control) eggs (closed circles) and 20-day (open squares) and 30-day (triangles) storage eggs.

membrane. As a result, the interposition of the albumen reduces the gas exchange of the chorioallantois, causing a pronounced fall in the arterialized blood P_{O_2} of late embryos, which is accompanied by an increase in hematocrit value.

The retardation of gas exchange due to lack of egg turning reflects on the developmental patterns of $\dot{M}_{O_2}$ during the last half of incubation (Pearson *et al.*, 1996). The $\dot{M}_{O_2}$ of unturned eggs decreases with incubation in comparison with turned eggs. Failure to turn eggs retards the growth of the embryo, but only after day 12 of incubation (Deeming, 1989a).

D. Ambient Temperatures and Embryonic Tolerance

While freshly laid chicken eggs can be stored at a low temperature, for maintaining dormancy of embryos, once incubation starts, the ambient temperature (T_a) must be kept within a certain range so that the embryo temperature is maintained adequately and cell proliferation may proceed. In artificial incubation of chicken eggs at constant temperature, the eggs are hatched at a temperature ranging from 35.5 to 39.5°C (Romanoff, 1960). The difference in incubation temperature (ambient temperature) causes different incubation periods and variations in embryo size and organ growth, changing the developmental patterns of $\dot{M}_{O_2}$ (Zhang and Whittow, 1992). The ambient temperature outside this range kills all embryos before hatching. In natural incubation, embryos are exposed to a variable T_a. Prolonged deviations from an adequate T_a can be fatal to developing embryos. The critical lethal ambient temperature and exposure time may be different among species and between developmental stages of embryos even in the same species. The tolerance limits of developing embryos to acutely lowered and increased T_a have been investigated in chickens in reference to their HR (Tazawa and Rahn, 1986; Ono *et al.*, 1994).

When 10-day-old embryos are exposed to a T_a of 28 or 18°C, the HR decreases in an exponential fashion to reach plateau values during the 2–3 hr of exposure (Tazawa and Rahn, 1986). The plateau values are about 100 bpm at 28°C and 30 bpm at 18°C, which are maintained until irreversible cardiac arrest occurs at about 100 and 60 hr after exposure to 28 and 18°C, respectively. Thus, 10-day-old embryos survive exposure to 28°C for no less than 4 days and to 18°C for about 2.5 days. Reduction of T_a by 30°C from the optimal incubation temperature forces the heartbeat of 10-day-old embryos to cease about 3 hr later. However, the cardiac arrest at this low T_a (8°C) does not mean the death of embryos. Ten-day-old embryos survive the low T_a, without a heartbeat, for 18 hr more and the heart begins to beat again after rewarming at 38°C. The survival time at 8°C becomes short as the embryos grow. While the hearts of 6-day embryos begin to beat even after 1-day exposure to 8°C, the hearts of 20-day embryos fail to beat after about 8 hr exposure (Tazawa and Rahn, 1986).

While chicken embryos can withstand a lowered T_a for a prolonged period without a heartbeat, at an increased T_a the cardiac arrest following arrhythmia is irreversible and the embryos cannot withstand exposure to an increased T_a for a prolonged period (Ono *et al.*, 1994). The HR of embryos increases in an exponential fashion at an increased T_a. The changes in HR are in parallel with those of egg temperature except for externally pipped embryos (Figure 14). When the internal egg temperature reaches 46–47°C, regardless of the developmental stage of embryos, the HR becomes arrhyth-

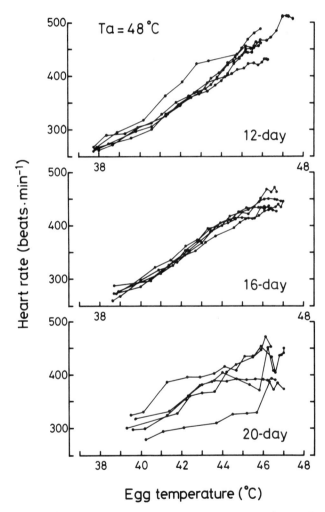

FIGURE 14 Changes in heart rate of 12-, 16-, and 20 (externally pipped)-day embryos exposed abruptly to an ambient temperature of 48°C, plotted against the simultaneously measured egg temperature. The heart rates of the 12- and 16-day embryos increase almost linearly with egg temperature, which increases in an exponential fashion (from Ono *et al.*, 1994, *Israel J. Zool.* **40**, 467–479. With permission).

mic and irreversible cardiac arrest follows. The lethal T_a and tolerance time of chicken embryos depend upon the time when the internal egg temperature takes to reach a lethal critical value of 46–47°C. At a T_a of 48°C, the tolerance time of 12-day-old embryos is about 100 min and that of 20-day-old embryos shortens by about half. The tolerance time to 48°C exposure is shortened as the embryos grow. This may be in part due to metabolic heat which is produced more by late embryos than young ones. The HR reaches about 450 bpm at the critical egg temperature (46–47°C) (Ono et al., 1994).

References

Ackerman, R. A., and Rahn, H. (1981). In vivo O_2 and water vapor permeability of the hen's eggshell during early development. Respir. Physiol. **45,** 1–8.

Ar, A., and Rahn, H. (1978). Interdependence of gas conductance, incubation length and weight of the avian egg. In "Respiratory Function in Birds, Adult and Embryonic" (J. Piiper, ed.), pp. 227–236. Springer-Verlag, New York.

Ar, A., and Rahn, H. (1985). Pores in avian eggshells: Gas conductance, gas exchange and embryonic growth rate. Respir. Physiol. **61,** 1–20.

Ar, A., Paganelli, C. V., Reeves, R. B., Greene, D. G., and Rahn, H. (1974). The avian egg: Water vapor conductance, shell thickness, and functional pore area. Condor **76,** 153–158.

Ar, A., Visschedijk, A. H. J., Rahn, H., and Piiper, J. (1980). Carbon dioxide in the chick embryo towards end of development: Effects of He and SF_6 in breathing mixtures. Respir. Physiol. **40,** 293–307.

Arora, K. L., and Kosin, I. L. (1966). Changes in the gross morphological appearance of chicken and turkey blastoderms during preincubation. Poult. Sci. **45,** 819–825.

Barnes, A. E., and Jensen, W. N. (1959). Blood volume and red cell concentration in the normal chick embryo. Am. J. Physiol. **197,** 403–405.

Biebach, H. (1986). Energetics of rewarming a clutch in starlings (Sturnus vulgaris). Physiol. Zool. **59,** 69–75.

Bissonnette, J. M., and Metcalfe, J. (1978). Gas exchange of the fertile hen's egg: Components of resistance. Respir. Physiol. **34,** 209–218.

Butler, D. E. (1991). Egg handling and storage at the farm and hatchery. In "Avian Incubation" (S. G. Tullett, ed.), pp. 195–203, London, Butterworth.

Cain, J. R., Abbott, U. K., and Rogallo, V. L. (1967). Heart rate of the developing chick embryo. Proc. Soc. Exp. Biol. Med. **126,** 507–510.

Clark, E. B., Hu, N., Dummett, J. L., Vandekieft, G. K., Olson, C., and Tomanek, R. (1986). Ventricular function and morphology in chick embryo from stages 18 and 29. Am. J. Physiol. **250,** H407–H413.

Decuypere, E., Nouwen, E. J., Kühn, E. R., Geers, R., and Michels, H. (1979). Differences in serum iodohormone concentration between chick embryos with and without the bill in the air chamber at different incubation temperatures. Gen. Comp. Endocrinol. **37,** 264–267.

Deeming, D. C. (1989a). Characteristics of unturned eggs: critical period, retarded embryonic growth and poor albumen utilisation. Br. Poult. Sci. **30,** 239–249.

Deeming, D. C. (1989b). Importance of sub-embryoic fluid and albumen in the embryo's response to turning of the egg during incubation. Br. Poult. Sci. **30,** 591–606.

Deeming, D. C. (1989c). Failure to turn eggs during incubation: development of the area vasculosa and embryonic growth. J. Morphol. **201,** 179–186.

Deeming, D. C., Rowlett, K., and Simkiss, K. (1987). Physical influences on embryo development. J. Exp. Zool. **1,** (Suppl.) 3341–345.

Erasmus, B. Dew., and Rahn, H. (1976). Effects of ambient pressures, He and SF_6 on O_2 and CO_2 transport in the avian egg. Respir. Physiol. **27,** 53–64.

Faber, J. J. (1968). Mechanical function of the septating embryonic heart. Am. J. Physiol. **214,** 475–481.

Faber, J. J., Green, T. J., and Thornburg, K. L. (1974). Embryonic stroke volume and cardiac output in the chick. Dev. Biol. **41,** 14–21.

Girard, H. (1973). Arterial pressure in the chick embryo. Am. J. Physiol. **224,** 454–460.

Haque, M. A., Pearson, J. T., Hou, P.-C. L., and Tazawa, H. (1996). Effects of pre–incubation egg storage on embryonic functions and growth. Respir. Physiol. **103,** 89–98.

Hoyt, D. F., and Rahn, H. (1980). Respiration of avian embryos—A comparative analysis. Respir. Physiol. **39,** 255–264.

Howe, R. S., Burggren, W. W., and Warburton, S. J. (1995). Fixed patterns of bradycardia during late embryonic development in domestic fowl with C locus mutations. Am. J. Physiol. **268,** H56–H60.

Hu, N., and Clark, E. B. (1989). Hemodynamics of the stage 12 to stage 29 chick embryo. Circ. Res. **65,** 1665–1670.

Hughes, A. F. W. (1949). The heart output of the chick embryo. J. Roy. Microsc. Soc. **69,** 145–152.

Kern, M. D., and Van Riper, C. (1984). Altitudinal variations in nests of the Hawaiian Honeycreeper Hemignathans virens virens. Condor **86,** 443–454.

Kind, C. (1975). The development of the circulating blood volume of the chick embryo. Anat. Embryol. **147,** 127–132.

Kuroda, O., Matsunaga, C., Whittow, G. C., and Tazawa, H. (1990). Comparative metabolic responses to prolonged cooling in precocial duck (Anas domestica) and altricial pigeon (Columba domestica) embryos. Comp. Biochem. Physiol. A **95,** 407–410.

Laughlin, K. F., Lundy, H., and Tait, J. A. (1976). Chick embryo heart rate during the last week of incubation: Population studies. Br. Poult. Sci. **17,** 293–301.

Lemez, L. (1972). Thrombocytes of chick embryos from the 2nd day of incubation till the 1st postembryonic day. Acta Univ. Carol. Ser. Med. Mono. 53–54, 365–371.

Mather, C. M., and Laughlin, K. F. (1979). Storage of hatching eggs: the interaction between parental age and early embryonic development. Br. Poult. Sci. **20,** 595–604.

Matsunaga, C., Mathiu, P. M., Whittow, G. C., and Tazawa, H. (1989). Oxygen consumption of Brown Noddy (Anous stolidus) embryos in a quasiequilibrium state at lowered ambient temperatures. Comp. Biochem. Physiol. A **93,** 707–710.

Mayer, A. A., Metcalfe, J., and Stock, M. K. (1995). Relocation during incubation of endothelial nuclei in the chick chorioallantois. Respir. Physiol. **100,** 171–176.

McNabb, F. M. A. (1987). Comparative thyroid development in precocial Japanese qauil and altricial ring doves. J. Exp. Zool. **1,** (Suppl.) 281–290.

Mitgard, V., Sejrsen, P., and Johansen, K. (1985). Blood flow in the brood patch of Bantam hens: evidence of cold vasodilation. J. Comp. Physiol. B **155,** 703–709.

New, D. A. T. (1957). A critical period for the turning of hens' eggs. J. Embryol. Exp. Morphol. **5,** 293–299.

Nice, M. (1962). Development of behavior in precocial birds. Trans. Linn. Soc. N. Y. **8,** 1–211.

Nichelmann, M., Lange, B., Pirow, R., Langbein, J., and Herrmann, S. (1994). Avian thermoregulation during the perinatal period. In "Advances in Pharmacological Sciences: Thermal Balance in Health and Disease" (E. Zeisberger, E. Schönbaum, and P. Lomax, eds.), pp. 167–173. Birkhäuser Verlag, Basel.

Okuda, A., and Tazawa, H. (1988). Gas exchange and development of chicken embryos with widely altered shell conductance from the beginning of incubation. *Respir. Physiol.* **74,** 187–198.

Ono, H., Hou, P.-C. L., and Tazawa, H. (1994). Responses of developing chicken embryos to acute changes in ambient temperature: Noninvasive study of heart rate. *Israel J. Zool.* **40,** 467–479.

Pearson, J. T., Haque, M. A., Hou, P.-C. L., and Tazawa, H. (1996). Developmental patterns of O_2 consumption, heart rate and O_2 pulse in unturned eggs. *Respir. Physiol.* **103,** 83–87.

Pettit, T. N., Whittow, G. C., and Ellis, H. I. (1984). Food and energetic requirements of seabirds at French Frigate Shoals, Hawaii. *In* "Proceedings of the 2nd Symposium on Resource Investigations in the Northwestern Hawaiian Islands" (R. W. Grigg and K. Y. Tanoue, eds.), pp. 265–282.

Prinzinger, R., and Dietz, V. (1995). Qualitative course of embryonic O_2 consumption in altricial and precocial birds. *Respir. Physiol.* **100,** 289–294.

Rahn, H. (1984). Factors controlling the rate of incubation water loss in bird eggs. *In* "Respiration and Metabolism of Embryonic Vertebrates" (R. S. Seymour, ed), pp. 271–288. Martinus Nijhoff, The Hague.

Rahn, H. (1991). Why birds lay eggs. *In* "Egg Incubation: Its Effects on Embryonic Development in Birds and Reptiles" (D. C. Deeming and M. W. J. Ferguson, eds.), pp. 345–360. Cambridge Univ. Press, Cambridge.

Rahn, H., and Ar, A. (1974). The avian egg: incubation time and water loss. *Condor* **76,** 147–152.

Rahn, H., Carey, C., Balmas, K., Bhatia, B., and Paganelli, C. (1977). Reduction of pore area of the avian eggshell as an adaptation to altitude. *Proc. Natl. Acad. Sci. USA.* **74,** 3095–3098.

Rahn, H., Matalon, S., and Sotherland, P. R. (1985). Circulatory changes and oxygen delivery in the chick embryo prior to hatching. *In* "Cardiovascular Shunts: Phylogeneic, Ontogenetic and Clinical Aspects" (K. Johansen and W. Burggren, ed), pp. 179–198, Munksggard, Copenhagen.

Rahn, H., Paganelli, C. V., and Ar, A. (1974). The avian egg: Air-cell gas tension, metabolism and incubation time. *Respir. Physiol.* **22,** 297–309.

Rahn, H., Paganelli, C. V., Nisbet, I. C. T., and Whittow, G. C. (1976). Regulation of incubation water loss in eggs of seven species of terns. *Physiol. Zool.* **49,** 245–259.

Romanoff, A. L. (1960). "The Avian Embryo." Macmillan, New York.

Romanoff, A. L. (1967). "Biochemistry of the Avian Embryo." Wiley, New York.

Skowron, C., and Kern, M. (1980). The insulation in nests of selected North American songbirds. *Auk* **97,** 816–824.

Sotherland, P. R., and Rahn, H. (1987). On the composition of bird eggs. *Condor* **89,** 48–65.

Sotherland, P. R., Spotila, J. R., and Paganelli, C. V. (1987). Avian eggs: Barriers to the exchange of heat and mass. *J. Exp. Zool.,* **1,** (Suppl.) 81–86.

Tazawa, H. (1980). Adverse effect of failure to turn the avian egg on embryo oxygen exchange. *Respir. Physiol.* **41,** 137–142.

Tazawa, H. (1981a). Compensation of diffusive respiratory disturbances of the acid–base balance in the chick embryo. *Comp. Biochem. Physiol. A* **69,** 333–336.

Tazawa, H. (1981b). Measurement of blood pressure of chick embryo with an implanted needle catheter. *J. Appl. Physiol.* **51,** 1023–1026.

Tazawa, H. (1982). Regulatory process of metabolic and respiratory acid–base disturbances in embryos. *J. Appl. Physiol.* **53,** 1449–1454.

Tazawa, H. (1986). Acid–base equilibrium in birds and eggs. *In* "Acid–Base Regulation in Animals" (N. Heisler, ed), pp. 203–233. Elsevier, Amsterdam.

Tazawa, H. (1987). Embryonic respiration. *In* "Bird Respiration" (T. J. Seller, ed), Vol. II, pp. 3–41. CRC Press, Boca Raton, FL.

Tazawa, H., and Mochizuki, M. (1976). Estimation of contact time and diffusing capacity for oxygen in the chorioallantoic vascular plexus. *Respir. Physiol.* **28,** 119–128.

Tazawa, H., and Mochizuki, M. (1977). Oxygen analysis of chicken embryo blood. *Respir. Physiol.* **31,** 203–215.

Tazawa, H., and Nakagawa, S. (1985). Response of egg temperature, heart rate and blood pressure in the chick embryo to hypothermal stress. *J. Comp. Physiol. B* **155,** 195–200.

Tazawa, H., and Ono, T. (1974). Microscopic observation of the chorioallantoic capillary bed of chicken embryo. *Respir. Physiol.* **20,** 81–90.

Tazawa, H., and Rahn, H. (1986). Tolerance of chick embryos to low temperatures in reference to the heart rate. *Comp. Biochem. Physiol. A* **85,** 531–534.

Tazawa, H., and Rahn, H. (1987). Temperature and metabolism of chick embryos and hatchlings after prolonged cooling. *J. Exp. Zool.,* **1,** (Suppl.) 105–109.

Tazawa, H., and Takenaka, H. (1985). Cardiovascular shunt and model analysis in the chick embryo. *In* "Cardiovascular Shunts. Phylogeneic, Ontogenetic and Clinical Aspects" (K. Johansen and W. Burggren, eds.), pp. 179–198, Munksggard, Copenhagen.

Tazawa, H., Hiraguchi, T., Kuroda, O., Tullett, S. G., and Deeming, O. C. (1991). Embryonic heart rate during development of domesticated birds. *Physiol. Zool.* **64,** 1002–1022.

Tazawa, H., Mikami, T., and Yoshimoto, C. (1971). Respiratory properties of chicken embryonic blood during development. *Respir. Physiol.* **13,** 160–170.

Tazawa, H., Nakazawa, S., Okuda, A., and Whittow, G. C. (1988b). Short-term effects of altered shell conductance on oxygen uptake and hematological variables of late chicken embryos. *Respir. Physiol.* **74,** 199–210.

Tazawa, H., Okuda, A., Nakazawa, S., and Whittow, G. C. (1989a). Metabolic responses of chicken embryos to graded, prolonged alterations in ambient temperature. *Comp. Biochem. Physiol. A* **92,** 613–617.

Tazawa, H., Piiper, J., Ar, A., and Rahn, H. (1981). Changes in acid–base balance of chick embryos exposed to a He and SF_6 atmosphere. *J. Appl. Physiol.* **50,** 819–823.

Tazawa, H., Takami, M., Kobayashi, K., Hasegawa, J., and Ar, A. (1992). Non-invasive determination of heart rate in newly hatched chicks. *Br. Poult. Sci.* **33,** 1111–1118.

Tazawa, H., Turner, J. S., and Paganelli, C. V. (1988a). Cooling rates of living and killed chicken and quail eggs in air and in helium-oxygen gas mixture. *Comp. Biochem. Physiol. A* **90,** 99–102.

Tazawa, H., Visschedijk, A. H. J., and Piiper, J. (1983). Blood gases and acid–base status in chicken embryos with naturally varying egg shell conductance. *Respir. Physiol.* **54,** 137–144.

Tazawa, H., Wakayama, H., Turner, J. S., and Paganelli, C. V. (1988c). Metabolic compensation for gradual cooling in developing chick embryos. *Comp. Biochem. Physiol. A* **89,** 125–129.

Tazawa, H., Whittow, G. C., Turner, J. S., and Paganelli, C. V. (1989b). Metabolic responses to gradual cooling in chicken eggs treated with thiourea and oxygen. *Comp. Biochem. Physiol. A* **92,** 619–622.

Tøien, O., Aulie, A., and Steen, J. B. (1986). Thermoregulatory responses to egg cooling in incubating bantam hens. *J. Comp. Physiol. B* **156,** 303–307.

Tullett, S. G., and Board, R. G. (1977). Determinants of avian egg shell porosity. *J. Zool. (London)* **183,** 203–211.

Tullett, S. G., and Deeming, D. C. (1987). Failure to turn eggs during incubation: effects on embryo weight, development of the chorioallantois and absorption of albumen. *Br. Poult. Sci.* **28,** 239–243.

Turner, J. S. (1986). Cooling rate and size of bird's eggs—A natural isomorphic body. *J. Therm. Biol.* **10,** 101–104.

Turner, J. S. (1987). Blood circulation and the flows of heat in an incubated egg. *J. Exp. Zool.,* **1,** (Suppl.) 99–104.

Turner, J. S. (1991). The thermal energetics of incubated bird eggs. In "Egg Incubation: Its Effects on Embryonic Development in Birds and Reptiles" (D. C. Deeming and M. W. J. Fergusson, eds.), pp. 117–145. Cambridge Univ. Press.

Van Mierop, L. H. S., and Bertuch, C. J., Jr. (1967). Development of arterial blood pressure in the chick embryo. *Am. J. Physiol.* **212,** 43–48.

Visschedijk, A. H. J., Tazawa, H., and Piiper, J. (1985). Variability of shell conductance and gas exchange of chicken eggs. *Respir. Physiol.* **59,** 339–345.

Vleck, C. M., and Vleck, D. (1987). Metabolism and energetics of avian embryos. *J. Exp. Zool.,* **1,** (Suppl.) 111–125.

Wangensteen, O. D., and Rahn, H. (1970/71). Respiratory gas exchange by the avian embryo. *Respir. Physiol.* **11,** 31–45.

Wangensteen, O. D., and Weibel, E. R. (1982). Morphometric evaluation of chorioallantoic oxygen transport in the chick embryo. *Respir. Physiol.* **47,** 1–20.

Wangensteen, O. D., Wilson, D., and Rahn, H. (1970/71). Diffusion of gases across the shell of the hen's egg. *Respir. Physiol.* **11,** 16–30.

White, P. T. (1974). Experimental studies on the circulatory system of the late chick embryo. *J. Exp. Biol.* **61,** 571–592.

Whittow, G. C. (1980). Physiological and ecological correlates of prolonged incubation in sea birds. *Am. Zool.* **20,** 427–436.

Whittow, G. C. (1984). Physiological ecology of incubation in tropical seabirds. *Studies Avian Biol.* **8,** 47–72.

Whittow, G. C., and Berger, A. J. (1977). Heat loss from the nest of the Hawaiian honeycreeper, "Amakihi." *Wilson Bull.* **89,** 480–483.

Whittow, G. C., and Tazawa, H. (1991). The early development of thermoregulation in birds. *Physiol. Zool.* **64,** 1371–1390.

Wilson, H. R. (1991). Physiological requirements of the developing embryo: temperature and turning. In "Avian Incubation" (S. G. Tullett, ed), pp. 145–156. London, Butterworth.

Yosphe-Purer, Y., Fendrich, J., and Davies, A. M. (1953). Estimation of the blood volumes of embryonated hen eggs at different ages. *Am. J. Physiol.* **175,** 178–180.

Zhang, Q., and Whittow, G. C. (1992). The effect of incubation temperature on oxygen consumption and organ growth in domestic-fowl embryos. *J. Therm. Biol.* **17,** 339–345.

CHAPTER 25

Physiology of Growth and Development

LARRY A. COGBURN AND JOAN BURNSIDE
*Department of Animal and Food Sciences
College of Agricultural Sciences
Delaware Agricultural Experiment Station
University of Delaware
Newark, Delaware 19717*

COLIN G. SCANES
*College of Agriculature
Iowa State University
Ames, Iowa 50011*

I. Introduction 635
  A. Hormonal Regulation of Growth and Development 635
  B. Molecular Mechanisms of Hormone Action 636
II. Somatotropic Axis 636
  A. Growth Hormone and Its Receptor 637
  B. Insulinlike Growth Factors, Type-I IGF Receptor, and IGF-Binding Proteins 638
III. Thyrotropic Axis 641
  A. Thyroid Hormones and Their Receptors 641
IV. Gonadotropic Axis 642
  A. Gonadal Steroids and Their Receptors 642
V. Lactotropic Axis 642
  A. Prolactin and Its Receptor 642
VI. Adrenocorticotropic Axis 644
  A. Adrenocorticotropin and Corticosterone 644
VII. Pancreatic Hormones 644
  A. Insulin and Its Receptor 644
VIII. Growth Factors 644
  A. Epidermal Growth Factor (EGF) and Transforming Growth Factor-α (TGF-α) 644
  B. Fibroblast Growth Factor (FGF) 645
  C. Platelet-Derived Growth Factor (PDGF) 646
  D. Nerve Growth Factor (NGF) 646
  E. Transforming Growth Factor-β (TGF-β) Superfamily 646
  F. Homeobox Genes 647
IX. Models in Avian Growth 648
  A. Sex-Linked Dwarf Chickens 648
  B. Growth-Selected Strains of Birds 649
  References 649

I. INTRODUCTION

A. Hormonal Regulation of Growth and Development

In this chapter, we review the current knowledge of hormonal regulation of avian growth (i.e., increase in body mass) and development (i.e., progression from fertilization toward sexual maturity). Growth is a complex process involving numerous interactions among the endocrine regulatory systems and other contributing (genetic, nutritional, and environmental) factors. Our discussion will be limited to the contribution of major pituitary axes (somatotropic, thyrotropic, gonadotropic, lactotropic, adrenocorticotropic), pancreatic hormones (insulin/glucagon), and certain tissue growth factors.

Avian development (Figure 1) follows five stages (embryonic, posthatch, juvenile, pubertal, and adult). Sigmoidal growth curves have been described for numerous species of domestic and feral birds (Ricklefs, 1983; Barbato, 1991; Hnizetova *et al.*, 1991). Most species of domestic birds (chicken, turkey, and Japanese quail) show similar patterns of embryonic and posthatching growth (Anthony *et al.*, 1991). The most rapid growth rate, or increase in body mass, of domestic birds occurs during the period encompassing late embryonic, posthatch, and juvenile stages of development. The growth rate of chickens and turkeys is sexually dimor-

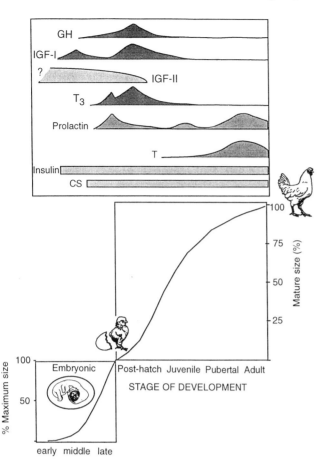

FIGURE 1 Relative importance of hormones during the five stages of growth and development of the chicken. The action of each hormone is based on both presence of ligand and expression of ligand-specific receptors in target tissue. The hormones presented are: growth hormone (GH), insulinlike growth factor-I (IGF-I), IGF-II, triiodothyronine (T_3), prolactin, testosterone (T), insulin, and corticosterone (CS). The importance of IGF-II during embryonic and posthatch development has not been established. Developmental profiles of insulin and corticosterone have not been determined since plasma levels of these hormones vary dramatically according to their respective metabolic and adaptive responses. However, both insulin and corticosterone are required for normal growth and development of birds. The growth curves for the embryo and posthatch chicken are expressed as body weight (percentage of mature size) versus developmental age.

phic with males reaching a greater body weight at sexual maturity, although adult body weight is achieved well in advance of sexual maturity. In contrast, mature body weight of the female Japanese quail exceeds that of the male (Marks, 1989). Our current understanding of endocrine factors which regulate growth and development acknowledges that may different regulatory axes participate in each stage of growth.

B. Molecular Mechanisms of Hormone Action

The ability of a cell to respond to a particular hormone (ligand) depends upon the presence of specific receptors for each hormone. Cell receptors fall into two general categories: those on the cell surface which interact with water-soluble (peptide/protein) hormones and the cytoplasmic or nuclear receptors that interact with lipid-soluble hormones (i.e., steroids, thyroid hormones, retinoids, and 1,25-dihydroxy vitamin D_3), which diffuse through the plasma membrane. A general scheme of hormone action (Figure 2) shows binding of a protein ligand (i.e., growth hormone, GH) to its membrane-bound receptor (i.e., the GHR). In this example, binding of GH to its cell surface receptor activates a signal transduction cascade, commonly used by several members of the GH/prolactin/cytokine receptor superfamily, which regulates transcription of genes (i.e., insulinlike growth factor-I (IGF-I). The lipophilic hormones (i.e., T_3) diffuse through the plasma membrane and bind to specific nuclear receptors (i.e., T_3-R), which recognize target sequences on specific genes and affect gene transcription (i.e., malic enzyme). Thus, activation of specific hormonal pathways depends upon the presence of both ligand and functional receptors on/in target cells.

II. SOMATOTROPIC AXIS

The somatotropic axis of birds (see Chapter 14) is controlled by hypothalamic-releasing factors (growth hormone-releasing hormone, thyrotropin-releasing factor, somatostatin), which regulate secretion of GH (somatotropin) from the anterior pituitary gland. Other components of the axis include the cell membrane-bound growth hormone receptor (GHR), the soluble GH-binding protein (GHBP), IGF-I, several IGF-binding proteins (IGFBPs) found in circulation, and the type-I IGF receptor which can be stimulated in an endocrine, paracrine, and/or autocrine manner (Figure 3). In general, GH exerts two types of action: the *direct* effects are on metabolism and the *indirect* effects that are mediated by IGF-I, which stimulates the differentiation and proliferation of target cells (i.e., muscle cells). Although IGF-II is not under direct control of the somatotropic axis, the chicken type-I IGF receptor has a very high affinity for IGF-II. The type-I IGF receptor also binds IGF-I with a somewhat lower affinity (Duclos and Goddard, 1990; Duclos *et al.*, 1991; Armstrong and Hogg, 1994). Thus, the avian type-I IGF receptor exerts endocrine, paracrine, and autocrine effects on differentiation, proliferation, and growth of cells by interacting

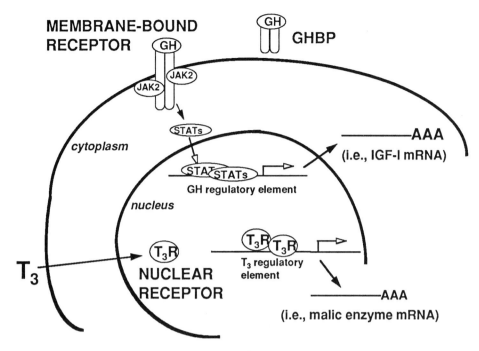

FIGURE 2 General mechanisms of hormone action. The action of peptide hormones (i.e., GH) on a target cell is initiated by binding and subsequent dimerization of intracellular domains of the cell-bound growth hormone receptor (GHR). Members of the GH/prolactin/cytokine receptor superfamily share a similar signal transduction mechanism that involves activation (via phosphorylation) of intracellular proteins (Janus kinase-2 (JAK2) and signal transducers and activators of transcription (STATs)). Binding of these activated signaling molecules to hormone-specific regulatory elements affects transcription of target genes (i.e., IGF-I). On the other hand, the lipid-soluble hormones (i.e., T_3, steroids, retinoids, and vitamin D_3) diffuse through the plasma membrane, bind to specific nuclear receptors (i.e., T_3-R), and affect transcription of target genes (i.e., malic enzyme).

at different affinities with the insulinlike ligands (IGF-II > IGF-I > insulin). The cross-reactivity of these ligands with the avian type-I IGF receptor indicates the importance of this regulatory pathway in embryonic and posthatching development (Bassas et al., 1987; De Pablo et al., 1993).

A. Growth Hormone and Its Receptor

1. GH and GHR Ontogeny

Ontogeny studies show that GH is found in the circulation of the chicken embryo as early as 11–13 days (McGuinness and Cogburn, 1990) and increases prior to hatching. After hatching, plasma GH levels increase sharply until a peak is reached between 3 and 4 weeks of age and then abruptly decline to the low levels found in pubertal and mature birds (Goddard et al., 1988; Johnson et al., 1990; McGuinness and Cogburn, 1990; Burnside and Cogburn, 1992; McCann-Levorse et al., 1993). Circulating GH levels reflect the developmental changes in pituitary GH mRNA (McCann-Levorse et al., 1993). Basal levels of GH and pituitary GH mRNA (McCann-Levorse et al., 1993) increase sharply from hatching until a peak is reached between 3 and 5 weeks of age. In the turkey, circulating GH reaches an earlier peak at 2 weeks of age (Vasilatos-Younken et al., 1988a), which declines to the low basal level found after 10 weeks of age. Japanese quail also show an early peak (1–2 weeks of age) in plasma GH levels, which fall to low adult levels by 3 weeks of age (Bacon et al., 1987). In the chicken, the hepatic GHR is transiently expressed in the embryo between days 17 and 19 (Figure 4A inset). After hatching, hepatic expression of the GHR mRNA progressively increases until a plateau is reached with the onset of sexual maturity (Burnside and Cogburn, 1992). Despite the transient expression of the hepatic GHR during late embryonic development, growth of the avian embryo is essentially GH independent, as it is in mammals. The onset of GH-dependent growth is marked by activation of the somatotropic axis as indicated by increases in circulating GH, hepatic GHR mRNA, and plasma IGF-I levels. However, hepatic GH-binding activity and expression of the GHR in tissue are inversely related to circulating GH levels after 10

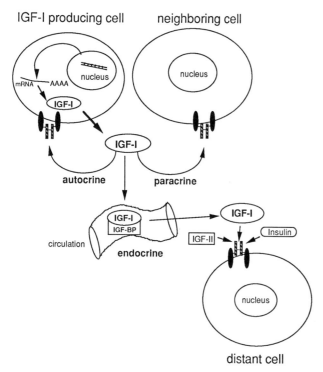

FIGURE 3 Mechanisms of IGF-I action in birds. IGF-I can affect the same cell (autocrine), adjacent cells (paracrine), or distant cells via the blood stream (endocrine). In circulation, IGF-I is complexed with the IGF binding proteins (IGFBP). In tissue, the type-I IGF receptor binds with different affinities to the IGFs (IGF-II > IGF-I > insulin).

weeks of age (Figure 4B). Mature broiler chickens (Leung et al., 1987) and male turkeys have higher hepatic GH binding activity than slower-growing hens (Krishnan et al., 1989). Turkeys exhibit a developmental increase in hepatic GH binding which is also inversely related to plasma GH levels (Vasilatos-Younken et al., 1990). The idea of down-regulation of the GHR by GH is further supported by the observation that hypophysectomized (hypox) chickens have greater hepatic GH binding than intact birds and that GH injections reduce hepatic GH binding in both normal and hypox chickens (Vanderpooten et al., 1991a). Genetic selection for rapid growth rate in the chicken (Vanderpooten et al., 1993), turkey (Vasilatos-Younken et al., 1988a), and Japanese quail (Bacon et al., 1987) is also associated with lower circulating levels of GH and higher GH-binding activity in liver (Vanderpooten et al., 1993). Thus, an inverse relationship is found between circulating GH and the abundance of GHR in tissue of pubertal and adult birds. The abundance of GHR mRNA is greater in liver and adipose tissue than in either breast or leg muscle in growth-selected strains of broiler chickens (Mao et al., 1998).

2. GH and Growth

Hypophysectomy reduces the growth rate of young chickens (King, 1969) and turkeys (Proudman et al., 1994), which suggests that growth of young birds is pituitary dependent. In hypox chickens, T_3 (or T_4) replacement therapy restores much of the growth deficit, albeit GH treatment alone has only a slight stimulatory effect (King, 1969; King and Scanes, 1986; Scanes et al., 1986). However, GH replacement has no discernable effect on growth rate of young hypox turkeys (Proudman et al., 1994). Thus, in young birds, the thyroid hormones appear to mediate much of pituitary-dependent growth.

Young rapidly growing broiler chickens (<7 weeks of age) do not respond to exogenous GH (Burke et al., 1987; Cogburn et al., 1989, Cogburn, 1991), while older chickens appear to be GH responsive (Vasilatos-Younken et al., 1988b; Scanes et al., 1990) presumably due to greater abundance of the hepatic GHR (Burnside and Cogburn, 1992). There is a strong synergism between exogenous GH and dietary T_3 in reducing body fat deposition in young broiler chickens (Cogburn, 1991). In 8-week-old female chickens, pulsatile, but not continuous, GH infusion stimulates growth (Vasilatos-Younken et al., 1988b), while in 12-week-old male chickens, continuous administration of GH increases growth rate (Scanes et al., 1990).

B. Insulinlike Growth Factors, Type-I IGF Receptor, and IGF-Binding Proteins

As in mammals, birds have two insulinlike growth factors (IGF-I and IGF-II) that exert insulinlike effects on metabolism and are important regulators of cellular differentiation, proliferation, and growth of tissue. Although the liver is the major source of IGFs found in circulation, these peptide growth factors are also produced by many different tissues and exert local (autocrine/paracrine) effects (Figure 3). A major distinction of the somatotropic axis in birds is the presence of a single receptor for IGF-I and IGF-II (i.e., the type-I IGF receptor). Thus, the IGFs share similarities in the structure of ligand (insulin, IGF-I, and IGF-II) and the cell surface receptors (i.e., insulin receptor and type-I IGF receptor). On target cells, IGF-I and IGF-II probably bind to different sites on the same receptor. Binding of ligand to the type-I IGF receptor activates a receptor tyrosine kinase which activates the phosphorylation cascade responsible for the biological actions of each ligand.

1. Insulinlike Growth Factors

Avian IGF-I (chicken and Japanese quail) and IGF-II (chicken) show strong nucleotide or amino acid sequence similarity with mammalian IGF-I and IGF-II

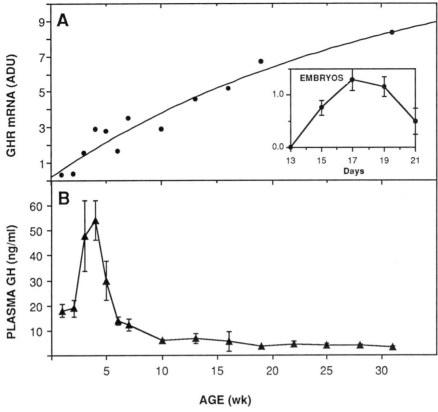

FIGURE 4 Developmental profiles of hepatic GHR mRNA (A) and plasma GH levels (B) in broiler chickens. Each value represents the mean ± SEM of four different birds. Hepatic GHR mRNA levels were determined by Northern blot analysis of total RNA and expressed in arbitrary densitometric units (ADU). (Reprinted from *Molecular and Cellular Endocrinology* **89,** J. Burnside and L. A. Cogburn, Developmental expression of hepatic growth hormone receptor and insulin-like growth factor-I mRNA in the chicken, pp. 91–96, Copyright (1992), with permission from Elsevier Science.)

(Dawe *et al.*, 1988; Kajimoto and Rotwein, 1989; Ballard *et al.*, 1990; Fawcett and Bulfield, 1990; Kallincos *et al.*, 1990; Kida *et al.*, 1994; Darling and Brickell, 1996). In mammals (i.e., the rat), IGF-II is more abundant than IGF-I in the fetus, while IGF-I predominates during postnatal development. Presently, there is no information on the ontogeny or tissue distribution of IGF-II mRNA in the chicken embryo. The expression of IGF-I in tissue of the chicken embryo increases dramatically between days 3 and 8 (Serrano *et al.*, 1990), which is the period of organogenesis (De Pablo *et al.*, 1993). IGF-I appears in circulation at day 6 (De Pablo *et al.*, 1993), reaches a peak between days 14 and 18, and then declines before hatching (Serrano *et al.*, 1990; Kikuchi *et al.*, 1991; De Pablo *et al.*, 1993). A similar development profile of IGF-I, with a midincubation peak, is found in the turkey embryo (McMurtry *et al.*, 1994). The absence of detectible hepatic IGF-I mRNA, using Northern blot analysis (inset Figure 5B), suggests that the low levels of the IGF-I found in circulation and in tissue of the chicken embryo could be derived from an extrahepatic origin (Serrano *et al.*, 1990; Kikuchi *et al.*, 1991; Burnside and Cogburn, 1992). After hatching, plasma (IGF-I levels and expression of IGF-I mRNA in liver (Figure 5) increase sharply, reach a broad plateau between 3 and 7 weeks of age, and then gradually decline during the juvenile and pubertal periods (McGuinness and Cogburn, 1990; Kikuchi *et al.*, 1991; Burnside and Cogburn, 1992). The developmental plateau of plasma IGF-I in the turkey is reached at the onset of the juvenile period (7 weeks of age) and declines during adolescence (Bacon *et al.*, 1993). Recent evidence indicates that IGF-I could be important for follicular development and egg production in the chicken (Hocking *et al.*, 1994; Roberts *et al.*, 1994) and Japanese quail (Kida *et al.*, 1994). The administration of IGF-I to posthatch chickens does not influence growth rate of normal chickens (McGuinness and Cogburn, 1991), but slightly stimulates growth of GHR-deficient dwarf chickens (Tixier-Boichard *et al.*, 1992).

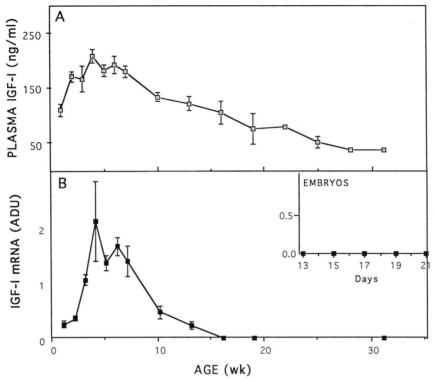

FIGURE 5 Developmental profile of plasma IGF-I (A) and hepatic expression of IGF-I mRNA (B) in broiler chickens. Each value represents the mean ± SEM of four different birds. Hepatic IGF-I mRNA levels are expressed in arbitrary densitometric units (ADU). (From J. Burnside and L. A. Cogburn, Developmental expression of hepatic growth hormone receptor and insulin-like growth factor-I mRNA in the chicken, *Molecular and Cellular Endocrinology,* **89,** pp. 91–96, 1992, © The Endocrine Society.)

2. Insulinlike Growth Factor Receptors

In mammals, three distinct receptors exist with respectively maximal affinity for insulin (the insulin receptor), IGF-I (the IGF-I receptor), and IGF-II (the IGF-II/cation independent mannose 6-phosphate receptor). It is still not clear whether IGF-II exerts its physiological effect in mammals through the IGF-II receptor or IGF-I receptor, which has a relatively high affinity for IGF-II. Based on studies of the chicken (De Pablo *et al.,* 1990), it is widely held that birds have only two receptors for IGFs (the insulin receptor and the type-I IGF receptor). The type-I IGF receptor mediates the effects of both IGF-I and IGF-II. Studies on the specificity of the insulin receptor and the type-I IGF receptor are complicated by the limited ability of ^{125}I-insulin to bind to the type-1 IGF receptor and of ^{125}I-IGF-I to bind to the insulin receptor. However, it appears that insulin is more potent than IGF-I in competing with binding of ^{125}I-insulin to the chicken insulin receptor, while IGF-I is considerably more potent than insulin in displacing ^{125}I-IGF-I from the type-I IGF receptor (Bassas *et al.,* 1987; Duclos and Goddard, 1990).

The chicken (cation-independent) mannose-6-phosphate receptor does not bind IGF-II (Yang *et al.,* 1991) and, hence, cannot function as a receptor for IGF-II. However, there is strong evidence that IGF-II is active in birds and mediates its action via the type-I IGF receptor. The specificity of the chicken type-I IGF receptor differs from its mammalian homolog since IGF-II appears to be equipotent or even more potent than IGF-I in radioreceptor assays with chick satellite cells (Duclos *et al.,* 1991) or *in vitro* bioassays with embryonic tissue (Kallincos *et al.,* 1990). In muscle, IGF-I stimulates the differentiation and proliferation of myoblasts into myotubes. In chicken muscle satellite cells, IGF-II is more potent than IGF-I in binding to the type-I IGF receptor (Duclos *et al.,* 1991). However, the type-I IGF receptor found in turkey muscle cells has a higher affinity for IGF-I than IGF-II (Minshall *et al.,* 1990; Sun *et al.,* 1992). The relative affinity of the chicken type-I IGF-I receptor for heterologous ligands (i.e., human IGF-I and IGF-II) must be reestablished by using recombinant-derived chicken IGF-I and IGF-II as ligand.

Developmental regulation of the hepatic type-I IGF receptor in chickens shows that receptor binding levels increase sharply after hatching, reach a peak at 1 week of age, and fall to low levels by 3 weeks of age (Bassas *et al.,* 1987; Duclos and Goddard, 1990; Duclos *et al.,*

1991). The chicken type-I IGF receptor cDNA has a very high sequence homology (85%) to that of the human IGF-I receptor (Holzenberger et al., 1996). The relative abundance of the type-I IGF receptor in different tissue (brain > muscle > liver) is similar for the chicken (Armstrong and Hogg, 1994) and turkey (McFarland et al., 1992). The number of type-I IGF receptors is greater than insulin receptors in most tissues during early embryonic development, except in the liver which predominately expresses the insulin receptor (De Pablo et al., 1993). Expression of type-I IGF receptor mRNA is higher in day-21 chicken embryos (brain > muscle > liver) than in tissue of the day 16 embryo or the 4-week-old chicken (Armstrong and Hogg, 1994). The extensive distribution of type-I IGF receptor transcripts in the central nervous system of the chicken embryo points to the importance of the IGFs in growth and development of the avian brain (Holzenberger et al., 1996). Similarly, the abundance of the type-I IGF receptor is greater in the brain than in either muscle or liver of the young turkey (McFarland et al., 1992). Thus, the developmental rise in IGF-I binding in the chicken (Duclos and Goddard, 1990) corresponds to peak levels of circulating IGF-I found after hatching.

3. Insulinlike Growth Factor Binding Proteins

Both IGF-I and IGF-II circulate bound to carrier proteins known as the IGF-binding proteins (IGFBPs). Six IGFBPs have been identified in mammals and three IGFBPs have been reported for the chicken (Armstrong et al., 1989; Francis et al., 1990). There is additional evidence of at least six IGFBPs in the serum of the broiler chicken (Goddard et al., 1993). The role of the different IGFBPs in mediating the action of the IGFs in birds is complex since one IGFBP can enhance the activity of IGF-I (Elgin et al., 1987), while binding to another IGFBP can reduce IGF-I activity (Burch et al., 1986; Delbe et al., 1991). Possible mechanisms by which each IGFBP functions include: (1) transport of IGFs to specific tissues or cells, (2) prolonging the half-life of IGFs by regulating metabolic clearance, or (3) controlling activity by modulating interaction with receptors. For example, most of the circulating IGF-I and IGF-II is bound to IGFBP-3 as part of a 150-kDa complex. In mammals, synthesis of IGFBP-3 is enhanced by GH and binding of IGF-I to IGFBP-3 prolongs its half-life. Alterations in levels of other IGFBPs during pathophysiological states, such as GH deficiency, protein restriction, or starvation, suggest that the IGFBPs determine the biological activity of IGFs (Ambler et al., 1996).

4. IGFs and Growth

The action of the IGFs at the cellular level depends on both their local production (autocrine/paracrine) and hepatic production which determines the IGF concentration in the circulatory system (endocrine) (Figure 3). Much of our knowledge of the biological effects of IGF-I and IGF-II in birds come from *in vitro* studies. In early chicken embryos, IGF-I stimulates growth as indicated by increased embryo weight, although the magnitude and potency of the effects are less than those found with insulin (Girbau et al., 1987). In chicken embryos incubated *ex ovo*, IGF-I increases protein synthesis as indicated by increasing phenylalanine extraction from the media (Munaim et al., 1988). It should be noted, however, that the administration of IGF-I into the chicken embryo *in ovo* does not stimulate growth (Spencer et al., 1990).

In contrast, IGF-I or IGF-II promote cell proliferation and differentiation of avian cells in culture. Examples of this include IGF-I- and IGF-II-enhanced proliferation of preadipocytes (Butterwith and Goddard, 1991), chondrocytes (Leach and Rosselot, 1992), fibroblasts (Cynober et al., 1985), and heart mesenchymal cells (Balk et al., 1982). Moreover, IGF-I and/or IGF-II increase net protein accretion in fibroblasts by decreasing protein breakdown (Kallincos et al., 1990) and by increasing amino acid uptake and protein synthesis (Cynober et al., 1985). In the chicken embryo, IGF-I induces the differentiation of the lens as well as epithelial cells (Caldes et al., 1991). There is also evidence that IGF-I is produced locally in some of these tissues. For instance, IGF-1 stimulates the growth of chicken embryo cartilage (increasing weight and/or chondroitin sulfate) *in vitro*, although cartilage is also capable of producing IGF-I (Burch et al., 1986). Similarly, IGF-I induces proliferation of chicken heart cells *in vitro*; IGF-I may exert the same effect *in vivo* since antisera to IGF-I blocks the proliferation of cultured myocardial cells (Balk et al., 1984).

III. THYROTROPIC AXIS

A. Thyroid Hormones and Their Receptors

The physiology of thyroid hormones is considered elsewhere (Chapter 15); however, it is appropriate to briefly review the role of thyroid hormones in regulation of growth and development. In birds and mammals, the predominant iodothyronine secreted by the thyroid glands is thyroxine (T_4), a pro-hormone which is converted to the active form triiodothyronine (T_3) or catabolized to metabolically inactive reverse T_3 (rT_3).

In birds, the thyroid hormones are required for normal growth, since thyroidectomy results in a considerable reduction in growth rate in young chickens (Snedecor, 1968; Raheja and Snedecor, 1971; King and King, 1973, 1976; Moore et al., 1984) or in the chick embryo (McNabb et al., 1984). In young chickens, the decrease in growth results in reductions in skeletal and bone masses (King and King, 1973, 1976; Moore et al., 1984). These effects of thyroidectomy can be overcome partially by T_3 or T_4 replacement therapy (King and King, 1973). Similarly, growth can be stimulated by T_3 or T_4 administration in hypophysectomized chicks (King, 1969; Scanes et al., 1986) and in sex-linked dwarf chickens, which are T_3 deficient (Scanes et al., 1983; Marsh et al., 1984; Leung et al., 1984c). The stimulatory effect of thyroid hormones in growing chicks can be mediated either by increases in the circulating concentrations of IGF-I and/or by direct effects of T_3 on the growing tissues. There is, indeed, evidence that T_3 can stimulate the growth of chick embryo cartilage *in vitro* (Burch and Lebovitz, 1982; Burch and Van Wyk, 1987).

While T_3 is required for the full manifestation of growth, dietary T_3 depresses growth rate when continuously administered to growing chickens (May, 1980; Decuypere et al., 1987). This could be due to the inhibitory effects of T_3 on GH release in chickens (Leung et al., 1984b; Marsh et al., 1984). However, short-term treatment of premarket broiler chickens with a low level of dietary T_3 reduces accumulation of excess body fat and increases accretion of muscle protein (Cogburn, 1991; Cogburn et al., 1995). Furthermore, there is a strong synergism between dietary T_3 and exogenous cGH in improving body composition of broiler chickens (Cogburn, 1991; Cogburn et al., 1995).

Not only do thyroid hormones appear to be required for growth in the avian ambryo but also they influence differentiation and the acquisition of full functioning by embryonic organs prior to hatching (see Chapter 15 for details). Effects of T_3 in the late-stage embryo are not surprising in view of the increases in circulating concentrations of T_3 observed before hatching (McNabb and Hughes, 1983). However, the reduction in growth rate in chick embryos receiving goitrogen treatment is perhaps surprising since circulating T_3 concentrations are low, although T_4 levels are high at mid-stage (days 9 to 11) (McNabb et al., 1984). There could be significant monodeiodination of T_4 to T_3 in the peripheral tissues or high levels of T_4 could exert effects by binding to the avian thyroid hormone receptor. The latter is unlikely based on the chicken hepatic nuclear binding studies (Bellabarba and Lehoux, 1981). Further evidence that the thyroid hormones are exerting a role in the midstage chick embryo comes from the reports of high levels of thyroid hormone receptor in the brain and liver (Bellabarba and Lehoux, 1981; Bellabarba et al., 1983). The thyroid hormone receptors are considered elsewhere in this volume (Chapter 15).

IV. GONADOTROPIC AXIS

A. Gonadal Steroids and Their Receptors

In many mammals, growth can be stimulated by androgens and/or estrogens. These effects are exerted either directly at the tissue level or mediated by the hypothalamo–pituitary axis. In either case, estrogen acts via binding to nuclear receptors while androgens can evoke their effect via either the nuclear androgen receptor or the estrogen receptor following aromatization to estrogen.

Sex steroids have been observed to influence growth but with remarkably large differences between species, although only a few avian species have been examined. In the chicken, the marked sexual dimorphism in growth rate does not appear to be related to either androgens or estrogens. The administration of androgens (either aromatizable (e.g., testosterone or 19-nortestosterone) or nonaromatizable (e.g., 5α-dihydrotestosterone)), at physiological concentrations, to male, female, or castrated male chicks does not stimulate growth (Fennell and Scanes, 1992a). Instead, skeletal growth, particularly of the long bones, is considerably reduced (Fennell and Scanes, 1992a). In contrast in the domestic turkey, administration of androgens is followed by considerable increases in growth (Fennell and Scanes, 1992b). In view of the ability of either aromatizable and nonaromatizable androgens to enhance growth of the turkey, it is probable that the effect is mediated by androgen receptors in the muscle and bone. While androgens can influence GH secretion in birds (Fennell et al., 1990), there is not evidence that circulating concentrations of IGF-I are affected by androgen treatment (M. J. Fennel, S. V. Radecki, and C. G. Scanes, unpublished observations).

V. LACTOTROPIC AXIS

A. Prolactin and Its Receptor

Prolactin (PRL) is most noted for its role in development of the mammary gland and stimulation of lactation in mammals. However, the biological activity of purified bovine PRL was originally assessed by measurement of its proliferative action on pigeon crop sac epithelium which yields "crop milk" (Riddle et al., 1935). It has become clear that PRL participates in a variety of physiological functions in birds including reproduction and maternal behavior, osmoregulation (Scances et al., 1976; Harvey et al., 1984; Ensor, 1978; Philips and Harvey, 1982), fat metabolism (Meier et al., 1965; Garrison and Scow, 1975), and immunomodulation (Bhat et al., 1983; Skwarlo-Sonta, 1990; Skwarlo-Sonta, 1992). It appears that PRL is required for normal gonadal development and the photosexual response of chickens (Sharp et al., 1989; Talbot et al., 1991) and turkeys (El Halawani et al., 1991; Wong et al., 1991; Youngren et al., 1991; Wong et al., 1992). In the chicken and turkey hen, enhanced PRL secretion is usually associated with initiation of incubation behavior and subsequent regression of the ovary (Proudman and Opel, 1981; Hall et al., 1986; El Halawani et al., 1988; Sharp et al., 1988; Talbot et al., 1991; Wong et al., 1992; El Halawani and Rozenboim, 1993). In male chickens and turkeys, plasma PRL levels are generally high for the first 2 or 3 weeks after hatching, decline during the juvenile period, increase with the onset of sexual maturity, and then decline in adults (Scanes et al., 1976; Harvey et al., 1979; Proudman and Opel, 1981; Burke and Marks, 1982; Sterling et al., 1984). Recently, the cDNA that encode turkey (Wong et al., 1991) and chicken (Hanks et al., 1989a; Watahiki et al., 1989) PRL have been isolated, sequenced, and characterized (Shimada et al., 1993). In the embryo, pituitary PRL mRNA levels increase dramatically just before hatching and remain high in the newly hatched chick (Ishida et al., 1991). As expected, pituitary levels of PRL mRNA are higher in incubating hens than in laying chickens (Talbot et al., 1991; Kansaku et al., 1994) and turkeys (Wong et al., 1991).

The administration of exogenous (ovine) PRL to laying hens arrests ovulation, causes ovarian involution, and initiates incubation behavior (Riddle et al., 1935; Nalbandov, 1945; Opel and Proudman, 1980; El Halawani et al., 1984; Sharp et al., 1988). A number of studies have shown that exogenous (mammalian) PRL causes involution of the testes in domestic and feral birds (Breneman, 1942; Nalbandov, 1945; Lofts and Marshall, 1956; Cramer and Harper, 1973; Opel and Proudman, 1980; Buntin and Tesch, 1985). Thus, PRL appears to be required for normal sexual development in domestic fowl, whereas hyperprolactinemia causes cessation of reproductive activity (i.e., incubation/brooding behavior, ovarian regression, and male impotence). However, considerable caution must be used in interpreting the responses of birds to ovine or bovine PRL treatment since these forms of PRL also bind with high affinity to the avian GHR (Leung et al., 1984a; Krishnan et al., 1989; Burnside and Cogburn, 1993). Thus, heterologous (ovine and bovine) PRL ligands do not discriminate between the somatotropic (GHR) and lactotropic (PRL-R) receptors of some birds. Homologous PRL does not bind to the abundant GHR found in liver of chickens (Leung et al., 1984a; Krishnan et al., 1989; Burnside and Cogburn, 1993; Kuhn et al., 1996) or turkeys (Krishnan et al., 1989); and, similarly, homologous GH does not bind to the PRL-R expressed in the kidney of these birds (Krishnan et al., 1991). The availability of biosynthetic cPRL (Hanks et al., 1989b; Ohkubo et al., 1993) and tPRL (Guemene et al., 1994) for in vitro and in vivo experiments will make an important contribution in resolving PRL action in avian species. A single in ovo injection of turkey embryos with recombinant-derived tPRL slightly depressed posthatching growth and diminished the endogenous developmental pattern of PRL secretion (Guemene et al., 1992). In chicken embryos, a single injection of biosynthetic cPRL alters peripheral metabolism of thyroid hormones and increases corticosterone secretion (Kuhn et al., 1996).

As a member of the cytokine/GH receptor gene superfamily, the PRL receptor (PRL-R) consists of an extracellular ligand binding domain, a single transmembrane domain, and an intracellular domain which is involved in signal transduction (see Figure 2). Mammalian PRL-R isoforms differ only in length of the intracellular domain (Kelly et al., 1993), whose heterogeneity could transduce different PRL signals (Ali et al., 1992; Horseman and Yu-Lee, 1994). The cDNA sequence of the chicken PRL-R (cPRL-R) (Tanaka et al., 1992), pigeon PRL-R (pPRL-R) (Chen and Horseman, 1994), and turkey PRL-R (tPRL-R) (Zhou et al., 1996) have been recently determined. The structure of avian PRL-Rs differs from that of mammalian PRL-Rs due to the presence of two conserved ligand-binding regions in the extracellular domain. Although the extracellular structure of the avian PRL-Rs is unusually large, tandem repeats of the extracellular domain are found among other members of this distinct gene family; e.g., interleukin-3 receptor (Gorman et al., 1990). The GHR/PRL-R family appears to use a common activation pathway that involves ligand-induced receptor homodimerization and tyrosine phosphorylation of intermediates (i.e., JAK2) for signal transduction (Fuh et al., 1992, 1993; Carter-Su et al., 1994; Horseman and Yu-Lee, 1994; Kelly et al., 1994; Rui et al., 1994). The PRL signal which is initiated by ligand binding to its receptor, activates transcription of specific milk protein genes in mammals (Doppler et al., 1990; Lesueur et al., 1990; Djiane et al., 1994) and of cropmilk genes in the pigeon (Horseman, 1989; Horseman et al., 1992; Horseman and Buntin, 1995).

Specific PRL-binding activity is found in the cropsac (Anderson et al., 1984), kidney (Krishnan et al., 1991), liver (Buntin et al., 1984), and brain (Buntin and Ruzychi, 1987) of birds. Unfortunately, these early binding studies were conducted with ovine PRL as the radioligand, which complicates the issue of whether the ligand binds to the lactogenic or somatotropic receptor. In the pigeon (Chen and Horseman, 1994) and chicken (Tanaka et al., 1992), PRL-R transcripts are found in cropsac, intestine, liver, and kidney. In the turkey, PRL-R mRNA levels are highest in the hypothalamus and kidney (Zhou et al., 1996). The wide distribution of

PRL-R transcripts in ovary, testes, kidney, bursa, and thymus tissue of chickens (J. N. C. Mao, J. Burnside, and L. A. Cogburn, unpublished observations) is consistent with the proposed functions of PRL in birds mentioned above. Also, we have recently identified unique truncated cPRL-R transcripts, which correspond only to the intracellular domain in the testes of sexually mature chickens. The function of these testis-specific PRL-R transcripts in the sexual development of the chicken is presently unknown. Our knowledge of PRL action in birds is rapidly advancing through use of molecular probes, for PRL and the PRL-R, and characterization of PRL-regulated genes (Horseman, 1994; Horseman and Buntin, 1995).

VI. ADRENOCORTICOTROPIC AXIS

A. Adrenocorticotropin and Corticosterone

Glucocorticoids, of which corticosterone (CS) is the most abundant in birds, play important roles in the control of metabolism (see Chapter 17). It is reasonable to infer that maximal growth rates require normal metabolism and hence adequate circulating concentrations of glucocorticoids. There is little evidence that insufficient levels of CS adversely affects growth in birds. Indeed, the administration of CS to hypophysectomized chicks failed to stimulate growth (King and Scanes, 1986). However, treatment of growing chickens with CS depresses growth rate while abdominal fat weight is increased (Buyse et al., 1987; Siegel et al., 1989; Taouis et al., 1993). Similarly, injection of glucocorticoids into growing Japanese quail with reduces fat-free body weight (Bray, 1993) while in adults, CS depresses nitrogen detention and increases nitrogen excretion (De La Cruz et al., 1988).

VII. PANCREATIC HORMONES

A. Insulin and Its Receptor

While insulin plays a major role in the control of metabolism (Simon, 1989, 1995; Simon and Taouis, 1993), particularly glucose metabolism, insulin is involved in the control of growth of avian embryos. Growth of chicken embryos, between days 2 and 5 of development, is increased by administration of low doses of insulin (De Pablo et al., 1985b; Girbau et al., 1987). Conversely, chick embryos treated with antisera to insulin either die or show marked growth retardation (De Pablo et al., 1985a). These reports coupled with the definitive observations of the presence of insulin in early chick embryos argue that insulin is involved in the control of early chick embryo development. Insulin can compete for binding to type-I IGF receptors, albeit to a less extent than IGF-I. Thus, the effects of insulin on chicken embryo growth are probably mediated by insulin receptors because insulin is more potent than IGF in stimulating embryo growth (Girbau et al., 1987) and there is evidence from binding studies for specific insulin receptors in chick embryos (Bassas et al., 1987). Moreover, the administration of antisera against the insulin receptor to chick embryos results in reduced growth (body weight, DNA, RNA, protein) (Girbau et al., 1988). However, the effects of insulin on muscle differentiation are probably mediated via type-I IGF receptors as antisera against the insulin receptor failed to influence embryo creatine-kinase MB isozyme (Girbau et al., 1988). Regulation of metabolism and body composition in broiler chickens by the insulin/glucagon molar ratio has been reviewed (Simon, 1989; Cogburn, 1991; Simon and Taouis, 1993). The interactions of chicken insulin and glucagon with their respective receptors has been recently described (Simon, 1995).

VIII. GROWTH FACTORS

It is now apparent that there are a series of growth factors, in addition to IGF-I and IGF-II, which influence embryonic and posthatch development. These factors act by stimulating cell proliferation or differentiation. Large numbers of genes, involved in regulation of embryogenesis and development, have been studied in the chicken. During early embryogenesis left/right asymmetry, and anterior/posterior and dorsal/ventral axes are formed. Later in development, structures like the limbs and digits are formed. The identification of tissue growth factors and their actions in birds will be briefly considered below.

A. Epidermal Growth Factor (EGF) and Transforming Growth Factor-α (TGF-α)

EGF has been identified in a wide variety of tissues and is mitogenic for cells of ectodermal and mesodermal origin. EGF has been characterized in a number of mammalian species as a protein with 53 amino acid residues (Johnson, 1992). In mammals, TGF-α is a peptide containing 50 amino acid residues that shows marked homologies to EGF. As yet, a cDNA encoding an avian EGF has not been isolated and sequenced. It is likely that an avian homolog to mammalian TGF-α exists, but the sequence has not been reported.

In birds, EGF stimulates the proliferation of heart mesenchymal cells (Balk et al., 1982); this effect is potentiated by IGF-I (Balk et al., 1984). In addition, not only do chick embryo epidermal cells respond to EGF, but this has been used as a bioassay for EGF (Cohen, 1965). Evidence that EGF itself (and not TGF-α) stimulates embryonic growth comes from measurement of EGF activity during development of the chick embryo (Mesiano et al., 1985). The existence of an avian TGF-α is supported by the immunocytochemical detection of TGF-α, but not of EGF-like immunoreactivity, in embryonic chick kidney tubules (both mesonephric and metanephric) (Diaz-Ruiz et al., 1993).

The sequence of chicken EGF receptor (EGF-R) has been established (Lax et al., 1988). Both EGF and TGF-α bind to the EGF receptor, suggesting that TGF-α exerts its biological effects via the EGF receptor (Lax et al., 1988). Moreover, the parallel demonstration of EGF-R immunoreactivity in the developing chick kidney provides strong support to the view that TGF-α acts via EGF receptors (Diaz-Ruiz et al., 1993). In addition, *in vitro* studies with chicken cells stimulated with mammalian EGF and TGF indicate that TGF-α is considerably more potent than EGF (Lax et al., 1988).

B. Fibroblast Growth Factor (FGF)

In mammals, fibroblast growth factors exist primarily in two forms, a ubiquitous basic form (bFGF/FGF2) and an acidic form (aFGF/FGF1) found primarily in neural tissue (i.e., brain and retina). These factors stimulate the proliferation of many mesoderm- and neuroectoderm-derived cells. Based on analogy with mammalian FGFs, two FGFs have been identified in birds. A 16-kDa protein extracted from chick brain cross reacts with antibodies to bovine bFGF (Risau et al., 1988). A putative chicken brain aFGF has also been isolated which shows very close homology to the N-terminal amino acid region of mammalian aFGF (Risau et al., 1988). Both FGFs bind to heparin in a manner similar to mammalian FGFs (Risau et al., 1988). In addition, the cDNA for chicken FGF4 has been isolated and characterized. The primary structure of the deduced amino acid sequence shows high homology to mammalian FGF4 (Niswander et al., 1994).

The chicken FGF receptor (FGF-R) has been partially characterized (Burrus and Olwin, 1989). Several mammalian FGF receptors have been cloned and sequenced including FGF-R1, FGF-R2 and FGF-R3. Avian homologs of these FGF-R exist (Pasquale and Singer, 1989; Lee et al., 1989; Pasquale, 1990). There is an additional member of the FGF-R family found in the chick embryo, namely fibroblast growth factor receptorlike embryonic kinase (FREK) (Marcelle and Eichmann, 1992; Marcelle et al., 1994). FREK mRNA is present in chick embryo chondrocytes and satellite cells (Halevy et al., 1994) with increased expression in the presence of bFGF (Halevy et al., 1994). Evidence that FREK is the physiological receptor for bFGF comes from the ability of retinoic acid to down-regulate the expression of FREK and to inhibit bFGF-induced proliferation of chick embryo muscle satellite cells (Halevy et al., 1994).

There is substantial evidence that FGF is involved in embryonic development of birds. bFGF stimulates angiogenesis in the chorioallantoic membrane of the chick egg (Gospodarowicz et al., 1987); this response has been employed as a biological assay for FGF activity. Both muscle and bone development appear to be affected by FGF, albeit both stimulatory and inhibitory effects on muscle differentiation have been found. The proliferation of both myogenic satellite cells and myoblasts from turkey embryo is stimulated by bFGF (Sun and McFarland, 1993). FGF can delay the differentiation of chick embryo skeletal muscle colony-forming cells. In addition, a subset of muscle cells require FGF for myogenic differentiation (Seed and Hauschka, 1988; Seed et al., 1988). In the case of chondrocytes, bFGF increases both proliferation (O'Keefe et al., 1988a,b) and differentiation (Leach and Rosselot 1992) of chick chondrocytes. In addition, bFGF stimulates production of another growth factor, TGF-β, from cultured chick chondrocytes (Gelb et al., 1990).

bFGF transforms chick embryo blastoderm cells into erythropoietic cells (Gospodarowicz et al., 1987). It also appears that FGF is produced by hypoblast cells which causes differentiation of blastoderm cells into erythropoietic cells, since the erythropoietic effect of the hypoblast cells blocked by antisera to bFGF (Gordon-Thomson and Fabian, 1994). The commitment to melanogenesis, in chick embryo neural crest cells, is stimulated by bFGF (Stocker et al., 1991). Although aFGF is more potent, both FGFs enhance proliferation of chick embryo preadipocytes (Butterwith et al., 1993). Moreover, the proliferative effects of IGF-I, PDGF, TGF-α and TGF-β 1 are greatly enhanced in the presence of either aFGF or bFGF (Butterwith et al., 1993). bFGF mRNA can be detected in both proliferating and differentiating chick preadipocytes (Burt et al., 1992), but bFGF has no effect on the differentiation of preadipocytes (Butterwith and Gilroy, 1991; Butterwith and Goddard, 1991).

The response of avian embryonic cells to FGF indicates that FGF and FGF receptor are expressed during embryogenesis (Munaim et al., 1988; Kalcheim and Neufeld, 1990). Furthermore, the presence of FGF in both the limb (Munaim et al., 1988) and nervous system (Kalcheim and Neufeld, 1990) of the bird would support physiological roles for FGF in the development of these organs/tissues. Recently, it has been shown that coadministration of FGF4 and retinoic acid induces the expression of sonic hedgehog (*Shh*) mRNA. In turn, *Shh* expression leads to *hoxd*-11 expression (one of the homeobox family) and ultimately supports formation of the embryonic skeleton (Niswander *et al.*, 1994). Moreover, a crucial piece of evidence for the role of FGF4 was the observation that an additional limb bud was formed after chick embryos were given implants of FGF4-expressing cells in the lateral plate mesoderm (Ohuchi *et al.*, 1995).

C. Platelet-Derived Growth Factor (PDGF)

PDGF is released during activation of platelets and accounts for much of the ability of serum to stimulate growth of cells in culture. There is relatively little information of the structure, tissue expression, or biological actions of PDGF in birds. There is evidence that PDGF may have a role in embryo development by stimulating the proliferation of preadipocytes (Butterwith and Goddard, 1991). Moreover, PDGF influences ovarian functioning by elevating plasminogen activity in chicken granulosa cells (Tilly and Johnson, 1990).

D. Nerve Growth Factor (NGF)

An avian NGF has been partially purified from chick embryos (Belew and Ebendal, 1986). The amino acid sequence of chicken NGF has been deduced from the cloned cDNA (Meier *et al.*, 1986; Ebendal *et al.*, 1986). The interaction of NGF with the NGF receptor mediates the effects of NGF on avian neural development. For instance, the volumes of both sympathetic and Remak's ganglia are increased following NGF administration to chick embryos with concomitant increases in the number of neurons per ganglion (Dimberg and Ebendal, 1987). Conversely, *in ovo* administration of antisera to mammalian NGF reduces the number of neurons found in the sympathetic ganglion (Dimberg and Ebendal, 1988). *In vitro*, NGF stimulates neural development including neurite formation by chick embryo neurons (Rush *et al.*, 1986). NGF also exerts a retrograde trophic action. NGF mRNA has been observed in the chick embryo (e.g., skin, eye) with peak levels around day 10 of embryonic development (Ebendal and Persson, 1988; Goedert, 1986; Ebendal *et al.*, 1982). There is evidence that there are growth factors distinct from NGF which exert neurotropic effects on chick embryo cells. For instance, avian muscle cells produce a factor which influences neurite formation in chick sympathetic neurons (Rush *et al.*, 1986); this effect was not blocked by antisera to NGF.

E. Transforming Growth Factor-β (TGF-β) Superfamily

Members of the transforming growth factor-β (TGF-β) superfamily exhibit a panoply of actions on growth and differentiation. This family includes several TGF-βs, the inhibin/activin subunits, Muellerian inhibitory substance, bone morphogenic proteins (BMP), and glial-derived neurotrophic factor. Studies in avian systems are not available for all members of this family, but information on avian homologs is forthcoming.

1. TGF-βs

A number of closely related forms of TGF-β have been characterized in mammals including TGF-β1, TGF-β2, and TGF-β3. Structure of chicken TGF-β1, TGF-β2, and TGF-β3 have been deduced from the cDNA sequences. An additional isoform TGF-β4 has been characterized in the chicken. TGF-β isoforms show 75–80% homology to each other in their amino acid sequence. Moreover, the TGF-βs isoforms have the same overall structure as homodimers of 114-amino-acid-residue-containing subunits (Johnson, 1992). TGF-β isoforms are known to be produced by many cell types in mammals. As yet, TGF-β synthesis and/or expression has been reported in only a few cell types in only one avian species, the chicken. In this case, chondrocytes, myocytes, and preadipocytes have been found to express TGF-β2, TGF-β3, and TGF-β4 mRNA (Jalowlew *et al.*, 1991). The biological effects of TGF-β have been extensively examined with both mammalian and avian (chicken) cells *in vitro*. Studies with cultured chicken cells have employed mammalian TGF-β which is biologically active in birds. The biological activity of the different TGF-β isoforms appear to be similar to that of mammals. *In vitro*, TGF-β influences the proliferation and differentiation of different cell types. For instance, TGF-β decreases both DNA synthesis (proliferation) and melanocyte formation of quail neural crest cells (differentiation) while increasing expression of fibronectin (differentiation) (Rogers *et al.*, 1992). Similarly, with chick embryo neural crest cells, TGF-β1 inhibits the basal rate of differentiation of pigmented cells and the stimulated rate of melanogenesis caused by bFGF (Stocker *et al.*, 1991). With chick embryo chondrocytes, TGF-β increases DNA synthesis (proliferation) (Crabb *et al.*, 1988; O'Keefe *et al.*, 1988a) and stimulates sulfate incorporation and synthesis of noncollagen protein, although collagen synthesis is reduced (O'Keefe *et al.*, 1988b). The evidence for the effects of TGF-β together with the local expression of TGF-β isoforms and their respective mRNAs would support a role for TGF-β in embryonic limb formation (Leonard et al., 1991). On the other hand, in chick adipocytes, TGF-β acts as a very potent agent which increases proliferation (Butterwith and Goddard, 1991) while suppressing differentiation (Butterwith and Gilroy, 1991). In tandem with the effects of TGF-β on cell proliferation and differentiation, TGF-β1 has been shown to increase the expression of heat shock protein (Hsp 70 and Hsp 90) mRNA by chick embryo cells (Takenaka and Hightower, 1993).

2. Inhibin/Activin

These proteins were originally discovered by measuring their effects on gonadotrophin secretion from the anterior pituitary (Gaddy-Kurten et al., 1995). As closely related members of the TGF-β superfamily, inhibin and activin share common subunits which are encoded by distinct genes (Kingsley, 1994). Inhibin blocks pituitary release of follicle-stimulating hormone (FSH) and is composed of an α-subunit (unique to inhibin) and a β-subunit. Conversely, activin stimulates FSH release and is composed of dimers of the β-subunit, designated as β_A and β_B. A cDNA encoding the α-subunit of chicken inhibin has been cloned, sequenced, and used to demonstrate abundant mRNA levels in the granulosa layer of preovulatory follicles, but not other tissues (i.e., brain, kidney, liver, or spleen), of the laying hen (Johnson and Wang, 1993; Wang and Johnson, 1993). Recently, a cDNA clone which encodes the β_A-subunit of chicken inhibin/activin has been isolated, sequenced, and used to establish high levels of β_A-subunit mRNA in the granulosa layer of preovulatory follicles (Chen and Johnson, 1996). Apparently, a shorter transcript of chicken inhibin α-subunit is only found in muscle and the lung of the laying hen. There is some new evidence that inhibin and activin may play antagonistic roles in paracrine regulation of gonadal steroidogenesis in the chicken embryo (Rombauts et al., 1996). A unique nonreproductive role has been described for activin or activinlike molecules in determining left–right asymmetry during early organogenesis of the chick embryo (Levin et al., 1995; see Section VIII,F below).

3. Bone Morphogenic Protein-4

Bone morphogenic protein-4 (BMP-4) is osteoinductive, important for gastrulation, and involved in neural development. In the chick embryo, BMP-4 is expressed in the presumptive neural fold of the hind brain in a manner consistent with the depletion of neural crest cells in the rhombomeres (Graham et al., 1994). Moreover, treatment of these neural crest cells with BMP-4 induces programmed cell death or apoptosis (Graham et al., 1994).

4. Glial Cell-Derived Nerve Growth Factor (GDNF)

In ovo, GDNF increases the survival of motor neurons (Oppenheim et al., 1995). In vitro studies with chick embryo motor neurons further demonstrate that the mechanism for the improve survival is a reduction in programmed cell death or apoptosis (Oppenheim et al., 1995).

F. Homeobox Genes

Among the genes involved in morphogenesis and patterning, the *Hox* gene family of transcription factors is the best characterized. Homeobox genes are master genes which function during patterning and morphogenesis of the developing embryo. These genes are DNA binding transcription factors with high sequence homology in the DNA binding domain, referred to as the homeobox. The *Hox* class of homeobox genes is a linked group of genes, originally identified in *Drosophila*, that shows a high degree of structural and functional conservation in vertebrates (Fainsod and Gruenbaum, 1995). The expression of different combinations of *Hox* genes in a given region of a developing embryo determines the resulting structure of the embryo. Recently, factors involved in the regulation of expression of the *Hox* genes have provided more details on mechanisms regulating early development. For example, Hensen's node orchestrates anteroposterior, dorsoventral, and left–right body axis formation. Four RNAs are asymmetrically expressed along the left–right axis during early development (Levin et al., 1995). An activinlike molecule on the right side activates signaling through its receptor (cAct-RIIa) and suppresses sonic hedgehog (*Shh*) on the right. The absence of activin activity on the left side allows expression of *Shh*, which induces chicken nodal-related gene 1 (cNR-1) on the left side. This eventually leads to asymmetric heart development. *Shh* also participates in limb bud development (Laufer et al., 1994). In this case, FGF-4 acts in concert with *Shh* to control expression of critical *Hox* genes involved in formation of the limb. Retinoids can also control *Hox* gene expression (Tabin, 1995). Retinoids act in a manner analogous to steroid hormones (with nuclear receptors, namely the retinoic acid receptors (RARs) and the retinoid X receptors (RXRs)). A gradient of retinoic acid originating in the zone of polarizing activity induces the expression of *Hoxb*-1 in mesoderm and neuroectoderm of the early chick embryo (Marshall et al., 1994); *Hoxb*-1 is one of the earliest expressed members of the *Hoxb* homeobox gene complex. Although the target genes which are activated by the *Hox* gene network are not known, studies in the chick embryo are providing clues to factors controlling vertebrate morphogenesis and development.

IX. MODELS IN AVIAN GROWTH

A. Sex-Linked Dwarf Chickens

The sex-linked dwarf chicken (gene symbol, *dw*) (Hutt, 1959; Guillaume, 1976) represents an interesting model for understanding the importance of a functional somatotropic axis for normal growth and development. In general, dwarf chickens grow to about two-thirds of normal stature (Guillaume, 1976) and plasma levels of GH and T_4 are elevated, while plasma T_3 and IGF-1 levels are depressed (Scanes *et al.*, 1983; Stewart *et al.*, 1984; Huybrechts *et al.*, 1987). The disruption of the somatotropic axis in dwarf chickens is evidenced by a virtual absence of GH-binding activity in liver (Leung *et al.*, 1987; Vanderpooten *et al.*, 1991b; Burnside *et al.*, 1992) and serum (Cogburn *et al.*, 1997) and low circulating IGF-I (Huybrechts *et al.*, 1987; Cogburn *et al.*, 1995). The abnormalities in peripheral thyroid hormone metabolism reflect impaired conversion of T_4 into metabolically active T_3 (May and Marks, 1983; Scanes *et al.*, 1983; Kuhn *et al.*, 1990; Bartha *et al.*, 1994), which appears to be regulated by GH (Kuhn *et al.*, 1987; Darras *et al.*, 1990). The hypothyroid state of dwarf chickens contributes to lowered basal metabolism (Guillaume, 1976; Tixier-Boichard *et al.*, 1989; Decuypere *et al.*, 1991a), increased insulin sensitivity (Gueritault *et al.*, 1990), and their mild obesity (Guillaume, 1976; Stewart *et al.*, 1984). These numerous endocrine and metabolic maladies have led to considerable speculation about the primary lesion of sex-linked dwarfism in chickens (Scanes *et al.*, 1983; Decuypere *et al.*, 1987; Tixier-Boichard *et al.*, 1989; Decuypere *et al.*, 1991a).

A critical step in resolving the molecular basis of the dwarfing phenotype in chickens was cloning of the chicken GHR cDNA (Burnside *et al.*, 1991). This enabled the examination of the GHR gene in different strains of dwarf chickens for molecular defects (Burnside *et al.*, 1992; Huang *et al.*, 1993; Agarwal *et al.*, 1994). A mutation in the GHR gene has been characterized in a strain of broiler chickens maintained at the University of Georgia. In this strain, a T-to-C point mutation of the GT in a splice-donor site results in improper splicing of the GHR mRNA and absence of full-length transcripts (Huang *et al.*, 1993). Exactly the same mutation was reported later in a Japanese strain of sex-linked dwarf chickens (Tanaka *et al.*, 1995). A deletion of a portion of the intracellular domain was found in the GHR gene of the Connecticut dwarf broiler strain (Agarwal *et al.*, 1994). In a Leghorn dwarf strain, a point mutation results in the substitution of an isoleucine for a conserved serine in the extracellular domain (Duriez *et al.*, 1993). The mutation in the Cornell dwarf strain of Leghorn chickens has not yet been identified (Hull *et al.*, 1993). Regardless of the specific mutation in the GHR gene of different strains of dwarf chickens (Cogburn *et al.*, 1997), the physiological outcome is similar due to the absence of a functional GHR protein which is required for transduction of the GH signal in target cells (Burnside *et al.*, 1997).

The growth rate of dwarf chickens is similar to that of normal chickens until about 4 weeks of age (i.e., GH-independent growth); thereafter, the growth of normal chickens proceeds at a higher rate which marks the period of GH-dependent growth (Figure 6A). Although plasma IGF-I levels in dwarf chickens fail to show the typical developmental rise, IGF-I levels are maintained at about one-third that of normal birds (Figure 6B). However, GH-dependent IGF-I production can be restored in cultured dwarf hepatocytes after transfection and transient expression of the full-length chicken GHR

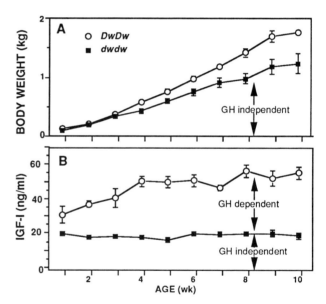

FIGURE 6 Body weight gain and plasma IGF-I levels of normal (*DwDw*) and sex-linked dwarf male (*dwdw*) broiler chickens from the Georgia strain (Huang *et al.*, 1993). Each value represents the mean ± SEM of six different birds. Plasma IGF-I levels were determined by a specific radioimmunoassay validated for chicken plasma (McGuinness and Cogburn, 1990, *Gen. Comp. Endocrinol.* with permission). Developmental changes in body weight and plasma IGF-I are considered to be GH independent since the Georgia dwarf does not express a functional GHR protein in tissue (Huang *et al.*, 1993). Adapted from Cogburn *et al.* (1997) with permission. (N. Huang, L. A. Cogburn, S. K. Agarwal, H. L. Marks, and J. Burnside; Over-expression of a truncated growth hormone receptor in the sex-linked dwarf chicken: Evidence for a splice mutation; *Molecular Endocrinology*; **7**; 1391–1398; 1993; © The Endocrine Society.)

cDNA (Huang et al., 1993). Thus, the normal balance between the somatotropic and thyrotropic axes is disrupted in the GHR-deficient dwarf chicken, where many of the endocrine and metabolic maladies reflect an absence of GH action in target tissue.

B. Growth-Selected Strains of Birds

Genetic selection for different growth traits represents another approach to understanding endocrine regulation of growth and development in domestic birds. In broiler chickens, selection for higher growth rate results in fatter chickens, while selection for improved feed efficiency results in leaner, slightly slower-growing chickens (Chambers et al., 1989; Pym et al., 1991; Leenstra et al., 1991; Tomas et al., 1991; Vanderpooten et al., 1993; Buyse et al., 1994). The endocrine profiles of different genetic lines of European broiler chickens has been recently reviewed (Decuypere et al., 1995). Faster growing chickens appear to have lower circulating GH levels (Goddard et al., 1988; Decuypere et al., 1991b; Leenstra et al., 1991; Decuypere et al., 1995) and higher hepatic GH-binding activity than their slower-growing counterparts (Vanderpooten et al., 1993). Divergent selection for abdominal fat results in slightly lower plasma cGH levels and higher hepatic GH-binding in the fat line (Buyse et al., 1994). Genetically fat chickens also have lower plasma T_3 (Buyse et al., 1994), a higher rate of hepatic lipogenesis, and increased insulin sensitivity when compared to lean chickens (Simon et al., 1991). While there is no direct effect of genetic selection on plasma IGF-I levels (Goddard et al., 1988; Leenstra et al., 1991; Pym et al., 1991), higher IGF-I appears to be correlated with a higher rate of protein accretion in leaner birds (Tomas et al., 1991). In growth-selected turkeys, plasma GH levels are higher in the random-bred population than in the fast-growing line (Vasilatos-Younken et al., 1988a). In contrast, plasma IGF-I levels were initially higher in the fast-growing line than in random-bred turkeys (Bacon et al., 1993). The strong inverse relationship between circulating GH and hepatic GH-binding activity could result from intensive selection pressure for rapid growth rate of modern broiler chickens. Examination of the developmental expression and tissue distribution of the GHR (Mao et al., 1998) and IGF-I genes in different lines of growth-selected broiler chickens could provide new insight to endocrine regulation of somatic growth and genetic selection for production traits.

References

Agarwal, S. K., Cogburn, L. A., and Burnside, J. (1994). Dysfunctional growth hormone receptor in a strain of sex-linked dwarf chicken: Evidence for a mutation in the intracellular domain. *J. Endocrinol.* **142**, 427–434.

Ali, S., Edery, M., Pellegrini, I., Lesueur, L., Paly, J., Djiane, J., and Kelly, P. A. (1992). The Nb2 form of prolactin receptor is able to activate a milk protein gene promoter. *Mol. Endocrinol.* **6**, 1242–1249.

Ambler, G. R., Butler, A. A., Padmanabhan, J., Breier, B. H., and Gluckman, P. D. (1996). The effects of octoreotide on GH receptor and IGF-I expression in the GH-deficient rat. *J. Endocrinol.* **149**, 223–231.

Anderson, T. R., Pitts, D. S., and Nicoll, C. S. (1984). Prolactin's mitogenic action on the pigeon crop-sac mucosal epithelium involves direct and indirect mechanisms. *Gen. Comp. Endocrinol.* **54**, 236–246.

Anthony, N. B., Emmerson, D. A., Nestor, K. E., and Bacon, W. L. (1991). Comparison of growth curves of weight selected populations of turkeys, quail and chickens. *Poult. Sci.* **70**, 13–19.

Armstrong, D. G., and Hogg, C. O. (1994). Type-I insulin-like growth factor receptor gene expression in the chick: Developmental changes and the effect of selection for increased growth on the amount of receptor mRNA. *J. Mol. Endocrinol.* **12**, 3–12.

Armstrong, D. G., McKay, C. O., Morrell, D. J., and Goddard, C. (1989). Insulin-like growth factor-I binding proteins in serum from the domestic fowl. *J. Endocrinol.* **120**, 373–378.

Bacon, W. L., Burke, W. H., Anthony, N. B., and Nestor, K. E. (1987). Growth hormone status and growth characteristics of Japanese quail divergently selected for four-week body weights. *Poult. Sci.* **66**, 1541–1544.

Bacon, W. L., Nestor, K. E., Emmerson, D. A., Vasilatos-Younken, R., and Long, D. W. (1993). Circulating IGF-I in plasma of growing male and female turkeys of medium and heavy weight lines. *Dom. Anim. Endocrinol.* **10**, 267–277.

Balk, S. D., Morisi, A., Gunther, H. S., Svoboda, M. F., Van Wyk, J. J., Nissley, S. P., and Scanes, C. G. (1984). Somatomedins (insulin-like growth factors) but not growth hormone, are mitogenic for chicken heart mesenchymal cells and act synergistically with epidermal growth factor and brain fibroblast growth factor. *Life Sci.* **35**, 335–346.

Balk, S. D., Shiu, R. P. C., La Fleur, M. M., and Young, L. L. (1982). Epidermal growth factor and insulin cause normal chicken heart mesenchymal cells to proliferate like their Rous sarcoma virus-infected counterparts. *Proc. Nat. Acad. Sci. USA* **79**, 1154–1157.

Ballard, F. J., Johnson, R. J., Owens, P. C., Francis, G. L., Upton, F. M., McMurtry, J. P., and Wallace, J. C. (1990). Chicken insulin-like growth factor-I: Amino acid sequence, radioimmunoassay, and plasma levels between strains and during growth. *Gen. Comp. Endocrinol.* **79**, 459–468.

Barbato, G. F. (1991). Genetic architecture of growth curve parameters in chickens. *Theor. Appl. Genet.* **83**, 24–32.

Bartha, T., Dewil, E., Rudas, P., Kuhn, E. R., Scanes, C. G., and Decuypere, E. (1994). Kinetic parameters of plasma thyroid hormone and thyroid hormome receptors in a dwarf and control line of chicken. *Gen. Comp. Endocrinol.* **96**, 140–148.

Bassas, L., De Pablo, F., Lesniak, M. A., and Roth, J. (1987). The insulin receptors of chick embryo show tissue-specific structural differences which parallel those of the insulin-like growth factor-I receptors. *Endocrinology* **121**, 1468–1476.

Baumann, G. (1990). Growth hormone-binding proteins. *Trends Endo. Metab.* **Sept./Oct.**, 342–347.

Belew, M., and Ebendal, T. (1986). Chick embryo nerve growth factor, fractionation and biological activity. *Exp. Cell Res.* **167**, 550–558.

Bellabarba, D., and Lehoux, J. G. (1981). Triiodothyronine nuclear receptor in chick embryo: Nature and properties of hepatic receptor. *Endocrinology* **109**, 1017–1025.

Bellabarba, D., Redard, S., Fortier, S., and Leoux, J. G. (1983). 3,5,3'-Triiodothyronine nuclear receptor in chick embryo: Properties and ontogeny of brain and lung receptors. *Endocrinology* **112**, 353–359.

Bhat, G., Gupta, S. K., and Maiti, B. R. (1983). Influence of prolactin on mitotic activity of the bursa of Fabricius of the chick. *Gen. Comp. Endocrinol.* **52,** 452–455.

Bray, M. M. (1993). Effect of ACTH and glucocorticoids on lipid metabolism in the Japanese quail, *Coturnix coturnix japonica. Comp. Biochem. Physiol. A* **105,** 689–696.

Breneman, W. R. (1942). Action of prolactin and estrone on weights of reproductive organs and viscera of the cockerel. *Endocrinology* **30,** 609–615.

Buntin, J. D., and Ruzycki, E. (1987). Characteristics of prolactin binding sites in the brain of the ring dove (*Streptopeliz risoria*). *Gen. Comp. Endocrinol.* **65,** 243–253.

Buntin, J. D., and Tesch, D. (1985). Effects of intracranial prolactin administration on maintenance of incubation readiness, ingestive behavior, and gonadal condition in ring doves. *Horm. Behav.* **19,** 188–203.

Buntin, J. D., Keskey, T. S., and Janik, D. S. (1984). Properties of hepatic binding sites for prolactin in ring dove (*Streptopelia risoria*). *Gen. Comp. Endocrinol.* **55,** 418–428.

Burch, W. M., and Lebovitz, H. E. (1982). Triiodothyronine stimulation of *in vitro* growth and maturation of embryonic chick cartilage. *Endocrinology* **111,** 462–468.

Burch, W. M., and Van Wyk, J. J. (1987). Triiodothyronine stimulates cartilage growth and maturation by different mechanisms. *Am. J. Physiol.* **252,** E176–E182.

Burch, W. M., Corda, G., Kopchick, J. J., and Leung, F. C. (1985). Homologous and heterologous growth hormones fail to stimulate avain cartilage growth in vitro. *J. Clin. Endocrinol. Metab.* **60,** 747–750.

Burch, W. M., Weir, S., and Van Wyk, J. J. (1986). Embryonic chick cartilage produces its own somatomedin-like peptide to stimulate cartilage growth *in vitro. Endocrinology* **119,** 1370–1376.

Burke, W. H., and Marks, H. L. (1982). Growth hormone and prolactin levels in nonselected and selected broiler lines of chickens from hatch to eight weeks of age. *Growth* **46,** 283–295.

Burke, W. H., Moore, J. A., Ogez, J. R., and Builder, S. E. (1987). The properties of recombinant chicken growth hormone and its effects on growth, body composition, feed efficiency, and other factors in broiler chickens. *Endocrinology* **120,** 651–658.

Burnside, J., and Cogburn, L. A. (1992). Developmental expression of hepatic growth hormone receptor and insulin-like growth factor-I mRNA in the chicken. *Mol. Cell. Endocrinol.* **89,** 91–96.

Burnside, J., and Cogburn, L. A. (1993). Molecular biology of the chicken growth hormone receptor. *In* "Avian Endocrinology" (P. J. Sharp, ed.), pp. 161–176. Journal of Endocrinology Ltd., Bristol.

Burnside, J., Agarwal, S. K., and Cogburn, L. A. (1997). Intracellular mechanism of growth hormone signaling. *In* "Perspectives in Avian Endocrinology" (S. Harvey and R. Etches, eds.), pp. 359–373. Journal of Endocrinology Press, Bristol.

Burnside, J., Liou, S. S., and Cogburn, L. A. (1991). Molecular cloning of the chicken growth hormone receptor complementary DNA: Mutation of the gene in sex-linked dwarf chickens *Endocrinology* **128,** 3183–3192.

Burnside, J., Liou, S. S., Zhong, C., and Cogburn, L. A. (1992). Abnormal growth hormone receptor gene expression in the sex-linked dwarf chicken. *Gen. Comp. Endocrinol.* **88,** 20–28.

Burrus, L. W., and Olwin, B. B. (1989). Isolation of a receptor for acidic and basic fibroblast growth factor from embryonic chick. *J. Biol. Chem.* **264,** 18647.

Burt, D. W., Boswell, J. M., Paton, I. R., and Butterwith, S. C. (1992). Multiple growth factor mRNAS are expressed in chicken adipocyte precursor cells. *Biochem. Biophys. Res. Commun.* **187,** 1298–1305.

Butterwith, S. C., and Gilroy, M. (1991). Effects of transforming growth factor-β1 and basic fibroblast growth factor on lipoprotein lipase activity in primary cultures of chicken (*Gallus domesticus*) adipocyte precursors. *Comp. Biochem. Physiol. A* **100,** 473–476.

Butterwith, S. C., and Goddard, C. (1991). Regulation of DNA synthesis in chicken adipocyte precursor cells by insulin-like growth factors, platelet-derived growth factor and transforming growth factor-β. *J. Endocrinol.* **131,** 203–209.

Butterwith, S. C., Peddie, C. D., and Goddard, C. (1993). Regulation of adipocyte precursor DNA synthesis by acidic and basic fibroblast growth factors: Interaction with heparin and other growth factors. *J. Endocrinol.* **137,** 369–374.

Buyse, J., Decuypere, E., Sharp, P. J., Huybrechts, L. M., Kuhn, E. R., and Whitehead, C. (1987). Effect of corticosterone on circulating concentrations of corticosterone, prolactin, thyroid hormones and somatomedin C and on fattening in broilers selected for high or low fat content. *J. Endocrinol.* **112,** 229–237.

Buyse, J., Vanderpooten, A., Leclercq, B., Berghman, L. R., and Decuypere, E. (1994). Pulsatility of plasma growth hormone and hepatic growth hormone receptor characteristics of broiler chickens divergently selected for abdominal fat content. *Br. Poult. Sci.* **35,** 145–152.

Caldes, J., Alemany, J., Robcis, H. L., and De Pablo, F. (1991). Expression of insulin-like growth factor I in developing lens is compartmentalized. *J. Biol. Chem.* **266,** 20786–20790.

Carter-Su, C., Argetsinger, L. S., Campbell, G. S., Wang, X., Ihle, J., and Witthuhn, B. (1994). The identification of JAK2 tyrosine kinase as a signaling molecule for growth hormone. *Proc. Soc. Exp. Biol. Med.* **206,** 210–215.

Chambers, J. R., Fortin, A., Mackie, D. A., and Larmond, E. (1989). Comparison of sensory properties of meat from broilers of modern stocks and experimental strains differing in growth and fatness. *Can. Inst. Food Sci. Technol. J.* **22,** 353–358.

Chen, C-C., and Johnson, P. A. (1996). Molecular cloning of inhibin/activin β_A-subunit complementary deoxyribonucleic acid and expression of inhibin/activin α- and β_A-subunits in the domestic hen. *Biol. Reprod.* **54,** 429–435.

Chen, X., and Horseman, N. D. (1994). Cloning, expression, and mutational analysis of the pigeon prolactin receptor. *Endocrinology* **135,** 269–276.

Cogburn, L. A. (1991). Endocrine manipulation of body composition in broiler chickens. *Crit. Rev. Poult. Biol.* **3,** 283–305.

Cogburn, L. A., Liou, S. S., Rand, A. L., and McMurtry, J. P. (1989). Growth, metabolic and endocrine responses of broiler cockerels given a daily subcutaneous injection of natural or biosynthetic chicken growth hormone. *J. Nutr.* **119,** 1213–1222.

Cogburn, L. A., Mao, J. N. C., and Burnside, J. (1997). The growth hormone receptor in growth and development. *In* "Perspectives in Avian Endocrinology" (S. Harvey and R. Etches, eds.), pp. 101–118. Journal of Endocrinology Press, Bristol.

Cogburn, L. A., Mao, N. C., Agarwal, S., and Burnside, J. (1995). Interaction between somatotropic and thyrotropic axes in regulation of growth and development of broiler chickens. *Arch. Geflugel. Sonderh.* **1,** 18–21.

Cohen, S. (1965). The stimulation of epidermal proliferation by a specific protein (EGF). *Dev. Biol.* **12,** 394.

Crabb, I. D., O'Keefe, R. J., Puzas, J. E., and Rosier, R. N. (1988). Stimulation of DNA synthesis in chick growth plate chondrocytes by transforming growth factor-β and fibroblast growth factor. *Orthop. Trans.* **12,** 533.

Cramer, J. L., and Harper, J. A. (1973). The effects of prolactin and other hormones on behavioral and physiological responses in turkey males. *Poult. Sci.* **52,** 2016.

Cynober, L., Aussel, C., Chatelain, P., Vaubourdolle, M., Agneray, J., and Ekindjian, O. G. (1985). Insulin-like growth factor I/somatomedin C action on 2-deoxyglucose and a-amino isobutyrate uptake in chick embryo fibroblasts. *Biochimie* **67**, 1185–1190.

Darling, D. C., and Brickell, P. M. (1996). Nucleotide sequence and genomic structure of the chicken insulin-like growth factor-II (IGF-II) coding region. *Gen. Comp. Endocrinol.* **102**, 283–287.

Darras, V. M., Huybrechts, L. M., Berghman, L., Kuhn, E. R., and Decuypere, E. (1990). Ontogeny of the effect of purified chicken growth hormone on the liver 5' monodeiodination activity in the chicken: Reversal of the activity after hatching. *Gen. Comp. Endocrinol.* **77**, 212–220.

Dawe, S. R., Francis, G. L., McNamara, P. J., Wallace, J. C., and Ballard, F. J. (1988). Purification, partial sequences and properties of chicken insulin-like growth factors. *J. Endocrinol.* **117**, 173–181.

De La Cruz, L. F., Mataix, F. J., and Illera, M. (1981). Effects of glucocorticoids on protein metabolism in laying quail (*Coturnix coturnix japonica*). *Comp. Biochem. Physiol. A* **70**, 649–652.

De Pablo, F., Girbau, M., Gomez, J. A., Hernandez, E., and Roth, J. (1985a). Insulin antibodies retard and insulin accelerates growth and differentiation in early embryos. *Diabetes* **34**, 1063.

De Pablo, F., Hernandez, E., Collia, F., and Gomez, J. A. (1985b). Untoward effects of pharmacological doses of insulin in early chick embryos: Through which receptor are they mediated? *Diabetologia* **28**, 308–313.

De Pablo, F., Perez, V. B., Serna, J., and Gonzalez, G. P. R. (1993). IGF-I and the IGF-I receptor in development of nonmammalian vertebrates. *Mol. Reprod. Dev.* **35**, 427–432.

De Pablo, F., Scott, L. A., and Roth, J. (1990). Insulin and insulin-like growth factor I in early development: Peptides, receptors and biological events. *Endocr. Rev.* **11**, 558–577.

Decuypere, E., Buyse, J., Scanes, C. G., Huybrechts, L., and Kuhn, E. R. (1987). Effects of hyper- or hypothyroid status on growth, adiposity and levels of growth hormone, somatomedin-C and thyroid metabolism in broiler chickens. *Reprod. Nutr. Dev.* **27**, 555–565.

Decuypere, E., Buyse, J., Rahimi, G., and Zeman, M. (1995). Comparative study of endocrinological parameters in the genetic lines of broilers: A review. *Arch. Geflugel. Sonderh.* **1**, 6–8.

Decuypere, E., Huybrechts, L. M., Kuhn, E. R., Tixier-Biochard, M., and Merat, P. (1991a). Physiological alterations associated with the chicken sex-linked dwarfing gene. *Crit. Rev. Poult. Biol.* **3**, 191–221.

Decuypere, E., Leenstra, F., Buyse, J., Beuving, G., and Berghman, L. (1991b). Temporal secretory patterns of growth hormone in male meat-type chickens of lines selected for body weight gain or food conversion. *Br. Poult. Sci.* **32**, 1121–1128.

Delbe, J., Blat, C., Desauty, G., and Harel, L. (1991). Presence of IDF45 (mIGFBP-3) binding sites on chick embryo fibroblasts. *Biochem. Biophys. Res. Commun.* **179**, 495–501.

Diaz-Ruiz, C., Perez-Tomas, R., Cullere, X., and Domingo, J. (1993). Immunohistochemical localization of transforming growth factor-α and epidermal growth factor-receptor in the mesonephros and metanephros of the chicken. *Cell Tissue Res.* **271**, 3–8.

Dimberg, Y., and Ebendal, T. (1987). Effects of nerve growth factor on autonomic neurons in the chick embryo: a stereological study. *Int. J. Dev. Neurosci.* **5**, 195–205.

Dimberg, Y., and Ebendal, T. (1988). Effects of injecting antibodies to mouse nerve growth factor into the chicken embryo. *Int. J. Dev. Neurosci.* **6**, 513–523.

Djiane, J., Daniel, N., Bignon, C., Paly, J., Waters, M., Vacher, P., and Duft, B. (1994). Prolactin receptor and signal transduction to milk protein genes. *Proc. Soc. Exp. Biol. Med.* **206**, 299–303.

Donoghue, D. J., Campbell, R. M., and Scanes, C. G. (1990). Effect of biosynthetic chicken growth hormone on egg production in white Leghorn hens. *Poult. Sci.* **69**, 1818–1821.

Doppler, W., Hock, W., Hofer, P., Groner, B., and Ball, R. K. (1990). Prolactin and glucocorticoid hormones control transcription of the beta-casein gene by kinetically distinct mechanisms. *Mol. Endocrinol.* **4**, 912–919.

Duclos, M. J., and Goddard, C. (1990). Insulin-like growth factor receptors in chicken liver membranes: Binding properties, specificity, developmental pattern and evidence for a single receptor type. *J. Endocrinol.* **125**, 199–206.

Duclos, M. J., Wilkie, R. S., and Goddard, C. (1991). Stimulation of DNA synthesis in chicken muscle satellite cells by insulin and insulin-like growth factors: Evidence for exclusive mediation by a type-I insulin-like growth factor receptor. *J. Endocrinol.* **128**, 35–42.

Duriez, B., Sobrier, M., Duquesnoy, P., Tixier-Boichard, M., Decuypere, E., Coquerelle, G., Zeman, M., Goossens, M., and Amselem, S. (1993). A naturally occurring growth hormone receptor mutation: *In vivo* and *in vitro* evidence for the functional importance of the WS motif common to all members of the cytokine receptor superfamily. *Mol. Endocrinol.* **7**, 806–814.

Ebendal, T., and Persson, H. (1988). Detection of nerve growth factor mRNA in the developing chicken embryo. *Development* **102**, 101–106.

Ebendal, T., Hedlung, K.-O., and Norrgren, G. (1982). Nerve growth factors in chick tissues. *J. Neurosci. Res.* **8**, 153–164.

Ebendal, T., Larhammar, D., and Persson, H. (1986). Structure and expression of the chicken β nerve growth factor gene. *EMBO J.* **5**, 1483–1487.

El Halawani, M. E., and Rozenboim, I. (1993). The ontogeny and control of incubation behavior in turkey. *Poult. Sci.* **72**, 906–911.

El Halawani, M. E., Burke, W. H., Millam, J. R., Fehrer, S. C., and Hargis, B. M. (1984). Regulation of prolactin and its role in Gallinaceous bird reproduction. *J. Exp. Zool.* **232**, 521–529.

El Halawani, M. E., Fehrer, S., Hargis, B. M., and Porter, T. E. (1988). Incubation behavior in the domestic turkey: Physiological correlates. *CRC Rev. Poult. Biol.* **1**, 285–314.

El Halawani, M. E., Silsby, J. L., Youngren, O. M., and Phillips, R. E. (1991). Exogenous prolactin delays photo-induced sexual maturity and suppresses ovariectomy-induced luteinizing hormone secretion in the turkey (*Meleagris gallopavo*). *Biol. Reprod.* **44**, 420–424.

Elgin, R. G., Rusby, W. H., and Clemmons, D. R. (1987). An insulin-like growth factor (IGF) binding protein enhances the biologic response to IGF-I. *Proc. Natl. Acad. Sci. USA* **84**, 3254–3258.

Ensor, D. M. (1978). "Comparative Endocrinology of Prolactin." Chapman and Hall Ltd., London.

Fainsod, A., Gruenbaum, Y. (1995). Homeobox genes in avian development. *Poult. Avian Biol. Rev.* **6**, 19–34.

Fawcett, D. H., and Bulfield, G. (1990). Molecular cloning, sequence analysis and expression of putative chicken insulin-like growth factor-I cDNAs. *J. Mol. Endocrinol.* **4**, 201–211.

Fennell, M. J., and Scanes, C. G. (1992a). Effects of androgen (testosterone, 5α-dihydrotestosterone, 19-nortestosterone) administration on growth in turkeys. *Poultry Sci.* **71**, 539–547.

Fennell, M. J., and Scanes, C. G. (1992b). Inhibition of growth in chickens by testosterone, 5α-dihydrotestosterone and 19-nortestosterone. *Poult. Sci.* **71**, 357–366.

Fennell, M. J., Johnson, A. L., and Scanes, C. G. (1990). Influence of androgens on plasma concentrations of growth hormone in growing castrated and intact chickens. *Gen. Comp. Endocrinol.* **77**, 466–475.

Francis, G. L., McMurtry, J. P., Johnson, R. J., and Ballard, F. J. (1990). Plasma clearance of chicken and human insulin-like growth factor-I and their association with circulating binding proteins in chickens. *J. Endocrinol.* **124**, 361–370.

Fuh, G., Cunningham, B. C., Fukunaga, R., Nagata, S., Goeddel, D. V., and Wells, J. A. (1992). Rational design of potent antagonists to the human growth hormone receptor. *Science* **256,** 1677–1680.

Fuh, G., Colosi, P., Wood, W. I., and Wells, J. A. (1993). Mechanism-based design of prolactin receptor antagonists. *J. Biol. Chem.* **268,** 5376–5381.

Gaddy-Kurten, D., Tsuchida, K., and Vale, W. (1995). Activins and the receptor serine kinase superfamily. *Rec. Prog. Horm. Res.* **50,** 109–129.

Garrison, M. M., and Scow, R. O. (1975). Effect of prolactin on lipoprotein lipase in crop sac and adipose tissue of pigeons. *Am. J. Physiol.* **228,** 1542–1544.

Gelb, D. E., Rosier, R. N., and Puzas, J. E. (1990). The production of transforming growth factor-β by chick growth plate chondrocytes in short term monolayer culture. *Endocrinology* **127,** 1941–1947.

Girbau, M., Gomez, J. A., Lesniak, M. A., and De Pablo, F. (1987). Insulin and insulin-like growth factor I both stimulate metabolism, growth, and differentiation in the postneurula chick embryo. *Endocrinology* **121,** 1477–1482.

Girbau, M., Lesniak, M. A., Gomez, J. A., and De Pablo, F. (1988). Insulin action in early embryonic life: Anti-insulin receptor antibodies retard chicken embryo growth and not muscle differentiation *in vivo*. *Biochem. Biophys. Res. Commun.* **153,** 142–148.

Goddard, C., Wilkie, R. S., and Dunn, I. C. (1988). The relationship between insulin-like growth factor-I, growth hormone, thyroid hormones, and insulin in chickens selected for growth. *Dom. Anim. Endocrinol.* **5,** 165–176.

Goddard, C., Butterwith, S. C., Roberts, R. D., and Duclos, M. (1993). Insulin-like growth factors and IGF binding proteins. *In* "Avian Endocrinology" (P. J. Sharp, ed.), pp. 161–176. Journal of Endocrinology Ltd., Bristol.

Goedert, M. (1986). Molecular cloning of the chicken nerve growth factor gene: mRNA distribution in developing and adult tissues. *Biochem. Biophys. Res. Commun.* **141,** 1116–1122.

Gordon-Thomson, C., and Fabian, B. C. (1994). Hypoblastic tissue and fibroblast growth factor in chicken blood tissue (haemoglobin) in the early chicken embryo. *Development* **120,** 3571–3579.

Gorman, D. M., Itoh, N., Kitamura, T., Schreurs, J., Yonehara, S., Yahara, I., Arai, K., and Miyajima, A. (1990). Cloning and expression of a gene encoding an interleukin 3 receptor-like protein: Identification of another member of the cytokine receptor gene family. *Proc. Natl. Acad. Sci. USA* **87,** 5459–5463.

Gospodarowicz, D., Ferrara, N., Schweigerer, L., and Neufeld, G. (1987). Structural characterization and biological functions of fibroblast growth factor. *Endocr. Rev.* **8,** 95–114.

Graham, A., Francis-West, P., Brickell, P., and Lumsden, A. (1994). The signalling molecule BMP4 mediates apoptosis in the rhonbencephalic neural crest. *Nature* **372,** 684–686.

Guemene, D., Bedecarrats, G., Karatzas, C. N., Garreau-Mills, M., Kuhnlein, U., Crisostomo-Pinto, S., and Zadworny, D. (1994). Development and validation of a homologous radioimmunoassay using a biologically active recombinant turkey prolactin. *Br. Poult. Sci.* **35,** 775–787.

Guemene, D., Karatzas, C. N., Kuhnlein, U., and Zadworny, D. (1992). Post-hatching effects of in ovo injection of recombinant prolactin in domestic turkeys. *Proc. XIV World Poult. Congr.* **1,** 783–787.

Gueritault, I., Simon, J., Chevalier, B., Derouet, M., Tixier-Boichard, M., and Merat, P. (1990). Increased *in vivo* insulin sensitivity but normal liver insulin receptor kinase activity in dwarf chickens. *J. Endocrinol.* **126,** 67–74.

Guillaume, J. (1976). The dwarfing gene *dw*: Its effects on anatomy, physiology, nutrition, management. *World's Poult. Sci. J.* **32,** 285–304.

Halevy, O., Monsonego, E., Marcelle, C., Hodik, V., Mett, A., and Pines, M. (1994). A new avian fibroblast growth factor receptor in myogenic and chondrogenic cell differentiation. *Exp. Cell Res.* **212,** 278–284.

Hall, T. R., Harvey, S., and Chadwick, A. (1986). Control of prolactin secretion in birds: A review. *Gen. Comp. Endocrinol.* **62,** 171–184.

Hanks, M. C., Alonzi, J. A., Sharp, P. J., and Sang, H. M. (1989a). Molecular cloning and sequence analysis of putative chicken prolactin cDNA. *J. Mol. Endocrinol.* **2,** 21–30.

Hanks, M. C., Talbot, R. T., and Sang, H. M. (1989b). Expression of biologically active recombinant-derived chicken prolactin in *Escherichia coli*. *J. Mol. Endocrinol.* **3,** 15–21.

Harvey, S., Hall, T. R., and Chadwick, A. (1984). Growth hormone and prolactin secretion in water deprived chickens. *Gen. Comp. Endocrinol.* **54,** 46–50.

Harvey, S., Scanes, C. G., Chadwick, A., and Bolton, N. J. (1979). Growth hormone and prolactin secretion in growing domestic fowl: Influence of sex and breed. *Br. Poult. Sci.* **20,** 9–17.

Hnizetova, H., Hyanek, J., Knize, B., and Roubicek, J. (1991). Analysis of growth curves of fowl. I. chickens. *Br. Poult. Sci.* **32,** 1027–1038.

Hocking, P. M., Bernard, R., Wilkie, R. S., and Goddard, C. (1994). Plasma growth hormone and insulin-like growth factor-I (IGF-I) concentrations at the onset of lay in ad libitum and restricted broiler breeder fowl. *Br. Poult. Sci.* **35,** 299–308.

Holzenberger, M., Lapointe, F., Leibovici, M., and Ayer-Le Lievre, C. (1996). The avian IGF type I receptor: cDNA analysis and *in situ* hybridization reveal conserved sequence elements and expression patterns relevant for the development of the nervous system. *Dev. Brain Res.* **97,** 76–87.

Horseman, N. D. (1989). A prolactin-inducible gene product which is a member of the calpactin/lipocortin family. *Mol. Endocrinol.* **3,** 773–779.

Horseman, N. D. (1994). Famine to feast: Growth hormone and prolactin signal transducers. *Endocrinology* **135,** 1289–1291.

Horseman, N., and Buntin, J. D. (1995). Regulation of pigeon cropmilk secretion and parental behaviors by prolactin. *Ann. Rev. Nutr.* **15,** 213–238.

Horseman, N. D., and Yu-Lee, L. Y. (1994). Transcriptional regulation by the helix bundle peptide hormones: Growth hormone, prolactin, and hematopoietic cytokines. *Endocr. Rev.* **15,** 627–649.

Horseman, N. D., Chen, X., Liu, L., Poyet, P., and Hitti, Y. (1992). Cell and species distribution of prolactin-inducible annexin I mRNA. *Gen. Comp Endocrinol.* **85,** 405–414.

Huang, N., Cogburn, L. A., Agarwal, S. K., Marks, H. L., and Burnside, J. (1993). Over-expression of a truncated growth hormone receptor in the sex-linked dwarf chicken: Evidence for a splice mutation. *Mol. Endocrinol.* **7,** 1391–1398.

Hutt, F. B. (1959). Sex-linked dwarfism in the fowl. *J. Hered.* **15,** 97–110.

Hull, K. L., Fraser, R. A., Marsh, J. A., and Harvey, S. (1993). Growth hormone receptor gene expression in sex-linked dwarf Leghorn chickens: Evidence against a gene deletion. *J. Endocrinol.* **137,** 91–98.

Huybrechts, L. M., King, D. B., Lauterio, T. J., Marsh, J., and Scanes, C. G. (1987). Plasma concentrations of growth hormone and somatomedin C in dwarf and normal chickens. *Reprod. Nutr. Dev.* **27,** 547–553.

Ishida, H., Shimada, K., Sato, K., Seo, H., Murata, Y., Matsui, N., and Zadworny, D. (1991). Developmental expression of the prolactin gene in the chicken. *Gen. Comp. Endocrinol.* **83,** 463–467.

Jalowlew, S. B., Dillard, P. J., Winokur, T. S., Flanders, K. C., Sporn, M. B., and Roberts, A. B. (1991). Expression of transforming growth factor-β 1-4 in chicken embryo chondrocytes and myocytes. *Dev. Biol.* **143,** 135.

Johnson, A. L. (1992). Non-IGF growth factors. In "The Endocrinology of Growth, Development and Metabolism" (M. P. Schreibman, C. G. Scanes, and P. K. T. Pang, eds.), pp. 220–248. Academic Press, San Diego.

Johnson, P. A., and Wang, S.-Y. (1993). Characterization and quantitation of mRNA for the inhibin α-subunit in the granulosa layer of the domestic hen. *Gen. Comp. Endocrinol.* **90,** 43–50.

Johnson, R. J., McMurtry, J. P., and Ballard, F. J. (1990). Ontogeny and secretory patterns of plasma insulin-like growth factor-I concentrations in meat-type chickens. *J. Endocrinol.* **124,** 81–87.

Kajimoto, Y., and Rotwein, P. (1989). Structure and expression of a chicken insulin-like growth factor I precursor. *J. Mol. Endocrinol.* **3,** 1907–1913.

Kalcheim, C., and Neufeld, G. (1990). Expression of basic fibroblast growth factor in the nervous system of early avian embryos. *Development* **109,** 203.

Kallincos, N. C., Wallace, J. C., Francis, G. L., and Ballard, F. J. (1990). Chemical and biological characterization of chicken insulin-like growth factor-II. *J. Endocrinol.* **124,** 89–97.

Kansaku, N., Shimada, K., Terada, O., and Saito, N. (1994). Prolactin, growth hormone, and luteinizing hormone-β subunit gene expression in the cephalic and caudal lobes of the anterior pituitary gland during embryogenesis and different reproductive stages in the chicken. *Gen. Comp. Endocrinol.* **96,** 197–205.

Kelly, P. A., Ali, S., Rozakis, M., Goujon, L., Nagano, M., Pellefrini, I., Gould, D., Djiane, J., Edery, M., Finidori, J., and Postel-Vinay, M. C. (1993). The growth hormone-prolactin receptor family. *Rec. Prog. Horm. Res.* **48,** 123–164.

Kelly, P. A., Edery, M., Finidori, J., Postel-Vinay, M. C., Gougon, L., Ali, S., Dinerstein, H., Sotiropoulos, A., Lochnan, H., Ferrag, F., Lebrun, J. J., Ormandy, C., Buteau, H., Esposito, N., Vincent, V., and Moldrup, A. (1994). Receptor domains involved in signal transduction of prolactin and growth hormone. *Proc. Soc. Exp. Biol. Med.* **206,** 280–283.

Kida, S., Iwaki, M., Nakamura, A., Miura, Y., Takenaka, A., Takahashi, S., and Noguchi, T. (1994). Insulin-like growth factor-I messenger RNA content in the oviduct of Japnese quail (*Coturnix coturnix japonica*): Changes during growth and development or after estrogen administration. *Comp. Biochem. Physiol. C* **109,** 191–204.

Kikuchi, K., Buonomo, F. C., Kajimoto, Y., and Rotwein, P. (1991). Expression of insulin-like growth factor-I during chicken development. *Endocrinology* **128,** 1323–1328.

King, D. B. (1969). Effect of hypophysectomy of young cockerels, with particular reference to body growth, liver weight, and liver glycogen level. *Gen. Comp. Endocrinol.* **12,** 242–255.

King, D. B., and King, C. R. (1973). Thyroidal influence on early muscle growth of chickens. *Gen. Comp. Endocrinol.* **21,** 517–529.

King, D. B., and King, C. R. (1976). Thyroidal influence on gastrocnemius and sartorius muscle growth in young White Leghorn cockerels. *Gen. Comp. Endocrinol.* **29,** 473–479.

King, D. B., and Scanes, C. G. (1986). Effect of mammalian growth hormone and prolactin on the growth of hypophysectomized chickens. *Proc. Soc. Exp. Biol. Med.* **182,** 201–207.

Krishnan, K. A., Proudman, J. A., and Bahr, J. M. (1989). Avian growth hormone receptor assay: Use of chicken and turkey liver membranes. *Mol. Cell. Endocrinol.* **66,** 125–134.

Krishnan, K. A., Proudman, J. A., and Bahr, J. M. (1991). Radioligand receptor assay for prolactin using chicken and turkey kidney membranes. *Comp. Biochem. Physiol. B* **100,** 769–774.

Kuhn, E. R., Huybrechts, L. M., Darras, V. M., Meeuwis, R., and Decuypere, E. (1990). Impaired peripheral T_3 production but normal induced thyroid hormone secretion in the sex-linked dwarf chick embryo. *Reprod. Nutr. Dev.* **30,** 193–201.

Kuhn, E. R., Shimada, K., Ohkubo, T., Vleurick, L. M., Berghman, L. R., and Darras, V. M. (1996). Influence of recombinant chicken prolactin on thyroid hormone metabolism in the chick embryo. *Gen. Comp. Endocrinol.* **103,** 349–358.

Kuhn, E. R., Verheyen, G., Chiasson, R. B., Huts, C., Huybrechts, L., Van den Steen, P., and Decuypere, E. (1987). Growth hormone stimulates the peripheral conversion of thyroxine into triiodothyronine by increasing the liver 5'-monodeiodinase activity in the fasted and normal fed chicken. *Horm. Metab. Res.* **19,** 304–308.

Laufer, E., Nelson, C. E., Johnson, R. L., Morgan, B. A., and Tabin, C. (1994). *Sonic hedgehog* and *FGF-4* act through a signaling cascade and feedback loop to integrate growth and patterning of the developing limb bud. *Cell* **79,** 993–1003.

Lax, I., Johnson, A., Howk, R., Sap, J., Bellot, F., Winkler, M., Ullrich, M., Vennstrom, B., Schlessinger, J., and Givol, D. (1988). Chicken epidermal growth factor (EGF) receptor: cDNA cloning, expression in mouse cells, and differential binding of EGF and transforming growth factor alpha. *Mol. Cell. Biol.* **8,** 1970–1978.

Leach, R. M., and Rosselot, G. E. (1992). The use of avian epiphyseal chondrocytes for in vitro studies of skeletal metabolism. *J. Nutr.* **122,** 802–805.

Lee, P. L., Johnson, D. E., Cousens, L. S., Fried, V. A., and Williams, L. T. (1989). Purification and complementary DNA cloning of a receptor for basic fibroblast growth factor. *Science* **245,** 57–60.

Leenstra, F. R., Decuypere, E., Beuving, G., Buyse, J., and Berghman, L. (1991). Concentration of hormones, glucose, triglycerides and free fatty acids in plasma of broiler chickens, selected for weight gain or feed conversion. *Br. Poult. Sci.* **32,** 619–632.

Leonard, C. M., Fuld, H. M., Frenz, D. A., Downie, S. A., Massague, J., and Newman, S. A. (1991). Role of transforming growth factor-β in chondrogenic pattern formation in the embryonic limb: Stimulation of mesenchymal condensation and fibronectin gene expression by exogenous TGF-β and evidence for endogenous TGF-β-like activity. *Dev. Biol.* **145,** 99–109.

Lesueur, L., Edery, M., Paly, J., Clark, J., Kelly, P. A., and Djiane, J. (1990). Prolactin stimulates milk protein promoter in CHO cells cotransfected with prolactin receptor cDNA. *Mol. Cell Endocrinol.* **71,** R7–R12.

Leung, F. C., Styles, W. J., Rosenblum, C. I., Lilburn, M. S., and Marsh, J. A. (1987). Diminished hepatic growth hormone receptor binding in sex-linked dwarf broiler and Leghorn chickens. *Proc. Soc. Exp. Biol. Med.* **184,** 234–238.

Leung, F. C., Taylor, J. E., Steelman, S. L., Bennett, C. D., Rodkey, J. A., Long, R. A., Serio, R., Weppelman, R. M., and Olson, G. (1984a). Purification and properties of chicken growth hormone and the development of a homologous radioimmunoassay. *Gen. Comp. Endocrinol.* **56,** 389–400.

Leung, F. C., Taylor, J. E., and Van Iderstine, A. (1984b). Thyrotropin-releasing hormone stimulates body weight gain and increases thyroid hormones and growth hormone in plasma of cockerels. *Endocrinology* **115,** 736–740.

Leung, F. C., Taylor, J. E., and Van Iderstine, A. (1984c). Effects of dietary thyroid hormones on growth and serum T_3, T_4 and growth hormone in sex-linked dwarf chickens. *Proc. Soc. Exp. Biol. Med.* **177,** 77–81.

Levin, M., Johnson, R. L., Stern, C. D., Kuehn, M., and Tabin, C. (1995). A molecular pathway determining left-right asymmetry in chick embryogenesis. *Cell* **82,** 803–814.

Lofts, B., and Marshall, A. J. (1956). The effects of prolactin administration on the internal rhythm of reproduction in male birds. *J. Endocrinol.* **13,** 101–106.

Mao, J. N. C., Burnside, J., Postel-Vinay, M. C., Chambers, J., Pesek, J., and Cogburn, L. A. (1998). Ontogeny of growth hormone receptor gene expression in tissue of growth-selected strains of broiler chickens. *J. Endocrinol.* **156,** 67–75.

Marcelle, C., and Eichmann, A. (1992). Molecular cloning of a family of protein kinase genes expressed in the avian embryo. *Oncogene* **7,** 2479–2487.

Marcelle, C., Eichmann, A., Halevy, O., Breant, C., and Douarin, N. M. (1994). Distinct developmental expression of a new avian fibroblast growth factor receptor. *Development* **120,** 683–694.

Marks, H. L. (1989). Long-term selection for four-week body weight in Japanese quail following modification of the selection environment. *Poult. Sci.* **68,** 455–459.

Marsh, J. A., Lauterio, T. J., and Scanes, C. G. (1984). Effects of triiodothyronine treatments on body and organ growth and the development of immune function in dwarf chickens. *Proc. Soc. Exp. Biol. Med.* **177,** 82–91.

Marshall, H., Studer, M., Poppert, H., Aparecio, S., Kuroiwa, A., Brenner, S., and Krumtauf, R. (1994). A conserved retinoic and response element required for early expression of the homeobox gene, Hoxb-1. *Nature* **370,** 567–571.

May, J. D. (1980). Effect of dietary thyroid hormone on growth and feed efficiency of broilers. *Poult. Sci.* **59,** 888–892.

May, J. D., and Marks, H. L. (1983). Thyroid activity of selected, nonselected, and dwarf broiler lines. *Poult. Sci.* **62,** 1721–1724.

McCann-Levorse, L. M., Radecki, S. V., Donoghue, D. J., Malamed, S., Foster, D. N., and Scanes, C. G. (1993). Ontogeny of pituitary growth hormone and growth hormone mRNA in the chicken. *Proc. Soc. Exp. Biol. Med.* **202,** 109–113.

McFarland, D. C., Ferrin, N. H., Gilkerson, K. K., and Pesall, J. E. (1992). Tissue distribution of insulin-like growth factor receptors in the turkey. *Comp. Biochem. Physiol. B* **103,** 601–607.

McGuinness, M. C., and Cogburn, L. A. (1990). Measurement of developmental changes in plasma insulin-like growth factor-I levels of broiler chickens by radioreceptor assay and radioimmunoassay. *Gen. Comp. Endocrinol.* **79,** 446–458.

McGuinness, M. C., and Cogburn, L. A. (1991). Response of young broiler chickens to chronic injection of recombinant-derived human insulin-like growth factor-1. *Dom. Anim. Endocrinol.* **8,** 611–620.

McNabb, F. M. A., and Hughes, T. E. (1983). The role of serum binding proteins in determining free thyroid hormone concentrations during development in quail. *Endocrinology* **113,** 957–963.

McNabb, F. M. A., Stanton, F. W., Weirich, R. T., and Hughes, T. E. (1984). Responses to thyrotropin during development in Japanese quail. *Endocrinology* **114,** 1238–1244.

McMurtry, J. P., Francis, G. L., Upton, F. Z., Rosselot, G., and Brocht, D. M. (1994). Developmental changes in chicken and turkey insulin-like growth factor-I (IGF-I) studied with a homologous radioimmunoassay for chicken IGF-I. *J. Endocrinol.* **142,** 225–234.

Meier, R., Becker-Andre, M., Gotz, R., Heumann, R., Shaw, A., and Thoenen, H. (1986). Molecular cloning of bovine and chick nerve growth factor (NGF): Delineation of conserved and unconserved domains and their relationship to the biological activity and antigenicity of NGF. *EMBO J.* **5,** 1489–1493.

Meier, A. H., Farner, D. S., and King, J. R. (1965). A possible endocrine basis for migratory behaviour in the white-crowned sparrow, *Zonotrichia Leucophrys Gambelii. Anim. Behav.* **4,** 453–465.

Mesiano, S., Browne, C. A., and Thorburn, G. D. (1985). Detection of endogenous epidermal growth factor-like activity in the developing chick embryo. *Dev. Biol.* **110,** 23–28.

Minshall, R. D., McFarland, D. C., and Doumit, M. E. (1990). Interaction of insulin-like growth factor I with turkey satellite cells and satellite cell-derived myotubes. *Dom. Anim. Endocrinol.* **7,** 413–424.

Moore, G. E., Harvey, S., Klandorf, H., and Goldspink, G. (1984). Muscle development in thyroidectomized chickens (*Gallus domesticus*). *Gen. Comp. Endocrinol.* **55,** 195–199.

Munaim, S. I., Klagsbrun, M., and Toole, B. P. (1988). Developmental changes in fibroblast growth factor in the chicken embryo limb bud. *Proc. Natl. Acad. Sci. USA* **85,** 8091–8093.

Nalbandov, A. V. (1945). A study of the effect of prolactin on broodiness and on the cock testes. *Endocrinology* **36,** 251–258.

Niswander, L., Jeffrey, S., Martin, G. R., and Tickle, C. (1994). A positive feedback loop coordinates growth and patterning in the vertebrate limb. *Nature* **371,** 609–612.

O'Keefe, R. J., Puzas, J. E., Brand, J. S., and Rosier, R. N. (1988a). Effect of transforming growth-factor-β on DNA synthesis by growth plate chondrocytes: Modulation by factors present in serum. *Calcif. Tissue Int.* **43,** 352–358.

O'Keefe, R. J., Puzas, J. E., Brand, J. S., and Rosier, R. N. (1988b). Effects of transforming growth factor-β on matrix synthesis by chick growth plate chondrocytes. *Endocrinology* **122,** 2953–2961.

Ohkubo, T., Tanaka, M., Nakashima, K., Shimada, K., Saito, N., and Sato, K. (1993). High-level expression of biologically active chicken prolactin in *E. coli. Comp. Biochem. Physiol. A* **105,** 123–128.

Ohuchi, H., Nakagawa, T., Yamauchi, M., Ohata, T., Yoshioka, H., Kuwana, T., Mima, T., Mikawa, T., Nohno, T., and Ngi, S. (1995). An additional limb bud can be induced from the flank of the chick embryo by FGF4. *Biochem. Biophys. Res. Commun.* **209,** 809–816.

Opel, H., and Proudman, J. A. (1980). Failure of mammalian prolactin to induce incubation behavior in chickens and turkeys. *Poult. Sci.* **59,** 2550–2558.

Oppenheim, R. W., Houenou, L. J., Johnson, J. E., Lin, L.-F. H., Li, L., Lo, A. C., Newsome, A. L., Prevette, D. M., and Wang, S. (1995). Developing motor neurons rescued from programmed and axotomy-induced cell death by GDNF. *Nature* **373,** 344–346.

Pasquale, E. B. (1990). A distinctive family of embryonic protein-tyrosine kinase receptors. *Proc. Natl. Acad. Sci. USA* **87,** 5812–5816.

Pasquale, E. B., and Singer, S. J. (1989). Identification of a developmentally regulated protein-tyrosine kinase by using anti-phophotyrosine antibodies to screen a cDNA expression library. *Proc. Natl. Acad. Sci. USA* **86,** 5449–5453.

Philips, J. G., and Harvey, S. (1982). A reappraisal of the role of prolactin in osmoregulation. *In* "Aspects of Avian Endocrinology" (C. G. Scanes, M. A. Ottinger, A. D. Kenny, J. Balthazart, J. Cronshaw, and I. Chester Jones, eds.), pp. 309–327. Texas Tech University, Lubbock.

Proudman, J. A., and Opel, H. (1981). Turkey prolactin: Validation of a radioimmunoassay and measurement of changes associated with broodiness. *Biol. Reprod.* **25,** 573–580.

Proudman, J. A., McGuinness, M. C., Krishnan, K. A., and Cogburn, L. A. (1994). Endocrine and metabolic responses of intact and hypophysectomized turkey poults given a daily injection of chicken growth hormone. *Comp. Biochem. Physiol. C* **109,** 47–56.

Pym, R. A. E., Johnson, J., Etse, D. B., and Eason, P. (1991). Inheritance of plasma insulin-like growth factor-I and growth rate, food intake, food efficiency and abdominal fatness in chickens. *Br. Poult. Sci.* **32,** 285–293.

Raheja, K. L., and Snedecor, J. G. (1970). Comparison of subnormal multiple doses of L-thyroxine and L-triiodothyronine in thyroidectomized chickens (*Gallus domesticus*). *Gen. Comp. Endocrinol.* **55,** 195–199.

Raheja, K. L., and Snedecor, J. G. (1971). Some effects of single doses of triiodothyronine and thyroxine in hypothyroid chicks. *Gen. Comp. Endocrinol.* **16,** 97–104.

Ricklefs, R. E. (1983). Avian postnatal development. *In* "Avian Biology" (D. S. Farner, J. R. King, and K. C. Parkes, eds.), Vol. VII, pp. 2–83. Academic Press, New York.

Riddle, O., Bates, R. W., and Lahr, E. L. (1935). Prolactin induces broodiness in fowl. *Am. J. Physiol.* **111,** 355–360.

Risau, W., Gantschi-Sova, P., and Bohlen, P. (1988). Endothelial cell growth factors in embryonic and adult chick brain are related to human acidic fibroblast growth factor. *EMBO J.* **7,** 959.

Roberts, R. D., Sharp, P. J., Burt, D. W., and Goddard, C. (1994). Insulin-like growth factor-I in the ovary of the laying hen: Gene expression and biological actions on granulosa and thecal cells. *Gen. Comp. Endocrinol.* **93,** 327–336.

Rogers, S. L., Gegick, P. J., Alexander, S. M., and McGuire, P. G. (1992). Transforming growth factor-β alters differentiation in cultures of avian neural crest-derived cells: Effects on cell morphology, proliferation, fibronectin expression and melanogenesis. *Dev. Biol.* **151,** 192–203.

Rombauts, L., Vanmountfort, D., Decuypere, E., and Verhoeven, G. (1996). Inhibin and activin have antagonistic paracrine effects on gonadal steroidogenesis of the chicken embryo. *Biol. Reprod.* **54,** 1229–1237.

Rui, H., Lebrun, J. J., Kirken, R. A., Kelly, P. A., and Farrar, W. L. (1994). JAK2 activation and cell proliferation induced by antibody-mediated prolactin receptor dimerization. *Endocrinology* **135,** 1299–1306.

Rush, R. A., Abrahamson, I. K., Belford, D. A., Murdoch, S. Y., and Wilson, P. A. (1986). Regulation of sympathetic trophic factors in smooth muscle. *Int. J. Dev. Neurosci.* **4,** 51–59.

Scanes, C. G., Chadwick, A., and Bolton, N. J. (1976). Radioimmunoassay of prolactin in the plasma of the domestic fowl. *Gen. Comp. Endocrinol.* **30,** 12–20.

Scanes, C. G., Duyka, D. R., Lauterio, T. J., Bowen, S. J., Huybrechts, L. M., Bacon, W. L., and King, D. B. (1986). Effect of chicken growth hormone, triiodothyronine and hypophysectomy in growing domestic fowl. *Growth* **50,** 12–31.

Scanes, C. G., Marsh, J., Decuypere, E., and Rudas, P. (1983). Abnormalities in the plasma concentrations of thyroxine, triiodothyronine and growth hormone in sex-linked dwarf and autosomal dwarf White Leghorn domestic fowl (*Gallus domesticus*). *J. Endocrinol.* **97,** 127–135.

Scanes, C. G., Peterla, T. A., Kantor, S., and Ricks, C. A. (1990). In vivo effects of biosynthetic chicken growth hormone in broiler-strain chickens. *Growth Dev. Aging* **54,** 95–101.

Seed, J., and Hauschka, S. D. (1988). Clonol analysis of vertebrate myogenesis. VIII. Fibroblast growth factor (FGF) dependent and FGF-independent muscle colony types during chick wing development. *Dev. Biol.* **128,** 40–49.

Seed, J., Olwin, B. B., and Houschka, S. (1988). Fibroblast growth factor levels in the whole embryo and limb bud during chick development. *Dev. Biol.* **128,** 50–57.

Serrano, J., Shuldiner, A. R., Roberts, C. T., LeRoith, D., and De Pablo, F. (1990). The insulin-like growth factor I (IGF-I) gene is expressed in chick embryos during early organogenesis. *Endocrinology* **127,** 1547–1549.

Sharp, P. J., Macnamee, M. C., Sterling, R. J., Lea, R. W., and Pedersen, H. C. (1988). Relationships between prolactin, LH and broody behaviour in bantam hens. *J. Endocrinol.* **118,** 279–286.

Sharp, P. J., Sterling, R. J., Talbot, R. T., and Huskisson, N. S. (1989). The role of hypothalamic vasoactive intestinal polypeptide in the maintenance of prolactin secretion in incubating bantam hens: Observations using passive immunization, radioimmunoassay and immunohistochemistry. *J. Endocrinol.* **122,** 5–13.

Shimada, K., Ohkubo, T., Saito, N., Talbot, R. T., and Sharp, P. J. (1993). The molecular biology of prolactin. In "Avian Endocrinology" (P. J. Sharp, ed.), pp. 135–148. Journal of Endocrinology Ltd., Bristol.

Siegel, P. B., Gross, W. B., and Dunnington, E. A. (1989). Effects of dietary corticosterone in young Leghorn and meat-type cockerels. *Br. Poult. Sci.* **30,** 185–192.

Simon, J. (1989). Chicken as a useful species for the comprehension of insulin action. *Crit. Rev. Poult. Biol.* **2,** 121–148.

Simon, J. (1995). Insulin-glucagon and growth in broilers. *Arch. Geflugel. Sonderh.* **1,** 14–17.

Simon, J., and Taouis, M. (1993). The insulin receptor in chicken tissues. In "Avian Endocrinology" (P. J. Sharp, ed.), pp. 177–188. Journal of Endocrinology Ltd., Bristol.

Simon, J., Chevalier, B., Derouet, M., and Leclercq, B. (1991). Normal number and kinase activity of insulin receptors in liver of genetically fat chickens. *J. Nutr.* **121,** 379–385.

Skwarlo-Sonta, K. (1990). Mitogenic effect of prolactin on chicken lymphocytes in vitro. *Immunol. Lett.* **24,** 171–178.

Skwarlo-Sonta, K. (1992). Prolactin as an immunoregulatory hormone in mammals and birds. *Immunol. Lett.* **33,** 105–121.

Spencer, G. S. G., Garsson, G. J., Gerrits, A. R., Spencer, E. M., and Kestin, S. C. (1990). Lack of effect of exogenous insulin-like growth factor-I (IGF-I) on chick embryo growth rate. *Reprod. Nutr. Dev.* **30,** 515–521.

Sterling, R. J., Sharp, P. J., Klandorf, H., Harvey, S., and Lea, R. W. (1984). Plasma concentrations of luteinising hormone, follicle stimulating hormone, androgen, growth hormone, prolactin, thyroxine and triiodothyronine during growth and sexual development in the cockerel. *Br. Poult. Sci.* **25,** 353–359.

Stewart, P. A., Washburn, K. W., and Marks, H. L. (1984). Effect of the dw gene on growth, plasma hormone concentrations and hepatic enzyme activity in a randombred population of chickens. *Growth* **48,** 59–73.

Stocker, K. M., Sherman, L., Rees, S., and Ciment, G. (1991). Basic FGF and TGF-β1 influence committment to melanogenesis in neural crest-derived cells of avian embryos. *Development* **111,** 635–645.

Sun, S. S., and McFarland, D. C. (1993). Interaction of fibroblast growth factor with turkey embryonic myoblasts and myogenic satellite cells. *Comp. Biochem. Physiol. A* **105,** 85–89.

Sun, S. S., McFarland, D. C., Ferrin, N. H., and Gilkerson, K. K. (1992). Comparison of insulin-like growth factor interaction with satellite cells and embryonic myoblasts derived from the turkey. *Comp. Biochem. Physiol. A* **102,** 235–243.

Tabin, C. (1995). The initiation of the limb bud: Growth factors, *Hox* genes, and retinoids. *Cell* **80,** 671–674.

Takenaka, I. M., and Hightower, L. E. (1993). Regulation of chicken *Hsp 70* and *Hsp 90* family gene expression by transforming growth factor-β1. *J. Cell Physiol.* **155,** 54–62.

Talbot, R. T., Hanks, M. C., Sterling, R. J., Sang, H. M., and Sharp, P. J. (1991). Pituitary prolactin messenger ribonucleic acid levels in incubating and laying hens: Effects of manipulating plasma levels of vasoactive intestinal polypeptide. *Endocrinology* **129,** 496–502.

Tanaka, M., Hayashida, Y., Wakita, M., Hoshino, S., and Nakashima, K. (1995). Expression of aberrantly spliced growth hormone receptor (GHR) mRNA in the sex-linked dwarf (SLD) chicken, *Gifu 20*. *Growth Regul.* **5,** 218–223.

Tanaka, M., Maeda, K., Okubo, T., and Nakashima, K. (1992). Double antenna structure of chicken prolactin receptor deduced from the cDNA sequence. *Biochem. Biophys. Res. Commun.* **188,** 490–496.

Taouis, M., Derouet, M., Chevalier, B., and Simon, J. (1993). Corticosterone effect on insulin receptor number and kinase activity in chicken muscle and liver. *Gen. Comp. Endocrinol.* **89,** 167–175.

Tilly, J. L., and Johnson, A. L. (1990). Effect of several growth factors on plasminogen activator activity in granulosa and theca cells of the domestic hen. *Poult. Sci.* **69,** 292–299.

Tixier-Boichard, M., Huybrechts, L. M., Decuypere, E., Kuhn, E. R., Monovoisin, J.-L., Coquerelle, G., Charrier, J., and Simon, J. (1992). Effects of insulin-like growth factor-I (IGF-I) infusion and dietary tri-iodothyronine (T_3) supplementation on growth, body composition and plasma hormone levels in sex-linked dwarf mutant and normal chickens. *J. Endocrinol.* **133,** 101–110.

Tixier-Boichard, M., Huybrechts, L. M., Kuhn, E., Decuypere, E., Charrier, J., and Mongin, P. (1989). Physiological studies on the sex-linked dwarfism of the fowl: A review on the search for the gene's primary effect. *Genet. Sel. Evol.* **21,** 217–234.

Tomas, F. M., Pym, R. A., and Johnson, R. J. (1991). Muscle protein turnover in chickens selected for increased growth rate, food consumption of efficiency of food utilisation: Effects of genotype and relationship to plasma IGF-I and growth hormone. *Br. Poult. Sci.* **32,** 363–376.

Vanderpooten, A., Darras, V. M., Huybrechts, L. M., Rudas, P., Decuypere, E., and Kuhn, E. R. (1991a). Effect of hypophysectomy and acute administration of growth hormone (GH) on GH-receptor binding in chick liver membranes. *J. Endocrinol.* **129,** 275–281.

Vanderpooten, A., Huybrechts, L. M., Decuypere, E., and Kuhn, E. R. (1991b). Differences in hepatic growth hormone receptor binding during development of normal and dwarf chickens. *Reprod. Nutr. Dev.* **31,** 47–55.

Vanderpooten, A., Janssens, W., Buyse, J., Leenstra, F., Berghman, L., Decuypere, E., and Kuhn, E. R. (1993). Study of the hepatic growth hormone (GH) receptor at different ages in chickens selected for a good feed conversion (FC) and a fast weight gain (GL.) *Dom. Anim. Endocrinol.* **10,** 199–206.

Vasilatos-Younken, R., Bacon, W. L., and Nestor, K. E. (1988a). Relationship of plasma growth hormone to growth within and between turkey lines selected for differential growth rates. *Poult. Sci.* **67,** 826–834.

Vasilatos-Younken, R., Cravener, T. L., Cogburn, L. A., Mast, M. G., and Wellenreiter, R. H. (1988b). Effect of pattern of administration on the response to exogenous, pituitary-derived chicken growth hormone by broiler-strain pullets. *Gen. Comp. Endocrinol.* **71,** 268–283.

Vasilatos-Younken, R., Gray, K. S., Bacon, W. L., Nestor, K. E., Long, D. W., and Rosenberger, J. L. (1990). Ontogeny of growth hormone (GH) binding in the domestic turkey: Evidence of sexual dimorphism and developmental changes in relationship to plasma GH. *J. Endocrinol.* **126,** 131–139.

Wang, S-Y., and Johnson, P. (1993). Complementary deoxyribonucleic acid cloning and sequence analysis of the α-subunit of inhibin from chicken ovarian granulosa cells. *Biol. Reprod.* **49,** 453–458.

Watahiki, M., Tanaka, M., Masuda, N., Sugisaki, K., Yamamoto, M., Yamakawa, M., Nagai, J., and Nakashima, K. (1989). Primary structure of chicken pituitary prolactin deduced from the cDNA sequence: Conserved and specific amino acid residues in the domain of the prolactins. *J. Biol. Chem.* **264,** 5535–5539.

Wong, E. A., Ferrin, N. H., Silsby, J. L., and El Halawani, M. E. (1991). Cloning of a turkey prolactin cDNA: Expression of prolactin mRNA throughout the reproductive cycle of the domestic turkey (*Meleagris gallopavo*). *Gen. Comp. Endocrinol.* **83,** 18–26.

Wong, E. A., Silsby, J. L., Ishii, S., and El Halawani, M. E. (1992). Pituitary luteinizing hormone and prolactin messenger ribonucleic acid levels are inversely related in laying and incubating turkey hens. *Biol. Reprod.* **47,** 598–602.

Yang, Y. W. H., Robbins, A. R., Nissley, S. P., and Rechler, M. M. (1991). The chick embryo fibroblast cation-independent mannose 6-phosphate receptor is functional and immunologically related to the mammalian insulin-growth factor-II (IGF-II)/man 6-p receptor but does not bind IGF-II. *Endocrinology* **128,** 1177–1189.

Youngren, O. M., El Halawani, M. E., Silsby, J. L., and Phillips, R. E. (1991). Intracranial prolactin perfusion induces incubation behavior in turkey hens. *Biol. Reprod.* **44,** 425–431.

Zhou, J. F., Zadworny, D., Guemene, D., and Kuhnlein, U. (1996). Molecular cloning, tissue distribution, and expression of the prolactin receptor during various reproductive states in *Meleagris gallopavo*. *Biol. Reprod.* **55,** 1081–1090.

CHAPTER 26

Immunophysiology

B. GLICK[1]
Department of Poultry Science
College of Agricultural Sciences
Clemson University
Clemson, South Carolina 29634

I. Introduction 657
II. Cytoarchitecture and Development of the Immune System 657
  A. Primary Immune Tissue 657
  B. Secondary Lymphoid Tissue 659
III. Regulation of Immune Response 662
  A. Major Histocompatibility Complex 662
  B. Cytokines 664
  C. Antibody-Mediated Immunity and B-Cell Repertoire 664
  D. Macrophages, Natural Killer Cells, Heterophils, and Thrombocytes 666
  References 667

I. INTRODUCTION

An understanding of avian immunology requires some incite into a history of immunology (Silverstein, 1989) and a comprehension of fundamental immunology (Paul, 1993). The primary purpose of this chapter will be to present a knowledge base for avian immunology by writing a brief narrative accompanied by limited illustrations and citing review articles and selected original research. Immunology, like the musical painting of Haydn's Sixth, Seventh, and Eighth Symphonies, begins with the sunrise in LeMatin ("new language and ideas"),

[1] Retired, June 1995, as Distinguished Emeritus Professor, Mississippi State University; Emeritus Professor, Clemson University; and Adjunct Professor Biomedical Cooperative Greenville Hospital and Clemson University.

continues to the calm ("assimilation of new concepts"), and is followed by the storm of LeSoir ("challenges leading back to LeMatin").

II. CYTOARCHITECTURE AND DEVELOPMENT OF THE IMMUNE SYSTEM

The immune system is dependent on specialized microenvironments that (1) offer a primary educational milieu where pluripotent precursors will differentiate into clones of lymphocytes endowed with the ability to respond to self or foreign antigens, and (2) offer a secondary educational milieu in which primary educated lymphocytes gather with various accessory cells to respond to specific cell-associated antigens (or noncell associated antigens) and clonally expand. These microenvironments will be discussed under primary immune tissue and secondary lymphoid tissue.

A. Primary Immune Tissue

The T- and B-cell concept entered the vocabulary of immunology only after basic research with the chicken model revealed an immunological role for the bursa of Fabricius (Glick *et al.*, 1956; Warner *et al.*, 1962; Cooper *et al.*, 1966) and the avian thymus (Cooper *et al.*, 1966; Figure 1). The B-lymphocyte of the concept was so named to identify its avian bursal origin and mammalian bone marrow origin and the T-lymphocyte of the concept identified its thymic origin (Roitt *et al.*, 1969).

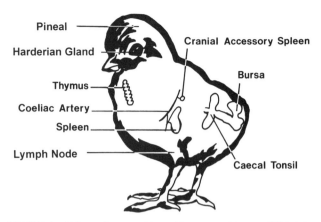

FIGURE 1 Primary immune tissue, thymus, and bursa of Fabricius and secondary lymphoid tissue, pineal, Harderian gland, accessory spleen, spleen, lymph node, and cecal tonsil.

1. Bursa of Fabricius: Morphology

The initial descriptive study of the bursa was by Hieronymus Fabricius after whom the gland was named (Adelman, 1942). Bursa growth studies revealed (1) a *rapid* growth from hatch to 3 or 4 weeks, (2) a *plateau* period for the next 5 or 6 weeks, and (3) regression occurring before sexual maturity (Glick, 1956). These observations set the stage for the functional studies since they directed that bursectomies be performed prior to 3 weeks of age or as close to hatch as possible. Serendipity then entered the picture when bursectomized birds were injected with *Salmonella pullorum* to satisfy a student laboratory and not as a part of a designed experiment. A more complete description of these events may be found in selected publications (Glick *et al.,* 1956; Glick, 1977, 1987).

The bursa is a dorsal diverticulum of the proctodaeal region of the cloaca (Jolly, 1915). The ovallike bursa of the chicken contrasts to the elongated bursae of the duck and starling (*Sternus vulgaris*) and to the ostrich (*Struthio camelus australis*) and emu (*Dromaius novaehollandio*) bursae, which are an integral part of the proctodaeal mucosa (Glick, 1986; von Rautenfeld and Budras, 1982). The bursal anlage appears between 3 and 5 days of embryonic development (DE) (Romanoff, 1960; Olah *et al.,* 1986). A major feature of embryonic development is the formation of buds, the forerunner of the bursal follicle. The bud develops into the medulla of the bursal follicle. Scanning electron microscopy revealed two types of surface epithelium: follicle-associated-epithelium (FAE), associated with the medulla, and the interfollicular epithelium (IFE), which is between the follicles (Bockman and Cooper, 1973; Holbrook *et al.,* 1974). The IFE and FAE morphologically may appear at 12 and 15 DE, respectively (Naukkarinen *et al.,* 1978). However, the pinocytotic ability of the FAE (Bockman and Cooper, 1973) was not evident until 19–21 DE. The observation that carbon, applied to the vent of the chick, would enter the bursal duct and be pinocytosed by the FAE (Bockman and Cooper, 1973) stimulated a carbon-vent study in a 4-week-old chicken that revealed an average of 800 FAE areas/fold or 8,000 to 12,000 bursal follicles per bursa considering the presence of 10 to 15 folds (Olah and Glick, 1978a).

2. Bursal B-Cell Markers

The hallmark of B-cells is the presence of membrane-associated immunoglobulin. Membrane Ig in mammals (rodents/humans) occurs after the appearance of cytoplasmic Ig (pre-B-cells) while in birds there is no such sequence. Gilmour *et al.* (1976) produced alloantisera that revealed two independent autosomal loci, Bu-1 and Th-1, recognizing an antigen of bursal lymphocyte/peripheral B-cells and thymic cells/peripheral T-cells, respectively. Allelic forms have been identified, Bu-1a (94 kDa) and Bu-1b (70 kDa), and utilized in bursal follicle colonization studies to suggest entry of no more than three precursors per follicle (Chen *et al.,* 1991). A Bu-2 antigen (66 kDa) was shown to be distinct from Bu-1. The Bu-2 antigen identified both Ig^+ and Ig^- cells and in control bursae identified lymphocytes in the cortex and medulla but not the epithelium. Other B-cell antigens, the CB antigens, have been reviewed by Chen *et al.* (1987, 1991).

3. Thymus, T-cell Receptors, and Cluster of Differentiation

The third and fourth pharyngeal pouches contributed to the formation of the thymus which consisted of seven lobes developed along each side of the jugular veins (Romanoff, 1960; Venzke, 1952). The thymus, like the bursa, possessed cortical and medullary regions. An isolated protein, the putative avian thymic hormone, located in the thymus and blood, stimulated bone marrow cells to express T-cell markers (Murthy *et al.,* 1984) and revealed an amino acid sequence similar to parvalbumin (Brewer *et al.,* 1990).

Antigen recognition in T-cells differs from B-cells in that the T-cell receptor (TCR) (1) is not an immunoglobulin molecule, (2) recognizes surface-bound antigen only, and (3) is not secreted but remains an integral part of the cell membrane. A cluster of differentiation (CD) identifies in a cell membrane specific groups of determinates which define stages of cellular differentiation and are detected by monoclonal antibodies (Table 1).

The human TCR complex includes an invariant five-polypeptide complex (γ, δ, ε, ζ, η,) molecule termed CD3. A similar CD3 molecule was identified in chickens by Chen *et al.* (1986). The chicken CD3 possessed three

TABLE 1 T-Cell Antigens[a]

Antigens	Molecular mass ($\times 10^{-3}$)	Antibody
CT1	65	CT1, Ct1-α
CD3	17, 19, and 20	CT3
γ/δ TCR	90 (subunits 40 and 50)	TCR1
α/β TCR	90 (subunits 40 and 50)	TCR2
α/β TCR	88 (subunits 40 and 48)	TCR3
CD2	40	2–4
CD4	64	CT4
CD8	64 (34 dimer)	CT8
CD5	56	CTLA-5, CTLA-8, 3–8
CD28	40	AV7
CD45	200	L-17, CL1
CD25 (IL-2 receptor)	50	INN-CH-16

[a] All cited in narrative with the exception of CT1, CD5, and CD25 (Chen et al., 1991).

polypeptides of M_r 20,000, 19,000, and 17,000 under nonreducing conditions.

Avian homologies to mammalian T-cell receptor γ/σ (TCR1) and TCR α/β (TCR2) have been identified (Chen et al., 1991). While TCR1- and TCR2-positive thymocytes appeared at 12 DE, only TCR1 was detected by surface staining and these were equivalent to the number of CD3 surface-stained thymocytes (Bucy et al., 1990). TCR1 and TCR2 were heterodimers of 50- and 40-kDa glycoproteins. A third TCR had a lower molecular weight (Chen et al., 1991). Cells positive for TCR1 and TCR2 migrate to the spleen by 15 and 19 DE, respectively, while the TCR3 positive cells do not appear in the spleen until after hatching. The range of TCR1, -2, and -3 in peripheral blood was 15–25, 45–55, and 10–15%, respectively (Chen et al., 1991). Precursor cells entered the thymus in three waves. All three lineages were present in the first and second arrivals of precursor cells (Chen et al., 1991).

A CD2 antigen (monomeric, 40 kDa), identified by mAb 2-4, appeared in the avian thymus by 11 DE and may influence T-cell growth and differentiation by way of cell adhesion. Avian thymocytes are 98% CD2 positive. CD2 is a coreceptor cooperating with the TCR and contributing to T-cell binding to antigen-presenting cells and/or T-cell signaling. There may be a question concerning the identification of avian CD2 (Young et al., 1994). Young et al. (1994) identified a 40-kDa molecule with a 50% amino acid sequence identical to mammalian CD28. In mammals, CD28 receptors of primed T-cells bound a B7 epitope present in B-cells, dendritic cells, or macrophages and signaled the induction of interleukin 2 and proliferation of the T-cell (Linsley and Ledbetter, 1993).

Avian T-cells possessed homologs to mammalian CD4 (64-kDa monomeric polypeptide) helper T-cells, and CD8 (a dimer, 31 and 34 kDa) cytotoxic T-cells (Chen et al., 1991). Veillette and Ratcliffe (1991) revealed that like the mammalian systems, the chicken homologies of CD4 and CD8 associated with a 56-kDa tyrosine-specific protein kinase. The thymic hormone, avian thymulin, influenced CD4/CD8 ratios and the expression of CD4 and CD8 based on fluorescent staining (Marsh, 1993).

Double-positive CD4 CD8 cells, single-positive CD8$^+$ cells, and single-positive CD4$^+$ cells appeared in the thymic cortex at 9 or 10 DE, 13 DE, and 15 DE, respectively (Bucy et al., 1990). The CD8$^+$ cells appeared earlier in the medulla (15 DE) than did the CD4$^+$ cells (17 DE).

4. Origin of Bursal and Thymic Lymphocytes

Jaffe and Fechheimer (1966) and Moore and Owen (1965, 1966) utilized sex chromosome techniques and suggested the possibility that immigrant cells were the progenitors of bursal lymphocytes. LeDouarin et al. (1984) took advantage of the one or two large clumps of heterochromatin associated with the cell nucleus of quail and their absence in chick cells to incisively reveal the blood-borne origin of bursal and thymic lymphocytes. Experiments revealed quail basophilic stem cell migration between days 7 and 11 of embryogenesis while the chick basophilic cell migrated into the bursa between 8 and 15 days DE (LeDouarin et al., 1984). Unlike the bursa, the thymic precursor lymphocytes entered the thymus in three waves (Le Douarin et al., 1984, 1990) between 6.5 and 8, 12 and 14, and 18 and 20 DE with a refractory period of 4 days between the first and second and second and third waves (Le Douarin et al. 1984, 1990; Figure 2).

With the acceptance that the bursal and thymic lymphocytes originated from a blood-borne stem cell, the origin of the blood-borne stem came into question. Chimeras were developed prior to circulation (<14 somite stage) by replacing a chick area pellucida (embryo proper) with that of a quail (Martin, 1990). The developing lymphocyte nuclear characteristics in this embryo chimera resembled those of a quail and were, therefore, of intraembryonic origin. The origin of the intraembryonic stem cells may be intraaortic, paraaortic, or from the coelomic epithelium (Dieterlen-Lievre et al., 1990; Olah et al., 1988).

B. Secondary Lymphoid Tissue

1. Spleen

Adjacent to the dorsal surface of the right lobe of the liver and dorsal to the proventriculus is the reddish-brown oval spleen (Nickel et al., 1977). Accessory

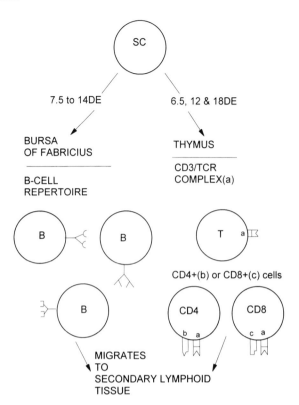

FIGURE 2 Stem cell (SC) migration to the chick bursa occurs between 7.5 and 14 days of embryogenesis (DE) and in the thymus occurs in three waves at 6.5, 12, and 18 DE. A B-cell repertoire able to recognize 10^6 different antigens forms in the microenvironment of the bursa while the thymus environment signals the formation of clonally specific T-cell receptors (a) and T-helper (CD4) (b) or T-cytotoxic (CD8) (c) cells. The differentiated B- and T-cells then migrate to the secondary lymphoid tissue.

spleens have been described cranial, adjacent, and caudal to the spleen. Subsequent to splenectomy, the cranial accessory spleen hypertrophied (Glick, 1986). The most rapid rate of splenic growth occurred during the first 6 weeks after hatching with maximum size (spleen-to-body weight ratio) attained by 10 weeks of age (Glick, 1986).

The avian spleen like the mammalian spleen possesses red and white pulp. Within the white pulp are located (1) the periarteriolar lymphatic sheath (PALS); (2) germinal centers; and (3) periellipsoid white pulp region (PWP). A central artery arising from the splenic artery was surrounded by the PALS, which contained lymphocytes, macrophages, and dendritic cells and was thymic dependent (Glick, 1986). Germinal centers, bursal-dependent regions, were located at the edge of the PALS. The central artery as it entered the PWP became the penicilliform capillary (PC). The midregion of the penicilliform capillary was surrounded by the capillary sleeve (CS) or ellipsoid (Olah and Glick, 1982; Figure 3). The CS was embroidered by the dendritic ellipsoid-associated cell (EAC) which bound diverse substances that entered the CS through stomata formed by the endothelial cells of the midregion of penicilliform capillaries (White et al., 1970; Olah and Glick, 1982; Figure 4). The EAC is activated following binding and migrates into the PWP, red pulp, PALS, and GC regions. The EAC appeared to be a messenger cell and may be lineage related to interdigitating dendritic cells of the PALS and follicular dendritic cells of the GC (Olah and Glick, 1982; Gallego et al., 1993). The mammalian spleen possessed a marginal zone (macrophages and lymphocytes) which surrounded the PALS separating this region from the red pulp (Weiss, 1972). An avian marginal zone has been suggested by Jeurissen et al. (1992) and Jeurissen (1993) to include the CS and the surrounding EAC, B-cells, and macrophages.

2. Cecal Tonsil

The cecal tonsil is an enlarged patch of tissue (4–18 mm) in the proximal region of each cecum (Muthmann, 1913; Glick, 1986). The cecal tonsil villi were longer and less broad than those from the remainder of the cecum's proximal region (Glick, 1986). The polycryptic cecal tissue was similar to the mammalian palatine tonsil (Glick et al., 1981). The location and continuous exposure of the tonsil villi to the fecal content suggested a sentinel role for this peripheral lymphoid tissue. The cecal tonsil possessed approximately 400 spherical units, each with a central crypt, diffuse lymphoid tissue, and germinal centers (Glick et al., 1981). The cecal tonsil possessed T- and B-cells (Albini and Wick, 1974) and IgM, IgG, and IgA plasma cells (Jeurissen et al., 1989b) and produced antibody to soluble antigens (Jankovic and Mitrovic, 1967; Orlans and Rose, 1970).

3. Peyer's Patches

Peyer's patches appeared in 10-day-old chickens along the intestine cranial to the ileocecal junction (Schat and Myers, 1991). They possessed lymphocytes beneath the epithelium but were not polycryptic like the cecal tonsil. There were approximately five or six peyer's patches, 5 mm in diameter, in the intestine of 12-week-old chickens (Schat and Myers, 1991). The majority of T-cells were TCR-1 (α/β) and CD4 (T-helper). In general, plasma cells produced each of the three Ig isotypes (Jeurissen et al., 1989a; Schat and Meyers, 1991). Hormonal bursectomy depopulated the lymphocytes in the subepithelial zone (B-dependent) and central zone.

4. Meckel's Diverticulum

The yolk stalk, Meckel's Diverticulum (MD), in 2-week-old chickens, is 3–6 mm long and 1.7 mm thick (Olah and Glick, 1984a). Its distal end continued as the

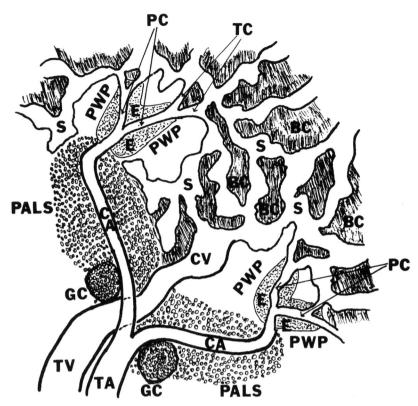

FIGURE 3 Chicken spleen: BC, billroth cord; E, ellipsoid or capillary sleeve; CA, central artery; CV, central vein; GC, germinal center; PALS, periarteriolar lymphatic sheath; PC, penicilliform capillary; PWP, periellipsoid white pulp; S, sinus; TA, trabecular artery; TV, trabecular vein. (From Olah and Glick, 1982, *Am. J. Anat.* **165**, 445–480. Copyright © 1982 Wiley-Liss, Inc. Reprinted by permission of Wiley-Liss, Inc., a subsidiary of John Wiley & Sons, Inc.)

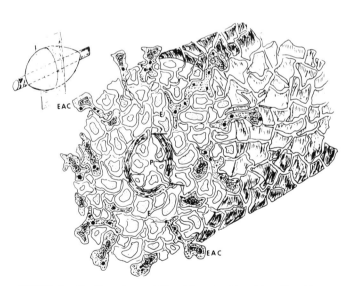

FIGURE 4 Capillary sleeve (ellipsoid) embroidered by ellipsoid associated cell (EAC); PC, penicilliform capillary. (From Olah and Glick, 1982, *Am. J. Anat.* **165**, 445–480. Copyright © 1982 Wiley-Liss, Inc. Reprinted by permission of Wiley-Liss, Inc., a subsidiary of John Wiley & Sons, Inc.)

yolk sac (Figure 5). The MD may contribute to the circulating pool of white blood cells and may be a site to isolate colony stimulating factors that lead to monocytic or granulocytic colonies. Olah *et al.* (1984) identified lymphoid accumulation in MD (yolk stalk) at 2 weeks of age and confirmed Calhoun's (1933) observation of its absence at 1 day old.

Jeurissen *et al.* (1989a, 1989b) identified leukocytes in MD of late embryos, IgM-positive cells underneath the epithelium at 5 days posthatch, and IgG- and IgA-positive cells between 2 and 6 weeks of age. Olah *et al.* (1984) identified dendritic cells (possibly secretory cells) and suggested that they may initiate germinal center formation and may be follicular dendritic cells. Jeurissen *et al.* (1989b) also reported the presence of dendriticlike cells, possibly follicular dendritic cells, in germinal centers.

5. Intestinal Lymphocytes

Intestinal lymphocytes generally resided in the epithelium or lamina propria (Schat and Meyers, 1991). The lamina propria contained IgM- and IgA-positive

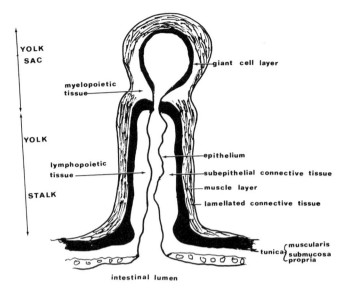

FIGURE 5 Schematic of Meckel's Diverticulum.

B-cells and plasma cells and CD4⁺ T-cells. These intraepithelial cells located between the epithelial cells and basement membrane. They appeared to immigrate from the lamina propria. The intraepithelial cells may be of thymic origin since they declined following thymectomy but not bursectomy (Schat and Myers, 1991). A functional separation of the intestinal lymphocytes occurred between the lamina propria, B- and CD4⁺cells, and epithelium CD8⁺cells (Schat and Myers, 1991). TCR2 predominated in the lamina propria and TCR1 in the epithelium. The presence of an antigen, similar to the expression of $\beta 7$ integrin of human and mice, on avian T-cells after their arrival in the intestine suggested that this antigen retained T-cells in the intestinal epithelium (Haury et al., 1993).

6. Lymph Node

The most developed lymphoid accumulations along the posterior tibiopopliteal and lower femoral veins were true lymph nodes possessing afferent and efferent lymphatics, T- and B-cells, germinal centers, and a prominent lymphatic sinus system (Olah and Glick, 1983, 1985). Kampmeier (1969) reported similar nodes, the cervicothoracic node in waterfowl. A footpad injection of sheep red blood cells stimulated enlargement of the nodes and generated more plaque forming cells (20×10^3 PFC/10^6 lymphoid cells) than in the spleen (4×10^3 PFC/10^6 lymphoid cells) (McCorkle et al., 1979).

7. Pineal

The pineal gland is a lymphopoietic tissue in the chicken (Romieu and Jullien, 1942; Cogburn and Glick, 1981; Olah and Glick, 1984b). By 9 days of age lymphocytes appeared in the pineal and attained maximum concentration by 32 days. Three to 5 days following a carotid injection of bovine serum albumin, pineal plasma cells revealed the presence of antibody to bovine serum albumin (Cogburn and Glick, 1983). Immunohistochemistry revealed an IgA-like substance on the luminal surface of the pineal follicles and in the perifollicular layer (Olah and Glick, 1991). Neonatal bursectomy/thymectomy or at-hatch administration of cyclophosphamide significantly reduced or eliminated T- and B-cells and germinal centers within the pineal (Cogburn and Glick, 1981). The pineal may contribute immunocompetent cells and their products for immune surveillance of the central nervous system.

8. Harderian Gland

The Harderian gland (HG), located ventral and posteromedial to the eyeball (Wight et al., 1971), performs a sentinel role in immune protection of the chicken (Glick and Olah, 1981). The identification of plasma cells in the HG by Bang and Bang (1968) was followed by the observation of Mueller et al. (1971) that the HG was capable of producing a specific antibody. The HG has been suggested to influence B-cell activation, proliferation, and differentiation (Gallego and Glick, 1988; Mansikka et al., 1989; Olah et al., 1992a; Savage et al., 1992; Scott et al., 1993; Scott and Savage, 1996; Tsuji et al., 1993). Later papers identified the response of the HG to a variety of pathogens (Darbyshire, 1987). Plasma cells within the HG (1) concentrated by 3 to 4 weeks posthatch (Niedford and Wolters, 1978; Gallego and Glick, 1988); (2) experienced a proliferation rate 2 to 3 times higher than in the spleen (Gallego and Glick, 1988); (3) peaked in the rate of S-phase by 6 or 8 weeks of age and declined thereafter (Savage et al., 1992); and (4) markedly declined in numbers following bursectomy (Mueller et al., 1971; Sundick et al., 1973). Immunoglobulin class switch from IgM to IgA or IgG occurred in HG B-cells (Mansikka et al., 1989). Immunohistochemical techniques revealed that IgA produced by plasma cells in the central canal and primary branches was secreted by the epithelium of secondary branches and entered the lumen of the primary branch and central canal (Olah et al., 1992a).

III. REGULATION OF IMMUNE RESPONSE

A. Major Histocompatibility Complex

The major histocompatibility complex (MHC) is a highly polymorphic cluster of genes that produce membrane-associated products that influence T-cell development (possibly B-cell development), recognition by T-cells of antigen (MHC restriction or allospecificity), graft rejection, disease resistance, and production traits.

1. Membrane-Associated Glycoproteins, Class I and Class II

The MHC of the chicken originally described by Briles led to the identification of three interesting MHC loci in birds, B-F (class I), B-L (class II), and B-G (class IV) (Crone and Simonsen, 1987). The B-F antigens, expressed on most chicken cells, and the B-L antigens present on B-cells, macrophage/dendritic cells, and activated T-cells are analogous to class I and class II mammalian antigens, respectively.

The B-F antigens possessed a membrane-bound glycosylated polymorphic heavy α-chain (40 to 45 kDa) noncovalently linked with the invariant β_2-microglobulin (12 kDa) (Crone and Simonsen, 1987; Chen et al., 1991). The α-chain possesses three domains, the most distal pair, α_1 and α_2, to form a cleft which presents the processed antigen to T-cytotoxic cells (CD8). The B-L antigen possesses a nonpolymorphic α-chain (32 kDa) and a polymorphic beta chain (27 kDa) (Pharr et al., 1993). The peptide cleft of MHC class II molecules is formed by β_1 and α_1, the most distal domains from the membrane. These peptide-associated class II molecules are presented to T-helper (CD4) cells (Figure 6).

2. MHC Restriction, Allospecific Response, and T-Cytotoxic Cells

Successful presentation of the peptide to a T-cell requires a similar MHC between the antigen-presenting cell and T-cell (MHC restriction). T-cytotoxic cells (CD8) have been identified in chickens (Glick, 1977). Numerous observations suggested that the cytotoxic response is, in part, dependent on T-cells since B-cells were absent in treated bird. A syngeneic-restricted T-cytotoxic lymphocyte response, MHC restriction, has been demonstrated in vitro with the Schmidt-Ruppin strain of avian sarcoma virus or Reticuloendotheliosis virus-infected cells (Schat and Myers, 1991).

Another example of MHC restriction is delayed type hypersensitivity. Several weeks after receiving a specific antigen, a subcutaneous injection of the same antigen will induce a T-cell mediated (CD4) inflammation response within 24 to 72 hr (Glick, 1986).

T-cytotoxic cells will react with peptides presented by different (allogeneic) MHC molecules to produce an allograft or graft-vs-host response (Glick, 1986). An allograft is a skin transplant to an animal of the same species but with different MHC and demonstrates the

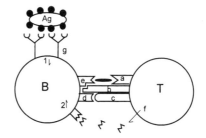

Ag = Antigen ● = Epitope — = Processed Antigen
a = TCR b = CD4 c = CD28 d = B7
e = MHCII f = cytokine release g = Ab Molecule
Σ = Il-4 1 = 1st signal 2 = 2nd signal

FIGURE 6 A model depicting events leading to activation of T- and B-cells. The recognition of a complementary epitope by the antigen receptor of the B-cell (g) transmits an initial signal (1) to the B-cell which then internalizes and processes the antigen for presentation (e) to T-cell receptors (a) of a clonally specific T-helper (CD4) cell. Coreceptor signals (e–b and d–c) help to activate the T-cell to release a cytokine (f) (IL-4) that binds to the B-cell supplying a second signal (2) which activates the B-cell to proliferate and differentiate into an antibody-secreting cell.

rejection of the skin by the host's T-cells. Like the allograft response, the graft-vs-host response is governed by differences at the BF and/or B-L loci. A synergistic effect on the graft-vs-host response occurred when bone marrow and thymic cells were combined (Glick, 1986).

3. Signaling and Coreceptors

MHC restriction depends on the TCR complex (see Section II,A,3) recognizing a peptide MHC complex on the surface of an antigen-presenting cell (monocyte/macrophage B-cell or dendritic cell). The CD8 or CD4 molecules associate with the TCR/CD3 complex and are considered coreceptors (Figure 6). In mammals, optimal T-cell activation depends on direct or indirect association of TCR/CD3 complex with the above coreceptors and a membrane-associated CD45 molecule, leukocyte common antigen. Signals received by the T-cell will aggregate the TCR/CD3 complex together with either coreceptor CD4 or CD8, activating the tyrosine protein kinase associated with coreceptors. CD45 (about 200 kDa) contributes to the activation of tyrosine protein kinase by way of its tyrosine-specific phosphatase (Janeway and Travers, 1992). The chicken model has revealed similar signaling and coreceptor activation. Avian CD4 and CD8, the only known nonmammalian

CD4 and CD8 homologs, have been shown to physically associate with a tyrosine phosphatase kinase homologous to mammalian p56^{lck} (Veillette and Ratcliffe, 1991). Taken together, the data of Veillette and Ratcliffe (1991) and Paramithiotis *et al.* (1991) suggest that signaling by way of the avian TCR/CD3 complex is modulated by the activation of the tyrosine protein kinase-p56^{lck} originating in the coreceptor molecules CD4 and CD8 and the tyrosine-specific phosphatase released by the CD45 surface molecule.

B. Cytokines

A variety of protein mediators or cytokines are produced by T-lymphocytes (Howard *et al.,* 1993). Of these cytokines only interferon (IFN), interleukin (Il)-I, Il-2, Il-3, and tumor necrosis factor (TNF-α) have been studied in birds (Lillehoj *et al.,* 1992; Klasing, 1994). In addition, a stem cell factor (SCF) and myelomonocytic growth factor (MGF) (Leutz *et al.,* 1989) have been studied. The MGF, originated from a transformed macrophage cell line, HDII, has been cloned by Leutz and shown to have some homology to mammalian granulocyte-colony stimulating factor (G-CSF). The recent work of Nicholas-Bolnet *et al.* (1991, correspondence) offers important steps in isolating SCF, Il-3, G-CSF, and M-CSF.

Interleukin-1 is produced by monocytes and stimulated thymocytes to release Il-2 (Schat and Myers, 1991). Limited or no functional cross-reactivity has been reported with chicken, human, and murine Il-1 or Il-2 (Schat and Myers, 1991).

The conditioned medium of a concanavalin-A-stimulated culture of adherent splenocytes and thymocytes stimulated proliferation of preactivated T-lymphoblasts and Il-2 activity (Schat and Myers, 1991). The molecular weight of between 15.5 and 30 kDa for avian Il-2 is similar to that of mammals.

Like the interleukin studies, IFN studies in chickens are in the very early stages of development (Schat and Myers, 1991; Lillehoj *et al.,* 1992). This is surprising since the first inference of IFN was from the chicken embryo data of Isaacs and Lindemann (1957). Transforming growth factor β which influences cell differentiation, wound repair and bone metabolism and growth and tumor necrotic factor-α have been identified in the chicken (Klasing, 1994).

A lymphocyte inhibitory factor (LyIF) was released from thymic or bursal cells sensitized to purified protein derivative (Glick, 1983). Sensitized thymic cells required macrophages for the release of LyIF and a chemotactic factor while the LyIF response of bursal cells was independent of macrophages (Joshi and Glick, 1990). These types of experiments illustrate an advanced stage of maturity of cells within the thymic and bursal microenvironments.

C. Antibody-Mediated Immunity and B-Cell Repertoire

Antibody production by B-cells prevents the spread of pathogens by (1) combining with the pathogen and neutralizing it, (2) facilitating uptake and digestion of the pathogen by phagocytic cells, and (3) facilitating cell lysis and death. Most antibody responses in birds are thymic dependent (TD) and require cooperation between T-, B-, and macrophage/dendritic cells (Weinbaum *et al.,* 1973; Thorbecke *et al.,* 1980). Thymic-independent (TI) antigens include pneumococcal polysaccharide and Brucella, examples of TI-type I, and ficoll and dextran, examples of a TI-type II (Golub and Green, 1991). A TI response has been demonstrated in birds by allogeneic bursal cell transfers to cyclophosphamide recipients (Toivanen *et al.,* 1974). Antibody response in these birds to *Brucella abortus* but not the TD antigen sheep red blood cells was restored. The allogeneic bursal stem cells induced B-cell chimerism and tolerance to donor MHC (Vainio and Toivanen, 1987). These types of allogeneic responses will reconstitute TI but not TD responses. The TD responses of cyclophosphamide birds required transfer of thymic and bursal cells syngeneic to host or possibly one MHC similar to donor B-cell and host T-cell (Vanio and Toivanen, 1987).

1. Immunoglobulins

B-cells and plasma cells synthesize and release immunoglobulin. Antibody, a glycoprotein, is structurally identical to Ig and is represented by five distinct classes or isotypes: IgM, IgD, IgG, IgA, and IgE. Each class exhibits at least one monomer of two-different-molecular-weight polypeptide chains: two heavy (H) and two light (L) chains. Beginning at the N-terminal of the H polypeptide class and extending to its C-terminal there is a variable (VH) domain and three to four constant domains (CH) depending on the isotype. The light chain possesses a VL and a single CL domain. The constant regions differentiate the distinctiveness of each heavy chain which are named μ (IgM), δ (IgD), γ (IgG), α (IgA), and ε (IgE). In the chicken there is a single light or κ-chain. The VH and VL (N terminal) domains contribute to the antigen binding site or F_{ab} while the CH domains (C-terminal) identify the constant or Fc portion of the antibody molecule (Figure 7).

Avian IgM resembled mammalian IgM [$(\mu_2 L_2)_5$] on the basis of physiochemical, antigenic characteristics; sedimentation coefficients (S, 16.7 to 16.9), and molecu-

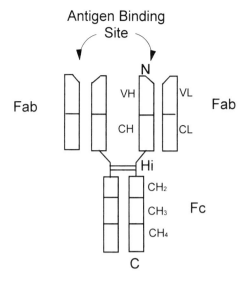

FIGURE 7 The immunoglobulin molecules possess two heavy (H) chains with a single variable domain (VH) and three or four constant domains (CH to CH4). Each light (L) chain possess a VL- and CL-domain. The Fab refers to the antigen binding position of the molecule (N-terminus) while F_c identifies the constant portion of the molecule (C-terminus).

lar weight (890 kDa) (Leslie and Clem, 1969). Sequence data for chicken μ identified a homology with mammalian μ chain (Dahan et al., 1983); there were five IgM CH allotypes (Benedict and Berestecky, 1987). On the other hand, disagreement exists concerning the name of the major avian 7.1- to 7.8-serum Ig IgG ($\gamma_2 L_2$). Leslie and Clem (1969) named this 7SIg IgY, in part, because it possessed 4CH and not 3CH as in human IgG and possessed a greater concentration of carbohydrate (6.0%) than did the human IgG. It has been suggested that avian serum 7SIg resembled mammalian IgA more than IgG (Tenenhouse and Deutch, 1966) and that human IgA cross-reacted with the avian serum 7SIg (Ambrosius and Hadge, 1987). The inference that the avian 7SIg could be a precursor of IgA was weakened with the identification of chicken secretory IgA (Benedict and Berestecky, 1987). The amino acid sequence of the IgY H chain placed it closest to mammalian IgE (Parvari et al., 1988). Functionally, the avian 7SIg (IgG or IgY) was similar to mammalian IgG (Benedict and Berestecky, 1987). Interestingly, the duck possessed two serum Igs, a 5.7-SIgG and a 7.8-SIgG, which may be influenced by a single gene (Magor et al., 1992; Higgins and Warr, 1993).

A third avian Ig was termed IgA because of its prevalence in lymphoepithelial tissue and structural similarity to mammalian IgA (Benedict and Berestecky, 1987). Cloning of the IgA H chain revealed the greatest structural similarity to mammalian αH chains and not μH or εH (IgE) chains (Mansikka, 1992). The heavy chain of chicken IgA contained four constant α domains or one more than human constant α chains. These results suggested the possible occurrence of a deletion during evolution. Avian IgA was polymorphic and highly concentrated in the bile (3–12 mg/ml) (Benedict and Berestecky, 1987). A J-component and secretory component were present in avian IgA. The secretory component (60 kDa) was synthesized by epithelial cells and attached to bile IgA before secretion (Benedict and Berestecky, 1987). While chicken and turkey IgA may be monomeric or polymeric, the secretory IgAs were generally trimeric or tetrameric. Polymeric IgA, which lacked the secretory component, may be transported from blood to bile by combining with membrane secretory components produced by hepatocytes (Schat and Myers, 1991).

Immunoglobulin G and IgM antibodies possess complement binding sites. Complement is a complex of 9 major components (Janeway and Travers, 1994). Avian complement 3 (C3) fragments to C3b after binding to an antibody–antigen complex (classical pathway) or in the presence of a pathogenic surface (alternate pathway, AP) and continues the cascade of complement factors that will remove the immune complexes, lyse the pathogens bound by the antibody, and enhance inflammation (Koch, 1987). It appeared that the B-like protein important in the AP existed in the chicken and was similar to mammalian C2 and component B (Kjalke et al., 1994).

2. V-Gene Repertoire and the Contributions of the Bursal Milieu

Antibody diversity in mammals depends on somatic-driven events within the B-cell leading to (1) heavy-chain rearrangement between D and J segments; (2) rearrangement of several hundred V-genes with the rearranged D-J segments (VDJ); (3) light-chain rearrangement of numerous V and J segments; (4) junctional diversity; (5) nucleotide additions; and (6) combinatorial joining between the rearranged H (VDJ)- and L (VJ)-chains (Golub and Green, 1991).

Since the chick H chain possesses a single V and 10 D (similar homologies) and a single J gene and the light chain possesses single V and J genes, rearrangement of the H- and L-chains will not contribute diversity as in mammals. Rather, the chicken depends on gene conversion, a transfer of nucleotide sequences from upstream pseudogenes to the rearranged VH (VDJ) and VL (VJ) genes, to produce antibody diversity (McCormack et al., 1991). Including the values of 10^4 follicles per bursa, 10^5 cells per follicle and other assumptions, mathematical models have been presented to support a B-cell repertoire (antibody diversity) of about 10^6 different specificities (Salanti et al., 1989; Langman and Cohn, 1993).

The rearrangement of H & L chains may occur in the bursa or at other sites (McCormack et al., 1991). These molecular observations supported the original observation of a bursal-independent site for Ig (Lerner et al., 1971; Glick, 1977; Ratcliffe et al., 1986; McCormack et al., 1991).

Schemes to explain the bursal-dependent roles of the expansion of in-frame B-cells (i.e., B-cells with rearranged H- and L-chains) and gene conversion (McCormack et al., 1991) have been discussed and conceptually approached by Langman and Cohn (1993). The microenvironmental milieu necessary for these events appears to include (1) the complex interaction of endodermal and mesodermal germ layers (Houssaint et al., 1976), (2) a singular role of the endodermal epithelium (LeDouarin et al., 1980), and (3) a dark mesenchymal cell that is the precursor of the bursal secretory dendritic cell (BSDC) (Olah et al., 1986; Olah and Glick, 1978b; Olah et al., 1992b). During late embryonic development the only cell to possess IgG on its membrane was the BSDC (Olah et al., 1991). We have proposed a receptor paracrine pathway that involved the interaction of the IgM of the in-frame B-cell with the IgG of the BSDC (Glick and Olah 1993a,b; Glick, 1995). This may lead to activation and proliferation of the B-cells and a BSDC secretion that signals gene conversion of the expanded in-frame B-cells. Alternatively, replication of in-frame B-cells might occur when the B-cell identified a bursal specific antigen (Masteller and Thompson, 1994), possibly expressed by the BSDC. In either case, apoptosis (programmed cell death) (Compton, 1993) of the in-frame B-cell would occur if the in-frame B-cell fails to make contact with a specialized cell.

The receptor–paracrine hypothesis has an analogy with the activity in mammalian germinal centers where follicular dendritic cells trap antigen–antibody complexes, present them to the B-cell, and then release a secretion that contributes to B-cell differentiation (Gordon et al., 1989). Kaufman and Salomonsen (1993), subsequent to several assumptions, proposed the B–G complex of the MHC as the antigen influencing the selection of germline Ig bursal cells. In their negative-selection model, the germline Ig B-cells bound to the B–G self molecule and were eliminated while the B-cells that experienced gene conversion would not bind and would emigrate. In the positive-selection model, the germline B-cells bind to the B–G self molecule, receive a signal to proliferate, and undergo gene conversion, eventually losing their recognition of B–G self and then emigrating (Kaufman et al., 1991). The influence of cell adhesion molecules on the cells of the bursal microenvironment has been briefly discussed (Glick, 1995) and experimental evidence offered by Masteller and Thompson (1994). Bu-1-positive cells in the embryonic spleen are committed to the B-cell lineage and express a cell adhesion ligand Sialyl Lewis x (Masteller and Thompson, 1994). By 10 to 12 days of embryonic development, Sialyl Lewis x-positive cells appear in the bursa, proliferate, and then contribute to the developing follicle. These B-cells lose Sialyl Lewis x in parallel with the initiation of gene conversion (15 to 18 days of embryogenesis). The loss of Sialyl Lewis x and the switch to Lewis x high in Ig-positive B-cells is followed by emigration to the spleen. This might suggest that migration from the bursa is influenced by a change in expression of Lewis x.

D. Macrophages, Natural Killer Cells, Heterophils, and Thrombocytes

The phagocytic avian macrophage, similar to the mammalian macrophage, performs the pivotal role of an antigen-presenting cell (Powell, 1987; Vainio et al., 1988). Several investigators have studied the avian macrophages' effector functions (Dietert et al., 1991; Qureshi et al., 1994) by employing peritoneal exudate macrophages and a malignantly transformed chicken macrophage cell line, MQ-NCSU. Optimum activation of avian macrophages required two signals, a lymphokine (IFN(?)) and lipopolysaccharide (LPS) (Qureshi et al., 1994). Activated macrophages will release Il-1, tumor necrotic factor (TNF), and colony stimulating factors (CSF). The cytotoxicity of TNF may be species specific (Qureshi et al., 1993). Chicken embryonic bone marrow cells exposed to supernatants from cultured MQ-NCSU macrophage cells produce colonies of granulocytes and macrophage-granulomonocytic cells, suggesting the ability of these macrophages to produce G-CSF and GM-CSF, respectively. Macrophage cell lines pulsed with LPS produce reactive nitrogen intermediates (e.g., nitric oxide, ·NO) (Qureshi et al., 1993, 1994). Unlike mammalian macrophages that synthesize arginine, avian macrophages require exogenous sources of L-arginine, which are converted to reactive nitrogen intermediates (e.g., ·NO) by way of oxidative enzymes (Qureshi et al., 1993). These reactive nitrogen intermediates have an antineoplastic and antimicrobial function. Nitric oxide may have an autocrine effect since it suppresses complexes I and II enzymes in the mitochondria of macrophages and tumor target cells (Sung and Dietert, 1994). Further challenges in understanding avian macrophage will be to identify a homolog to the mammalian differentiation antigen CD14 and the LPS–LBP (lipopolysaccharide binding protein) complex that binds to the membrane-associated CD14 to trigger the release of macrophage cytokines (Martin et al., 1994). Also, will

·NO inhibit the expression of avian macrophage Ia (class II MHC), which in mammals results in reduced ·NO production (Sicher et al., 1994)?

Natural killer (NK) cells are large granular leukocytes exhibiting natural cytotoxicity to a variety of tumor cells. The NK cells found in peripheral blood, intestinal epithelium, and spleen tend to be nonadherent cells. Cells exhibiting spontaneous cytotoxicity, NK cells, were identified by two mAb, K-108 and K-4 (Chung and Lillehoj, 1991). NK cells increase with age and in the presence of viruses. Disease and suppressor cells depress NK cells. Mammalian NK cells respond positively to interferon while avian NK cells may not (Sharma and Schat, 1991).

The nonlymphoid heterophil and thrombocyte were active in immunity and infection (Powell, 1987). Heterophils are the early cells of inflammation and are phagocytic. Serotonin inhibited the cytotoxicity of granulocytes while pretreatment with interferon impeded the inhibitory effects of serotonin (Garssadi et al., 1994). The thrombocytes role in hemostasis and phagocytosis was shown initially by Stalsberg and Prydz (1963) and Glick et al.(1964), respectively. Monoclonal antibodies raised against platelets will identify thrombocytes (Kunicki and Newman, 1985) while Kaspers et al. (1993) have described a mAb that appears to identify monocyte/thrombocytes.

References

Adelmann, H. B. (1942). "The Embryological Treatises of Hieronymus Fabricius of Aquapendente," Vol. 1, p. 376, Cornell Univ. Press, Ithaca, NY.
Albini, B., and Wick, G. (1974). Delineation of B and T–lymphoid cells in the chicken. *J. Immunol.* **112**, 444–450.
Ambrosius, H., and Hadge, D. (1987). Chicken immunoglobulins. *Vet. Immunol. Immunopathol.* **17**, 57–67.
Bang, B. G., and Bang, F. B. (1968). Localized lymphoid tissues and plasma cells in paraocular and paranasal organ systems in chickens. *Am. J. Pathol.* **33**, 735–751.
Benedict, A. A., and Berestecky, J. M. (1987). Special features of avian immunoglobulins. In "Avian Immunology: Basis and Practice" (A. Toivanen and P. Toivanen, eds.), Vol. 1, pp. 113–126. CRC Press, Boca Raton, FL.
Bockman, D. E., and M. D. Cooper (1973). Pinocytosis by epithelium associated with lymphoid follicles in the bursa of Fabricius, appendix, and Peyer's patches. An electron microscopic study. *Am. J. Anat.* **136**, 455–478.
Brewer, J. M., Wunderlich, J. K., and Ragland, W. L. (1990). The amino acid sequence of avian thymic hormone, a parvalbumin. *Biochimie* **72**, 653–660.
Bucy, R. P., Chen, Chen-Lo H., and Cooper, M. D. (1990). Ontogeny of T–cell receptors in the chicken thymus. *J. Immunol.* **144**, 1161–1608.
Calhoun, M. L. (1933). The microscopic anatomy of the digestive tract of *Gallus domesticus. Iowa St. Coll. J. Sci.* **7**, 261–381.
Chen, Chen-Lo H., Ager, L. L., Gartland, E. L., and Cooper, M. D. (1986). Identification of a T3/T cell receptor complex in chickens. *J. Exp. Med.* **164**, 375–380.
Chen, Chen-Lo H., and Cooper, M. D. (1987). Identification of cell surface molecules on chicken lymphocytes with monoclonal antibodies. In "Avian Immunology: Basis and Practice" (A. Toivanen and P. Toivanen, eds.), Vol. 1, pp. 138–154. CRC Press, Boca Raton, FL.
Chen, Chen-Lo H., Pickel, J. M., Lahti, J. M., and Cooper, M. D. (1991). Surface markers on avian immune cells: In "Avian Cellular Immunology" (J. M. Sharma, ed.), pp. 1–22. CRC Press, Boca Raton, FL.
Chung, K. S., and Lillehoj, H. S. (1991). Developmental and functional characterization of monoclonal antibodies recognizing chicken lymphocytes with natural killer cell activity. *Vet. Immunol. Immunopathol.* **28**, 351–363.
Cogburn, L. A., and Glick, B. (1981). Lymphopoiesis in the chicken pineal gland. *Am. J. Anat.* **102**, 131–142.
Cogburn, L. A., and Glick, B. (1983). Functional lymphocytes in the chicken pineal gland. *J. Immunol.* **130**, 2109–2112.
Compton, M. M. (1993). Programmed cell death in avian thymocytes: Role of the apoptotic endonuclease. *Poult. Sci.* **72**, 1267–1272.
Cooper, M. D., Peterson, R. D. A., South, M. A., and Good, R. A. (1966). The functions of the thymus system and bursa system in the chicken. *J. Exp. Med.* **123**, 75–106.
Crone, M., and Simonsen, M. (1987). Avian major histocompatibility complex. In "Avian Immunology: Basis and Practice" (A. Toivanen and P. Toivanen, eds.), Vol. II, pp. 25–42. CRC Press, Boca Raton, FL.
Dahan, A., Raynaud, C.-A., and Weill, J.-C. (1983). Nucleotide sequence of the constant region of a chicken μ heavy chain immunoglobulin mRNA. *Nucleic Acids Res.* **11**, 5381–5389.
Darbyshire, J. H. (1987). Immunity and resistance in respiratory tract diseases. In "Avian Immunology: Basis and Practice" (A. Toivanen and P. Toivanen, eds.), pp. 129–141. CRC Press, Boca Raton, FL.
Dieterlen-Lievre, F., Pardanaud, L., Bolnet, C., and Cormier, F. (1990). Development of the hemopoietic and vascular systems studied in the avian embryo. In "The Avian Model in Developmental Biology: From Organism to Genes" (N. LeDouarin, F. Dieterlen–Lievre, and J. Smith, eds.), pp. 181–196. Edition Du Centre National De La Recherche Scientifique, Paris, France.
Dietert, R. R., Golemboski, K. A., Bloom, S. E., and Qureshi, M. A. (1991). The avian macrophage in cellular immunity. In "Avian Cellular Immunology" (J. A. Sharma, ed.), pp. 71–95. CRC Press, Boca Raton, FL.
Gallego, M., and Glick, B. (1988). The proliferative capacity of the avian Harderian gland. *Dev. Comp. Immunol.* **12**, 157–166.
Gallego, M., Olah, I., Del Cacho, E., and Glick, B. (1993). Anti S-100 antibody recognizes ellipsoid–associated cells and other dendritic cells in the chicken spleen. *Dev. Comp. Immunol.* **17**, 77–83.
Garssadi, S. E., Regely, K., Mandi, Y., and Beladi, I. (1994). Inhibition of cytotoxicity of chicken granulocytes by serotonin and ketamine. *Vet. Immunol. Immunopathol.* **41**, 101–112.
Gilmour, D. G., Brand, A., Donnelly, N., and Stone, H. A. (1976). Bu-1 and Th-1, two loci determining surface antigens of B and T lymphocytes in chickens. *Immunogenetics* **3**, 549–563.
Glick, B. (1956). Normal growth of the bursa of Fabricius in chickens. *Poult. Sci.* **35**, 843–851.
Glick, B. (1977). The bursa of Fabricius and immunoglobulin synthesis. *Int. Rev. Cytol.* **48**, 345–402.
Glick, B. (1983). Bursa of Fabricius. In "Avian Biology" (D. S. Farner, R. King, and K. C. Parkes, eds.), Vol. 7, pp. 443–500. Academic Press, New York.
Glick, B. (1986). Immunophysiology. In "Avian Physiology" (P. D. Sturkie, ed.), pp. 87–101. Springer-Verlag, New York.

Glick, B. (1987). How it all began: A continuing story of the bursa of Fabricius. In "Avian Immunology: Basis and Practice" (A. Toivanen and P. Toivanen, eds.), pp. 1–8. CRC Press, Boca Raton, FL.

Glick, B. (1995). Embryogenesis in the bursa of Fabricius: Stem cell, microenvironment and receptor–paracrine pathways. *Poult. Sci.* **74**, 419–426.

Glick, B., and Olah, I. (1981). Gut–associated lymphoid tissue of the chicken. In "Scanning Electron Microscopy III" (O. Jahari, ed.), pp. 95–108. SEM, Inc., Chicago, IL.

Glick, B., and Olah, I. (1993a). Bursal secretory dendritic-like cell: A microenvironment issue. *Poult. Sci.* **72**, 1262–1266.

Glick, B., and Olah, I. (1993b). A bursal secretory dendritic cell and its contributions to the microenvironment of the developing bursal follicle. *Res. Immunol.* **144**, 446–448.

Glick, B., Chang, T. S., and Jaap, R. G. (1956). The bursa of Fabricius and antibody production in the domestic fowl. *Poult. Sci.* **35**, 224–225.

Glick, B., Sato, K., and Cohenour, F. (1964). Comparison of the phagocytic ability of normal and bursectomized birds. *J. Reticuloendothel. Soc.* **1**, 442–449.

Glick, B., Holbrook, K. A., Olah, I., Perkins, W. D., and Stinson, R. A. (1981). An electron and light microscope study of the caecal tonsil: The basic unit of the caecal tonsil. *J. Dev. Comp. Immunol.* **5**, 95–104.

Golub, E. S., and Green, D. R. (1991). "Immunology: A Synthesis," 2nd Edition. Sinauer Associates, Sumderland, MA.

Gordon, J., Flores-Romo, L., Cairns, J., Millsum, M., Lane, P., Johnson, G., and MacLennan, I. C. M. (1989). CD23: A multifunctional receptor/lymphokine? *Immunol. Today* **10**, 153–157.

Haury, M., Kasahara, Y., Schaal, S., Bucy, R., and Cooper, M. D. (1993). Intestinal T-lymphocytes in the chicken express an integrin–like antigen. *Eur. J. Immunol.* **23**, 313–319.

Higgins, D. A., and Warr, G. W. (1993). Duck immunoglobulins: Structure, functions and molecular genetics. *Avian Pathol.* **22**, 211–230.

Holbrook, K. A., Perkins, W. D., and Glick, B. (1974). The fine structure of the bursa of Fabricius: "B" cell surface configuration and lymphoepithelial organization revealed by scanning and transmission electron microscopy. *J. Reticuloendothel. Soc.* **16**, 300–311.

Houssaint, E., Belo, M., and LeDouarin, N. M. (1976). Investigations on cell lineage and tissue interactions in the developing bursa of Fabricius through interspecific chimeras. *Dev. Biol.* **53**, 200–264.

Howard, M. C., Miyajima, A., and Coffman, R. (1993). T-cell derived cytokines and their receptors. In "Fundamentals of Immunology" (W. E. Paul, ed.) pp. 763–800. Raven, New York.

Issaacs, A., and Lindenmann, J. (1957). Virus interference. I. The interferon. *Proc. Roy Soc. London (Biol.)* **147**, 258–267.

Jaffe, W. P., and Fechheimer, N. S. (1966). Cell transport and the bursa of Fabricius. *Nature (London)* **212**, 92.

Janeway, C. A., Jr., and Travers, P. (1994). "Immunobiology, The Immune System in Health and Disease." Garland Publishing, New York.

Jeurissen, S. H. M. (1993). The role of various compartments in the chicken spleen during an antigen–specific humoral response. *Immunology* **80**, 29–33.

Jankovic, B. D., and Mitrovic, K. (1967). Antibody producing cells in the chicken, as observed by fluorescent antibody technique. *Folia. Biol.* **13**, 406–410.

Jeurissen, S. H. M., Janse, E. M., and Koch, G. (1989a). Meckel's diverticle: A gut-associated lymphoid organ in chickens. *Adv. Exp. Med. Biol.* **237**, 599–606.

Jeurissen, S. H. M., Janse, E. M., Koch, G., and deBoer, G.F. (1989b). Post natal development of mucosa–associated lymphoid tissues in chickens. *Cell Tissue Res.* **258**, 119–124.

Jeurissen, S. H. M., Claassen, E., and Janse, E. M. (1992). Histological and functional differentiation of non-lymphoid cells in the chicken spleen. *Immunology* **77**, 75–80.

Jolly, J. (1915). Le bourse de Fabricius et les organes lympho-epitheliaux. *Arch. Anat. Microsc. Morphol. Exp.* **16**, 363–546.

Joshi, P., and Glick, B. (1990). The role of avian macrophages in the production of avian lymphokines. *Dev. Comp. Immunol.* **14**, 319–325.

Kampmeier, O. F. (1969). "Evolution and Comparative Morphology of the Lymphatic System." Charles C. Thomas, Springfield, IL.

Kaspers, B., Lillehoj, H. S., and Lillehoj, E. P. (1993). Chicken macrophages and thrombocytes share a common cell surface antigen defined by a monoclonal antibody. *Vet. Immunol. Immunopathol.* **36**, 333–346.

Kaufman, J., and Salomonsen, J. (1993). What in the dickens is with those chickens? An only slightly silly response to the first draft of Langman and Cohn. *Res. Immunol.* **144**, 495–502.

Kaufman, J., Skjodt, K., and Salomonsen, J. (1991). The B-G multigene family of the chicken major histocompatibility complex. *Crit. Rev. Immunol.* **11**, 113–143.

Kjalke, M., Welender, K. G., and Koch, C. (1994). Structural analysis of chicken factor B-like protease and comparison with mammalian complement proteins factor B and C2. *J. Immunol.* **151**, 4147–4152.

Klasing, K. C. (1994). Avian leukocytic cytokines. *Poult. Sci.* **73**, 1035–1043.

Koch, C. (1987). Complement system in avian species. In "Avian Immunology Basics and Practice" (A. Toivanen and P. Toivanen, eds.), Vol. II, pp. 43–55. CRC Press, Boca Raton, FL.

Kunicki, T., and Newman, P., (1985). Synthesis of analog of human platelet membrane glycoprotein 11b-111a complex by chicken peripheral blood monocytes. *Proc. Natl. Acad. Sci. USA* **82**, 7319–7323.

Langman, R. E., and Cohn, M. (1993). A theory of the ontogeny of the chicken immune system: The consequences of diversification by gene conversion hyperconversion and its extension to the rabbit. *Res. Immunol.* **144**, 422–496.

Le Douarin, N. M., Michel, G., and Baulieu, E. E. (1980). Studies of testosterone-induced involation of the bursa of Fabricius. *Dev. Biol.* **75**, 288–302.

Le Douarin, N. M., Dieterlen-Lievre, F., and Oliver, P. D. (1984). Ontogeny of primary lymphoid organs and lymphoid stem cells. *Am. J. Anat.* **170**, 261–299.

Le Douarin, N. M., Martin, C., Ohki-Hamazaki, H., Belo, M., Coltey, M. D., and Corbel, C. (1990). Development of the immune system and self/non-self recognition studied in the avian embryo. In "The Avian Model in Developmental Biology: From Organism to Genes" (N. Le Douarin, F. Dieterlen-Lievre, and J. Smith, eds.), pp. 219–237. Edition Du Centre National De La Recherche Scientifique, Paris, France.

Lerner, K. G., Glick, B., and McDuffie, F. C. (1971). Role of the bursa of Fabricius in IgG and IgM production in the chicken: Evidence for the role of a non-bursal site in the development of humoral immunity. *J. Immunol.* **107**, 493–503.

Leslie, G. A., and Clem, L. W. (1969). Phylogeny of immunoglobulin structure and function. III. Immunoglobulins of the chicken. *J. Exp. Med.* **130**, 1337–1352.

Leutz, A., Damm, K., Sterneck, E., Kowenz, E., Ness, S., Frank, R., Gauseponl, H., Pan, Y. C., Smart, J., Haiman, M., and Graft, T. (1989). Molecular cloning of the chicken myelomonocytic growth factor (cMGF) reveals relationship to interleukin-6 and granulocyte colony stimulating factor. *EMBO J.* **8**, 175–181.

Lillehoj, H. S., Kaoporo, B., Jenkins, M. O., and Lillehoj, E. P. (1992). Avian interleukin-2: A review by comparison with mammalian homologues. *Poult. Sci. Rev.* **4**, 67–86.

Linsley, P. S., and Ledbetter, J. A. (1993). The role of the CD28 receptor during T-cell responses to antigen. *Annu. Rev. Immunol.* **11**, 191–212.

Magor, K. E., Warr, G. W., Middleton, D., Wilson, M. R., and Higgins, J. D. A. (1992). Structural relationship between the two IgY of the duck, *Anas platyrhynchos:* Molecular genetic evidence. *J. Immunol.* **149**, 2627–2633.

Mansikka, A. (1992). Chicken IgA H chains: Implications concerning the evolution of H chain genes. *J. Immunol.* **149**, 855–861.

Mansikka, A., Sandberg, M., Veromaa, T., Vainio, O., Granfors, K., and Toivanen, P. (1989). B-cell maturation in the chicken Harderian Gland. *J. Immunol.* **142**, 1826–1833.

Marsh, J. (1993). The humoral activity of the avian thymic microenvironment. *Poult. Sci.* **72**, 1294–1300.

Martin, C. (1990). Quail chick chimeras, a tool for developmental immunology. *In* "The Avian Model in Developmental Biology: From Organisms to Genes" (N. Le Douarin, F. Dieterlen-Lievre, and J. Smith, eds), pp. 207–217. Editions Du Centre Nationale De La Recherche Scientifique, Paris, France.

Martin, T. R., Mongovin, S. M., Tobias, P. S., Mathison, J. C., Moriarty, A. M., Leturcq, D. J., and Ulevitch, R. J., (1994). The CD14 differentiation antigen mediates the development of endotoxin responsiveness during differentiation of mononuclear phagocytes. *J. Leuk. Biol.* **56**, 1–9.

Masteller, E. L., and Thompson, C. B. (1994). B-cell development in the chicken. *Poult. Sci.* **73**, 998–1011.

McCorkle, F. M., Stinson, R. S., Olah, I., and Glick, B. (1979). The chicken's femoral lymph nodules: T&B cells and the immune response. *J. Immunol.* **123**, 667–669.

McCormack, W. T., Tjoelker, L. W., and Thompson, C. B. (1991). Avian B-cell development: Generation of an immunoglobulin repertoire by gene conversion. *Annu. Rev. Immunol.* **9**, 219–241.

Moore, M. A. S., and Owen, J. J. T. (1965). Chromosome marker studies on the development of the hematopoietic system in the chick embryo. *Nature (London)* **708**, 956,989–990.

Moore, M. A. S., and Owen, J. J. T. (1966). Experimental studies in the development of the bursa of Fabricius. *Dev. Biol.* **14**, 40–51.

Mueller, A., Sato, K., and Glick, B. (1971). The chicken lacrimal gland, gland of Harder, caecal tonsil, and accessory spleens as sources of antibody producing cells. *Cell. Immunol.* **2**, 140–152.

Murthy, K. K., Pace, J. L., Barger, B. O., Dawe, D. L., and Ragland, W. L. (1984). Localization and distribution by age and species of a thymus-specific antigen. *Thymus* **6**, 43–56.

Muthmann, E., (1913). Beitrage Zur vergleichenden anatomie des Blinddarmes und der lymphoiden organe des darmkanals bei saugetieren und vogeln. *Anat. Hefte* **48**, 67–114.

Naukkarinen, A., Arstela, A. U., and Sorvari, T. E. (1978). Morphological and functional differentiation of the surface epithelium of the bursa of Fabricii in chicken. *Anat. Rec.* **191**, 415–432.

Nicholas-Bolnet, C., Yassine, F., Cormier, F., and Dieterlen-Lievre, F. (1991). Developmental kinetics of hemopoietic progenitors in the avian embryo spleen. *Exp. Cell Res.* **196**, 294–301.

Nickel, R., Schummer, A., Seiferle, E., Siller, W. G., and Wight, P. A. L., (1977). "Anatomy of the Domestic Birds." Springer-Verlag, New York.

Niedorf, H. R., and Wolters, B. (1978). Development of the Harderian gland in the chicken: Light and electron microscopic investigations. *Invest. Cell. Pathol.* **1**, 205–215.

Olah, I., and Glick, B. (1978a). The number and size of the follicular epithelium (FE) and follicles in the bursa of Fabricius. *Poult. Sci.* **57**, 1445–1450.

Olah, I., and Glick, B. (1978b). Secretory cell in the medulla of the bursa of Fabricius. *Experientia* **34**, 1642–1643.

Olah, I., and Glick, B. (1982). Splenic white pulp and associated vascular channels in chicken spleen. *Am. J. Anat.* **165**, 445–480.

Olah, I., and Glick, B. (1983). Avian lymph node: Light and electron microscopic study. *Anat. Rec.* **205**, 287–299.

Olah, I., and Glick, B. (1984a). Meckel's diverticulum. I. Extra medullary myelopoiesis in the yolk sac of hatched chickens (*Gallus domesticus*). *Anat. Rec.* **208**, 243–252.

Olah, I., and Glick, B. (1984b). Lymphopineal tissue in the chicken. *Dev. Comp. Immunol.* **8**, 855–862.

Olah, I., and Glick, B. (1985). Lymphocyte migration through the lymphatic sinuses of the chicken's lymph node. *Poult. Sci.* **64**, 159–168.

Olah, I., and Glick, B. (1991). An Ig-A-like substance in the chicken's pineal. *Experientia* **147**, 202–205.

Olah, I., Glick, B., and Taylor, R. L., Jr. (1984). Meckel's diverticulum. II. A novel lymphoepithelial organ in the chicken. *Anat. Rec.* **208**, 253–263.

Olah, I., Glick, B., and Toro, I. (1986). Bursal development in normal and testosterone-treated chick embryos. *Poult. Sci.* **65**, 574–588.

Olah, I., Medgyes, T., and Glick, B. (1988). Origin of aortic cell clusters in the chicken embryo. *Anat. Rec.* **222**, 60–68.

Olah, I., Kendall, C., and Glick, B. (1991). Bursal secretory-dendritic cell express a homologue of IgGFc. *Fed. Am. Soc. Exp. Biol. J.* **5**, A1064. [Abstract]

Olah, I., Scott, T. R., Gallego, M. Kendall, C., and Glick, B. (1992a). Plasma cells expressing immunoglobulins M and A but not immunoglobulin B develop an intimate relationship with central canal epithelium in the Harderian gland of the chicken. *Poult. Sci.* **71**, 664–676.

Olah, I., Kendall, C., and Glick, B. (1992b). Anti-vimentin monoclonal antibody recognizes a cell with dendritic appearance in the chick's bursa of Fabricius. *Anat. Rec.* **232**, 121–125.

Orlans, E., and Rose, M. E. (1970). Antibody formation by transferred cells in inbred fowls. *Immunology* **18**, 473–482.

Paramithiotis, E., Tkalec, L., and Ratcliff, M. J. H. (1991). High levels of CD45 are coordinately expressed with CD4 and CD8 in avian thymocytes. *J. Immunol.* **147**, 3710–3717.

Parvari, R., Avivi, A., Lentner, F., Ziv, E., Tel-or, S., Burstein, Y., and Schechter, I. (1988). Chicken immunoglobulin γ heavy chains: limited VH gene repertoire, combinatorial diversification by D gene segments and evolution of the heavy chicken locus. *EMBO J.* **7**, 739–744.

Paul, W. E. (1993). "Fundamental Immunology." Raven, New York.

Pharr, G. T., Hunt, H. D., Bacon, L. D., and Dodgson, J. B. (1993). Identification of class II major histocompatibility complex Polymorphisms predicted to be important in peptide antigen presentation. *Poult. Sci.* **72**, 1312–1317.

Powell, P. C. (1987). Macrophages and other non-lymphoid cells contributing to immunity. *In* "Avian Immunology: Basis and Practice" (A. Toivanen and P. Toivanen, eds.), Vol. 1, pp. 195–212. CRC Press, Boca Raton, FL.

Qureshi, M. A., Petitte, J. H., Laster, S. M., and Dietert, R. R. (1993). Avian macrophages: Contribution to cellular microenvironment and changes in effector functions following activation. *Poult. Sci.* **72**, 1280–1284.

Qureshi, M. A., Marsh, J. A., Dietert, R. R., Sung, Y.-J. Nicholas-Bolnet, C., and Petitte, J. H. (1994). Profiles of chicken macrophage effector functions. *Poult. Sci.* **73**, 1027–1034.

Ratcliffe, M. Lassila, O., Pink, J. R. L., and Vainio, O. (1986). Avian B-cell precursors: Surface immunoglobulin expression is an early, possibly bursa-independent event. *Eur. J. Immunol.* **16**, 129–133.

Roitt, I. M., Torrigiani, G., Greaves, M. F., Brostaff, J., and Playfair, J. H. (1969). The cellular basis of immunological responses. *Lancet* **2**, 367–371.

Romanoff, A. (1960). "The Avian Embryo: Structural and Functional Development." Macmillan, New York.

Romieu, M., and Jullien, G. (1942). Sur l'existence d'une formation lymphoid dans l'epiphyse des Gallinces. *C. R. Soc. Biol.* **136**, 626–628.

Salanti, E. P., Pink, J., Richard, L., and Steinberg, C. M. (1989). A model of bursal colonization. *Theor. Biol.* **139**, 1–6.

Savage, M. L., Olah, I., and Scott, T. R. (1992). Plasma cell proliferation in the chicken Harderian gland. *Cell Prolif.* **25**, 337–344.

Schat, K. A., and Myers, T. J. (1991). Avian intestinal immunity. *Crit. Rev. Poult. Biol.* **3**, 19–34.

Scott, T. R., and Savage, M. L. (1996). Immune cell proliferation in the Harderian gland: An avian model. *Microscopy Res. Tech.* **29**, 149–155.

Scott, T. R., Savage, M. L., and Olah, I. (1993). Plasma cells of the chicken Harderian gland. *Poult. Sci.* **72**, 1273–1279.

Sharma, J. M., and Schat, J. A. (1991). Natural immune functions. *In* "Avian Cellular Immunology" (J. M. Sharma, ed.), pp. 51–70. CRC Press, Boca Raton, FL.

Sicher, S. C., Vazquez, M. A., and Christopher, Y. Lu (1994). Inhibition of macrophage Ia expression by nitric oxide. *J. Immunol.* **153**, 1293–1300.

Silverstein, A. M. (1989). "History of Immunology." Academic Press, Orlando, FL.

Stalsberg, H., and Prydz, H. (1963). Studies on chick thrombocytes II. Function in primary hemostasis. *Thromb. Diath. Haemorr.* **9**, 291–299.

Sundick, Roy S., Albini, B., and Wick, G. (1973). Chicken Harderian's gland: Evidence for a relatively pure bursa dependent lymphoid cell population. *Cell Immunol.* **7**, 332–335.

Sung, Y.-J., and Dietert, R. R. (1994). Nitric oxide (·No)-induced mitochondrial injury among chickens ·NO-generating and target leukocytes. *J. Leuk. Biol.* **56**, 52–58.

Tenenhouse, H. S., and Deutsch, H. F. (1966). Some physical chemical properties of chicken γ-globulins and their pepsin and papain digestion products. *Immunochemistry* **3**, 11–20.

Thorbecke, G. J., Palladino, M. A., and Lerman, S. P. (1980). Lymphoid-cell cooperation in immune responses of the chicken. *Contemp. Top. Immunol.* **9**, 91–107.

Toivanen, P., Toivanen, A., and Vanio, D. (1974). Complete restoration of bursa-dependent immune system after transplantation of semi-allogenic stem cells into immunodeficient chickens. *J. Exp. Med.* **139**, 1344–1349.

Tsuji, S., Baba, T., Kawata, T., and Kajikawa, T. (1993). Role of Harderian gland on differentiation and proliferation of immunoglobulin A-bearing lymphocytes in chickens. *Vet. Immunol. Immunopathol.* **37**, 276–283.

Vainio, O., and Toivanen, A. (1987). Cellular cooperation in immunity. *In* "Avian Immunology: Basis and Practice" (A. Toivanen and P. Toivanen, eds.), Vol. II, pp. 1–12. CRC Press, Boca Raton, FL.

Vainio, O., Veromaa, T., Eerola, E., Toivanen, P., and Radcliffe, M. J. H. (1988). Antigen presenting cell T-cell interaction in the chicken is MHC Class II antigen restricted. *J. Immunol.* **140**, 2864–2868.

Venzke, W. G. (1952). Morphogenesis of the thymus of chicken embryos. *Am. J. Vet. Res.* **13**, 395–404.

Veillette, A., and Ratcliffe, M. J. H. (1991). Avian CD4 and CD8 interact with a cellular tyrosine protein kinase homologous to mammalian p56[lck]. *J. Immunol.* **21**, 397–401.

von Rautenfeld, D. B., and Budras, K. D. (1982). The bursa cloacae (Fabricii) of struthioniformes in comparison with the bursa of other birds. *J. Morphol.* **172**, 123–138.

Warner, N. L., Szenberg, A. J., and Burnet, F. M. (1962). The immunological role of different lymphoid organs in the chicken. I. Dissociation of immunological responsiveness. *Aust. J. Exp. Biol. Med. Sci.* **40**, 373–388.

Weinbaum, F. I., Glimour, D. G., and Thorbecke, G. J. (1973). Immunocompetent cells of the chicken. III. Cooperation of carrier sensitized T-cells from agammaglobulinemic donors with Hapten immune B-cells. *J. Immunol.* **110**, 1434–1436.

Weiss, L. (1972). "The cells and tissues of the immune system." Prentice-Hall, Englewood Cliffs, NJ.

White, R. G., French, V. I., and Stark, J. M. (1970). A study of the localization of a protein integrin in the chicken spleen and its relation to the formation of germinal centres. *J. Med. Microbiol.* **3**, 65–83.

Wight, P. A. L., Burns, R. B., Rothwell, B., and MacKenzie, G. M. (1971). The Harderian gland of the domestic fowl. I. Histology, with reference to the genesis of plasma cells and Russell bodies. *J. Anat.* **120**, 307–315.

Young, J. R., Davison, T. F., Tregaskes, C. A., Rennie, M. C., and Vainio, O. (1994). Monomeric homologue of mammalian CD28 is expressed on chicken T-cells. *J. Immunol.* **152**, 3848–3851.

Index

A

Acclimatization and acclimation, thermoregulation
  body temperature, 347
  evaporative heat loss, 363
  heat production, 353–355
  heat transfer within the body, 355–356
Acetylcholine
  corticosteroid function regulation, 504
  gastrointestinal system motility control, 310–313
  neuromuscular transmission role, 130–133
  parasympathetic heart control, 205–210
Acid–base equilibrium
  incubation, 624–625
  renal contribution, 280–281
  respiratory system, 248–249
Activin, development role, 647
Adrenal glands
  adrenal chromaffin hormones, 519–522
    catecholamine synthesis, 519–520
    developmental changes, 521–522
    epinephrine physiology, 520–521
    maturational changes, 522
    norepinephrine physiology, 520–521
    senescencal changes, 522
  adrenocortical hormones, 492–514
    circulating concentrations, 496–497
    clearance, 497–498
    developmental changes, 513–514
    function, 500–502
    hypothalamo–pituitary–adrenal axis, 498–500
    maturational changes, 514
    metabolism, 497–498
    physiological and ecological contexts, 509–513
      diurnal rhythms, 509–511
      reproductive parameters, 511–512
      seasonal parameters, 511–512
    physiology, 514–519
      action in target cells, 515–516
      behavior, 517–518
      corticosteroid receptors, 514–515
      electrolyte balance, 518–519
      immune function, 517
      intermediary metabolism, 516–517
    regulation
      adrenocortical hormone, 502–503, 508
      aldosterone secretion, 506–509
      angiotensins, 503, 506–507
      biphasic modulators, 505–506
      inhibitors, 505, 508–509
      negative modulators, 505
      positive modulators, 503–505, 508
      renin–angiotensin system, 506–507
      stimulators, 503–505
    secretion, 497–498
    secretory products, 492–493
    senescencal changes, 514
    stress, 500–502
    synthesis, 493–496
    transport, 496
  anatomy, 489–492
Adrenocorticotropic hormone
  aldosterone secretion regulation, 508
  characteristics, 449–450
  corticosteroid secretion regulation, 492, 502–503, 508
  development role, 644
  diurnal rhythm effects, 509–511
  growth hormone interactions, 445
  hypothalamo–pituitary–adrenal axis, 498–500
  inhibin secretion regulation, 498
Aerodynamic models, flight study, 394–396
Aerodynamic valving, respiratory flow patterns, 241, 243
Age, gastrointestinal function effects, 320
Air flow visualization, flight study, 396
Albumen, characteristics, 586–587
Aldosterone
  osmoregulation role, 278
  secretion regulation, 506–509
    adrenocorticotropic hormone, 508
    angiotensin II action mechanisms, 507–508
    inhibitors, 508–509
    negative modulators, 508–509
    positive modulators, 508
    renin–angiotensin system, 506–507
Alloxan, pancreatic toxicity, 552–553
Altitude
  air density effects, 428
  blood changes, 180, 428
  circulation control, 217–219, 429
  gas exchange, 254
  hypobaria, 428–429
  navigation, 428
  oxygen partial pressure, 428–429
Amino acids, absorption, 318
Amylin, calcium regulation role, 480–481
Androgens
  development role, 642
  ovulation regulation, 581–582
  testicular excurrent duct formation, 603
Angiotensins
  aldosterone secretion regulation, 506–508
  corticosteroid function regulation, 503
  osmoregulation role, 277–278
Ansa lenticularis, motor control role, 97
Antibodies, immune response regulation, 664–666
  immunoglobulins, 664–665
  macrophage, 666–667
  V-genes, 665–666
Anting, olfactory response, 50
Archistriatum tract, motor control role, 95–97
Arginine vasotocin
  characteristics, 451–452, 498
  osmoregulation role, 273, 277
  oviposition induction, 584
  prolactin release control, 449
Arteries, see also Cardiovascular system
  baroreflexes, 217–219
  chemoreceptors, 256–257
  coronary circulation, 145–147
  gross anatomy, 157–158
  hemodynamics, 154–157
  osmoregulation role, 268

Arteries (*continued*)
　peripheral blood flow control, systemic
　　innervation, 186–188
　pressure–flow relationship, 162–165
　vascular impedance, 165
　wall morphology, 158–162
Ascites, pulmonary circulation effects, 244
Ascorbic acid, pancreatic toxicity, 552
Atretic follicles, anatomy, 570–572
Atrial natriuretic peptide
　aldosterone secretion regulation, 509
　osmoregulation role, 278
Auditory system, 21–35
　auditory nerve fibers, 27–28
　behavioral aspects, 33–35
　　audiograms, 33
　　echolocation, 35
　　frequency discrimination, 33
　　intensity discrimination, 33
　　sound localization, 34–35
　　time resolution, 33–34
　central pathways, 28–33
　　anatomy, 28–29
　　electrophysiology, 29–33
　　　cochlear nuclei, 30–31
　　　lateral lemniscus nuclei, 31
　　　midbrain auditory nucleus, 31–32
　　　nucleus laminaris, 31
　　　telencephalic auditory field L,
　　　　32–33
　　　thalamic nucleus ovoidalis, 32
　ear function, 24–27
　　basilar membrane oscillations, 24
　　hair cell activation, 24–27
　　middle ear transfer of function, 24
　ear structure, 21–24
　　external ear, 21–22
　　inner ear, 22–24
　　middle ear, 22
　overview, 21, 35
Autonomic nervous system, *see also*
　　Central nervous system
　components, 105–112
　　circumventricular organs, 102, 111–112
　　hypothalamopituitary system, 108–111
　　parasympathetic nervous system,
　　　105–108
　　sympathetic nervous system, 105
　functional neural pathways, 112–117
　　food intake regulation, 112–114
　　limbic system, 115–117
　　visceral forebrain system, 114–115
　genetic studies, 118
　heart control
　　control mechanisms, 205–210
　　integrative control, 221–223
　　neuron anatomy, 198–205
　　sympathetic innervation, 191–198
　homeostasis shifts, 117–118
　imbalance consequences, 118
　migration regulation, 117–118
　overview, 102
　peripheral blood flow control
　　autonomic pathways, 190

　　oviduct innervation, 575
　　pulmonary vessel innervation, 189–190
　　systemic arterial innervation, 186–188
　　systemic venous innervation, 188–189
　peripheral motor components, 102–104
　　parasympathetic nervous system, 104
　　sympathetic nervous system, 102–104

B

Baroreflexes, circulation control, 217–219
Basilar membrane, oscillations, 24
Beak
　anatomy, 299–302
　jaw and tongue motor control, 87–91
　　bulbar premotor centers, 89–91
　　jaw opening and closing, 87–89
　　proprioceptive control, 89–91
　　tongue muscles, 89
　mechanoreception, 64–65
　mouth digestion and secretion, 313
Behavior, *see also specific behaviors*
　annual cycles, autonomic hypothesis,
　　117–118
　auditory aspects, 33–35
　　audiograms, 33
　　echolocation, 35
　　frequency discrimination, 33
　　intensity discrimination, 33
　　sound localization, 34–35
　　time resolution, 33–34
　chemical stimuli responses, 40–42
　corticosteroid effects, 517–518
　energy costs, 331–334
　　foraging, 333
　　locomotion, 332
　　reproduction, 333–334
　migration, *see* Migration
　olfactory response, 50
　somatosensory effects, 65–66
　　mechanoreception, 65–66
　　pain, 66
　　thermoreception, 66
　taste response, 43
　thermoregulation, 364–365, 370–371
Bile
　absorption, 318
　secretion, 317
Blood
　altitude effects, 180
　capillary bed
　　flow distribution, 171–173
　　gas exchange, 165–170
　　pulmonary volume, 243
　carbohydrate metabolism, 551–552
　circulation, *see* Cardiovascular system
　components, 176–177
　heat exchange, 355
　hemodynamics, 154–157
　microvascular fluid exchange, 170–171
　osmoregulation, *see* Osmoregulation
　oxygenation, *see* Respiratory system
　peripheral flow control
　　autonomic pathways, 190

　　autoregulation, 181–182
　　chemical factors, 182
　　circulating agents, 183–186
　　humoral factors, 182–190
　　neural control, 186–190
　　oviduct innervation, 575
　　pulmonary vessel innervation, 189–190
　　systemic arterial innervation, 186–188
　　systemic venous innervation, 188–189
　　vasoactive agents, 182–183
　pulmonary circulation
　　ascites role, 244
　　blood flow distribution, 244
　　capillary volume, 243
　　circulation anatomy, 243
　　fluid balance, 244
　　gas exchange, 249–254
　　　blood–gas barrier, 251–252
　　　cross-current exchange, 250–251
　　　lung diffusing capacity, 251–252
　　　lung heterogeneity, 252–254
　　　oxygen transport principles, 250
　　gas transport, 244–249
　　　acid–base equilibrium, 248–249
　　　blood–CO_2 equilibrium curves, 248
　　　carbon dioxide, 247–248
　　　forms in blood, 248
　　　hemoglobin, 245, 247
　　　Henderson–Hasselbalch equation,
　　　　248–249
　　　measurement, 249
　　　O_2–blood equilibrium curves, 245
　　　O_2 capacity, 247
　　　O_2–hemoglobin affinity, 246–247
　　vascular resistance, 244
　rheology, 177–180
　　flow velocity, 178
　　hematocrit effect, 177–178
　　vessel radius, 178–180
　temperature effects, 180–181
B-lymphocytes
　development
　　bursa of fabricius morphology, 658
　　cluster of differentiation, 658–659
　　markers, 658
　　origin, 659
　　overview, 657–658
　immune response regulation, repertoire,
　　664–666
　　immunoglobulins, 664–665
　　macrophage, 666–667
　　V-genes, 665–666
　origin, 659
Body size, energy exchange measurement,
　331
Body temperature, *see* Thermoregulation
Bone morphogenic protein-4, development
　role, 647
Bones
　eggshell formation, medullary bone
　　calcium mobilization, 590
　parathyroid hormone actions, 475–477
　vitamin D effects, 482–483
Box temperature, *see* Thermoregulation

Index

Brain, *see also* Central nervous system
  autonomic nervous system pathways, 114–115
  gas exchange, 255
  motor control role, 92–94
Breeding, *see* Reproduction
Brooding
  induction, 585–586
  thermoregulation
    brood patch, 355, 585
    hatchling responses, 375
    nesting, 378
Bursal antisteroidogenic peptide,
  corticosteroid function regulation, 505
Bursa of fabricius, morphology, 658

C

Calcitonin
  actions, 480
  amylin, 480–481
  calcitonin gene-related peptide, 480–481
  characteristics, 479
  circulating levels, 479–480
  extracellular calcium role, 480
  overview, 473–474, 484
  ultimobranchial glands, 479
Calcitonin gene-related peptide,
  characteristics, 480–481
Calcium
  absorption, 319, 590–591
  eggshell metabolism, 589–591
    carbonate formation and deposition, 590–591
    medullary bone calcium mobilization, 590
    shell gland, 590–591
    sources, 589–590
    vitamin D metabolism role, 590
  excretion, 278, 590–591
  regulation
    calcitonin role
      actions, 480
      amylin, 480–481
      calcitonin gene-related peptide, 480–481
      characteristics, 479
      circulating levels, 479–480
      extracellular calcium role, 480
      overview, 473–474, 484
      ultimobranchial glands, 479
    parathyroid hormone role
      chemistry, 474–475
      circulating hormone, 475
      overview, 473–474, 484
      parathyroidectomy, 474
      parathyroid glands, 474
      parathyroid hormone-related peptide, 477–479
      renal actions, 477
      skeletal actions, 475–477
    prostaglandin role, 483–484
    vitamin D role, 482–483
      bone actions, 482–483
      circulating vitamin D metabolite levels, 482
      eggshell metabolism, 590
      intestine actions, 482
      overview, 473–474, 484
      oviduct actions, 482
      renal metabolism, 481–482
Capillaries, *see also* Cardiovascular system
  blood flow distribution, 171–173
  flight effects, 172–173
  fluid exchange, 170–171
  gas exchange, 165–170
  glomerular filtration, 272–273
  hemodynamics, 154–157
  ovulation effects, 173
  pulmonary volume, 243
  resting output distribution, 171
  swimming and submergence effects, 171–172
Carbohydrates
  absorption, 317–318
  glucose metabolism
    altered metabolism, 552–554
      diabetes mellitus, 552
      growth hormone effects, 554
      pancreatic chemocytotoxins, 552–553
      pancreatropic agents, 553–554
      prolactin effects, 554
      surgical extirpation of pancreas, 553
    birds and mammals compared, 541–542
    liver role, 540
    mechanisms, 539–540
    pancreas function, 540–541
    pancreatic-enteric regulation, 548–551
      adipose tissue, 550
      erythrocytes, 551–552
      glycogen body, 552
      heart, 551
      insulin/glucagon molar ratio role, 550–551
      liver role, 549–550
      muscle tissue, 550
      plasma effects, 549
      tissue effects, 549–552
Carbon dioxide
  blood–CO_2 equilibrium curves, 248
  blood measurement, 249
  forms in blood, 248
  pulmonary exchange, 250–251
  thermoregulation control, 371–372
  ventilatory reflexes, 258–259
Cardiovascular system
  blood, 176–181
    altitude effects, 180, 428–429
    components, 176–177
    rheology, 177–180
      flow velocity, 178
      hematocrit effect, 177–178
      vessel radius, 178–180
    temperature effects, 180–181
  control, 181–223
    circulation controlling reflexes
      baroreflexes, 217–219, 429
      cardiac receptor reflexes, 219–220
      chemoreflexes, 215–217
      skeletal muscle afferents, 220–221
    heart control, 190–215
      catecholamine effects, 190–191
      chronotropic effects, 207–209
      dromotropic effects, 207
      heart rate–output relationship, 210–212
      inotropic effects, 209–210
      neural control, 191–215, 221–223
      parasympathetic innervation, 198–210
      stroke volume–output relationship, 212–215
      sympathetic innervation, 191–198
    integrative neural control, 221–223
    overview, 181
    peripheral blood flow, 181–190
      autonomic pathways, 190
      autoregulation, 181–182
      chemical factors, 182
      circulating agents, 183–186
      humoral factors, 182–190
      neural control, 186–190
      pulmonary vessel innervation, 189–190
      systemic arterial innervation, 186–188
      systemic venous innervation, 188–189
      vasoactive agents, 182–183
  flight effects
    adjustments, 416–417
    blood flow changes, 172–173, 429
    cardiac muscles, 417–419
    cardiovascular function modeling, 399–400
  heart
    carbohydrate metabolism, 551
    cardiac variables, 147–148
    control mechanisms, 190–215
      catecholamine effects, 190–191
      chronotropic effects, 207–209
      dromotropic effects, 207
      heart rate–output relationship, 210–212
      inotropic effects, 209–210
      integrative neural control, 221–223
      neural control, 191–215
      parasympathetic innervation, 198–210
      stroke volume–output relationship, 212–215
      sympathetic innervation, 191–198
    electrophysiology, 148–154
      conduction system, 150–151
      excitation–contraction coupling, 149–150
      fine structure, 148–149
      mechanisms, 151–154
    fine structure, 148–149
    gross structure and function, 142–147
      cardiac chambers, 143
      coronary circulation, 145–147

Cardiovascular system (*continued*)
　functional anatomy, 142
　　heart size, 142–143
　　valves, 143–145
　hemodynamics, 154–157
　overview, 141–142
　peripheral flow control, 181–190
　　autonomic pathways, 190
　　autoregulation, 181–182
　　chemical factors, 182
　　circulating agents, 183–186
　　humoral factors, 182–190
　　neural control, 186–190
　　pulmonary vessel innervation, 189–190
　　systemic arterial innervation, 186–188
　　systemic venous innervation, 188–189
　　vasoactive agents, 182–183
　pulmonary circulation, 243–244
　　ascites role, 244
　　blood flow distribution, 244
　　capillary volume, 243
　　circulation anatomy, 243
　　fluid balance, 244
　　vascular resistance, 244
　rheology, 177–180
　　flow velocity, 178
　　hematocrit effect, 177–178
　　vessel radius, 178–180
　temperature effects, 180–181
　vascular tree, 157–176
　　arterial system, 157–165
　　　arterial wall morphology, 158–162
　　　gross anatomy, 157–158
　　　pressure–flow relationship, 162–165
　　　vascular impedance, 165
　　capillary beds
　　　blood flow distribution, 171–173
　　　flight effects, 172–173
　　　fluid exchange, 170–171, 272–273
　　　gas exchange, 165–170, 243
　　　ovulation effects, 173
　　　resting output distribution, 171
　　　swimming and submergence effects, 171–172
　　venous system, 173–176
　　　capacitance function, 174
　　　development, 173–174
　　　exercise physiology, 174–175
　　　renal portal system, 175–176
　　　submersion physiology, 174–175
Catecholamines
　corticosteroid function regulation, 504
　heart control, 190–191
　physiological actions, 520–521
　senescence role, 522
　synthesis and release regulation, 519–520
Ceca
　anatomy, 304
　cecal tonsil, 660
　motility, 309
　osmoregulation role
　　transport properties, 286–288
　　ureteral urine postrenal modifications, 288

Central nervous system, *see also* Autonomic nervous system
　auditory pathways, 28–33
　　anatomy, 28–29
　　electrophysiology, 29–33
　　　cochlear nuclei, 30–31
　　　lateral lemniscus nuclei, 31
　　　midbrain auditory nucleus, 31–32
　　　nucleus laminaris, 31
　　　telencephalic auditory field L, 32–33
　　　thalamic nucleus ovoidalis, 32
　motor control, *see* Motor control system
　neurons, *see* Neurons
　somatosensory processing
　　electrophysiology, 65
　　pathways, 63–64
　spinal cord
　　organization, 71–80
　　　ascending pathways, 78–79
　　　Clarke's column, 76–77
　　　column of terni, 77
　　　descending pathways, 79
　　　gross anatomy, 71–72
　　　laminae, 75–76
　　　magnocellular column, 76–77
　　　motoneurons, 76
　　　nucleus proprius, 75
　　　overview, 71, 79–80
　　　paragriseal cells, 77–78
　　　peripheral input, 74
　　　spinal gray cytoarchitectonic organization, 72–74
　　thermoregulatory role, 370
Centrifugal pathway, central visual processing, 7–9
Cerebellum, motor control role, 92–94
Chemesthesis, *see also* Gustation; Olfaction
　behavioral responses, 40–42
　breathing control, 42, 256–258
　　arterial chemoreceptors, 256–257
　　central chemoreceptors, 256
　　intrapulmonary chemoreceptors, 257–258
　chemesthetic receptor innervation, 40
　nasal irritation, 42–43
　overview, 39–40, 43
　respiratory irritation, 42–43
　structure–activity relationships for aromatic stimuli, 42
　trigeminal chemoreceptors, 40
　trigeminal–olfactory interactions, 42–43
Chemoreflexes, circulation control, 215–217
Chloride
　absorption, 319–320
　osmoregulation role
　　excretion regulation in kidneys, 277–278
　　intestinal transport, 283–284, 286–287
　　ureteral urine postrenal modifications
　　　dehydration/depletion, 287
　　　hydration/loading, 287
Chromaffin cells
　adrenal hormones, 519–522
　　catecholamine synthesis, 519–520

　　developmental changes, 521–522
　　epinephrine physiology, 520–521
　　maturational changes, 522
　　norepinephrine physiology, 520–521
　　senescencal changes, 522
　innervation, 490
Chromogranin A, characteristics, 450–451
Circadian rhythms
　pineal gland physiological, 560–564
　　diversity, 563
　　pacemaker function, 560–563
　　pineal–melatonin synchronization, 563–564
　thermoregulation role
　　body temperature, 345–347, 565
　　heat production, 350
Circumventricular organs, autonomic nervous system components, 102, 111–112
Clarke's column, organization, 76–77
Cloaca, anatomy, 304–305, 600
Cobalt chloride, pancreatic toxicity, 552
Cochlea, structure, 22–24
Cochlear nuclei, central auditory pathway function, electrophysiology, 30–31
Cold response, *see* Hypothermia; Thermoregulation
Cold vasodilation, thermoregulation, 355
Colon
　digestion and secretion, 316
　osmoregulation role
　　coprodeum compared, 285
　　transport properties, 283–285
Column of terni, organization, 77
Convection, pulmonary gas exchange, 250
Coprodeum, osmoregulation role
　colon compared, 285
　transport properties, 283–285
Corticosteroids
　circulating concentrations, 496–497
　clearance, 497–498
　developmental changes, 513–514
　development role, 644
　female reproduction role, 582
　function, 500–502
　hypothalamo–pituitary–adrenal axis, 449–450, 498–500
　maturational changes, 514
　metabolism, 497–498
　physiological and ecological contexts, 509–513
　　diurnal rhythms, 509–511
　　reproductive parameters, 511–512
　　seasonal parameters, 511–512
　physiology, 514–519
　　action in target cells, 515–516
　　behavior, 517–518
　　corticosteroid receptors, 514–515
　　electrolyte balance, 518–519
　　immune function, 517
　　intermediary metabolism, 516–517
　regulation, 502–506
　　adrenocortical hormone, 502–503, 508
　　aldosterone secretion, 506–509

angiotensins, 503, 506–507
biphasic modulators, 505–506
inhibitors, 505, 508–509
negative modulators, 505
positive modulators, 503–505, 508
renin–angiotensin system, 506–507
stimulators, 503–505
secretion, 497–498
secretory products, 492–493
senescencal changes, 514
stress, 500–502
synthesis, 493–496
transport, 496
Crop
anatomy, 302–303
digestion and secretion, 313–314
Cytokines, immune response regulation, 664

D

Data logging, flight study, 397–398
Dehydration
sodium chloride depletion, 287
thermoregulation, 347
Development
adrenocorticotropic axis
adrenocorticotropic hormones, 513–514, 644
corticosterone, 513–514, 644
chromaffin cells, 521–522
gonadotropic axis, gonadal steroids and receptors, 642
growth factors, 644–648
activin, 647
bone morphogenic protein-4, 647
epidermal growth factor, 644–645
fibroblast growth factor, 645–646
glial cell-derived nerve growth factor, 647
inhibin, 647
nerve growth factor, 646
platelet-derived growth factor, 646
transforming growth factor-α, 644–645
transforming growth factor-β, 646–647
growth models, 648–649
growth-selected strains, 649
sex-linked dwarf chickens, 648–649
homeobox genes, 647–648
hormonal regulation
molecular mechanisms, 636
overview, 635–636
incubation physiology, 621–629
acid–base regulation, 624–625
cardiovascular function, 625–627
respiration, 621–624
thermoregulation, 627–629
lactotropic axis, prolactin and receptor, 642–644
pancreatic hormones, insulin and receptor, 644
pancreatic organ role, 542–543
skeletal muscles, 123–126
preflight muscles, 412–414

somatotropic axis, 636–641
growth hormone and receptor, 637–638
insulinlike growth factor binding proteins, 641
insulinlike growth factor receptors, 640–641
insulinlike growth factors, 638–639, 641
thermoregulation
embryo, 375
hatchling, 375–376
brooding responses, 375
cold responses, 375
heat responses, 376
nesting, 376–379
altricial species, 377–378
brooding, 378
growth rates, 376–377
heat loss, 378
hypothermia, 378–379
incubation, 620–621, 627–629
precocial species, 377–378
solar heat gain, 378
thyrotropic axis, thyroid hormones and receptors, 467–468, 641–642
venous system, 173–174
Diabetes mellitus
characteristics, 552
pancreatropic agents, 553–554
Diffusion, pulmonary gas exchange, 250–252
air capillary diffusion, 251
blood–gas barrier diffusion, 251–252
mechanisms, 250
Digestion, see Gastrointestinal system
Diurnal effects
gastrointestinal system motility, 310
plasma corticosterone concentrations, 509–511
Dopamine
corticosteroid function regulation, 504
prolactin release control, 449
Doubly-labeled water, flight energetics study, 397
Drinking, see Osmoregulation

E

Ears, see Auditory system
Echolocation, auditory behavioral aspects, 35
Egg production, see also Embryos
energy costs, 333, 619–620
formation
albumen, 586–587
calcium metabolism, 589–591
carbonate formation and deposition, 591
crystallization layers, 588–589
mammillary knob layer, 588
palisade layer, 588
shell respiration, 589, 621–624
surface layer, 588–589

freshly laid egg composition, 617–618
organic matrix, 587–588
cuticle, 587–588
mammillary cores, 587
shell matrix, 587
shell membranes, 587
respiration, 589, 621–624
vitamin D metabolism, 590
yolk, 586
incubation, see Incubation
preincubation storage, 629–630
Electrolyte balance
corticosteroid effects, 518–519
gastrointestinal absorption, 267, 319
osmoregulation, 267, 518–519
Electrophysiology
auditory pathways, 29–33
cochlear nuclei, 30–31
lateral lemniscus nuclei, 31
midbrain auditory nucleus, 31–32
nucleus laminaris, 31
telencephalic auditory field L, 32–33
thalamic nucleus ovoidalis, 32
heart, 148–154
conduction system, 150–151
excitation–contraction coupling, 149–150
fine structure, 148–149
mechanisms, 151–154
mechanoreception
investigations, 64–65
mechanoreceptor characteristics, 59–61
muscle fibers, 129
somatosensory processing, 65
Embryos, see also Egg production
ambient temperature tolerance, 631–632
embryology, see Development
growth, energy costs, 336–337
incubation, see Incubation
pancreatic organ role, 542
thermoregulation, 375, 627–629
thyroid gland histology, 461–462
Eminentia sagittalis, motor control role, 94–95
Energy balance, see also Thermoregulation
activity costs, 331–334
foraging, 333
locomotion, 332
reproduction, 333–334
daily budgets, 337–338
energy exchange measurement, 327–331
basal metabolism, 328–329
body size energetics, 331
demand, 327–328
existence energy, 329–330
metabolized energy, 330
resting metabolism, 329
flight
energy costs, 332
power input data, 400–405
flight muscle efficiency scaling, 403–405

Energy balance (*continued*)
　　study techniques, 394–400
　　　aerodynamic models, 394–396
　　　air flow visualization, 396
　　　cardiovascular function modeling, 399–400
　　　data logging, 397–398
　　　direct force measurement, 396
　　　doubly-labeled water, 397
　　　mass loss, 396–397
　　　telemetry, 397–398
　　　wind tunnel, 398–399
　　geographic variation, 338
　　overview, 327, 338–339
　　population requirements, 338
　　production and storage
　　　egg production, 335
　　　embryo growth, 336–337
　　　freshly laid egg, 619–620
　　　gonad growth, 335
　　　molt, 335–336
　　　newly hatched young, 336–337
　　　overwinter fattening, 334–335
　　　premigratory fattening, 334
　　　starvation, 334
Environmental influences, thyroid glands, 469
Epidermal growth factor, development role, 644–645
Epididymis, characteristics, 599–600
Epinephrine
　　chromaffin cell innervation, 490
　　corticosteroid function regulation, 504
　　physiological actions, 520–521
　　senescence role, 522
　　synthesis and release regulation, 519–520
Esophagus
　　anatomy, 302–303
　　digestion and secretion, 313–314
　　motility, 305–306
Estrogens
　　development role, 642
　　female reproduction role, 582
Evaporative water loss
　　osmoregulation, 291
　　thermoregulation, 356–363
　　　acclimatization and acclimation, 363
　　　cooling site, 361–362
　　　cutaneous heat loss, 357–359
　　　gular flutter, 361
　　　heat conservation, 362–363
　　　nonevaporative heat loss, 356
　　　osmotic state, 362
　　　partition of heat loss, 363
　　　respiratory heat loss, 359–363
　　　thermal tachypnea, 359–361
　　　water conservation, 362–363
Exercise, *see also* Flight
　　gas exchange effects, 255
　　venous system physiology, 174–175
　　ventilatory response, 259
Eyes, *see* Vision

F

Fat
　　follicle growth, deposition, 572–574
　　metabolism, growth hormone effects, 445
　　overwinter fattening, 334–335
　　premigratory fattening, 334, 414–416
Fatty acids, absorption, 318–319
Feathers, molt
　　energy costs, 335–336
　　regulation, autonomic hypothesis, 117–118
Fibroblast growth factor, development role, 645–646
Flight
　　altitude effects
　　　air density effects, 428
　　　blood changes, 180, 428
　　　circulation control, 217–219, 429
　　　gas exchange, 254
　　　hypobaria, 428–429
　　　navigation, 428
　　　oxygen partial pressure, 428–429
　　cardiovascular system
　　　adjustments, 416–417
　　　blood flow changes, 172–173, 429
　　　cardiac muscles, 417–419
　　energetics
　　　power input data, 400–405
　　　　energy costs, 332
　　　　flight muscle efficiency scaling, 403–405
　　　study techniques, 394–400
　　　　aerodynamic models, 394–396
　　　　air flow visualization, 396
　　　　cardiovascular function modeling, 399–400
　　　　data logging, 397–398
　　　　direct force measurement, 396
　　　　doubly-labeled water, 397
　　　　mass loss, 396–397
　　　　telemetry, 397–398
　　　　wind tunnel, 398–399
　　long distance performance, 424–428
　　　migration preparation, 424–427
　　　migratory behavior, 427–428
　　　metabolic substrates and fuel deposits, 414–416
　　muscles
　　　cardiac muscles, 417–419
　　　flight muscles, 405–412
　　　　biochemistry, 408–410
　　　　efficiency scaling, 403–405
　　　　fiber types, 405–408
　　　　function, 410–412
　　　　morphology, 405–408
　　　　neurophysiology, 410–412
　　　forward flapping, 401–403
　　　gliding, 400–401
　　　hovering, 403
　　　locomotor muscles, 412–414
　　　preflight muscle development, 412–414
　　　soaring, 400–401
　　overview, 391–392
　　respiratory system
　　　temperature control, 421–424
　　　ventilatory adjustments, 259, 419–421
　　　ventilatory/locomotor coupling, 419–421
　　　water loss, 424
　　scaling, 392–393
　　vein physiology, 174–175
Follicles
　　anatomy
　　　atretic follicles, 570–572
　　　ovarian follicle, 570
　　　postovulatory follicles, 570–572
　　growth, 572–574
　　hormone effects
　　　atretic follicles, 579–580
　　　hierarchal follicles, 578–579
　　　postovulatory follicles, 579–580
　　　prehierarchal follicles, 579
　　　stromal follicles, 579
　　oviposition
　　　postovulatory follicle, 585
　　　preovulatory follicle, 585
Follicle-stimulating hormone, action
　　females, 439, 581
　　males, 441
Food
　　absorption, 317–319
　　deprivation, thermoregulation, 347–348
　　digestion, *see* Gastrointestinal system
　　foraging, 333
　　growth hormone secretion control, 446–447
　　intake regulation, 112–114, 320–321
　　metabolism, *see* Energy Balance
Foraging, energy costs, 333
Forebrain, autonomic nervous system pathways, 114–115

G

Galanin, corticosteroid function regulation, 504
Gas transport, *see* Cardiovascular system; Respiratory system
Gastrointestinal system
　　absorption, 317–320
　　　amino acids and peptides, 318
　　　bile acids, 318
　　　calcium, 482
　　　carbohydrates, 317–318
　　　chloride, 319–320
　　　electrolytes, 319
　　　fatty acids, 318–319
　　　sodium, 319–320
　　　vitamins, 320
　　　water, 319–320
　　accessory organ anatomy
　　　liver, 305
　　　pancreas, 305
　　age-related effects, 320
　　digestion and secretion, 313–317
　　　bile, 317

colon, 316
crop, 313–314
esophagus, 313–314
intestines, 315–316
mouth, 313
pancreas, 316–317, 548–551
salivary glands, 313
stomach, 314–315
digestive tract anatomy, 299–305
beak, 299–302
ceca, 304
cloaca, 304–305
crop, 302–303
esophagus, 302–303
mouth, 299–302
pharynx, 299–302
rectum, 304–305
small intestine, 303–304
stomach, 303
food intake regulation, 320–321
intestinal lymphocytes, 661–662
motility, 305–313
ceca, 309
diurnal variation, 310
esophagus, 305–306
gastrointestinal cycle, 306–308
hormonal control, 310–313
neural control, 310–313
passage rate, 313
rectum, 309–310
small intestine, 308–309
osmoregulation role, lower intestines, 283–288
caeca transport properties, 286–288
colon transport properties, 283–285
coprodeum transport properties, 283–285
coprodeum versus colon role compared, 285
dietary regulation mechanisms, 285
hormonal regulation mechanisms, 285
molecular induction, 285–286
potassium transport, 284–285
sodium transport, 283–284, 286–287
ultrastructural adaptation, 285–286
ureteral urine postrenal modifications, 287–288
caeca quantitative role, 288
dehydration/sodium chloride depletion, 287
hydration/sodium chloride loading, 287
ratite special cases, 288
salt glands, 287–288
water transport, 283–284, 286–287
Genetic studies, autonomic nervous system imbalance, 118
Geographic variations
energy balance, 338
thermoregulation, 353–355
Glial cell-derived nerve growth factor, development role, 647
Gliding, power input, 400–401

Glomerular filtration, osmoregulation role, 272–273
Glucagon
action mechanisms, 544–546
insulin/glucagon molar ratio
pancreas function, 544
pancreatic–enteric regulation, 550–551
secretogenesis, 543–544
Glucose, metabolism
altered metabolism, 552–554
diabetes mellitus, 552
growth hormone effects, 554
pancreatic chemocytotoxins, 552–553
pancreatropic agents, 553–554
prolactin effects, 554
surgical extirpation of pancreas, 553
birds and mammals compared, 541–542
liver role, 540
mechanisms, 539–540
pancreas function, 540–541
pancreatic–enteric regulation, 548–551
adipose tissue, 550
erythrocytes, 551–552
glycogen body, 552
heart, 551
insulin/glucagon molar ratio role, 550–551
liver role, 549–550
muscle tissue, 550
plasma effects, 549
tissue effects, 549–552
Glycoproteins, major histocompatibility complex immune response regulation, 663
Gonadotropins, 438–443
female action, 438–439
male action, 439–441, 607
release control, 441–443
female reproductive cycle, 442
luteinizing hormone releasing hormone, 441–442
negative and positive feedback, 442
photoperiodism, 442
seasonal breeding, 442–443
structure, 438
Gonads
differentiation, 602
growth, energy costs, 335
undifferentiated formation, 601–602
Grandry corpuscles, morphology and distribution, 58–59
Growth factors
development role, 644–648
activin, 647
bone morphogenic protein-4, 647
epidermal growth factor, 644–645
fibroblast growth factor, 645–646
glial cell-derived nerve growth factor, 647
inhibin, 647
nerve growth factor, 646
platelet-derived growth factor, 646
transforming growth factor-α, 644–645
transforming growth factor-β, 646–647
female reproduction role, 580

Growth hormone, 444–447
actions
adrenocortical hormone interactions, 445
development role, 637–638
growth, 444–445
immune function, 445
lipid metabolism, 445
thyroid hormone interactions, 445, 468
assay, 444
carbohydrate metabolism role, 554
chemistry, 444
corticosteroid function regulation, 503–504
growth hormone receptors, 445–446, 637–638
pituitary origin, 447
secretion control, 446–447
growth hormone releasing factor, 446
hypophysiotropic peptide release, 446
nutrition effects, 446–447
ontogeny, 447
somatostatin, 446
thyrotropin releasing hormone, 446
variants, 444
Growth hormone releasing factor, growth hormone secretion control, 446
Gular flutter, evaporative thermoregulation, 361
Gustation
bitter response, 44–46
complex taste response, 46
overview, 39–40
salt response, 44
sour response, 44
sweet response, 43–44
taste behavior, 43
taste receptor innervation, 43
taste receptors, 43
temperature effects, 46

H

Hair cells, activation, 24–27
Harderian gland, immunophysiology, 662
Hatchlings
energy costs, 336–337
thermoregulation, 375–376
brooding responses, 375
cold responses, 375
heat responses, 376
thyroid hormone effects, 468
Hearing, see Auditory system
Heart, see Cardiovascular system
Heat balance, see Thermoregulation
Hemoglobin
lung diffusing capacity, 252
oxygen transport
characteristics, 245
molecular engineering, 247
O_2–hemoglobin affinity, 246–247
Henderson–Hasselbalch equation, gas transport, 248–249

Herbst corpuscles, morphology and distribution, 58
Hierarchal follicles, steroid production, 578–579
Hippocampus, food storage behavior, 116–117
Histidine isoleucine peptide, prolactin release control, 449
Homeobox genes, development role, 647–648
Hormones, *see specific hormones*
Hovering, power input, 403
Hypophysiotropic peptide, growth hormone secretion control, 446
Hypothalamopituitary system, *see also* Adrenal glands; Pituitary gland
   adrenocorticotropic hormone, 449–450, 498–500, 513–514
   arginine vasotocin
      action, 451–452
      chemistry, 451
      prolactin release, 449
   autonomic nervous system components, 108–111
   chromogranin A, 450–451
   gonadotropins, 438–443
      female action, 438–439
      male action, 439–441
      release control, 441–443
         female reproductive cycle, 442
         luteinizing hormone releasing hormone, 441–442
         negative and positive feedback, 442
         photoperiodism, 442
         seasonal breeding, 442–443
      structure, 438
   growth hormone, 444–447
      actions, 444–445
         adrenocortical hormone interactions, 445
         growth, 444–445
         immune function, 445
         lipid metabolism, 445
         thyroid hormone interactions, 445
      assay, 444
      chemistry, 444
      growth hormone receptors, 445–446
      pituitary origin, 447
      secretion control, 446–447
         growth hormone releasing factor, 446
         hypophysiotropic peptide release, 446
         nutrition effects, 446–447
         ontogeny, 447
         somatostatin, 446
         thyrotropin releasing hormone, 446
      variants, 444
   hypothalamic–hypophyseal complex, 438, 606–607
   hypothalamic–pituitary–thyroid axis, 466
   mesotocin, 451–452
   neurohypophysis, 451
   overview, 437–438
   prolactin, 447–449
      actions, 448
      assay, 447–448
      chemistry, 447
      pituitary origin, 447
      receptor, 448–449
      release control, 449
   thermoregulation, 368–369
   thyrotropin, 443–444
Hypothermia
   arousal, 374–375
   development in nesting, 378–379
   hypothermic state, 373–374
   patterns, 372–373
Hypoxia
   gas exchange effects, 255
   ventilatory response, 259

I

Immune system
   corticosteroid effects, 517
   cytoarchitecture, 657–662
   development, 657–662
      primary immune tissue
         bursal B-cell markers, 658
         bursal lymphocyte origin, 659
         bursa of fabricius morphology, 658
         cluster of differentiation, 658–659
         T-cell receptors, 658–659
         thymic lymphocyte origin, 659
      secondary lymphoid tissue
         cecal tonsil, 660
         Harderian gland, 662
         lymph node, 662
         Meckel's diverticulum, 660–661
         Peyer's patches, 660
         pineal gland, 662
         spleen, 659–660
   growth hormone effects, 445
   immune response regulation, 662–667
      antibody-mediated immunity, 664–666
         immunoglobulins, 664–665
         macrophage, 666–667
         V-genes, 665–666
      cytokines, 664
      major histocompatibility complex, 662–664
   overview, 657
Immunoglobulins
   bursal B-cell markers, 658
   immune response regulation, 664–665
Incubation, *see also* Development
   physiology
      ambient temperature effects, 631–632
      development, 621–629
         acid–base regulation, 624–625
         cardiovascular function, 625–627
         respiration, 621–624
         thermoregulation, 627–629
      egg composition changes, 618–620
         energy content, 619–620
         water content, 618–619
      egg turning, 630–631
      embryonic tolerance, 631–632
      energy costs, 333, 619–620
      freshly laid egg composition, 617–618
      heat transfer, 620–621
      incubation period, 629
      overview, 617
      preincubation egg storage, 629–630
Infundibulum, anatomy, 574
Inhibin
   adrenal origin, 498
   development role, 647
Insulin
   action mechanisms, 544
   development role, 644
   insulin/glucagon molar ratio
      pancreas function, 544
      pancreatic–enteric regulation, 550–551
   secretogenesis, 543–544
Insulinlike growth factor binding proteins, development role, 641
Insulinlike growth factor receptors, development role, 640–641
Insulinlike growth factors, development role, 638–639, 641
Interleukin-1, corticosteroid function regulation, 504–505
Intestines, *see* Gastrointestinal system
Isthmus, anatomy, 574

J

Jaw, motor control, 87–91
   bulbar premotor centers, 89–91
   jaw opening and closing, 87–89
   proprioceptive control, 89–91
   tongue muscles, 89

K

Kidneys, osmoregulation, 267–282
   anatomy, 267–270
      arterial supply, 268
      blood flow, 267–270
      gross anatomy, 267
      nephron types and numbers, 267
      renal portal system, 268–270
      venous drainage, 270
   physiology, 270–282
      acid/base regulation, 280–281
      aldosterone, 278
      arginine vasotocin, 273, 277
      atrial natriuretic peptide, 278
      calcium excretion regulation, 278
      collecting duct, 275
      descending thin limb, 274
      glomerular filtration, 272–273
      nitrogen excretion regulation, 278–280
      phosphate excretion regulation, 278
      renal blood flow, 270–272
      renin/angiotensin system, 277–278

sodium excretion regulation, 277–278
thick ascending limb, 275
ureter function, 282
urinary concentrating mechanisms, 273–277
urine composition and flow, 281–282
water excretion regulation, 273–277
water reabsorption, 273–274

L

Laminae, organization, 75–76
Lateral lemniscus nuclei, central auditory pathway function, electrophysiology, 31
Lateral septal organ, characteristics, 111–112
Lateral spinal nucleus, organization, 77
Limbic system, autonomic nervous system pathways, 115–117
Lipids
  follicle growth, deposition, 572–574
  metabolism, growth hormone effects, 445
Lissauer tract neurons, organization, 77
Liver
  anatomy, 305
  bile secretion, 317
  glucoregulation
    mechanisms, 540
    pancreatic–enteric regulation, 549–550
Locomotion, see also specific types
  control centers, 91
  energy costs, 332
Lungs, see Respiratory system
Luteinizing hormone, action
  females, 438–439, 442, 580–581
  males, 439–441
Luteinizing hormone releasing hormone, release control, 441–442, 580
Lymph nodes, immunophysiology, 662

M

Macrophage, immune response regulation, 666–667
Magnocellular column, organization, 76–77
Magnum, anatomy, 574
Major histocompatibility complex, immune response regulation, 662–664
  allospecific response, 663
  coreceptors, 663–664
  membrane-associated glycoproteins, 663
  signaling, 663–664
  T-cytotoxic cells, 663
Marginal nuclei, organization, 77–78
Mass, loss, flight study, 396–397
Mechanoreception
  behavioral aspects, 65–66
  electrophysiological investigations, 64–65
  mechanoreceptor characteristics, 58–61
    cutaneous mechanoreceptors, 58–59
    electrophysiology, 59–61
    Grandry corpuscles, 58–59
    herbst corpuscles, 58
    Merkel cell receptors, 58
    Ruffini endings, 59

Meckel's diverticulum, immunophysiology, 660–661
Median eminence, characteristics, 112
Meiosis, spermatocyte formation, 605
Melatonin
  characteristics, 558–559
  circadian rhythm synchronization, 563–564
  extrapineal production, 559–560
  sleep regulation, 565
Merkel cell receptors, morphology and distribution, 58
Mesonephros, testicular excurrent duct formation, 602–603
Mesotocin, characteristics, 451–452
Metabolism, see Energy balance; Thermoregulation
Midbrain auditory nucleus, central auditory pathway function, electrophysiology, 31–32
Migration
  coronary circulation, 147
  long distance performance, 424–428
    migratory behavior, 427–428
    preparation, 424–427
  metabolic substrates and fuel deposits, 414–416
  premigratory fattening, 334, 414–416
  regulation, autonomic hypothesis, 117–118
Molt
  energy costs, 335–336
  female reproduction role, 583–584
  regulation, autonomic hypothesis, 117–118
  thyroid hormone effects, 468
Motor control system
  autonomic nervous system components, 102–104
    parasympathetic nervous system, 104
    sympathetic nervous system, 102–104
  cerebellum structure and function, 92–94
  flight, 410–412
  jaw and tongue movements, 87–91
    bulbar premotor centers, 89–91
    jaw opening and closing, 87–89
    proprioceptive control, 89–91
    tongue muscles, 89
  locomotion centers, 91, 412–414
  oculomotor control, 84–87
    eye muscles, 84–87
    head and neck movements, 87
    oculomotor nuclei, 84–87
    optokinetic nystagmus, 84–86
    tectobulbospinal control, 87
    vestibulocollic control, 87
    vestibulooculomotor reflex, 86–87
  overview, 83–84, 97
  respiration centers, 91–92
  telencephalic centers, 94–97
    ansa lenticularis, 97
    archistriatum tract, 95–97
    eminentia sagittalis, 94–95
    occipitomesencephalic tract, 95–97

paleostriatal complex, 97
septomesencephalic tract, 94–95
songbird vocalization, 96–97
vocalization centers, 91–92
Mouth, see also Beak; Gustation
  anatomy, 299–302
  digestion and secretion, 313
Müllerian ducts, regression, 602
Muscles, see Motor control system; Skeletal muscles
Myoglobin, gas exchange role, 255

N

Nephrons, osmoregulation role, types and numbers, 267
Nerve growth factor, development role, 646
Nesting
  induction, 585–586
  thermoregulation
    altricial species, 377–378
    brooding, 378
    growth rates, 376–377
    heat loss, 378
    hypothermia, 378–379
    incubation, 620–621
    precocial species, 377–378
    solar heat gain, 378
Neurons, see also Autonomic nervous system; Central nervous system
  auditory nerve fibers, 27–28
  central auditory pathways, 28–33
    anatomy, 28–29
    electrophysiology, 29–33
      cochlear nuclei, 30–31
      lateral lemniscus nuclei, 31
      midbrain auditory nucleus, 31–32
      nucleus laminaris, 31
      telencephalic auditory field L, 32–33
      thalamic nucleus ovoidalis, 32
  muscle innervation, 128–129
  olfactory neuronal response, 48
  retina structure and function, 4–7
  spinal cord organization, see Central nervous system
Nitrogen, excretion regulation, 278–280
Nociception
  electrophysiological investigations, 64
  receptor characteristics, 62
Nonsteroidal hormones, female reproduction role, 580
Norepinephrine
  chromaffin cell innervation, 490
  corticosteroid function regulation, 504
  female reproduction role, 580
  physiological actions, 520–521
  senescence role, 522
  synthesis and release regulation, 519–520
Nucleus laminaris, central auditory pathway function, electrophysiology, 31
Nucleus proprius, organization, 75
Nutrition, see also Food
  growth hormone secretion control, 446–447

O

Occipitomesencephalic tract, motor control role, 95–97
Oculomotor system, see Vision
Oil droplets, vision role, 4
Olfaction
　discrimination capabilities, 48–49
　field performance, 49–50
　laboratory detection, 48–49
　neuronal response, 48
　overview, 39–40, 50–51
　receptor innervation, 47–48
　system morphology, 46–47
　trigeminal–olfactory interactions, 42–43
Optokinetic nystagmus, oculomotor control, 84–86
Osmoregulation
　evaporative water loss, 291, 362, 424
　extrarenal organs, 282–283
　intestine, lower regions, 283–288
　　caeca transport properties, 286–288
　　colon transport properties, 283–285
　　coprodeum transport properties, 283–285
　　coprodeum versus colon role compared, 285
　　dietary regulation mechanisms, 285
　　hormonal regulation mechanisms, 285
　　molecular induction, 285–286
　　potassium transport, 284–285
　　sodium transport, 283–284, 286–287
　　ultrastructural adaptation, 285–286
　　ureteral urine postrenal modifications, 287–288
　　　caeca quantitative role, 288
　　　dehydration/sodium chloride depletion, 287
　　　hydration/sodium chloride loading, 287
　　　ratite special cases, 288
　　　salt glands, 287–288
　　water transport, 283–284, 286–287
　kidneys, 267–282
　　anatomy, 267–270
　　　arterial supply, 268
　　　blood flow, 267–270
　　　gross anatomy, 267
　　　nephron types and numbers, 267
　　　renal portal system, 268–270
　　　venous drainage, 270
　　physiology
　　　acid/base regulation, 280–281
　　　aldosterone, 278
　　　arginine vasotocin, 273, 277, 451
　　　atrial natriuretic peptide, 278
　　　calcium excretion regulation, 278
　　　collecting duct, 275
　　　descending thin limb, 274
　　　glomerular filtration, 272–273
　　　nitrogen excretion regulation, 278–280
　　　phosphate excretion regulation, 278
　　　renal blood flow, 270–272
　　　renin/angiotensin system, 277–278
　　　sodium excretion regulation, 277–278
　　　thick ascending limb, 275
　　　ureter function, 282
　　　urinary concentrating mechanisms, 273–277
　　　urine composition and flow, 281–282
　　　water excretion regulation, 273–277
　　　water reabsorption, 273–274
　overview, 265–266
　salt glands
　　anatomy, 288–289
　　function, 289–290
　　　fluid composition, 289–290
　　　regulatory mediators, 290
　　　secretion mechanisms, 289–290
　　　secretion stimulus, 289
　　postrenal urine modification, 287–288
　　quantitative contribution, 290–291
　solute intake, 267
　water intake, 266–267
Ovary, anatomy, 569–574
Oviduct
　anatomy, 574–575
　blood flow, 575
　calcium regulation, vitamin D action, 482
　innervation, 575
　motility, 575
Oviposition
　broodiness and nesting behavior, 585–586
　hormone effects, 580–586
　lay rate, 575–577
　light effects, 583–584
　parthenogenesis, 577
　postovulatory follicle, 585
　preovulatory follicle, 585
Ovulation
　arginine vasotocin role, 451–452
　blood flow changes, 173
　hormone effects, 580–583
　lay rate, 575–577
　light effects, 583–584
　parthenogenesis, 577
Oxygen
　high altitude flight, 428–429
　oxygen cascade, 234
　transport
　　hemoglobin, 245
　　hemoglobin engineering, 247
　　measurement, 249
　　O_2–blood equilibrium curves, 245
　　O_2 capacity, 247
　　O_2–hemoglobin affinity, 246–247
　　pulmonary exchange, 250–251
Oxytocin, oviposition induction, 584

P

Pain, behavioral effects, 66
Paleostriatal complex, motor control role, 97
Pancreas
　anatomy, 305
　function
　　cell distribution, 542–543
　　cytodifferentiation, 542–543
　　digestion and secretion, 316–317
　　embryogenesis, 542
　　glucoregulation, 540–541
　　hormonogenesis, 543–544
　　insulin/glucagon molar ratio, 544
　　secretogenesis, 543–544
　glucoregulation
　　altered metabolism, 552–554
　　　diabetes mellitus, 552
　　　growth hormone effects, 554
　　　pancreatic chemocytotoxins, 552–553
　　　pancreatropic agents, 553–554
　　　prolactin effects, 554
　　　surgical extirpation of pancreas, 553
　　birds and mammals compared, 541–542
　　liver role, 540
　　mechanisms, 539–540
　　pancreas function, 540–541
　　pancreatic–enteric regulation, 548–551
　　　adipose tissue, 550
　　　erythrocytes, 551–552
　　　glycogen body, 552
　　　heart, 551
　　　insulin/glucagon molar ratio role, 550–551
　　　liver role, 549–550
　　　muscle tissue, 550
　　　plasma effects, 549
　　　tissue effects, 549–552
　　pancreatic hormone action mechanisms, 544–546
　　　glucagon, 544–546
　　　insulin, 544
　　　pancreatic polypeptide, 546
　　　receptor considerations, 546–548
　　　somatostatin, 546
Paragriseal cells, organization, 77–78
Parasympathetic nervous system
　central components, 105–108
　heart control, 198–210
　　control mechanisms, 205–210
　　　chronotropic effects, 207–209
　　　dromotropic effects, 207
　　　inotropic effects, 209–210
　　neuron anatomy, 198–205
　peripheral motor components, 104
Parathyroid glands, characteristics, 474
Parathyroid hormone
　actions, 475–477
　　renal actions, 477
　　skeletal actions, 475–477
　chemistry, 474–475
　circulating hormone, 475
　corticosteroid function regulation, 504
　overview, 473–474, 484
　parathyroidectomy, 474
　parathyroid glands, 474
　parathyroid hormone-related peptide, 477–479

Parathyroid hormone-related peptide, characteristics, 477–479
Parthenogenesis, ovulation and oviposition cycles, 577
Peptides, see also specific types
  absorption, 318
Peyer's patches, immunophysiology, 660
Pharynx, anatomy, 299–302
Phosphate
  absorption, 319
  excretion regulation, 278
Photoperiodism
  pineal gland role, 564–565
  seasonal breeding role, 442, 583–584
Photoreceptors, function, 4
Pineal gland
  anatomy, 557–558
  annual cycles, 564–565
  melatonin
    characteristics, 558–559
    circadian rhythm synchronization, 563–564
    extrapineal production, 559–560
    photoperiodism, 564–565
    physiological effects, 560–565
      circadian system, 560–564
        diversity, 563
          pacemaker function, 560–563
      immunophysiology, 662
    pineal hormones, 558–560, 563–564
    sleep, 565
    thermoregulation, 565
Pituitary gland
  adrenocorticotropic hormone, 449–450
  arginine vasotocin
    action, 451–452
    chemistry, 451
    prolactin release, 449
  autonomic nervous system components, 108–111
  chromogranin A, 450–451
  gonadotropins, 438–443
    female action, 438–439
    male action, 439–441
    release control, 441–443
      female reproductive cycle, 442
      luteinizing hormone releasing hormone, 441–442
      negative and positive feedback, 442
      photoperiodism, 442
      seasonal breeding, 442–443
    structure, 438
  growth hormone, 444–447
    actions, 444–445
      adrenocortical hormone interactions, 445
      growth, 444–445
      immune function, 445
      lipid metabolism, 445
      thyroid hormone interactions, 445
    assay, 444
    chemistry, 444
    growth hormone receptors, 445–446
    pituitary origin, 447

  secretion control, 446–447
    growth hormone releasing factor, 446
    hypophysiotropic peptide release, 446
    nutrition effects, 446–447
    ontogeny, 447
    somatostatin, 446
    thyrotropin releasing hormone, 446
    variants, 444
  hypothalamic–hypophyseal complex, 438
  mesotocin, 451–452
  neurohypophysis, 451
  overview, 437–438
  prolactin, 447–449
    actions, 448
    assay, 447–448
    chemistry, 447
    pituitary origin, 447
    receptor, 448–449
    release control, 449
  thermoregulation, 368–369
  thyrotropin, 443–444
Platelet-derived growth factor, development role, 646
Populations, energy balance requirements, 338
Prehierarchal follicles, steroid production, 579
Progesterone, female reproduction role, 581
Prolactin
  actions, 448
  assay, 447–448
  carbohydrate metabolism role, 554
  chemistry, 447
  corticosteroid function regulation, 503
  development role, 642–644
  female reproduction role, 582–583
  pituitary origin, 447
  receptor, 448–449
  release control, 449
Prostaglandins
  characteristics, 483–484
  corticosteroid function regulation, 506
  female reproduction role, 580, 583–585
Protein, see specific proteins

R

Ratites, intestinal osmoregulation, 288
Rectum
  anatomy, 304–305
  motility, 309–310
Renal function, see Osmoregulation
Renal portal system
  characteristics, 175–176
  osmoregulation role, 268–270
Renin
  aldosterone secretion regulation, 506–507
  osmoregulation role, 277–278
Reproduction
  energy costs
    egg production, 335
    embryo growth, 336–337

    gonad growth, 335
    incubation, 333
    newly hatched young, 336–337
    rearing young, 333
    territorial defense, 333
  females
    anatomy, 569–575
      atretic follicles, 570–572
      blood flow, 575
      follicle growth, 572–574
      infundibulum, 574
      isthmus, 574
      lipid deposition, 572–574
      magnum, 574
      ovarian follicle, 570
      ovary, 569–574
      oviduct, 574–575
      postovulatory follicles, 570–572
      shell gland, 574
      sperm storage glands, 574–575
      vagina, 574–575
    breeding cycles, 575–577
    egg formation, 586–591
      albumen, 586–587
      calcium metabolism, 589–591
      carbonate formation and deposition, 591
      crystallization layers, 588–589
      cuticle, 587–588
      mammillary cores, 587
      mammillary knob layer, 588
      organic matrix, 587–588
      palisade layer, 588
      shell matrix, 587
      shell membranes, 587
      shell respiration, 589
      vitamin D metabolism, 590
      yolk, 586
    follicle-stimulating hormone role, 439
    hormones
      atretic follicles, 579–580
      corticosterone, 511–512, 582
      estrogens, 582
      follicle-stimulating hormone, 581
      growth factors, 580
      hierarchal follicles, 578–579
      luteinizing hormone, 580–581
      luteinizing hormone releasing hormone, 580
      molt, 583–584
      nonsteroidal hormones, 580
      oxytocin, 584
      postovulatory follicles, 579–580
      prehierarchal follicles, 579
      progesterone, 581
      prolactin, 582–583
      prostaglandins, 580, 583–585
      serotonin, 583
      sexual maturation, 583
      steroids, 578–580
      stromal follicles, 579
    luteinizing hormone role, 438–439, 442
    oviposition
      broodiness and nesting behavior, 585–586

Reproduction (*continued*)
    hormone effects, 580–586
    lay rate, 575–577
    light effects, 583–584
    parthenogenesis, 577
    postovulatory follicle, 585
    preovulatory follicle, 585
  ovulation cycles
    hormone effects, 580–583
    lay rate, 575–577
    light effects, 583–584
    parthenogenesis, 577
  males
    anatomy, 597–600
      accessory organs, 600
      excurrent ducts, 599–600
      testis, 597–599
    follicle-stimulating hormone role, 441
    luteinizing hormone role, 439–441
    ontogeny, 600–603
      excurrent duct formation, 602–603
      gonad differentiation, 602
      Müllerian duct regression, 602
      undifferentiated gonad formation, 601–602
    overview, 597
    sperm
      extragonadal sperm transport, 609–612
      spermatogenesis, 607–609
    testis
      adenohypophyseal function control, 606–607
      anatomy, 597–599
      central function control, 605–606
      gonadotropin effects, 607
      growth pattern alteration, 605
      meiosis initiation, 605
      somatic cell proliferation and differentiation, 603–605
      stem cell proliferation, 603
  seasonal breeding, 442–443
  thyroid hormone role, 468
Respiratory system
  anatomy, 234–238
    air sacs, 236–238
    air volumes, 238
    lungs, 235–236
      conducting airways, 235–236
      parabronchi, 236
    upper airways, 235
  breathing control, 256–259
    respiratory rhythm-generation, 256
    sensory inputs, 256–258
      arterial chemoreceptors, 256–257
      central chemoreceptors, 256
      intrapulmonary chemoreceptors, 257–258
    ventilatory reflexes, 258–259
      CO_2 response, 258–259
      exercise response, 259
      hypoxic response, 259
  capillary gas exchange, 165–170
  chemical stimuli responses, 42
  eggshell ventilation, 589–590, 621–624
  evaporative heat loss, 359–363
    cooling site, 361–362
    gular flutter, 361
    heat conservation, 362–363
    osmotic state, 362
    thermal tachypnea, 359–361
    water conservation, 362–363
  flight effects, 419–424
    temperature control, 421–424
    ventilatory adjustments, 419–421
    ventilatory/locomotor coupling, 419–421
    water loss, 424
  gas transport, 244–249
    acid–base equilibrium, 248–249
    carbon dioxide, 247–248
      blood–CO_2 equilibrium curves, 248
      forms in blood, 248
      Henderson–Hasselbalch equation, 248–249
    measurement, 249
    oxygen, 245–247
      hemoglobin, 245
      hemoglobin molecular engineering, 247
      O_2–blood equilibrium curves, 245
      O_2 capacity, 247
      O_2–hemoglobin affinity, 246–247
  mechanics, 238–243
    flow patterns, 240–243
      aerodynamic valving, 241, 243
      air sac Po_2 and Pco_2, 241–242
      air sac ventilation, 240
      artificial ventilation, 242
      parabronchial ventilation, 242
      pulmonary ventilation, 240–241
    properties, 239–240
      air capillary surface forces, 240
      compliance, 239
      resistance, 239–240
    respiratory muscles, 238–239
  motor control centers, 91–92
  overview, 233–234
  oxygen cascade, 234
  pulmonary circulation, 243–244
    ascites role, 244
    blood flow distribution, 244
    capillary volume, 243
    circulation anatomy, 243
    fluid balance, 244
    vascular resistance, 244
  pulmonary gas exchange, 249–254
    cross-current gas exchange, 250–251
      CO_2 exchange, 251
      O_2 exchange, 250–251
    high-altitude flight, 254
    lung diffusing capacity, 251–252
      air capillary diffusion, 251
      blood–gas barrier diffusion, 251–252
      O_2–hemoglobin reaction rates, 252
      O_{L2} physiological estimates, 252
    lung heterogeneity, 252–254
      dead space physiology, 252
      temporal heterogeneity, 253–254
      $\dot{V}/\dot{Q}$ mismatching, 253
    overview, 254
    oxygen transport principles, 250
      convection, 250
      diffusion, 250
  pulmonary vessel innervation, 189–190
  symbols and units, 234
  tissue gas exchange, 254–255
    exercise effects, 255
    hypoxia effects, 255
    microcirculation, 255
      cerebral circulation, 255
      skeletal muscle, 255
    myoglobin, 255
Ruffini endings, morphology and distribution, 59

S

Salt glands, osmoregulation role
  anatomy, 288–289
  function, 289–290
    fluid composition, 289–290
    regulatory mediators, 290
    secretion mechanisms, 289–290
    secretion stimulus, 289
  postrenal urine modification, 287–288
  quantitative contribution, 290–291
Seasonal patterns
  breeding
    gonadotropin release control, 442–443
    photoperiodism role, 442, 583–584
  corticosteroid concentration effects, 511–512
  migration, *see* Migration
Senescence
  catecholamine effects, 522
  hypothalamo–pituitary–adrenal axis role, 514
Senses, *see specific senses*
Septomesencephalic tract, motor control role, 94–95
Serotonin
  aldosterone secretion regulation, 508–509
  female reproduction role, 583
Sexual maturation, female reproduction role, 583
Shell
  formation
    matrix, 587
    membranes, 587
    respiration, 589, 621–624
  shell gland anatomy, 574
Sight, *see* Vision
Size, energy exchange measurement, 331
Skeletal muscles
  circulation controlling reflexes, 220–221
  contractile properties, 129–130
  development, 123–126

electrical properties, 129
flight effects
  cardiac muscles, 417–419
  flight muscles, 405–412
    biochemistry, 408–410
    efficiency scaling, 403–405
    fiber types, 405–408
    function, 410–412
    morphology, 405–408
    neurophysiology, 410–412
  forward flapping, 401–403
  gliding, 400–401
  hovering, 403
  locomotor muscles, 412–414
  preflight muscle development, 412–414
  soaring, 400–401
gas exchange, 255
innervation, 128–129
insulin effects, 550
motor control, see Motor control system
neuromuscular pharmacology, 133–135
neuromuscular transmission, 130–133
overview, 123
respiratory muscles, 238–239
Skeleton
  eggshell formation, medullary bone calcium mobilization, 590
  parathyroid hormone actions, 475–477
  vitamin D effects, 482–483
Skin, sensory receptors, see Somatosensory system
Sleep, melatonin role, 565
Small intestine
  anatomy, 303–304
  motility, 308–309
Smell, see Olfaction
Soaring, power input, 400–401
Sodium
  absorption, 319–320
  osmoregulation role
    excretion regulation in kidneys, 277–278
    intestinal transport, 283–284, 286–287
    ureteral urine postrenal modifications
      dehydration/depletion, 287
      hydration/loading, 287
Solute balance, see Osmoregulation; specific solutes
Somatic cells, proliferation and differentiation in testis, 603–605
Somatosensory system, 57–67
  afferent fibers, 57–62
  behavioral aspects, 65–66
    mechanoreception, 65–66
    pain, 66
    thermoreception, 66
  central processing, 62–65
    electrophysiological investigations, 64–65
      mechanoreception, 64–65
      nociception, 64
      spinal system and body representation, 65
      thermoreception, 64

trigeminal system and beak representation, 64–65
    somatosensory pathways, 62–64
      spinal system, 63–64
      trigeminal system, 62–63
  overview, 57, 66–67
  receptor types, 57–62
    mechanoreceptors, 58–61
      cutaneous mechanoreceptors, 58–59
      electrophysiology, 59–61
      Grandry corpuscles, 58–59
      herbst corpuscles, 58
      Merkel cell receptors, 58
      Ruffini endings, 59
    nociceptors, 62
    thermoreceptors, 61–62
Somatostatin
  action mechanisms, 546
  growth hormone secretion control, 446
  secretogenesis, 543–544
Songbirds, vocalization, motor control centers, 96–97
Sound, see Auditory system; Vocalization
Sperm
  extragonadal sperm transport, 609–612
  female storage glands, 574–575
  spermatogenesis, 607–609
Spinal cord, see Central nervous system
Spleen, immunophysiology, 659–660
Starvation, energy balance, 334
Stem cells, proliferation in testis, 603
Stereopsis, eye structure and function, 2–4
Steroids, female reproduction role, 578–580
Stomach
  anatomy, 303
  digestion and secretion, 314–315
Streptozotocin, pancreatic toxicity, 552–553
Stress, adrenocortical function, 500–502
Swimming
  energy costs, 332
  tissue blood flow effects, 171–172
  venous system physiology, 174–175
Sympathetic nervous system
  central components, 105
  heart control, 191–198
  peripheral motor components, 102–104
Synthalin-A, pancreatic toxicity, 552

T

Tachypnea, evaporative water loss, 359–361
Taste, see Gustation
T-cell receptors, development, 658–659
Tectobulbospinal control center, oculomotor control, 87
Tectofugal pathway, central visual processing, 9–12
Telemetry, flight study, 397–398
Telencephalic centers
  central auditory pathway function, electrophysiology, 32–33
  motor control role, 94–97
    ansa lenticularis, 97
    archistriatum tract, 95–97

eminentia sagittalis, 94–95
    occipitomesencephalic tract, 95–97
    paleostriatal complex, 97
    septomesencephalic tract, 94–95
    songbird vocalization, 96–97
Temperature, ambient, see also Thermoreception; Thermoregulation
  blood effects, 180–181
  body temperature regulation, 347
  embryonic tolerance, 631–632
  gustation sensitivity, 46
  metabolic rate effects, 350–353
Territorial defense, energy costs, 333
Testis
  adenohypophyseal function control, 606–607
  anatomy, 597–599
  central function control, 605–606
  gonadotropin effects, 607
  growth pattern alteration, 605
  meiosis initiation, 605
  somatic cell proliferation and differentiation, 603–605
  stem cell proliferation, 603
Thalamic nucleus ovoidalis, central auditory pathway function, electrophysiology, 32
Thalamofugal pathway, central visual processing, 12–14
Thermal tachypnea
  afferent inputs, 371
  evaporative water loss, 359–361
Thermoreception
  behavioral aspects, 66
  electrophysiological investigations, 64
  thermoreceptor characteristics, 61–62
  thermoregulation control mechanisms, 367–368
Thermoregulation, see also Energy balance
  behavioral thermoregulation, 364–367, 370–371
  body heat content changes, 348–349
  body temperature, 344–348
    acclimatization and acclimation, 347
    ambient temperature, 347
    circadian rhythm, 345–347, 565
    core temperature, 344
    dehydration, 347
    food deprivation, 347–348
    interspecific variation, 345
    measurement, 344
  control mechanisms, 367–372
    behavioral thermoregulation, 370–371
    carbon dioxide effects, 371–372
    deep body receptors role, 370
    hypothalamus role, 368–369
    peripheral thermoreception, 367–368
    pineal gland role, 565
    spinal cord role, 370
    thermosensitivity, 367–370
  development
    embryo, 375
    hatchling, 375–376
      brooding responses, 375

Thermoregulation (continued)
    cold responses, 375
    heat responses, 376
      nesting, 376–379
        altricial species, 377–378
        brooding, 378
        growth rates, 376–377
        heat loss, 378
        hypothermia, 378–379
        incubation, 620–621, 627–629
        precocial species, 377–378
        solar heat gain, 378
    heat balance, 348
    heat exchange, 363–364
    heat loss, 356–363
      evaporative heat loss, 356–363
        acclimatization and acclimation, 363
        cooling site, 361–362
        cutaneous heat loss, 357–359
        flight effects, 421–424
        gular flutter, 361
        heat conservation, 362–363
        osmotic state, 362
        partition of heat loss, 363
        thermal tachypnea, 359–361
        water conservation, 362–363
      nonevaporative heat loss, 356
      total thermal conductance, 356
    heat production, 349–353
      acclimatization and acclimation, 353–355
      ambient temperature influence, 350–353
      circadian rhythm, 350
      geographical variation, 353–355
      interspecific variation, 350, 353–355
      measurement, 349
      metabolic rates, 349–350
    heat transfer within the body, 355–356
      acclimatization and acclimation, 355–356
      brood patch, 355
      cold vasodilation, 355
      tissues, 355
      vascular heat exchange, 355
    overview, 344, 379
    reduced body temperatures, 372–375
      hypothermia arousal, 374–375
      hypothermia patterns, 372–373
      hypothermic state, 373–374
Thymus, development, 658–659
Thyroid glands
    environmental influences, 469
    hypothalamic–pituitary–thyroid axis, 466
    overview, 461–462
    thyroid hormones
      action mechanisms, 466–467
      activation, 464–466
      circulating hormone, 463–464
      degradation, 464–466
      development role, 467–468, 641–642
      effects, 467–468
      hatching role, 468
      hormone interactions, 445, 468–469
      molting role, 468
      reproduction role, 468
      synthesis and release, 462–463
Thyroid hormone, corticosteroid function regulation, 505
Thyrotropin, characteristics, 443–444
Thyrotropin releasing hormone
    growth hormone secretion control, 446
    prolactin release control, 449
Time resolution, auditory behavioral aspects, 33–34
Tissues, *see also specific tissues*
    blood flow effects to swimming, 171–172
    gas exchange mechanisms, 254–255
      exercise effects, 255
      hypoxia effects, 255
      microcirculation, 255
        cerebral circulation, 255
        skeletal muscle, 255
      myoglobin, 255
    pancreatic hormone effects, 549–550
      adipose tissue, 550
      liver, 549–550
      muscle tissue, 550
      plasma effects, 549
    thermoregulation, 355
T-lymphocytes, origin, 659
Tolbutamide, hypoglycemic effects, 553–554
Tongue
    gustation
      bitter response, 44–46
      complex taste response, 46
      overview, 39–40
      salt response, 44
      sour response, 44
      sweet response, 43–44
      taste behavior, 43
      taste receptor innervation, 43
      taste receptors, 43
      temperature effects, 46
    motor control, 87–91
      bulbar premotor centers, 89–91
      jaw opening and closing, 87–89
      proprioceptive control, 89–91
      tongue muscles, 89
Touch, *see* Mechanoreception
Transforming growth factor-α, development role, 644–645
Transforming growth factor-β, development role, 646–647
Trigeminal receptors
    function, 40
    olfactory interactions, 42–43
    somatosensory central processing
      electrophysiological investigations, 64–65
      pathways, 62–63

U

Ultimobranchial glands, characteristics, 479

Urine
    kidney physiology
      urinary concentrating mechanisms, 273–277
      urine composition and flow, 281–282
    postrenal modifications, 287–288
      caeca quantitative role, 288
      dehydration/sodium chloride depletion, 287
      hydration/sodium chloride loading, 287
      ratite special cases, 288
      salt glands, 287–288

V

Vagina, anatomy, 574–575
Vascular system, *see* Cardiovascular system
Vasoactive intestinal peptide
    female reproduction role, 580
    prolactin release control, 449
Vasodilation
    peripheral blood flow control, vasoactive agents, 182–183
    thermoregulation, 355
Venous system, *see also* Cardiovascular system
    capacitance function, 174
    coronary circulation, 146–147
    development, 173–174
    exercise physiology, 174–175
    hemodynamics, 154–157
    kidney drainage, 270
    peripheral blood flow control, systemic innervation, 188–189
    renal portal system, 175–176
    submersion physiology, 174–175
Ventilation/perfusion ratio, mismatching, 253
Vestibulocollic control center, oculomotor control, 87
Vestibuloocculomotor reflex, oculomotor control, 86–87
V gene, immunophysiology, 665–666
Visceral forebrain, autonomic nervous system pathways, 114–115
Vision
    central processing, 7–14
      centrifugal pathway, 7–9
      tectofugal pathway, 9–12
      thalamofugal pathway, 12–14
    eye structure and function, 1–7
      acuity, 2–4
      retina, 4–7
        color vision, 4
        neuronal wiring, 4–7
        oil droplets, 4
        photoreceptors, 4
      shape, 2–4
      stereopsis, 2–4
    oculomotor control
      eye muscles, 84–87
      head and neck movements, 87
      oculomotor nuclei, 84–87
      optokinetic nystagmus, 84–86
      parasympathetic nervous system, 104
      tectobulbospinal control, 87

vestibulocollic control, 87
vestibulooculomotor reflex, 86–87
overview, 1
Vitamin B_6, absorption, 320
Vitamin C, pancreatic toxicity, 552
Vitamin D
    actions, 482–483
        bone actions, 482–483
        intestine, 482
        oviduct, 482
    circulating vitamin D metabolite levels, 482
    eggshell metabolism, 590
    overview, 473–474, 484
    renal metabolism, 481–482
Vocalization, motor control centers, 91–92, 96–97

W

Walking, energy costs, 332
Water balance, *see* Osmoregulation
Wind tunnel, flight study, 398–399
Wulst, motor control role, 94–95

Y

Yolk, characteristics, 586